POWER SYSTEM ANALYSIS AND DESIGN

FIFTH EDITION, SI

J. DUNCAN GLOVER
FAILURE ELECTRICAL, LLC

MULUKUTLA S. SARMA
NORTHEASTERN UNIVERSITY

THOMAS J. OVERBYE
UNIVERSITY OF ILLINOIS

Australia • Brazil • Japan • Korea • Mexico • Singapore • Spain • United Kingdom • United States

Power System Analysis and Design, Fifth Edition, SI
J. Duncan Glover, Mulukutla S. Sarma, and Thomas J. Overbye

Publisher, Global Engineering: Christopher M. Shortt
Acquisitions Editor: Swati Meherishi
Senior Developmental Editor: Hilda Gowans
Editorial Assistant: Tanya Altieri
Team Assistant: Carly Rizzo
Marketing Manager: Lauren Betsos
Media Editor: Chris Valentine
Content Project Manager: Jennifer Ziegler
Production Service: RPK Editorial Services
Copyeditor: Shelly Gerger-Knechtl
Proofreader: Becky Taylor
Indexer: Glyph International
Compositor: Glyph International
Senior Art Director: Michelle Kunkler
Internal Designer: Carmela Periera
Cover Designer: Andrew Adams
Cover Image: ©Shebeko/Shutterstock
Senior Rights Acquisitions Specialists: Mardell Glinski-Schultz and John Hill
Text and Image Permissions Researcher: Kristiina Paul
First Print Buyer: Arethea L. Thomas

Library of Congress Control Number: 2011924686

ISBN-13: 978-1-111-42579-1

ISBN-10: 1-111-42579-5

Cengage Learning
200 First Stamford Place, Suite 400
Stamford, CT 06902
USA

Printed in the United States of America
1 2 3 4 5 6 7 13 12 11

CONVERSIONS BETWEEN U.S. CUSTOMARY UNITS AND SI UNITS (Continued)

U.S. Customary unit		Times conversion factor		Equals SI unit	
		Accurate	Practical		
Moment of inertia (area)					
inch to fourth power	in.^{4}	416,231	416,000	millimeter to fourth power	mm^4
inch to fourth power	in.^{4}	0.416231×10^{-6}	0.416×10^{-6}	meter to fourth power	m^4
Moment of inertia (mass)					
slug foot squared	slug-ft^2	1.35582	1.36	kilogram meter squared	kg·m^2
Power					
foot-pound per second	ft-lb/s	1.35582	1.36	watt (J/s or N·m/s)	W
foot-pound per minute	ft-lb/min	0.0225970	0.0226	watt	W
horsepower (550 ft-lb/s)	hp	745.701	746	watt	W
Pressure; stress					
pound per square foot	psf	47.8803	47.9	pascal (N/m^2)	Pa
pound per square inch	psi	6894.76	6890	pascal	Pa
kip per square foot	ksf	47.8803	47.9	kilopascal	kPa
kip per square inch	ksi	6.89476	6.89	megapascal	MPa
Section modulus					
inch to third power	in.^{3}	16,387.1	16,400	millimeter to third power	mm^3
inch to third power	in.^{3}	16.3871×10^{-6}	16.4×10^{-6}	meter to third power	m^3
Velocity (linear)					
foot per second	ft/s	0.3048*	0.305	meter per second	m/s
inch per second	in./s	0.0254*	0.0254	meter per second	m/s
mile per hour	mph	0.44704*	0.447	meter per second	m/s
mile per hour	mph	1.609344*	1.61	kilometer per hour	km/h
Volume					
cubic foot	ft^3	0.0283168	0.0283	cubic meter	m^3
cubic inch	in.^{3}	16.3871×10^{-6}	16.4×10^{-6}	cubic meter	m^3
cubic inch	in.^{3}	16.3871	16.4	cubic centimeter (cc)	cm^3
gallon (231 in.^{3})	gal.	3.78541	3.79	liter	L
gallon (231 in.^{3})	gal.	0.00378541	0.00379	cubic meter	m^3

*An asterisk denotes an *exact* conversion factor

Note: **To convert from SI units to USCS units, *divide* by the conversion factor**

Temperature Conversion Formulas

$$T(°C) = \frac{5}{9}[T(°F) - 32] = T(K) - 273.15$$

$$T(K) = \frac{5}{9}[T(°F) - 32] + 273.15 = T(°C) + 273.15$$

$$T(°F) = \frac{9}{5}T(°C) + 32 = \frac{9}{5}T(K) - 459.67$$

TO LOUISE, TATIANA & BRENDAN, ALISON & JOHN, LEAH, OWEN, ANNA, EMILY & BRIGID

Dear Lord! Kind Lord!
Gracious Lord! I pray
Thou wilt look on all I love,
Tenderly to-day!
Weed their hearts of weariness;
Scatter every care
Down a wake of angel-wings
Winnowing the air.

Bring unto the sorrowing
All release from pain;
Let the lips of laughter
Overflow again;
And with all the needy
O divide, I pray,
This vast treasure of content
That is mine to-day!

James Whitcomb Riley

CONTENTS

CHAPTER 10 System Protection 516

CHAPTER 11 Transient Stability 579

CHAPTER 12 Power System Controls 639

CHAPTER 13 Transmission Lines: Transient Operation 690

PREFACE TO THE SI EDITION

This edition of ***Power System Analysis and Design*** has been adapted to incorporate the International System of Units (*Le Système International d'Unités* or SI) throughout the book.

LE SYSTÈME INTERNATIONAL D'UNITÉS

The United States Customary System (USCS) of units uses FPS (foot–pound–second) units (also called English or Imperial units). SI units are primarily the units of the MKS (meter–kilogram–second) system. However, CGS (centimeter–gram–second) units are often accepted as SI units, especially in textbooks.

USING SI UNITS IN THIS BOOK

In this book, we have used both MKS and CGS units. USCS units or FPS units used in the US Edition of the book have been converted to SI units throughout the text and problems. However, in case of data sourced from handbooks, government standards, and product manuals, it is not only extremely difficult to convert all values to SI, it also encroaches upon the intellectual property of the source. Also, some quantities such as the ASTM grain size number and Jominy distances are generally computed in FPS units and would lose their relevance if converted to SI. Some data in figures, tables, examples, and references, therefore, remains in FPS units. For readers unfamiliar with the relationship between the FPS and the SI systems, conversion tables have been provided inside the front and back covers of the book.

To solve problems that require the use of sourced data, the sourced values can be converted from FPS units to SI units just before they are to be used in a calculation. To obtain standardized quantities and manufacturers' data in SI units, the readers may contact the appropriate government agencies or authorities in their countries/regions.

INSTRUCTOR RESOURCES

A Printed Instructor's Solution Manual in SI units is available on request. An electronic version of the Instructor's Solutions Manual, and PowerPoint slides of the figures from the SI text are available through http://login.cengage.com.

The readers' feedback on this SI Edition will be highly appreciated and will help us improve subsequent editions.

The Publishers

PREFACE

The objective of this book is to present methods of power system analysis and design, particularly with the aid of a personal computer, in sufficient depth to give the student the basic theory at the undergraduate level. The approach is designed to develop students' thinking processes, enabling them to reach a sound understanding of a broad range of topics related to power system engineering, while motivating their interest in the electrical power industry. Because we believe that fundamental physical concepts underlie creative engineering and form the most valuable and permanent part of an engineering education, we highlight physical concepts while giving due attention to mathematical techniques. Both theory and modeling are developed from simple beginnings so that they can be readily extended to new and complex situations.

This edition of the text features new Chapter 14 entitled, *Power Distribution.* During the last decade, major improvements in distribution reliability have come through automated distribution and more recently through the introduction of "smart grids." Chapter 14 introduces the basic features of primary and secondary distribution systems as well as basic distribution components including distribution substation transformers, distribution transformers, and shunt capacitors. We list some of the major distribution software vendors followed by an introduction to distribution reliability, distribution automation, and smart grids.

This edition also features the following: (1) wind-energy systems modeling in the chapter on transient stability; (2) discussion of reactive/pitch control of wind generation in the chapter on powers system controls; (3) updated case studies for nine chapters along with four case studies from the previous edition describing present-day, practical applications and new technologies; (4) an updated PowerWorld Simulator package; and (5) updated problems at the end of chapters.

One of the most challenging aspects of engineering education is giving students an intuitive feel for the systems they are studying. Engineering systems are, for the most part, complex. While paper-and-pencil exercises can be quite useful for highlighting the fundamentals, they often fall short in imparting the desired intuitive insight. To help provide this insight, the book uses PowerWorld Simulator to integrate computer-based examples, problems, and design projects throughout the text.

PowerWorld Simulator was originally developed at the University of Illinois at Urbana–Champaign to teach the basics of power systems to nontechnical people involved in the electricity industry, with version 1.0 introduced in June 1994. The program's interactive and graphical design made it an immediate hit as an educational tool, but a funny thing happened—its interactive and graphical design also appealed to engineers doing analysis of real power systems. To meet the needs of a growing group of users, PowerWorld Simulator was commercialized in 1996 by the formation of PowerWorld Corporation. Thus while retaining its appeal for education, over the years PowerWorld Simulator has evolved into a top-notch analysis package, able to handle power systems of any size. PowerWorld Simulator is now used throughout the power industry, with a range of users encompassing universities, utilities of all sizes, government regulators, power marketers, and consulting firms.

In integrating PowerWorld Simulator with the text, our design philosophy has been to use the software to extend, rather than replace, the fully worked examples provided in previous editions. Therefore, except when the problem size makes it impractical, each PowerWorld Simulator example includes a fully worked hand solution of the problem along with a PowerWorld Simulator case. This format allows students to simultaneously see the details of how a problem is solved and a computer implementation of the solution. The added benefit from PowerWorld Simulator is its ability to easily extend the example. Through its interactive design, students can quickly vary example parameters and immediately see the impact such changes have on the solution. By reworking the examples with the new parameters, students get immediate feedback on whether they understand the solution process. The interactive and visual design of PowerWorld Simulator also makes it an excellent tool for instructors to use for in-class demonstrations. With numerous examples utilizing PowerWorld Simulator instructors can easily demonstrate many of the text topics. Additional PowerWorld Simulator functionality is introduced in the text problems and design projects.

The text is intended to be fully covered in a two-semester or three-quarter course offered to seniors and first-year graduate students. The organization of chapters and individual sections is flexible enough to give the instructor sufficient latitude in choosing topics to cover, especially in a one-semester course. The text is supported by an ample number of worked examples covering most of the theoretical points raised. The many problems to be worked with a calculator as well as problems to be worked using a personal computer have been expanded in this edition.

As background for this course, it is assumed that students have had courses in electric network theory (including transient analysis) and ordinary differential equations and have been exposed to linear systems, matrix algebra, and computer programming. In addition, it would be helpful, but not necessary, to have had an electric machines course.

After an introduction to the history of electric power systems along with present and future trends, Chapter 2 on fundamentals orients the students to the terminology and serves as a brief review. The chapter reviews phasor concepts, power, and single-phase as well as three-phase circuits.

Chapters 3 through 6 examine power transformers, transmission-line parameters, steady-state operation of transmission lines, and power flows including the Newton–Raphson method. These chapters provide a basic understanding of power systems under balanced three-phase, steady-state, normal operating conditions.

Chapters 7 through 10, which cover symmetrical faults, symmetrical components, unsymmetrical faults, and system protection, come under the general heading of power system short-circuit protection. Chapter 11 (previously Chapter 13) examines transient stability, which includes the swing equation, the equal-area criterion, and multi-machine stability with modeling of wind-energy systems as a new feature. Chapter 12 (previously Chapter 11) covers power system controls, including turbine-generator controls, load-frequency control, economic dispatch, and optimal power flow, with reactive/pitch control of wind generation as a new feature. Chapter 13 (previously Chapter 12) examines transient operation of transmission lines including power system overvoltages and surge protection. The final and new Chapter 14 introduces power distribution.

ADDITIONAL RESOURCES

Companion websites for this book are available for both students and instructors. These websites provide useful links, figures, and other support material. The **Student Companion Site** includes a link to download the free student version of PowerWorld. The **Instructor Companion Site** includes access to the solutions manual and PowerPoint slides. Through the Instructor Companion Site, instructors can also request access to additional support material, including a printed solutions manual.

To access the support material described here along with all additional course materials, please visit www.cengagebrain.com. At the cengagebrain.com home page, search for the ISBN of your title (from the back cover of your book) using the search box at the top of the page. This will take you to the product page where these resources can be found.

ACKNOWLEDGMENTS

The material in this text was gradually developed to meet the needs of classes taught at universities in the United States and abroad over the past 30 years. The original 13 chapters were written by the first author, J. Duncan Glover, *Failure Electrical LLC*, who is indebted to many people who helped during the planning and writing of this book. The profound influence of earlier texts written on power systems, particularly by W. D. Stevenson, Jr., and the developments made by various outstanding engineers are gratefully acknowledged. Details of sources can only be made through references at the end of each chapter, as they are otherwise too numerous to mention.

Chapter 14 (*Power Distribution*) was a collaborative effort between Dr. Glover (Sections 14.1–14.7) and Co-author Thomas J. Overbye (Sections 14.8 & 14.9). Professor Overbye, *University of Illinois at Urbana-Champaign*, updated Chapter 6 (*Power Flows*), Chapter 11 (*Transient Stability*), and Chapter 12 (*Power System Controls*) for this edition of the text. He also provided the examples and problems using PowerWorld Simulator as well as three design projects. Co-author Mulukutla Sarma, *Northeastern University*, contributed to end-of-chapter multiple-choice questions and problems.

We commend the following Cengage Learning professionals: Chris Shortt, Publisher, Global Engineering; Hilda Gowans, Senior Developmental Editor; Swati Meherishi, Acquisitions Editor; and Kristiina Paul, Permissions Researcher; as well as Rose Kernan of RPK Editorial Services, Inc., for their broad knowledge, skills, and ingenuity in publishing this edition.

The reviewers for the fifth edition are as follows: Thomas L. Baldwin, *Florida State University*; Ali Emadi, *Illinois Institute of Technology*; Reza Iravani, *University of Toronto*; Surya Santoso, *University of Texas at Austin*; Ali Shaban, *California Polytechnic State University, San Luis Obispo*; and Dennis O. Wiitanen, *Michigan Technological University,* and Hamid Jaffari, *Danvers Electric.*

Substantial contributions to prior editions of this text were made by a number of invaluable reviewers, as follows:

Fourth Edition: Robert C. Degeneff, *Rensselaer Polytechnic Institute*; Venkata Dinavahi, *University of Alberta*; Richard G. Farmer, *Arizona State University*; Steven M. Hietpas, *South Dakota State University*; M. Hashem Nehrir, *Montana State University*; Anil Pahwa, *Kansas State University*; and Ghadir Radman, *Tennessee Technical University.*

Third Edition: Sohrab Asgarpoor, *University of Nebraska–Lincoln*; Mariesa L. Crow, *University of Missouri–Rolla*; Ilya Y. Grinberg, *State University of New York, College at Buffalo*; Iqbal Husain, *The University of Akron*; W. H. Kersting, *New Mexico State University*; John A. Palmer, *Colorado School of Mines*; Satish J. Ranada, *New Mexico State University*; and Shyama C. Tandon, *California Polytechnic State University*.

Second Edition: Max D. Anderson, *University of Missouri–Rolla*; Sohrab Asgarpoor, *University of Nebraska–Lincoln*; Kaveh Ashenayi, *University of Tulsa*; Richard D. Christie, Jr., *University of Washington*; Mariesa L. Crow, *University of Missouri–Rolla*; Richard G. Farmer, *Arizona State University*; Saul Goldberg, *California Polytechnic University*; Clifford H. Grigg, *Rose-Hulman Institute of Technology*; Howard B. Hamilton, *University of Pittsburgh*; Leo Holzenthal, Jr., *University of New Orleans*; Walid Hubbi, *New Jersey Institute of Technology*; Charles W. Isherwood, *University of Massachusetts–Dartmouth*; W. H. Kersting, *New Mexico State University*; Wayne E. Knabach, *South Dakota State University*; Pierre-Jean Lagace, *IREQ Institut de Reserche d'Hydro–Quebec*; James T. Lancaster, *Alfred University*; Kwang Y. Lee, *Pennsylvania State University*; Mohsen Lotfalian, *University of Evansville*; Rene B. Marxheimer, *San Francisco State University*, Lamine Mili, *Virginia Polytechnic Institute and State University*; Osama A. Mohammed, *Florida International University*; Clifford C. Mosher, *Washington State University*, Anil Pahwa, *Kansas State University*; M. A. Pai, *University of Illinois at Urbana–Champaign*; R. Ramakumar, *Oklahoma State University*; Teodoro C. Robles, *Milwaukee School of Engineering*, Ronald G. Schultz, *Cleveland State University*; Stephen A. Sebo, *Ohio State University*; Raymond Shoults, *University of Texas at Arlington*, Richard D. Shultz, *University of Wisconsin at Platteville*; Charles Slivinsky, *University of Missouri–Columbia*; John P. Stahl, *Ohio Northern University*; E. K. Stanek, *University of Missouri–Rolla*; Robert D. Strattan, *University of Tulsa*; Tian-Shen Tang, *Texas A&M University–Kingsville*; S. S. Venkata, *University of Washington*; Francis M. Wells, *Vanderbilt University*; Bill Wieserman, *University of Pennsylvania–Johnstown*; Stephen Williams, *U.S. Naval Postgraduate School*; and Salah M. Yousif, *California State University–Sacramento*.

First Edition: Frederick C. Brockhurst, *Rose-Hulman Institute of Technology*; Bell A. Cogbill. *Northeastern University*; Saul Goldberg, *California Polytechnic State University*; Mack Grady, *University of Texas at Austin*; Leonard F. Grigsby, *Auburn University*; Howard Hamilton, *University of Pittsburgh*; William F. Horton, *California Polytechnic State University*; W. H. Kersting, *New Mexico State University*; John Pavlat, *Iowa State University*; R. Ramakumar, *Oklahoma State University*; B. Don Russell, *Texas A&M*; Sheppard Salon, *Rensselaer Polytechnic Institute*; Stephen A. Sebo, *Ohio State University*; and Dennis O. Wiitanen, *Michigan Technological University*.

In conclusion, the objective in writing this text and the accompanying software package will have been fulfilled if the book is considered to be student-oriented, comprehensive, and up to date, with consistent notation and necessary detailed explanation at the level for which it is intended.

J. Duncan Glover

Mulukutla S. Sarma

Thomas J. Overbye

LIST OF SYMBOLS, UNITS, AND NOTATION

Symbol	Description
a	operator $1\underline{/120^\circ}$
a_t	transformer turns ratio
A	area
A	transmission line parameter
$\boldsymbol{A}$	symmetrical components transformation matrix
B	loss coefficient
B	frequency bias constant
B	phasor magnetic flux density
B	transmission line parameter
C	capacitance
C	transmission line parameter
D	distance
D	transmission line parameter
E	phasor source voltage
E	phasor electric field strength
f	frequency
G	conductance
G	conductance matrix
H	normalized inertia constant
H	phasor magnetic field intensity
$i(t)$	instantaneous current
I	current magnitude (rms unless otherwise indicated)
I	phasor current
$\boldsymbol{I}$	vector of phasor currents
j	operator $1\underline{/90^\circ}$
J	moment of inertia
l	length
l	length
L	inductance
L	inductance matrix
N	number (of buses, lines, turns, etc.)
p.f.	power factor
$p(t)$	instantaneous power
P	real power
q	charge
Q	reactive power
r	radius
R	resistance
R	turbine-governor regulation constant
R	resistance matrix
s	Laplace operator
S	apparent power
S	complex power
t	time
T	period
T	temperature
T	torque
$v(t)$	instantaneous voltage
V	voltage magnitude (rms unless otherwise indicated)
V	phasor voltage
$\boldsymbol{V}$	vector of phasor voltages
X	reactance
X	reactance matrix
Y	phasor admittance
$\boldsymbol{Y}$	admittance matrix
Z	phasor impedance
$\boldsymbol{Z}$	impedance matrix
α	angular acceleration
α	transformer phase shift angle
β	current angle
β	area frequency response characteristic
δ	voltage angle
δ	torque angle
ε	permittivity
Γ	reflection or refraction coefficient

Symbol	Description
λ	magnetic flux linkage
λ	penalty factor
Φ	magnetic flux
ρ	resistivity
τ	time in cycles
τ	transmission line transit time
θ	impedance angle
θ	angular position
μ	permeability
v	velocity of propagation
ω	radian frequency

SI Units

A	ampere
C	coulomb
F	farad
H	henry
Hz	hertz
J	joule
kg	kilogram
m	meter
N	newton
rad	radian
s	second
S	siemen
VA	voltampere
var	voltampere reactive
W	watt
Wb	weber
Ω	ohm

English Units

BTU	British thermal unit
cmil	circular mil
ft	foot
hp	horsepower
in	inch
mi	mile

Notation

Lowercase letters such as v(t) and i(t) indicate instantaneous values.
Uppercase letters such as V and I indicate rms values.
Uppercase letters in italic such as *V* and *I* indicate rms phasors.
Matrices and vectors with real components such as **R** and **I** are indicated by boldface type.
Matrices and vectors with complex components such as ***Z*** and ***I*** are indicated by boldface italic type.
Superscript T denotes vector or matrix transpose.
Asterisk (*) denotes complex conjugate.
■ indicates the end of an example and continuation of text.
PW highlights problems that utilize PowerWorld Simulator.

1300 MW coal-fired power plant (Courtesy of American Electric Power Company)

1 INTRODUCTION

Electrical engineers are concerned with every step in the process of generation, transmission, distribution, and utilization of electrical energy. The electric utility industry is probably the largest and most complex industry in the world. The electrical engineer who works in that industry will encounter challenging problems in designing future power systems to deliver increasing amounts of electrical energy in a safe, clean, and economical manner.

The objectives of this chapter are to review briefly the history of the electric utility industry, to discuss present and future trends in electric power systems, to describe the restructuring of the electric utility industry, and to introduce PowerWorld Simulator—a power system analysis and simulation software package.

CASE STUDY The following article describes the restructuring of the electric utility industry that has been taking place in the United States and the impacts on an aging transmission infrastructure. Independent power producers, increased competition in the generation sector, and open access for generators to the U.S. transmission system have changed the way the transmission system is utilized. The need for investment in new transmission and transmission technologies, for further refinements in restructuring, and for training and education systems to replenish the workforce are discussed [8].

The Future Beckons: Will the Electric Power Industry Heed the Call?

CHRISTOPHER E. ROOT

Over the last four decades, the U.S. electric power industry has undergone unprecedented change. In the 1960s, regulated utilities generated and delivered power within a localized service area. The decade was marked by high load growth and modest price stability. This stood in sharp contrast to the wild increases in the price of fuel oil, focus on energy conservation, and slow growth of the 1970s. Utilities quickly put the brakes on generation expansion projects, switched to coal or other nonoil fuel sources, and significantly cut back on the expansion of their networks as load growth slowed to a crawl. During the 1980s, the economy in many regions of the country began to rebound. The 1980s also brought the emergence of independent power producers and the deregulation of the natural gas wholesale markets and pipelines. These developments resulted in a significant increase in natural gas transmission into the northeastern United States and in the use of natural gas as the preferred fuel for new generating plants.

During the last ten years, the industry in many areas of the United States has seen increased competition in the generation sector and a fundamental shift in the role of the nation's electric transmission system, with the 1996 enactment of the Federal Energy Regulatory Commission (FERC) Order No. 888, which mandated open access for generators to the nation's transmission system. And while prices for distribution and transmission of electricity remained regulated, unregulated energy commodity markets have developed in several regions. FERC has supported these changes with rulings leading to the formation of independent system operators (ISOs) and regional transmission organizations (RTOs) to administer the electricity markets in several regions of the United States, including New England, New York, the Mid-Atlantic, the Midwest, and California.

The transmission system originally was built to deliver power from a utility's generator across town to its distribution company. Today, the transmission system is being used to deliver power across states or entire regions. As market forces increasingly determine the location of generation sources, the transmission grid is being asked to play an even more important role in markets and the reliability of the system. In areas where markets have been restructured, customers have begun to see significant benefits. But full delivery of restructuring's benefits is being impeded by an inadequate, underinvested transmission system.

If the last 30 years are any indication, the structure of the industry and the increasing demands placed on the nation's transmission infrastructure and the people who operate and manage it are likely to continue unabated. In order to meet the challenges of the future, to continue to maintain the stable, reliable, and efficient system we have known for more than a century and to support the continued development of efficient competitive markets, U.S. industry leaders must address three significant issues:

- an aging transmission system suffering from substantial underinvestment, which is exacerbated by an out-of-date industry structure
- the need for a regulatory framework that will spur independent investment, ownership, and management of the nation's grid
- an aging workforce and the need for a succession plan to ensure the existence of the next generation of technical expertise in the industry.

ARE WE SPENDING ENOUGH?

In areas that have restructured power markets, substantial benefits have been delivered to customers in the form of lower prices, greater supplier choice, and environmental benefits, largely due to the development and operation of new, cleaner generation. There is, however, a growing recognition that the delivery of the full value of restructuring to customers has been stalled by an inadequate transmission system that was not designed for the new demands being placed on it. In fact, investment in the nation's electricity infrastructure has been declining for decades. Transmission investment has been falling for a quarter century at an average rate of almost US$50 million a year (in constant 2003 U.S. dollars), though there has been a small upturn in the last few years. Transmission investment has not kept up with load growth or generation investment in recent years, nor has it been sufficiently expanded to accommodate the advent of regional power markets (see Figure 1).

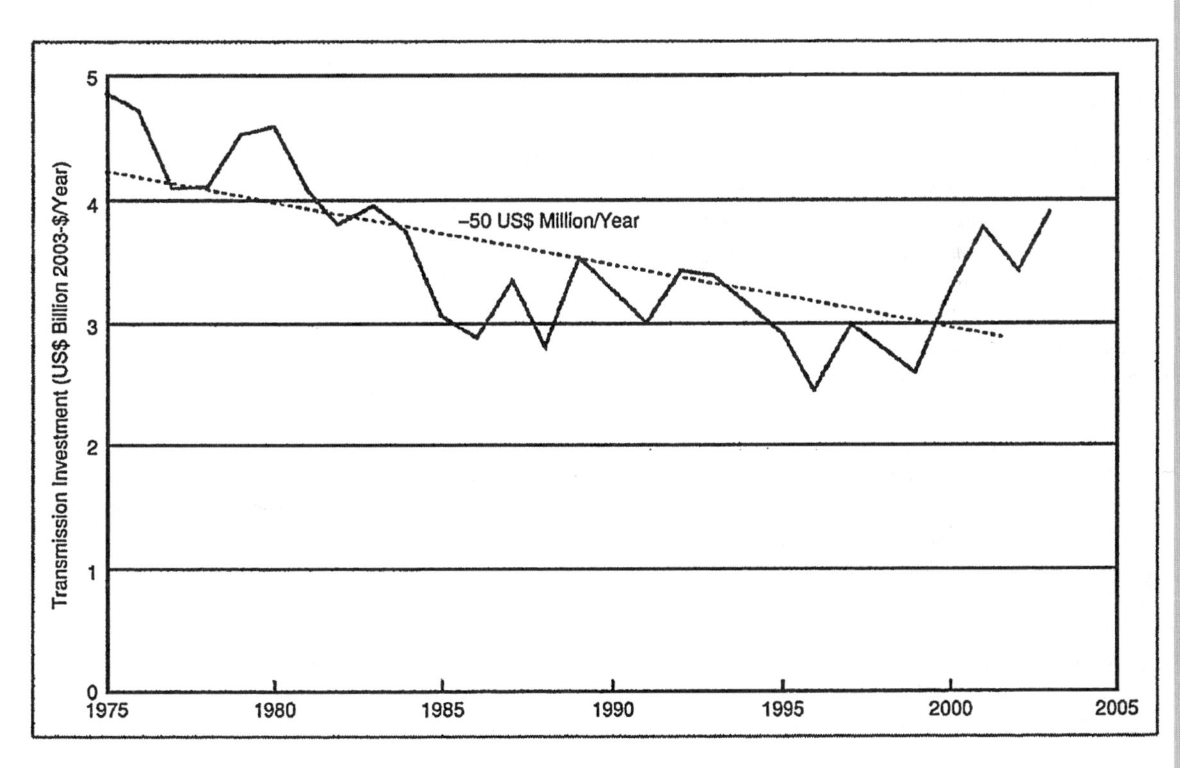

Figure 1
Annual transmission investments by investor-owned utilities, 1975–2003 (Source: Eric Hirst, "U.S. Transmission Capacity: Present Status and Future Prospects," 2004. Graph used with permission from the Edison Electric Institute, 2004. All rights reserved)

TABLE 1 Transmission investment in the United States and in international competitive markets

Country	Investment in High Voltage Transmission (>230 kV) Normalized by Load for 2004–2008 (in US$M/GW/year)	Number of Transmission-Owning Entities
New Zealand	22.0	1
England & Wales (NGT)	16.5	1
Denmark	12.5	2
Spain	12.3	1
The Netherlands	12.0	1
Norway	9.2	1
Poland	8.6	1
Finland	7.2	1
United States	4.6 (based on representative data from EEI)	450 (69 in EEI)

Outlooks for future transmission development vary, with Edison Electric Institute (EEI) data suggesting a modest increase in expected transmission investment and other sources forecasting a continued decline. Even assuming EEI's projections are realized, this level of transmission investment in the United States is dwarfed by that of other international competitive electricity markets, as shown in Table 1, and is expected to lag behind what is needed.

The lack of transmission investment has led to a high (and increasing in some areas) level of congestion-related costs in many regions. For instance, total uplift for New England is in the range of US$169 million per year, while locational installed capacity prices and reliability must-run charges are on the rise. In New York, congestion costs have increased substantially, from US$310 million in 2001 to US$525 million in 2002, US$688 million in 2003, and US$629 million in 2004. In PJM Interconnection (PJM), an RTO that administers electricity markets for all or parts of 14 states in the Northeast, Midwest, and Mid-Atlantic, congestion costs have continued to increase, even when adjusted to reflect PJM's expanding footprint into western and southern regions.

Because regions do not currently quantify the costs of constraints in the same way, it is difficult to make direct comparisons from congestion data between regions. However, the magnitude and upward trend of available congestion cost data indicates a significant and growing problem that is increasing costs to customers.

THE SYSTEM IS AGING

While we are pushing the transmission system harder, it is not getting any younger. In the northeastern United States, the bulk transmission system operates primarily at 345 kV. The majority of this system originally was constructed during the 1960s and into the early 1970s, and its substations, wires, towers, and poles are, on average, more than 40 years old. (Figure 2 shows the age of National Grid's U.S. transmission structures.) While all utilities have maintenance plans in place for these systems, ever-increasing congestion levels in many areas are making it increasingly difficult to schedule circuit outages for routine upgrades.

The combination of aging infrastructure, increased congestion, and the lack of significant expansion in transmission capacity has led to the need to carefully prioritize maintenance and construction, which in turn led to the evolution of the science of asset management, which many utilities have adopted. Asset management entails quantifying the risks of not doing work as a means to ensure that the highest priority work is performed. It has significantly helped the industry in maintaining reliability. As the assets continue to age, this combination of engineering, experience, and business risk will grow in importance to the industry. If this is not done well, the impact on utilities in terms of reliability and asset replacement will be significant.

And while asset management techniques will help in managing investment, the age issue undoubtedly will require substantial reinvestment at some point to replace the installed equipment at the end of its lifetime.

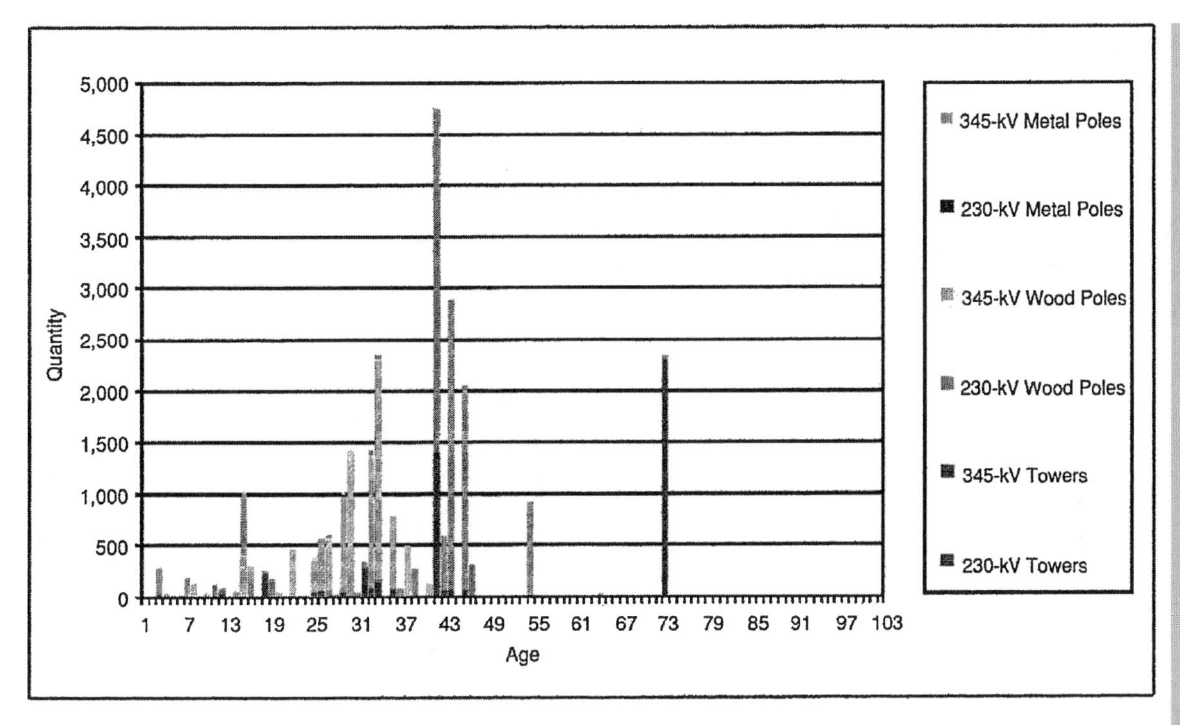

Figure 2
Age of National Grid towers and poles

TECHNOLOGY WILL HAVE A ROLE

The expansion of the transmission network in the United States will be very difficult, if not impossible, if the traditional approach of adding new overhead lines continues. Issues of land availability, concerns about property values, aesthetics, and other licensing concerns make siting new lines a difficult proposition in many areas of the United States. New approaches to expansion will be required to improve the transmission networks of the future.

Where new lines are the only answer, more underground solutions will be chosen. In some circumstances, superconducting cable will become a viable option. There are several companies, including National Grid, installing short superconducting lines to gain experience with this newly available technology and solve real problems. While it is reasonable to expect this solution to become more prevalent, it is important to recognize that it is not inexpensive.

Technology has an important role to play in utilizing existing lines and transmission corridors to increase capacity. Lightweight, high-temperature overhead conductors are now becoming available for line upgrades without significant tower modifications. Monitoring systems for real-time ratings and better computer control schemes are providing improved information to control room operators to run the system at higher load levels. The development and common use of static var compensators for voltage and reactive control, and the general use of new solid-state equipment to solve real problems are just around the corner and should add a new dimension to the traditional wires and transformers approach to addressing stability and short-term energy storage issues.

These are just a few examples of some of the exciting new technologies that will be tools for the future. It is encouraging that the development of new and innovative solutions to existing problems continues. In the future, innovation must take a leading role in developing solutions to transmission problems, and it will be important for the regulators to encourage the use of new techniques and technologies. Most of these new technologies have a higher cost than traditional solutions, which will place increasing pressure on capital investment. It will be important to ensure that appropriate cost recovery mechanisms are developed to address this issue.

INDUSTRY STRUCTURE

Another factor contributing to underinvestment in the transmission system is the tremendous fragmentation that exists in the U.S. electricity industry. There are literally hundreds of entities that own and operate transmission. The United States has more than 100 separate control areas and more than 50 regulators that oversee the nation's grid. The patchwork of ownership and operation lies in stark contrast to the interregional delivery demands that are being placed on the nation's transmission infrastructure.

Federal policymakers continue to encourage transmission owners across the nation to join RTOs. Indeed, RTO/ISO formation was intended to occupy a central role in carrying forward FERC's vision of restructuring, and an extraordinary amount of effort has been expended in making this model work. While RTOs/ISOs take a step toward an independent, coordinated transmission system, it remains unclear whether they are the best long-term solution to deliver efficient transmission system operation while ensuring reliability and delivering value to customers.

Broad regional markets require policies that facilitate and encourage active grid planning, management, and the construction of transmission upgrades both for reliability and economic needs. A strong transmission infrastructure or network platform would allow greater fuel diversity, more stable and competitive energy prices, and the relaxation (and perhaps ultimate removal) of administrative mechanisms to mitigate market power. This would also allow for common asset management approaches to the transmission system. The creation of independent transmission companies (ITCs), i.e., companies that focus on the investment in and operation of transmission independent of generation interests, would be a key institutional step toward an industry structure that appropriately views transmission as a facilitator of robust competitive electricity markets. ITCs recognize transmission as an enabler of competitive electricity markets. Policies that provide a more prominent role for such companies would align the interests of transmission owners/operators with those of customers, permitting the development of well-designed and enduring power markets that perform the function of any market, namely, to drive the efficient allocation of resources for the benefit of customers. In its policy statement released in June 2005, FERC reiterated its commitment to ITC formation to support improving the performance and efficiency of the grid.

Having no interest in financial outcomes within a power market, the ITC's goal is to deliver maximum value to customers through transmission operation and investment. With appropriate incentives, ITCs will pursue opportunities to leverage relatively small expenditures on transmission construction and management to create a healthy market and provide larger savings in the supply portion of customer's bills. They also offer benefits over nonprofit RTO/ISO models, where the incentives for efficient operation and investment may be less focused.

An ideal industry structure would permit ITCs to own, operate, and manage transmission assets over a wide area. This would allow ITCs to access economies of scale in asset investment, planning, and operations to increase throughout and enhance reliability in the most cost-effective manner. This structure would also avoid ownership fragmentation within a single market, which is a key obstacle to the introduction of performance-based rates that benefit customers by aligning the interests of transmission companies and customers in reducing congestion. This approach to "horizontal integration" of the transmission sector under a single regulated for-profit entity is key to establishing an industry structure that recognizes the transmission system as a market enabler and provider of infrastructure to support effective competitive markets. Market administration would be contracted out to another (potentially nonprofit) entity while generators, other suppliers, demand response providers, and load serving entities (LSEs) would all compete and innovate in fully functioning markets, delivering still-increased efficiency and more choices for customers.

REGULATORY ISSUES

The industry clearly shoulders much of the responsibility for determining its own future and for taking the steps necessary to ensure the robustness of the nation's transmission system. However, the industry also operates within an environment governed by substantial regulatory controls. Therefore, policymakers also will have a significant role in helping to remove the obstacles to the delivery of the full benefits of industry restructuring to customers. In order to ensure adequate transmission investment and the expansion of the system as appropriate, the following policy issues must be addressed:

- *Regional planning:* Because the transmission system is an integrated network, planning for system needs should occur on a regional basis. Regional planning recognizes that transmission investment and the benefits transmission can deliver to customers are regional in nature rather than bounded by state or service area lines. Meaningful regional planning processes also take into account the fact that transmission provides both reliability and economic benefits. Comprehensive planning processes provide for mechanisms to pursue regulated transmission solutions for reliability and economic needs in the event that the market fails to respond or is identified as unlikely to respond to these needs in a timely manner. In areas where regional system planning processes have been implemented, such as New England and PJM, progress is being made towards identifying and building transmission projects that will address regional needs and do so in a way that is cost effective for customers.
- *Cost recovery and allocation:* Comprehensive regional planning processes that identify needed transmission projects must be accompanied by cost recovery and allocation mechanisms that recognize the broad benefits of transmission and its role in supporting and enabling regional electricity markets. Mechanisms that allocate the costs of transmission investment broadly view transmission as the regional market enabler it is and should be, provide greater certainty and reduce delays in cost recovery, and, thus, remove obstacles to provide further incentives for the owners and operators of transmission to make such investment.
- *Certainty of rate recovery and state cooperation:* It is critical that transmission owners are assured certain and adequate rate recovery under a regional planning process. Independent administration of the planning processes will assure that transmission enhancements required for reliability and market efficiency do not unduly burden retail customers with additional costs. FERC and the states must work together to provide for certainty in rate recovery from ultimate customers through federal and state jurisdictional rates.
- *Incentives to encourage transmission investment, independence, and consolidation:* At a time when a significant increase in transmission investment is needed to ensure reliability, produce an adequate platform for competitive power markets and regional electricity commerce, and to promote fuel diversity and renewable sources of supply, incentives not only for investment but also for independence and consolidation of transmission are needed and warranted. Incentives should be designed to promote transmission organizations that acknowledge the benefits to customers of varying degrees of transmission independence and reward that independence accordingly. These incentives may take the form of enhanced rates of return or other financial incentives for assets managed, operated, and/or owned by an ITC.

The debate about transmission regulation will continue. Ultimately, having the correct mixture of incentives and reliability standards will be a critical factor that will determine whether or not the nation's grid can successfully tie markets together and improve the overall reliability of the bulk transmission system in the United States. The future transmission system must be able to meet the needs of customers reliably and support competitive markets that provide them with electricity efficiently. Failure to invest in the transmission system now will mean an increased likelihood of reduced reliability and higher costs to customers in the future.

WORKFORCE OF THE FUTURE

Clearly, the nation's transmission system will need considerable investment and physical work due to age, growth of the use of electricity, changing markets, and how the networks are used. As previously noted, this will lead to a required significant increase in capital spending. But another critical resource is beginning to become a concern to many in the industry, specifically the continued availability of qualified power system engineers.

Utility executives polled by the Electric Power Research Institute in 2003 estimated that 50% of the technical workforce will reach retirement in the next 5–10 years. This puts the average age near 50, with many utilities still hiring just a few college graduates each year. Looking a few years ahead, at the same time when a significant number of power engineers will be considering retirement, the need for them will be significantly increasing. The supply of power engineers will have to be great enough to replace the large numbers of those retiring in addition to the number required to respond to the anticipated increase in transmission capital spending.

Today, the number of universities offering power engineering programs has decreased. Some universities, such as Rensselaer Polytechnic Institute, no longer have separate power system engineering departments. According to the IEEE, the number of power system engineering graduates has dropped from approximately 2,000 per year in the 1980s to 500 today. Overall, the number of engineering graduates has dropped 50% in the last 15 years. Turning this situation around will require a long-term effort by many groups working together, including utilities, consultants, manufacturers, universities, and groups such as the IEEE Power Engineering Society (PES).

Part of the challenge is that utilities are competing for engineering students against other industries, such as telecommunications or computer software development, that are perceived as being more glamorous or more hip than the power industry and have no problem attracting large numbers of new engineers.

For the most part, the power industry has not done a great job of selling itself. Too often, headlines focus on negatives such as rate increases, power outages, and community relations issues related to a proposed new generation plant or transmission line. To a large extent, the industry also has become a victim of its own success by delivering electricity so reliably that the public generally takes it for granted, which makes the good news more difficult to tell. It is incumbent upon the industry to take a much more proactive role in helping its public—including talented engineering students—understand the dedication, commitment, ingenuity, and innovation that is required to keep the nation's electricity system humming. PES can play an important role in this.

On a related note, as the industry continues to develop new, innovative technologies, they should be documented and showcased to help generate excitement about the industry among college-age engineers and help attract them to power system engineering.

The utilities, consultants, and manufacturers must strengthen their relationships with strong technical institutions to continue increasing support for electrical engineering departments to offer power systems classes at the undergraduate level. In some cases, this may even require underwriting a class. Experience at National Grid has shown that when support for a class is guaranteed, the number of students who sign up typically is greater than expected. The industry needs to further support these efforts by offering presentations to students on the complexity of the power system, real problems that need to be solved, and the impact that a reliable, cost-efficient power system has on society. Sponsoring more student internships and research projects will introduce additional students and faculty to the unique challenges of the industry. In the future, the industry will have to hire more nonpower engineers and train them in the specifics of power system engineering or rely on hiring from overseas.

Finally, the industry needs to cultivate relationships with universities to assist in developing professors who are knowledgeable about the industry. This can take the form of research work, consulting, and teaching custom programs for the industry. National Grid has developed relationships with several northeastern U.S. institutions that are offering courses for graduate engineers who may not have power backgrounds. The courses can be offered online, at the university, or on site at the utility.

This problem will only get worse if industry leaders do not work together to resolve it. The industry's future depends on its ability to anticipate what lies ahead and the development of the necessary human resources to meet the challenges.

CONCLUSIONS

The electric transmission system plays a critical role in the lives of the people of the United States. It is an ever-changing system both in physical terms and how it is operated and regulated. These changes must be recognized and actions developed accordingly. Since the industry is made up of many organizations that share the system, it can be difficult to agree on action plans.

There are a few points on which all can agree. The first is that the transmission assets continue to get older and investment is not keeping up with needs when looking over a future horizon. The issue will only get worse as more lines and substations exceed the 50-year age mark. Technology development and application undoubtedly will increase as engineers look for new and creative ways to combat the congestion issues and increased electrical demand—and new overhead transmission lines will be only one of the solutions considered.

The second is that it will be important for further refinement in the restructuring of the industry to occur. The changes made since the late 1990s have delivered benefits to customers in the Northeast in the form of lower energy costs and access to greater competitive electric markets. Regulators and policymakers should recognize that independently owned, operated, managed, and widely planned networks are important to solving future problems most efficiently. Having a reliable, regional, uncongested transmission system will enable a healthy competitive marketplace.

The last, but certainly not least, concern is with the industry's future workforce. Over the last year, there has been significant discussion of the issue, but it will take a considerable effort by many to guide the future workforce into a position of appreciating the electricity industry and desiring to enter it and to ensure that the training and education systems are in place to develop the new engineers who will be required to upgrade and maintain the electric power system.

The industry has many challenges, but it also has great resources and a good reputation. Through the efforts of many and by working together through organizations such as PES, the industry can move forward to the benefit of the public and the United States as a whole.

ACKNOWLEDGMENTS

The following National Grid staff members contributed to this article: Jackie Barry, manager, transmission communications; Janet Gail Besser, vice president, regulatory affairs, U.S. Transmission; Mary Ellen Paravalos, director, regulatory policy, U.S. Transmission; Joseph Rossignoli, principal analyst, regulatory policy, U.S. Transmission.

FOR FURTHER READING

National Grid, "Transmission: The critical link. Delivering the promise of industry restructuring to customers," June 2005 [Online]. Available: http://www.nationalgridus.com/transmission_the_critical_link/

E. Hirst, "U.S. transmission capacity: Present status and future prospects," Edison Electric Inst. and U.S. Dept. Energy, Aug. 2004.

Consumer Energy Council of America, "Keeping the power flowing: Ensuring a strong transmission system to support consumer needs for cost-effectiveness, security and reliability," Jan. 2005 [Online]. Available: http://www.cecarf.org

"Electricity sector framework for the future," Electric Power Res. Inst., Aug. 2003.

J. R. Borland, "A shortage of talent," *Transmission Distribution World*, Sep. 1, 2002.

BIOGRAPHY

Christopher E. Root is senior vice president of Transmission and Distribution (T&D) Technical Services of National Grid's U.S. business. He oversees the T&D technical services organization in New England and New York. He received a B.S. in electrical engineering from Northeastern University, Massachusetts, and a master's in engineering from Rensselaer Polytechnic Institute, New York. In 1997, he completed the Program for Management Development from the Harvard University Graduate School of Business. He is a registered Professional Engineer in the states of Massachusetts and Rhode Island and is a Senior Member of the IEEE.

1.1 HISTORY OF ELECTRIC POWER SYSTEMS

In 1878, Thomas A. Edison began work on the electric light and formulated the concept of a centrally located power station with distributed lighting serving a surrounding area. He perfected his light by October 1879, and the opening of his historic Pearl Street Station in New York City on September 4, 1882, marked the beginning of the electric utility industry (see Figure 1.1). At Pearl Street, dc generators, then called dynamos, were driven by steam engines to supply an initial load of 30 kW for 110-V incandescent lighting to 59 customers in a one-square-mile (2.5-square-km) area. From this beginning in 1882 through 1972, the electric utility industry grew at a remarkable pace—a growth based on continuous reductions in the price of electricity due primarily to technological acomplishment and creative engineering.

The introduction of the practical dc motor by Sprague Electric, as well as the growth of incandescent lighting, promoted the expansion of Edison's dc systems. The development of three-wire 220-V dc systems allowed load to increase somewhat, but as transmission distances and loads continued to increase, voltage problems were encountered. These limitations of maximum distance and load were overcome in 1885 by William Stanley's development of a commercially practical transformer. Stanley installed an ac distribution system in Great Barrington, Massachusetts, to supply 150 lamps. With the transformer, the ability to transmit power at high voltage with corresponding lower current and lower line-voltage drops made ac more attractive than dc. The first single-phase ac line in the United States operated in 1889 in Oregon, between Oregon City and Portland—21 km at 4 kV.

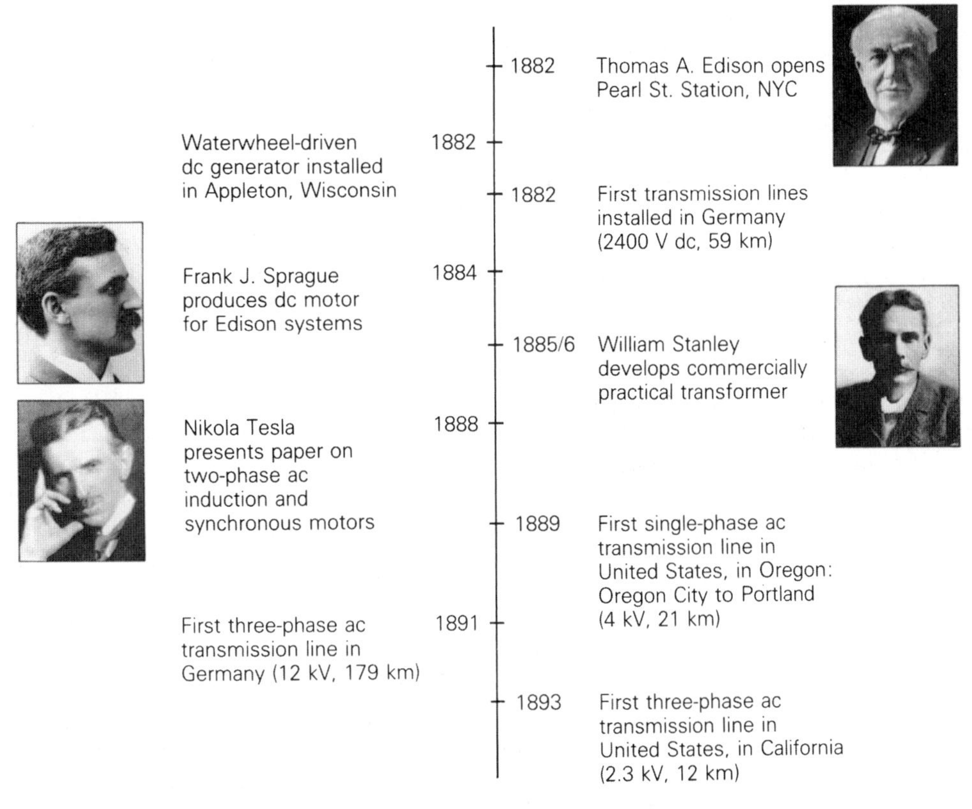

FIGURE 1.1 Milestones of the early electric utility industry [1] (H.M. Rustebakke et al., Electric Utility Systems Practice, 4th Ed. (New York: Wiley, 1983). Reprinted with permission of John Wiley & Sons, Inc. Photos courtesy of Westinghouse Historical Collection)

The growth of ac systems was further encouraged in 1888 when Nikola Tesla presented a paper at a meeting of the American Institute of Electrical Engineers describing two-phase induction and synchronous motors, which made evident the advantages of polyphase versus single-phase systems. The first three-phase line in Germany became operational in 1891, transmitting power 179 km at 12 kV. The first three-phase line in the United States (in California) became operational in 1893, transmitting power 12 km at 2.3 kV. The three-phase induction motor conceived by Tesla went on to become the workhorse of the industry.

In the same year that Edison's steam-driven generators were inaugurated, a waterwheel-driven generator was installed in Appleton, Wisconsin. Since then, most electric energy has been generated in steam-powered and in water-powered (called hydro) turbine plants. Today, steam turbines account for more than 85% of U.S. electric energy generation, whereas hydro turbines account for about 6%. Gas turbines are used in some cases to meet peak loads. Also, the addition of wind turbines into the bulk power system is expected to grow considerably in the near future.

Steam plants are fueled primarily by coal, gas, oil, and uranium. Of these, coal is the most widely used fuel in the United States due to its abundance in the country. Although many of these coal-fueled power plants were converted to oil during the early 1970s, that trend has been reversed back to coal since the 1973–74 oil embargo, which caused an oil shortage and created a national desire to reduce dependency on foreign oil. In 2008, approximately 48% of electricity in the United States was generated from coal [2].

In 1957, nuclear units with 90-MW steam-turbine capacity, fueled by uranium, were installed, and today nuclear units with 1312-MW steam-turbine capacity are in service. In 2008, approximately 20% of electricity in the United States was generated from uranium from 104 nuclear power plants. However, the growth of nuclear capacity in the United States has been halted by rising construction costs, licensing delays, and public opinion. Although there are no emissions associated with nuclear power generation, there are safety issues and environmental issues, such as the disposal of used nuclear fuel and the impact of heated cooling-tower water on aquatic habitats. Future technologies for nuclear power are concentrated on safety and environmental issues [2, 3].

Starting in the 1990s, the choice of fuel for new power plants in the United States has been natural gas due to its availability and low cost as well as the higher efficiency, lower emissions, shorter construction-lead times, safety, and lack of controversy associated with power plants that use natural gas. Natural gas is used to generate electricity by the following processes: (1) gas combustion turbines use natural gas directly to fire the turbine; (2) steam turbines burn natural gas to create steam in a boiler, which is then run through the steam turbine; (3) combined cycle units use a gas combustion turbine by burning natural gas, and the hot exhaust gases from the combustion turbine are used to boil water that operates a steam turbine; and (4) fuel cells powered by natural gas generate electricity using electrochemical reactions by passing streams of natural gas and oxidants over electrodes that are separated by an electrolyte. In 2008, approximately 21% of electricity in the United States was generated from natural gas [2, 3].

In 2008, in the United States, approximately 9% of electricity was generated by renewable sources and 1% by oil [2, 3]. Renewable sources include conventional hydroelectric (water power), geothermal, wood, wood waste, all municipal waste, landfill gas, other biomass, solar, and wind power. Renewable sources of energy cannot be ignored, but they are not expected to supply a large percentage of the world's future energy needs. On the other hand, nuclear fusion energy just may. Substantial research efforts have shown nuclear fusion energy to be a promising technology for producing safe, pollution-free, and economical electric energy later in the 21st century and beyond. The fuel consumed in a nuclear fusion reaction is deuterium, of which a virtually inexhaustible supply is present in seawater.

The early ac systems operated at various frequencies including 25, 50, 60, and 133 Hz. In 1891, it was proposed that 60 Hz be the standard frequency in the United States. In 1893, 25-Hz systems were introduced with the

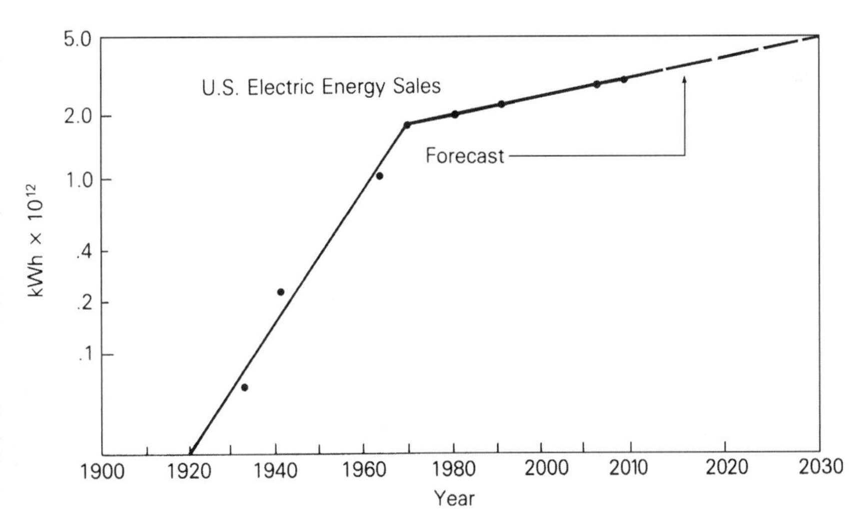

FIGURE 1.2 Growth of U.S. electric energy consumption [1, 2, 3, 5] (H. M. Rustebakke et al., Electric Utility Systems Practice, 4th ed. (New York: Wiley, 1983); U.S. Energy Information Administration, Existing Capacity by Energy Source—2008, www.eia.gov; U.S. Energy Information Administration, Annual Energy Outlook 2010 Early Release Overview, www.eia.gov; M.P. Bahrman and B.K. Johnson, "The ABCs of HVDC Transmission Technologies," IEEE Power & Energy Magazine, 5, 2 (March/April 2007), pp. 33–44)

synchronous converter. However, these systems were used primarily for railroad electrification (and many are now retired) because they had the disadvantage of causing incandescent lights to flicker. In California, the Los Angeles Department of Power and Water operated at 50 Hz, but converted to 60 Hz when power from the Hoover Dam became operational in 1937. In 1949, Southern California Edison also converted from 50 to 60 Hz. Today, the two standard frequencies for generation, transmission, and distribution of electric power in the world are 60 Hz (in the United States, Canada, Japan, Brazil) and 50 Hz (in Europe, the former Soviet republics, South America except Brazil, and India). The advantage of 60-Hz systems is that generators, motors, and transformers in these systems are generally smaller than 50-Hz equipment with the same ratings. The advantage of 50-Hz systems is that transmission lines and transformers have smaller reactances at 50 Hz than at 60 Hz.

As shown in Figure 1.2, the rate of growth of electric energy in the United States was approximately 7% per year from 1902 to 1972. This corresponds to a doubling of electric energy consumption every 10 years over the 70-year period. In other words, every 10 years the industry installed a new electric system equal in energy-producing capacity to the total of what it had built since the industry began. The annual growth rate slowed after the oil embargo of 1973–74. Kilowatt-hour consumption in the United States increased by 3.4% per year from 1972 to 1980, and by 2.1% per year from 1980 to 2008.

Along with increases in load growth, there have been continuing increases in the size of generating units (Table 1.1). The principal incentive to build larger units has been economy of scale—that is, a reduction in installed cost per kilowatt of capacity for larger units. However, there have also been steady improvements in generation efficiency. For example, in 1934 the average heat rate for steam generation in the U.S. electric industry was

TABLE 1.1

Growth of generator sizes in the United States [1] (H. M. Rustebakke et al., Electric Utility Systems Practice, 4th Ed. (New York: Wiley, 1983). Reprinted with permission of John Wiley & Sons, Inc.)

Hydroelectric Generators		Generators Driven by Single-Shaft, 3600 r/min Fossil-Fueled Steam Turbines	
Size (MVA)	Year of Installation	Size (MVA)	Year of Installation
4	1895	5	1914
108	1941	50	1937
158	1966	216	1953
232	1973	506	1963
615	1975	907	1969
718	1978	1120	1974

18,938 kJ/kWh, which corresponds to 19% efficiency. By 1991, the average heat rate was 10,938 kJ/kWh, which corresponds to 33% efficiency. These improvements in thermal efficiency due to increases in unit size and in steam temperature and pressure, as well as to the use of steam reheat, have resulted in savings in fuel costs and overall operating costs.

There have been continuing increases, too, in transmission voltages (Table 1.2). From Edison's 220-V three-wire dc grid to 4-kV single-phase and 2.3-kV three-phase transmission, ac transmission voltages in the United States have risen progressively to 150, 230, 345, 500, and now 765 kV. And ultra-high voltages (UHV) above 1000 kV are now being studied. The incentives for increasing transmission voltages have been: (1) increases in transmission distance and transmission capacity, (2) smaller line-voltage drops, (3) reduced line losses, (4) reduced right-of-way requirements per MW transfer, and (5) lower capital and operating costs of transmission. Today, one 765-kV three-phase line can transmit thousands of megawatts over hundreds of kilometers.

TABLE 1.2

History of increases in three-phase transmission voltages in the United States [1] (H. M. Rustebakke et al., Electric Utility Systems Practice, 4th Ed. (New York: Wiley, 1983). Reprinted with permission of John Wiley & Sons, Inc.)

Voltage (kV)	Year of Installation
2.3	1893
44	1897
150	1913
165	1922
230	1923
287	1935
345	1953
500	1965
765	1969

The technological developments that have occurred in conjunction with ac transmission, including developments in insulation, protection, and control, are in themselves important. The following examples are noteworthy:

1. The suspension insulator
2. The high-speed relay system, currently capable of detecting short-circuit currents within one cycle (0.017 s)
3. High-speed, extra-high-voltage (EHV) circuit breakers, capable of interrupting up to 63-kA three-phase short-circuit currents within two cycles (0.033 s)
4. High-speed reclosure of EHV lines, which enables automatic return to service within a fraction of a second after a fault has been cleared
5. The EHV surge arrester, which provides protection against transient overvoltages due to lightning strikes and line-switching operations

6. Power-line carrier, microwave, and fiber optics as communication mechanisms for protecting, controlling, and metering transmission lines

7. The principle of insulation coordination applied to the design of an entire transmission system

8. Energy control centers with supervisory control and data acquisition (SCADA) and with automatic generation control (AGC) for centralized computer monitoring and control of generation, transmission, and distribution

9. Automated distribution features, including advanced metering infrastructure (AMI), reclosers and remotely controlled sectionalizing switches with fault-indicating capability, along with automated mapping/facilities management (AM/FM) and geographic information systems (GIS) for quick isolation and identification of outages and for rapid restoration of customer services

10. Digital relays capable of circuit breaker control, data logging, fault locating, self-checking, fault analysis, remote query, and relay event monitoring/recording.

In 1954, the first modern high-voltage dc (HVDC) transmission line was put into operation in Sweden between Vastervik and the island of Gotland in the Baltic sea; it operated at 100 kV for a distance of 100 km. The first HVDC line in the United States was the $\pm$400-kV (now $\pm$500 kV), 1360-km Pacific Intertie line installed between Oregon and California in 1970. As of 2008, seven other HVDC lines up to 500 kV and eleven back-to-back ac-dc links had been installed in the United States, and a total of 57 HVDC lines up to 600 kV had been installed worldwide [4].

For an HVDC line embedded in an ac system, solid-state converters at both ends of the dc line operate as rectifiers and inverters. Since the cost of an HVDC transmission line is less than that of an ac line with the same capacity, the additional cost of converters for dc transmission is offset when the line is long enough. Studies have shown that overhead HVDC transmission is economical in the United States for transmission distances longer than about 600 km. However, HVDC also has the advantage that it may be the only feasible method to:

1. interconnect two asynchronous networks;
2. utilize long underground or underwater cable circuits;
3. bypass network congestion;
4. reduce fault currents;
5. share utility rights-of-way without degrading reliability; and
6. mitigate environmental concerns [5].

In the United States, electric utilities grew first as isolated systems, with new ones continuously starting up throughout the country. Gradually, however,

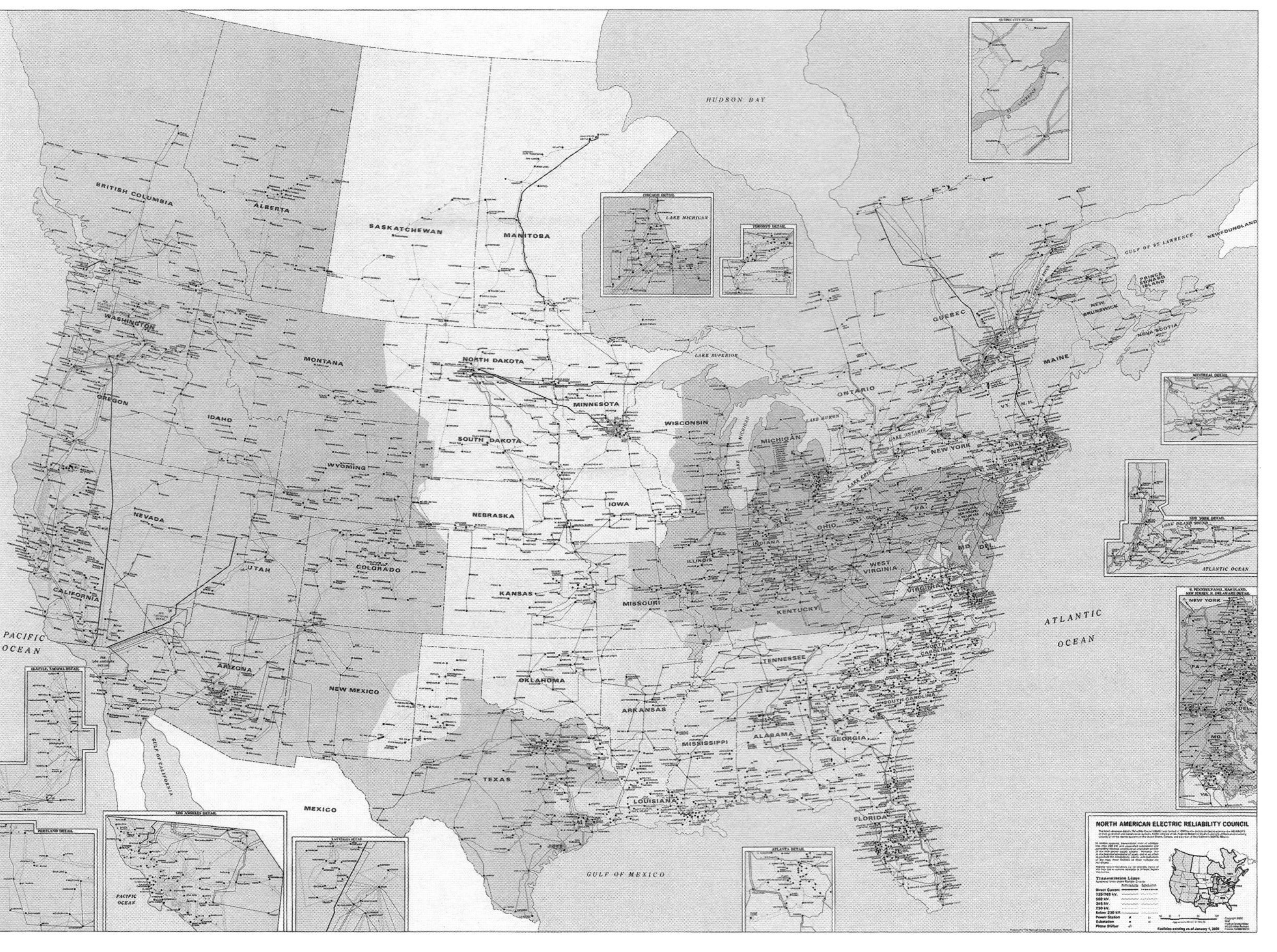

FIGURE 1.3 Major transmission in the United States—2000 [8] (© North American Electric Reliability Council. Reprinted with permission)

neighboring electric utilities began to interconnect, to operate in parallel. This improved both reliability and economy. Figure 1.3 shows major 230-kV and higher-voltage, interconnected transmission in the United States in 2000. An interconnected system has many advantages. An interconnected utility can draw upon another's rotating generator reserves during a time of need (such as a sudden generator outage or load increase), thereby maintaining continuity of service, increasing reliability, and reducing the total number of generators that need to be kept running under no-load conditions. Also, interconnected utilities can schedule power transfers during normal periods to take advantage of energy-cost differences in respective areas, load diversity, time zone differences, and seasonal conditions. For example, utilities whose generation is primarily hydro can supply low-cost power during high-water periods in spring/summer, and can receive power from the interconnection during low-water periods in fall/winter. Interconnections also allow shared ownership of larger, more efficient generating units.

While sharing the benefits of interconnected operation, each utility is obligated to help neighbors who are in trouble, to maintain scheduled intertie transfers during normal periods, and to participate in system frequency regulation.

In addition to the benefits/obligations of interconnected operation, there are disadvantages. Interconnections, for example, have increased fault currents that occur during short circuits, thus requiring the use of circuit breakers with higher interrupting capability. Furthermore, although overall system reliability and economy have improved dramatically through interconnection, there is a remote possibility that an initial disturbance may lead to a regional blackout, such as the one that occurred in August 2003 in the northeastern United States and Canada.

1.2 PRESENT AND FUTURE TRENDS

Present trends indicate that the United States is becoming more electrified as it shifts away from a dependence on the direct use of fossil fuels. The electric power industry advances economic growth, promotes business development and expansion, provides solid employment opportunities, enhances the quality of life for its users, and powers the world. Increasing electrification in the United States is evidenced in part by the ongoing digital revolution. Today the United States electric power industry is a robust, $342-billion-plus industry that employs nearly 400,000 workers. In the United States economy, the industry represents 3% of real gross domestic product (GDP) [6].

As shown in Figure 1.2, the growth rate in the use of electricity in the United States is projected to increase by about 1% per year from 2008 to 2030 [2]. Although electricity forecasts for the next ten years are based on economic and social factors that are subject to change, 1% annual growth rate is considered necessary to generate the GDP anticipated over that period. Variations in longer-term forecasts of 0.5 to 1.5% annual growth from 2008 to 2030 are based on low-to-high ranges in economic growth. Following a recent rapid decline in natural gas prices, average delivered electricity prices are projected to fall sharply from 9.8 cents per kilowatt-hour in 2008 to 8.6 cents per kilowatt-hour in 2011 and remain below 9.0 cents per kilowatt-hour through 2020 [2, 3].

Figure 1.4 shows the percentages of various fuels used to meet U.S. electric energy requirements for 2008 and those projected for 2015 and 2030. Several trends are apparent in the chart. One is the continuing use of coal. This trend is due primarily to the large amount of U.S. coal reserves, which, according to some estimates, is sufficient to meet U.S. energy needs for the next 500 years. Implementation of public policies that have been proposed to reduce carbon dioxide emissions and air pollution could reverse this trend. Another trend is the continuing consumption of natural gas in the long term with gas-fired turbines that are safe, clean, and more efficient than competing technologies. Regulatory policies to lower greenhouse gas emissions could accelerate a switchover from coal to gas, but that would require an increasing supply of deliverable natural gas. A slight percentage decrease in nuclear fuel consumption is also evident. No new nuclear plant has been

FIGURE 1.4

Electric energy generation in the United States, by principal fuel types [2, 3] (U.S. Energy Information Administration, Existing Capacity by Energy Source—2008, www.eia.gov; U.S. Energy Information Administration, Annual Energy Outlook 2010 Early Release Overview, www.eia.gov)

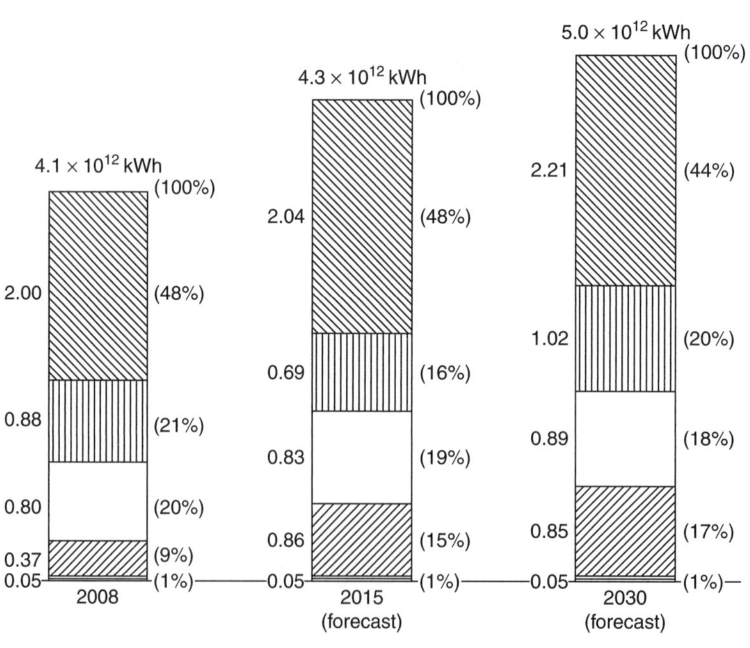

ordered in the United States for more than 30 years. The projected growth from 0.80×10^{12} kWh in 2008 to 0.89×10^{12} kWh in 2030 in nuclear generation is based on uprates at existing plants and some new nuclear capacity that is cost competitive. Safety concerns will require passive or inherently safe reactor designs with standardized, modular construction of nuclear units. Also shown in Figure 1.4 is an accelerating increase in electricity generation from renewable resources in response to federal subsidies supported by many state requirements for renewable generation.

Figure 1.5 shows the 2008 and projected 2015 U.S. generating capability by principal fuel type. As shown, total U.S. generating capacity is projected to reach 1,069 GW (1 GW = 1000 MW) by the year 2015. This represents a 0.8% annual projected growth in generating capacity, which is slightly above the 0.7% annual projected growth in electric energy production. The projected increase in generating capacity together with lowered load forecasts have contributed to generally improved generating capacity reserve margins for most of the United States and North America [2, 3, 7].

As of 2008, there were 584,093 circuit km of existing transmission (above 100 kV) in the United States, with an additional 50,265 circuit km (already under construction, planned, and conceptual) projected for the ten-year period from 2008 to 2018. The North American Electric Reliability Council (NERC) has identified bulk power system reliability and the integration of variable renewable generation (particularly wind and solar generation)

FIGURE 1.5

Installed generating capability in the United States by principal fuel types [2] (U.S. Energy Information Administration, Existing Capacity by Energy Source—2008, www.eia.gov)

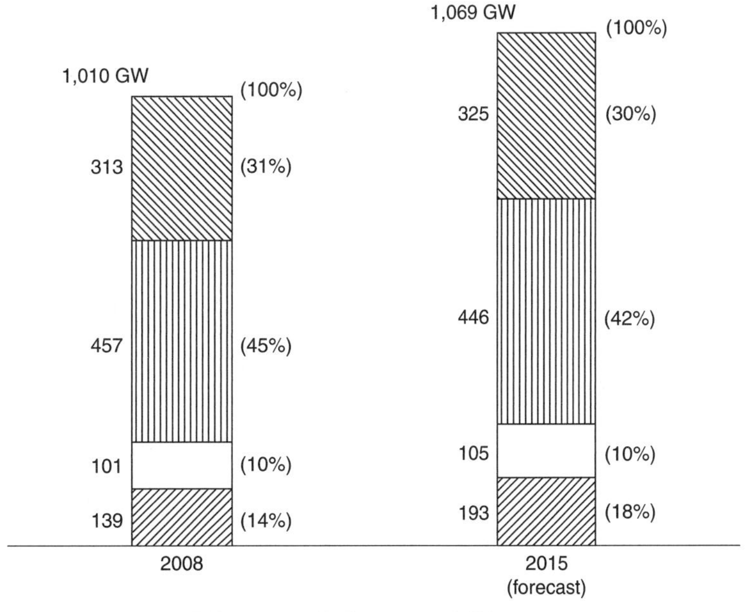

as the predominant reasons for projected transmission additions and upgrades. NERC has concluded that while recent progress has been made in the development of transmission, much work will be required to ensure that planned and conceptual transmission is sited and built. NERC also concludes that significant transmission will be required to "unlock" projected renewable generation resources. Without this transmission, the integration of variable generation resources could be limited [7].

Siting of new bulk power transmission lines has unique challenges due to their high visibility, their span through multiple states, and potentially the amount of coordination and cooperation required among multiple regulating agencies and authorities. A recent court decision to limit the Federal Energy Regulatory Commission's (FERC's) siting authority will lengthen the permit issuing process and cause new transmission projects, particularly multiple-state or regional projects from moving forward in timely manner. This creates a potential transmission congestion issue and challenges the economic viability of new generation projects [7].

Growth in distribution construction roughly correlates with growth in electric energy construction. During the last two decades, many U.S. utilities converted older 2.4-, 4.1-, and 5-kV primary distribution systems to 12 or 15 kV. The 15-kV voltage class is widely preferred by U.S. utilities for new installations; 25 kV, 34.5 kV, and higher primary distribution voltages are also utilized. Secondary distribution reduces the voltage for utilization by commercial and residential customers. Common secondary distribution voltages in the United States are 240/120 V, single-phase, three-wire; 208Y/120 V, three-phase, four-wire; and 480Y/277 V, three-phase, four-wire.

Transmission and distribution grids in the United States as well as other industrialized countries are aging and being stressed by operational uncertainties and challenges never envisioned when they were developed many decades ago. There is a growing consensus in the power industry and among many governments that smart grid technology is the answer to the uncertainties and challenges. A smart grid is characterized by the follolwing attributes:

1. Self-healing from power system disturbances;
2. Enables active participation by consumers in demand response;
3. Operates resiliently against both physical and cyber attacks;
4. Provides quality power that meets 21st century needs;
5. Accommodates all generation and energy storage technologies;
6. Enables new products, services, and markets; and
7. Optimizes asset utilization and operating efficiency.

The objective of a smart grid is to provide reliable, high-quality electric power to digital societies in an environmentally friendly and sustainable manner [9].

Utility executives polled by the Electric Power Research Institute (EPRI) in 2003 estimated that 50% of the electric-utility technical workforce in the United States will reach retirement in the next five to ten years. And according to the Institute of Electrical and Electronics Engineers (IEEE), the number of

U.S. power system engineering graduates has dropped from approximately 2,000 per year in the 1980s to 500 in 2006. The continuing availability of qualified power system engineers is a critical resource to ensure that transmission and distribution systems are maintained and operated efficiently and reliably [8].

1.3 ELECTRIC UTILITY INDUSTRY STRUCTURE

The case study at the beginning of this chapter describes the restructuring of the electric utility industry that has been ongoing in the United States. The previous structure of large, vertically integrated monopolies that existed until the last decade of the twentieth century is being replaced by a horizontal structure with generating companies, transmission companies, and distribution companies as separate business facilities.

In 1992, the United States Congress passed the Energy Policy Act, which has shifted and continues to further shift regulatory power from the state level to the federal level. The 1992 Energy Policy Act mandates the Federal Energy Regulatory Commission (FERC) to ensure that adequate transmission and distribution access is available to Exempt Wholesale Generators (EWGs) and nonutility generation (NUG). In 1996, FERC issued the "MegaRule," which regulates Transmission Open Access (TOA).

TOA was mandated in order to facilitate competition in wholesale generation. As a result, a broad range of Independent Power Producers (IPPs) and cogenerators now submit bids and compete in energy markets to match electric energy supply and demand. In the future, the retail structure of power distribution may resemble the existing structure of the telephone industry; that is, consumers would choose which generator to buy power from. Also, with demand-side metering, consumers would know the retail price of electric energy at any given time and choose when to purchase it.

Overall system reliability has become a major concern as the electric utility industry adapts to the new horizontal structure. The North American Electric Reliability Council (NERC), which was created after the 1965 Northeast blackout, is responsible for maintaining system standards and reliability. NERC coordinates its efforts with FERC and other organizations such as the Edison Electric Institute (EEI) [10].

As shown in Figure 1.3, the transmission system in North America is interconnected in a large power grid known as the North American Power Systems Interconnection. NERC divides this grid into ten geographic regions known as coordinating councils (such as WSCC, the Western Systems Coordinating Council) or power pools (such as MAPP, the Mid-Continent Area Power Pool). The councils or pools consist of several neighboring utility companies that jointly perform regional planning studies and operate jointly to schedule generation.

The basic premise of TOA is that transmission owners treat all transmission users on a nondiscriminatory and comparable basis. In December 1999, FERC issued Order 2000, which calls for companies owning transmission systems to put transmission systems under the control of Regional Transmission Organizations (RTOs). Several of the NERC regions have either established Independent System Operators (ISOs) or planned for ISOs to operate the transmission system and facilitate transmission services. Maintenance of the transmission system remains the responsibility of the transmission owners.

At the time of the August 14, 2003 blackout in the northeastern United States and Canada, NERC reliability standards were voluntary. In August 2005, the U.S. Federal government passed the Energy Policy Act of 2005, which authorizes the creation of an electric reliability organization (ERO) with the statutory authority to enforce compliance with reliability standards among all market participants. As of June 18, 2007, FERC granted NERC the legal authority to enforce reliability standards with all users, owners, and operators of the bulk power system in the United States, and made compliance with those standards mandatory and enforceable. Reliability standards are also mandatory and enforceable in Ontario and New Brunswick, and NERC is seeking to achieve comparable results in the other Canadian provinces.

The objectives of electric utility restructuring are to increase competition, decrease regulation, and in the long run lower consumer prices. There is a concern that the benefits from breaking up the old vertically integrated utilities will be unrealized if the new unbundled generation and transmission companies are able to exert market power. Market power refers to the ability of one seller or group of sellers to maintain prices above competitive levels for a significant period of time, which could be done via collusion or by taking advantage of operational anomalies that create and exploit transmission congestion. Market power can be eliminated by independent supervision of generation and transmission companies, by ensuring that there are an ample number of generation companies, by eliminating transmission congestion, and by creating a truly competitive market, where the spot price at each node (bus) in the transmission system equals the marginal cost of providing energy at that node, where the energy provider is any generator bidding into the system [11].

1.4 COMPUTERS IN POWER SYSTEM ENGINEERING

As electric utilities have grown in size and the number of interconnections has increased, planning for future expansion has become increasingly complex. The increasing cost of additions and modifications has made it imperative that utilities consider a range of design options, and perform detailed studies of the effects on the system of each option, based on a number of assumptions:

normal and abnormal operating conditions, peak and off-peak loadings, and present and future years of operation. A large volume of network data must also be collected and accurately handled. To assist the engineer in this power system planning, digital computers and highly developed computer programs are used. Such programs include power-flow, stability, short-circuit, and transients programs.

Power-flow programs compute the voltage magnitudes, phase angles, and transmission-line power flows for a network under steady-state operating conditions. Other results, including transformer tap settings and generator reactive power outputs, are also computed. Today's computers have sufficient storage and speed to efficiently compute power-flow solutions for networks with 100,000 buses and 150,000 transmission lines. High-speed printers then print out the complete solution in tabular form for analysis by the planning engineer. Also available are interactive power-flow programs, whereby power-flow results are displayed on computer screens in the form of single-line diagrams; the engineer uses these to modify the network with a mouse or from a keyboard and can readily visualize the results. The computer's large storage and high-speed capabilities allow the engineer to run the many different cases necessary to analyze and design transmission and generation-expansion options.

Stability programs are used to study power systems under disturbance conditions to determine whether synchronous generators and motors remain in synchronism. System disturbances can be caused by the sudden loss of a generator or transmission line, by sudden load increases or decreases, and by short circuits and switching operations. The stability program combines power-flow equations and machine-dynamic equations to compute the angular swings of machines during disturbances. The program also computes critical clearing times for network faults, and allows the engineer to investigate the effects of various machine parameters, network modifications, disturbance types, and control schemes.

Short-circuits programs are used to compute three-phase and line-to-ground faults in power system networks in order to select circuit breakers for fault interruption, select relays that detect faults and control circuit breakers, and determine relay settings. Short-circuit currents are computed for each relay and circuit-breaker location, and for various system-operating conditions such as lines or generating units out of service, in order to determine minimum and maximum fault currents.

Transients programs compute the magnitudes and shapes of transient overvoltages and currents that result from lightning strikes and line-switching operations. The planning engineer uses the results of a transients program to determine insulation requirements for lines, transformers, and other equipment, and to select surge arresters that protect equipment against transient overvoltages.

Other computer programs for power system planning include relay-coordination programs and distribution-circuits programs. Computer programs for generation-expansion planning include reliability analysis and loss-of-load probability (LOLP) programs, production cost programs, and investment cost programs.

1.5 POWERWORLD SIMULATOR

PowerWorld Simulator (PowerWorld) version 15 is a commercial-grade power system analysis and simulation package that accompanies this text. The purposes of integrating PowerWorld with the text are to provide computer solutions to examples in the text, to extend the examples, to demonstrate topics covered in the text, to provide a software tool for more realistic design projects, and to provide the readers with experience using a commercial grade power system analysis package. To use this software package, you must first install PowerWorld, along with all of the necessary case files onto your computer. The PowerWorld software and case files can be downloaded by going to the www.powerworld.com/GloverSarmaOverbye webpage, and clicking on the **DownLoad PowerWorld Software and Cases for the 5th Edition** button. The remainder of this section provides the necessary details to get up and running with PowerWorld.

EXAMPLE 1.1 Introduction to PowerWorld Simulator

After installing PowerWorld, double-click on the PW icon to start the program. Power system analysis requires, of course, that the user provide the program with a model of the power system. With PowerWorld, you can either build a new case (model) from scratch or start from an existing case. Initially, we'll start from an existing case. PowerWorld uses the common Ribbon user interface in which common commands, such as opening or saving a case, are available by clicking on the blue and white PowerWorld icon in the upper left-hand corner. So to open a case click on the icon and select **Open Case.** This displays the Open Dialog. Select the Example 1.1 case in the Chapter 1 directory, and then click Open. The display should look similar to Figure 1.6.

For users familiar with electric circuit schematics it is readily apparent that Figure 1.6 does NOT look like a traditional schematic. This is because the system is drawn in what is called one-line diagram form. A brief explanation is in order. Electric power systems range in size from small dc systems with peak power demands of perhaps a few milliwatts (mW) to large continent-spanning interconnected ac systems with peak demands of hundreds of Gigawatts (GW) of demand (1 GW $= 1 \times 10^9$ Watt). The subject of this book and also PowerWorld are the high voltage, high power, interconnected ac systems. Almost without exception these systems operate using three-phase ac power at either 50 or 60 Hz. As discussed in Chapter 2, a full analysis of an arbitrary three-phase system requires consideration of each of the three phases. Drawing such systems in full schematic form quickly gets excessively complicated. Thankfully, during normal operation three-phase systems are usually balanced. This permits the system to be accurately modeled as an equivalent single-phase system (the details are discussed in Chapter 8, *Symmetrical Components*). Most power system analysis packages, including PowerWorld,

FIGURE 1.6

Example power system

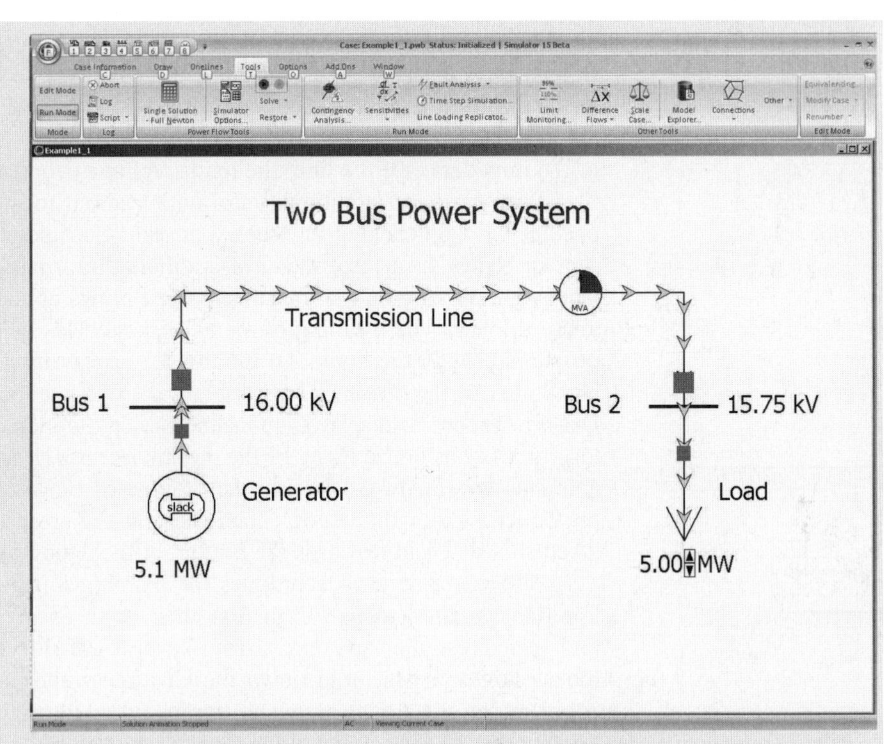

use this approach. Then connections between devices are then drawn with a single line joining the system devices, hence the term "one-line" diagram. However, do keep in mind that the actual systems are three phase.

Figure 1.6 illustrates how the major power system components are represented in PowerWorld. Generators are shown as a circle with a "dog-bone" rotor, large arrows represent loads, and transmission lines are simply drawn as lines. In power system terminology, the nodes at which two or more devices join are called buses. In PowerWorld thicker lines usually represent buses; the bus voltages are shown in kilovolts (kV) in the fields immediately to the right of the buses. In addition to voltages, power engineers are also concerned with how power flows through the system (the solution of the power flow problem is covered in Chapter 6, *Power Flows*). In PowerWorld, power flows can be visualized with arrows superimposed on the generators, loads, and transmission lines. The size and speed of the arrows indicates the direction of flow. One of the unique aspects of PowerWorld is its ability to animate power systems. To start the animation, select the **Tools** tab on the Ribbon and then click on the green and black arrow button above **Solve** (i.e., the "Play" button). The one-line should spring to life! While the one-line is being animated you can interact with the system. Figure 1.6 represents a simple power system in which a generator is supplying power to a load through a 16 kV distribution system feeder. The solid red blocks on the line and load represent circuit breakers. To open, a circuit breaker simply click on it. Since the load is series connected to the generator, clicking on any of the circuit breakers isolates the load from the generator resulting in a blackout. To restore the system click again on the circuit breaker to close it and then again select the button on the **Tools** ribbon. To vary the load click on the up or down arrows between the load value and the "MW" field. Note that because of the impedance of the line, the load's voltage drops as its value is increased.

You can view additional information about most of the elements on the one-line by right-clicking on them. For example right-clicking on the generator symbol brings up a local menu of additional information about the generator, while right-clicking on the transmission line brings up local menu of information about the line. The meaning of many of these fields will become clearer as you progress through the book. To modify the display itself simply right-click on a blank area of the one-line. This displays the one-line local menu. Select **Oneline Display Options** to display the Oneline Display Options Dialog. From this dialog you can customize many of the display features. For example, to change the animated flow arrow color select the "Animated Flows" from the options shown on the left side of the dialog. Then click on the green colored box next to the "Actual MW" field (towards the bottom of the dialog) to change its color.

There are several techniques for panning and/or zooming on the one-line. One method to pan is to first click in an empty portion of the display and then press the keyboard arrow keys in the direction you would like to move. To zoom just hold down the Ctrl key while pressing the up arrow to zoom in, or the down arrow to zoom out. Alternatively you can drag the one-line by clicking and holding the left mouse button down and then moving the mouse–the one-line should follow. To go to a favorite view from the one-line local menu select the **Go To View** to view a list of saved views.

If you would like to retain your changes after you exit PowerWorld you need to save the results. To do this, select the PowerWorld icon in the upper left portion of the Ribbon and then **Save Case As**; enter a different file name so as to not overwrite the initial case. One important note: PowerWorld actually saves the information associated with the power system model itself in a different file from the information associated with the one-line. The power system model is stored in *.pwb files (PowerWorld binary file) while the one-line display information is stored in *.pwd files (PowerWorld display file). For all the cases discussed in this book, the names of both files should be the same (except the different extensions). The reason for the dual file system is to provide flexibility. With large system models, it is quite common for a system to be displayed using multiple one-line diagrams. Furthermore, a single one-line diagram might be used at different times to display information about different cases. ■

EXAMPLE 1.2 PowerWorld Simulator—Edit Mode

PowerWorld has two major modes of operations. The Run Mode, which was just introduced, is used for running simulations and performing analysis. The Edit Mode, which is used for modifying existing cases and building new cases, is introduced in this example. To switch to the Edit Mode click on the

Edit Mode button, which is located in the upper left portion of the display immediately below the PowerWorld icon. We'll use the edit mode to add an additional bus and load as well as two new lines to the Example 1.1 system.

When switching to the Edit Mode notice that the Ribbon changes slightly, with several of the existing buttons and icons disabled and others enabled. Also, the one-line now has a superimposed grid to help with alignment (the grid can be customized using the Grid/Highlight Unlinked options category on the One-line Display Options Dialog). In the Edit Mode, we will first add a new bus to the system. This can be done graphically by first selecting the **Draw** tab, then clicking on the **Network** button and selecting **Bus**. Once this is done, move the mouse to the desired one-line location and click (note the **Draw** tab is only available in the Edit Mode). The Bus Options dialog then appears. This dialog is used to set the bus parameters. For now leave all the bus fields at their default values, except set Bus Name to "Bus 3" and set the nominal voltage to 16.0; note that the number for this new bus was automatically set to the one greater than the highest bus number in the case. The one-line should look similar to Figure 1.7. You may wish to save your case now to avoid losing your changes.

By default, when a new bus is inserted a "bus field" is also inserted. Bus fields are used to show information about buses on the one-lines. In this case the new field shows the bus name, although initially in rather small fonts. To change the field's font size click on the field to select it, and then select the **Format** button (on the Draw Ribbon) to display the Format dialog. Click on the **Font** tab and change the font's size to a larger value to make it easier to see.

FIGURE 1.7

Example 1.2—Edit Mode view with new bus

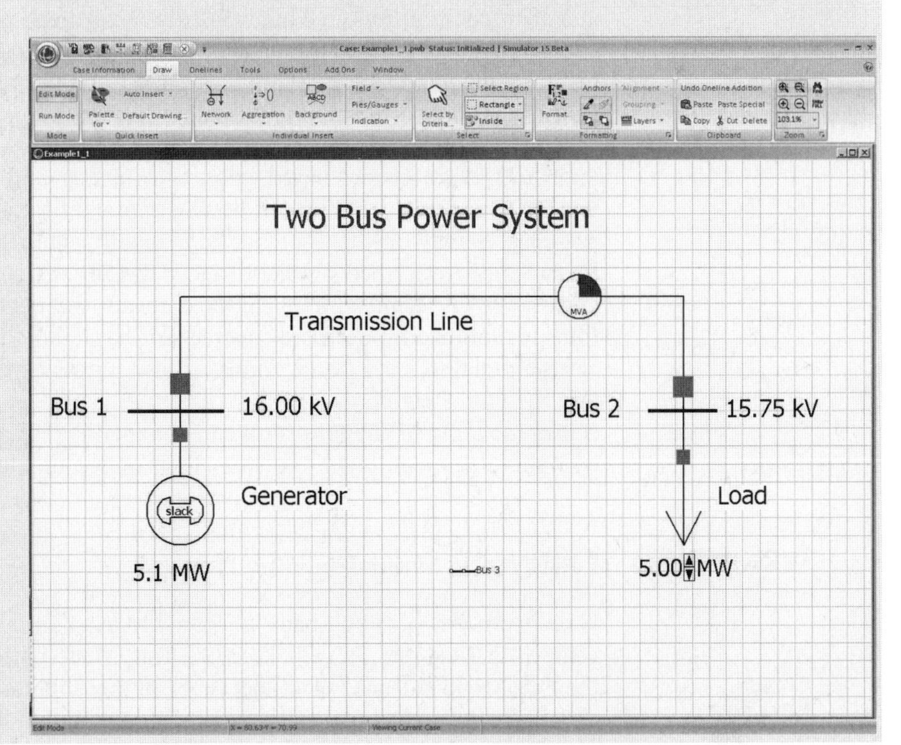

You can also change the size of the bus itself using the Format dialog, Display/Size tab. Since we would also like to see the bus voltage magnitude, we need to add an additional bus field. On the Draw ribbon select **Field, Bus Field**, and then click near bus 3. This displays the Bus Field Options dialog. Make sure the bus number is set to 3, and that the "Type of Field" is Bus Voltage. Again, resize with the **Format, Font** dialog.

Next, we'll insert some load at bus 3. This can be done graphically by selecting **Network, Load**, and then clicking on bus 3. The Load Options dialog appears, allowing you to set the load parameters. Note that the load was automatically assigned to bus 3. Leave all the fields at their default values, except set the orientation to "Down," and enter 10.0 in the Constant Power column MW Value field. As the name implies, a constant power load treats the load power as being independent of bus voltage; constant power load models are commonly used in power system analysis. By default PowerWorld "anchors" each load symbol to its bus. This is a handy feature when changing a drawing since when you drag the bus the load and all associated fields move as well. Note that two fields showing the load's real (MW) and reactive (Mvar) power were also auto-inserted with the load. Since we won't be needing the reactive field right now, select this field and then select click **Delete** (located towards the right side of the **Tools** Ribbon) to remove it. You should also resize the MW field using the **Format, Font** command.

Now we need to join the bus 3 load to the rest of the system. We'll do this by adding a line from bus 2 to bus 3. Select **Network, Transmission Line** and then click on bus 2. This begins the line drawing. During line drawing PowerWorld adds a new line segment for each mouse click. After adding several segments place the cursor on bus 3 and double-click. The Transmission Line/Transformer Options dialog appears allowing you to set the line parameters. Note that PowerWorld should have automatically set the "from" and "to" bus numbers based upon the starting and ending buses (buses 2 and 3). If these values have not been set automatically then you probably did not click exactly on bus 2 or bus 3; manually enter the values. Next, set the line's Series Resistance (R) field to 0.3, the Series Reactance (X) field to 0.6, and the MVA Limits Limit (A) field to 20 (the details of transformer and transmission line modeling is covered in Chapters 3 through 5). Select OK to close the dialog. Note that Simulator also auto-inserted two circuit breakers and a round "pie chart" symbol. The pie charts are used to show the percentage loading of the line. You can change the display size for these objects by right-clicking on them to display their option dialogs. ■

EXAMPLE 1.3 PowerWorld Simulator—Run Mode

Next, we need to switch back to Run Mode to animate the new system developed in Example 1.2. Click on the **Run Mode** button (immediately below the **Edit Mode** button), select the **Tools** on the ribbon and then click the green and black button above **Solve** to start the simulation. You should see the arrows flow from bus 1 to bus 2 to bus 3. Note that the total generation is now about 16.2 MW, with 15 MW flowing to the two loads and 1.2 MW lost to the wire resistance. To add the load variation arrows to the bus 3 load right click on the load MW field (not the load arrow itself) to display the field's local menu. Select **Load Field Information Dialog** to view the Load Field Options dialog. Set the "Delta per Mouse Click" field to "1.0," which will change the load by one MW per click on the up/down arrows. You may also like to set the "Digits to Right of Decimal" to 2 to see more digits in the load field. Be sure to save your case. The new system now has one generator and two loads. The system is still radial, meaning that a break anywhere on the wire joining bus 1 to bus 2 would result in a blackout of all the loads. Radial power systems are quite common in the lower voltage distribution systems. At higher voltage levels, networked systems are typically used. In a networked system, each load has at least two possible sources of power. We can convert our system to a networked system simply by adding a new line from bus 1 to bus 3. To do this switch back to Edit Mode and then repeat the previous line insertion process except you should start at bus 1 and end at bus 3; use the same line parameters as for the bus 2 to 3 line. Also before returning to Run Mode, right click on the blue "Two Bus Power System" title and change it to "Three Bus Power System." Return to Run Mode and again solve. Your final system should look similar to the system shown in Figure 1.8. Note that now you can open any single line and still supply both loads—a nice increase in reliability!

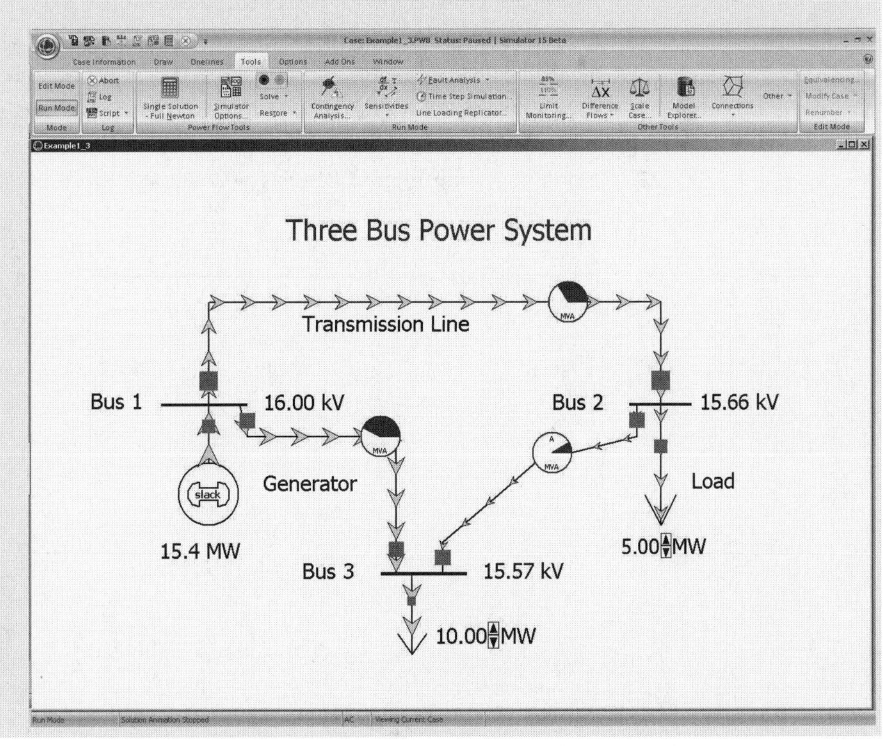

FIGURE 1.8

Example 1.3—new three-bus system

With this introduction you now have the skills necessary to begin using PowerWorld to interactively learn about power systems. If you'd like to take a look at some of the larger systems you'll be studying, open PowerWorld case Example 6.13. This case models a power system with 37 buses. Notice that when you open any line in the system the flow of power immediately redistributes to continue to meet the total load demand. ■

REFERENCES

1. H. M. Rustebakke et al., *Electric Utility Systems Practice*, 4th ed. (New York: Wiley, 1983). Photos courtesy of Westinghouse Historical Collection.
2. U.S. Energy Information Administration, *Existing Capacity by Energy Source—2008*, www.eia.gov.
3. U.S. Energy Information Administration, *Annual Energy Outlook 2010 Early Release Overview*, www.eia.gov.
4. Wikipedia Encyclopedia, *List of HVDC Projects*, en.wikipedia.org.
5. M.P. Bahrman and B.K. Johnson, "The ABCs of HVDC Transmission Technologies," *IEEE Power & Energy Magazine*, 5,2 (March/April 2007), pp. 33–44.
6. Edison Electric Institute, *About the Industry*, www.eei.org.
7. North American Electric Reliability Council (NERC), *2009 Long-Term Reliability Assessment* (Princeton, NJ: www.nerc.com, October 2009).
8. C.E. Root, "The Future Beckons," *Supplement to IEEE Power & Energy Magazine*, 4,3 (May/June 2006), pp. 58–65.
9. E. Santacana, G. Rackliffe, L. Tang & X. Feng, "Getting Smart," *IEEE Power & Energy Magazine*, 8,2 (March/April 2010), pp. 41–48.
10. North American Electric Reliability Council (NERC), *About NERC* (Princeton, NJ: www.nerc.com).
11. T.J. Overbye and J. Weber, "Visualizing the Electric Grid", *IEEE Spectrum*, 38,2 (February 2001), pp. 52–58.

Fossil-fuel (oil/gas) power plant with two 850-MVA generating units (Courtesy of PacifiCorp)

2

FUNDAMENTALS

The objective of this chapter is to review basic concepts and establish terminology and notation. In particular, we review phasors, instantaneous power, complex power, network equations, and elementary aspects of balanced three-phase circuits. Students who have already had courses in electric network theory and basic electric machines should find this chapter to be primarily refresher material.

CASE STUDY Throughout most of the 20th-century, electric utility companies built increasingly larger generation plants, primarily hydro or thermal (using coal, gas, oil, or nuclear fuel). At the end of the twentieth century, following the ongoing deregulation of the electric utility industry with increased competition in the United States and in other countries, smaller generation sources that connect directly to distribution systems have emerged. Distributed energy resources are sources of energy including generation and storage devices that are located near local loads. Distributed generation sources include renewable technologies (including geothermal, ocean tides, solar and wind) and nonrenewable technologies (including internal combustion engines, combustion turbines, combined cycle, microturbines, and fuel cells). Microgrids are systems that have distributed energy resources and associated loads that can form intentional islands in distribution systems. The following article describes the benefits of microgrids and several microgrid technologies under development in the United States and other countries. [5].

Making Microgrids Work

Distributed energy resources (DER), including distributed generation (DG) and distributed storage (DS), are sources of energy located near local loads and can provide a variety of benefits including improved reliability if they are properly operated in the electrical distribution system. Microgrids are systems that have at least one distributed energy resource and associated loads and can form intentional islands in the electrical distribution systems. Within microgrids, loads and energy sources can be disconnected from and reconnected to the area or local electric power system with minimal disruption to the local loads. Any time a microgrid is implemented in an electrical distribution system, it needs to be well planned to avoid causing problems.

For microgrids to work properly, an upstream switch must open (typically during an unacceptable power quality condition), and the DER must be able to carry the load on the islanded section. This includes maintaining suitable voltage and frequency levels for all islanded loads. Depending on switch technology, momentary interruptions may occur during transfer from grid-connected to islanded mode. In this case, the DER assigned to carry the island loads should be able to restart and pick up the island load after the switch has opened. Power flow analysis of island scenarios should be performed to insure that proper voltage regulation is maintained and to establish that the DER can handle inrush during "starting" of the island. The DER must be able to supply the real and reactive power requirements during islanded operation and to sense if a fault current has occurred downstream of the switch location. When power is restored on the utility side, the switch must not close unless the utility and "island" are synchronized. This requires measuring the voltage on both sides of the switch to allow synchronizing the island and the utility.

Microgrids' largest impact will be in providing higher reliability electric service and better power quality to the end customers. Microgrids can also provide additional benefits to the local utility by providing dispatchable power for use during peak power conditions and alleviating or postponing distribution system upgrades.

MICROGRID TECHNOLOGIES

Microgrids consist of several basic technologies for operation. These include DG, DS, interconnection switches, and control systems. One of the technical challenges is the design, acceptance, and availability of low-cost technologies for installing and using microgrids. Several technologies are under development to allow the safe interconnection and use of microgrids (see Figure 1).

DISTRIBUTED GENERATION

DG units are small sources of energy located at or near the point of use. DG technologies (Figures 2–5) typically include photovoltaic (PV), wind, fuel cells, microturbines, and reciprocating internal combustion engines with generators. These systems may be powered by either fossil or renewable fuels.

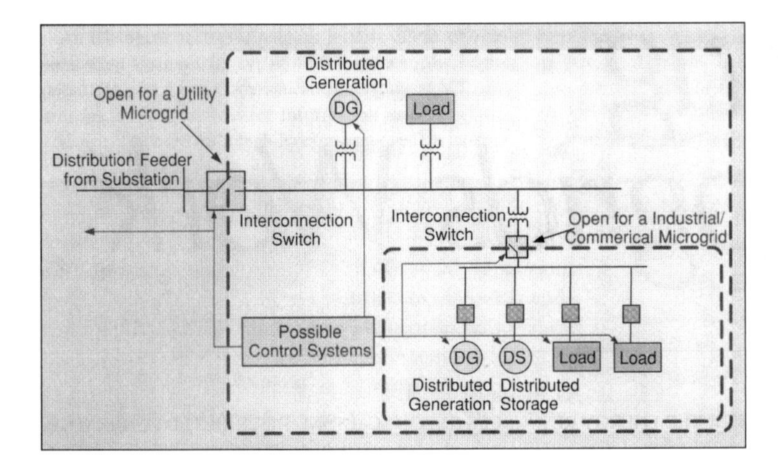

Figure 1
Microgrids and components

Some types of DG can also provide combined heat and power by recovering some of the waste heat generated by the source such as the microturbine in Figure 2. This can significantly increase the efficiency of the DG unit. Most of the DG technologies require a power electronics interface in order to convert the energy into grid-compatible ac power. The power electronics interface contains the necessary circuitry to convert power from one form to another. These converters may include both a rectifier and an inverter or just an inverter. The converter is compatible in voltage and frequency with the electric power system to which it will be connected and contains the necessary output filters. The power electronics interface can also contain protective functions for both the distributed energy system and the local electric power system that allow paralleling and disconnection from the electric power system. These power electronic interfaces provide a unique capability to the DG units and can enhance the operations of a microgrid.

Figure 2
Microturbines with heat recovery

DISTRIBUTED STORAGE

DS technologies are used in microgrid applications where the generation and loads of the microgrid cannot be exactly matched. Distributed storage provides a bridge in meeting the power and energy requirements of the microgrid. Storage capacity is defined in terms of the time that the nominal energy capacity can cover the load at rated power. Storage capacity can be then categorized in terms of energy density requirements (for medium- and long-term needs) or in terms of power density requirements (for short- and very short-term needs). Distributed storage enhances the overall performance of microgrid systems in three ways. First, it stabilizes and permits DG units to run at a constant and stable output, despite load fluctuations. Second, it provides the ride-through capability when there are dynamic variations of primary energy (such as those of sun, wind, and hydropower sources). Third, it permits DG to seamlessly operate as a dispatchable unit. Moreover, energy storage can benefit power systems by damping peak surges in electricity demand, countering momentary power disturbances, providing outage ride-through while backup generators respond, and reserving energy for future demand.

Figure 3
Wind turbine

Figure 4
Fuel cell

Figure 5
PV array

Figure 6
Large lead-acid battery bank

There are several forms of energy storage available that can be used in microgrids; these include batteries, supercapacitors, and flywheels. Battery systems store electrical energy in the form of chemical energy (Figure 6). Batteries are dc power systems that require power electronics to convert the energy to and from ac power. Many utility connections for batteries have bidirectional converters, which allow energy to be stored and taken from the batteries. Supercapacitors, also known as ultracapacitors, are electrical energy storage devices that offer high power density and extremely high cycling capability. Flywheel systems have recently regained consideration as a viable means of supporting critical load during grid power interruption because of their fast response compared to electrochemical energy storage. Advances in power electronics and digitally controlled fields have led to better flywheel designs that deliver a cost-effective alternative in the power quality market. Typically, an electric motor supplies mechanical energy to the flywheel and a generator is coupled on the same shaft that outputs the energy, when needed, through a converter. It is also possible to design a bidirectional system with one machine that is capable of motoring and regenerating operations.

INTERCONNECTION SWITCH

The interconnection switch (Figure 7) ties the point of connection between the microgrid and the rest of the distribution system. New technology in this area consolidates the various power and switching functions (e.g., power switching, protective relaying, metering, and communications) traditionally provided by relays, hardware, and other components at the utility interface into a single system with a digital signal processor (DSP). Grid conditions are measured both on the utility and microgrid sides of the switch through current transformers (CTs) and potential transformers (PTs) to determine operational conditions (Figure 8). The interconnection switches are designed to meet grid interconnection standards (IEEE 1547 and UL 1741 for North America) to minimize custom engineering and site-specific approval processes and lower cost. To maximize applicability and functionality, the controls are also designed to be technology neutral and can be used with a circuit breaker as well as faster semiconductor-based static switches like thyristors and integrated gate bipolar transistor technologies and are applicable to a variety of DG assets with conventional generators or power converters.

Figure 7
Interconnection switch and control board

CONTROL SYSTEMS

The control system of a microgrid is designed to safely operate the system in grid-connected and stand-alone modes. This system may be based on a central controller or imbedded as autonomous parts of each distributed generator. When the utility is disconnected the control system must control the local voltage and frequency, provide (or absorb) the instantaneous real power difference between generation and loads, provide the difference between generated reactive power and the actual reactive power consumed by the load; and protect the internal microgrid.

In stand-alone mode, frequency control is a challenging problem. The frequency response of larger systems is based on rotating masses and these are regarded as essential for the inherent stability of these systems. In contrast, microgrids are inherently converter-dominated grids without or with very little directly connected rotating masses, like flywheel energy storage coupled through a converter. Since microturbines and fuel cells have slow response to control signals and are inertia-less, isolated operation is technically demanding and raises load-tracking problems. The converter control systems must be adapted to provide the response previously obtained from directly connected rotating masses. The frequency control strategy should exploit, in a cooperative way, the capabilities of the micro sources to change their active power, through frequency control droops, the response of the storage devices, and load shedding.

Appropriate voltage regulation is necessary for local reliability and stability. Without effective local voltage control, systems with high penetration of distributed energy resources are likely to experience voltage and/or reactive power excursions and oscillations. Voltage control requires that there are no large circulating reactive currents between sources. Since the voltage control is inherently a local problem, voltage regulation faces the same problems in both modes of operation; i.e., isolated or interconnected. In the grid-interconnected mode, it is conceivable to consider that DG units can provide ancillary services in the form of local voltage support. The capability of modern power electronic interfaces offers solutions to the provision of reactive power locally by the adoption of a voltage versus reactive current droop controller, similar to the droop controller for frequency control.

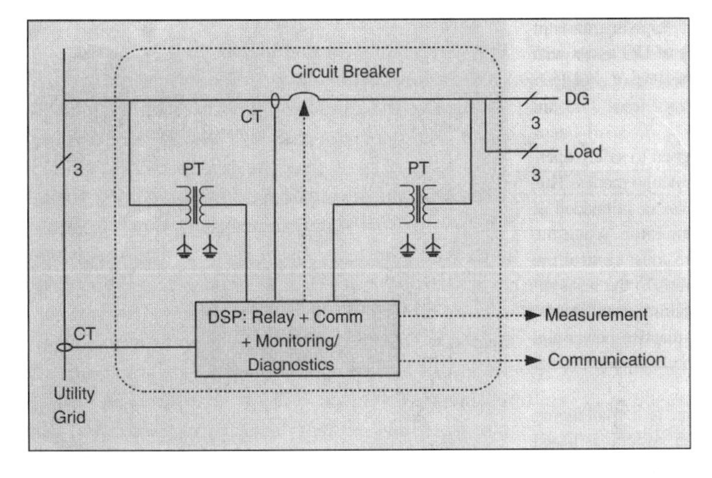

Figure 8
Schematic diagram of a circuit breaker-based interconnection switch

MICROGRID TESTING EXPERIENCE

Around the world, there are several active experiments in the microgrid area covering an array of technologies. As part of this research, microgrid topologies and operational configurations are being defined and design criteria established for all possibilities of microgrid applications.

TESTING EXPERIENCE IN THE UNITED STATES

Consortium for Electric Reliability Solutions (CERTS) Testbed

The objective of the CERTS microgrid testbed is to demonstrate a mature system approach that allows for high penetration of DER equipment by providing a resilient platform for plug-and-play operation, use of waste heat and intermittent sources, and enhancement of the robustness and reliability of the customers' electrical supply. The CERTS microgrid has two main components: a static switch and autonomous sources. The static switch has the ability to autonomously island the microgrid from disturbances such as faults, IEEE 1547 events, or power quality events. After islanding, the reconnection of the microgrid is achieved autonomously after the tripping event is no longer present. This synchronization is achieved by using the frequency difference between the islanded microgrid and the utility grid. Each source can seamlessly balance the power on the islanded microgrid using real power versus frequency droop and maintain voltage using the reactive power versus voltage droop. The coordination between sources is through frequency, and the voltage controller provides local stability. Without local voltage control, systems with high penetrations of DG could experience voltage and/or reactive power oscillations. Voltage control must also insure that there are no large circulating reactive currents between sources. This requires a voltage versus reactive power droop controller so that, as the reactive power generated by the source becomes more capacitive, the local voltage set point is reduced. Conversely, as reactive power becomes more inductive, the voltage set point is increased.

The CERTS microgrid has no "master" controller or source. Each source is connected in a peer-to-peer fashion with a localized control scheme implemented with each component. This arrangement increases the reliability of the system in comparison to a master–slave or centralized control scheme. In the case of a master–slave architecture, the failure of the master controller could compromise the operation of the whole system. The CERTS testbed uses a central communication system to dispatch DG set points as needed to improve overall system operation. However, this communication network is not used for the dynamic operation of the microgrid. This plug-and-play approach allows expansion of the microgrid to meet the requirements of the site without extensive re-engineering.

The CERTS testbed (Figure 9) is located at American Electric Power's Walnut test site in Columbus, Ohio. It consists of three 60-kW converter based sources and a thyristor based static switch. The prime mover in this case is an automobile internal combustion engine converted to run on natural gas. It drives a synchronous generator at variable speeds to achieve maximum efficiencies over a wide range of loads. The output is rectified and inverted to insure a constant ac frequency

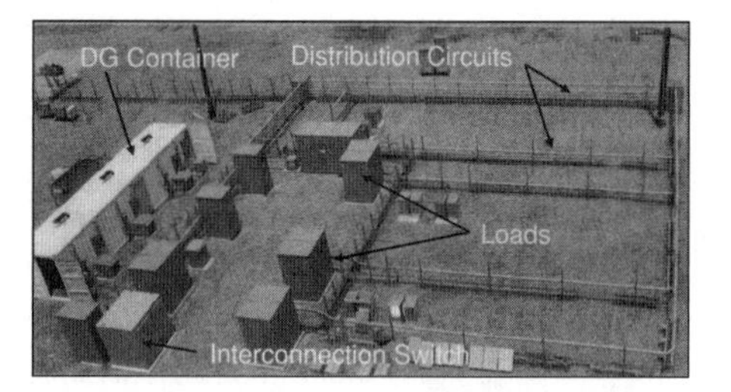

Figure 9
CERTS/AEP microgrid testbed

at the microgrid. To insure that the converter can provide the necessary energy demanded by the CERTS controls there is storage on the dc bus. This also insures that the dynamics of the permanent magnet and generator are decoupled from the dynamics of the converter. This insures that a variety of energy sources can have the same dynamic response as the sources used at the testbed.

The testbed has three feeders, two of which have DG units connected and can be islanded. One of these feeders has two sources separated by 170 m of cable. The other feeder has a single source, which allows for testing parallel operation of sources. The third feeder stays connected to the utility but can receive power from the micro sources when the static switch is closed without injecting power into the utility. The objective of the testing is to demonstrate the system dynamics of each component of the CERTS microgrid. This includes smooth transitions from grid-connected to islanded operation and back, high power quality, system protection, speed of response of the sources, operation under difficult loads, and autonomous load tracking.

Figure 10 is an example of islanding dynamics between two sources on a single feeder at the CERTS testbed. Initially, the microgrid is utility connected with unit A and unit B output at 6 kW and 54 kW, respectively. The load is such that the grid provides 42 kW. Upon islanding, unit B exceeds 60 kW and quickly settles at its maximum steady-state operating point of 60 kW with a reduced frequency of 59.8 Hz due to the power versus frequency droop. Unit A increases to 42 kW and converges to the same islanded frequency. The smoothness and speed of the transition is seen in the invert currents and the microgrid voltages. The loads do not see the islanding event.

Figure 11 shows voltage across the switch and the phase currents through the static switch during autonomous synchronization. This synchronization is achieved by using the frequency difference between the islanded

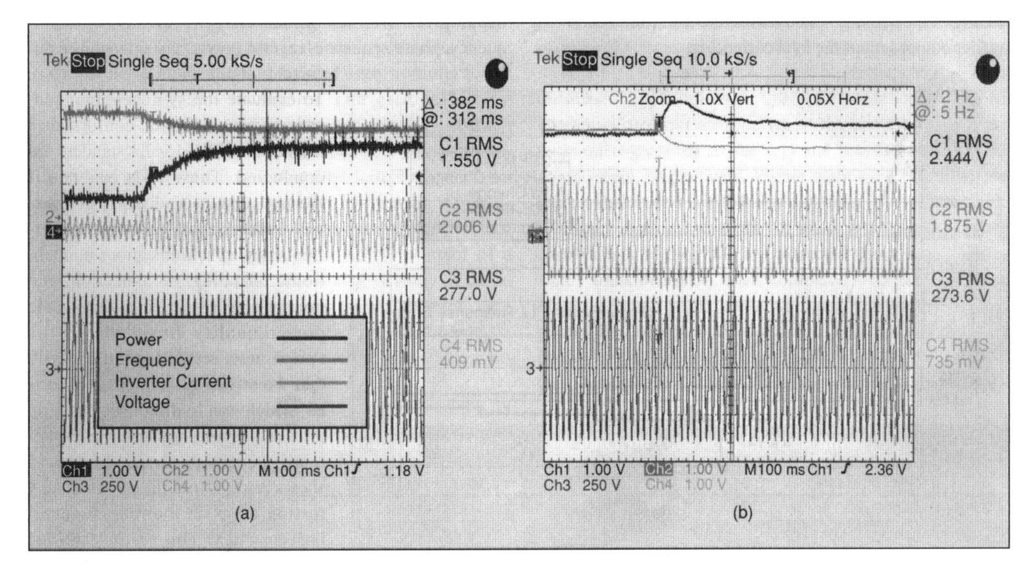

Figure 10
Operation of two 60-kW sources using CERTS autonomous controls during an islanding event

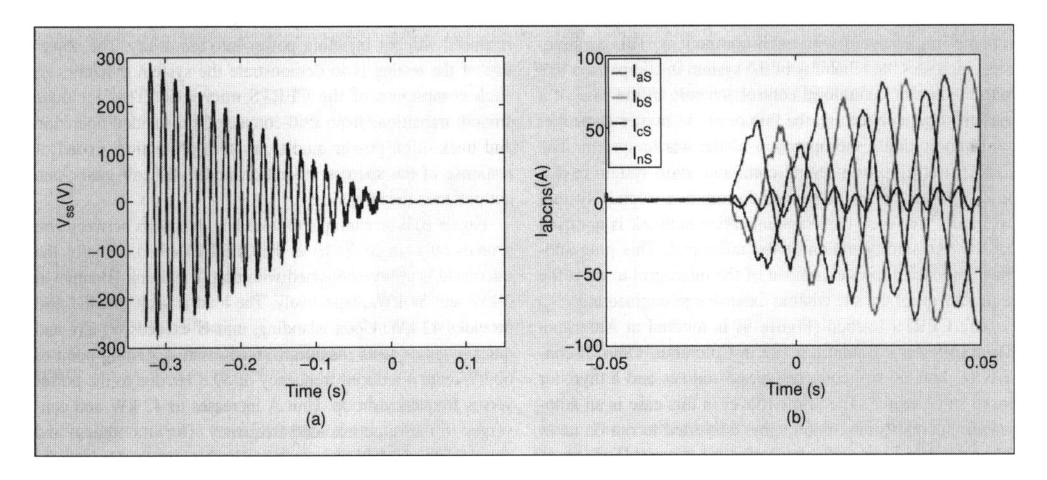

Figure 11
Synchronization of the microgrid to the utility

microgrid and the utility grid. This results in a low-frequency beat voltage across the switch. When the two voltages come in phase due to this frequency difference the switch will close. The phase currents display a smooth transition due to closing at zero voltage phase difference. The unbalanced currents are driven by a utility voltage unbalance of around 1% and a balanced voltage created by the DG source. All loads see balanced voltages provided by the DG sources. The neutral third harmonic current and phase current distortion are due to transformer magnetization currents.

The fundamental and third-harmonic frequency component from the transformer magnetization is apparent. As the loading of the transformer increases, the distortion becomes a smaller component of the total current.

Interconnection Switch Testing

The National Renewable Energy Laboratory has worked with a variety of U.S. interconnection switch manufacturers on the development of advanced interconnection technologies that allow paralleling of distributed generators with the utility for uninterrupted electrical service and the ability to parallel and sell electricity back to the utility. This research promotes the development of new products and technologies that enable faster switching, greater reliability, and lower fault currents on the electrical grids, thereby providing fewer disruptions for customers while expanding capabilities as an energy-intensive world becomes more energy efficient in the future.

Testing of the various switch technologies includes typical protective relay function tests such as detection and tripping for over- and undervoltage, over- and underfrequency, phase sequence, reverse power, instantaneous over-current, and discrete event trip tests. To evaluate the switches' interconnection requirements, conformance tests to the IEEE 1547.1 standard are conducted. These tests evaluate if the unit detects and trips for over- and undervoltage, over- and underfrequency, synchronization, unintentional islanding, reconnection, and open-phase tests. To evaluate the power quality functions of the switch, tests are performed to verify that the switch responded as expected, which was to disconnect the grid and DG terminals when a power quality event occurred. Figure 12 shows results from the power quality testing done on a circuit-breaker-based switch. This testing showed that there is a minimum trip time for the breaker (0.005 s) and that the control logic for the breaker needs to be more accurately tuned to stay within the Information Technology Industry (ITI) Council curve.

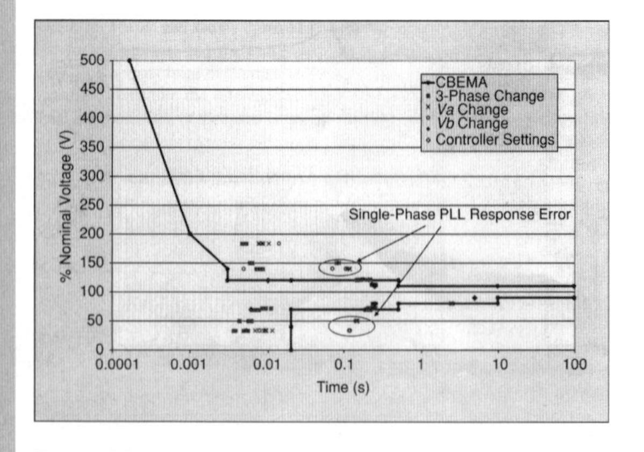

Figure 12
Testing of a circuit breaker-based microgrid switch versus the ITI curve

TESTING EXPERIENCE IN JAPAN

The New Energy and Industrial Technology Development Organization (NEDO) is currently supporting a variety of microgrid demonstration projects applying renewable and distributed generation. The first group of projects, called Regional Power Grids with Various New Energies, was implemented at three locations in Japan: Expo 2005 Aichi, recently moved to the Central Japan Airport City (Aichi project), Kyoto Eco-Energy project (Kyotango project), and Regional Power Grid with Renewable Energy Resources in Hachinohe City (Hachinohe project). In these three projects, control systems capable of matching energy demand and supply for microgrid operation were established. An important target in all of the projects is achieving a matched supply and demand of electricity. In each project, a standard for the margin of error between supplied energy and consumed energy over a certain period was set as a control target.

In the Aichi project, a power supply system utilizing fuel cells, PV, and a battery storage system, all equipped with converters, was constructed. A block diagram of the supply system for the project is shown in Figure 13. The fuel cells adopted for the system include two molten carbonate fuel cells (MCFCs) with capacities of 270 kW and 300 kW, one 25-kW solid oxide fuel cell (SOFC), and four 200-kW phosphoric acid fuel cells (PAFCs). The total capacity of the installed PV systems is 330 kW, and the adopted cell types include multicrystalline silicon, amorphous silicon, and a single crystalline silicon bifacial type. A sodium-sulfur (NaS) battery is used to store energy within the supply system and it plays an important role in matching supply and demand. In the Aichi project, the load-generation balancing has been maintained at 3% for as short as ten-minute intervals. The Aichi project experienced a second grid-independent operation mode in September 2007. In this operational mode, the NaS battery converter controls voltage and balancing of the load.

In the Kyotango project, the energy supply facilities and demand sites are connected to a utility grid and are integrated by a master control system. The energy supply system functions as a "virtual microgrid." A management system for matching the demand and supply of electricity is being demonstrated and a reduction in imbalances to within 3% of expected demand for five-minute intervals was achieved. Several criteria related to power quality (outages, voltage fluctuations, and frequency fluctuations) are being

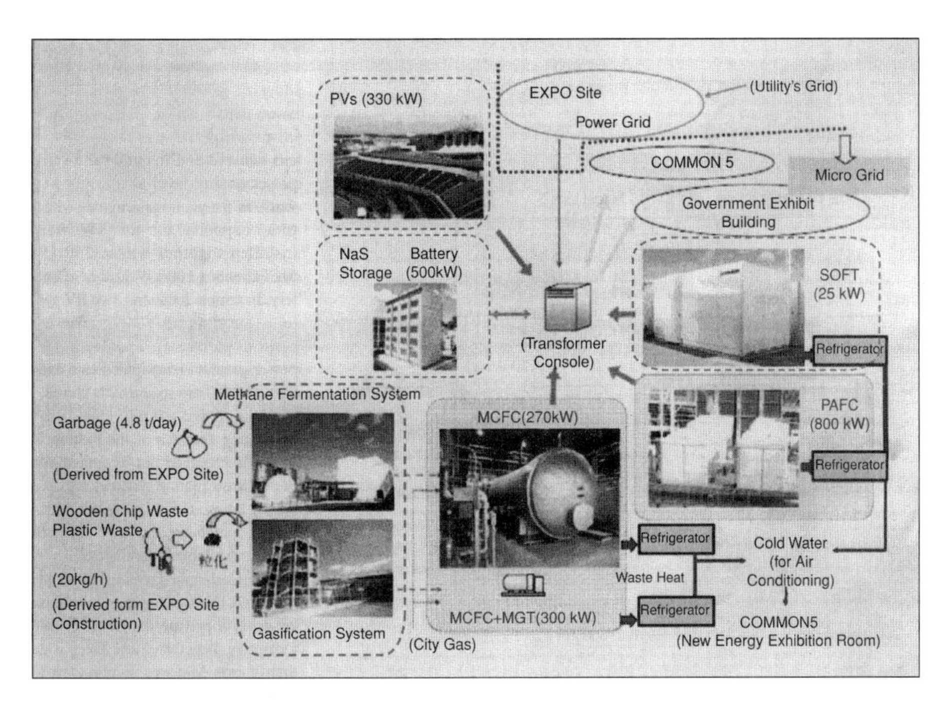

Figure 13
Diagram of Aichi Microgrid project

monitored during the demonstration period to determine if the system can achieve and maintain the same power quality level as a utility network. In the plant, gas engines with a total capacity of 400 kW were installed together with a 250-kW MCFC and a 100-kW lead-acid battery. In remote locations, two PV systems and one 50-kW small wind turbine were also installed. The power generation equipment and end-user demand are managed by remote monitoring and control. One of the interesting features of the system is that it is managed not by a state-of-the-art information network system but by conventional information networks, which are the only network systems available in rural areas.

The Hachinohe project (Figure 14) features a microgrid system constructed using a private distribution line measuring more than 5 km. The private distribution line was constructed to transmit electricity primarily generated by the gas engine system. Several PV systems and small wind turbines are also connected to the microgrid. At the sewage plant, three 170-kW gas engines and a 50-kW PV system have been installed. To support the creation of digestion gas by the sewage plant, a wood-waste steam boiler was also installed due to a shortage of thermal heat to safeguard the bacteria. Between the sewage plant and city office, four schools and a water supply authority office are connected to the private distribution line. At the school sites, renewable energy resources are used to create a power supply that fluctuates according to weather conditions in order to prove the microgrid's control system's capabilities to match demand and supply. The control system used to balance supply and demand consists of three facets: weekly supply and demand planning, economic dispatch control once every three minutes, and second-by-second power flow control at interconnection points. The control target is a margin of error between supply and demand of less than 3% for every six-minute interval. During testing, a margin of error rate of less than 3% was achieved during 99.99% of the system's operational time. The Hachinohe project experienced one week of grid-independent operation in November 2007. In this operational mode, imbalance among the three phases was compensated by the PV converter.

The New Power Network Systems project is evaluating new test equipment installed on a test distribution network (Figure 15) constructed at the Akagi Test Center of the Central Research Institute of the Electric Power Industry (CRIEPI). This equipment includes a static var compensator (SVC), a step voltage regulator (SVR), and loop balance controllers (LBCs). The SVC and SVR are

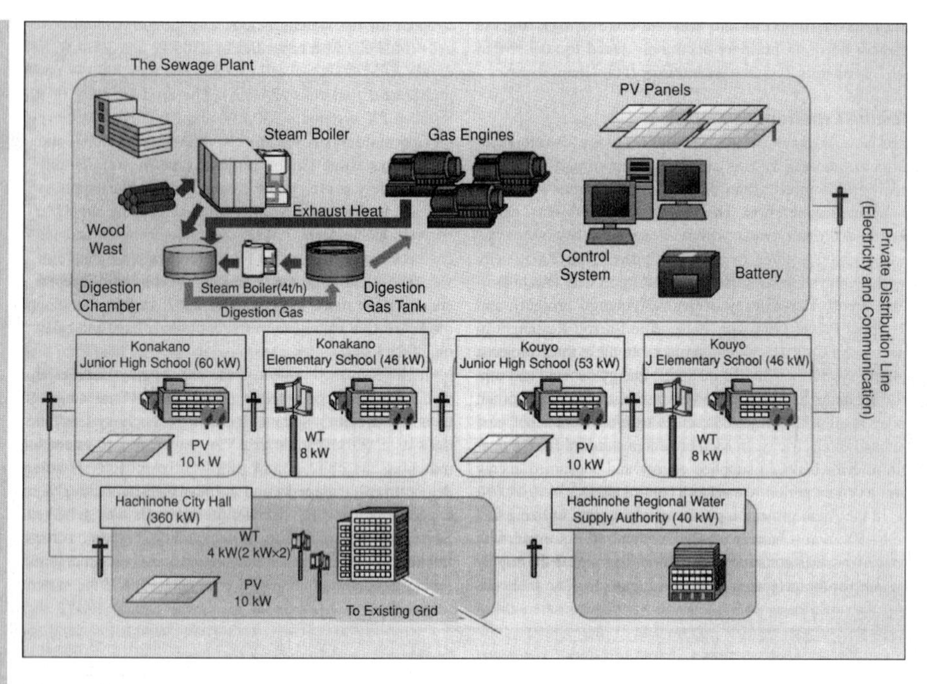

Figure 14
Overview of the Hachinohe project

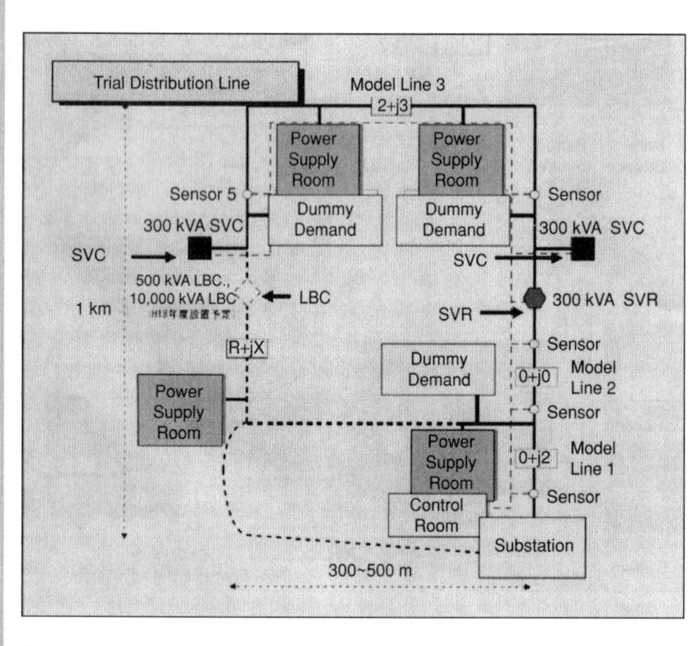

Figure 15
Structure of test network at CRIEPI

used for controlling the voltage on a distribution line, and they are sometimes applied on an actual utility network. In this project, the effects of integrated control of this equipment are being examined. LBCs are a new type of distribution network equipment that can control the power flow between two distribution feeders by means of a back-to-back (BTB) type converter. The LBCs allow connections of two sources with different voltages, frequencies, and phase angles by providing a dc link.

A final microgrid project is evaluating the possibility that grid technology can create value for consumers and various energy service levels. In Sendai City a microgrid consisting of two 350-kW gas engine generators, one 250-kW MCFC, and various types of compensating equipment is being evaluated to demonstrate four levels of customer power. Two of the service levels will have compensating equipment that includes an integrated power quality backup system that supplies high-quality power that reduces interruptions and voltage drops. In one of these cases, the wave pattern is guaranteed. Two additional lower service levels have only short-term voltage drops compensated by a series compensator. This work will evaluate the possibility of providing various service levels to customers located in the same area. Since summer of 2007, the Sendai system has been in operation and has improved the power quality at the site. Before starting actual operation, the compensation equipment was tested by using a BTB power supply system to create artificial voltage sag.

In addition to the NEDO-sponsored projects, there are several private microgrid projects. Tokyo Gas has been evaluating a 100-kW microgrid test facility since September 2006 at the Yokohama Research Institute, consisting of gas-engine combined heat and power (CHP), PV, wind power, and battery-incorporated power electronics. Shimizu Corp. has developed a microgrid control system with a small microgrid that consists of gas engines, gas turbines, PV, and batteries. The system is designed for load following and includes load forecasting and integrated control for heat and power.

TESTING EXPERIENCE IN CANADA

Planned microgrid islanding application, also known as intentional islanding, is an early utility adaptation of the microgrid concept that has been implemented by BC Hydro and Hydro Quebec, two of the major utility companies in Canada. The main objective of planned islanding projects is to enhance customer-based power supply reliability on rural feeders by utilizing an appropriately located independent power producer (IPP), which is, for instance, located on the same or adjacent feeder of a distribution substation. In one case, the customers in Boston Bar town, part of the BC Hydro rural areas, which is supplied by three 25-kV medium-voltage distribution feeders, had been exposed to power outages of 12 to 20 hrs two or three times per year. This area, as shown in Figure 16, is supplied by a 69/25-kV distribution substation and is connected to the BC Hydro high-voltage system through 60 km of 69-kV line. Most of the line is built off a highway in a canyon that is difficult to access with high potential of rock/mud/snow slides. The implemented option to reduce sustained power-outage durations is based on utilizing a local IPP to operate in an intentional island mode and supply the town load on one or more feeders of the substation. The Boston Bar IPP has two 3.45-MW hydro power generators and is connected to one of the three feeders with a peak load of 3.0 MW. Depending on the water level, the Boston Bar IPP can supply the community load on one or more of the feeders during the islanding operation. If the water level is not sufficient, the load on one feeder can be sectioned to adequate portions.

Based on the BC Hydro islanding guideline, to perform planned islanding, an IPP should be equipped with additional equipment and control systems for voltage regulation, frequency stabilization, and fault protection. In addition, the island-load serving capability of an IPP needs to be tested prior to and during the project commissioning to ensure

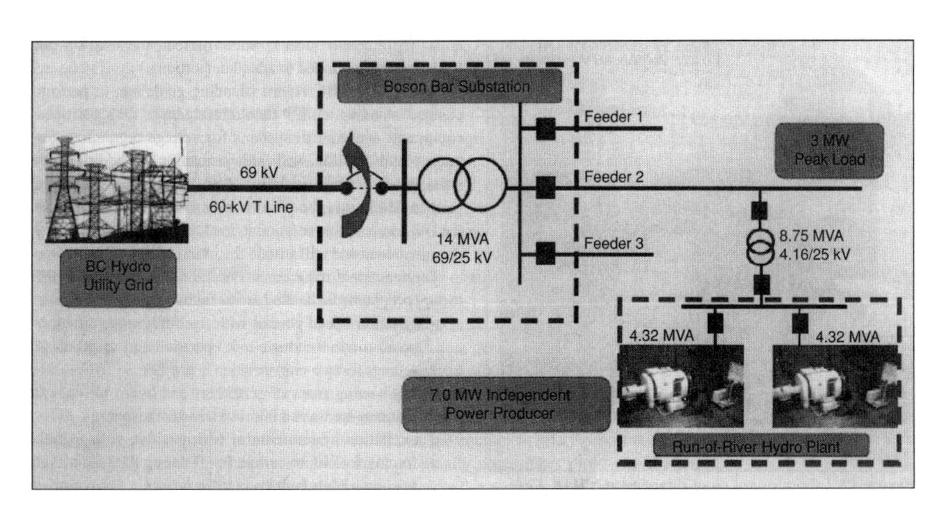

Figure 16
System configuration for the Boston Bar IPP and BC Hydro planned islanding site

that the IPP can properly respond to load transients such as a step change in load and still sustain the island.

The functional requirements added to the Boston Bar IPP to support planned islanding are as follows:

- governor speed control with fixed-frequency (isochronous) mode for single-unit operation and speed-droop settings for two-unit operation in parallel
- engineering mass of generators and hydro turbines to increase inertia and improve transient response
- excitation system control with positive voltage field forcing for output current boost during the feeder fault to supply high fault current for proper coordination of protection relays
- automatic voltage regulation control to regulate voltages at the point of common coupling
- two sets of overcurrent protection set-points for the grid-connected and the islanding operating modes
- real-time data telemetry via a leased telephone line between the IPP remote control site and the utility area control center
- black start capability via an onsite 55-kW diesel generator.

In addition to the above upgrades, the auto-recloser on the connecting IPP feeder is equipped with a secondary voltage supervision function for voltage supervisory close and blocking of the auto-reclosing action. Remote auto-synchronization capability was also added at the substation level to synchronize and connect the island area to the 69-kV feeder without causing load interruption. When a sustain power outage event, such as a permanent fault or line breakdown, occurs on the utility side of the substation, the main circuit breaker and feeder reclosers are opened (Figure 16). Then, the substation breaker open position is telemetered to the IPP operator. Subsequently, the IPP changes the control and protection settings to the island mode and attempts to hold the island downstream of the feeder 2 recloser. If the IPP fails to sustain the island, the IPP activates a black-start procedure and picks up the dead feeder load under the utility supervision. The island load may be supplied by one generator or both generators in parallel.

Two sets of tests were performed during the generator commissioning as follows:

1) grid parallel operation tests including a) the automatic and manual synchronization, and b) output load, voltage and frequency controls, and load rejection tests
2) island operation tests comprising a) load pick-up and drop-off tests in 350-kW increments, b) dead load pick-up of 1.2 MW when only one of the two generators is in operation, and c) islanded operation and load following capability when one unit is generating and/or both units are operating in parallel.

The planned islanding operation of the Boston Bar IPP has been successfully demonstrated and performed several times during power outages caused by adverse environmental effects. Building on the knowledge and experience gained from this project, BC Hydro has recently completed a second case of planned islanding and is presently assessing a third project.

TESTING IN EUROPE

At the international level, the European Union has supported two major research efforts devoted exclusively to microgrids: the Microgrids and More Microgrids projects. The Microgrids project focused on the operation of a single microgrid, has successfully investigated appropriate control techniques, and demonstrated the feasibility of microgrid operation through laboratory experiments. The Microgrids project investigated a microgrid central controller (MCC) that promotes technical and economical operation, interfaces with loads and micro sources and demand-side management, and provides set points or supervises local control to interruptible loads and micro sources. A pilot installation was installed in Kythnos Island, Greece, that evaluated a variety of DER to create a microgrid.

Continuing microgrid projects in Greece include a laboratory facility (Figure 17) that has been set up at the National Technical University of Athens (NTUA), with the objective to test small-scale equipment and control strategies for micro-grid operation. The system comprises two poles, each equipped with local (PV and wind) generation and battery storage, connected to each other via a low-voltage line as well as to the main grid. Each pole may operate as a micro-grid via its own connection to the grid, or both poles may be connected via the low-voltage line to form a two-bus micro-grid connected to the main grid at one end. The battery converters are the main regulating units in island mode, regulated via active power-frequency and reactive power-voltage droops. Multi-agent technology has been implemented for the control of the sources and the loads.

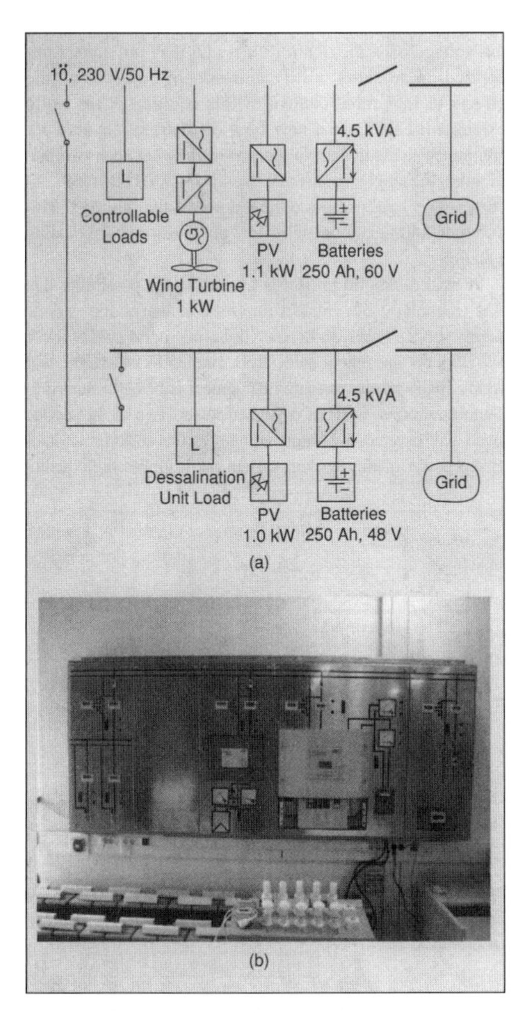

Figure 17
Laboratory microgrid facility at NTUA, Greece: (a) single-line diagram and (b) view of one pole

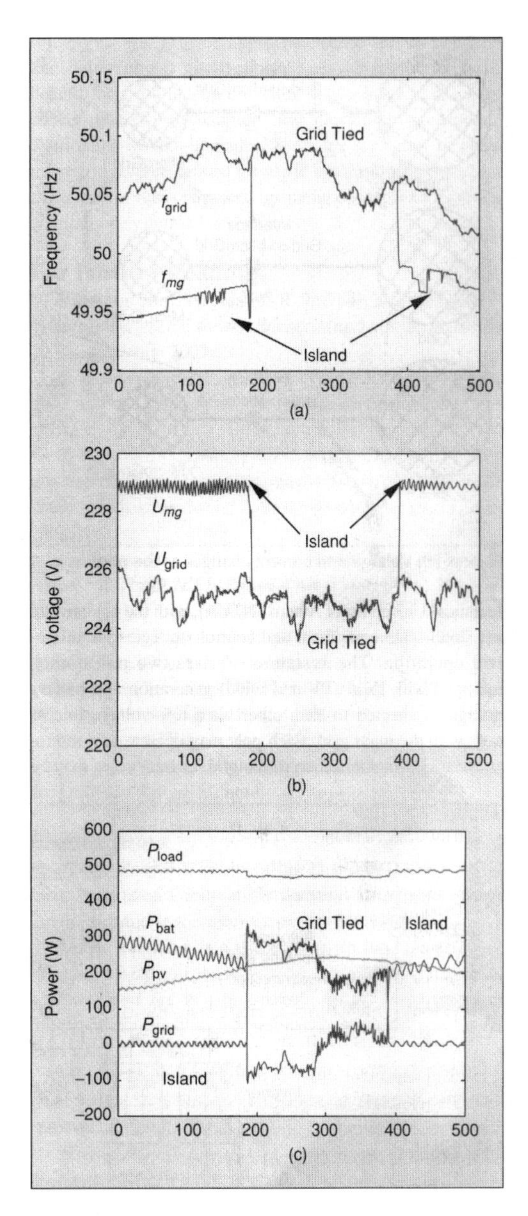

Figure 18
Changes of the microgrid operating mode from island to interconnected mode and vice-versa: (a) frequency, (b) voltage, and (c) component powers

Figure 18 shows indicative test results demonstrating the seamless transition of the microgrid from grid-connected to island mode and vice-versa (one-pole microgrid operation). The first diagram illustrates the variation of the frequency and the second of the voltage. The change of the component power flows is shown in the third illustration. While the load and the PV continue operating at the same power, the output of the battery converter and the power flow from the grid change to maintain the power equilibrium in the microgrid.

Testing on microgrid components has also been extensively conducted by ISET in Germany. Figure 19 shows testing conducted to examine voltage and current transient when microgrids transfer from grid-connected to islanded mode. This figure shows that with proper design, there can be minimal load disruption during the transfer.

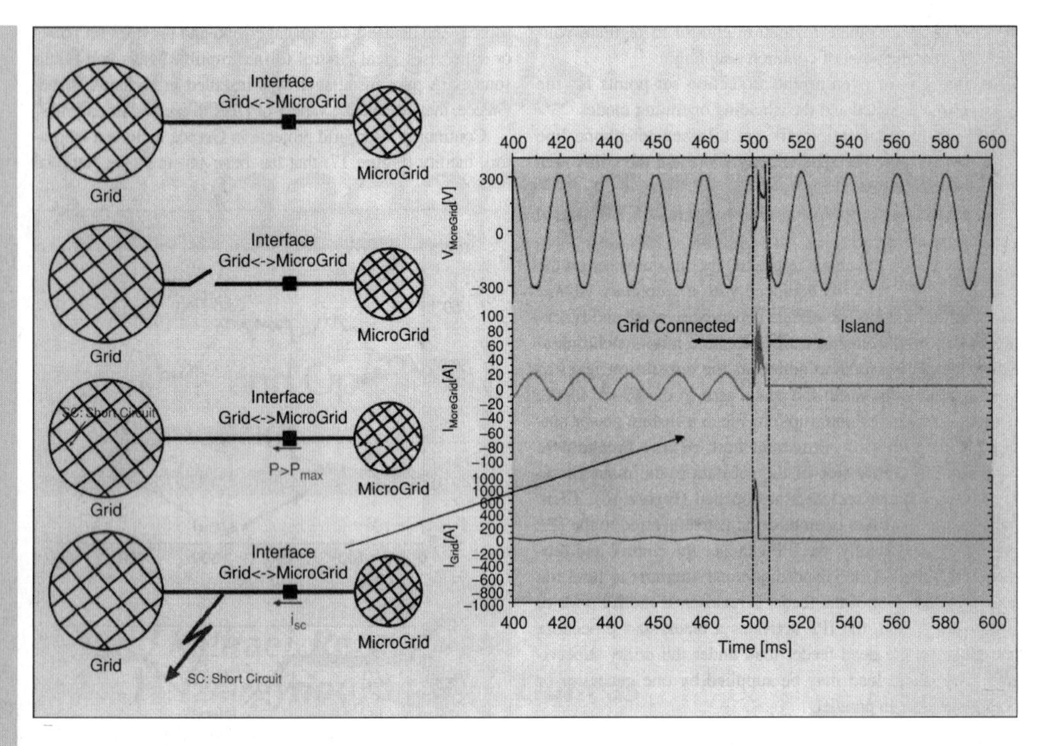

Figure 19
Voltage and current changes as the microgrid switches to islanded mode

The More Microgrids project aims at the increase of penetration of microgeneration in electrical networks through the exploitation and extension of the Microgrids concept, involving the investigation of alternative micro-generator control strategies and alternative network designs, development of new tools for multimicrogrid management operation and standardization of technical and commercial protocols, and field trials on actual microgrids and evaluation of the system performance on power system operation.

One of the More Microgrids projects is located at Bronsbergen Holiday Park, located near Zutphen in the Netherlands. It comprises 210 cottages, 108 of which are equipped with grid-connected PV systems. The park is electrified by a traditional three-phase 400-V network, which is connected to a 10-kV medium-voltage network via a distribution transformer located on the premises (Figure 20). The distribution transformer does not feed any low-voltage loads outside of the holiday park. Internally in the park, the 400-V supply from the distribution transformer is distributed over four cables, each protected by 200-A fuses on the three phases. The peak load is approximately 90 kW. The installed power of all the PV systems together is 315 kW. The objective of this project is experimental validation of islanded microgrids by means of smart storage (coupled by a flexible ac distribution system) including evaluation of islanded operation, automatic isolation and re-connection, fault level of the microgrid, harmonic voltage distortion, energy management and lifetime optimization of the storage system, and parallel operation of converters.

Another More Microgrids project involves field test on the transfer between interconnected and islanding mode with German utility MVV Energie. MVV Energie is planning to develop an efficient solution to cope with the expected future high penetration of renewable energy sources and distributed generation in the low-voltage distribution grid. If integrated in an intelligent way, these new players in the distribution grid will improve independence from energy imports, reliability, and power quality at lower cost than the "business as usual" regarding replacement or reinforcement of the regional energy infrastructure. A successful transfer between interconnected and islanding

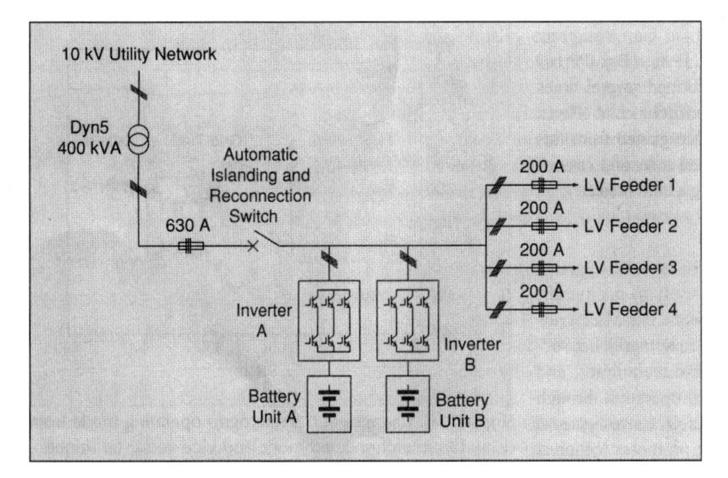

Figure 20
Schematic for the Bronsbergen Holiday Park microgrid

mode would provide a substantial benefit for the grid operator.

This project will evaluate decentralized control in a residential site in the ecological settlement in Mannheim-Wallstadt. The new control structures for the decentralized control with agents will be tested and allow the transition from grid connection to islanding operation without interruptions. This would improve reliability of the grid and support for black start after failure of the grid.

The CESI RICERCA test facility in Italy will also be used to experiment, demonstrate, and validate the operation of an actual microgrid field test of different microgrid topologies at steady and transient state and power quality analysis. During a transient state, the behavior during short-duration voltage variation for single/three-phase ac faults, or dynamic response to sudden load changes and to conditions of phase imbalance or loss of phase, the islanding conditions following interruption of the supply will be analyzed.

CONCLUSIONS

Microgrids will provide improved electric service reliability and better power quality to end customers and can also benefit local utilities by providing dispatchable load for use during peak power conditions and alleviating or postponing distribution system upgrades. There are a number of active microgrid projects around the world involved with testing and evaluation of these advanced operating concepts for electrical distribution systems.

FOR FURTHER READING

N. Hatziargyriou, A. Asano, R. Iravani, and C. Marnay, "Microgrids," *IEEE Power Energy Mag.*, vol. 5, no. 4, pp. 78–94, July/Aug. 2007.

R. Lasseter, and P. Piagi, "MicroGrids: A conceptual solution," in *Proc. IEEE PESC'04*, Aachen, Germany, June 2004, pp. 4285–4290.

B. Kroposki, C. Pink, T. Basso, and R. DeBlasio, "Microgrid standards and technology development," in *Proc. IEEE Power Engineering Society General Meeting*, Tampa, FL, June 2007, pp. 1–4.

S. Morozumi, "Micro-grid demonstration projects in Japan," in *Proc. IEEE Power Conversion Conf.*, Nagoya, Japan, Apr. 2007, pp. 635–642.

C. Abby, F. Katiraei, C. Brothers, L. Dignard-Bailey, and G. Joos, "Integration of distributed generation and wind energy in Canada," in *Proc. IEEE Power Engineering General Meeting*, Montreal, Canada, June 2006.

BC Hydro (2006, June), "Distribution power generator islanding guidelines," [Online]. Available: http://www.bchydro.com/info/ipp/ipp992.html

BIOGRAPHIES

Benjamin Kroposki manages the Distributed Energy Systems Integration Group at the National Renewable Energy Laboratory and serves as chairman for IEEE P1547.4.

Robert Lasseter is a professor with the University of Wisconsin-Madison and leads the CERTS Microgrid project.

Toshifumi Ise is a professor with the Department of Electrical Engineering, Faculty of Engineering, at Osaka University in Japan.

Satoshi Morozumi leads research activities in microgrids for the New Energy and Industrial Technology Development Organization in Japan.

Stavros Papathanassiou is an assistant professor with the National Technical University of Athens, Greece.

Nikos Hatziargyriou is a professor with the National Technical University of Athens and executive vice-chair and deputy CEO of the Public Power Corporation of Greece.

2.1 PHASORS

A sinusoidal voltage or current at constant frequency is characterized by two parameters: a maximum value and a phase angle. A voltage

$$v(t) = V_{max}\cos(\omega t + \delta) \tag{2.1.1}$$

has a maximum value V_{max} and a phase angle δ when referenced to $\cos(\omega t)$. The root-mean-square (rms) value, also called *effective value*, of the sinusoidal voltage is

$$V = \frac{V_{max}}{\sqrt{2}} \tag{2.1.2}$$

Euler's identity, $e^{j\phi} = \cos\phi + j\sin\phi$, can be used to express a sinusoid in terms of a phasor. For the above voltage,

$$\begin{aligned} v(t) &= \mathrm{Re}[V_{max}e^{j(\omega t+\delta)}] \\ &= \mathrm{Re}[\sqrt{2}(Ve^{j\delta})e^{j\omega t}] \end{aligned} \tag{2.1.3}$$

where $j = \sqrt{-1}$ and Re denotes "real part of." The rms phasor representation of the voltage is given in three forms—exponential, polar, and rectangular:

$$V = \underbrace{Ve^{j\delta}}_{\text{exponential}} = \underbrace{V\angle\delta}_{\text{polar}} = \underbrace{V\cos\delta + jV\sin\delta}_{\text{rectangular}} \tag{2.1.4}$$

A phasor can be easily converted from one form to another. Conversion from polar to rectangular is shown in the phasor diagram of Figure 2.1. Euler's identity can be used to convert from exponential to rectangular form. As an example, the voltage

$$v(t) = 169.7\cos(\omega t + 60^\circ) \quad \text{volts} \tag{2.1.5}$$

has a maximum value $V_{max} = 169.7$ volts, a phase angle $\delta = 60^\circ$ when referenced to $\cos(\omega t)$, and an rms phasor representation in polar form of

$$V = 120\angle 60^\circ \quad \text{volts} \tag{2.1.6}$$

Also, the current

$$i(t) = 100\cos(\omega t + 45^\circ) \quad \text{A} \tag{2.1.7}$$

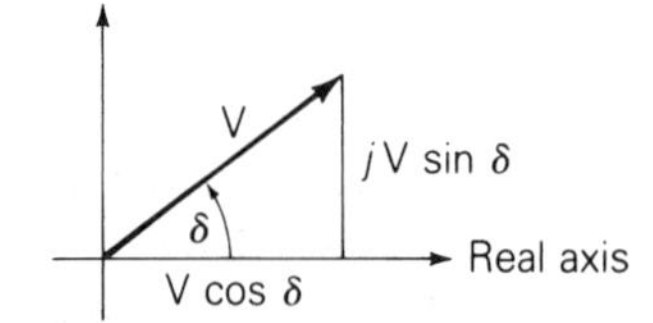

FIGURE 2.1 Phasor diagram for converting from polar to rectangular form

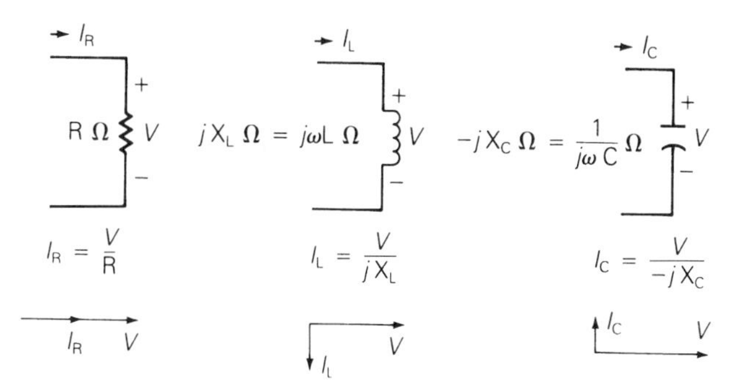

FIGURE 2.2 Summary of relationships between phasors V and I for constant R, L, and C elements with sinusoidal-steady-state excitation

has a maximum value $I_{max} = 100$ A, an rms value $I = 100/\sqrt{2} = 70.7$ A, a phase angle of 45°, and a phasor representation

$$I = 70.7\underline{/45^\circ} = 70.7e^{j45} = 50 + j50 \quad \text{A} \tag{2.1.8}$$

The relationships between the voltage and current phasors for the three passive elements—resistor, inductor, and capacitor—are summarized in Figure 2.2, where sinusoidal-steady-state excitation and constant values of R, L, and C are assumed.

When voltages and currents are discussed in this text, lowercase letters such as $v(t)$ and $i(t)$ indicate instantaneous values, uppercase letters such as V and I indicate rms values, and uppercase letters in italics such as V and I indicate rms phasors. When voltage or current values are specified, they shall be rms values unless otherwise indicated.

2.2 INSTANTANEOUS POWER IN SINGLE-PHASE AC CIRCUITS

Power is the rate of change of energy with respect to time. The unit of power is a watt, which is a joule per second. Instead of saying that a load absorbs energy at a rate given by the power, it is common practice to say that a load absorbs power. The instantaneous power in watts absorbed by an electrical load is the product of the instantaneous voltage across the load in volts and the instantaneous current into the load in amperes. Assume that the load voltage is

$$v(t) = V_{max} \cos(\omega t + \delta) \quad \text{volts} \tag{2.2.1}$$

We now investigate the instantaneous power absorbed by purely resistive, purely inductive, purely capacitive, and general RLC loads. We also introduce the concepts of real power, power factor, and reactive power. The physical significance of real and reactive power is also discussed.

PURELY RESISTIVE LOAD

For a purely resistive load, the current into the load is in phase with the load voltage, $I = V/\mathrm{R}$, and the current into the resistive load is

$$i_{\mathrm{R}}(t) = \mathrm{I}_{\mathrm{Rmax}} \cos(\omega t + \delta) \quad \mathrm{A} \tag{2.2.2}$$

where $\mathrm{I}_{\mathrm{Rmax}} = \mathrm{V}_{\mathrm{max}}/\mathrm{R}$. The instantaneous power absorbed by the resistor is

$$\begin{aligned} p_{\mathrm{R}}(t) &= v(t) i_{\mathrm{R}}(t) = \mathrm{V}_{\mathrm{max}} \mathrm{I}_{\mathrm{Rmax}} \cos^2(\omega t + \delta) \\ &= \tfrac{1}{2} \mathrm{V}_{\mathrm{max}} \mathrm{I}_{\mathrm{Rmax}} \{1 + \cos[2(\omega t + \delta)]\} \\ &= \mathrm{VI}_{\mathrm{R}} \{1 + \cos[2(\omega t + \delta)]\} \quad \mathrm{W} \end{aligned} \tag{2.2.3}$$

As indicated by (2.2.3), the instantaneous power absorbed by the resistor has an average value

$$\mathrm{P}_{\mathrm{R}} = \mathrm{VI}_{\mathrm{R}} = \frac{\mathrm{V}^2}{\mathrm{R}} = \mathrm{I}_{\mathrm{R}}^2 \mathrm{R} \quad \mathrm{W} \tag{2.2.4}$$

plus a double-frequency term $\mathrm{VI}_{\mathrm{R}} \cos[2(\omega t + \delta)]$.

PURELY INDUCTIVE LOAD

For a purely inductive load, the current lags the voltage by 90°, $I_{\mathrm{L}} = V/(j\mathrm{X}_{\mathrm{L}})$, and

$$i_{\mathrm{L}}(t) = \mathrm{I}_{\mathrm{Lmax}} \cos(\omega t + \delta - 90^\circ) \quad \mathrm{A} \tag{2.2.5}$$

where $\mathrm{I}_{\mathrm{Lmax}} = \mathrm{V}_{\mathrm{max}}/\mathrm{X}_{\mathrm{L}}$, and $\mathrm{X}_{\mathrm{L}} = \omega \mathrm{L}$ is the inductive reactance. The instantaneous power absorbed by the inductor is*

$$\begin{aligned} p_{\mathrm{L}}(t) &= v(t) i_{\mathrm{L}}(t) = \mathrm{V}_{\mathrm{max}} \mathrm{I}_{\mathrm{Lmax}} \cos(\omega t + \delta) \cos(\omega t + \delta - 90^\circ) \\ &= \tfrac{1}{2} \mathrm{V}_{\mathrm{max}} \mathrm{I}_{\mathrm{Lmax}} \cos[2(\omega t + \delta) - 90^\circ] \\ &= \mathrm{VI}_{\mathrm{L}} \sin[2(\omega t + \delta)] \quad \mathrm{W} \end{aligned} \tag{2.2.6}$$

As indicated by (2.2.6), the instantaneous power absorbed by the inductor is a double-frequency sinusoid with *zero* average value.

PURELY CAPACITIVE LOAD

For a purely capacitive load, the current leads the voltage by 90°, $I_{\mathrm{C}} = V/(-j\mathrm{X}_{\mathrm{C}})$, and

$$i_{\mathrm{C}}(t) = \mathrm{I}_{\mathrm{Cmax}} \cos(\omega t + \delta + 90^\circ) \quad \mathrm{A} \tag{2.2.7}$$

where $I_{Cmax} = V_{max}/X_C$, and $X_C = 1/(\omega C)$ is the capacitive reactance. The instantaneous power absorbed by the capacitor is

$$\begin{aligned} p_C(t) = v(t)i_C(t) &= V_{max}I_{Cmax}\cos(\omega t+\delta)\cos(\omega t+\delta+90°) \\ &= \tfrac{1}{2}V_{max}I_{Cmax}\cos[2(\omega t+\delta)+90°)] \\ &= -VI_C\sin[2(\omega t+\delta)] \quad \text{W} \end{aligned} \tag{2.2.8}$$

The instantaneous power absorbed by a capacitor is also a double-frequency sinusoid with *zero* average value.

GENERAL RLC LOAD

For a general load composed of RLC elements under sinusoidal-steady-state excitation, the load current is of the form

$$i(t) = I_{max}\cos(\omega t+\beta) \quad \text{A} \tag{2.2.9}$$

The instantaneous power absorbed by the load is then*

$$\begin{aligned} p(t) = v(t)i(t) &= V_{max}I_{max}\cos(\omega t+\delta)\cos(\omega t+\beta) \\ &= \tfrac{1}{2}V_{max}I_{max}\{\cos(\delta-\beta)+\cos[2(\omega t+\delta)-(\delta-\beta)]\} \\ &= VI\cos(\delta-\beta)+VI\cos(\delta-\beta)\cos[2(\omega t+\delta)] \\ &\quad + VI\sin(\delta-\beta)\sin[2(\omega t+\delta)] \end{aligned}$$

$$p(t) = VI\cos(\delta-\beta)\{1+\cos[2(\omega t+\delta)]\}+VI\sin(\delta-\beta)\sin[2(\omega t+\delta)]$$

Letting $I\cos(\delta-\beta) = I_R$ and $I\sin(\delta-\beta) = I_X$ gives

$$p(t) = \underbrace{VI_R\{1+\cos[2(\omega t+\delta)]\}}_{p_R(t)} + \underbrace{VI_X\sin[2(\omega t+\delta)]}_{p_X(t)} \tag{2.2.10}$$

As indicated by (2.2.10), the instantaneous power absorbed by the load has two components: One can be associated with the power $p_R(t)$ absorbed by the resistive component of the load, and the other can be associated with the power $p_X(t)$ absorbed by the reactive (inductive or capacitive) component of the load. The first component $p_R(t)$ in (2.2.10) is identical to (2.2.3), where $I_R = I\cos(\delta-\beta)$ is the component of the load current in phase with the load voltage. The phase angle $(\delta-\beta)$ represents the angle between the voltage and current. The second component $p_X(t)$ in (2.2.10) is identical to (2.2.6) or (2.2.8), where $I_X = I\sin(\delta-\beta)$ is the component of load current 90° out of phase with the voltage.

* Use the identity: $\cos A\cos B = \frac{1}{2}[\cos(A-B)+\cos(A+B)]$.

REAL POWER

Equation (2.2.10) shows that the instantaneous power $p_R(t)$ absorbed by the resistive component of the load is a double-frequency sinusoid with average value P given by

$$P = VI_R = VI\cos(\delta - \beta) \quad W \tag{2.2.11}$$

The *average power* P is also called *real power* or *active power*. All three terms indicate the same quantity P given by (2.2.11).

POWER FACTOR

The term $\cos(\delta - \beta)$ in (2.2.11) is called the *power factor*. The phase angle $(\delta - \beta)$, which is the angle between the voltage and current, is called the *power factor angle*. For dc circuits, the power absorbed by a load is the product of the dc load voltage and the dc load current; for ac circuits, the average power absorbed by a load is the product of the rms load voltage V, rms load current I, and the power factor $\cos(\delta - \beta)$, as shown by (2.2.11). For inductive loads, the current lags the voltage, which means β is less than δ, and the power factor is said to be *lagging*. For capacitive loads, the current leads the voltage, which means β is greater than δ, and the power factor is said to be *leading*. By convention, the power factor $\cos(\delta - \beta)$ is positive. If $|\delta - \beta|$ is greater than 90°, then the reference direction for current may be reversed, resulting in a positive value of $\cos(\delta - \beta)$.

REACTIVE POWER

The instantaneous power absorbed by the reactive part of the load, given by the component $p_X(t)$ in (2.2.10), is a double-frequency sinusoid with zero average value and with amplitude Q given by

$$Q = VI_X = VI\sin(\delta - \beta) \quad \text{var} \tag{2.2.12}$$

The term Q is given the name *reactive power*. Although it has the same units as real power, the usual practice is to define units of reactive power as volt-amperes reactive, or var.

EXAMPLE 2.1 Instantaneous, real, and reactive power; power factor

The voltage $v(t) = 141.4\cos(\omega t)$ is applied to a load consisting of a 10-Ω resistor in parallel with an inductive reactance $X_L = \omega L = 3.77\ \Omega$. Calculate the instantaneous power absorbed by the resistor and by the inductor. Also calculate the real and reactive power absorbed by the load, and the power factor.

FIGURE 2.3

Circuit and phasor diagram for Example 2.1

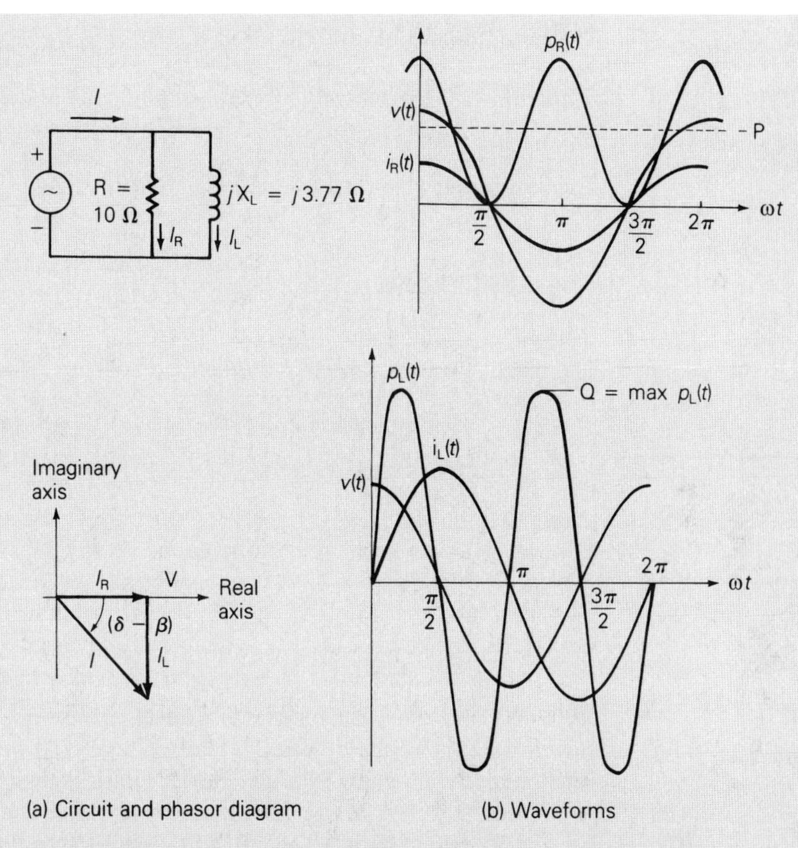

SOLUTION The circuit and phasor diagram are shown in Figure 2.3(a). The load voltage is

$$V = \frac{141.4}{\sqrt{2}}\underline{/0^\circ} = 100\underline{/0^\circ} \quad \text{volts}$$

The resistor current is

$$I_R = \frac{V}{R} = \frac{100}{10}\underline{/0^\circ} = 10\underline{/0^\circ} \quad \text{A}$$

The inductor current is

$$I_L = \frac{V}{jX_L} = \frac{100}{(j3.77)}\underline{/0^\circ} = 26.53\underline{/-90^\circ} \quad \text{A}$$

The total load current is

$$I = I_R + I_L = 10 - j26.53 = 28.35\underline{/-69.34^\circ} \quad \text{A}$$

The instantaneous power absorbed by the resistor is, from (2.2.3),

$$\begin{aligned} p_R(t) &= (100)(10)[1 + \cos(2\omega t)] \\ &= 1000[1 + \cos(2\omega t)] \quad \text{W} \end{aligned}$$

The instantaneous power absorbed by the inductor is, from (2.2.6),

$$p_L(t) = (100)(26.53)\sin(2\omega t)$$
$$= 2653\sin(2\omega t) \quad \text{W}$$

The real power absorbed by the load is, from (2.2.11),

$$P = VI\cos(\delta - \beta) = (100)(28.53)\cos(0° + 69.34°)$$
$$= 1000 \quad \text{W}$$

(*Note*: P is also equal to $VI_R = V^2/R$.)
The reactive power absorbed by the load is, from (2.2.12),

$$Q = VI\sin(\delta - \beta) = (100)(28.53)\sin(0° + 69.34°)$$
$$= 2653 \quad \text{var}$$

(*Note*: Q is also equal to $VI_L = V^2/X_L$.)
The power factor is

$$\text{p.f.} = \cos(\delta - \beta) = \cos(69.34°) = 0.3528 \quad \text{lagging}$$

Voltage, current, and power waveforms are shown in Figure 2.3(b).

As shown for this parallel RL load, the resistor absorbs real power (1000 W) and the inductor absorbs reactive power (2653 var). The resistor current $i_R(t)$ is in phase with the load voltage, and the inductor current $i_L(t)$ lags the load voltage by 90°. The power factor is lagging for an RL load.

Note that $p_R(t)$ and $p_X(t)$, given by (2.2.10), are strictly valid only for a parallel R-X load. For a general RLC load, the voltages across the resistive and reactive components may not be in phase with the source voltage $v(t)$, resulting in additional phase shifts in $p_R(t)$ and $p_X(t)$ (see Problem 2.13). However, (2.2.11) and (2.2.12) for P and Q are valid for a general RLC load. ■

PHYSICAL SIGNIFICANCE OF REAL AND REACTIVE POWER

The physical significance of real power P is easily understood. The total energy absorbed by a load during a time interval T, consisting of one cycle of the sinusoidal voltage, is PT watt-seconds (Ws). During a time interval of n cycles, the energy absorbed is $P(nT)$ watt-seconds, all of which is absorbed by the resistive component of the load. A kilowatt-hour meter is designed to measure the energy absorbed by a load during a time interval $(t_2 - t_1)$, consisting of an integral number of cycles, by integrating the real power P over the time interval $(t_2 - t_1)$.

The physical significance of reactive power Q is not as easily understood. Q refers to the maximum value of the instantaneous power absorbed by the reactive component of the load. The instantaneous reactive power, given by the second term $p_X(t)$ in (2.2.10), is alternately positive and negative, and it expresses the reversible flow of energy to and from the reactive component of the load. Q may be positive or negative, depending on the sign of $(\delta - \beta)$ in (2.2.12). Reactive power Q is a useful quantity when describing the operation of power systems (this will become evident in later chapters). As one example, shunt capacitors can be used in transmission systems to deliver reactive power and thereby increase voltage magnitudes during heavy load periods (see Chapter 5).

2.3 COMPLEX POWER

For circuits operating in sinusoidal-steady-state, real and reactive power are conveniently calculated from complex power, defined below. Let the voltage across a circuit element be $V = \mathrm{V}\underline{/\delta}$, and the current into the element be $I = \mathrm{I}\underline{/\beta}$. Then the complex power S is the product of the voltage and the conjugate of the current:

$$\begin{aligned} S = VI^* &= [\mathrm{V}\underline{/\delta}][\mathrm{I}\underline{/\beta}]^* = \mathrm{VI}\underline{/\delta - \beta} \\ &= \mathrm{VI}\cos(\delta - \beta) + j\mathrm{VI}\sin(\delta - \beta) \end{aligned} \tag{2.3.1}$$

where $(\delta - \beta)$ is the angle between the voltage and current. Comparing (2.3.1) with (2.2.11) and (2.2.12), S is recognized as

$$S = \mathrm{P} + j\mathrm{Q} \tag{2.3.2}$$

The magnitude $\mathrm{S} = \mathrm{VI}$ of the complex power S is called the *apparent power*. Although it has the same units as P and Q, it is common practice to define the units of apparent power S as voltamperes or VA. The real power P is obtained by multiplying the apparent power $\mathrm{S} = \mathrm{VI}$ by the power factor p.f. $= \cos(\delta - \beta)$.

The procedure for determining whether a circuit element absorbs or delivers power is summarized in Figure 2.4. Figure 2.4(a) shows the *load*

FIGURE 2.4 Load and generator conventions

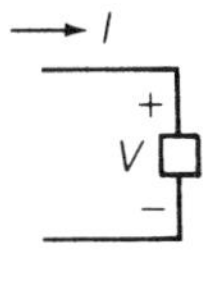

(a) *Load convention.* Current *enters* positive terminal of circuit element. If P is positive, then positive real power is *absorbed.* If Q is positive, then positive reactive power is *absorbed.* If P (Q) is negative, then positive real (reactive) power is *delivered.*

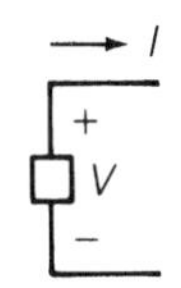

(b) *Generator convention.* Current *leaves* positive terminal of the circuit element. If P is positive, then positive real power is *delivered.* If Q is positive, then positive reactive power is *delivered.* If P (Q) is negative, then positive real (reactive) power is *absorbed.*

convention, where the current *enters* the positive terminal of the circuit element, and the complex power *absorbed* by the circuit element is calculated from (2.3.1). This equation shows that, depending on the value of $(\delta - \beta)$, P may have either a positive or negative value. If P is positive, then the circuit element absorbs positive real power. However, if P is negative, the circuit element absorbs negative real power, or alternatively, it delivers positive real power. Similarly, if Q is positive, the circuit element in Figure 2.4(a) absorbs positive reactive power. However, if Q is negative, the circuit element absorbs negative reactive power, or it delivers positive reactive power.

Figure 2.4(b) shows the *generator convention*, where the current *leaves* the positive terminal of the circuit element, and the complex power *delivered* is calculated from (2.3.1). When P is positive (negative) the circuit element *delivers* positive (negative) real power. Similarly, when Q is positive (negative), the circuit element *delivers* positive (negative) reactive power.

EXAMPLE 2.2 Real and reactive power, delivered or absorbed

A single-phase voltage source with $V = 100\underline{/130^\circ}$ volts delivers a current $I = 10\underline{/10^\circ}$ A, which leaves the positive terminal of the source. Calculate the source real and reactive power, and state whether the source delivers or absorbs each of these.

SOLUTION Since I leaves the positive terminal of the source, the generator convention is assumed, and the complex power delivered is, from (2.3.1),

$$S = VI^* = [100\underline{/130^\circ}][10\underline{/10^\circ}]^*$$

$$S = 1000\underline{/120^\circ} = -500 + j866$$

$$\text{P} = \text{Re}[S] = -500 \quad \text{W}$$

$$\text{Q} = \text{Im}[S] = +866 \quad \text{var}$$

where Im denotes "imaginary part of." The source absorbs 500 W and delivers 866 var. Readers familiar with electric machines will recognize that one example of this source is a synchronous motor. When a synchronous motor operates at a leading power factor, it absorbs real power and delivers reactive power. ■

The *load convention* is used for the RLC elements shown in Figure 2.2. Therefore, the complex power *absorbed* by any of these three elements can be calculated as follows. Assume a load voltage $V = \text{V}\underline{/\delta}$. Then, from (2.3.1),

$$\text{resistor: } S_\text{R} = VI_\text{R}^* = [\text{V}\underline{/\delta}]\left[\frac{\text{V}}{\text{R}}\underline{/-\delta}\right] = \frac{\text{V}^2}{\text{R}} \tag{2.3.3}$$

$$\text{inductor: } S_\text{L} = VI_\text{L}^* = [\text{V}\underline{/\delta}]\left[\frac{\text{V}}{-j\text{X}_\text{L}}\underline{/-\delta}\right] = +j\frac{\text{V}^2}{\text{X}_\text{L}} \tag{2.3.4}$$

$$\text{capacitor: } S_\text{C} = VI_\text{C}^* = [\text{V}\underline{/\delta}]\left[\frac{\text{V}}{j\text{X}_\text{C}}\underline{/-\delta}\right] = -j\frac{\text{V}^2}{\text{X}_\text{C}} \tag{2.3.5}$$

FIGURE 2.5

Power triangle

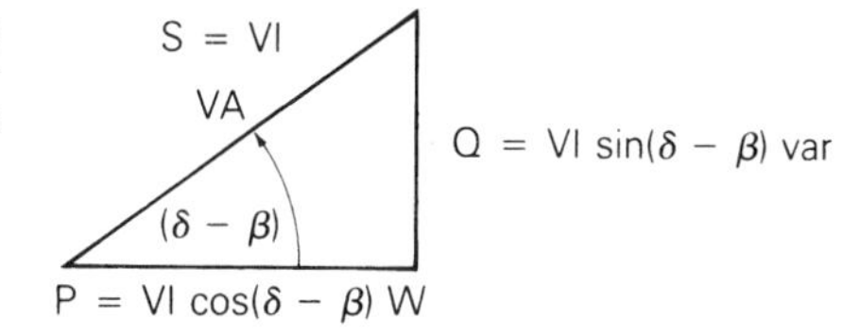

From these complex power expressions, the following can be stated:

A (positive-valued) resistor absorbs (positive) real power, $P_R = V^2/R$ W, and zero reactive power, $Q_R = 0$ var.

An inductor absorbs zero real power, $P_L = 0$ W, and positive reactive power, $Q_L = V^2/X_L$ var.

A capacitor *absorbs* zero real power, $P_C = 0$ W, and *negative* reactive power, $Q_C = -V^2/X_C$ var. Alternatively, a capacitor *delivers positive* reactive power, $+V^2/X_C$.

For a general load composed of RLC elements, complex power S is also calculated from (2.3.1). The real power $P = \text{Re}(S)$ absorbed by a passive load is always positive. The reactive power $Q = \text{Im}(S)$ absorbed by a load may be either positive or negative. When the load is inductive, the current lags the voltage, which means β is less than δ in (2.3.1), and the reactive power absorbed is positive. When the load is capacitive, the current leads the voltage, which means β is greater than δ, and the reactive power absorbed is negative; or, alternatively, the capacitive load delivers positive reactive power.

Complex power can be summarized graphically by use of the power triangle shown in Figure 2.5. As shown, the apparent power S, real power P, and reactive power Q form the three sides of the power triangle. The power factor angle $(\delta - \beta)$ is also shown, and the following expressions can be obtained:

$$S = \sqrt{P^2 + Q^2} \tag{2.3.6}$$

$$(\delta - \beta) = \tan^{-1}(Q/P) \tag{2.3.7}$$

$$Q = P\tan(\delta - \beta) \tag{2.3.8}$$

$$\text{p.f.} = \cos(\delta - \beta) = \frac{P}{S} = \frac{P}{\sqrt{P^2 + Q^2}} \tag{2.3.9}$$

EXAMPLE 2.3 Power triangle and power factor correction

A single-phase source delivers 100 kW to a load operating at a power factor of 0.8 lagging. Calculate the reactive power to be delivered by a capacitor connected in parallel with the load in order to raise the source power factor

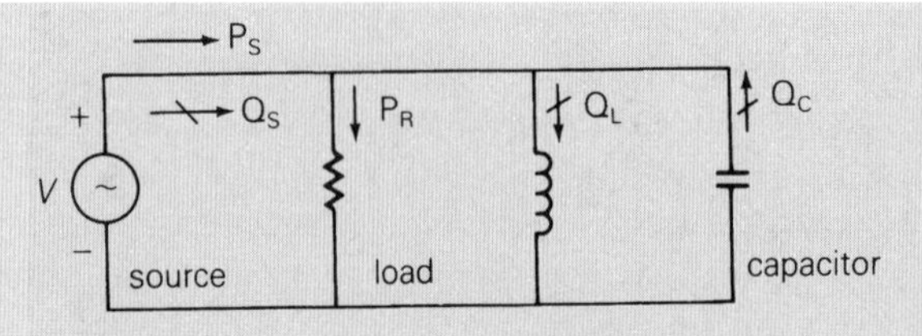

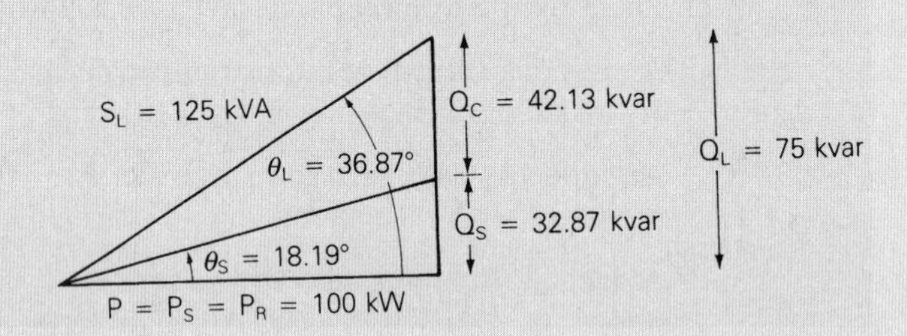

FIGURE 2.6 Circuit and power triangle for Example 2.3

to 0.95 lagging. Also draw the power triangle for the source and load. Assume that the source voltage is constant, and neglect the line impedance between the source and load.

SOLUTION The circuit and power triangle are shown in Figure 2.6. The real power $P = P_S = P_R$ delivered by the source and absorbed by the load is not changed when the capacitor is connected in parallel with the load, since the capacitor delivers only reactive power Q_C. For the load, the power factor angle, reactive power absorbed, and apparent power are

$$\theta_L = (\delta - \beta_L) = \cos^{-1}(0.8) = 36.87^\circ$$

$$Q_L = P \tan\theta_L = 100\tan(36.87^\circ) = 75 \quad \text{kvar}$$

$$S_L = \frac{P}{\cos\theta_L} = 125 \quad \text{kVA}$$

After the capacitor is connected, the power factor angle, reactive power delivered, and apparent power of the source are

$$\theta_S = (\delta - \beta_S) = \cos^{-1}(0.95) = 18.19^\circ$$

$$Q_S = P \tan\theta_S = 100\tan(18.19^\circ) = 32.87 \quad \text{kvar}$$

$$S_S = \frac{P}{\cos\theta_S} = \frac{100}{0.95} = 105.3 \quad \text{kVA}$$

The capacitor delivers

$$Q_C = Q_L - Q_S = 75 - 32.87 = 42.13 \quad \text{kvar}$$

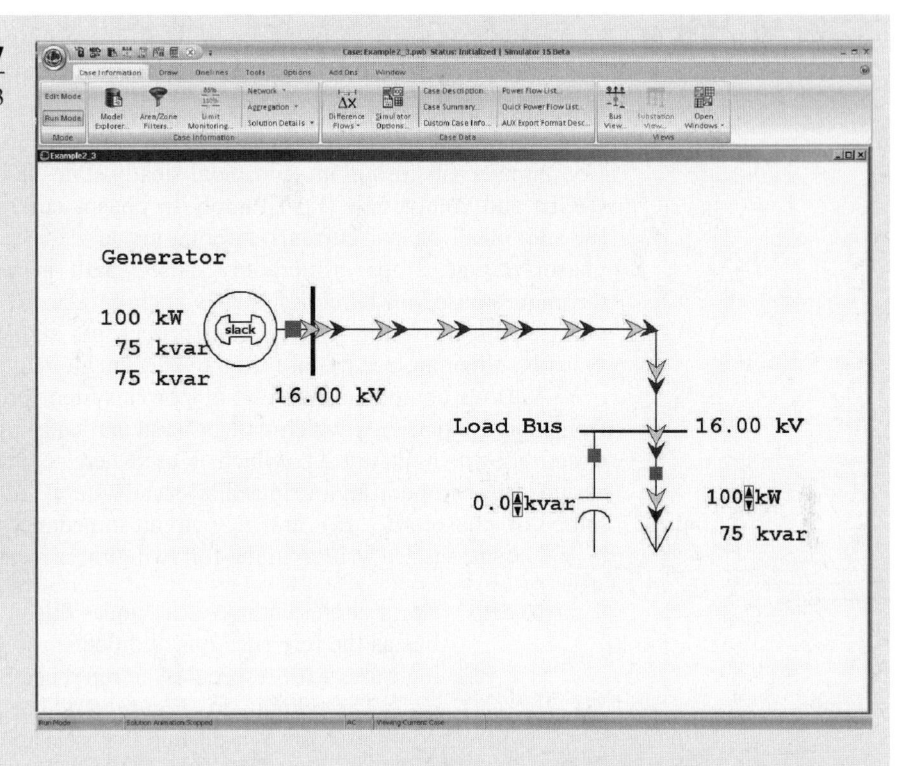

FIGURE 2.7

Screen for Example 2.3

The method of connecting a capacitor in parallel with an inductive load is known as *power factor correction*. The effect of the capacitor is to increase the power factor of the source that delivers power to the load. Also, the source apparent power S_S decreases. As shown in Figure 2.6, the source apparent power for this example decreases from 125 kVA without the capacitor to 105.3 kVA with the capacitor. The source current $I_S = S_S/V$ also decreases. When line impedance between the source and load is included, the decrease in source current results in lower line losses and lower line-voltage drops. The end result of power factor correction is improved efficiency and improved voltage regulation.

To see an animated view of this example, open PowerWorld Simulator case Example 2.3 (see Figure 2.7). From the Ribbon select the green and black "Play" button to begin the simulation. The speed and size of the green arrows are proportional to the real power supplied to the load bus, and the blue arrows are proportional to the reactive power. Here reactive compensation can be supplied in discrete 20-kVar steps by clicking on the arrows in the capacitor's kvar field, and the load can be varied by clicking on the arrows in the load field. Notice that increasing the reactive compensation decreases both the reactive power flow on the supply line and the kVA power supplied by the generator; the real power flow is unchanged. ■

2.4

NETWORK EQUATIONS

For circuits operating in sinusoidal-steady-state, Kirchhoff's current law (KCL) and voltage law (KVL) apply to phasor currents and voltages. Thus the sum of all phasor currents entering any node is zero and the sum of the phasor-voltage drops around any closed path is zero. Network analysis techniques based on Kirchhoff's laws, including nodal analysis, mesh or loop analysis, superposition, source transformations, and Thévenin's theorem or Norton's theorem, are useful for analyzing such circuits.

Various computer solutions of power system problems are formulated from nodal equations, which can be systematically applied to circuits. The circuit shown in Figure 2.8, which is used here to review nodal analysis, is assumed to be operating in sinusoidal-steady-state; source voltages are represented by phasors E_{S1}, E_{S2}, and E_{S3}; circuit impedances are specified in ohms. Nodal equations are written in the following three steps:

STEP 1 For a circuit with $(N+1)$ nodes (also called buses), select one bus as the reference bus and define the voltages at the remaining buses with respect to the reference bus.

The circuit in Figure 2.8 has four buses—that is, $N+1=4$ or $N=3$. Bus 0 is selected as the reference bus, and bus voltages V_{10}, V_{20}, and V_{30} are then defined with respect to bus 0.

STEP 2 Transform each voltage source in series with an impedance to an equivalent current source in parallel with that impedance. Also, show admittance values instead of impedance values on the circuit diagram. Each current source is equal to the voltage source divided by the source impedance.

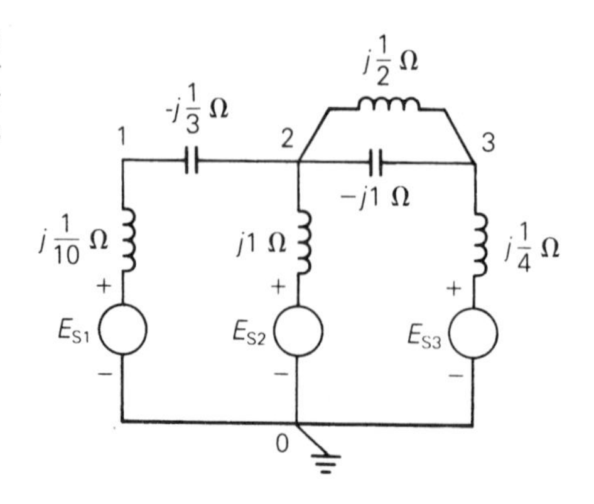

FIGURE 2.8 Circuit diagram for reviewing nodal analysis

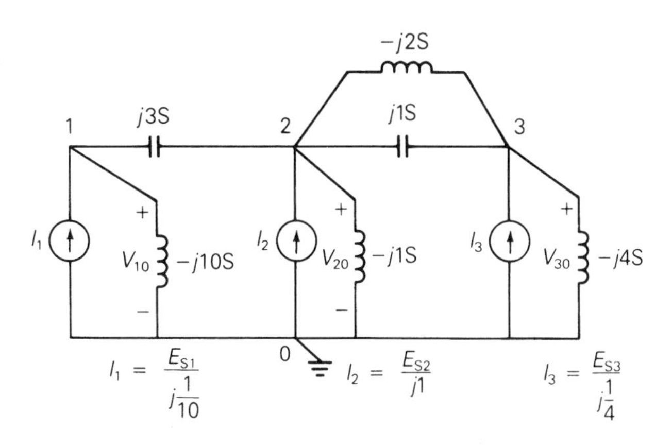

FIGURE 2.9

Circuit of Figure 2.8 with equivalent current sources replacing voltage sources. Admittance values are also shown

In Figure 2.9 equivalent current sources I_1, I_2, and I_3 are shown, and all impedances are converted to corresponding admittances.

STEP 3 Write nodal equations in matrix format as follows:

$$\begin{bmatrix} Y_{11} & Y_{12} & Y_{13} & \cdots & Y_{1N} \\ Y_{21} & Y_{22} & Y_{23} & \cdots & Y_{2N} \\ Y_{31} & Y_{32} & Y_{33} & \cdots & Y_{3N} \\ \vdots & \vdots & \vdots & & \vdots \\ Y_{N1} & Y_{N2} & Y_{N3} & \cdots & Y_{NN} \end{bmatrix} \begin{bmatrix} V_{10} \\ V_{20} \\ V_{30} \\ \vdots \\ V_{N0} \end{bmatrix} = \begin{bmatrix} I_1 \\ I_2 \\ I_3 \\ \vdots \\ I_N \end{bmatrix} \tag{2.4.1}$$

Using matrix notation, (2.4.1) becomes

$$\boldsymbol{YV} = \boldsymbol{I} \tag{2.4.2}$$

where $\boldsymbol{Y}$ is the $N \times N$ bus admittance matrix, $\boldsymbol{V}$ is the column vector of N bus voltages, and $\boldsymbol{I}$ is the column vector of N current sources. The elements Y_{kn} of the bus admittance matrix $\boldsymbol{Y}$ are formed as follows:

diagonal elements: Y_{kk} = sum of admittances connected to bus k $(k = 1, 2, \ldots, N)$ (2.4.3)

off-diagonal elements: $Y_{kn} = -$(sum of admittances connected between buses k and n) $(k \neq n)$ (2.4.4)

The diagonal element Y_{kk} is called the *self-admittance* or the *driving-point admittance* of bus k, and the off-diagonal element Y_{kn} for $k \neq n$ is called the *mutual admittance* or the *transfer admittance* between buses k and n. Since $Y_{kn} = Y_{nk}$, the matrix $\boldsymbol{Y}$ is symmetric.

For the circuit of Figure 2.9, (2.4.1) becomes

$$\begin{bmatrix} (j3-j10) & -(j3) & 0 \\ -(j3) & (j3-j1+j1-j2) & -(j1-j2) \\ 0 & -(j1-j2) & (j1-j2-j4) \end{bmatrix} \begin{bmatrix} V_{10} \\ V_{20} \\ V_{30} \end{bmatrix} = \begin{bmatrix} I_1 \\ I_2 \\ I_3 \end{bmatrix}$$

$$j\begin{bmatrix} -7 & -3 & 0 \\ -3 & 1 & 1 \\ 0 & 1 & -5 \end{bmatrix} \begin{bmatrix} V_{10} \\ V_{20} \\ V_{30} \end{bmatrix} = \begin{bmatrix} I_1 \\ I_2 \\ I_3 \end{bmatrix} \tag{2.4.5}$$

The advantage of this method of writing nodal equations is that a digital computer can be used both to generate the admittance matrix $\boldsymbol{Y}$ and to solve (2.4.2) for the unknown bus voltage vector $\boldsymbol{V}$. Once a circuit is specified with the reference bus and other buses identified, the circuit admittances and their bus connections become computer input data for calculating the elements Y_{kn} via (2.4.3) and (2.4.4). After $\boldsymbol{Y}$ is calculated and the current source vector $\boldsymbol{I}$ is given as input, standard computer programs for solving simultaneous linear equations can then be used to determine the bus voltage vector $\boldsymbol{V}$.

When double subscripts are used to denote a voltage in this text, the voltage shall be that at the node identified by the first subscript with respect to the node identified by the second subscript. For example, the voltage V_{10} in Figure 2.9 is the voltage at node 1 with respect to node 0. Also, a current I_{ab} shall indicate the current from node a to node b. Voltage polarity marks $(+/-)$ and current reference arrows $(\rightarrow)$ are not required when double subscript notation is employed. The polarity marks in Figure 2.9 for V_{10}, V_{20}, and V_{30}, although not required, are shown for clarity. The reference arrows for sources I_1, I_2, and I_3 in Figure 2.9 are required, however, since single subscripts are used for these currents. Matrices and vectors shall be indicated in this text by boldface type (for example, $\boldsymbol{Y}$ or $\boldsymbol{V}$).

2.5 BALANCED THREE-PHASE CIRCUITS

In this section we introduce the following topics for balanced three-phase circuits: Y connections, line-to-neutral voltages, line-to-line voltages, line currents, Δ loads, Δ–Y conversions, and equivalent line-to-neutral diagrams.

FIGURE 2.10

Circuit diagram of a three-phase Y-connected source feeding a balanced-Y load

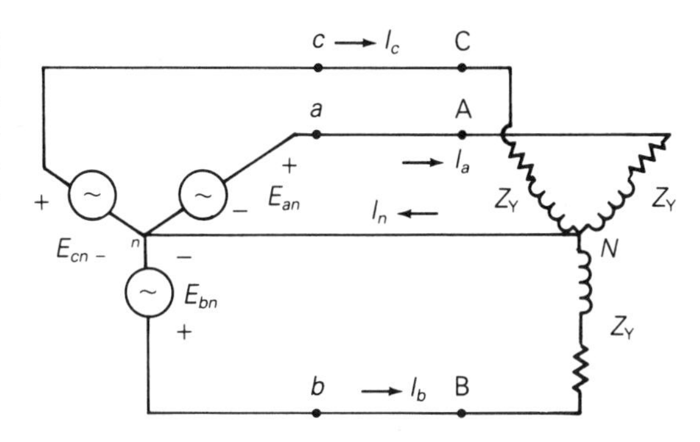

BALANCED-Y CONNECTIONS

Figure 2.10 shows a three-phase Y-connected (or "wye-connected") voltage source feeding a balanced-Y-connected load. For a Y connection, the neutrals of each phase are connected. In Figure 2.10 the source neutral connection is labeled bus n and the load neutral connection is labeled bus N. The three-phase source is assumed to be ideal since source impedances are neglected. Also neglected are the line impedances between the source and load terminals, and the neutral impedance between buses n and N. The three-phase load is *balanced*, which means the load impedances in all three phases are identical.

BALANCED LINE-TO-NEUTRAL VOLTAGES

In Figure 2.10, the terminal buses of the three-phase source are labeled a, b, and c, and the source line-to-neutral voltages are labeled E_{an}, E_{bn}, and E_{cn}. The source is *balanced* when these voltages have equal magnitudes and an equal 120°-phase difference between any two phases. An example of balanced three-phase line-to-neutral voltages is

$$
\begin{aligned}
E_{an} &= 10\underline{/0^\circ} \\
E_{bn} &= 10\underline{/-120^\circ} = 10\underline{/+240^\circ} \\
E_{cn} &= 10\underline{/+120^\circ} = 10\underline{/-240^\circ} \quad \text{volts}
\end{aligned}
\tag{2.5.1}
$$

FIGURE 2.11

Phasor diagram of balanced positive-sequence line-to-neutral voltages with E_{an} as the reference

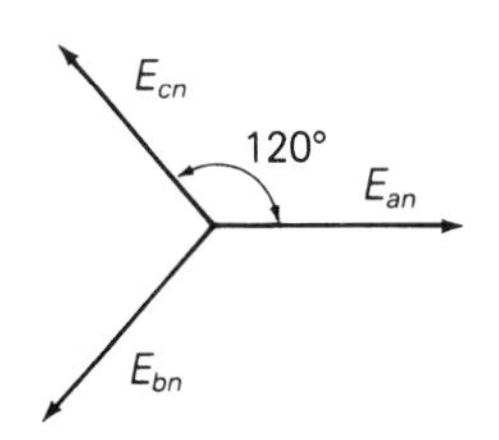

where the line-to-neutral voltage magnitude is 10 volts and E_{an} is the reference phasor. The phase sequence is called *positive sequence* or *abc* sequence when E_{an} leads E_{bn} by 120° and E_{bn} leads E_{cn} by 120°. The phase sequence is called *negative sequence* or *acb* sequence when E_{an} leads E_{cn} by 120° and E_{cn} leads E_{bn} by 120°. The voltages in (2.5.1) are positive-sequence voltages, since E_{an} leads E_{bn} by 120°. The corresponding phasor diagram is shown in Figure 2.11.

BALANCED LINE-TO-LINE VOLTAGES

The voltages E_{ab}, E_{bc}, and E_{ca} between phases are called line-to-line voltages. Writing a KVL equation for a closed path around buses a, b, and n in Figure 2.10,

$$E_{ab} = E_{an} - E_{bn} \tag{2.5.2}$$

For the line-to-neutral voltages of (2.5.1),

$$\begin{aligned} E_{ab} &= 10\underline{/0^\circ} - 10\underline{/-120^\circ} = 10 - 10\left[\frac{-1 - j\sqrt{3}}{2}\right] \\ E_{ab} &= \sqrt{3}(10)\left(\frac{\sqrt{3} + j1}{2}\right) = \sqrt{3}(10\underline{/30^\circ}) \quad \text{volts} \end{aligned} \tag{2.5.3}$$

Similarly, the line-to-line voltages E_{bc} and E_{ca} are

$$\begin{aligned} E_{bc} &= E_{bn} - E_{cn} = 10\underline{/-120^\circ} - 10\underline{/+120^\circ} \\ &= \sqrt{3}(10\underline{/-90^\circ}) \quad \text{volts} \end{aligned} \tag{2.5.4}$$

$$\begin{aligned} E_{ca} &= E_{cn} - E_{an} = 10\underline{/+120^\circ} - 10\underline{/0^\circ} \\ &= \sqrt{3}(10\underline{/150^\circ}) \quad \text{volts} \end{aligned} \tag{2.5.5}$$

The line-to-line voltages of (2.5.3)–(2.5.5) are also balanced, since they have equal magnitudes of $\sqrt{3}(10)$ volts and 120° displacement between any two phases. Comparison of these line-to-line voltages with the line-to-neutral voltages of (2.5.1) leads to the following conclusion:

> In a balanced three-phase Y-connected system with positive-sequence sources, the line-to-line voltages are $\sqrt{3}$ times the line-to-neutral voltages and lead by 30°. That is,
>
> $$\begin{aligned} E_{ab} &= \sqrt{3}E_{an}\underline{/+30^\circ} \\ E_{bc} &= \sqrt{3}E_{bn}\underline{/+30^\circ} \\ E_{ca} &= \sqrt{3}E_{cn}\underline{/+30^\circ} \end{aligned} \tag{2.5.6}$$

This very important result is summarized in Figure 2.12. In Figure 2.12(a) each phasor begins at the origin of the phasor diagram. In Figure 2.12(b) the line-to-line voltages form an equilateral triangle with vertices labeled a, b, c corresponding to buses a, b, and c of the system; the line-to-neutral voltages begin at the vertices and end at the center of the triangle, which is labeled n for neutral bus n. Also, the clockwise sequence of the vertices abc in Figure 2.12(b) indicates positive-sequence voltages. In both diagrams, E_{an} is the reference. However, the diagrams could be rotated to align with any other reference.

Since the balanced line-to-line voltages form a closed triangle in Figure 2.12, their sum is zero. In fact, the sum of line-to-line voltages $(E_{ab} + E_{bc} + E_{ca})$

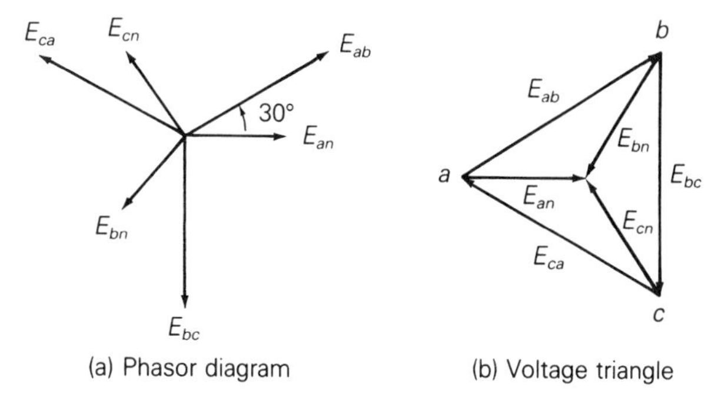

FIGURE 2.12 Positive-sequence line-to-neutral and line-to-line voltages in a balanced three-phase Y-connected system

is *always* zero, even if the system is unbalanced, since these voltages form a closed path around buses a, b, and c. Also, in a balanced system the sum of the line-to-neutral voltages $(E_{an} + E_{bn} + E_{cn})$ equals zero.

BALANCED LINE CURRENTS

Since the impedance between the source and load neutrals in Figure 2.10 is neglected, buses n and N are at the same potential, $E_{nN} = 0$. Accordingly, a separate KVL equation can be written for each phase, and the line currents can be written by inspection:

$$\begin{aligned} I_a &= E_{an}/Z_Y \\ I_b &= E_{bn}/Z_Y \\ I_c &= E_{cn}/Z_Y \end{aligned} \tag{2.5.7}$$

For example, if each phase of the Y-connected load has an impedance $Z_Y = 2\underline{/30^\circ}\ \Omega$, then

$$\begin{aligned} I_a &= \frac{10\underline{/0^\circ}}{2\underline{/30^\circ}} = 5\underline{/-30^\circ} \quad \text{A} \\ I_b &= \frac{10\underline{/-120^\circ}}{2\underline{/30^\circ}} = 5\underline{/-150^\circ} \quad \text{A} \\ I_c &= \frac{10\underline{/+120^\circ}}{2\underline{/30^\circ}} = 5\underline{/90^\circ} \quad \text{A} \end{aligned} \tag{2.5.8}$$

The line currents are also balanced, since they have equal magnitudes of 5 A and 120° displacement between any two phases. The neutral current I_n is determined by writing a KCL equation at bus N in Figure 2.10.

$$I_n = I_a + I_b + I_c \tag{2.5.9}$$

Using the line currents of (2.5.8),

$$I_n = 5\underline{/-30^\circ} + 5\underline{/-150^\circ} + 5\underline{/90^\circ}$$
$$I_n = 5\left(\frac{\sqrt{3} - j1}{2}\right) + 5\left(\frac{-\sqrt{3} - j1}{2}\right) + j5 = 0 \tag{2.5.10}$$

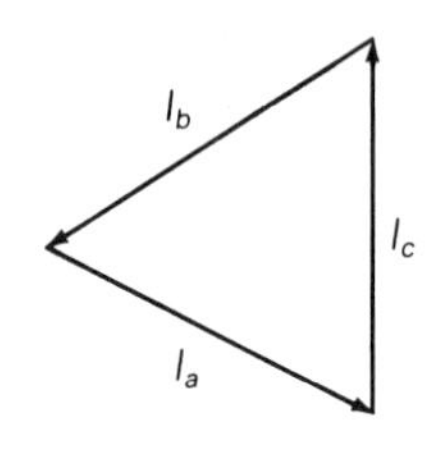

FIGURE 2.13 Phasor diagram of line currents in a balanced three-phase system

The phasor diagram of the line currents is shown in Figure 2.13. Since these line currents form a closed triangle, their sum, which is the neutral current I_n, is zero. In general, the sum of any balanced three-phase set of phasors is zero, since balanced phasors form a closed triangle. Thus, although the impedance between neutrals n and N in Figure 2.10 is assumed to be zero, the neutral current will be zero for *any* neutral impedance ranging from short circuit (0 Ω) to open circuit (∞ Ω), as long as the system is balanced. If the system is not balanced—which could occur if the source voltages, load impedances, or line impedances were unbalanced—then the line currents will not be balanced and a neutral current I_n may flow between buses n and N.

BALANCED Δ LOADS

Figure 2.14 shows a three-phase Y-connected source feeding a balanced-Δ-connected (or "delta-connected") load. For a balanced-Δ connection, equal load impedances Z_Δ are connected in a triangle whose vertices form the buses, labeled A, B, and C in Figure 2.14. The Δ connection does not have a neutral bus.

Since the line impedances are neglected in Figure 2.14, the source line-to-line voltages are equal to the load line-to-line voltages, and the Δ-load currents I_{AB}, I_{BC}, and I_{CA} are

$$I_{AB} = E_{ab}/Z_\Delta$$
$$I_{BC} = E_{bc}/Z_\Delta \tag{2.5.11}$$
$$I_{CA} = E_{ca}/Z_\Delta$$

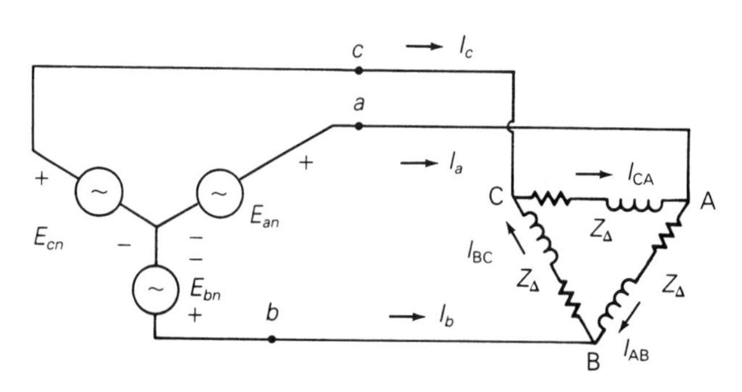

FIGURE 2.14 Circuit diagram of a Y-connected source feeding a balanced-Δ load

For example, if the line-to-line voltages are given by (2.5.3)–(2.5.5) and if $Z_\Delta = 5\underline{/30^\circ}\ \Omega$, then the Δ-load currents are

$$I_{AB} = \sqrt{3}\left(\frac{10\underline{/30^\circ}}{5\underline{/30^\circ}}\right) = 3.464\underline{/0^\circ} \quad \text{A}$$

$$I_{BC} = \sqrt{3}\left(\frac{10\underline{/-90^\circ}}{5\underline{/30^\circ}}\right) = 3.464\underline{/-120^\circ} \quad \text{A} \tag{2.5.12}$$

$$I_{CA} = \sqrt{3}\left(\frac{10\underline{/150^\circ}}{5\underline{/30^\circ}}\right) = 3.464\underline{/+120^\circ} \quad \text{A}$$

Also, the line currents can be determined by writing a KCL equation at each bus of the Δ load, as follows:

$$I_a = I_{AB} - I_{CA} = 3.464\underline{/0^\circ} - 3.464\underline{/120^\circ} = \sqrt{3}(3.464\underline{/-30^\circ})$$

$$I_b = I_{BC} - I_{AB} = 3.464\underline{/-120^\circ} - 3.464\underline{/0^\circ} = \sqrt{3}(3.464\underline{/-150^\circ}) \tag{2.5.13}$$

$$I_c = I_{CA} - I_{BC} = 3.464\underline{/120^\circ} - 3.464\underline{/-120^\circ} = \sqrt{3}(3.464\underline{/+90^\circ})$$

Both the Δ-load currents given by (2.5.12) and the line currents given by (2.5.13) are balanced. Thus the sum of balanced Δ-load currents $(I_{AB} + I_{BC} + I_{CA})$ equals zero. The sum of line currents $(I_a + I_b + I_c)$ is always zero for a Δ-connected load even if the system is unbalanced, since there is no neutral wire. Comparison of (2.5.12) and (2.5.13) leads to the following conclusion:

For a balanced-Δ load supplied by a balanced positive-sequence source, the line currents into the load are $\sqrt{3}$ times the Δ-load currents and lag by 30°. That is,

$$I_a = \sqrt{3} I_{AB}\underline{/-30^\circ}$$

$$I_b = \sqrt{3} I_{BC}\underline{/-30^\circ} \tag{2.5.14}$$

$$I_c = \sqrt{3} I_{CA}\underline{/-30^\circ}$$

This result is summarized in Figure 2.15.

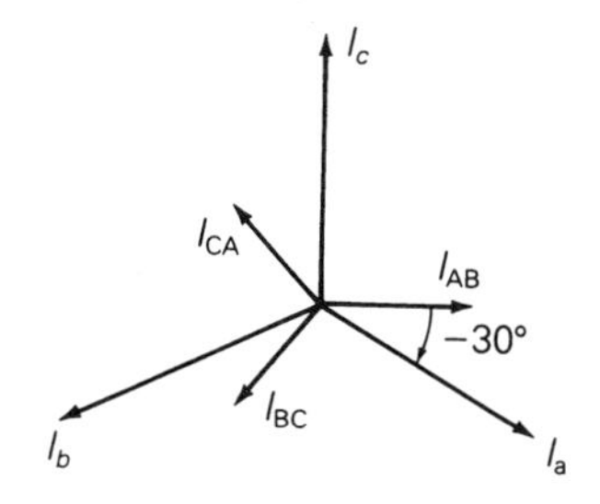

FIGURE 2.15

Phasor diagram of line currents and load currents for a balanced-Δ load

FIGURE 2.16

Δ–Y conversion for balanced loads

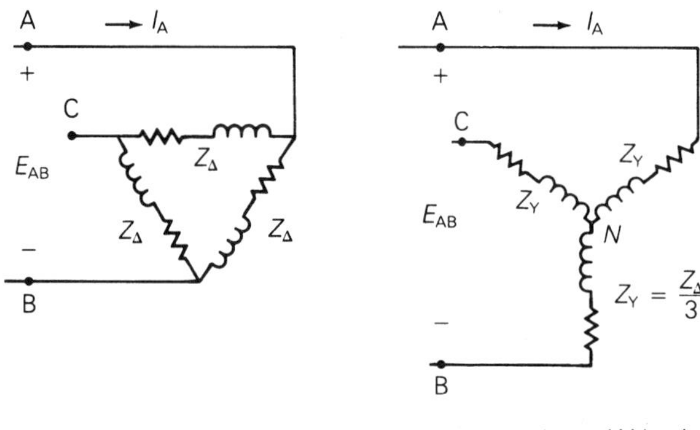

(a) Balanced-Δ load (b) Equivalent balanced-Y load

Δ–Y CONVERSION FOR BALANCED LOADS

Figure 2.16 shows the conversion of a balanced-Δ load to a balanced-Y load. If balanced voltages are applied, then these loads will be equivalent as viewed from their terminal buses A, B, and C when the line currents into the Δ load are the same as the line currents into the Y load. For the Δ load,

$$I_A = \sqrt{3} I_{AB} \underline{/-30^\circ} = \frac{\sqrt{3} E_{AB} \underline{/-30^\circ}}{Z_\Delta} \tag{2.5.15}$$

and for the Y load,

$$I_A = \frac{E_{AN}}{Z_Y} = \frac{E_{AB} \underline{/-30^\circ}}{\sqrt{3} Z_Y} \tag{2.5.16}$$

Comparison of (2.5.15) and (2.5.16) indicates that I_A will be the same for both the Δ and Y loads when

$$Z_Y = \frac{Z_\Delta}{3} \tag{2.5.17}$$

Also, the other line currents I_B and I_C into the Y load will equal those into the Δ load when $Z_Y = Z_\Delta/3$, since these loads are balanced. Thus a balanced-Δ load can be converted to an equivalent balanced-Y load by dividing the Δ-load impedance by 3. The angles of these Δ- and equivalent Y-load impedances are the same. Similarly, a balanced-Y load can be converted to an equivalent balanced-Δ load using $Z_\Delta = 3Z_Y$.

EXAMPLE 2.4 **Balanced Δ and Y loads**

A balanced, positive-sequence, Y-connected voltage source with $E_{ab} = 480\underline{/0^\circ}$ volts is applied to a balanced-Δ load with $Z_\Delta = 30\underline{/40^\circ}$ Ω. The line impedance between the source and load is $Z_L = 1\underline{/85^\circ}$ Ω for each phase. Calculate the line currents, the Δ-load currents, and the voltages at the load terminals.

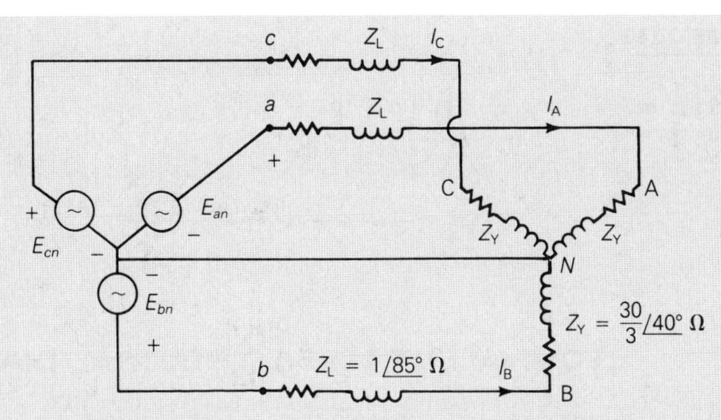

FIGURE 2.17 Circuit diagram for Example 2.4

SOLUTION The solution is most easily obtained as follows. First, convert the Δ load to an equivalent Y. Then connect the source and Y-load neutrals with a zero-ohm neutral wire. The connection of the neutral wire has no effect on the circuit, since the neutral current $I_n = 0$ in a balanced system. The resulting circuit is shown in Figure 2.17. The line currents are

$$I_A = \frac{E_{an}}{Z_L + Z_Y} = \frac{\frac{480}{\sqrt{3}}\underline{/-30°}}{1\underline{/85°} + \frac{30}{3}\underline{/40°}}$$

$$= \frac{277.1\underline{/-30°}}{(0.0872 + j0.9962) + (7.660 + j6.428)} \tag{2.5.18}$$

$$= \frac{277.1\underline{/-30°}}{(7.748 + j7.424)} = \frac{277.1\underline{/-30°}}{10.73\underline{/43.78°}} = 25.83\underline{/-73.78°}\quad \text{A}$$

$$I_B = 25.83\underline{/166.22°}\quad \text{A}$$

$$I_C = 25.83\underline{/46.22°}\quad \text{A}$$

The Δ-load currents are, from (2.5.14),

$$I_{AB} = \frac{I_a}{\sqrt{3}}\underline{/+30°} = \frac{25.83}{\sqrt{3}}\underline{/-73.78° + 30°} = 14.91\underline{/-43.78°}\quad \text{A}$$

$$I_{BC} = 14.91\underline{/-163.78°}\quad \text{A} \tag{2.5.19}$$

$$I_{CA} = 14.91\underline{/+76.22°}\quad \text{A}$$

The voltages at the load terminals are

$$E_{AB} = Z_\Delta I_{AB} = (30\underline{/40°})(14.91\underline{/-43.78°}) = 447.3\underline{/-3.78°}$$

$$E_{BC} = 447.3\underline{/-123.78°} \tag{2.5.20}$$

$$E_{CA} = 447.3\underline{/116.22°}\quad \text{volts}$$

■

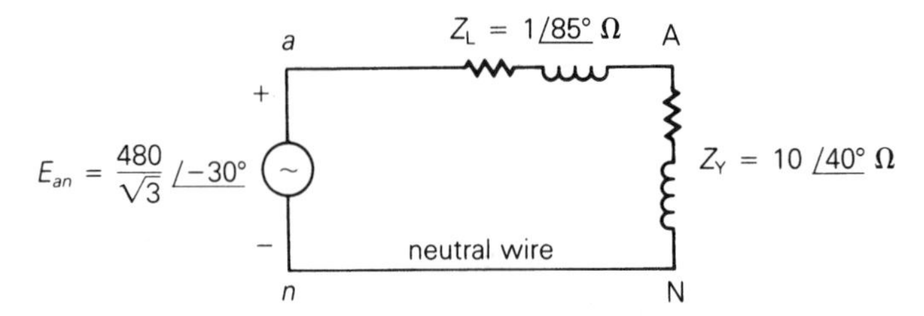

FIGURE 2.18 Equivalent line-to-neutral diagram for the circuit of Example 2.4

EQUIVALENT LINE-TO-NEUTRAL DIAGRAMS

When working with balanced three-phase circuits, only one phase need be analyzed. Δ loads can be converted to Y loads, and all source and load neutrals can be connected with a zero-ohm neutral wire without changing the solution. Then one phase of the circuit can be solved. The voltages and currents in the other two phases are equal in magnitude to and $\pm 120°$ out of phase with those of the solved phase. Figure 2.18 shows an equivalent line-to-neutral diagram for one phase of the circuit in Example 2.4.

When discussing three-phase systems in this text, voltages shall be rms line-to-line voltages unless otherwise indicated. This is standard industry practice.

2.6 POWER IN BALANCED THREE-PHASE CIRCUITS

In this section, we discuss instantaneous power and complex power for balanced three-phase generators and motors and for balanced-Y and Δ-impedance loads.

INSTANTANEOUS POWER: BALANCED THREE-PHASE GENERATORS

Figure 2.19 shows a Y-connected generator represented by three voltage sources with their neutrals connected at bus n and by three identical generator impedances Z_g. Assume that the generator is operating under balanced steady-state conditions with the instantaneous generator terminal voltage given by

$$v_{an}(t) = \sqrt{2} V_{LN} \cos(\omega t + \delta) \quad \text{volts} \tag{2.6.1}$$

and with the instantaneous current leaving the positive terminal of phase a given by

$$i_a(t) = \sqrt{2} I_L \cos(\omega t + \beta) \quad \text{A} \tag{2.6.2}$$

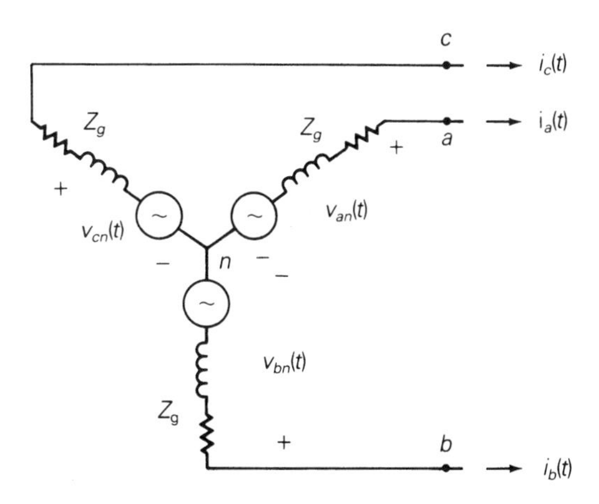

FIGURE 2.19

Y-connected generator

where V_{LN} is the rms line-to-neutral voltage and I_L is the rms line current. The instantaneous power $p_a(t)$ delivered by phase a of the generator is

$$\begin{aligned} p_a(t) &= v_{an}(t)i_a(t) \\ &= 2V_{LN}I_L \cos(\omega t + \delta)\cos(\omega t + \beta) \\ &= V_{LN}I_L \cos(\delta - \beta) + V_{LN}I_L \cos(2\omega t + \delta + \beta) \quad \text{W} \end{aligned} \tag{2.6.3}$$

Assuming balanced operating conditions, the voltages and currents of phases b and c have the same magnitudes as those of phase a and are $\pm 120°$ out of phase with phase a. Therefore the instantaneous power delivered by phase b is

$$\begin{aligned} p_b(t) &= 2V_{LN}I_L \cos(\omega t + \delta - 120°)\cos(\omega t + \beta - 120°) \\ &= V_{LN}I_L \cos(\delta - \beta) + V_{LN}I_L \cos(2\omega t + \delta + \beta - 240°) \quad \text{W} \end{aligned} \tag{2.6.4}$$

and by phase c,

$$\begin{aligned} p_c(t) &= 2V_{LN}I_L \cos(\omega t + \delta + 120°)\cos(\omega t + \beta + 120°) \\ &= V_{LN}I_L \cos(\delta - \beta) + V_{LN}I_L \cos(2\omega t + \delta + \beta + 240°) \quad \text{W} \end{aligned} \tag{2.6.5}$$

The total instantaneous power $p_{3\phi}(t)$ delivered by the three-phase generator is the sum of the instantaneous powers delivered by each phase. Using (2.6.3)–(2.6.5):

$$\begin{aligned} p_{3\phi}(t) &= p_a(t) + p_b(t) + p_c(t) \\ &= 3V_{LN}I_L \cos(\delta - \beta) + V_{LN}I_L[\cos(2\omega t + \delta + \beta) \\ &\quad + \cos(2\omega t + \delta + \beta - 240°) \\ &\quad + \cos(2\omega t + \delta + \beta + 240°)] \quad \text{W} \end{aligned} \tag{2.6.6}$$

The three cosine terms within the brackets of (2.6.6) can be represented by a *balanced* set of three phasors. Therefore, the sum of these three terms is zero for any value of δ, for any value of β, and for all values of t. Equation (2.6.6) then reduces to

$$p_{3\phi}(t) = P_{3\phi} = 3V_{LN}I_L \cos(\delta - \beta) \quad W \tag{2.6.7}$$

Equation (2.6.7) can be written in terms of the line-to-line voltage V_{LL} instead of the line-to-neutral voltage V_{LN}. Under balanced operating conditions,

$$V_{LN} = V_{LL}/\sqrt{3} \qquad \text{and} \qquad P_{3\phi} = \sqrt{3}V_{LL}I_L \cos(\delta - \beta) \quad W \tag{2.6.8}$$

Inspection of (2.6.8) leads to the following conclusion:

> The total instantaneous power delivered by a three-phase generator under balanced operating conditions is not a function of time, but a *constant*, $p_{3\phi}(t) = P_{3\phi}$.

INSTANTANEOUS POWER: BALANCED THREE-PHASE MOTORS AND IMPEDANCE LOADS

The total instantaneous power absorbed by a three-phase motor under balanced steady-state conditions is also a constant. Figure 2.19 can be used to represent a three-phase motor by reversing the line currents to enter rather than leave the positive terminals. Then (2.6.1)–(2.6.8), valid for power *delivered* by a generator, are also valid for power *absorbed* by a motor. These equations are also valid for the instantaneous power absorbed by a balanced three-phase impedance load.

COMPLEX POWER: BALANCED THREE-PHASE GENERATORS

The phasor representations of the voltage and current in (2.6.1) and (2.6.2) are

$$V_{an} = V_{LN}\underline{/\delta} \quad \text{volts} \tag{2.6.9}$$

$$I_a = I_L\underline{/\beta} \quad A \tag{2.6.10}$$

where I_a leaves positive terminal "a" of the generator. The complex power S_a delivered by phase a of the generator is

$$\begin{aligned} S_a &= V_{an}I_a^* = V_{LN}I_L\underline{/(\delta - \beta)} \\ &= V_{LN}I_L \cos(\delta - \beta) + jV_{LN}I_L \sin(\delta - \beta) \end{aligned} \tag{2.6.11}$$

Under balanced operating conditions, the complex powers delivered by phases b and c are identical to S_a, and the total complex power $S_{3\phi}$ delivered by the generator is

$$S_{3\phi} = S_a + S_b + S_c = 3S_a$$
$$= 3V_{LN}I_L\angle(\delta - \beta)$$
$$= 3V_{LN}I_L \cos(\delta - \beta) + j3V_{LN}I_L \sin(\delta - \beta) \quad (2.6.12)$$

In terms of the total real and reactive powers,

$$S_{3\phi} = P_{3\phi} + jQ_{3\phi} \quad (2.6.13)$$

where

$$P_{3\phi} = \mathrm{Re}(S_{3\phi}) = 3V_{LN}I_L \cos(\delta - \beta)$$
$$= \sqrt{3}V_{LL}I_L \cos(\delta - \beta) \quad \mathrm{W} \quad (2.6.14)$$

and

$$Q_{3\phi} = \mathrm{Im}(S_{3\phi}) = 3V_{LN}I_L \sin(\delta - \beta)$$
$$= \sqrt{3}V_{LL}I_L \sin(\delta - \beta) \quad \mathrm{var} \quad (2.6.15)$$

Also, the total apparent power is

$$S_{3\phi} = |S_{3\phi}| = 3V_{LN}I_L = \sqrt{3}V_{LL}I_L \quad \mathrm{VA} \quad (2.6.16)$$

COMPLEX POWER: BALANCED THREE-PHASE MOTORS

The preceding expressions for complex, real, reactive, and apparent power *delivered* by a three-phase generator are also valid for the complex, real, reactive, and apparent power *absorbed* by a three-phase motor.

COMPLEX POWER: BALANCED-Y AND BALANCED-Δ IMPEDANCE LOADS

Equations (2.6.13)–(2.6.16) are also valid for balanced-Y and -Δ impedance loads. For a balanced-Y load, the line-to-neutral voltage across the phase *a* load impedance and the current entering the positive terminal of that load impedance can be represented by (2.6.9) and (2.6.10). Then (2.6.11)–(2.6.16) are valid for the power absorbed by the balanced-Y load.

For a balanced-Δ load, the line-to-line voltage across the phase *a–b* load impedance and the current into the positive terminal of that load impedance can be represented by

$$V_{ab} = V_{LL}\angle\delta \quad \text{volts} \quad (2.6.17)$$

$$I_{ab} = I_{\Delta}\angle\beta \quad \mathrm{A} \quad (2.6.18)$$

where V_{LL} is the rms line-to-line voltage and I_Δ is the rms Δ-load current. The complex power S_{ab} absorbed by the phase a–b load impedance is then

$$S_{ab} = V_{ab}I_{ab}^* = V_{LL}I_\Delta \underline{/(\delta - \beta)} \tag{2.6.19}$$

The total complex power absorbed by the Δ load is

$$\begin{aligned} S_{3\phi} &= S_{ab} + S_{bc} + S_{ca} = 3S_{ab} \\ &= 3V_{LL}I_\Delta \underline{/(\delta - \beta)} \\ &= 3V_{LL}I_\Delta \cos(\delta - \beta) + j3V_{LL}I_\Delta \sin(\delta - \beta) \end{aligned} \tag{2.6.20}$$

Rewriting (2.6.19) in terms of the total real and reactive power,

$$S_{3\phi} = P_{3\phi} + jQ_{3\phi} \tag{2.6.21}$$

$$\begin{aligned} P_{3\phi} &= \text{Re}(S_{3\phi}) = 3V_{LL}I_\Delta \cos(\delta - \beta) \\ &= \sqrt{3}V_{LL}I_L \cos(\delta - \beta) \quad \text{W} \end{aligned} \tag{2.6.22}$$

$$\begin{aligned} Q_{3\phi} &= \text{Im}(S_{3\phi}) = 3V_{LL}I_\Delta \sin(\delta - \beta) \\ &= \sqrt{3}V_{LL}I_L \sin(\delta - \beta) \quad \text{var} \end{aligned} \tag{2.6.23}$$

where the Δ-load current I_Δ is expressed in terms of the line current $I_L = \sqrt{3}I_\Delta$ in (2.6.22) and (2.6.23). Also, the total apparent power is

$$S_{3\phi} = |S_{3\phi}| = 3V_{LL}I_\Delta = \sqrt{3}V_{LL}I_L \quad \text{VA} \tag{2.6.24}$$

Equations (2.6.21)–(2.6.24) developed for the balanced-Δ load are identical to (2.6.13)–(2.6.16).

EXAMPLE 2.5 Power in a balanced three-phase system

Two balanced three-phase motors in parallel, an induction motor drawing 400 kW at 0.8 power factor lagging and a synchronous motor drawing 150 kVA at 0.9 power factor leading, are supplied by a balanced, three-phase 4160-volt source. Cable impedances between the source and load are neglected, (a) Draw the power triangle for each motor and for the combined-motor load. (b) Determine the power factor of the combined-motor load. (c) Determine the magnitude of the line current delivered by the source. (d) A delta-connected capacitor bank is now installed in parallel with the combined-motor load. What value of capacitive reactance is required in each leg of the capacitor bank to make the source power factor unity? (e) Determine the magnitude of the line current delivered by the source with the capacitor bank installed.

SOLUTION **(a)** For the induction motor, P = 400 kW and:

$$S = P/\text{p.f.} = 400/0.8 = 500 \text{ kVA}$$

$$Q = \sqrt{S^2 - P^2} = \sqrt{(500)^2 - (400)^2} = 300 \text{ kvar absorbed}$$

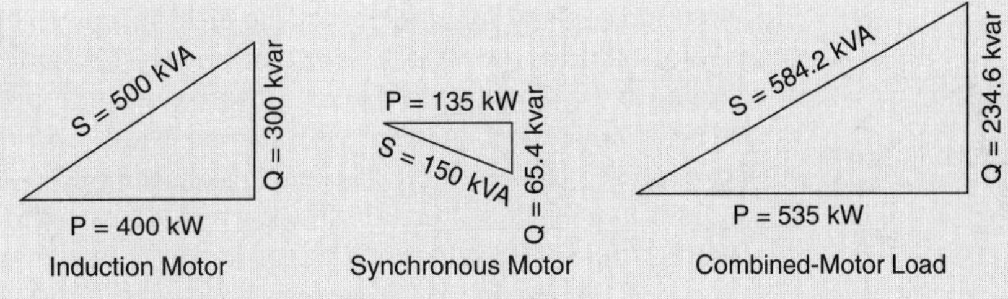

FIGURE 2.20 Power triangles for Example 2.5

For the synchronous motor, S = 150 kVA and

$$P = S(\text{p.f.}) = 150(0.9) = 135 \text{ kW}$$

$$Q = \sqrt{S^2 - P^2} = \sqrt{(150)^2 - (135)^2} = 65.4 \text{ kvar delivered}$$

For the combined-motor load:

$$P = 400 + 135 = 535 \text{ kW} \quad Q = 300 - 65.4 = 234.6 \text{ kvar absorbed}$$

$$S = \sqrt{P^2 + Q^2} = \sqrt{(535)^2 + (234.6)^2} = 584.2 \text{ kVA}$$

(a) The power triangles for each motor and the combined-motor load are shown in Figure 2.20.

(b) The power factor of the combined-motor load is p.f. = P/S = 535/584.2 = 0.916 lagging.

(c) The line current delivered by the source is $I = S/(\sqrt{3}\,V)$, where S is the three-phase apparent power of the combined-motor load and V is the magnitude of the line-to-line load voltage, which is the same as the source voltage for this example. $I = 584.2/(\sqrt{3}\ 4160\ V) = 0.0811$ kA = 81.1 per phase.

(d) For unity power factor, the three-phase reactive power supplied by the capacitor bank should equal the three-phase reactive power absorbed by the combined-motor load. That is, $Q_c = 234.6$ kvar. For a delta-connected capacitor bank, $Q_c = 3V^2/X_\Delta$ where V is the line-to-line voltage across the bank and X_Δ the capacitive reactance of each leg of the bank. The capacitive reactance of each leg is

$$X_\Delta = 3V^2/Q_c = 3(4160^2)/234.6 \times 10^3 = 221.3\ \Omega$$

(e) With the capacitor bank installed, the source power factor is unity and the apparent power S delivered by the source is the same as the real power P delivered by the source. The line current magnitude is

$$I = S/(\sqrt{3}\,V) = P/(\sqrt{3}\,V) = 535/(\sqrt{3}\ 4160) = 0.0743 \text{ kA} = 74.3 \text{ A per phase}$$

In this example, the source voltage of 4160 V is not specified as a line-to-line voltage or line-to-neutral voltage, rms or peak. Therefore, it is assumed to be an rms line-to-line voltage, which is the convention throughout this text and a standard practice in the electric power industry. The combined-motor load absorbs 535 kW of real power. The induction motor, which operates at lagging power factor, absorbs reactive power (300 kvar) and the synchronous motor, which operates at leading power factor, delivers reactive power (65.4 kvar). The capacitor bank also delivers reactive power (234.6 kvar). Note that the line current delivered by the source is reduced from 81.1 A without the capacitor bank to 74.3 A with the capacitor bank. Any I^2R losses due to cable resistances and voltage drops due to cable reactances between the source and loads (not included in this example) would also be reduced. ■

2.7 ADVANTAGES OF BALANCED THREE-PHASE VERSUS SINGLE-PHASE SYSTEMS

Figure 2.21 shows three separate single-phase systems. Each single-phase system consists of the following identical components: (1) a generator represented by a voltage source and a generator impedance Z_g; (2) a forward and return conductor represented by two series line impedances Z_L; (3) a load represented by an impedance Z_Y. The three single-phase systems, although completely separated, are drawn in a Y configuration in the figure to illustrate two advantages of three-phase systems.

Each separate single-phase system requires that *both* the forward and return conductors have a current capacity (or *ampacity*) equal to or greater than the load current. However, if the source and load neutrals in Figure 2.21 are connected to form a three-phase system, and if the source voltages are

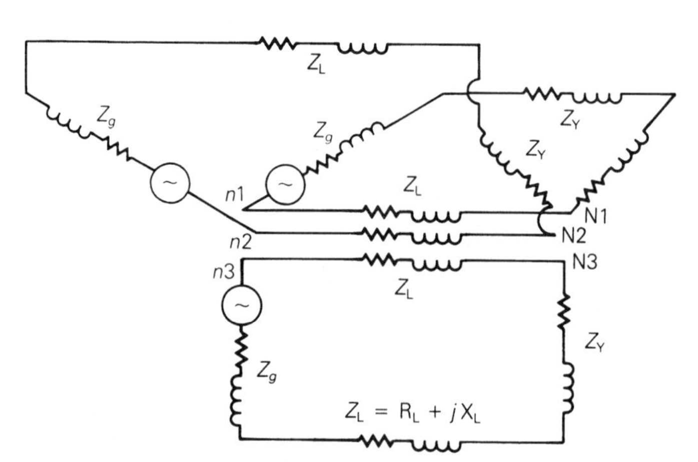

FIGURE 2.21 Three single-phase systems

balanced with equal magnitudes and with 120° displacement between phases, then the neutral current will be zero [see (2.5.10)] and the three neutral conductors can be removed. Thus, the balanced three-phase system, while delivering the same power to the three load impedances Z_Y, requires only half the number of conductors needed for the three separate single-phase systems. Also, the total I^2R line losses in the three-phase system are only half those of the three separate single-phase systems, and the line-voltage drop between the source and load in the three-phase system is half that of each single-phase system. Therefore, one advantage of balanced three-phase systems over separate single-phase systems is reduced capital and operating costs of transmission and distribution, as well as better voltage regulation.

Some three-phase systems such as Δ-connected systems and three-wire Y-connected systems do not have any neutral conductor. However, the majority of three-phase systems are four-wire Y-connected systems, where a grounded neutral conductor is used. Neutral conductors are used to reduce transient overvoltages, which can be caused by lightning strikes and by line-switching operations, and to carry unbalanced currents, which can occur during unsymmetrical short-circuit conditions. Neutral conductors for transmission lines are typically smaller in size and ampacity than the phase conductors because the neutral current is nearly zero under normal operating conditions. Thus, the cost of a neutral conductor is substantially less than that of a phase conductor. The capital and operating costs of three-phase transmission and distribution systems with or without neutral conductors are substantially less than those of separate single-phase systems.

A second advantage of three-phase systems is that the total instantaneous electric power delivered by a three-phase generator under balanced steady-state conditions is (nearly) constant, as shown in Section 2.6. A three-phase generator (constructed with its field winding on one shaft and with its three-phase windings equally displaced by 120° on the stator core) will also have a nearly constant mechanical input power under balanced steady-state conditions, since the mechanical input power equals the electrical output power plus the small generator losses. Furthermore, the mechanical shaft torque, which equals mechanical input power divided by mechanical radian frequency ($T_{mech} = P_{mech}/\omega_m$) is nearly constant.

On the other hand, the equation for the instantaneous electric power delivered by a single-phase generator under balanced steady-state conditions is the same as the instantaneous power delivered by one phase of a three-phase generator, given by $p_a(t)$ in (2.6.3). As shown in that equation, $p_a(t)$ has two components: a constant and a double-frequency sinusoid. Both the mechanical input power and the mechanical shaft torque of the single-phase generator will have corresponding double-frequency components that create shaft vibration and noise, which could cause shaft failure in large machines. Accordingly, most electric generators and motors rated 5 kVA and higher are constructed as three-phase machines in order to produce nearly constant torque and thereby minimize shaft vibration and noise.

MULTIPLE CHOICE QUESTIONS

SECTION 2.1

2.1 The rms value of $v(t) = \mathrm{V}_{\max} \cos(\omega t + \delta)$ is given by
(a) $\mathrm{V}_{\max}$ (b) $\mathrm{V}_{\max}/\sqrt{2}$ (c) 2 $\mathrm{V}_{\max}$ (d) $\sqrt{2}\ \mathrm{V}_{\max}$

2.2 If the rms phasor of a voltage is given by $\mathrm{V} = 120\underline{/60^\circ}$ volts, then the corresponding $v(t)$ is given by
(a) $120\sqrt{2}\cos(\omega t + 60^\circ)$ (b) $120\cos(\omega t + 60^\circ)$
(c) $120\sqrt{2}\sin(\omega t + 60^\circ)$

2.3 If a phasor representation of a current is given by $I = 70.7\underline{/45^\circ}$ A, it is equivalent to
(a) $100\ e^{j45^\circ}$ (b) $100 + j100$
(c) $50 + j50$

2.4 With sinusoidal steady-state excitation, for a purely resistive circuit, the voltage and current phasors are
(a) in phase
(b) perpendicular with each other with V leading I
(c) perpendicular with each other with I leading V.

2.5 For a purely inductive circuit, with sinusoidal steady-state excitation, the voltage and current phasors are
(a) in phase
(b) perpendicular to each other with V leading I
(c) perpendicular to each other with I leading V.

2.6 For a purely capacitive circuit, with sinusoidal steady-state excitation, the voltage and current phasors are
(a) in phase
(b) perpendicular to each other with V leading I
(c) perpendicular to each other with I leading V.

SECTION 2.2

2.7 With sinusoidal steady-state excitation, the average power in a single-phase ac circuit with a purely resistive load is given by
(a) $\mathrm{I}^2_{\mathrm{rms}}\ R$ (b) $\mathrm{V}^2_{\max}/R$ (c) Zero

2.8 The average power in a single-phase ac circuit with a purely inductive load, for sinusoidal steady-state excitation, is
(a) $\mathrm{I}^2_{\mathrm{rms}}\ X_L$ (b) $\mathrm{V}^2_{\max}/X_L$ (c) Zero

[**Note:** $X_L - \omega L$ is the inductive reactance]

2.9 The average power in a single-phase ac circuit with a purely capacitive load, for sinusoidal steady-state excitation, is
(a) zero (b) $\mathrm{V}^2_{\max}/X_C$ (c) $\mathrm{I}^2_{\mathrm{rms}} X_C$

[**Note:** $X_C = 1/(\omega L_c)$ is the capacitive reactance]

2.10 The average value of a double-frequency sinusoid, $\sin 2(\omega t + \delta)$, is given by
(a) 1 (b) δ (c) zero

2.11 The power factor for an inductive circuit (*R-L* load), in which the current lags the voltage, is said to be
(a) Lagging (b) Leading (c) Zero

2.12 The power factor for a capacitive circuit (*R-C* load), in which the current leads the voltage, is said to be
(a) Lagging (b) Leading (c) One

SECTION 2.3

2.13 In a single-phase ac circuit, for a general load composed of *RLC* elements under sinusoidal-steady-state excitation, the average reactive power is given by
(a) $V_{rms} I_{rms} \cos \phi$ (b) $V_{rms} I_{rms} \sin \phi$
(c) zero

[**Note:** ϕ is the power-factor angle]

2.14 The instantaneous power absorbed by the load in a single-phase ac circuit, for a general *RLC* load under sinusoidal-steady-state excitation, is
(a) Nonzero constant (b) zero
(c) containing double-frequency components

2.15 With load convention, where the current enters the positive terminal of the circuit element, if Q is positive then positive reactive power is absorbed.
(a) True (b) False

2.16 With generator convention, where the current leaves the positive terminal of the circuit element, if P is positive then positive real power is delivered.
(a) False (b) True

2.17 Consider the load convention that is used for the *RLC* elements shown in Figure 2.2 of the text.
A. If one says that an inductor absorbs zero real power and positive reactive power, is it
(a) True (b) False

B. If one says that a capacitor absorbs zero real power and negative reactive power (or delivers positive reactive power), is it
(a) False (b) True

C. If one says that a (positive-valued) resistor absorbs (positive) real power and zero reactive power, is it
(a) True (b) False

2.18 In an ac circuit, power factor connection or improvement is achieved by
(a) connecting a resistor in parallel with the inductive load.
(b) connecting an inductor in parallel with the inductive load.
(c) connecting a capacitor in parallel with the inductive load.

SECTION 2.4

2.19 The admittance of the impedance $-j\frac{1}{2}\,\Omega$ is given by
(a) $-j2$ S (b) $j2$ S (c) $-j4$ S

2.20 Consider Figure 2.9 of the text. Let the nodal equations in matrix form be given by Eq. (2.4) of the text.
A. The element Y_{11} is given by
(a) 0 (b) $j\,13$ (c) $-j\,7$

B. The element Y_{31} is given by
(a) 0 (b) $-j\,5$ (c) $j\,1$

C. The admittance matrix is always symmetric square.
(a) False (b) True

SECTION 2.5 AND 2.6

2.21 The three-phase source line-to-neutral voltages are given by
$E_{an} = 10\angle 0°$, $E_{bn} = 10\angle +240°$, and $E_{cn} = 10\angle -240°$ volts.

Is the source balanced?
(a) Yes (b) No

2.22 In a balanced 3-phase wye-connected system with positive-sequence source, the line-to-line voltages are $\sqrt{3}$ times the line-to-neutral voltages and lend by 30°.
(a) True (b) False

2.23 In a balanced system, the phasor sum of line-to-line voltages and the phasor sum of line-to-neutral voltages are always equal to zero.
(a) False (b) True

2.24 Consider a three-phase Y-connected source feeding a balanced-Y load. The phasor sum of the line currents as well as the neutral current are always zero.
(a) True (b) False

2.25 For a balanced-Δ load supplied by a balanced positive-sequence source, the line currents into the load are $\sqrt{3}$ times the Δ-load currents and lag by 30°.
(a) True (b) False

2.26 A balanced Δ-load can be converted to an equivalent balanced-Y load by dividing the Δ-load impedance by
(a) $\sqrt{3}$ (b) 3 (c) 1/3

2.27 When working with balanced three-phase circuits, per-phase analysis is commonly done after converting Δ loads to Y loads, thereby solving only one phase of the circuit.
(a) True (b) False

2.28 The total instantaneous power delivered by a three-phase generator under balanced operating conditions is
(a) a function of time (b) a constant

2.29 The total instantaneous power absorbed by a three-phase motor (under balanced steady-state conditions) as well as a balanced three-phase impedance load is
(a) a constant (b) a function of time

2.30 Under balanced operating conditions, consider the 3-phase complex power delivered by the 3-phase source to the 3-phase load. Match the following expressions, those on the left to those on the right.

(i) Real power, $P_{3\phi}$ (a) $(\sqrt{3}\,V_{LL}\,I_L)$VA
(ii) Reactive power, $Q_{3\phi}$ (b) $(\sqrt{3}\,V_{LL}\,I_L \sin\phi)$ var
(iii) Total apparent power $S_{3\phi}$ (c) $(\sqrt{3}\,V_{LL}\,I_L \cos\phi)$ W
(iv) Complex power, $S_{3\phi}$ (d) $P_{3\phi} + jQ_{3\phi}$

Note that VLL is the rms line-to-line voltage, I_L is the rms line current, and ϕ is the power-factor angle.

2.31 One advantage of balanced three-phase systems over separate single-phase systems is reduced capital and operating costs of transmission and distribution.
(a) True (b) False

2.32 While the instantaneous electric power delivered by a single-phase generator under balanced steady-state conditions is a function of time having two components of a constant and a double-frequency sinusoid, the total instantaneous electric power delivered by a three-phase generator under balanced steady-state conditions is a constant.
(a) True (b) False

PROBLEMS

SECTION 2.1

2.1 Given the complex numbers $A_1 = 5\underline{/30^\circ}$ and $A_2 = -3 + j4$, (a) convert A_1 to rectangular form; (b) convert A_2 to polar and exponential form; (c) calculate $A_3 = (A_1 + A_2)$, giving your answer in polar form; (d) calculate $A_4 = A_1 A_2$, giving your answer in rectangular form; (e) calculate $A_5 = A_1/(A_2^*)$, giving your answer in exponential form.

2.2 Convert the following instantaneous currents to phasors, using $\cos(\omega t)$ as the reference. Give your answers in both rectangular and polar form.
(a) $i(t) = 400\sqrt{2}\cos(\omega t - 30^\circ)$;
(b) $i(t) = 5\sin(\omega t + 15^\circ)$;
(c) $i(t) = 4\cos(\omega t - 30^\circ) + 5\sqrt{2}\sin(\omega t + 15^\circ)$.

2.3 The instantaneous voltage across a circuit element is $v(t) = 359.3\sin(\omega t + 15^\circ)$ volts, and the instantaneous current entering the positive terminal of the circuit element is $i(t) = 100\cos(\omega t + 5^\circ)$ A. For both the current and voltage, determine (a) the maximum value, (b) the rms value, (c) the phasor expression, using $\cos(\omega t)$ as the reference.

2.4 For the single-phase circuit shown in Figure 2.22, $I = 10\underline{/0^\circ}$ A. (a) Compute the phasors I_1, I_2, and V. (b) Draw a phasor diagram showing I, I_1, I_2, and V.

FIGURE 2.22
Circuit for Problem 2.4

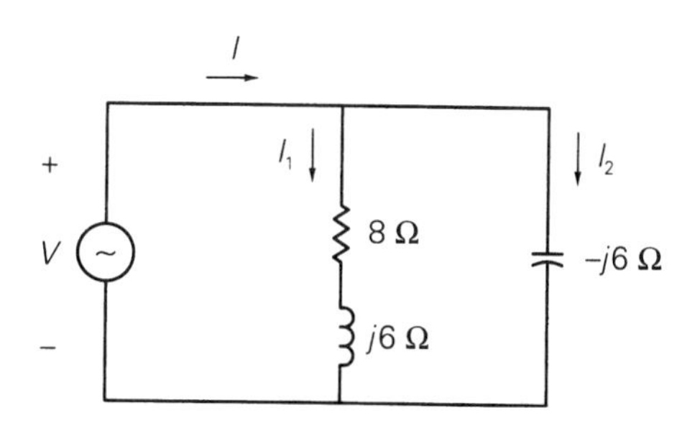

2.5 A 60-Hz, single-phase source with $V = 277\underline{/30^\circ}$ volts is applied to a circuit element. (a) Determine the instantaneous source voltage. Also determine the phasor and instantaneous currents entering the positive terminal if the circuit element is (b) a 20-Ω resistor, (c) a 10-mH inductor, (d) a capacitor with 25-Ω reactance.

2.6 (a) Transform $v(t) = 100\cos(377t - 30^\circ)$ to phasor form. Comment on whether $\omega = 377$ appears in your answer. (b) Transform $V = 100\underline{/20^\circ}$ to instantaneous form. Assume that $\omega = 377$. (c) Add the two sinusoidal functions $a(t)$ and $b(t)$ of the same frequency given as follows: $a(t) = A\sqrt{2}\cos(\omega t + \alpha)$ and $b(t) = B\sqrt{2}\cos(\omega t + \beta)$. Use phasor methods and obtain the resultant $c(t)$. Does the resultant have the same frequency?

2.7 Let a 100-V sinusoidal source be connected to a series combination of a 3-Ω resistor, an 8-Ω inductor, and a 4-Ω capacitor. (a) Draw the circuit diagram. (b) Compute the series impedance. (c) Determine the current I delivered by the source. Is the current lagging or leading the source voltage? What is the power factor of this circuit?

2.8 Consider the circuit shown in Figure 2.23 in time domain. Convert the entire circuit into phasor domain.

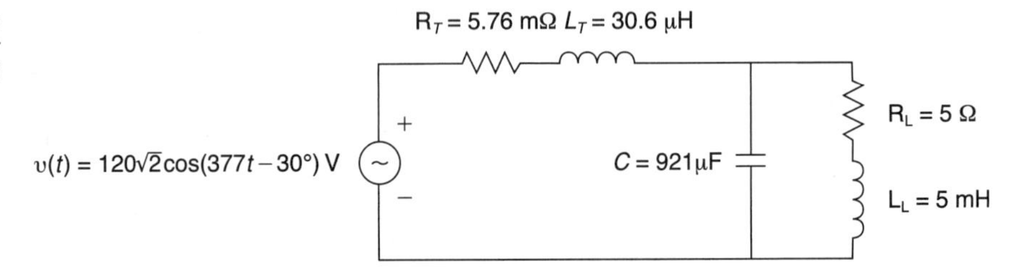

FIGURE 2.23 Circuit for Problem 2.8

2.9 For the circuit shown in Figure 2.24, compute the voltage across the load terminals.

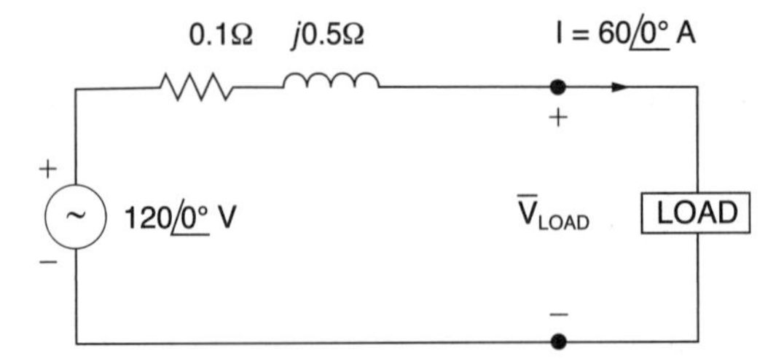

FIGURE 2.24 Circuit for Problem 2.9

SECTION 2.2

2.10 For the circuit element of Problem 2.3, calculate (a) the instantaneous power absorbed, (b) the real power (state whether it is delivered or absorbed), (c) the reactive power (state whether delivered or absorbed), (d) the power factor (state whether lagging or leading).

[*Note*: By convention the power factor $\cos(\delta - \beta)$ is positive. If $|\delta - \beta|$ is greater than 90°, then the reference direction for current may be reversed, resulting in a positive value of $\cos(\delta - \beta)$].

2.11 Referring to Problem 2.5, determine the instantaneous power, real power, and reactive power absorbed by: (a) the 20-Ω resistor, (b) the 10-mH inductor, (c) the capacitor with 25-Ω reactance. Also determine the source power factor and state whether lagging or leading.

2.12 The voltage $v(t) = 359.3\cos(\omega t)$ volts is applied to a load consisting of a 10-Ω resistor in parallel with a capacitive reactance $X_C = 25\ \Omega$. Calculate (a) the instantaneous power absorbed by the resistor, (b) the instantaneous power absorbed by the capacitor, (c) the real power absorbed by the resistor, (d) the reactive power delivered by the capacitor, (e) the load power factor.

2.13 Repeat Problem 2.12 if the resistor and capacitor are connected in series.

2.14 A single-phase source is applied to a two-terminal, passive circuit with equivalent impedance $Z = 2.0\underline{/-45°}\ \Omega$ measured from the terminals. The source current is $i(t) = 4\sqrt{2}\cos(\omega t)$ kA. Determine the (a) instantaneous power, (b) real power, and (c) reactive power delivered by the source. (d) Also determine the source power factor.

2.15 Let a voltage source $v(t) = 4\cos(\omega t + 60°)$ be connected to an impedance $Z = 2\underline{/30°}\ \Omega$. (a) Given the operating frequency to be 60 Hz, determine the expressions for the current and instantaneous power delivered by the source as functions of time. (b) Plot these functions along with $v(t)$ on a single graph for comparison. (c) Find the frequency and average value of the instantaneous power.

2.16 A single-phase, 120-V (rms), 60-Hz source supplies power to a series R-L circuit consisting of R = 10 Ω and L = 40 mH. (a) Determine the power factor of the circuit and state whether it is lagging or leading. (b) Determine the real and reactive power absorbed by the load. (c) Calculate the peak magnetic energy W_{int} stored in the inductor by using the expression $W_{int} = L(I_{rms})^2$ and check whether the reactive power $Q = \omega W$ is satisfied. (*Note*: The instantaneous magnetic energy storage fluctuates between zero and the peak energy. This energy must be sent twice each cycle to the load from the source by means of reactive power flows.)

SECTION 2.3

2.17 Consider a load impedance of $Z = j\omega L$ connected to a voltage V let the current drawn be I.
(a) Develop an expression for the reactive power Q in terms of ω, L, and I, from complex power considerations.
(b) Let the instantaneous current be $i(t) = \sqrt{2}I\cos(\omega t + \theta)$. Obtain an expression for the instantaneous power $p(t)$ into L, and then express it in terms of Q.
(c) Comment on the average real power P supplied to the inductor and the instantaneous power supplied.

2.18 Let a series R-L-C network be connected to a source voltage V, drawing a current I.
(a) In terms of the load impedance $Z = Z < Z$, find expressions for P and Q, from complex power considerations.
(b) Express $p(t)$ in terms of P and Q, by choosing $i(t) = \sqrt{2}I\ cos\ \omega t$.
(c) For the case of $Z = R + j\omega L + 1/j\omega c$, interpret the result of part (b) in terms of P, Q_L, and Q_C. In particular, if $\omega^2 LC = 1$, when the inductive and capacitive reactances cancel, comment on what happens.

2.19 Consider a single-phase load with an applied voltage $v(t) = 150\cos(\omega t + 10°)$ volts and load current $i(t) = 5\cos(\omega t - 50°)$ A. (a) Determine the power triangle. (b) Find the power factor and specify whether it is lagging or leading. (c) Calculate the reactive power supplied by capacitors in parallel with the load that correct the power factor to 0.9 lagging.

2.20 A circuit consists of two impedances, $Z_1 = 20\underline{/30°}\ \Omega$ and $Z_2 = 25\underline{/60°}\ \Omega$, in parallel, supplied by a source voltage $V = 100\underline{/60°}$ volts. Determine the power triangle for each of the impedances and for the source.

2.21 An industrial plant consisting primarily of induction motor loads absorbs 500 kW at 0.6 power factor lagging. (a) Compute the required kVA rating of a shunt capacitor to improve the power factor to 0.9 lagging. (b) Calculate the resulting power factor if a synchronous motor rated 500 hp with 90% efficiency operating at rated load and at unity power factor is added to the plant instead of the capacitor. Assume constant voltage. (1 hp = 0.746 kW)

2.22 The real power delivered by a source to two impedances, $Z_1 = 3 + j4\ \Omega$ and $Z_2 = 10\ \Omega$, connected in parallel, is 1100 W. Determine (a) the real power absorbed by each of the impedances and (b) the source current.

2.23 A single-phase source has a terminal voltage $V = 120\underline{/0°}$ volts and a current $I = 10\underline{/30°}$ A, which leaves the positive terminal of the source. Determine the real and reactive power, and state whether the source is delivering or absorbing each.

2.24 A source supplies power to the following three loads connected in parallel: (1) a lighting load drawing 10 kW, (2) an induction motor drawing 10 kVA at 0.90 power factor lagging, and (3) a synchronous motor operating at 10 hp, 85% efficiency and 0.95 power factor leading (1 hp = 0.746 kW). Determine the real, reactive, and apparent power delivered by the source. Also, draw the source power triangle.

2.25 Consider the series R-L-C circuit of Problem 2.7 and calculate the complex power absorbed by each of the elements R, L, and C, as well as the complex power absorbed by the total load. Draw the resultant power triangle. Check whether the complex power delivered by the source equals the total complex power absorbed by the load.

2.26 A small manufacturing plant is located 2 km down a transmission line, which has a series reactance of 0.5 Ω/km. The line resistance is negligible. The line voltage at the plant is $480\underline{/0°}$ V (rms), and the plant consumes 120 kW at 0.85 power factor lagging. Determine the voltage and power factor at the sending end of the transmission line by using (a) a complex power approach and (b) a circuit analysis approach.

2.27 An industrial load consisting of a bank of induction motors consumes 50 kW at a power factor of 0.8 lagging from a 220-V, 60-Hz, single-phase source. By placing a bank of capacitors in parallel with the load, the resultant power factor is to be raised to 0.95 lagging. Find the net capacitance of the capacitor bank in μF that is required.

2.28 Three loads are connected in parallel across a single-phase source voltage of 240 V (rms).
Load 1 absorbs 12 kW and 6.667 kvar;
Load 2 absorbs 4 kVA at 0.96 p.f. leading;
Load 3 absorbs 15 kW at unity power factor.
Calculate the equivalent impedance, Z, for the three parallel loads, for two cases:
(i) Series combination of R and X, and (ii) parallel combination of R and X.

2.29 Modeling the transmission lines as inductors, with $Sij = S_{ji}^*$,
Compute S_{13}, S_{31}, S_{23}, S_{32}, and S_{G3}, in Figure 2.25. (*Hint*: complex power balance holds good at each bus, statisfying KCL.)

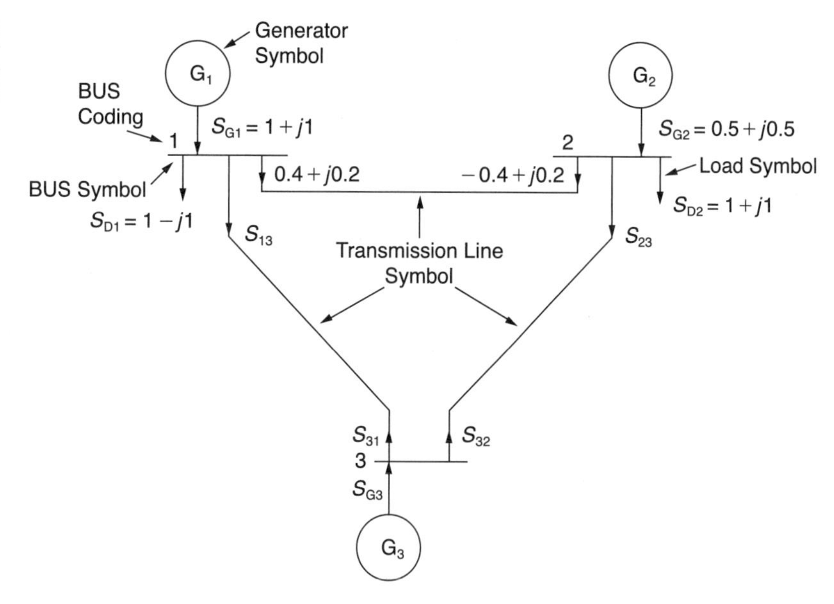

FIGURE 2.25 System diagram for Problem 2.29

2.30 Figure 2.26 shows three loads connected in parallel across a 1000-V (rms), 60-Hz single-phase source.
Load 1: Inductive load, 125 kVA, 0.28 p.f. lagging
Load 2: Capacitive load, 10 kW, 40 kvar
Load 3: Resistive load, 15 kW
(a) Determine the total kW, kvar, kva, and supply power factor.
(b) In order to improve the power factor to 0.8 lagging, a capacitor of negligible resistance is connected in parallel with the above loads. Find the KVAR rating of that capacitor and the capacitance in μf.
Comment on the magnitude of the supply current after adding the capacitor.

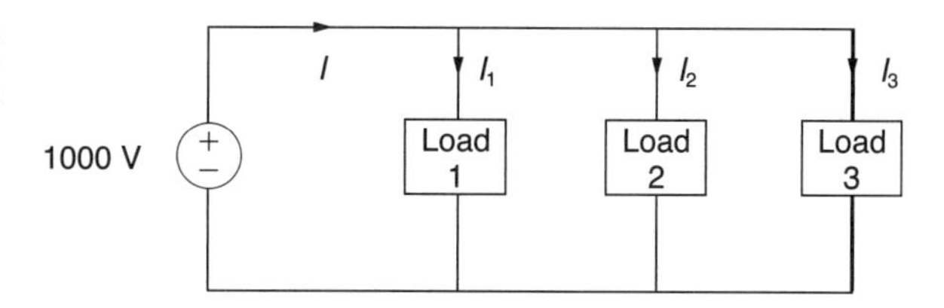

FIGURE 2.26 Circuit for Problem 2.30

2.31 Consider two interconnected voltage sources connected by a line of impedance $Z = jx\ \Omega$, as shown in Figure 2.27.
(a) Obtain expressions for P_{12} and Q_{12}.
(b) Determine the maximum power transfer and the condition for it to occur.

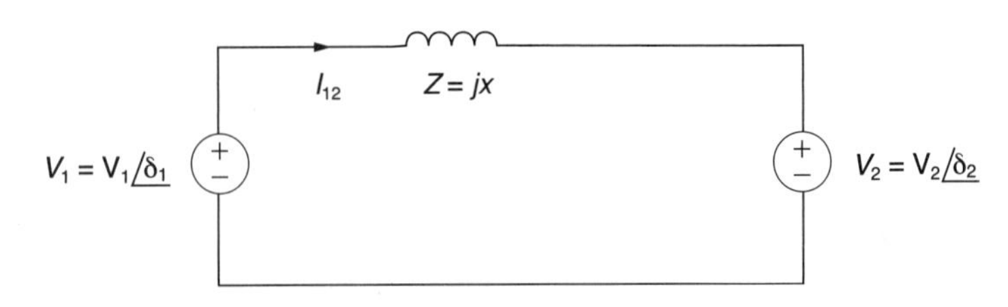

FIGURE 2.27 Circuit for Problem 2.31

PW **2.32** In PowerWorld Simulator Problem 2.32 (see Figure 2.28) a 8 MW/4 Mvar load is supplied at 13.8 kV through a feeder with an impedance of $1 + j2\ \Omega$. The load is compensated with a capacitor whose output, Q_{cap}, can be varied in 0.5 Mvar steps between 0 and 10.0 Mvar. What value of Q_{cap} minimizes the real power line losses? What value of Q_{cap} minimizes the MVA power flow into the feeder?

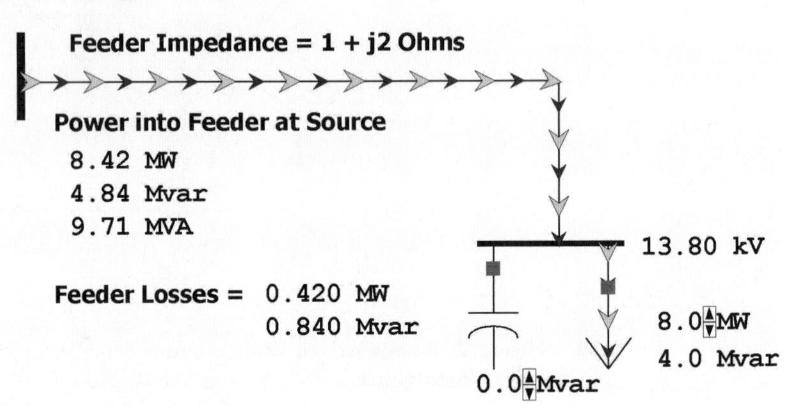

FIGURE 2.28 Screen for Problem 2.32

PW **2.33** For the system from Problem 2.32, plot the real and reactive line losses as Q_{cap} is varied between 0 and 10.0 Mvar.

PW **2.34** For the system from Problem 2.32, assume that half the time the load is 10 MW/5 Mvar, and for the other half it is 20 MW/10 Mvar. What single value of Q_{cap} would minimize the average losses? Assume that Q_{cap} can only be varied in 0.5 Mvar steps.

SECTION 2.4

2.35 For the circuit shown in Figure 2.29, convert the voltage sources to equivalent current sources and write nodal equations in matrix format using bus 0 as the reference bus. Do not solve the equations.

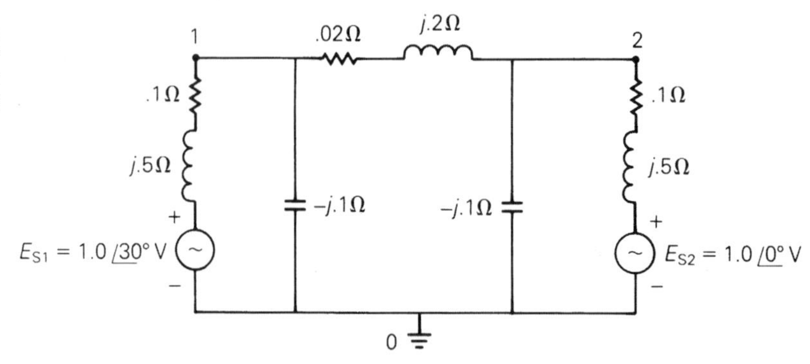

FIGURE 2.29 Circuit diagram for Problems 2.35 and 2.36

2.36 For the circuit shown in Figure 2.29, write a computer program that uses the sources, impedances, and bus connections as input data to (a) compute the 2×2 bus admittance matrix $\boldsymbol{Y}$, (b) convert the voltage sources to current sources and compute the vector of source currents into buses 1 and 2.

2.37 Determine the 4×4 bus admittance matrix and write nodal equations in matrix format for the circuit shown in Figure 2.30. Do not solve the equations.

FIGURE 2.30

Circuit for Problem 2.37

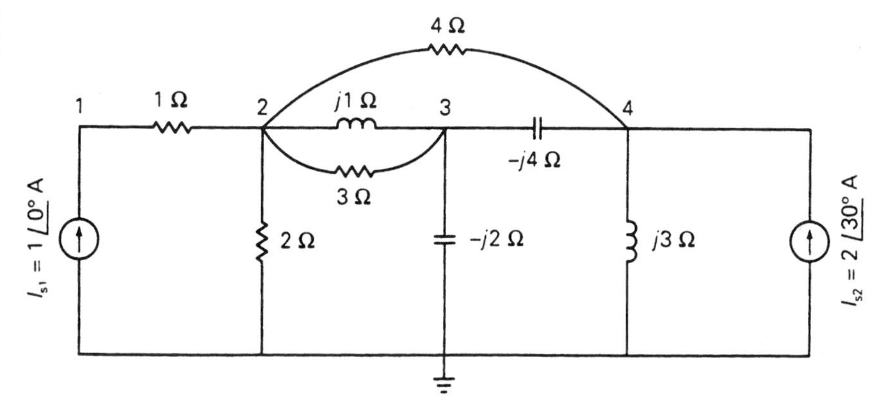

2.38 Given the impedance diagram of a simple system as shown in Figure 2.31, draw the admittance diagram for the system and develop the 4×4 bus admittance matrix $\boldsymbol{Y}_{\text{bus}}$ by inspection.

FIGURE 2.31

System diagram for Problem 2.38

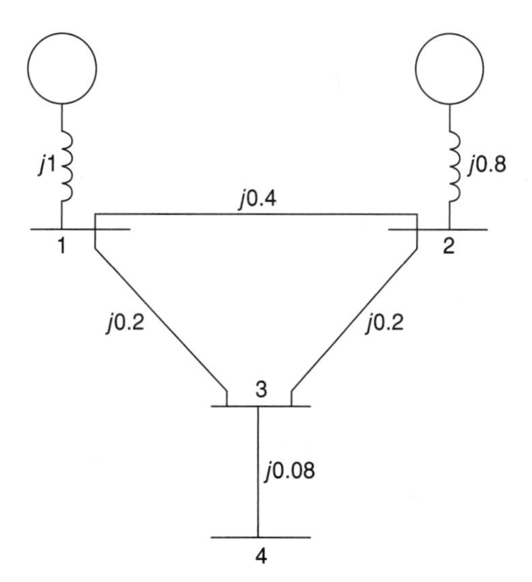

2.39 (a) Given the circuit diagram in Figure 2.32 showing admittances and current sources at nodes 3 and 4, set up the nodal equations in matrix format. (b) If the parameters are given by: $Y_a = -j0.8$ S, $Y_b = -j4.0$ S, $Y_c = -j4.0$ S, $Y_d = -j8.0$ S, $Y_e = -j5.0$ S, $Y_f = -j2.5$ S, $Y_g = -j0.8$ S, $I_3 = 1.0\underline{/-90^\circ}$ A, and $I_4 = 0.62\underline{/-135^\circ}$ A, set up the nodal equations and suggest how you would go about solving for the voltages at the nodes.

FIGURE 2.32

Circuit diagram for Problem 2.39

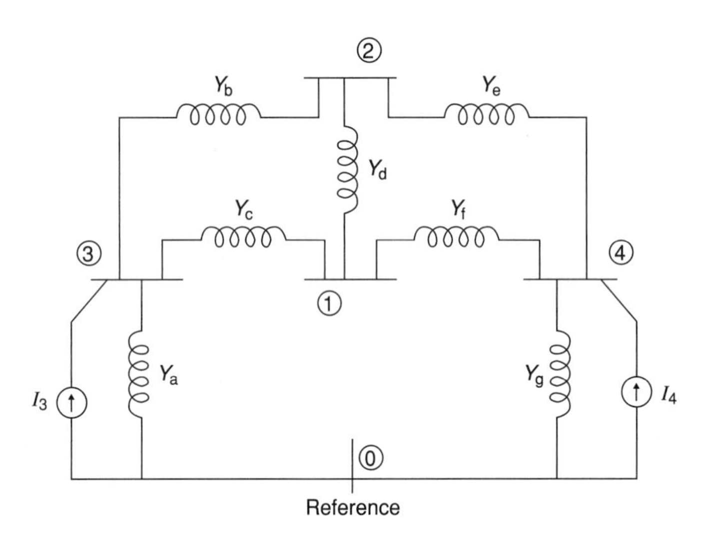

SECTIONS 2.5 AND 2.6

2.40 A balanced three-phase 208-V source supplies a balanced three-phase load. If the line current I_A is measured to be 10 A and is in phase with the line-to-line voltage V_{BC}, find the per-phase load impedance if the load is (a) Y-connected, (b) Δ-connected.

2.41 A three-phase 25-kVA, 480-V, 60-Hz alternator, operating under balanced steady-state conditions, supplies a line current of 20 A per phase at a 0.8 lagging power factor and at rated voltage. Determine the power triangle for this operating condition.

2.42 A balanced Δ-connected impedance load with $(12 + j9)\ \Omega$ per phase is supplied by a balanced three-phase 60-Hz, 208-V source. (a) Calculate the line current, the total real and reactive power absorbed by the load, the load power factor, and the apparent load power. (b) Sketch a phasor diagram showing the line currents, the line-to-line source voltages, and the Δ-load currents. Assume positive sequence and use V_{ab} as the reference.

2.43 A three-phase line, which has an impedance of $(2 + j4)\ \Omega$ per phase, feeds two balanced three-phase loads that are connected in parallel. One of the loads is Y-connected with an impedance of $(30 + j40)\ \Omega$ per phase, and the other is Δ-connected with an impedance of $(60 - j45)\ \Omega$ per phase. The line is energized at the sending end from a 60-Hz, three-phase, balanced voltage source of $120\sqrt{3}$ V (rms, line-to-line). Determine (a) the current, real power, and reactive power delivered by the sending-end source; (b) the line-to-line voltage at the load; (c) the current per phase in each load; and (d) the total three-phase real and reactive powers absorbed by each load and by the line. Check that the total three-phase complex power delivered by the source equals the total three-phase power absorbed by the line and loads.

2.44 Two balanced three-phase loads that are connected in parallel are fed by a three-phase line having a series impedance of $(0.4 + j2.7)\ \Omega$ per phase. One of the loads absorbs 560 kVA at 0.707 power factor lagging, and the other 132 kW at unity power factor. The line-to-line voltage at the load end of the line is $2200\sqrt{3}$ V. Compute (a) the line-to-line voltage at the source end of the line, (b) the total real and reactive power losses in the three-phase line, and (c) the total three-phase real and reactive power supplied at the sending end of the line. Check that the total three-phase complex power delivered by the source equals the total three-phase complex power absorbed by the line and loads.

2.45 Two balanced Y-connected loads, one drawing 10 kW at 0.8 power factor lagging and the other 15 kW at 0.9 power factor leading, are connected in parallel and supplied by a balanced three-phase Y-connected, 480-V source. (a) Determine the source current. (b) If the load neutrals are connected to the source neutral by a zero-ohm neutral wire through an ammeter, what will the ammeter read?

2.46 Three identical impedances $Z_\Delta = 30\underline{/30^\circ}\ \Omega$ are connected in Δ to a balanced three-phase 208-V source by three identical line conductors with impedance $Z_L = (0.8 + j0.6)\ \Omega$ per line. (a) Calculate the line-to-line voltage at the load terminals. (b) Repeat part (a) when a Δ-connected capacitor bank with reactance $(-j60)\ \Omega$ per phase is connected in parallel with the load.

2.47 Two three-phase generators supply a three-phase load through separate three-phase lines. The load absorbs 30 kW at 0.8 power factor lagging. The line impedance is $(1.4 + j1.6)\ \Omega$ per phase between generator G1 and the load, and $(0.8 + j1)\ \Omega$ per phase between generator G2 and the load. If generator G1 supplies 15 kW at 0.8 power factor lagging, with a terminal voltage of 460 V line-to-line, determine (a) the voltage at the load terminals, (b) the voltage at the terminals of generator G2, and (c) the real and reactive power supplied by generator G2. Assume balanced operation.

2.48 Two balanced Y-connected loads in parallel, one drawing 15 kW at 0.6 power factor lagging and the other drawing 10 kVA at 0.8 power factor leading, are supplied by a balanced, three-phase, 480-volt source. (a) Draw the power triangle for each load and for the combined load. (b) Determine the power factor of the combined load and state whether lagging or leading. (c) Determine the magnitude of the line current from the source. (d) Δ-connected capacitors are now installed in parallel with the combined load. What value of capacitive reactance is needed in each leg of the Δ to make the source power factor unity? Give your answer in Ω. (e) Compute the magnitude of the current in each capacitor and the line current from the source.

2.49 Figure 2.33 gives the general Δ–Y transformation. (a) Show that the general transformation reduces to that given in Figure 2.16 for a balanced three-phase load. (b) Determine the impedances of the equivalent Y for the following Δ impedances: $Z_{AB} = j10$, $Z_{BC} = j20$, and $Z_{CA} = -j25\ \Omega$.

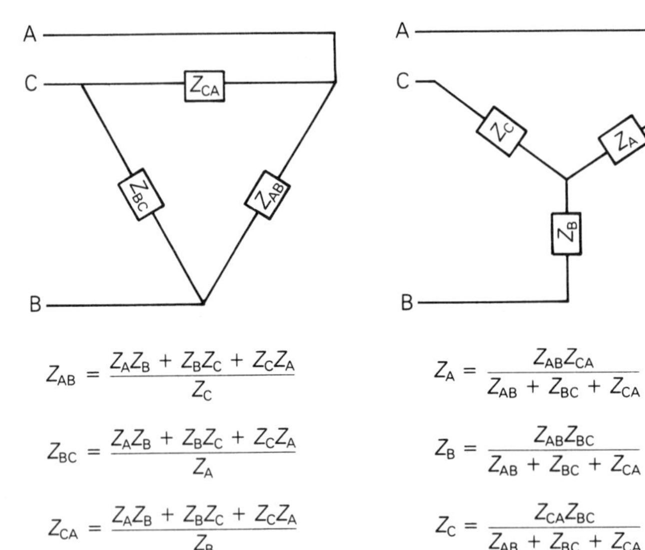

FIGURE 2.33 General Δ–Y transformation

2.50 Consider the balanced three-phase system shown in Figure 2.34. Determine $v_1(t)$ and $i_2(t)$. Assume positive phase sequence.

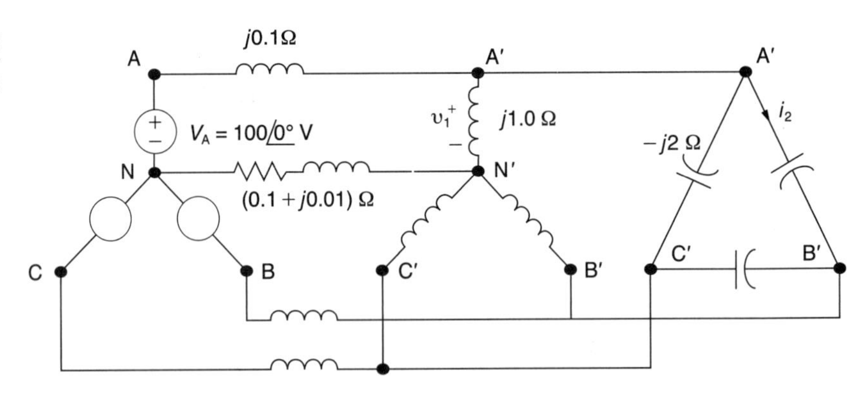

FIGURE 2.34 Circuit for Problem 2.50

2.51 A three-phase line with an impedance of $(0.2 + j1.0)$ Ω/phase feeds three balanced three-phase loads connected in parallel.
Load 1: Absorbs a total of 150 kW and 120 kvar; Load 2: Delta connected with an impedance of $(150 - j48)$ Ω/phase; Load 3: 120 kVA at 0.6 p.f. leading. If the line-to-neutral voltage at the load end of the line is 2000 V (rms), determine the magnitude of the line-to-line voltage at the source end of the line.

2.52 A balanced three-phase load is connected to a 4.16-kV, three-phase, four-wire, grounded-wye dedicated distribution feeder. The load can be modeled by an impedance of $Z_L = (4.7 + j9)$ Ω/phase, wye-connected. The impedance of the phase

conductors is $(0.3 + j1)\ \Omega$. Determine the following by using the phase A to neutral voltage as a reference and assume positive phase sequence:
(a) Line currents for phases A, B, and C.
(b) Line-to-neutral voltages for all three phases at the load.
(c) Apparent, active, and reactive power dissipated per phase, and for all three phases in the load.
(d) Active power losses per phase and for all three phases in the phase conductors.

CASE STUDY QUESTIONS

A. What is a microgrid?
B. What is an island in an interconnected power system?
C. Why is a microgrid designed to be able to operate in both grid-connected and stand-alone modes?
D. When operating in the stand-alone mode, what control features should be associated with a microgrid?

REFERENCES

1. W. H. Hayt, Jr., and J. E. Kemmerly, *Engineering Circuit Analysis*, 7th ed. (New York: McGraw-Hill, 2006).
2. W. A. Blackwell and L. L. Grigsby, *Introductory Network Theory* (Boston: PWS, 1985).
3. A. E. Fitzgerald, D. E. Higginbotham, and A. Grabel, *Basic Electrical Engineering* (New York: McGraw-Hill, 1981).
4. W. D. Stevenson, Jr., *Elements of Power System Analysis*, 4th ed. (New York: McGraw-Hill, 1982).
5. B. Kroposki, R. Lasseter, T. Ise, S. Morozumi, S. Papathanassiou & N. Hatziargyriou, "Making Microgrids Work," *IEEE Power & Energy Magazine*, 6,3 (May/June 2008), pp. 40–53.

Core and coil assemblies of a three-phase 20.3 kVΔ/345 kVY step-up transformer. This oil-immersed transformer is rated 325 MVA self-cooled (OA)/542 MVA forced oil, forced air-cooled (FOA)/607MVA forced oil, forced air-cooled (FOA) (Courtesy of General Electric)

3 POWER TRANSFORMERS

The power transformer is a major power system component that permits economical power transmission with high efficiency and low series-voltage drops. Since electric power is proportional to the product of voltage and current, low current levels (and therefore low I^2R losses and low IZ voltage drops) can be maintained for given power levels via high voltages. Power transformers transform ac voltage and current to optimum levels for generation, transmission, distribution, and utilization of electric power.

The development in 1885 by William Stanley of a commercially practical transformer was what made ac power systems more attractive than dc power systems. The ac system with a transformer overcame voltage problems encountered in dc systems as load levels and transmission distances increased. Today's modern power transformers have nearly 100% efficiency, with ratings up to and beyond 1300 MVA.

In this chapter, we review basic transformer theory and develop equivalent circuits for practical transformers operating under sinusoidal-steady-state conditions. We look at models of single-phase two-winding, three-phase two-winding, and three-phase three-winding transformers, as well as auto-transformers and regulating transformers. Also, the per-unit system, which simplifies power system analysis by eliminating the ideal transformer winding in transformer equivalent circuits, is introduced in this chapter and used throughout the remainder of the text.

CASE STUDY

The following article describes how transmission transformers are managed in the Pennsylvania–New Jersey (PJM) Interconnection. PJM is a regional transmission organization (RTO) that operates approximately 19% of the transmission infrastructure of the U.S. Eastern Interconnection. As of 2007, there were 188 transmission transformers (500/230 kV) and 29 dedicated spare transformers in the PJM system. A Probabilistic Risk assessment (PRA) model is applied to PJM transformer asset management [8].

PJM Manages Aging Transformer Fleet: Risk-based tools enable regional transmission owner to optimize asset service life and manage spares.

BY DAVID EGAN AND KENNETH SEILER
PJM INTERCONNECTION

The PJM interconnection system has experienced both failures and degradation of older transmission transformers (Fig. 1). Steps required to mitigate potential system reliability issues, such as operation of out-of-merit generation, have led to higher operating costs of hundreds of millions of dollars for transmission system users over the last several years.

The PJM (Valley Forge, Pennsylvania, U.S.) system has 188 transmission transformers (500 kV/230 kV) in service and 29 dedicated spares. Figure 2 shows the age distribution of this transformer fleet. Note that 113 transformers are more than 30 years old and will reach or exceed their design life over the course of the next 10 years. To address increasing

Figure 1
PJM is evaluating the risk of older transformers. The Probabilistic Risk Assessment also considers the effectiveness of alternative spares strategies

(*"PJM Manages Aging Transformer Fleet" by David Egan and Kenneth Seiler,* Transmission & Distribution World Magazine, *March 2007)*

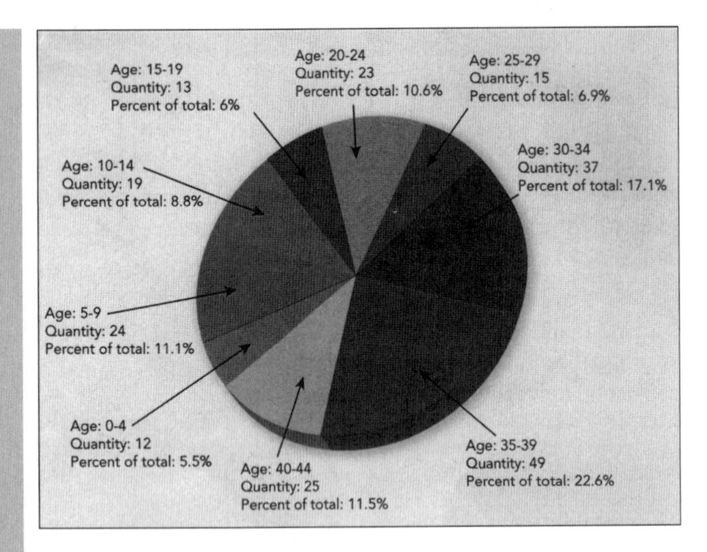

Figure 2
Age distribution of the PJM 500-kV/230-kV transformer fleet. Note that more than half of this population is over 30 years old

concerns regarding potential reliability impacts and the ability to replace failed transformer units in a timely fashion, PJM and its transmission-owning members are establishing a systematic, proactive transformer replacement program to mitigate negative impacts on PJM stakeholders, operations and ultimately the consumers. PJM now assesses the risk exposure from an aging 500-kV/230-kV transformer fleet through its Probabilistic Risk Assessment (PRA) model.

CONGESTION

Generally PJM's backbone high-voltage transmission system delivers lower-cost power from sources in the western side of the regional transmission organization (RTO) to serve load centers in the eastern side. Delivery of power in PJM includes transformation from 500-kV lines to 230-kV facilities for further delivery to and consumption by customers.

Congestion on the electric system can occur when a transmission transformer unit must be removed from service and the redirected electricity flow exceeds the capabilities of parallel transmission facilities. When congestion occurs, higher-cost generation on the restricted side of the constraint must operate to keep line flows under specified limits and to meet customer demand. The cost of congestion results from the expense of operating higher-cost generators. Congestion and its related costs exist on all electric power systems. However, in a RTO such as PJM, the cost of congestion is readily knowable and identified.

The failure impact of certain 500-kV/230-kV transformers on the PJM system can mean annual congestion costs of hundreds of millions of dollars if the failure cannot be addressed with a spare. Lead times for replacement transformer units at this voltage class can take up to 18 months, and each replacement unit cost is several million dollars. These costly transformer-loss consequences, coupled with the age distribution of the transformer population, have raised PJM's concern that the existing system spare quantities could be deficient and locations of existing spares suboptimal.

DEVELOPING PRA

PJM reviewed existing methods for determining transformer life expectancy, assessing failure impacts, mitigating transformer failures, ensuring spare-quantity adequacy and locating spares. Each of these methodologies has weaknesses when applied to an RTO scenario. In addition, no existing method identified the best locations for spare transformers on the system.

Transformer condition assessments are the primary means for predicting failures. Although technology advancements have improved condition-monitoring data, unless a transformer exhibits signs of imminent failure, predicting when a transformer will fail based on a condition assessment is still mostly guesswork. Traditional methods have quantified the impacts of transformer failure based on reliability criteria; they have not typically included economic considerations. Also, while annual failure rate analysis is used to determine the number of spares required, assuming a constant failure rate may be a poor assumption if a large portion of the transformer fleet is entering the wear-out stage of asset life.

Recognizing the vulnerabilities of existing methods, PJM proceeded to develop a risk-based approach to transformer asset management. The PJM PRA model couples the loss consequence of a transformer with its loss likelihood (Fig. 3). The product of these inputs, risk, is expressed in terms of annual risk-exposure dollars.

PRA requires a detailed understanding of failure consequences. PJM projects the dollar value of each transformer's failure consequence, including cost estimates for replacement, litigation, environmental impact and congestion. PJM's PRA also permits the assessment of various spare-unit and replacement policies based on sensitivity analysis of these four cost drivers.

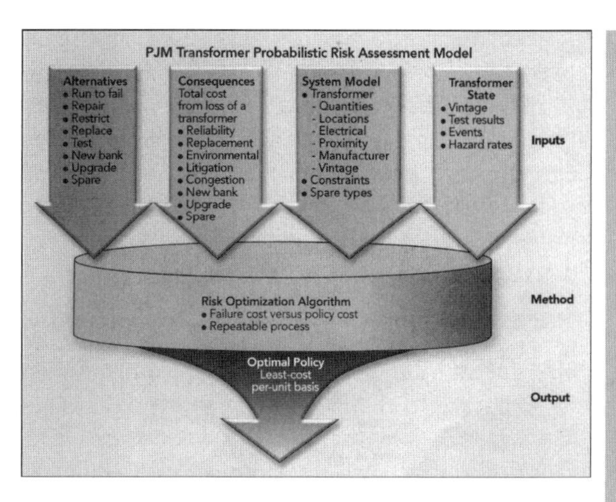

Figure 3
The PJM Probabilistic Risk Assessment model uses drivers to represent overall failure consequences: costs of replacement, litigation, environmental and congestion

PRA MODEL INPUTS

The PRA model depends on several inputs to determine the likelihood of asset failure. One key input is the number of existing fleet transformers. Individual utilities within PJM may not have enough transformers to develop statistically significant assessment results. However, PJM's region-wide perspective permits evaluation of the entire transformer population within its footprint.

Second, rather than applying the annual failure rate of the aggregate transformer population, each transformer's failure rate is determined as a function of its effective age. PJM developed its own method for determining this effective age-based failure rate, or hazard rate. Effective age combines condition data with age-based failure history. By way of analogy, consider a 50-year-old person who smokes and has high cholesterol and high blood pressure (condition data). This individual may have the same risk of death as a healthy 70-year-old non-smoker. Thus, while the individual's actual age is 50 years, his effective age could be as high as 70 years.

Third, the PRA model inputs also include transformers' interactions with each other in terms of the probabilities of cascading events and large-impact, low-likelihood events. For example, transformers are cooled with oil, which, if a transformer ruptures, can become a fuel source for fire. Such a fire can spread to neighboring units causing them to fail as well. PJM determined cascading event probability by reviewing industry events and consulting industry subject-matter experts. Further, the impacts of weather events also are considered. For example, a tornado could damage multiple transformer units at a substation. PJM uses National Oceanic and Atmospheric Administration statistical data for probabilities of such weather-related phenomena.

The remaining PRA model inputs include the possible risk-mitigation alternatives and transformer groupings. The possible risk-mitigation alternatives include running to failure, overhauling or retrofitting, restricting operations, replacing in-kind or with an upgraded unit, increasing test frequency to better assess condition, adding redundant transformers or purchasing a spare. The PRA model objective is to select the appropriate alternative commensurate with risk. To accomplish this objective, the PRA model also requires inputs of the cost and time to implement each alternative. The time to implement an alternative is important because failure consequences accumulate until restoration is completed.

Also, transformers must be grouped by spare applicability. Design parameters can limit the number of in-service transformers that can be served by a designated spare. Additionally, without executed sharing agreements in place between transmission owners, PJM cannot recognize transformer spare sharing beyond the owner's service territory.

THE QUESTION OF SPARES

PRA determines the amount of transformer-loss risk exposure to the PJM system and to PJM members. To calculate the total risk exposure from transformer loss, each transformer's risk is initially determined assuming no available spare. This initial total-system transformer-loss risk is a baseline for comparing potential mitigation approaches. For this baseline, with no spares available, US$553 million of annual risk exposure was identified.

A spare's value is equal to the cumulative risk reduction, across all facilities that can be served by a given spare. The existing system spares were shown to mitigate $396 million of the annual risk, leaving $157 million of annual exposure. The PRA showed that planned projects would further mitigate $65 million, leaving $92 million of exposed annual risk.

With the value of existing spares and planned reliability upgrade projects known, the PRA can then assess the value of additional spares in reducing this risk exposure. As long as the risk mitigated by an additional spare exceeds the payback value of a new transformer, purchasing a spare is justified. The PRA identified $75 million of justifiable risk mitigation from seven additional spares.

PRA also specifies the best spare type. If a spare can be cost justified, asset owners can use two types of spare transformers: used or new. As an in-service unit begins to show signs of failure, it can be replaced. Since the unit removed has not yet failed, it can be stored as an emergency spare. However, the downsides of this approach are the expense, work efforts and congestion associated with handling the spare twice. Also, the likelihood of a used spare unit's success is lower than that of a new unit because of its preexisting degradation.

PJM's PRA analysis revealed that it is more cost-effective to purchase a new unit as a spare. In this case, when a failure occurs, the spare transformer can be installed permanently to remedy the failure and a replacement spare purchased. This process allows expedient resolution of a failure and reduces handling.

Existing spares may not be located at optimal sites. PRA also reveals ideal locations for storing spares. A spare can be located on-site or at a remote location. An on-site spare provides the benefit of expedient installation. A remote spare requires added transportation and handling. Ideally, spares would be located at the highest risk sites. Remote spares serve lower risk sites. The PRA both identifies the best locations to position spares on the system to minimize risk and evaluates relocation of existing spares by providing the cost/benefit analysis of moving a spare to a higher risk site.

The PRA has shown that the type of spare (no spare, old spare or new spare) and a transformer's loss consequence strongly influence the most cost-effective retirement age. High-consequence transformers should be replaced at younger ages due to the risk they impose on the system as their effective age increases. PRA showed that using new spares maximizes a transformer's effective age for retirement.

STANDARDIZATION IMPACT

Approximately one-third of the number of current spares would be required if design standardization and sharing between asset owners were achieved. This allows a single spare to reduce the loss consequence for a larger number of in-service units. Increasing the number of transformers covered by a spare improves the spare's risk-mitigation value. Having more transformers covered by spares reduces the residual risk exposure that accumulates with having many spare subgroups.

PJM transmission asset owners have finalized a standardized 500-kV/230-kV transformer design to apply to future purchase decisions. For the benefits of standardization to be achieved, PJM asset owners

PJM BACKGROUND

Formally established on Sept. 16, 1927, the Pennsylvania-New Jersey Interconnection allowed Philadelphia Electric, Pennsylvania Power & Light, and Public Service Electric & Gas of New Jersey to share their electric loads and receive power from the huge new hydroelectric plant at Conowingo, Maryland, U.S. Throughout the years, neighboring utilities also connected into the system. Today, the interconnection, now called the PJM Interconnection, has far exceeded its original footprint.

PJM is the operator of the world's largest centrally dispatched grid, serving about 51 million people in 13 states and the District of Columbia. A regional transmission organization that operates 19% of the transmission infrastructure of the U.S. Eastern Interconnection on behalf of transmission system owners, PJM dispatches 164,634 MW of generating capacity over 56,000 miles (91,800 km) of transmission. Within PJM, 12 utilities individually own the 500-kV/230-kV transformer assets.

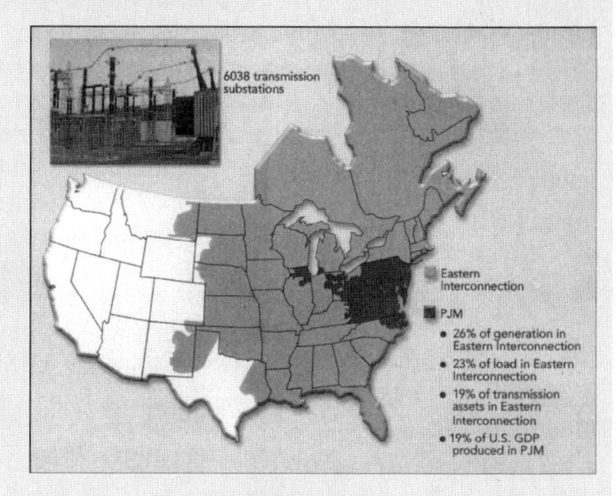

PJM system information breakdown and location

also are developing a spare-sharing agreement. Analysis showed that $50 million of current spare transformer requirements could be avoided by standardization and sharing.

The PRA model is a useful tool for managing PJM's aging 500-kV/230-kV transformer infrastructure. While creating the PRA model was challenging, system planners and asset owners have gained invaluable insights from both the development process and the model use. Knowing and understanding risk has better prepared PJM and its members to proactively and economically address their aging transformer fleet. PRA results have been incorporated into PJM's regional transmission-expansion planning process. PRA will be performed annually to ensure minimum transformer fleet risk exposure. PJM is also investigating the use of this risk quantification approach for other power-system assets.

Kenneth Seiler is manager of power system coordination at PJM Interconnection. He is responsible for the interconnection coordination of generation, substation and transmission projects, and outage planning. He has been actively involved in the PJM Planning Committee and the development of the PJM's aging infrastructure initiatives. Prior to working for PJM, he was with GPU Energy for nearly 15 years in the Electrical Equipment Construction and Maintenance and System Operation departments. Seiler earned his BSEE degree from Pennsylvania State University and MBA from Lebanon Valley College, **seilek@pjm.com**

David Egan is a senior engineer in PJM's Interconnection Planning department, where he has worked for three years. He earned his BSME degree from Binghamton University. Previously he worked at Oyster Creek Generating Station for 13 years. During this time, he worked as a thermal performance engineer and turbine-generator systems' manager, and coordinated implementation of the site's Maintenance Rule program, **egand@pjm.com**

3.1

THE IDEAL TRANSFORMER

Figure 3.1 shows a basic single-phase two-winding transformer, where the two windings are wrapped around a magnetic core [1, 2, 3]. It is assumed here that the transformer is operating under sinusoidal-steady-state excitation. Shown in the figure are the phasor voltages E_1 and E_2 across the windings, and the phasor currents I_1 entering winding 1, which has N_1 turns, and I_2 leaving winding 2, which has N_2 turns. A phasor flux Φ_c set up in the core and a magnetic field intensity phasor H_c are also shown. The core has a cross-sectional area denoted A_c, a mean length of the magnetic circuit l_c, and a magnetic permeability μ_c, assumed constant.

For an ideal transformer, the following are assumed:

1. The windings have zero resistance; therefore, the I^2R losses in the windings are zero.
2. The core permeability μ_c is infinite, which corresponds to zero core reluctance.
3. There is no leakage flux; that is, the entire flux Φ_c is confined to the core and links both windings.
4. There are no core losses.

A schematic representation of a two-winding transformer is shown in Figure 3.2. Ampere's and Faraday's laws can be used along with the preceding assumptions to derive the ideal transformer relationships. Ampere's law states that the tangential component of the magnetic field intensity vector

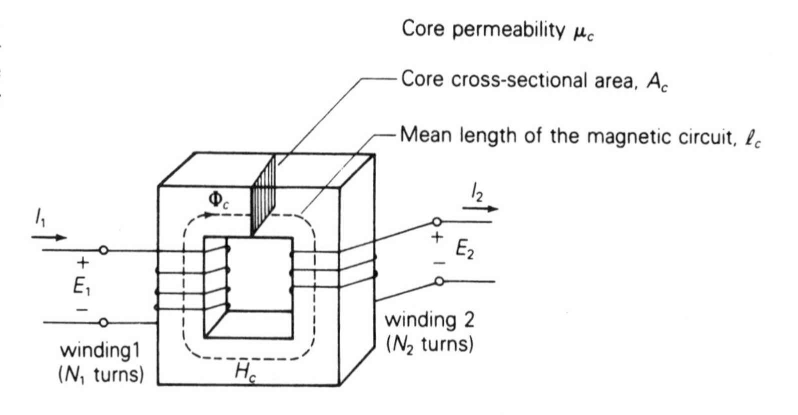

FIGURE 3.1 Basic single-phase two-winding transformer

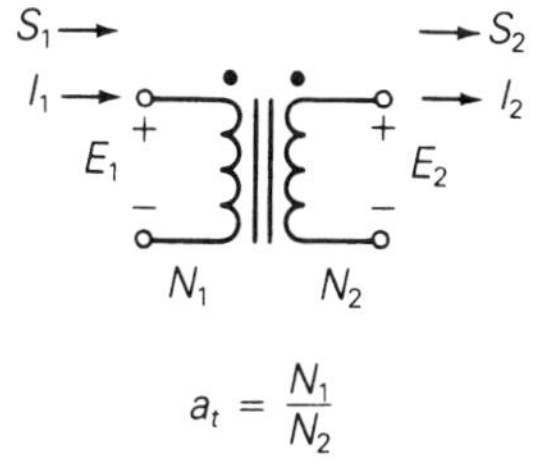

FIGURE 3.2 Schematic representation of a single-phase two-winding transformer

integrated along a closed path equals the net current enclosed by that path; that is,

$$\oint H_{\tan}\, dl = I_{\text{enclosed}} \tag{3.1.1}$$

If the core center line shown in Figure 3.1 is selected as the closed path, and if H_c is constant along the path as well as tangent to the path, then (3.1.1) becomes

$$H_c l_c = N_1 I_1 - N_2 I_2 \tag{3.1.2}$$

Note that the current I_1 is enclosed N_1 times and I_2 is enclosed N_2 times, one time for each turn of the coils. Also, using the right-hand rule*, current I_1 contributes to clockwise flux but current I_2 contributes to counterclockwise flux. Thus, in (3.1.2) the net current enclosed is $N_1 I_1 - N_2 I_2$. For constant core permeability μ_c, the magnetic flux density B_c within the core, also constant, is

$$B_c = \mu_c H_c \quad \text{Wb/m}^2 \tag{3.1.3}$$

and the core flux Φ_c is

$$\Phi_c = B_c \mathrm{A}_c \quad \text{Wb} \tag{3.1.4}$$

Using (3.1.3) and (3.1.4) in (3.1.2) yields

$$N_1 I_1 - N_2 I_2 = l_c B_c / \mu_c = \left(\frac{l_c}{\mu_c \mathrm{A}_c} \right) \Phi_c \tag{3.1.5}$$

We define core reluctance R_c as

$$\mathrm{R}_c = \frac{l_c}{\mu_c \mathrm{A}_c} \tag{3.1.6}$$

Then (3.1.5) becomes

$$N_1 I_1 - N_2 I_2 = \mathrm{R}_c \Phi_c \tag{3.1.7}$$

*The right-hand rule for a coil is as follows: Wrap the fingers of your right hand around the coil in the direction of the current. Your right thumb then points in the direction of the flux.

Equation (3.1.7) can be called "Ohm's law" for the magnetic circuit, wherein the net magnetomotive force mmf $= N_1I_1 - N_2I_2$ equals the product of the core reluctance R_c and the core flux Φ_c. Reluctance R_c, which impedes the establishment of flux in a magnetic circuit, is analogous to resistance in an electric circuit. For an ideal transformer, μ_c is assumed infinite, which, from (3.1.6), means that R_c is 0, and (3.1.7) becomes

$$N_1I_1 = N_2I_2 \tag{3.1.8}$$

In practice, power transformer windings and cores are contained within enclosures, and the winding directions are not visible. One way of conveying winding information is to place a dot at one end of each winding such that when current enters a winding at the dot, it produces an mmf acting in the *same* direction. This dot convention is shown in the schematic of Figure 3.2. The dots are conventionally called *polarity marks.*

Equation (3.1.8) is written for current I_1 *entering* its dotted terminal and current I_2 *leaving* its dotted terminal. As such, I_1 and I_2 are *in phase,* since $I_1 = (N_2/N_1)I_2$. If the direction chosen for I_2 were reversed, such that both currents entered their dotted terminals, then I_1 would be 180° *out of phase* with I_2.

Faraday's law states that the voltage $e(t)$ induced across an N-turn winding by a time-varying flux $\phi(t)$ linking the winding is

$$e(t) = N\frac{d\phi(t)}{dt} \tag{3.1.9}$$

Assuming a sinusoidal-steady-state flux with constant frequency ω, and representing $e(t)$ and $\phi(t)$ by their phasors E and Φ, (3.1.9) becomes

$$E = N(j\omega)\Phi \tag{3.1.10}$$

For an ideal transformer, the entire flux is assumed to be confined to the core, linking both windings. From Faraday's law, the induced voltages across the windings of Figure 3.1 are

$$E_1 = N_1(j\omega)\Phi_c \tag{3.1.11}$$

$$E_2 = N_2(j\omega)\Phi_c \tag{3.1.12}$$

Dividing (3.1.11) by (3.1.12) yields

$$\frac{E_1}{E_2} = \frac{N_1}{N_2} \tag{3.1.13}$$

or

$$\frac{E_1}{N_1} = \frac{E_2}{N_2} \tag{3.1.14}$$

The dots shown in Figure 3.2 indicate that the voltages E_1 and E_2, both of which have their + polarities at the dotted terminals, are in phase. If the polarity chosen for one of the voltages in Figure 3.1 were reversed, then E_1 would be 180° out of phase with E_2.

The turns ratio a_t is defined as follows:

$$a_t = \frac{N_1}{N_2} \tag{3.1.15}$$

Using a_t in (3.1.8) and (3.1.14), the basic relations for an ideal single-phase two-winding transformer are

$$E_1 = \left(\frac{N_1}{N_2}\right) E_2 = a_t E_2 \tag{3.1.16}$$

$$I_1 = \left(\frac{N_2}{N_1}\right) I_2 = \frac{I_2}{a_t} \tag{3.1.17}$$

Two additional relations concerning complex power and impedance can be derived from (3.1.16) and (3.1.17) as follows. The complex power entering winding 1 in Figure 3.2 is

$$S_1 = E_1 I_1^* \tag{3.1.18}$$

Using (3.1.16) and (3.1.17),

$$S_1 = E_1 I_1^* = (a_t E_2)\left(\frac{I_2}{a_t}\right)^* = E_2 I_2^* = S_2 \tag{3.1.19}$$

As shown by (3.1.19), the complex power S_1 entering winding 1 equals the complex power S_2 leaving winding 2. That is, an ideal transformer has no real or reactive power loss.

If an impedance Z_2 is connected across winding 2 of the ideal transformer in Figure 3.2, then

$$Z_2 = \frac{E_2}{I_2} \tag{3.1.20}$$

This impedance, when measured from winding 1, is

$$Z_2' = \frac{E_1}{I_1} = \frac{a_t E_2}{I_2/a_t} = a_t^2 Z_2 = \left(\frac{N_1}{N_2}\right)^2 Z_2 \tag{3.1.21}$$

Thus, the impedance Z_2 connected to winding 2 is referred to winding 1 by multiplying Z_2 by a_t^2, the square of the turns ratio.

EXAMPLE 3.1 Ideal, single-phase two-winding transformer

A single-phase two-winding transformer is rated 20 kVA, 480/120 V, 60 Hz. A source connected to the 480-V winding supplies an impedance load connected to the 120-V winding. The load absorbs 15 kVA at 0.8 p.f. lagging

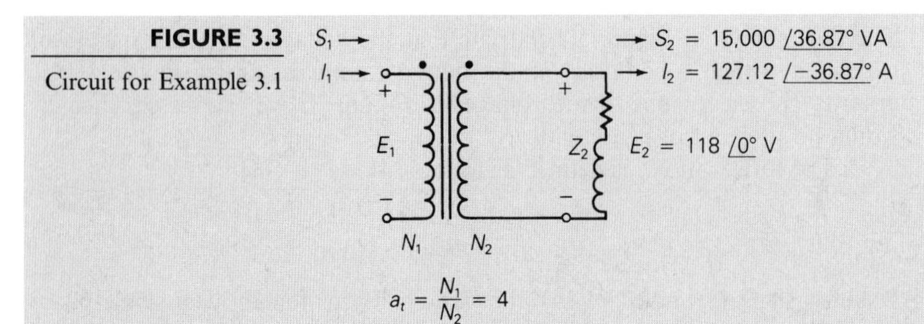

FIGURE 3.3 Circuit for Example 3.1

when the load voltage is 118 V. Assume that the transformer is ideal and calculate the following:

a. The voltage across the 480-V winding.

b. The load impedance.

c. The load impedance referred to the 480-V winding.

d. The real and reactive power supplied to the 480-V winding.

SOLUTION

a. The circuit is shown in Figure 3.3, where winding 1 denotes the 480-V winding and winding 2 denotes the 120-V winding. Selecting the load voltage E_2 as the reference,

$$E_2 = 118\underline{/0^\circ} \quad \text{V}$$

The turns ratio is, from (3.1.13),

$$a_t = \frac{N_1}{N_2} = \frac{\text{E}_{1\text{rated}}}{\text{E}_{2\text{rated}}} = \frac{480}{120} = 4$$

and the voltage across winding 1 is

$$E_1 = a_t E_2 = 4(118\underline{/0^\circ}) = 472\underline{/0^\circ} \quad \text{V}$$

b. The complex power S_2 absorbed by the load is

$$S_2 = E_2 I_2^* = 118 I_2^* = 15{,}000\underline{/\cos^{-1}(0.8)} = 15{,}000\underline{/36.87^\circ} \quad \text{VA}$$

Solving, the load current I_2 is

$$I_2 = 127.12\underline{/-36.87^\circ} \quad \text{A}$$

The load impedance Z_2 is

$$Z_2 = \frac{E_2}{I_2} = \frac{118\underline{/0^\circ}}{127.12\underline{/-36.87^\circ}} = 0.9283\underline{/36.87^\circ} \quad \Omega$$

c. From (3.1.21), the load impedance referred to the 480-V winding is

$$Z_2' = a_t^2 Z_2 = (4)^2(0.9283\underline{/36.87^\circ}) = 14.85\underline{/36.87^\circ} \quad \Omega$$

d. From (3.1.19)

$$S_1 = S_2 = 15{,}000\underline{/36.87^\circ} = 12{,}000 + j9000$$

Thus, the real and reactive powers supplied to the 480-V winding are

$$\mathrm{P}_1 = \mathrm{Re}\, S_1 = 12{,}000 \text{ W} = 12 \text{ kW}$$

$$\mathrm{Q}_1 = \mathrm{Im}\, S_1 = 9000 \text{ var} = 9 \text{ kvar}$$ ■

Figure 3.4 shows a schematic of a conceptual single-phase, phase-shifting transformer. This transformer is not an idealization of an actual transformer since it is physically impossible to obtain a complex turns ratio. It will be used later in this chapter as a mathematical model for representing phase shift of three-phase transformers. As shown in Figure 3.4, the complex turns ratio a_t is defined for the phase-shifting transformer as

$$a_t = \frac{e^{j\phi}}{1} = e^{j\phi} \tag{3.1.22}$$

where ϕ is the phase-shift angle. The transformer relations are then

$$E_1 = a_t E_2 = e^{j\phi} E_2 \tag{3.1.23}$$

$$I_1 = \frac{I_2}{a_t^*} = e^{j\phi} I_2 \tag{3.1.24}$$

Note that the phase angle of E_1 leads the phase angle of E_2 by ϕ. Similarly, I_1 leads I_2 by the angle ϕ. However, the magnitudes are unchanged; that is, $|E_1| = |E_2|$ and $|I_1| = |I_2|$.

FIGURE 3.4 Schematic representation of a conceptual single-phase, phase-shifting transformer

S_1 → S_2, I_1 → I_2, E_1, E_2, $e^{j\phi}$: 1

$a_t = e^{j\phi}$

$E_1 = a_t E_2 = e^{j\phi} E_2$

$I_1 = \frac{I_2}{a_t^*} = e^{j\phi} I_2$

$S_1 = S_2$

$Z_2' = Z_2$

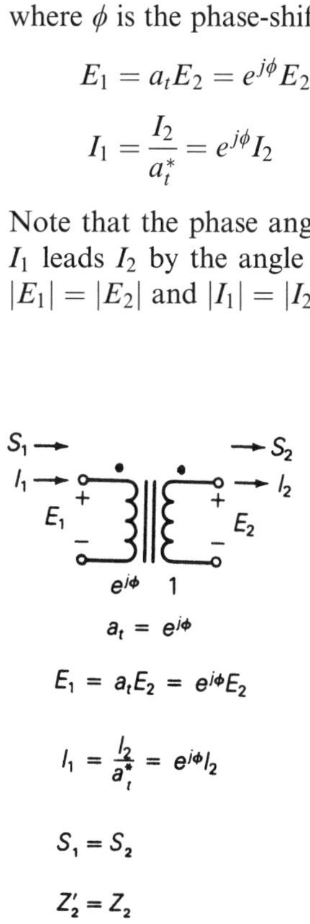

From these two relations, the following two additional relations are derived:

$$S_1 = E_1 I_1^* = (a_t E_2)\left(\frac{I_2}{a_t^*}\right)^* = E_2 I_2^* = S_2 \tag{3.1.25}$$

$$Z_2' = \frac{E_1}{I_1} = \frac{a_t E_2}{\dfrac{1}{a_t^*} I_2} = |a_t|^2 Z_2 = Z_2 \tag{3.1.26}$$

Thus, impedance is unchanged when it is referred from one side of an ideal phase-shifting transformer to the other. Also, the ideal phase-shifting transformer has no real or reactive power losses since $S_1 = S_2$.

Note that (3.1.23) and (3.1.24) for the phase-shifting transformer are the same as (3.1.16) and (3.1.17) for the ideal physical transformer except for the complex conjugate (*) in (3.1.24). The complex conjugate for the phase-shifting transformer is required to make $S_1 = S_2$ (complex power into winding 1 equals complex power out of winding 2), as shown in (3.1.25).

3.2 EQUIVALENT CIRCUITS FOR PRACTICAL TRANSFORMERS

Figure 3.5 shows an equivalent circuit for a practical single-phase two-winding transformer, which differs from the ideal transformer as follows:

1. The windings have resistance.
2. The core permeability μ_c is finite.
3. The magnetic flux is not entirely confined to the core.
4. There are real and reactive power losses in the core.

The resistance R_1 is included in series with winding 1 of the figure to account for I^2R losses in this winding. A reactance X_1, called the leakage

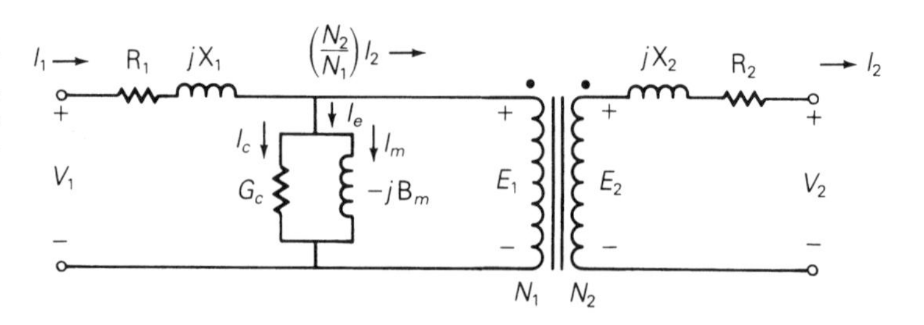

FIGURE 3.5 Equivalent circuit of a practical single-phase two-winding transformer

reactance of winding 1, is also included in series with winding 1 to account for the leakage flux of winding 1. This leakage flux is the component of the flux that links winding 1 but does not link winding 2; it causes a voltage drop $I_1(jX_1)$, which is proportional to I_1 and leads I_1 by 90°. There is also a reactive power loss $I_1^2X_1$ associated with this leakage reactance. Similarly, there is a resistance R_2 and a leakage reactance X_2 in series with winding 2.

Equation (3.1.7) shows that for finite core permeability μ_c, the total mmf is not 0. Dividing (3.1.7) by N_1 and using (3.1.11), we get

$$I_1 - \left(\frac{N_2}{N_1}\right)I_2 = \frac{R_c}{N_1}\Phi_c = \frac{R_c}{N_1}\left(\frac{E_1}{j\omega N_1}\right) = -j\left(\frac{R_c}{\omega N_1^2}\right)E_1 \tag{3.2.1}$$

Defining the term on the right-hand side of (3.2.1) to be I_m, called *magnetizing* current, it is evident that I_m lags E_1 by 90°, and can be represented by a shunt inductor with susceptance $B_m = \left(\frac{R_c}{\omega N_1^2}\right)$ mhos.* However, in reality there is an additional shunt branch, represented by a resistor with conductance G_c mhos, which carries a current I_c, called the *core loss* current. I_c is in phase with E_1. When the core loss current I_c is included, (3.2.1) becomes

$$I_1 - \left(\frac{N_2}{N_1}\right)I_2 = I_c + I_m = (G_c - jB_m)E_1 \tag{3.2.2}$$

The equivalent circuit of Figure 3.5, which includes the shunt branch with admittance $(G_c - jB_m)$ mhos, satisfies the KCL equation (3.2.2). Note that when winding 2 is open ($I_2 = 0$) and when a sinusoidal voltage V_1 is applied to winding 1, then (3.2.2) indicates that the current I_1 will have two components: the core loss current I_c and the magnetizing current I_m. Associated with I_c is a real power loss $I_c^2/G_c = E_1^2G_c$ W. This real power loss accounts for both hysteresis and eddy current losses within the core. Hysteresis loss occurs because a cyclic variation of flux within the core requires energy dissipated as heat. As such, hysteresis loss can be reduced by the use of special high grades of alloy steel as core material. Eddy current loss occurs because induced currents called eddy currents flow within the magnetic core perpendicular to the flux. As such, eddy current loss can be reduced by constructing the core with laminated sheets of alloy steel. Associated with I_m is a reactive power loss $I_m^2/B_m = E_1^2B_m$ var. This reactive power is required to magnetize the core. The phasor sum $(I_c + I_m)$ is called the *exciting* current I_e.

Figure 3.6 shows three alternative equivalent circuits for a practical single-phase two-winding transformer. In Figure 3.6(a), the resistance R_2 and leakage reactance X_2 of winding 2 are referred to winding 1 via (3.1.21).

---

*The units of admittance, conductance, and susceptance, which in the SI system are siemens (with symbol S), are also called mhos (with symbol ℧) or ohms^{-1} (with symbol Ω^{-1}).

FIGURE 3.6

Equivalent circuits for a practical single-phase two-winding transformer

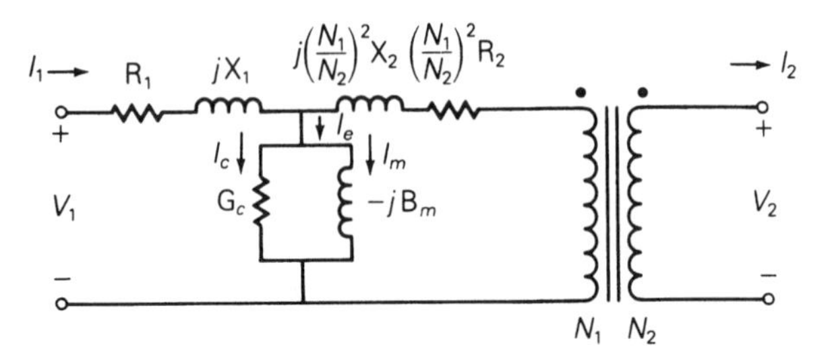

(a) R_2 and X_2 are referred to winding 1

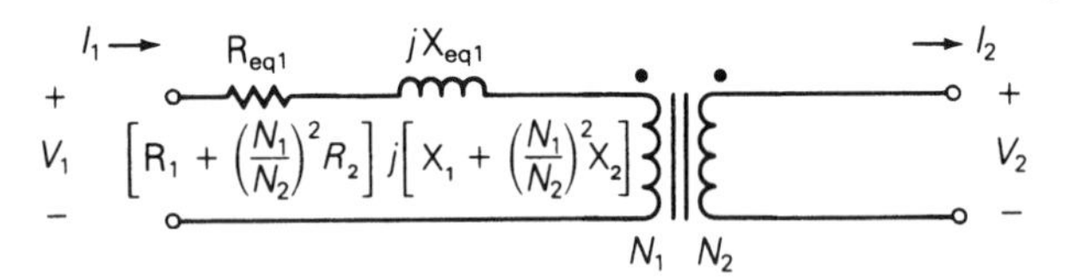

(b) Neglecting exciting current

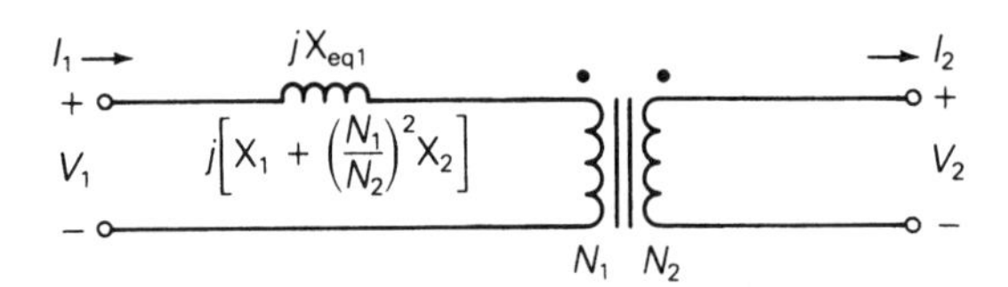

(c) Neglecting exciting current and I^2R winding loss

In Figure 3.6(b), the shunt branch is omitted, which corresponds to neglecting the exciting current. Since the exciting current is usually less than 5% of rated current, neglecting it in power system studies is often valid unless transformer efficiency or exciting current phenomena are of particular concern. For large power transformers rated more than 500 kVA, the winding resistances, which are small compared to the leakage reactances, can often be neglected, as shown in Figure 3.6(c).

Thus, a practical transformer operating in sinusoidal steady state is equivalent to an ideal transformer with external impedance and admittance branches, as shown in Figure 3.6. The external branches can be evaluated from short-circuit and open-circuit tests, as illustrated by the following example.

EXAMPLE 3.2 Transformer short-circuit and open-circuit tests

A single-phase two-winding transformer is rated 20 kVA, 480/120 volts, 60 Hz. During a short-circuit test, where rated current at rated frequency is applied to the 480-volt winding (denoted winding 1), with the 120-volt winding (winding 2) shorted, the following readings are obtained: $V_1 = 35$ volts, $P_1 = 300$ W. During an open-circuit test, where rated voltage is applied to winding 2, with winding 1 open, the following readings are obtained: $I_2 = 12$ A, $P_2 = 200$ W.

a. From the short-circuit test, determine the equivalent series impedance $Z_{eq1} = R_{eq1} + jX_{eq1}$ referred to winding 1. Neglect the shunt admittance.

b. From the open-circuit test, determine the shunt admittance $Y_m = G_c - jB_m$ referred to winding 1. Neglect the series impedance.

SOLUTION

a. The equivalent circuit for the short-circuit test is shown in Figure 3.7(a), where the shunt admittance branch is neglected. Rated current for winding 1 is

$$I_{1\text{rated}} = \frac{S_{\text{rated}}}{V_{1\text{rated}}} = \frac{20 \times 10^3}{480} = 41.667 \quad \text{A}$$

R_{eq1}, Z_{eq1}, and X_{eq1} are then determined as follows:

$$R_{eq1} = \frac{P_1}{I^2_{1\text{rated}}} = \frac{300}{(41.667)^2} = 0.1728 \quad \Omega$$

$$|Z_{eq1}| = \frac{V_1}{I_{1\text{rated}}} = \frac{35}{41.667} = 0.8400 \quad \Omega$$

$$X_{eq1} = \sqrt{Z^2_{eq1} - R^2_{eq1}} = 0.8220 \quad \Omega$$

$$Z_{eq1} = R_{eq1} + jX_{eq1} = 0.1728 + j0.8220 = 0.8400\underline{/78.13^\circ} \quad \Omega$$

FIGURE 3.7

Circuits for Example 3.2

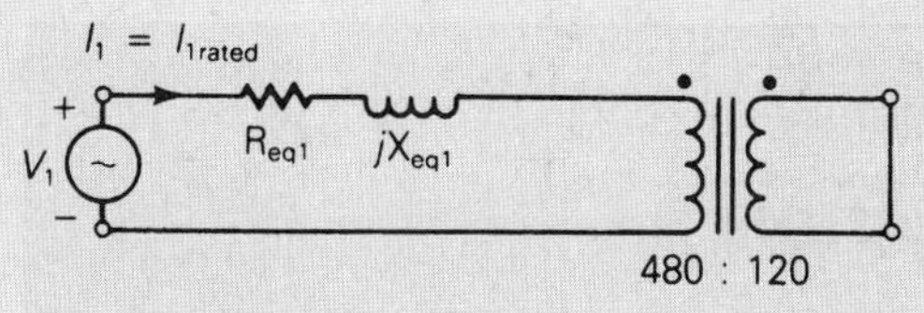

(a) Short-circuit test (neglecting shunt admittance)

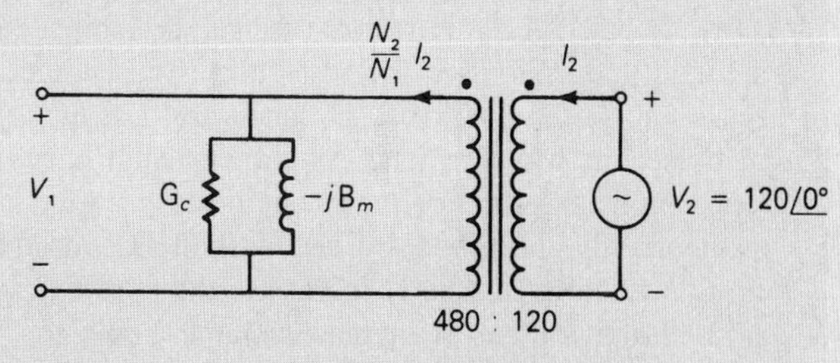

(b) Open-circuit test (neglecting series impedance)

b. The equivalent circuit for the open-circuit test is shown in Figure 3.7(b), where the series impedance is neglected. From (3.1.16),

$$\mathrm{V}_1 = \mathrm{E}_1 = a_t\mathrm{E}_2 = \frac{N_1}{N_2}\mathrm{V}_{2\text{rated}} = \frac{480}{120}(120) = 480 \text{ volts}$$

G_c, Y_m, and B_m are then determined as follows:

$$\mathrm{G}_c = \frac{\mathrm{P}_2}{\mathrm{V}_1^2} = \frac{200}{(480)^2} = 0.000868 \quad \mathrm{S}$$

$$|Y_m| = \frac{\left(\frac{N_2}{N_1}\right)\mathrm{I}_2}{\mathrm{V}_1} = \frac{\left(\frac{120}{480}\right)(12)}{480} = 0.00625 \quad \mathrm{S}$$

$$\mathrm{B}_m = \sqrt{\mathrm{Y}_m^2 - \mathrm{G}_c^2} = \sqrt{(0.00625)^2 - (0.000868)^2} = 0.00619 \quad \mathrm{S}$$

$$\mathrm{Y}_m = \mathrm{G}_c - j\mathrm{B}_m = 0.000868 - j0.00619 = 0.00625\underline{/-82.02^\circ} \quad \mathrm{S}$$

Note that the equivalent series impedance is usually evaluated at rated current from a short-circuit test, and the shunt admittance is evaluated at rated voltage from an open-circuit test. For small variations in transformer operation near rated conditions, the impedance and admittance values are often assumed constant. ■

The following are not represented by the equivalent circuit of Figure 3.5:

1. Saturation
2. Inrush current
3. Nonsinusoidal exciting current
4. Surge phenomena

They are briefly discussed in the following sections.

SATURATION

In deriving the equivalent circuit of the ideal and practical transformers, we have assumed constant core permeability μ_c and the linear relationship $B_c = \mu_c H_c$ of (3.1.3). However, the relationship between B and H for ferromagnetic materials used for transformer cores is nonlinear and multivalued. Figure 3.8 shows a set of B–H curves for a grain-oriented electrical steel typically used in transformers. As shown, each curve is multivalued, which is caused by hysteresis. For many engineering applications, the B–H curves can be adequately described by the dashed line drawn through the curves in Figure 3.8. Note that as H increases, the core becomes saturated; that is, the curves flatten out as B increases above 1 Wb/m^2. If the magnitude of the voltage applied to a transformer is too large, the core will saturate and a high

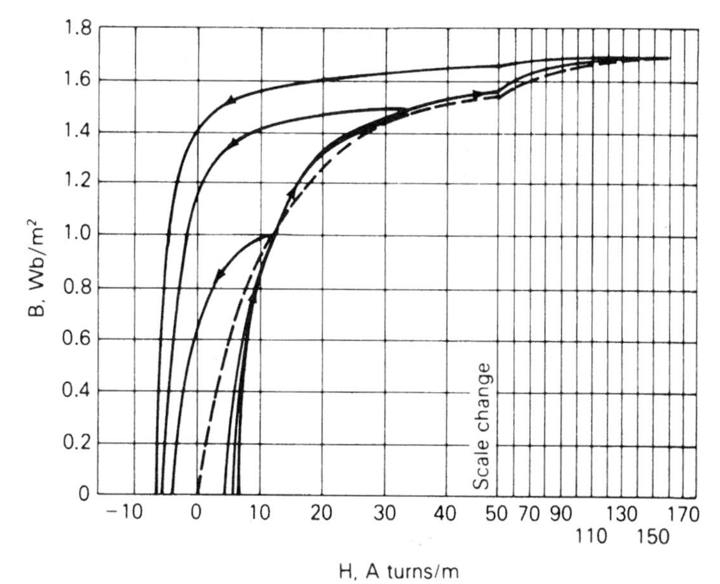

FIGURE 3.8 B–H curves for M-5 grain-oriented electrical steel 0.012 in. (0.305 mm) thick (Reprinted with permission of AK Steel Corporation)

magnetizing current will flow. In a well-designed transformer, the applied peak voltage causes the peak flux density in steady state to occur at the knee of the B–H curve, with a corresponding low value of magnetizing current.

INRUSH CURRENT

When a transformer is first energized, a transient current much larger than rated transformer current can flow for several cycles. This current, called *inrush current*, is nonsinusoidal and has a large dc component. To understand the cause of inrush, assume that before energization, the transformer core is magnetized with a residual flux density $B(0) = 1.5\ \text{Wb/m}^2$ (near the knee of the dotted curve in Figure 3.8). If the transformer is then energized when the source voltage is positive and increasing, Faraday's law, (3.1.9), will cause the flux density $B(t)$ to increase further, since

$$B(t) = \frac{\phi(t)}{A} = \frac{1}{NA}\int_0^t e(t)\,dt + B(0)$$

As $B(t)$ moves into the saturation region of the B–H curve, large values of $H(t)$ will occur, and, from Ampere's law, (3.1.1), corresponding large values of current $i(t)$ will flow for several cycles until it has dissipated. Since normal inrush currents can be as large as abnormal short-circuit currents in transformers, transformer protection schemes must be able to distinguish between these two types of currents.

NONSINUSOIDAL EXCITING CURRENT

When a sinusoidal voltage is applied to one winding of a transformer with the other winding open, the flux $\phi(t)$ and flux density $B(t)$ will, from Faraday's law, (3.1.9), be very nearly sinusoidal in steady state. However, the magnetic field intensity $H(t)$ and the resulting exciting current will not be sinusoidal in steady state, due to the nonlinear B–H curve. If the exciting current is measured and analyzed by Fourier analysis techniques, one finds that it has a fundamental component and a set of odd harmonics. The principal harmonic is the third, whose rms value is typically about 40% of the total rms exciting current. However, the nonsinusoidal nature of exciting current is usually neglected unless harmonic effects are of direct concern, because the exciting current itself is usually less than 5% of rated current for power transformers.

SURGE PHENOMENA

When power transformers are subjected to transient overvoltages caused by lightning or switching surges, the capacitances of the transformer windings have important effects on transient response. Transformer winding capacitances and response to surges are discussed in Chapter 12.

3.3 THE PER-UNIT SYSTEM

Power-system quantities such as voltage, current, power, and impedance are often expressed in per-unit or percent of specified base values. For example, if a base voltage of 20 kV is specified, then the voltage 18 kV is $(18/20) = 0.9$ per unit or 90%. Calculations can then be made with per-unit quantities rather than with the actual quantities.

One advantage of the per-unit system is that by properly specifying base quantities, the transformer equivalent circuit can be simplified. The ideal transformer winding can be eliminated, such that voltages, currents, and external impedances and admittances expressed in per-unit do not change when they are referred from one side of a transformer to the other. This can be a significant advantage even in a power system of moderate size, where hundreds of transformers may be encountered. The per-unit system allows us to avoid the possibility of making serious calculation errors when referring quantities from one side of a transformer to the other. Another advantage of the per-unit system is that the per-unit impedances of electrical equipment of similar type usually lie within a narrow numerical range when the equipment ratings are used as base values. Because of this, per-unit impedance data can be checked rapidly for gross errors by someone familiar with per-unit quantities. In addition, manufacturers usually specify the impedances of machines and transformers in per-unit or percent of nameplate rating.

Per-unit quantities are calculated as follows:

$$\text{per-unit quantity} = \frac{\text{actual quantity}}{\text{base value of quantity}} \tag{3.3.1}$$

where *actual quantity* is the value of the quantity in the actual units. The base value has the same units as the actual quantity, thus making the per-unit quantity dimensionless. Also, the base value is always a real number. Therefore, the angle of the per-unit quantity is the same as the angle of the actual quantity.

Two independent base values can be arbitrarily selected at one point in a power system. Usually the base voltage V_{baseLN} and base complex power $S_{base1\phi}$ are selected for either a single-phase circuit or for one phase of a three-phase circuit. Then, in order for electrical laws to be valid in the per-unit system, the following relations must be used for other base values:

$$P_{base1\phi} = Q_{base1\phi} = S_{base1\phi} \tag{3.3.2}$$

$$I_{base} = \frac{S_{base1\phi}}{V_{baseLN}} \tag{3.3.3}$$

$$Z_{base} = R_{base} = X_{base} = \frac{V_{baseLN}}{I_{base}} = \frac{V_{baseLN}^2}{S_{base1\phi}} \tag{3.3.4}$$

$$Y_{base} = G_{base} = B_{base} = \frac{1}{Z_{base}} \tag{3.3.5}$$

In (3.3.2)–(3.3.5) the subscripts LN and 1ϕ denote "line-to-neutral" and "per-phase," respectively, for three-phase circuits. These equations are also valid for single-phase circuits, where subscripts can be omitted.

By convention, we adopt the following two rules for base quantities:

1. The value of $S_{base1\phi}$ is the same for the entire power system of concern.
2. The ratio of the voltage bases on either side of a transformer is selected to be the same as the ratio of the transformer voltage ratings.

With these two rules, a per-unit impedance remains unchanged when referred from one side of a transformer to the other.

EXAMPLE 3.3 Per-unit impedance: single-phase transformer

A single-phase two-winding transformer is rated 20 kVA, 480/120 volts, 60 Hz. The equivalent leakage impedance of the transformer referred to the 120-volt winding, denoted winding 2, is $Z_{eq2} = 0.0525\underline{/78.13^\circ}\ \Omega$. Using the transformer ratings as base values, determine the per-unit leakage impedance referred to winding 2 and referred to winding 1.

SOLUTION The values of S_{base}, V_{base1}, and V_{base2} are, from the transformer ratings,

$$S_{base} = 20 \text{ kVA}, \qquad V_{base1} = 480 \text{ volts}, \qquad V_{base2} = 120 \text{ volts}$$

Using (3.3.4), the base impedance on the 120-volt side of the transformer is

$$Z_{base2} = \frac{V_{base2}^2}{S_{base}} = \frac{(120)^2}{20{,}000} = 0.72 \quad \Omega$$

Then, using (3.3.1), the per-unit leakage impedance referred to winding 2 is

$$Z_{eq2p.u.} = \frac{Z_{eq2}}{Z_{base2}} = \frac{0.0525\underline{/78.13^\circ}}{0.72} = 0.0729\underline{/78.13^\circ} \quad \text{per unit}$$

If Z_{eq2} is referred to winding 1,

$$Z_{eq1} = a_t^2 Z_{eq2} = \left(\frac{N_1}{N_2}\right)^2 Z_{eq2} = \left(\frac{480}{120}\right)^2 (0.0525\underline{/78.13^\circ})$$

$$= 0.84\underline{/78.13^\circ} \quad \Omega$$

The base impedance on the 480-volt side of the transformer is

$$Z_{base1} = \frac{V_{base1}^2}{S_{base}} = \frac{(480)^2}{20{,}000} = 11.52 \quad \Omega$$

and the per-unit leakage reactance referred to winding 1 is

$$Z_{eq1p.u.} = \frac{Z_{eq1}}{Z_{base1}} = \frac{0.84\underline{/78.13^\circ}}{11.52} = 0.0729\underline{/78.13^\circ} \text{ per unit} = Z_{eq2p.u.}$$

Thus, the *per-unit* leakage impedance remains unchanged when referred from winding 2 to winding 1. This has been achieved by specifying

$$\frac{V_{base1}}{V_{base2}} = \frac{V_{rated1}}{V_{rated2}} = \left(\frac{480}{120}\right)$$

■

Figure 3.9 shows three per-unit circuits of a single-phase two-winding transformer. The ideal transformer, shown in Figure 3.9(a), satisfies the per-unit relations $E_{1p.u.} = E_{2p.u.}$, and $I_{1p.u.} = I_{2p.u.}$, which can be derived as follows. First divide (3.1.16) by V_{base1}:

$$E_{1p.u.} = \frac{E_1}{V_{base1}} = \frac{N_1}{N_2} \times \frac{E_2}{V_{base1}} \tag{3.3.6}$$

Then, using $V_{base1}/V_{base2} = V_{rated1}/V_{rated2} = N_1/N_2$,

$$E_{1p.u.} = \frac{N_1}{N_2} \frac{E_2}{\left(\frac{N_1}{N_2}\right) V_{base2}} = \frac{E_2}{V_{base2}} = E_{2p.u.} \tag{3.3.7}$$

FIGURE 3.9

Per-unit equivalent circuits of a single-phase two-winding transformer

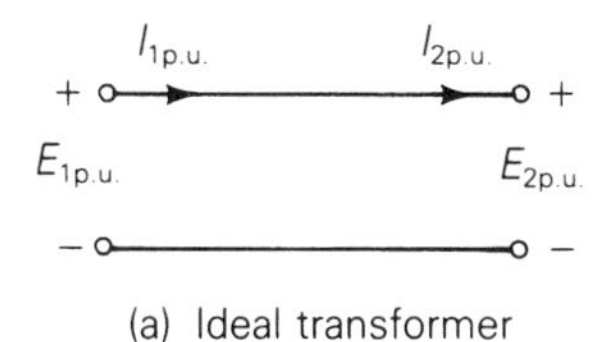

(a) Ideal transformer

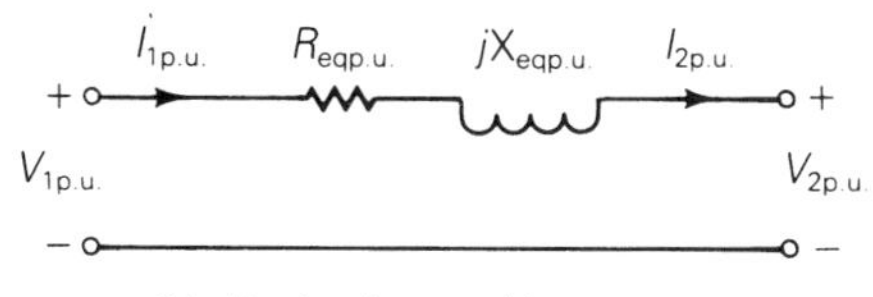

(b) Neglecting exciting current

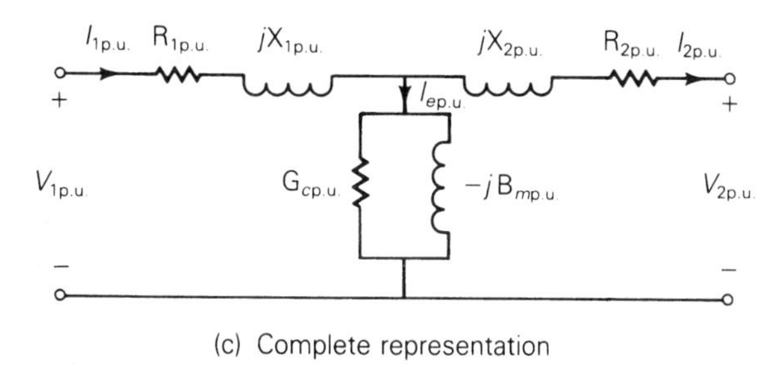

(c) Complete representation

Similarly, divide (3.1.17) by I_{base1}:

$$I_{1\text{p.u.}} = \frac{I_1}{\text{I}_{\text{base1}}} = \frac{N_2}{N_1}\frac{I_2}{\text{I}_{\text{base1}}} \tag{3.3.8}$$

Then, using $\text{I}_{\text{base1}} = \text{S}_{\text{base}}/\text{V}_{\text{base1}} = \text{S}_{\text{base}}/[(N_1/N_2)\text{V}_{\text{base2}}] = (N_2/N_1)\text{I}_{\text{base2}}$,

$$I_{1\text{p.u.}} = \frac{N_2}{N_1}\frac{I_2}{\left(\frac{N_2}{N_1}\right)\text{I}_{\text{base2}}} = \frac{I_2}{\text{I}_{\text{base2}}} = I_{2\text{p.u.}} \tag{3.3.9}$$

Thus, the ideal transformer winding in Figure 3.2 is eliminated from the per-unit circuit in Figure 3.9(a). The per-unit leakage impedance is included in Figure 3.9(b), and the per-unit shunt admittance branch is added in Figure 3.9(c) to obtain the complete representation.

When only one component, such as a transformer, is considered, the nameplate ratings of that component are usually selected as base values. When several components are involved, however, the system base values may be different from the nameplate ratings of any particular device. It is then necessary to convert the per-unit impedance of a device from its nameplate ratings to the system base values. To convert a per-unit impedance from "old" to "new" base values, use

$$Z_{\text{p.u.new}} = \frac{Z_{\text{actual}}}{Z_{\text{basenew}}} = \frac{Z_{\text{p.u.old}} Z_{\text{baseold}}}{Z_{\text{basenew}}} \tag{3.3.10}$$

or, from (3.3.4),

$$Z_{\text{p.u.new}} = Z_{\text{p.u.old}} \left(\frac{V_{\text{baseold}}}{V_{\text{basenew}}}\right)^2 \left(\frac{S_{\text{basenew}}}{S_{\text{baseold}}}\right) \tag{3.3.11}$$

EXAMPLE 3.4 Per-unit circuit: three-zone single-phase network

Three zones of a single-phase circuit are identified in Figure 3.10(a). The zones are connected by transformers T_1 and T_2, whose ratings are also shown. Using base values of 30 kVA and 240 volts in zone 1, draw the per-unit circuit and

FIGURE 3.10

Circuits for Example 3.4

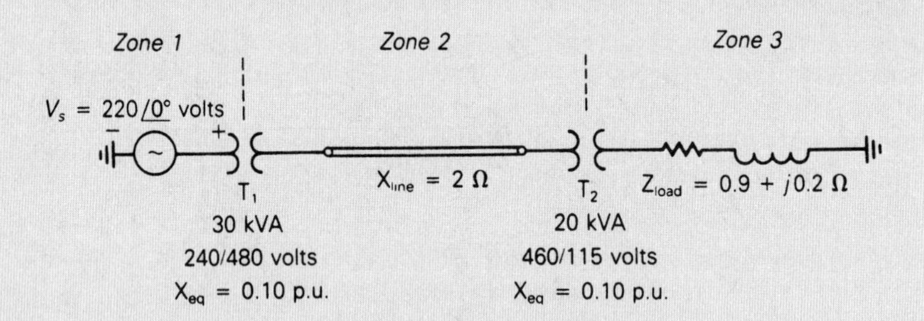

(a) Single-phase circuit

(b) Per-unit circuit

determine the per-unit impedances and the per-unit source voltage. Then calculate the load current both in per-unit and in amperes. Transformer winding resistances and shunt admittance branches are neglected.

SOLUTION First the base values in each zone are determined. $S_{base} = 30$ kVA is the same for the entire network. Also, $V_{base1} = 240$ volts, as specified for zone 1. When moving across a transformer, the voltage base is changed in proportion to the transformer voltage ratings. Thus,

$$V_{base2} = \left(\frac{480}{240}\right)(240) = 480 \quad \text{volts}$$

and

$$V_{base3} = \left(\frac{115}{460}\right)(480) = 120 \quad \text{volts}$$

The base impedances in zones 2 and 3 are

$$Z_{base2} = \frac{V_{base2}^2}{S_{base}} = \frac{480^2}{30{,}000} = 7.68 \quad \Omega$$

and

$$Z_{base3} = \frac{V_{base3}^2}{S_{base}} = \frac{120^2}{30{,}000} = 0.48 \quad \Omega$$

and the base current in zone 3 is

$$I_{base3} = \frac{S_{base}}{V_{base3}} = \frac{30{,}000}{120} = 250 \quad \text{A}$$

Next, the per-unit circuit impedances are calculated using the system base values. Since $S_{base} = 30$ kVA is the same as the kVA rating of transformer T_1, and $V_{base1} = 240$ volts is the same as the voltage rating of the zone 1 side of transformer T_1, the per-unit leakage reactance of T_1 is the same as its nameplate value, $X_{T1p.u.} = 0.1$ per unit. However, the per-unit leakage reactance of transformer T_2 must be converted from its nameplate rating to the system base. Using (3.3.11) and $V_{base2} = 480$ volts,

$$X_{T2p.u.} = (0.10)\left(\frac{460}{480}\right)^2\left(\frac{30{,}000}{20{,}000}\right) = 0.1378 \quad \text{per unit}$$

Alternatively, using $V_{base3} = 120$ volts,

$$X_{T2p.u.} = (0.10)\left(\frac{115}{120}\right)^2\left(\frac{30{,}000}{20{,}000}\right) = 0.1378 \quad \text{per unit}$$

which gives the same result. The line, which is located in zone 2, has a per-unit reactance

$$X_{linep.u.} = \frac{X_{line}}{Z_{base2}} = \frac{2}{7.68} = 0.2604 \quad \text{per unit}$$

and the load, which is located in zone 3, has a per-unit impedance

$$Z_{\text{loadp.u.}} = \frac{Z_{\text{load}}}{Z_{\text{base3}}} = \frac{0.9 + j0.2}{0.48} = 1.875 + j0.4167 \quad \text{per unit}$$

The per-unit circuit is shown in Figure 3.10(b), where the base values for each zone, per-unit impedances, and the per-unit source voltage are shown. The per-unit load current is then easily calculated from Figure 3.10(b) as follows:

$$\begin{aligned} I_{\text{loadp.u.}} = I_{\text{sp.u.}} &= \frac{V_{\text{sp.u.}}}{j(X_{\text{T1p.u.}} + X_{\text{linep.u.}} + X_{\text{T2p.u.}}) + Z_{\text{loadp.u.}}} \\ &= \frac{0.9167\angle 0^\circ}{j(0.10 + 0.2604 + 0.1378) + (1.875 + j0.4167)} \\ &= \frac{0.9167\angle 0^\circ}{1.875 + j0.9149} = \frac{0.9167\angle 0^\circ}{2.086\angle 26.01^\circ} \\ &= 0.4395\angle -26.01^\circ \quad \text{per unit} \end{aligned}$$

The actual load current is

$$I_{\text{load}} = (I_{\text{loadp.u.}})I_{\text{base3}} = (0.4395\angle -26.01^\circ)(250) = 109.9\angle -26.01^\circ \quad \text{A}$$

Note that the per-unit equivalent circuit of Figure 3.10(b) is relatively easy to analyze, since ideal transformer windings have been eliminated by proper selection of base values. ■

Balanced three-phase circuits can be solved in per-unit on a per-phase basis after converting Δ-load impedances to equivalent Y impedances. Base values can be selected either on a per-phase basis or on a three-phase basis. Equations (3.3.1)–(3.3.5) remain valid for three-phase circuits on a per-phase basis. Usually $S_{\text{base}3\phi}$ and V_{baseLL} are selected, where the subscripts 3ϕ and LL denote "three-phase" and "line-to-line," respectively. Then the following relations must be used for other base values:

$$S_{\text{base}1\phi} = \frac{S_{\text{base}3\phi}}{3} \tag{3.3.12}$$

$$V_{\text{baseLN}} = \frac{V_{\text{baseLL}}}{\sqrt{3}} \tag{3.3.13}$$

$$S_{\text{base}3\phi} = P_{\text{base}3\phi} = Q_{\text{base}3\phi} \tag{3.3.14}$$

$$I_{\text{base}} = \frac{S_{\text{base}1\phi}}{V_{\text{baseLN}}} = \frac{S_{\text{base}3\phi}}{\sqrt{3}V_{\text{baseLL}}} \tag{3.3.15}$$

$$Z_{\text{base}} = \frac{V_{\text{baseLN}}}{I_{\text{base}}} = \frac{V^2_{\text{baseLN}}}{S_{\text{base}1\phi}} = \frac{V^2_{\text{baseLL}}}{S_{\text{base}3\phi}} \tag{3.3.16}$$

$$R_{\text{base}} = X_{\text{base}} = Z_{\text{base}} = \frac{1}{Y_{\text{base}}} \tag{3.3.17}$$

EXAMPLE 3.5 Per-unit and actual currents in balanced three-phase networks

As in Example 2.5, a balanced-Y-connected voltage source with $E_{ab} = 480\underline{/0^\circ}$ volts is applied to a balanced-Δ load with $Z_\Delta = 30\underline{/40^\circ}\ \Omega$. The line impedance between the source and load is $Z_L = 1\underline{/85^\circ}\ \Omega$ for each phase. Calculate the per-unit and actual current in phase *a* of the line using $S_{base3\phi} = 10$ kVA and $V_{baseLL} = 480$ volts.

SOLUTION First, convert Z_Δ to an equivalent Z_Y; the equivalent line-to-neutral diagram is shown in Figure 2.17. The base impedance is, from (3.3.16),

$$Z_{base} = \frac{V_{baseLL}^2}{S_{base3\phi}} = \frac{(480)^2}{10{,}000} = 23.04 \quad \Omega$$

The per-unit line and load impedances are

$$Z_{Lp.u.} = \frac{Z_L}{Z_{base}} = \frac{1\underline{/85^\circ}}{23.04} = 0.04340\underline{/85^\circ} \quad \text{per unit}$$

and

$$Z_{Yp.u.} = \frac{Z_Y}{Z_{base}} = \frac{10\underline{/40^\circ}}{23.04} = 0.4340\underline{/40^\circ} \quad \text{per unit}$$

Also,

$$V_{baseLN} = \frac{V_{baseLL}}{\sqrt{3}} = \frac{480}{\sqrt{3}} = 277 \quad \text{volts}$$

and

$$E_{anp.u.} = \frac{E_{an}}{V_{baseLN}} = \frac{277\underline{/-30^\circ}}{277} = 1.0\underline{/-30^\circ} \quad \text{per unit}$$

The per-unit equivalent circuit is shown in Figure 3.11. The per-unit line current in phase *a* is then

FIGURE 3.11
Circuit for Example 3.5

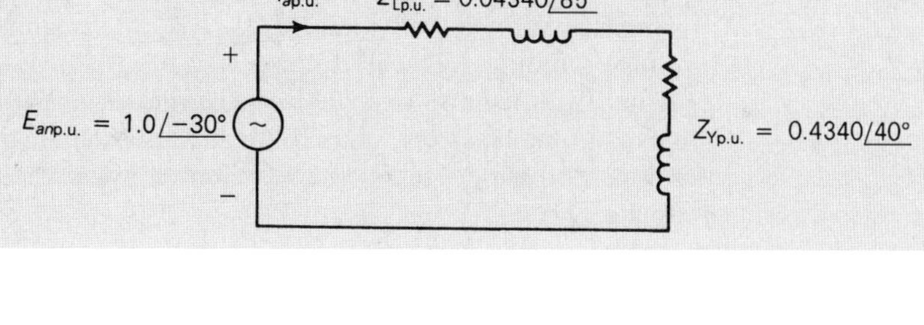

$$I_{ap.u.} = \frac{E_{anp.u.}}{Z_{Lp.u.} + Z_{Yp.u.}} = \frac{1.0\angle{-30^\circ}}{0.04340\angle{85^\circ} + 0.4340\angle{40^\circ}}$$

$$= \frac{1.0\angle{-30^\circ}}{(0.00378 + j0.04323) + (0.3325 + j0.2790)}$$

$$= \frac{1.0\angle{-30^\circ}}{0.3362 + j0.3222} = \frac{1.0\angle{-30^\circ}}{0.4657\angle{43.78^\circ}}$$

$$= 2.147\angle{-73.78^\circ} \quad \text{per unit}$$

The base current is

$$I_{base} = \frac{S_{base3\phi}}{\sqrt{3}V_{baseLL}} = \frac{10{,}000}{\sqrt{3}(480)} = 12.03 \quad \text{A}$$

and the actual phase a line current is

$$I_a = (2.147\angle{-73.78^\circ})(12.03) = 25.83\angle{-73.78^\circ} \quad \text{A}$$

■

3.4 THREE-PHASE TRANSFORMER CONNECTIONS AND PHASE SHIFT

Three identical single-phase two-winding transformers may be connected to form a three-phase bank. Four ways to connect the windings are Y–Y, Y–Δ, Δ–Y, and Δ–Δ. For example, Figure 3.12 shows a three-phase Y–Y bank. Figure 3.12(a) shows the core and coil arrangements. The American standard for marking three-phase transformers substitutes H1, H2, and H3 on the high-voltage terminals and X1, X2, and X3 on the low-voltage terminals in place of the polarity dots. Also, in this text, we will use uppercase letters *ABC* to identify phases on the high-voltage side of the transformer and lowercase letters *abc* to identify phases on the low-voltage side of the transformer. In Figure 3.12(a) the transformer high-voltage terminals H1, H2, and H3 are connected to phases *A*, *B*, and *C*, and the low-voltage terminals X1, X2, and X3 are connected to phases *a*, *b*, and *c*, respectively.

Figure 3.12(b) shows a schematic representation of the three-phase Y–Y transformer. Windings on the same core are drawn in parallel, and the phasor relationship for balanced positive-sequence operation is shown. For example, high-voltage winding H1–N is on the same magnetic core as low-voltage winding X1–n in Figure 3.12(b). Also, V_{AN} is in phase with V_{an}. Figure 3.12(c) shows a single-line diagram of a Y–Y transformer. A single-line diagram shows one phase of a three-phase network with the neutral wire omitted and with components represented by symbols rather than equivalent circuits.

FIGURE 3.12

Three-phase two-winding Y–Y transformer bank

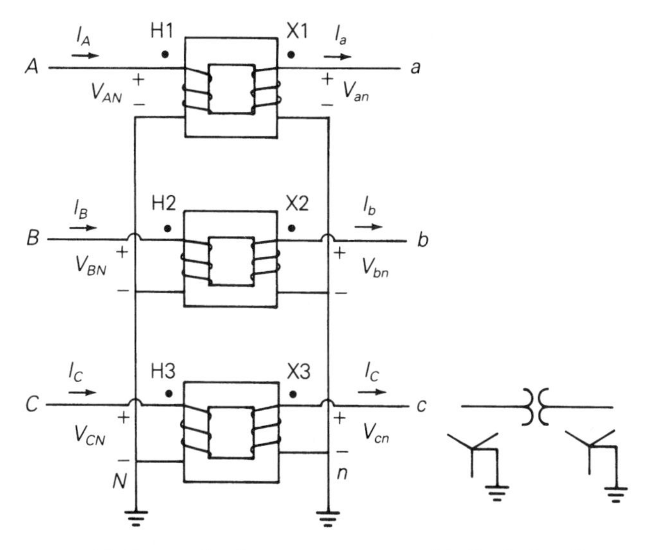

(a) Core and coil arrangements (c) Single-line diagram

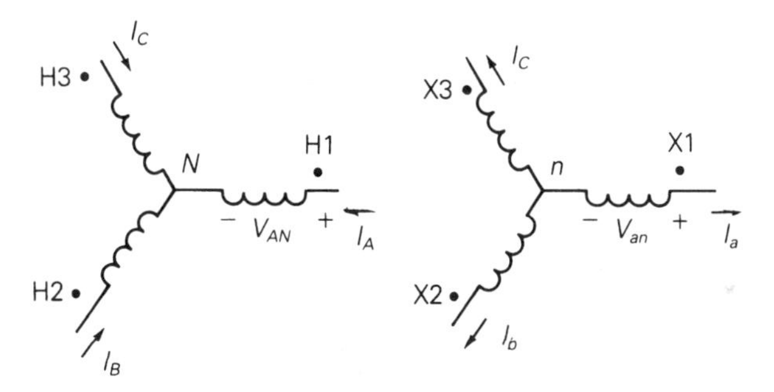

(b) Schematic representation showing phasor relationship for positive sequence operation

The phases of a Y–Y or a Δ–Δ transformer can be labeled so there is no phase shift between corresponding quantities on the low- and high-voltage windings. However, for Y–Δ and Δ–Y transformers, there is always a phase shift. Figure 3.13 shows a Y–Δ transformer. The labeling of the windings and the schematic representation are in accordance with the American standard, which is as follows:

> In either a Y–Δ or Δ–Y transformer, positive-sequence quantities on the high-voltage side shall lead their corresponding quantities on the low-voltage side by 30°.

As shown in Figure 3.13(b), V_{AN} leads V_{an} by 30°.

The positive-sequence phasor diagram shown in Figure 3.13(b) can be constructed via the following five steps, which are also indicated in Figure 3.13:

FIGURE 3.13 Three-phase two-winding Y–Δ transformer bank

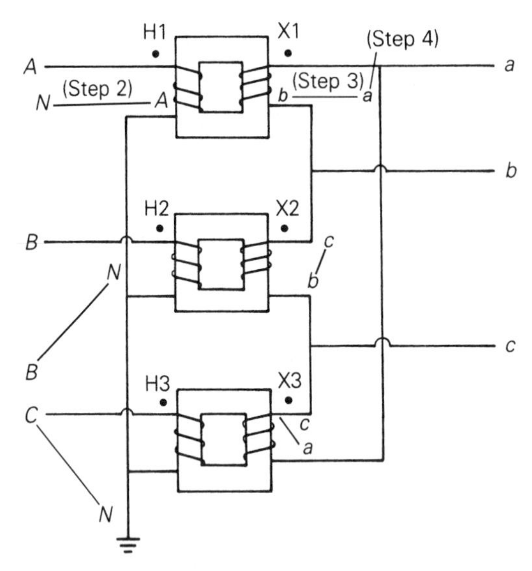

(a) Core and coil arrangement

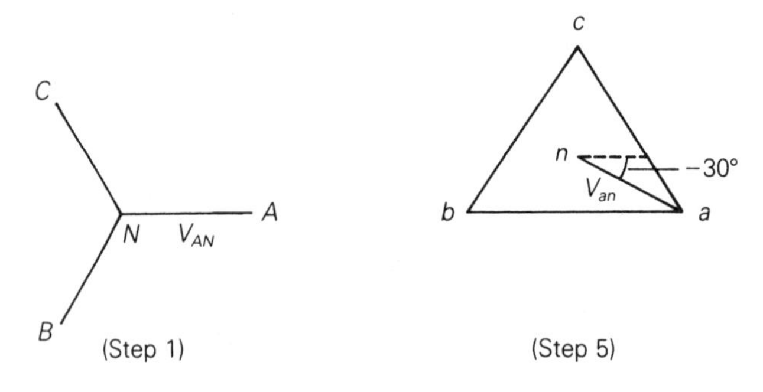

(b) Positive-sequence phasor diagram

STEP 1 Assume that balanced positive-sequence voltages are applied to the Y winding. Draw the positive-sequence phasor diagram for these voltages.

STEP 2 Move phasor *A–N* next to terminals *A–N* in Figure 3.13(a). Identify the ends of this line in the same manner as in the phasor diagram. Similarly, move phasors *B–N* and *C–N* next to terminals *B–N* and *C–N* in Figure 3.13(a).

STEP 3 For each single-phase transformer, the voltage across the low-voltage winding must be in phase with the voltage across the high-voltage winding, assuming an ideal transformer. Therefore, draw a line next to each low-voltage winding parallel to the corresponding line already drawn next to the high-voltage winding.

STEP 4 Label the ends of the lines drawn in Step 3 by inspecting the polarity marks. For example, phase A is connected to dotted terminal H1, and A appears on the *right* side of line A–N. Therefore, phase a, which is connected to dotted terminal X1, must be on the *right* side, and b on the left side of line a–b. Similarly, phase B is connected to dotted terminal H2, and B is *down* on line B–N. Therefore, phase b, connected to dotted terminal X2, must be *down* on line b–c. Similarly, c is *up* on line c–a.

STEP 5 Bring the three lines labeled in Step 4 together to complete the phasor diagram for the low-voltage Δ winding. Note that V_{AN} leads V_{an} by 30° in accordance with the American standard.

EXAMPLE 3.6 Phase shift in Δ–Y transformers

Assume that balanced negative-sequence voltages are applied to the high-voltage windings of the Y–Δ transformer shown in Figure 3.13. Determine the negative-sequence phase shift of this transformer.

SOLUTION The negative-sequence diagram, shown in Figure 3.14, is constructed from the following five steps, as outlined above:

STEP 1 Draw the phasor diagram of balanced negative-sequence voltages, which are applied to the Y winding.

STEP 2 Move the phasors A–N, B–N, and C–N next to the high-voltage Y windings.

STEP 3 For each single-phase transformer, draw a line next to the low-voltage winding that is parallel to the line drawn in Step 2 next to the high-voltage winding.

STEP 4 Label the lines drawn in Step 3. For example, phase B, which is connected to dotted terminal H2, is shown *up* on line B–N; therefore phase b, which is connected to dotted terminal X2, must be *up* on line b–c.

STEP 5 Bring the lines drawn in Step 4 together to form the negative-sequence phasor diagram for the low-voltage Δ winding.

As shown in Figure 3.14, the high-voltage phasors *lag* the low-voltage phasors by 30°. Thus the negative-sequence phase shift is the reverse of the positive-sequence phase shift. ■

The Δ–Y transformer is commonly used as a generator step-up transformer, where the Δ winding is connected to the generator terminals and the

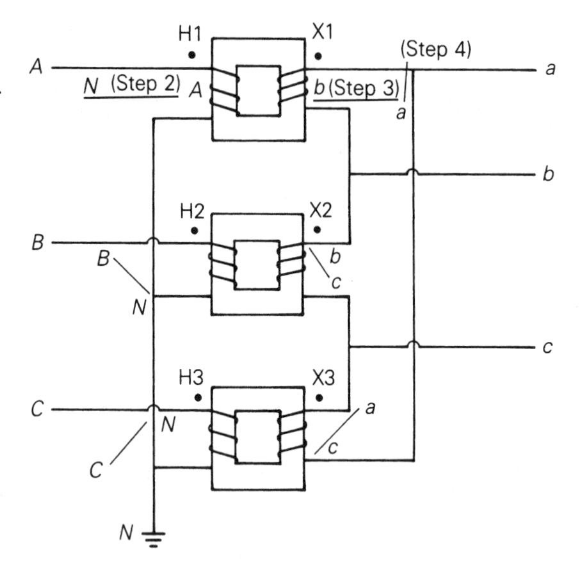

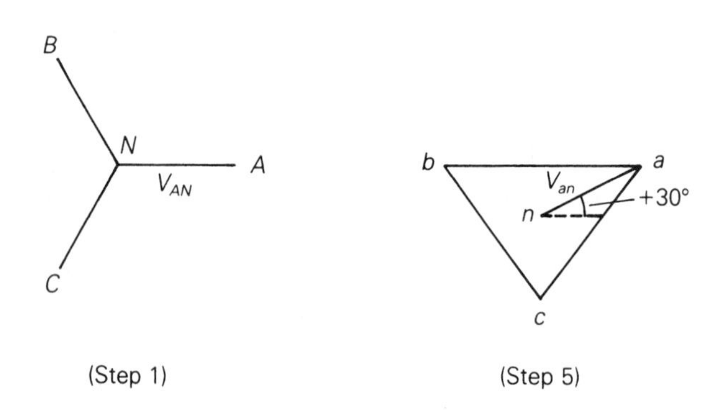

FIGURE 3.14 Example 3.6—Construction of negative-sequence phasor diagram for Y–Δ transformer bank

Y winding is connected to a transmission line. One advantage of a high-voltage Y winding is that a neutral point N is provided for grounding on the high-voltage side. With a permanently grounded neutral, the insulation requirements for the high-voltage transformer windings are reduced. The high-voltage insulation can be graded or tapered from maximum insulation at terminals ABC to minimum insulation at grounded terminal N. One advantage of the Δ winding is that the undesirable third harmonic magnetizing current, caused by the nonlinear core B–H characteristic, remains trapped inside the Δ winding. Third harmonic currents are (triple-frequency) zero-sequence currents, which cannot enter or leave a Δ connection, but can flow within the Δ. The Y–Y transformer is seldom used because of difficulties with third harmonic exciting current.

The Δ–Δ transformer has the advantage that one phase can be removed for repair or maintenance while the remaining phases continue to operate as

FIGURE 3.15

Transformer core configurations

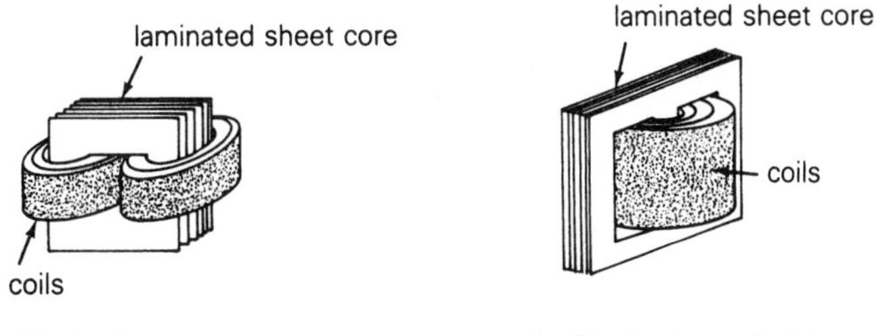

(a) Single-phase core type

(b) Single-phase shell type

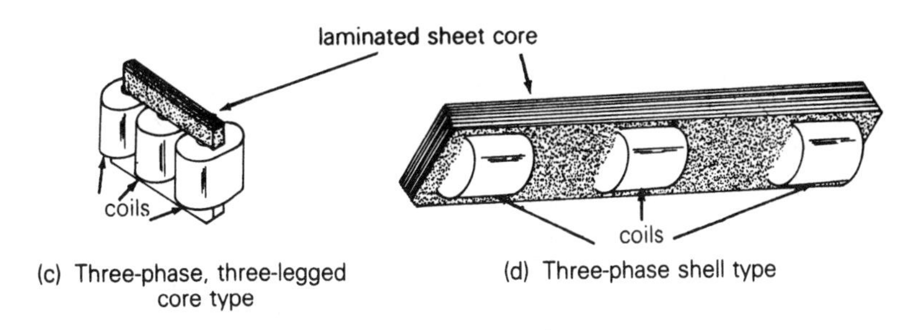

(c) Three-phase, three-legged core type

(d) Three-phase shell type

a three-phase bank. This *open*-Δ connection permits balanced three-phase operation with the kVA rating reduced to 58% of the original bank (see Problem 3.36).

Instead of a bank of three single-phase transformers, all six windings may be placed on a common three-phase core to form a three-phase transformer, as shown in Figure 3.15. The three-phase core contains less iron than the three single-phase units; therefore it costs less, weighs less, requires less floor space, and has a slightly higher efficiency. However, a winding failure would require replacement of an entire three-phase transformer, compared to replacement of only one phase of a three-phase bank.

3.5 PER-UNIT EQUIVALENT CIRCUITS OF BALANCED THREE-PHASE TWO-WINDING TRANSFORMERS

Figure 3.16(a) is a schematic representation of an ideal Y–Y transformer grounded through neutral impedances Z_N and Z_n. Figure 3.16(b) shows the per-unit equivalent circuit of this ideal transformer for balanced three-phase operation. Throughout the remainder of this text, per-unit quantities will be used unless otherwise indicated. Also, the subscript "p.u.," used to indicate a per-unit quantity, will be omitted in most cases.

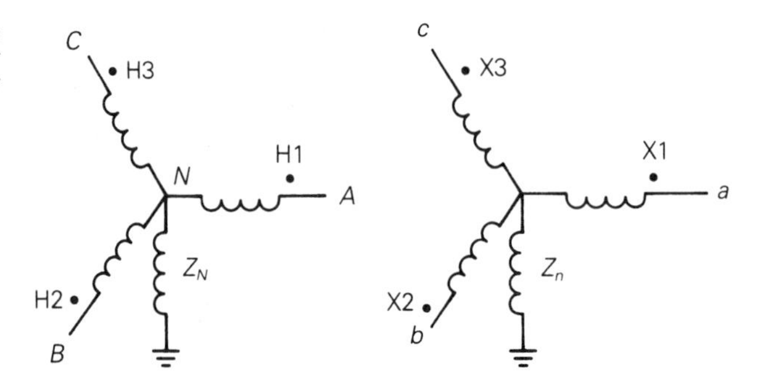

(a) Schematic representation

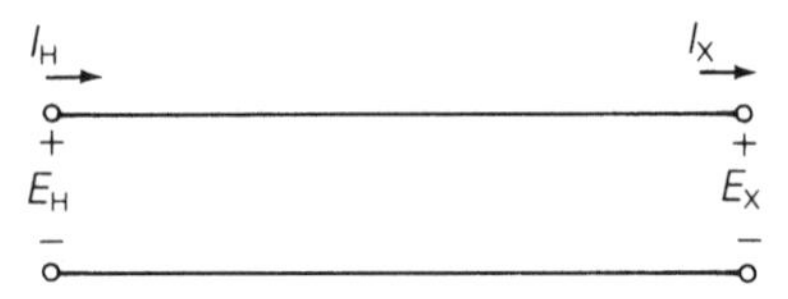

(b) Per-unit equivalent circuit for balanced three-phase operation

FIGURE 3.16 Ideal Y–Y transformer

By convention, we adopt the following two rules for selecting base quantities:

1. A common S_{base} is selected for both the H and X terminals.
2. The ratio of the voltage bases V_{baseH}/V_{baseX} is selected to be equal to the ratio of the rated line-to-line voltages $V_{ratedHLL}/V_{ratedXLL}$.

When balanced three-phase currents are applied to the transformer, the neutral currents are zero and there are no voltage drops across the neutral impedances. Therefore, the per-unit equivalent circuit of the ideal Y–Y transformer, Figure 3.16(b), is the same as the per-unit single-phase ideal transformer, Figure 3.9(a).

The per-unit equivalent circuit of a practical Y–Y transformer is shown in Figure 3.17(a). This network is obtained by adding external impedances to the equivalent circuit of the ideal transformer, as in Figure 3.9(c).

The per-unit equivalent circuit of the Y–Δ transformer, shown in Figure 3.17(b), includes a phase shift. For the American standard, the positive-sequence voltages and currents on the high-voltage side of the Y–Δ transformer lead the corresponding quantities on the low-voltage side by 30°. The phase shift in the equivalent circuit of Figure 3.17(b) is represented by the phase-shifting transformer of Figure 3.4.

The per-unit equivalent circuit of the Δ–Δ transformer, shown in Figure 3.17(c), is the same as that of the Y–Y transformer. It is assumed that the windings are labeled so there is no phase shift. Also, the per-unit impedances do not depend on the winding connections, but the base voltages do.

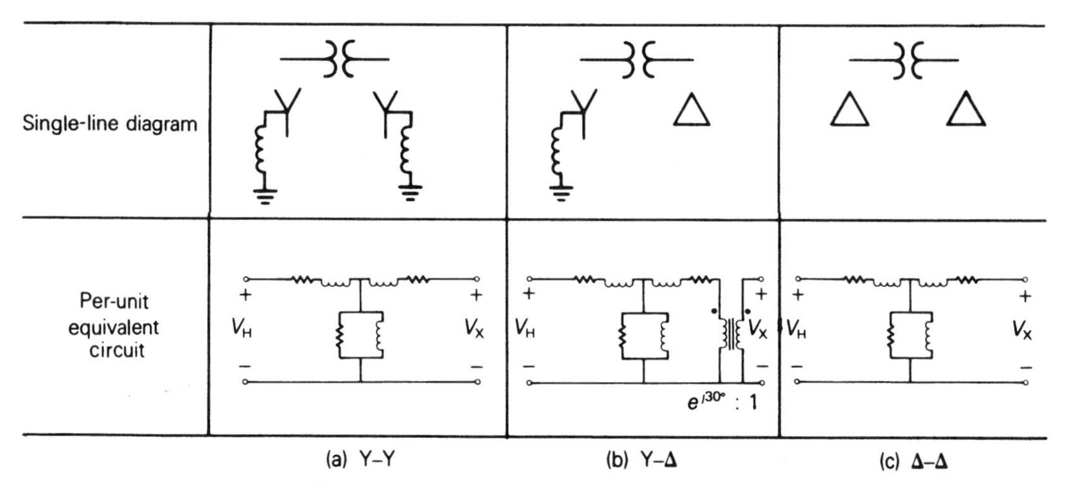

FIGURE 3.17 Per-unit equivalent circuits of practical Y–Y, Y–Δ, and Δ–Δ transformers for balanced three-phase operation

EXAMPLE 3.7 Voltage calculations: balanced Y–Y and Δ–Y transformers

Three single-phase two-winding transformers, each rated 400 MVA, 13.8/199.2 kV, with leakage reactance $X_{eq} = 0.10$ per unit, are connected to form a three-phase bank. Winding resistances and exciting current are neglected. The high-voltage windings are connected in Y. A three-phase load operating under balanced positive-sequence conditions on the high-voltage side absorbs 1000 MVA at 0.90 p.f. lagging, with $V_{AN} = 199.2\underline{/0°}$ kV. Determine the voltage V_{an} at the low-voltage bus if the low-voltage windings are connected (a) in Y, (b) in Δ.

SOLUTION The per-unit network is shown in Figure 3.18. Using the transformer bank ratings as base quantities, $S_{base3\phi} = 1200$ MVA, $V_{baseHLL} = 345$ kV, and $I_{baseH} = 1200/(345\sqrt{3}) = 2.008$ kA. The per-unit load voltage and load current are then

$$V_{AN} = 1.0\underline{/0°} \quad \text{per unit}$$

$$I_A = \frac{1000/(345\sqrt{3})}{2.008}\underline{/-\cos^{-1}0.9} = 0.8333\underline{/-25.84°} \quad \text{per unit}$$

a. For the Y–Y transformer, Figure 3.18(a),

$$I_a = I_A = 0.8333\underline{/-25.84°} \quad \text{per unit}$$

$$\begin{aligned} V_{an} &= V_{AN} + (jX_{eq})I_A \\ &= 1.0\underline{/0°} + (j0.10)(0.8333\underline{/-25.84°}) \\ &= 1.0 + 0.08333\underline{/64.16°} = 1.0363 + j0.0750 = 1.039\underline{/4.139°} \\ &= 1.039\underline{/4.139°} \quad \text{per unit} \end{aligned}$$

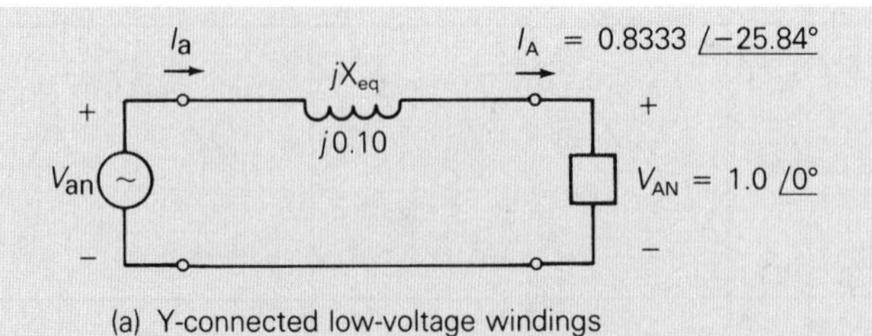

(a) Y-connected low-voltage windings

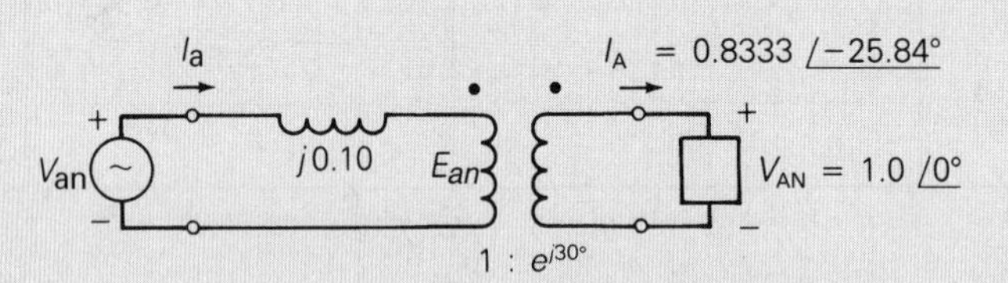

(b) Δ-connected low-voltage windings

FIGURE 3.18 Per-unit network for Example 3.7

Further, since $V_{baseXLN} = 13.8$ kV for the low-voltage Y windings, $V_{an} = 1.039(13.8) = 14.34$ kV, and

$$V_{an} = 14.34\underline{/4.139°} \quad \text{kV}$$

b. For the Δ–Y transformer, Figure 3.18(b),

$$E_{an} = e^{-j30°} V_{AN} = 1.0\underline{/-30°} \quad \text{per unit}$$

$$I_a = e^{-j30°} I_A = 0.8333\underline{/-25.84° - 30°} = 0.8333\underline{/-55.84°} \quad \text{per unit}$$

$$V_{an} = E_{an} + (jX_{eq})I_a = 1.0\underline{/-30°} + (j0.10)(0.8333\underline{/-55.84°})$$

$$V_{an} = 1.039\underline{/-25.861°} \quad \text{per unit}$$

Further, since $V_{baseXLN} = 13.8/\sqrt{3} = 7.967$ kV for the low-voltage Δ windings, $V_{an} = (1.039)(7.967) = 8.278$ kV, and

$$V_{an} = 8.278\underline{/-25.861°} \quad \text{kV}$$

■

EXAMPLE 3.8 Per-unit voltage drop and per-unit fault current: balanced three-phase transformer

A 200-MVA, 345-kVΔ/34.5-kV Y substation transformer has an 8% leakage reactance. The transformer acts as a connecting link between 345-kV transmission and 34.5-kV distribution. Transformer winding resistances and exciting current are neglected. The high-voltage bus connected to the transformer is assumed to be an ideal 345-kV positive-sequence source with negligible source impedance. Using the transformer ratings as base values, determine:

a. The per-unit magnitudes of transformer voltage drop and voltage at the low-voltage terminals when rated transformer current at 0.8 p.f. lagging enters the high-voltage terminals

b. The per-unit magnitude of the fault current when a three-phase-to-ground bolted short circuit occurs at the low-voltage terminals

SOLUTION In both parts (a) and (b), only balanced positive-sequence current will flow, since there are no imbalances. Also, because we are interested only in voltage and current magnitudes, the Δ–Y transformer phase shift can be omitted.

a. As shown in Figure 3.19(a),

$$V_{\text{drop}} = \mathrm{I}_{\text{rated}}\mathrm{X}_{\text{eq}} = (1.0)(0.08) = 0.08 \quad \text{per unit}$$

and

$$\begin{aligned} V_{an} &= V_{\text{AN}} - (j\mathrm{X}_{\text{eq}})I_{\text{rated}} \\ &= 1.0\underline{/0^\circ} - (j0.08)(1.0\underline{/-36.87^\circ}) \\ &= 1.0 - (j0.08)(0.8 - j0.6) = 0.952 - j0.064 \\ &= 0.954\underline{/-3.85^\circ} \quad \text{per unit} \end{aligned}$$

b. As shown in Figure 3.19(b),

$$\mathrm{I}_{\text{SC}} = \frac{\mathrm{V}_{\text{AN}}}{\mathrm{X}_{\text{eq}}} = \frac{1.0}{0.08} = 12.5 \quad \text{per unit}$$

Under rated current conditions [part (a)], the 0.08 per-unit voltage drop across the transformer leakage reactance causes the voltage at the low-voltage terminals to be 0.954 per unit. Also, under three-phase short-circuit conditions

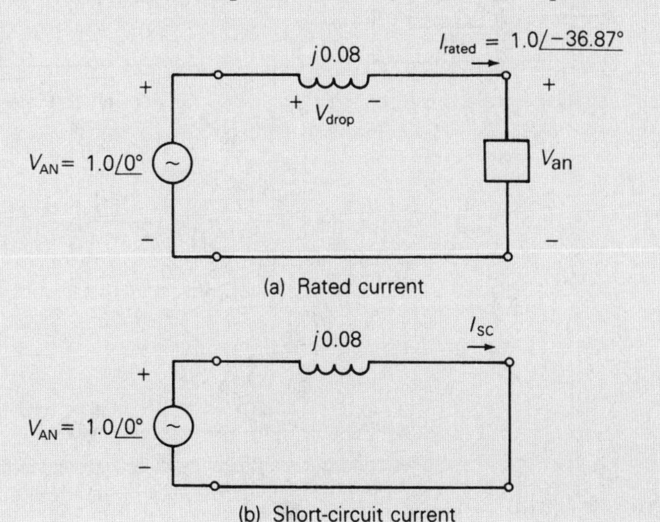

FIGURE 3.19

Circuits for Example 3.8

[part (b)], the fault current is 12.5 times the rated transformer current. This example illustrates a compromise in the design or specification of transformer leakage reactance. A low value is desired to minimize voltage drops, but a high value is desired to limit fault currents. Typical transformer leakage reactances are given in Table A.2 in the Appendix. ■

3.6 THREE-WINDING TRANSFORMERS

Figure 3.20(a) shows a basic single-phase three-winding transformer. The ideal transformer relations for a two-winding transformer, (3.1.8) and (3.1.14), can easily be extended to obtain corresponding relations for an ideal three-winding transformer. In actual units, these relations are

$$N_1 I_1 = N_2 I_2 + N_3 I_3 \tag{3.6.1}$$

$$\frac{E_1}{N_1} = \frac{E_2}{N_2} = \frac{E_3}{N_3} \tag{3.6.2}$$

where I_1 enters the dotted terminal, I_2 and I_3 leave dotted terminals, and E_1, E_2, and E_3 have their + polarities at dotted terminals. In per-unit, (3.6.1) and (3.6.2) are

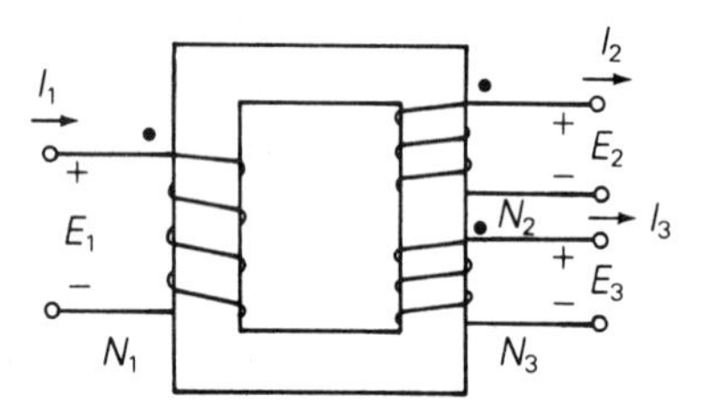

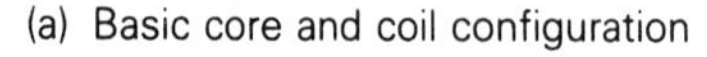
(a) Basic core and coil configuration

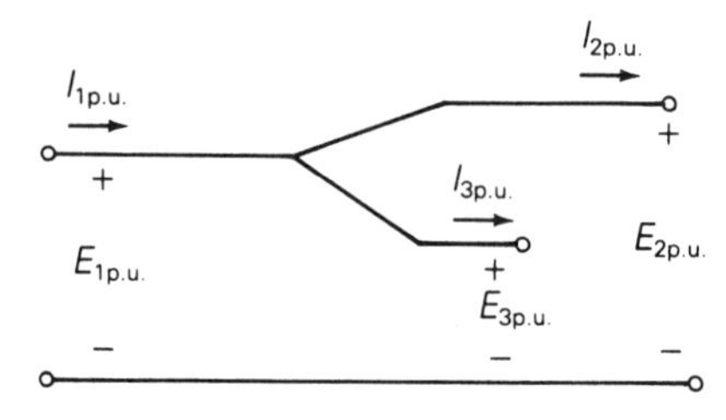

(b) Per-unit equivalent circuit—ideal transformer

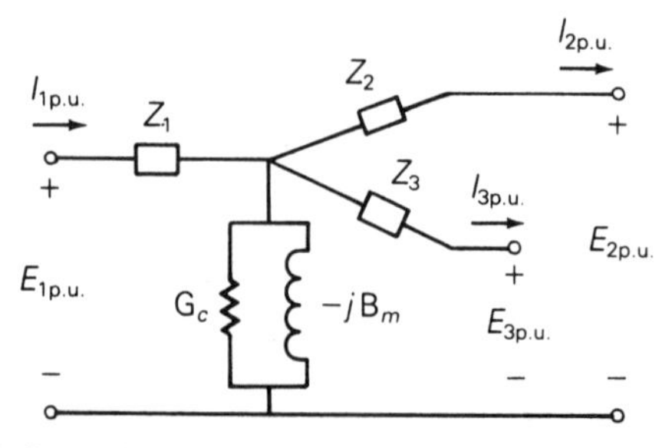

(c) Per-unit equivalent circuit—practical transformer

FIGURE 3.20 Single-phase three-winding transformer

$$I_{1\text{p.u.}} = I_{2\text{p.u.}} + I_{3\text{p.u.}} \tag{3.6.3}$$

$$E_{1\text{p.u.}} = E_{2\text{p.u.}} = E_{3\text{p.u.}} \tag{3.6.4}$$

where a common S_{base} is selected for all three windings, and voltage bases are selected in proportion to the rated voltages of the windings. These two perunit relations are satisfied by the per-unit equivalent circuit shown in Figure 3.20(b). Also, external series impedance and shunt admittance branches are included in the practical three-winding transformer circuit shown in Figure 3.20(c). The shunt admittance branch, a core loss resistor in parallel with a magnetizing inductor, can be evaluated from an open-circuit test. Also, when one winding is left open, the three-winding transformer behaves as a two-winding transformer, and standard short-circuit tests can be used to evaluate per-unit leakage impedances, which are defined as follows:

Z_{12} = per-unit leakage impedance measured from winding 1, with winding 2 shorted and winding 3 open

Z_{13} = per-unit leakage impedance measured from winding 1, with winding 3 shorted and winding 2 open

Z_{23} = per-unit leakage impedance measured from winding 2, with winding 3 shorted and winding 1 open

From Figure 3.20(c), with winding 2 shorted and winding 3 open, the leakage impedance measured from winding 1 is, neglecting the shunt admittance branch,

$$Z_{12} = Z_1 + Z_2 \tag{3.6.5}$$

Similarly,

$$Z_{13} = Z_1 + Z_3 \tag{3.6.6}$$

and

$$Z_{23} = Z_2 + Z_3 \tag{3.6.7}$$

Solving (3.6.5)–(3.6.7),

$$Z_1 = \tfrac{1}{2}(Z_{12} + Z_{13} - Z_{23}) \tag{3.6.8}$$

$$Z_2 = \tfrac{1}{2}(Z_{12} + Z_{23} - Z_{13}) \tag{3.6.9}$$

$$Z_3 = \tfrac{1}{2}(Z_{13} + Z_{23} - Z_{12}) \tag{3.6.10}$$

Equations (3.6.8)–(3.6.10) can be used to evaluate the per-unit series impedances Z_1, Z_2, and Z_3 of the three-winding transformer equivalent circuit from the per-unit leakage impedances Z_{12}, Z_{13}, and Z_{23}, which, in turn, are determined from short-circuit tests.

Note that each of the windings on a three-winding transformer may have a *different* kVA rating. If the leakage impedances from short-circuit tests are expressed in per-unit based on winding ratings, they must first be converted to per-unit on a common S_{base} *before* they are used in (3.6.8)–(3.6.10).

EXAMPLE 3.9 Three-winding single-phase transformer: per-unit impedances

The ratings of a single-phase three-winding transformer are

winding 1: 300 MVA, 13.8 kV

winding 2: 300 MVA, 199.2 kV

winding 3: 50 MVA, 19.92 kV

The leakage reactances, from short-circuit tests, are

$X_{12} = 0.10$ per unit on a 300-MVA, 13.8-kV base

$X_{13} = 0.16$ per unit on a 50-MVA, 13.8-kV base

$X_{23} = 0.14$ per unit on a 50-MVA, 199.2-kV base

Winding resistances and exciting current are neglected. Calculate the impedances of the per-unit equivalent circuit using a base of 300 MVA and 13.8 kV for terminal 1.

SOLUTION $S_{base} = 300$ MVA is the same for all three terminals. Also, the specified voltage base for terminal 1 is $V_{base1} = 13.8$ kV. The base voltages for terminals 2 and 3 are then $V_{base2} = 199.2$ kV and $V_{base3} = 19.92$ kV, which are the rated voltages of these windings. From the data given, $X_{12} = 0.10$ per unit was measured from terminal 1 using the same base values as those specified for the circuit. However, $X_{13} = 0.16$ and $X_{23} = 0.14$ per unit on a 50-MVA base are first converted to the 300-MVA circuit base.

$$X_{13} = (0.16)\left(\frac{300}{50}\right) = 0.96 \quad \text{per unit}$$

$$X_{23} = (0.14)\left(\frac{300}{50}\right) = 0.84 \quad \text{per unit}$$

Then, from (3.6.8)–(3.6.10),

$$X_1 = \tfrac{1}{2}(0.10 + 0.96 - 0.84) = 0.11 \quad \text{per unit}$$

$$X_2 = \tfrac{1}{2}(0.10 + 0.84 - 0.96) = -0.01 \quad \text{per unit}$$

$$X_3 = \tfrac{1}{2}(0.84 + 0.96 - 0.10) = 0.85 \quad \text{per unit}$$

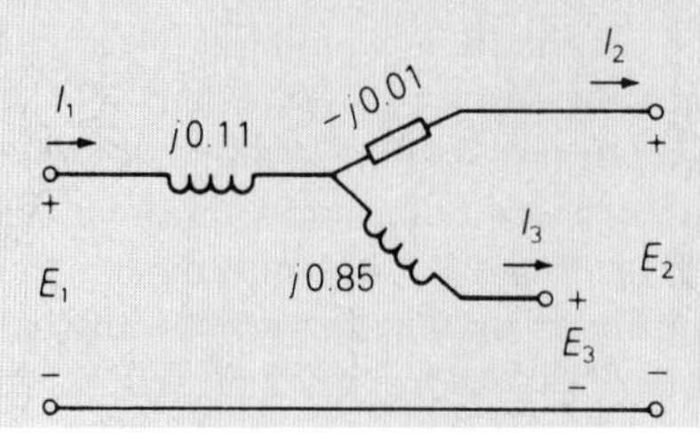

FIGURE 3.21
Circuit for Example 3.9

The per-unit equivalent circuit of this three-winding transformer is shown in Figure 3.21. Note that X_2 is negative. This illustrates the fact that X_1, X_2, and X_3 are *not* leakage reactances, but instead are equivalent reactances derived from the leakage reactances. Leakage reactances are always positive.

Note also that the node where the three equivalent circuit reactances are connected does not correspond to any physical location within the transformer. Rather, it is simply part of the equivalent circuit representation. ■

EXAMPLE 3.10 Three-winding three-phase transformer: balanced operation

Three transformers, each identical to that described in Example 3.9, are connected as a three-phase bank in order to feed power from a 900-MVA, 13.8-kV generator to a 345-kV transmission line and to a 34.5-kV distribution line. The transformer windings are connected as follows:

13.8-kV windings (X): Δ, to generator
199.2-kV windings (H): solidly grounded Y, to 345-kV line
19.92-kV windings (M): grounded Y through $Z_n = j0.10\ \Omega$, to 34.5-kV line

The positive-sequence voltages and currents of the high- and medium-voltage Y windings lead the corresponding quantities of the low-voltage Δ winding by 30°. Draw the per-unit network, using a three-phase base of 900 MVA and 13.8 kV for terminal X. Assume balanced positive-sequence operation.

SOLUTION The per-unit network is shown in Figure 3.22. $V_{baseX} = 13.8$ kV, which is the rated line-to-line voltage of terminal X. Since the M and H windings are Y-connected, $V_{baseM} = \sqrt{3}(19.92) = 34.5$ kV, and $V_{baseH} = \sqrt{3}(199.2) = 345$ kV, which are the rated line-to-line voltages of the M and H windings. Also, a phase-shifting transformer is included in the network. The neutral impedance is not included in the network, since there is no neutral current under balanced operation. ■

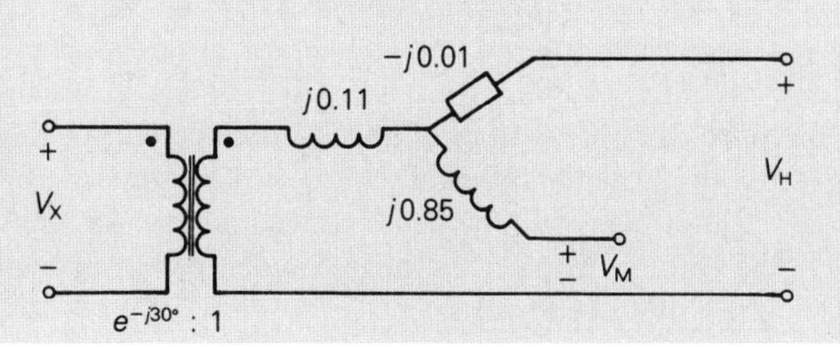

FIGURE 3.22 Per-unit network for Example 3.10

FIGURE 3.23

Ideal single-phase transformers

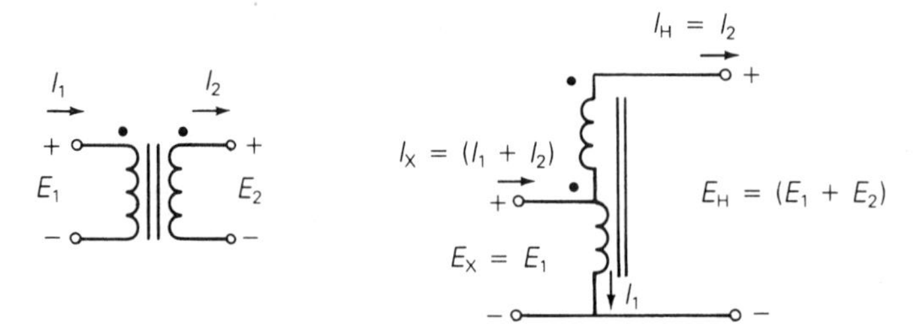

3.7 AUTOTRANSFORMERS

A single-phase two-winding transformer is shown in Figure 3.23(a) with two separate windings, which is the usual two-winding transformer; the same transformer is shown in Figure 3.23(b) with the two windings connected in series, which is called an *autotransformer*. For the usual transformer [Figure 3.23(a)] the two windings are coupled magnetically via the mutual core flux. For the autotransformer [Figure 3.23(b)] the windings are both electrically and magnetically coupled. The autotransformer has smaller per-unit leakage impedances than the usual transformer; this results in both smaller series-voltage drops (an advantage) and higher short-circuit currents (a disadvantage). The autotransformer also has lower per-unit losses (higher efficiency), lower exciting current, and lower cost if the turns ratio is not too large. The electrical connection of the windings, however, allows transient overvoltages to pass through the autotransformer more easily.

EXAMPLE 3.11 Autotransformer: single-phase

The single-phase two-winding 20-kVA, 480/120-volt transformer of Example 3.3 is connected as an autotransformer, as in Figure 3.23(b), where winding 1 is the 120-volt winding. For this autotransformer, determine (a) the voltage ratings E_X and E_H of the low- and high-voltage terminals, (b) the kVA rating, and (c) the per-unit leakage impedance.

SOLUTION

a. Since the 120-volt winding is connected to the low-voltage terminal, $E_X = 120$ volts. When $E_X = E_1 = 120$ volts is applied to the low-voltage terminal, $E_2 = 480$ volts is induced across the 480-volt winding, neglecting the voltage drop across the leakage impedance. Therefore, $E_H = E_1 + E_2 = 120 + 480 = 600$ volts.

b. As a normal two-winding transformer rated 20 kVA, the rated current of the 480-volt winding is $I_2 = I_H = 20{,}000/480 = 41.667$ A. As an autotransformer, the 480-volt winding can carry the same current. Therefore, the kVA rating $S_H = E_H I_H = (600)(41.667) = 25$ kVA. Note also that when $I_H = I_2 = 41.667$ A, a current $I_1 = 480/120(41.667) = 166.7$ A is induced in the 120-volt winding. Therefore, $I_X = I_1 + I_2 = 208.3$ A (neglecting exciting current) and $S_X = E_X I_X = (120)(208.3) = 25$ kVA, which is the same rating as calculated for the high-voltage terminal.

c. From Example 3.3, the leakage impedance is $0.0729\underline{/78.13^\circ}$ per unit as a normal, two-winding transformer. As an autotransformer, the leakage impedance *in ohms* is the same as for the normal transformer, since the core and windings are the same for both (only the external winding connections are different). However, the base impedances are different. For the high-voltage terminal, using (3.3.4),

$$Z_{\text{baseHold}} = \frac{(480)^2}{20{,}000} = 11.52 \quad \Omega \quad \text{as a normal transformer}$$

$$Z_{\text{baseHnew}} = \frac{(600)^2}{25{,}000} = 14.4 \quad \Omega \quad \text{as an autotransformer}$$

Therefore, using (3.3.10),

$$Z_{\text{p.u.new}} = (0.0729\underline{/78.13^\circ})\left(\frac{11.52}{14.4}\right) = 0.05832\underline{/78.13^\circ} \quad \text{per unit}$$

For this example, the rating is 25 kVA, 120/600 volts as an autotransformer versus 20 kVA, 120/480 volts as a normal transformer. The autotransformer has both a larger kVA rating and a larger voltage ratio for the same cost. Also, the per-unit leakage impedance of the autotransformer is smaller. However, the increased high-voltage rating as well as the electrical connection of the windings may require more insulation for both windings. ■

3.8 TRANSFORMERS WITH OFF-NOMINAL TURNS RATIOS

It has been shown that models of transformers that use per-unit quantities are simpler than those that use actual quantities. The ideal transformer winding is eliminated when the ratio of the selected voltage bases equals the ratio of the voltage ratings of the windings. In some cases, however, it is impossible to select voltage bases in this manner. For example, consider the two transformers connected in parallel in Figure 3.24. Transformer T_1 is rated 13.8/345 kV and T_2 is rated 13.2/345 kV. If we select $V_{\text{baseH}} = 345$ kV, then

FIGURE 3.24

Two transformers connected in parallel

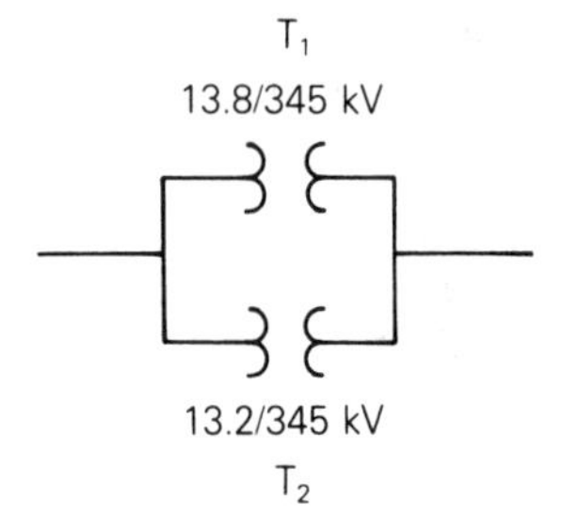

transformer T_1 requires $V_{baseX} = 13.8$ kV and T_2 requires $V_{baseX} = 13.2$ kV. It is clearly impossible to select the appropriate voltage bases for both transformers.

To accommodate this situation, we will develop a per-unit model of a transformer whose voltage ratings are not in proportion to the selected base voltages. Such a transformer is said to have an "off-nominal turns ratio." Figure 3.25(a) shows a transformer with rated voltages V_{1rated} and V_{2rated}, which satisfy

$$V_{1rated} = a_t V_{2rated} \tag{3.8.1}$$

where a_t is assumed, in general, to be either real or complex. Suppose the selected voltage bases satisfy

$$V_{base1} = bV_{base2} \tag{3.8.2}$$

Defining $c = \dfrac{a_t}{b}$, (3.8.1) can be rewritten as

$$V_{1rated} = b\left(\frac{a_t}{b}\right)V_{2rated} = bcV_{2rated} \tag{3.8.3}$$

Equation (3.8.3) can be represented by two transformers in series, as shown in Figure 3.25(b). The first transformer has the same ratio of rated winding voltages as the ratio of the selected base voltages, b. Therefore, this transformer has a standard per-unit model, as shown in Figure 3.9 or 3.17. We will assume that the second transformer is ideal, and all real and reactive losses are associated with the first transformer. The resulting per-unit model is shown in Figure 3.25(c), where, for simplicity, the shunt-exciting branch is neglected. Note that if $a_t = b$, then the ideal transformer winding shown in this figure can be eliminated, since its turns ratio $c = (a_t/b) = 1$.

The per-unit model shown in Figure 3.25(c) is perfectly valid, but it is not suitable for some of the computer programs presented in later chapters because these programs do not accommodate ideal transformer windings. An alternative representation can be developed, however, by writing nodal equations for this figure as follows:

FIGURE 3.25

Transformer with off-nominal turns ratio

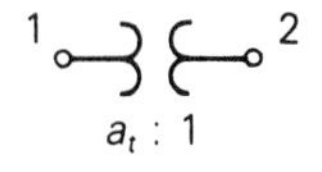

(a) Single-line diagram

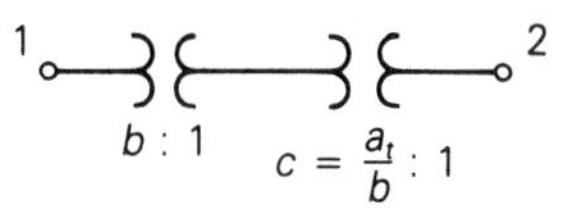

(b) Represented as two transformers in series

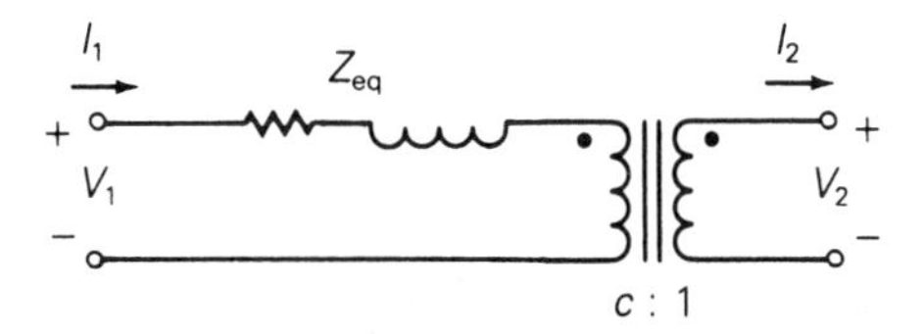

(c) Per-unit equivalent circuit (Per-unit impedance is shown)

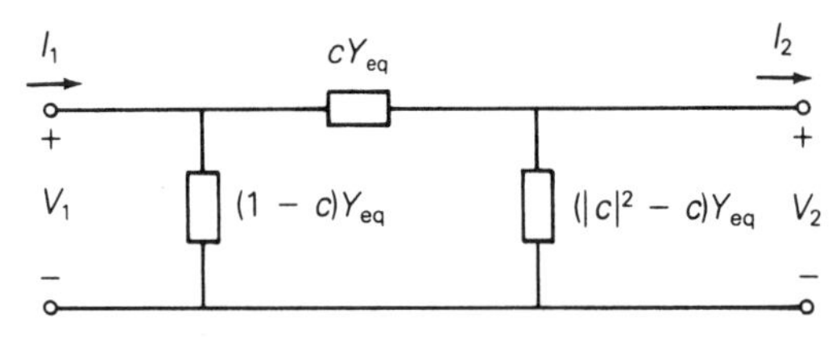

(d) π circuit representation for real c $\left(\text{Per-unit admittances are shown; } Y_{eq} = \frac{1}{Z_{eq}}\right)$

$$\begin{bmatrix} I_1 \\ -I_2 \end{bmatrix} = \begin{bmatrix} Y_{11} & Y_{12} \\ Y_{21} & Y_{22} \end{bmatrix} \begin{bmatrix} V_1 \\ V_2 \end{bmatrix} \tag{3.8.4}$$

where both I_1 and $-I_2$ are referenced *into* their nodes in accordance with the nodal equation method (Section 2.4). Recalling two-port network theory, the admittance parameters of (3.8.4) are, from Figure 3.23(c)

$$Y_{11} = \left.\frac{I_1}{V_1}\right|_{V_2 = 0} = \frac{1}{Z_{eq}} = Y_{eq} \tag{3.8.5}$$

$$Y_{22} = \left.\frac{-I_2}{V_2}\right|_{V_1 = 0} = \frac{1}{Z_{eq}/|c|^2} = |c|^2 Y_{eq} \tag{3.8.6}$$

$$Y_{12} = \left.\frac{I_1}{V_2}\right|_{V_1 = 0} = \frac{-cV_2/Z_{eq}}{V_2} = -cY_{eq} \tag{3.8.7}$$

$$Y_{21} = \left.\frac{-I_2}{V_1}\right|_{V_2 = 0} = \frac{-c^* I_1}{V_1} = -c^* Y_{eq} \tag{3.8.8}$$

Equations (3.8.4)–(3.8.8) with real or complex c are convenient for representing transformers with off-nominal turns ratios in the computer programs presented later. Note that when c is complex, Y_{12} is not equal to Y_{21}, and the preceding admittance parameters cannot be synthesized with a passive RLC circuit. However, the π network shown in Figure 3.25(d), which has the same admittance parameters as (3.8.4)–(3.8.8), can be synthesized for real c. Note also that when $c = 1$, the shunt branches in this figure become open circuits (zero per unit mhos), and the series branch becomes Y_{eq} per unit mhos (or Z_{eq} per unit ohms).

EXAMPLE 3.12 Tap-changing three-phase transformer: per-unit positive-sequence network

A three-phase generator step-up transformer is rated 1000 MVA, 13.8 kV Δ/345 kV Y with $Z_{eq} = j0.10$ per unit. The transformer high-voltage winding has $\pm 10\%$ taps. The system base quantities are

$$S_{base3\phi} = 500 \quad \text{MVA}$$

$$V_{baseXLL} = 13.8 \quad \text{kV}$$

$$V_{baseHLL} = 345 \quad \text{kV}$$

Determine the per-unit equivalent circuit for the following tap settings:

a. Rated tap

b. -10% tap (providing a 10% voltage decrease for the high-voltage winding)

Assume balanced positive-sequence operation. Neglect transformer winding resistance, exciting current, and phase shift.

SOLUTION

a. Using (3.8.1) and (3.8.2) with the low-voltage winding denoted winding 1,

$$a_t = \frac{13.8}{345} = 0.04 \qquad b = \frac{V_{baseXLL}}{V_{baseHLL}} = \frac{13.8}{345} = a_t \qquad c = 1$$

From (3.3.11)

$$Z_{p.u.new} = (j0.10)\left(\frac{500}{1000}\right) = j0.05 \quad \text{per unit}$$

The per-unit equivalent circuit, not including winding resistance, exciting current, and phase shift is:

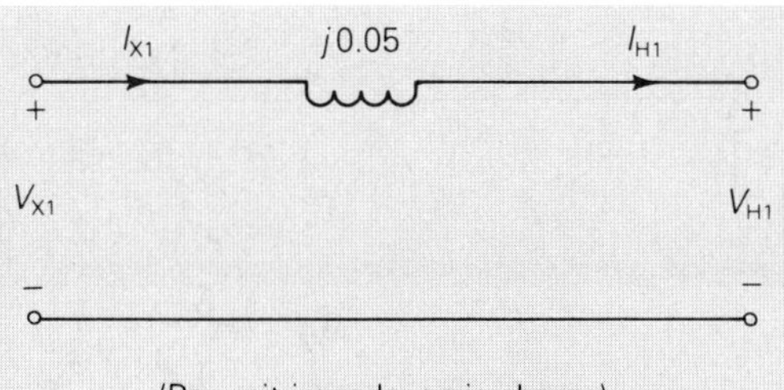

(Per-unit impedance is shown)

b. Using (3.8.1) and (3.8.2),

$$a_t = \frac{13.8}{345(0.9)} = 0.04444 \qquad b = \frac{13.8}{345} = 0.04$$

$$c = \frac{a_t}{b} = \frac{0.04444}{0.04} = 1.1111$$

From Figure 3.23(d),

$$cY_{\text{eq}} = 1.1111\left(\frac{1}{j0.05}\right) = -j22.22 \quad \text{per unit}$$

$$(1-c)Y_{\text{eq}} = (-0.11111)(-j20) = +j2.222 \quad \text{per unit}$$

$$(|c|^2 - c)Y_{\text{eq}} = (1.2346 - 1.1)(-j20) = -j2.469 \quad \text{per unit}$$

The per-unit positive-sequence network is:

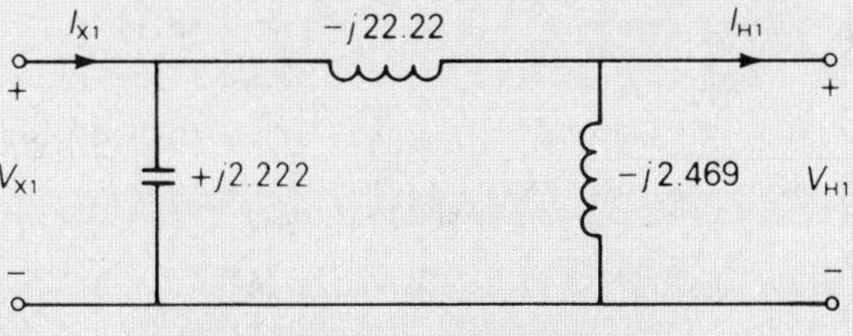

(Per-unit admittances are shown)

Open PowerWorld Simulator case Example 3.12 (see Figure 3.26) and select **Tools, Play** to see an animated view of this LTC transformer example. Initially the generator/step-up transformer feeds a 500 MW/100 Mvar load. As is typical in practice, the transformer's taps are adjusted in discrete steps, with each step changing the tap ratio by 0.625% (hence a 10% change requires 16 steps). Click on arrows next to the transformer's tap to manually adjust the tap by one step. Note that changing the tap directly changes the load voltage.

Because of the varying voltage drops caused by changing loads, LTCs are often operated to automatically regulate a bus voltage. This is particularly true when they are used as step-down transformers. To place the example transformer on automatic control, click on the "Manual" field. This toggles the transformer control mode to automatic. Now the transformer manually

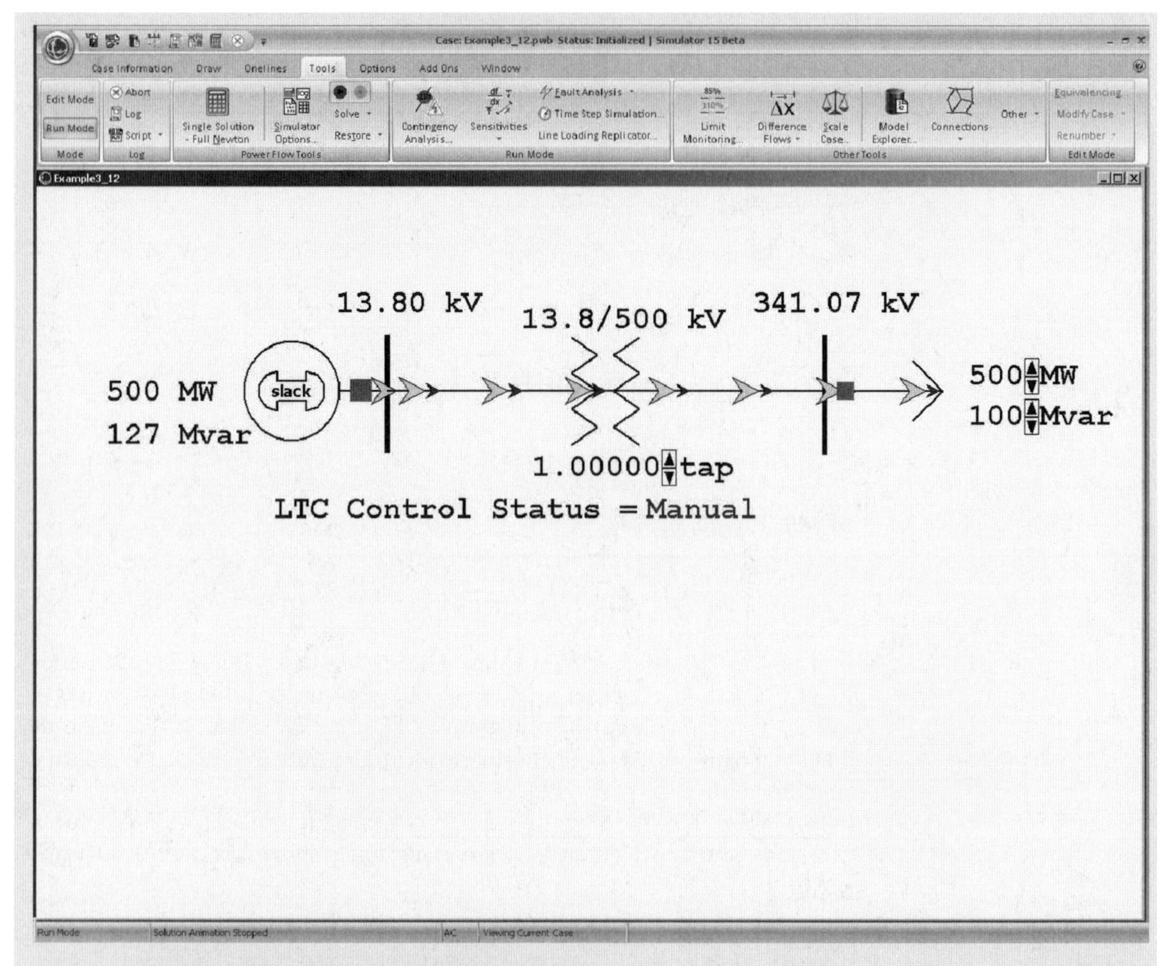

FIGURE 3.26 Screen for Example 3.12

changes its tap ratio to maintain the load voltage within a specified voltage range, between 0.995 and 1.005 per unit (343.3 to 346.7 kV) in this case. To see the LTC in automatic operation use the load arrows to vary the load, particularly the Mvar field, noting that the LTC changes to keep the load's voltage within the specified deadband. ■

The three-phase regulating transformers shown in Figures 3.27 and 3.28 can be modeled as transformers with off-nominal turns ratios. For the voltage-magnitude-regulating transformer shown in Figure 3.27, adjustable voltages ΔV_{an}, ΔV_{bn}, and ΔV_{cn}, which have equal magnitudes ΔV and which are in phase with the phase voltages V_{an}, V_{bn}, and V_{cn}, are placed in the series link between buses a–a', b–b', and c–c'. Modeled as a transformer with an off-nominal turns ratio (see Figure 3.25), $c = (1 + \Delta \mathrm{V})$ for a voltage-magnitude increase toward bus abc, or $c = (1 + \Delta \mathrm{V})^{-1}$ for an increase toward bus $a'b'c'$.

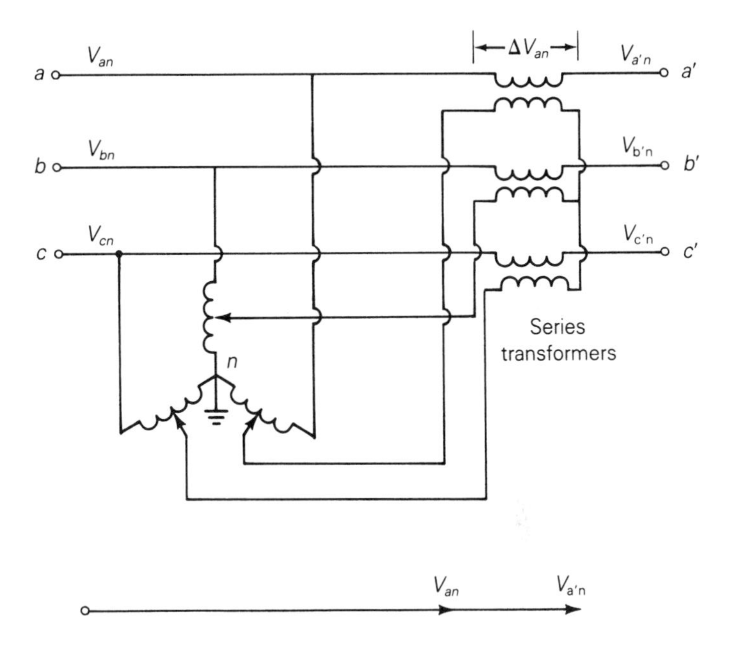

FIGURE 3.27 An example of a voltage-magnitude-regulating transformer

For the phase-angle-regulating transformer in Figure 3.28, the series voltages ΔV_{an}, ΔV_{bn}, and ΔV_{cn} are $\pm 90°$ out of phase with the phase voltages V_{an}, V_{bn}, and V_{cn}. The phasor diagram in Figure 3.28 indicates that each of the bus voltages $V_{a'n}$, $V_{b'n}$, and $V_{c'n}$ has a phase shift that is approximately proportional to the magnitude of the added series voltage. Modeled as a transformer with an off-nominal turns ratio (see Figure 3.25), $c \approx 1\underline{/\alpha}$ for a phase increase toward bus abc or $c \approx 1\underline{/-\alpha}$ for a phase increase toward bus $a'b'c'$.

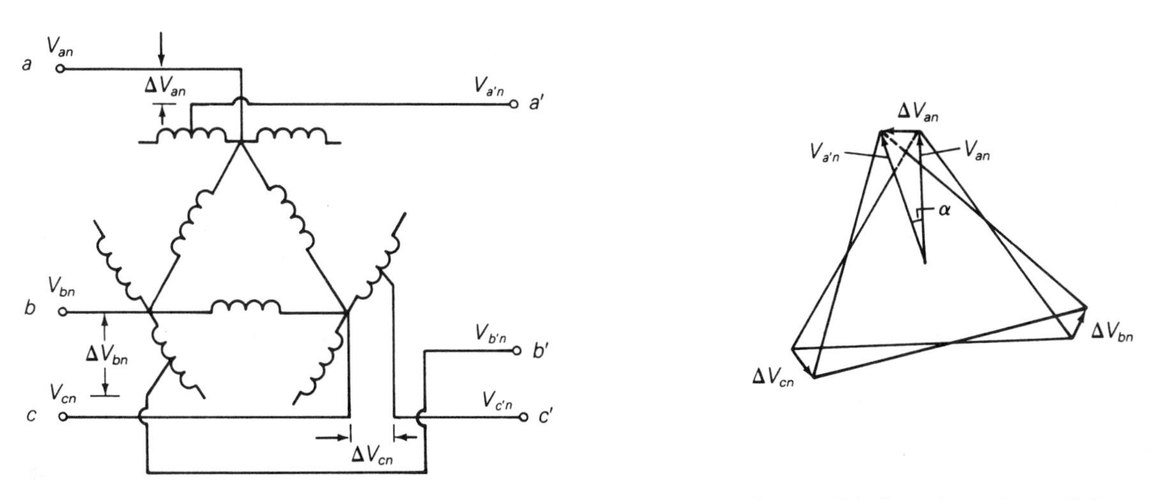

FIGURE 3.28 An example of a phase-angle-regulating transformer. Windings drawn in parallel are on the same core

EXAMPLE 3.13 Voltage-regulating and phase-shifting three-phase transformers

Two buses *abc* and *a′b′c′* are connected by two parallel lines L1 and L2 with positive-sequence series reactances $X_{L1} = 0.25$ and $X_{L2} = 0.20$ per unit. A regulating transformer is placed in series with line L1 at bus *a′b′c′*. Determine the 2×2 bus admittance matrix when the regulating transformer (a) provides a 0.05 per-unit increase in voltage magnitude toward bus *a′b′c′* and (b) advances the phase 3° toward bus *a′b′c′*. Assume that the regulating transformer is ideal. Also, the series resistance and shunt admittance of the lines are neglected.

SOLUTION The circuit is shown in Figure 3.29.

a. For the voltage-magnitude-regulating transformer, $c = (1 + \Delta V)^{-1} = (1.05)^{-1} = 0.9524$ per unit. From (3.7.5)–(3.7.8), the admittance parameters of the regulating transformer in series with line L1 are

$$Y_{11L1} = \frac{1}{j0.25} = -j4.0$$

$$Y_{22L1} = (0.9524)^2(-j4.0) = -j3.628$$

$$Y_{12L1} = Y_{21L1} = (-0.9524)(-j4.0) = j3.810$$

For line L2 alone,

$$Y_{11L2} = Y_{22L2} = \frac{1}{j0.20} = -j5.0$$

$$Y_{12L2} = Y_{21L2} = -(-j5.0) = j5.0$$

Combining the above admittances in parallel,

$$Y_{11} = Y_{11L1} + Y_{11L2} = -j4.0 - j5.0 = -j9.0$$

$$Y_{22} = Y_{22L1} + Y_{22L2} = -j3.628 - j5.0 = -j8.628$$

$$Y_{12} = Y_{21} = Y_{12L1} + Y_{12L2} = j3.810 + j5.0 = j8.810 \quad \text{per unit}$$

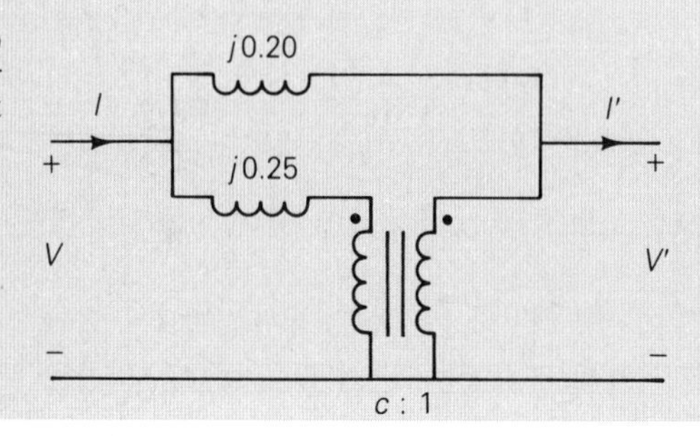

FIGURE 3.29 Positive-sequence circuit for Example 3.13

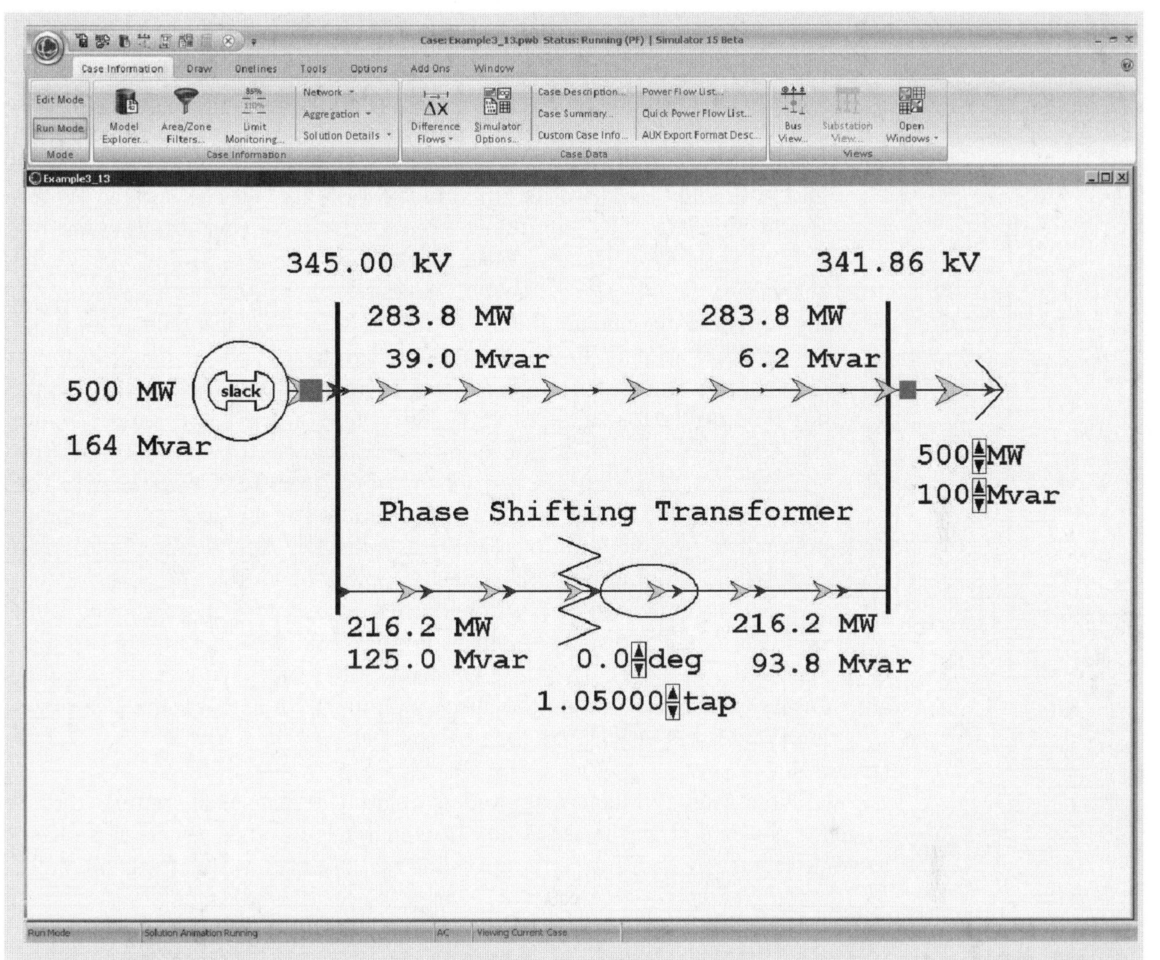

FIGURE 3.30 Screen for Example 3.13

b. For the phase-angle-regulating transformer, $c = 1\underline{/-\alpha} = 1\underline{/-3^\circ}$. Then, for this regulating transformer in series with line L1,

$$Y_{11\text{L1}} = \frac{1}{j0.25} = -j4.0$$

$$Y_{22\text{L1}} = |1.0\underline{/-3^\circ}|^2(-j4.0) = -j4.0$$

$$Y_{12\text{L1}} = -(1.0\underline{/-3^\circ})(-j4.0) = 4.0\underline{/87^\circ} = 0.2093 + j3.9945$$

$$Y_{21\text{L1}} = -(1.0\underline{/-3^\circ})^*(-j4.0) = 4.0\underline{/93^\circ} = -0.2093 + j3.9945$$

The admittance parameters for line L2 alone are given in part (a) above. Combining the admittances in parallel,

$$Y_{11} = Y_{22} = -j4.0 - j5.0 = -j9.0$$

$$Y_{12} = 0.2093 + j3.9945 + j5.0 = 0.2093 + j8.9945$$

$$Y_{21} = -0.2093 + j3.9945 + j5.0 = -0.2093 + j8.9945 \quad \text{per unit}$$

To see this example in PowerWorld Simulator open case Example 3.13 (see Figure 3.30). In this case, the transformer and a parallel transmission line are assumed to be supplying power from a 345-kV generator to a 345-kV load. Initially, the off-nominal turns ratio is set to the value in part (a) of the example (PowerWorld has the off-nominal turns ratio on the load side [right-hand] so its tap value of 1.05 = c^{-1}). To view the PowerWorld Simulator bus admittance matrix, select the **Case Information** ribbon, then **Solution Details, Ybus**. To see how the system flows vary with changes to the tap, select **Tools, Play,** and then click on the arrows next to the tap field to change the LTC tap in 0.625% steps. Next, to verify the results from part (b), change the tap field to 1.0 and the deg field to 3.0 degrees, and then again look at the bus admittance matrix. Click on the deg field arrow to vary the phase shift angle in one-degree steps. Notice that changing the phase angle primarily changes the real power flow, whereas changing the LTC tap changes the reactive power flow. In this example, the line flow fields show the absolute value of the real or reactive power flow; the direction of the flow is indicated with arrows. Traditional power flow programs usually indicate power flow direction using a convention that flow into a transmission line or transformer is assumed to be positive. You can display results in PowerWorld Simulator using this convention by first clicking on the **Onelines** ribbon and then selecting **Oneline Display Options.** Then on the Display Options tab uncheck the Use Absolute Values for MW/Mvar Line Flows" fields. ■

Note that a voltage-magnitude-regulating transformer controls the *reactive* power flow in the series link in which it is installed, whereas a phase-angle-regulating transformer controls the *real* power flow (see Problem 3.59).

MULTIPLE CHOICE QUESTIONS

SECTION 3.1

3.1 The "Ohm's law" for the magnetic circuit states that the net magnetomotive force (mmf) equals the product of the core reluctance and the core flux.
(a) True (b) False

3.2 For an ideal transformer, the efficiency is
(a) 0% (b) 100% (c) 50%

3.3 For an ideal 2-winding transformer, the ampere-turns of the primary winding, N_1I_1, is equal to the ampere-turns of the secondary winding, N_2I_2.
(a) True (b) False

3.4 An ideal transformer has no real or reactive power loss.
(a) True (b) False

3.5 For an ideal 2-winding transformer, an impedance Z_2 connected across winding 2 (secondary) is referred to winding 1 (primary) by multiplying Z_2 by
(a) The turns ratio (N_1/N_2)
(b) The square of the turns ratio $(N_1/N_2)^2$
(c) The cubed turns ratio $(N_1/N_2)^3$

3.6 Consider Figure 3.4 of the text. For an ideal phase-shifting transformer, the impedance is unchanged when it is referred from one side to the other.
(a) True (b) False

SECTION 3.2

3.7 Consider Figure 3.5 of the text. Match the following, those on the left to those on the right.

(i) I_m	(a) Exciting current
(ii) I_C	(b) Magnetizing current
(iii) I_e	(c) Core loss current

3.8 The units of admittance, conductance, and susceptance are siemens.
(a) True (b) False

3.9 Match the following:

(i) Hysteresis loss	(a) Can be reduced by constructing the core with laminated sheets of alloy steel
(ii) Eddy current loss	(b) Can be reduced by the use of special high grades of alloy steel as core material.

3.10 For large power transformers rated more than 500 kVA, the winding resistances, which are small compared with the leakage reactances, can often be neglected.
(a) True (b) False

3.11 For a short-circuit test on a 2-winding transformer, with one winding shorted, can you apply the rated voltage on the other winding?
(a) Yes (b) No

SECTION 3.3

3.12 The per-unit quantity is always dimensionless.
(a) True (b) False

3.13 Consider the adopted per-unit system for the transformers. Specify true or false for each of the following statements:
(a) For the entire power system of concern, the value of S_{base} is not the same.
(b) The ratio of the voltage bases on either side of a transformer is selected to be the same as the ratio of the transformer voltage ratings.
(c) Per-unit impedance remains unchanged when referred from one side of a transformer to the other.

3.14 The ideal transformer windings are eliminated from the per-unit equivalent circuit of a transformer.
(a) True (b) False

3.15 To convert a per-unit impedance from "old" to "new" base values, the equation to be used is

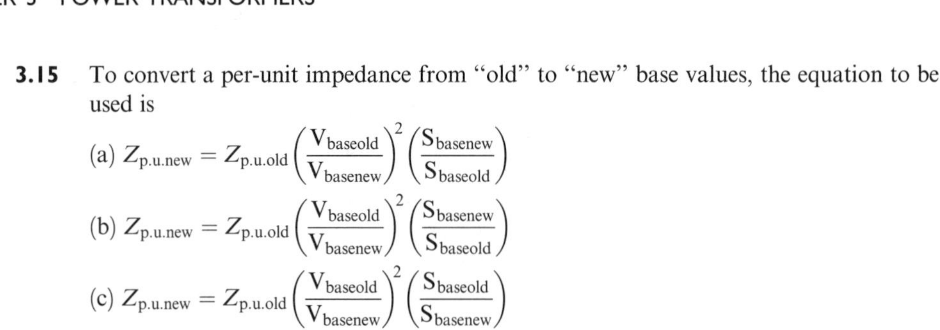

(a) $Z_{\text{p.u.new}} = Z_{\text{p.u.old}} \left(\frac{V_{\text{baseold}}}{V_{\text{basenew}}}\right)^2 \left(\frac{S_{\text{basenew}}}{S_{\text{baseold}}}\right)$

(b) $Z_{\text{p.u.new}} = Z_{\text{p.u.old}} \left(\frac{V_{\text{baseold}}}{V_{\text{basenew}}}\right)^2 \left(\frac{S_{\text{basenew}}}{S_{\text{baseold}}}\right)$

(c) $Z_{\text{p.u.new}} = Z_{\text{p.u.old}} \left(\frac{V_{\text{baseold}}}{V_{\text{basenew}}}\right)^2 \left(\frac{S_{\text{baseold}}}{S_{\text{basenew}}}\right)$

3.16 In developing per-unit circuits of systems such as the one shown in Figure 3.10 of the text, when moving across a transformer, the voltage base is changed in proportion to the transformer voltage ratings.
(a) True (b) False

3.17 Consider Figure 3.10 of the text. The per-unit leakage reactance of transformer T_1, given as 0.1 p.u., is based on the name plate ratings of transformer T_1.
(a) True (b) False

3.18 For balanced three-phase systems, Z_{base} is given by

$$Z_{\text{base}} = \frac{V_{\text{baseLL}}^2}{S_{\text{base3}}}$$

(a) True (b) False

SECTION 3.4

3.19 With the American Standard notation, in either Y–Δ OR Δ–Y transformer, positive-sequence quantities on the high-voltage side shall lead their corresponding quantities on the low-voltage side by 30°.
(a) True (b) False

3.20 In either Y–Δ or Δ–Y transformer, as per the American Standard notation, the negative-sequence phase shift is the reverse of the positive-sequence phase shift.
(a) True (b) False

3.21 In order to avoid difficulties with third-harmonic exciting current, which three-phase transformer connection is seldom used for step-up transformers between a generator and a transmission line in power systems.
(a) Y–Δ (b) Δ–Y (c) Y–Y

3.22 Does open –Δ connection permit balanced three-phase operation?
(a) Yes (b) No

3.23 With the open –Δ operation, the kVA rating compared to that of the original three-phase bank is
(a) 2/3 (b) 58% (c) 1

SECTION 3.5

3.24 It is stated that
(i) balanced three-phase circuits can be solved in per unit on a per-phase basis after converting Δ-load impedances to equivalent Y impedances.

(ii) Base values can be selected either on a per-phase basis or on a three-phase basis.
(a) Both statements are true.
(b) Neither is true.
(c) Only one of the above is true.

3.25 In developing per-unit equivalent circuits for three-phase transformers, under balanced three-phase operation,
(i) A common S_{base} is selected for both the H and X terminals.
(ii) The ratio of the voltage bases V_{baseH}/V_{baseX} is selected to be equal to the ratio of the rated line-to-line voltages $V_{ratedHLL}/V_{ratedXLL}$.
(a) Only one of the above is true.
(b) Neither is true.
(c) Both statements are true.

3.26 In per-unit equivalent circuits of practical three-phase transformers, under balanced three-phase operation, in which of the following connections would a phase-shifting transformer come up?
(a) Y–Y (b) Y–Δ (c) Δ–Δ

3.27 A low value of transformer leakage reactance is desired to minimize the voltage drop, but a high value is derived to limit the fault current, thereby leading to a compromise in the design specification.
(a) True (b) False

SECTION 3.6

3.28 Consider a single-phase three-winding transformer with the primary excited winding of N_1 turns carrying a current I_1 and two secondary windings of N_2 and N_3 turns, delivering currents of I_2 and I_3 respectively. For an ideal case, how are the ampere-turns balanced?
(a) $N_1I_1 = N_2I_2 - N_3I_3$ (b) $N_1I_1 = N_2I_2 + N_3I_3$
(c) $N_1I_1 = -(N_2I_2 - N_3I_3)$

3.29 For developing per-unit equivalent circuits of single-phase three-winding transformer, a common S_{base} is selected for all three windings, and voltage bases are selected in proportion to the rated voltage of the windings.
(a) True (b) False

3.30 Consider the equivalent circuit of Figure 3.20 (c) in the text. After neglecting the winding resistances and exciting current, could X_1, X_2, or X_3 become negative, even though the leakage reactance are always positive?
(a) Yes (b) No

SECTION 3.7

3.31 Consider an ideal single-phase 2-winding transformer of turns ratio N_1/N_2 = a. If it is converted to an autotransformer arrangement with a transformation ratio of $V_H/V_X = 1 + a$, (the autotransformer rating/two-winding transformer rating) would then be
(a) $1 + a$ (b) $1 + \frac{1}{a}$ (c) a

3.32 For the same output, the autotransformer (with not too large turns ratio) is smaller in size than a two-winding transformer and has high efficiency as well as superior voltage regulation.
(a) True (b) False

3.33 The direct electrical connection of the windings allows transient over voltages to pass through the autotransformer more easily, and that is an important disadvantage of the autotransformer.
(a) True (b) False

SECTION 3.8

3.34 Consider Figure 3.25 of the text for a transformer with off-nominal turns ratio.
(i) The per-unit equivalent circuit shown in Part (c) contains an ideal transformer which cannot be accommodated by some computer programs.
(a) True (b) False
(ii) In the π-circuit representation for real C in Part (d), the admittance parameters Y_{12} and Y_{21} would be unequal.
(a) True (b) False
(iii) For complex C, can the admittance, parameters the synthesized with a passive RLC circuit?
(a) Yes (b) No

PROBLEMS

SECTION 3.1

3.1 (a) An ideal single-phase two-winding transformer with turns ratio $a_t = N_1/N_2$ is connected with a series impedance Z_2 across winding 2. If one wants to replace Z_2, with a series impedance Z_1 across winding 1 and keep the terminal behavior of the two circuits to be identical, find Z_1 in terms of Z_2.
(b) Would the above result be true if instead of a series impedance there is a shunt impedance?
(c) Can one refer a ladder network on the secondary (2) side to the primary (1) side simply by multiplying every impendance by a_t^2?

3.2 An ideal transformer with $N_1 = 2000$ and $N_2 = 500$ is connected with an impedance Z_{2_2} across winding 2, called secondary. If $V_1 = 1000 \angle 0°$ V and $I_1 = 5 \angle -30°$ A, determine V_2, I_2, Z_2, and the impedance Z_2', which is the value of Z_2 referred to the primary side of the transformer.

3.3 Consider an ideal transformer with $N_1 = 3000$ and $N_2 = 1000$ turns. Let winding 1 be connected to a source whose voltage is $e_1(t) = 100(1 - |t|)$ volts for $-1 \le t \le 1$ and $e_1(t) = 0$ for $|t| > 1$ second. A 2-farad capacitor is connected across winding 2. Sketch $e_1(t)$, $e_2(t)$, $i_1(t)$, and $i_2(t)$ versus time t.

3.4 A single-phase 100-kVA, 2400/240-volt, 60-Hz distribution transformer is used as a step-down transformer. The load, which is connected to the 240-volt secondary winding, absorbs 80 kVA at 0.8 power factor lagging and is at 230 volts. Assuming an ideal transformer, calculate the following: (a) primary voltage, (b) load impedance, (c) load impedance referred to the primary, and (d) the real and reactive power supplied to the primary winding.

3.5 Rework Problem 3.4 if the load connected to the 240-V secondary winding absorbs 110 kVA under short-term overload conditions at 0.85 power factor leading and at 230 volts.

3.6 For a conceptual single-phase, phase-shifting transformer, the primary voltage leads the secondary voltage by 30°. A load connected to the secondary winding absorbs 100 kVA at 0.9 power factor leading and at a voltage $E_2 = 277\underline{/0^\circ}$ volts. Determine (a) the primary voltage, (b) primary and secondary currents, (c) load impedance referred to the primary winding, and (d) complex power supplied to the primary winding.

3.7 Consider a source of voltage $v(t) = 10\sqrt{2}\sin(2t)$ V, with an internal resistance of 1800 Ω. A transformer that can be considered as ideal is used to couple a 50-Ω resistive load to the source. (a) Determine the transformer primary-to-secondary turns ratio required to ensure maximum power transfer by matching the load and source resistances. (b) Find the average power delivered to the load, assuming maximum power transfer.

3.8 For the circuit shown in Figure 3.31, determine $v_{out}(t)$.

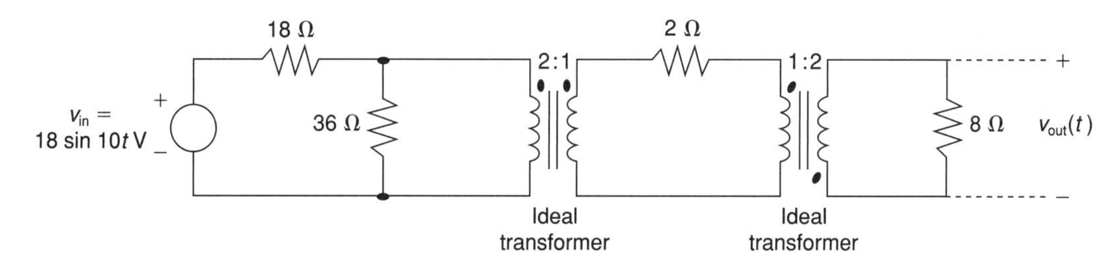

FIGURE 3.31 Problem 3.8

SECTION 3.2

3.9 A single-phase transformer has 2000 turns on the primary winding and 500 turns on the secondary. Winding resistances are $R_1 = 2\ \Omega$ and $R_2 = 0.125\ \Omega$; leakage reactances are $X_1 = 8\ \Omega$ and $X_2 = 0.5\ \Omega$. The resistance load on the secondary is 12 Ω.
(a) If the applied voltage at the terminals of the primary is 1000 V, determine V_2 at the load terminals of the transformer, neglecting magnetizing current.
(b) If the voltage regulation is defined as the difference between the voltage magnitude at the load terminals of the transformer at full load and at no load in percent of full-load voltage with input voltage held constant, compute the percent voltage regulation.

3.10 A single-phase step-down transformer is rated 15 MVA, 66 kV/11.5 kV. With the 11.5 kV winding short-circuited, rated current flows when the voltage applied to the primary is 5.5 kV. The power input is read as 100 kW. Determine R_{eq1} and X_{eq1} in ohms referred to the high-voltage winding.

3.11 For the transformer in Problem 3.10, the open-circuit test with 11.5 kV applied results in a power input of 65 kW and a current of 30 A. Compute the values for G_c and B_m

in siemens referred to the high-voltage winding. Compute the efficiency of the transformer for a load of 10 MW at 0.8 p.f. lagging at rated voltage.

3.12 The following data are obtained when open-circuit and short-circuit tests are performed on a single-phase, 50-kVA, 2400/240-volt, 60-Hz distribution transformer.

	VOLTAGE (volts)	CURRENT (amperes)	POWER (watts)
Measurements on low-voltage side with high-voltage winding open	240	4.85	173
Measurements on high-voltage side with low-voltage winding shorted	52.0	20.8	650

(a) Neglecting the series impedance, determine the exciting admittance referred to the high-voltage side. (b) Neglecting the exciting admittance, determine the equivalent series impedance referred to the high-voltage side. (c) Assuming equal series impedances for the primary and referred secondary, obtain an equivalent T-circuit referred to the high-voltage side.

3.13 A single-phase 50-kVA, 2400/240-volt, 60-Hz distribution transformer has a 1-ohm equivalent leakage reactance and a 5000-ohm magnetizing reactance referred to the high-voltage side. If rated voltage is applied to the high-voltage winding, calculate the open-circuit secondary voltage. Neglect I^2R and G_c^2V losses. Assume equal series leakage reactances for the primary and referred secondary.

3.14 A single-phase 50-kVA, 2400/240-volt, 60-Hz distribution transformer is used as a step-down transformer at the load end of a 2400-volt feeder whose series impedance is $(1.0 + j2.0)$ ohms. The equivalent series impedance of the transformer is $(1.0 + j2.5)$ ohms referred to the high-voltage (primary) side. The transformer is delivering rated load at 0.8 power factor lagging and at rated secondary voltage. Neglecting the transformer exciting current, determine (a) the voltage at the transformer primary terminals, (b) the voltage at the sending end of the feeder, and (c) the real and reactive power delivered to the sending end of the feeder.

3.15 Rework Problem 3.14 if the transformer is delivering rated load at rated secondary voltage and at (a) unity power factor, (b) 0.8 power factor leading. Compare the results with those of Problem 3.14.

3.16 A single-phase, 50-kVA, 2400/240-V, 60-Hz distribution transformer has the following parameters:

Resistance of the 2400-V winding: $R_1 = 0.75\ \Omega$

Resistance of the 240-V winding: $R_2 = 0.0075\ \Omega$

Leakage reactance of the 2400-V winding: $X_1 = 1.0\ \Omega$

Leakage reactance of the 240-V winding: $X_2 = 0.01\ \Omega$

Exciting admittance on the 240-V side $= 0.003 - j0.02$ S

(a) Draw the equivalent circuit referred to the high-voltage side of the transformer. (b) Draw the equivalent circuit referred to the low-voltage side of the transformer. Show the numerical values of impedances on the equivalent circuits.

3.17 The transformer of Problem 3.16 is supplying a rated load of 50 kVA at a rated secondary voltage of 240 V and at 0.8 power factor lagging. Neglecting the transformer exciting current, (a) Determine the input terminal voltage of the transformer on the high-voltage side. (b) Sketch the corresponding phasor diagram. (c) If the transformer is used as a step-down transformer at the load end of a feeder whose impedance is $0.5 + j2.0\ \Omega$, find the voltage V_S and the power factor at the sending end of the feeder.

SECTION 3.3

3.18 Using the transformer ratings as base quantities, work Problem 3.13 in per-unit.

3.19 Using the transformer ratings as base quantities, work Problem 3.14 in per-unit.

3.20 Using base values of 20 kVA and 115 volts in zone 3, rework Example 3.4.

3.21 Rework Example 3.5, using $S_{base3\phi} = 100$ kVA and $V_{baseLL} = 600$ volts.

3.22 A balanced Y-connected voltage source with $E_{ag} = 277\underline{/0^\circ}$ volts is applied to a balanced-Y load in parallel with a balanced-Δ load, where $Z_Y = 20 + j10$ and $Z_\Delta = 30 - j15$ ohms. The Y load is solidly grounded. Using base values of $S_{base1\phi} = 10$ kVA and $V_{baseLN} = 277$ volts, calculate the source current I_a in per-unit and in amperes.

3.23 Figure 3.32 shows the one-line diagram of a three-phase power system. By selecting a common base of 100 MVA and 22 kV on the generator side, draw an impedance diagram showing all impedances including the load impedance in per-unit. The data are given as follows:

G:	90 MVA	22 kV	x = 0.18 per unit
*T*1:	50 MVA	22/220 kV	x = 0.10 per unit
*T*2:	40 MVA	220/11 kV	x = 0.06 per unit
*T*3:	40 MVA	22/110 kV	x = 0.064 per unit
*T*4:	40 MVA	110/11 kV	x = 0.08 per unit
M:	66.5 MVA	10.45 kV	x = 0.185 per unit

Lines 1 and 2 have series reactances of 48.4 and 65.43 Ω, respectively. At bus 4, the three-phase load absorbs 57 MVA at 10.45 kV and 0.6 power factor lagging.

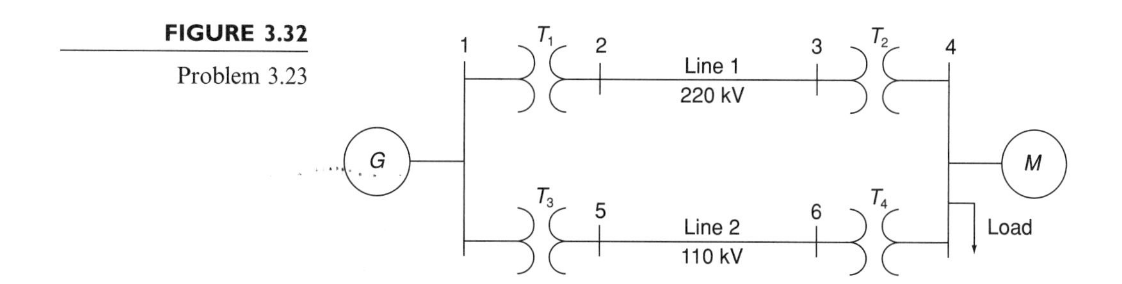

FIGURE 3.32
Problem 3.23

3.24 For Problem 3.18, the motor operates at full load, at 0.8 power factor leading, and at a terminal voltage of 10.45 kV. Determine (a) the voltage at bus 1, the generator bus, and (b) the generator and motor internal EMFs.

3.25 Consider a single-phase electric system shown in Figure 3.33.
Transformers are rated as follows:
X–Y 15 MVA, 13.8/138 kV, leakage reactance 10%
Y–Z 15 MVA, 138/69 kV, leakage reactance 8%
With the base in circuit Y chosen as 15 MVA, 138 kV, determine the per-unit impedance of the 500 Ω resistive load in circuit Z, referred to circuits Z, Y, and X. Neglecting magnetizing currents, transformer resistances, and line impedances, draw the impedance diagram in per unit.

FIGURE 3.33

Single-phase electric system for Problem 3.25

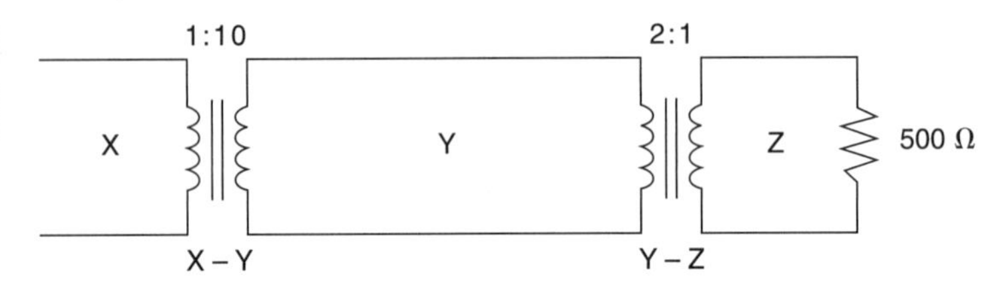

3.26 A bank of three single-phase transformers, each rated 30 MVA, 38.1/3.81 kV, are connected in Y–Δ with a balanced load of three 1-Ω, wye-connected resistors. Choosing a base of 90 MVA, 66 kV for the high-voltage side of the three-phase transformer, specify the base for the low-voltage side. Compute the per-unit resistance of the load on the base for the low-voltage side. Also, determine the load resistance in ohms referred to the high-voltage side and the per-unit value on the chosen base.

3.27 A three-phase transformer is rated 500 MVA, 220 Y/22 Δ kV. The wye-equivalent short-circuit impedance, considered equal to the leakage reactance, measured on the low-voltage side is 0.1 Ω. Compute the per-unit reactance of the transformer. In a system in which the base on the high-voltage side of the transformer is 100 MVA, 230 kV, what value of the per-unit reactance should be used to represent this transformer?

3.28 For the system shown in Figure 3.34, draw an impedance diagram in per unit, by choosing 100 kVA to be the base kVA and 2400 V as the base voltage for the generators.

FIGURE 3.34

System for Problem 3.28

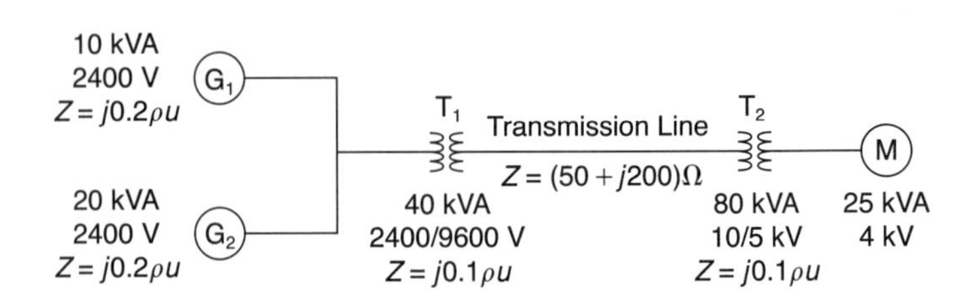

3.29 Consider three ideal single-phase transformers (with a voltage gain of η) put together as a delta-wye three-phase bank as shown in Figure 3.35. Assuming positive-sequence voltages for V_{an}, V_{bn}, and V_{cn}, find $V_{a'n'}$, $V_{b'n'}$, and $V_{c'n'}$ in terms of V_{an}, V_{bn}, and V_{cn}, respectively.
(a) Would such relationships hold for the line voltages as well?
(b) Looking into the current relationships, express I'_a, I'_b, and I'_c in terms of I_a, I_b, and I_c, respectively.
(c) Let S' and S be the per-phase complex power output and input, respectively. Find S' in terms of S.

FIGURE 3.35

Δ–Y connection for Problem 3.29

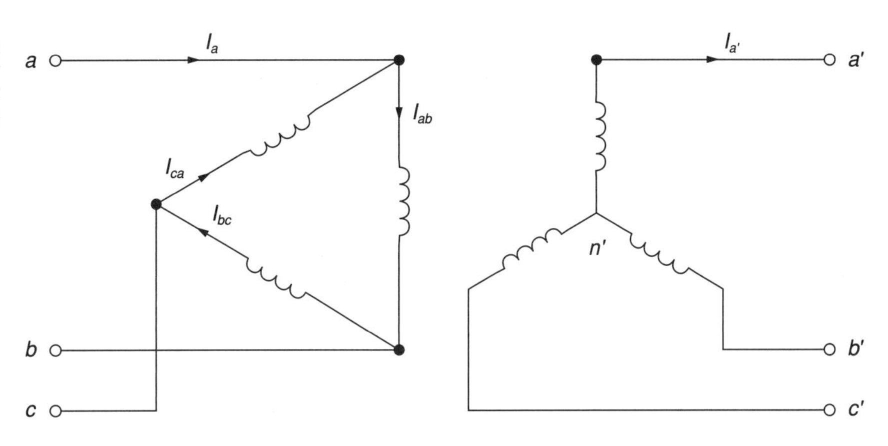

3.30 Reconsider Problem 3.29. If V_{an}, V_{bn}, and V_{cn} are a negative-sequence set, how would the voltage and current relationships change?
(a) If C_1 is the complex positive-sequence voltage gain in Problem 3.29, and C_2 is the negative sequence complex voltage gain, express the relationship between C_1 and C_2.

3.31 If positive-sequence voltages are assumed and the wye-delta connection is considered, again with ideal transformers as in Problem 3.29, find the complex voltage gain C_3.
(a) What would the gain be for a negative-sequence set?
(b) Comment on the complex power gain.
(c) When terminated in a symmetric wye-connected load, find the referred impedance Z'_L, the secondary impedance Z_L referred to primary (i.e., the per-phase driving-point impedance on the primary side), in terms of Z_L and the complex voltage gain C.

SECTION 3.4

3.32 Determine the positive- and negative-sequence phase shifts for the three-phase transformers shown in Figure 3.36.

3.33 Consider the three single-phase two-winding transformers shown in Figure 3.37. The high-voltage windings are connected in Y. (a) For the low-voltage side, connect the windings in Δ, place the polarity marks, and label the terminals a, b, and c in accordance with the American standard. (b) Relabel the terminals a', b', and c' such that V_{AN} is 90° out of phase with $V_{a'n}$ for positive sequence.

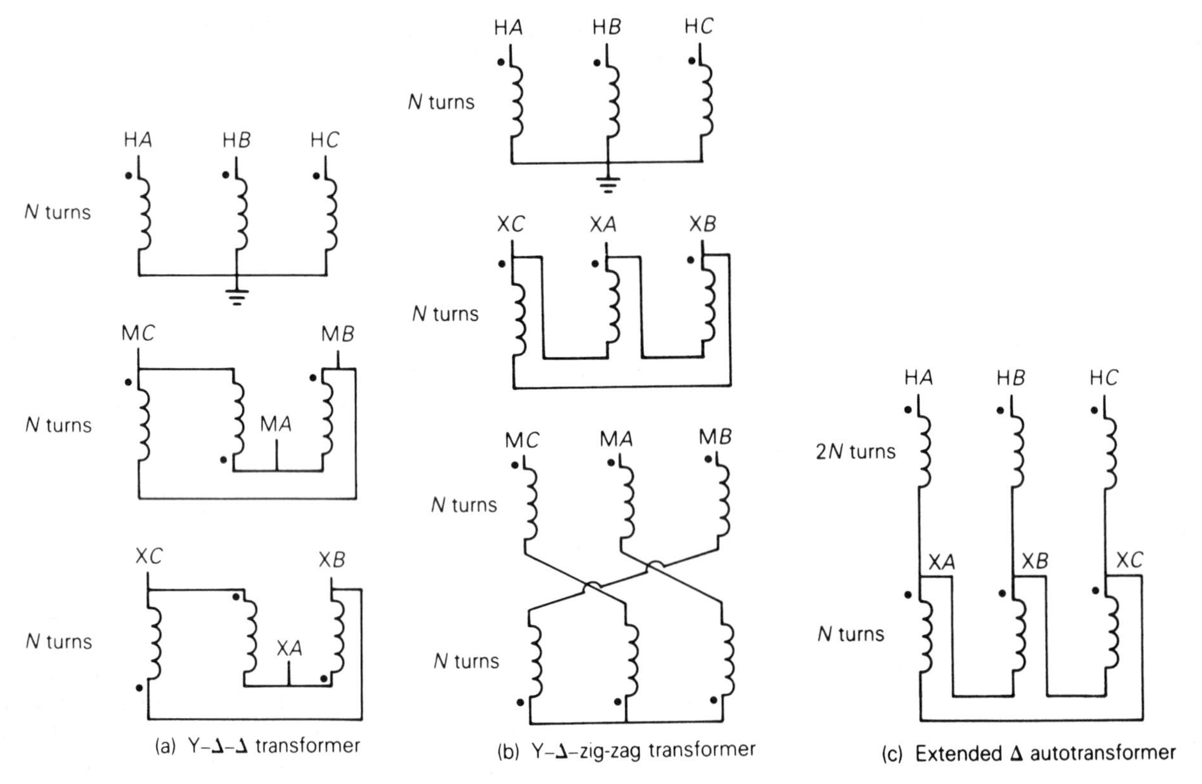

(a) Y–Δ–Δ transformer (b) Y–Δ–zig-zag transformer (c) Extended Δ autotransformer

FIGURE 3.36 Problems 3.32 and 3.52 (Coils drawn on the same vertical line are on the same core)

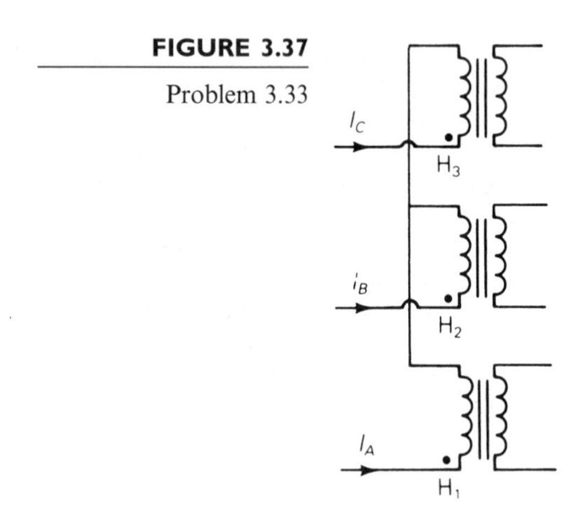

FIGURE 3.37 Problem 3.33

3.34 Three single-phase, two-winding transformers, each rated 450 MVA, 20 kV/288.7 kV, with leakage reactance $X_{eq} = 0.10$ per unit, are connected to form a three-phase bank. The high-voltage windings are connected in Y with a solidly grounded neutral. Draw the per-unit equivalent circuit if the low-voltage windings are connected (a) in Δ with American standard phase shift, (b) in Y with an open neutral. Use the transformer ratings as base quantities. Winding resistances and exciting current are neglected.

3.35 Consider a bank of three single-phase two-winding transformers whose high-voltage terminals are connected to a three-phase, 13.8-kV feeder. The low-voltage terminals are connected to a three-phase substation load rated 2.1 MVA and 2.3 kV. Determine the required voltage, current, and MVA ratings of both windings of each transformer, when the high-voltage/low-voltage windings are connected (a) Y–Δ, (b) Δ–Y, (c) Y–Y, and (d) Δ–Δ.

3.36 Three single-phase two-winding transformers, each rated 25 MVA, 34.5/13.8 kV, are connected to form a three-phase Δ–Δ bank. Balanced positive-sequence voltages are applied to the high-voltage terminals, and a balanced, resistive Y load connected to the low-voltage terminals absorbs 75 MW at 13.8 kV. If one of the single-phase transformers is removed (resulting in an open-Δ connection) and the balanced load is simultaneously reduced to 43.3 MW (57.7% of the original value), determine (a) the load voltages V_{an}, V_{bn}, and V_{cn}; (b) load currents I_a, I_b, and I_c; and (c) the MVA supplied by each of the remaining two transformers. Are balanced voltages still applied to the load? Is the open-Δ transformer overloaded?

3.37 Three single-phase two-winding transformers, each rated 25 MVA, 38.1/3.81 kV, are connected to form a three-phase Y–Δ bank with a balanced Y-connected resistive load of 0.6 Ω per phase on the low-voltage side. By choosing a base of 75 MVA (three phase) and 66 kV (line-to-line) for the high voltage side of the transformer bank, specify the base quantities for the low-voltage side. Determine the per-unit resistance of the load on the base for the low-voltage side. Then determine the load resistance R_L in ohms referred to the high-voltage side and the per-unit value of this load resistance on the chosen base.

3.38 Consider a three-phase generator rated 300 MVA, 23 kV, supplying a system load of 240 MVA and 0.9 power factor lagging at 230 kV through a 330 MVA, 23 Δ/ 230 Y-kV step-up transformer with a leakage reactance of 0.11 per unit. (a) Neglecting the exciting current and choosing base values at the load of 100 MVA and 230 kV, find the phasor currents I_A, I_B, and I_C supplied to the load in per unit. (b) By choosing the load terminal voltage V_A as reference, specify the proper base for the generator circuit and determine the generator voltage V as well as the phasor currents I_a, I_b, and I_c, from the generator. (*Note:* Take into account the phase shift of the transformer.) (c) Find the generator terminal voltage in kV and the real power supplied by the generator in MW. (d) By omitting the transformer phase shift altogether, check to see whether you get the same magnitude of generator terminal voltage and real power delivered by the generator.

SECTION 3.5

3.39 The leakage reactance of a three-phase, 300-MVA, 230 Y/23 Δ-kV transformer is 0.06 per unit based on its own ratings. The Y winding has a solidly grounded neutral. Draw the per-unit equivalent circuit. Neglect the exciting admittance and assume American standard phase shift.

3.40 Choosing system bases to be 240/24 kV and 100 MVA, redraw the per-unit equivalent circuit for Problem 3.39.

3.41 Consider the single-line diagram of the power system shown in Figure 3.38. Equipment ratings are:

Generator 1:	1000 MVA, 18 kV, $X'' = 0.2$ per unit
Generator 2:	1000 MVA, 18 kV, $X'' = 0.2$
Synchronous motor 3:	1500 MVA, 20 kV, $X'' = 0.2$
Three-phase Δ–Y transformers T_1, T_2, T_3, T_4:	1000 MVA, 500 kV Y/20 kV Δ, $X = 0.1$
Three-phase Y–Y transformer T_5:	1500 MVA, 500 kV Y/20 kV Y, $X = 0.1$

Neglecting resistance, transformer phase shift, and magnetizing reactance, draw the equivalent reactance diagram. Use a base of 100 MVA and 500 kV for the 50-ohm line. Determine the per-unit reactances.

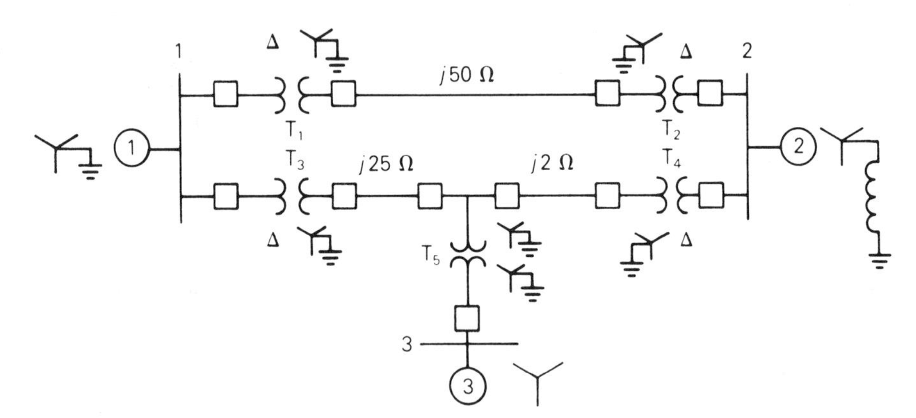

FIGURE 3.38 Problems 3.41 and 3.42

3.42 For the power system in Problem 3.41, the synchronous motor absorbs 1500 MW at 0.8 power factor leading with the bus 3 voltage at 18 kV. Determine the bus 1 and bus 2 voltages in kV. Assume that generators 1 and 2 deliver equal real powers and equal reactive powers. Also assume a balanced three-phase system with positive-sequence sources.

3.43 Three single-phase transformers, each rated 10 MVA, 66.4/12.5 kV, 60 Hz, with an equivalent series reactance of 0.1 per unit divided equally between primary and secondary, are connected in a three-phase bank. The high-voltage windings are Y connected and their terminals are directly connected to a 115-kV three-phase bus. The secondary terminals are all shorted together. Find the currents entering the high-voltage terminals and leaving the low-voltage terminals if the low-voltage windings are (a) Y connected, (b) Δ connected.

3.44 A 130-MVA, 13.2-kV three-phase generator, which has a positive-sequence reactance of 1.5 per unit on the generator base, is connected to a 135-MVA, 13.2 Δ/115 Y-kV step-up transformer with a series impedance of $(0.005 + j0.1)$ per unit on its own base. (a) Calculate the per-unit generator reactance on the transformer base. (b) The load at the transformer terminals is 15 MW at unity power factor and at 115 kV. Choosing the transformer high-side voltage as the reference phasor, draw a phasor diagram for this condition. (c) For the condition of part (b), find the transformer low-side voltage and the generator internal voltage behind its reactance. Also compute the generator output power and power factor.

3.45 Figure 3.39 shows a one-line diagram of a system in which the three-phase generator is rated 300 MVA, 20 kV with a subtransient reactance of 0.2 per unit and with its neutral grounded through a 0.4-Ω reactor. The transmission line is 64 km long with a series reactance of 0.5 Ω/km. The three-phase transformer T_1 is rated 350 MVA, 230/20 kV with a leakage reactance of 0.1 per unit. Transformer T_2 is composed of three single-phase transformers, each rated 100 MVA, 127/13.2 kV with a leakage reactance of 0.1 per unit. Two 13.2-kV motors M_1 and M_2 with a subtransient reactance of 0.2 per unit for each motor represent the load. M_1 has a rated input of 200 MVA with its neutral grounded through a 0.4-Ω current-limiting reactor. M_2 has a rated input of 100 MVA with its neutral not connected to ground. Neglect phase shifts associated with the transformers. Choose the generator rating as base in the generator circuit and draw the positive-sequence reactance diagram showing all reactances in per unit.

FIGURE 3.39 Problems 3.45 and 3.46

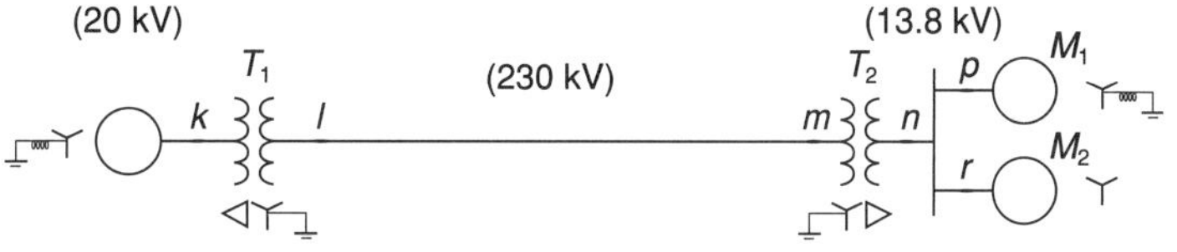

3.46 The motors M_1 and M_2 of Problem 3.45 have inputs of 120 and 60 MW, respectively, at 13.2 kV, and both operate at unity power factor. Determine the generator terminal voltage and voltage regulation of the line. Neglect transformer phase shifts.

3.47 Consider the one-line diagram shown in Figure 3.40. The three-phase transformer bank is made up of three identical single-phase transformers, each specified by $X_l = 0.24\ \Omega$ (on the low-voltage side), negligible resistance and magnetizing current, and turns ratio $\eta = N_2/N_1 = 10$. The transformer bank is delivering 100 MW at 0.8 p.f. lagging to a substation bus whose voltage is 230 kV.
(a) Determine the primary current magnitude, primary voltage (line-to-line) magnitude, and the three-phase complex power supplied by the generator. Choose the line-to-neutral voltage at the bus, $V_{a'n'}$, as the reference. Account for the phase shift, and assume positive-sequence operation.
(b) Find the phase shift between the primary and secondary voltages.

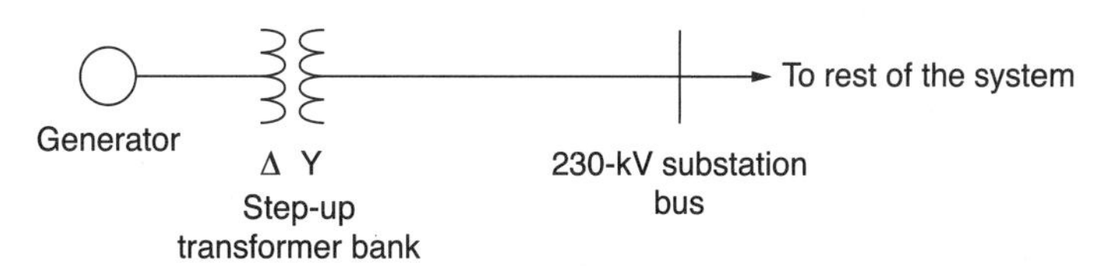

FIGURE 3.40 One-line diagram for Problem 3.47

3.48 With the same transformer banks as in Problem 3.47, Figure 3.41 shows the one-line diagram of a generator, a step-up transformer bank, a transmission line, a step-down

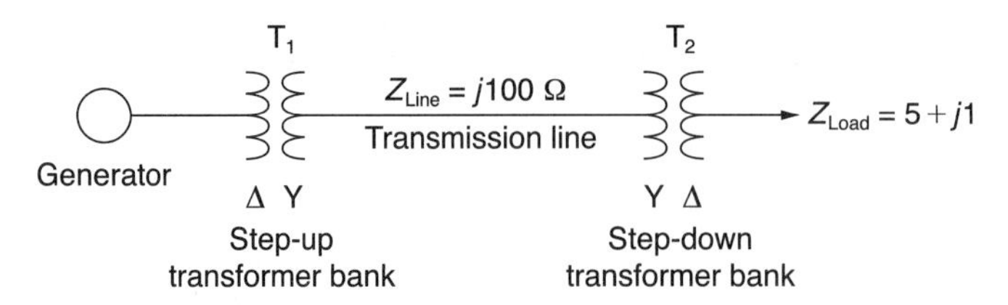

FIGURE 3.41 One-line diagram for Problem 3.48

transformer bank, and an impedance load. The generator terminal voltage is 15 kV (line-to-line).

(a) Draw the per-phase equivalent circuit, accounting for phase shifts for positive-sequence operation.

(b) By choosing the line-to-neutral generator terminal voltage as the reference, determine the magnitudes of the generator current, transmission-line current, load current, and line-to-line load voltage. Also, find the three-phase complex power delivered to the load.

3.49 Consider the single-line diagram of a power system shown in Figure 3.42 with equipment ratings given below:

Generator G_1:	50 MVA, 13.2 kV, $x = 0.15$ *pu*
Generator G_2:	20 MVA, 13.8 kV, $x = 0.15$ *pu*
three-phase Δ–Y transformer T_1:	80 MVA, 13.2 Δ/165 Y kV, $X = 0.1$ *pu*
three-phase Y–Δ transformer T_2:	40 MVA, 165 Y/13.8 Δ kV, $X = 0.1$ *pu*
Load:	40 MVA, 0.8 p.f. lagging, operating at 150 kV

Choose a base of 100 MVA for the system and 132-kV base in the transmission-line circuit. Let the load be modeled as a parallel combination of resistance and inductance. Neglect transformer phase shifts.

Draw a per-phase equivalent circuit of the system showing all impedances in per unit.

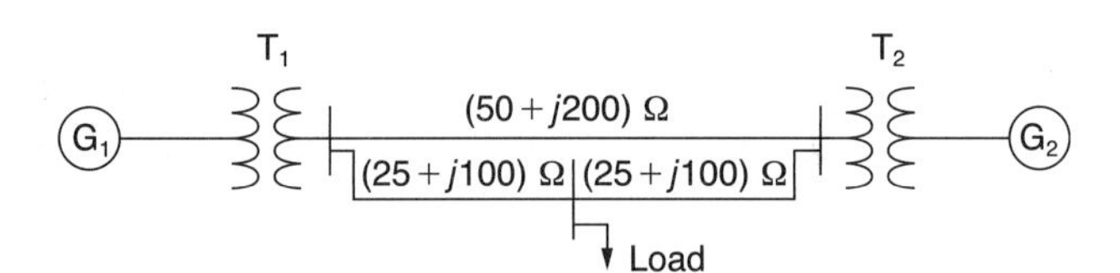

FIGURE 3.42 One-line diagram for Problem 3.49

SECTION 3.6

3.50 A single-phase three-winding transformer has the following parameters: $Z_1 = Z_2 = Z_3 = 0 + j0.05$, $G_c = 0$, and $B_m = 0.2$ per unit. Three identical transformers, as described, are connected with their primaries in Y (solidly grounded neutral) and with their secondaries and tertiaries in Δ. Draw the per-unit sequence networks of this transformer bank.

3.51 The ratings of a three-phase three-winding transformer are:

Primary (1):	Y connected, 66 kV, 15 MVA
Secondary (2):	Y connected, 13.2 kV, 10 MVA
Tertiary (3):	Δ connected, 2.3 kV, 5 MVA

Neglecting winding resistances and exciting current, the per-unit leakage reactances are:

$$X_{12} = 0.08 \text{ on a 15-MVA, 66-kV base}$$

$$X_{13} = 0.10 \text{ on a 15-MVA, 66-kV base}$$

$$X_{23} = 0.09 \text{ on a 10-MVA, 13.2-kV base}$$

(a) Determine the per-unit reactances X_1, X_2, X_3 of the equivalent circuit on a 15-MVA, 66-kV base at the primary terminals. (b) Purely resistive loads of 7.5 MW at 13.2 kV and 5 MW at 2.3 kV are connected to the secondary and tertiary sides of the transformer, respectively. Draw the per-unit impedance diagram, showing the per-unit impedances on a 15-MVA, 66-kV base at the primary terminals.

3.52 Draw the per-unit equivalent circuit for the transformers shown in Figure 3.34. Include ideal phase-shifting transformers showing phase shifts determined in Problem 3.32. Assume that all windings have the same kVA rating and that the equivalent leakage reactance of any two windings with the third winding open is 0.10 per unit. Neglect the exciting admittance.

3.53 The ratings of a three-phase, three-winding transformer are:

Primary:	Y connected, 66 kV, 15 MVA
Secondary:	Y connected, 13.2 kV, 10 MVA
Tertiary:	Δ connected, 2.3 kV, 5 MVA

Neglecting resistances and exciting current, the leakage reactances are:

$$X_{PS} = 0.07 \text{ per unit on a 15-MVA, 66-kV base}$$

$$X_{PT} = 0.09 \text{ per unit on a 15-MVA, 66-kV base}$$

$$X_{ST} = 0.08 \text{ per unit on a 10-MVA, 13.2-kV base}$$

Determine the per-unit reactances of the per-phase equivalent circuit using a base of 15 MVA and 66 kV for the primary.

3.54 An infinite bus, which is a constant voltage source, is connected to the primary of the three-winding transformer of Problem 3.53. A 7.5-MVA, 13.2-kV synchronous motor with a subtransient reactance of 0.2 per unit is connected to the transformer secondary. A 5-MW, 2.3-kV three-phase resistive load is connected to the tertiary. Choosing a base of 66 kV and 15 MVA in the primary, draw the impedance diagram of the system showing per-unit impedances. Neglect transformer exciting current, phase shifts, and all resistances except the resistive load.

SECTION 3.7

3.55 A single-phase 10-kVA, 2300/230-volt, 60-Hz two-winding distribution transformer is connected as an autotransformer to step up the voltage from 2300 to 2530 volts. (a) Draw a schematic diagram of this arrangement, showing all voltages and currents when delivering full load at rated voltage. (b) Find the permissible kVA rating of the autotransformer if the winding currents and voltages are not to exceed the rated values as a two-winding transformer. How much of this kVA rating is transformed by magnetic induction? (c) The following data are obtained from tests carried out on the transformer when it is connected as a two-winding transformer:

Open-circuit test with the low-voltage terminals excited:
Applied voltage = 230 V, Input current = 0.45 A, Input power = 70 W.

Short-circuit test with the high-voltage terminals excited:
Applied voltage = 120 V, Input current = 4.5 A, Input power = 240 W.

Based on the data, compute the efficiency of the autotransformer corresponding to full load, rated voltage, and 0.8 power factor lagging. Comment on why the efficiency is higher as an autotransformer than as a two-winding transformer.

3.56 Three single-phase two-winding transformers, each rated 3 kVA, 220/110 volts, 60 Hz, with a 0.10 per-unit leakage reactance, are connected as a three-phase extended Δ autotransformer bank, as shown in Figure 3.31(c). The low-voltage Δ winding has a 110 volt rating. (a) Draw the positive-sequence phasor diagram and show that the high-voltage winding has a 479.5 volt rating. (b) A three-phase load connected to the low-voltage terminals absorbs 6 kW at 110 volts and at 0.8 power factor lagging. Draw the per-unit impedance diagram and calculate the voltage and current at the high-voltage terminals. Assume positive-sequence operation.

3.57 A two-winding single-phase transformer rated 60 kVA, 240/1200 V, 60 Hz, has an efficiency of 0.96 when operated at rated load, 0.8 power factor lagging. This transformer is to be utilized as a 1440/1200-V step-down autotransformer in a power distribution system. (a) Find the permissible kVA rating of the autotransformer if the winding currents and voltages are not to exceed the ratings as a two-winding transformer. Assume an ideal transformer. (b) Determine the efficiency of the autotransformer with the kVA loading of part (a) and 0.8 power factor leading.

3.58 A single-phase two-winding transformer rated 90 MVA, 80/120 kV is to be connected as an autotransformer rated 80/200 kV. Assume that the transformer is ideal. (a) Draw a schematic diagram of the ideal transformer connected as an autotransformer, showing the voltages, currents, and dot notation for polarity. (b) Determine the permissible kVA rating of the autotransformer if the winding currents and voltages are not to exceed the rated values as a two-winding transformer. How much of the kVA rating is transferred by magnetic induction?

SECTION 3.8

3.59 The two parallel lines in Example 3.13 supply a balanced load with a load current of $1.0\angle{-30^\circ}$ per unit. Determine the real and reactive power supplied to the load bus from each parallel line with (a) no regulating transformer, (b) the voltage-magnitude-regulating transformer in Example 3.13(a), and (c) the phase-angle-regulating transformer in Example 3.13(b). Assume that the voltage at bus *abc* is adjusted so that the voltage at bus $a'b'c'$ remains constant at $1.0\angle{0^\circ}$ per unit. Also assume positive sequence. Comment on the effects of the regulating transformers.

PW **3.60** PowerWorld Simulator case Problem 3.60 duplicates Example 3.13 except that a resistance term of 0.06 per unit has been added to the transformer and 0.05 per unit to the transmission line. Since the system is no longer lossless, a field showing the real power losses has also been added to the one-line. With the LTC tap fixed at 1.05, plot the real power losses as the phase shift angle is varied from -10 to $+10$ degrees. What value of phase shift minimizes the system losses?

PW **3.61** Repeat Problem 3.60, except keep the phase-shift angle fixed at 3.0 degrees, while varying the LTC tap between 0.9 and 1.1. What tap value minimizes the real power losses?

3.62 Rework Example 3.12 for a +10% tap, providing a 10% increase for the high-voltage winding.

3.63 A 23/230-kV step-up transformer feeds a three-phase transmission line, which in turn supplies a 150-MVA, 0.8 lagging power factor load through a step-down 230/23-kV transformer. The impedance of the line and transformers at 230 kV is $18 + j60\ \Omega$. Determine the tap setting for each transformer to maintain the voltage at the load at 23 kV.

3.64 The per-unit equivalent circuit of two transformers T_a and T_b connected in parallel, with the same nominal voltage ratio and the same reactance of 0.1 per unit on the same base, is shown in Figure 3.43. Transformer T_b has a voltage-magnitude step-up toward the load of 1.05 times that of T_b (that is, the tap on the secondary winding of T_a is set to 1.05). The load is represented by $0.8 + j0.6$ per unit at a voltage $V_2 = 1.0\angle 0^\circ$ per unit. Determine the complex power in per unit transmitted to the load through each transformer. Comment on how the transformers share the real and reactive powers.

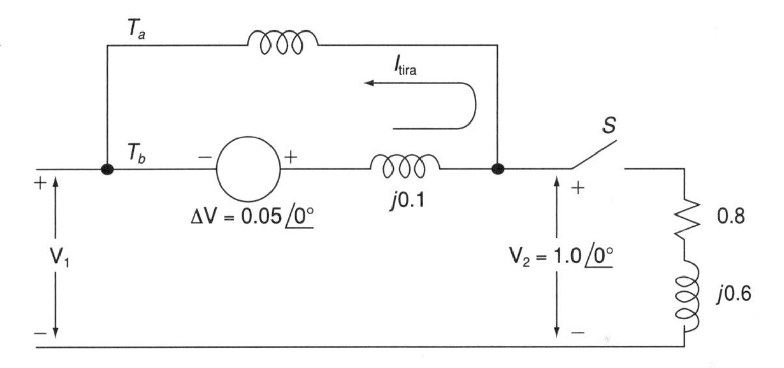

FIGURE 3.43 Problem 3.64

3.65 Reconsider Problem 3.64 with the change that now T_b includes both a transformer of the same turns ratio as T_a and a regulating transformer with a 3° phase shift. On the base of T_a, the impedance of the two components of T_b is $j0.1$ per unit. Determine the complex power in per unit transmitted to the load through each transformer. Comment on how the transformers share the real and reactive powers.

CASE STUDY QUESTIONS

A. What are the potential consequences of running a transmission transformer to failure with no available spare to replace it?

B. What are the benefits of sharing spare transmission transformers among utility companies?

C. Where should spare transmission be located?

REFERENCES

1. R. Feinberg, *Modern Power Transformer Practice* (New York: Wiley, 1979).
2. A. C. Franklin and D. P. Franklin, *The J & P Transformer Book*, 11th ed. (London: Butterworths, 1983).
3. W. D. Stevenson, Jr., *Elements of Power System Analysis*, 4th ed. (New York: McGraw-Hill, 1982).
4. J. R. Neuenswander, *Modern Power Systems* (Scranton, PA: International Textbook Company, 1971).
5. M. S. Sarma, *Electric Machines* (Dubuque, IA: Brown, 1985).
6. A. E. Fitzgerald, C. Kingsley, and S. Umans, *Electric Machinery*, 4th ed. (New York: McGraw-Hill, 1983).
7. O. I. Elgerd, *Electric Energy Systems: An Introduction* (New York: McGraw-Hill, 1982).
8. D. Egan and K. Seiler, "PJM Manages Aging Transformer Fleet," *Transmission & Distribution World Magazine*, March 2007, pp. 42–45.

765-kV transmission line with aluminum guyed-V towers (Courtesy of American Electric Power Company)

4 TRANSMISSION LINE PARAMETERS

In this chapter, we discuss the four basic transmission-line parameters: series resistance, series inductance. shunt capacitance, and shunt conductance. We also investigate transmission-line electric and magnetic fields.

Series resistance accounts for ohmic (I^2R) line losses. Series impedance, including resistance and inductive reactance, gives rise to series-voltage drops along the line. Shunt capacitance gives rise to line-charging currents. Shunt conductance accounts for V^2G line losses due to leakage currents between conductors or between conductors and ground. Shunt conductance of overhead lines is usually neglected.

Although the ideas developed in this chapter can be applied to underground transmission and distribution, the primary focus here is on overhead lines. Underground transmission in the United States presently accounts for less than 1% of total transmission, and is found mostly in large cities or under waterways. There is, however, a large application for underground cable in distribution systems.

CASE STUDY Two transmission articles are presented here. The first article covers transmission conductor technologies including conventional conductors, high-temperature conductors, and emerging conductor technologies [10]. Conventional conductors include the aluminum conductor steel reinforced (ACSR), the homogeneous all aluminum alloy conductor (AAAC), the aluminum conductor alloy reinforced (ACAR), and others. High-temperature conductors are based on aluminum-zirconium alloys that resist the annealing effects of high temperatures. Emerging conductor designs make use of composite material technology. The second article describes trends in transmission and distribution line insulators for six North American electric utilities [12]. Insulator technologies include porcelain, toughened glass, and polymer (also known as composite or non-ceramic). All three technologies are widely used. Current trends favor polymer insulators for distribution (less than 69 kV) because they are lightweight, easy to handle, and economical. Porcelain remains in wide use for bulk power transmission lines, but maintenance concerns associated with management and inspection of aging porcelain insulators are driving some utilities to question their use. Life-cycle cost considerations and ease of inspection for toughened glass insulators are steering some utilities toward glass technology.

Transmission Line Conductor Design Comes of Age

ART J. PETERSON JR. AND SVEN HOFFMANN

Deregulation and competition have changed power flows across transmission networks significantly. Meanwhile, demand for electricity continues to grow, as do the increasing challenges of building new transmission circuits. As a result, utilities need innovative ways to increase circuit capacities to reduce congestion and maintain reliability.

National Grid is monitoring transmission conductor technologies with the intent of testing and deploying innovative conductor technologies within the United States over the next few years. In the UK, National Grid has been using conductor replacement as a means of increasing circuit capacity since the mid 1980s, most recently involving the high-temperature, low-sag "Gap-type" conductor. As a first step in developing a global conductor deployment strategy, National Grid embarked on an overall assessment of overhead transmission line conductor technologies, examining innovative, and emerging technologies.

("Transmission Line Conductor Design Comes of Age" by Art J. Peterson Jr. and Sven Hoffmann, Transmission & Distribution World Magazine, (Aug/2006). Reprinted with permission of Penton Media)

About National Grid

National Grid USA is a subsidiary of National Grid Transco, an international energy-delivery business with principal activities in the regulated electric and gas industries. National Grid is the largest transmission business in the northeast United States, as well as one of the 10 largest electric utilities in the United States National Grid achieved this by combining New England Electric System, Eastern Utilities Associates and Niagara Mohawk between March 2000 and January 2002. Its electricity-delivery network includes 9000 miles (14,484 km) of transmission lines and 72,000 miles (115,872 km) of distribution lines.

National Grid UK is the owner, operator and developer of the high-voltage electricity transmission network in England and Wales, comprising approximately 9000* circuit-miles of overhead line and 600* circuit-miles of underground cable at 275 and 400 kV, connecting more than 300 substations.

**9000 circuit-miles = 14,500 circuit-km*
**600 circuit-miles = 1000 circuit-km*

De-stranding the Gap conductor for field installation

Re-stranding of conductor

CONVENTIONAL CONDUCTORS

The reality is that there is no single "wonder material." As such, the vast majority of overhead line conductors are nonhomogeneous (made up of more than one material). Typically, this involves a high-strength core material surrounded by a high-conductivity material. The most common conductor type is the aluminum conductor steel reinforced (ACSR), which has been in use for more than 80 years. By varying the relative cross-sectional areas of steel and aluminum, the conductor can be made stronger at the expense of conductivity (for areas with high ice loads, for example), or it can be made more conductive at the expense of strength where it's not required.

More recently, in the last 15 to 20 years, the homogeneous all-aluminum alloy conductor (AAAC) has become quite popular, especially for National Grid in the UK where it is now the standard conductor type employed for new and refurbished lines. Conductors made up of this alloy (a heat treatable aluminum-magnesium-silicon alloy) are, for the same diameter as an ACSR, stronger, lighter, and more conductive although they are a little more expensive and have a higher expansion coefficient. However, their high strength-to-weight ratio allows them to be strung to much lower initial sags, which allows higher operating temperatures. The resulting tension levels are relatively high, which could result in increased vibration and early fatigue of the conductors. In the UK, with favorable terrain, wind conditions and dampers, these tensions are acceptable and have allowed National Grid to increase the capacities of some lines by up to 50%.

For the purpose of this article, the three materials mentioned so far—steel, aluminum and aluminum alloy—are considered to be the materials from which conventional conductors are made. The ACSR and AAAC are two examples of such conductors. Other combinations available include aluminum conductor alloy reinforced (ACAR), aluminum alloy conductor steel reinforced (AACSR) and the less common all-aluminum conductor (AAC).

Conductors of these materials also are available in other forms, such as compacted conductors, where the strands are shaped so as not to leave any voids within the conductor's cross section (a standard conductor uses round strands), increasing the amount of conducting material without increasing the diameter. These conductors are designated trapezoidal-wire (TW) or, for example, ACSR/TW and AACSR/TW. Other shaped conductors are available that have noncircular cross sections designed to minimize the effects of wind-induced motions and vibrations.

HIGH-TEMPERATURE CONDUCTORS

Research in Japan in the 1960s produced a series of aluminum-zirconium alloys that resisted the annealing effects of high temperatures. These alloys can retain their strength at temperatures up to 230 °C (446 °F). The most common of these alloys—TA1, ZTA1 and XTA1—are the basis of a variety of high-temperature conductors.

Clamp used for Gap conductor

The thermal expansion coefficients of all the conventional steel-cored conductors are governed by both materials together, resulting in a value between that of the steel and that of the aluminum. This behavior relies on the fact that both components are carrying mechanical stress.

However, because the expansion coefficient of aluminum is twice that of steel, stress will be increasingly transferred to the steel core as the conductor's temperature rises. Eventually the core bears all the stress in the conductor. From this point on, the conductor as a whole essentially takes on the expansion coefficient of the core. For a typical 54/7 ACSR (54 aluminum strands, 7 steel) this transition point (also known as the "knee-point") occurs around 100 °C (212 °F).

For lines built to accommodate relatively large sags, the T-aluminum conductor, steel reinforced (TACSR) conductor was developed. (This is essentially identical to ACSR but uses the heat-resistant aluminum alloy designated TA1). Because this conductor can be used at high temperatures with no strength loss, advantage can be taken of the low-sag behavior above the knee-point.

If a conductor could be designed with a core that exhibited a lower expansion coefficient than steel, or that exhibited a lower knee-point temperature, more advantage could be taken of the high-temperature alloys. A conductor that exhibits both of these properties uses Invar, an alloy of iron and nickel. Invar has an expansion coefficient about one-third of steel (2.8 microstrain per Kelvin up to 100 °C, and 3.6 over 100 °C, as opposed to 11.5 for steel). T-aluminum conductor Invar reinforced (TACIR) is capable of operation up to 150 °C (302 °F), with ZTACIR and XTACIR capable of 210 °C (410 °F) and 230 °C (446 °F), respectively.

Further, the transition temperature, although dependent on many factors, is typically lower than that for an ACSR, allowing use of the high temperatures within lower sag limits than required for the TACSR conductors. One disadvantage of this conductor is that Invar is considerably weaker than steel. Therefore, for high-strength applications (to resist ice loading, for example), the core needs to make up a greater proportion of the conductor's area, reducing or even negating the high-temperature benefits. As a result, the ACIR-type conductors are used in favorable areas in Japan and Asia, but are not commonly used in the United States or Europe.

There will still be instances, however, where insufficient clearance is available to take full advantage of the transitional behavior of the ACIR conductors. A conductor more suitable for uprating purposes would exhibit a knee-point at much lower temperatures. Two conductors are available that exhibit this behavior: the Gap-type conductor and a variant of the ACSR that uses fully annealed aluminum.

Developed in Japan during the 1970s, Gap-type ZT-aluminum conductor steel reinforced (GZTACSR) uses heat-resistant aluminum over a steel core. It has been used in Japan, Saudi Arabia, and Malaysia, and is being extensively implemented by National Grid in the UK. The principle of the Gap-type conductor is that it can be tensioned on the steel core alone during erection. A small annular Gap exists between a high-strength steel core and the first layer of trapezoidal-shaped aluminum strands, which allows this to be achieved. The result is a conductor with a knee-point at the erection temperature. Above this, thermal expansion is that of steel (11.5 microstrain per Kelvin), while below it is that of a comparable ACSR (approximately 18). This construction allows for low-sag properties above the erection temperature and good strength below it as the aluminum alloy can take up significant load.

For example, the application of GZTACSR by National Grid in the UK allowed a 90 °C (194 °F) rated 570 mm^2 AAAC to be replaced with a 620 mm^2 GZTACSR (Matthew). The Gap-type conductor, being of compacted construction, actually had a smaller diameter than the AAAC, despite having a larger nominal area. The low-sag properties allowed

Semi-strain assembly installed on line in a rural area of the UK

a rated temperature of 170 °C (338 °F) and gave a 30% increase in rating for the same sag.

The principal drawback of the Gap-type conductor is its complex installation procedure, which requires destranding the aluminum alloy to properly install on the joints. There is also the need for "semi-strain" assemblies for long line sections (typically every five spans). Experience in the UK has shown that a Gap-type conductor requires about 25% more time to install than an ACSR.

A semi-strain assembly is, in essence, a pair of back-to-back compression anchors at the bottom of a suspension insulator set. It is needed to avoid potential problems caused by the friction that developes between the steel core and the aluminum layers when using running blocks. This helps to prevent the steel core from hanging up within the conductor.

During 1999 and 2000, in the UK, National Grid installed 8 km (single circuit) of Matthew GZTACSR. Later this year and continuing through to next year, National Grid will be refurbishing a 60 km (37-mile) double-circuit (120 circuit-km) route in the UK with Matthew.

A different conductor of a more standard construction is aluminum conductor steel supported (ACSS), formerly known as SSAC. Introduced in the 1980s, this conductor uses fully annealed aluminum around a steel core. The steel core provides the entire conductor support. The aluminum strands are "dead soft," thus the conductor may be operated at temperatures in excess of 200 °C without loss of strength. The maximum operating temperature of the conductor is limited by the coating used on the steel core. Conventional galvanized coatings deteriorate rapidly at temperatures above 245 °C (473 °F). If a zinc-5% aluminum mischmetal alloy coated steel core is used, temperatures of 250 °C are possible.

Since the fully annealed aluminum cannot support significant stress, the conductor has a thermal expansion similar to that of steel. Tension in the aluminum strands is normally low. This helps to improve the conductor's self-damping characteristics and helps to reduce the need for dampers.

For some applications there will be concern over the lack of strength in the aluminum, as well as the possibility of damage to the relatively soft outer layers. However, ACSS is available as ACSS/TW, improving, its strength. ACSS requires special care when installing. The soft annealed aluminum wires can be easily damaged and "bird-caging" can occur. As with the other high-temperature conductors, the heat requires the use of special suspension clamps, high-temperature deadends, and high-temperature splices to avoid hardware damage.

EMERGING CONDUCTOR TECHNOLOGIES

Presently, all the emerging designs have one thing in common—the use of composite material technology.

Aluminum conductor carbon fiber reinforced (ACFR) from Japan makes use of the very-low-expansion coefficient of carbon fiber, resulting in a conductor with a lower knee-point of around 70 °C (158 °F). The core is a resin-matrix composite containing carbon fiber. This composite is capable of withstanding temperatures up to 150 °C. The ACFR is about 30% lighter and has an expansion coefficient (above the knee-point) that is 8% that of an ACSR of the same stranding, giving a rating increase of around 50% with no structural work required.

Meanwhile, in the United States, 3M has developed the Aluminum Conductor Composite Reinforced (ACCR). The core is an aluminum-matrix

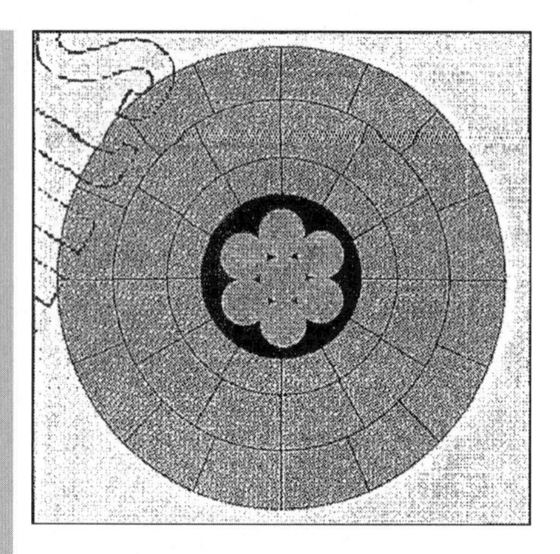

A cross section of the Gap conductor

composite containing alumina fibers, with the outer layers made from a heat-resistant aluminum alloy. As with the ACFR, the low-expansion coefficient of the core contributes to a fairly low knee-point, allowing the conductor to make full use of the heat resistant alloy within existing sag constraints. Depending on the application, rating increases between 50% and 200% are possible as the conductor can be rated up to 230 °C.

Also in the United States, two more designs based on glass-fiber composites are emerging. Composite Technology Corp. (CTC; Irvine, California, U.S.) calls it the aluminum conductor composite core (ACCC), and W. Brandt Goldsworthy and Associates (Torrance, California) are developing composite reinforced aluminum conductor (CRAC). These conductors are expected to offer between 40% and 100% increases in ratings.

Over the next few years, National Grid plans to install ACSS and the Gap conductor techology within its U.S. transmission system. Even a test span of one or more of the new composite conductors is being considered.

Art J. Peterson Jr. is a senior engineer in National Grid's transmission line engineering and project management department in Syracuse, New York. Peterson received a BS degree in physics from Le Moyne College in Syracuse; a MS degree in physics from Clarkson University in Potsdam, New York; a M. Eng. degree in nuclear engineering from Pennsylvania State University in State College, Pennsylvania; and a Ph.D. in organization and management from Capella University in Minneapolis, Minnesota. He has 20 years of experience in electric generation and transmission.

Art.Peterson@us.ngrid.com

Sven Hoffmann is the circuits forward policy team leader in National Grid's asset strategy group in Coventry, United Kingdom. Hoffmann has a bachelor's in engineering degree from the University of Birmingham in England. He is a chartered engineer with the Institution of Electrical Engineers, and the UK Regular Member for CIGRE Study Committee B2. Hoffmann has been working at National Grid, specializing in thermal and mechanical aspects of overhead lines for eight years.

sven.hoffmann@uk.ngrid.com

Six Utilities Share Their Perspectives on Insulators

APR 1, 2010 12:00 PM

BY RAVI S. GORUR, ARIZONA STATE UNIVERSITY

Trends in the changing landscape of high-voltage insulators are revealed through utility interviews.

("Six Utilities Share Their Perspectives on Insulators" by Ravi S. Gorur, *Transmission & Distribution World Magazine* (April/2010). *Reprinted with permission of Penton Media)*

The high-voltage transmission system in North America is the result of planning and execution initiated soon after World War II. Ambitious goals, sound engineering and the vertically integrated structure of utilities at that time all contributed to high reliability and good quality of electric power.

The high-voltage transmission infrastructure development peaked in the 1970s. From then on until the turn of the century, load growth was not as high as anticipated, resulting in a drastic reduction in transmission activity. Consequently, the system was pushed to its limits, which led to a few large-scale blackouts. The consensus is that the existing system is bursting at its seams, continuing to age and needs refurbishment; at the same time, new lines are needed to handle load growth and transfer massive amounts of power from remote regions to load centers.

Today, several thousand kilometers of transmission lines at voltages from 345 kV ac to 765 kV ac and high-voltage dc lines are either in the planning or construction stages. A catalyst for this renewed interest in transmission line construction is renewable energy. It is clear that in order to reap the benefits of green and clean energy (mostly solar and wind), there is an urgent need to build more lines to transfer power from locations rich in these resources to load centers quite distant from them.

For this upcoming surge of new high-voltage projects and refurbishment of older lines, insulators play a critical and often grossly underestimated role in power delivery. Over many decades, the utility perspective regarding insulation technologies has changed in several ways.

INSULATOR TYPES

When the *original transmission system* was built, the porcelain insulator industry was strong in North America and utilities preferred to use domestic products. Toughened glass insulators were introduced in Europe in the 1950s and gained worldwide acceptance. In the United States, many users adopted the new technology in the 1960s and 1970s, while others were reluctant to use them because of perceived concerns with vandalism. However, the use of glass insulators in the United States continued to expand.

Polymer (also known as composite or nonceramic) insulators were introduced in the 1970s and have been widely used in North America since the 1980s. With the advent of polymers, it seemed the use of glass and porcelain suspension insulators started to decline. Polymers are particularly suited for compact line construction. Such compact lines minimized right-of-way requirements and facilitated the permitting of new transmission corridors in congested and urban areas.

With the growing number of high-voltage lines now reaching their life expectancy, many utilities are turning their attention to the fast-growing population of aging porcelain insulators. Deterioration of porcelain insulators typically stems from impurities or voids in the porcelain dielectric and expansion of the cement in the pin region, which leads to radial cracks in the shell. As internal cracks or punctures in porcelain cannot be visually detected and require tools, the labor-intensive process is expensive and requires special training of the work force.

SUPPLY CHAIN

Today, there is no domestic supplier of porcelain suspension insulators in North America. However, there are quite a few suppliers of porcelain insulators in several other countries, but most of them have limited or no experience in North America. This naturally has raised concerns among many utilities in North America about the quality and consistency of such productions.

Polymer insulators have been widely used at all voltages but largely in the 230-kV and below range. There are still unresolved issues with degradation, life expectancy and live-line working—all of which are hindering large-scale acceptance at higher voltages. The *Electric Power Research Institute (EPRI)* recently suggested that composite insulators for voltages in the range of 115 kV to 161 kV may require corona rings, which would not only increase the cost of composites but could create possible confusion as the corona rings offered vary from one manufacturer to another. With respect to toughened glass, not much has been published or discussed in the United States.

SALT RIVER PROJECT, ARIZONA

Salt River Project (SRP) serves the central and eastern parts of Arizona. Except for small pockets in the eastern parts, which are subject to contamination from the mining industry, SRP's service territory is fairly clean and dry. Its bulk transmission and distribution networks are based largely on porcelain insulators. The utility began to use polymer insulators in the early 1980s and has successfully used them at all voltages. Polymers are favored for line post construction and account for the majority of 69-kV through 230-kV constructions in the last 30 years. The 500-kV ac Mead-Phoenix line, operational since 1990, was one of the first long transmission lines in the country to use silicone rubber composite insulators. The utility's service experience with these has been excellent.

The need for corona rings for composite insulators at 230 kV and higher voltage was recognized in the early 1980s by many users that experience fairly high wet periods in addition to contamination. This was not a concern for *SRP;* consequently, the first batch of composite insulators installed in the 1980s on several 230-kV lines had no corona rings. These insulators were inspected visually and with a corona camera about 10 years ago and most recently in 2009.

Some 230-kV lines are constructed with polymer insulators and no corona rings, and the insulators are in remarkably good condition. The relatively clean and dry environment in Arizona creates a corona-free setting most of the time, and this contributes greatly to SRP's problem-free experience with all types of insulators. In keeping with industry practice, all 230-kV suspension composite insulators subsequently installed by SRP have a corona ring at the line end, and those installed on 500-kV lines have rings at the line and tower ends.

SRP performs helicopter inspections of its transmission lines annually. Insulators with visual damage are replaced. Like many utilities, SRP trains and equips its linemen to perform line maintenance under energized (live or hot) conditions. Even though most maintenance is done with the lines de-energized, it is considered essential to preserve the ability to work on energized 500-kV lines. Future conditions may make outages unobtainable or unreasonably expensive.

Because there is no industry standard on live-line working with composite insulators and because of the difficulty in getting an outage on its 500-kV lines, which are co-owned by several utilities, SRP decided not to use polymer insulators at 500 kV. After reviewing the service experience of toughened glass insulators, SRP decided to consider them equal to porcelain in bid processes. This has resulted in the installation of toughened glass insulators on a portion of the utility's recent 500-kV line construction. The ease of detection of damaged glass bells was a factor, although not the most important one as its service experience with porcelain has been excellent.

PUBLIC SERVICE ELECTRIC & GAS, NEW JERSEY

Public Service Electric & Gas (PSE&G) has experienced problems with loss of dielectric strength and punctures on porcelain insulators from some suppliers. Lines with such insulators are being examined individually using a buzzer or electric field probe, but the results are not always reliable.

The utility has used composite insulators extensively on compact lines (line post configuration) up to 69 kV, and the experience has been good. It has experienced degradation (erosion, corona cutting) on some composite suspension insulators at 138 kV. These insulators were installed without a corona ring as is common practice. In one instance, PSE&G was fortunate to remove a composite insulator with part of the fiberglass core exposed before any mechanical failure (brittle fracture) could occur.

In the last five years, the utility has been using toughened glass insulators on new construction and as a replacement of degraded porcelain insulators on 138-kV and higher lines. Since many of these lines are shared with other utilities, PSE&G needs to have the ability to maintain them live; it calls itself a live-line utility. A major factor for using glass was the ease of spotting damaged bells. For example, the utility flies about 6 miles (10 km) per day and inspects roughly 30 towers; in contrast, a ground crew climbing and inspecting averages about three towers per day. In many cases, the entire circuit using glass insulators can be inspected in a single day with helicopters. PSE&G has estimated the maintenance of porcelain insulators can be up to 25 times more than that of glass insulators.

The utility is working to make its specifications for porcelain insulators more stringent than dictated by present ANSI standards, so that only good-quality insulators can be selected.

PACIFIC GAS AND ELECTRIC CO., CALIFORNIA

Pacific Gas and Electric Co. (PG&E) operates its extra-high-voltage lines at either 230 kV or 500 kV. The primary insulator type used is ceramic or glass. The exceptions are in vandalism-prone locations and areas with high insulator wash cycles, where composite insulators are used. Composite insulators are also used at lower voltages. However, fairly recently, corona cutting and cracks have been found on some 115-kV composite insulators installed without corona rings, which was the normal practice.

PG&E has reduced the use of composite insulators somewhat at all voltages in the last five years. In addition to aging-related issues, the utility has experienced damage by birds, specifically crows.

The utility has approved two offshore suppliers of porcelain insulators and expects several more vying for acceptance. While PG&E does not differentiate between porcelain and glass in the specification, design and installation, it is seeing an increase in the use of toughened glass insulators at all voltages in the 69-kV to 500-kV range. The utility attributes this to better education of the work force and performance characteristics associated with glass insulators.

The utility performs an aerial helicopter inspection annually, wherein insulators with visible damage are noted. A detailed ground inspection is done every five years. Climbing inspections are performed only if triggered by a specific condition.

XCEL ENERGY, MINNESOTA

Xcel Energy recently updated its standard designs by voltage. All technologies—porcelain, toughened glass and polymer—may be used for voltages below 69 kV. For 69 kV to 345 kV, polymers are used for suspension, braced and unbraced line post applications. For deadend application in this range and higher voltages, only toughened glass insulators are used. This change was driven by problems encountered with porcelain and early generation polymers.

For example, several porcelain suspension and deadend insulators on 115-kV and 345-kV lines in critical locations failed mechanically, attributable to cement growth. The age of these insulators was in excess of 20 years. As most of the porcelain insulators in the system are of this vintage or older, the utility has instituted a rigorous maintenance procedure where lines are examined regularly by fixed wing, helicopter and foot patrols. Those identified for detailed inspection are worked by linemen from buckets using the buzz technique. Needless to say, this is a very expensive undertaking and adds to the life-cycle cost of porcelain insulators.

Xcel has also experienced failures (brittle fracture) with early generation composites, primarily on 115-kV and 345-kV installations, and is concerned about longevity in 345-kV and higher applications. The utility evaluated life-cycle costs with the three insulator technologies before proceeding with revisions to its design philosophy.

HYDRO ONE, ONTARIO

Hydro One has excellent experience with all three insulator technologies for lines up to 230 kV. For higher voltages, it uses porcelain and glass, and does not use polymer insulators because issues with live-line working, bird damage, corona and aging have not been fully resolved. Porcelain and glass insulators on Hydro One's system in many places are 60 years or older for some porcelain. The porcelain insulators are tested on a regular basis for punctures and cracks attributed to cement expansion, which have caused primarily mechanical failures of several strings across the high-voltage system.

The utility has success using thermovision equipment, and punctured bells show a temperature difference of up to 10 °C (18 °F) under damp conditions. In the last two years, Hydro One has examined more than 3000 porcelain strings at 230-kV and 500-kV lines. With a five-man crew, the utility can inspect five towers per day. Indeed, this is a time- and labor-intensive, not to mention expensive, undertaking.

Owing to the ease of visually detecting damaged units on toughened glass insulator strings, Hydro

One will be using such insulators on its new construction of 230-kV and 500-kV lines.

NORTHWESTERN ENERGY, MONTANA

NorthWestern Energy has been using toughened glass insulators on its 500-kV lines since the 1980s. It has had very good experience with them and will continue this practice on its new construction of a 430-mile (692-km) 500-kV line being built for the Mountain State Transmission Intertie project. The utility performs much of its maintenance under live conditions; it calls itself a live-line-friendly utility. Since most of North Western's lines are in remote locations, routine inspections by helicopter occur four times a year on the 500-kV lines and once per year for all other lines. More detailed inspections are done on a five- to 10-year cycle. The utility has experienced problems due to vandalism in some pockets, but since the damaged glass insulators are easy to spot, it finds that glass is advantageous over other options.

NorthWestern has had good experience with porcelain at 230-kV and lower voltage lines. It inspects these insulators under de-energized conditions. Owing to the relatively dry climate in Montana, the utility has many thousands of porcelain insulators well in excess of 60 years old. Composite insulators are the preferred choice for lines of 115 kV and below. At 161 kV and 230 kV, composites are used on a limited basis for project-specific needs.

Porcelain is still the preferred choice for the bulk transmission lines. NorthWestern has experienced problems with many of the early vintage composite insulators due to corona cutting and moisture ingress. One severe example of this was a 161-kV line built in the early 1990s with composite horizontal line post insulators. The line has only been operated at 69 kV since construction, yet moisture ingress failures, believed to occur during manufacturing, have occurred on the 161-kV insulators, forcing NorthWestern to replace them recently.

OVERALL PERSPECTIVE

It seems a shift is occurring in the use of the various insulator technologies for high-voltage lines in North America. Users pointed out that, for distribution (less than 69 kV), polymers are favored, because they are lightweight, easy to handle and low cost; however, several utilities are limiting the use of polymers at higher voltages. Polymers seem to be established as the technological choice for compact line applications (line posts and braced posts). Maintenance concerns associated with the management of aging porcelain insulators and associated inspection costs are driving some utilities to question the use of porcelain insulators, while life-cycle cost considerations and ease of inspection associated with toughened glass insulators are steering other utilities toward this latter technology.

Clearly, all three insulation technologies are still very much alive, and decisions made with regard to insulation systems for the refurbishment of older lines and the upcoming surge of new high-voltage projects will depend on past experience and the expected performance and life-cycle cost criteria utilities set for the operation of their systems.

Ravi Gorur (ravi.gorur@asu.edu) is a professor in the school of electrical, computer and energy engineering at Arizona State University, Tempe. He has authored a textbook and more than 150 publications on the subject of outdoor insulators. He is the U.S. representative to CIGRÉ Study Committee DI (Materials and Emerging Technologies) and is actively involved in various IEEE working groups and task forces related to insulators. Gorur is a fellow of the IEEE.

The purpose of this article is to provide a current review of the trends in insulator technologies through interviews with several utilities, all familiar with and having experience in the three technologies. The utilities selected for soliciting input cover a wide range of geographic and climatic conditions from the U.S. West Coast to the East Coast, including one major Canadian utility. The author gratefully acknowledges input from the following:

- J. Hunt, Salt River Project
- G. Giordanella, Public Service Electric and Gas
- D.H. Shaffner, Pacific Gas and Electric
- D. Berklund, Xcel Energy
- H. Crockett, Hydro One
- T. Pankratz, North Western Energy.

4.1

TRANSMISSION LINE DESIGN CONSIDERATIONS

An overhead transmission line consists of conductors, insulators, support structures, and, in most cases, shield wires.

CONDUCTORS

Aluminum has replaced copper as the most common conductor metal for overhead transmission. Although a larger aluminum cross-sectional area is required to obtain the same loss as in a copper conductor, aluminum has a lower cost and lighter weight. Also, the supply of aluminum is abundant, whereas that of copper is limited.

One of the most common conductor types is aluminum conductor, steel-reinforced (ACSR), which consists of layers of aluminum strands surrounding a central core of steel strands (Figure 4.1). Stranded conductors are easier to manufacture, since larger conductor sizes can be obtained by simply adding successive layers of strands. Stranded conductors are also easier to handle and more flexible than solid conductors, especially in larger sizes. The use of steel strands gives ACSR conductors a high strength-to-weight ratio. For purposes of heat dissipation, overhead transmission-line conductors are bare (no insulating cover).

Other conductor types include the all-aluminum conductor (AAC), all-aluminum-alloy conductor (AAAC), aluminum conductor alloy-reinforced (ACAR), and aluminum-clad steel conductor (Alumoweld). Higher-temperature conductors capable of operation in excess of 150 °C include the aluminum conductor steel supported (ACSS), which uses fully annealed aluminum around a steel core, and the gap-type ZT-aluminum conductor (GTZACSR) which uses heat-resistant aluminum over a steel core with a small annular gap between the steel and first layer of aluminum strands. Emerging technologies use composite materials, including the aluminum conductor carbon reinforced (ACFR), whose core is a resinmatrix composite containing carbon fiber, and the aluminum conductor composite reinforced (ACCR), whose core is an aluminum-matrix containing aluminum fibers [10].

FIGURE 4.1 Typical ACSR conductor

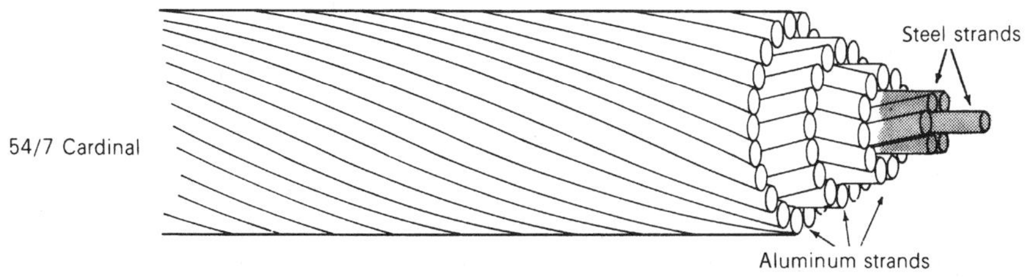

FIGURE 4.2 A 765-kV transmission line with self-supporting lattice steel towers (Courtesy of the American Electric Power Company)

FIGURE 4.3 A 345-kV double-circuit transmission line with self-supporting lattice steel towers (Courtesy of NSTAR, formerly Boston Edison Company)

EHV lines often have more than one conductor per phase; these conductors are called a *bundle*. The 765-kV line in Figure 4.2 has four conductors per phase, and the 345-kV double-circuit line in Figure 4.3 has two conductors per phase. Bundle conductors have a lower electric field strength at the conductor surfaces, thereby controlling corona. They also have a smaller series reactance.

INSULATORS

Insulators for transmission lines above 69 kV are typically suspension-type insulators, which consist of a string of discs constructed porcelain, toughened glass, or polymer. The standard disc (Figure 4.4) has a 0.254-m (10-in.) diameter, 0.146-m ($5\frac{3}{4}$-in.) spacing between centers of adjacent discs, and a mechanical strength of 7500 kg. The 765-kV line in Figure 4.2 has two strings

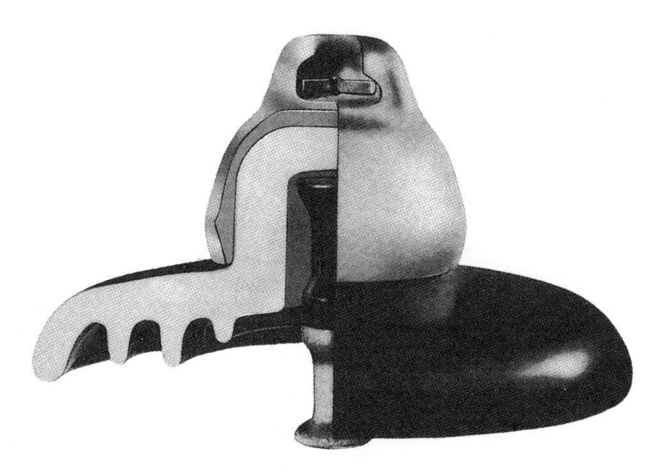

FIGURE 4.4

Cut-away view of a standard porcelain insulator disc for suspension insulator strings (Courtesy of Ohio Brass)

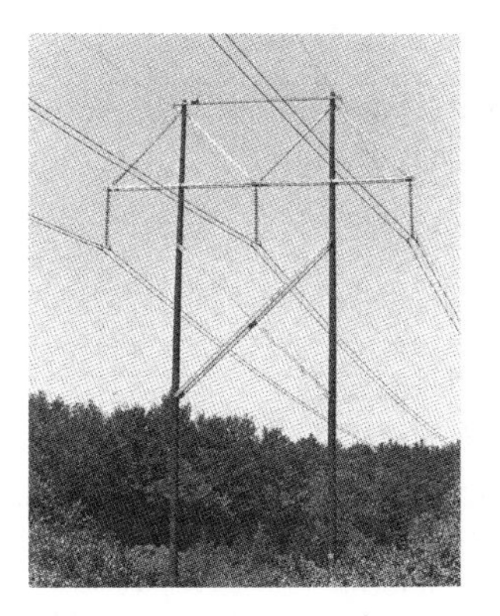

FIGURE 4.5

Wood frame structure for a 345-kV line (Courtesy of NSTAR, formerly Boston Edison Company)

per phase in a V-shaped arrangement, which helps to restrain conductor swings. The 345-kV line in Figure 4.5 has one vertical string per phase. The number of insulator discs in a string increases with line voltage (Table 4.1). Other types of discs include larger units with higher mechanical strength and fog insulators for use in contaminated areas.

SUPPORT STRUCTURES

Transmission lines employ a variety of support structures. Figure 4.2 shows a self-supporting, lattice steel tower typically used for 500- and 765-kV lines. Double-circuit 345-kV lines usually have self-supporting steel towers with the phases arranged either in a triangular configuration to reduce tower height or in a vertical configuration to reduce tower width (Figure 4.3). Wood frame configurations are commonly used for voltages of 345 kV and below (Figure 4.5).

TABLE 4.1

Typical transmission-line characteristics [1, 2] (Electric Power Research Institute (EPRI), EPRI AC Transmission Line Reference Book—200 kV and Above (Palo Alto, CA: EPRI, www.epri.com, December 2005); Westinghouse Electric Corporation, Electrical Transmission and Distribution Reference Book, 4th ed. (East Pittsburgh, PA, 1964))

Nominal Voltage	Phase Conductors				
				Minimum Clearances	
(kV)	Number of Conductors per Bundle	Aluminum Cross-Section Area per Conductor (ACSR) (kcmil)*	Bundle Spacing (cm)	Phase-to-Phase (m)	Phase-to-Ground (m)
69	1	—	—	—	—
138	1	300–700	—	4 to 5	—
230	1	400–1000	—	6 to 9	—
345	1	2000–2500	—	6 to 9	7.6 to 11
345	2	800–2200	45.7	6 to 9	7.6 to 11
500	2	2000–2500	45.7	9 to 11	9 to 14
500	3	900–1500	45.7	9 to 11	9 to 14
765	4	900–1300	45.7	13.7	12.2

*1 kcmil = 0.5 mm^2

Nominal Voltage	Suspension Insulator String		Shield Wires		
(kV)	Number of Strings per Phase	Number of Standard Insulator Discs per Suspension String	Type	Number	Diameter (cm)
69	1	4 to 6	Steel	0, 1 or 2	—
138	1	8 to 11	Steel	0, 1 or 2	—
230	1	12 to 21	Steel or ACSR	1 or 2	1.1 to 1.5
345	1	18 to 21	Alumoweld	2	0.87 to 1.5
345	1 and 2	18 to 21	Alumoweld	2	0.87 to 1.5
500	2 and 4	24 to 27	Alumoweld	2	0.98 to 1.5
500	2 and 4	24 to 27	Alumoweld	2	0.98 to 1.5
765	2 and 4	30 to 35	Alumoweld	2	0.98

SHIELD WIRES

Shield wires located above the phase conductors protect the phase conductors against lightning. They are usually high- or extra-high-strength steel, Alumoweld, or ACSR with much smaller cross section than the phase conductors. The number and location of the shield wires are selected so that almost all lightning strokes terminate on the shield wires rather than on the phase conductors. Figures 4.2, 4.3, and 4.5 have two shield wires. Shield wires are grounded to the tower. As such, when lightning strikes a shield wire, it flows harmlessly to ground, provided the tower impedance and tower footing resistance are small.

The decision to build new transmission is based on power-system planning studies to meet future system requirements of load growth and new generation. The points of interconnection of each new line to the system, as well as the power and voltage ratings of each, are selected based on these studies. Thereafter, transmission-line design is based on optimization of electrical, mechanical, environmental, and economic factors.

ELECTRICAL FACTORS

Electrical design dictates the type, size, and number of bundle conductors per phase. Phase conductors are selected to have sufficient thermal capacity to meet continuous, emergency overload, and short-circuit current ratings. For EHV lines, the number of bundle conductors per phase is selected to control the voltage gradient at conductor surfaces, thereby reducing or eliminating corona.

Electrical design also dictates the number of insulator discs, vertical or V-shaped string arrangement, phase-to-phase clearance, and phase-to-tower clearance, all selected to provide adequate line insulation. Line insulation must withstand transient overvoltages due to lightning and switching surges, even when insulators are contaminated by fog, salt, or industrial pollution. Reduced clearances due to conductor swings during winds must also be accounted for.

The number, type, and location of shield wires are selected to intercept lightning strokes that would otherwise hit the phase conductors. Also, tower footing resistance can be reduced by using driven ground rods or a buried conductor (called *counterpoise*) running parallel to the line. Line height is selected to satisfy prescribed conductor-to-ground clearances and to control ground-level electric field and its potential shock hazard.

Conductor spacings, types, and sizes also determine the series impedance and shunt admittance. Series impedance affects line-voltage drops, I^2R losses, and stability limits (Chapters 5, 13). Shunt admittance, primarily capacitive, affects line-charging currents, which inject reactive power into the power system. Shunt reactors (inductors) are often installed on lightly loaded EHV lines to absorb part of this reactive power, thereby reducing overvoltages.

MECHANICAL FACTORS

Mechanical design focuses on the strength of the conductors, insulator strings, and support structures. Conductors must be strong enough to support a specified thickness of ice and a specified wind in addition to their own weight. Suspension insulator strings must be strong enough to support the phase conductors with ice and wind loadings from tower to tower (span length). Towers that satisfy minimum strength requirements, called suspension towers, are designed to support the phase conductors and shield wires with ice and wind loadings, and, in some cases, the unbalanced pull due to breakage of one or two conductors. Dead-end towers located every mile or so satisfy the maximum strength requirement of breakage of all conductors on one side of the tower. Angles in the line employ angle towers with intermediate strength. Conductor vibrations, which can cause conductor fatigue failure and damage to towers, are also of concern. Vibrations are controlled by adjustment of conductor tensions, use of vibration dampers, and—for bundle conductors—large bundle spacing and frequent use of bundle spacers.

ENVIRONMENTAL FACTORS

Environmental factors include land usage and visual impact. When a line route is selected, the effect on local communities and population centers, land values, access to property, wildlife, and use of public parks and facilities must all be considered. Reduction in visual impact is obtained by aesthetic tower design and by blending the line with the countryside. Also, the biological effects of prolonged exposure to electric and magnetic fields near transmission lines is of concern. Extensive research has been and continues to be done in this area.

ECONOMIC FACTORS

The optimum line design meets all the technical design criteria at lowest overall cost, which includes the total installed cost of the line as well as the cost of line losses over the operating life of the line. Many design factors affect cost. Utilities and consulting organizations use digital computer programs combined with specialized knowledge and physical experience to achieve optimum line design.

4.2 RESISTANCE

The dc resistance of a conductor at a specified temperature T is

$$R_{dc,T} = \frac{\rho_T l}{A} \quad \Omega \tag{4.2.1}$$

where ρ_T = conductor resistivity at temperature T

l = conductor length

A = conductor cross-sectional area

Two sets of units commonly used for calculating resistance, SI and English units, are summarized in Table 4.2. In this text we will use SI units throughout except where manufacturers' data is in English units. To interpret American manufacturers' data, it is useful to learn the use of English units in resistance calculations. In English units, conductor cross-sectional area is

TABLE 4.2

Comparison of SI and English units for calculating conductor resistance

Quantity	Symbol	SI Units	English Units
Resistivity	ρ	Ωm	Ω-cmil/ft
Length	ℓ	m	ft
Cross-sectional area	A	m^2	cmil
dc resistance	$R_{dc} = \frac{\rho \ell}{A}$	Ω	Ω

expressed in circular mils (cmil). One inch (2.54 cm) equals 1000 mils and 1 cmil equals $\pi/4$ sq mil. A circle with diameter D inches, or (D in.) (1000 mil/in.) = 1000 D mil = d mil, has an area

$$A = \left(\frac{\pi}{4} D^2 \text{ in.}^2\right)\left(1000 \frac{\text{mil}}{\text{in.}}\right)^2 = \frac{\pi}{4}(1000 \text{ D})^2 = \frac{\pi}{4} d^2 \quad \text{sq mil}$$

or

$$A = \left(\frac{\pi}{4} d^2 \text{ sq mil}\right)\left(\frac{1 \text{ cmil}}{\pi/4 \text{ sq mil}}\right) = d^2 \quad \text{cmil} \tag{4.2.2}$$

1000 cmil or 1 kcmil is equal to 0.506 mm^2, often approximated to 0.5 mm^2.

Resistivity depends on the conductor metal. Annealed copper is the international standard for measuring resistivity ρ (or conductivity σ, where $\sigma = 1/\rho$). Resistivity of conductor metals is listed in Table 4.3. As shown, hard-drawn aluminum, which has 61% of the conductivity of the international standard, has a resistivity at 20 °C of 2.83×10^{-8} Ωm.

Conductor resistance depends on the following factors:

1. Spiraling
2. Temperature
3. Frequency ("skin effect")
4. Current magnitude—magnetic conductors

These are described in the following paragraphs.

TABLE 4.3

% Conductivity, resistivity, and temperature constant of conductor metals

Material	% Conductivity	$\rho_{20\,^\circ C}$ Resistivity at 20 °C $\Omega m \times 10^{-8}$	T Temperature Constant °C
Copper:			
Annealed	100%	1.72	234.5
Hard-drawn	97.3%	1.77	241.5
Aluminum			
Hard-drawn	61%	2.83	228.1
Brass	20–27%	6.4–8.4	480
Iron	17.2%	10	180
Silver	108%	1.59	243
Sodium	40%	4.3	207
Steel	2–14%	12–88	180–980

For stranded conductors, alternate layers of strands are spiraled in opposite directions to hold the strands together. Spiraling makes the strands 1 or 2% longer than the actual conductor length. As a result, the dc resistance of a stranded conductor is 1 or 2% larger than that calculated from (4.2.1) for a specified conductor length.

Resistivity of conductor metals varies linearly over normal operating temperatures according to

$$\rho_{T2} = \rho_{T1}\left(\frac{T_2 + T}{T_1 + T}\right) \tag{4.2.3}$$

where ρ_{T2} and ρ_{T1} are resistivities at temperatures T_2 and T_1 °C, respectively. T is a temperature constant that depends on the conductor material, and is listed in Table 4.3.

The ac resistance or *effective* resistance of a conductor is

$$R_{ac} = \frac{P_{loss}}{|I|^2} \quad \Omega \tag{4.2.4}$$

where P_{loss} is the conductor real power loss in watts and I is the rms conductor current. For dc, the current distribution is uniform throughout the conductor cross section, and (4.2.1) is valid. However, for ac, the current distribution is nonuniform. As frequency increases, the current in a solid cylindrical conductor tends to crowd toward the conductor surface, with smaller current density at the conductor center. This phenomenon is called *skin effect*. A conductor with a large radius can even have an oscillatory current density versus the radial distance from the conductor center.

With increasing frequency, conductor loss increases, which, from (4.2.4), causes the ac resistance to increase. At power frequencies (60 Hz), the ac resistance is at most a few percent higher than the dc resistance. Conductor manufacturers normally provide dc, 50-Hz, and 60-Hz conductor resistance based on test data (see Appendix Tables A.3 and A.4).

For magnetic conductors, such as steel conductors used for shield wires, resistance depends on current magnitude. The internal flux linkages, and therefore the iron or magnetic losses, depend on the current magnitude. For ACSR conductors, the steel core has a relatively high resistivity compared to the aluminum strands, and therefore the effect of current magnitude on ACSR conductor resistance is small. Tables on magnetic conductors list resistance at two current levels (see Table A.4).

EXAMPLE 4.1 Stranded conductor: dc and ac resistance

Table A.3 lists a 4/0 copper conductor with 12 strands. Strand diameter is 0.3373 cm (0.1328 in.). For this conductor:

a. Verify the total copper cross-sectional area of 107.2 mm^2 (211,600 cmil in the table).

b. Verify the dc resistance at 50 °C of 0.1876 Ω/km or 0.302 Ω/mi. Assume a 2% increase in resistance due to spiraling.

c. From Table A.3, determine the percent increase in resistance at 60 Hz versus dc.

SOLUTION

a. The strand diameter is $d = (0.3373\ \text{cm})\ (10\ \text{mm/cm}) = 3.373\ \text{mm}$, and, from (4.2.2), the strand area is

$$A = \frac{12\pi d^2}{4} = 3\pi(3.373)^2 = 107.2 \quad \text{mm}^2$$

which agrees with the value given in Table A.3.

b. Using (4.2.3) and hard-drawn copper data from Table 4.3,

$$\rho_{50\,^\circ\text{C}} = 1.77 \times 10^{-8}\left(\frac{50 + 241.5}{20 + 241.5}\right) = 1.973 \times 10^{-8} \quad \Omega\text{-m}$$

From (4.2.1), the dc resistance at 50 °C for a conductor length of 1 km is

$$R_{\text{dc},\,50\,^\circ\text{C}} = \frac{(1.973 \times 10^{-8})(10^3 \times 1.02)}{107.2 \times 10^{-6}} = 0.1877 \quad \Omega/\text{km}$$

which agrees with the value listed in Table A.3.

c. From Table A.3,

$$\frac{R_{60\ \text{Hz},\,50\,^\circ\text{C}}}{R_{\text{dc},\,50\,^\circ\text{C}}} = \frac{0.1883}{0.1877} = 1.003 \qquad \frac{R_{60\ \text{Hz},\,25\,^\circ\text{C}}}{R_{\text{dc},\,25\,^\circ\text{C}}} = \frac{0.1727}{0.1715} = 1.007$$

Thus, the 60-Hz resistance of this conductor is about 0.3–0.7% higher than the dc resistance. The variation of these two ratios is due to the fact that resistance in Table A.3 is given to only three significant figures. ■

4.3 CONDUCTANCE

Conductance accounts for real power loss between conductors or between conductors and ground. For overhead lines, this power loss is due to leakage currents at insulators and to corona. Insulator leakage current depends on the amount of dirt, salt, and other contaminants that have accumulated on insulators, as well as on meteorological factors, particularly the presence of moisture. Corona occurs when a high value of electric field strength at a conductor surface causes the air to become electrically ionized and to conduct. The real power loss due to corona, called *corona loss*, depends on meteorological conditions, particularly rain, and on conductor surface irregularities. Losses due to insulator leakage and corona are usually small compared to conductor I^2R loss. Conductance is usually neglected in power system studies because it is a very small component of the shunt admittance.

4.4

INDUCTANCE: SOLID CYLINDRICAL CONDUCTOR

The inductance of a magnetic circuit that has a constant permeability μ can be obtained by determining the following:

1. Magnetic field intensity H, from Ampere's law
2. Magnetic flux density B ($B = \mu H$)
3. Flux linkages λ
4. Inductance from flux linkages per ampere ($L = \lambda/I$)

As a step toward computing the inductances of more general conductors and conductor configurations, we first compute the internal, external, and total inductance of a solid cylindrical conductor. We also compute the flux linking one conductor in an array of current-carrying conductors.

Figure 4.6 shows a 1-meter section of a solid cylindrical conductor with radius r, carrying current I. For simplicity, assume that the conductor (1) is sufficiently long that end effects are neglected, (2) is nonmagnetic ($\mu = \mu_0 = 4\pi \times 10^{-7}$ H/m), and (3) has a uniform current density (skin effect is neglected). From (3.1.1), Ampere's law states that

$$\oint H_{\tan}\, dl = I_{\text{enclosed}} \tag{4.4.1}$$

To determine the magnetic field inside the conductor, select the dashed circle of radius $x < r$ shown in Figure 4.6 as the closed contour for Ampere's law. Due to symmetry, H_x is constant along the contour. Also, there is no radial component of H_x, so H_x is tangent to the contour. That is, the conductor has a concentric magnetic field. From (4.4.1), the integral of H_x around the selected contour is

$$H_x(2\pi x) = I_x \qquad \text{for } x < r \tag{4.4.2}$$

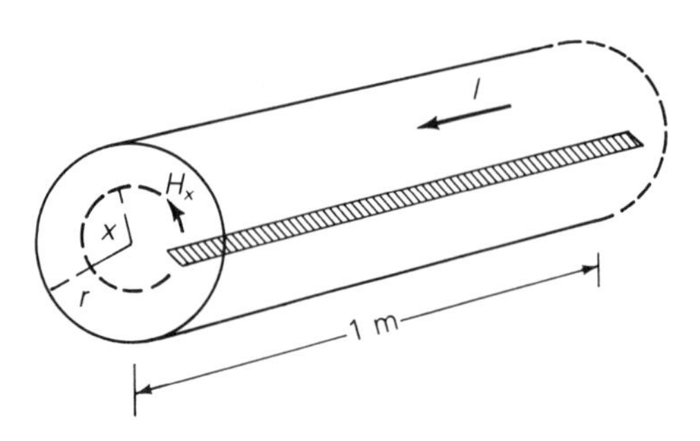

FIGURE 4.6 Internal magnetic field of a solid cylindrical conductor

where I_x is the portion of the total current enclosed by the contour. Solving (4.4.2)

$$H_x = \frac{I_x}{2\pi x} \quad \text{A/m} \tag{4.4.3}$$

Now assume a uniform current distribution within the conductor, that is

$$I_x = \left(\frac{x}{r}\right)^2 I \qquad \text{for } x < r \tag{4.4.4}$$

Using (4.4.4) in (4.4.3)

$$H_x = \frac{xI}{2\pi r^2} \quad \text{A/m} \tag{4.4.5}$$

For a nonmagnetic conductor, the magnetic flux density B_x is

$$B_x = \mu_0 H_x = \frac{\mu_0 x I}{2\pi r^2} \quad \text{Wb/m}^2 \tag{4.4.6}$$

The differential flux $d\Phi$ per-unit length of conductor in the cross-hatched rectangle of width dx shown in Figure 4.6 is

$$d\Phi = B_x \, dx \quad \text{Wb/m} \tag{4.4.7}$$

Computation of the differential flux linkage $d\lambda$ in the rectangle is tricky since only the fraction $(x/r)^2$ of the total current I is linked by the flux. That is,

$$d\lambda = \left(\frac{x}{r}\right)^2 d\Phi = \frac{\mu_0 I}{2\pi r^4} x^3 \, dx \quad \text{Wb-t/m} \tag{4.4.8}$$

Integrating (4.4.8) from $x = 0$ to $x = r$ determines the total flux linkages λ_{int} inside the conductor

$$\lambda_{\text{int}} = \int_0^r d\lambda = \frac{\mu_0 I}{2\pi r^4} \int_0^r x^3 \, dx = \frac{\mu_0 I}{8\pi} = \frac{1}{2} \times 10^{-7} I \quad \text{Wb-t/m} \tag{4.4.9}$$

The internal inductance L_{int} per-unit length of conductor due to this flux linkage is then

$$L_{\text{int}} = \frac{\lambda_{\text{int}}}{I} = \frac{\mu_0}{8\pi} = \frac{1}{2} \times 10^{-7} \quad \text{H/m} \tag{4.4.10}$$

Next, in order to determine the magnetic field outside the conductor, select the dashed circle of radius $x > r$ shown in Figure 4.7 as the closed contour for Ampere's law. Noting that this contour encloses the entire current I, integration of (4.4.1) yields

$$H_x(2\pi x) = I \tag{4.4.11}$$

which gives

$$H_x = \frac{I}{2\pi x} \quad \text{A/m} \qquad x > r \tag{4.4.12}$$

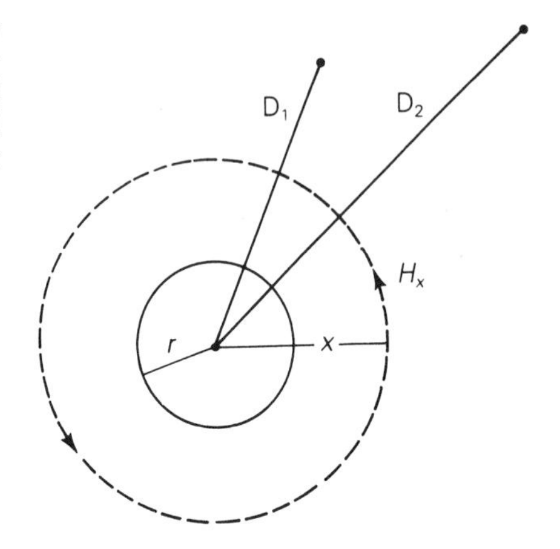

FIGURE 4.7

External magnetic field of a solid cylindrical conductor

Outside the conductor, $\mu = \mu_0$ and

$$B_x = \mu_0 H_x = (4\pi \times 10^{-7})\frac{I}{2\pi x} = 2 \times 10^{-7}\frac{I}{x} \quad \text{Wb/m}^2 \tag{4.4.13}$$

$$d\Phi = B_x\,dx = 2 \times 10^{-7}\frac{I}{x}\,dx \quad \text{Wb/m} \tag{4.4.14}$$

Since the entire current I is linked by the flux outside the conductor,

$$d\lambda = d\Phi = 2 \times 10^{-7}\frac{I}{x}\,dx \quad \text{Wb-t/m} \tag{4.4.15}$$

Integrating (4.4.15) between two external points at distances D_1 and D_2 from the conductor center gives the external flux linkage λ_{12} between D_1 and D_2:

$$\lambda_{12} = \int_{D_1}^{D_2} d\lambda = 2 \times 10^{-7} I \int_{D_1}^{D_2} \frac{dx}{x}$$

$$= 2 \times 10^{-7} I \ln\left(\frac{D_2}{D_1}\right) \quad \text{Wb-t/m} \tag{4.4.16}$$

The external inductance L_{12} per-unit length due to the flux linkages between D_1 and D_2 is then

$$L_{12} = \frac{\lambda_{12}}{I} = 2 \times 10^{-7} \ln\left(\frac{D_2}{D_1}\right) \quad \text{H/m} \tag{4.4.17}$$

The total flux λ_P linking the conductor out to external point P at distance D is the sum of the internal flux linkage, (4.4.9), and the external flux linkage, (4.4.16) from $D_1 = r$ to $D_2 = D$. That is

$$\lambda_P = \frac{1}{2} \times 10^{-7} I + 2 \times 10^{-7} I \ln\frac{D}{r} \tag{4.4.18}$$

Using the identity $\frac{1}{2} = 2 \ln e^{1/4}$ in (4.4.18), a more convenient expression for λ_P is obtained:

$$\lambda_P = 2 \times 10^{-7} I \left(\ln e^{1/4} + \ln \frac{D}{r} \right)$$

$$= 2 \times 10^{-7} I \ln \frac{D}{e^{-1/4} r}$$

$$= 2 \times 10^{-7} I \ln \frac{D}{r'} \quad \text{Wb-t/m} \tag{4.4.19}$$

where

$$r' = e^{-1/4} r = 0.7788 r \tag{4.4.20}$$

Also, the total inductance L_P due to both internal and external flux linkages out to distance D is

$$L_P = \frac{\lambda_P}{I} = 2 \times 10^{-7} \ln \left(\frac{D}{r'} \right) \quad \text{H/m} \tag{4.4.21}$$

Finally, consider the array of M solid cylindrical conductors shown in Figure 4.8. Assume that each conductor m carries current I_m referenced out of the page. Also assume that the sum of the conductor currents is zero—that is,

$$I_1 + I_2 + \cdots + I_M = \sum_{m=1}^{M} I_m = 0 \tag{4.4.22}$$

The flux linkage λ_{kPk}, which links conductor k out to point P due to current I_k, is, from (4.4.19),

$$\lambda_{kPk} = 2 \times 10^{-7} I_k \ln \frac{D_{Pk}}{r'_k} \tag{4.4.23}$$

FIGURE 4.8

Array of M solid cylindrical conductors

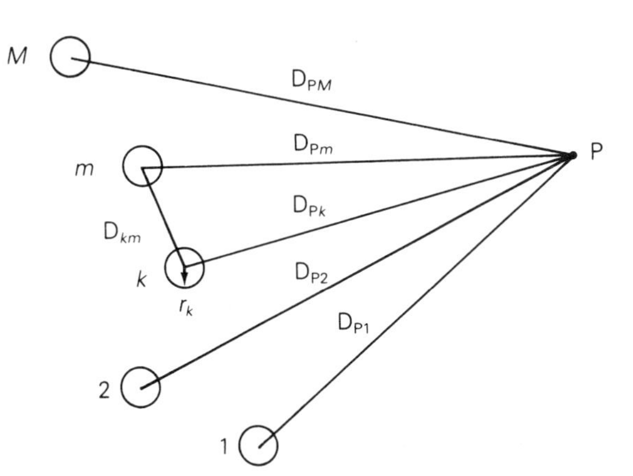

Note that λ_{kPk} includes both internal and external flux linkages due to I_k. The flux linkage λ_{kPm}, which links conductor k out to P due to I_m, is, from (4.4.16),

$$\lambda_{kPm} = 2 \times 10^{-7} I_m \ln \frac{D_{Pm}}{D_{km}} \tag{4.4.24}$$

In (4.4.24) we use D_{km} instead of $(D_{km} - r_k)$ or $(D_{km} + r_k)$, which is a valid approximation when D_{km} is much greater than r_k. It can also be shown that this is a good approximation even when D_{km} is small. Using superposition, the total flux linkage λ_{kP}, which links conductor k out to P due to all the currents, is

$$\begin{aligned} \lambda_{kP} &= \lambda_{kP1} + \lambda_{kP2} + \cdots + \lambda_{kPM} \\ &= 2 \times 10^{-7} \sum_{m=1}^{M} I_m \ln \frac{D_{Pm}}{D_{km}} \end{aligned} \tag{4.4.25}$$

where we define $D_{kk} = r'_k = e^{-1/4} r_k$ when $m = k$ in the above summation. Equation (4.4.25) is separated into two summations:

$$\lambda_{kP} = 2 \times 10^{-7} \sum_{m=1}^{M} I_m \ln \frac{1}{D_{km}} + 2 \times 10^{-7} \sum_{m=1}^{M} I_m \ln D_{Pm} \tag{4.4.26}$$

Removing the last term from the second summation we get:

$$\lambda_{kP} = 2 \times 10^{-7} \left[\sum_{m=1}^{M} I_m \ln \frac{1}{D_{km}} + \sum_{m=1}^{M-1} I_m \ln D_{Pm} + I_M \ln D_{PM} \right] \tag{4.4.27}$$

From (4.4.22),

$$I_M = -(I_1 + I_2 + \cdots + I_{M-1}) = -\sum_{m=1}^{M-1} I_m \tag{4.4.28}$$

Using (4.4.28) in (4.4.27)

$$\begin{aligned} \lambda_{kP} &= 2 \times 10^{-7} \left[\sum_{m=1}^{M} I_m \ln \frac{1}{D_{km}} + \sum_{m=1}^{M-1} I_m \ln D_{Pm} - \sum_{m=1}^{M-1} I_m \ln D_{PM} \right] \\ &= 2 \times 10^{-7} \left[\sum_{m=1}^{M} I_m \ln \frac{1}{D_{km}} + \sum_{m=1}^{M-1} I_m \ln \frac{D_{Pm}}{D_{PM}} \right] \end{aligned} \tag{4.4.29}$$

Now, let λ_k equal the total flux linking conductor k out to infinity. That is, $\lambda_k = \lim_{P \to \infty} \lambda_{kP}$. As $P \to \infty$, all the distances D_{Pm} become equal, the ratios D_{Pm}/D_{PM} become unity, and $\ln(D_{Pm}/D_{PM}) \to 0$. Therefore, the second summation in (4.4.29) becomes zero as $P \to \infty$, and

$$\lambda_k = 2 \times 10^{-7} \sum_{m=1}^{M} I_m \ln \frac{1}{D_{km}} \quad \text{Wb-t/m} \tag{4.4.30}$$

Equation (4.4.30) gives the total flux linking conductor k in an array of M conductors carrying currents $I_1, I_2, \ldots, I_M$, whose sum is zero. This equation is valid for either dc or ac currents. λ_k is a dc flux linkage when the currents are dc, and λ_k is a phasor flux linkage when the currents are phasor representations of sinusoids.

4.5 INDUCTANCE: SINGLE-PHASE TWO-WIRE LINE AND THREE-PHASE THREE-WIRE LINE WITH EQUAL PHASE SPACING

The results of the previous section are used here to determine the inductances of two relatively simple transmission lines: a single-phase two-wire line and a three-phase three-wire line with equal phase spacing.

Figure 4.9(a) shows a single-phase two-wire line consisting of two solid cylindrical conductors x and y. Conductor x with radius r_x carries phasor current $I_x = I$ referenced out of the page. Conductor y with radius r_y carries return current $I_y = -I$. Since the sum of the two currents is zero, (4.4.30) is valid, from which the total flux linking conductor x is

$$\begin{aligned}\lambda_x &= 2 \times 10^{-7}\left(I_x \ln \frac{1}{\mathrm{D}_{xx}} + I_y \ln \frac{1}{\mathrm{D}_{xy}}\right) \\ &= 2 \times 10^{-7}\left(I \ln \frac{1}{r'_x} - I \ln \frac{1}{\mathrm{D}}\right) \\ &= 2 \times 10^{-7} I \ln \frac{\mathrm{D}}{r'_x} \quad \text{Wb-t/m} \qquad (4.5.1)\end{aligned}$$

where $r'_x = e^{-1/4} r_x = 0.7788 r_x$.

The inductance of conductor x is then

$$\mathrm{L}_x = \frac{\lambda_x}{I_x} = \frac{\lambda_x}{I} = 2 \times 10^{-7} \ln \frac{\mathrm{D}}{r'_x} \quad \text{H/m per conductor} \qquad (4.5.2)$$

FIGURE 4.9 Single-phase two-wire line

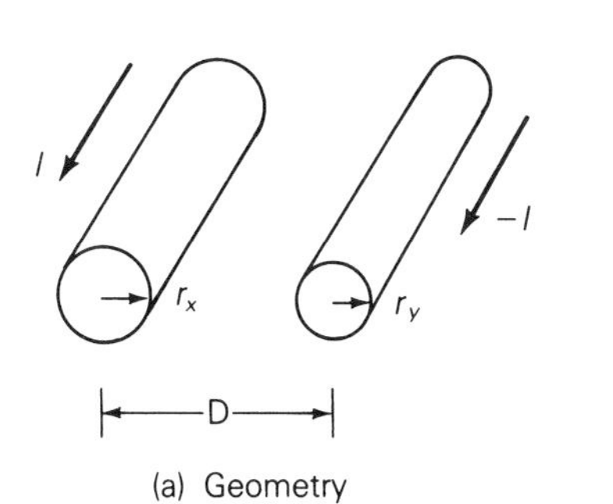

(a) Geometry

(b) Inductances

Similarly, the total flux linking conductor y is

$$\lambda_y = 2 \times 10^{-7}\left(I_x \ln \frac{1}{D_{yx}} + I_y \ln \frac{1}{D_{yy}}\right)$$

$$= 2 \times 10^{-7}\left(I \ln \frac{1}{D} - I \ln \frac{1}{r'_y}\right)$$

$$= -2 \times 10^{-7} I \ln \frac{D}{r'_y} \tag{4.5.3}$$

and

$$L_y = \frac{\lambda_y}{I_y} = \frac{\lambda_y}{-I} = 2 \times 10^{-7} \ln \frac{D}{r'_y} \quad \text{H/m per conductor} \tag{4.5.4}$$

The total inductance of the single-phase circuit, also called *loop inductance*, is

$$L = L_x + L_y = 2 \times 10^{-7}\left(\ln \frac{D}{r'_x} + \ln \frac{D}{r'_y}\right)$$

$$= 2 \times 10^{-7} \ln \frac{D^2}{r'_x r'_y}$$

$$= 4 \times 10^{-7} \ln \frac{D}{\sqrt{r'_x r'_y}} \quad \text{H/m per circuit} \tag{4.5.5}$$

Also, if $r'_x = r'_y = r'$, the total circuit inductance is

$$L = 4 \times 10^{-7} \ln \frac{D}{r'} \quad \text{H/m per circuit} \tag{4.5.6}$$

The inductances of the single-phase two-wire line are shown in Figure 4.9(b).

Figure 4.10(a) shows a three-phase three-wire line consisting of three solid cylindrical conductors a, b, c, each with radius r, and with equal phase spacing D between any two conductors. To determine inductance, assume balanced positive-sequence currents I_a, I_b, I_c that satisfy $I_a + I_b + I_c = 0$. Then (4.4.30) is valid and the total flux linking the phase a conductor is

FIGURE 4.10 Three-phase three-wire line with equal phase spacing

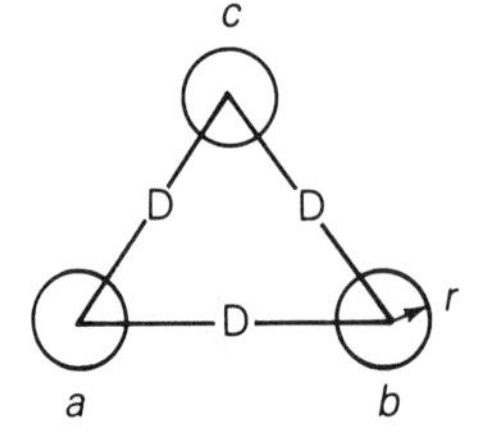

(a) Geometry

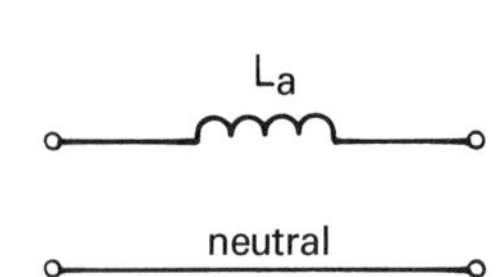

(b) Phase inductance

$$\lambda_a = 2 \times 10^{-7}\left(I_a \ln \frac{1}{r'} + I_b \ln \frac{1}{D} + I_c \ln \frac{1}{D}\right)$$

$$= 2 \times 10^{-7}\left[I_a \ln \frac{1}{r'} + (I_b + I_c) \ln \frac{1}{D}\right] \tag{4.5.7}$$

Using $(I_b + I_c) = -I_a$,

$$\lambda_a = 2 \times 10^{-7}\left(I_a \ln \frac{1}{r'} - I_a \ln \frac{1}{D}\right)$$

$$= 2 \times 10^{-7} I_a \ln \frac{D}{r'} \quad \text{Wb-t/m} \tag{4.5.8}$$

The inductance of phase a is then

$$L_a = \frac{\lambda_a}{I_a} = 2 \times 10^{-7} \ln \frac{D}{r'} \quad \text{H/m per phase} \tag{4.5.9}$$

Due to symmetry, the same result is obtained for $L_b = \lambda_b/I_b$ and for $L_c = \lambda_c/I_c$. However, only one phase need be considered for balanced three-phase operation of this line, since the flux linkages of each phase have equal magnitudes and 120° displacement. The phase inductance is shown in Figure 4.10(b).

4.6 INDUCTANCE: COMPOSITE CONDUCTORS, UNEQUAL PHASE SPACING, BUNDLED CONDUCTORS

The results of Section 4.5 are extended here to include composite conductors, which consist of two or more solid cylindrical subconductors in parallel. A stranded conductor is one example of a composite conductor. For simplicity we assume that for each conductor, the subconductors are identical and share the conductor current equally.

Figure 4.11 shows a single-phase two-conductor line consisting of two composite conductors x and y. Conductor x has N identical subconductors, each with radius r_x and with current (I/N) referenced out of the page. Similarly, conductor y consists of M identical subconductors, each with radius r_y and with return current $(-I/M)$. Since the sum of all the currents is zero, (4.4.30) is valid and the total flux Φ_k linking subconductor k of conductor x is

$$\Phi_k = 2 \times 10^{-7}\left[\frac{I}{N}\sum_{m=1}^{N} \ln \frac{1}{D_{km}} - \frac{I}{M}\sum_{m=1'}^{M} \ln \frac{1}{D_{km}}\right] \tag{4.6.1}$$

Since only the fraction $(1/N)$ of the total conductor current I is linked by this flux, the flux linkage λ_k of (the current in) subconductor k is

FIGURE 4.11

Single-phase two-conductor line with composite conductors

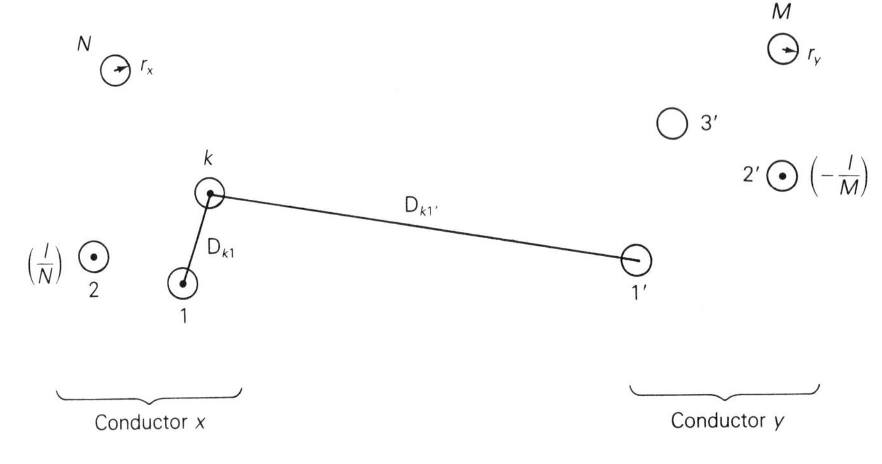

$$\lambda_k = \frac{\Phi_k}{N} = 2 \times 10^{-7} I \left[\frac{1}{N^2} \sum_{m=1}^{N} \ln \frac{1}{D_{km}} - \frac{1}{NM} \sum_{m=1'}^{M} \ln \frac{1}{D_{km}} \right] \tag{4.6.2}$$

The total flux linkage of conductor x is

$$\lambda_x = \sum_{k=1}^{N} \lambda_k = 2 \times 10^{-7} I \sum_{k=1}^{N} \left[\frac{1}{N^2} \sum_{m=1}^{N} \ln \frac{1}{D_{km}} - \frac{1}{NM} \sum_{m=1'}^{M} \ln \frac{1}{D_{km}} \right] \tag{4.6.3}$$

Using $\ln A^{\alpha} = \alpha \ln A$ and $\sum \ln A_k = \ln \prod A_k$ (sum of $\ln$ s $= \ln$ of products), (4.6.3) can be rewritten in the following form:

$$\lambda_x = 2 \times 10^{-7} I \ln \prod_{k=1}^{N} \frac{\left(\prod_{m=1'}^{M} D_{km} \right)^{1/NM}}{\left(\prod_{m=1}^{N} D_{km} \right)^{1/N^2}} \tag{4.6.4}$$

and the inductance of conductor x, $L_x = \frac{\lambda_x}{I}$, can be written as

$$L_x = 2 \times 10^{-7} \ln \frac{D_{xy}}{D_{xx}} \quad \text{H/m per conductor} \tag{4.6.5}$$

where

$$D_{xy} = \sqrt[MN]{\prod_{k=1}^{N} \prod_{m=1'}^{M} D_{km}} \tag{4.6.6}$$

$$D_{xx} = \sqrt[N^2]{\prod_{k=1}^{N} \prod_{m=1}^{N} D_{km}} \tag{4.6.7}$$

D_{xy}, given by (4.6.6), is the MNth root of the product of the MN distances from the subconductors of conductor x to the subconductors of conductor y. Associated with each subconductor k of conductor x are the M distances $D_{k1'}, D_{k2'}, \ldots, D_{kM}$ to the subconductors of conductor y. For N subconductors in conductor x, there are therefore MN of these distances. D_{xy} is called the *geometric mean distance* or GMD between conductors x and y.

Also, D_{xx}, given by (4.6.7), is the N^2 root of the product of the N^2 distances between the subconductors of conductor x. Associated with each subconductor k are the N distances $D_{k1}, D_{k2}, \ldots, D_{kk} = r', \ldots, D_{kN}$. For N subconductors in conductor x, there are therefore N^2 of these distances. D_{xx} is called the *geometric mean radius* or GMR of conductor x.

Similarly, for conductor y,

$$L_y = 2 \times 10^{-7} \ln \frac{D_{xy}}{D_{yy}} \quad \text{H/m per conductor} \tag{4.6.8}$$

where

$$D_{yy} = \sqrt[M^2]{\prod_{k=1'}^{M} \prod_{m=1'}^{M} D_{km}} \tag{4.6.9}$$

D_{yy}, the GMR of conductor y, is the M^2 root of the product of the M^2 distances between the subconductors of conductor y. The total inductance L of the single-phase circuit is

$$L = L_x + L_y \quad \text{H/m per circuit} \tag{4.6.10}$$

EXAMPLE 4.2 GMR, GMD, and inductance: single-phase two-conductor line

Expand (4.6.6), (4.6.7), and (4.6.9) for $N = 3$ and $M = 2'$. Then evaluate L_x, L_y, and L in H/m for the single-phase two-conductor line shown in Figure 4.12.

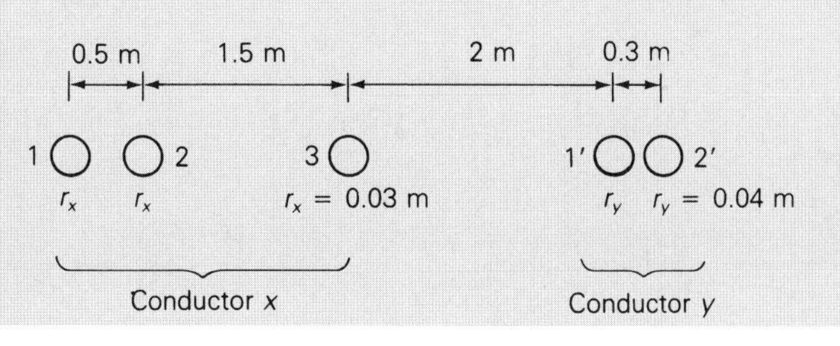

FIGURE 4.12 Single-phase two-conductor line for Example 4.2

SOLUTION For $N = 3$ and $M = 2'$, (4.6.6) becomes

$$\mathrm{D}_{xy} = \sqrt[6]{\prod_{k=1}^{3} \prod_{m=1'}^{2'} \mathrm{D}_{km}}$$

$$= \sqrt[6]{\prod_{k=1}^{3} \mathrm{D}_{k1'}\mathrm{D}_{k2'}}$$

$$= \sqrt[6]{(\mathrm{D}_{11'}\mathrm{D}_{12'})(\mathrm{D}_{21'}\mathrm{D}_{22'})(\mathrm{D}_{31'}\mathrm{D}_{32'})}$$

Similarly, (4.6.7) becomes

$$\mathrm{D}_{xx} = \sqrt[9]{\prod_{k=1}^{3} \prod_{m=1}^{3} \mathrm{D}_{km}}$$

$$= \sqrt[9]{\prod_{k=1}^{3} \mathrm{D}_{k1}\mathrm{D}_{k2}\mathrm{D}_{k3}}$$

$$= \sqrt[9]{(\mathrm{D}_{11}\mathrm{D}_{12}\mathrm{D}_{13})(\mathrm{D}_{21}\mathrm{D}_{22}\mathrm{D}_{23})(\mathrm{D}_{31}\mathrm{D}_{32}\mathrm{D}_{33})}$$

and (4.6.9) becomes

$$\mathrm{D}_{yy} = \sqrt[4]{\prod_{k=1'}^{2'} \prod_{m=1'}^{2'} \mathrm{D}_{km}}$$

$$= \sqrt[4]{\prod_{k=1'}^{2'} \mathrm{D}_{k1'}\mathrm{D}_{k2'}}$$

$$= \sqrt[4]{(\mathrm{D}_{1'1'}\mathrm{D}_{1'2'})(\mathrm{D}_{2'1'}\mathrm{D}_{2'2'})}$$

Evaluating D_{xy}, D_{xx}, and D_{yy} for the single-phase two-conductor line shown in Figure 4.12,

$$\mathrm{D}_{11'} = 4 \text{ m} \qquad \mathrm{D}_{12'} = 4.3 \text{ m} \qquad \mathrm{D}_{21'} = 3.5 \text{ m}$$

$$\mathrm{D}_{22'} = 3.8 \text{ m} \qquad \mathrm{D}_{31'} = 2 \text{ m} \qquad \mathrm{D}_{32'} = 2.3 \text{ m}$$

$$\mathrm{D}_{xy} = \sqrt[6]{(4)(4.3)(3.5)(3.8)(2)(2.3)} = 3.189 \text{ m}$$

$$\mathrm{D}_{11} = \mathrm{D}_{22} = \mathrm{D}_{33} = r'_x = e^{-1/4} r_x = (0.7788)(0.03) = 0.02336 \text{ m}$$

$$\mathrm{D}_{21} = \mathrm{D}_{12} = 0.5 \text{ m}$$

$$\mathrm{D}_{23} = \mathrm{D}_{32} = 1.5 \text{ m}$$

$$\mathrm{D}_{31} = \mathrm{D}_{13} = 2.0 \text{ m}$$

$$D_{xx} = \sqrt[9]{(0.02336)^3(0.5)^2(1.5)^2(2.0)^2} = 0.3128 \text{ m}$$

$$D_{1'1'} = D_{2'2'} = r'_y = e^{-1/4}r_y = (0.7788)(0.04) = 0.03115 \text{ m}$$

$$D_{1'2'} = D_{2'1'} = 0.3 \text{ m}$$

$$D_{yy} = \sqrt[4]{(0.03115)^2(0.3)^2} = 0.09667 \text{ m}$$

Then, from (4.6.5), (4.6.8), and (4.6.10):

$$L_x = 2 \times 10^{-7} \ln\left(\frac{3.189}{0.3128}\right) = 4.644 \times 10^{-7} \quad \text{H/m per conductor}$$

$$L_y = 2 \times 10^{-7} \ln\left(\frac{3.189}{0.09667}\right) = 6.992 \times 10^{-7} \quad \text{H/m per conductor}$$

$$L = L_x + L_y = 1.164 \times 10^{-6} \quad \text{H/m per circuit}$$

■

It is seldom necessary to calculate GMR or GMD for standard lines. The GMR of standard conductors is provided by conductor manufacturers and can be found in various handbooks (see Appendix Tables A.3 and A.4). Also, if the distances between conductors are large compared to the distances between subconductors of each conductor, then the GMD between conductors is approximately equal to the distance between conductor centers.

EXAMPLE 4.3 Inductance and inductive reactance: single-phase line

A single-phase line operating at 60 Hz consists of two 4/0 12-strand copper conductors with 1.5 m spacing between conductor centers. The line length is 32 km. Determine the total inductance in H and the total inductive reactance in Ω.

SOLUTION The GMD between conductor centers is $D_{xy} = 1.5$ m. Also, from Table A.3, the GMR of a 4/0 12-strand copper conductor is $D_{xx} = D_{yy} =$ 0.01750 ft or 0.5334 cm. From (4.6.5) and (4.6.8),

$$L_x = L_y = 2 \times 10^{-7} \ln\left(\frac{150}{0.5334}\right) \frac{\text{H}}{\text{m}} \times 32 \times 10^3$$

$$= 0.03609 \quad \text{H per conductor}$$

The total inductance is

$$L = L_x + L_y = 2 \times 0.03609 = 0.07218 \quad \text{H per circuit}$$

and the total inductive reactance is

$$X_L = 2\pi f L = (2\pi)(60)(0.07218) = 27.21 \quad \Omega \text{ per circuit}$$

■

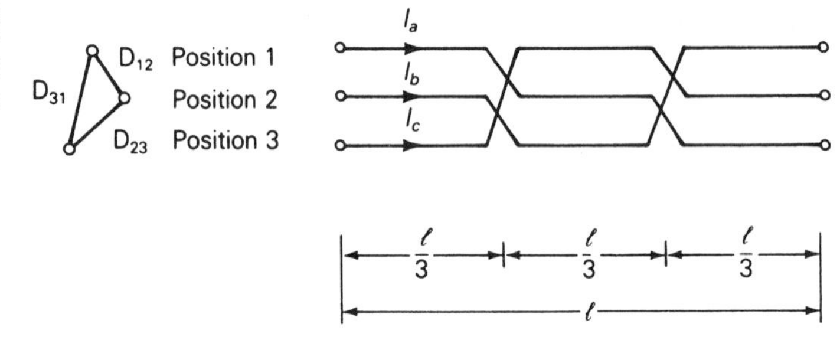

FIGURE 4.13 Completely transposed three-phase line

To calculate inductance for three-phase lines with stranded conductors and equal phase spacing, r' is replaced by the conductor GMR in (4.5.9). If the spacings between phases are unequal, then balanced positive-sequence flux linkages are not obtained from balanced positive-sequence currents. Instead, unbalanced flux linkages occur, and the phase inductances are unequal. However, balance can be restored by exchanging the conductor positions along the line, a technique called *transposition*.

Figure 4.13 shows a completely transposed three-phase line. The line is transposed at two locations such that each phase occupies each position for one-third of the line length. Conductor positions are denoted 1, 2, 3 with distances D_{12}, D_{23}, D_{31} between positions. The conductors are identical, each with GMR denoted D_S. To calculate inductance of this line, assume balanced positive-sequence currents I_a, I_b, I_c, for which $I_a + I_b + I_c = 0$. Again, (4.4.30) is valid, and the total flux linking the phase a conductor while it is in position 1 is

$$\lambda_{a1} = 2 \times 10^{-7}\left[I_a \ln\frac{1}{D_S} + I_b \ln\frac{1}{D_{12}} + I_c \ln\frac{1}{D_{31}}\right] \quad \text{Wb-t/m} \tag{4.6.11}$$

Similarly, the total flux linkage of this conductor while it is in positions 2 and 3 is

$$\lambda_{a2} = 2 \times 10^{-7}\left[I_a \ln\frac{1}{D_S} + I_b \ln\frac{1}{D_{23}} + I_c \ln\frac{1}{D_{12}}\right] \quad \text{Wb-t/m} \tag{4.6.12}$$

$$\lambda_{a3} = 2 \times 10^{-7}\left[I_a \ln\frac{1}{D_S} + I_b \ln\frac{1}{D_{31}} + I_c \ln\frac{1}{D_{23}}\right] \quad \text{Wb-t/m} \tag{4.6.13}$$

The average of the above flux linkages is

$$\lambda_a = \frac{\lambda_{a1}\left(\frac{l}{3}\right) + \lambda_{a2}\left(\frac{l}{3}\right) + \lambda_{a3}\left(\frac{l}{3}\right)}{l} = \frac{\lambda_{a1} + \lambda_{a2} + \lambda_{a3}}{3}$$

$$= \frac{2 \times 10^{-7}}{3}\left[3I_a \ln\frac{1}{D_S} + I_b \ln\frac{1}{D_{12}D_{23}D_{31}} + I_c \ln\frac{1}{D_{12}D_{23}D_{31}}\right] \tag{4.6.14}$$

Using $(I_b + I_c) = -I_a$ in (4.6.14),

$$\lambda_a = \frac{2 \times 10^{-7}}{3}\left[3I_a \ln \frac{1}{D_S} - I_a \ln \frac{1}{D_{12}D_{23}D_{31}}\right]$$

$$= 2 \times 10^{-7} I_a \ln \frac{\sqrt[3]{D_{12}D_{23}D_{31}}}{D_S} \quad \text{Wb-t/m} \tag{4.6.15}$$

and the average inductance of phase a is

$$L_a = \frac{\lambda_a}{I_a} = 2 \times 10^{-7} \ln \frac{\sqrt[3]{D_{12}D_{23}D_{31}}}{D_S} \quad \text{H/m per phase} \tag{4.6.16}$$

The same result is obtained for $L_b = \lambda_b/I_b$ and for $L_c = \lambda_c/I_c$. However, only one phase need be considered for balanced three-phase operation of a completely transposed three-phase line. Defining

$$D_{eq} = \sqrt[3]{D_{12}D_{23}D_{31}} \tag{4.6.17}$$

we have

$$L_a = 2 \times 10^{-7} \ln \frac{D_{eq}}{D_S} \quad \text{H/m} \tag{4.6.18}$$

D_{eq}, the cube root of the product of the three-phase spacings, is the geometric mean distance between phases. Also, D_S is the conductor GMR for stranded conductors, or r' for solid cylindrical conductors.

EXAMPLE 4.4 Inductance and inductive reactance: three-phase line

A completely transposed 60-Hz three-phase line has flat horizontal phase spacing with 10 m between adjacent conductors. The conductors are 806 mm^2 (1,590,000 cmil) ACSR with 54/3 stranding. Line length is 200 km. Determine the inductance in H and the inductive reactance in Ω.

SOLUTION From Table A.4, the GMR of a 806 mm^2 (1,590,000 cmil) 54/3 ACSR conductor is

$$D_S = 0.0520 \text{ ft } \frac{1 \text{ m}}{3.28 \text{ ft}} = 0.0159 \text{ m}$$

Also, from (4.6.17) and (4.6.18),

$$D_{eq} = \sqrt[3]{(10)(10)(20)} = 12.6 \text{ m}$$

$$L_a = 2 \times 10^{-7} \ln\left(\frac{12.6}{0.0159}\right) \frac{\text{H}}{\text{m}} \times \frac{1000 \text{ m}}{\text{km}} \times 200 \text{ km}$$

$$= 0.267 \quad \text{H}$$

The inductive reactance of phase a is

$$X_a = 2\pi f L_a = 2\pi(60)(0.267) = 101 \quad \Omega$$

■

FIGURE 4.14 Bundle conductor configurations

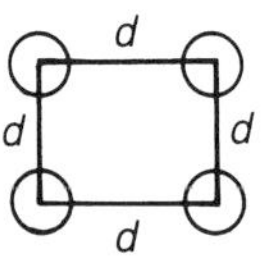

It is common practice for EHV lines to use more than one conductor per phase, a practice called *bundling*. Bundling reduces the electric field strength at the conductor surfaces, which in turn reduces or eliminates corona and its results: undesirable power loss, communications interference, and audible noise. Bundling also reduces the series reactance of the line by increasing the GMR of the bundle.

Figure 4.14 shows common EHV bundles consisting of two, three, or four conductors. The three-conductor bundle has its conductors on the vertices of an equilateral triangle, and the four-conductor bundle has its conductors on the corners of a square. To calculate inductance, D_S in (4.6.18) is replaced by the GMR of the bundle. Since the bundle constitutes a composite conductor, calculation of bundle GMR is, in general, given by (4.6.7). If the conductors are stranded and the bundle spacing d is large compared to the conductor outside radius, each stranded conductor is first replaced by an equivalent solid cylindrical conductor with GMR $= D_S$. Then the bundle is replaced by one equivalent conductor with GMR $= D_{SL}$, given by (4.6.7) with $n = 2$, 3, or 4 as follows:

Two-conductor bundle:

$$D_{SL} = \sqrt[4]{(D_S \times d)^2} = \sqrt{D_S d} \tag{4.6.19}$$

Three-conductor bundle:

$$D_{SL} = \sqrt[9]{(D_S \times d \times d)^3} = \sqrt[3]{D_S d^2} \tag{4.6.20}$$

Four-conductor bundle:

$$D_{SL} = \sqrt[16]{(D_S \times d \times d \times d\sqrt{2})^4} = 1.091 \sqrt[4]{D_S d^3} \tag{4.6.21}$$

The inductance is then

$$L_a = 2 \times 10^{-7} \ln \frac{D_{eq}}{D_{SL}} \quad \text{H/m} \tag{4.6.22}$$

If the phase spacings are large compared to the bundle spacing, then sufficient accuracy for D_{eq} is obtained by using the distances between bundle centers.

EXAMPLE 4.5 Inductive reactance: three-phase line with bundled conductors

Each of the 806 mm^2 conductors in Example 4.4 is replaced by two 403 mm^2 ACSR 26/2 conductors, as shown in Figure 4.15. Bundle spacing is 0.40 m.

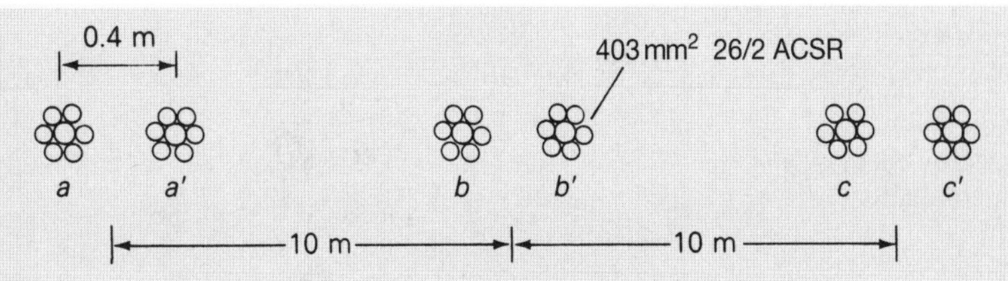

FIGURE 4.15 Three-phase bundled conductor line for Example 4.5

Flat horizontal spacing is retained, with 10 m between adjacent bundle centers. Calculate the inductive reactance of the line and compare it with that of Example 4.4.

SOLUTION From Table A.4, the GMR of a 403 mm^2 (795,000 cmil) 26/2 ACSR conductor is

$$D_S = 0.0375 \text{ ft} \times \frac{1 \text{ m}}{3.28 \text{ ft}} = 0.0114 \text{ m}$$

From (4.6.19), the two-conductor bundle GMR is

$$D_{SL} = \sqrt{(0.0114)(0.40)} = 0.0676 \text{ m}$$

Since $D_{eq} = 12.6$ m is the same as in Example 4.4,

$$L_a = 2 \times 10^{-7} \ln\left(\frac{12.6}{0.0676}\right)(1000)(200) = 0.209 \text{ H}$$

$$X_a = 2\pi f L_1 = (2\pi)(60)(0.209) = 78.8 \ \Omega$$

The reactance of the bundled line, 78.8 Ω, is 22% less than that of Example 4.4, even though the two-conductor bundle has the same amount of conductor material (that is, the same cmil per phase). One advantage of reduced series line reactance is smaller line-voltage drops. Also, the loadability of medium and long EHV lines is increased (see Chapter 5). ■

4.7 SERIES IMPEDANCES: THREE-PHASE LINE WITH NEUTRAL CONDUCTORS AND EARTH RETURN

In this section, we develop equations suitable for computer calculation of the series impedances, including resistances and inductive reactances, for the three-phase overhead line shown in Figure 4.16. This line has three phase conductors a, b, and c, where bundled conductors, if any, have already been replaced by equivalent conductors, as described in Section 4.6. The line also has N neutral conductors denoted $n1, n2, \ldots, nN$.* All the neutral conductors

* Instead of *shield wire* we use the term *neutral conductor*, which applies to distribution as well as transmission lines.

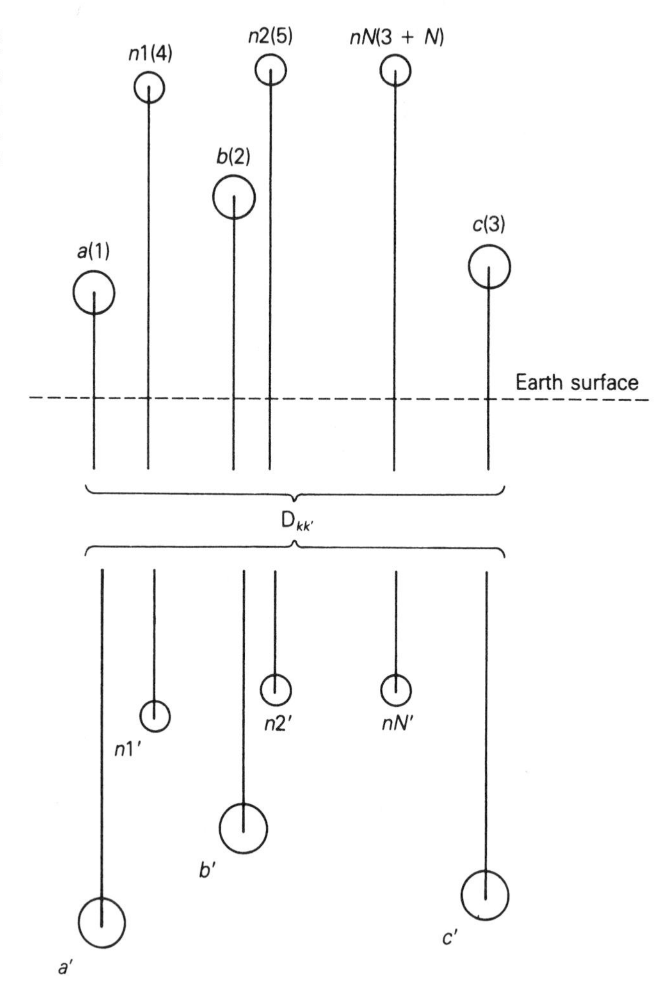

FIGURE 4.16 Three-phase transmission line with earth replaced by earth return conductors

are connected in parallel and are grounded to the earth at regular intervals along the line. Any isolated neutral conductors that carry no current are omitted. The phase conductors are insulated from each other and from earth.

If the phase currents are not balanced, there may be a return current in the grounded neutral wires and in the earth. The earth return current will spread out under the line, seeking the lowest impedance return path. A classic paper by Carson [4], later modified by others [5, 6], shows that the earth can be replaced by a set of "earth return" conductors located directly under the overhead conductors, as shown in Figure 4.16. Each earth return conductor carries the negative of its overhead conductor current, has a GMR denoted $\text{D}_{k'k'}$, distance $D_{kk'}$ from its overhead conductor, and resistance $\text{R}_{k'}$ given by:

TABLE 4.4

Earth resistivities and 60-Hz equivalent conductor distances

Type of Earth	Resistivity (Ωm)	$D_{kk'}$ (m)
Sea water	0.01–1.0	8.50–85.0
Swampy ground	10–100	269–850
Average damp earth	100	850
Dry earth	1000	2690
Pure slate	10^7	269,000
Sandstone	10^9	2,690,000

$$D_{k'k'} = D_{kk} \quad \text{m} \tag{4.7.1}$$

$$D_{kk'} = 658.5\sqrt{\rho/f} \quad \text{m} \tag{4.7.2}$$

$$R_{k'} = 9.869 \times 10^{-7} f \quad \Omega/\text{m} \tag{4.7.3}$$

where ρ is the earth resistivity in ohm-meters and f is frequency in hertz. Table 4.4 lists earth resistivities and 60-Hz equivalent conductor distances for various types of earth. It is common practice to select $\rho = 100\ \Omega$m when actual data are unavailable.

Note that the GMR of each earth return conductor, $D_{k'k'}$, is the same as the GMR of its corresponding overhead conductor, D_{kk}. Also, all the earth return conductors have the same distance $D_{kk'}$ from their overhead conductors and the same resistance $R_{k'}$.

For simplicity, we renumber the overhead conductors from 1 to $(3+N)$, beginning with the phase conductors, then overhead neutral conductors, as shown in Figure 4.16. Operating as a transmission line, the sum of the currents in all the conductors is zero. That is,

$$\sum_{k=1}^{(6+2N)} I_k = 0 \tag{4.7.4}$$

Equation (4.4.30) is therefore valid, and the flux linking overhead conductor k is

$$\lambda_k = 2 \times 10^{-7} \sum_{m=1}^{(3+N)} I_m \ln \frac{D_{km'}}{D_{km}} \quad \text{Wb-t/m} \tag{4.7.5}$$

In matrix format, (4.7.5) becomes

$$\boldsymbol{\lambda} = \mathbf{L}\boldsymbol{I} \tag{4.7.6}$$

where

$\boldsymbol{\lambda}$ is a $(3+N)$ vector

$\boldsymbol{I}$ is a $(3+N)$ vector

$\mathbf{L}$ is a $(3+N) \times (3+N)$ matrix whose elements are:

$$L_{km} = 2 \times 10^{-7} \ln \frac{D_{km'}}{D_{km}} \tag{4.7.7}$$

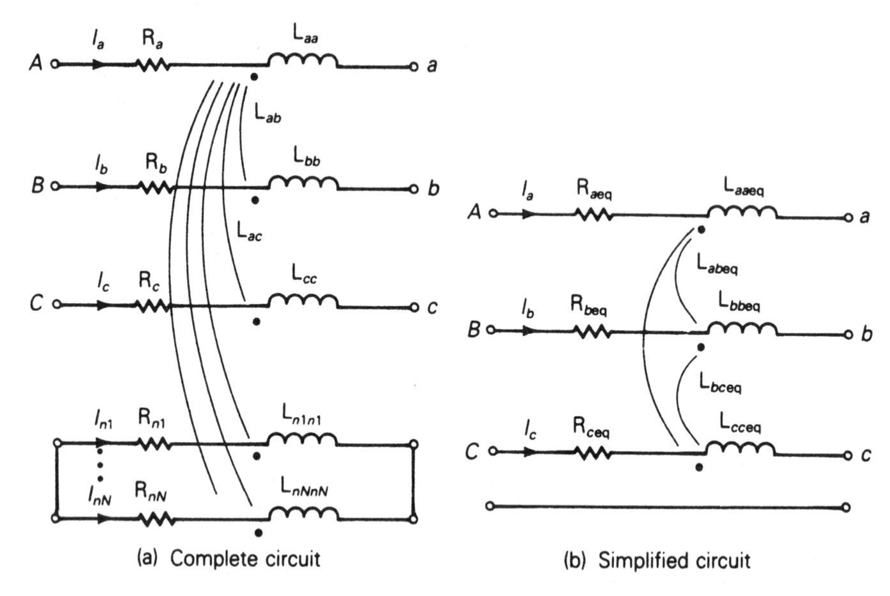

FIGURE 4.17 Circuit representation of series-phase impedances

When $k = m$, D_{kk} in (4.7.7) is the GMR of (bundled) conductor k. When $k \neq m$, D_{km} is the distance between conductors k and m.

A circuit representation of a 1-meter section of the line is shown in Figure 4.17(a). Using this circuit, the vector of voltage drops across the conductors is:

$$\begin{bmatrix} E_{Aa} \\ E_{Bb} \\ E_{Cc} \\ 0 \\ 0 \\ \vdots \\ 0 \end{bmatrix} = (\mathbf{R} + j\omega\mathbf{L}) \begin{bmatrix} I_a \\ I_b \\ I_c \\ I_{n1} \\ \vdots \\ I_{nN} \end{bmatrix} \tag{4.7.8}$$

where $\mathbf{L}$ is given by (4.7.7) and $\mathbf{R}$ is a $(3+N) \times (3+N)$ matrix of conductor resistances.

$$\mathbf{R} = \begin{bmatrix} (R_a + R_{k'})R_{k'} \cdots & & & & R_{k'} \\ R_{k'}(R_b + R_{k'})R_{k'} \cdots & & & & \vdots \\ & (R_c + R_{k'})R_{k'} \cdots & & & \\ & & (R_{n1} + R_{k'})R_{k'} \cdots & & \\ & & \ddots & & \\ R_{k'} & & & (R_{nN} + R_{k'}) & \end{bmatrix} \Omega/\text{m} \tag{4.7.9}$$

The resistance matrix of (4.7.9) includes the resistance R_k of each overhead conductor and a mutual resistance $R_{k'}$ due to the image conductors. R_k of each overhead conductor is obtained from conductor tables such as Appendix Table A.3 or A.4, for a specified frequency, temperature, and current. $R_{k'}$ of all the image conductors is the same, as given by (4.7.3).

Our objective now is to reduce the $(3 + N)$ equations in (4.7.8) to three equations, thereby obtaining the simplified circuit representations shown in Figure 4.17(b). We partition (4.7.8) as follows:

$$\left[\begin{array}{c} E_{Aa} \\ E_{Bb} \\ E_{Cc} \\ \hline 0 \\ \cdots \\ 0 \end{array}\right] \left[\begin{array}{c|c} \overbrace{\begin{matrix} Z_{11} & Z_{12} & Z_{13} \\ Z_{21} & Z_{22} & Z_{23} \\ Z_{31} & Z_{32} & Z_{33} \end{matrix}}^{Z_A} & \overbrace{\begin{matrix} Z_{14} & \cdots & Z_{1(3+N)} \\ Z_{24} & \cdots & Z_{2(3+N)} \\ Z_{34} & \cdots & Z_{3(3+N)} \end{matrix}}^{Z_B} \\ \hline \underbrace{\begin{matrix} Z_{41} & Z_{42} & Z_{43} \\ & & \\ Z_{(3+N)1} & Z_{(3+N)2} & Z_{(3+N)3} \end{matrix}}_{Z_C} & \underbrace{\begin{matrix} Z_{44} & \cdots & Z_{4(3+N)} \\ & & \\ Z_{(3+N)4} & \cdots & Z_{(3+N)(3+N)} \end{matrix}}_{Z_D} \end{array}\right] \left[\begin{array}{c} I_a \\ I_b \\ I_c \\ \hline I_{n1} \\ I_{nN} \\ \vdots \end{array}\right] \tag{4.7.10}$$

The diagonal elements of this matrix are

$$Z_{kk} = R_k + R_{k'} + j\omega 2 \times 10 \ln \frac{D_{kk'}}{D_{kk}} \quad \Omega/\text{m} \tag{4.7.11}$$

And the off-diagonal elements, for $k \neq m$, are

$$Z_{km} = R_{k'} + j\omega 2 \times 10 \ln \frac{D_{km'}}{D_{km}} \quad \Omega/\text{m} \tag{4.7.12}$$

Next, (4.7.10) is partitioned as shown above to obtain

$$\left[\begin{array}{c} \boldsymbol{E}_P \\ \hline \boldsymbol{0} \end{array}\right] = \left[\begin{array}{c|c} \boldsymbol{Z}_A & \boldsymbol{Z}_B \\ \hline \boldsymbol{Z}_C & \boldsymbol{Z}_D \end{array}\right] \left[\begin{array}{c} \boldsymbol{I}_P \\ \hline \boldsymbol{I}_n \end{array}\right] \tag{4.7.13}$$

where

$$\boldsymbol{E}_P = \begin{bmatrix} E_{Aa} \\ E_{Bb} \\ E_{Cc} \end{bmatrix}; \quad \boldsymbol{I}_P = \begin{bmatrix} I_a \\ I_b \\ I_c \end{bmatrix}; \quad \boldsymbol{I}_n = \begin{bmatrix} I_{n1} \\ \vdots \\ I_{nN} \end{bmatrix}$$

$\boldsymbol{E}_P$ is the three-dimensional vector of voltage drops across the phase conductors (including the neutral voltage drop). $\boldsymbol{I}_P$ is the three-dimensional vector of phase currents and $\boldsymbol{I}_n$ is the N vector of neutral currents. Also, the $(3 + N) \times (3 + N)$ matrix in (4.7.10) is partitioned to obtain the following matrices:

$\boldsymbol{Z}_A$ with dimension 3×3

$\boldsymbol{Z}_B$ with dimension $3 \times N$

$\boldsymbol{Z}_C$ with dimension $N \times 3$

$\boldsymbol{Z}_D$ with dimension $N \times N$

Equation (4.7.13) is rewritten as two separate matrix equations:

$$\boldsymbol{E}_{\mathrm{P}} = \boldsymbol{Z}_A \boldsymbol{I}_{\mathrm{P}} + \boldsymbol{Z}_B \boldsymbol{I}_n \tag{4.7.14}$$

$$\boldsymbol{0} = \boldsymbol{Z}_C \boldsymbol{I}_{\mathrm{P}} + \boldsymbol{Z}_D \boldsymbol{I}_n \tag{4.7.15}$$

Solving (4.7.15) for $\boldsymbol{I}_n$,

$$\boldsymbol{I}_n = -\boldsymbol{Z}_D^{-1} \boldsymbol{Z}_C \boldsymbol{I}_{\mathrm{P}} \tag{4.7.16}$$

Using (4.7.16) in (4.7.14):

$$\boldsymbol{E}_{\mathrm{P}} = [\boldsymbol{Z}_A - \boldsymbol{Z}_B \boldsymbol{Z}_D^{-1} \boldsymbol{Z}_C] \boldsymbol{I}_{\mathrm{P}} \tag{4.7.17}$$

or

$$\boldsymbol{E}_{\mathrm{P}} = \boldsymbol{Z}_{\mathrm{P}} \boldsymbol{I}_{\mathrm{P}} \tag{4.7.18}$$

where

$$\boldsymbol{Z}_{\mathrm{P}} = \boldsymbol{Z}_A - \boldsymbol{Z}_B \boldsymbol{Z}_D^{-1} \boldsymbol{Z}_C \tag{4.7.19}$$

Equation (4.7.17), the desired result, relates the phase-conductor voltage drops (including neutral voltage drop) to the phase currents. $\boldsymbol{Z}_{\mathrm{P}}$ given by (4.7.19) is the 3×3 series-phase impedance matrix, whose elements are denoted

$$\boldsymbol{Z}_{\mathrm{P}} = \begin{bmatrix} Z_{aaeq} & Z_{abeq} & Z_{aceq} \\ Z_{abeq} & Z_{bbeq} & Z_{bceq} \\ Z_{aceq} & Z_{bceq} & Z_{cceq} \end{bmatrix} \quad \Omega/\mathrm{m} \tag{4.7.20}$$

If the line is completely transposed, the diagonal and off-diagonal elements are averaged to obtain

$$\hat{\boldsymbol{Z}}_{\mathrm{P}} = \begin{bmatrix} \hat{Z}_{aaeq} & \hat{Z}_{abeq} & \hat{Z}_{abeq} \\ \hat{Z}_{abeq} & \hat{Z}_{aaeq} & \hat{Z}_{abeq} \\ \hat{Z}_{abeq} & \hat{Z}_{abeq} & \hat{Z}_{aaeq} \end{bmatrix} \quad \Omega/\mathrm{m} \tag{4.7.21}$$

where

$$\hat{Z}_{aaeq} = \tfrac{1}{3}(Z_{aaeq} + Z_{bbeq} + Z_{cceq}) \tag{4.7.22}$$

$$\hat{Z}_{abeq} = \tfrac{1}{3}(Z_{abeq} + Z_{aceq} + Z_{bceq}) \tag{4.7.23}$$

4.8

ELECTRIC FIELD AND VOLTAGE: SOLID CYLINDRICAL CONDUCTOR

The capacitance between conductors in a medium with constant permittivity ε can be obtained by determining the following:

1. Electric field strength E, from Gauss's law
2. Voltage between conductors
3. Capacitance from charge per unit volt $(C = q/V)$

As a step toward computing capacitances of general conductor configurations, we first compute the electric field of a uniformly charged, solid cylindrical conductor and the voltage between two points outside the conductor. We also compute the voltage between two conductors in an array of charged conductors.

Gauss's law states that the total electric flux leaving a closed surface equals the total charge within the volume enclosed by the surface. That is, the normal component of electric flux density integrated over a closed surface equals the charge enclosed:

$$\oiint D_\perp \, ds = \oiint \varepsilon E_\perp \, ds = Q_{\text{enclosed}} \tag{4.8.1}$$

where $D_\perp$ denotes the normal component of electric flux density, $E_\perp$ denotes the normal component of electric field strength, and ds denotes the differential surface area. From Gauss's law, electric charge is a source of electric fields. Electric field lines originate from positive charges and terminate at negative charges.

Figure 4.18 shows a solid cylindrical conductor with radius r and with charge q coulombs per meter (assumed positive in the figure), uniformly distributed on the conductor surface. For simplicity, assume that the conductor is (1) sufficiently long that end effects are negligible, and (2) a perfect conductor (that is, zero resistivity, $\rho = 0$).

Inside the perfect conductor, Ohm's law gives $E_{\text{int}} = \rho J = 0$. That is, the internal electric field E_{int} is zero. To determine the electric field outside the conductor, select the cylinder with radius $x > r$ and with 1-meter length, shown in Figure 4.18, as the closed surface for Gauss's law. Due to the uniform charge distribution, the electric field strength E_x is constant on the cylinder. Also, there is no tangential component of E_x, so the electric field is radial to the conductor. Then, integration of (4.8.1) yields

$$\varepsilon E_x(2\pi x)(1) = q(1)$$

$$E_x = \frac{q}{2\pi\varepsilon x} \quad \text{V/m} \tag{4.8.2}$$

where, for a conductor in free space, $\varepsilon = \varepsilon_0 = 8.854 \times 10^{-12}$ F/m.

FIGURE 4.18

Perfectly conducting solid cylindrical conductor with uniform charge distribution

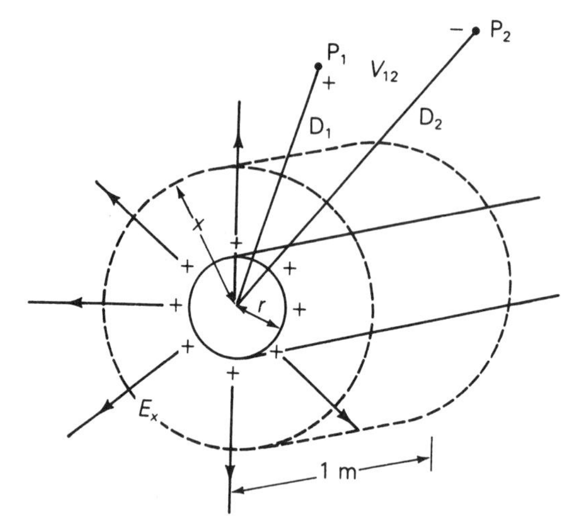

A plot of the electric field lines is also shown in Figure 4.18. The direction of the field lines, denoted by the arrows, is from the positive charges where the field originates, to the negative charges, which in this case are at infinity. If the charge on the conductor surface were negative, then the direction of the field lines would be reversed.

Concentric cylinders surrounding the conductor are constant potential surfaces. The potential difference between two concentric cylinders at distances D_1 and D_2 from the conductor center is

$$V_{12} = \int_{D_1}^{D_2} E_x \, dx \tag{4.8.3}$$

Using (4.8.2) in (4.8.1),

$$V_{12} = \int_{D_1}^{D_2} \frac{q}{2\pi\varepsilon x} \, dx = \frac{q}{2\pi\varepsilon} \ln \frac{D_2}{D_1} \quad \text{volts} \tag{4.8.4}$$

Equation (4.8.4) gives the voltage V_{12} between two points, P_1 and P_2, at distances D_1 and D_2 from the conductor center, as shown in Figure 4.18. Also, in accordance with our notation, V_{12} is the voltage at P_1 with respect to P_2. If q is positive and D_2 is greater than D_1, as shown in the figure, then V_{12} is positive; that is, P_1 is at a higher potential than P_2. Equation (4.8.4) is also valid for either dc or ac. For ac, V_{12} is a phasor voltage and q is a phasor representation of a sinusoidal charge.

Now apply (4.8.4) to the array of M solid cylindrical conductors shown in Figure 4.19. Assume that each conductor m has an ac charge q_m C/m uniformly distributed along the conductor. The voltage V_{kim} between conductors k and i due to the charge q_m acting alone is

$$V_{kim} = \frac{q_m}{2\pi\varepsilon} \ln \frac{D_{im}}{D_{km}} \quad \text{volts} \tag{4.8.5}$$

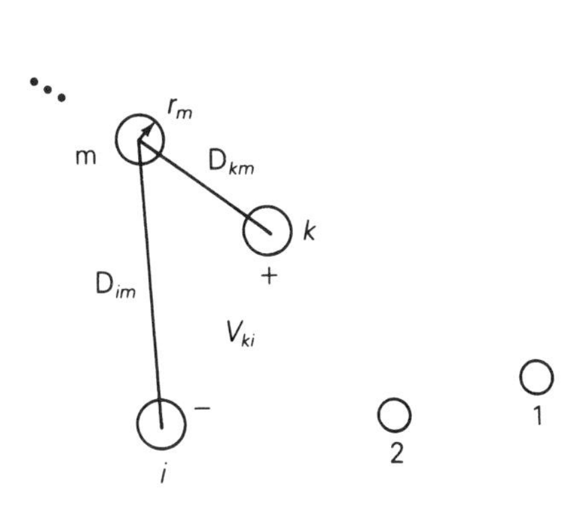

FIGURE 4.19 Array of M solid cylindrical conductors

where $D_{mm} = r_m$ when $k = m$ or $i = m$. In (4.8.5) we have neglected the distortion of the electric field in the vicinity of the other conductors, caused by the fact that the other conductors themselves are constant potential surfaces. V_{kim} can be thought of as the voltage between cylinders with radii D_{km} and D_{im} concentric to conductor m at points on the cylinders remote from conductors, where there is no distortion.

Using superposition, the voltage V_{ki} between conductors k and i due to all the changes is

$$V_{ki} = \frac{1}{2\pi\varepsilon} \sum_{m=1}^{M} q_m \ln \frac{D_{im}}{D_{km}} \quad \text{volts} \tag{4.8.6}$$

4.9 CAPACITANCE: SINGLE-PHASE TWO-WIRE LINE AND THREE-PHASE THREE-WIRE LINE WITH EQUAL PHASE SPACING

The results of the previous section are used here to determine the capacitances of the two relatively simple transmission lines considered in Section 4.5, a single-phase two-wire line and a three-phase three-wire line with equal phase spacing.

First we consider the single-phase two-wire line shown in Figure 4.9. Assume that the conductors are energized by a voltage source such that conductor x has a uniform charge q C/m and, assuming conservation of charge, conductor y has an equal quantity of negative charge $-q$. Using (4.8.6) with $k = x$, $i = y$, and $m = x, y$,

$$V_{xy} = \frac{1}{2\pi\varepsilon}\left[q \ln \frac{D_{yx}}{D_{xx}} - q \ln \frac{D_{yy}}{D_{xy}}\right]$$

$$= \frac{q}{2\pi\varepsilon} \ln \frac{D_{yx}D_{xy}}{D_{xx}D_{yy}} \tag{4.9.1}$$

Using $D_{xy} = D_{yx} = D$, $D_{xx} = r_x$, and $D_{yy} = r_y$, (4.9.1) becomes

$$V_{xy} = \frac{q}{\pi\varepsilon} \ln \frac{D}{\sqrt{r_x r_y}} \quad \text{volts} \tag{4.9.2}$$

For a 1-meter line length, the capacitance between conductors is

$$C_{xy} = \frac{q}{V_{xy}} = \frac{\pi\varepsilon}{\ln\left(\frac{D}{\sqrt{r_x r_y}}\right)} \quad \text{F/m line-to-line} \tag{4.9.3}$$

and if $r_x = r_y = r$,

$$C_{xy} = \frac{\pi\varepsilon}{\ln(D/r)} \quad \text{F/m line-to-line} \tag{4.9.4}$$

If the two-wire line is supplied by a transformer with a grounded center tap, then the voltage between each conductor and ground is one-half that given by (4.9.2). That is,

$$V_{xn} = V_{yn} = \frac{V_{xy}}{2} \tag{4.9.5}$$

and the capacitance from either line to the grounded neutral is

$$C_n = C_{xn} = C_{yn} = \frac{q}{V_{xn}} = 2C_{xy}$$

$$= \frac{2\pi\varepsilon}{\ln(D/r)} \quad \text{F/m line-to-neutral} \tag{4.9.6}$$

Circuit representations of the line-to-line and line-to-neutral capacitances are shown in Figure 4.20. Note that if the neutral is open in Figure 4.20(b), the two line-to-neutral capacitances combine in series to give the line-to-line capacitance.

Next consider the three-phase line with equal phase spacing shown in Figure 4.10. We shall neglect the effect of earth and neutral conductors here. To determine the positive-sequence capacitance, assume positive-sequence charges q_a, q_b, q_c such that $q_a + q_b + q_c = 0$. Using (4.8.6) with $k = a$, $i = b$, and $m = a, b, c$, the voltage V_{ab} between conductors a and b is

$$V_{ab} = \frac{1}{2\pi\varepsilon}\left[q_a \ln \frac{D_{ba}}{D_{aa}} + q_b \ln \frac{D_{bb}}{D_{ab}} + q_c \ln \frac{D_{bc}}{D_{ac}}\right] \tag{4.9.7}$$

Using $D_{aa} = D_{bb} = r$, and $D_{ab} = D_{ba} = D_{ca} = D_{cb} = D$, (4.9.7) becomes

(a) Line-to-line capacitance (b) Line-to-neutral capacitances

FIGURE 4.20 Circuit representation of capacitances for a single-phase two-wire line

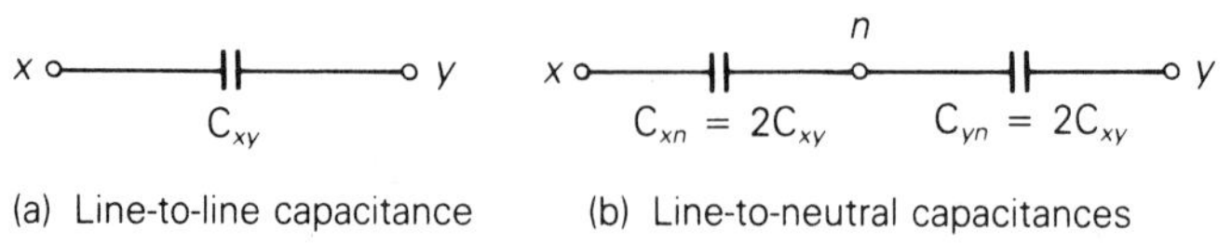

$$V_{ab} = \frac{1}{2\pi\varepsilon}\left[q_a \ln\frac{D}{r} + q_b \ln\frac{r}{D} + q_c \ln\frac{D}{D}\right]$$

$$= \frac{1}{2\pi\varepsilon}\left[q_a \ln\frac{D}{r} + q_b \ln\frac{r}{D}\right] \quad \text{volts} \tag{4.9.8}$$

Note that the third term in (4.9.8) is zero because conductors a and b are equidistant from conductor c. Thus, conductors a and b lie on a constant potential cylinder for the electric field due to q_c.

Similarly, using (4.8.6) with $k = a$, $i = c$, and $m = a, b, c$, the voltage V_{ac} is

$$V_{ac} = \frac{1}{2\pi\varepsilon}\left[q_a \ln\frac{D_{ca}}{D_{aa}} + q_b \ln\frac{D_{cb}}{D_{ab}} + q_c \ln\frac{D_{cc}}{D_{ac}}\right]$$

$$= \frac{1}{2\pi\varepsilon}\left[q_a \ln\frac{D}{r} + q_b \ln\frac{D}{D} + q_c \ln\frac{r}{D}\right]$$

$$= \frac{1}{2\pi\varepsilon}\left[q_a \ln\frac{D}{r} + q_c \ln\frac{r}{D}\right] \quad \text{volts} \tag{4.9.9}$$

Recall that for balanced positive-sequence voltages,

$$V_{ab} = \sqrt{3}V_{an}\underline{/+30^\circ} = \sqrt{3}V_{an}\left[\frac{\sqrt{3}}{2} + j\frac{1}{2}\right] \tag{4.9.10}$$

$$V_{ac} = -V_{ca} = \sqrt{3}V_{an}\underline{/-30^\circ} = \sqrt{3}V_{an}\left[\frac{\sqrt{3}}{2} - j\frac{1}{2}\right] \tag{4.9.11}$$

Adding (4.9.10) and (4.9.11) yields

$$V_{ab} + V_{ac} = 3V_{an} \tag{4.9.12}$$

Using (4.9.8) and (4.9.9) in (4.9.12),

$$V_{an} = \frac{1}{3}\left(\frac{1}{2\pi\varepsilon}\right)\left[2q_a \ln\frac{D}{r} + (q_b + q_c)\ln\frac{r}{D}\right] \tag{4.9.13}$$

and with $q_b + q_c = -q_a$,

$$V_{an} = \frac{1}{2\pi\varepsilon} q_a \ln\frac{D}{r} \quad \text{volts} \tag{4.9.14}$$

The capacitance-to-neutral per line length is

$$C_{an} = \frac{q_a}{V_{an}} = \frac{2\pi\varepsilon}{\ln\left(\frac{D}{r}\right)} \quad \text{F/m line-to-neutral} \tag{4.9.15}$$

Due to symmetry, the same result is obtained for $C_{bn} = q_b/V_{bn}$ and $C_{cn} = q_c/V_{cn}$. For balanced three-phase operation, however, only one phase need be considered. A circuit representation of the capacitance-to-neutral is shown in Figure 4.21.

FIGURE 4.21 Circuit representation of the capacitance-to-neutral of a three-phase line with equal phase spacing

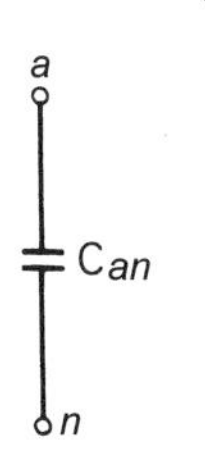

4.10

CAPACITANCE: STRANDED CONDUCTORS, UNEQUAL PHASE SPACING, BUNDLED CONDUCTORS

Equations (4.9.6) and (4.9.15) are based on the assumption that the conductors are solid cylindrical conductors with zero resistivity. The electric field inside these conductors is zero, and the external electric field is perpendicular to the conductor surfaces. Practical conductors with resistivities similar to those listed in Table 4.3 have a small internal electric field. As a result, the external electric field is slightly altered near the conductor surfaces. Also, the electric field near the surface of a stranded conductor is not the same as that of a solid cylindrical conductor. However, it is normal practice when calculating line capacitance to replace a stranded conductor by a perfectly conducting solid cylindrical conductor whose radius equals the outside radius of the stranded conductor. The resulting error in capacitance is small since only the electric field near the conductor surfaces is affected.

Also, (4.8.2) is based on the assumption that there is uniform charge distribution. But conductor charge distribution is nonuniform in the presence of other charged conductors. Therefore (4.9.6) and (4.9.15), which are derived from (4.8.2), are not exact. However, the nonuniformity of conductor charge distribution can be shown to have a negligible effect on line capacitance.

For three-phase lines with unequal phase spacing, balanced positive-sequence voltages are not obtained with balanced positive-sequence charges. Instead, unbalanced line-to-neutral voltages occur, and the phase-to-neutral capacitances are unequal. Balance can be restored by transposing the line such that each phase occupies each position for one-third of the line length. If equations similar to (4.9.7) for V_{ab} as well as for V_{ac} are written for each position in the transposition cycle, and are then averaged and used in (4.9.12)–(4.9.14), the resulting capacitance becomes

$$C_{an} = \frac{2\pi\varepsilon}{\ln(D_{eq}/r)} \quad \text{F/m} \tag{4.10.1}$$

where

$$D_{eq} = \sqrt[3]{D_{ab} D_{bc} D_{ac}} \tag{4.10.2}$$

Figure 4.22 shows a bundled conductor line with two conductors per bundle. To determine the capacitance of this line, assume balanced positive-sequence charges q_a, q_b, q_c for each phase such that $q_a + q_b + q_c = 0$. Assume that the conductors in each bundle, which are in parallel, share the charges equally. Thus conductors a and a' each have the charge $q_a/2$. Also assume that the phase spacings are much larger than the bundle spacings so that D_{ab} may be used instead of $(D_{ab} - d)$ or $(D_{ab} + d)$. Then, using (4.8.6) with $k = a$, $i = b$, $m = a, a', b, b', c, c'$,

FIGURE 4.22 Three-phase line with two conductors per bundle

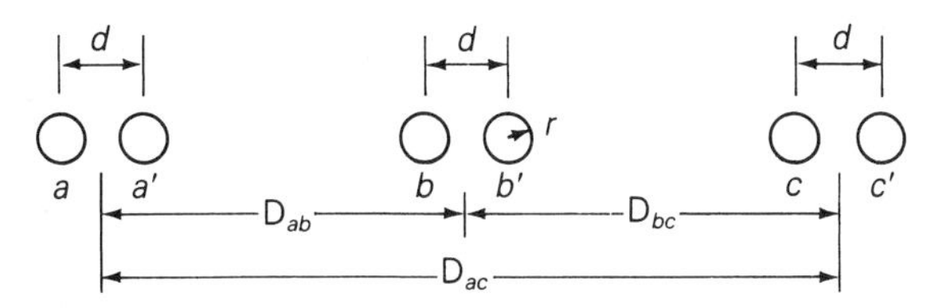

$$V_{ab} = \frac{1}{2\pi\varepsilon}\left[\frac{q_a}{2}\ln\frac{D_{ba}}{D_{aa}} + \frac{q_a}{2}\ln\frac{D_{ba'}}{D_{aa'}} + \frac{q_b}{2}\ln\frac{D_{bb}}{D_{ab}}\right.$$
$$\left. + \frac{q_b}{2}\ln\frac{D_{bb'}}{D_{ab'}} + \frac{q_c}{2}\ln\frac{D_{bc}}{D_{ac}} + \frac{q_c}{2}\ln\frac{D_{bc'}}{D_{ac'}}\right]$$
$$= \frac{1}{2\pi\varepsilon}\left[\frac{q_a}{2}\left(\ln\frac{D_{ab}}{r} + \ln\frac{D_{ab}}{d}\right) + \frac{q_b}{2}\left(\ln\frac{r}{D_{ab}} + \ln\frac{d}{D_{ab}}\right)\right.$$
$$\left. + \frac{q_c}{2}\left(\ln\frac{D_{bc}}{D_{ac}} + \ln\frac{D_{bc}}{D_{ac}}\right)\right]$$
$$= \frac{1}{2\pi\varepsilon}\left[q_a\ln\frac{D_{ab}}{\sqrt{rd}} + q_b\ln\frac{\sqrt{rd}}{D_{ab}} + q_c\ln\frac{D_{bc}}{D_{ac}}\right] \tag{4.10.3}$$

Equation (4.10.3) is the same as (4.9.7), except that D_{aa} and D_{bb} in (4.9.7) are replaced by $\sqrt{rd}$ in this equation. Therefore, for a transposed line, derivation of the capacitance would yield

$$C_{an} = \frac{2\pi\varepsilon}{\ln(D_{eq}/D_{SC})} \quad \text{F/m} \tag{4.10.4}$$

where

$$D_{SC} = \sqrt{rd} \quad \text{for a two-conductor bundle} \tag{4.10.5}$$

Similarly,

$$D_{SC} = \sqrt[3]{rd^2} \quad \text{for a three-conductor bundle} \tag{4.10.6}$$

$$D_{SC} = 1.091\sqrt[4]{rd^3} \quad \text{for a four-conductor bundle} \tag{4.10.7}$$

Equation (4.10.4) for capacitance is analogous to (4.6.22) for inductance. In both cases D_{eq}, given by (4.6.17) or (4.10.2), is the geometric mean of the distances between phases. Also, (4.10.5)–(4.10.7) for D_{SC} are analogous to (4.6.19)–(4.6.21) for D_{SL}, except that the conductor outside radius r replaces the conductor GMR D_S.

The current supplied to the transmission-line capacitance is called *charging current*. For a single-phase circuit operating at line-to-line voltage $V_{xy} = V_{xy}\underline{/0°}$, the charging current is

$$I_{chg} = Y_{xy}V_{xy} = j\omega C_{xy}V_{xy} \quad \text{A} \tag{4.10.8}$$

As shown in Chapter 2, a capacitor delivers reactive power. From (2.3.5), the reactive power delivered by the line-to-line capacitance is

$$Q_C = \frac{V_{xy}^2}{X_c} = Y_{xy}V_{xy}^2 = \omega C_{xy}V_{xy}^2 \quad \text{var} \tag{4.10.9}$$

For a completely transposed three-phase line that has balanced positive-sequence voltages with $V_{an} = V_{LN}\angle 0°$, the phase a charging current is

$$I_{chg} = YV_{an} = j\omega C_{an}V_{LN} \quad \text{A} \tag{4.10.10}$$

and the reactive power delivered by phase a is

$$Q_{C1\phi} = YV_{an}^2 = \omega C_{an}V_{LN}^2 \quad \text{var} \tag{4.10.11}$$

The total reactive power supplied by the three-phase line is

$$Q_{C3\phi} = 3Q_{C1\phi} = 3\omega C_{an}V_{LN}^2 = \omega C_{an}V_{LL}^2 \quad \text{var} \tag{4.10.12}$$

EXAMPLE 4.6 Capacitance, admittance, and reactive power supplied: single-phase line

For the single-phase line in Example 4.3, determine the line-to-line capacitance in F and the line-to-line admittance in S. If the line voltage is 20 kV, determine the reactive power in kvar supplied by this capacitance.

SOLUTION From Table A.3, the outside radius of a 4/0 12-strand copper conductor is

$$r = \frac{0.552}{2} \text{ in.} = 0.7 \text{ cm}$$

and from (4.9.4),

$$C_{xy} = \frac{\pi(8.854 \times 10^{-12})}{\ln\left(\frac{150}{0.7}\right)} = 5.182 \times 10^{-12} \quad \text{F/m}$$

or

$$C_{xy} = 5.182 \times 10^{-12} \frac{\text{F}}{\text{m}} \times 32 \times 10^3 \text{ m} = 1.66 \times 10^{-7} \quad \text{F}$$

and the shunt admittance is

$$Y_{xy} = j\omega C_{xy} = j(2\pi 60)(1.66 \times 10^{-7})$$

$$= j6.27 \times 10^{-5} \quad \text{S line-to-line}$$

From (4.10.9),

$$Q_C = (6.27 \times 10^{-5})(20 \times 10^3)^2 = 25.1 \quad \text{kvar}$$

■

EXAMPLE 4.7 Capacitance and shunt admittance; charging current and reactive power supplied: three-phase line

For the three-phase line in Example 4.5, determine the capacitance-to-neutral in F and the shunt admittance-to-neutral in S. If the line voltage is 345 kV, determine the charging current in kA per phase and the total reactive power in Mvar supplied by the line capacitance. Assume balanced positive-sequence voltages.

SOLUTION From Table A.4, the outside radius of a 403 mm^2 26/2 ACSR conductor is

$$r = \frac{1.108}{2} \text{ in.} \times 0.0254 \frac{\text{m}}{\text{in.}} = 0.0141 \quad \text{m}$$

From (4.10.5), the equivalent radius of the two-conductor bundle is

$$D_{SC} = \sqrt{(0.0141)(0.40)} = 0.0750 \quad \text{m}$$

$D_{eq} = 12.6$ m is the same as in Example 4.5. Therefore, from (4.10.4),

$$C_{an} = \frac{(2\pi)(8.854 \times 10^{-12})}{\ln\left(\dfrac{12.6}{0.0750}\right)} \frac{\text{F}}{\text{m}} \times 1000 \frac{\text{m}}{\text{km}} \times 200 \text{ km}$$

$$= 2.17 \times 10^{-6} \quad \text{F}$$

The shunt admittance-to-neutral is

$$Y_{an} = j\omega C_{an} = j(2\pi 60)(2.17 \times 10^{-6})$$

$$= j8.19 \times 10^{-4} \quad \text{S}$$

From (4.10.10),

$$I_{chg} = |I_{chg}| = (8.19 \times 10^{-4})\left(\frac{345}{\sqrt{3}}\right) = 0.163 \quad \text{kA/phase}$$

and from (4.10.12),

$$Q_{C3\phi} = (8.19 \times 10^{-4})(345)^2 = 97.5 \quad \text{Mvar}$$

■

4.11 SHUNT ADMITTANCES: LINES WITH NEUTRAL CONDUCTORS AND EARTH RETURN

In this section, we develop equations suitable for computer calculation of the shunt admittances for the three-phase overhead line shown in Figure 4.16. We approximate the earth surface as a perfectly conducting horizontal plane,

FIGURE 4.23

Method of images

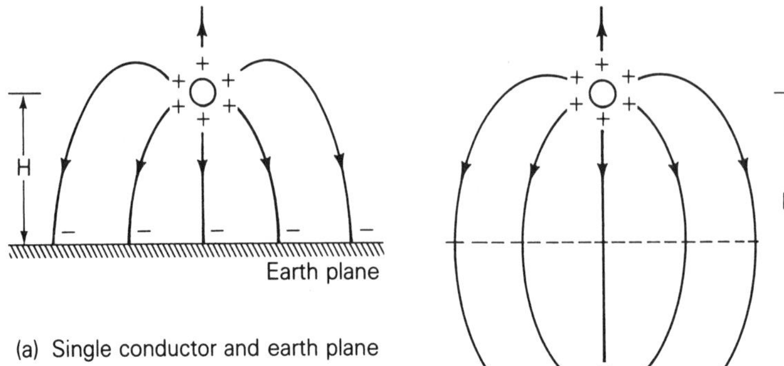

(a) Single conductor and earth plane

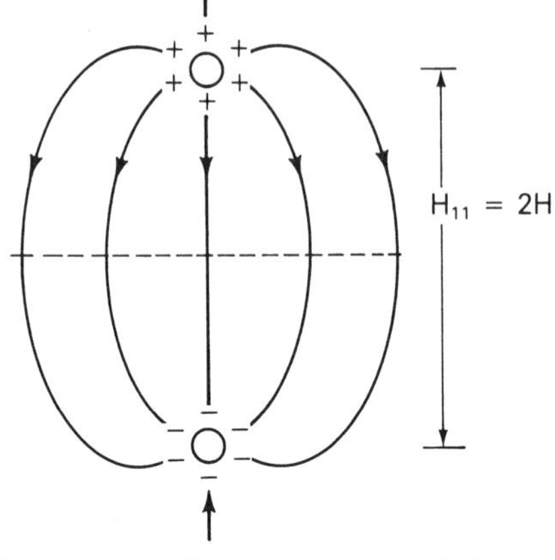

(b) Earth plane replaced by image conductor

even though the earth under the line may have irregular terrain and resistivities as shown in Table 4.4.

The effect of the earth plane is accounted for by the *method of images*, described as follows. Consider a single conductor with uniform charge distribution and with height H above a perfectly conducting earth plane, as shown in Figure 4.23(a). When the conductor has a positive charge, an equal quantity of negative charge is induced on the earth. The electric field lines will originate from the positive charges on the conductor and terminate at the negative charges on the earth. Also, the electric field lines are perpendicular to the surfaces of the conductor and earth.

Now replace the earth by the image conductor shown in Figure 4.23(b), which has the same radius as the original conductor, lies directly below the original conductor with conductor separation $H_{11} = 2H$, and has an equal quantity of negative charge. The electric field above the dashed line representing the location of the removed earth plane in Figure 4.23(b) is identical to the electric field above the earth plane in Figure 4.23(a). Therefore, the voltage between any two points above the earth is the same in both figures.

EXAMPLE 4.8 Effect of earth on capacitance: single-phase line

If the single-phase line in Example 4.6 has flat horizontal spacing with 5.49 m average line height, determine the effect of the earth on capacitance. Assume a perfectly conducting earth plane.

SOLUTION The earth plane is replaced by a separate image conductor for each overhead conductor, and the conductors are charged as shown in Figure 4.24. From (4.8.6), the voltage between conductors x and y is

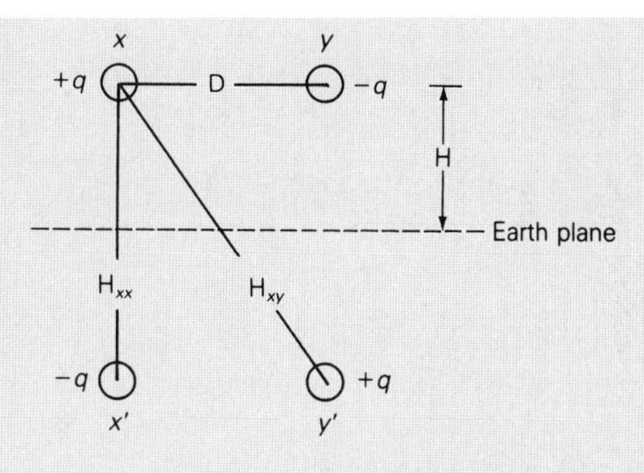

FIGURE 4.24 Single-phase line for Example 4.8

$$V_{xy} = \frac{q}{2\pi\varepsilon}\left[\ln\frac{D_{yx}}{D_{xx}} - \ln\frac{D_{yy}}{D_{xy}} - \ln\frac{H_{yx}}{H_{xx}} + \ln\frac{H_{yy}}{H_{xy}}\right]$$

$$= \frac{q}{2\pi\varepsilon}\left[\ln\frac{D_{yx}D_{xy}}{D_{xx}D_{yy}} - \ln\frac{H_{yx}H_{xy}}{H_{xx}H_{yy}}\right]$$

$$= \frac{q}{\pi\varepsilon}\left[\ln\frac{D}{r} - \ln\frac{H_{xy}}{H_{xx}}\right]$$

The line-to-line capacitance is

$$C_{xy} = \frac{q}{V_{xy}} = \frac{\pi\varepsilon}{\ln\dfrac{D}{r} - \ln\dfrac{H_{xy}}{H_{xx}}} \quad \text{F/m}$$

Using $D = 1.5$ m, $r = 0.7$ cm, $H_{xx} = 2H = 10.98$, and $H_{xy} = \sqrt{(10.98)^2 + (1.5)^2} = 11.08$ m,

$$C_{xy} = \frac{\pi(8.854 \times 10^{-12})}{\ln\dfrac{150}{0.7} - \ln\dfrac{11.08}{11}} = 5.189 \times 10^{-12} \quad \text{F/m}$$

compared with 5.182×10^{-12} F/m in Example 4.6. The effect of the earth plane is to slightly increase the capacitance. Note that as the line height H increases, the ratio H_{xy}/H_{xx} approaches 1, $\ln(H_{xy}/H_{xx}) \to 0$, and the effect of the earth becomes negligible. ■

For the three-phase line with N neutral conductors shown in Figure 4.25, the perfectly conducting earth plane is replaced by a separate image conductor for each overhead conductor. The overhead conductors a, b, c, $n1$, $n2, \ldots, nN$ carry charges $q_a, q_b, q_c, q_{n1}, \ldots, q_{nN}$, and the image conductors a', b', c', $n1', \ldots, nN'$ carry charges $-q_a$, $-q_b$, $-q_c$, $-q_{n1}, \ldots, -q_{nN}$. Applying (4.8.6) to determine the voltage $V_{kk'}$ between any conductor k and its image conductor k',

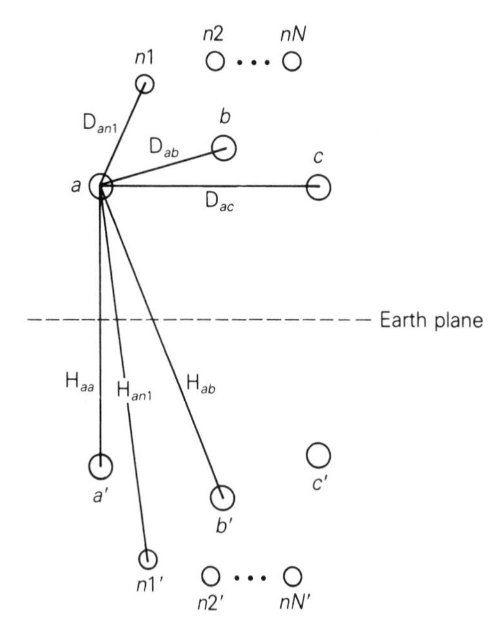

FIGURE 4.25 Three-phase line with neutral conductors and with earth plane replaced by image conductors

$$V_{kk'} = \frac{1}{2\pi\varepsilon}\left[\sum_{m=a}^{nN} q_m \ln\frac{\mathrm{H}_{km}}{\mathrm{D}_{km}} - \sum_{m=a}^{nN} q_m \ln\frac{\mathrm{D}_{km}}{\mathrm{H}_{km}}\right]$$

$$= \frac{2}{2\pi\varepsilon}\sum_{m=a}^{nN} q_m \ln\frac{\mathrm{H}_{km}}{\mathrm{D}_{km}} \tag{4.11.1}$$

where $\mathrm{D}_{kk} = r_k$ and D_{km} is the distance between overhead conductors k and m. H_{km} is the distance between overhead conductor k and image conductor m. By symmetry, the voltage V_{kn} between conductor k and the earth is one-half of $V_{kk'}$.

$$V_{kn} = \frac{1}{2}V_{kk'} = \frac{1}{2\pi\varepsilon}\sum_{m=a}^{nN} q_m \ln\frac{\mathrm{H}_{km}}{\mathrm{D}_{km}} \tag{4.11.2}$$

where

$$k = a, b, c, n1, n2, \ldots, nN$$

$$m = a, b, c, n1, n2, \ldots, nN$$

Since all the neutral conductors are grounded to the earth,

$$V_{kn} = 0 \qquad \text{for } k = n1, n2, \ldots, nN \tag{4.11.3}$$

In matrix format, (4.11.2) and (4.11.3) are

$$\begin{bmatrix} V_{an} \\ V_{bn} \\ V_{cn} \\ \hline 0 \\ \vdots \\ 0 \end{bmatrix} = \left[\begin{array}{ccc|ccc} \overbrace{\mathrm{P}_{aa} \quad \mathrm{P}_{ab} \quad \mathrm{P}_{ac}}^{\mathbf{P}_A} & & & \overbrace{\mathrm{P}_{an1} \quad \cdots \quad \mathrm{P}_{anN}}^{\mathbf{P}_B} & & \\ \mathrm{P}_{ba} & \mathrm{P}_{bb} & \mathrm{P}_{bc} & \mathrm{P}_{bn1} & \cdots & \mathrm{P}_{bnN} \\ \mathrm{P}_{ca} & \mathrm{P}_{cb} & \mathrm{P}_{cc} & \mathrm{P}_{cn1} & \cdots & \mathrm{P}_{cnN} \\ \hline \mathrm{P}_{n1a} & \mathrm{P}_{n1b} & \mathrm{P}_{n1c} & \mathrm{P}_{n1n1} & \cdots & \mathrm{P}_{n1nN} \\ \vdots & & & & & \\ \underbrace{\mathrm{P}_{nNa} \quad \mathrm{P}_{nNb} \quad \mathrm{P}_{nNc}}_{\mathbf{P}_C} & & & \underbrace{\mathrm{P}_{nNn1} \quad \cdots \quad \mathrm{P}_{nNnN}}_{\mathbf{P}_D} & & \end{array}\right] \begin{bmatrix} q_a \\ q_b \\ q_c \\ \hline q_{n1} \\ \vdots \\ q_{nN} \end{bmatrix} \tag{4.11.4}$$

The elements of the $(3+N) \times (3+N)$ matrix $\mathbf{P}$ are

$$\mathrm{P}_{km} = \frac{1}{2\pi\varepsilon} \ln \frac{\mathrm{H}_{km}}{\mathrm{D}_{km}} \quad \text{m/F} \tag{4.11.5}$$

where

$$k = a, b, c, n1, \ldots, nN$$

$$m = a, b, c, n1, \ldots, nN$$

Equation (4.11.4) is now partitioned as shown above to obtain

$$\begin{bmatrix} \boldsymbol{V}_\mathrm{P} \\ \hline \mathbf{0} \end{bmatrix} = \left[\begin{array}{c|c} \mathbf{P}_A & \mathbf{P}_B \\ \hline \mathbf{P}_C & \mathbf{P}_D \end{array}\right] \begin{bmatrix} \boldsymbol{q}_\mathrm{P} \\ \hline \boldsymbol{q}_n \end{bmatrix} \tag{4.11.6}$$

$\boldsymbol{V}_\mathrm{P}$ is the three-dimensional vector of phase-to-neutral voltages. $\boldsymbol{q}_\mathrm{P}$ is the three-dimensional vector of phase-conductor charges and $\boldsymbol{q}_n$ is the N vector of neutral conductor charges. The $(3+N) \times (3+N)\mathbf{P}$ matrix is partitioned as shown in (4.11.4) to obtain:

$\mathbf{P}_A$ with dimension 3×3

$\mathbf{P}_B$ with dimension $3 \times N$

$\mathbf{P}_C$ with dimension $N \times 3$

$\mathbf{P}_D$ with dimension $N \times N$

Equation (4.11.6) is rewritten as two separate equations:

$$\boldsymbol{V}_\mathrm{P} = \mathbf{P}_A \boldsymbol{q}_\mathrm{P} + \mathbf{P}_B \boldsymbol{q}_n \tag{4.11.7}$$

$$\mathbf{0} = \mathbf{P}_C \boldsymbol{q}_\mathrm{P} + \mathbf{P}_D \boldsymbol{q}_n \tag{4.11.8}$$

Then (4.11.8) is solved for $\boldsymbol{q}_n$, which is used in (4.11.7) to obtain

$$\boldsymbol{V}_\mathrm{P} = (\mathbf{P}_A - \mathbf{P}_B \mathbf{P}_D^{-1} \mathbf{P}_C) \boldsymbol{q}_\mathrm{P} \tag{4.11.9}$$

or

$$\boldsymbol{q}_\mathrm{P} = \mathbf{C}_\mathrm{P} \boldsymbol{V}_\mathrm{P} \tag{4.11.10}$$

where

$$\mathbf{C}_P = (\mathbf{P}_A - \mathbf{P}_B \mathbf{P}_D^{-1} \mathbf{P}_C)^{-1} \quad \text{F/m} \tag{4.11.11}$$

Equation (4.11.10), the desired result, relates the phase-conductor charges to the phase-to-neutral voltages. $\mathbf{C}_P$ is the 3×3 matrix of phase capacitances whose elements are denoted

$$\mathbf{C}_P = \begin{bmatrix} C_{aa} & C_{ab} & C_{ac} \\ C_{ab} & C_{bb} & C_{bc} \\ C_{ac} & C_{bc} & C_{cc} \end{bmatrix} \quad \text{F/m} \tag{4.11.12}$$

It can be shown that $\mathbf{C}_P$ is a symmetric matrix whose diagonal terms C_{aa}, C_{bb}, C_{cc} are positive, and whose off-diagonal terms C_{ab}, C_{bc}, C_{ac} are negative. This indicates that when a positive line-to-neutral voltage is applied to one phase, a positive charge is induced on that phase and negative charges are induced on the other phases, which is physically correct.

If the line is completely transposed, the diagonal and off-diagonal elements of $\mathbf{C}_P$ are averaged to obtain

$$\hat{\mathbf{C}}_P = \begin{bmatrix} \hat{C}_{aa} & \hat{C}_{ab} & \hat{C}_{ab} \\ \hat{C}_{ab} & \hat{C}_{aa} & \hat{C}_{ab} \\ \hat{C}_{ab} & \hat{C}_{ab} & \hat{C}_{aa} \end{bmatrix} \quad \text{F/m} \tag{4.11.13}$$

where

$$\hat{C}_{aa} = \tfrac{1}{3}(C_{aa} + C_{bb} + C_{cc}) \quad \text{F/m} \tag{4.11.14}$$

$$\hat{C}_{ab} = \tfrac{1}{3}(C_{ab} + C_{bc} + C_{ac}) \quad \text{F/m} \tag{4.11.15}$$

$\hat{\mathbf{C}}_P$ is a symmetrical capacitance matrix.

The shunt phase admittance matrix is given by

$$\boldsymbol{Y}_P = j\omega \mathbf{C}_P = j(2\pi f)\mathbf{C}_P \quad \text{S/m} \tag{4.11.16}$$

or, for a completely transposed line,

$$\hat{\boldsymbol{Y}}_P = j\omega \hat{\mathbf{C}}_P = j(2\pi f)\hat{\mathbf{C}}_P \quad \text{S/m} \tag{4.11.17}$$

4.12 ELECTRIC FIELD STRENGTH AT CONDUCTOR SURFACES AND AT GROUND LEVEL

When the electric field strength at a conductor surface exceeds the breakdown strength of air, current discharges occur. This phenomenon, called corona, causes additional line losses (corona loss), communications interference, and audible noise. Although breakdown strength depends on many factors, a rough value is 30 kV/cm in a uniform electric field for dry air at atmospheric pressure. The presence of water droplets or rain can lower this value significantly. To control corona, transmission lines are usually designed to maintain calculated values of conductor surface electric field strength below 20 $\mathrm{kV_{rms}/cm}$.

When line capacitances are determined and conductor voltages are known, the conductor charges can be calculated from (4.9.3) for a single-phase line or from (4.11.10) for a three-phase line. Then the electric field strength at the surface of one phase conductor, neglecting the electric fields due to charges on other phase conductors and neutral wires, is, from (4.8.2),

$$E_r = \frac{q}{2\pi\varepsilon r} \quad \text{V/m} \tag{4.12.1}$$

where r is the conductor outside radius.

For bundled conductors with N_b conductors per bundle and with charge q C/m per phase, the charge per conductor is q/N_b and

$$E_{r\text{ave}} = \frac{q/N_b}{2\pi\varepsilon r} \quad \text{V/m} \tag{4.12.2}$$

Equation (4.12.2) represents an average value for an individual conductor in a bundle. The maximum electric field strength at the surface of one conductor due to all charges in a bundle, obtained by the vector addition of electric fields (as shown in Figure 4.26), is as follows:

Two-conductor bundle ($N_b = 2$):

$$E_{r\max} = \frac{q/2}{2\pi\varepsilon r} + \frac{q/2}{2\pi\varepsilon d} = \frac{q/2}{2\pi\varepsilon r}\left(1 + \frac{r}{d}\right)$$

$$= \mathrm{E}_{r\text{ave}}\left(1 + \frac{r}{d}\right) \tag{4.12.3}$$

Three-conductor bundle ($N_b = 3$):

$$E_{r\max} = \frac{q/3}{2\pi\varepsilon}\left(\frac{1}{r} + \frac{2\cos 30^\circ}{d}\right) = E_{r\text{ave}}\left(1 + \frac{r\sqrt{3}}{d}\right) \tag{4.12.4}$$

Four-conductor bundle ($N_b = 4$):

$$E_{r\max} = \frac{q/4}{2\pi\varepsilon}\left(\frac{1}{r} + \frac{1}{d\sqrt{2}} + \frac{2\cos 45^\circ}{d}\right) = E_{r\text{ave}}\left[1 + \frac{r}{d}(2.1213)\right] \tag{4.12.5}$$

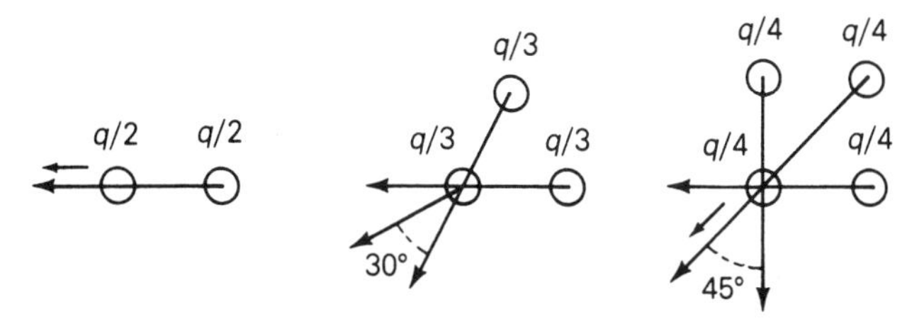

FIGURE 4.26 Vector addition of electric fields at the surface of one conductor in a bundle

TABLE 4.5

Examples of maximum ground-level electric field strength versus transmission-line voltage [1] (© Copyright 1987. Electric Power Research Institute (EPRI), Publication Number EL-2500. *Transmission Line Reference Book, 345-kV and Above, Second Edition, Revised.* Reprinted with permission)

Line Voltage (kV_{rms})	Maximum Ground-Level Electric Field Strength (kV_{rms}/m)
23 (1ϕ)	0.01–0.025
23 (3ϕ)	0.01–0.05
115	0.1–0.2
345	2.3–5.0
345 (double circuit)	5.6
500	8.0
765	10.0

Although the electric field strength at ground level is much less than at conductor surfaces where corona occurs, there are still capacitive coupling effects. Charges are induced on ungrounded equipment such as vehicles with rubber tires located near a line. If a person contacts the vehicle and ground, a discharge current will flow to ground. Transmission-line heights are designed to maintain discharge currents below prescribed levels for any equipment that may be on the right-of-way. Table 4.5 shows examples of maximum ground-level electric field strength.

As shown in Figure 4.27, the ground-level electric field strength due to charged conductor k and its image conductor is perpendicular to the earth plane, with value

$$E_k(w) = \left(\frac{q_k}{2\pi\varepsilon}\right)\frac{2\cos\theta}{\sqrt{y_k^2 + (w - x_k)^2}}$$

$$= \left(\frac{q_k}{2\pi\varepsilon}\right)\frac{2y_k}{y_k^2 + (w - x_k)^2} \quad \text{V/m} \tag{4.12.6}$$

where (x_k, y_k) are the horizontal and vertical coordinates of conductor k with respect to reference point R, w is the horizontal coordinate of the ground-level point where the electric field strength is to be determined, and q_k is the charge on conductor k. The total ground-level electric field is the phasor sum of terms $E_k(w)$ for all overhead conductors. A lateral profile of ground-level

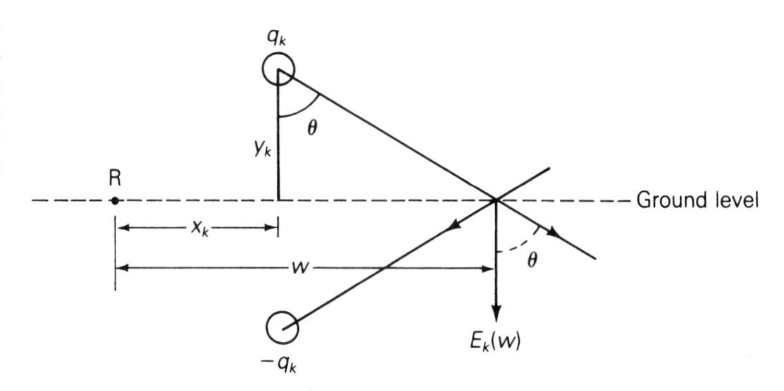

FIGURE 4.27

Ground-level electric field strength due to an overhead conductor and its image

electric field strength is obtained by varying w from the center of the line to the edge of the right-of-way.

EXAMPLE 4.9 Conductor surface and ground-level electric field strengths: single-phase line

For the single-phase line of Example 4.8, calculate the conductor surface electric field strength in kV_{rms}/cm. Also calculate the ground-level electric field in kV_{rms}/m directly under conductor x. The line voltage is 20 kV.

SOLUTION From Example 4.8, $C_{xy} = 5.189 \times 10^{-12}$ F/m. Using (4.9.3) with $V_{xy} = 20\angle 0^\circ$ kV,

$$q_x = -q_y = (5.189 \times 10^{-12})(20 \times 10^3\angle 0^\circ) = 1.038 \times 10^{-7}\angle 0^\circ \quad \text{C/m}$$

From (4.12.1), the conductor surface electric field strength is, with $r = 0.023$ ft $= 0.00701$ m,

$$E_r = \frac{1.036 \times 10^{-7}}{(2\pi)(8.854 \times 10^{-12})(0.00701)} \frac{\text{V}}{\text{m}} \times \frac{\text{kV}}{1000\ \text{V}} \times \frac{\text{m}}{100\ \text{cm}}$$

$$= 2.66 \quad \text{kV}_{\text{rms}}/\text{cm}$$

Selecting the center of the line as the reference point R, the coordinates (x_x, y_x) for conductor x are (−0.75 m, 5.49 m) and (+0.75 m, 5.49 m) for conductor y. The ground-level electric field directly under conductor x, where $w = -0.75$ m, is, from (4.12.6),

$$E(-0.762) = E_x(-0.75) + E_y(-0.75)$$

$$= \frac{1.036 \times 10^{-7}}{(2\pi)(8.85 \times 10^{-12})} \left[\frac{(2)(5.49)}{(5.49)^2} - \frac{(2)(5.49)}{(5.49)^2 + (0.75 + 0.75)^2} \right]$$

$$= 1.862 \times 10^3(0.364 - 0.338) = 48.5\angle 0^\circ \text{ V/m} = 0.0485 \text{ kV/m}$$

For this 20-kV line, the electric field strengths at the conductor surface and at ground level are low enough to be of relatively small concern. For EHV lines, electric field strengths and the possibility of corona and shock hazard are of more concern. ■

4.13 PARALLEL CIRCUIT THREE-PHASE LINES

If two parallel three-phase circuits are close together, either on the same tower as in Figure 4.3, or on the same right-of-way, there are mutual inductive and capacitive couplings between the two circuits. When calculating the equivalent series impedance and shunt admittance matrices, these couplings should not be neglected unless the spacing between the circuits is large.

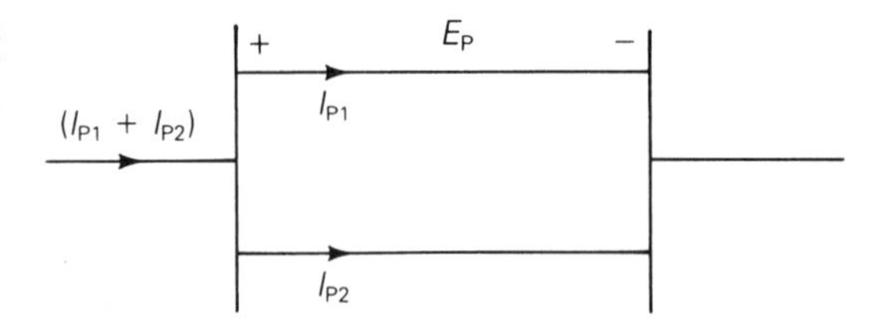

FIGURE 4.28 Single-line diagram of a double-circuit line

Consider the double-circuit line shown in Figure 4.28. For simplicity, assume that the lines are not transposed. Since both are connected in parallel, they have the same series-voltage drop for each phase. Following the same procedure as in Section 4.7, we can write $2(6+N)$ equations similar to (4.7.6)–(4.7.9): six equations for the overhead phase conductors, N equations for the overhead neutral conductors, and $(6+N)$ equations for the earth return conductors. After lumping the neutral voltage drop into the voltage drops across the phase conductors, and eliminating the neutral and earth return currents, we obtain

$$\begin{bmatrix} \boldsymbol{E}_{\mathrm{P}} \\ \boldsymbol{E}_{\mathrm{P}} \end{bmatrix} = \boldsymbol{Z}_{\mathrm{P}} \begin{bmatrix} \boldsymbol{I}_{\mathrm{P1}} \\ \boldsymbol{I}_{\mathrm{P2}} \end{bmatrix} \tag{4.13.1}$$

where $\boldsymbol{E}_{\mathrm{P}}$ is the vector of phase-conductor voltage drops (including the neutral voltage drop), and $\boldsymbol{I}_{\mathrm{P1}}$ and $\boldsymbol{I}_{\mathrm{P2}}$ are the vectors of phase currents for lines 1 and 2. $\boldsymbol{Z}_{\mathrm{P}}$ is a 6×6 impedance matrix. Solving (4.13.1)

$$\begin{bmatrix} \boldsymbol{I}_{\mathrm{P1}} \\ \hline \boldsymbol{I}_{\mathrm{P2}} \end{bmatrix} = \boldsymbol{Z}_{\mathrm{P}}^{-1} \begin{bmatrix} \boldsymbol{E}_{\mathrm{P}} \\ \hline \boldsymbol{E}_{\mathrm{P}} \end{bmatrix} = \left[\begin{array}{c|c} \boldsymbol{Y}_{A} & \boldsymbol{Y}_{B} \\ \hline \boldsymbol{Y}_{C} & \boldsymbol{Y}_{D} \end{array}\right] \begin{bmatrix} \boldsymbol{E}_{\mathrm{P}} \\ \hline \boldsymbol{E}_{\mathrm{P}} \end{bmatrix} = \begin{bmatrix} (\boldsymbol{Y}_{A} + \boldsymbol{Y}_{B}) \\ (\boldsymbol{Y}_{C} + \boldsymbol{Y}_{D}) \end{bmatrix} \boldsymbol{E}_{\mathrm{P}} \tag{4.13.2}$$

where $\boldsymbol{Y}_A$, $\boldsymbol{Y}_B$, $\boldsymbol{Y}_C$, and $\boldsymbol{Y}_D$ are obtained by partitioning $\boldsymbol{Z}_{\mathrm{P}}^{-1}$ into four 3×3 matrices. Adding $\boldsymbol{I}_{\mathrm{P1}}$ and $\boldsymbol{I}_{\mathrm{P2}}$,

$$(\boldsymbol{I}_{\mathrm{P1}} + \boldsymbol{I}_{\mathrm{P2}}) = (\boldsymbol{Y}_{A} + \boldsymbol{Y}_{B} + \boldsymbol{Y}_{C} + \boldsymbol{Y}_{D})\boldsymbol{E}_{\mathrm{P}} \tag{4.13.3}$$

and solving for $\boldsymbol{E}_{\mathrm{P}}$,

$$\boldsymbol{E}_{\mathrm{P}} = \boldsymbol{Z}_{\mathrm{Peq}}(\boldsymbol{I}_{\mathrm{P1}} + \boldsymbol{I}_{\mathrm{P2}}) \tag{4.13.4}$$

where

$$\boldsymbol{Z}_{\mathrm{Peq}} = (\boldsymbol{Y}_{A} + \boldsymbol{Y}_{B} + \boldsymbol{Y}_{C} + \boldsymbol{Y}_{D})^{-1} \tag{4.13.5}$$

$\boldsymbol{Z}_{\mathrm{Peq}}$ is the equivalent 3×3 series phase impedance matrix of the double-circuit line. Note that in (4.13.5) the matrices $\boldsymbol{Y}_B$ and $\boldsymbol{Y}_C$ account for the inductive coupling between the two circuits.

An analogous procedure can be used to obtain the shunt admittance matrix. Following the ideas of Section 4.11, we can write $(6+N)$ equations similar to (4.11.4). After eliminating the neutral wire charges, we obtain

$$\begin{bmatrix} \boldsymbol{q}_{\mathrm{P1}} \\ \hline \boldsymbol{q}_{\mathrm{P2}} \end{bmatrix} = \mathbf{C}_{\mathrm{P}} \begin{bmatrix} \boldsymbol{V}_{\mathrm{P}} \\ \hline \boldsymbol{V}_{\mathrm{P}} \end{bmatrix} = \left[\begin{array}{c|c} \mathbf{C}_{A} & \mathbf{C}_{B} \\ \hline \mathbf{C}_{C} & \mathbf{C}_{D} \end{array}\right] \begin{bmatrix} \boldsymbol{V}_{\mathrm{P}} \\ \hline \boldsymbol{V}_{\mathrm{P}} \end{bmatrix} = \begin{bmatrix} (\mathbf{C}_{A} + \mathbf{C}_{B}) \\ (\mathbf{C}_{C} + \mathbf{C}_{D}) \end{bmatrix} \boldsymbol{V}_{\mathrm{P}} \tag{4.13.6}$$

where V_P is the vector of phase-to-neutral voltages, and q_{P1} and q_{P2} are the vectors of phase-conductor charges for lines 1 and 2. $\mathbf{C}_P$ is a 6×6 capacitance matrix that is partitioned into four 3×3 matrices $\mathbf{C}_A$, $\mathbf{C}_B$, $\mathbf{C}_C$, and $\mathbf{C}_D$. Adding q_{P1} and q_{P2}

$$(q_{P1} + q_{P2}) = \mathbf{C}_{Peq} V_P \tag{4.13.7}$$

where

$$\mathbf{C}_{Peq} = (\mathbf{C}_A + \mathbf{C}_B + \mathbf{C}_C + \mathbf{C}_D) \tag{4.13.8}$$

Also,

$$Y_{Peq} = j\omega \mathbf{C}_{Peq} \tag{4.13.9}$$

Y_{Peq} is the equivalent 3×3 shunt admittance matrix of the double-circuit line. The matrices $\mathbf{C}_B$ and $\mathbf{C}_C$ in (4.13.8) account for the capacitive coupling between the two circuits.

These ideas can be extended in a straightforward fashion to more than two parallel circuits.

MULTIPLE CHOICE QUESTIONS

SECTION 4.1

4.1 ACSR stands for
(a) Aluminum-clad steel conductor
(b) Aluminum conductor steel supported
(c) Aluminum conductor steel reinforced

4.2 Overhead transmission-line conductors arc barc with no insulating cover.
(a) True (b) False

4.3 Alumoweld is an aluminum-clad steel conductor.
(a) True (b) False

4.4 EHV lines often have more than one conductor per phase; these conductors are called a ________. Fill in the Blank.

4.5 Shield wires located above the phase conductors protect the phase conductors against lightning.
(a) True (b) False

4.6 Conductor spacings, types, and sizes do have an impact on the series impedance and shunt admittance.
(a) True (b) False

SECTION 4.2

4.7 A circle with diameter D in = 1000 D mil = d mil has an area of _______ cmil. Fill in the Blank.

4.8 AC resistance is higher than dc resistance.
(a) True (b) False

4.9 Match the following for the current distribution throughout the conductor cross section:
(i) For dc (a) uniform
(ii) For ac (b) nonuniform

SECTION 4.3

4.10 Transmission line conductance is usually neglected in power system studies.
(a) True (b) False

SECTION 4.4

4.11 The internal inductance L_{int} per unit-length of a solid cylindrical conductor is a constant, given by $\frac{1}{2} \times 10^{-7}$H/m in SI system of units.
(a) True (b) False

4.12 The total inductance L_P of a solid cylindrical conductor (of radius r) due to both internal and external flux linkages out of distance D is given by (in H/m)
(a) 2×10^{-7} (b) $2 \times 10^{-7} \ln(\frac{D}{r})$
(c) $2 \times 10^{-7} \ln(\frac{D}{r'})$
where $r' = e^{-\frac{1}{4}} r = 0.778\, r$

SECTION 4.5

4.13 For a single-phase, two-wire line consisting of two solid cylindrical conductors of same radius, r, the total circuit inductance, also called loop inductance, is given by (in H/m)
(a) $2 \times 10^{-7} \ln(\frac{D}{r'})$ (b) $4 \times 10^{-7} \ln(\frac{D}{r'})$
where $r' = e^{-\frac{1}{4}} r = 0.778r$

4.14 For a three-phase, three-wire line consisting of three solid cylindrical conductors, each with radius r, and with equal phase spacing D between any two conductors, the inductance in H/m per phase is given by
(a) $2 \times 10^{-7} \ln(\frac{D}{r'})$ (b) $4 \times 10^{-7} \ln(\frac{D}{r'})$
(c) $6 \times 10^{-7} \ln(\frac{D}{r'})$
where $r' = e^{-\frac{1}{4}} r = 0.778\, r$

4.15 For a balanced three-phase, positive-sequence currents I_a, I_b, I_c, does the equation $I_a + I_b + I_c = 0$ hold good?
(a) Yes (b) No

SECTION 4.6

4.16 A stranded conductor is an example of a composite conductor.
(a) True (b) False

4.17 $\Sigma \ln A_k = \ln \Pi A_k$
(a) True (b) False

4.18 Is Geometric Mean Distance (GMD) the same as Geometric Mean Radius (GMR)?
(a) Yes (b) No

4.19 Expand $6\sqrt{\Pi_{k=1}^{3} \Pi_{m=1'}^{2'} D_{km}}$

4.20 If the distance between conductors are large compared to the distances between sub-conductors of each conductor, then the GMD between conductors is approximately equal to the distance between conductor centers.
(a) True (b) False

4.22 For a single-phase, two-conductor line with composite conductors x and y, express the inductance of conductor x in terms of GMD and its GMR.

4.23 In a three-phase line, in order to avoid unequal phase inductances due to unbalanced flux linkages, what technique is used?

4.24 For a completely transposed three-phase line identical conductors, each with GMR denoted D_S, with conductor distance D_{12}, D_{23}, and D_{31} give expressions for GMD between phases, and the average per-phase inductance.

4.25 For EHV lines, a common practice of conductor bundling is used. Why?

4.26 Does bundling reduce the series reactance of the line?
(a) Yes (b) No

4.27 Does $r' = e^{-\frac{1}{4}} r = 0.788\,r$, that comes in calculation of inductance, play a role in capacitance computations?
(a) Yes (b) No

4.28 In terms of line-to-line capacitance, the line-to-neutral capacitance of a single-phase transmission line is
(a) same (b) twice (c) one-half

4.29 For either single-phase two-wire line or balanced three-phase three-wire line, with equal phase spacing D and with conductor radius r, the capacitance (line-to-neutral) in F/m is given by C_{an} = __________. Fill in the Blank.

4.30 In deriving expressions for capacitance for a balanced three-phase, three-wire line with equal phase spacing, the following relationships may have been used.
(i) Sum of positive-sequence charges, $q_a + q_b + q_c = 0$
(ii) The sum of the two line-to-line voltages $V_{ab} + V_{ac}$, is equal to three-times the line-to-neutral voltage V_{an}.

Which of the following is true?
(a) both (b) only (i) (c) only (ii) (d) None

SECTION 4.10

4.31 When calculating line capacitance, it is normal practice to replace a stranded conductor by a perfectly conducting solid cylindrical conductor whose radius equals the outside radius of the stranded conductor.
(a) True (b) False

4.32 For bundled-conductor configurations, the expressions for calculating D_{SL} in inductance calculations and D_{SC} in capacitance calculations are analogous, except that the conductor outside radius r replaces the conductor GMR, D_S.
(a) True (b) False

4.33 The current supplied to the transmission-line capacitance is called __________. Fill in the Blank.

4.34 For a completely transposed three-phase line that has balanced positive-sequence voltages, the total reactive power supplied by the three-phase line, in var, is given by Q_{C3} = ____________, in terms of frequency ω, line-to-neutral capacitance C_{an}, and line-to-line voltage V_{LL}.

SECTION 4.11

4.35 Considering lines with neutral conductors and earth return, the effect of earth plane is accounted for by the method of __________ with a perfectly conducting earth plane.

4.36 The affect of the earth plane is to slightly increase the capacitance, an as the line height increases, the effect of earth becomes negligible.
(a) True (b) False

SECTION 4.12

4.37 When the electric field strength at a conductor surface exceeds the breakdown strength of air, current, discharges occur. This phenomenon is called ____________. Fill in the Blank.

4.38 To control corona, transmission lines are usually designed to maintain the calculated conductor surface electric field strength below ________ kV_{rms}/cm. Fill in the Blank.

4.39 Along with limiting corona and its effects, particularly for EHV lines, the maximum ground level electric field strength needs to be controlled to avoid the shock hazard.
(a) True (b) False

SECTION 4.13

4.40 Considering two parallel three-phase circuits that are close together, when calculating the equivalent series-impedance and shunt-admittance matrices, mutual inductive and capacitive couplings between the two circuits can be neglected.
(a) True (b) False

PROBLEMS

SECTION 4.2

4.1 The *Aluminum Electrical Conductor Handbook* lists a dc resistance of 0.01558 ohm per 1000 ft (or 0.05112 ohm per km) at 20 °C and a 60-Hz resistance of 0.0956 ohm per mile (or 0.0594 ohm per km) at 50 °C for the all-aluminum Marigold conductor, which has 61 strands and whose size is 564 mm^2 or 1113 kcmil. Assuming an increase in resistance of 2% for spiraling, calculate and verify the dc resistance. Then calculate the dc resistance at 50 °C, and determine the percentage increase due to skin effect.

4.2 The temperature dependence of resistance is also quantified by the relation $R_2 = R_1[1 + \alpha(T_2 - T_1)]$ where R_1 and R_2 are the resistances at temperatures T_1 and T_2, respectively, and α is known as the temperature coefficient of resistance. If a copper wire has a resistance of 50 Ω at 20 °C, find the maximum permissible operating temperature of the wire if its resistance is to increase by at most 10%. Take the temperature coefficient at 20 °C to be $\alpha = 0.00382$.

4.3 A transmission-line cable, of length 3 km, consists of 19 strands of identical copper conductors, each 1.5 mm in diameter. Because of the twist of the strands, the actual length of each conductor is increased by 5%. Determine the resistance of the cable, if the resistivity of copper is 1.72 μΩ·cm at 20 °C.

4.4 One thousand circular mils or 1 kcmil is sometimes designated by the abbreviation MCM. Data for commercial bare aluminum electrical conductors lists a 60-Hz resistance of 0.0880 ohm per kilometer at 75 °C for a 793-MCM AAC conductor. (a) Determine the cross-sectional conducting area of this conductor in square meters. (b) Find the 60-Hz resistance of this conductor in ohms per kilometer at 50 °C.

4.5 A 60-Hz, 765-kV three-phase overhead transmission line has four ACSR 900 kcmil 54/3 conductors per phase. Determine the 60-Hz resistance of this line in ohms per kilometer per phase at 50 °C.

4.6 A three-phase overhead transmission line is designed to deliver 190.5 MVA at 220 kV over a distance of 63 km, such that the total transmission line loss is not to exceed 2.5% of the rated line MVA. Given the resistivity of the conductor material to be 2.84×10^{-8} Ω-m, determine the required conductor diameter and the conductor size in circular mils. Neglect power losses due to insulator leakage currents and corona.

4.7 If the per-phase line loss in a 60-km-long transmission line is not to exceed 60 kW while it is delivering 100 A per phase, compute the required conductor diameter, if the resistivity of the conductor material is 1.72×10^{-8} Ωm.

SECTIONS 4.4 AND 4.5

4.8 A 60-Hz single-phase, two-wire overhead line has solid cylindrical copper conductors with 1.5 cm diameter. The conductors are arranged in a horizontal configuration with 0.5 m spacing. Calculate in mH/km (a) the inductance of each conductor due to internal flux linkages only, (b) the inductance of each conductor due to both internal and external flux linkages, and (c) the total inductance of the line.

4.9 Rework Problem 4.8 if the diameter of each conductor is: (a) increased by 20% to 1.8 cm, (b) decreased by 20% to 1.2 cm, without changing the phase spacing. Compare the results with those of Problem 4.8.

4.10 A 60-Hz three-phase, three-wire overhead line has solid cylindrical conductors arranged in the form of an equilateral triangle with 4 ft conductor spacing. Conductor diameter is 0.5 in. Calculate the positive-sequence inductance in H/m and the positive-sequence inductive reactance in Ω/km.

4.11 Rework Problem 4.10 if the phase spacing is: (a) increased by 20% to 4.8 ft, (b) decreased by 20% to 3.2 ft. Compare the results with those of Problem 4.10.

4.12 Find the inductive reactance per mile of a single-phase overhead transmission line operating at 60 Hz, given the conductors to be *Partridge* and the spacing between centers to be 20 ft.

4.13 A single-phase overhead transmission line consists of two solid aluminum conductors having a radius of 2.5 cm, with a spacing 3.6 m between centers. (a) Determine the total line inductance in mH/m. (b) Given the operating frequency to be 60 Hz, find the total inductive reactance of the line in Ω/km and in Ω/mi. (c) If the spacing is doubled to 7.2 m, how does the reactance change?

4.14 (a) In practice, one deals with the inductive reactance of the line per phase per mile and use the logarithm to the base 10.
Show that Eq. (4.5.9) of the text can be rewritten as

$$x = k \log \frac{D}{r'} \text{ ohms per mile per phase}$$
$$= x_d + x_a$$

where $x_d = k \log D$ is the inductive reactance spacing factor in ohms per km

$x_a = k \log \frac{1}{r'}$ is the inductive reactance at 1-m spacing in ohms per km

$k = 2.893 \times 10^{-6} f = 1.736$ at 60 Hz.

(b) Determine the inductive reactance per km per phase at 60 Hz for a single-phase line with phase separation of 3 m and conductor radius of 2 cm.
If the spacing is doubled, how does the reactance change?

SECTION 4.6

4.15 Find the GMR of a stranded conductor consisting of six outer strands surrounding and touching one central strand, all strands having the same radius r.

4.16 A bundle configuration for UHV lines (above 1000 kV) has identical conductors equally spaced around a circle, as shown in Figure 4.29. N_b is the number of conductors in the bundle, A is the circle radius, and D_S is the conductor GMR. Using the distance D_{1n} between conductors 1 and n given by $D_{1n} = 2A \sin[(n-1)\pi/N_b]$ for $n = 1, 2, \ldots, N_b$, and the following trigonometric identity:

$$[2 \sin(\pi/N_b)][2 \sin(2\pi/N_b)][2 \sin(3\pi/N_b)] \cdots [2 \sin\{(N_b - 1)\pi/N_b\}] = N_b$$

show that the bundle GMR, denoted D_{SL}, is

$$D_{SL} = [N_b D_S A^{(N_b-1)}]^{(1/N_b)}$$

Also show that the above formula agrees with (4.6.19)–(4.6.21) for EHV lines with $N_b = 2, 3$, and 4.

FIGURE 4.29 Bundle configuration for Problem 4.16

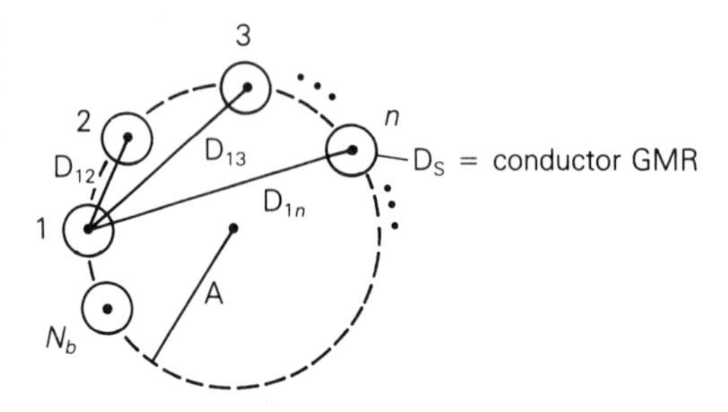

4.17 Determine the GMR of each of the unconventional stranded conductors shown in Figure 4.30. All strands have the same radius r.

FIGURE 4.30 Unconventional stranded conductors for Problem 4.17

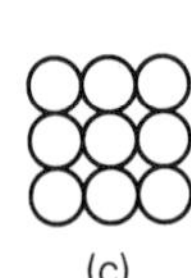

4.18 A 230-kV, 60-Hz, three-phase completely transposed overhead line has one ACSR 954-kcmil (or 564 mm^2) conductor per phase and flat horizontal phase spacing, with 8 m between adjacent conductors. Determine the inductance in H/m and the inductive reactance in Ω/km.

4.19 Rework Problem 4.18 if the phase spacing between adjacent conductors is: (a) increased by 10% to 8.8 m, (b) decreased by 10% to 7.2 m. Compare the results with those of Problem 4.18.

4.20 Calculate the inductive reactance in Ω/km of a bundled 500-kV, 60-Hz, three-phase completely transposed overhead line having three ACSR 1113-kcmil (556.50 mm^2) conductors per bundle, with 0.5 m between conductors in the bundle. The horizontal phase spacings between bundle centers are 10, 10, and 20 m.

4.21 Rework Problem 4.20 if the bundled line has: (a) three ACSR, 1351-kcmil (675.5-mm^2) conductors per phase, (b) three ACSR, 900-kcmil (450-mm^2) conductors per phase, without changing the bundle spacing or the phase spacings between bundle centers. Compare the results with those of Problem 4.20.

4.22 The conductor configuration of a bundled single-phase overhead transmission line is shown in Figure 4.31. Line X has its three conductors situated at the corners of an equilateral triangle with 10-cm spacing. Line Y has its three conductors arranged in a horizontal configuration with 10-cm spacing. All conductors are identical, solid-cylindrical conductors, each with a radius of 2 cm. (a) Find the equivalent representation in terms of the geometric mean radius of each bundle and a separation that is the geometric mean distance.

FIGURE 4.31

Problem 4.22

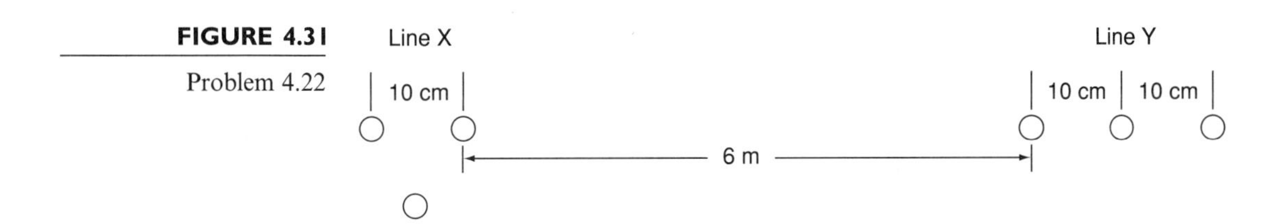

4.23 Figure 4.32 shows the conductor configuration of a completely transposed three-phase overhead transmission line with bundled phase conductors. All conductors have a radius of 0.74 cm with a 30-cm bundle spacing. (a) Determine the inductance per phase in mH/km. (b) Find the inductive line reactance per phase in Ω/km at 60 Hz.

FIGURE 4.32

Problem 4.23

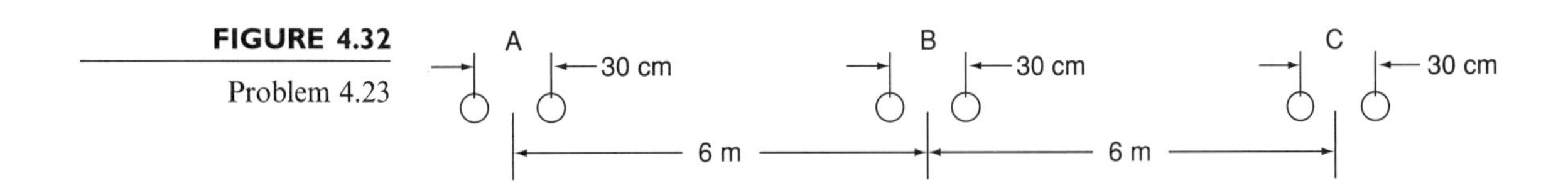

4.24 Consider a three-phase overhead line made up of three phase conductors, Linnet, 336.4 kcmil (170 mm^2), ACSR 26/7. The line configuration is such that the horizontal separation between center of C and that of A is 102 cm, and between that of A and B is also 102 cm in the same line; the vertical separation of A from the line of C–B is 41 cm. If the line is operated at 60 Hz at a conductor temperature of 75 °C, determine the inductive reactance per phase in Ω/km,
(a) By using the formula given in Problem 4.14 (a), and
(b) By using (4.6.18) of the text.

4.25 For the overhead line of configuration shown in Figure 4.33, operating at 60 Hz, and a conductor temperature of 70 °C, determine the resistance per phase, inductive reactance in ohms/km/phase and the current carrying capacity of the overhead line. Each conductor is ACSR Cardinal of Table A.4.

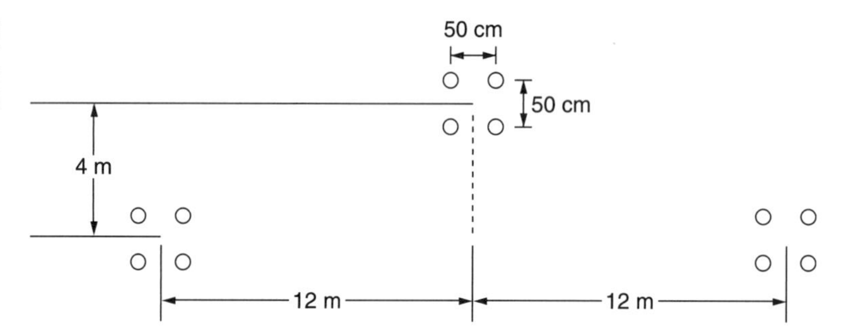

FIGURE 4.33 Line configuration for Problem 4.25

4.26 Consider a symmetrical bundle with N subconductors arranged in a circle of radius A. The inductance of a single-phase symmetrical bundle-conductor line is given by

$$L = 2 \times 10^{-7} \ln \frac{\text{GMD}}{\text{GMR}} \text{ H/m}$$

where GMR is given by $[Nr'(\text{A})^{N-1}]^{1/N}$

$r' = (e^{-1/4}r)$, r being the subconductor radius, and GMD is approximately the distance D between the bundle centers. Note that A is related to the subconductor spacing S in the bundle circle by $S = 2A \sin(\Pi/N)$
Now consider a 965-kV, single-phase, bundle-conductor line with eight subconductors per phase, with phase spacing D = 17 m, and the subconductor spacing S = 45.72 cm. Each subconductor has a diameter of 4.572 cm. Determine the line inductance in H/m.

4.27 Figure 4.34 shows double-circuit conductors' relative positions in Segment 1 of transposition of a completely transposed three-phase overhead transmission line. The inductance is given by

$$L = 2 \times 10^{-7} \ln \frac{\text{GMD}}{\text{GMR}} \text{ H/m/phase}$$

where $\text{GMD} = (D_{AB_{eq}} D_{BC_{eq}} D_{AC_{eq}})^{1/3}$, with mean distances defined by equivalent spacings

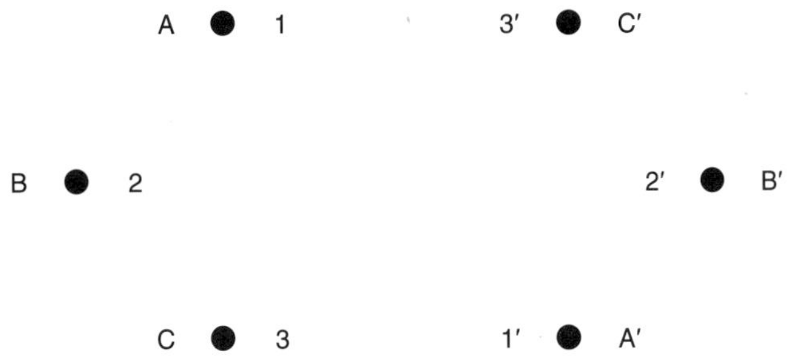

FIGURE 4.34

For Problem 4.27 (Double-circuit conductor configuration)

$$D_{AB_{eq}} = (D_{12}D_{1'2'}D_{12'}D_{1'2})^{1/4}$$

$$D_{BC_{eq}} = (D_{23}D_{2'3'}D_{2'3}D_{23'})^{1/4}$$

$$D_{AC_{eq}} = (D_{13}D_{1'3'}D_{13'}D_{1'3})^{1/4}$$

and GMR $= [(GMR)_A(GMR)_B(GMR)_C]^{1/3}$, with phase GMRs defined by

$$(GMR)_A = [r'D_{11'}]^{1/2}; \quad (GMR)_B = [r'D_{22'}]^{1/2}; \quad (GMR)_C = [r'D_{33'}]^{1/2}$$

and r' is the GMR of phase conductors.

Now consider A 345-kV, three-phase, double-circuit line with phase-conductor's GMR of 1.8 cm, and the horizontal conductor configuration shown in Figure 4.35.
(a) Determine the inductance per meter per phase in henries.
(b) Calculate the inductance of just one circuit and then divide by 2 to obtain the inductance of the double circuit.

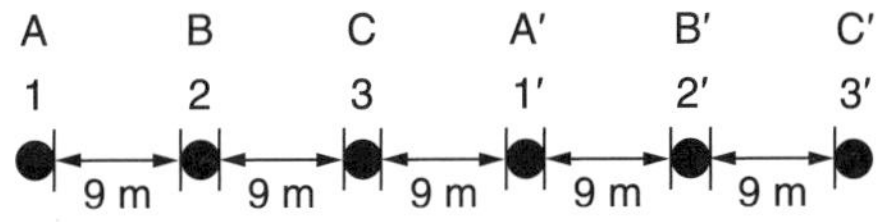

FIGURE 4.35

For Problem 4.27

Find the relative error involved

4.28 For the case of double-circuit, bundle-conductor lines, the same method indicated in Problem 4.27 applies with r' replaced by the bundle's GMR in the calculation of the overall GMR.

Now consider a double-circuit configuration shown in Figure 4.36, which belongs to a 500-kV, three-phase line with bundle conductors of three subconductors at 53-cm spacing. The GMR of each subconductor is given to be 1.5 cm.

Determine the inductive reactance of the line in ohms per km per phase. You may use

$$X_L = 0.1786 \log \frac{GMD}{GMR} \ \Omega/\text{km/phase}$$

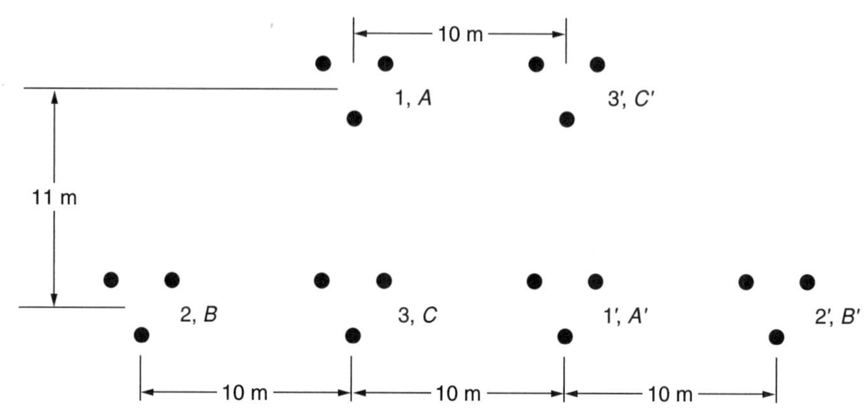

FIGURE 4.36 Configuration for Problem 4.28

4.29 Reconsider Problem 4.28 with an alternate phase placement given below:

	Physical Position					
	1	2	3	1′	2′	3′
Phase Placement	A	B	B'	C	C'	A'

Calculate the inductive reactance of the line in Ω/km/phase.

4.30 Reconsider Problem 4.28 with still another alternate phase placement shown below.

	Physical Position					
	1	2	3	1′	2′	3′
Phase Placement	C	A	B	B'	A'	C'

Find the inductive reactance of the line in Ω/km/phase.

4.31 Figure 4.37 shows the conductor configuration of a three-phase transmission line and a telephone line supported on the same towers. The power line carries a balanced

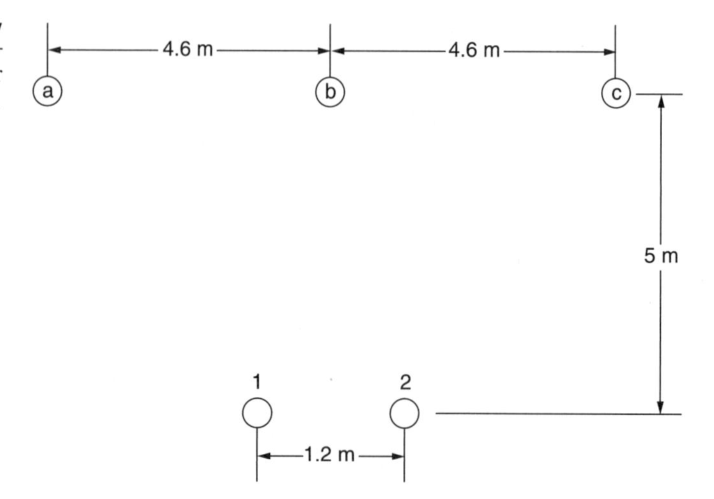

FIGURE 4.37 Conductor layout for Problem 4.31

current of 250 A/phase at 60 Hz, while the telephone line is directly located below phase b. Assume balanced three-phase currents in the power line. Calculate the voltage per kilometer induced in the telephone line.

SECTION 4.9

4.32 Calculate the capacitance-to-neutral in F/m and the admittance-to-neutral in S/km for the single-phase line in Problem 4.8. Neglect the effect of the earth plane.

4.33 Rework Problem 4.32 if the diameter of each conductor is: (a) increased by 20% to 1.8 cm, (b) decreased by 20% to 1.2 cm. Compare the results with those of Problem 4.32.

4.34 Calculate the capacitance-to-neutral in F/m and the admittance-to-neutral in S/km for the three-phase line in Problem 4.10. Neglect the effect of the earth plane.

4.35 Rework Problem 4.34 if the phase spacing is: (a) increased by 20% to 146.4 cm, (b) decreased by 20% to 97.6 cm. Compare the results with those of Problem 4.34.

4.36 The line of Problem 4.23 as shown in Figure 4.32 is operating at 60 Hz. Determine (a) the line-to-neutral capacitance in nF/km per phase; (b) the capacitive reactance in Ω-km per phase; and (c) the capacitive reactance in Ω per phase for a line length of 160 km.

4.37 (a) In practice, one deals with the capacitive reactance of the line in ohms-km to neutral. Show that Eq. (4.9.15) of the text can be rewritten as

$$\begin{aligned} X_C &= k' \log \frac{D}{r} \text{ ohms-km to neutral} \\ &= x'_d + x'_a \end{aligned}$$

where $x'_d = k' \log D$ is the capacitive reactance spacing factor

$x'_a = k' \log \frac{1}{r}$ is the capacitive reactance at 1-m spacing

$k' = (21.65 \times 10^6)/f = 0.36 \times 10^6$ at $f = 60$ Hz.

(b) Determine the capacitive reactance in Ω-km for a single-phase line of Problem 4.14. If the spacing is doubled, how does the reactance change?

4.38 The capacitance per phase of a balanced three-phase overhead line is given by

$$C = \frac{0.04217}{\log(\text{GMD}/r)} \; \mu f/\text{km/phase}$$

For the line of Problem 4.24, determine the capacitive reactance per phase in Ω-km.

SECTION 4.10

4.39 Calculate the capacitance-to-neutral in F/m and the admittance-to-neutral in S/km for the three-phase line in Problem 4.18. Also calculate the line-charging current in kA/phase if the line is 100 km in length and is operated at 230 kV. Neglect the effect of the earth plane.

4.40 Rework Problem 4.39 if the phase spacing between adjacent conductors is: (a) increased by 10% to 8.8 m, (b) decreased by 10% to 7.2 m. Compare the results with those of Problem 4.39.

4.41 Calculate the capacitance-to-neutral in F/m and the admittance-to-neutral in S/km for the line in Problem 4.20. Also calculate the total reactive power in Mvar/km supplied by the line capacitance when it is operated at 500 kV. Neglect the effect of the earth plane.

4.42 Rework Problem 4.41 if the bundled line has: (a) three ACSR, 1351-kcmil (685-mm^2) conductors per phase, (b) three ACSR, 900-kcmil (450 mm^2) conductors per phase, without changing the bundle spacing or the phase spacings between bundle centers.

4.43 Three ACSR *Drake* conductors are used for a three-phase overhead transmission line operating at 60 Hz. The conductor configuration is in the form of an isosceles triangle with sides of 6 m, 6 m, and 12 m. (a) Find the capacitance-to-neutral and capacitive reactance-to-neutral for each 1-km length of line. (b) For a line length of 280 km and a normal operating voltage of 220 kV, determine the capacitive reactance-to-neutral for the entire line length as well as the charging current per km and total three-phase reactive power supplied by the line capacitance.

4.44 Consider the line of Problem 4.25. Calculate the capacitive reactance per phase in Ω-km.

SECTION 4.11

4.45 For an average line height of 10 m, determine the effect of the earth on capacitance for the single-phase line in Problem 4.32. Assume a perfectly conducting earth plane.

4.46 A three-phase 60-Hz, 125-km overhead transmission line has flat horizontal spacing with three identical conductors. The conductors have an outside diameter of 3.28 cm with 12 m between adjacent conductors. (a) Determine the capacitive reactance-to-neutral in Ω-m per phase and the capacitive reactance of the line in Ω per phase. Neglect the effect of the earth plane. (b) Assuming that the conductors are horizontally placed 20 m above ground, repeat (a) while taking into account the effect of ground. Consider the earth plane to be a perfect conductor.

4.47 For the single-phase line of Problem 4.14 (b), if the height of the conductor above ground is 24 m, determine the line-to-line capacitance in F/m. Neglecting earth effect, evaluate the relative error involved. If the phase separation is doubled, repeat the calculations.

4.48 The capacitance of a single-circuit, three-phase transposed line, and with configuration shown in Figure 4.38 including ground effect, with conductors not equilaterally spaced, is given by

$$C_{an} \frac{2\pi\varepsilon_0}{\ln \frac{D_{eq}}{r} - \ln \frac{H_m}{H_s}} \text{ F/m Line-to-neutral}$$

where $D_{eq} = \sqrt[3]{D_{12}D_{23}D_{13}} = \text{GMD}$

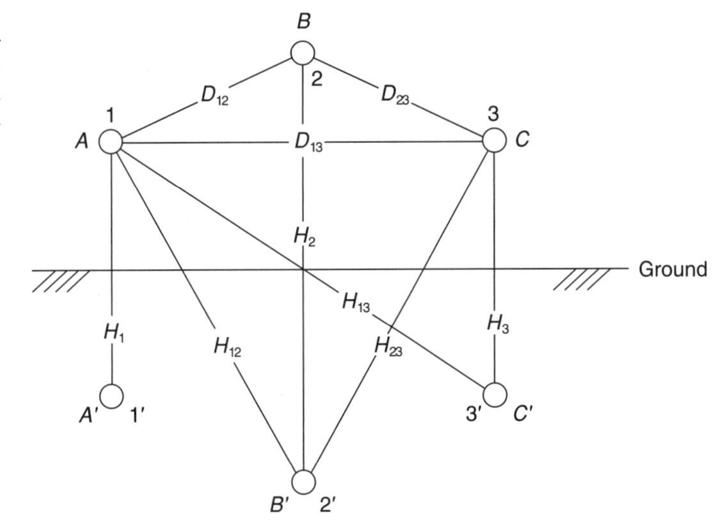

FIGURE 4.38

Three-phase single-circuit line configuration including ground effect for Problem 4.48

$$r = \text{conductor's outside radius}$$

$$H_m = (H_{12}H_{23}H_{13})^{1/3}$$

$$H_s = (H_1H_2H_3)^{1/3}$$

(a) Now consider Figure 4.39 in which the configuration of a three-phase, single circuit, 345-kV line, with conductors having an outside diameter of 27.051 mm (or 1.065 in.), is shown. Determine the capacitance to neutral in F/m, including the ground effect. (b) Next, neglecting the effect of ground, see how the value changes.

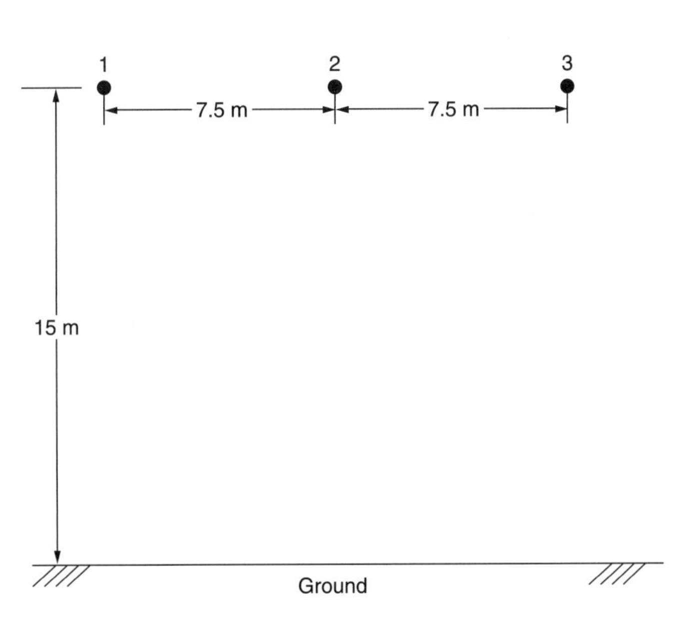

FIGURE 4.39

Configuration for Problem 4.48 (a)

4.49 The capacitance to neutral, neglecting the ground effect, for the three-phase, single-circuit, bundle-conductor line is given by

$$C_{an} = \frac{2\pi\varepsilon_0}{\ell n\left(\frac{\text{GMD}}{\text{GMR}}\right)} \text{ F/m Line-to-neutral}$$

where $\text{GMD} = (D_{AB}D_{BC}D_{AC})^{1/3}$

$$\text{GMR} = [rN(A)^{N-1}]^{1/N}$$

in which N is the number of subconductors of the bundle conductor on a circle of radius A, and each subconductor has an outside radius of r.
The capacitive reactance in mega-ohms for 1 km of line, at 60 Hz, can be shown to be

$$X_C = 0.11 \log\left(\frac{\text{GMD}}{\text{GMR}}\right) = X_a' + X_d'$$

where $X_a' = 0.11 \log\left(\frac{1}{\text{GMR}}\right)$ and $X_d' = 0.11 \log(\text{GMD})$

Note that A is related to the bundle spacing S given by

$$A = \frac{S}{2 \sin\left(\frac{\pi}{N}\right)} \quad \text{for } N > 1$$

Using the above information, for the configuration shown in Figure 4.40, compute the capacitance to neutral in F/m, and the capacitive reactance in Ω-km to neutral, for the three-phase, 765-kV, 60-Hz, single-circuit, bundle-conductor line ($N = 4$), with subconductor's outside diameter of 3 cm and subconductor spacing (S) of 46 cm.

FIGURE 4.40 Configuration for Problem 4.49

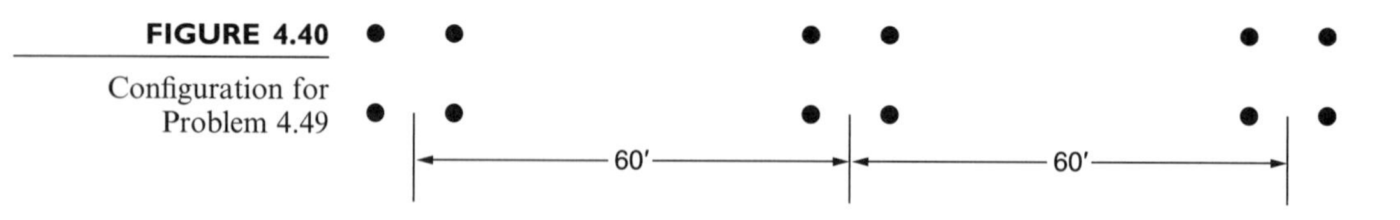

SECTION 4.12

4.50 Calculate the conductor surface electric field strength in kV_{rms}/cm for the single-phase line in Problem 4.32 when the line is operating at 20 kV. Also calculate the ground-level electric field strength in kV_{rms}/m directly under one conductor. Assume a line height of 10 m.

4.51 Rework Problem 4.50 if the diameter of each conductor is: (a) increased by 25% to 1.875 cm, (b) decreased by 25% to 1.125 cm, without changing the phase spacings. Compare the results with those of Problem 4.50.

CASE STUDY QUESTIONS

a. Why is aluminum today's choice of metal for overhead transmission line conductors versus copper or some other metal? How does the use of steel together with aluminum as well as aluminum alloys and composite materials improve conductor performance?

b. What is a high-temperature conductor? What are its advantages over conventional ACSR and AAC conductors? What are its drawbacks?

c. What are the concerns among utilities about porcelain insulators used for overhead transmission lines in the United States?

d. What are the advantages of toughened glass insulators versus porcelain? What are the advantages of polymer insulators versus porcelain? What are the disadvantages of polymer insulators?

REFERENCES

1. Electric Power Research Institute (EPRI), *EPRI AC Transmission Line Reference Book—200 kV and Above* (Palo Alto, CA: EPRI, www.epri.com, December 2005).
2. Westinghouse Electric Corporation, *Electrical Transmission and Distribution Reference Book*, 4th ed. (East Pittsburgh, PA, 1964).
3. General Electric Company, *Electric Utility Systems and Practices*, 4th ed. (New York: Wiley, 1983).
4. John R. Carson, "Wave Propagation in Overhead Wires with Ground Return," *Bell System Tech. J.* 5 (1926): 539–554.
5. C. F. Wagner and R. D. Evans, *Symmetrical Components* (New York: McGraw-Hill, 1933).
6. Paul M. Anderson, *Analysis of Faulted Power Systems* (Ames, IA: Iowa State Press, 1973).
7. M. H. Hesse, "Electromagnetic and Electrostatic Transmission Line Parameters by Digital Computer," *Trans. IEEE* PAS-82 (1963): 282–291.
8. W. D. Stevenson, Jr., *Elements of Power System Analysis*, 4th ed. (New York: McGraw-Hill, 1982).
9. C. A. Gross, *Power System Analysis* (New York: Wiley, 1979).

10. A. J. Peterson, Jr. and S. Hoffmann, "Transmission Line Conductor Design Comes of Age," *Transmission & Distribution World Magazine* (www.tdworld.com, June 2003).

11. ANCI C2. *National Electrical Safety Code*, 2007 edition (New York: Institute of Electrical and Electronics Engineers).

12. R. S. Gorur, "Six Utilities Share Their Perspectives on Insulators," *Transmission & Distribution World Magazine*, (www.tdworld.com, April 1, 2010).

Series capacitor installation at Goshen Substation, Goshen, Idaho, USA rated at 395 kV, 965 Mvar (Courtesy of PacifiCorp)

5 TRANSMISSION LINES: STEADY-STATE OPERATION

In this chapter, we analyze the performance of single-phase and balanced three-phase transmission lines under normal steady-state operating conditions. Expressions for voltage and current at any point along a line are developed, where the distributed nature of the series impedance and shunt admittance is taken into account. A line is treated here as a two-port network for which the *ABCD* parameters and an equivalent π circuit are derived. Also, approximations are given for a medium-length line lumping the shunt admittance, for a short line neglecting the shunt admittance, and for a lossless line assuming zero series resistance and shunt conductance. The concepts of *surge impedance loading* and transmission-line *wavelength* are also presented.

An important issue discussed in this chapter is *voltage regulation*. Transmission-line voltages are generally high during light load periods and low during heavy load periods. Voltage regulation, defined in Section 5.1, refers to the change in line voltage as line loading varies from no-load to full load.

Another important issue discussed here is line loadability. Three major line-loading limits are: (1) the thermal limit, (2) the voltage-drop limit, and (3) the steady-state stability limit. Thermal and voltage-drop limits are discussed in Section 5.1. The theoretical steady-state stability limit, discussed in Section 5.4 for lossless lines and in Section 5.5 for lossy lines, refers to the ability of synchronous machines at the ends of a line to remain in synchronism. Practical line loadability is discussed in Section 5.6.

In Section 5.7 we discuss line compensation techniques for improving voltage regulation and for raising line loadings closer to the thermal limit.

CASE STUDY High Voltage Direct Current (HVDC) applications embedded within ac power system grids have many benefits. A bipolar HVDC transmission line has only two insulated sets of conductors versus three for an ac transmission line. As such, HVDC transmission lines have smaller transmission towers, narrower rights-of-way, and lower line losses compared to ac lines with similar capacity. The resulting cost savings can offset the higher converter station costs of HVDC. Further, HVDC may be the only feasible method to: (1) interconnect two asynchronous ac networks; (2) utilize long underground or underwater cable circuits; (3) bypass network congestion; (4) reduce fault currents; (5) share utility rights-of-way without degrading reliability; and (6) mitigate environmental concerns. The following article provides an overview of HVDC along with HVDC applications [6].

The ABCs of HVDC Transmission Technologies: An Overview of High Voltage Direct Current Systems and Applications

BY MICHAEL P. BAHRMAN
AND BRIAN K. JOHNSON

High voltage direct current (HVDC) technology has characteristics that make it especially attractive for certain transmission applications. HVDC transmission is widely recognized as being advantageous for long-distance bulk-power delivery, asynchronous interconnections, and long submarine cable crossings. The number of HVDC projects committed or under consideration globally has increased in recent years reflecting a renewed interest in this mature technology. New converter designs have broadened the potential range of HVDC transmission to include applications for underground, offshore, economic replacement of reliability-must-run generation, and voltage stabilization. This broader range of applications has contributed to the recent growth of HVDC transmission. There are approximately ten new HVDC projects under construction or active consideration in North America along with many more projects underway globally. Figure 1 shows the Danish terminal for Skagerrak's pole 3, which is rated 440 MW. Figure 2 shows the ±500-kV HVDC transmission line for the 2,000 MW Intermountain Power Project between Utah and California. This article discusses HVDC technologies, application areas where HVDC is favorable compared to ac transmission, system configuration, station design, and operating principles.

Figure 1
HVDC converter station with ac filters in the foreground and valve hall in the background

Figure 2
A ±500-kV HVDC transmission line

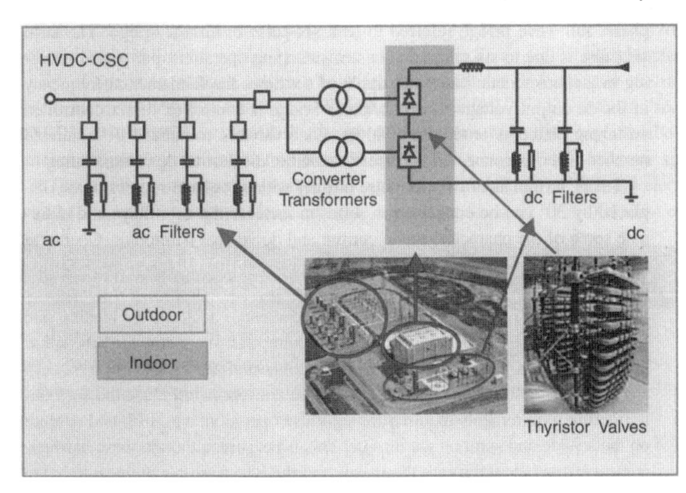

Figure 3
Conventional HVDC with current source converters

CORE HVDC TECHNOLOGIES

Two basic converter technologies are used in modem HVDC transmission systems. These are conventional line-commutated current source converters (CSCs) and self-commutated voltage source converters (VSCs). Figure 3 shows a conventional HVDC converter station with CSCs while Figure 4 shows a HVDC converter station with VSCs.

LINE-COMMUTATED CURRENT SOURCE CONVERTER

Conventional HVDC transmission employs line-commutated CSCs with thyristor valves. Such converters require a synchronous voltage source in order to operate. The basic building block used for HVDC conversion is the three-phase, full-wave bridge referred to as a six-pulse or Graetz bridge. The term six-pulse is due to six commutations or switching operations per period resulting in a characteristic harmonic ripple of six times the fundamental frequency in the dc output voltage. Each six-pulse bridge is comprised of six controlled switching elements or thyristor valves. Each valve is comprised of a suitable number of series-connected thyristors to achieve the desired dc voltage rating.

The dc terminals of two six-pulse bridges with ac voltage sources phase displaced by 30° can be connected in series to increase the dc voltage and eliminate some of the characteristic ac current and dc voltage harmonics. Operation in this manner is referred to as 12-pulse operation. In 12-pulse operation, the characteristic ac current and dc voltage harmonics have frequencies of $12n \pm 1$ and $12n$, respectively. The 30° phase displacement is achieved by feeding one bridge through a transformer with a wye-connected secondary and the other bridge through a transformer with a delta-connected secondary. Most modern HVDC transmission schemes utilize 12-pulse converters to reduce the harmonic filtering requirements required for six-pulse operation; e.g., fifth and seventh on the ac side and sixth on the dc side. This is because, although these harmonic currents still flow through the valves and the transformer windings, they are 180° out of phase and cancel out on the primary side of the converter transformer. Figure 5 shows the thyristor valve arrangement for a 12-pulse converter with three quadruple valves, one for each phase. Each thyristor valve is built up with series-connected thyristor modules.

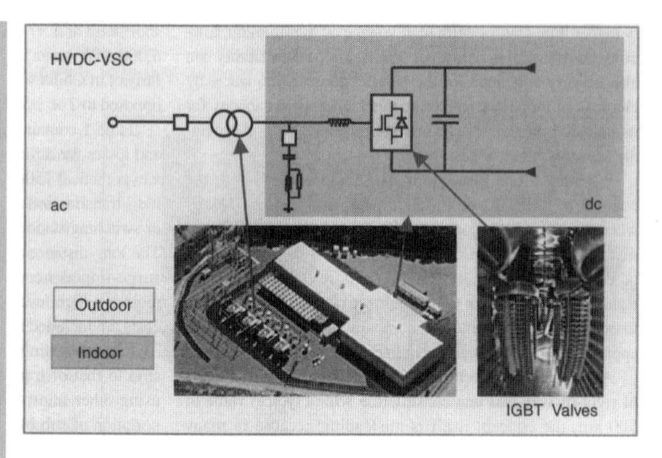

Figure 4
HVDC with voltage source converters

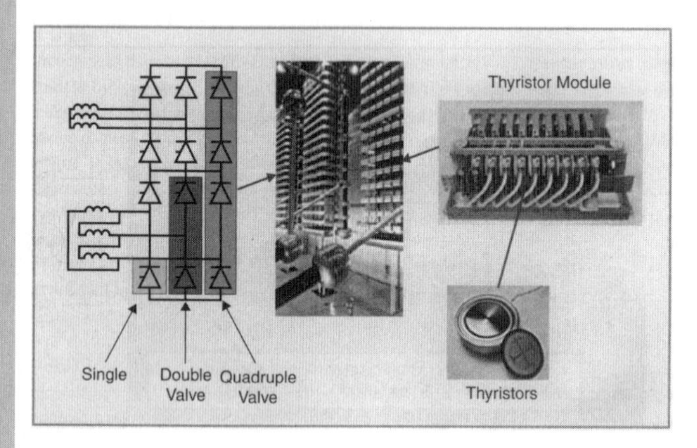

Figure 5
Thyristor valve arrangement for a 12-pulse converter with three quadruple valves, one for each phase

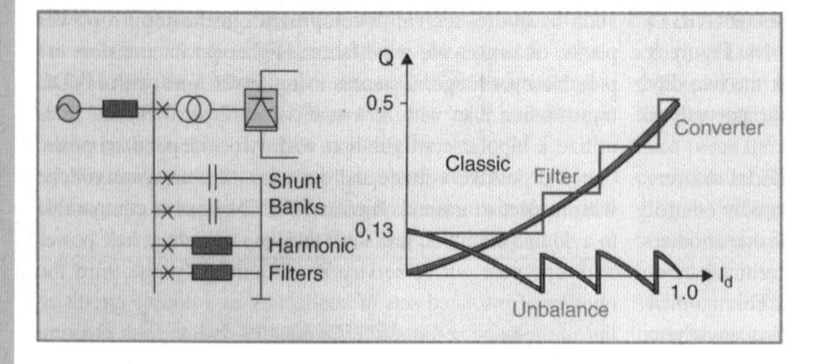

Figure 6
Reactive power compensation for conventional HVDC converter station

Line-commutated converters require a relatively strong synchronous voltage source in order to commutate. Commutation is the transfer of current from one phase to another in a synchronized firing sequence of the thyristor valves. The three-phase symmetrical short circuit capacity available from the network at the converter connection point should be at least twice the converter rating for converter operation. Line-commutated CSCs can only operate with the ac current lagging the voltage, so the conversion process demands reactive power. Reactive power is supplied from the ac filters, which look capacitive at the fundamental frequency, shunt banks, or series capacitors that are an integral part of the converter station. Any surplus or deficit in reactive power from these local sources must be accommodated by the ac system. This difference in reactive power needs to be kept within a given band to keep the ac voltage within the desired tolerance. The weaker the ac system or the further the converter is away from generation, the tighter the reactive power exchange must be to stay within the desired voltage tolerance. Figure 6 illustrates the reactive power demand, reactive power compensation, and reactive power exchange with the ac network as a function of dc load current.

Converters with series capacitors connected between the valves and the transformers were introduced in the late 1990s for weak-system, back-to-back applications. These converters are referred to as capacitor-commutated converters (CCCs). The series capacitor provides some of the converter reactive power compensation requirements automatically with load current and provides part of the commutation voltage, improving voltage stability. The overvoltage protection of the series capacitors is simple since the capacitor is not exposed to line faults, and the fault current for internal converter faults is limited by the impedance of the converter transformers. The CCC configuration allows higher power ratings in areas were the ac network is close to its voltage stability limit. The asynchronous Garabi interconnection between Brazil and Argentina consists of 4 × 550 MW parallel CCC links. The Rapid City Tie between the Eastern and Western interconnected systems consists of 2 × 10 MW parallel CCC links (Figure 7). Both installations use a modular design with

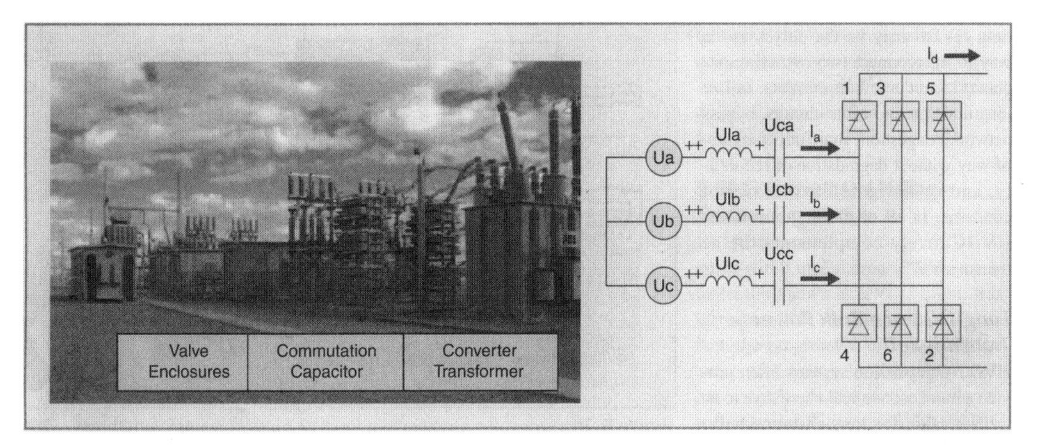

Figure 7
Asynchronous back-to-back tie with capacitor-commutated converter near Rapid City, South Dakota

converter valves located within prefabricated electrical enclosures rather than a conventional valve hall.

SELF-COMMUTATED VOLTAGE SOURCE CONVERTER

HVDC transmission using VSCs with pulse-width modulation (PWM), commercially known as HVDC Light, was introduced in the late 1990s. Since then the progression to higher voltage and power ratings for these converters has roughly paralleled that for thyristor valve converters in the 1970s. These VSC-based systems are self-commutated with insulated-gate bipolar transistor (IGBT) valves and solid-dielectric extruded HVDC cables. Figure 8 illustrates solid-state converter development for the two different types of converter technologies using thyristor valves and IGBT valves.

HVDC transmission with VSCs can be beneficial to overall system performance. VSC technology can rapidly control both active and reactive power independently of one another. Reactive power can also be controlled at each terminal independent of the dc transmission voltage level. This control capability gives total flexibility to place converters anywhere in the ac network since there is no restriction on minimum network short-circuit capacity. Self-commutation with VSC even permits black start; i.e., the converter can be used to synthesize a balanced set of three phase voltages like a virtual synchronous generator. The dynamic support of the ac voltage at each converter terminal improves the voltage stability and can increase the transfer capability of the sending- and receiving-end ac systems, thereby leveraging the transfer capability of the dc link. Figure 9 shows the IGBT converter valve arrangement for a VSC station. Figure 10 shows the active and reactive power operating range for a converter station with a VSC. Unlike conventional HVDC transmission, the converters themselves have

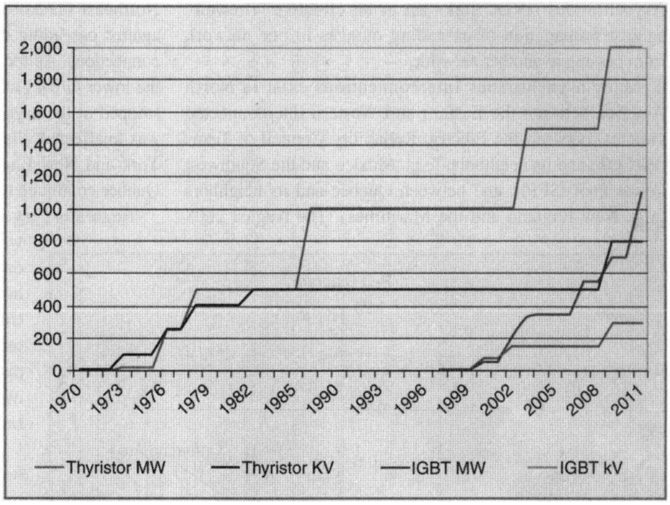

Figure 8
Solid-state converter development

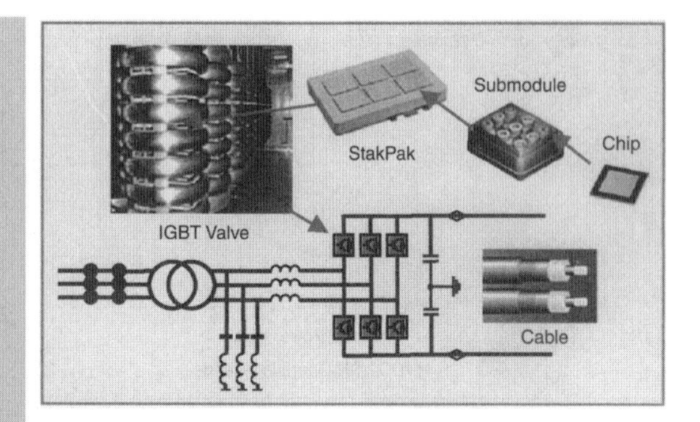

Figure 9
HVDC IGBT valve converter arrangement

no reactive power demand and can actually control their reactive power to regulate ac system voltage just like a generator.

HVDC APPLICATIONS

HVDC transmission applications can be broken down into different basic categories. Although the rationale for selection of HVDC is often economic, there may be other reasons for its selection. HVDC may be the only feasible way to interconnect two asynchronous networks, reduce fault currents, utilize long underground cable circuits, bypass network congestion, share utility rights-of-way without degradation of reliability, and to mitigate environmental concerns. In all of these applications, HVDC nicely complements the ac transmission system.

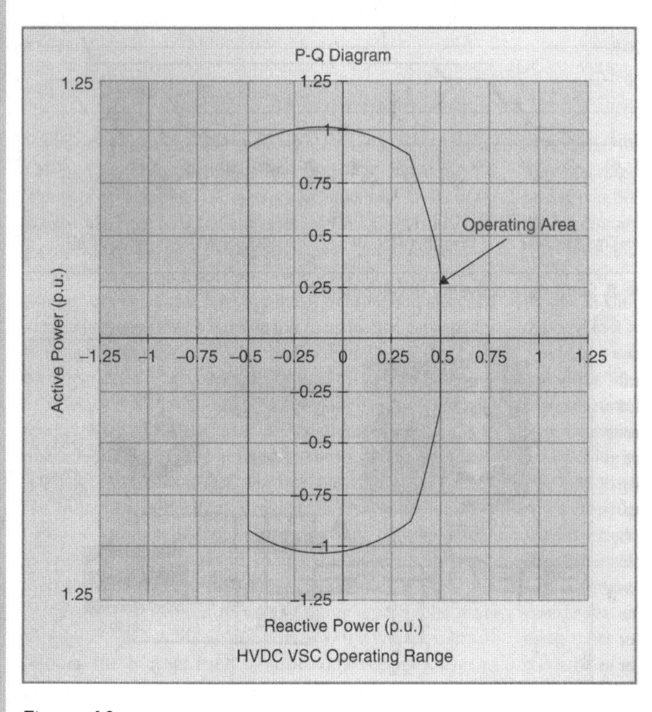

Figure 10
Operating range for voltage source converter HVDC transmission

LONG-DISTANCE BULK POWER TRANSMISSION

HVDC transmission systems often provide a more economical alternative to ac transmission for long-distance bulk-power delivery from remote resources such as hydroelectric developments, mine-mouth power plants, or large-scale wind farms. Higher power transfers are possible over longer distances using fewer lines with HVDC transmission than with ac transmission. Typical HVDC lines utilize a bipolar configuration with two independent poles, one at a positive voltage and the other at a negative voltage with respect to ground. Bipolar HVDC lines are comparable to a double circuit ac line since they can operate at half power with one pole out of service but require only one-third the number of insulated sets of conductors as a double circuit ac line. Automatic restarts from temporary dc line fault clearing sequences are routine even for generator outlet transmission. No synchro-checking is required as for automatic reclosures following ac line faults since the dc restarts do not expose turbine generator units to high risk of transient torque amplification from closing into faults or across high phase angles. The controllability of HVDC links offer firm transmission capacity without limitation due to network congestion or loop flow on parallel paths. Controllability allows the HVDC to "leap-frog" multiple "choke-points" or bypass sequential path limits in the ac network. Therefore, the utilization of HVDC links is usually higher than that for extra high voltage ac transmission, lowering the transmission cost per MWh. This controllability can also be very beneficial for the parallel transmission since, by eliminating loop flow, it frees up this transmission capacity for its intended purpose of serving intermediate load and providing an outlet for local generation.

Whenever long-distance transmission is discussed, the concept of "break-even distance" frequently arises. This is where the savings in line costs offset the higher converter station costs. A bipolar HVDC line uses only two insulated sets of conductors rather than three. This results in narrower rights-of-way, smaller transmission towers, and lower line losses than with ac lines of comparable capacity. A rough approximation of the savings in line construction is 30%.

Although break-even distance is influenced by the costs of right-of-way and line construction with a typical value of 500 km, the concept itself is misleading because in many cases more ac lines are needed to deliver the same power over the same distance due to system stability limitations. Furthermore, the long-distance ac lines usually require intermediate switching stations and reactive power compensation. This can increase the substation costs for ac transmission to the point where it is comparable to that for HVDC transmission.

For example, the generator outlet transmission alternative for the ±250-kV, 500-MW Square Butte Project was two 345-kV series-compensated ac transmission lines. The 12,600-MW Itaipu project has half its power delivered on three 800-kV series-compensated ac lines (three circuits) and the other half delivered on two ±600-kV bipolar HVDC lines (four circuits). Similarly, the ±500-kV, 1,600-MW Intermountain Power Project (IPP) ac alternative comprised two 500-kV ac lines. The IPP takes advantage of the double-circuit nature of the bipolar line and includes a 100% short-term and 50% continuous monopolar overload. The first 6,000-MW stage of the transmission for the Three Gorges Project in China would have required 5 × 500-kV ac lines as opposed to 2 × ±500-kV, 3,000-MW bipolar HVDC lines.

Table 1 contains an economic comparison of capital costs and losses for different ac and dc transmission alternatives for a hypothetical 750-mile (1200-km), 3,000-MW transmission system. The long transmission distance requires intermediate substations or switching stations and shunt reactors for the ac alternatives. The long distance and heavy power transfer, nearly twice the surge-impedance loading on the 500-kV ac alternatives, require a high level of series compensation. These ac station costs are included in the cost estimates for the ac alternatives.

It is interesting to compare the economics for transmission to that of transporting an equivalent amount of energy using other transport methods, in this case using rail transportation of sub-bituminous western coal with a heat content of 8,500 Btu/lb (19.8 MJ/kg) to support a 3,000-MW base load power plant with heat rate of 8,500 Btu/kWh (9 MJ/kWh) operating at an 85% load factor. The rail route is assumed to be longer than the more direct transmission route; i.e., 900 miles (1400 km). Each unit train is comprised of 100 cars each carrying 100 tons (90 tonnes) of coal. The plant requires three unit trains per day. The annual coal transportation costs are about US$560 million per year at an assumed rate of US$50/ton ($55/tonne). This works out to be US$186 kW/year and US$25 per MWh. The annual diesel fuel consumed in the process is in excess of 20 million gallons (76 million Liters) at 500 net ton-miles per gallon (193 net tonne-km per liter). The rail transportation costs are subject to escalation and congestion whereas the transmission costs are fixed. Furthermore, transmission is the only way to deliver remote renewable resources.

UNDERGROUND AND SUBMARINE CABLE TRANSMISSION

Unlike the case for ac cables, there is no physical restriction limiting the distance or power level for HVDC underground or submarine cables. Underground cables can be used on shared rights-of-way with other utilities without impacting reliability concerns over use of common corridors. For underground or submarine cable systems there is considerable savings in installed cable costs and cost of losses when using HVDC transmission. Depending on the power level to be transmitted, these savings can offset the higher converter station costs at distances of 40 km or more. Furthermore, there is a drop-off in cable capacity with ac transmission over distance due to its reactive component of charging current since cables have higher capacitances and lower inductances than ac overhead lines. Although this can be compensated by intermediate shunt compensation for underground cables at increased expense, it is not practical to do so for submarine cables.

For a given cable conductor area, the line losses with HVDC cables can be about half those of ac cables. This is due to ac cables requiring more conductors (three phases), carrying the reactive component of current, skin-effect, and induced currents in the cable sheath and armor.

With a cable system, the need to balance unequal loadings or the risk of postcontingency overloads often necessitates use of a series-connected reactors or phase shifting transformers. These potential problems do not exist with a controlled HVDC cable system.

Extruded HVDC cables with prefabricated joints used with VSC-based transmission are lighter, more flexible, and easier to splice than the mass-impregnated oil-paper cables (MINDs) used for conventional HVDC transmission, thus making them more conducive for land cable applications where transport limitations and extra splicing

TABLE 1 Comparative costs of HVDC and EHV AC transmission alternatives

	DC Alternatives				AC Alternatives			Hybrid AC/DC Alternative		
Alternative	+ 500 Kv Bipole	2 × +500 kV 2 bipoles	+600 kV Bipole	+800 kV Bipole	500 kV 2 Single Ckt	500 kV Double Ckt	765 kV 2 Singl Ckt	+500 kV Bipole	500 kV Single Ckt	Total AC + DC
Capital Cost										
Rated Power (MW)	3000	4000	3000	3000	3000	3000	3000	3000	1500	4500
Station costs including reactive compensation (M$)	$420	$680	$465	$510	$542	$542	$630	$420	$302	$722
Transmission line cost (M$/mile)*	$1.60	$1.60	$1.80	$1.95	$2.00	$3.20	$2.80	$1.60	$2.00	
Distance in miles*	750	1,500	750	750	1,500	750	1,500	750	750	1,500
Transmission Line Cost (M$)	$1,200	$2,400	$1,350	$1,463	$3,000	$2,400	$4,200	$1,200	$1,500	$2,700
Total Cost (M$)	**$1,620**	**$3,080**	**$1,815**	**$1,973**	**$3,542**	**$2,942**	**$4,830**	**$1,620**	**$1,802**	**$3,422**
Annual Payment, 30 years @ 10%	$172	$327	$193	$209	$376	$312	$512	$172	$191	$363
Cost per kW-Yr	$57.28	$81.68	$64.18	$69.75	$125.24	$104.03	$170.77	$57.28	$127.40	$80.66
Cost per MWh @ 85% Utilization Factor	$7.69	$10.97	$8.62	$9.37	$16.82	$13.97	$22.93	$7.69	$17.11	$10.83
Losses @ full load	193	134	148	103	208	208	139	106	48	154
Losses at full load in %	6.44%	3.35%	4.93%	3.43%	6.93%	6.93%	4.62%	5.29%	4.79%	5.12%
Capitalized cost of losses @ $1500 kW (M$)	$246	$171	$188	$131	$265	$265	$177	$135	$61	$196
Parameters:										
Interest rate %	10%									
Capitalized cost of losses $/kW	$1,500									

Note:
AC current assumes 94% pf
Full load converter station losses = 9.75% per station
Total substation losses (transformers, reactors) assumed = 0.5% of rated power

*1 mile = 1.6 km

costs can drive up installation costs. The lower-cost cable installations made possible by the extruded HVDC cables and prefabricated joints makes long-distance underground transmission economically feasible for use in areas with rights-of-way constraints or subject to permitting difficulties or delays with overhead lines.

ASYNCHRONOUS TIES

With HVDC transmission systems, interconnections can be made between asynchronous networks for more economic or reliable system operation. The asynchronous interconnection allows interconnections of mutual benefit while providing a buffer between the two systems. Often these interconnections use back-to-back converters with no transmission line. Asynchronous HVDC links act as an effective "firewall" against propagation of cascading outages in one network from passing to another network.

Many asynchronous interconnections exist in North America between the Eastern and Western interconnected systems, between the Electric Reliability Council of Texas (ERCOT) and its neighbors, [e.g., Mexico and the Southwest Power Pool (SPP)], and between Quebec and its neighbors (e.g., New England and the Maritimes). The August 2003 Northeast blackout provides an example of the "firewall" against cascading outages provided by asynchronous interconnections. As the outage expanded and propagated around the lower Great Lakes and through Ontario and New York, it stopped at the asynchronous interface with Quebec. Quebec was unaffected; the weak ac interconnections between New York and New England tripped, but the HVDC links from Quebec continued to deliver power to New England.

Regulators try to eliminate "seams" in electrical networks because of their potential restriction on power markets. Electrical "seams," however, serve as natural points of separation by acting as "shear-pins," thereby reducing the impact of large-scale system disturbances. Asynchronous ties can eliminate market "seams" while retaining natural points of separation.

Interconnections between asynchronous networks are often at the periphery of the respective systems where the networks tend to be weak relative to the desired power transfer. Higher power transfers can be achieved with improved voltage stability in weak system applications using CCCs. The dynamic voltage support and improved voltage stability offered by VSC-based converters permits even higher power transfers without as much need for ac system reinforcement. VSCs do not suffer commutation failures, allowing fast recoveries from nearby ac faults. Economic power schedules that reverse power direction can be made without any restrictions since there is no minimum power or current restrictions.

OFFSHORE TRANSMISSION

Self-commutation, dynamic voltage control, and black-start capability allow compact VSC HVDC transmission to serve isolated loads on islands or offshore production platforms over long-distance submarine cables. This capability can eliminate the need for running expensive local generation or provide an outlet for offshore generation such as that from wind. The VSCs can operate at variable frequency to more efficiently drive large compressor or pumping loads using high-voltage motors. Figure 11 shows the Troll A production platform in the North Sea where power to drive compressors is delivered from shore to reduce the higher carbon emissions and higher O&M costs associated with less efficient platform-based generation.

Large remote wind generation arrays require a collector system, reactive power support, and outlet transmission. Transmission for wind generation must often traverse scenic or environmentally sensitive areas or bodies of water. Many of the better wind sites with higher capacity factors are located offshore. VSC-based HVDC transmission allows efficient use of long-distance land or submarine cables and provides reactive support to the wind generation complex. Figure 12 shows a design for an

Figure 11
VSC power supply to Troll A production platform

Figure 12
VSC converter for offshore wind generation

offshore converter station designed to transmit power from offshore wind generation.

MULTITERMINAL SYSTEMS

Most HVDC systems are for point-to-point transmission with a converter station at each end. The use of intermediate taps is rare. Conventional HVDC transmission uses voltage polarity reversal to reverse the power direction. Polarity reversal requires no special switching arrangement for a two-terminal system where both terminals reverse polarity by control action with no switching to reverse power direction. Special dc-side switching arrangements are needed for polarity reversal in a multiterminal system, however, where it may be desired to reverse the power direction at a tap while maintaining the same power direction on the remaining terminals. For a bipolar system this can be done by connecting the converter to the opposite pole. VSC HVDC transmission, however, reverses power through reversal of the current direction rather than voltage polarity. Thus, power can be reversed at an intermediate tap independently of the main power flow direction without switching to reverse voltage polarity.

POWER DELIVERY TO LARGE URBAN AREAS

Power supply for large cities depends on local generation and power import capability. Local generation is often older and less efficient than newer units located remotely. Often, however, the older, less-efficient units located near the city center must be dispatched out-of-merit because they must be run for voltage support or reliability due to inadequate transmission. Air quality regulations may limit the availability of these units. New transmission into large cities is difficult to site due to right-of-way limitations and land-use constraints.

Compact VSC-based underground transmission circuits can be placed on existing dual-use rights-of-way to bring in power as well as to provide voltage support, allowing a more economical power supply without compromising reliability. The receiving terminal acts like a virtual generator delivering power and supplying voltage regulation and dynamic reactive power reserve. Stations are compact and housed mainly indoors, making siting in urban areas somewhat easier. Furthermore, the dynamic voltage support offered by the VSC can often increase the capability of the adjacent ac transmission.

SYSTEM CONFIGURATIONS AND OPERATING MODES

Figure 13 shows the different common system configurations and operating modes used for HVDC transmission. Monopolar systems are the simplest and least expensive systems for moderate power transfers since only two converters and one high-voltage insulated cable or line conductor are required. Such systems have been used with low-voltage electrode lines and sea electrodes to carry the return current in submarine cable crossings.

In some areas conditions are not conducive to monopolar earth or sea return. This could be the case in heavily congested areas, fresh water cable crossings, or areas with high earth resistivity. In such cases a metallic neutral- or low-voltage cable is used for the return path and the dc circuit uses a simple local ground connection for potential reference only. Back-to-back stations are used for interconnection of asynchronous networks and use ac lines to connect on either side. In such systems power transfer is limited by the relative capacities of the adjacent ac systems at the point of connection.

As an economic alternative to a monopolar system with metallic return, the midpoint of a 12-pulse converter can be connected to earth directly or through an impedance and two half-voltage cables or line conductors can be used. The converter is only operated in 12-pulse mode so there is never any stray earth current.

VSC-based HVDC transmission is usually arranged with a single converter connected pole-to-pole rather than pole-to-ground. The center point of the converter is connected to ground through a high impedance to provide a reference for the dc voltage. Thus, half the converter dc

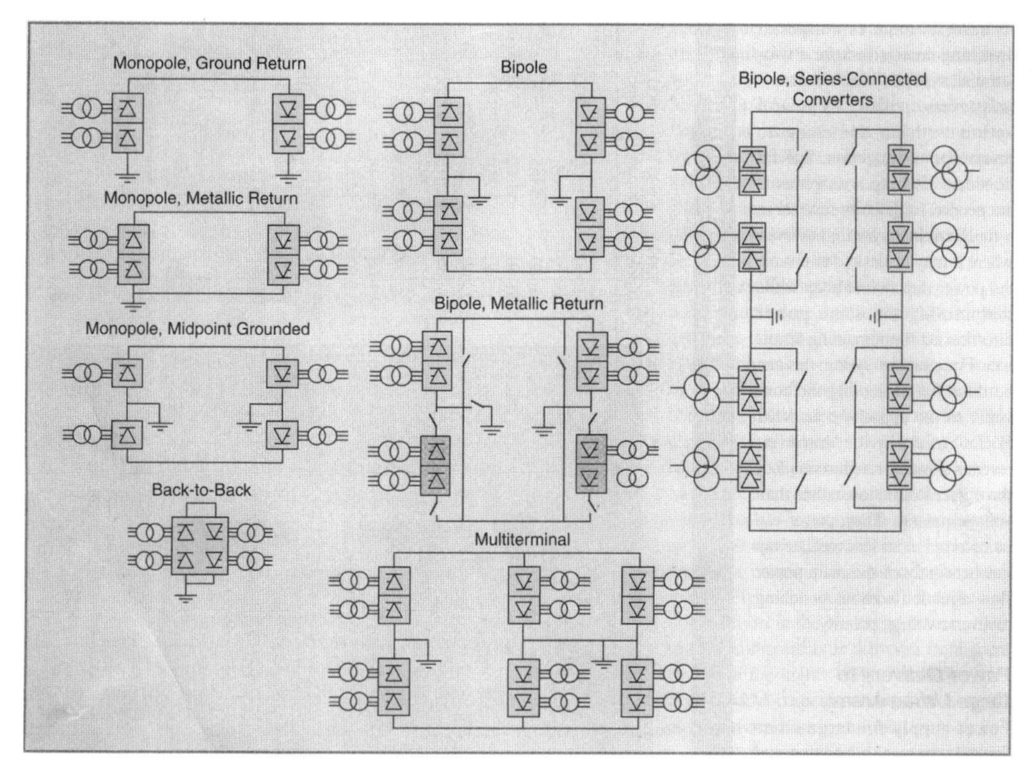

Figure 13
HVDC configurations and operating modes

voltage appears across the insulation on each of the two dc cables, one positive the other negative.

The most common configuration for modern overhead HVDC transmission lines is bipolar with a single 12-pulse converter for each pole at each terminal. This gives two independent dc circuits each capable of half capacity. For normal balanced operation there is no earth current. Monopolar earth return operation, often with overload capacity, can be used during outages of the opposite pole.

Earth return operation can be minimized during monopolar outages by using the opposite pole line for metallic return via pole/converter bypass switches at each end. This requires a metallic-return transfer breaker in the ground electrode line at one of the dc terminals to commutate the current from the relatively low resistance of the earth into that of the dc line conductor. Metallic return operation capability is provided for most dc transmission systems. This not only is effective during converter outages but also during line insulation failures where the remaining insulation strength is adequate to withstand the low resistive voltage drop in the metallic return path.

For very-high-power HVDC transmission, especially at dc voltages above ±500 kV (i.e., ±600 kV or ±800 kV), series-connected converters can be used to reduce the energy unavailability for individual converter outages or partial line insulation failure. By using two series-connected converters per pole in a bipolar system, only one quarter of the transmission capacity is lost for a converter outage or if the line insulation for the affected pole is degraded to where it can only support half the rated dc line voltage. Operating in this mode also avoids the need to transfer to monopolar metallic return to limit the duration of emergency earth return.

STATION DESIGN AND LAYOUT

CONVENTIONAL HVDC

The converter station layout depends on a number of factors such as the dc system configuration (i.e., monopolar, bipolar, or back-to-back), ac filtering, and reactive power compensation requirements. The thyristor valves are air-insulated, water-cooled, and enclosed in a converter building often referred to as a valve hall. For back-to-back ties with their characteristically low dc voltage, thyristor valves can be housed in prefabricated electrical enclosures, in which case a valve hall is not required.

To obtain a more compact station design and reduce the number of insulated high-voltage wall bushings, converter transformers are often placed adjacent to the valve hall with valve winding bushings protruding through the building walls for connection to the valves. Double or quadruple valve structures housing valve modules are used within the valve hall. Valve arresters are located immediately adjacent to the valves. Indoor motor-operated grounding switches are used for personnel safety during maintenance. Closed-loop valve cooling systems are used to circulate the cooling medium, deionized water or water-glycol mix, through the indoor thyristor valves with heat transfer to dry coolers located outdoors. Area requirements for conventional HVDC converter stations are influenced by the ac system voltage and reactive power compensation requirements where each individual bank rating may be limited by such system requirements as reactive power exchange and maximum voltage step on bank switching. The ac yard with filters and shunt compensation can take up as much as three quarters of the total area requirements of the converter station. Figure 14 shows a typical arrangement for an HVDC converter station.

VSC-BASED HVDC

The transmission circuit consists of a bipolar two-wire HVDC system with converters connected pole-to-pole. DC capacitors are used to provide a stiff dc voltage source. The dc capacitors are grounded at their electrical center point to establish the earth reference potential for the transmission system. There is no earth return operation. The converters are coupled to the ac system through ac phase reactors and power transformers. Unlike most conventional HVDC systems, harmonic filters are located between the phase reactors and power transformers. Therefore, the transformers are exposed to no dc voltage stresses or harmonic loading, allowing use of ordinary power transformers. Figure 15 shows the station arrangement for a ±150-kV, 350 to 550-MW VSC converter station.

The IGBT valves used in VSC converters are comprised of series-connected IGBT positions. The IGBT is a hybrid device exhibiting the low forward drop of a bipolar transistor as a conducting device. Instead of the regular current-controlled base, the IGBT has a

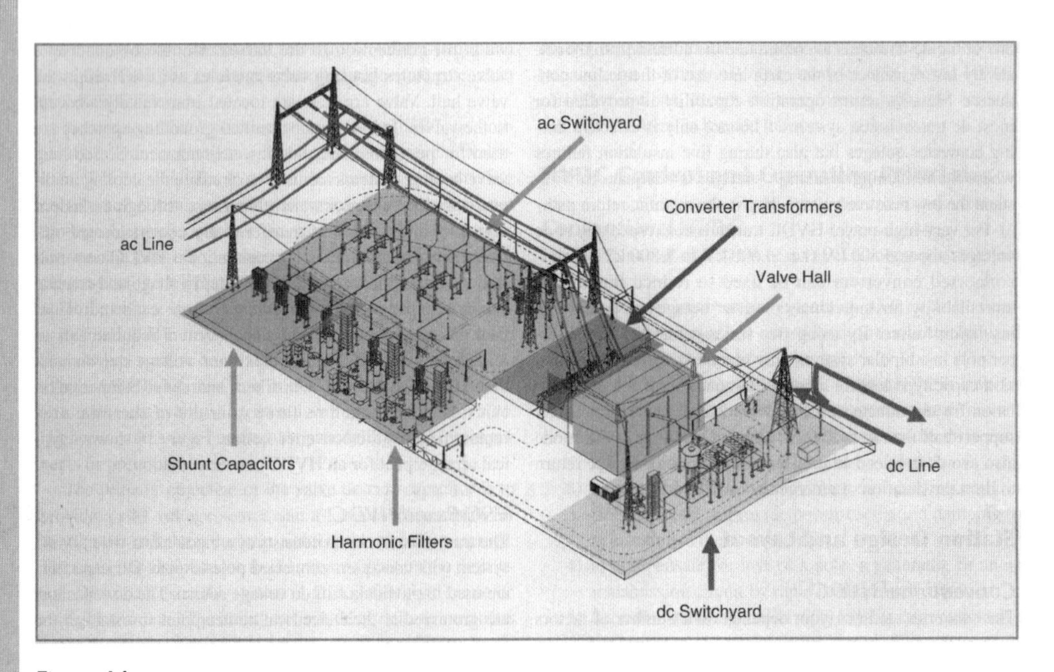

Figure 14
Monopolar HVDC converter station

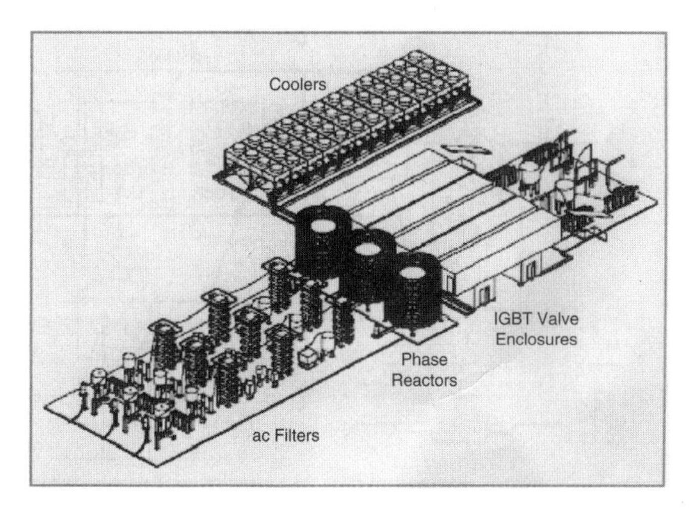

Figure 15
VSC HVDC converter station

voltage-controlled capacitive gate, as in the MOSFET device.

A complete IGBT position consists of an IGBT, an anti-parallel diode, a gate unit, a voltage divider, and a water-cooled heat sink. Each gate unit includes gate-driving circuits, surveillance circuits, and optical interface. The gate-driving electronics control the gate voltage and current at turn-on and turn-off to achieve optimal turn-on and turn-off processes of the IGBTs.

To be able to switch voltages higher than the rated voltage of one IGBT, many positions arc connected in series in each valve similar to thyristors in conventional HVDC valves. All IGBTs must turn on and off at the same moment to achieve an evenly distributed voltage across the valve. Higher currents are handled by paralleling IGBT components or press packs.

The primary objective of the valve dc-side capacitor is to provide a stiff voltage source and a low-inductance path for the turn-off switching currents and to provide energy storage. The capacitor also reduces the harmonic ripple on the dc voltage. Disturbances in the system (e.g., ac faults) will cause dc voltage variations. The ability to limit these voltage variations depends on the size of the dc-side capacitor. Since the dc capacitors are used indoors, dry capacitors are used.

AC filters for VSC HVDC converters have smaller ratings than those for conventional converters and are not required for reactive power compensation. Therefore, these filters are always connected to the converter bus and not switched with transmission loading. All equipment for VSC-based HVDC converter stations, except the transformer, high-side breaker, and valve coolers, is located indoors.

HVDC CONTROL AND OPERATING PRINCIPLES

CONVENTIONAL HVDC

The fundamental objectives of an HVDC control system are as follows:

1) to control basic system quantities such as dc line current, dc voltage, and transmitted power accurately and with sufficient speed of response
2) to maintain adequate commutation margin in inverter operation so that the valves can recover their forward blocking capability after conduction before their voltage polarity reverses
3) to control higher-level quantities such as frequency in isolated mode or provide power oscillation damping to help stabilize the ac network
4) to compensate for loss of a pole, a generator, or an ac transmission circuit by rapid readjustment of power
5) to ensure stable operation with reliable commutation in the presence of system disturbances
6) to minimize system losses and converter reactive power consumption
7) to ensure proper operation with fast and stable recoveries during ac system faults and disturbances.

For conventional HVDC transmission, one terminal sets the dc voltage level while the other terminal(s) regulates the (its) dc current by controlling its output voltage relative to that maintained by the voltage-setting terminal. Since the dc line resistance is low, large changes in current and hence power can be made with relatively small changes in firing angle (alpha). Two independent methods exist for controlling the converter dc output voltage. These are 1) by changing the ratio between the direct voltage and the ac voltage by varying the delay angle or 2) by changing the converter ac voltage via load tap changers (LTCs) on the converter transformer. Whereas the former method is rapid the latter method is slow due to the limited speed of response of the LTC. Use of high delay angles to achieve a larger dynamic range, however, increases the converter reactive power

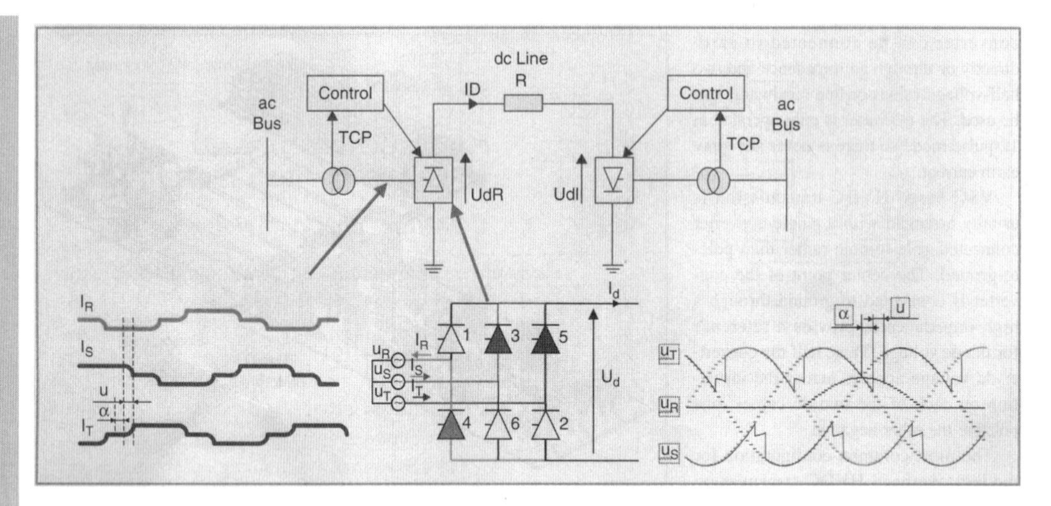

Figure 16
Conventional HVDC control

consumption. To minimize the reactive power demand while still providing adequate dynamic control range and commutation margin, the LTC is used at the rectifier terminal to keep the delay angle within its desired steady-slate range (e.g., 13–18°) and at the inverter to keep the extinction angle within its desired range (e.g., 17–20°), if the angle is used for dc voltage control or to maintain rated dc voltage if operating in minimum commutation margin control mode. Figure 16 shows the characteristic transformer current and dc bridge voltage waveforms along with the controlled items Ud, Id, and tap changer position (TCP).

VSC-BASED HVDC

Power can be controlled by changing the phase angle of the converter ac voltage with respect to the filter bus voltage, whereas the reactive power can be controlled by changing the magnitude of the fundamental component of the converter ac voltage with respect to the filter bus voltage. By controlling these two aspects of the converter voltage, operation in all four quadrants is possible. This means that the converter can be operated in the middle of its reactive power range near unity power factor to maintain dynamic reactive power reserve for contingency voltage support similar to a static var compensator. It also means that the real power transfer can be changed rapidly without altering the reactive power exchange with the ac network or waiting for switching of shunt compensation.

Being able to independently control ac voltage magnitude and phase relative to the system voltage allows use of separate active and reactive power control loops for HVDC system regulation. The active power control loop can be set to control either the active power or the dc-side voltage. In a dc link, one station will then be selected to control the active power while the other must be set to control the dc-side voltage. The reactive power control loop can be set to control either the reactive power or the ac-side voltage. Either of these two modes can be selected independently at either end of the dc link. Figure 17 shows the characteristic ac voltage waveforms before and after the ac filters along with the controlled items Ud, Id, Q, and Uac.

CONCLUSIONS

The favorable economics of long-distance bulk-power transmission with HVDC together with its controllability make it an interesting alternative or complement to ac transmission. The higher voltage levels, mature technology, and new converter designs have significantly increased the interest in HVDC transmission and expanded the range of applications.

FOR FURTHER READING

B. Jacobson, Y. Jiang-Hafner, P. Rey, and G. Asplund, "HVDC with voltage source converters and extruded

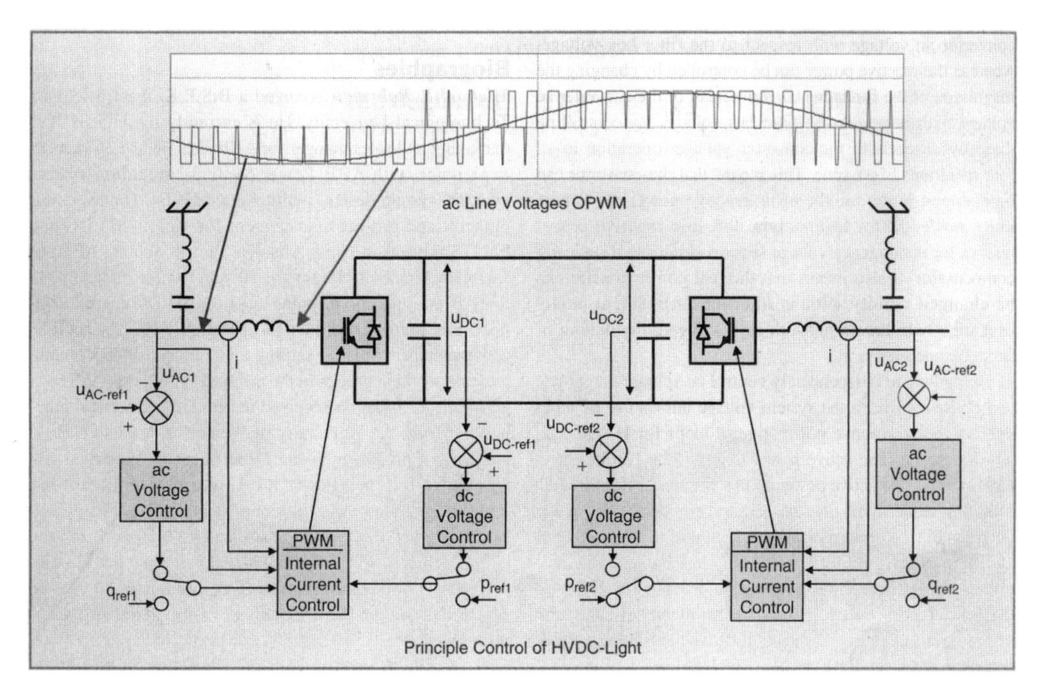

Figure 17
Control of VSC HVDC transmission

cables for up to ±300 kV and 1000 MW," in *Proc. CIGRÉ 2006*, Paris, France, pp. B4–105.

L. Ronstrom, B.D. Railing, J.J. Miller, P. Steckley, G. Moreau, P. Bard, and J. Lindberg, "Cross sound cable project second generation VSC technology for HVDC," *Proc. CIGRÉ 2006*, Paris, France, pp. B4–102.

M. Bahrman, D. Dickinson, P. Fisher, and M. Stoltz, "The Rapid City Tie—New technology tames the East-West interconnection," in *Proc. Minnesota Power Systems Conf.*, St. Paul, MN, Nov. 2004.

D. McCallum, G. Moreau, J. Primeau, D. Soulier, M. Bahrman, and B. Ekehov, "Multiterminal integration of the Nicolet Converter Station into the Quebec-New England Phase II transmission system," in *Proc. CIGRÉ 1994*, Paris, France.

A. Ekstrom and G. Liss, "A refined HVDC control system," *IEEE Trans. Power Systems*, vol. PAS-89, pp. 723–732, May–June 1970.

BIOGRAPHIES

Michael P. Bahrman received a B.S.E.E. from Michigan Technological University. He is currently the U.S. HVDC marketing and sales manger for ABB Inc. He has 24 years of experience with ABB Power Systems including system analysis, system design, multiterminal HVDC control development, and project management for various HVDC and FACTS projects in North America. Prior to joining ABB, he was with Minnesota Power for 10 years where he held positions as transmission planning engineer, HVDC control engineer, and manager of system operations. He has been an active member of IEEE, serving on a number of subcommittees and working groups in the area of HVDC and FACTS.

Brian K. Johnson received the Ph.D. in electrical engineering from the University of Wisconsin-Madison. He is currently a professor in the Department of Electrical and Computer Engineering at the University of Idaho. His interests include power system protection and the application of power electronics to utility systems, security and survivability of ITS systems and power systems, distributed sensor and control networks, and real-time simulation of traffic systems. He is a member of the Board of Governors of the IEEE Intelligent Transportation Systems Society and the Administrative Committee of the IEEE Council on Superconductivity.

5.1

MEDIUM AND SHORT LINE APPROXIMATIONS

In this section, we present short and medium-length transmission-line approximations as a means of introducing *ABCD* parameters. Some readers may prefer to start in Section 5.2, which presents the exact transmission-line equations.

It is convenient to represent a transmission line by the two-port network shown in Figure 5.1, where V_S and I_S are the sending-end voltage and current, and V_R and I_R are the receiving-end voltage and current.

The relation between the sending-end and receiving-end quantities can be written as

$$V_S = AV_R + BI_R \quad \text{volts} \tag{5.1.1}$$

$$I_S = CV_R + DI_R \quad \text{A} \tag{5.1.2}$$

or, in matrix format,

$$\begin{bmatrix} V_S \\ I_S \end{bmatrix} = \left[\begin{array}{c|c} A & B \\ \hline C & D \end{array}\right] \begin{bmatrix} V_R \\ I_R \end{bmatrix} \tag{5.1.3}$$

where *A*, *B*, *C*, and *D* are parameters that depend on the transmission-line constants R, L, C, and G. The *ABCD* parameters are, in general, complex numbers. *A* and *D* are dimensionless. *B* has units of ohms, and *C* has units of siemens. Network theory texts [5] show that *ABCD* parameters apply to linear, passive, bilateral two-port networks, with the following general relation:

$$AD - BC = 1 \tag{5.1.4}$$

The circuit in Figure 5.2 represents a short transmission line, usually applied to overhead 60-Hz lines less than 80 km long. Only the series resistance and reactance are included. The shunt admittance is neglected. The circuit applies to either single-phase or completely transposed three-phase lines operating under balanced conditions. For a completely transposed

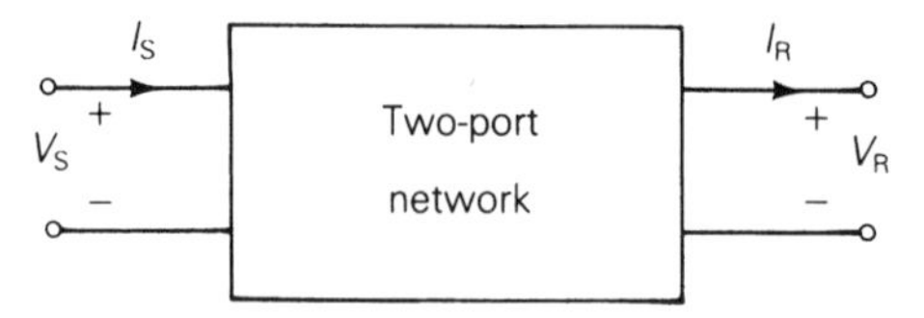

FIGURE 5.1 Representation of two-port network

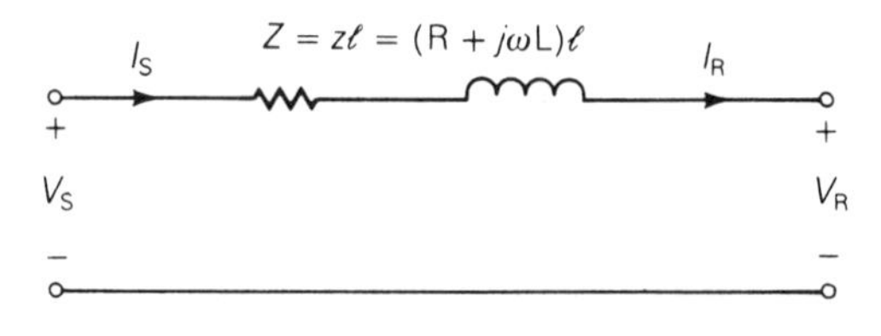

FIGURE 5.2 Short transmission line

three-phase line, Z is the series impedance, V_S and V_R are positive-sequence line-to-neutral voltages, and I_S and I_R are positive-sequence line currents.

To avoid confusion between total series impedance and series impedance per unit length, we use the following notation:

$z = R + j\omega L \quad \Omega/m$, series impedance per unit length

$y = G + j\omega C \quad S/m$, shunt admittance per unit length

$Z = zl \quad \Omega$, total series impedance

$Y = yl \quad S$, total shunt admittance

$l =$ line length m

Recall that shunt conductance G is usually neglected for overhead transmission.

The *ABCD* parameters for the short line in Figure 5.2 are easily obtained by writing a KVL and KCL equation as

$$V_S = V_R + ZI_R \tag{5.1.5}$$

$$I_S = I_R \tag{5.1.6}$$

or, in matrix format,

$$\begin{bmatrix} V_S \\ I_S \end{bmatrix} = \left[\begin{array}{c|c} 1 & Z \\ \hline 0 & 1 \end{array}\right] \begin{bmatrix} V_R \\ I_R \end{bmatrix} \tag{5.1.7}$$

Comparing (5.1.7) and (5.1.3), the *ABCD* parameters for a short line are

$$A = D = 1 \quad \text{per unit} \tag{5.1.8}$$

$$B = Z \quad \Omega \tag{5.1.9}$$

$$C = 0 \quad S \tag{5.1.10}$$

For medium-length lines, typically ranging from 80 to 250 km at 60 Hz, it is common to lump the total shunt capacitance and locate half at each end of the line. Such a circuit, called a *nominal π circuit*, is shown in Figure 5.3.

To obtain the *ABCD* parameters of the nominal π circuit, note first that the current in the series branch in Figure 5.3 equals $I_R + \frac{V_R Y}{2}$. Then, writing a KVL equation,

$$V_S = V_R + Z\left(I_R + \frac{V_R Y}{2}\right)$$

$$= \left(1 + \frac{YZ}{2}\right)V_R + ZI_R \tag{5.1.11}$$

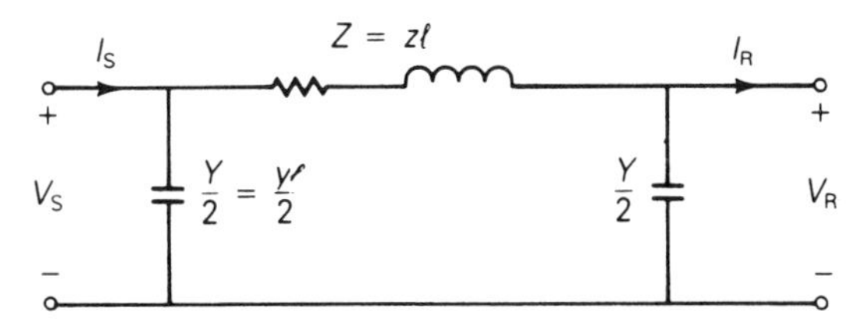

FIGURE 5.3 Medium-length transmission line—nominal π circuit

Also, writing a KCL equation at the sending end,

$$I_S = I_R + \frac{V_R Y}{2} + \frac{V_S Y}{2} \tag{5.1.12}$$

Using (5.1.11) in (5.1.12),

$$\begin{aligned} I_S &= I_R + \frac{V_R Y}{2} + \left[\left(1 + \frac{YZ}{2}\right) V_R + Z I_R\right] \frac{Y}{2} \\ &= Y\left(1 + \frac{YZ}{4}\right) V_R + \left(1 + \frac{YZ}{2}\right) I_R \end{aligned} \tag{5.1.13}$$

Writing (5.1.11) and (5.1.13) in matrix format,

$$\begin{bmatrix} V_S \\ \\ I_S \end{bmatrix} = \left[\begin{array}{c|c} \left(1 + \frac{YZ}{2}\right) & Z \\ \hline Y\left(1 + \frac{YZ}{4}\right) & \left(1 + \frac{YZ}{2}\right) \end{array}\right] \begin{bmatrix} V_R \\ \\ I_R \end{bmatrix} \tag{5.1.14}$$

Thus, comparing (5.1.14) and (5.1.3)

$$A = D = 1 + \frac{YZ}{2} \quad \text{per unit} \tag{5.1.15}$$

$$B = Z \quad \Omega \tag{5.1.16}$$

$$C = Y\left(1 + \frac{YZ}{4}\right) \quad \text{S} \tag{5.1.17}$$

Note that for both the short and medium-length lines, the relation $AD - BC = 1$ is verified. Note also that since the line is the same when viewed from either end, $A = D$.

Figure 5.4 gives the *ABCD* parameters for some common networks, including a series impedance network that approximates a short line and a π circuit that approximates a medium-length line. A medium-length line could also be approximated by the T circuit shown in Figure 5.4, lumping half of the series impedance at each end of the line. Also given are the *ABCD* parameters for networks in series, which are conveniently obtained by multiplying the *ABCD* matrices of the individual networks.

ABCD parameters can be used to describe the variation of line voltage with line loading. *Voltage regulation* is the change in voltage at the receiving end of the line when the load varies from no-load to a specified full load at a specified power factor, while the sending-end voltage is held constant. Expressed in percent of full-load voltage,

$$\text{percent VR} = \frac{|V_{RNL}| - |V_{RFL}|}{|V_{RFL}|} \times 100 \tag{5.1.18}$$

Circuit	*ABCD* Matrix
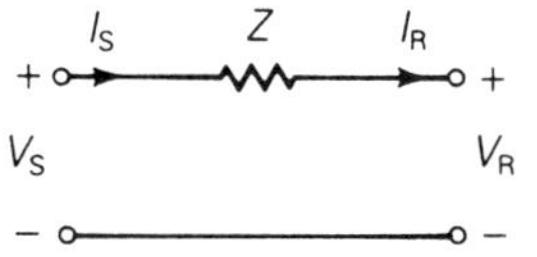 Series impedance (I_S, Z, I_R; V_S, V_R)	$\begin{bmatrix} 1 & Z \\ 0 & 1 \end{bmatrix}$
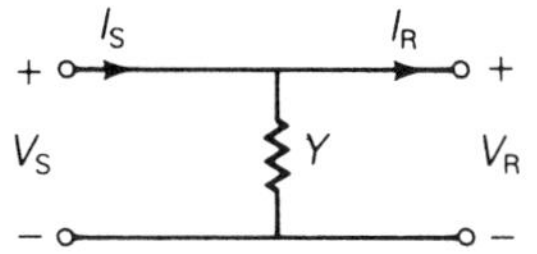 Shunt admittance (I_S, Y, I_R; V_S, V_R)	$\begin{bmatrix} 1 & 0 \\ Y & 1 \end{bmatrix}$
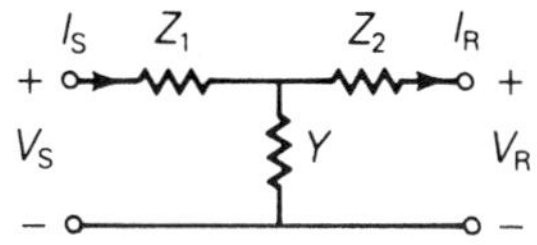 T circuit (I_S, Z_1, Z_2, I_R, Y; V_S, V_R)	$\begin{bmatrix} (1 + YZ_1) & (Z_1 + Z_2 + YZ_1Z_2) \\ Y & (1 + YZ_2) \end{bmatrix}$
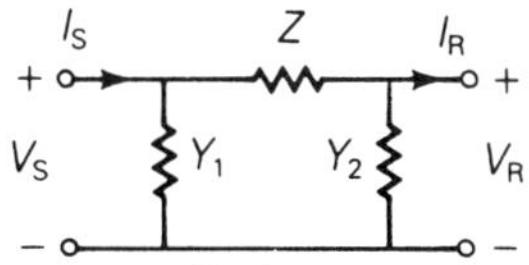 Π circuit (I_S, Z, I_R, Y_1, Y_2; V_S, V_R)	$\begin{bmatrix} (1 + Y_2Z) & Z \\ (Y_1 + Y_2 + Y_1Y_2Z) & (1 + Y_1Z) \end{bmatrix}$
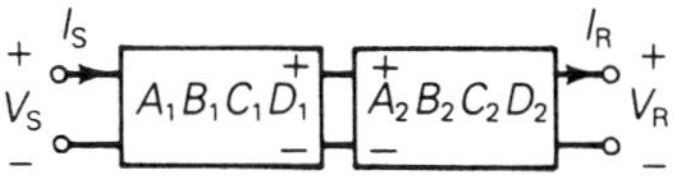 Series networks (I_S, $A_1B_1C_1D_1$, $A_2B_2C_2D_2$, I_R; V_S, V_R)	$\begin{bmatrix} A_1 & B_1 \\ C_1 & D_1 \end{bmatrix} \begin{bmatrix} A_2 & B_2 \\ C_2 & D_2 \end{bmatrix} = \begin{bmatrix} (A_1A_2 + B_1C_2) & (A_1B_2 + B_1D_2) \\ (C_1A_2 + D_1C_2) & (C_1B_2 + D_1D_2) \end{bmatrix}$

FIGURE 5.4 *ABCD* parameters of common networks

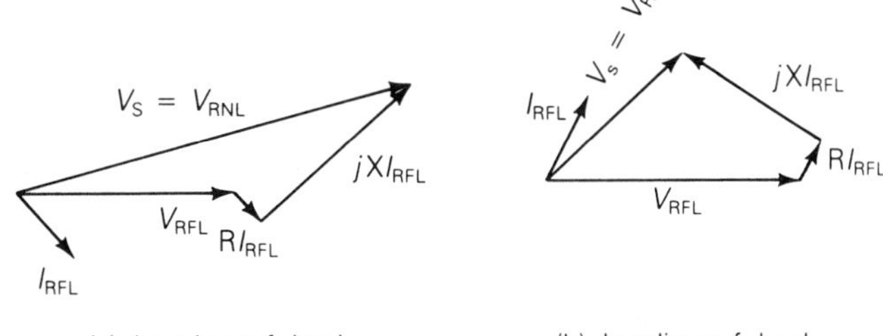

(a) Lagging p.f. load (b) Leading p.f. load

FIGURE 5.5 Phasor diagrams for a short transmission line

where percent VR is the percent voltage regulation, $|V_{RNL}|$ is the magnitude of the no-load receiving-end voltage, and $|V_{RFL}|$ is the magnitude of the full-load receiving-end voltage.

The effect of load power factor on voltage regulation is illustrated by the phasor diagrams in Figure 5.5 for short lines. The phasor diagrams are graphical representations of (5.1.5) for lagging and leading power factor loads. Note that, from (5.1.5) at no-load, $I_{RNL} = 0$ and $V_S = V_{RNL}$ for a short line. As shown, the higher (worse) voltage regulation occurs for the lagging p.f. load, where V_{RNL} exceeds V_{RFL} by the larger amount. A smaller or even negative voltage regulation occurs for the leading p.f. load. In general, the no-load voltage is, from (5.1.1), with $I_{RNL} = 0$,

$$V_{RNL} = \frac{V_S}{A} \tag{5.1.19}$$

which can be used in (5.1.18) to determine voltage regulation.

In practice, transmission-line voltages decrease when heavily loaded and increase when lightly loaded. When voltages on EHV lines are maintained within $\pm 5\%$ of rated voltage, corresponding to about 10% voltage regulation, unusual operating problems are not encountered. Ten percent voltage regulation for lower voltage lines including transformer-voltage drops is also considered good operating practice.

In addition to voltage regulation, line loadability is an important issue. Three major line-loading limits are: (1) the thermal limit, (2) the voltage-drop limit, and (3) the steady-state stability limit.

The maximum temperature of a conductor determines its thermal limit. Conductor temperature affects the conductor sag between towers and the loss of conductor tensile strength due to annealing. If the temperature is too high, prescribed conductor-to-ground clearances may not be met, or the elastic limit of the conductor may be exceeded such that it cannot shrink to its original length when cooled. Conductor temperature depends on the current magnitude and its time duration, as well as on ambient temperature, wind velocity, and conductor surface conditions. Appendix Tables A.3 and A.4 give approximate current-carrying capacities of copper and ACSR conductors. The loadability of short transmission lines (less than 80 km in length for 60-Hz overhead lines) is usually determined by the conductor thermal limit or by ratings of line terminal equipment such as circuit breakers.

For longer line lengths (up to 300 km), line loadability is often determined by the voltage-drop limit. Although more severe voltage drops may be tolerated in some cases, a heavily loaded line with $V_R/V_S \geqslant 0.95$ is usually considered safe operating practice. For line lengths over 300 km, steady-state stability becomes a limiting factor. Stability, discussed in Section 5.4, refers to the ability of synchronous machines on either end of a line to remain in synchronism.

EXAMPLE 5.1 *ABCD* parameters and the nominal π circuit: medium-length line

A three-phase, 60-Hz, completely transposed 345-kV, 200-km line has two 795,000-cmil (403-mm^2) 26/2 ACSR conductors per bundle and the following positive-sequence line constants:

$$z = 0.032 + j0.35 \quad \Omega/\text{km}$$

$$y = j4.2 \times 10^{-6} \quad \text{S/km}$$

Full load at the receiving end of the line is 700 MW at 0.99 p.f. leading and at 95% of rated voltage. Assuming a medium-length line, determine the following:

a. *ABCD* parameters of the nominal π circuit

b. Sending-end voltage V_S, current I_S, and real power P_S

c. Percent voltage regulation

d. Thermal limit, based on the approximate current-carrying capacity listed in Table A.4

e. Transmission-line efficiency at full load

SOLUTION

a. The total series impedance and shunt admittance values are

$$Z = zl = (0.032 + j0.35)(200) = 6.4 + j70 = 70.29\underline{/84.78^\circ} \quad \Omega$$

$$Y = yl = (j4.2 \times 10^{-6})(200) = 8.4 \times 10^{-4}\underline{/90^\circ} \quad \text{S}$$

From (5.1.15)–(5.1.17),

$$A = D = 1 + (8.4 \times 10^{-4}\underline{/90^\circ})(70.29\underline{/84.78^\circ})\left(\tfrac{1}{2}\right)$$

$$= 1 + 0.02952\underline{/174.78^\circ}$$

$$= 0.9706 + j0.00269 = 0.9706\underline{/0.159^\circ} \quad \text{per unit}$$

$$B = Z = 70.29\underline{/84.78^\circ} \quad \Omega$$

$$C = (8.4 \times 10^{-4}\underline{/90^\circ})(1 + 0.01476\underline{/174.78^\circ})$$

$$= (8.4 \times 10^{-4}\underline{/90^\circ})(0.9853 + j0.00134)$$

$$= 8.277 \times 10^{-4}\underline{/90.08^\circ} \quad \text{S}$$

b. The receiving-end voltage and current quantities are

$$\text{V}_\text{R} = (0.95)(345) = 327.8 \quad \text{kV}_\text{LL}$$

$$V_\text{R} = \frac{327.8}{\sqrt{3}}\underline{/0^\circ} = 189.2\underline{/0^\circ} \quad \text{kV}_\text{LN}$$

$$I_\text{R} = \frac{700\underline{/\cos^{-1} 0.99}}{(\sqrt{3})(0.95 \times 345)(0.99)} = 1.246\underline{/8.11^\circ} \quad \text{kA}$$

From (5.1.1) and (5.1.2), the sending-end quantities are

$$V_\text{S} = (0.9706\underline{/0.159^\circ})(189.2\underline{/0^\circ}) + (70.29\underline{/84.78^\circ})(1.246\underline{/8.11^\circ})$$

$$= 183.6\underline{/0.159^\circ} + 87.55\underline{/92.89^\circ}$$

$$= 179.2 + j87.95 = 199.6\underline{/26.14^\circ} \quad \text{kV}_\text{LN}$$

$$V_S = 199.6\sqrt{3} = 345.8 \text{ kV}_{LL} \approx 1.00 \quad \text{per unit}$$

$$I_S = (8.277 \times 10^{-4}\underline{/90.08^\circ})(189.2\underline{/0^\circ}) + (0.9706\underline{/0.159^\circ})(1.246\underline{/8.11^\circ})$$
$$= 0.1566\underline{/90.08^\circ} + 1.209\underline{/8.27^\circ}$$
$$= 1.196 + j0.331 = 1.241\underline{/15.5^\circ} \quad \text{kA}$$

and the real power delivered to the sending end is

$$P_S = (\sqrt{3})(345.8)(1.241)\cos(26.14^\circ - 15.5^\circ)$$
$$= 730.5 \quad \text{MW}$$

c. From (5.1.19), the no-load receiving-end voltage is

$$V_{RNL} = \frac{V_S}{A} = \frac{345.8}{0.9706} = 356.3 \quad \text{kV}_{LL}$$

and, from (5.1.18),

$$\text{percent VR} = \frac{356.3 - 327.8}{327.8} \times 100 = 8.7\%$$

d. From Table A.4, the approximate current-carrying capacity of two 795,000-cmil (403-mm^2) 26/2 ACSR conductors is $2 \times 0.9 = 1.8$ kA.

e. The full-load line losses are $P_S - P_R = 730.5 - 700 = 30.5$ MW and the full-load transmission efficiency is

$$\text{percent EFF} = \frac{P_R}{P_S} \times 100 = \frac{700}{730.5} \times 100 = 95.8\%$$

Since $V_S = 1.00$ per unit, the full-load receiving-end voltage of 0.95 per unit corresponds to $V_R/V_S = 0.95$, considered in practice to be about the lowest operating voltage possible without encountering operating problems. Thus, for this 345-kV 200-km uncompensated line, voltage drop limits the full-load current to 1.246 kA at 0.99 p.f. leading, well below the thermal limit of 1.8 kA. ■

5.2 TRANSMISSION-LINE DIFFERENTIAL EQUATIONS

The line constants R, L, and C are derived in Chapter 4 as per-length values having units of Ω/m, H/m, and F/m. They are not lumped, but rather are uniformly distributed along the length of the line. In order to account for the distributed nature of transmission-line constants, consider the circuit shown in Figure 5.6, which represents a line section of length Δx. $V(x)$ and $I(x)$ denote the voltage and current at position x, which is measured in meters from the right, or receiving end of the line. Similarly, $V(x + \Delta x)$ and $I(x + \Delta x)$ denote the voltage and current at position $(x + \Delta x)$. The circuit constants are

FIGURE 5.6
Transmission-line section of length Δx

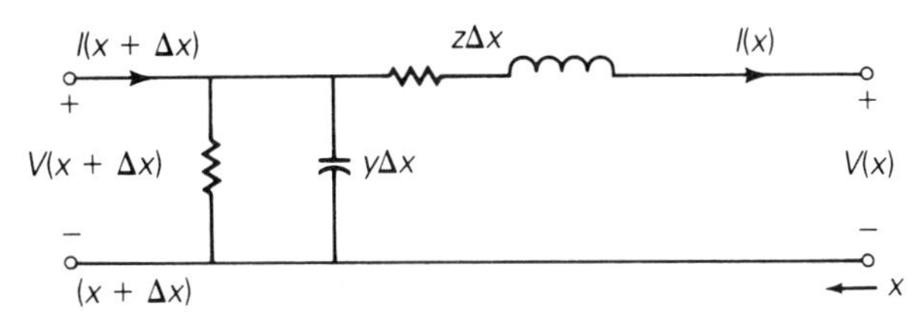

$$z = \mathrm{R} + j\omega \mathrm{L} \quad \Omega/\mathrm{m} \tag{5.2.1}$$

$$y = \mathrm{G} + j\omega \mathrm{C} \quad \mathrm{S/m} \tag{5.2.2}$$

where G is usually neglected for overhead 60-Hz lines. Writing a KVL equation for the circuit

$$V(x + \Delta x) = V(x) + (z\Delta x)I(x) \quad \text{volts} \tag{5.2.3}$$

Rearranging (5.2.3),

$$\frac{V(x + \Delta x) - V(x)}{\Delta x} = zI(x) \tag{5.2.4}$$

and taking the limit as Δx approaches zero,

$$\frac{dV(x)}{dx} = zI(x) \tag{5.2.5}$$

Similarly, writing a KCL equation for the circuit,

$$I(x + \Delta x) = I(x) + (y\Delta x)V(x + \Delta x) \quad \mathrm{A} \tag{5.2.6}$$

Rearranging,

$$\frac{I(x + \Delta x) - I(x)}{\Delta x} = yV(x) \tag{5.2.7}$$

and taking the limit as Δx approaches zero,

$$\frac{dI(x)}{dx} = yV(x) \tag{5.2.8}$$

Equations (5.2.5) and (5.2.8) are two linear, first-order, homogeneous differential equations with two unknowns, $V(x)$ and $I(x)$. We can eliminate $I(x)$ by differentiating (5.2.5) and using (5.2.8) as follows:

$$\frac{d^2V(x)}{dx^2} = z\frac{dI(x)}{dx} = zyV(x) \tag{5.2.9}$$

or

$$\frac{d^2V(x)}{dx^2} - zyV(x) = 0 \tag{5.2.10}$$

Equation (5.2.10) is a linear, second-order, homogeneous differential equation with one unknown, $V(x)$. By inspection, its solution is

$$V(x) = A_1 e^{\gamma x} + A_2 e^{-\gamma x} \quad \text{volts} \tag{5.2.11}$$

where A_1 and A_2 are integration constants and

$$\gamma = \sqrt{zy} \quad \text{m}^{-1} \tag{5.2.12}$$

γ, whose units are m^{-1}, is called the *propagation constant*. By inserting (5.2.11) and (5.2.12) into (5.2.10), the solution to the differential equation can be verified.

Next, using (5.2.11) in (5.2.5),

$$\frac{dV(x)}{dx} = \gamma A_1 e^{\gamma x} - \gamma A_2 e^{-\gamma x} = zI(x) \tag{5.2.13}$$

Solving for $I(x)$,

$$I(x) = \frac{A_1 e^{\gamma x} - A_2 e^{-\gamma x}}{z/\gamma} \tag{5.2.14}$$

Using (5.2.12), $z/\gamma = z/\sqrt{zy} = \sqrt{z/y}$, (5.2.14) becomes

$$I(x) = \frac{A_1 e^{\gamma x} - A_2 e^{-\gamma x}}{Z_c} \tag{5.2.15}$$

where

$$Z_c = \sqrt{\frac{z}{y}} \quad \Omega \tag{5.2.16}$$

Z_c, whose units are Ω, is called the *characteristic impedance*.

Next, the integration constants A_1 and A_2 are evaluated from the boundary conditions. At $x = 0$, the receiving end of the line, the receiving-end voltage and current are

$$V_R = V(0) \tag{5.2.17}$$

$$I_R = I(0) \tag{5.2.18}$$

Also, at $x = 0$, (5.2.11) and (5.2.15) become

$$V_R = A_1 + A_2 \tag{5.2.19}$$

$$I_R = \frac{A_1 - A_2}{Z_c} \tag{5.2.20}$$

Solving for A_1 and A_2,

$$A_1 = \frac{V_R + Z_c I_R}{2} \tag{5.2.21}$$

$$A_2 = \frac{V_R - Z_c I_R}{2} \tag{5.2.22}$$

Substituting A_1 and A_2 into (5.2.11) and (5.2.15),

$$V(x) = \left(\frac{V_{\rm R} + Z_c I_{\rm R}}{2}\right)e^{\gamma x} + \left(\frac{V_{\rm R} - Z_c I_{\rm R}}{2}\right)e^{-\gamma x} \tag{5.2.23}$$

$$I(x) = \left(\frac{V_{\rm R} + Z_c I_{\rm R}}{2Z_c}\right)e^{\gamma x} - \left(\frac{V_{\rm R} - Z_c I_{\rm R}}{2Z_c}\right)e^{-\gamma x} \tag{5.2.24}$$

Rearranging (5.2.23) and (5.2.24),

$$V(x) = \left(\frac{e^{\gamma x} + e^{-\gamma x}}{2}\right)V_{\rm R} + Z_c\left(\frac{e^{\gamma x} - e^{-\gamma x}}{2}\right)I_{\rm R} \tag{5.2.25}$$

$$I(x) = \frac{1}{Z_c}\left(\frac{e^{\gamma x} - e^{-\gamma x}}{2}\right)V_{\rm R} + \left(\frac{e^{\gamma x} + e^{-\gamma x}}{2}\right)I_{\rm R} \tag{5.2.26}$$

Recognizing the hyperbolic functions cosh and sinh,

$$V(x) = \cosh(\gamma x)V_{\rm R} + Z_c \sinh(\gamma x)I_{\rm R} \tag{5.2.27}$$

$$I(x) = \frac{1}{Z_c}\sinh(\gamma x)V_{\rm R} + \cosh(\gamma x)I_{\rm R} \tag{5.2.28}$$

Equations (5.2.27) and (5.2.28) give the *ABCD* parameters of the distributed line. In matrix format,

$$\begin{bmatrix} V(x) \\ I(x) \end{bmatrix} = \left[\begin{array}{c|c} A(x) & B(x) \\ \hline C(x) & D(x) \end{array}\right] \begin{bmatrix} V_{\rm R} \\ I_{\rm R} \end{bmatrix} \tag{5.2.29}$$

where

$$A(x) = D(x) = \cosh(\gamma x) \quad \text{per unit} \tag{5.2.30}$$

$$B(x) = Z_c \sinh(\gamma x) \quad \Omega \tag{5.2.31}$$

$$C(x) = \frac{1}{Z_c}\sinh(\gamma x) \quad \text{S} \tag{5.2.32}$$

Equation (5.2.29) gives the current and voltage at any point x along the line in terms of the receiving-end voltage and current. At the sending end, where $x = l$, $V(l) = V_{\rm S}$ and $I(l) = I_{\rm S}$. That is,

$$\begin{bmatrix} V_{\rm S} \\ I_{\rm S} \end{bmatrix} = \left[\begin{array}{c|c} A & B \\ \hline C & D \end{array}\right] \begin{bmatrix} V_{\rm R} \\ I_{\rm R} \end{bmatrix} \tag{5.2.33}$$

where

$$A = D = \cosh(\gamma l) \quad \text{per unit} \tag{5.2.34}$$

$$B = Z_c \sinh(\gamma l) \quad \Omega \tag{5.2.35}$$

$$C = \frac{1}{Z_c}\sinh(\gamma l) \quad \text{S} \tag{5.2.36}$$

Equations (5.2.34)–(5.2.36) give the *ABCD* parameters of the distributed line. In these equations, the propagation constant γ is a complex quantity with real and imaginary parts denoted α and β. That is,

$$\gamma = \alpha + j\beta \quad \text{m}^{-1} \tag{5.2.37}$$

The quantity γl is dimensionless. Also

$$e^{\gamma l} = e^{(\alpha l + j\beta l)} = e^{\alpha l} e^{j\beta l} = e^{\alpha l} \underline{/\beta l} \tag{5.2.38}$$

Using (5.2.38) the hyperbolic functions cosh and sinh can be evaluated as follows:

$$\cosh(\gamma l) = \frac{e^{\gamma l} + e^{-\gamma l}}{2} = \frac{1}{2}(e^{\alpha l} \underline{/\beta l} + e^{-\alpha l} \underline{/-\beta l}) \tag{5.2.39}$$

and

$$\sinh(\gamma l) = \frac{e^{\gamma l} - e^{-\gamma l}}{2} = \frac{1}{2}(e^{\alpha l} \underline{/\beta l} - e^{-\alpha l} \underline{/-\beta l}) \tag{5.2.40}$$

Alternatively, the following identities can be used:

$$\cosh(\alpha l + j\beta l) = \cosh(\alpha l)\cos(\beta l) + j\sinh(\alpha l)\sin(\beta l) \tag{5.2.41}$$

$$\sinh(\alpha l + j\beta l) = \sinh(\alpha l)\cos(\beta l) + j\cosh(\alpha l)\sin(\beta l) \tag{5.2.42}$$

Note that in (5.2.39)–(5.2.42), the dimensionless quantity βl is in radians, not degrees.

The *ABCD* parameters given by (5.2.34)–(5.2.36) are exact parameters valid for any line length. For accurate calculations, these equations must be used for overhead 60-Hz lines longer than 250 km. The *ABCD* parameters derived in Section 5.1 are approximate parameters that are more conveniently used for hand calculations involving short and medium-length lines. Table 5.1 summarizes the *ABCD* parameters for short, medium, long, and lossless (see Section 5.4) lines.

TABLE 5.1 Summary: Transmission-line *ABCD* parameters

Parameter	A = D	B	C
Units	per Unit	Ω	S
Short line (less than 80 km)	1	Z	0
Medium line—nominal π circuit (80 to 250 km)	$1 + \frac{YZ}{2}$	Z	$Y\left(1 + \frac{YZ}{4}\right)$
Long line—equivalent π circuit (more than 250 km)	$\cosh(\gamma\ell) = 1 + \frac{Y'Z'}{2}$	$Z_c \sinh(\gamma\ell) = Z'$	$(1/Z_c)\sinh(\gamma\ell) = Y'\left(1 + \frac{Y'Z'}{4}\right)$
Lossless line (R = G = 0)	$\cos(\beta\ell)$	$jZ_c \sin(\beta\ell)$	$\frac{j\sin(\beta\ell)}{Z_c}$

EXAMPLE 5.2 Exact *ABCD* parameters: long line

A three-phase 765-kV, 60-Hz, 300-km, completely transposed line has the following positive-sequence impedance and admittance:

$$z = 0.0165 + j0.3306 = 0.3310\underline{/87.14^\circ} \quad \Omega/\text{km}$$

$$y = j4.674 \times 10^{-6} \quad \text{S/km}$$

Assuming positive-sequence operation, calculate the exact *ABCD* parameters of the line. Compare the exact *B* parameter with that of the nominal π circuit.

SOLUTION From (5.2.12) and (5.2.16):

$$Z_c = \sqrt{\frac{0.3310\underline{/87.14^\circ}}{4.674 \times 10^{-6}\underline{/90^\circ}}} = \sqrt{7.082 \times 10^4\underline{/-2.86^\circ}}$$

$$= 266.1\underline{/-1.43^\circ} \quad \Omega$$

and

$$\gamma l = \sqrt{(0.3310\underline{/87.14^\circ})(4.674 \times 10^{-6}\underline{/90^\circ})} \times (300)$$

$$= \sqrt{1.547 \times 10^{-6}\underline{/177.14^\circ}} \times (300)$$

$$= 0.3731\underline{/88.57^\circ} = 0.00931 + j0.3730 \quad \text{per unit}$$

From (5.2.38),

$$e^{\gamma l} = e^{0.00931}e^{+j0.3730} = 1.0094\underline{/0.3730} \quad \text{radians}$$

$$= 0.9400 + j0.3678$$

and

$$e^{-\gamma l} = e^{-0.00931}e^{-j0.3730} = 0.9907\underline{/-0.3730} \quad \text{radians}$$

$$= 0.9226 - j0.3610$$

Then, from (5.2.39) and (5.2.40),

$$\cosh(\gamma l) = \frac{(0.9400 + j0.3678) + (0.9226 - j0.3610)}{2}$$

$$= 0.9313 + j0.0034 = 0.9313\underline{/0.209^\circ}$$

$$\sinh(\gamma l) = \frac{(0.9400 + j0.3678) - (0.9226 - j0.3610)}{2}$$

$$= 0.0087 + j0.3644 = 0.3645\underline{/88.63^\circ}$$

Finally, from (5.2.34)–(5.2.36),

$$A = D = \cosh(\gamma l) = 0.9313\angle 0.209^\circ \quad \text{per unit}$$

$$B = (266.1\angle -1.43^\circ)(0.3645\angle 88.63^\circ) = 97.0\angle 87.2^\circ \quad \Omega$$

$$C = \frac{0.3645\angle 88.63^\circ}{266.1\angle -1.43^\circ} = 1.37 \times 10^{-3}\angle 90.06^\circ \quad \text{S}$$

Using (5.1.16), the B parameter for the nominal π circuit is

$$B_{\text{nominal }\pi} = Z = (0.3310\angle 87.14^\circ)(300) = 99.3\angle 87.14^\circ \quad \Omega$$

which is 2% larger than the exact value. ■

5.3 EQUIVALENT π CIRCUIT

Many computer programs used in power system analysis and design assume circuit representations of components such as transmission lines and transformers (see the power-flow program described in Chapter 6 as an example). It is therefore convenient to represent the terminal characteristics of a transmission line by an equivalent circuit instead of its *ABCD* parameters.

The circuit shown in Figure 5.7 is called an *equivalent π circuit*. It is identical in structure to the nominal π circuit of Figure 5.3, except that Z' and Y' are used instead of Z and Y. Our objective is to determine Z' and Y' such that the equivalent π circuit has the same *ABCD* parameters as those of the distributed line, (5.2.34)–(5.2.36). The *ABCD* parameters of the equivalent π circuit, which has the same structure as the nominal π, are

$$A = D = 1 + \frac{Y'Z'}{2} \quad \text{per unit} \tag{5.3.1}$$

$$B = Z' \quad \Omega \tag{5.3.2}$$

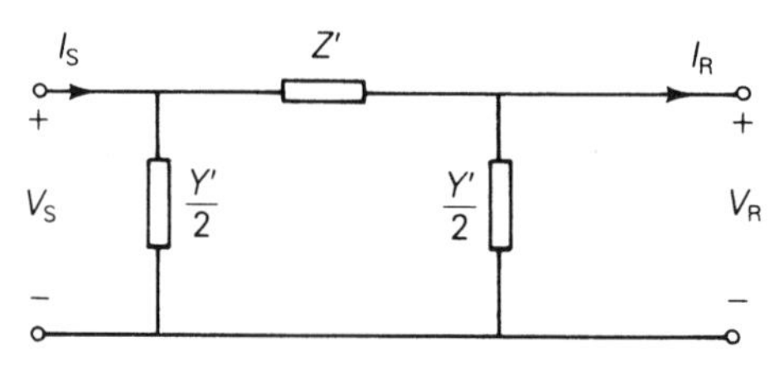

FIGURE 5.7 Transmission-line equivalent π circuit

$$Z' = Z_c \sinh(\gamma\ell) = ZF_1 = Z\frac{\sinh(\gamma\ell)}{\gamma\ell}$$

$$\frac{Y'}{2} = \frac{\tanh(\gamma\ell/2)}{Z_c} = \frac{Y}{2}F_2 = \frac{Y}{2}\frac{\tanh(\gamma\ell/2)}{(\gamma\ell/2)}$$

$$C = Y'\left(1 + \frac{Y'Z'}{4}\right) \quad \text{S} \tag{5.3.3}$$

where we have replaced Z and Y in (5.1.15)–(5.1.17) with Z' and Y' in (5.3.1)–(5.3.3). Equating (5.3.2) to (5.2.35),

$$Z' = Z_c \sinh(\gamma l) = \sqrt{\frac{z}{y}} \sinh(\gamma l) \tag{5.3.4}$$

Rewriting (5.3.4) in terms of the nominal π circuit impedance $Z = zl$,

$$\begin{aligned} Z' &= zl\left[\sqrt{\frac{z}{y}}\,\frac{\sinh(\gamma l)}{zl}\right] = zl\left[\frac{\sinh(\gamma l)}{\sqrt{zy}\,l}\right] \\ &= ZF_1 \quad \Omega \end{aligned} \tag{5.3.5}$$

where

$$F_1 = \frac{\sinh(\gamma l)}{\gamma l} \quad \text{per unit} \tag{5.3.6}$$

Similarly, equating (5.3.1) to (5.2.34),

$$\begin{aligned} 1 + \frac{Y'Z'}{2} &= \cosh(\gamma l) \\ \frac{Y'}{2} &= \frac{\cosh(\gamma l) - 1}{Z'} \end{aligned} \tag{5.3.7}$$

Using (5.3.4) and the identity $\tanh\left(\frac{\gamma l}{2}\right) = \frac{\cosh(\gamma l) - 1}{\sinh(\gamma l)}$, (5.3.7) becomes

$$\frac{Y'}{2} = \frac{\cosh(\gamma l) - 1}{Z_c \sinh(\gamma l)} = \frac{\tanh(\gamma l/2)}{Z_c} = \frac{\tanh(\gamma l/2)}{\sqrt{\frac{z}{y}}} \tag{5.3.8}$$

Rewriting (5.3.8) in terms of the nominal π circuit admittance $Y = yl$,

$$\begin{aligned} \frac{Y'}{2} &= \frac{yl}{2}\left[\frac{\tanh(\gamma l/2)}{\sqrt{\frac{z}{y}}\,\frac{\gamma l}{2}}\right] = \frac{yl}{2}\left[\frac{\tanh(\gamma l/2)}{\sqrt{zy}\,l/2}\right] \\ &= \frac{Y}{2}F_2 \quad \text{S} \end{aligned} \tag{5.3.9}$$

where

$$F_2 = \frac{\tanh(\gamma l/2)}{\gamma l/2} \quad \text{per unit} \tag{5.3.10}$$

Equations (5.3.6) and (5.3.10) give the correction factors F_1 and F_2 to convert Z and Y for the nominal π circuit to Z' and Y' for the equivalent π circuit.

EXAMPLE 5.3 Equivalent π circuit: long line

Compare the equivalent and nominal π circuits for the line in Example 5.2.

SOLUTION For the nominal π circuit,

$$Z = zl = (0.3310\underline{/87.14^\circ})(300) = 99.3\underline{/87.14^\circ} \quad \Omega$$

$$\frac{Y}{2} = \frac{yl}{2} = \left(\frac{j4.674 \times 10^{-6}}{2}\right)(300) = 7.011 \times 10^{-4}\underline{/90^\circ} \quad \text{S}$$

From (5.3.6) and (5.3.10), the correction factors are

$$F_1 = \frac{0.3645\underline{/88.63^\circ}}{0.3731\underline{/88.57^\circ}} = 0.9769\underline{/0.06^\circ} \quad \text{per unit}$$

$$F_2 = \frac{\tanh(\gamma l/2)}{\gamma l/2} = \frac{\cosh(\gamma l) - 1}{(\gamma l/2)\sinh(\gamma l)}$$

$$= \frac{0.9313 + j0.0034 - 1}{\left(\frac{0.3731}{2}\underline{/88.57^\circ}\right)(0.3645\underline{/88.63^\circ})}$$

$$= \frac{-0.0687 + j0.0034}{0.06800\underline{/177.20^\circ}}$$

$$= \frac{0.06878\underline{/177.17^\circ}}{0.06800\underline{/177.20^\circ}} = 1.012\underline{/-0.03^\circ} \quad \text{per unit}$$

Then, from (5.3.5) and (5.3.9), for the equivalent π circuit,

$$Z' = (99.3\underline{/87.14^\circ})(0.9769\underline{/0.06^\circ}) = 97.0\underline{/87.2^\circ} \quad \Omega$$

$$\frac{Y'}{2} = (7.011 \times 10^{-4}\underline{/90^\circ})(1.012\underline{/-0.03^\circ}) = 7.095 \times 10^{-4}\underline{/89.97^\circ} \quad \text{S}$$

$$= 3.7 \times 10^{-7} + j7.095 \times 10^{-4} \quad \text{S}$$

Comparing these nominal and equivalent π circuit values, Z' is about 2% smaller than Z, and $Y'/2$ is about 1% larger than $Y/2$. Although the circuit values are approximately the same for this line, the equivalent π circuit should be used for accurate calculations involving long lines. Note the small shunt conductance, $G' = 3.7 \times 10^{-7}$ S, introduced in the equivalent π circuit. G' is often neglected. ■

5.4

LOSSLESS LINES

In this section, we discuss the following concepts for lossless lines: surge impedance, $ABCD$ parameters, equivalent π circuit, wavelength, surge impedance loading, voltage profiles, and steady-state stability limit.

When line losses are neglected, simpler expressions for the line parameters are obtained and the above concepts are more easily understood. Since transmission and distribution lines for power transfer generally are designed to have low losses, the equations and concepts developed here can be used for quick and reasonably accurate hand calculations leading to seat-of-the-pants analyses and to initial designs. More accurate calculations can then be made with computer programs for follow-up analysis and design.

SURGE IMPEDANCE

For a lossless line, $R = G = 0$, and

$$z = j\omega L \quad \Omega/\text{m} \tag{5.4.1}$$

$$y = j\omega C \quad \text{S/m} \tag{5.4.2}$$

From (5.2.12) and (5.2.16),

$$Z_c = \sqrt{\frac{z}{y}} = \sqrt{\frac{j\omega L}{j\omega C}} = \sqrt{\frac{L}{C}} \quad \Omega \tag{5.4.3}$$

and

$$\gamma = \sqrt{zy} = \sqrt{(j\omega L)(j\omega C)} = j\omega\sqrt{LC} = j\beta \quad \text{m}^{-1} \tag{5.4.4}$$

where

$$\beta = \omega\sqrt{LC} \quad \text{m}^{-1} \tag{5.4.5}$$

The characteristic impedance $Z_c = \sqrt{L/C}$, commonly called *surge* impedance for a lossless line, is pure real—that is, resistive. The propagation constant $\gamma = j\beta$ is pure imaginary.

ABCD PARAMETERS

The *ABCD* parameters are, from (5.2.30)–(5.2.32),

$$A(x) = D(x) = \cosh(\gamma x) = \cosh(j\beta x)$$

$$= \frac{e^{j\beta x} + e^{-j\beta x}}{2} = \cos(\beta x) \quad \text{per unit} \tag{5.4.6}$$

$$\sinh(\gamma x) = \sinh(j\beta x) = \frac{e^{j\beta x} - e^{-j\beta x}}{2} = j\sin(\beta x) \quad \text{per unit} \tag{5.4.7}$$

$$B(x) = Z_c \sinh(\gamma x) = jZ_c \sin(\beta x) = j\sqrt{\frac{L}{C}}\sin(\beta x) \quad \Omega \tag{5.4.8}$$

$$C(x) = \frac{\sinh(\gamma x)}{Z_c} = \frac{j\sin(\beta x)}{\sqrt{\frac{L}{C}}} \quad \text{S} \tag{5.4.9}$$

$A(x)$ and $D(x)$ are pure real; $B(x)$ and $C(x)$ are pure imaginary.

A comparison of lossless versus lossy *ABCD* parameters is shown in Table 5.1.

EQUIVALENT π CIRCUIT

For the equivalent π circuit, using (5.3.4),

$$Z' = jZ_c \sin(\beta l) = jX' \quad \Omega \tag{5.4.10}$$

or, from (5.3.5) and (5.3.6),

$$Z' = (j\omega L l)\left(\frac{\sin(\beta l)}{\beta l}\right) = jX' \quad \Omega \tag{5.4.11}$$

Also, from (5.3.9) and (5.3.10),

$$\begin{aligned}\frac{Y'}{2} &= \frac{Y}{2}\frac{\tanh(j\beta l/2)}{j\beta l/2} = \frac{Y}{2}\frac{\sinh(j\beta l/2)}{(j\beta l/2)\cosh(j\beta l/2)} \\ &= \left(\frac{j\omega Cl}{2}\right)\frac{j\sin(\beta l/2)}{(j\beta l/2)\cos(\beta l/2)} = \left(\frac{j\omega Cl}{2}\right)\frac{\tan(\beta l/2)}{\beta l/2} \\ &= \left(\frac{j\omega C' l}{2}\right) \quad \text{S}\end{aligned} \tag{5.4.12}$$

Z' and Y' are both pure imaginary. Also, for βl less than π radians, Z' is pure inductive and Y' is pure capacitive. Thus the equivalent π circuit for a lossless line, shown in Figure 5.8, is also lossless.

WAVELENGTH

A *wavelength* is the distance required to change the phase of the voltage or current by 2π radians or 360°. For a lossless line, using (5.2.29),

$$\begin{aligned}V(x) &= A(x)V_R + B(x)I_R \\ &= \cos(\beta x)V_R + jZ_c\sin(\beta x)I_R\end{aligned} \tag{5.4.13}$$

FIGURE 5.8 Equivalent π circuit for a lossless line ($\beta\ell$ less than π)

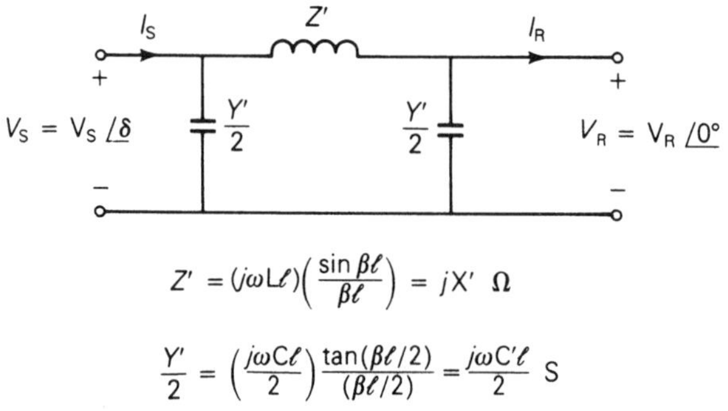

and

$$I(x) = C(x)V_R + D(x)I_R$$
$$= \frac{j\sin(\beta x)}{Z_c}V_R + \cos(\beta x)I_R \tag{5.4.14}$$

From (5.4.13) and (5.4.14), $V(x)$ and $I(x)$ change phase by 2π radians when $x = 2\pi/\beta$. Denoting wavelength by λ, and using (5.4.5),

$$\lambda = \frac{2\pi}{\beta} = \frac{2\pi}{\omega\sqrt{LC}} = \frac{1}{f\sqrt{LC}} \quad \text{m} \tag{5.4.15}$$

or

$$f\lambda = \frac{1}{\sqrt{LC}} \tag{5.4.16}$$

We will show in Chapter 12 that the term $(1/\sqrt{LC})$ in (5.4.16) is the velocity of propagation of voltage and current waves along a lossless line. For overhead lines, $(1/\sqrt{LC}) \approx 3 \times 10^8$ m/s, and for $f = 60$ Hz, (5.4.14) gives

$$\lambda \approx \frac{3 \times 10^8}{60} = 5 \times 10^6 \text{ m} = 5000 \text{ km}$$

Typical power-line lengths are only a small fraction of the above 60-Hz wavelength.

SURGE IMPEDANCE LOADING

Surge impedance loading (SIL) is the power delivered by a lossless line to a load resistance equal to the surge impedance $Z_c = \sqrt{L/C}$. Figure 5.9 shows a lossless line terminated by a resistance equal to its surge impedance. This line represents either a single-phase line or one phase-to-neutral of a balanced three-phase line. At SIL, from (5.4.13),

$$V(x) = \cos(\beta x)V_R + jZ_c\sin(\beta x)I_R$$
$$= \cos(\beta x)V_R + jZ_c\sin(\beta x)\left(\frac{V_R}{Z_c}\right)$$
$$= (\cos\beta x + j\sin\beta x)V_R$$
$$= e^{j\beta x}V_R \quad \text{volts} \tag{5.4.17}$$

$$|V(x)| = |V_R| \quad \text{volts} \tag{5.4.18}$$

FIGURE 5.9 Lossless line terminated by its surge impedance

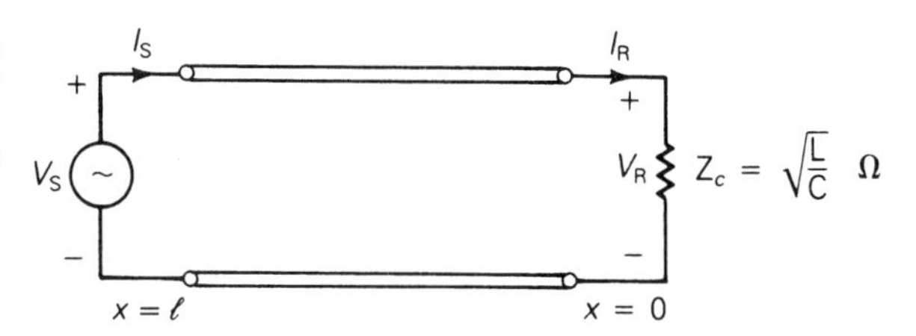

TABLE 5.2 Surge impedance and SIL values for typical 60-Hz overhead lines [1, 2] (Electric Power Research Institute (EPRI), EPRI AC Transmission Line Reference Book—200 kV and Above (Palo Alto, CA: EPRI, www.epri.com, December 2005); Westinghouse Electric Corporation, Electrical Transmission and Distribution Reference Book, 4th ed. (East Pittsburgh, PA, 1964))

V_{rated} (kV)	$Z_c = \sqrt{L/C}$ (Ω)	SIL$= V_{rated}^2/Z_c$ (MW)
69	366–400	12–13
138	366–405	47–52
230	365–395	134–145
345	280–366	325–425
500	233–294	850–1075
765	254–266	2200–2300

Thus, at SIL, the voltage profile is flat. That is, the voltage magnitude at any point x along a lossless line at SIL is constant.

Also from (5.4.14) at SIL,

$$\begin{aligned} I(x) &= \frac{j\sin(\beta x)}{Z_c} V_R + (\cos\beta x)\frac{V_R}{Z_c} \\ &= (\cos\beta x + j\sin\beta x)\frac{V_R}{Z_c} \\ &= (e^{j\beta x})\frac{V_R}{Z_c} \quad \text{A} \end{aligned} \tag{5.4.19}$$

Using (5.4.17) and (5.4.19), the complex power flowing at any point x along the line is

$$\begin{aligned} S(x) &= P(x) + jQ(x) = V(x)I^*(x) \\ &= (e^{j\beta x}V_R)\left(\frac{e^{j\beta x}V_R}{Z_c}\right)^* \\ &= \frac{|V_R|^2}{Z_c} \end{aligned} \tag{5.4.20}$$

Thus the real power flow along a lossless line at SIL remains constant from the sending end to the receiving end. The reactive power flow is zero.

At rated line voltage, the real power delivered, or SIL, is, from (5.4.20),

$$\text{SIL} = \frac{V_{rated}^2}{Z_c} \tag{5.4.21}$$

where rated voltage is used for a single-phase line and rated line-to-line voltage is used for the total real power delivered by a three-phase line. Table 5.2 lists surge impedance and SIL values for typical overhead 60-Hz three-phase lines.

VOLTAGE PROFILES

In practice, power lines are not terminated by their surge impedance. Instead, loadings can vary from a small fraction of SIL during light load conditions

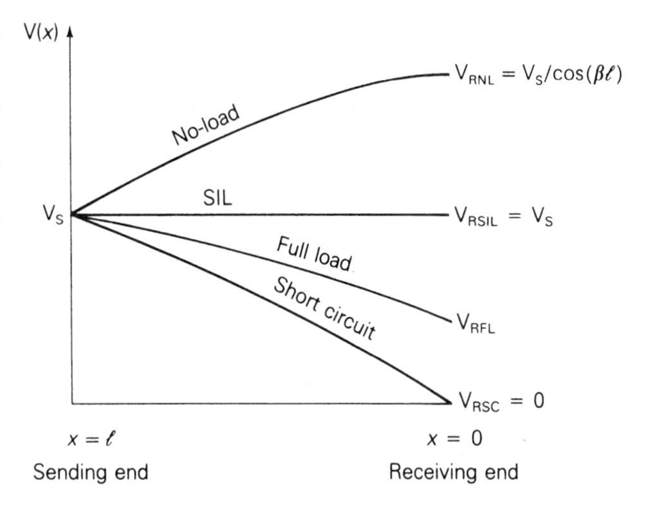

FIGURE 5.10 Voltage profiles of an uncompensated lossless line with fixed sending-end voltage for line lengths up to a quarter wavelength

up to multiples of SIL, depending on line length and line compensation, during heavy load conditions. If a line is not terminated by its surge impedance, then the voltage profile is not flat. Figure 5.10 shows voltage profiles of lines with a fixed sending-end voltage magnitude V_S for line lengths l up to a quarter wavelength. This figure shows four loading conditions: (1) no-load, (2) SIL, (3) short circuit, and (4) full load, which are described as follows:

1. At no-load, $I_{RNL} = 0$ and (5.4.13) yields

$$V_{NL}(x) = (\cos \beta x)V_{RNL} \tag{5.4.22}$$

The no-load voltage increases from $V_S = (\cos \beta l)V_{RNL}$ at the sending end to V_{RNL} at the receiving end (where $x = 0$).

2. From (5.4.18), the voltage profile at SIL is flat.

3. For a short circuit at the load, $V_{RSC} = 0$ and (5.4.13) yields

$$V_{SC}(x) = (Z_c \sin \beta x)I_{RSC} \tag{5.4.23}$$

The voltage decreases from $V_S = (\sin \beta l)(Z_c I_{RSC})$ at the sending end to $V_{RSC} = 0$ at the receiving end.

4. The full-load voltage profile, which depends on the specification of full-load current, lies above the short-circuit voltage profile.

Figure 5.10 summarizes these results, showing a high receiving-end voltage at no-load and a low receiving-end voltage at full load. This voltage regulation problem becomes more severe as the line length increases. In Section 5.6, we discuss shunt compensation methods to reduce voltage fluctuations.

STEADY-STATE STABILITY LIMIT

The equivalent π circuit of Figure 5.8 can be used to obtain an equation for the real power delivered by a lossless line. Assume that the voltage magnitudes V_S and V_R at the ends of the line are held constant. Also, let δ denote the voltage-phase angle at the sending end with respect to the receiving end. From KVL, the receiving-end current I_R is

$$\begin{aligned} I_R &= \frac{V_S - V_R}{Z'} - \frac{Y'}{2} V_R \\ &= \frac{V_S e^{j\delta} - V_R}{jX'} - \frac{j\omega C' l}{2} V_R \end{aligned} \tag{5.4.24}$$

and the complex power S_R delivered to the receiving end is

$$\begin{aligned} S_R = V_R I_R^* &= V_R \left(\frac{V_S e^{j\delta} - V_R}{jX'} \right)^* + \frac{j\omega C' l}{2} V_R^2 \\ &= V_R \left(\frac{V_S e^{-j\delta} - V_R}{-jX'} \right) + \frac{j\omega C l}{2} V_R^2 \\ &= \frac{jV_R V_S \cos\delta + V_R V_S \sin\delta - jV_R^2}{X'} + \frac{j\omega C l}{2} V_R^2 \end{aligned} \tag{5.4.25}$$

The real power delivered is

$$P = P_S = P_R = \mathrm{Re}(S_R) = \frac{V_R V_S}{X'} \sin\delta \quad \mathrm{W} \tag{5.4.26}$$

Note that since the line is lossless, $P_S = P_R$.

Equation (5.4.26) is plotted in Figure 5.11. For fixed voltage magnitudes V_S and V_R, the phase angle δ increases from 0 to 90° as the real power delivered increases. The maximum power that the line can deliver, which occurs when $\delta = 90°$, is given by

$$P_{max} = \frac{V_S V_R}{X'} \quad \mathrm{W} \tag{5.4.27}$$

FIGURE 5.11 Real power delivered by a lossless line versus voltage angle across the line

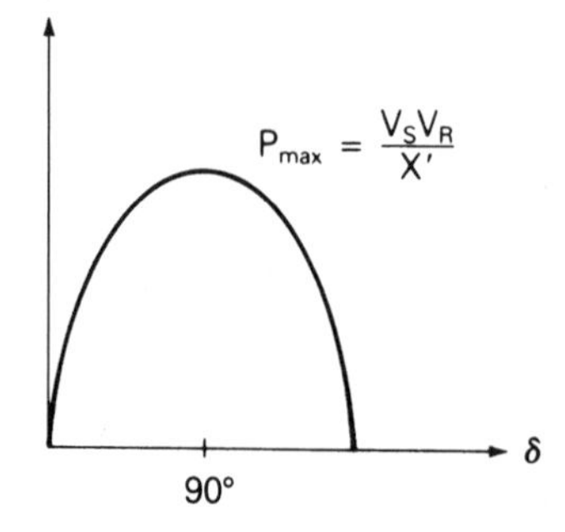

P_{max} represents the theoretical *steady-state stability* limit of a lossless line. If an attempt were made to exceed this steady-state stability limit, then synchronous machines at the sending end would lose synchronism with those at the receiving end. Stability is further discussed in Chapter 13.

It is convenient to express the steady-state stability limit in terms of SIL. Using (5.4.10) in (5.4.26),

$$P = \frac{V_S V_R \sin\delta}{Z_c \sin\beta l} = \left(\frac{V_S V_R}{Z_c}\right)\frac{\sin\delta}{\sin\left(\frac{2\pi l}{\lambda}\right)} \tag{5.4.28}$$

Expressing V_S and V_R in per-unit of rated line voltage,

$$P = \left(\frac{V_S}{V_{rated}}\right)\left(\frac{V_R}{V_{rated}}\right)\left(\frac{V_{rated}^2}{Z_c}\right)\frac{\sin\delta}{\sin\left(\frac{2\pi l}{\lambda}\right)}$$

$$= V_{S.p.u.} V_{R.p.u.} (\text{SIL}) \frac{\sin\delta}{\sin\left(\frac{2\pi l}{\lambda}\right)} \quad \text{W} \tag{5.4.29}$$

And for $\delta = 90°$, the theoretical steady-state stability limit is

$$P_{max} = \frac{V_{S.p.u.} V_{R.p.u.} (\text{SIL})}{\sin\left(\frac{2\pi l}{\lambda}\right)} \quad \text{W} \tag{5.4.30}$$

Equations (5.4.27)–(5.4.30) reveal two important factors affecting the steady-state stability limit. First, from (5.4.27), it increases with the square of the line voltage. For example, a doubling of line voltage enables a fourfold increase in maximum power flow. Second, it decreases with line length. Equation (5.4.30) is plotted in Figure 5.12 for $V_{S.p.u.} = V_{R.p.u.} = 1$, $\lambda = 5000$ km, and line lengths up to 1100 km. As shown, the theoretical steady-state stability limit decreases from 4(SIL) for a 200-km line to about 2(SIL) for a 400-km line.

FIGURE 5.12

Transmission-line loadability curve for 60-Hz overhead lines—no series or shunt compensation

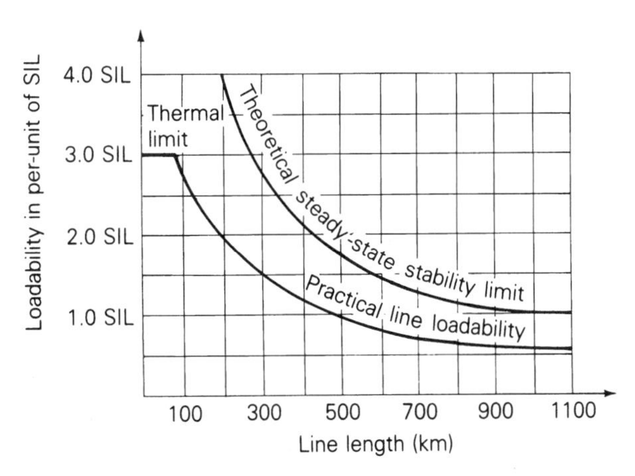

EXAMPLE 5.4 Theoretical steady-state stability limit: long line

Neglecting line losses, find the theoretical steady-state stability limit for the 300-km line in Example 5.2. Assume a 266.1-Ω surge impedance, a 5000-km wavelength, and $V_S = V_R = 765$ kV.

SOLUTION From (5.4.21),

$$\mathrm{SIL} = \frac{(765)^2}{266.1} = 2199 \quad \mathrm{MW}$$

From (5.4.30) with $l = 300$ km and $\lambda = 5000$ km,

$$P_{max} = \frac{(1)(1)(2199)}{\sin\left(\dfrac{2\pi \times 300}{5000}\right)} = (2.716)(2199) = 5974 \quad \mathrm{MW}$$

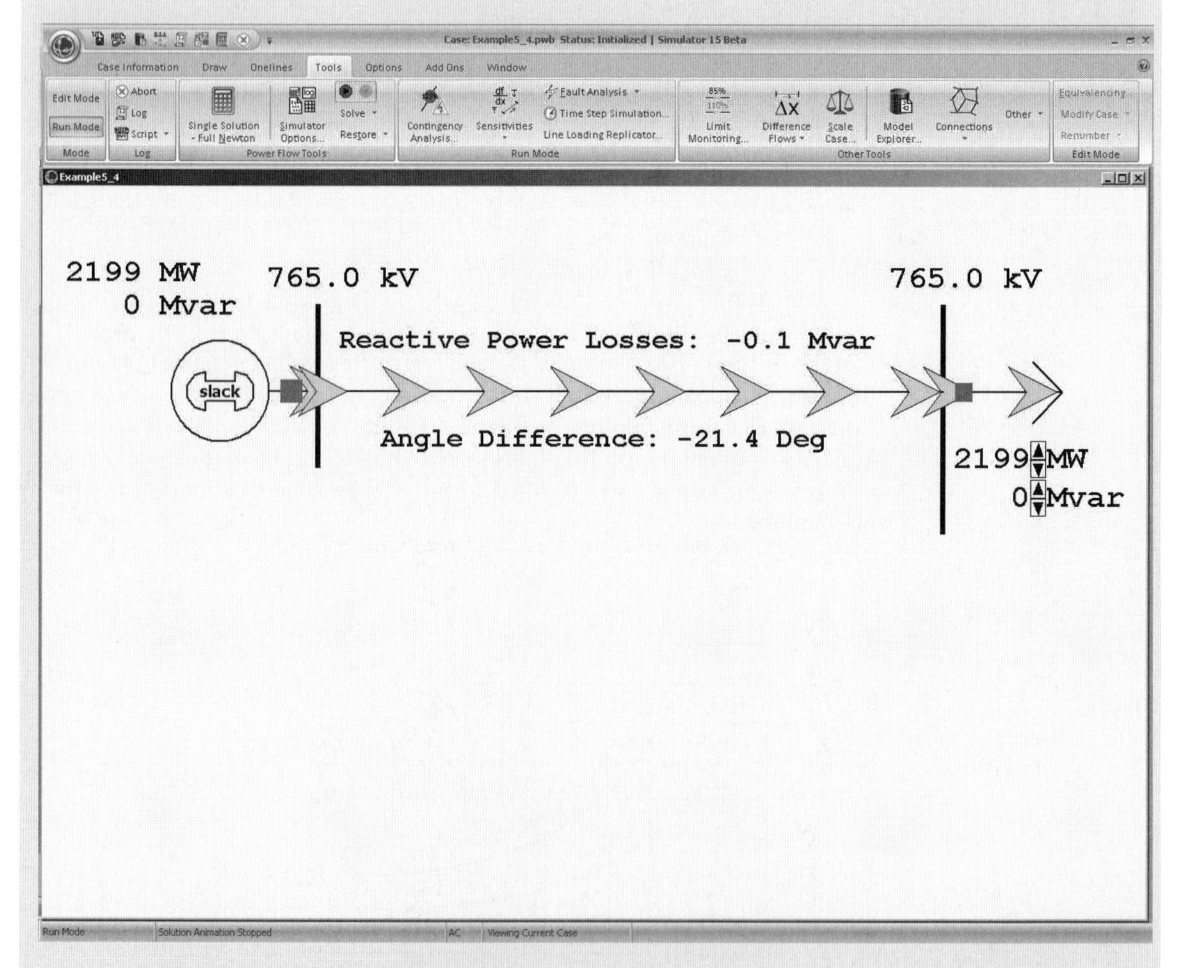

FIGURE 5.13 Screen for Example 5.4

Alternatively, from Figure 5.12, for a 300-km line, the theoretical steady-state stability limit is (2.72)SIL = (2.72)(2199) = 5980 MW, about the same as the above result (see Figure 5.13).

Open PowerWorld Simulator case Example 5_4 and select **Tools Play** to see an animated view of this example. When the load on a line is equal to the SIL, the voltage profile across the line is flat and the line's net reactive power losses are zero. For loads above the SIL, the line consumes reactive power and the load's voltage magnitude is below the sending-end value. Conversely, for loads below the SIL, the line actually generates reactive power and the load's voltage magnitude is above the sending-end value. Use the load arrow button to vary the load to see the changes in the receiving-end voltage and the line's reactive power consumption. ■

5.5 MAXIMUM POWER FLOW

Maximum power flow, discussed in Section 5.4 for lossless lines, is derived here in terms of the $ABCD$ parameters for lossy lines. The following notation is used:

$$A = \cosh(\gamma l) = \mathrm{A}\underline{/\theta_{\mathrm{A}}}$$

$$B = Z' = \mathrm{Z}'\underline{/\theta_{\mathrm{Z}}}$$

$$V_{\mathrm{S}} = \mathrm{V_S}\underline{/\delta} \qquad V_{\mathrm{R}} = \mathrm{V_R}\underline{/0^\circ}$$

Solving (5.2.33) for the receiving-end current,

$$I_{\mathrm{R}} = \frac{V_{\mathrm{S}} - AV_{\mathrm{R}}}{B} = \frac{\mathrm{V_S}e^{j\delta} - \mathrm{AV_R}e^{j\theta_{\mathrm{A}}}}{\mathrm{Z}'e^{j\theta_{\mathrm{Z}}}} \tag{5.5.1}$$

The complex power delivered to the receiving end is

$$S_{\mathrm{R}} = \mathrm{P_R} + j\mathrm{Q_R} = V_{\mathrm{R}}I_{\mathrm{R}}^* = \mathrm{V_R}\left[\frac{\mathrm{V_S}e^{j(\delta-\theta_{\mathrm{Z}})} - \mathrm{AV_R}e^{j(\theta_{\mathrm{A}}-\theta_{\mathrm{Z}})}}{\mathrm{Z}'}\right]^*$$

$$= \frac{\mathrm{V_R V_S}}{\mathrm{Z}'}e^{j(\theta_{\mathrm{Z}}-\delta)} - \frac{\mathrm{AV_R^2}}{\mathrm{Z}'}e^{j(\theta_{\mathrm{Z}}-\theta_{\mathrm{A}})} \tag{5.5.2}$$

The real and reactive power delivered to the receiving end are thus

$$\mathrm{P_R} = \mathrm{Re}(S_{\mathrm{R}}) = \frac{\mathrm{V_R V_S}}{\mathrm{Z}'}\cos(\theta_{\mathrm{Z}} - \delta) - \frac{\mathrm{AV_R^2}}{\mathrm{Z}'}\cos(\theta_{\mathrm{Z}} - \theta_{\mathrm{A}}) \tag{5.5.3}$$

$$\mathrm{Q_R} = \mathrm{Im}(S_{\mathrm{R}}) = \frac{V_{\mathrm{R}} V_{\mathrm{S}}}{Z'}\sin(\theta_{\mathrm{Z}} - \delta) - \frac{\mathrm{AV_R^2}}{\mathrm{Z}'}\sin(\theta_{\mathrm{Z}} - \theta_{\mathrm{A}}) \tag{5.5.4}$$

Note that for a lossless line, $\theta_A = 0°$, $B = Z' = jX'$, $Z' = X'$, $\theta_Z = 90°$, and (5.5.3) reduces to

$$\begin{aligned} P_R &= \frac{V_R V_S}{X'} \cos(90 - \delta) - \frac{A V_R^2}{X'} \cos(90°) \\ &= \frac{V_R V_S}{X'} \sin \delta \end{aligned} \tag{5.5.5}$$

which is the same as (5.4.26).

The theoretical maximum real power delivered (or steady-state stability limit) occurs when $\delta = \theta_Z$ in (5.5.3):

$$P_{Rmax} = \frac{V_R V_S}{Z'} - \frac{A V_R^2}{Z'} \cos(\theta_Z - \theta_A) \tag{5.5.6}$$

The second term in (5.5.6), and the fact that Z' is larger than X', reduce P_{Rmax} to a value somewhat less than that given by (5.4.27) for a lossless line.

EXAMPLE 5.5 Theoretical maximum power delivered: long line

Determine the theoretical maximum power, in MW and in per-unit of SIL, that the line in Example 5.2 can deliver. Assume $V_S = V_R = 765$ kV.

SOLUTION From Example 5.2,

$$A = 0.9313 \quad \text{per unit;} \qquad \theta_A = 0.209°$$

$$B = Z' = 97.0 \quad \Omega; \qquad \theta_Z = 87.2°$$

$$Z_c = 266.1 \quad \Omega$$

From (5.5.6) with $V_S = V_R = 765$ kV,

$$\begin{aligned} P_{Rmax} &= \frac{(765)^2}{97} - \frac{(0.9313)(765)^2}{97} \cos(87.2° - 0.209°) \\ &= 6033 - 295 = 5738 \quad \text{MW} \end{aligned}$$

From (5.4.20),

$$\text{SIL} = \frac{(765)^2}{266.1} = 2199 \quad \text{MW}$$

Thus

$$P_{Rmax} = \frac{5738}{2199} = 2.61 \quad \text{per unit}$$

This value is about 4% less than that found in Example 5.4, where losses were neglected. ■

5.6

LINE LOADABILITY

In practice, power lines are not operated to deliver their theoretical maximum power, which is based on rated terminal voltages and an angular displacement $\delta = 90°$ across the line. Figure 5.12 shows a practical line loadability curve plotted below the theoretical steady-state stability limit. This curve is based on the voltage-drop limit $V_R/V_S \geqslant 0.95$ and on a maximum angular displacement of 30 to 35° across the line (or about 45° across the line and equivalent system reactances), in order to maintain stability during transient disturbances [1, 3]. The curve is valid for typical overhead 60-Hz lines with no compensation. Note that for short lines less than 80 km long, loadability is limited by the thermal rating of the conductors or by terminal equipment ratings, not by voltage drop or stability considerations. In Section 5.7, we investigate series and shunt compensation techniques to increase the loadability of longer lines toward their thermal limit.

EXAMPLE 5.6 Practical line loadability and percent voltage regulation: long line

The 300-km uncompensated line in Example 5.2 has four 1,272,000-cmil (644.5-mm^2) 54/3 ACSR conductors per bundle. The sending-end voltage is held constant at 1.0 per-unit of rated line voltage. Determine the following:

a. The practical line loadability. (Assume an approximate receiving-end voltage $V_R = 0.95$ per unit and $\delta = 35°$ maximum angle across the line.)

b. The full-load current at 0.986 p.f. leading based on the above practical line loadability

c. The exact receiving-end voltage for the full-load current found in part (b)

d. Percent voltage regulation for the above full-load current

e. Thermal limit of the line, based on the approximate current-carrying capacity given in Table A.4

SOLUTION

a. From (5.5.3), with $V_S = 765$, $V_R = 0.95 \times 765$ kV, and $\delta = 35°$, using the values of Z', θ_Z, A, and θ_A from Example 5.5,

$$P_R = \frac{(765)(0.95 \times 765)}{97.0}\cos(87.2° - 35°)$$

$$- \frac{(0.9313)(0.95 \times 765)^2}{97.0}\cos(87.2° - 0.209°)$$

$$= 3513 - 266 = 3247 \quad \text{MW}$$

$P_R = 3247$ MW is the practical line loadability, provided the thermal and voltage-drop limits are not exceeded. Alternatively, from Figure 5.12 for a 300-km line, the practical line loadability is $(1.49)\text{SIL} = (1.49)(2199) = 3277$ MW, about the same as the above result.

b. For the above loading at 0.986 p.f. leading and at 0.95×765 kV, the full-load receiving-end current is

$$I_{RFL} = \frac{P}{\sqrt{3}V_R(\text{p.f.})} = \frac{3247}{(\sqrt{3})(0.95 \times 765)(0.986)} = 2.616 \quad \text{kA}$$

c. From (5.1.1) with $I_{RFL} = 2.616\underline{/\cos^{-1} 0.986} = 2.616\underline{/9.599^\circ}$ kA, using the A and B parameters from Example 5.2,

$$V_S = AV_{RFL} + BI_{RFL}$$

$$\frac{765}{\sqrt{3}}\underline{/\delta} = (0.9313\underline{/0.209^\circ})(V_{RFL}\underline{/0^\circ}) + (97.0\underline{/87.2^\circ})(2.616\underline{/9.599^\circ})$$

$$441.7\underline{/\delta} = (0.9313V_{RFL} - 30.04) + j(0.0034V_{RFL} + 251.97)$$

Taking the squared magnitude of the above equation,

$$(441.7)^2 = 0.8673V_{RFL}^2 - 54.24V_{RFL} + 64{,}391$$

Solving,

$$V_{RFL} = 420.7 \quad \text{kV}_{LN}$$

$$= 420.7\sqrt{3} = 728.7 \quad \text{kV}_{LL} = 0.953 \quad \text{per unit}$$

d. From (5.1.19), the receiving-end no-load voltage is

$$V_{RNL} = \frac{V_S}{A} = \frac{765}{0.9313} = 821.4 \quad \text{kV}_{LL}$$

And from (5.1.18),

$$\text{percent VR} = \frac{821.4 - 728.7}{728.7} \times 100 = 12.72\%$$

e. From Table A.4, the approximate current-carrying capacity of four 1,272,000-cmil (644.5-mm^2) 54/3 ACSR conductors is $4 \times 1.2 = 4.8$ kA.

Since the voltages $V_S = 1.0$ and $V_{RFL} = 0.953$ per unit satisfy the voltage-drop limit $V_R/V_S \geqslant 0.95$, the factor that limits line loadability is steady-state stability for this 300-km uncompensated line. The full-load current of 2.616 kA corresponding to loadability is also well below the thermal limit of 4.8 kA. The 12.7% voltage regulation is too high because the no-load voltage is too high. Compensation techniques to reduce no-load voltages are discussed in Section 5.7. ■

EXAMPLE 5.7 Selection of transmission line voltage and number of lines for power transfer

From a hydroelectric power plant 9000 MW are to be transmitted to a load center located 500 km from the plant. Based on practical line loadability criteria, determine the number of three-phase, 60-Hz lines required to transmit this power, with one line out of service, for the following cases: (a) 345-kV lines with $Z_c = 297\ \Omega$; (b) 500-kV lines with $Z_c = 277\ \Omega$; (c) 765-kV lines with $Z_c = 266\ \Omega$. Assume $V_S = 1.0$ per unit, $V_R = 0.95$ per unit, and $\delta = 35°$. Also assume that the lines are uncompensated and widely separated such that there is negligible mutual coupling between them.

SOLUTION

a. For 345-kV lines, (5.4.21) yields

$$\text{SIL} = \frac{(345)^2}{297} = 401 \quad \text{MW}$$

Neglecting losses, from (5.4.29), with $l = 500$ km and $\delta = 35°$,

$$P = \frac{(1.0)(0.95)(401)\sin(35°)}{\sin\left(\dfrac{2\pi \times 500}{5000}\right)} = (401)(0.927) = 372 \quad \text{MW/line}$$

Alternatively, the practical line loadability curve in Figure 5.12 can be used to obtain P = (0.93)SIL for typical 500-km overhead 60-Hz uncompensated lines.

In order to transmit 9000 MW with one line out of service,

$$\#345\text{-kV lines} = \frac{9000 \text{ MW}}{372 \text{ MW/line}} + 1 = 24.2 + 1 \approx 26$$

b. For 500-kV lines,

$$\text{SIL} = \frac{(500)^2}{277} = 903 \quad \text{MW}$$

$$P = (903)(0.927) = 837 \quad \text{MW/line}$$

$$\#500\text{-kV lines} = \frac{9000}{837} + 1 = 10.8 + 1 \approx 12$$

c. For 765-kV lines,

$$\text{SIL} = \frac{(765)^2}{266} = 2200 \quad \text{MW}$$

$$P = (2200)(0.927) = 2039 \quad \text{MW/line}$$

$$\#765\text{-kV lines} = \frac{9000}{2039} + 1 = 4.4 + 1 \approx 6$$

Increasing the line voltage from 345 to 765 kV, a factor of 2.2, reduces the required number of lines from 26 to 6, a factor of 4.3. ■

EXAMPLE 5.8 Effect of intermediate substations on number of lines required for power transfer

Can five instead of six 765-kV lines transmit the required power in Example 5.7 if there are two intermediate substations that divide each line into three 167-km line sections, and if only one line section is out of service?

SOLUTION The lines are shown in Figure 5.14. For simplicity, we neglect line losses. The equivalent π circuit of one 500-km, 765-kV line has a series reactance, from (5.4.10) and (5.4.15),

$$X' = (266)\sin\left(\frac{2\pi \times 500}{5000}\right) = 156.35 \quad \Omega$$

Combining series/parallel reactances in Figure 5.14, the equivalent reactance of five lines with one line section out of service is

$$X_{eq} = \frac{1}{5}\left(\frac{2}{3}X'\right) + \frac{1}{4}\left(\frac{X'}{3}\right) = 0.2167X' = 33.88 \quad \Omega$$

Then, from (5.4.26) with $\delta = 35°$,

$$P = \frac{(765)(765 \times 0.95)\sin(35°)}{33.88} = 9412 \quad \text{MW}$$

Inclusion of line losses would reduce the above value by 3 or 4% to about 9100 MW. Therefore, the answer is yes. Five 765-kV, 500-km uncompensated lines with two intermediate substations and with one line section out of service will transmit 9000 MW. Intermediate substations are often economical if their costs do not outweigh the reduction in line costs.

This example is modeled in PowerWorld Simulator case Example 5_8 (see Figure 5.15). Each line segment is represented with the lossless line model from Example 5.4 with the π circuit parameters modified to exactly match those for a 167 km distributed line. The pie charts on each line segment show the percentage loading of the line, assuming a rating of 3500 MVA. The solid red squares on the lines represent closed circuit breakers,

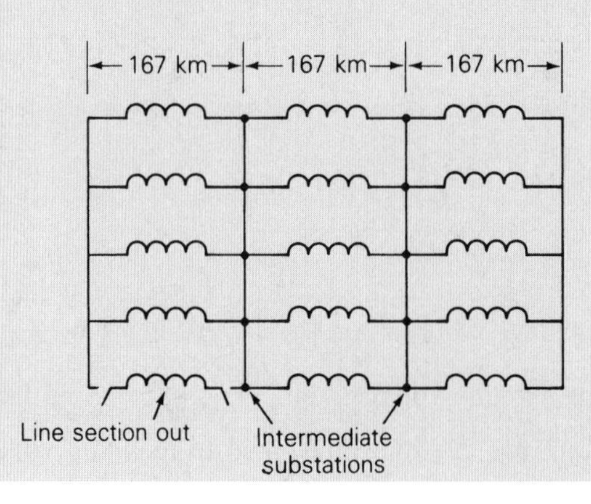

FIGURE 5.14 Transmission-line configuration for Example 5.8

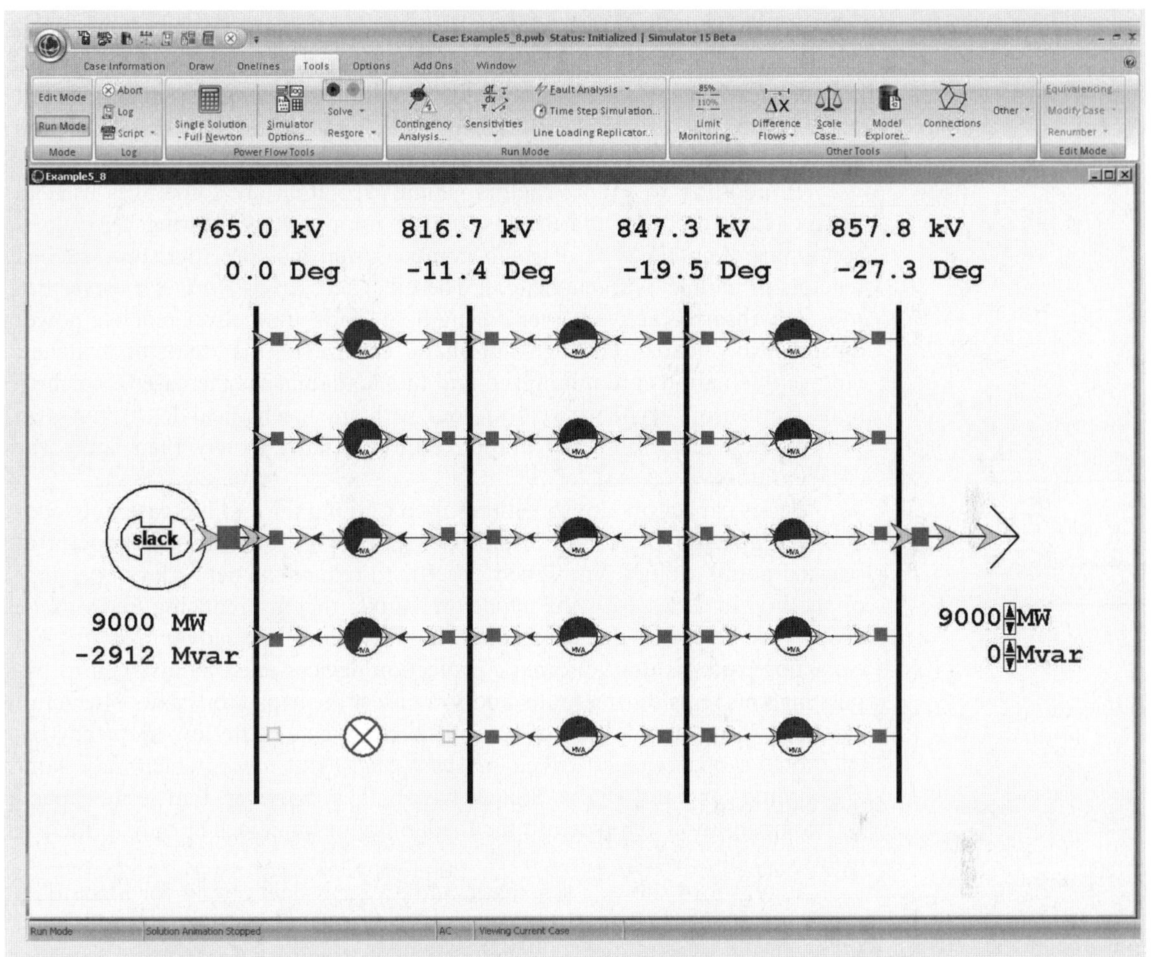

FIGURE 5.15 Screen for Example 5.8

and the green squares correspond to open circuit breakers. Clicking on a circuit breaker toggles its status. The simulation results differ slightly from the simplified analysis done earlier in the example because the simulation includes the charging capacitance of the transmission lines. With all line segments in-service, use the load's arrow to verify that the SIL for this system is 11,000 MW, five times that of the single circuit line in Example 5.4. ■

5.7 REACTIVE COMPENSATION TECHNIQUES

Inductors and capacitors are used on medium-length and long transmission lines to increase line loadability and to maintain voltages near rated values.

Shunt reactors (inductors) are commonly installed at selected points along EHV lines from each phase to neutral. The inductors absorb reactive power and reduce overvoltages during light load conditions. They also reduce transient overvoltages due to switching and lightning surges. However, shunt reactors can reduce line loadability if they are not removed under full-load conditions.

In addition to shunt reactors, shunt capacitors are sometimes used to deliver reactive power and increase transmission voltages during heavy load conditions. Another type of shunt compensation includes thyristor-switched reactors in parallel with capacitors. These devices, called *static var compensators*, can absorb reactive power during light loads and deliver reactive power during heavy loads. Through automatic control of the thyristor switches, voltage fluctuations are minimized and line loadability is increased. Synchronous condensors (synchronous motors with no mechanical load) can also control their reactive power output, although more slowly than static var compensators.

Series capacitors are sometimes used on long lines to increase line loadability. Capacitor banks are installed in series with each phase conductor at selected points along a line. Their effect is to reduce the net series impedance of the line in series with the capacitor banks, thereby reducing line-voltage drops and increasing the steady-state stability limit. A disadvantage of series capacitor banks is that automatic protection devices must be installed to bypass high currents during faults and to reinsert the capacitor banks after fault clearing. Also, the addition of series capacitors can excite low-frequency oscillations, a phenomenon called *subsynchronous resonance*, which may damage turbine-generator shafts. Studies have shown, however, that series capacitive compensation can increase the loadability of long lines at only a fraction of the cost of new transmission [1].

Figure 5.16 shows a schematic and an equivalent circuit for a compensated line section, where N_C is the amount of series capacitive compensation

FIGURE 5.16

Compensated transmission-line section

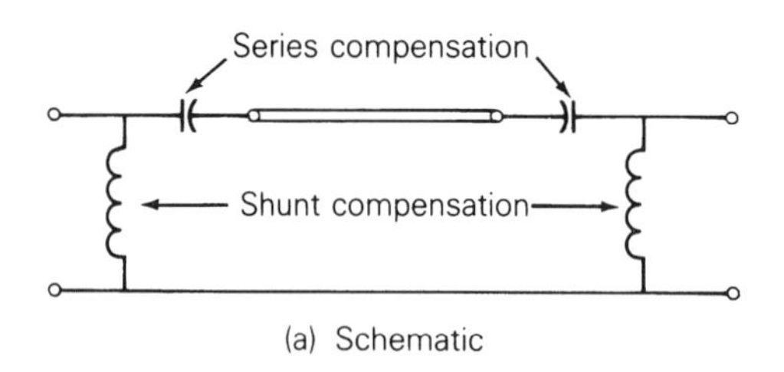

(a) Schematic

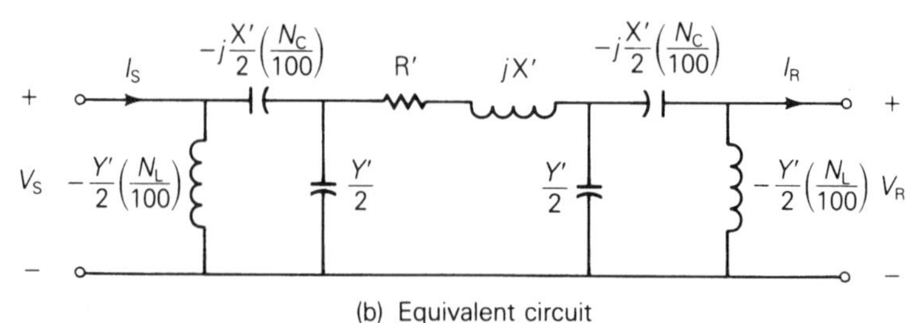

(b) Equivalent circuit

expressed in percent of the positive-sequence line impedance and N_L is the amount of shunt reactive compensation in percent of the positive-sequence line admittance. It is assumed in Figure 5.16 that half of the compensation is installed at each end of the line section. The following two examples illustrate the effect of compensation.

EXAMPLE 5.9 **Shunt reactive compensation to improve transmission-line voltage regulation**

Identical shunt reactors (inductors) are connected from each phase conductor to neutral at both ends of the 300-km line in Example 5.2 during light load conditions, providing 75% compensation. The reactors are removed during heavy load conditions. Full load is 1.90 kA at unity p.f. and at 730 kV. Assuming that the sending-end voltage is constant, determine the following:

a. Percent voltage regulation of the uncompensated line

b. The equivalent shunt admittance and series impedance of the compensated line

c. Percent voltage regulation of the compensated line

SOLUTION

a. From (5.1.1) with $I_{RFL} = 1.9\underline{/0^\circ}$ kA, using the A and B parameters from Example 5.2,

$$V_S = AV_{RFL} + BI_{RFL}$$

$$= (0.9313\underline{/0.209^\circ})\left(\frac{730}{\sqrt{3}}\underline{/0^\circ}\right) + (97.0\underline{/87.2^\circ})(1.9\underline{/0^\circ})$$

$$= 392.5\underline{/0.209^\circ} + 184.3\underline{/87.2^\circ}$$

$$= 401.5 + j185.5$$

$$= 442.3\underline{/24.8^\circ} \quad \text{kV}_{LN}$$

$$\text{V}_S = 442.3\sqrt{3} = 766.0 \quad \text{kV}_{LL}$$

The no-load receiving-end voltage is, from (5.1.19),

$$\text{V}_{RNL} = \frac{766.0}{0.9313} = 822.6 \quad \text{kV}_{LL}$$

and the percent voltage regulation for the uncompensated line is, from (5.1.18),

$$\text{percent VR} = \frac{822.6 - 730}{730} \times 100 = 12.68\%$$

b. From Example 5.3, the shunt admittance of the equivalent π circuit without compensation is

$$Y' = 2(3.7 \times 10^{-7} + j7.094 \times 10^{-4})$$
$$= 7.4 \times 10^{-7} + j14.188 \times 10^{-4} \quad \text{S}$$

With 75% shunt compensation, the equivalent shunt admittance is

$$Y_{eq} = 7.4 \times 10^{-7} + j14.188 \times 10^{-4}\left(1 - \tfrac{75}{100}\right)$$
$$= 3.547 \times 10^{-4}\underline{/89.88^\circ} \quad \text{S}$$

Since there is no series compensation, the equivalent series impedance is the same as without compensation:

$$Z_{eq} = Z' = 97.0\underline{/87.2^\circ} \quad \Omega$$

c. The equivalent A parameter for the compensated line is

$$A_{eq} = 1 + \frac{Y_{eq}Z_{eq}}{2}$$
$$= 1 + \frac{(3.547 \times 10^{-4}\underline{/89.88^\circ})(97.0\underline{/87.2^\circ})}{2}$$
$$= 1 + 0.0172\underline{/177.1^\circ}$$
$$= 0.9828\underline{/0.05^\circ} \quad \text{per unit}$$

Then, from (5.1.19),

$$V_{RNL} = \frac{766}{0.9828} = 779.4 \quad \text{kV}_{LL}$$

Since the shunt reactors are removed during heavy load conditions, $V_{RFL} = 730$ kV is the same as without compensation. Therefore

$$\text{percent VR} = \frac{779.4 - 730}{730} \times 100 = 6.77\%$$

The use of shunt reactors at light loads improves the voltage regulation from 12.68% to 6.77% for this line. ■

EXAMPLE 5.10 Series capacitive compensation to increase transmission-line loadability

Identical series capacitors are installed in each phase at both ends of the line in Example 5.2, providing 30% compensation. Determine the theoretical maximum power that this compensated line can deliver and compare with that of the uncompensated line. Assume $V_S = V_R = 765$ kV.

SOLUTION From Example 5.3, the equivalent series reactance without compensation is

$$X' = 97.0 \sin 87.2^\circ = 96.88 \quad \Omega$$

Based on 30% series compensation, half at each end of the line, the impedance of each series capacitor is

$$Z_{cap} = -jX_{cap} = -j(\tfrac{1}{2})(0.30)(96.88) = -j14.53 \quad \Omega$$

From Figure 5.4, the *ABCD* matrix of this series impedance is

$$\left[\begin{array}{c|c} 1 & -j14.53 \\ \hline 0 & 1 \end{array}\right]$$

As also shown in Figure 5.4, the equivalent *ABCD* matrix of networks in series is obtained by multiplying the *ABCD* matrices of the individual networks. For this example there are three networks: the series capacitors at the sending end, the line, and the series capacitors at the receiving end. Therefore the equivalent *ABCD* matrix of the compensated line is, using the *ABCD* parameters, from Example 5.2,

$$\left[\begin{array}{c|c} 1 & -j14.53 \\ \hline 0 & 1 \end{array}\right]\left[\begin{array}{c|c} 0.9313\underline{/0.209^\circ} & 97.0\underline{/87.2^\circ} \\ \hline 1.37 \times 10^{-3}\underline{/90.06^\circ} & 0.9313\underline{/0.209^\circ} \end{array}\right]\left[\begin{array}{c|c} 1 & -j14.53 \\ \hline 0 & 1 \end{array}\right]$$

After performing these matrix multiplications, we obtain

$$\left[\begin{array}{c|c} A_{eq} & B_{eq} \\ \hline C_{eq} & D_{eq} \end{array}\right] = \left[\begin{array}{c|c} 0.9512\underline{/0.205^\circ} & 69.70\underline{/86.02^\circ} \\ \hline 1.37 \times 10^{-3}\underline{/90.06^\circ} & 0.9512\underline{/0.205^\circ} \end{array}\right]$$

Therefore

$$A_{eq} = 0.9512 \quad \text{per unit} \qquad \theta_{Aeq} = 0.205^\circ$$

$$B_{eq} = Z'_{eq} = 69.70 \quad \Omega \qquad \theta_{Zeq} = 86.02^\circ$$

From (5.5.6) with $V_S = V_R = 765$ kV,

$$P_{Rmax} = \frac{(765)^2}{69.70} - \frac{(0.9512)(765)^2}{69.70}\cos(86.02^\circ - 0.205^\circ)$$

$$= 8396 - 583 = 7813 \quad \text{MW}$$

which is 36.2% larger than the value of 5738 MW found in Example 5.5 without compensation. We note that the practical line loadability of this series compensated line is also about 35% larger than the value of 3247 MW found in Example 5.6 without compensation.

This example is modeled in PowerWorld Simulator case Example 5_10 (see Figure 5.17). When opened, both of the series capacitors are bypassed (i.e., they are modeled as short circuits) meaning this case is initially identical to the Example 5.4 case. Click on the blue "Bypassed" field to place each of the series capacitors into the circuit. This decreases the angle across the line, resulting in more net power transfer.

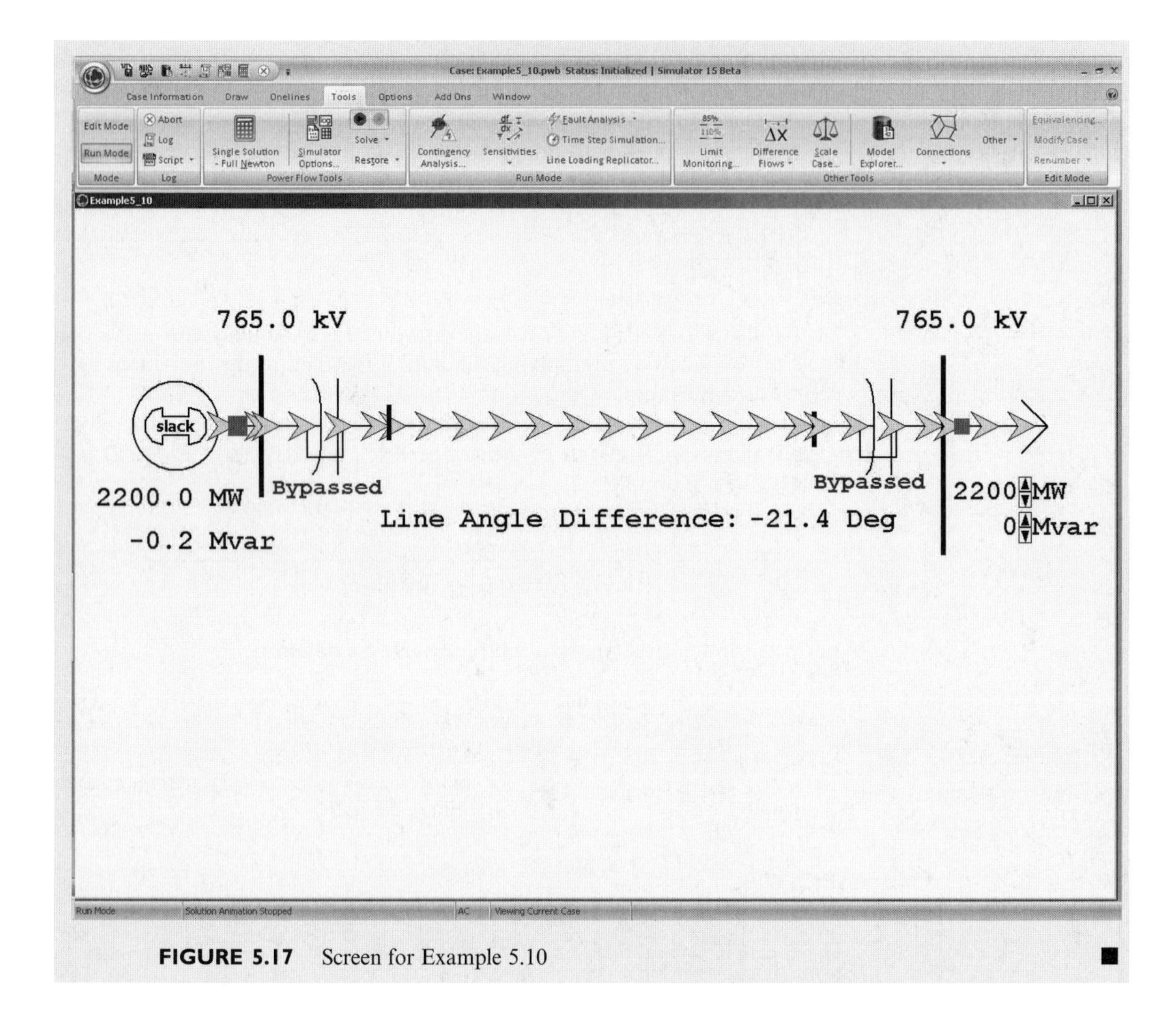

FIGURE 5.17 Screen for Example 5.10

MULTIPLE CHOICE QUESTIONS

SECTION 5.1

5.1 Representing a transmission line by the two-port network, in terms of $ABCD$ parameters, (a) express V_S, the sending-end voltage, in terms of V_R, the receiving-end voltage, and I_R, the receiving-end current, and (b) express I_S, the sending-end current, in terms of V_R and I_R.

(a) $V_S =$ __________ (b) $I_S =$ __________

5.2 As applied to linear, passive, bilateral two-port networks, the $ABCD$ parameters satisfy $AD - BC = 1$.

(a) True (b) False

5.3 Express the no-load receiving-end voltage V_{RNL} in terms of the sending-end voltage, V_S, and the $ABCD$ parameters.

$V_{RNL} =$__________

5.4 The *ABCD* parameters, which are in general complex numbers, have the units of __________, __________, __________, __________, respectively. Fill in the Blanks.

5.5 The loadability of short transmission lines (less than 80 km, represented by including only series resistance and reactance) is determined by __________; that of medium lines (less than 250 km, represented by nominal π circuit) is determined by __________; and that of long lines (more than 250 km, represented by equivalent π circuit) is determined by __________. Fill in the Blanks.

5.6 Can the voltage regulation, which is proportional to $(V_{\text{RNL}} - V_{\text{RFL}})$, be negative?
(a) Yes (b) No

SECTION 5.2

5.7 The propagation constant, which is a complex quantity in general, has the units of __________, and the characteristic impedance has the units of __________.

5.8 Express hyperbolic functions cosh $\sqrt{x}$ and sinh $\sqrt{x}$ in terms of exponential functions.

5.9 e^{γ}, where $\gamma = \alpha + j\beta$, can be expressed as $e^{\alpha l}\underline{/\beta l}$, in which αl is dimensionless and βl is in radians (also dimensionless).
(a) True (b) False

SECTION 5.3

5.10 The equivalent π circuit is identical in structure to the nominal π circuit.
(a) True (b) False

5.11 The correction factors $F_1 = \sinh(\gamma l)/\gamma l$ and $F_2 = \tanh(\gamma l/2)/(\gamma l/2)$, which are complex numbers, have the units of _______. Fill in the Blank.

SECTION 5.4

5.12 For a lossless line, the surge impedance is purely resistive and the propagation constant is pure imaginary.
(a) True (b) False

5.13 For equivalent π circuits of lossless lines, the *A* and *D* parameters are pure ______. whereas *B* and *C* parameters are pure __________. Fill in the Blanks.

5.14 In equivalent π circuits of lossless lines, Z' is pure __________, and Y' is pure __________. Fill in the Blanks.

5.15 Typical power-line lengths are only a small fraction of the 60-Hz wavelength.
(a) True (b) False

5.16 The velocity of propagation of voltage and current waves along a lossless overhead line is the same as speed of light.
(a) True (b) False

5.17 Surge Impedance Loading (SIL) is the power delivered by a lossless line to a load resistance equal to __________. Fill in the Blank.

5.18 For a lossless line, at SIL, the voltage profile is __________, and the real power delivered, in terms of rated line voltage V and surge impedance Z_C, is given by __________. Fill in the Blanks.

5.19 The maximum power that a lossless line can deliver, in terms of the voltage magnitudes V_S and V_R (in volts) at the ends of the line held constant, and the series reactance X' of the corresponding equivalent π circuit, is given by __________, in Watts. Fill in the Blank.

SECTION 5.5

5.20 The maximum power flow for a lossy line will be somewhat less than that for a lossless line.
(a) True (b) False

SECTION 5.6

5.21 For short lines less than 80 km long, loadability is limited by the thermal rating of the conductors or by terminal equipment ratings, not by voltage drop or stability considerations.
(a) True (b) False

5.22 Increasing the transmission line voltage reduces the required number of lines for the same power transfer.
(a) True (b) False

5.23 Intermediate substations are often economical from the viewpoint of the number of lines required for power transfer, if their costs do not outweigh the reduction in line costs.
(a) True (b) False

SECTION 5.7

5.24 Shunt reactive compensation improves transmission-line __________, whereas series capacitive compensation increases transmission-line __________. Fill in the Blanks.

5.25 Static-var-compensators can absorb reactive power during light loads, and deliver reactive power during heavy loads.
(a) True (b) False

PROBLEMS

SECTION 5.1

5.1 A 25-km, 34.5-kV, 60-Hz three-phase line has a positive-sequence series impedance $z = 0.19 + j0.34\ \Omega$/km. The load at the receiving end absorbs 10 MVA at 33 kV. Assuming a short line, calculate: (a) the *ABCD* parameters, (b) the sending-end voltage for a load power factor of 0.9 lagging, (c) the sending-end voltage for a load power factor of 0.9 leading.

5.2 A 200-km, 230-kV, 60-Hz three-phase line has a positive-sequence series impedance $z = 0.08 + j0.48\ \Omega$/km and a positive-sequence shunt admittance $y = j3.33 \times 10^{-6}$ S/km. At full load, the line delivers 250 MW at 0.99 p.f. lagging and at 220 kV. Using the nominal π circuit, calculate: (a) the *ABCD* parameters, (b) the sending-end voltage and current, and (c) the percent voltage regulation.

5.3 Rework Problem 5.2 in per-unit using 100-MVA (three-phase) and 230-kV (line-to-line) base values. Calculate: (a) the per-unit *ABCD* parameters, (b) the per-unit sending-end voltage and current, and (c) the percent voltage regulation.

5.4 Derive the *ABCD* parameters for the two networks in series, as shown in Figure 5.4.

5.5 Derive the *ABCD* parameters for the T circuit shown in Figure 5.4.

5.6 (a) Consider a medium-length transmission line represented by a nominal π circuit shown in Figure 5.3 of the text. Draw a phasor diagram for lagging power-factor condition at the load (receiving end).

(b) Now consider a nominal T-circuit of the medium-length transmission line shown in Figure 5.18.

(i) Draw the corresponding phasor diagram for lagging power-factor load condition

(ii) Determine the *ABCD* parameters in terms of Y and Z, for the nominal T-circuit and for the nominal π-circuit of part (a).

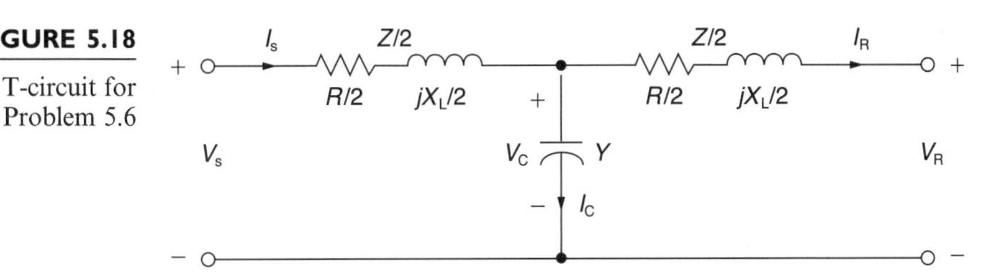

FIGURE 5.18 Nominal T-circuit for Problem 5.6

5.7 The per-phase impedance of a short three—phase transmission line is $0.5\angle 53.15°\,\Omega$. The three-phase load at the receiving end is 900 kW at 0.8 p.f. lagging. If the line-to-line sending-end voltage is 3.3 kV, determine (a) the receiving-end line-to-line voltage in kV, and (b) the line current.
Draw the phasor diagram with the line current I, as reference.

5.8 Reconsider Problem 5.7 and find the following: (a) sending-end power factor, (b) sending-end three-phase power, and (c) the three-phase line loss.

5.9 The 100-km, 230-kV, 60-Hz three-phase line in Problems 4.18 and 4.39 delivers 300 MVA at 218 kV to the receiving end at full load. Using the nominal π circuit, calculate the: *ABCD* parameters, sending-end voltage, and percent voltage regulation when the receiving-end power factor is (a) 0.9 lagging, (b) unity, and (c) 0.9 leading. Assume a 50 °C conductor temperature to determine the resistance of this line.

5.10 The 500-kV, 60-Hz three-phase line in Problems 4.20 and 4.41 has a 180-km length and delivers 1600 MW at 475 kV and at 0.95 power factor leading to the receiving end at full load. Using the nominal π circuit, calculate the: (a) *ABCD* parameters, (b) sending-end voltage and current, (c) sending-end power and power factor, (d) full-load line losses and efficiency, and (e) percent voltage regulation. Assume a 50 °C conductor temperature to determine the resistance of this line.

5.11 A 40-km, 220-kV, 60-Hz three-phase overhead transmission line has a per-phase resistance of 0.15 Ω/km, a per-phase inductance of 1.3263 mH/km, and negligible shunt capacitance. Using the short line model, find the sending-end voltage, voltage regulation, sending-end power, and transmission line efficiency when the line is supplying a three-phase load of: (a) 381 MVA at 0.8 power factor lagging and at 220 kV, (b) 381 MVA at 0.8 power factor leading and at 220 kV.

5.12 A 60-Hz, 100-km, three-phase overhead transmission line, constructed of ACSR conductors, has a series impedance of $(0.1826 + j0.784)$ Ω/km per phase and a shunt capacitive reactance-to-neutral of $185.5 \times 10^3 \angle -90^\circ$ Ω-km per phase. Using the nominal π circuit for a medium-length transmission line, (a) determine the total series impedance and shunt admittance of the line. (b) Compute the voltage, the current, and the real and reactive power at the sending end if the load at the receiving end draws 200 MVA at unity power factor and at a line-to-line voltage of 230 kV. (c) Find the percent voltage regulation of the line.

SECTION 5.2

5.13 Evaluate $\cosh(\gamma l)$ and $\tanh(\gamma l/2)$ for $\gamma l = 0.40\angle 85^\circ$ per unit.

5.14 A 400-km, 500-kV, 60-Hz uncompensated three-phase line has a positive-sequence series impedance $z = 0.03 + j0.35$ Ω/km and a positive-sequence shunt admittance $y = j4.4 \times 10^{-6}$ S/km. Calculate: (a) Z_c, (b) (γl), and (c) the exact $ABCD$ parameters for this line.

5.15 At full load the line in Problem 5.14 delivers 1000 MW at unity power factor and at 475 kV. Calculate: (a) the sending-end voltage, (b) the sending-end current, (c) the sending-end power factor, (d) the full-load line losses, and (e) the percent voltage regulation.

5.16 The 500-kV, 60-Hz three-phase line in Problems 4.20 and 4.41 has a 300-km length. Calculate: (a) Z_c, (b) (γl), and (c) the exact $ABCD$ parameters for this line. Assume a 50 °C conductor temperature.

5.17 At full load, the line in Problem 5.16 delivers 1500 MVA at 480 kV to the receiving-end load. Calculate the sending-end voltage and percent voltage regulation when the receiving-end power factor is (a) 0.9 lagging, (b) unity, and (c) 0.9 leading.

5.18 A 60-Hz, 230-km, three-phase overhead transmission line has a series impedance $z = 0.8431\angle 79.04^\circ$ Ω/km and a shunt admittance $y = 5.105 \times 10^{-6}\angle 90^\circ$ S/km. The load at the receiving end is 125 MW at unity power factor and at 215 kV. Determine the voltage, current, real and reactive power at the sending end and the percent voltage regulation of the line. Also find the wavelength and velocity of propagation of the line.

5.19 Using per-unit calculations, rework Problem 5.18 to determine the sending-end voltage and current.

5.20 (a) The series expansions of the hyperbolic functions are given by

$$\cosh\theta = 1 + \frac{\theta^2}{2} + \frac{\theta^4}{24} + \frac{\theta^6}{720} + \cdots$$

$$\sinh\theta = 1 + \frac{\theta^2}{6} + \frac{\theta^4}{120} + \frac{\theta^6}{5040} + \cdots$$

For the $ABCD$ parameters of a long transmission line represented by an equivalent π circuit, apply the above expansion and consider only the first two terms, and express the result in terms of Y and Z.

(b) For the nominal π and equivalent π circuits shown in Figures 5.3 and 5.7 of the text, show that

$$\frac{A-1}{B} = \frac{Y}{2} \quad \text{and} \quad \frac{A-1}{B} = \frac{Y'}{2}$$

hold good, respectively.

5.21 Starting with (5.1.1) of the text, show that

$$A = \frac{V_S I_S + V_R I_R}{V_R I_S + V_S I_R} \quad \text{and} \quad B = \frac{V_S^2 - V_R^2}{V_R I_S + V_S I_R}$$

5.22 Consider the A parameter of the long line given by $\cosh\theta$, where $\theta = \sqrt{ZY}$. With $x = e^{-\theta} = x_1 + jx_2$, and $A = A_1 + jA_2$, show that x_1 and x_2 satisfy the following:

$$x_1^2 - x_2^2 - 2(A_1x_1 - A_2x_2) + 1 = 0$$

and $x_1x_2 - (A_2x_1 + A_1x_2) = 0.$

SECTION 5.3

5.23 Determine the equivalent π circuit for the line in Problem 5.14 and compare it with the nominal π circuit.

5.24 Determine the equivalent π circuit for the line in Problem 5.16. Compare the equivalent π circuit with the nominal π circuit.

5.25 Let the transmission line of Problem 5.12 be extended to cover a distance of 200 km. Assume conditions at the load to be the same as in Problem 5.12. Determine the: (a) sending-end voltage, (b) sending-end current, (c) sending-end real and reactive powers, and (d) percent voltage regulation.

SECTION 5.4

5.26 A 300-km, 500-kV, 60-Hz three-phase uncompensated line has a positive-sequence series reactance $x = 0.34$ Ω/km and a positive-sequence shunt admittance $y = j4.5 \times 10^{-6}$ S/km. Neglecting losses, calculate: (a) Z_c, (b) (γl), (c) the $ABCD$ parameters, (d) the wavelength λ of the line, in kilometers, and (e) the surge impedance loading in MW.

5.27 Determine the equivalent π circuit for the line in Problem 5.26.

5.28 Rated line voltage is applied to the sending end of the line in Problem 5.26. Calculate the receiving-end voltage when the receiving end is terminated by (a) an open circuit, (b) the surge impedance of the line, and (c) one-half of the surge impedance. (d) Also calculate the theoretical maximum real power that the line can deliver when rated voltage is applied to both ends of the line.

5.29 Rework Problems 5.9 and 5.16 neglecting the conductor resistance. Compare the results with and without losses.

5.30 From (4.6.22) and (4.10.4), the series inductance and shunt capacitance of a three-phase overhead line are

$$L_a = 2 \times 10^{-7} \ln(D_{eq}/D_{SL}) = \frac{\mu_0}{2\pi} \ln(D_{eq}/D_{SL}) \quad \text{H/m}$$

$$C_{an} = \frac{2\pi\varepsilon_0}{\ln(D_{eq}/D_{SC})} \quad \text{F/m}$$

where $\mu_0 = 4\pi \times 10^{-7}$ H/m and $\varepsilon_0 = \left(\frac{1}{36\pi}\right) \times 10^{-9}$ F/m

Using these equations, determine formulas for surge impedance and velocity of propagation of an overhead lossless line. Then determine the surge impedance and velocity of propagation for the three-phase line given in Example 4.5. Assume positive-sequence operation. Neglect line losses as well as the effects of the overhead neutral wires and the earth plane.

5.31 A 500-kV, 300-km, 60-Hz three-phase overhead transmission line, assumed to be lossless, has a series inductance of 0.97 mH/km per phase and a shunt capacitance of 0.0115 μF/km per phase. (a) Determine the phase constant β, the surge impedance Z_C, velocity of propagation v, and the wavelength λ of the line. (b) Determine the voltage, current, real and reactive power at the sending end, and the percent voltage regulation of the line if the receiving-end load is 800 MW at 0.8 power factor lagging and at 500 kV.

5.32 The following parameters are based on a preliminary line design: $V_S = 1.0$ per unit, $V_R = 0.9$ per unit, $\lambda = 5000$ km, $Z_C = 320\ \Omega$, $\delta = 36.8°$. A three-phase power of 700 MW is to be transmitted to a substation located 315 km from the source of power. (a) Determine a nominal voltage level for the three-phase transmission line, based on the practical line-loadability equation. (b) For the voltage level obtained in (a), determine the theoretical maximum power that can be transferred by the line.

5.33 Consider a long radial line terminated in its characteristic impedance Z_C. Determine the following:

(a) V_1/I_1, known as the driving point impedance.

(b) $|V_2|/|V_1|$, known as the voltage gain, in terms of $\alpha\ell$.

(c) $|I_2|/|I_1|$, known as the current gain, in terms of $\alpha\ell$.

(d) The complex power gain, $-S_{21}/S_{12}$, in terms of $\alpha\ell$.

(e) The real power efficiency, $(-P_{21}/P_{12}) = \eta$, in terms of $\alpha\ell$.

[Note: 1 refers to sending end and 2 refers to receiving end. (S_{21}) is the complex power received at 2; S_{12} is sent from 1.]

5.34 For the case of a lossless line, how would the results of Problem 5.33 change? In terms of Z_C, which will be a real quantity for this case, express P_{12} in terms $|I_1|$ and $|V_1|$.

5.35 For a lossless open-circuited line, express the sending-end voltage, V_1, in terms of the receiving-end voltage, V_2, for the three cases of short-line model, medium-length line model, and long-line model. Is it true that the voltage at the open receiving end of a long line is higher than that at the sending end, for small $\beta\ell$.

5.36 For a short transmission line of impedance $(R + jX)$ ohms per phase, show that the maximum power that can be transmitted over the line is

$$P_{max} = \frac{V_R^2}{Z^2}\left(\frac{ZV_S}{V_R} - R\right) \quad \text{where } Z = \sqrt{R^2 + X^2}$$

when the sending-end and receiving-end voltages are fixed, and for the condition

$$Q = \frac{-V_R^2 X}{R^2 + X^2} \quad \text{when } dP/dQ = 0$$

5.37 (a) Consider complex power transmission via the three-phase short line for which the per-phase circuit is shown in Figure 5.19. Express S_{12}, the complex power sent by bus 1 (or V_1), and $(-S_{21})$, the complex power received by bus 2 (or V_2), in terms of V_1, V_2, Z, $\underline{/Z}$, and $\theta_{12} = \theta_1 - \theta_2$, the power angle.

(b) For a balanced three-phase transmission line, in per-unit notation, with $Z = 1\underline{/85°}$, $\theta_{12} = 10°$, determine S_{12} and $(-S_{21})$ for

(i) $V_1 = V_2 = 1.0$

(ii) $V_1 = 1.1$ and $V_2 = 0.9$

Comment on the changes of real and reactive powers from (i) to (ii).

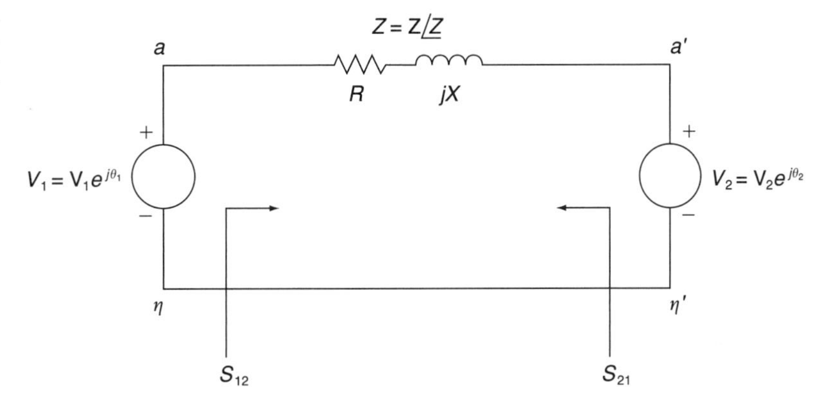

FIGURE 5.19 Per-phase circuit for Problem 5.37

SECTION 5.5

5.38 The line in Problem 5.14 has three ACSR 1113-kcmil (564-mm^2) conductors per phase. Calculate the theoretical maximum real power that this line can deliver and compare with the thermal limit of the line. Assume $V_S = V_R = 1.0$ per unit and unity power factor at the receiving end.

5.39 Repeat Problems 5.14 and 5.38 if the line length is (a) 200 km, (b) 600 km.

5.40 For the 500-kV line given in Problem 5.16, (a) calculate the theoretical maximum real power that the line can deliver to the receiving end when rated voltage is applied to both ends. (b) Calculate the receiving-end reactive power and power factor at this theoretical loading.

5.41 A 230-kV, 100-km, 60-Hz three-phase overhead transmission line with a rated current of 900 A/phase has a series impedance $z = 0.088 + j0.465\ \Omega/\text{km}$ and a shunt admittance $y = j3.524\ \mu\text{S/km}$. (a) Obtain the nominal π equivalent circuit in normal units and in per unit on a base of 100 MVA (three phase) and 230 kV (line-to-line). (b) Determine the three-phase rated MVA of the line. (c) Compute the ABCD parameters. (d) Calculate the SIL.

5.42 A three-phase power of 460 MW is to the transmitted to a substation located 500 km from the source of power. With $V_S = 1$ per unit, $V_R = 0.9$ per unit, $\lambda = 5000$ km, $Z_C = 500\ \Omega$, and $\delta = 36.87°$, determine a nominal voltage level for the lossless transmission line, based on Eq. (5.4.29) of the text.
Using this result, find the theoretical three-phase maximum power that can be transferred by the lossless transmission line.

PW **5.43** Open PowerWorld Simulator case Example 5_4 and graph the load bus voltage as a function of load real power (assuming unity power factor at the load). What is the maximum amount of real power that can be transferred to the load at unity power factor if we require the load voltage always be greater than 0.9 per unit?

PW **5.44** Repeat Problem 5.43, but now vary the load reactive power, assuming the load real power is fixed at 1000 MW.

SECTION 5.6

5.45 For the line in Problems 5.14 and 5.38, determine: (a) the practical line loadability in MW, assuming $V_S = 1.0$ per unit, $V_R \approx 0.95$ per unit, and $\delta_{max} = 35°$; (b) the full-load current at 0.99 p.f. leading, based on the above practical line loadability; (c) the exact receiving-end voltage for the full-load current in (b) above; and (d) the percent voltage regulation. For this line, is loadability determined by the thermal limit, the voltage-drop limit, or steady-state stability?

5.46 Repeat Problem 5.45 for the 500-kV line given in Problem 5.10.

5.47 Determine the practical line loadability in MW and in per-unit of SIL for the line in Problem 5.14 if the line length is (a) 200 km, (b) 600 km. Assume $V_S = 1.0$ per unit, $V_R = 0.95$ per unit, $\delta_{max} = 35°$, and 0.99 leading power factor at the receiving end.

5.48 It is desired to transmit 2000 MW from a power plant to a load center located 300 km from the plant. Determine the number of 60-Hz three-phase, uncompensated transmission lines required to transmit this power with one line out of service for the following cases: (a) 345-kV lines, $Z_c = 300\ \Omega$, (b) 500-kV lines, $Z_c = 275\ \Omega$, (c) 765-kV lines, $Z_c = 260\ \Omega$. Assume that $V_S = 1.0$ per unit, $V_R = 0.95$ per unit, and $\delta_{max} = 35°$.

5.49 Repeat Problem 5.48 if it is desired to transmit: (a) 3200 MW to a load center located 300 km from the plant, (b) 2000 MW to a load center located 400 km from the plant.

5.50 A three-phase power of 3600 MW is to be transmitted through four identical 60-Hz overhead transmission lines over a distance of 300 km. Based on a preliminary design, the phase constant and surge impedance of the line are $\beta = 9.46 \times 10^{-4}$ rad/km and $Z_C = 343\ \Omega$, respectively. Assuming $V_S = 1.0$ per unit, $V_R = 0.9$ per unit, and a power angle $\delta = 36.87°$, determine a suitable nominal voltage level in kV, based on the practical line-loadability criteria.

5.51 The power flow at any point on a transmission line can be calculated in terms of the *ABCD* parameters. By letting $A = |A|\angle\alpha$, $B = |B|\angle\beta$, $V_R = |V_R|\angle 0°$, and $V_S = |V_S|\angle\delta$, the complex power at the receiving end can be shown to be

$$P_R + jQ_R = \frac{|V_R|\,|V_S|\angle\beta - \alpha}{|B|} - \frac{|\delta|\,|V_R^2|\angle\beta - \alpha}{|B|}$$

(a) Draw a phasor diagram corresponding to the above equation. Let it be represented by a triangle O′OA with O′ as the origin and OA representing $P_R + jQ_R$.

(b) By shifting the origin from O′ to O, turn the result of (a) into a power diagram, redrawing the phasor diagram. For a given fixed value of $|V_R|$ and a set of values for $|V_S|$, draw the loci of point A, thereby showing the so-called receiving-end circles.

(c) From the result of (b) for a given load with a lagging power factor angle θ_R, determine the amount of reactive power that must be supplied to the receiving end to maintain a constant receiving-end voltage, if the sending-end voltage magnitude decreases from $|V_{S1}|$ to $|V_{S2}|$.

5.52 (a) Consider complex power transmission via the three-phase long line for which the per-phase circuit is shown in Figure 5.20. See Problem 5.37 in which the short-line case was considered. Show that

$$\text{sending-end power} = S_{12} = \frac{Y'^*}{2}\mathrm{V}_1^2 + \frac{\mathrm{V}_1^2}{Z'^*} - \frac{\mathrm{V}_1\mathrm{V}_2}{Z'^*}e^{j\theta_{12}}$$

$$\text{and received power} = -S_{21} = -\frac{Y'^*}{2}\mathrm{V}_2^2 - \frac{\mathrm{V}_2^2}{Z'^*} + \frac{\mathrm{V}_1\mathrm{V}_2}{Z'^*}e^{-j\theta_{12}}$$

where $\theta_{12} = \theta_1 - \theta_2$.

(b) For a lossless line with equal voltage magnitudes at each end, show that

$$\mathrm{P}_{12} = -\mathrm{P}_{21} = \frac{\mathrm{V}_1^2 \sin\theta_{12}}{Z_C \sin\beta\ell} = \mathrm{P}_{\mathrm{SIL}}\frac{\sin\theta_{12}}{\sin\beta\ell}$$

(c) For $\theta_{12} = 45°$, and $\beta = 0.002$ rad/km, find $(\mathrm{P}_{12}/\mathrm{P}_{\mathrm{SIL}})$ as a function of line length in km, and sketch it.

(d) If a thermal limit of $(\mathrm{P}_{12}/\mathrm{P}_{\mathrm{SIL}}) = 2$ is set, which limit governs for short lines and long lines?

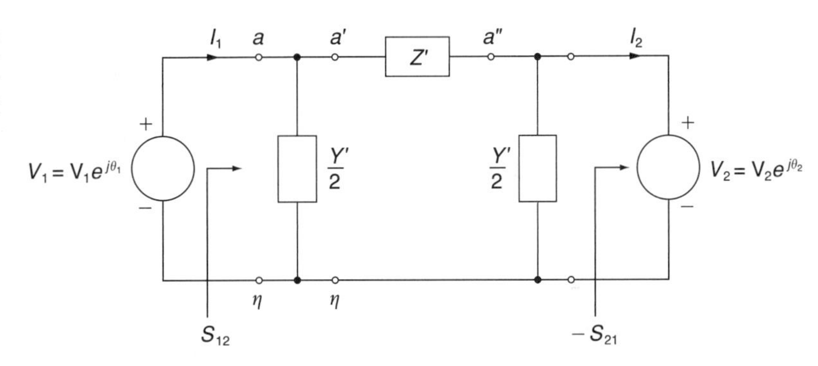

FIGURE 5.20 Per-phase circuit for Problem 5.52

PW **5.53** Open PowerWorld Simulator case Example 5_8. If we require the load bus voltage to be greater than or equal to 730 kV even with any line segment out of service, what is the maximum amount of real power that can be delivered to the load?

PW **5.54** Repeat Problem 5.53, but now assume any two line segments may be out of service.

SECTION 5.7

5.55 Recalculate the percent voltage regulation in Problem 5.15 when identical shunt reactors are installed at both ends of the line during light loads, providing 65% total shunt compensation. The reactors are removed at full load. Also calculate the impedance of each shunt reactor.

5.56 Rework Problem 5.17 when identical shunt reactors are installed at both ends of the line, providing 50% total shunt compensation. The reactors are removed at full load.

5.57 Identical series capacitors are installed at both ends of the line in Problem 5.14, providing 40% total series compensation. Determine the equivalent *ABCD* parameters of this compensated line. Also calculate the impedance of each series capacitor.

5.58 Identical series capacitors are installed at both ends of the line in Problem 5.16, providing 30% total series compensation. (a) Determine the equivalent *ABCD* parameters for this compensated line. (b) Determine the theoretical maximum real power that this series-compensated line can deliver when $V_S = V_R = 1.0$ per unit. Compare your result with that of Problem 5.40.

5.59 Determine the theoretical maximum real power that the series-compensated line in Problem 5.57 can deliver when $V_S = V_R = 1.0$ per unit. Compare your result with that of Problem 5.38.

5.60 What is the minimum amount of series capacitive compensation N_C in percent of the positive-sequence line reactance needed to reduce the number of 765-kV lines in Example 5.8 from five to four. Assume two intermediate substations with one line section out of service. Also, neglect line losses and assume that the series compensation is sufficiently distributed along the line so as to effectively reduce the series reactance of the equivalent π circuit to $X'(1 - N_C/100)$.

5.61 Determine the equivalent *ABCD* parameters for the line in Problem 5.14 if it has 70% shunt reactive (inductors) compensation and 40% series capacitive compensation. Half of this compensation is installed at each end of the line, as in Figure 5.14.

5.62 Consider the transmission line of Problem 5.18. (a) Find the *ABCD* parameters of the line when uncompensated. (b) For a series capacitive compensation of 70% (35% at the sending end and 35% at the receiving end), determine the *ABCD* parameters. Comment on the relative change in the magnitude of the *B* parameter with respect to the relative changes in the magnitudes of the *A*, *C*, and *D* parameters. Also comment on the maximum power that can be transmitted when series compensated.

5.63 Given the uncompensated line of Problem 5.18, let a three-phase shunt reactor (inductor) that compensates for 70% of the total shunt admittance of the line be connected at the receiving end of the line during no-load conditions. Determine the effect of voltage regulation with the reactor connected at no load. Assume that the reactor is removed under full-load conditions.

5.64 Let the three-phase lossless transmission line of Problem 5.31 supply a load of 1000 MVA at 0.8 power factor lagging and at 500 kV. (a) Determine the capacitance/phase and total three-phase Mvars supplied by a three-phase, Δ-connected shunt-capacitor bank at the receiving end to maintain the receiving-end voltage at 500 kV when the sending end of the line is energized at 500 kV. (b) If series capacitive compensation of 40% is installed at the midpoint of the line, without the shunt capacitor bank at the receiving end, compute the sending-end voltage and percent voltage regulation.

PW **5.65** Open PowerWorld Simulator case Example 5_10 with the series capacitive compensation at both ends of the line in service. Graph the load bus voltage as a function of load real power (assuming unity power factor at the load). What is the maximum amount of real power that can be transferred to the load at unity power factor if we require the load voltage always be greater than 0.85 per unit?

PW **5.66** Open PowerWorld Simulator case Example 5_10 with the series capacitive compensation at both ends of the line in service. With the reactive power load fixed at 500 Mvar, graph the load bus voltage as the MW load is varied between 0 and 2600 MW in 200 MW increments. Then repeat with both of the series compensation elements out of service.

CASE STUDY QUESTIONS

A. For underground and underwater transmission, why are line losses for HVDC cables lower than those of ac cables with similar capacity?

B. Where are back-to-back HVDC converters (back-to-back HVDC links) currently located in North America? What are the characteristics of those locations that prompted the installation of back-to-back HVDC links?

C. Which HVDC technology can independently control both active (real) power flow and reactive power flow to and from the interconnected ac system?

REFERENCES

1. Electric Power Research Institute (EPRI), *EPRI AC Transmission Line Reference Book—200 kV and Above* (Palo Alto, CA: EPRI, www.epri.com, December 2005).
2. Westinghouse Electric Corporation, *Electrical Transmission and Distribution Reference Book*, 4th ed. (East Pittsburgh, PA, 1964).
3. R. D. Dunlop, R. Gutman, and P. P. Marchenko, "Analytical Development of Loadability Characteristics for EHV and UHV Lines," *IEEE Trans. PAS*, Vol. PAS-98, No. 2 (March/April 1979): pp. 606–607.
4. W. D. Stevenson, Jr., *Elements of Power System Analysis*, 4th ed. (New York: McGraw-Hill, 1982).
5. W. H. Hayt, Jr., and J. E. Kemmerly, *Engineering Circuit Analysis*, 7th ed. (New York: McGraw-Hill, 2006).
6. M. P. Bahrman and B. K. Johnson, "The ABCs of HVDC Transmission Technologies," *IEEE Power & Energy Magazine*, 5, 2 (March/April 2007): pp. 32–44.

Tennessee Valley Authority (TVA) Regional Operations Center (Courtesy of TVA)

6

POWER FLOWS

Successful power system operation under normal balanced three-phase steady-state conditions requires the following:

1. Generation supplies the demand (load) plus losses.
2. Bus voltage magnitudes remain close to rated values.
3. Generators operate within specified real and reactive power limits.
4. Transmission lines and transformers are not overloaded.

The power-flow computer program (sometimes called *load flow*) is the basic tool for investigating these requirements. This program computes the voltage magnitude and angle at each bus in a power system under balanced three-phase steady-state conditions. It also computes real and reactive power flows for all equipment interconnecting the buses, as well as equipment losses.

Both existing power systems and proposed changes including new generation and transmission to meet projected load growth are of interest.

Conventional nodal or loop analysis is not suitable for power-flow studies because the input data for loads are normally given in terms of power, not impedance. Also, generators are considered as power sources, not voltage or current sources. The power-flow problem is therefore formulated as a set of nonlinear algebraic equations suitable for computer solution.

In Sections 6.1–6.3 we review some basic methods, including direct and iterative techniques for solving algebraic equations. Then in Sections 6.4–6.6 we formulate the power-flow problem, specify computer input data, and present two solution methods, Gauss–Seidel and Newton–Raphson. Means for controlling power flows are discussed in Section 6.7. Sections 6.8 and 6.9 introduce sparsity techniques and a fast decoupled power-flow method, while Section 6.10 discusses the dc power flow, and Section 6.11 considers the power-flow representation of wind turbine generators.

Since balanced three-phase steady-state conditions are assumed, we use only positive-sequence networks in this chapter. Also, all power-flow equations and input/output data are given in per-unit.

CASE STUDY

Power-flow programs are used to analyze large transmission grids and the complex interaction between transmission grids and the power markets. Historically, these transmission grids were designed primarily by local utilities to meet the needs of their own customers. But increasingly there is a need for coordinated transmission system planning to create coordinated, continent-spanning grids. The following article details some of the issues associated with such large-scale system planning.

Future Vision: The Challenge of Effective Transmission Planning

BY DONALD J. MORROW
AND RICHARD E. BROWN

Exceptional forces are changing the use of the transmission infrastructure in the United States. There are high expectations that the transmission system will support and enable national-level economic, renewable energy, and other emerging policy issues.

The U. S. transmission system was developed in a piecemeal fashion. Originally, transmission systems connected large generation facilities in remote areas to users of the electricity they produced. Shortly thereafter, utilities started to interconnect their systems in order to realize the benefits of improved reliability that larger systems offer and to get access to lower cost energy in other systems. Subsequent transmission lines were typically added incrementally to the network, primarily driven by the needs of the local utility and without wide-area planning considerations.

Opportunistic usage of the transmission system beyond its design occurred early in the U. S. electric system. The need for coordinated transmission planning among utilities soon followed. As early as 1925, small power pools formed to take advantage of the economies of developing larger, more cost-effective power plants that were made possible by the expanding transmission network. By today's standards, these power pools were rather simple affairs made up of localized pockets of utilities that shared the expenses of fuel and operation and maintenance of shared units.

Today, the transmission system is increasingly being called upon to serve as the platform to enable sophisticated and complex energy and financial transactions. New market systems have been developed that allow transactions interconnection-wide. Today, a utility can purchase power without knowing the seller. These same market systems have the ability to enable transactions to be interconnection-wide and will soon accommodate the ability of load-serving entities to bid in their loads.

As the barriers to participate in electricity markets start to disappear, the U. S. electric system starts to look small from the perspective of market participants. In his book *The World is Flat*, author Thomas Friedman states, "The world is flat." That is, the location of producers and consumers no longer matters in the world. It is the expectation of wholesale electricity market participants that they can soon claim, "The transmission system is flat." That is, the transmission system is such that the location of power producers and power purchasers does not matter in terms of participation in national electricity markets.

Unfortunately, the vast majority of transmission infrastructure was not designed for this purpose. The existing transmission infrastructure is aging, and new transmission investment hasn't kept pace with other development. This article discusses these challenges and then presents a vision for the future where effective planning can address the transmission expectations of today.

BENEFITS OF TRANSMISSION

The primary function of transmission is to transport bulk power from sources of desirable generation to bulk power delivery points. Benefits have traditionally included lower electricity costs, access to renewable energy such as wind and hydro, locating power plants away from large population centers, and access to alternative generation sources when primary sources are not available.

Historically, transmission planning has been done by individual utilities with a focus on local benefits. However, proponents of nationwide transmission policies now view the transmission system as an "enabler" of energy policy objectives at even the national level. This is an understandable expectation since a well-planned transmission grid has the potential to enable the following:

- **Efficient bulk power markets.** Bulk power purchasers should almost always be able to purchase from the lowest cost generation. Today, purchasers are often forced to buy higher-cost electricity to avoid violating transmission loading constraints. The difference between the actual price of electricity at the point of consumption and the lowest price on the grid is called the "congestion" cost.
- **Hedge against generation outages.** The transmission system should typically allow access to alternative economic energy sources to replace lost resources. This is especially critical when long-term, unplanned outages of large generation units occur.
- **Hedge against fuel price changes.** The transmission system should allow purchasers to economically access generation from diversified fuel resources as a hedge against fuel disruptions that may occur from strikes, natural disasters, rail interruptions, or natural fuel price variation.
- **Low-cost access to renewable energy.** Many areas suitable for producing electricity from renewable resources are not near transmission with spare capacity. The transmission system should usually allow developers to build renewable sources of energy without the need for expensive transmission upgrades (Figure 1).
- **Operational flexibility.** The transmission system should allow for the economic scheduling of maintenance outages and for the economic reconfiguration of the grid when unforeseen events occur.

Many of these benefits are available on a local level, since transmission systems have been planned by the local utility with these objectives in mind. However, these benefits are not fully realized on a regional or national level, since planning has traditionally been focused on providing these benefits at the local level.

AGING TRANSMISSION SYSTEM

Even at a local level, transmission benefits are in jeopardy. For the past 20 years, the growth of electricity demand has far outpaced the growth of transmission capacity. With limited new transmission capacity available, the loading of existing transmission lines has dramatically increased (Figure 2). North American Reliability Corporation (NERC) reliability criteria have still been maintained for the most part, but the transmission system is far more vulnerable to multiple contingencies and cascading events.

A large percentage of transmission equipment was installed in the postwar period between the mid-1950s and

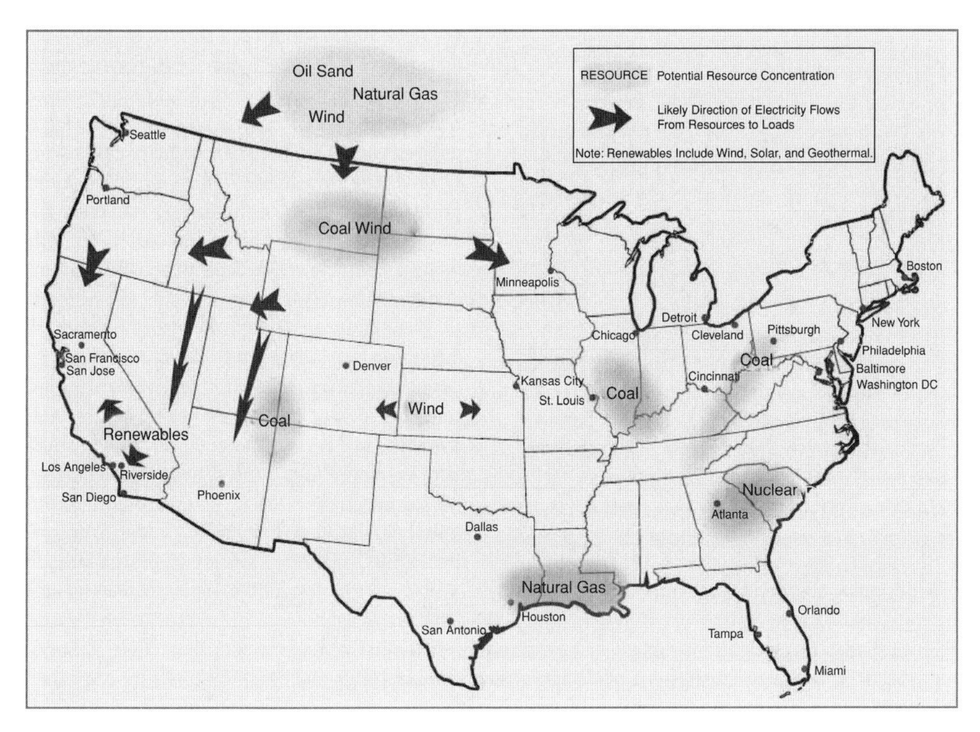

Figure 1
Potential sources of renewable energy concentrations (U.S. Department of Energy, *National Electric Transmission Congestion Study,* 2006)

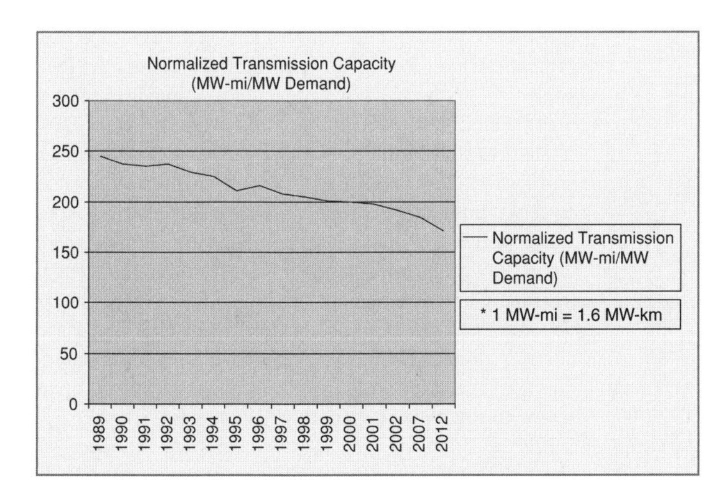

Figure 2
Transmission capacity normalized over MW demand (E. Hurst, *U.S. Transmission Capacity: Present Status and Future Prospects,* prepared for EEI and DOE, Aug. 2004)

the mid-1970s, with limited construction in the past 20 years. The equipment installed in the postwar period is now between 30 and 50 years old and is at the end of its expected life (Figure 3). Having a large amount of old and aging equipment typically results in higher probabilities of failure, higher maintenance costs, and higher replacement costs. Aging equipment will eventually have to be replaced, and this replacement should be planned and coordinated with capacity additions.

According to Fitch Ratings, 70% of transmission lines and power transformers in the United States are 25 years old or older. Their report also states that 60% of high-voltage circuit breakers are 30 years old or older. It is this aging infrastructure that is being asked to bear the burden of increased market activity and to support policy developments such as massive wind farm deployment.

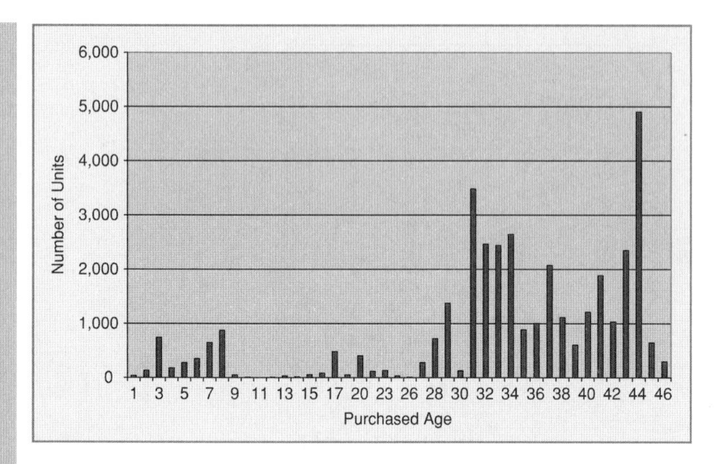

Figure 3
The age distribution of wood transmission poles for a Midwestern utility. Most of these structures are over 30 years old

Today, the industry is beginning to spend more money on new transmission lines and on upgrading existing transmission lines. It is critical that this new transmission construction be planned well, so that the existing grid can be systematically transformed into a desired future state rather than becoming a patchwork of incremental decisions and uncoordinated projects.

PLANNING CHALLENGES

As the transmission system becomes flatter, the processes to analyze and achieve objectives on a regional or interconnection-wide basis have lagged. Current planning processes simply do not have the perspective necessary to keep pace with the scope of the economic and policy objectives being faced today. While the planners of transmission owners often recognize these needs, addressing these needs exceeds the scope of their position. Regional transmission organizations exist today, but these organizations do not have the ability to effectively plan for interconnect-wide objectives.

PLANNING BEFORE OPEN ACCESS

Before access to the electric system was required by the Federal Energy Regulatory Commission (FERC) in 1996, a vertically integrated utility would plan for generation and transmission needs within its franchise territory. This allowed for a high degree of certainty because the decisions regarding the timing and location of new generation and transmission were controlled by the utility. These projects were developed to satisfy the utility's reliability and economic needs.

Transmission interconnections to neighboring utilities for the purposes of importing and exporting bulk power and the development of transmission projects that spanned multiple utilities were also the responsibility of the vertically integrated utility. They were negotiated projects that often took years of effort to ensure that ownership shares and cost allocations were acceptable to each party and that no undue burden was placed on the affected systems.

Planning coordination eventually emerged, facilitated through the regional reliability councils (RRCs). Committees were formed that performed aggregate steady-state and dynamic analysis on the total set of transmission owner (TO) plans. These studies were performed under the direction of committee members, facilitated by RRC staff, to ensure that NERC planning policies (the predecessor to today's NERC standards) and regional planning guidelines were satisfied. Insights from these studies were used by planners to adjust their projects if necessary. Some regions still follow this process for their coordinated planning activities.

PLANNING AFTER OPEN ACCESS

The Open Access Tariff of 1996 (created through FERC Order 888) requires functional separation of generation and transmission within a vertically integrated utility. A generation queue process is now required to ensure that generation interconnection requests are processed in a nondiscriminatory fashion and in a first-come, first-served order. FERC Order 889, the companion to Order 888, establishes the OASIS (Open Access Same-time Information System) process that requires transmission service requests, both external and internal, to be publicly posted and processed in the order in which they arc entered. Order 889 requires each utility to ensure nonpreferential treatment of its own generation plan. Effectively, generation and transmission planning, even within the same utility, are not allowed to be coordinated and integrated. This has been done to protect nondiscriminatory, open access to the electric system for all parties.

These landmark orders have removed barriers to market participation by entities such as independent power producers (IPPs) and power marketers. They force utilities to follow standardized protocols to address their needs and allow, for the most part, market forces to drive the addition of new generation capacity.

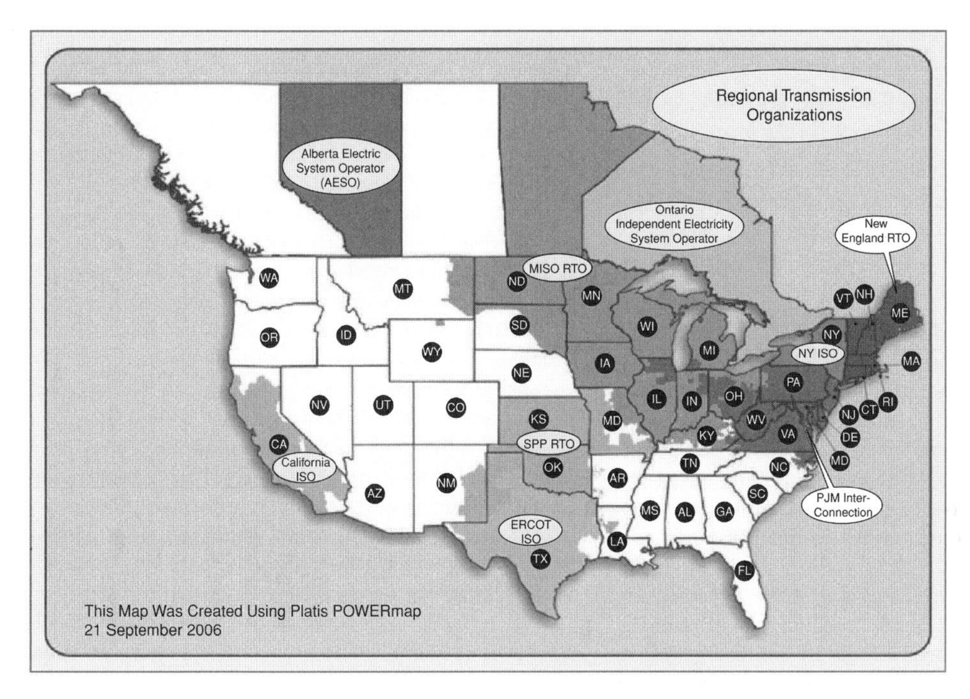

Figure 4
Regional transmission organizations in the United Stated and Canada

These orders also complicated the planning process, since information flow within planning departments becomes one-directional. Transmission planners know all the details of proposed generation planners through the queue process, but not vice versa. A good transmission plan is now supposed to address the economic objectives of all users of the transmission grid by designing plans to accommodate generation entered into the generation queue and to ensure the viability of long-term firm transmission service requests entered through OASIS. However, utility transmission planners continue to design their transmission systems largely to satisfy their own company's reliability objectives.

These planning processes designed the electric system in the Eastern United States and Canada that existed on 13 August 2003. The blackout that occurred that day which interrupted more the 50 million customers made it clear what planners were beginning to suspect-that the margins within the system were becoming dangerously small. The comprehensive report performed by the U. S.—Canada Power System Outage Task Force summarizes the situation as follows:

> A smaller transmission margin for reliability makes the preservation of system reliability a harder job than it used to be. The system is being operated closer to the edge of reliability than it was just a few years ago.

PLANNING IN THE ERA OF THE RTO

Well before the 2003 blackout, FERC realized that better coordination among transmission owners is required for efficient national electricity markets. FERC Order 2000 issued in December 1999 established the concept of the regional transmission operator (RTO) and requires transmission operators to make provisions to form and participate in these organizations.

In this order, FERC establishes the authority of an RTO to perform regional planning and gives it the ultimate responsibility for planning within its region. Order 2000 allowed a 3-year phase-in to allow the RTO to develop the processes and capabilities to perform this function. For the first time in its history, the U. S. electric system has the potential for a coordinated, comprehensive regional planning process (Figure 4 shows the existing RTOs in the United States and Canada).

Despite the advance of developing planning organizations that aligned with the scope of the reliability and economic needs of a region, a significant gap was introduced between planning a system and implementing the plan. Order 2000 recognizes this gap with the following statement:

> We also note that the RTO's implementation of this general standard requires addressing many specific design questions, including who decides which projects should be built and how the costs and benefits of the project should be allocated.

Determining who decides which project should be built is a difficult problem. Does the RTO decide which projects are to be built since it has planned the system? Does the TO decide which projects are to be built since it bears the project development risks such as permitting, regulatory approval, right-of-way acquisition financing, treatment of allowance for funds used during construction (AFUDC), construction, cost escalation, and prudency reviews?

If the issue of project approval is not properly addressed, it is easy to envision a situation where planners spend significant efforts and costs to design a grid that satisfies critical economic and policy objectives. This plan ultimately languishes on the table because no TO wants to build it, no TO has the ability to build it, or no state regulator will approve it. To their credit, RTOs and their member transmission owners recognize this gap and have begun to take steps to resolve it.

TECHNICAL CHALLENGES

The main technical criteria that should drive transmission planning are reliability and congestion. Reliability relates to unexpected transmission contingencies (such as faults) and the ability of the system to respond to these contingencies without interrupting load. Congestion occurs when transmission reliability limitations result in the need to use higher-cost generation than would be the case without any reliability constraints. Both reliability and congestion are of critical importance and present difficult technical challenges.

Transmission reliability is tracked and managed by NERC, which as of 20 July 2006 now serves as the federal electric reliability organization (ERO) under the jurisdiction of FERC. For decades, the primary reliability consideration used by NERC for transmission planning has been "*N*-1." For a system consisting of N major components, the *N*-1 criterion is satisfied if the system can perform properly with only *N*-1 components in service. An *N*-1 analysis consists of a steady-state and a dynamic component. The steady-state analysis checks to see if the transmission system can withstand the loss of any single major piece of equipment (such as a transmission line or a transformer) without violating voltage or equipment loading limits. The dynamic analysis checks to see if the system can retain synchronism after all potential faults.

N-1 has served the industry well but has several challenges when applied to transmission planning today. The first is its deterministic nature; all contingencies are treated equal regardless of how likely they are to occur or the severity of consequences. The second, and more insidious, is the inability of *N*-1 (and *N*-2) to account for the increased risk associated with a more heavily interconnected system and a more heavily loaded system.

When a system is able to withstand any single major contingency, it is termed "*N*-1 secure." For a moderately loaded *N*-1 secure system, most single contingencies can be handled even if the system response to the contingency is not perfect. When many components of a transmission system are operated close to their thermal or stability limits, a single contingency can significantly stress the system and can lead to problems unless all protection systems and remedial actions operate perfectly. In this sense, moderately loaded systems are "resilient" and can often absorb multiple contingencies and/or cascading events. Heavily loaded systems are brittle and run the risk of widespread outages if an initiating event is followed by a protection system failure or a mistake in remedial actions. Since blackouts invariably involve multiple contingencies and/or cascading events, *N*-1 and *N*-2 are not able to effectively plan for wide-area events.

N-1 secure systems are, by design, not able to withstand certain multiple contingencies. When equipment failure rates are low, this is a minor problem. When equipment failure rates increase due to aging and higher loading, this problem becomes salient. Consider the likelihood of two pieces of equipment experiencing outages that overlap. If the outages are independent, the probability of overlap increases with the square of outage rate. Similarly, the probability of three outages overlapping (exceeding *N*-2) increases with the cube of outage rate. Blackouts typically result from three or more simultaneous contingencies. If transmission failure rates double due to aging and higher loading, the likelihood of a third-order event increases by a factor of eight or more. Today's transmission systems may remain *N*-1 or *N*-2 secure, but the risk of wide-area events is much higher than a decade ago.

Computationally it is difficult to plan for wide-area events. This is due to large system models, a high number of potential contingencies, and convergence difficulties.

Consider the eastern interconnected system, which would require over 150,000 major components in a power flow model. This size exceeds the useful capabilities of present planning software, even when exploring only a few cases. To plan for all triple contingencies, more than 3 sextillion (thousand trillion) cases must be considered. Even if only one out of every million of cases is considered, more than 3 billion simulations must be performed. Each simulation is also at risk for nonconvergence, since a system under multiple contingencies will often have a solution very different from the base case.

In addition to reliability planning, it is becoming increasingly important to plan for congestion (the 2006 Department of Energy congestion study reports that two constraints alone in PJM Interconnection resulted in congestion costs totaling US$1. 2 billion in 2005). Basic congestion planning tools work as follows. First, hourly loads for an entire year are assigned to each bulk power delivery point. Second, a load flow is performed for each hour (accounting for scheduled generation and transmission maintenance). If transmission reliability criteria are violated, remedial actions such as generation re-dispatch is performed until the constraints are relieved. The additional energy costs resulting from these remedial actions is assigned to congestion cost (sophisticated tools will also incorporate generation bidding strategies and customer demand curves). Each case examined in a congestion study is computationally intensive.

There are many ways to address existing congestion problems, but it difficult from a technical perspective to combine congestion planning with reliability planning. Imagine a tool with the capability to compute both the reliability and congestion characteristics of a system. A congestion simulation is still required, but unplanned contingencies must now be considered. To do this, each transmission component is checked in each hour of the simulation to see if a random failure occurs. If so, this component is removed from the system until it is repaired, potentially resulting in increased congestion costs. Since each simulated year will only consider a few random transmission failures, many years must be simulated (typically 1,000 or more) for each case under consideration. These types of tools are useful when only the existing transmission system is of interest, such as for energy traders or for dealing with existing congestion problems. For transmission planners that need to consider many scenarios and many project alternatives, these types of tools are insufficient at this time.

The last major technical challenge facing transmission planning is the application of new technologies such as phasor measurements units, real-time conductor ratings, and power electronic devices. Proper application of these devices to address a specific problem already requires a specialist familiar with the technology. Considering each new technology as part of an overall proactive planning process would require new tools, new processes, and transmission planners familiar with the application of all new technologies.

Perhaps the biggest technical challenge to transmission planning is overcoming the traditional mindset of planners. Traditionally a utility transmission planner was primarily concerned with the transport of bulk generation to load centers without violation of local constraints. In today's environment, effective transmission planning requires a wide-area perspective, aging infrastructure awareness, a willingness to coordination extensively, an economic mindset, and an ability to effectively integrate new technologies with traditional approaches.

INFRASTRUCTURE DEVELOPMENT CHALLENGES

Developing transmission projects has been a daunting affair in recent years, and significant roadblocks still exist. A partial list of these roadblocks includes:

- NIMBY mentality (Not In My Back Yard)
- organized public opposition
- environmental concerns
- lack of institutional knowledge
- regulatory risk
- uncontrolled cost increases
- political pressures
- financing risks.

Perhaps the biggest impediment to transmission infrastructure development is the risk of cost recovery. AFUDC rate treatment is the present norm for transmission project financing. This allows the accrued cost of financing for development of a utility project to be included in rates for cost recovery. Recovery is typically only allowed after a project is completed and after state regulatory prudency review on the project. The effect is a substantial risk of cost nonrecovery that discourages transmission investment. If a project fails during development or is judged to be imprudent, AFUDC recovery may not be allowed and the shareholders then bear the financial risk. Without assurances for cost recovery, it will be very difficult to build substantial amounts of new transmission. Minimizing development risks becomes of paramount importance when developing the types of projects necessary for regional and national purposes.

VISION FOR THE FUTURE

The challenges facing effective transmission planning are daunting, but pragmatic steps can be taken today to help the industry move toward a future vision capable of meeting these challenges. The following are suggestions that address the emerging economic and policy issues of today and can help to plan for a flexible transmission system that can effectively serve a variety of different future scenarios.

DEVELOP AN ALIGNED PLANNING PROCESS

Effective planning requires processes and methodologies that align well with the specific objectives being addressed. A good process should "de-clutter" a planning problem and align planning activity with the geographic scope of the goals. The process should push down the planning problem to the lowest possible level to reduce analytical requirements and organizational burden to a manageable size.

If the planning goal is to satisfy the reliability needs for communities in a tight geographic area, planning efforts should be led by the associated TO. This type of planning can be considered "bottom up" planning since it starts with the specific needs of specific customers. If the planning goal is to address regional market issues, planning efforts should be led by the associated RTO. This type of planning can be considered "top down" since it addresses the general requirements of the transmission system itself (in this case the ability to be an efficient market maker).

Typically, RTOs have drawn a demarcation line at an arbitrary voltage level (100 kV is typical). Below this line, TOs are responsible for the transmission plan. Above the line, RTOs are responsible for the transmission plan. This criterion can run counter to the "de-cluttering" principle. Very often, local planning requires solutions that go above 100 kV, and regional solutions may require the need to reach below 100 kV.

TOs and RTOs can effectively address planning issues corresponding to local and regional areas, respectively, but what about issues of national scope? Consider the current issue of renewable energy. For example, many states in the Northeast are beginning to set renewable energy portfolio targets that will require access to renewable energy concentrations in other parts of the country. Access to these resources will require crossing multiple RTO boundaries and/or transmission systems currently without RTO oversight.

Individual RTOs and TOs do not have the geographic perspective necessary to effectively address these types of broader issues. Who then should play this national role? RTOs working together could potentially be effective if the process is perceived as fair and equitable for all regions. However, if it is perceived that one region's objectives are beginning to take precedence over others, then a new national organization may be required.

If such a national step were taken, the role of the RTO must shift toward integrating member TO plans necessary to meet local load serving needs, integrating the EHV plan to address the national policy, and creating the regional plan that necessarily results to accommodate the regional objectives. The role would implement the strategic national plan and enables the tactical at the regional and local levels.

ADDRESSING THE REGULATORY NEED

The gap between planning a system and getting it developed needs to be closed. Planners should recognize that regulators are the ultimate decision makers. They decide whether or not a project is developed, not the planner. Therefore, planners must perform their work in a way that maximizes the probability of regulatory approval for their projects.

The regulatory oversight role is to ensure that transmission investment is prudent. It also ensures that public impacts are minimized. Planners need to recognize these roles and address these concerns early in and throughout the planning processes.

To address the prudency question, transmission planning processes should be open to stakeholder participation and permit stakeholders to have influence on a project. This ensures that a broadly vetted set of goals and objectives are being addressed by the process.

The objectives of an open planning process are:

- Transparency: the ability of affected stakeholders to observe and influence the planning processes and decisions
- Traceable: the ability for all parties to track the flow of planning effort throughout the life cycle of a project or overall plan
- Defendable: the appropriateness and completeness of the process from the perspective of key decision

makers such as RTO management, TOs, and regulators
- Dynamic: the ability to adjust the process for good reasons.

For planning at the regional or national levels, regulators expect that plans balance the benefits across the footprint and that stakeholder needs are addressed in an unbiased way. By design, RTOs do not own the facilities they plan and operate. By de-coupling the financial benefit of the transmission plan from the RTO, FERC hoped to ensure that plans were forwarded only driven by the needs of the stakeholders and designed in such a way as to minimize the overall cost regardless of the ownership boundaries. This independence is used by regulators to help make the prudency assessment since a project will, at least theoretically, only be approved for the "right" reasons.

To address the impacts on the public, planning processes need to encourage public involvement preferably early on in the process. Use of techniques such as press releases, community meetings, public planning meetings, open houses, and interactions with community development groups, economic development commissions, and regional planning commissions are extremely effective in addressing the public concerns in a meaningful way.

The effect is significant. First, the feedback provided can significantly aid in route selection and allow the planner and ultimate developer to better predict the costs of a project. Second, and equally important, if the public feels it has been heard and has had a meaningful chance to influence the results, the opposition is significantly muted. If not, the opposition is empowered and is able to recruit support from a much wider audience. The public tends to fear the unknown more than the known.

Many TOs know that these efforts are critical to the success of their projects, and some have successfully incorporated this outreach into their planning and infrastructure development efforts.

However, RTOs seem less aware of the importance of the public outreach step. A search of RTO Web sites shows significant efforts expended to bring certain stakeholders into their processes (highly commendable and necessary) but little efforts to bring in the public. There is a need for the public to be appropriately involved in the process. If regional and national transmission projects are to be planned in a way that maximizes the likelihood of approval, then the public input must be meaningfully provided. While difficult, creative thought needs to be applied to determine how to meaningfully bring the public into the regional and national forums.

ADDRESSING THE NEEDS OF THE DEVELOPER

At the RTO level, the regulatory need for an independent plan makes it more difficult to incorporate the needs of developers. The perception of independence needs to be protected to ensure the RTO appropriately plays its FERC-appointed role.

However, by bringing the stakeholders and the public into the planning process, developers have greater assurance that a project will be approved, that costs have been more accurately estimated, and that opposition has been minimized. Meaningfully addressing these issues in the RTO process are significant steps in encouraging developers to come forward.

ENHANCED PROJECT JUSTIFICATION

The advent of electricity markets illustrates the need for a richer understanding of the economic benefits of transmission projects. New facilities can have significant energy price impacts and, therefore, affect the underlying value of financial transmission rights. The evolving electricity markets are creating new winners and losers. As a result, it has become more critical to understand the economic benefits of transmission projects, especially at regional and national levels.

Project justification during the planning process needs to incorporate the pricing information available from these developing markets. Energy price history is now available to calibrate the analysis (Figure 5). Analysis tools that merge production cost analysis with transmission system constraints now exist to aid the planning in getting insights into the economic value of projects. As discussed above, these tools are difficult to use when considering myriads of alternative projects. However, they can be extremely effective in selecting between a narrowed-down set of alternatives.

For planning on a regional or national level, probabilistic methods show promise in managing the scope of studies necessary to perform *N*-2 or higher contingency analysis. At the regional or national level, decluttering still results in a network of significant scope. At the national level, the dynamics of an interconnect-wide system are poorly understood by any one planning entity.

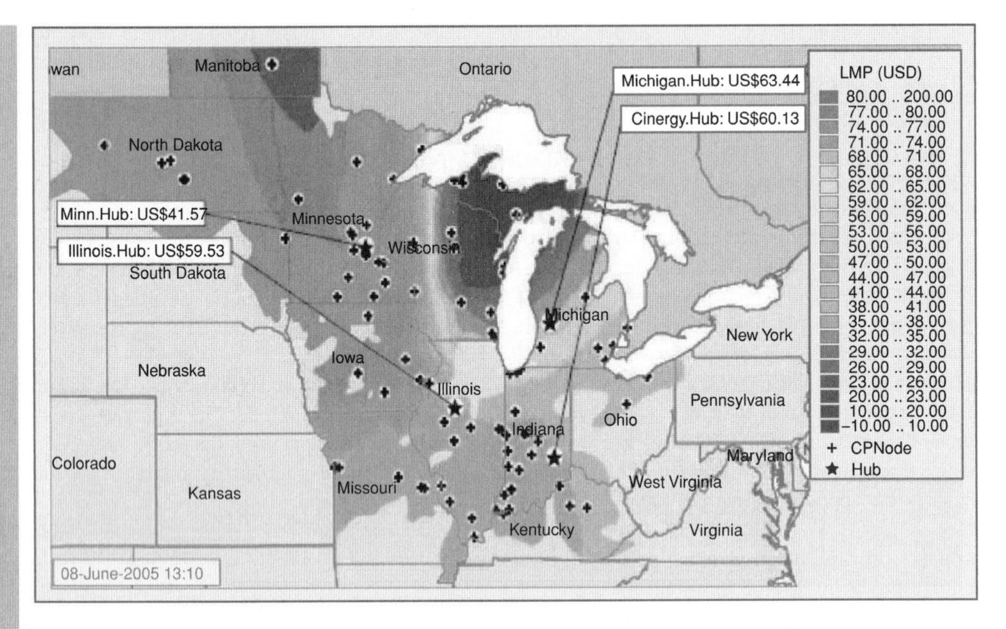

Figure 5
Examples of locational marginal price (LMP) information

Two things are certain; the United States needs to build more transmission capacity and it needs to begin to deal with aging transmission infrastructure. There are many challenges, but better transmission planning is needed to effectively address these issues in an integrated and cost-effective manner.

FOR FURTHER READING

B. Beck, *Interconnections: The History of the Mid-Continent Area Power Pool.* Minneapolis, MN: The Pool, 1988.

U.S. Department of Energy, *National Electric Transmission Congestion Study,* 2006, [Online]. Available: http://nietc.anl.gov/documents/docs/Congestion_Study_2006-9MB.pdf

E. Hirst, *U.S. Transmission Capacity: Present Status and Future Prospects,* EEI and DOE, Aug. 2004. [Online]. Available: http://www.oe.energy.gov/DocumentsandMedia/transmission_capacity.pdf

Fitch Ratings, *Frayed Wires: U.S. Transmission System Shows Its Age,* Oct. 25, 2006.

U.S.-Canada Power System Outage Task Force, *Causes of the August 14 2003 Blackout in the United States and Canada,* 2003.

BIOGRAPHIES

Donald. J. Morrow is vice president with the Technology Division of InfraSource. Morrow has extensive experience in transmission system planning, system operations, and transmission development. In his previous role, he was director of system planning and protection at American Transmission Company, a stand-alone transmission company in the upper Midwest. In this capacity he managed a US$3 billion/year capital budget portfolio. Morrow has been actively involved in many industry organizations including NERC and MISO. He has a B.S.E.E. and an executive M.B.A., both from the University of Wisconsin, Madison. He is a registered professional engineer in Wisconsin and a member of the IEEE.

Richard E. Brown is a vice president with the Technology Division of InfraSource. Brown has published more than 70 technical papers related to power system reliability and asset management, is author of the book *Electric Power Distribution Reliability,* and has provided consulting services to most major utilities in the United States. He is an IEEE Fellow and vice-chair of the Planning and Implementation Committee. Dr. Brown has a B.S.E.E., M.S.E.E., and Ph.D. from the University of Washington, Seattle, and an M.B.A. from the University of North Carolina, Chapel Hill.

Characteristics of Wind Turbine Generators for Wind Power Plants: IEEE PES Wind Plant Collector System Design Working Group

CONTRIBUTING MEMBERS: E. H. CAMM, M. R. BEHNKE, O. BOLADO, M. BOLLEN, M. BRADT, C. BROOKS, W. DILLING, M. EDDS, W. J. HEJDAK, D. HOUSEMAN, S. KLEIN, F. LI, J. LI, P. MAIBACH, T. NICOLAI, J. PATIÑO, S. V. PASUPULATI, N. SAMAAN, S. SAYLORS, T. SIEBERT, T. SMITH, M. STARKE, R. WALLING

Abstract—This paper presents a summary of the most important characteristics of wind turbine generators applied in modern wind power plants. Various wind turbine generator designs, based on classification by machine type and speed control capabilities, are discussed along with their operational characteristics, voltage, reactive power, or power factor control capabilities, voltage ride-through characteristics, behavior during short circuits, and reactive power capabilities.

Index Terms—Wind turbine generator, voltage ride-through, wind power plants.

I. INTRODUCTION

Modern wind power plants (WPPs), comprised of a large number of wind turbine generators (WTGs), a collector system, collector and/or interconnect substation utilize machines that are designed to optimize the generation of power using the energy in the wind. WTGs have developed from small machines with output power ratings on the order of kilowatts to several megawatts, and from machines with limited speed control and other capabilities to machines with variable speed control capabilities over a wide speed range and sophisticated control capabilities using modern power electronics [1].

The application of WTGs in modern WPPs requires an understanding of a number of different aspects related to the design and capabilities of the machines involved. This paper, authored by members of the Wind Plant Collector Design Working Group of the IEEE, is intended to provide insight into the various wind turbine generator designs, based on classification by machine type and speed control capabilities, along with their operational characteristics, voltage, reactive power, or power factor control capabilities, voltage ride-through characteristics, behavior during short circuits, and reactive power capabilities.

II. TURBINE CHARACTERISTICS

The principle of wind turbine operation is based on two well-known processes. The first one involves the conversion of kinetic energy of moving air into mechanical energy. This is accomplished by using aerodynamic rotor blades and a variety of methodologies for mechanical power control. The second process is the electromechanical energy conversion through a generator that is transmitted to the electrical grid.

Wind turbines can be classified by their mechanical power control, and further divided by their speed control. All turbine blades convert the motion of air across the air foils to torque, and then regulate that torque in an attempt to capture as much energy as possible, yet prevent damage. At the top level turbines can be classified as either stall regulated (with active stall as an improvement) or pitch regulated.

Stall regulation is achieved by shaping the turbine blades such that the airfoil generates less aerodynamic force at high wind speed, eventually stalling, thus reducing the turbine's torque-this is a simple, inexpensive and robust mechanical system. Pitch regulation, on the other hand, is achieved through the use of pitching devices in the turbine hub, which twist the blades around their own axes. As the wind speed changes, the blade quickly pitches to the optimum angle to control torque in order to capture the maximum energy or self-protect, as needed. Some turbines now are able to pitch each blade independently to achieve more balanced torques on the rotor shaft given wind speed differences at the top and bottom of the blade arcs.

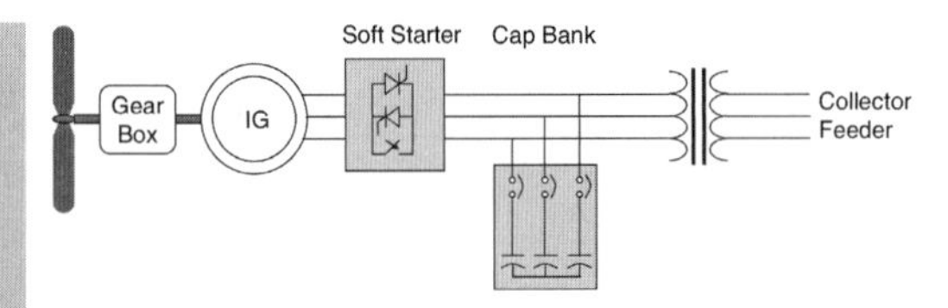

Figure 1
Typical Configuration of a Type 1 WTG

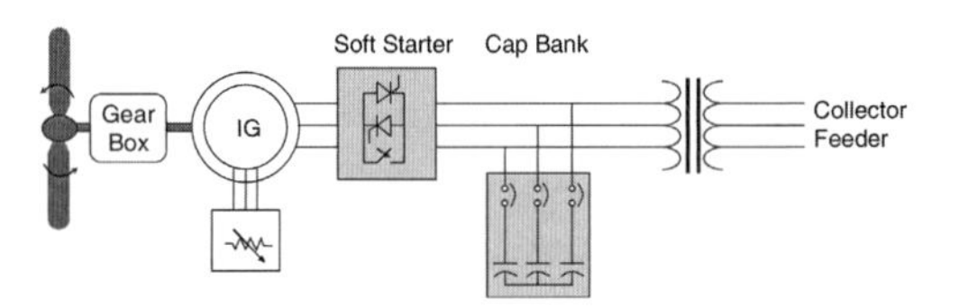

Figure 3
Typical Configuration of a Type 2 WTG

Beyond mechanical power regulation, turbines are further divided into fixed speed (Type 1), limited variable speed (Type 2), or variable speed with either partial (Type 3) or full (Type 4) power electronic conversion. The different speed control types are implemented via different rotating ac machines and the use of power electronics. There is one other machine type that will be referred to as Type 5 in which a mechanical torque converter between the rotor's low-speed shaft and the generator's high-speed shaft controls the generator speed to the electrical synchronous speed. This type of machine then uses a synchronous machine directly connected to the medium voltage grid.

The Type 1 WTG is implemented with a squirrel-cage induction generator (SCIG) and is connected to the step-up transformer directly. See Figure 1. The turbine speed is fixed (or nearly fixed) to the electrical grid's frequency, and generates real power (P) when the turbine shaft rotates faster than the electrical grid frequency creating a negative slip (positive slip and power is motoring convention).

Figure 2 shows the power flow at the SCIG terminals. While there is a bit of variability in output with the slip of the machine, Type 1 turbines typically operate at or very close to a rated speed. A major drawback of the induction machine is the reactive power that it consumes for its excitation field and the large currents the machine can draw when started "across-the-line." To ameliorate these effects the turbine typically employs a soft starter and discrete steps of capacitor banks within the turbine.

In Type 2 turbines, wound rotor induction generators arc connected directly to the WTG step-up transformer in a fashion similar to Type 1 with regards to the machines stator circuit, but also include a variable resistor in the rotor circuit. See Figure 3. This can be accomplished with a set of resistors and power electronics external to the rotor with currents flowing between the resistors and rotor via slip rings. Alternately, the resistors and electronics can be mounted on the rotor, eliminating the slip rings—this is the Weier design. The variable resistors are connected into the rotor circuit softly and can control the rotor currents quite rapidly so as to keep constant power even during gusting conditions, and can influence the machine's dynamic response during grid disturbances.

By adding resistance to the rotor circuit, the real power curve, which was shown in Figure 2, can be "stretched" to the higher slip and higher speed ranges. See Figure 4. That is to say that the turbine would have

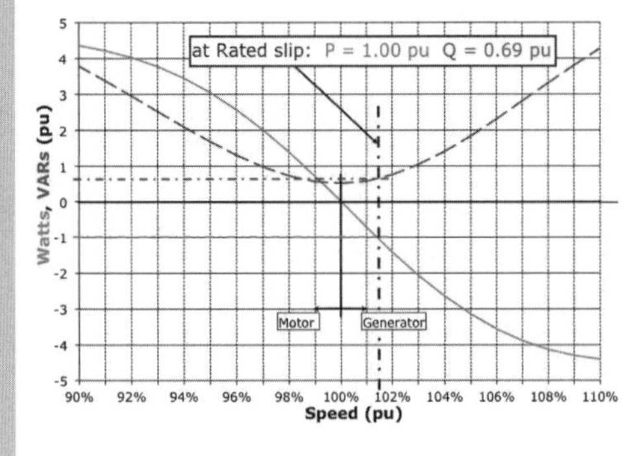

Figure 2
Variation of Real and Reactive Power for SCIG

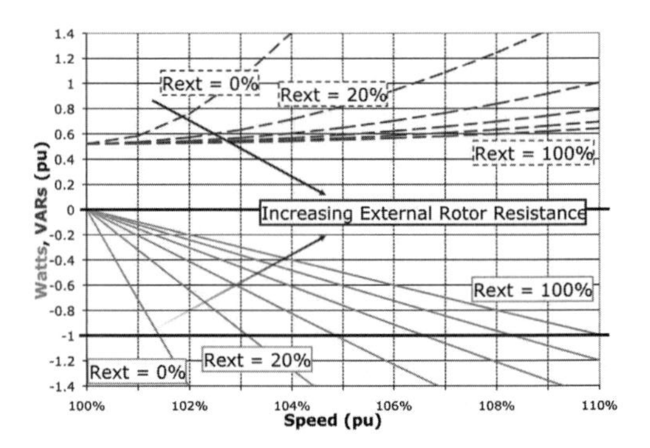

Figure 4
Variation of Real and Reactive Power with External Rotor Resitor in a Type 2 WTG

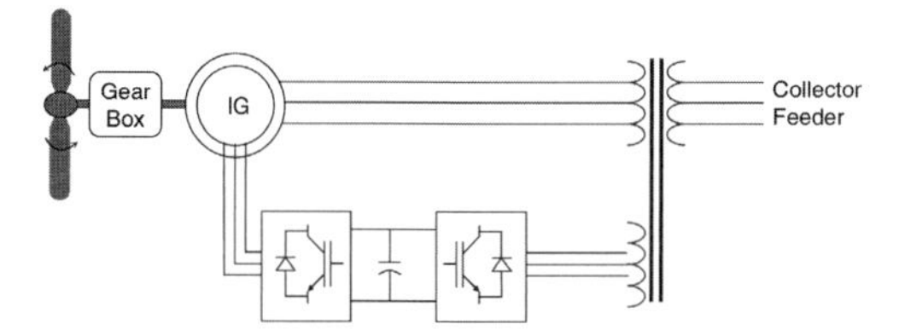

Figure 5
Typical Configuration of a Type 3 WTG

to spin faster to create the same output power, for an added rotor resistance. This allows some ability to control the speed, with the blades' pitching mechanisms and move the turbines operation to a tip speed ratio (ration of tip speed to the ambient wind speed) to achieve the best energy capture. It is typical that speed variations of up to 10% are possible, allowing for some degree of freedom in energy capture and self protective torque control.

The Type 3 turbine, known commonly as the Doubly Fed Induction Generator (DFIG) or Doubly Fed Asynchronous Generator (DFAG), takes the Type 2 design to the next level, by adding variable frequency ac excitation (instead of simply resistance) to the rotor circuit. The additional rotor excitation is supplied via slip rings by a current regulated, voltage-source converter, which can adjust the rotor currents' magnitude and phase nearly instantaneously. This rotor-side converter is connected back-to-back with a grid side converter, which exchanges power directly with the grid. See Figure 5.

A small amount power injected into the rotor circuit can effect a large control of power in the stator circuit. This is a major advantage of the DFIG—a great deal of control of the output is available with the presence of a set of converters that typically are only 30% of the rating of the machine. In addition to the real power that is delivered to the grid from the generator's stator circuit, power is delivered to the grid through the grid-connected inverter when the generator is moving faster than synchronous speed. When the generator is moving slower than synchronous speed, real power flows from the grid, through both converters, and from rotor to stator. These two modes, made possible by the four-quadrant nature of the two converters, allows a much wider speed range, both above and *below* synchronous speed by up to 50%, although narrower ranges are more common.

The greatest advantage of the DFIG, is that it offers the benefits of separate real and reactive power control, much like a traditional synchronous generator, while being able to run asynchronously. The field of industrial drives has produced and matured the concepts of vector or field oriented control of induction machines. Using these control schemes, the torque producing components of the rotor flux can be made to respond fast enough that the machine remains under relative control, even during significant grid disturbances. Indeed, while more expensive than the Type 1 or 2 machines, the Type 3 is becoming popular due to its advantages.

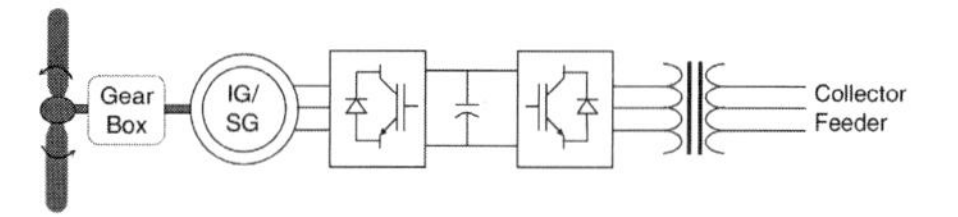

Figure 6
Typical Configuration of a Type 4 WTG

The Type 4 turbine (Figure 6) offers a great deal of flexibility in design and operation as the output of the rotating machine is sent to the grid through a full-scale back-to-back frequency converter. The turbine is allowed to rotate at its optimal aerodynamic speed, resulting in a "wild" ac output from the machine. In addition, the gearbox may be eliminated, such that the machine spins at the slow turbine speed and generates an electrical frequency well below that of the grid. This is no problem for a Type 4 turbine, as the inverters convert the power, and offer the possibility of reactive power supply to the grid, much like a STATCOM. The rotating machines of this type have been constructed as wound rotor synchronous machines, similar to conventional generators found in hydroelectric plants with control of the field current and high pole numbers, as permanent magnet synchronous machines, or as squirrel cage induction machines. However, based upon the ability of the machine side inverter to control real and reactive power flow, any type of machine could be used. Advances in power electronic devices and controls in the last decade have made the converters both responsive and efficient. It does bear mentioning, however, that the power electronic converters have to be sized to pass the full rating of the rotating machine, plus any capacity to be used for reactive compensation.

Type 5 turbines (Figure 7) consist of a typical WTG variable-speed drive train connected to a torque/speed converter coupled with a synchronous generator. The torque/speed converter changes the variable speed of the rotor shaft to a constant output shaft speed. The closely coupled synchronous generator,

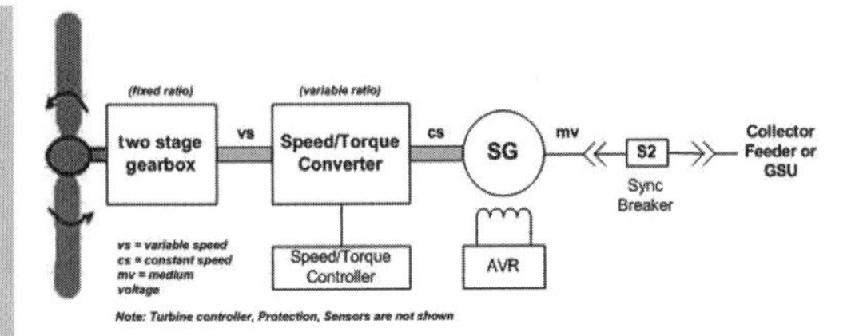

Figure 7
Typical Configuration of a Type 5 WTG

operating at a fixed speed (corresponding to grid frequency), can then be directly connected to the grid through a synchronizing circuit breaker. The synchronous generator can be designed appropriately for any desired speed (typically 6 pole or 4 pole) and voltage (typically medium voltage for higher capacities). This approach requires speed and torque control of the torque/speed converter along with the typical voltage regulator (AVR), synchronizing system, and generator protection system inherent with a grid-connected synchronous generator.

III. VOLTAGE, REACTIVE POWER, AND POWER FACTOR CONTROL CAPABILITIES

The voltage control capabilities of a WTG depend on the wind turbine type. Type I and Type 2 WTGs can typically not control voltage. Instead, these WTGs typically use power factor correction capacitors (PFCCs) to maintain the power factor or reactive power output on the low-voltage terminals of the machine to a setpoint. Types 3 through 5 WTGs can control voltage. These WTGs are capable of varying the reactive power at a given active power and terminal voltage, which enables voltage control [2]. In a Type 3 WTG voltage is controlled by changing the direct component of the rotor current (this is the component of the current that is in-line with the stator flux). In a Type 4 WTG voltage control is achieved by varying the quadrature (reactive) component of current at the grid-side converter. To allow voltage control capability, the grid-side converter must be rated above the rated MW of the machine. Since a synchronous generator is used in a Type 5 WTG, an automatic voltage regulator (AVR) is typically needed. Modern AVRs can be programmed to control reactive power, power factor and voltage.

The voltage control capabilities of individual WTGs are typically used to control the voltage at the collector bus or on the high side of the main power transformer. Usually a centralized wind farm controller will manage the control of the voltage through communication with the individual WTGs. A future companion Working Group paper is planned to discuss the WPP SCADA and control capabilities.

IV. REACTIVE POWER CAPABILITIES

The reactive power capabilities of modern WTGs are significant as most grid codes require the WPP to have reactive power capability at the point of interconnect over a specified power factor range, for example 0.95 leading (inductive) to 0.95 lagging (capacitive). Typical interconnect requirements related to total WPP reactive power capabilities are discussed in [3].

As stated earlier, Type I and Type 2 WTGs typically use PFCCs to maintain the power factor or reactive power of the machine to a specified setpoint. The PFCCs may be sized to maintain a slightly leading (inductive) power factor of around 0.98 at rated power output. This is often referred to as no-load compensation. With full-load compensation, the PFCCs are sized to maintain unity power factor or, in some cases, a slightly lagging (capacitive) power factor at the machine's rated power output. The PFCCs typically consists of multiple stages of capacitors switched with a low-voltage ac contactor.

Type 3 (DFIG) WTGs typically have a reactive power capability corresponding to a power factor of 0.95 lagging (capacitive) to 0.90 leading (inductive) at the terminals of the machines. Options for these machines include an expanded reactive power capability of 0.90 lagging to 0.90 leading. Some Type 3 WTGs can deliver reactive power even when the turbine is not operating mechanically, while no real power is generated.

As previously stated, Type 4 WTGs can vary the grid-side converter current, allowing control of the effective power factor of the machines over a wide range. Reactive power limit curves for different terminal voltage levels are typically provided. Some Type 4 WTGs can deliver reactive power even when the turbine is not operating mechanically, while no real power is generated.

The synchronous generator in a Type 5 WTG has inherent dynamic reactive power capabilities similar to that of Type 3 and 4 machines. See Figure 8. Depending on the design of the generator, operating power factor ranges at rated output can vary from 0.8 leading to 0.8 lagging.

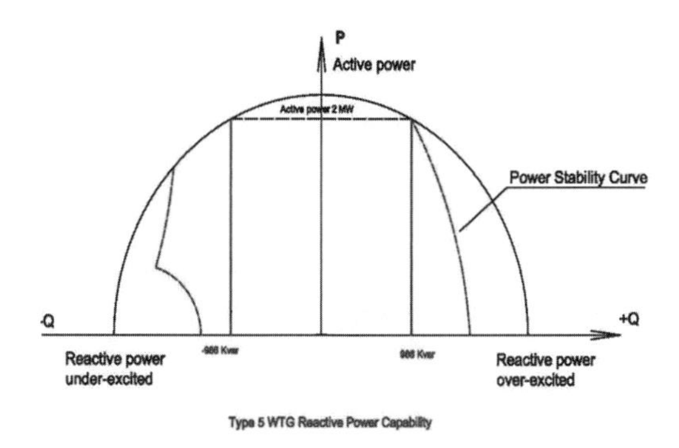

Figure 8
Reactive Power Capabilities of a 2 MW Type 5 WTG

A range of 0.9 leading and lagging is more typical. At power outputs below rated power, the reactive power output is only limited by rotor or stator heating, stability concerns, and local voltage conditions and it is unlikely that PFCCs would be required. As with some Type 3 and 4 WTGs, it is also possible to operate the machine as a synchronous condenser, requiring minimal active power output with adjustable reactive power output levels.

V. VOLTAGE RIDE-THROUOH

The voltage ride-through (VRT) capabilities of WTGs vary widely and have evolved based on requirements in various grid codes. In the United States, low voltage ride-through (LVRT) requirements specified in FERC Order 661-A [5] calls for wind power plants to ride-through a three-phase fault on the high side of the substation transformer for up to 9 cycles, depending on the primary fault clearing time of the fault interrupting circuit breakers at the location. There is no high voltage ride-through (HVRT) requirement in FERC order 661-A, but NERC and some ISO/RTOs are in the process of imposing such requirement. In many European countries WPP are required not to trip for a high voltage level up to 110% of the nominal voltage at the POI [4].

Some of the Type I WTGs have limited VRT capability and may require a central reactive power compensation system [4] to meet wind power plant VRT capability. Many of the Types 2, 3, and 4 WTGs have VRT capabilities that may meet the requirements of FERC Order 661, which was issued before FERC Order 661-A (i.e., withstand a three-phase fault for 9 cycles at a voltage as low as 0.15 p.u measured on the high side of the substation transformer). Most WTGs are expected to ultimately meet the FERC 661-A requirements.

The VRT of a Type 5 WTG is very similar to that of standard grid-connected synchronous generators, which are well understood. The capabilities of the excitation system (AVR) and physical design of the generator (machine constants, time constants) will determine the basic performance of a synchronous generator during transient conditions. In order to meet utility VRT requirements, the settings and operation of the turbine control system, excitation system and protection systems must be generally coordinated and then fine-tuned for a specific site.

VI. WTG BEHAVIOR DURING GRID SHORT CIRCUITS

The response of WTGs to short circuits on the grid depends largely on the type of WTG. While the response of Type I and Type 2 WTGs are essentially similar to that of large induction machines used in industrial applications, the response of Type 3, 4, and 5 WTGs is dictated by the WTG controls. In short circuit calculations, a Type I WTG can be represented as a voltage source in series with the direct axis sub-transient inductance X_d''. This practice is used to consider the maximum short-circuit contribution from the induction generator as it determines the symmetrical current magnitude during the first few cycles after the fault. A Type I WTG can contribute short circuit current up to the value of its locked rotor current which is usually on the order of 5 to 6 p.u [6].

Type 2 WTGs employing limited speed control via controlled external rotor resistance are fundamentally induction generators. If, during the fault, the external resistance control were to result in short-circuiting of the generator rotor, the short-circuit behavior would be similar to Type 1. On the other hand, if the control action at or shortly after fault inception were to result in insertion of the full external resistance, the equivalent voltage source-behind-Thevenin impedance representation for the WTG should be modified to include this significant resistance value in series with the equivalent turbine inductance.

Other wind turbine topologies employ some type of power electronic control. Consequently, the behavior during short-circuit conditions cannot be ascertained directly from the physical structure of the electrical generator. Algorithms which control the power electronic switches can have significant influence on the short-circuit currents contributed by the turbine, and the details of these controllers are generally held closely by the turbine manufacturers.

For Type 3 WTGs (DFIG), if during the fault, the rotor power controller remains active, the machine stator currents would be limited between 1.1 to 2.5 p.u. of the machine rated current. Under conditions where protective functions act to "crowbar" the rotor circuit, the short-circuit behavior defaults to 5 to 6 p.u. in the case of a fault applied directly to the WTG terminals. [7]

In turbines employing full-rated power converters as the interface to the grid (Type 4), currents during network faults will be limited to slightly above rated current. This limitation is affected by the power converter control, and is generally necessary to protect the power semiconductor switches.

Type 5 WTGs exhibit typical synchronous generator behavior during grid short circuits. Generator contribution to grid faults can be calculated from the machine constants, obtainable from the generator manufacturer. Fault current contribution for line to ground faults will depend on the type of generator grounding used. Typical generator fault current contribution can range from 4 to more times rated current for close-in bolted three-phase faults. Fault current contribution for single-line to ground faults can range from near zero amps (ungrounded neutral) to more than the three-phase bolted level (depending on the zero sequence impedance of solidly grounded generators.)

A joint Working Group sponsored by the Power Systems Relaying Committee (PSRC) and the T&D Committee on short-circuit contributions from WTGs is currently discussing this topic. It is expected that more specific guidelines on considerations in determining short-circuit contributions from different types of WTGs will be forthcoming.

VII. REFERENCES

[1] Robert Zavadil, Nicholas Miller, Abraham Ellis, and Eduard Muljadi, "Making Connections [Wind Generation Facilities]," *IEEE Power & Energy Magazine,* vol. 3, no. 6, pp. 26-37, Nov.–Dec. 2005.

[2] W.L. Kling, J.G. Slootweg, "Wind Turbines as Power Plants" IEEE/Cigré Workshop on Wind Power and the Impacts on Power Systems, June 2002, Oslo, Norway.

[3] Wind Plant Collector System Design Working Group, "Wind Power Plant Collector System Design Considerations," in *Proc. 2009 IEEE Power and Energy Society General Meeting,* Calgary, Canada, July 2009.

[4] Wind Plant Collector SystemDesign Working Group, "Reactive Power Compensation for Wind Power Plants," in *Proc. 2009 IEEE Power and Energy Society General Meeting,* Calgary, Canada, July 2009.

[5] FERC Order no. 661-A, "Interconnection for Wind Energy," Docket No. RM05-4-001, December 2005.

[6] Nader Samaan, Robert Zavadil. J. Charles Smith and Jose Conto, "Modeling of Wind Power Plants for Short Circuit Analysis in the Transmission Network," in *Proc. of IEEE/PES Transmission and Distribution Conference,* Chicago, USA, April 2008.

[7] J. Morreu, S.W.H. de Haan, "Ridethrough of Wind Turbines with Doubly-Fed Induction Generator During a Voltage Dip," IEEE Transactions on Energy Conversion, vol. 20, no. 2, June 2005.

[8] Ackermann, Thomas, ed. Wind Power in Power Systems. West Sussex, UK: John Wiley & Sons, 2005. ISBN 13: 978-0-470-85508-9.

[9] Hau, Erich. Wind Turbines: Fundamentals. Technologies. Application, Economics. 2nd Edition. Trans. Horst von Renouard. Sidcup, Kent, UK: Springer, 2006. ISBN 13: 978-3-540-24240-6.

6.1

DIRECT SOLUTIONS TO LINEAR ALGEBRAIC EQUATIONS: GAUSS ELIMINATION

Consider the following set of linear algebraic equations in matrix format:

$$\begin{bmatrix} A_{11} & A_{12} & \cdots & A_{1N} \\ A_{21} & A_{22} & \cdots & A_{2N} \\ \vdots & & \vdots & \\ A_{N1} & A_{N2} & \cdots & A_{NN} \end{bmatrix} \begin{bmatrix} x_1 \\ x_2 \\ \vdots \\ x_N \end{bmatrix} = \begin{bmatrix} y_1 \\ y_2 \\ \vdots \\ y_N \end{bmatrix} \tag{6.1.1}$$

or

$$\mathbf{Ax} = \mathbf{y} \tag{6.1.2}$$

where $\mathbf{x}$ and $\mathbf{y}$ are N vectors and $\mathbf{A}$ is an $N \times N$ square matrix. The components of $\mathbf{x}$, $\mathbf{y}$, and $\mathbf{A}$ may be real or complex. Given $\mathbf{A}$ and $\mathbf{y}$, we want to solve for $\mathbf{x}$. We assume the det($\mathbf{A}$) is nonzero, so a unique solution to (6.1.1) exists.

The solution $\mathbf{x}$ can easily be obtained when $\mathbf{A}$ is an upper triangular matrix with nonzero diagonal elements. Then (6.1.1) has the form

$$\begin{bmatrix} A_{11} & A_{12}\ldots & & A_{1N} \\ 0 & A_{22}\ldots & & A_{2N} \\ \vdots & & & \\ 0 & 0\ldots & A_{N-1,N-1} & A_{N-1,N} \\ 0 & 0\ldots 0 & & A_{NN} \end{bmatrix} \begin{bmatrix} x_1 \\ x_2 \\ \vdots \\ x_{N-1} \\ x_N \end{bmatrix} = \begin{bmatrix} y_1 \\ y_2 \\ \vdots \\ y_{N-1} \\ y_N \end{bmatrix} \tag{6.1.3}$$

Since the last equation in (6.1.3) involves only x_N,

$$x_N = \frac{y_N}{A_{NN}} \tag{6.1.4}$$

After x_N is computed, the next-to-last equation can be solved:

$$x_{N-1} = \frac{y_{N-1} - A_{N-1,N}x_N}{A_{N-1,N-1}} \tag{6.1.5}$$

In general, with $x_N, x_{N-1}, \ldots, x_{k+1}$ already computed, the kth equation can be solved

$$x_k = \frac{y_k - \sum_{n=k+1}^{N} A_{kn}x_n}{A_{kk}} \qquad k = N, N-1, \ldots, 1 \tag{6.1.6}$$

This procedure for solving (6.1.3) is called *back substitution.*

If $\mathbf{A}$ is not upper triangular, (6.1.1) can be transformed to an equivalent equation with an upper triangular matrix. The transformation, called *Gauss elimination*, is described by the following $(N-1)$ steps. During Step 1, we use the first equation in (6.1.1) to eliminate x_1 from the remaining equations. That is, Equation 1 is multiplied by A_{n1}/A_{11} and then subtracted from equation n, for $n = 2, 3, \ldots, N$. After completing Step 1, we have

$$\begin{bmatrix} A_{11} & A_{12} & \cdots & A_{1N} \\ 0 & \left(A_{22} - \frac{A_{21}}{A_{11}}A_{12}\right) & \cdots & \left(A_{2N} - \frac{A_{21}}{A_{11}}A_{1N}\right) \\ 0 & \left(A_{32} - \frac{A_{31}}{A_{11}}A_{12}\right) & \cdots & \left(A_{3N} - \frac{A_{31}}{A_{11}}A_{1N}\right) \\ \vdots & \vdots & & \vdots \\ 0 & \left(A_{N2} - \frac{A_{N1}}{A_{11}}A_{12}\right) & \cdots & \left(A_{NN} - \frac{A_{N1}}{A_{11}}A_{1N}\right) \end{bmatrix} \begin{bmatrix} x_1 \\ x_2 \\ x_3 \\ \vdots \\ x_N \end{bmatrix}$$

$$= \begin{bmatrix} y_1 \\ y_2 - \frac{A_{21}}{A_{11}}y_1 \\ y_3 - \frac{A_{31}}{A_{11}}y_1 \\ \vdots \\ y_N - \frac{A_{N1}}{A_{11}}y_1 \end{bmatrix} \tag{6.1.7}$$

Equation (6.1.7) has the following form:

$$\begin{bmatrix} A_{11}^{(1)} & A_{12}^{(1)} & \cdots & A_{1N}^{(1)} \\ 0 & A_{22}^{(1)} & \cdots & A_{2N}^{(1)} \\ 0 & A_{32}^{(1)} & \cdots & A_{3N}^{(1)} \\ \vdots & \vdots & & \vdots \\ 0 & A_{N2}^{(1)} & \cdots & A_{NN}^{(1)} \end{bmatrix} \begin{bmatrix} x_1 \\ x_2 \\ x_3 \\ \vdots \\ x_N \end{bmatrix} = \begin{bmatrix} y_1^{(1)} \\ y_2^{(1)} \\ y_3^{(1)} \\ \vdots \\ y_N^{(1)} \end{bmatrix} \tag{6.1.8}$$

where the superscript (1) denotes Step 1 of Gauss elimination.

During Step 2 we use the second equation in (6.1.8) to eliminate x_2 from the remaining (third, fourth, fifth, and so on) equations. That is, Equation 2 is multiplied by $A_{n2}^{(1)}/A_{22}^{(1)}$ and subtracted from equation n, for $n = 3, 4, \ldots, N$.

After Step 2, we have

$$\begin{bmatrix} A_{11}^{(2)} & A_{12}^{(2)} & A_{13}^{(2)} & \cdots & A_{1N}^{(2)} \\ 0 & A_{22}^{(2)} & A_{23}^{(2)} & \cdots & A_{2N}^{(2)} \\ 0 & 0 & A_{33}^{(2)} & \cdots & A_{3N}^{(2)} \\ 0 & 0 & A_{43}^{(2)} & \cdots & A_{4N}^{(2)} \\ \vdots & \vdots & \vdots & & \vdots \\ 0 & 0 & A_{N3}^{(2)} & \cdots & A_{NN}^{(2)} \end{bmatrix} \begin{bmatrix} x_1 \\ x_2 \\ x_3 \\ x_4 \\ \vdots \\ x_N \end{bmatrix} = \begin{bmatrix} y_1^{(2)} \\ y_2^{(2)} \\ y_3^{(2)} \\ y_4^{(2)} \\ \vdots \\ y_N^{(2)} \end{bmatrix} \tag{6.1.9}$$

During step k, we start with $\mathbf{A}^{(k-1)}\mathbf{x} = \mathbf{y}^{(k-1)}$. The first k of these equations, already triangularized, are left unchanged. Also, equation k is multiplied by $A_{nk}^{(k-1)}/A_{kk}^{(k-1)}$ and then subtracted from equation n, for $n = k+1, k+2, \ldots, N$.

After $(N-1)$ steps, we arrive at the equivalent equation $\mathbf{A}^{(N-1)}\mathbf{x} = \mathbf{y}^{(N-1)}$, where $\mathbf{A}^{(N-1)}$ is upper triangular.

EXAMPLE 6.1 Gauss elimination and back substitution: direct solution to linear algebraic equations

Solve

$$\left[\begin{array}{c|c} 10 & 5 \\ \hline 2 & 9 \end{array}\right] \begin{bmatrix} x_1 \\ x_2 \end{bmatrix} = \begin{bmatrix} 6 \\ 3 \end{bmatrix}$$

using Gauss elimination and back substitution.

SOLUTION Since $N = 2$ for this example, there is $(N-1) = 1$ Gauss elimination step. Multiplying the first equation by $A_{21}/A_{11} = 2/10$ and then subtracting from the second,

$$\left[\begin{array}{c|c} 10 & 5 \\ \hline 0 & 9 - \frac{2}{10}(5) \end{array}\right] \begin{bmatrix} x_1 \\ x_2 \end{bmatrix} = \begin{bmatrix} 6 \\ 3 - \frac{2}{10}(6) \end{bmatrix}$$

or

$$\left[\begin{array}{c|c} 10 & 5 \\ \hline 0 & 8 \end{array}\right] \begin{bmatrix} x_1 \\ x_2 \end{bmatrix} = \begin{bmatrix} 6 \\ 1.8 \end{bmatrix}$$

which has the form $\mathbf{A}^{(1)}\mathbf{x} = \mathbf{y}^{(1)}$, where $\mathbf{A}^{(1)}$ is upper triangular. Now, using back substitution, (6.1.6) gives, for $k = 2$:

$$x_2 = \frac{y_2^{(1)}}{A_{22}^{(1)}} = \frac{1.8}{8} = 0.225$$

and, for $k = 1$,

$$x_1 = \frac{y_1^{(1)} - A_{12}^{(1)} x_2}{A_{11}^{(1)}} = \frac{6 - (5)(0.225)}{10} = 0.4875$$ ■

EXAMPLE 6.2 Gauss elimination: triangularizing a matrix

Use Gauss elimination to triangularize

$$\begin{bmatrix} 2 & 3 & -1 \\ -4 & 6 & 8 \\ 10 & 12 & 14 \end{bmatrix} \begin{bmatrix} x_1 \\ x_2 \\ x_3 \end{bmatrix} = \begin{bmatrix} 5 \\ 7 \\ 9 \end{bmatrix}$$

SOLUTION There are $(N - 1) = 2$ Gauss elimination steps. During Step 1, we subtract $A_{21}/A_{11} = -4/2 = -2$ times Equation 1 from Equation 2, and we subtract $A_{31}/A_{11} = 10/2 = 5$ times Equation 1 from Equation 3, to give

$$\begin{bmatrix} 2 & 3 & -1 \\ 0 & 6-(-2)(3) & 8-(-2)(-1) \\ 0 & 12-(5)(3) & 14-(5)(-1) \end{bmatrix} \begin{bmatrix} x_1 \\ x_2 \\ x_3 \end{bmatrix} = \begin{bmatrix} 5 \\ 7-(-2)(5) \\ 9-(5)(5) \end{bmatrix}$$

or

$$\begin{bmatrix} 2 & 3 & -1 \\ 0 & 12 & 6 \\ 0 & -3 & 19 \end{bmatrix} \begin{bmatrix} x_1 \\ x_2 \\ x_3 \end{bmatrix} = \begin{bmatrix} 5 \\ 17 \\ -16 \end{bmatrix}$$

which is $\mathbf{A}^{(1)}\mathbf{x} = \mathbf{y}^{(1)}$. During Step 2, we subtract $A_{32}^{(1)}/A_{22}^{(1)} = -3/12 = -0.25$ times Equation 2 from Equation 3, to give

$$\begin{bmatrix} 2 & 3 & -1 \\ 0 & 12 & 6 \\ 0 & 0 & 19-(-.25)(6) \end{bmatrix} \begin{bmatrix} x_1 \\ x_2 \\ x_3 \end{bmatrix} = \begin{bmatrix} 5 \\ 17 \\ -16-(-.25)(17) \end{bmatrix}$$

or

$$\begin{bmatrix} 2 & 3 & -1 \\ 0 & 12 & 6 \\ 0 & 0 & 20.5 \end{bmatrix} \begin{bmatrix} x_1 \\ x_2 \\ x_3 \end{bmatrix} = \begin{bmatrix} 5 \\ 17 \\ -11.75 \end{bmatrix}$$

which is triangularized. The solution $\mathbf{x}$ can now be easily obtained via back substitution. ■

Computer storage requirements for Gauss elimination and back substitution include N^2 memory locations for $\mathbf{A}$ and N locations for $\mathbf{y}$. If there is no further need to retain $\mathbf{A}$ and $\mathbf{y}$, then $\mathbf{A}^{(k)}$ can be stored in the location of $\mathbf{A}$, and $\mathbf{y}^{(k)}$, as well as the solution $\mathbf{x}$, can be stored in the location of $\mathbf{y}$. Additional memory is also required for iterative loops, arithmetic statements, and working space.

Computer time requirements can be evaluated by determining the number of arithmetic operations required for Gauss elimination and back substitution. One can show that Gauss elimination requires $(N^3 - N)/3$ multiplications, $(N)(N-1)/2$ divisions, and $(N^3 - N)/3$ subtractions. Also, back substitution requires $(N)(N-1)/2$ multiplications, N divisions, and $(N)(N-1)/2$ subtractions. Therefore, for very large N, the approximate computer time for solving (6.1.1) by Gauss elimination and back substitution is the time required to perform $N^3/3$ multiplications and $N^3/3$ subtractions.

For example, consider a digital computer with a 2×10^{-9} s multiplication time and 1×10^{-9} s addition or subtraction time. Solving $N = 10{,}000$ equations would require approximately

$$\tfrac{1}{3}N^3(2 \times 10^{-9}) + \tfrac{1}{3}N^3(1 \times 10^{-9}) = \tfrac{1}{3}(10{,}000)^3(3 \times 10^{-9}) = 1000 \quad \text{s}$$

plus some additional bookkeeping time for indexing and managing loops.

Since the power-flow problem often involves solving power systems with tens of thousands of equations, by itself Gauss elimination would not be a good solution. However, for matrixes that have relatively few nonzero elements, known as sparse matrices, special techniques can be employed to significantly reduce computer storage and time requirements. Since all large power systems can be modeled using sparse matrices, these techniques are briefly introduced in Section 6.8.

6.2 ITERATIVE SOLUTIONS TO LINEAR ALGEBRAIC EQUATIONS: JACOBI AND GAUSS–SEIDEL

A general iterative solution to (6.1.1) proceeds as follows. First select an initial guess $\mathbf{x}(0)$. Then use

$$\mathbf{x}(i+1) = \mathbf{g}[\mathbf{x}(i)] \qquad i = 0, 1, 2, \ldots \tag{6.2.1}$$

where $\mathbf{x}(i)$ is the ith guess and $\mathbf{g}$ is an N vector of functions that specify the iteration method. Continue the procedure until the following stopping condition is satisfied:

$$\left| \frac{x_k(i+1) - x_k(i)}{x_k(i)} \right| < \varepsilon \qquad \text{for all } k = 1, 2, \ldots, N \tag{6.2.2}$$

where $x_k(i)$ is the kth component of $\mathbf{x}(i)$ and ε is a specified *tolerance level.*

The following questions are pertinent:

1. Will the iteration procedure converge to the unique solution?
2. What is the convergence rate (how many iterations are required)?
3. When using a digital computer, what are the computer storage and time requirements?

These questions are addressed for two specific iteration methods: *Jacobi* and *Gauss–Seidel*.* The Jacobi method is obtained by considering the kth equation of (6.1.1), as follows:

$$y_k = A_{k1}x_1 + A_{k2}x_2 + \cdots + A_{kk}x_k + \cdots + A_{kN}x_N \tag{6.2.3}$$

Solving for x_k,

$$\begin{aligned} x_k &= \frac{1}{A_{kk}}[y_k - (A_{k1}x_1 + \cdots + A_{k,k-1}x_{k-1} + A_{k,k+1}x_{k+1} + \cdots + A_{kN}x_N)] \\ &= \frac{1}{A_{kk}}\left[y_k - \sum_{n=1}^{k-1} A_{kn}x_n - \sum_{n=k+1}^{N} A_{kn}x_n\right] \end{aligned} \tag{6.2.4}$$

The Jacobi method uses the "old" values of $\mathbf{x}(i)$ at iteration i on the right side of (6.2.4) to generate the "new" value $x_k(i+1)$ on the left side of (6.2.4). That is,

$$x_k(i+1) = \frac{1}{A_{kk}}\left[y_k - \sum_{n=1}^{k-1} A_{kn}x_n(i) - \sum_{n=k+1}^{N} A_{kn}x_n(i)\right] \qquad k = 1, 2, \ldots, N \tag{6.2.5}$$

The Jacobi method given by (6.2.5) can also be written in the following matrix format:

$$\mathbf{x}(i+1) = \mathbf{M}\mathbf{x}(i) + \mathbf{D}^{-1}\mathbf{y} \tag{6.2.6}$$

where

$$\mathbf{M} = \mathbf{D}^{-1}(\mathbf{D} - \mathbf{A}) \tag{6.2.7}$$

and

$$\mathbf{D} = \begin{bmatrix} A_{11} & 0 & 0 & \cdots & 0 \\ 0 & A_{22} & 0 & \cdots & 0 \\ 0 & \vdots & \vdots & & \vdots \\ \vdots & & & & 0 \\ 0 & 0 & 0 & \cdots & A_{NN} \end{bmatrix} \tag{6.2.8}$$

For Jacobi, $\mathbf{D}$ consists of the diagonal elements of the $\mathbf{A}$ matrix.

---

*The Jacobi method is also called the Gauss method.

EXAMPLE 6.3 Jacobi method: iterative solution to linear algebraic equations

Solve Example 6.1 using the Jacobi method. Start with $x_1(0) = x_2(0) = 0$ and continue until (6.2.2) is satisfied for $\varepsilon = 10^{-4}$.

SOLUTION From (6.2.5) with $N = 2$,

$$k = 1 \qquad x_1(i+1) = \frac{1}{A_{11}}[y_1 - A_{12}x_2(i)] = \frac{1}{10}[6 - 5x_2(i)]$$

$$k = 2 \qquad x_2(i+1) = \frac{1}{A_{22}}[y_2 - A_{21}x_1(i)] = \frac{1}{9}[3 - 2x_1(i)]$$

Alternatively, in matrix format using (6.2.6)–(6.2.8),

$$\mathbf{D}^{-1} = \begin{bmatrix} 10 & 0 \\ 0 & 9 \end{bmatrix}^{-1} = \begin{bmatrix} \frac{1}{10} & 0 \\ 0 & \frac{1}{9} \end{bmatrix}$$

$$\mathbf{M} = \begin{bmatrix} \frac{1}{10} & 0 \\ 0 & \frac{1}{9} \end{bmatrix} \begin{bmatrix} 0 & -5 \\ -2 & 0 \end{bmatrix} = \begin{bmatrix} 0 & -\frac{5}{10} \\ -\frac{2}{9} & 0 \end{bmatrix}$$

$$\begin{bmatrix} x_1(i+1) \\ x_2(i+1) \end{bmatrix} = \begin{bmatrix} 0 & -\frac{5}{10} \\ -\frac{2}{9} & 0 \end{bmatrix} \begin{bmatrix} x_1(i) \\ x_2(i) \end{bmatrix} + \begin{bmatrix} \frac{1}{10} & 0 \\ 0 & \frac{1}{9} \end{bmatrix} \begin{bmatrix} 6 \\ 3 \end{bmatrix}$$

The above two formulations are identical. Starting with $x_1(0) = x_2(0) = 0$, the iterative solution is given in the following table:

JACOBI

i	0	1	2	3	4	5	6	7	8	9	10
$x_1(i)$	0	0.60000	0.43334	0.50000	0.48148	0.48889	0.48683	0.48766	0.48743	0.48752	0.48749
$x_2(i)$	0	0.33333	0.20000	0.23704	0.22222	0.22634	0.22469	0.22515	0.22496	0.22502	0.22500

As shown, the Jacobi method converges to the unique solution obtained in Example 6.1. The convergence criterion is satisfied at the 10th iteration, since

$$\left|\frac{x_1(10) - x_1(9)}{x_1(9)}\right| = \left|\frac{0.48749 - 0.48752}{0.48749}\right| = 6.2 \times 10^{-5} < \varepsilon$$

and

$$\left|\frac{x_2(10) - x_2(9)}{x_2(9)}\right| = \left|\frac{0.22500 - 0.22502}{0.22502}\right| = 8.9 \times 10^{-5} < \varepsilon$$

■

The Gauss–Seidel method is given by

$$x_k(i+1) = \frac{1}{A_{kk}}\left[y_k - \sum_{n=1}^{k-1} A_{kn}x_n(i+1) - \sum_{n=k+1}^{N} A_{kn}x_n(i)\right] \tag{6.2.9}$$

Comparing (6.2.9) with (6.2.5), note that Gauss–Seidel is similar to Jacobi except that during each iteration, the "new" values, $x_n(i+1)$, for $n < k$ are used on the right side of (6.2.9) to generate the "new" value $x_k(i+1)$ on the left side.

The Gauss–Seidel method of (6.2.9) can also be written in the matrix format of (6.2.6) and (6.2.7), where

$$\mathbf{D} = \begin{bmatrix} A_{11} & 0 & 0 & \cdots & 0 \\ A_{21} & A_{22} & 0 & \cdots & 0 \\ \vdots & \vdots & & & \vdots \\ A_{N1} & A_{N2} & \cdots & & A_{NN} \end{bmatrix} \tag{6.2.10}$$

For Gauss–Seidel, **D** in (6.2.10) is the lower triangular portion of **A**, whereas for Jacobi, **D** in (6.2.8) is the diagonal portion of **A**.

EXAMPLE 6.4 Gauss–Seidel method: iterative solution to linear algebraic equations

Rework Example 6.3 using the Gauss–Seidel method.

SOLUTION From (6.2.9),

$$k=1 \qquad x_1(i+1) = \frac{1}{A_{11}}[y_1 - A_{12}x_2(i)] = \frac{1}{10}[6 - 5x_2(i)]$$

$$k=2 \qquad x_2(i+1) = \frac{1}{A_{22}}[y_2 - A_{21}x_1(i+1)] = \frac{1}{9}[3 - 2x_1(i+1)]$$

Using this equation for $x_1(i+1)$, $x_2(i+1)$ can also be written as

$$x_2(i+1) = \frac{1}{9}\left\{3 - \frac{2}{10}[6 - 5x_2(i)]\right\}$$

Alternatively, in matrix format, using (6.2.10), (6.2.6), and (6.2.7):

$$\mathbf{D}^{-1} = \left[\begin{array}{c|c} 10 & 0 \\ \hline 2 & 9 \end{array}\right]^{-1} = \left[\begin{array}{c|c} \frac{1}{10} & 0 \\ \hline -\frac{2}{90} & \frac{1}{9} \end{array}\right]$$

$$\mathbf{M} = \begin{bmatrix} \frac{1}{10} & 0 \\ -\frac{2}{90} & \frac{1}{9} \end{bmatrix} \begin{bmatrix} 0 & -5 \\ 0 & 0 \end{bmatrix} = \begin{bmatrix} 0 & -\frac{1}{2} \\ 0 & \frac{1}{9} \end{bmatrix}$$

$$\begin{bmatrix} x_1(i+1) \\ x_2(i+1) \end{bmatrix} = \begin{bmatrix} 0 & -\frac{1}{2} \\ 0 & \frac{1}{9} \end{bmatrix} \begin{bmatrix} x_1(i) \\ x_2(i) \end{bmatrix} + \begin{bmatrix} \frac{1}{10} & 0 \\ -\frac{2}{90} & \frac{1}{9} \end{bmatrix} \begin{bmatrix} 6 \\ 3 \end{bmatrix}$$

These two formulations are identical. Starting with $x_1(0) = x_2(0) = 0$, the solution is given in the following table:

GAUSS–SEIDEL

i	0	1	2	3	4	5	6
$x_1(i)$	0	0.60000	0.50000	0.48889	0.48765	0.48752	0.48750
$x_2(i)$	0	0.20000	0.22222	0.22469	0.22497	0.22500	0.22500

For this example, Gauss–Seidel converges in 6 iterations, compared to 10 iterations with Jacobi. ■

The convergence rate is faster with Gauss–Seidel for some **A** matrices, but faster with Jacobi for other **A** matrices. In some cases, one method diverges while the other converges. In other cases both methods diverge, as illustrated by the next example.

EXAMPLE 6.5 Divergence of Gauss–Seidel method

Using the Gauss–Seidel method with $x_1(0) = x_2(0) = 0$, solve

$$\begin{bmatrix} 5 & 10 \\ 9 & 2 \end{bmatrix} \begin{bmatrix} x_1 \\ x_2 \end{bmatrix} = \begin{bmatrix} 6 \\ 3 \end{bmatrix}$$

SOLUTION Note that these equations are the same as those in Example 6.1, except that x_1 and x_2 are interchanged. Using (6.2.9),

$$k = 1 \qquad x_1(i+1) = \frac{1}{A_{11}}[y_1 - A_{12}x_2(i)] = \frac{1}{5}[6 - 10x_2(i)]$$

$$k = 2 \qquad x_2(i+1) = \frac{1}{A_{22}}[y_2 - A_{21}x_1(i+1)] = \frac{1}{2}[3 - 9x_1(i+1)]$$

Successive calculations of x_1 and x_2 are shown in the following table:

GAUSS–SEIDEL

i	0	1	2	3	4	5
$x_1(i)$	0	1.2	9	79.2	711	6397
$x_2(i)$	0	−3.9	−39	−354.9	−3198	−28786

The unique solution by matrix inversion is

$$\begin{bmatrix} x_1 \\ x_2 \end{bmatrix} = \begin{bmatrix} 5 & 10 \\ 9 & 2 \end{bmatrix}^{-1} \begin{bmatrix} 6 \\ 3 \end{bmatrix} = \frac{-1}{80} \begin{bmatrix} 2 & -10 \\ -9 & 5 \end{bmatrix} \begin{bmatrix} 6 \\ 3 \end{bmatrix} = \begin{bmatrix} 0.225 \\ 0.4875 \end{bmatrix}$$

As shown, Gauss–Seidel does not converge to the unique solution; instead it diverges. We could show that Jacobi also diverges for this example. ■

If any diagonal element A_{kk} equals zero, then Jacobi and Gauss–Seidel are undefined, because the right-hand sides of (6.2.5) and (6.2.9) are divided by A_{kk}. Also, if any one diagonal element has too small a magnitude, these methods will diverge. In Examples 6.3 and 6.4, Jacobi and Gauss–Seidel converge, since the diagonals (10 and 9) are both large; in Example 6.5, however, the diagonals (5 and 2) are small compared to the off-diagonals, and the methods diverge.

In general, convergence of Jacobi or Gauss–Seidel can be evaluated by recognizing that (6.2.6) represents a digital filter with input $\mathbf{y}$ and output $\mathbf{x}(i)$. The z-transform of (6.2.6) may be employed to determine the filter transfer function and its poles. The output $\mathbf{x}(i)$ converges if and only if all the filter poles have magnitudes less than 1 (see Problems 6.16 and 6.17).

Rate of convergence is also established by the filter poles. Fast convergence is obtained when the magnitudes of all the poles are small. In addition, experience with specific $\mathbf{A}$ matrices has shown that more iterations are required for Jacobi and Gauss–Seidel as the dimension N increases.

Computer storage requirements for Jacobi include N^2 memory locations for the $\mathbf{A}$ matrix and $3N$ locations for the vectors $\mathbf{y}$, $\mathbf{x}(i)$, and $\mathbf{x}(i+1)$. Storage space is also required for loops, arithmetic statements, and working space to compute (6.2.5). Gauss–Seidel requires N fewer memory locations, since for (6.2.9) the new value $x_k(i+1)$ can be stored in the location of the old value $x_k(i)$.

Computer time per iteration is relatively small for Jacobi and Gauss–Seidel. Inspection of (6.2.5) or (6.2.9) shows that N^2 multiplications/divisions and $N(N-1)$ subtractions per iteration are required [one division, $(N-1)$ multiplications, and $(N-1)$ subtractions for each $k = 1, 2, \ldots, N$]. But as was the case with Gauss elimination, if the matrix is sparse (i.e., most of the elements are zero), special sparse matrix algorithms can be used to substantially decrease both the storage requirements and the computation time.

6.3

ITERATIVE SOLUTIONS TO NONLINEAR ALGEBRAIC EQUATIONS: NEWTON–RAPHSON

A set of nonlinear algebraic equations in matrix format is given by

$$\mathbf{f}(\mathbf{x}) = \begin{bmatrix} f_1(\mathbf{x}) \\ f_2(\mathbf{x}) \\ \vdots \\ f_N(\mathbf{x}) \end{bmatrix} = \mathbf{y} \tag{6.3.1}$$

where $\mathbf{y}$ and $\mathbf{x}$ are N vectors and $\mathbf{f}(\mathbf{x})$ is an N vector of functions. Given $\mathbf{y}$ and $\mathbf{f}(\mathbf{x})$, we want to solve for $\mathbf{x}$. The iterative methods described in Section 6.2 can be extended to nonlinear equations as follows. Rewriting (6.3.1),

$$\mathbf{0} = \mathbf{y} - \mathbf{f}(\mathbf{x}) \tag{6.3.2}$$

Adding $\mathbf{Dx}$ to both sides of (6.3.2), where $\mathbf{D}$ is a square $N \times N$ invertible matrix,

$$\mathbf{Dx} = \mathbf{Dx} + \mathbf{y} - \mathbf{f}(\mathbf{x}) \tag{6.3.3}$$

Premultiplying by $\mathbf{D}^{-1}$,

$$\mathbf{x} = \mathbf{x} + \mathbf{D}^{-1}[\mathbf{y} - \mathbf{f}(\mathbf{x})] \tag{6.3.4}$$

The old values $\mathbf{x}(i)$ are used on the right side of (6.3.4) to generate the new values $\mathbf{x}(i+1)$ on the left side. That is,

$$\mathbf{x}(i+1) = \mathbf{x}(i) + \mathbf{D}^{-1}\{\mathbf{y} - \mathbf{f}[\mathbf{x}(i)]\} \tag{6.3.5}$$

For linear equations, $\mathbf{f}(\mathbf{x}) = \mathbf{Ax}$ and (6.3.5) reduces to

$$\mathbf{x}(i+1) = \mathbf{x}(i) + \mathbf{D}^{-1}[\mathbf{y} - \mathbf{Ax}(i)] = \mathbf{D}^{-1}(\mathbf{D} - \mathbf{A})\mathbf{x}(i) + \mathbf{D}^{-1}\mathbf{y} \tag{6.3.6}$$

which is identical to the Jacobi and Gauss–Seidel methods of (6.2.6). For nonlinear equations, the matrix $\mathbf{D}$ in (6.3.5) must be specified.

One method for specifying $\mathbf{D}$, called *Newton–Raphson*, is based on the following Taylor series expansion of $\mathbf{f}(\mathbf{x})$ about an operating point $\mathbf{x}_0$.

$$\mathbf{y} = \mathbf{f}(\mathbf{x}_0) + \left.\frac{d\mathbf{f}}{d\mathbf{x}}\right|_{\mathbf{x}=\mathbf{x}_0} (\mathbf{x} - \mathbf{x}_0) \cdots \tag{6.3.7}$$

Neglecting the higher order terms in (6.3.7) and solving for $\mathbf{x}$,

$$\mathbf{x} = \mathbf{x}_0 + \left[\left. \frac{d\mathbf{f}}{d\mathbf{x}} \right|_{\mathbf{x}=\mathbf{x}_0} \right]^{-1} [\mathbf{y} - \mathbf{f}(\mathbf{x}_0)] \tag{6.3.8}$$

The Newton–Raphson method replaces $\mathbf{x}_0$ by the old value $\mathbf{x}(i)$ and $\mathbf{x}$ by the new value $\mathbf{x}(i+1)$ in (6.3.8). Thus,

$$\mathbf{x}(i+1) = \mathbf{x}(i) + \mathbf{J}^{-1}(i)\{\mathbf{y} - \mathbf{f}[\mathbf{x}(i)]\} \tag{6.3.9}$$

where

$$\mathbf{J}(i) = \left. \frac{d\mathbf{f}}{d\mathbf{x}} \right|_{\mathbf{x}=\mathbf{x}(i)} = \begin{bmatrix} \frac{\partial f_1}{\partial x_1} & \frac{\partial f_1}{\partial x_2} & \cdots & \frac{\partial f_1}{\partial x_N} \\ \frac{\partial f_2}{\partial x_1} & \frac{\partial f_2}{\partial x_2} & \cdots & \frac{\partial f_2}{\partial x_N} \\ \vdots & \vdots & & \vdots \\ \frac{\partial f_N}{\partial x_1} & \frac{\partial f_N}{\partial x_2} & \cdots & \frac{\partial f_N}{\partial x_N} \end{bmatrix}_{\mathbf{x}=\mathbf{x}(i)} \tag{6.3.10}$$

The $N \times N$ matrix $\mathbf{J}(i)$, whose elements are the partial derivatives shown in (6.3.10), is called the Jacobian matrix. The Newton–Raphson method is similar to extended Gauss–Seidel, except that $\mathbf{D}$ in (6.3.5) is replaced by $\mathbf{J}(i)$ in (6.3.9).

EXAMPLE 6.6 Newton–Raphson method: solution to polynomial equations

Solve the scalar equation $f(x) = y$, where $y = 9$ and $f(x) = x^2$. Starting with $x(0) = 1$, use (a) Newton–Raphson and (b) extended Gauss–Seidel with $D = 3$ until (6.2.2) is satisfied for $\varepsilon = 10^{-4}$. Compare the two methods.

SOLUTION

a. Using (6.3.10) with $f(x) = x^2$,

$$\mathbf{J}(i) = \left. \frac{d}{dx}(x^2) \right|_{x=x(i)} = \left. 2x \right|_{x=x(i)} = 2x(i)$$

Using $\mathbf{J}(i)$ in (6.3.9),

$$x(i+1) = x(i) + \frac{1}{2x(i)}[9 - x^2(i)]$$

Starting with $x(0) = 1$, successive calculations of the Newton–Raphson equation are shown in the following table:

NEWTON–RAPHSON

i	0	1	2	3	4	5
$x(i)$	1	5.00000	3.40000	3.02353	3.00009	3.00000

b. Using (6.3.5) with D = 3, the Gauss–Seidel method is

$$x(i+1) = x(i) + \tfrac{1}{3}[9 - x^2(i)]$$

The corresponding Gauss–Seidel calculations are as follows:

GAUSS–SEIDEL (D = 3)

i	0	1	2	3	4	5	6
$x(i)$	1	3.66667	2.18519	3.59351	2.28908	3.54245	2.35945

As shown, Gauss–Seidel oscillates about the solution, slowly converging, whereas Newton–Raphson converges in five iterations to the solution $x = 3$. Note that if $x(0)$ is negative, Newton–Raphson converges to the negative solution $x = -3$. Also, it is assumed that the matrix inverse $\mathbf{J}^{-1}$ exists. Thus the initial value $x(0) = 0$ should be avoided for this example. ■

EXAMPLE 6.7 Newton–Raphson method: solution to nonlinear algebraic equations

Solve

$$\begin{bmatrix} x_1 + x_2 \\ x_1 x_2 \end{bmatrix} = \begin{bmatrix} 15 \\ 50 \end{bmatrix} \qquad \mathbf{x}(0) = \begin{bmatrix} 4 \\ 9 \end{bmatrix}$$

Use the Newton–Raphson method starting with the above $\mathbf{x}(0)$ and continue until (6.2.2) is satisfied with $\varepsilon = 10^{-4}$.

SOLUTION Using (6.3.10) with $f_1 = (x_1 + x_2)$ and $f_2 = x_1 x_2$,

$$\mathbf{J}(i)^{-1} = \begin{bmatrix} \dfrac{\partial f_1}{\partial x_1} & \dfrac{\partial f_1}{\partial x_2} \\[2ex] \dfrac{\partial f_2}{\partial x_1} & \dfrac{\partial f_2}{\partial x_2} \end{bmatrix}^{-1}_{\mathbf{x}=\mathbf{x}(i)} = \begin{bmatrix} 1 & 1 \\ x_2(i) & x_1(i) \end{bmatrix}^{-1} = \frac{\begin{bmatrix} x_1(i) & -1 \\ -x_2(i) & 1 \end{bmatrix}}{x_1(i) - x_2(i)}$$

Using $\mathbf{J}(i)^{-1}$ in (6.3.9),

$$\begin{bmatrix} x_1(i+1) \\ x_2(i+1) \end{bmatrix} = \begin{bmatrix} x_1(i) \\ x_2(i) \end{bmatrix} + \frac{\begin{bmatrix} x_1(i) & -1 \\ -x_2(i) & 1 \end{bmatrix}}{x_1(i) - x_2(i)} \begin{bmatrix} 15 - x_1(i) - x_2(i) \\ 50 - x_1(i) x_2(i) \end{bmatrix}$$

Writing the preceding as two separate equations,

$$x_1(i+1) = x_1(i) + \frac{x_1(i)[15 - x_1(i) - x_2(i)] - [50 - x_1(i)x_2(i)]}{x_1(i) - x_2(i)}$$

$$x_2(i+1) = x_2(i) + \frac{-x_2(i)[15 - x_1(i) - x_2(i)] + [50 - x_1(i)x_2(i)]}{x_1(i) - x_2(i)}$$

Successive calculations of these equations are shown in the following table:

NEWTON–RAPHSON

i	0	1	2	3	4
$x_1(i)$	4	5.20000	4.99130	4.99998	5.00000
$x_2(i)$	9	9.80000	10.00870	10.00002	10.00000

Newton–Raphson converges in four iterations for this example. ■

Equation (6.3.9) contains the matrix inverse $\mathbf{J}^{-1}$. Instead of computing $\mathbf{J}^{-1}$, (6.3.9) can be rewritten as follows:

$$\mathbf{J}(i)\Delta\mathbf{x}(i) = \Delta\mathbf{y}(i) \tag{6.3.11}$$

where

$$\Delta\mathbf{x}(i) = \mathbf{x}(i+1) - \mathbf{x}(i) \tag{6.3.12}$$

and

$$\Delta\mathbf{y}(i) = \mathbf{y} - \mathbf{f}[\mathbf{x}(i)] \tag{6.3.13}$$

Then, during each iteration, the following four steps are completed:

STEP 1 Compute $\Delta\mathbf{y}(i)$ from (6.3.13).

STEP 2 Compute $\mathbf{J}(i)$ from (6.3.10).

STEP 3 Using Gauss elimination and back substitution, solve (6.3.11) for $\Delta\mathbf{x}(i)$.

STEP 4 Compute $\mathbf{x}(i+1)$ from (6.3.12).

EXAMPLE 6.8 Newton–Raphson method in four steps

Complete the above four steps for the first iteration of Example 6.7.

SOLUTION

STEP 1 $\Delta\mathbf{y}(0) = \mathbf{y} - \mathbf{f}[\mathbf{x}(0)] = \begin{bmatrix} 15 \\ 50 \end{bmatrix} - \begin{bmatrix} 4+9 \\ (4)(9) \end{bmatrix} = \begin{bmatrix} 2 \\ 14 \end{bmatrix}$

STEP 2 $\mathbf{J}(0) = \left[\begin{array}{c|c} 1 & 1 \\ \hline x_2(0) & x_1(0) \end{array}\right] = \left[\begin{array}{c|c} 1 & 1 \\ \hline 9 & 4 \end{array}\right]$

STEP 3 Using $\Delta\mathbf{y}(0)$ and $\mathbf{J}(0)$, (6.3.11) becomes

$$\left[\begin{array}{c|c} 1 & 1 \\ \hline 9 & 4 \end{array}\right]\begin{bmatrix} \Delta x_1(0) \\ \Delta x_2(0) \end{bmatrix} = \begin{bmatrix} 2 \\ 14 \end{bmatrix}$$

Using Gauss elimination, subtract $\mathbf{J}_{21}/\mathbf{J}_{11} = 9/1 = 9$ times the first equation from the second equation, giving

$$\left[\begin{array}{c|c} 1 & 1 \\ \hline 0 & -5 \end{array}\right]\begin{bmatrix} \Delta x_1(0) \\ \Delta x_2(0) \end{bmatrix} = \begin{bmatrix} 2 \\ -4 \end{bmatrix}$$

Solving by back substitution,

$$\Delta x_2(0) = \frac{-4}{-5} = 0.8$$

$$\Delta x_1(0) = 2 - 0.8 = 1.2$$

STEP 4 $\mathbf{x}(1) = \mathbf{x}(0) + \Delta\mathbf{x}(0) = \begin{bmatrix} 4 \\ 9 \end{bmatrix} + \begin{bmatrix} 1.2 \\ 0.8 \end{bmatrix} = \begin{bmatrix} 5.2 \\ 9.8 \end{bmatrix}$

This is the same as computed in Example 6.7. ■

Experience from power-flow studies has shown that Newton–Raphson converges in many cases where Jacobi and Gauss–Seidel diverge. Furthermore, the number of iterations required for convergence is independent of the dimension N for Newton–Raphson, but increases with N for Jacobi and Gauss–Seidel. Most Newton–Raphson power-flow problems converge in fewer than 10 iterations [1].

6.4 THE POWER-FLOW PROBLEM

The power-flow problem is the computation of voltage magnitude and phase angle at each bus in a power system under balanced three-phase steady-state conditions. As a by-product of this calculation, real and reactive power flows in equipment such as transmission lines and transformers, as well as equipment losses, can be computed.

The starting point for a power-flow problem is a single-line diagram of the power system, from which the input data for computer solutions can be obtained. Input data consist of bus data, transmission line data, and transformer data.

As shown in Figure 6.1, the following four variables are associated with each bus k: voltage magnitude V_k, phase angle δ_k, net real power P_k, and reactive power Q_k supplied to the bus. At each bus, two of these variables are specified as input data, and the other two are unknowns to be computed by

FIGURE 6.1

Bus variables V_k, δ_k, P_k, and Q_k

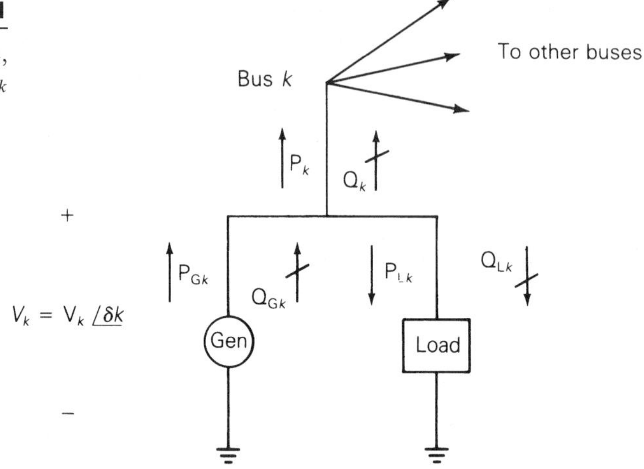

the power-flow program. For convenience, the power delivered to bus k in Figure 6.1 is separated into generator and load terms. That is,

$$P_k = P_{Gk} - P_{Lk}$$

$$Q_k = Q_{Gk} - Q_{Lk} \tag{6.4.1}$$

Each bus k is categorized into one of the following three bus types:

1. Swing bus (or slack bus)—There is only one swing bus, which for convenience is numbered bus 1 in this text. The swing bus is a reference bus for which $V_1\underline{/\delta_1}$, typically $1.0\underline{/0^\circ}$ per unit, is input data. The power-flow program computes P_1 and Q_1.

2. Load (PQ) bus—P_k and Q_k are input data. The power-flow program computes V_k and δ_k. Most buses in a typical power-flow program are load buses.

3. Voltage controlled (PV) bus—P_k and V_k are input data. The power-flow program computes Q_k and δ_k. Examples are buses to which generators, switched shunt capacitors, or static var systems are connected. Maximum and minimum var limits $Q_{Gk\max}$ and $Q_{Gk\min}$ that this equipment can supply are also input data. If an upper or lower reactive power limit is reached, then the reactive power output of the generator is held at the limit, and the bus is modeled as a PQ bus. Another example is a bus to which a tap-changing transformer is connected; the power-flow program then computes the tap setting.

Note that when bus k is a load bus with no generation, $P_k = -P_{Lk}$ is negative; that is, the real power supplied to bus k in Figure 6.1 is negative. If the load is inductive, $Q_k = -Q_{Lk}$ is negative.

Transmission lines are represented by the equivalent π circuit, shown in Figure 5.7. Transformers are also represented by equivalent circuits, as shown in Figure 3.9 for a two-winding transformer, Figure 3.20 for a three-winding transformer, or Figure 3.25 for a tap-changing transformer.

Input data for each transmission line include the per-unit equivalent π circuit series impedance Z' and shunt admittance Y', the two buses to which the line is connected, and maximum MVA rating. Similarly, input data for each transformer include per-unit winding impedances Z, the per-unit exciting branch admittance Y, the buses to which the windings are connected, and maximum MVA ratings. Input data for tap-changing transformers also include maximum tap settings.

The bus admittance matrix $\boldsymbol{Y}_{\text{bus}}$ can be constructed from the line and transformer input data. From (2.4.3) and (2.4.4), the elements of $\boldsymbol{Y}_{\text{bus}}$ are:

$$\begin{aligned} &\text{Diagonal elements:} \quad Y_{kk} = \text{sum of admittances connected to bus } k \\ &\text{Off-diagonal elements:} \quad Y_{kn} = -(\text{sum of admittances connected between buses } k \text{ and } n) \quad k \neq n \end{aligned} \tag{6.4.2}$$

EXAMPLE 6.9 Power-flow input data and $\boldsymbol{Y}_{\text{bus}}$

Figure 6.2 shows a single-line diagram of a five-bus power system. Input data are given in Tables 6.1, 6.2, and 6.3. As shown in Table 6.1, bus 1, to which a generator is connected, is the swing bus. Bus 3, to which a generator and a load are connected, is a voltage-controlled bus. Buses 2, 4, and 5 are load buses. Note that the loads at buses 2 and 3 are inductive since $Q_2 = -Q_{L2} = -2.8$ and $-Q_{L3} = -0.4$ are negative.

For each bus k, determine which of the variables V_k, δ_k, P_k, and Q_k are input data and which are unknowns. Also, compute the elements of the second row of $\boldsymbol{Y}_{\text{bus}}$.

SOLUTION The input data and unknowns are listed in Table 6.4. For bus 1, the swing bus, P_1 and Q_1 are unknowns. For bus 3, a voltage-controlled bus,

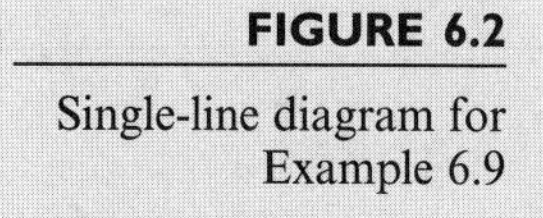
FIGURE 6.2 Single-line diagram for Example 6.9

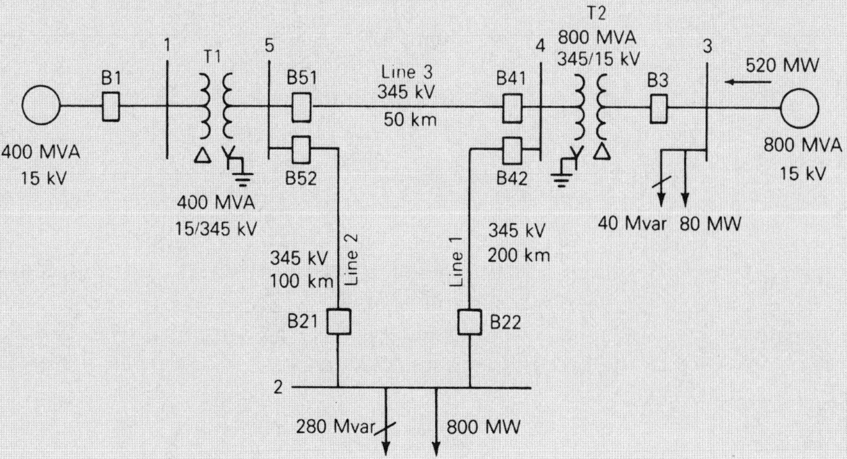

TABLE 6.1

Bus input data for Example 6.9*

Bus	Type	V per unit	δ degrees	P_G per unit	Q_G per unit	P_L per unit	Q_L per unit	Q_{Gmax} per unit	Q_{Gmin} per unit
1	Swing	1.0	0	—	—	0	0	—	—
2	Load	—	—	0	0	8.0	2.8	—	—
3	Constant voltage	1.05	—	5.2	—	0.8	0.4	4.0	−2.8
4	Load	—	—	0	0	0	0	—	—
5	Load	—	—	0	0	0	0	—	—

* $S_{base} = 100$ MVA, $V_{base} = 15$ kV at buses 1, 3, and 345 kV at buses 2, 4, 5

TABLE 6.2

Line input data for Example 6.9

Bus-to-Bus	R′ per unit	X′ per unit	G′ per unit	B′ per unit	Maximum MVA per unit
2–4	0.0090	0.100	0	1.72	12.0
2–5	0.0045	0.050	0	0.88	12.0
4–5	0.00225	0.025	0	0.44	12.0

TABLE 6.3

Transformer input data for Example 6.9

Bus-to-Bus	R per unit	X per unit	G_c per unit	B_m per unit	Maximum MVA per unit	Maximum TAP Setting per unit
1–5	0.00150	0.02	0	0	6.0	—
3–4	0.00075	0.01	0	0	10.0	—

TABLE 6.4

Input data and unknowns for Example 6.9

Bus	Input Data	Unknowns
1	$V_1 = 1.0, \delta_1 = 0$	P_1, Q_1
2	$P_2 = P_{G2} - P_{L2} = -8$ $Q_2 = Q_{G2} - Q_{L2} = -2.8$	V_2, δ_2
3	$V_3 = 1.05$ $P_3 = P_{G3} - P_{L3} = 4.4$	Q_3, δ_3
4	$P_4 = 0, Q_4 = 0$	V_4, δ_4
5	$P_5 = 0, Q_5 = 0$	V_5, δ_5

Q_3 and δ_3 are unknowns. For buses 2, 4, and 5, load buses, V_2, V_4, V_5 and δ_2, δ_4, δ_5 are unknowns.

The elements of Y_{bus} are computed from (6.4.2). Since buses 1 and 3 are not directly connected to bus 2,

$$Y_{21} = Y_{23} = 0$$

Using (6.4.2),

$$Y_{24} = \frac{-1}{R'_{24} + jX'_{24}} = \frac{-1}{0.009 + j0.1} = -0.89276 + j9.91964 \quad \text{per unit}$$

$$= 9.95972\underline{/95.143^\circ} \quad \text{per unit}$$

$$Y_{25} = \frac{-1}{R'_{25} + jX'_{25}} = \frac{-1}{0.0045 + j0.05} = -1.78552 + j19.83932 \quad \text{per unit}$$

$$= 19.9195\underline{/95.143^\circ} \quad \text{per unit}$$

$$Y_{22} = \frac{1}{R'_{24} + jX'_{24}} + \frac{1}{R'_{25} + jX'_{25}} + j\frac{B'_{24}}{2} + j\frac{B'_{25}}{2}$$

$$= (0.89276 - j9.91964) + (1.78552 - j19.83932) + j\frac{1.72}{2} + j\frac{0.88}{2}$$

$$= 2.67828 - j28.4590 = 28.5847\underline{/-84.624^\circ} \quad \text{per unit}$$

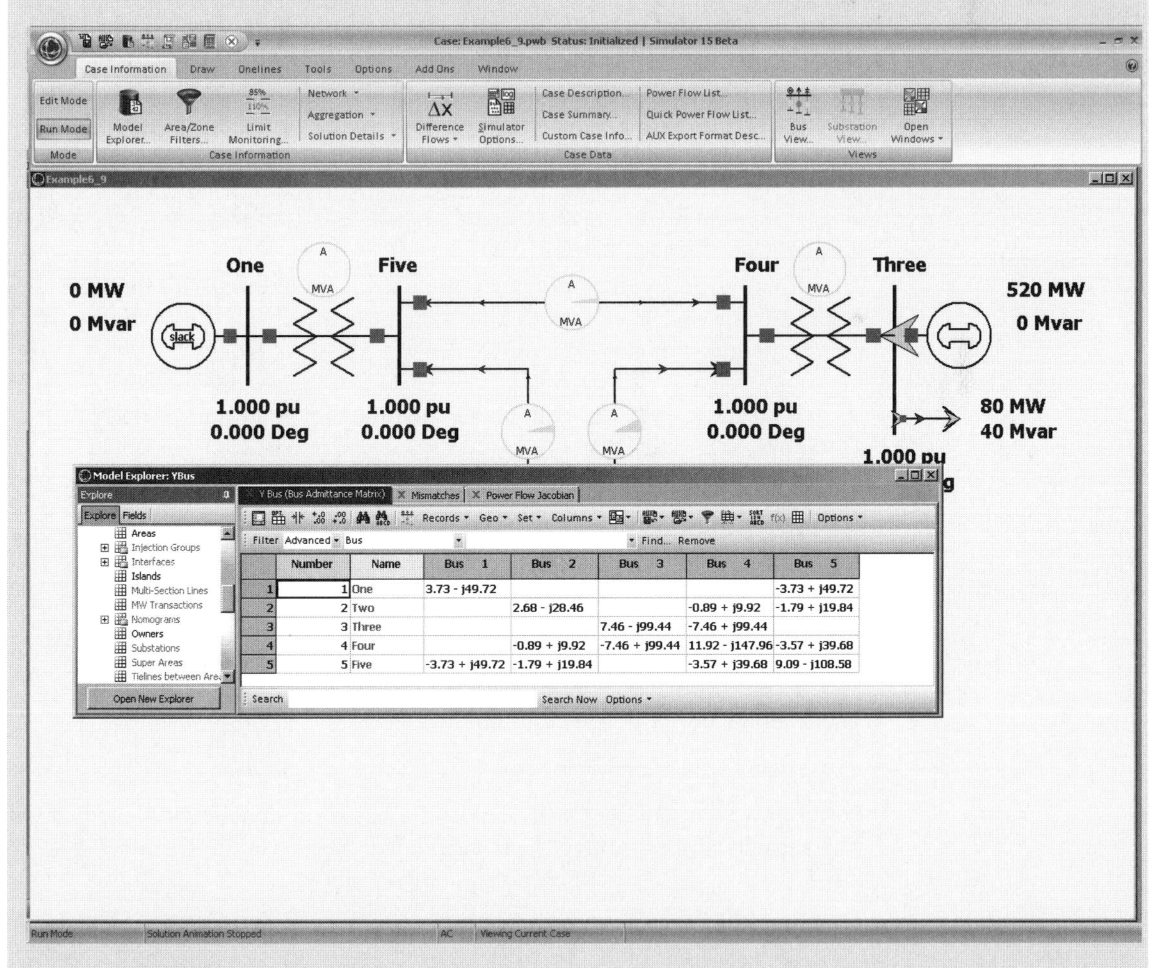

FIGURE 6.3 Screen for Example 6.9

where half of the shunt admittance of each line connected to bus 2 is included in Y_{22} (the other half is located at the other ends of these lines).

This five-bus power system is modeled in PowerWorld Simulator case Example 6_9 (see Figure 6.3). To view the input data, first click on the **Edit Mode** button (on the far left-hand side of the ribbon) to switch into the Edit mode (the Edit mode is used for modifying system parameters). Then by selecting the **Case Information** tab you can view tabular displays showing the various parameters for the system. For example, use **Network, Buses** to view the parameters for each bus, and **Network, Lines and Transformers** to view the parameters for the transmission lines and transformers. Fields shown in blue can be directly changed simply by typing over them, and those shown in green can be toggled by clicking on them. Note that the values shown on these displays match the values from Tables 6.1 to 6.3, except the power values are shown in actual MW/Mvar units.

The elements of Y_{bus} can also be displayed by selecting **Solution Details, $\mathbf{Y_{bus}}$**. Since the Y_{bus} entries are derived from other system parameters, they cannot be changed directly. Notice that several of the entries are blank, indicating that there is no line directly connecting these two buses (a blank entry is equivalent to zero). For larger networks most of the elements of the Y_{bus} are zero since any single bus usually only has a few incident lines (such sparse matrices are considered in Section 6.8). The elements of the Y_{bus} can be saved in a Matlab compatible format by first right-clicking within the Y_{bus} matrix to display the local menu, and then selecting **Save $\mathbf{Y_{bus}}$ in Matlab Format** from the local menu.

Finally, notice that no flows are shown on the one-line because the nonlinear power-flow equations have not yet been solved. We cover the solution of these equations next. ■

Using $\boldsymbol{Y}_{\text{bus}}$, we can write nodal equations for a power system network, as follows:

$$\boldsymbol{I} = \boldsymbol{Y}_{\text{bus}} \boldsymbol{V} \tag{6.4.3}$$

where I is the N vector of source currents injected into each bus and V is the N vector of bus voltages. For bus k, the kth equation in (6.4.3) is

$$I_k = \sum_{n=1}^{N} Y_{kn} V_n \tag{6.4.4}$$

The complex power delivered to bus k is

$$S_k = \text{P}_k + j\text{Q}_k = V_k I_k^* \tag{6.4.5}$$

Power-flow solutions by Gauss–Seidel are based on nodal equations, (6.4.4), where each current source I_k is calculated from (6.4.5). Using (6.4.4) in (6.4.5),

$$\text{P}_k + j\text{Q}_k = V_k \left[\sum_{n=1}^{N} Y_{kn} V_n \right]^* \qquad k = 1, 2, \ldots, N \tag{6.4.6}$$

With the following notation,

$$V_n = \mathrm{V}_n e^{j\delta_n} \tag{6.4.7}$$

$$Y_{kn} = \mathrm{Y}_{kn} e^{j\theta_{kn}} = G_{kn} + jB_{kn} \qquad k, n = 1, 2, \ldots, N \tag{6.4.8}$$

(6.4.6) becomes

$$\mathrm{P}_k + j\mathrm{Q}_k = \mathrm{V}_k \sum_{n=1}^{N} \mathrm{Y}_{kn} \mathrm{V}_n e^{j(\delta_k - \delta_n - \theta_{kn})} \tag{6.4.9}$$

Taking the real and imaginary parts of (6.4.9), we can write the power balance equations as either

$$\mathrm{P}_k = \mathrm{V}_k \sum_{n=1}^{N} \mathrm{Y}_{kn} \mathrm{V}_n \cos(\delta_k - \delta_n - \theta_{kn}) \tag{6.4.10}$$

$$\mathrm{Q}_k = \mathrm{V}_k \sum_{n=1}^{N} \mathrm{Y}_{kn} \mathrm{V}_n \sin(\delta_k - \delta_n - \theta_{kn}) \qquad k = 1, 2, \ldots, N \tag{6.4.11}$$

or when the Y_{kn} is expressed in rectangular coordinates by

$$\mathrm{P}_K = \mathrm{V}_K \sum_{n=1}^{N} V_n [G_{kn} \cos(\delta_k - \delta_n) + B_{kn} \sin(\delta_k - \delta_n)] \tag{6.4.12}$$

$$\mathrm{Q}_K = \mathrm{V}_K \sum_{n=1}^{N} V_n [G_{kn} \sin(\delta_k - \delta_n) - B_{kn} \cos(\delta_k - \delta_n)] \qquad k = 1, 2, \ldots, N \tag{6.4.13}$$

Power-flow solutions by Newton–Raphson are based on the nonlinear power-flow equations given by (6.4.10) and (6.4.11) [or alternatively by (6.4.12) and (6.4.13)].

6.5 POWER-FLOW SOLUTION BY GAUSS–SEIDEL

Nodal equations $\boldsymbol{I} = \boldsymbol{Y}_{\text{bus}} \boldsymbol{V}$ are a set of linear equations analogous to $\boldsymbol{y} = \boldsymbol{Ax}$, solved in Section 6.2 using Gauss–Seidel. Since power-flow bus data consists of P_k and Q_k for load buses or P_k and V_k for voltage-controlled buses, nodal equations do not directly fit the linear equation format; the current source vector $\boldsymbol{I}$ is unknown and the equations are actually nonlinear. For each load bus, I_k can be calculated from (6.4.5), giving

$$I_k = \frac{\mathrm{P}_k - j\mathrm{Q}_k}{V_k^*} \tag{6.5.1}$$

Applying the Gauss–Seidel method, (6.2.9), to the nodal equations, with I_k given above, we obtain

$$V_k(i+1) = \frac{1}{Y_{kk}}\left[\frac{P_k - jQ_k}{V_k^*(i)} - \sum_{n=1}^{k-1} Y_{kn}V_n(i+1) - \sum_{n=k+1}^{N} Y_{kn}V_n(i)\right] \quad (6.5.2)$$

Equation (6.5.2) can be applied twice during each iteration for load buses, first using $V_k^*(i)$, then replacing $V_k^*(i)$, by $V_k^*(i+1)$ on the right side of (6.5.2).

For a voltage-controlled bus, Q_k is unknown, but can be calculated from (6.4.11), giving

$$Q_k = V_k(i)\sum_{n=1}^{N} Y_{kn}V_n(i)\sin[\delta_k(i) - \delta_n(i) - \theta_{kn}] \quad (6.5.3)$$

Also,

$$Q_{Gk} = Q_k + Q_{Lk}$$

If the calculated value of Q_{Gk} does not exceed its limits, then Q_k is used in (6.5.2) to calculate $V_k(i+1) = V_k(i+1)\underline{/\delta_k(i+1)}$. Then the magnitude $V_k(i+1)$ is changed to V_k, which is input data for the voltage-controlled bus. Thus we use (6.5.2) to compute only the angle $\delta_k(i+1)$ for voltage-controlled buses.

If the calculated value exceeds its limit $Q_{Gk\max}$ or $Q_{Gk\min}$ during any iteration, then the bus type is changed from a voltage-controlled bus to a load bus, with Q_{Gk} set to its limit value. Under this condition, the voltage-controlling device (capacitor bank, static var system, and so on) is not capable of maintaining V_k as specified by the input data. The power-flow program then calculates a new value of V_k.

For the swing bus, denoted bus 1, V_1 and δ_1 are input data. As such, no iterations are required for bus 1. After the iteration process has converged, one pass through (6.4.10) and (6.4.11) can be made to compute P_1 and Q_1.

EXAMPLE 6.10 Power-flow solution by Gauss–Seidel

For the power system of Example 6.9, use Gauss–Seidel to calculate $V_2(1)$, the phasor voltage at bus 2 after the first iteration. Use zero initial phase angles and 1.0 per-unit initial voltage magnitudes (except at bus 3, where $V_3 = 1.05$) to start the iteration procedure.

SOLUTION Bus 2 is a load bus. Using the input data and bus admittance values from Example 6.9 in (6.5.2),

$$V_2(1) = \frac{1}{Y_{22}}\left\{\frac{P_2 - jQ_2}{V_2^*(0)} - [Y_{21}V_1(1) + Y_{23}V_3(0) + Y_{24}V_4(0) + Y_{25}V_5(0)]\right\}$$

$$= \frac{1}{28.5847\underline{/-84.624^\circ}}\left\{\frac{-8 - j(-2.8)}{1.0\underline{/0^\circ}}\right.$$

$$\left. - [(-1.78552 + j19.83932)(1.0) + (-0.89276 + j9.91964)(1.0)]\right\}$$

$$= \frac{(-8 + j2.8) - (-2.67828 + j29.7589)}{28.5847\underline{/-84.624^\circ}}$$

$$= 0.96132\underline{/-16.543^\circ} \quad \text{per unit}$$

Next, the above value is used in (6.5.2) to recalculate $V_2(1)$:

$$V_2(1) = \frac{1}{28.5847\underline{/-84.624^\circ}}\left\{\frac{-8 + j2.8}{0.96132\underline{/16.543^\circ}}\right.$$

$$\left. - [-2.67828 + j29.75829]\right\}$$

$$= \frac{-4.4698 - j24.5973}{28.5847\underline{/-84.624^\circ}} = 0.87460\underline{/-15.675^\circ} \quad \text{per unit}$$

Computations are next performed at buses 3, 4, and 5 to complete the first Gauss–Seidel iteration.

To see the complete convergence of this case, open PowerWorld Simulator case Example 6_10. By default, PowerWorld Simulator uses the Newton–Raphson method described in the next section. However, the case can be solved with the Gauss–Seidel approach by selecting **Tools, Solve, Gauss–Seidel Power Flow**. To avoid getting stuck in an infinite loop if a case does not converge, PowerWorld Simulator places a limit on the maximum number of iterations. Usually for a Gauss–Seidel procedure this number is quite high, perhaps equal to 100 iterations. However, in this example to demonstrate the convergence characteristics of the Gauss–Seidel method it has been set to a single iteration, allowing the voltages to be viewed after each iteration. To step through the solution one iteration at a time, just repeatedly select **Tools, Solve, Gauss–Seidel Power Flow**.

A common stopping criteria for the Gauss–Seidel is to use the scaled difference in the voltage from one iteration to the next (6.2.2). When this difference is below a specified convergence tolerance ε for each bus, the problem is considered solved. An alternative approach, implemented in PowerWorld Simulator, is to examine the real and reactive mismatch equations, defined as the difference between the right- and left-hand sides of (6.4.10) and (6.4.11). PowerWorld Simulator continues iterating until all the bus mismatches are below an MVA (or kVA) tolerance. When single-stepping through the solution, the bus mismatches can be viewed after each iteration on the **Case**

Information, Mismatches display. The solution mismatch tolerance can be changed on the Power Flow Solution page of the PowerWorld Simulator Options dialog (select **Tools, Simulator Options**, then select the **Power Flow Solution** category to view this dialog); the maximum number of iterations can also be changed from this page. A typical convergence tolerance is about 0.5 MVA. ■

6.6 POWER-FLOW SOLUTION BY NEWTON–RAPHSON

Equations (6.4.10) and (6.4.11) are analogous to the nonlinear equation $\mathbf{y} = \mathbf{f}(\mathbf{x})$, solved in Section 6.3 by Newton–Raphson. We define the $\mathbf{x}$, $\mathbf{y}$, and $\mathbf{f}$ vectors for the power-flow problem as

$$\mathbf{x} = \begin{bmatrix} \boldsymbol{\delta} \\ \mathbf{V} \end{bmatrix} = \begin{bmatrix} \delta_2 \\ \vdots \\ \delta_N \\ V_2 \\ \vdots \\ V_N \end{bmatrix}; \quad \mathbf{y} = \begin{bmatrix} \mathbf{P} \\ \mathbf{Q} \end{bmatrix} = \begin{bmatrix} P_2 \\ \vdots \\ P_N \\ Q_2 \\ \vdots \\ Q_N \end{bmatrix};$$

$$\mathbf{f}(\mathbf{x}) = \begin{bmatrix} \mathbf{P}(\mathbf{x}) \\ \mathbf{Q}(\mathbf{x}) \end{bmatrix} = \begin{bmatrix} P_2(\mathbf{x}) \\ \vdots \\ P_N(\mathbf{x}) \\ Q_2(\mathbf{x}) \\ \vdots \\ Q_N(\mathbf{x}) \end{bmatrix} \tag{6.6.1}$$

where all V, P, and Q terms are in per-unit and δ terms are in radians. The swing bus variables δ_1 and V_1 are omitted from (6.6.1), since they are already known. Equations (6.4.10) and (6.4.11) then have the following form:

$$y_k = P_k = P_k(\mathbf{x}) = V_k \sum_{n=1}^{N} Y_{kn} V_n \cos(\delta_k - \delta_n - \theta_{kn}) \tag{6.6.2}$$

$$y_{k+N} = Q_k = Q_k(\mathbf{x}) = V_k \sum_{n=1}^{N} Y_{kn} V_n \sin(\delta_k - \delta_n - \theta_{kn})$$

$$k = 2, 3, \ldots, N \tag{6.6.3}$$

TABLE 6.5 Elements of the Jacobian matrix

$n \neq k$	$J1_{kn} = \dfrac{\partial P_k}{\partial \delta_n} = V_k Y_{kn} V_n \sin(\delta_k - \delta_n - \theta_{kn})$ $J2_{kn} = \dfrac{\partial P_k}{\partial V_n} = V_k Y_{kn} \cos(\delta_k - \delta_n - \theta_{kn})$ $J3_{kn} = \dfrac{\partial Q_k}{\partial \delta_n} = -V_k Y_{kn} V_n \cos(\delta_k - \delta_n - \theta_{kn})$ $J4_{kn} = \dfrac{\partial Q_k}{\partial V_n} = V_k Y_{kn} \sin(\delta_k - \delta_n - \theta_{kn})$
$n = k$	$J1_{kk} = \dfrac{\partial P_k}{\partial \delta_k} = -V_k \sum_{n=1, n\neq k}^{N} Y_{kn} V_n \sin(\delta_k - \delta_n - \theta_{kn})$ $J2_{kk} = \dfrac{\partial P_k}{\partial V_k} = V_k Y_{kk} \cos\theta_{kk} + \sum_{n=1}^{N} Y_{kn} V_n \cos(\delta_k - \delta_n - \theta_{kn})$ $J3_{kk} = \dfrac{\partial Q_k}{\partial \delta_k} = V_k \sum_{n=1, n\neq k}^{N} Y_{kn} V_n \cos(\delta_k - \delta_n - \theta_{kn})$ $J4_{kk} = \dfrac{\partial Q_k}{\partial V_k} = -V_k Y_{kk} \sin\theta_{kk} + \sum_{n=1}^{N} Y_{kn} V_n \sin(\delta_k - \delta_n - \theta_{kn})$
	$k, n = 2, 3, \ldots, N$

The Jacobian matrix of (6.3.10) has the form

$$
\mathbf{J} = \left[\begin{array}{ccc|ccc}
\multicolumn{3}{c|}{\mathbf{J1}} & \multicolumn{3}{c}{\mathbf{J2}} \\
\dfrac{\partial P_2}{\partial \delta_2} & \cdots & \dfrac{\partial P_2}{\partial \delta_N} & \dfrac{\partial P_2}{\partial V_2} & \cdots & \dfrac{\partial P_2}{\partial V_N} \\
\vdots & & & \vdots & & \\
\dfrac{\partial P_N}{\partial \delta_2} & \cdots & \dfrac{\partial P_N}{\partial \delta_N} & \dfrac{\partial P_N}{\partial V_2} & \cdots & \dfrac{\partial P_N}{\partial V_N} \\
\hline
\dfrac{\partial Q_2}{\partial \delta_2} & \cdots & \dfrac{\partial Q_2}{\partial \delta_N} & \dfrac{\partial Q_2}{\partial V_2} & \cdots & \dfrac{\partial Q_2}{\partial V_N} \\
\vdots & & & \vdots & & \\
\dfrac{\partial Q_N}{\partial \delta_2} & \cdots & \dfrac{\partial Q_N}{\partial \delta_N} & \dfrac{\partial Q_N}{\partial V_2} & \cdots & \dfrac{\partial Q_N}{\partial V_N} \\
\multicolumn{3}{c|}{\mathbf{J3}} & \multicolumn{3}{c}{\mathbf{J4}}
\end{array}\right] \tag{6.6.4}
$$

Equation (6.6.4) is partitioned into four blocks. The partial derivatives in each block, derived from (6.6.2) and (6.6.3), are given in Table 6.5.

We now apply to the power-flow problem the four Newton–Raphson steps outlined in Section 6.3, starting with $\mathbf{x}(i) = \begin{bmatrix} \boldsymbol{\delta}(i) \\ \mathbf{V}(i) \end{bmatrix}$ at the ith iteration.

STEP 1 Use (6.6.2) and (6.6.3) to compute

$$\Delta \mathbf{y}(i) = \begin{bmatrix} \Delta \mathbf{P}(i) \\ \Delta \mathbf{Q}(i) \end{bmatrix} = \begin{bmatrix} \mathbf{P} - \mathbf{P}[\mathbf{x}(i)] \\ \mathbf{Q} - \mathbf{Q}[\mathbf{x}(i)] \end{bmatrix} \tag{6.6.5}$$

STEP 2 Use the equations in Table 6.5 to calculate the Jacobian matrix.

STEP 3 Use Gauss elimination and back substitution to solve

$$\left[\begin{array}{c|c} \mathbf{J1}(i) & \mathbf{J2}(i) \\ \hline \mathbf{J3}(i) & \mathbf{J4}(i) \end{array}\right] \begin{bmatrix} \Delta \boldsymbol{\delta}(i) \\ \Delta \mathbf{V}(i) \end{bmatrix} = \begin{bmatrix} \Delta \mathbf{P}(i) \\ \Delta \mathbf{Q}(i) \end{bmatrix} \tag{6.6.6}$$

STEP 4 Compute

$$\mathbf{x}(i+1) = \begin{bmatrix} \boldsymbol{\delta}(i+1) \\ \mathbf{V}(i+1) \end{bmatrix} = \begin{bmatrix} \boldsymbol{\delta}(i) \\ \mathbf{V}(i) \end{bmatrix} + \begin{bmatrix} \Delta \boldsymbol{\delta}(i) \\ \Delta \mathbf{V}(i) \end{bmatrix} \tag{6.6.7}$$

Starting with initial value $\mathbf{x}(0)$, the procedure continues until convergence is obtained or until the number of iterations exceeds a specified maximum. Convergence criteria are often based on $\Delta \mathbf{y}(i)$ (called *power mismatches*) rather than on $\Delta \mathbf{x}(i)$ (phase angle and voltage magnitude mismatches).

For each voltage-controlled bus, the magnitude V_k is already known, and the function $Q_k(\mathbf{x})$ is not needed. Therefore, we could omit V_k from the $\mathbf{x}$ vector and Q_k from the $\mathbf{y}$ vector. We could also omit from the Jacobian matrix the column corresponding to partial derivatives with respect to V_k and the row corresponding to partial derivatives of $Q_k(\mathbf{x})$. Alternatively, rows and corresponding columns for voltage-controlled buses can be retained in the Jacobian matrix. Then during each iteration, the voltage magnitude $V_k(i+1)$ of each voltage-controlled bus is reset to V_k, which is input data for that bus.

At the end of each iteration, we compute $Q_k(\mathbf{x})$ from (6.6.3) and $Q_{Gk} = Q_k(\mathbf{x}) + Q_{Lk}$ for each voltage-controlled bus. If the computed value of Q_{Gk} exceeds its limits, then the bus type is changed to a load bus with Q_{Gk} set to its limit value. The power-flow program also computes a new value for V_k.

EXAMPLE 6.11 Jacobian matrix and power-flow solution by Newton–Raphson

Determine the dimension of the Jacobian matrix for the power system in Example 6.9. Also calculate $\Delta P_2(0)$ in Step 1 and $J1_{24}(0)$ in Step 2 of the first Newton–Raphson iteration. Assume zero initial phase angles and 1.0 per-unit initial voltage magnitudes (except $V_3 = 1.05$).

SOLUTION Since there are $N = 5$ buses for Example 6.9, (6.6.2) and (6.6.3) constitute $2(N-1) = 8$ equations, for which $\mathbf{J}(i)$ has dimension 8×8.

However, there is one voltage-controlled bus, bus 3. Therefore, V_3 and the equation for $Q_3(\mathbf{x})$ could be eliminated, with $\mathbf{J}(i)$ reduced to a 7×7 matrix.

From Step 1 and (6.6.2),

$$\begin{aligned}
\Delta P_2(0) = P_2 - P_2(\mathbf{x}) = P_2 - V_2(0)\{&Y_{21}V_1 \cos[\delta_2(0) - \delta_1(0) - \theta_{21}] \\
&+ Y_{22}V_2 \cos[-\theta_{22}] + Y_{23}V_3 \cos[\delta_2(0) - \delta_3(0) - \theta_{23}] \\
&+ Y_{24}V_4 \cos[\delta_2(0) - \delta_4(0) - \theta_{24}] \\
&+ Y_{25}V_5 \cos[\delta_2(0) - \delta_5(0) - \theta_{25}]\}
\end{aligned}$$

$$\begin{aligned}
\Delta P_2(0) &= -8.0 - 1.0\{28.5847(1.0) \cos(84.624^\circ) \\
&\quad + 9.95972(1.0) \cos(-95.143^\circ) \\
&\quad + 19.9159(1.0) \cos(-95.143^\circ)\} \\
&= -8.0 - (-2.89 \times 10^{-4}) = -7.99972 \quad \text{per unit}
\end{aligned}$$

From Step 2 and J1 given in Table 6.5

$$\begin{aligned}
J1_{24}(0) &= V_2(0)Y_{24}V_4(0) \sin[\delta_2(0) - \delta_4(0) - \theta_{24}] \\
&= (1.0)(9.95972)(1.0) \sin[-95.143^\circ] \\
&= -9.91964 \quad \text{per unit}
\end{aligned}$$

To see the complete convergence of this case, open PowerWorld Simulator case Example 6_11 (see Figure 6.4). Select **Case Information, Network, Mismatches** to see the initial mismatches, and **Case Information, Solution Details, Power Flow Jacobian** to view the initial Jacobian matrix. As is common in commercial power flows, PowerWorld Simulator actually includes rows in the Jacobian for voltage-controlled buses. When a generator is regulating its terminal voltage, this row corresponds to the equation setting the bus voltage magnitude equal to the generator voltage setpoint. However, if the generator hits a reactive power limit, the bus type is switched to a load bus.

To step through the New-Raphson solution, from the **Tools** Ribbon select **Solve, Single Solution—Full Newton**. Ordinarily this selection would perform a complete Newton-Raphson iteration, stopping only when all the mismatches are less than the desired tolerance. However, for this case, in order to allow you to see the solution process, the maximum number of iterations has been set to 1, allowing the voltages, mismatches and the Jacobian to be viewed after each iteration. To complete the solution, continue to select **Single Solution—Full Newton** until the solution convergence to the values shown in Tables 6.6, 6.7 and 6.8 (in about three iterations).

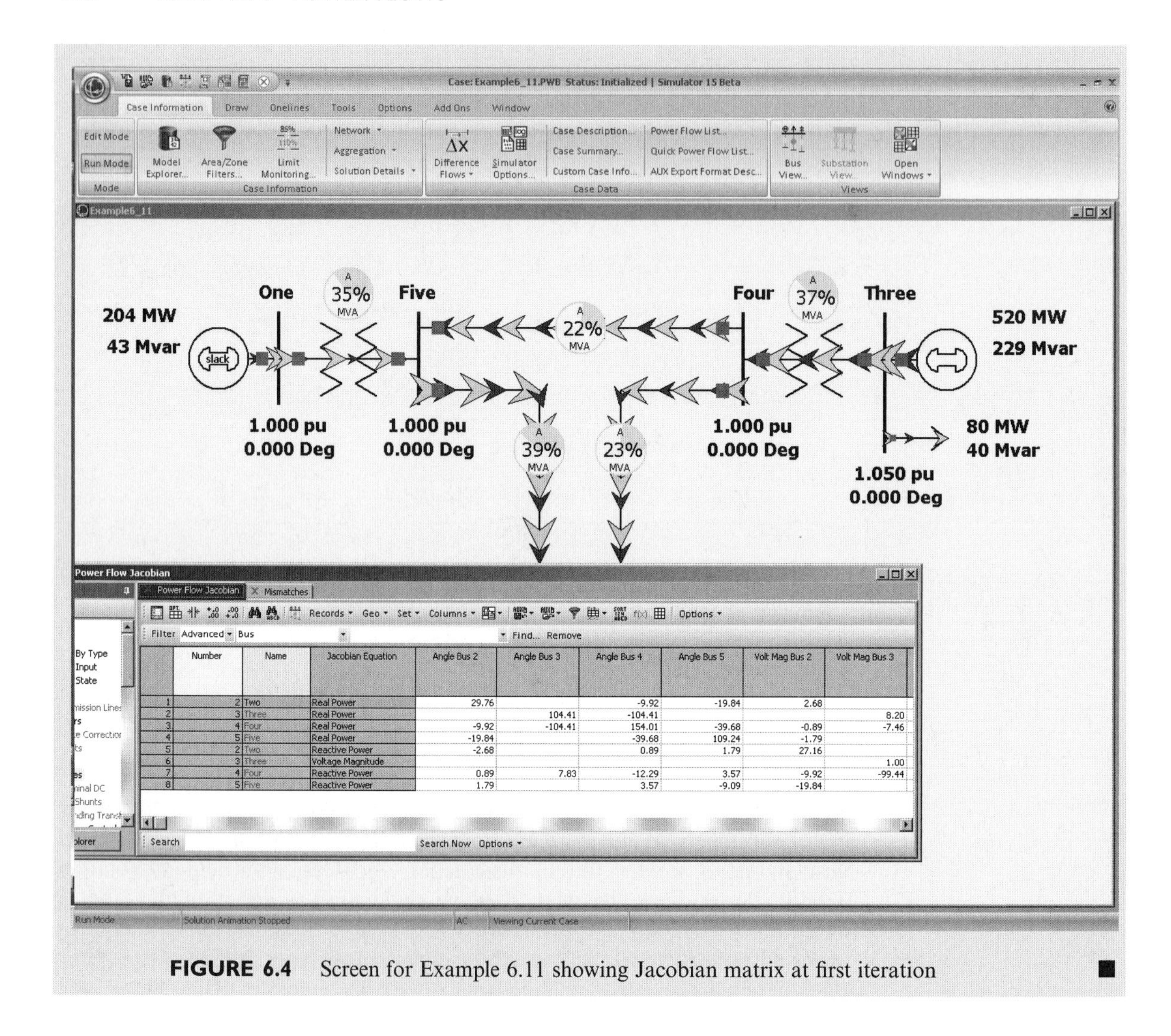

FIGURE 6.4 Screen for Example 6.11 showing Jacobian matrix at first iteration ■

TABLE 6.6

Bus output data for the power system given in Example 6.9

Bus #	Voltage Magnitude (per unit)	Phase Angle (degrees)	Generation PG (per unit)	Generation QG (per unit)	Load PL (per unit)	Load QL (per unit)
1	1.000	0.000	3.948	1.144	0.000	0.000
2	0.834	−22.407	0.000	0.000	8.000	2.800
3	1.050	−0.597	5.200	3.376	0.800	0.400
4	1.019	−2.834	0.000	0.000	0.000	0.000
5	0.974	−4.548	0.000	0.000	0.000	0.000
		TOTAL	9.148	4.516	8.800	3.200

TABLE 6.7 Line output data for the power system given in Example 6.9

Line #	Bus to Bus		P	Q	S
1	2	4	−2.920	−1.392	3.232
	4	2	3.036	1.216	3.272
2	2	5	−5.080	−1.408	5.272
	5	2	5.256	2.632	5.876
3	4	5	1.344	1.504	2.016
	5	4	−1.332	−1.824	2.260

TABLE 6.8 Transformer output data for the power system given in Example 6.9

Tran. #	Bus to Bus		P	Q	S
1	1	5	3.948	1.144	4.112
	5	1	−3.924	−0.804	4.004
2	3	4	4.400	2.976	5.312
	4	3	−4.380	−2.720	5.156

EXAMPLE 6.12 Power-flow program: change in generation

Using the power-flow system given in Example 6.9, determine the acceptable generation range at bus 3, keeping each line and transformer loaded at or below 100% of its MVA limit.

SOLUTION Load PowerWorld Simulator case Example 6.9. Select **Single Solution-Full Newton** to perform a single power-flow solution using the Newton–Raphson approach. Then view the **Case Information** displays to verify that the PowerWorld Simulator solution matches the solution shown in Tables 6.6, 6.7, and 6.8. Additionally, the pie charts on the one-lines show the percentage line and transformer loadings. Initially transformer T1, between buses 1 and 5, is loaded at about 68% of its maximum MVA limit, while transformer T2, between buses 3 and 4, is loaded at about 53%.

Next, the bus 3 generation needs to be varied. This can be done a number of different ways in PowerWorld Simulator. The easiest (for this example) is to use the bus 3 generator MW one-line field to manually change the generation (see Figure 6.5). Right-click on the "520 MW" field to the right of the bus 3 generator and select '**Generator Field Information**' dialog to view the '**Generator Field Options**' dialog. Set the "Delta Per Mouse Click" field to 10 and select OK. Small arrows are now visible next to this field on the one-line; clicking on the up arrow increases the generator's MW output by 10 MW, while clicking on the down arrow decreases the generation by 10 MW. Select **Tools, Play** to begin the simulation. Increase the generation until the pie chart for the transformer from bus 3 to 4 is loaded to 100%. This occurs at about 1000 MW. Notice that as the bus 3 generation is increased the bus 1 slack generation decreases by a similar amount. Repeat the process, except

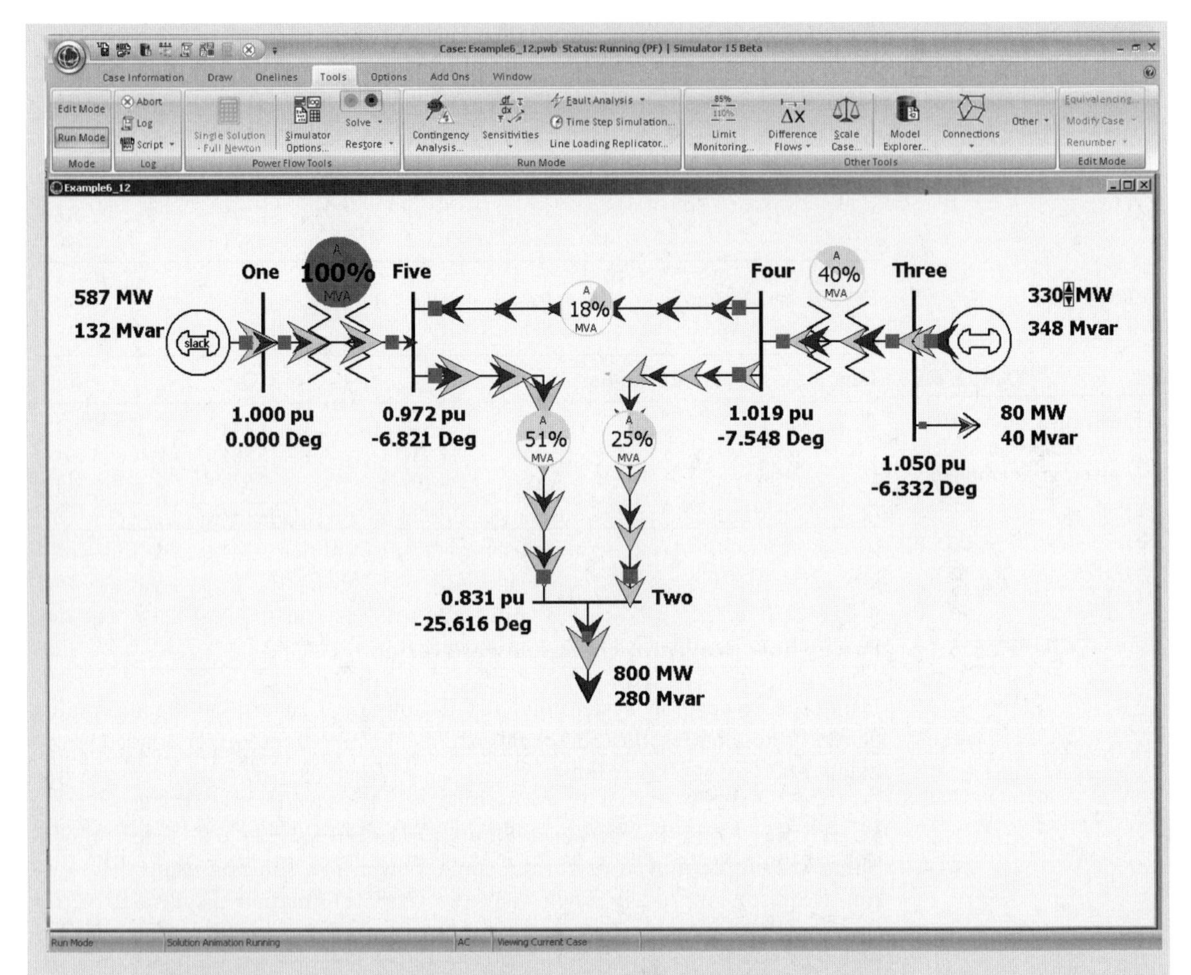

FIGURE 6.5 Screen for Example 6.12, Minimum Bus 3 Generator Loading

now decreasing the generation. This unloads the transformer from bus 3 to 4, but increases the loading on the transformer from bus 1 to bus 5. The bus 1 to 5 transformer should reach 100% loading with the bus 3 generation equal to about 330 MW. ■

Voltage-controlled buses to which tap-changing or voltage-regulating transformers are connected can be handled by various methods. One method is to treat each of these buses as a load bus. The equivalent π circuit parameters (Figure 3.25) are first calculated with tap setting $c = 1.0$ for starting. During each iteration, the computed bus voltage magnitude is compared with the desired value specified by the input data. If the computed voltage is low (or high), c is increased (or decreased) to its next setting, and the parameters of the equivalent π circuit as well as $\boldsymbol{Y}_{\text{bus}}$ are recalculated. The procedure continues until the computed bus voltage magnitude equals the desired value within a specified tolerance, or until the high or low tap-setting limit is reached. Phase-shifting transformers can be handled in a similar way by using a complex turns ratio $c = 1.0\underline{/\alpha}$, and by varying the phase-shift angle α.

A method with faster convergence makes c a variable and includes it in the $\mathbf{x}$ vector of (6.6.1). An equation is then derived to enter into the Jacobian matrix [4].

In comparing the Gauss-Seidel and Newton-Raphson algorithms, experience from power-flow studies has shown that Newton-Raphson converges in many cases where Jacobi and Gauss-Seidel diverge. Furthermore, the number of iterations required for convergence is independent of the number of buses N for Newton-Raphson, but increases with N for Jacobi and Gauss-Seidel. The principal advantage of the Jacobi and Gauss-Seidel methods had been their more modest memory storage requirements and their lower computational requirements per iteration. However, with the vast increases in low-cost computer memory over the last several decades, coupled with the need to solve power-flow problems with tens of thousands of buses, these advantages have been essentially eliminated. Therefore the Newton-Raphson, or one of the derivative methods discussed in Sections 6.9 and 6.10, are the preferred power-flow solution approaches.

EXAMPLE 6.13 Power-flow program: 37-bus system

To see a power-flow example of a larger system, open PowerWorld Simulator case Example 6_13 (see Figure 6.6). This case models a 37-bus, 9-generator power system containing three different voltage levels (345 kV, 138 kV, and 69 kV) with 57 transmission lines or transformers. The one-line can be panned by pressing the arrow keys, and it can be zoomed by pressing the ⟨ctrl⟩ with the up arrow key to zoom in or with the down arrow key to zoom out. Use **Tools, Play** to animate the one-line and **Tools, Pause** to stop the animation.

Determine the lowest per-unit voltage and the maximum line/transformer loading both for the initial case and for the case with the line from bus TIM69 to HANNAH69 out of service.

SOLUTION Use single solution to initially solve the power flow, and then **Case Information, Network, Buses...** to view a listing of all the buses in the case. To quickly determine the lowest per-unit voltage magnitude, left-click on the PU Volt column header to sort the column (clicking a second time reverses the sort). The lowest initial voltage magnitude is 0.9902 at bus DEMAR69. Next, select **Case Information, Network, Lines and Transformers...** to view the Line and Transformer Records display. Left-click on % of Max Limit to sort the lines by percentage loading. Initially the highest percentage loading is 64.9% on the line from UIUC69 to BLT69 circuit 1.

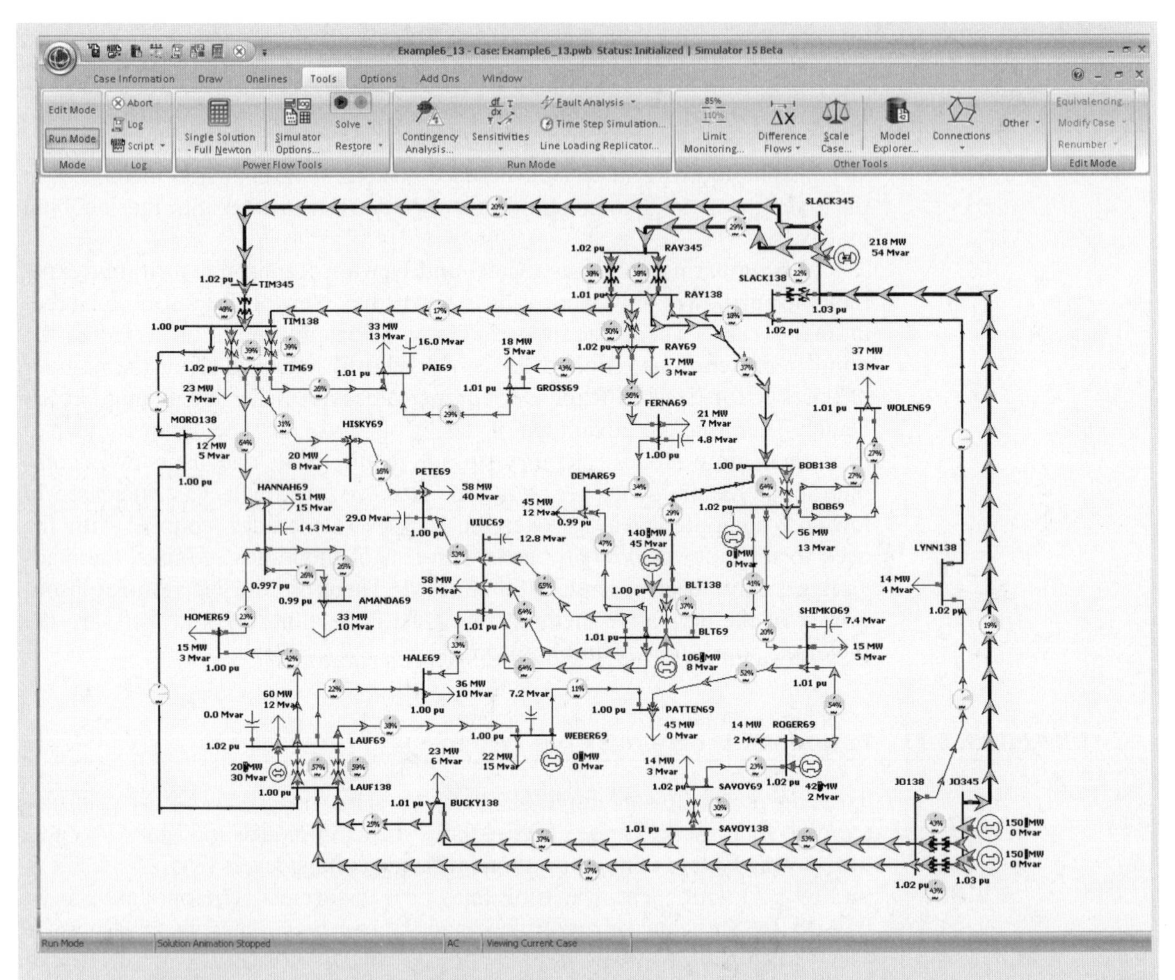

FIGURE 6.6 Screen for Example 6.13 showing the initial flows

There are several ways to remove the TIM69 to HANNAH69 line. One approach is to locate the line on the Line and Transformer Records display and then double-click on the Status field to change its value. An alternative approach is to find the line on the one-line (it is in the upper-lefthand portion) and then click on one of its circuit breakers. Once the line is removed, use single solution to resolve the power flow. The lowest per-unit voltage is now 0.9104 at AMANDA69 and the highest percentage line loading is 134.8%, on the line from HOMER69 to LAUF69. Since there are now several bus and line violations, the power system is no longer at a secure operating point. Control actions and/or design improvements are needed to correct these problems. Design Project 1 discusses these options. ■

6.7

CONTROL OF POWER FLOW

The following means are used to control system power flows:

1. Prime mover and excitation control of generators.
2. Switching of shunt capacitor banks, shunt reactors, and static var systems.
3. Control of tap-changing and regulating transformers.

A simple model of a generator operating under balanced steady-state conditions is the Thévenin equivalent shown in Figure 6.7. V_t is the generator terminal voltage, E_g is the excitation voltage, δ is the power angle, and X_g is the positive-sequence synchronous reactance. From the figure, the generator current is

$$I = \frac{E_g e^{j\delta} - V_t}{jX_g} \tag{6.7.1}$$

and the complex power delivered by the generator is

$$S = P + jQ = V_t I^* = V_t \left(\frac{E_g e^{-j\delta} - V_t}{-jX_g} \right)$$

$$= \frac{V_t E_g (j \cos \delta + \sin \delta) - jV_t^2}{X_g} \tag{6.7.2}$$

The real and reactive powers delivered are then

$$P = \text{Re } S = \frac{V_t E_g}{X_g} \sin \delta \tag{6.7.3}$$

$$Q = \text{Im } S = \frac{V_t}{X_g} (E_g \cos \delta - V_t) \tag{6.7.4}$$

Equation (6.7.3) shows that the real power P increases when the power angle δ increases. From an operational standpoint, when the prime mover increases the power input to the generator while the excitation voltage is held constant, the rotor speed increases. As the rotor speed increases, the power angle δ also increases, causing an increase in generator real power output P. There is also a decrease in reactive power output Q, given by (6.7.4). However, when δ is

FIGURE 6.7

Generator Thévenin equivalent

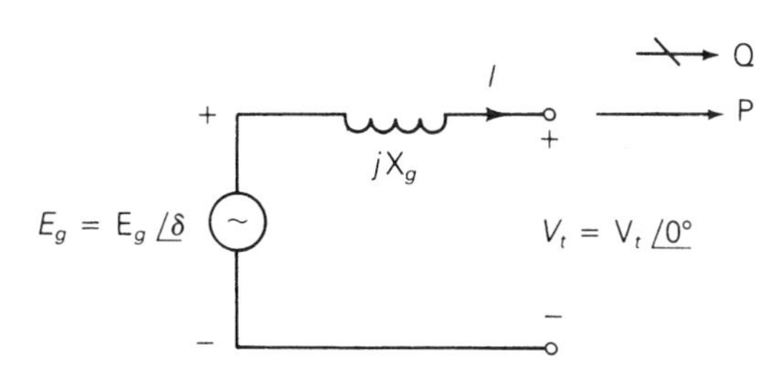

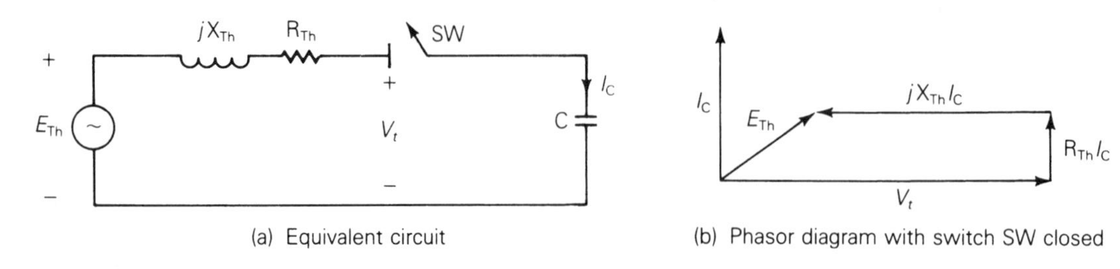

FIGURE 6.8 Effect of adding a shunt capacitor bank to a power system bus

less than 15°, the increase in P is much larger than the decrease in Q. From the power-flow standpoint, an increase in prime-move power corresponds to an increase in P at the constant-voltage bus to which the generator is connected. The power-flow program computes the increase in δ along with the small change in Q.

Equation (6.7.4) shows that reactive power output Q increases when the excitation voltage E_g increases. From the operational standpoint, when the generator exciter output increases while holding the prime-mover power constant, the rotor current increases. As the rotor current increases, the excitation voltage E_g also increases, causing an increase in generator reactive power output Q. There is also a small decrease in δ required to hold P constant in (6.7.3). From the power-flow standpoint, an increase in generator excitation corresponds to an increase in voltage magnitude at the constant-voltage bus to which the generator is connected. The power-flow program computes the increase in reactive power Q supplied by the generator along with the small change in δ.

Figure 6.8 shows the effect of adding a shunt capacitor bank to a power system bus. The system is modeled by its Thévenin equivalent. Before the capacitor bank is connected, the switch SW is open and the bus voltage equals E_{Th}. After the bank is connected, SW is closed, and the capacitor current I_C leads the bus voltage V_t by 90°. The phasor diagram shows that V_t is larger than E_{Th} when SW is closed. From the power-flow standpoint, the addition of a shunt capacitor bank to a load bus corresponds to the addition of a negative reactive load, since a capacitor absorbs negative reactive power. The power-flow program computes the increase in bus voltage magnitude along with the small change in δ. Similarly, the addition of a shunt reactor corresponds to the addition of a positive reactive load, wherein the power-flow program computes the decrease in voltage magnitude.

Tap-changing and voltage-magnitude-regulating transformers are used to control bus voltages as well as reactive power flows on lines to which they are connected. Similarly, phase-angle regulating transformers are used to control bus angles as well as real power flows on lines to which they are connected. Both tap-changing and regulating transformers are modeled by a transformer with an off-nominal turns ratio c (Figure 3.25). From the power-flow standpoint, a change in tap setting or voltage regulation corresponds to a change in c. The power-flow program computes the changes in $\boldsymbol{Y}_{bus}$, bus voltage magnitudes and angles, and branch flows.

Besides the above controls, the power-flow program can be used to investigate the effect of switching in or out lines, transformers, loads, and generators. Proposed system changes to meet future load growth, including new transmission, new transformers, and new generation can also be investigated. Power-flow design studies are normally conducted by trial and error. Using engineering judgment, adjustments in generation levels and controls are made until the desired equipment loadings and voltage profile are obtained.

EXAMPLE 6.14 Power-flow program: effect of shunt capacitor banks

Determine the effect of adding a 200-Mvar shunt capacitor bank at bus 2 on the power system in Example 6.9.

SOLUTION Open PowerWorld Simulator case Example 6_14 (see Figure 6.9). This case is identical to Example 6.9 except that a 200-Mvar shunt capacitor

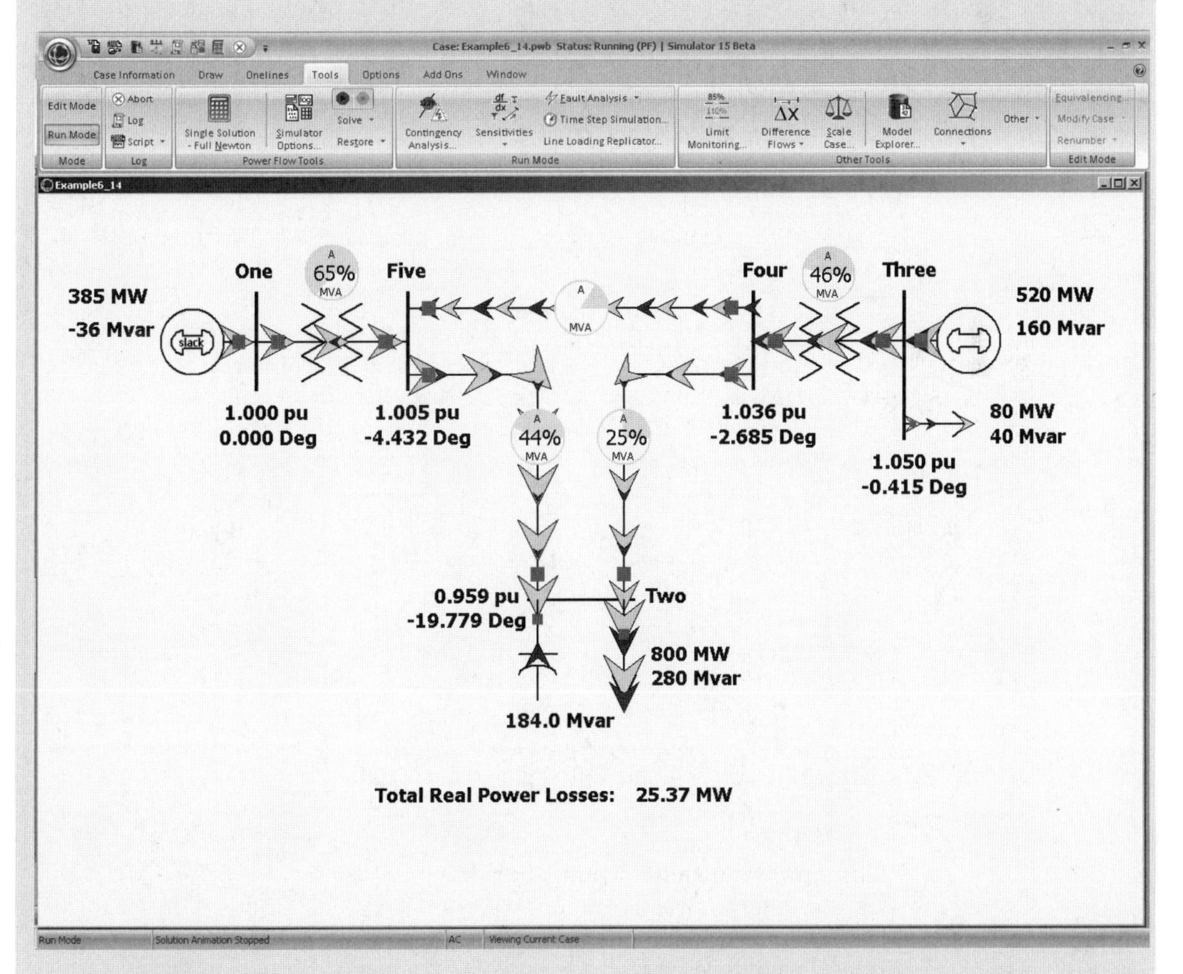

FIGURE 6.9 Screen for Example 6.14

bank has been added at bus 2. Initially this capacitor is open. Click on the capacitor's circuit to close the capacitor and then solve the case. The capacitor increases the bus 2 voltage from 0.834 per unit to a more acceptable 0.959 per unit. The insertion of the capacitor has also substantially decreased the losses, from 34.84 to 25.37 MW.

Notice that the amount of reactive power actually supplied by the capacitor is only 184 Mvar. This discrepancy arises because a capacitor's reactive output varies with the square of the terminal voltage, $Q_{cap} = V_{cap}{}^2/X_c$ (see 2.3.5). A capacitor's Mvar rating is based on an assumed voltage of 1.0 per unit. ■

EXAMPLE 6.15

PowerWorld Simulator Case Example 6_15 (see Figure 6.10), which modifies the Example 6.13 case by (1) opening one of the 138/69 kV transformers at the LAUF substation, and (2) opening the 69 kV transmission line between

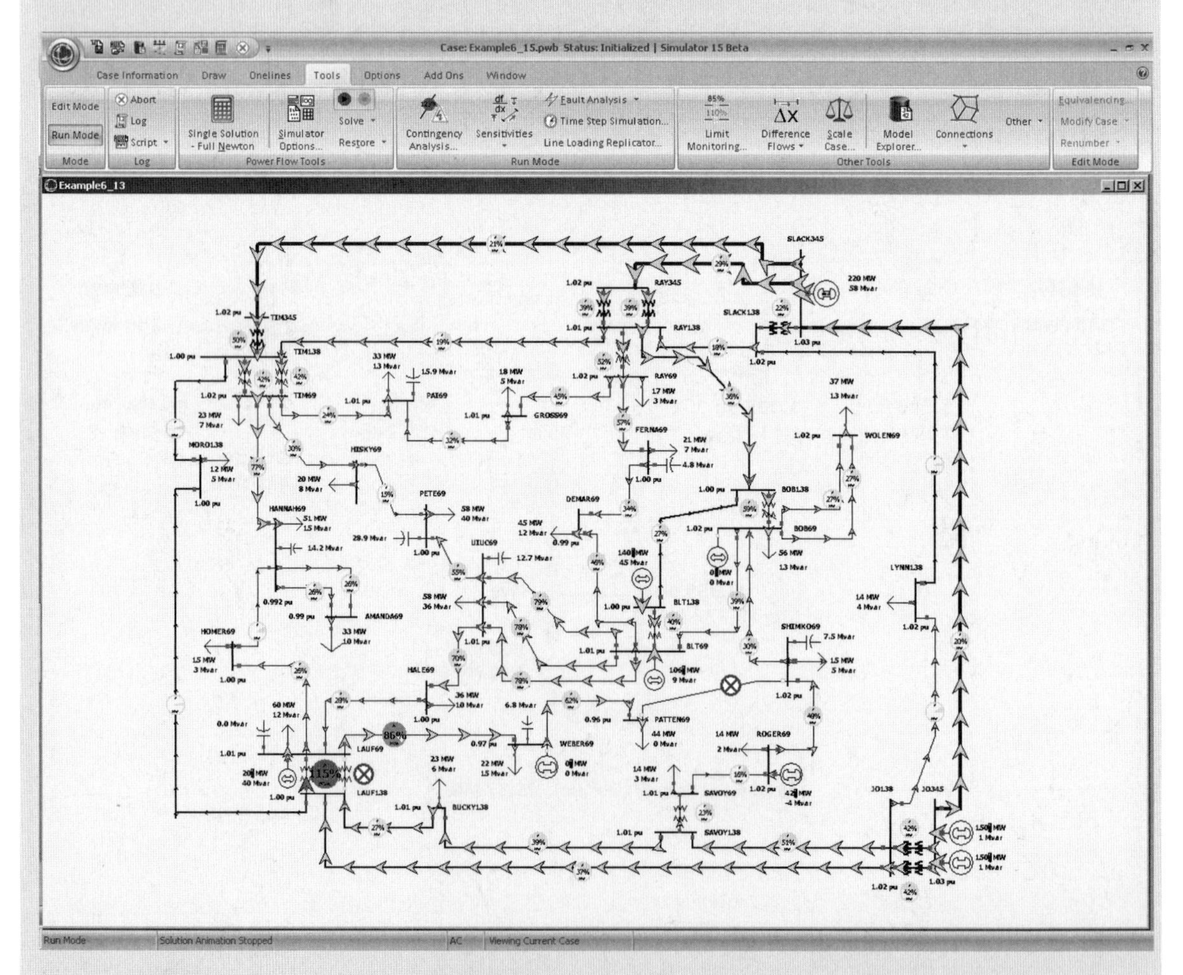

FIGURE 6.10 Screen for Example 6.15

PATTEN69 and SHIMKO69. This causes a flow of 116.2 MVA on the remaining 138/69 kV transformer at LAUF. Since this transformer has a limit of 101 MVA, it results in an overload at 115%. Redispatch the generators in order to remove this overload.

SOLUTION There are a number of solutions to this problem, and several solution techniques. One solution technique would be to use engineering intuition, along with a trial and error approach (see Figure 6.11). Since the overload is from the 138 kV level to the 69 kV level, and there is a generator directly connected to at the LAUF 69 kV bus, it stands to reason that increasing this generation would decrease the overload. Using this approach, we can remove the overload by increasing the Lauf generation until the transformer flow is reduced to 100%. This occurs when the generation is increased from 20 MW to 51 MW. Notice that as the generation is increased, the swing bus (SLACK345) generation automatically decreases in order to satisfy the requirement that total system load plus losses must be equal to total generation.

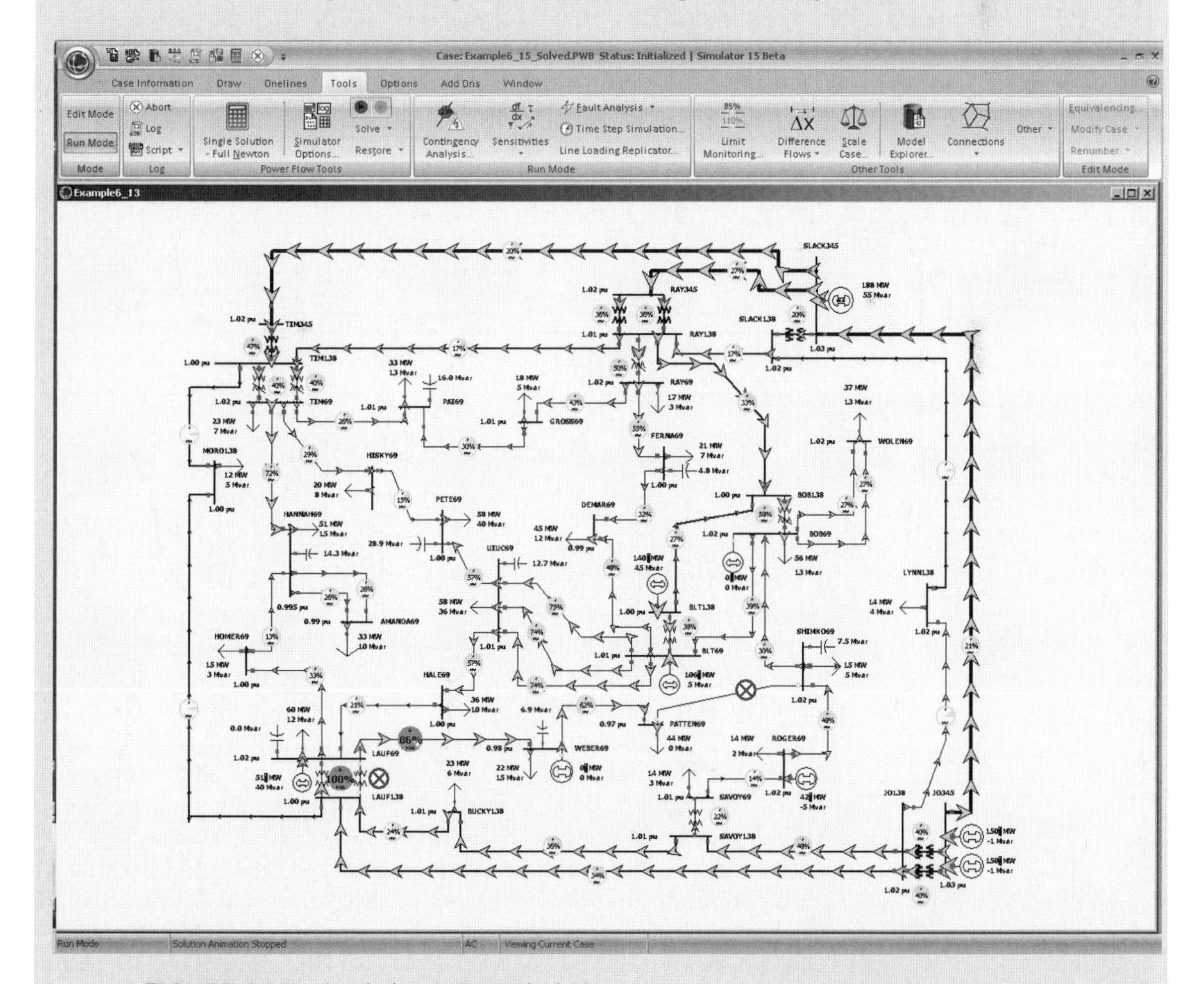

FIGURE 6.11 A solution to Example 6.15

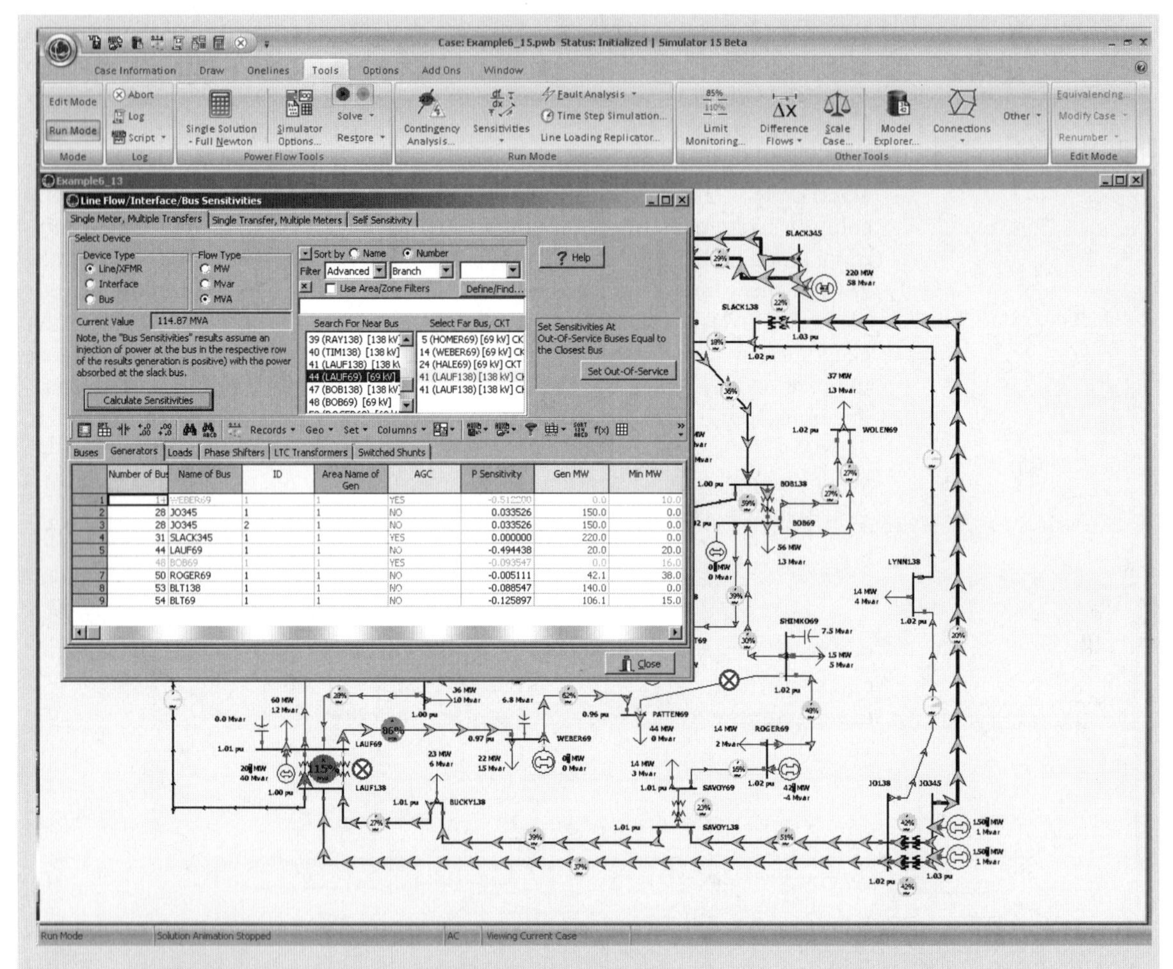

FIGURE 6.12 Example 6.15 Flow Sensitivities Dialog

An alternative possible solution is seen by noting that since the overload is caused by power flowing from the 138 kV bus, decreasing the generation at JO345 might also decrease this flow. This is indeed the case, but now the trial and error approach requires a substantial amount of work, and ultimately doesn't solve the problem. Even when we decrease the total JO345 generation from 300 MW to 0 MW, the overload is still present, albeit with its percentage decreased to 105%.

An alternative solution approach would be to first determine the generators with the most sensitivity to this violation and then adjust these (see Figure 6.12). This can be done in PowerWorld Simulator by selecting **Tools, Sensitivities, Flows and Voltage Sensitivities**. Select the LAUF 138/69 kV transformer, click on the **Calculate Sensitivities** button, and select the Generator Sensitivities tab towards the bottom of the dialog. The "P Sensitivity" field tells how increasing the output of each generator by one MW would affect the MVA flow on this transformer. Note that the sensitivity for the Lauf generator is −0.494, indicating that if we increase this generation by 1 MW the transformer MVA flow would decrease by 0.494 MVA. Hence, in order to decrease the flow by 15.2 MVA we would expect to increase the LAUF69 generator by 31 MW, exactly what we got by the trial and error approach. It is also clear that the JO345 generators, with a sensitivity of just 0.0335, would be relatively ineffective. In actual power system operation these sensitivities, known as generator shift factors, are used extensively. These sensitivities are also used in the Optimal Power Flow (introduced in Section 11.5). ■

6.8 SPARSITY TECHNIQUES

A typical power system has an average of fewer than three lines connected to each bus. As such, each row of Y_{bus} has an average of fewer than four nonzero elements, one off-diagonal for each line and the diagonal. Such a matrix, which has only a few nonzero elements, is said to be *sparse*.

Newton–Raphson power-flow programs employ sparse matrix techniques to reduce computer storage and time requirements [2]. These techniques include compact storage of Y_{bus} and $\mathbf{J}(i)$ and reordering of buses to avoid fill-in of $\mathbf{J}(i)$ during Gauss elimination steps. Consider the following matrix:

$$\mathbf{S} = \begin{bmatrix} 1.0 & -1.1 & -2.1 & -3.1 \\ -4.1 & 2.0 & 0 & -5.1 \\ -6.1 & 0 & 3.0 & 0 \\ -7.1 & 0 & 0 & 4.0 \end{bmatrix} \tag{6.8.1}$$

One method for compact storage of **S** consists of the following four vectors:

$$\mathbf{DIAG} = [1.0 \quad 2.0 \quad 3.0 \quad 4.0] \tag{6.8.2}$$

$$\mathbf{OFFDIAG} = [-1.1 \quad -2.1 \quad -3.1 \quad -4.1 \quad -5.1 \quad -6.1 \quad -7.1] \tag{6.8.3}$$

$$\mathbf{COL} = [2 \quad 3 \quad 4 \quad 1 \quad 4 \quad 1 \quad 1] \tag{6.8.4}$$

$$\mathbf{ROW} = [3 \quad 2 \quad 1 \quad 1] \tag{6.8.5}$$

DIAG contains the ordered diagonal elements and **OFFDIAG** contains the nonzero off-diagonal elements of **S**. **COL** contains the column number of each off-diagonal element. For example, the *fourth* element in **COL** is 1, indicating that the *fourth* element of **OFFDIAG**, −4.1, is located in column 1. **ROW** indicates the number of off-diagonal elements in each row of **S**. For example, the *first* element of **ROW** is 3, indicating the *first* three elements of **OFFDIAG**, −1.1, −2.1, and −3.1, are located in the *first* row. The *second* element of **ROW** is 2, indicating the next two elements of **OFFDIAG**, −4.1 and −5.1, are located in the *second* row. The **S** matrix can be completely reconstructed from these four vectors. Note that the dimension of **DIAG** and

ROW equals the number of diagonal elements of **S**, whereas the dimension of **OFFDIAG** and **COL** equals the number of nonzero off-diagonals.

Now assume that computer storage requirements are 4 bytes to store each magnitude and 4 bytes to store each phase of Y_{bus} in an N-bus power system. Also assume Y_{bus} has an average of $3N$ nonzero off-diagonals (three lines per bus) along with its N diagonals. Using the preceding compact storage technique, we need $(4+4)3N = 24N$ bytes for **OFFDIAG** and $(4+4)N = 8N$ bytes for **DIAG**. Also, assuming 2 bytes to store each integer, we need $6N$ bytes for **COL** and $2N$ bytes for **ROW**. Total computer memory required is then $(24+8+6+2)N = 40N$ bytes with compact storage of Y_{bus}, compared to $8N^2$ bytes without compact storage. For a 1000-bus power system, this means 40 instead of 8000 kilobytes to store Y_{bus}. Further storage reduction could be obtained by storing only the upper triangular portion of the symmetric Y_{bus} matrix.

The Jacobian matrix is also sparse. From Table 6.5, whenever $Y_{kn} = 0$, $J1_{kn} = J2_{kn} = J3_{kn} = J4_{kn} = 0$. Compact storage of **J** for a 30,000-bus power system requires less than 10 megabytes with the above assumptions.

The other sparsity technique is to reorder buses. Suppose Gauss elimination is used to triangularize **S** in (6.8.1). After one Gauss elimination step, as described in Section 6.1, we have

$$\mathbf{S}^{(1)} = \begin{bmatrix} 1.0 & -1.1 & -2.1 & -3.1 \\ 0 & -2.51 & -8.61 & -7.61 \\ 0 & -6.71 & -9.81 & -18.91 \\ 0 & -7.81 & -14.91 & -18.01 \end{bmatrix} \tag{6.8.6}$$

We can see that the zeros in columns 2, 3, and 4 of **S** are filled in with nonzero elements in $\mathbf{S}^{(1)}$. The original degree of sparsity is lost.

One simple reordering method is to start with those buses having the fewest connected branches and to end with those having the most connected branches. For example, **S** in (6.8.1) has three branches connected to bus 1 (three off-diagonals in row 1), two branches connected to bus 2, and one branch connected to buses 3 and 4. Reordering the buses 4, 3, 2, 1 instead of 1, 2, 3, 4 we have

$$\mathbf{S}_{\text{reordered}} = \begin{bmatrix} 4.0 & 0 & 0 & -7.1 \\ 0 & 3.0 & 0 & -6.1 \\ -5.1 & 0 & 2.0 & -4.1 \\ -3.1 & -2.1 & -1.1 & 1.0 \end{bmatrix} \tag{6.8.7}$$

Now, after one Gauss elimination step,

$$\mathbf{S}^{(1)}_{\text{reordered}} = \begin{bmatrix} 4.0 & 0 & 0 & -7.1 \\ 0 & 3.0 & 0 & -6.1 \\ 0 & 0 & 2.0 & -13.15 \\ 0 & -2.1 & -1.1 & -4.5025 \end{bmatrix} \tag{6.8.8}$$

Note that the original degree of sparsity is not lost in (6.8.8).

Reordering buses according to the fewest connected branches can be performed once, before the Gauss elimination process begins. Alternatively, buses can be renumbered during each Gauss elimination step in order to account for changes during the elimination process.

Sparsity techniques similar to those described in this section are a standard feature of today's Newton–Raphson power-flow programs. As a result of these techniques, typical 30,000-bus power-flow solutions require less than 10 megabytes of storage, less than one second per iteration of computer time, and less than 10 iterations to converge.

EXAMPLE 6.16 Sparsity in a 37-bus system

To see a visualization of the sparsity of the power-flow Ybus and Jacobian matrices in a 37-bus system, open PowerWorld Simulator case Example 6_13.

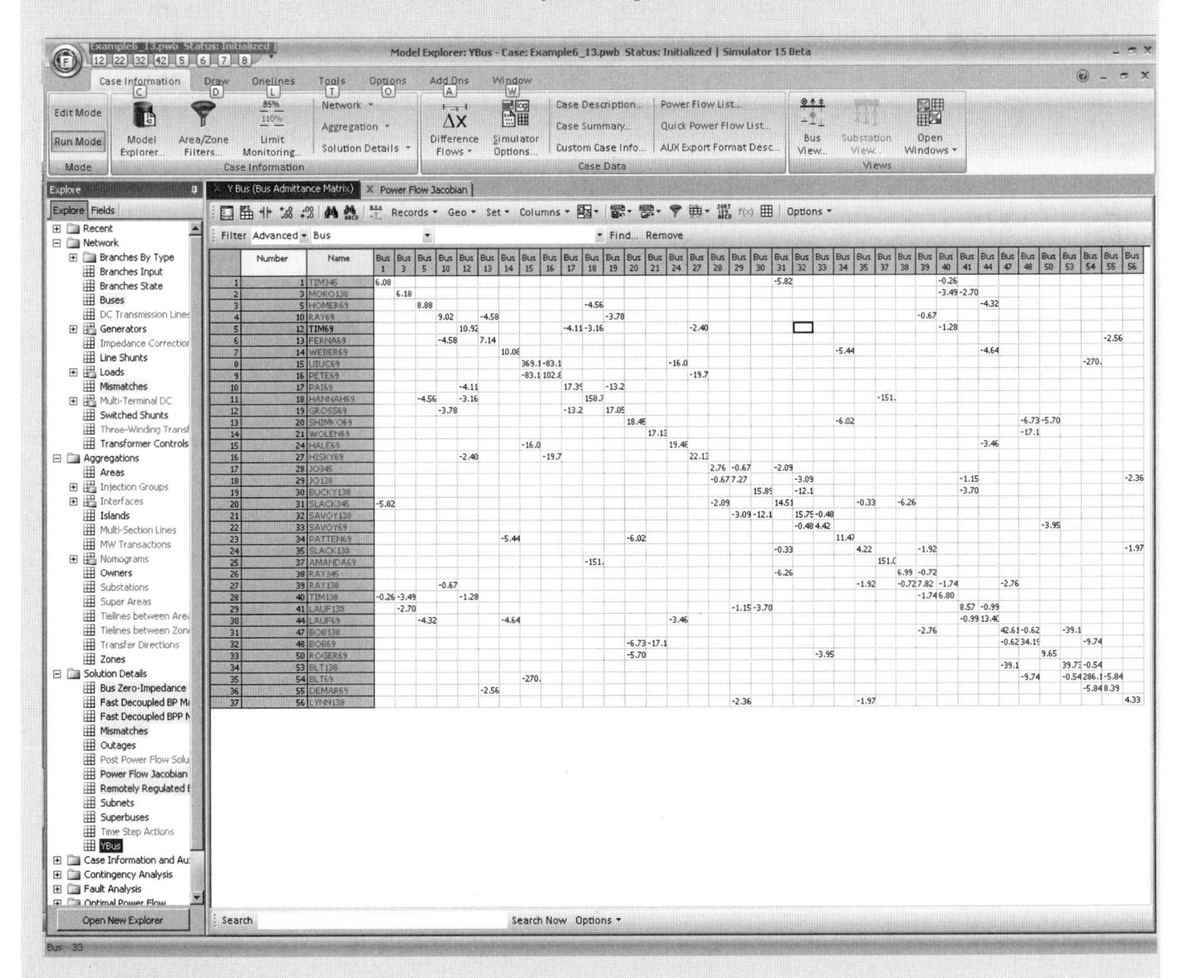

FIGURE 6.13 Screen for Example 6.16

Select **Case Information, Solution Details, Ybus** to view the bus admittance matrix. Then press ⟨ctrl⟩ Page Down to zoom the display out. Blank entries in the matrix correspond to zero entries. The 37×37 Ybus has a total of 1369 entries, with only about 10% nonzero (see Figure 6.13). Select **Case Information, Solution Details, Power Flow Jacobian** to view the Jacobian matrix. ■

6.9 FAST DECOUPLED POWER FLOW

Contingencies are a major concern in power system operations. For example, operating personnel need to know what power-flow changes will occur due to a particular generator outage or transmission-line outage. Contingency information, when obtained in real time, can be used to anticipate problems caused by such outages, and can be used to develop operating strategies to overcome the problems.

Fast power-flow algorithms have been developed to give power-flow solutions in seconds or less [8]. These algorithms are based on the following simplification of the Jacobian matrix. Neglecting $\mathbf{J}_2(i)$ and $\mathbf{J}_3(i)$, (6.6.6) reduces to two sets of decoupled equations:

$$\mathbf{J}_1(i)\Delta\boldsymbol{\delta}(i) = \Delta\mathbf{P}(i) \tag{6.9.1}$$

$$\mathbf{J}_4(i)\Delta\mathbf{V}(i) = \Delta\mathbf{Q}(i) \tag{6.9.2}$$

The computer time required to solve (6.9.1) and (6.9.2) is significantly less than that required to solve (6.6.6). Further reduction in computer time can be obtained from additional simplification of the Jacobian matrix. For example, assume $V_k \approx V_n \approx 1.0$ per unit and $\delta_k \approx \delta_n$. Then $\mathbf{J}_1$ and $\mathbf{J}_4$ are constant matrices whose elements in Table 6.5 are the negative of the imaginary components of $\boldsymbol{Y}_{\text{bus}}$. As such, $\mathbf{J}_1$ and $\mathbf{J}_4$ do not have to be recalculated during successive iterations.

The above simplifications can result in rapid power-flow solutions for most systems. While the fast decoupled power flow usually takes more iterations to converge, it is usually significantly faster then the Newton-Raphson algorithm since the Jacobian does not need to be recomputed each iteration. And since the mismatch equations themselves have not been modified, the solution obtained by the fast decoupled algorithm is the same as that found with the Newton-Raphson algorithm. However, in some situations in which only an approximate power-flow solution is needed the fast decoupled approach can be used with a fixed number of iterations (typically one) to give an extremely fast, albeit approximate solution.

6.10

THE "DC" POWER FLOW

The power-flow problem can be further simplified by extending the fast decoupled power flow to completely neglect the Q-V equation, assuming that the voltage magnitudes are constant at 1.0 per unit. With these simplifications the power flow on the line from bus j to bus k with reactive X_{jk} becomes

$$P_{jk} = \frac{\delta_j - \delta_k}{X_{jk}} \tag{6.10.1}$$

and the real power balance equations reduce to a completely linear problem

$$-\mathbf{B}\delta = \mathbf{P} \tag{6.10.2}$$

where $\mathbf{B}$ is the imaginary component of the of $\mathbf{Y}_{\text{bus}}$ calculated neglecting line resistance and excepting the slack bus row and column.

Because (6.10.2) is a linear equation with a form similar to that found in solving dc resistive circuits, this technique is referred to as the dc power flow. However, in contrast to the previous power-flow algorithms, the dc power flow only gives an approximate solution, with the degree of approximation system dependent. Nevertheless, with the advent of power system restructuring the dc power flow has become a commonly used analysis technique.

EXAMPLE 6.17

Determine the dc power-flow solution for the five bus system from Example 6.9.

SOLUTION With bus 1 as the system slack, the $\mathbf{B}$ matrix and $\mathbf{P}$ vector for this system are

$$\mathbf{B} = \begin{bmatrix} -30 & 0 & 10 & 20 \\ 0 & -100 & 100 & 0 \\ 10 & 100 & -150 & 40 \\ 20 & 0 & 40 & -110 \end{bmatrix} \qquad \mathbf{P} = \begin{bmatrix} -8.0 \\ 4.4 \\ 0 \\ 0 \end{bmatrix}$$

$$\delta = -\mathbf{B}^{-1}\mathbf{P} = \begin{bmatrix} -0.3263 \\ 0.0091 \\ -0.0349 \\ -0.0720 \end{bmatrix} \text{radians} = \begin{bmatrix} -18.70 \\ 0.5214 \\ -2.000 \\ -4.125 \end{bmatrix} \text{degrees}$$

To view this example in PowerWorld Simulator open case Example 6_17 which has this example solved using the dc power flow (see Figure 6.14). To view the dc power flow options select **Options, Simulator Options** to show the PowerWorld Simulator Options dialog. Then select the Power Flow Solution category, and the DC Options page.

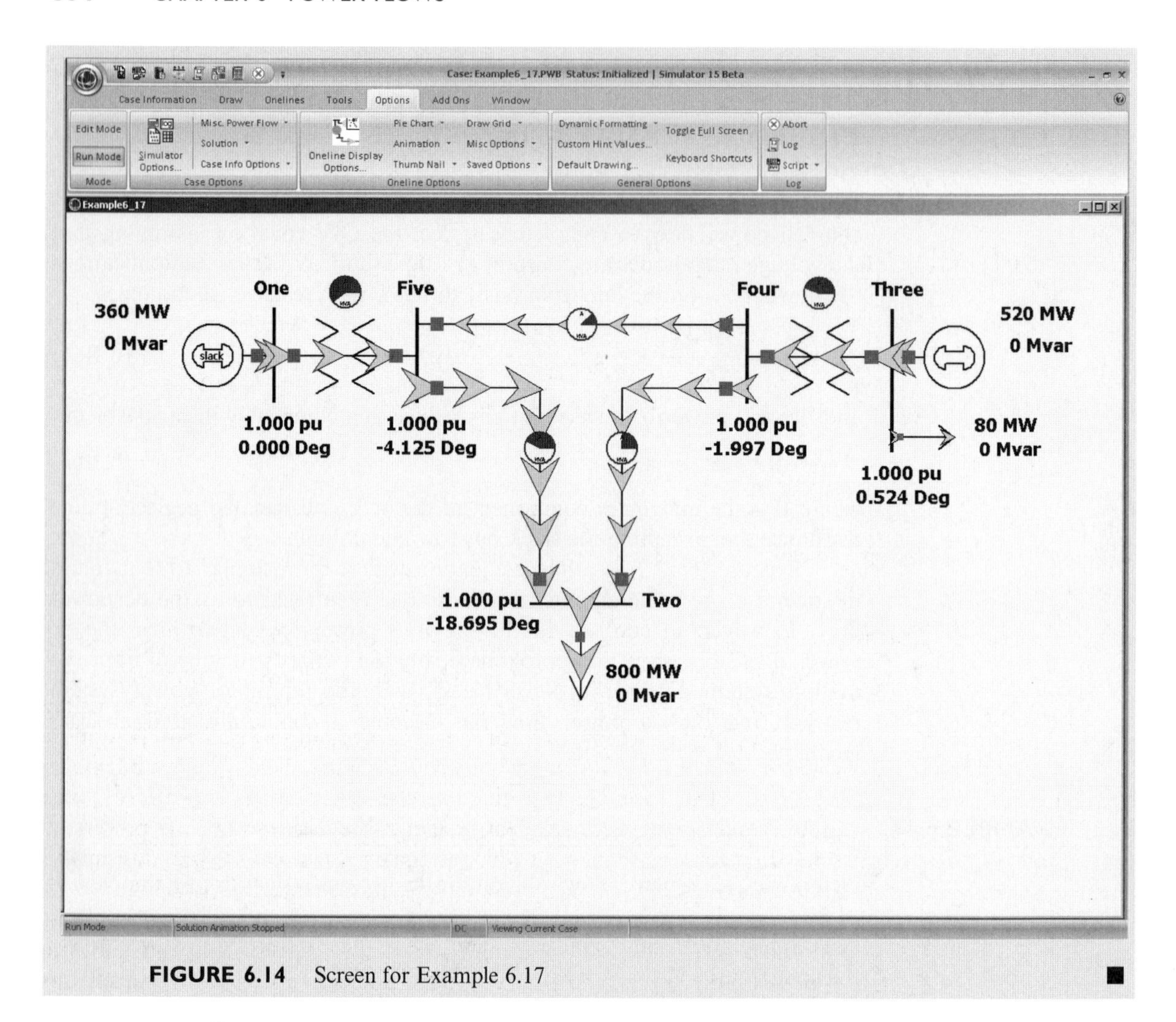

FIGURE 6.14 Screen for Example 6.17 ■

6.11 POWER-FLOW MODELING OF WIND GENERATION

As was mentioned in Chapter 1, the amount of renewable generation, particularly wind, being integrated into electric grids around the world is rapidly growing. For example, in 2008 Denmark obtained almost 20% of their total electric energy from wind while Spain was over 10%. In the United States that amount of wind capacity has been rapidly escalating from less than 2.5 GW in 2000 to more than 35 GW in 2009 (out of a total generation capacity of about 1000 GW).

Whereas most energy from traditional synchronous generators comes from large units with ratings of hundreds of MWs, comparatively speaking, individual wind turbine generator (WTG) power ratings are quite low, with

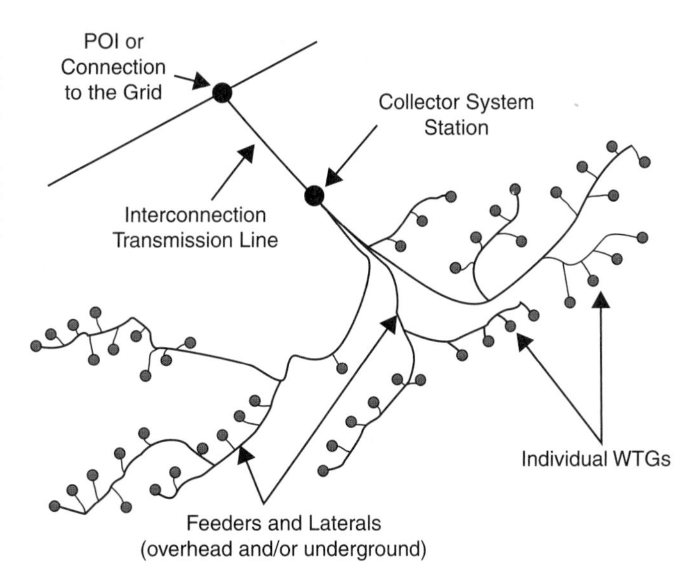

FIGURE 6.15 Wind power plant collector system topology [14] (Figure 1 from WECC Wind Generation Modeling Group, "WECC Wind Power Plant Power Flow Model Guide," WECC, May 2008, p. 2)

common values for new WTGs between one to three MWs. This power is generated at low voltage (e.g., 600 V) and then usually stepped-up with a pad-mounted transformer at the base of the turbine to a distribution-level voltage (e.g., 34.5 kV). Usually dozens or even hundreds of individual WTGs are located in wind "farms" or "parks" that cover an area of many square kilometers, with most of the land still available for other uses such as farming. An underground and/or overhead collector system is used to transmit the power to a single interconnection point at which its voltage is stepped-up to a transmission level voltage (> 100 kV). The layout of such a system is shown in Figure 6.15.

From a power system analysis perspective for large-scale studies the entire wind farm can usually be represented as a single equivalent generator which is either directly connected at the interconnection point transmission system bus, or connected to this bus through an equivalent impedance that represents the impedance of the collector system and the step-up transformers. The parameters associated with the equivalent generator are usually just scaled values of the parameters for the individual WTGs.

There are four main types of WTGs [13], with more details on each type provided in Chapter 11—here the focus is on their power-flow characteristics. As is the case with traditional synchronous generators, the real power outputs for all the WTG types are considered to be a constant value in power-flow studies. Of course how much real power a wind farm can actually produce at any moment depends upon the wind speed, with a typical wind speed versus power curve shown in Figure 6.16.

Type 1 WTGs are squirrel-cage induction machines. Since induction machines consume reactive power and their reactive power output cannot be independently controlled, typically these machines are modeled as a constant power factor PQ bus. By themselves these machines have under-excited

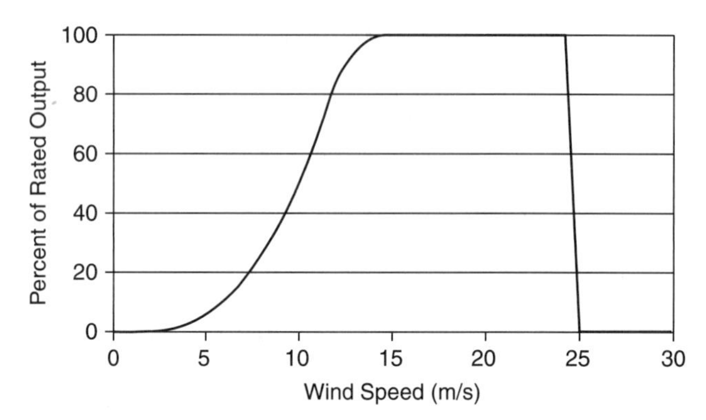

FIGURE 6.16 Typical wind speed versus power curve

(consuming reactive power) power factors of between 0.85 and 0.9, but banks of switched capacitors are often used to correct the wind farm power factor. Type 2 WTGs are wound rotor induction machines in which the rotor resistance can be controlled. The advantages of this approach are discussed in Chapter 11; from a power-flow perspective, they perform like Type 1 WTGs.

Most new WTGs are either Type 3 or Type 4. Type 3 wind turbines are used to represent doubly-fed asynchronous generators (DFAGs), also sometimes referred to as doubly-fed induction generators (DFIGs). This type models induction machines in which the rotor circuit is also connected to the ac network through an ac-dc-ac converter allowing for much greater control of the WTG. Type 4 wind turbines are fully asynchronous machines in which the full power output of the machine is coupled to the ac network through an ac-dc-ac converter. From a power-flow perspective both types are capable of full voltage control like a traditional PV bus generator with reactive power control between a power factor of up to $\pm$ 0.9. However, like traditional synchronous generators, how their reactive power is actually controlled depends on commercial considerations, with many generator owners desiring to operate at unity power factor to maximize their real power outputs.

MULTIPLE CHOICE QUESTIONS

SECTION 6.1

6.1 For a set of linear algebraic equations in matrix format, $\mathbf{Ax} = \mathbf{y}$, for a unique solution to exist, det ($\mathbf{A}$) should be ________. Fill in the Blank.

6.2 For an $N \times N$ square matrix $\mathbf{A}$, in $(N - 1)$ steps, the technique of gauss elimination can transform into an ________ matrix. Fill in the Blank.

SECTION 6.2

6.3 For the iterative solution to linear algebraic equations $\mathbf{Ax} = \mathbf{y}$, the $\mathbf{D}$ matrix in the Jacobi method is the ________ portion of $\mathbf{A}$, whereas $\mathbf{D}$ for Gauss-Siedel is the ________ portion of $\mathbf{A}$.

6.4 Is convergence guaranteed always with Jacobi and Gauss-Siedel methods, as applied to iterative solutions of linear algebraic equations?
(a) Yes (b) No

SECTION 6.3

6.5 For the iterative solutions to nonlinear algebraic equations with Newton-Raphson Method, the Jacobian Matrix $\mathbf{J}$ (i) consists of the partial derivatives. Write down the elements of first row of $\mathbf{J}$ (i).

6.6 For the Newton-Raphson method to work, one should make sure that $\mathbf{J}^{-1}$ exists.
(a) True (b) False

6.7 The Newton-Raphson method in four steps makes use of Gauss elimination and Back Substitution.
(a) True (b) False

6.8 The number of iterations required for convergence is dependent/independent of the dimension N for Newton-Raphson method. Choose one.

SECTION 6.4

6.9 The swing bus or slack bus is a reference bus for which $V_1\underline{/\delta_1}$, typically $1.0\underline{/0^\circ}$ per unit, is input data. The power-flow program computes ________. Fill in the Blank.

6.10 Most buses in a typical power-flow program are load buses, for which P_k and Q_k are input data. The power-flow program computes ________. Fill in the Blank.

6.11 For a voltage-controlled bus k, ________ are input data, while the power-flow program computes ________. Fill in the Blanks.

6.12 When the bus k is a load bus with no generation and inductive load, in terms of generation and load, $P_k =$ ________, and $Q_k =$ ________. Fill in the Blanks.

6.13 Starting from a single-line diagram of a power system, the input data for a power-flow problem consists of ________, ________, and ________. Fill in the Blanks.

SECTION 6.5

6.14 Nodal equations $\boldsymbol{I} = \boldsymbol{Y}_{\text{bus}} \boldsymbol{V}$ are a set of linear equations analogous to $\boldsymbol{y} = \boldsymbol{Ax}$.
(a) True (b) False

6.15 Because of the nature of the power-flow bus data, nodal equations do not directly fit the linear-equation format, and power-flow equations are actually nonlinear. However, Gauss-Siedel method can be used for the power-flow solution.
(a) True (b) False

SECTION 6.6

6.16 The Newton-Raphson method is most well suited for solving the nonlinear power-flow equations.
(a) True (b) False

6.17 By default, PowerWorld Simulator uses ________ method for the power-flow solution. Fill in the Blank.

SECTION 6.7

6.18 Prime-mover control of a generator is responsible for a significant change in ________, whereas excitation control significantly changes ________. Fill in the Blanks.

6.19 From the power-flow standpoint, the addition of a shunt-capacitor bank to a load bus corresponds to the addition of a positive/negative reactive load. Choose the right word.

6.20 Tap-changing and voltage-magnitude-regulating transformers are used to control bus voltages and reactive power flows on lines to which they are connected.
(a) True (b) False

SECTION 6.8

6.21 A matrix, which has only a few nonzero elements, is said to be ________. Fill in the Blank.

6.22 Sparse-matrix techniques are used in Newton-Raphson power-flow programs in order to reduce computer ________ and ________ requirements. Fill in the Blanks.

6.23 Reordering buses can be an effective sparsity technique, in power-flow solution.
(a) True (b) False

SECTION 6.9

6.24 While the fast decoupled power flow usually takes more iterations to converge, it is usually significantly faster than the Newton-Raphson method.
(a) True (b) False

SECTION 6.10

6.25 The "dc" power-flow solution, giving approximate answers, is based on completely neglecting the Q–V equation, and solving the linear real-power balance equations.
(a) True (b) False

PROBLEMS

SECTION 6.1

6.1 Using Gauss elimination, solve the following linear algebraic equations:

$$-25x_1 + 5x_2 + 10x_3 + 10x_4 = 0$$
$$5x_1 - 10x_2 + 5x_3 = 2$$
$$10x_1 + 5x_2 - 10x_3 + 10x_4 = 1$$
$$10x_1 - 20x_4 = -2$$

6.2 Using Gauss elimination and back substitution, solve

$$\begin{bmatrix} 6 & 2 & 1 \\ 4 & 10 & 2 \\ 3 & 4 & 14 \end{bmatrix} \begin{bmatrix} x_1 \\ x_2 \\ x_3 \end{bmatrix} = \begin{bmatrix} 3 \\ 4 \\ 2 \end{bmatrix}$$

6.3 Rework Problem 6.2 with the value of A_{11} changed to 4.

6.4 What is the difficulty in applying Gauss elimination to the following linear algebraic equations?

$$-10x_1 + 10x_2 = 10$$
$$5x_1 - 5x_2 = -10$$

6.5 Show that, after triangularizing $\mathbf{Ax} = \mathbf{y}$, the back substitution method of solving $\mathbf{A}^{(N-1)}\mathbf{x} = \mathbf{y}^{(N-1)}$ requires N divisions, $N(N-1)/2$ multiplications, and $N(N-1)/2$ subtractions. Assume that all the elements of $\mathbf{A}^{(N-1)}$ and $\mathbf{y}^{(N-1)}$ are nonzero and real.

SECTION 6.2

6.6 Solve Problem 6.2 using the Jacobi iterative method. Start with $x_1(0) = x_2(0) = x_3(0) = 0$, and continue until (6.2.2) is satisfied with $\varepsilon = 0.01$.

6.7 Repeat Problem 6.6 using the Gauss–Seidel iterative method. Which method converges more rapidly?

6.8 Express the below set of equations in the form of (6.2.6), and then solve using the Jacobi iterative method with $\varepsilon = 0.05$, and $x_1(0), = 1, x_2(0) = 1, x_3(0) = 0$.

$$\begin{bmatrix} 10 & -2 & -4 \\ -2 & 6 & -2 \\ -4 & -2 & 10 \end{bmatrix} \begin{bmatrix} x_1 \\ x_2 \\ x_3 \end{bmatrix} = \begin{bmatrix} -2 \\ 3 \\ -1 \end{bmatrix}$$

6.9 Solve for x_1 and x_2 in the system of equations given by

$$x_2 - 3x_1 + 1.9 = 0$$
$$x_2 + x_1^2 - 3.0 = 0$$

by Gauss method with an initial guess of $x_1 = 1$ and $x_2 = 1$.

6.10 Solve $x^2 - 4x + 1 = 0$ using the Jacobi iterative method with $x(0) = 1$. Continue until (Eq. 6.2.2) is satisfied with $\varepsilon = 0.01$. Check using the quadratic formula.

6.11 Try to solve Problem 6.2 using the Jacobi and Gauss–Seidel iterative methods with the value of A_{33} changed from 14 to 0.14 and with $x_1(0) = x_2(0) = x_3(0) = 0$. Show that neither method converges to the unique solution.

6.12 Using the Jacobi method (also known as the Gauss method), solve for x_1 and x_2 in the system of equations.

$$x_2 - 3x_1 + 1.9 = 0$$
$$x_2 + {x_1}^2 - 1.8 = 0$$

Use an initial guess $x_1(0) = 1.0 = x_2(0) = 1.0$. Also, see what happens when you choose an uneducated initial guess $x_1(0) = x_2(0) = 100$.

6.13 Use the Gauss-Seidel method to solve the following equations that contain terms that are often found in power-flow equations.

$$x_1 = (1/(-20j)) * [(-1 + 0.5j)/(x_1)^* - (j10) * x_2 - (j10)]$$
$$x_2 = (1/(-20j)) * [(-3 + j)/(x_2)^* - (j10) * x_1 - (j10)]$$

Use an initial estimate of $x_1(0) = 1$ and $x_2(0) = 1$, and a stopping of $\varepsilon = 0.05$.

6.14 Find a root of the following equation by using the Gauss-Seidel method: (use an initial estimate of $x = 2$) $f(x) = x^3 - 6x^2 + 9x - 4 = 0$.

6.15 Use the Jacobi method to find a solution to $x^2 \cos x - x + 0.5 = 0$. Use $x(0) = 1$ and $\varepsilon = 0.01$. Experimentally determine the range of initial values that results in convergence.

6.16 Take the z-transform of (6.2.6) and show that $\mathbf{X}(z) = \mathbf{G}(z)\mathbf{Y}(z)$, where $\mathbf{G}(z) = (z\mathbf{U} - \mathbf{M})^{-1}\mathbf{D}^{-1}$ and $\mathbf{U}$ is the unit matrix.

$\mathbf{G}(z)$ is the matrix transfer function of a digital filter that represents the Jacobi or Gauss–Seidel methods. The filter poles are obtained by solving $\det(z\mathbf{U} - \mathbf{M}) = 0$. The filter is stable if and only if all the poles have magnitudes less than 1.

6.17 Determine the poles of the Jacobi and Gauss–Seidel digital filters for the general two-dimensional problem ($N = 2$):

$$\left[\begin{array}{c|c} A_{11} & A_{12} \\ \hline A_{21} & A_{22} \end{array}\right]\begin{bmatrix} x_1 \\ x_2 \end{bmatrix} = \begin{bmatrix} y_1 \\ y_2 \end{bmatrix}$$

Then determine a necessary and sufficient condition for convergence of these filters when $N = 2$.

SECTION 6.3

6.18 Use Newton–Raphson to find a solution to the polynomial equation $f(x) = y$ where $y = 0$ and $f(x) = x^3 + 8x^2 + 2x - 50$. Start with $x(0) = 1$ and continue until (6.2.2) is satisfied with $\varepsilon = 0.001$.

6.19 Repeat 6.19 using $x(0) = -2$.

6.20 Use Newton–Raphson to find one solution to the polynomial equation $f(x) = y$, where $y = 7$ and $f(x) = x^4 + 3x^3 - 15x^2 - 19x + 30$. Start with $x(0) = 0$ and continue until (6.2.2) is satisfied with $\varepsilon = 0.001$.

6.21 Repeat Problem 6.20 with an initial guess of $x(1) = 4$.

6.22 For Problem 6.20 plot the function $f(x)$ between $x = 0$ and 4. Then provide a graphical interpretation why points close to $x = 2.2$ would be poorer initial guesses.

6.23 Use Newton–Raphson to find a solution to

$$\begin{bmatrix} e^{x_1 x_2} \\ \cos(x_1 + x_2) \end{bmatrix} = \begin{bmatrix} 1.2 \\ 0.5 \end{bmatrix}$$

where x_1 and x_2 are in radians. (a) Start with $x_1(0) = 1.0$ and $x_2(0) = 0.5$ and continue until (6.2.2) is satisfied with $\varepsilon = 0.005$. (b) Show that Newton–Raphson diverges for this example if $x_1(0) = 1.0$ and $x_2(0) = 2.0$.

6.24 Solve the following equations by the Newton–Raphson method:

$$2x_1^2 + x_2^2 - 10 = 0$$

$$x_1^2 - x_2^2 + x_1 x_2 - 4 = 0$$

Start with an initial guess of $x_1 = 1$ and $x_2 = 1$.

6.25 The following nonlinear equations contain terms that are often found in the power-flow equations:

$$f_1(x) = 10x_1 \sin x_2 + 2 = 0$$
$$f_2(x) = 10(x_1)^2 - 10x_1 \cos x_2 + 1 = 0$$

Solve using the Newton–Raphson method starting with an initial guess of $x_1(0) = 1$ and $x_2(0) = 0$ radians, and a stopping criteria of $\varepsilon = 10^{-4}$.

6.26 Repeat 6.25 except using $x_1(0) = 0.25$ and $x_2(0) = 0$ radians as an initial guess.

6.27 For the Newton–Raphson method the *region of attraction* (or *basin of attraction*) for a particular solution is the set of all initial guesses that converge to that solution. Usually initial guesses close to a particular solution will converge to that solution. However, for all but the simplest of multi-dimensional, nonlinear problems the region of attraction boundary is often fractal. This makes it impossible to quantify the region of attraction, and hence to guarantee convergence. Problem 6.25 has two solutions when x_2 is restricted to being between $-\pi$ and π. With the x_2 initial guess fixed at 0 radians, numerically determine the values of the x_1 initial guesses that converge to the Problem 6.25 solution. Restrict your search to values of x_1 between 0 and 1.

SECTION 6.4

6.28 Consider the simplified electric power system shown in Figure 6.17 for which the power-flow solution can be obtained without resorting to iterative techniques. (a) Compute the elements of the bus admittance matrix Y_{bus}. (b) Calculate the phase angle δ_2 by using the real power equation at bus 2 (voltage-controlled bus). (c) Determine $|V_3|$ and δ_3 by using both the real and reactive power equations at bus 3 (load bus). (d) Find the real power generated at bus 1 (swing bus). (e) Evaluate the total real power losses in the system.

6.29 In Example 6.9, double the impedance on the line from bus 2 to bus 5. Determine the new values for the second row of Y_{bus}. Verify your result using PowerWorld Simulator case Example 6.9.

6.30 Determine the bus admittance matrix (Y_{bus}) for the following power three phase system (note that some of the values have already been determined for you). Assume a three-phase 100 MVA per unit base.

6.31 For the system from Problem 6.30, assume that a 75 Mvar shunt capacitance (three phase assuming one per unit bus voltage) is added at bus 4. Calculate the new value of Y_{44}.

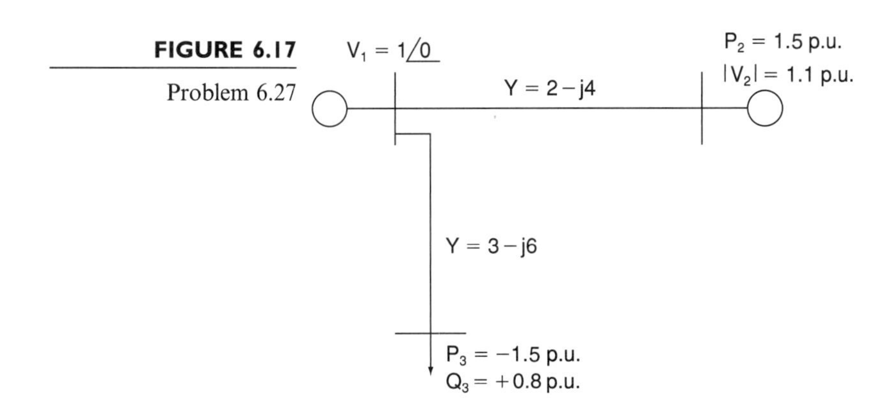

FIGURE 6.17 Problem 6.27

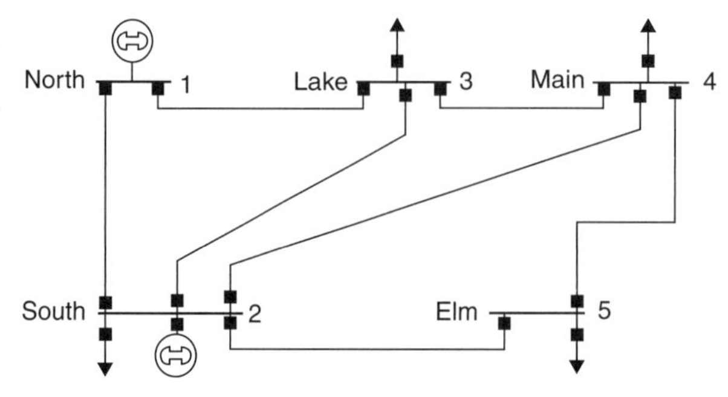

FIGURE 6.18 Sample System Diagram

TABLE 6.9 Bus input data for Problem 6.30

Bus-to-Bus	R per unit	X per unit	B per unit
1-2	0.02	0.06	0.06
1-3	0.08	0.24	0.05
2-3	0.06	0.18	0.04
2-4	0.08	0.24	0.05
2-5	0.02	0.06	0.02
3-4	0.01	0.04	0.01
4-5	0.03	0.10	0.04

TABLE 6.10 Partially Completed Bus Admittance Matrix (Y_{bus})

$6.25 - j18.695$	$-5.00 + j15.00$	$-1.25 + j3.75$	0	0
$-5.00 + j15.00$				

SECTION 6.5

6.32 Assume a $0.8 + j0.4$ per unit load at bus 2 is being supplied by a generator at bus 1 through a transmission line with series impedance of $0.05 + j0.1$ per unit. Assuming bus 1 is the swing bus with a fixed per unit voltage of $1.0\underline{/0}$, use the Gauss-Seidel method to calculate the voltage at bus 2 after three iterations.

6.33 Repeat the above problem with the swing bus voltage changed to $1.0\underline{/30^\circ}$ per unit.

6.34 For the three bus system whose Y_{bus} is given below, calculate the second iteration value of V_3 using the Gauss-Seidel method. Assume bus 1 as the slack (with $V_1 = 1.0\underline{/0^\circ}$), and buses 2 and 3 are load buses with a per unit load of $S_2 = 1 + j0.5$ and $S_3 = 1.5 + j0.75$. Use voltage guesses of $1.0\underline{/0^\circ}$ at both buses 2 and 3. The bus admittance matrix for a three-bus system is

$$Y_{bus} = \begin{bmatrix} -j10 & j5 & j5 \\ j5 & -j10 & j5 \\ j5 & j2 & -j10 \end{bmatrix}$$

6.35 Repeat Problem 6.34 except assume the bus 1 (slack bus) voltage of $V_1 = 1.05\underline{/0^\circ}$.

FIGURE 6.19

Problem 6.36

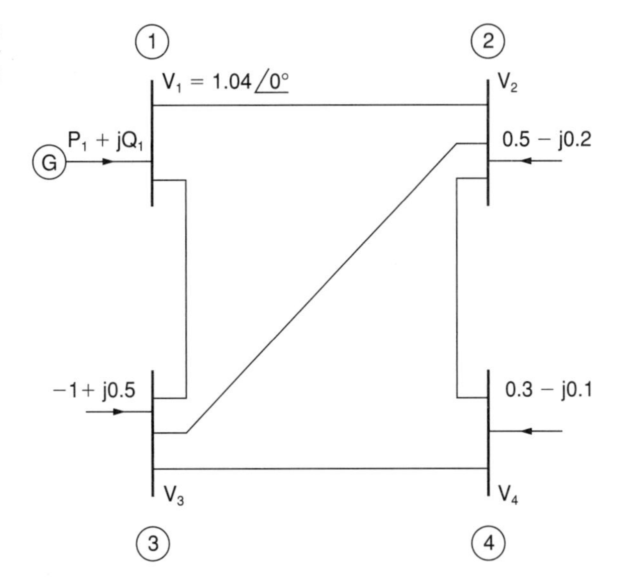

6.36 The bus admittance matrix for the power system shown in Figure 6.19 is given by

$$Y_{bus} = \begin{bmatrix} 3-j9 & -2+j6 & -1+j3 & 0 \\ -2+j6 & 3.666-j11 & -0.666+j2 & -1+j3 \\ -1+j3 & -0.666+j2 & 3.666-j11 & -2+j6 \\ 0 & -1+j3 & -2+j6 & 3-j9 \end{bmatrix} \text{ per unit}$$

With the complex powers on load buses 2, 3, and 4 as shown in Figure 6.19, determine the value for V_2 that is produced by the first and second iterations of the Gauss–Seidel procedure. Choose the initial guess $V_2(0) = V_3(0) = V_4(0) = 1.0\underline{/0°}$ per unit.

6.37 The bus admittance matrix of a three-bus power system is given by

$$Y_{bus} = -j\begin{bmatrix} 7 & -2 & -5 \\ -2 & 6 & -4 \\ -5 & -4 & 9 \end{bmatrix} \text{ per unit}$$

with $V_1 = 1.0\underline{/0°}$ per unit; $V_2 = 1.0$ per unit; $P_2 = 60$ MW; $P_3 = -80$ MW; $Q_3 = -60$ MVAR (lagging) as a part of the power-flow solution of the system, find V_2 and V_3 within a tolerance of 0.01 per unit, by using Gauss-Seidel iteration method. Start with $\delta_2 = 0$, $V_3 = 1.0$ per unit, and $\delta_3 = 0$.

SECTION 6.6

6.38 A generator bus (with a 1.0 per unit voltage) supplies a 150 MW, 50 Mvar load through a lossless transmission line with per unit (100 MVA base) impedance of $j0.1$ and no line charging. Starting with an initial voltage guess of $1.0\underline{/0°}$, iterate until converged using the Newton–Raphson power flow method. For convergence criteria use a maximum power flow mismatch of 0.1 MVA.

6.39 Repeat Problem 6.37 except use an initial voltage guess of $1.0\underline{/30°}$.

6.40 Repeat Problem 6.37 except use an initial voltage guess of $0.25\underline{/0°}$.

6.41 Determine the initial Jacobian matrix for the power system described in Problem 6.33.

6.42 Use the Newton–Raphson power flow to solve the power system described in Problem 6.34. For convergence criteria use a maximum power flow mismatch of 0.1 MVA.

6.43 For a three bus power system assume bus 1 is the swing with a per unit voltage of $1.0\underline{/0^\circ}$, bus 2 is a PQ bus with a per unit load of $2.0 + j0.5$, and bus 3 is a PV bus with 1.0 per unit generation and a 1.0 voltage setpoint. The per unit line impedances are $j0.1$ between buses 1 and 2, $j0.4$ between buses 1 and 3, and $j0.2$ between buses 2 and 3. Using a flat start, use the Newton–Raphson approach to determine the first iteration phasor voltages at buses 2 and 3.

6.44 Repeat Problem 6.42 except with the bus 2 real power load changed to 1.0 per unit.

PW **6.45** Load PowerWorld Simulator case Example 6.11; this case is set to perform a single iteration of the Newton–Raphson power flow each time **Single Solution** is selected. Verify that initially the Jacobian element J_{33} is 104.41. Then, give and verify the value of this element after each of the next three iterations (until the case converges).

PW **6.46** Load PowerWorld Simulator case Problem 6_46. Using a 100 MVA base, each of the three transmission lines have an impedance of $0.05 + j0.1$ pu. There is a single 180 MW load at bus 3, while bus 2 is a PV bus with generation of 80 MW and a voltage setpoint of 1.0 pu. Bus 1 is the system slack with a voltage setpoint of 1.0 pu. Manually solve this case using the Newton–Raphson approach with a convergence criteria of 0.1 MVA. Show all your work. Then verify your solution by solving the case with PowerWorld Simulator.

PW **6.47** As was mentioned in Section 6.4, if a generator's reactive power output reaches its limit, then it is modeled as though it were a PQ bus. Repeat Problem 6.46, except assume the generator at bus 2 is operating with its reactive power limited to a maximum of 50 Mvar. Then verify your solution by solving the case with PowerWorld Simulator. To increase the reactive power output of the bus 2 generator, select **Tools, Play** to begin the power flow simulation, then click on the up arrow on the bus 2 magenta voltage setpoint field until the reactive power output reaches its maximum.

PW **6.48** Load PowerWorld Simulator case Problem 6_46. Plot the reactive power output of the generator at bus 2 as a function of its voltage setpoint value in 0.005 pu voltage steps over the range between its lower limit of −50 Mvar and its upper limit of 50 Mvar. To change the generator 2 voltage set point first select **Tools, Play** to begin the power flow simulation, and then click on the up/down arrows on the bus 2 magenta voltage setpoint field.

SECTION 6.7

PW **6.49** Open PowerWorld Simulator case Problem 6_49. This case is identical to Example 6.9 except that the transformer between buses 1 and 5 is now a tap-changing transformer with a tap range between 0.9 and 1.1 and a tap step size of 0.00625. The tap is on the high side of the transformer. As the tap is varied between 0.975 and 1.1, show the variation in the reactive power output of generator 1, V_5, V_2, and the total real power losses.

PW **6.50** Use PowerWorld Simulator to determine the Mvar rating of the shunt capacitor bank in the Example 6_14 case that increases V_2 to 1.0 per unit. Also determine the effect of this capacitor bank on line loadings and the total real power losses (shown immediately below bus 2 on the one-line). To vary the capacitor's nominal Mvar rating, right-click on the capacitor symbol to view the Switched Shunt Dialog, and then change Nominal Mvar field.

PW **6.51** Use PowerWorld Simulator to modify the Example 6.9 case by inserting a second line between bus 2 and bus 5. Give the new line a circuit identifier of "2" to distinguish it from the existing line. The line parameters of the added line should be identical to those of the existing lines 2–5. Determine the new line's effect on V_2, the line loadings, and on the total real power losses.

PW **6.52** Open PowerWorld Simulator case Problem 6_52. Open the 69 kV line between buses HOMER69 and LAUF69 (shown toward the bottom-left). With the line open, determine the amount of Mvar (to the nearest 1 Mvar) needed from the HANNAH69 capacitor bank to correct the HANNAH69 voltage to at least 1.0 pu.

PW **6.53** Open PowerWorld Simulator case Problem 6_53. Plot the variation in the total system real power losses as the generation at bus BLT138 is varied in 20-MW blocks between 0 MW and 400 MW. What value of BLT138 generation minimizes the total system losses?

PW **6.54** Repeat Problem 6.53, except first remove the 138-69 kV transformer between BLT138 and BLT69.

SECTION 6.8

6.55 Using the compact storage technique described in Section 6.8, determine the vectors **DIAG**, **OFFDIAG**, **COL**, and **ROW** for the following matrix:

$$\mathbf{S} = \begin{bmatrix} 17 & -9.1 & 0 & 0 & -2.1 & -7.1 \\ -9.1 & 25 & -8.1 & -1.1 & -6.1 & 0 \\ 0 & -8.1 & 9 & 0 & 0 & 0 \\ 0 & -1.1 & 0 & 2 & 0 & 0 \\ -2.1 & -6.1 & 0 & 0 & 14 & -5.1 \\ -7.1 & 0 & 0 & 0 & -5.1 & 15 \end{bmatrix}$$

6.56 For the triangular factorization of the corresponding $\boldsymbol{Y}_{\text{bus}}$, number the nodes of the graph shown in Figure 6.9 in an optimal order.

SECTION 6.10

6.57 Compare the angles and line flows between the Example 6.17 case and results shown in Tables 6.6, 6.7, and 6.8.

6.58 Redo Example 6.17 with the assumption that the per unit reactance on the line between buses 2 and 5 is changed from 0.05 to 0.03.

PW **6.59** Open PowerWorld Simulator case Problem 6.58, which models a seven bus system using the dc power flow approximation. Bus 7 is the system slack. The real power generation/load at each bus is as shown, while the per unit reactance of each of the lines (on a 100 MVA base) is as shown in yellow on the one-line. (a) Determine the six by six **B** matrix for this system and the **P** vector. (b) Use a matrix package such as Matlab to verify the angles as shown on the one-line.

PW **6.60** Using the PowerWorld Simulator case from Problem 6.59, if the rating on the line between buses 1 and 3 is 65 MW, the current flow is 59 MW (from one to three), and the current bus one generation is 160 MW, analytically determine the amount this generation can increase until this line reaches 100% flow. Assume any change in the bus 1 generation is absorbed at the system slack.

SECTION 6.11

6.61 PowerWorld Simulator cases Problem 6_61_PQ and 6_61_PV model a seven bus power system in which the generation at bus 4 is modeled as a Type 1 or 2 wind turbine in the first case, and as a Type 3 or 4 wind turbine in the second. A shunt capacitor is used to make the net reactive power injection at the bus the same in both cases. Compare the bus 4 voltage between the two cases for a contingency in which the line between buses 2 and 4 is opened. What is an advantage of a Type 3 or 4 wind turbine with respect to voltage regulation following a contingency? What is the variation in the Mvar output of a shunt capacitor with respect to bus voltage magnitude?

CASE STUDY QUESTIONS

A. What are some of the benefits of a high voltage electric transmission system?

B. Why is transmission capacity in the U.S. decreasing?

C. How has transmission planning changed since the mid 1990s?

D. How is the power flow used in the transmission planning process?

DESIGN PROJECT 1: A NEW WIND FARM

You've just been hired as a new power engineer with Kyle and Weber Wind (KWW), one of the country's leading wind energy developers. KWW has identified the rolling hills to the northwest of the Metropolis urban area as an ideal location for a new 200 MW wind farm. The local utility, Metropolis Light and Power (MLP), seems amenable to this new generation development taking place within their service territory. However, they are also quite adamant that any of the costs associated with transmission system upgrades necessary to site this new generation be funded by KWW. Therefore, your supervisor at KWW has requested that you do a preliminary transmission planning assessment to determine the least cost design.

Hence, your job is to make recommendations on the least cost design for the construction of new lines and transformers to ensure that the transmission system in the MLP system is adequate for any base case or first contingency loading situation when the KWW wind farm is installed and operating at its maximum output of 200 MW. Since the wind farm will be built with Type 3 DFAG wind turbines, you can model the wind farm in the power flow as a single, equivalent traditional PV bus generator with an output of 200 MW, a voltage setpoint of 1.05 per unit, and with reactive power limits of ± 100 Mvar. In keeping with KWW tradition, the wind interconnection point will be at 69 kV, and for reliability purposes your supervisor requests that there be two separate feeds into the interconnection substation.

The following table shows the available right-of-way distances for the construction of new 69 kV and/or new 138 kV lines. All existing 69 kV only substations are large enough to accommodate 138 kV as well.

Design Procedure

1. Load DesignCase1 into PowerWorld Simulator. This case contains the initial system power flow case, and the disconnected KWW generator and its interconnection bus. Perform an initial power-flow solution to determine the initial system operating point. From this solution you should find that all the line flows and bus voltage magnitudes are within their limits. Assume all line MVA flows must be at or below 100% of their limit values, and all voltages must be between 0.95 and 1.10 per unit.

2. Repeat the above analysis considering the impact of any single transmission line or transformer outage. This is known as n-1 contingency analysis. To simplify this analysis, PowerWorld Simulator has the ability to automatically perform a contingency analysis study. Select **Tools, Contingency Analysis** to show the Contingency Analysis display. Note that the 57 single line/transformer contingencies are already defined. Select **Start Run** (toward the bottom right corner of the display) to automatically see the impact of removing any single element. Without the KWW generation the system has no contingency (n-1) violations.

3. Using the available rights-of-ways and the transmission line parameters/costs given in the table, iteratively determine the least expensive system additions so that the base case and all the contingences result in reliable operation points with the KWW generation connected with an output of 200 MW. The parameters of the new transmission lines(s) need to be derived using the tower configurations and conductor types provided by the instructor. In addition, the transmission changes you propose will modify the total system losses, indicated by the yellow field on the one-line. While the system losses are not KWW's responsibility, your supervisor has asked you to consider the impact your design changes will have on the total system losses assuming the system operates in the studied condition for the next five years. Hence, you should minimize the total construction costs minus the savings associated with any decrease in system losses over the next five years.

4. Write a detailed report including the justification for your final recommendation.

Simplifying Assumptions

To simplify the analysis, several assumptions are made:

1. You need only consider the base case loading level given in DesignCase1. In a real design, typically a number of different operating points/loading levels must be considered.

2. You should consider all the generator real power outputs, including that of the new KWW generation, as fixed values. The change in the total system generation due to the addition of the 200 MW in KWW generation and any changes in the system losses are always picked up by the system slack.
3. You should not modify the status of the capacitors or the transformer taps.
4. You should assume that the system losses remain constant over the five-year period, and you need only consider the impact and new design has on the base case losses. The price for losses can be assumed to be $50/MWh.
5. You do not need to consider contingencies involving the new transmission lines and possibly any transformers you may be adding.

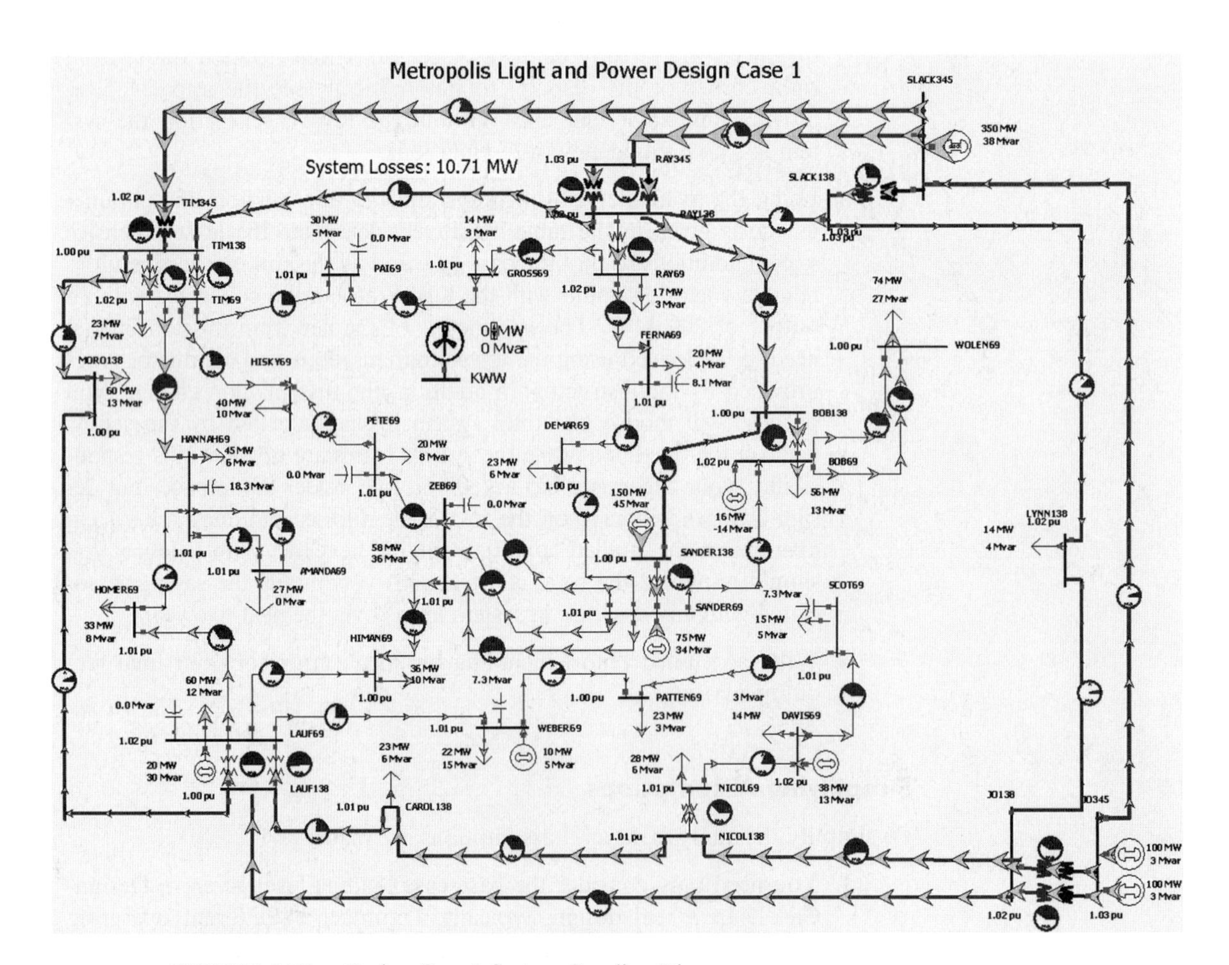

FIGURE 6.20 Design Case 1 System One-line Diagram

6. While an appropriate control response to a contingency might be to decrease the KWW wind farm output (by changing the pitch on the wind turbine blades), your supervisor has specifically asked you not to consider this possibility. Therefore the KWW generator should always be assumed to have a 200 MW output.

Available New Rights-of-Ways for Design Case 1

Right-of-Way/Substation	Right-of-Way Mileage(km)
KWW to PAI	9.66
KWW to PETE	11.91
KKWW to DEMAR	19.31
KKWW to GROSS	7.24
KKWW to HISKY	18.02
KKWW to TIM	20.92
KKWW to RAY	24.14
KWW to ZEB	17.7

DESIGN PROJECT 2: SYSTEM PLANNING FOR GENERATION RETIREMENT

After more than 70 years of supplying downtown Metropolis with electricity it is time to retire the SANDERS69 power plant. The city's downtown revitalization plan, coupled with a desire for more green space, make it impossible to build new generation in the downtown area. At the same time, a booming local economy means that the city-wide electric demand is still as high as ever, so this impending plant retirement is going to have some adverse impacts on the electric grid. As a planning engineer for the local utility, Metropolis Light and Power (MLP), your job is to make recommendations on the construction of new lines and transformers to ensure that the transmission system in the MLP system is adequate for any base case or first contingency loading situation. The below table shows the right-of-way distances that are available for the construction of new 69 kV and/or new 138 kV lines. All existing 69 kV only substations are large enough to accommodate 138 kV as well.

Design Procedure

1. Load DesignCase2 into PowerWorld Simulator which contains the system dispatch without the SANDERS69 generator. Perform an initial power flow solution to determine the initial system operating point. From this solution you should find that all the line flows and bus voltage magnitudes are within their limits. Assume all line MVA

flows must be at or below 100% of their limit values, and all voltages must be between 0.95 and 1.10 per unit.

2. Repeat the above analysis considering the impact of any single transmission line or transformer outage. This is known as n-1 contingency analysis. To simplify this analysis, PowerWorld Simulator has the ability to automatically perform a contingency analysis study. Select **Tools, Contingency Analysis** to show the Contingency Analysis display. Note that the 57 single line/transformer contingencies are already defined. Select **Start Run** (toward the bottom right corner of the display) to automatically see the impact of removing any single element. Without the SANDERS69 generation this system is insecure for several contingencies, including at least one that has nothing to do with the power plant retirement (but it still needs to be fixed).

3. Using the rights-of-way and the transmission line parameters/costs given in the table, iteratively determine the least expensive system additions so that the base case and all the contingences result in secure operation points. The parameters of the new transmission lines(s) need to be derived using the tower configurations and conductor types provided by the instructor. The total cost of an addition is defined as the construction costs minus the savings associated with any decrease in system losses over the next five years.

4. Write a detailed report discussing the initial system problems, your approach to optimally solving the system problems and the justification for your final recommendation.

Simplifying Assumptions

To simplify the analysis, several assumptions are made:

1. You need only consider the base case loading level given in DesignCase2. In a real design, typically a number of different operating points/loading levels must be considered.

2. You should consider the generator outputs as fixed values; any changes in the losses are always picked up by the system slack.

3. You should not modify the status of the capacitors or the transformer taps.

4. You should assume that the system losses remain constant over the five-year period and need only consider the impact and new design has on the base case losses. The price for losses can be assumed to be $50/MWh.

Available New Rights-of-Ways

Right-of-Way/Substation	Right-of-Way Mileage (km)
BOB to SCOT	13.68
BOB to WOLEN	7.72
FERNA to RAY	9.66
LYNN to SCOT	19.31
LYNN to WOLEN	24.14
SANDER to SCOTT	9.66
SLACK to WOLEN	18.51
JO to SCOT	24.14

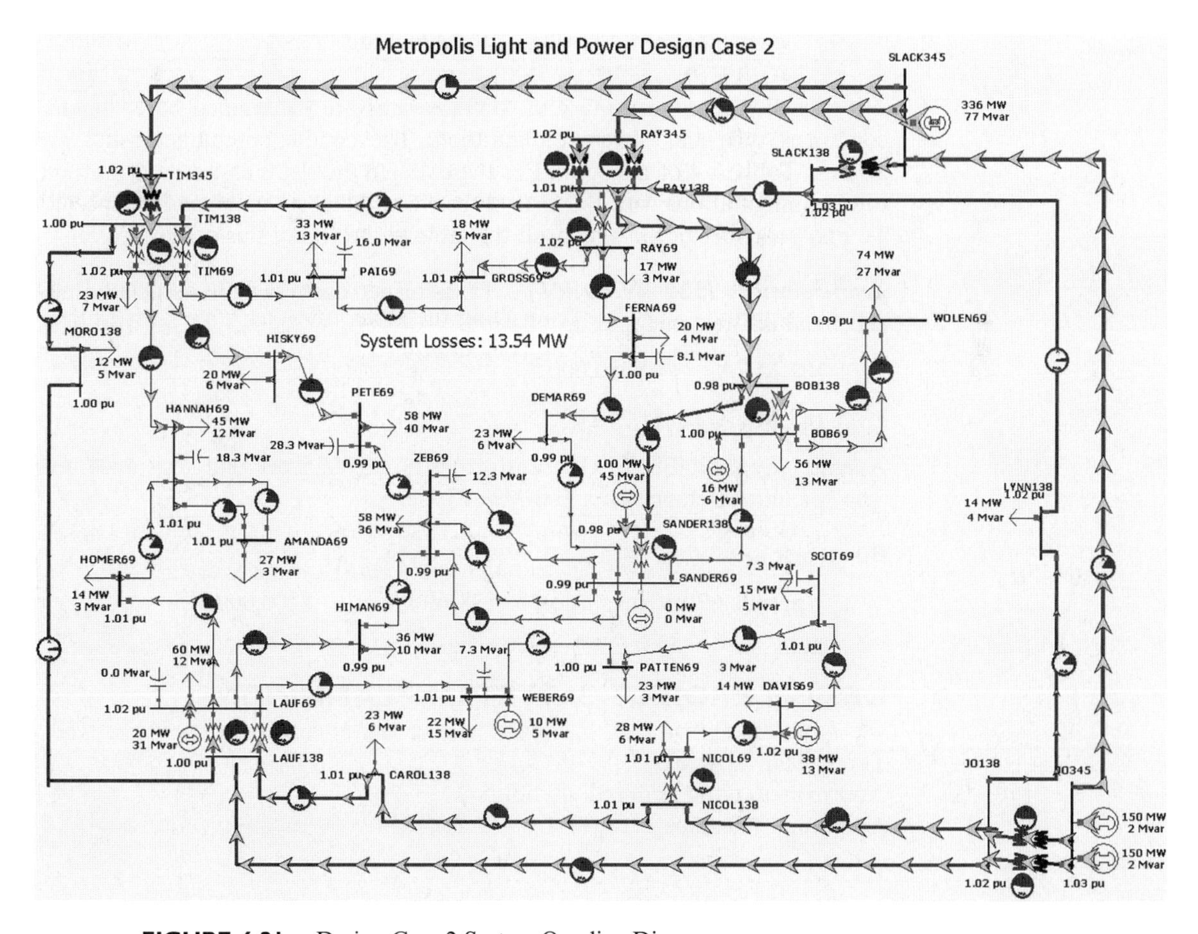

FIGURE 6.21 Design Case 2 System One-line Diagram

DESIGN PROJECTS 1 AND 2: SAMPLE TRANSMISSION SYSTEM DESIGN COSTS

Transmission lines (69 kV and 138 kV) New transmission lines include a fixed cost and a variable cost. The fixed cost is for the design work, the purchase/installation of the three-phase circuit breakers, associated relays, and changes to the substation bus structure. The fixed costs are **$200,000** for a 138-kV line and **$125,000** for a 69-kV line.

The variable costs depend on the type of conductor and the length of the line. The assumed cost in $/km are given here.

Conductor Type	Current Rating (Amps)	138-kV Lines	69-kV Lines
Rook	770	$250,000/km	$200,000/km
Crow	830	$270,000/km	$220,000/km
Condor	900	$290,000/km	$240,000/km
Cardinal	1110	$310,000/km	

Lined impedance data and MVA ratings are determined based on the conductor type and tower configuration. The conductor characteristics are given in Table A.4 of the book. For these design problems assume a symmetric tower configurations with the spacing between the conductors student specific. To find your specific value consult the table at the end of this design project.

Transformers (138 kV/69 kV) Transformer costs include associated circuit breakers, relaying and installation.

101 MVA $950,000

187 MVA $1,200,000

Assume any new 138/69 kV transformer has 0.0025 per unit resistance and 0.04 per unit reactance on a 100-MVA base.

Bus work

Upgrade 69-kV substation to 138/69 kV $200,000

DESIGN PROJECT 3: SYSTEM PLANNING*

Time given: 11 weeks
Approximate time required: 40 hours
Additional references: [10, 11]

*This case is based on a project assigned by Adjunct Professor Leonard Dow at Northeastern University, Boston, Massachusetts.

FIGURE 6.22 Design Project 3: Single-line diagram for 31-bus interconnected power system

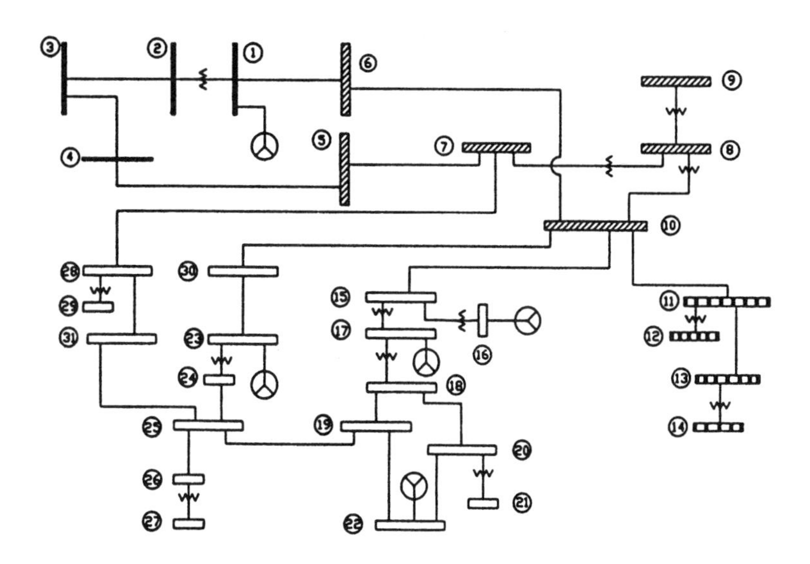

Figure 6.22 shows a single-line diagram of four interconnected power systems identified by different graphic bus designations. The following data are given:

1. There are 31 buses, 21 lines, and 13 transformers.
2. Generation is present at buses 1, 16, 17, 22, and 23.
3. Total load of the four systems is 400 MW.
4. Bus 1 is the swing bus.
5. The system base is 100 MVA.
6. Additional information on transformers and transmission lines is provided in [10, 11].

Based on the data given:

1. Allocate the total 400-MW system load among the four systems.
2. For each system, allocate the load to buses that you want to represent as load buses. Select reasonable load power factors.
3. Taking into consideration the load you allocated above, select appropriate transmission-line voltage ratings, MVA ratings, and distances necessary to supply these loads. Then determine per-unit transmission-line impedances for the lines shown on the single-line diagram (show your calculations).
4. Also select appropriate transformer voltage and MVA ratings, and determine per-unit transformer leakage impedances for the transformers shown on the single-line diagram.
5. Develop a generation schedule for the 5 generator buses.

6. Show on a copy of the single-line diagram per-unit line impedances, transformer impedances, generator outputs, and loads that you selected above.

7. Using PowerWorld Simulator, run a base case power flow. In addition to the printed input/output data files, show on a separate copy of the single-line diagram per-unit bus voltages as well as real and reactive line flows, generator outputs, and loads. Flag any high/low bus voltages for which $0.95 \leq V \leq 1.05$ per unit and any line or transformer flows that exceed normal ratings.

8. If the base case shows any high/low voltages or ratings exceeded, then correct the base case by making changes. Explain the changes you have made.

9. Repeat (7). Rerun the power-flow program and show your changes on a separate copy of the single-line diagram.

10. Provide a typed summary of your results along with your above calculations, printed power-flow input/output data files, and copies of the single-line diagram.

DESIGN PROJECT 4: POWER FLOW/SHORT CIRCUITS

Time given: 3 weeks
Approximate time required: 15 hours

Each student is assigned one of the single-line diagrams shown in Figures 6.23 and 6.24. Also, the length of line 2 in these figures is varied for each student.

Assignment 1: Power-Flow Preparation

For the single-line diagram that you have been assigned (Figure 6.23 or 6.24), convert all positive-sequence impedance, load, and voltage data to per unit using the given system base quantities. Then using PowerWorld Simulator, create three input data files: bus input data, line input data, and transformer input data. Note that bus 1 is the swing bus. Your output for this assignment consists of three power-flow input data files.

The purpose of this assignment is to get started and to correct errors before going to the next assignment. It requires a knowledge of the per-unit system, which was covered in Chapter 3, but may need review.

Assignment 2: Power Flow

Case 1. Run the power flow program and obtain the bus, line, and transformer input/output data files that you prepared in Assignment 1.

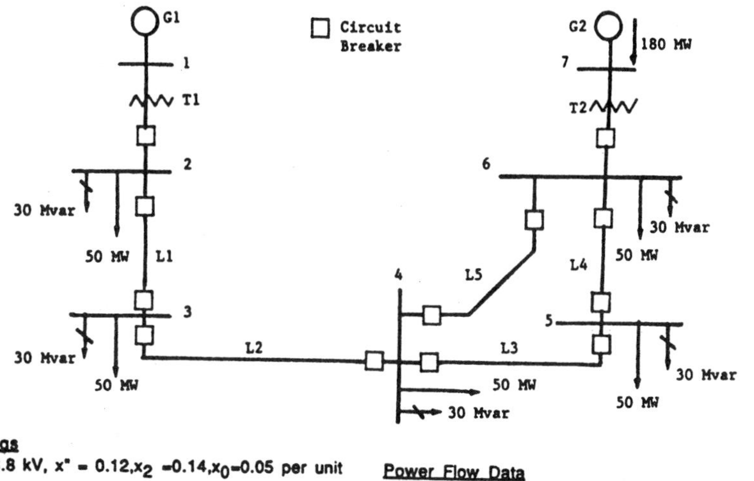

<u>Generator Ratings</u>
G1: 100 MVA, 13.8 kV, $x'' = 0.12$, $x_2 = 0.14$, $x_0 = 0.05$ per unit
G2: 200 MVA, 15.0 kV, $x'' = 0.12$, $x_2 = 0.14$, $x_0 = 0.05$ per unit
The generator neutrals are solidly grounded
<u>Transformer Ratings</u>
T1: 100 MVA, 13.8 kVΔ/230 kVY, $x = 0.1$ per unit
T2: 200 MVA, 15 kVΔ/230 kVY, $x = 0.1$ per unit
The transformer neutrals are solidly grounded
<u>Transmission Line Ratings</u>
All Lines: 230 kV, $z_1 = 0.08 + j0.5$ Ω/km,
$z_0 = 0.2 + j1.5$ Ω/km, $y_1 = j3.3$ E-6 S/km,
Maximum MVA = 400
Line Lengths: $L_1 = 15$ km, L_2 assigned by the instructor (20 to 50 km), $L_3 = 40$ km, $L_4 = 15$ km, $L_5 = 50$ km.

<u>Power Flow Data</u>
Bus 1 : Swing bus, $V_1 = 13.8$ kV, $\delta_1 = 0°$
Bus 2,3,4,5,6 : Load buses
Bus 7 : Constant voltage magnitude bus, $V_7 = 15$ kV,
$P_{G7} = 180$ MW, -87 Mvar $< Q_{G7} < +87$ Mvar
<u>System Base Quantities</u>
$S_{base} = 100$ MVA (three-phase)
$V_{base} = 13.8$ kV (line-to-line) in the zone of G1

FIGURE 6.23 Single-line diagram for Design Project 4—transmission loop

Case 2. Suggest one method of increasing the voltage magnitude at bus 4 by 5%. Demonstrate the effectiveness of your method by making appropriate changes to the input data of case 1 and by running the power flow program.

Your output for this assignment consists of 12 data files, 3 input and 3 output data files for each case, along with a one-paragraph explanation of your method for increasing the voltage at bus 4 by 5%.

During this assignment, course material contains voltage control methods, including use of generator excitation control, tap changing and regulating transformers, static capacitors, static var systems, and parallel transmission lines.

This project continues in Chapters 7 and 9.

DESIGN PROJECT 5: POWER FLOW*

Time given: 4 weeks
Approximate time required: 25 hours

*This case is based on a project assigned by Adjunct Professor Richard Farmer at Arizona State University, Tempe, Arizona.

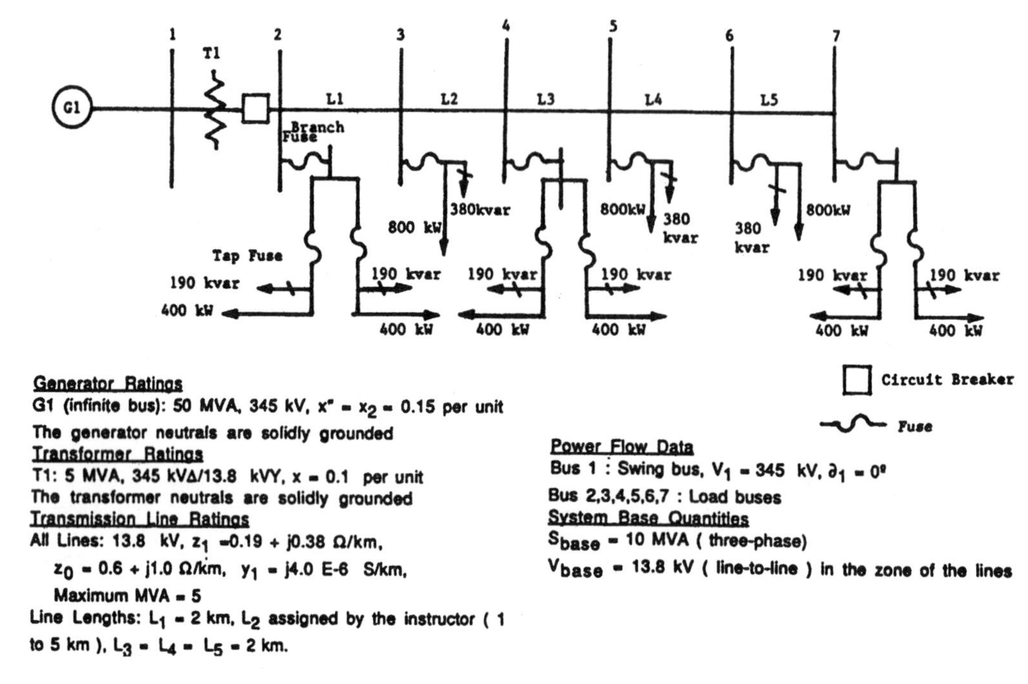

1
2
3
4
5
6
7
T1
G1
L1
L2
L3
L4
L5
Branch Fuse
Tap Fuse
190 kvar
400 kW
800 kW
380kvar
190 kvar
190 kvar
400 kW
400 kW
800kW
380 kvar
190 kvar
400 kW
380 kvar
800kW
190 kvar
190 kvar
400 kW
400 kW
Circuit Breaker
Fuse
Generator Ratings
G1 (infinite bus): 50 MVA, 345 kV, x" = x2 = 0.15 per unit
The generator neutrals are solidly grounded
Transformer Ratings
T1: 5 MVA, 345 kVΔ/13.8 kVY, x = 0.1 per unit
The transformer neutrals are solidly grounded
Transmission Line Ratings
All Lines: 13.8 kV, z1 = 0.19 + j0.38 Ω/km,
z0 = 0.6 + j1.0 Ω/km, y1 = j4.0 E-6 S/km,
Maximum MVA = 5
Line Lengths: L1 = 2 km, L2 assigned by the instructor (1 to 5 km), L3 = L4 = L5 = 2 km.
Power Flow Data
Bus 1 : Swing bus, V1 = 345 kV, ∂1 = 0°
Bus 2,3,4,5,6,7 : Load buses
System Base Quantities
Sbase = 10 MVA (three-phase)
Vbase = 13.8 kV (line-to-line) in the zone of the lines

FIGURE 6.24 Single-line diagram for Design Project 4—radial distribution feeder

Figure 6.25 shows the single-line diagram of a 10-bus power system with 7 generating units, 2 345-kV lines, 7 230-kV lines, and 5 transformers. Per-unit transformer leakage reactances, transmission-line series impedances and shunt susceptances, real power generation, and real and reactive loads during heavy load periods, all on a 100-MVA system base, are given on the diagram. Fixed transformer tap settings are also shown. During light load periods, the real and reactive loads (and generation) are 25% of those shown. Note that bus 1 is the swing bus.

Design Procedure

Using PowerWorld Simulator (convergence can be achieved by changing load buses to constant voltage magnitude buses with wide var limits), determine:

1. The amount of shunt compensation required at 230- and 345-kV buses such that the voltage magnitude $0.99 \leq V \leq 1.02$ per unit at all buses during both light and heavy loads. Find two settings for the compensation, one for light and one for heavy loads.

2. The amount of series compensation required during heavy loads on each 345-kV line such that there is a maximum of 40° angular displacement between bus 4 and bus 10. Assume that one 345-kV line is

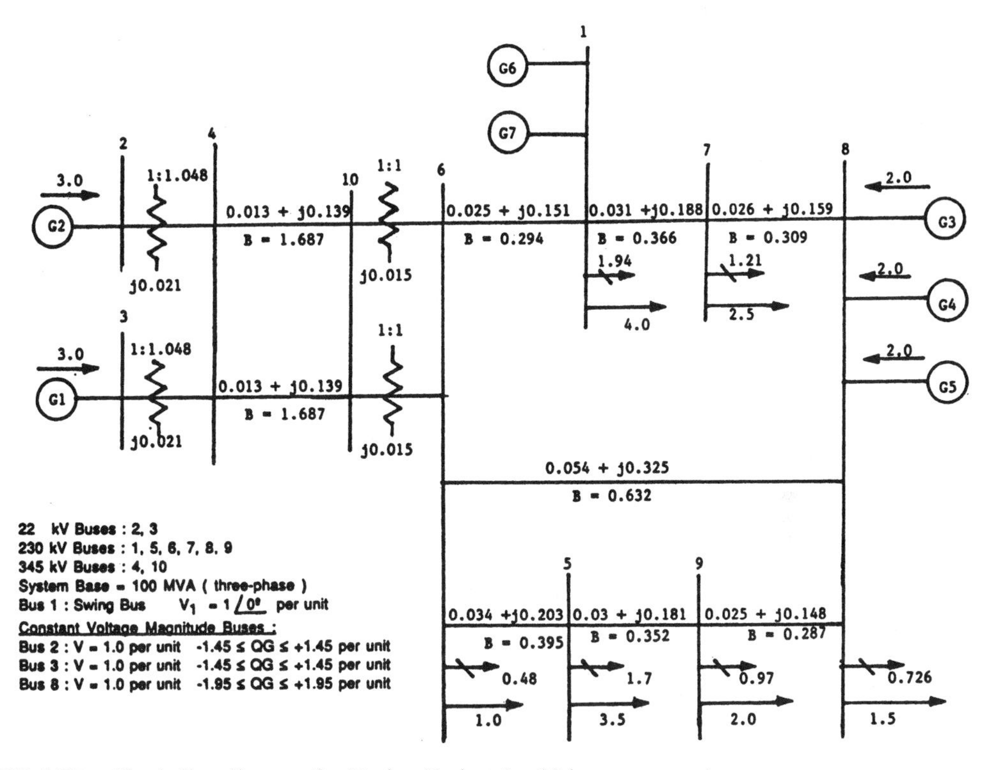

FIGURE 6.25 Single-line diagram for Design Project 5—10-bus power system

out of service. Also assume that the series compensation is effectively distributed such that the net series reactance of each 345-kV line is reduced by the percentage compensation. Determine the percentage series compensation to within $\pm 10\%$.

REFERENCES

1. W. F. Tinney and C. E. Hart, "Power Flow Solutions by Newton's Method," *IEEE Trans. PAS*, *86* (November 1967), p. 1449.
2. W. F. Tinney and J. W. Walker, "Direct Solution of Sparse Network Equations by Optimally Ordered Triangular Factorization," *Proc. IEEE*, *55* (November 1967), pp. 1801–1809.
3. Glenn W. Stagg and Ahmed H. El-Abiad, *Computer Methods in Power System Analysis* (New York: McGraw-Hill, 1968).
4. N. M. Peterson and W. S. Meyer, "Automatic Adjustment of Transformer and Phase Shifter Taps in Newton Power Flow," *IEEE Trans. PAS*, *90* (January–February 1971), pp. 103–108.

5. W. D. Stevenson, Jr., *Elements of Power Systems Analysis*, 4th ed. (New York: McGraw-Hill, 1982).
6. A. Bramellar and R. N. Allan, *Sparsity* (London: Pitman, 1976).
7. C. A. Gross, *Power Systems Analysis* (New York: Wiley, 1979).
8. B. Stott, "Fast Decoupled Load Flow," *IEEE Trans. PAS*, Vol. PAS 91 (September–October 1972), pp. 1955–1959.
9. T. Overbye and J. Weber, "Visualizing the Electric Grid," *IEEE Spectrum*, 38, 2 (February 2001), pp. 52–58.
10. Westinghouse Electric Corporation, *Transmission and Distribution Reference Book*, 4th ed. (Pittsburgh: Westinghouse, 1964).
11. Aluminum Association, *The Aluminum Electrical Conductor Handbook* (Washington, D.C.: Aluminum Association).
12. A. J. Wood and B. F. Wollenberg, *Power Generation, Operation and Control*, 2nd ed. (New York: John Wiley & Sons, 1996).
13. A. Ellis, "Wind Power Plant Models for System Studies," Tutorial on Fundamentals of Wind Energy, Section V, *IEEE PES GM* (Calgary, AB: July 2009).
14. WECC Wind Generator Modeling Group, "WECC Wind Power Plant Power Flow Modeling Guide," *WECC*, May 2008.
15. E.H. Camm, et. al., "Characteristics of Wind Turbine Generators for Wind Power Plants," *Proc. IEEE* 2009 General Meeting (Calgary, AB: July 2009).

345-kV SF6 circuit breaker installation at Goshen Substation, Idaho Falls, Idaho, USA. This circuit breaker has a continuous current rating of 2,000A and an interrupting current rating of 40 kA (Courtesy of PacifiCorp)

7 SYMMETRICAL FAULTS

Short circuits occur in power systems when equipment insulation fails due to system overvoltages caused by lightning or switching surges, to insulation contamination (salt spray or pollution), or to other mechanical causes. The resulting short circuit or "fault" current is determined by the internal voltages of the synchronous machines and by the system impedances between the machine voltages and the fault. Short-circuit currents may be several orders of magnitude larger than normal operating currents and, if allowed to persist, may cause thermal damage to equipment. Windings and busbars may also suffer mechanical damage due to high magnetic forces during faults. It is therefore necessary to remove faulted sections of a power system from service as soon as possible. Standard EHV protective equipment is designed to clear faults within 3 cycles (50 ms at 60 Hz). Lower voltage protective equipment operates more slowly (for example, 5 to 20 cycles).

We begin this chapter by reviewing series R–L circuit transients in Section 7.1, followed in Section 7.2 by a description of three-phase short-circuit currents at unloaded synchronous machines. We analyze both the ac component, including subtransient, transient, and steady-state currents, and the dc component of fault current. We then extend these results in Sections 7.3 and 7.4 to power system three-phase short circuits by means of the superposition principle. We observe that the bus impedance matrix is the key to calculating fault currents. The SHORT CIRCUITS computer program that accompanies this text may be utilized in power system design to select, set, and coordinate protective equipment such as circuit breakers, fuses, relays, and instrument transformers. We discuss circuit breaker and fuse selection in Section 7.5.

Balanced three-phase power systems are assumed throughout this chapter. We also work in per-unit.

CASE STUDY Short circuits can cause severe damage when not interrupted promptly. In some cases, high-impedance fault currents may be insufficient to operate protective relays or blow fuses. Standard overcurrent protection schemes utilized on secondary distribution at some industrial, commercial, and large residential buildings may not detect high-impedance faults, commonly called arcing faults. In these cases, more careful design techniques, such as the use of ground fault circuit interruption, are required to detect arcing faults and prevent burndown. The following case histories [11] give examples of the destructive effects of arcing faults.

The Problem of Arcing Faults in Low-Voltage Power Distribution Systems

FRANCIS J. SHIELDS

ABSTRACT

Many cases of electrical equipment burndown arising from low-level arcing-fault currents have occurred in recent years in low-voltage power distribution systems. Burndown, which is the severe damage or complete destruction of conductors, insulation systems, and metallic enclosures, is caused by the concentrated release of energy in the fault arc. Both grounded and ungrounded electrical distribution systems have experienced burndown, and the reported incidents have involved both industrial and commercial building distribution equipment, without regard to manufacturer, geographical location, or operating environment.

BURNDOWN CASE HISTORIES

The reported incidents of equipment burndown are many. One of the most publicized episodes involved a huge apartment building complex in New York City (Fig. 1), in which two main 480Y/277-volt switchboards were completely destroyed, and two 5000-ampere service entrance buses were burned-off right back to the utility vault. This arcing fault blazed and sputtered for over an hour, and inconvenienced some 10,000 residents of the development through loss of service to building water

Figure 1
Burndown damage caused by arcing fault. View shows low-voltage cable compartments of secondary unit substation

Figure 2
Service entrance switch and current-limiting fuses completely destroyed by arcing fault in main low-voltage switchboard

Figure 3
Fused feeder switch consumed by arcing fault in high-rise apartment main switchboard. No intermediate segregating barriers had been used in construction

pumps, hall and stair lighting, elevators, appliances, and apartment lights. Several days elapsed before service resembling normal was restored through temporary hookups. Illustrations of equipment damage in this burndown are shown in Figs. 2 and 3.

Another example of burndown occurred in the Midwest, and resulted in completely gutting a service entrance switchboard and burning up two 1000-kVA supply transformers. This burndown arc current flowed for about 15 minutes.

In still other reported incidents, a Maryland manufacturer experienced four separate burndowns of secondary unit substations in a little over a year; on the West Coast a unit substation at an industrial process plant burned for more than eight minutes, resulting in destruction of the low-voltage switchgear equipment; and this year [1966] several burndowns have occurred in government office buildings at scattered locations throughout the country.

An example of the involvement of the latter type of equipment in arcing-fault burndowns is shown in Fig. 4. The arcing associated with this fault continued for over 20 minutes, and the fault was finally extinguished only when the relays on the primary system shut down the whole plant.

The electrical equipment destruction shown in the sample photographs is quite startling, but it is only one aspect of this type of fault. Other less graphic but no less serious effects of electrical

Figure 4
Remains of main secondary circuit breaker burned down during arcing fault in low-voltage switchgear section of unit substation

equipment burndown may include personnel fatalities or serious injury, contingent fire damage, loss of vital services (lighting, elevators, ventilation, fire pumps, etc.), shutdown of critical loads, and loss of product revenue. It should be pointed out that the cases reported have involved both industrial and commercial building distribution equipment, without regard to manufacturer, geographical location, operating environment, or the presence or absence of electrical system neutral grounding. Also, the reported burndowns have included a variety of distribution equipment—load center unit substations, switchboards, busway, panelboards, service-entrance equipment, motor control centers, and cable in conduit, for example.

It is obvious, therefore, when all the possible effects of arcing-fault burndowns are taken into consideration, that engineers responsible for electrical power system layout and operation should be anxious both to minimize the probability of arcing faults in electrical systems and to alleviate or mitigate the destructive effects of such faults if they should inadvertently occur despite careful design and the use of quality equipment.

7.1

SERIES R–L CIRCUIT TRANSIENTS

Consider the series R–L circuit shown in Figure 7.1. The closing of switch SW at $t = 0$ represents to a first approximation a three-phase short circuit at the terminals of an unloaded synchronous machine. For simplicity, assume zero fault impedance; that is, the short circuit is a solid or "bolted" fault. The current is assumed to be zero before SW closes, and the source angle α determines the source voltage at $t = 0$. Writing a KVL equation for the circuit,

$$\frac{\mathrm{L}di(t)}{dt} + \mathrm{R}i(t) = \sqrt{2}\mathrm{V}\sin(\omega t + \alpha) \quad t \geqslant 0 \tag{7.1.1}$$

The solution to (7.1.1) is

$$\begin{aligned} i(t) &= i_{\mathrm{ac}}(t) + i_{\mathrm{dc}}(t) \\ &= \frac{\sqrt{2}\mathrm{V}}{\mathrm{Z}}[\sin(\omega t + \alpha - \theta) - \sin(\alpha - \theta)e^{-t/\mathrm{T}}] \quad \mathrm{A} \end{aligned} \tag{7.1.2}$$

FIGURE 7.1

Current in a series R–L circuit with ac voltage source

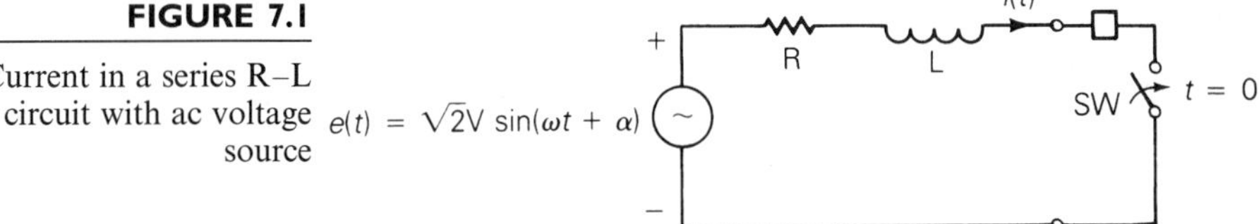

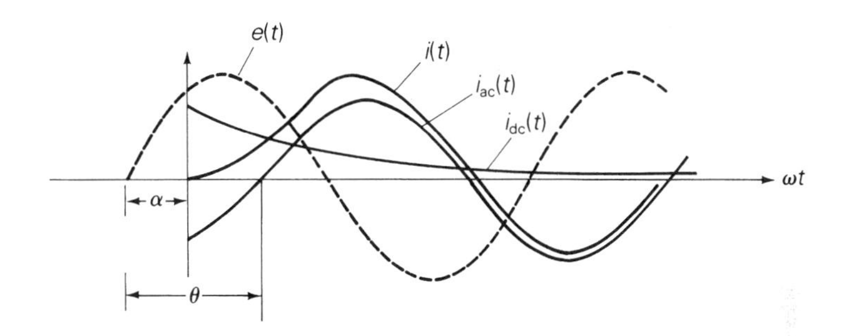

where

$$i_{ac}(t) = \frac{\sqrt{2}V}{Z}\sin(\omega t + \alpha - \theta) \quad A \tag{7.1.3}$$

$$i_{dc}(t) = -\frac{\sqrt{2}V}{Z}\sin(\alpha - \theta)e^{-t/T} \quad A \tag{7.1.4}$$

$$Z = \sqrt{R^2 + (\omega L)^2} = \sqrt{R^2 + X^2} \quad \Omega \tag{7.1.5}$$

$$\theta = \tan^{-1}\frac{\omega L}{R} = \tan^{-1}\frac{X}{R} \tag{7.1.6}$$

$$T = \frac{L}{R} = \frac{X}{\omega R} = \frac{X}{2\pi f R} \quad s \tag{7.1.7}$$

The total fault current in (7.1.2), called the *asymmetrical fault current*, is plotted in Figure 7.1 along with its two components. The ac fault current (also called *symmetrical* or *steady-state fault current*), given by (7.1.3), is a sinusoid. The *dc offset current*, given by (7.1.4), decays exponentially with time constant $T = L/R$.

The rms ac fault current is $I_{ac} = V/Z$. The magnitude of the dc offset, which depends on α, varies from 0 when $\alpha = \theta$ to $\sqrt{2}I_{ac}$ when $\alpha = (\theta \pm \pi/2)$. Note that a short circuit may occur at any instant during a cycle of the ac source; that is, α can have any value. Since we are primarily interested in the largest fault current, we choose $\alpha = (\theta - \pi/2)$. Then (7.1.2) becomes

$$i(t) = \sqrt{2}I_{ac}[\sin(\omega t - \pi/2) + e^{-t/T}] \quad A \tag{7.1.8}$$

TABLE 7.1 Short-circuit current—series R–L circuit*

Component	Instantaneous Current (A)	rms Current (A)
Symmetrical (ac)	$i_{ac}(t) = \frac{\sqrt{2}V}{Z}\sin(\omega t + \alpha - \theta)$	$I_{ac} = \frac{V}{Z}$
dc offset	$i_{dc}(t) = \frac{-\sqrt{2}V}{Z}\sin(\alpha - \theta)e^{-t/T}$	
Asymmetrical (total)	$i(t) = i_{ac}(t) + i_{dc}(t)$	$I_{rms}(t) = \sqrt{I_{ac}^2 + i_{dc}(t)^2}$ with maximum dc offset: $I_{rms}(\tau) = K(\tau)I_{ac}$

* See Figure 7.1 and (7.1.1)–(7.1.12).

where

$$I_{ac} = \frac{V}{Z} \quad A \tag{7.1.9}$$

The rms value of $i(t)$ is of interest. Since $i(t)$ in (7.1.8) is not strictly periodic, its rms value is not strictly defined. However, treating the exponential term as a constant, we stretch the rms concept to calculate the rms asymmetrical fault current with maximum dc offset, as follows:

$$\begin{aligned} I_{rms}(t) &= \sqrt{[I_{ac}]^2 + [I_{dc}(t)]^2} \\ &= \sqrt{[I_{ac}]^2 + [\sqrt{2}I_{ac}e^{-t/T}]^2} \\ &= I_{ac}\sqrt{1 + 2e^{-2t/T}} \quad A \end{aligned} \tag{7.1.10}$$

It is convenient to use $T = X/(2\pi fR)$ and $t = \tau/f$, where τ is time in cycles, and write (7.1.10) as

$$I_{rms}(\tau) = K(\tau)I_{ac} \quad A \tag{7.1.11}$$

where

$$K(\tau) = \sqrt{1 + 2e^{-4\pi\tau/(X/R)}} \quad \text{per unit} \tag{7.1.12}$$

From (7.1.11) and (7.1.12), the rms asymmetrical fault current equals the rms ac fault current times an "asymmetry factor," $K(\tau)$. $I_{rms}(\tau)$ decreases from $\sqrt{3}I_{ac}$ when $\tau = 0$ to I_{ac} when τ is large. Also, higher X to R ratios (X/R) give higher values of $I_{rms}(\tau)$. The above series R–L short-circuit currents are summarized in Table 7.1.

EXAMPLE 7.1 Fault currents: R–L circuit with ac source

A bolted short circuit occurs in the series R–L circuit of Figure 7.1 with $V = 20$ kV, $X = 8\ \Omega$, $R = 0.8\ \Omega$, and with maximum dc offset. The circuit breaker opens 3 cycles after fault inception. Determine (a) the rms ac fault current, (b) the rms "momentary" current at $\tau = 0.5$ cycle, which passes through the breaker before it opens, and (c) the rms asymmetrical fault current that the breaker interrupts.

SOLUTION

a. From (7.1.9),

$$I_{ac} = \frac{20 \times 10^3}{\sqrt{(8)^2 + (0.8)^2}} = \frac{20 \times 10^3}{8.040} = 2.488 \quad \text{kA}$$

b. From (7.1.11) and (7.1.12) with $(X/R) = 8/(0.8) = 10$ and $\tau = 0.5$ cycle,

$$K(0.5 \text{ cycle}) = \sqrt{1 + 2e^{-4\pi(0.5)/10}} = 1.438$$

$$I_{momentary} = K(0.5 \text{ cycle}) I_{ac} = (1.438)(2.488) = 3.576 \quad \text{kA}$$

c. From (7.1.11) and (7.1.12) with $(X/R) = 10$ and $\tau = 3$ cycles,

$$K(3 \text{ cycles}) = \sqrt{1 + 2e^{-4\pi(3)/10}} = 1.023$$

$$I_{rms}(3 \text{ cycles}) = (1.023)(2.488) = 2.544 \quad \text{kA}$$

■

7.2 THREE-PHASE SHORT CIRCUIT—UNLOADED SYNCHRONOUS MACHINE

One way to investigate a three-phase short circuit at the terminals of a synchronous machine is to perform a test on an actual machine. Figure 7.2 shows an oscillogram of the ac fault current in one phase of an unloaded synchronous machine during such a test. The dc offset has been removed

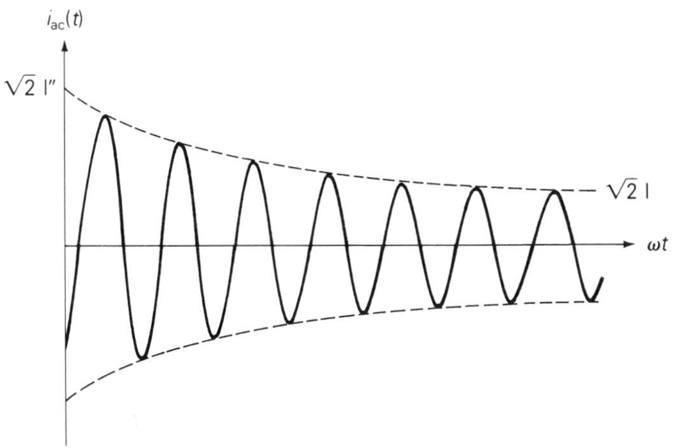

FIGURE 7.2 The ac fault current in one phase of an unloaded synchronous machine during a three-phase short circuit (the dc offset current is removed)

from the oscillogram. As shown, the amplitude of the sinusoidal waveform decreases from a high initial value to a lower steady-state value.

A physical explanation for this phenomenon is that the magnetic flux caused by the short-circuit armature currents (or by the resultant armature MMF) is initially forced to flow through high reluctance paths that do not link the field winding or damper circuits of the machine. This is a result of the theorem of constant flux linkages, which states that the flux linking a closed winding cannot change instantaneously. The armature inductance, which is inversely proportional to reluctance, is therefore initially low. As the flux then moves toward the lower reluctance paths, the armature inductance increases.

The ac fault current in a synchronous machine can be modeled by the series R–L circuit of Figure 7.1 if a time-varying inductance $L(t)$ or reactance $X(t) = \omega L(t)$ is employed. In standard machine theory texts [3, 4], the following reactances are defined:

X''_d = direct axis subtransient reactance

X'_d = direct axis transient reactance

X_d = direct axis synchronous reactance

where $X''_d < X'_d < X_d$. The subscript d refers to the direct axis. There are similar quadrature axis reactances X''_q, X'_q, and X_q [3, 4]. However, if the armature resistance is small, the quadrature axis reactances do not significantly affect the short-circuit current. Using the above direct axis reactances, the instantaneous ac fault current can be written as

$$i_{ac}(t) = \sqrt{2}E_g\left[\left(\frac{1}{X''_d} - \frac{1}{X'_d}\right)e^{-t/T''_d} + \left(\frac{1}{X'_d} - \frac{1}{X_d}\right)e^{-t/T'_d} + \frac{1}{X_d}\right]\sin\left(\omega t + \alpha - \frac{\pi}{2}\right) \tag{7.2.1}$$

where E_g is the rms line-to-neutral prefault terminal voltage of the unloaded synchronous machine. Armature resistance is neglected in (7.2.1). Note that at $t = 0$, when the fault occurs, the rms value of $i_{ac}(t)$ in (7.2.1) is

$$I_{ac}(0) = \frac{E_g}{X''_d} = I'' \tag{7.2.2}$$

which is called the rms *subtransient fault current*, I''. The duration of I'' is determined by the time constant T''_d, called the *direct axis short-circuit subtransient time constant*.

At a later time, when t is large compared to T''_d but small compared to the *direct axis short-circuit transient time constant* T'_d, the first exponential term in (7.2.1) has decayed almost to zero, but the second exponential has not decayed significantly. The rms ac fault current then equals the rms *transient fault current*, given by

$$I' = \frac{E_g}{X'_d} \tag{7.2.3}$$

TABLE 7.2 Short-circuit current—unloaded synchronous machine*

Component	Instantaneous Current (A)	rms Current (A)
Symmetrical (ac)	(7.2.1)	$I_{ac}(t) = E_g\left[\left(\frac{1}{X''_d} - \frac{1}{X'_d}\right)e^{-t/T''_d} + \left(\frac{1}{X'_d} - \frac{1}{X_d}\right)e^{-t/T'_d} + \frac{1}{X_d}\right]$
Subtransient		$I'' = E_g/X''_d$
Transient		$I' = E_g/X'_d$
Steady-state		$I = E_g/X_d$
Maximum dc offset	$i_{dc}(t) = \sqrt{2}I''e^{-t/T_A}$	
Asymmetrical (total)	$i(t) = i_{ac}(t) + i_{dc}(t)$	$I_{rms}(t) = \sqrt{I_{ac}(t)^2 + i_{dc}(t)^2}$ with maximum dc offset: $I_{rms}(t) = \sqrt{I_{ac}(t)^2 + [\sqrt{2}I''e^{-t/T_A}]^2}$

*See Figure 7.2 and (7.2.1)–(7.2.5).

When t is much larger than T'_d, the rms ac fault current approaches its steady-state value, given by

$$I_{ac}(\infty) = \frac{E_g}{X_d} = I \tag{7.2.4}$$

Since the three-phase no-load voltages are displaced 120° from each other, the three-phase ac fault currents are also displaced 120° from each other. In addition to the ac fault current, each phase has a different dc offset. The maximum dc offset in any one phase, which occurs when $\alpha = 0$ in (7.2.1), is

$$i_{dcmax}(t) = \frac{\sqrt{2}E_g}{X''_d}e^{-t/T_A} = \sqrt{2}I''e^{-t/T_A} \tag{7.2.5}$$

where T_A is called the *armature time constant*. Note that the magnitude of the maximum dc offset depends only on the rms subtransient fault current I''. The above synchronous machine short-circuit currents are summarized in Table 7.2.

Machine reactances X''_d, X'_d, and X_d as well as time constants T''_d, T'_d, and T_A are usually provided by synchronous machine manufacturers. They can also be obtained from a three-phase short-circuit test, by analyzing an oscillogram such as that in Figure 7.2 [2]. Typical values of synchronous machine reactances and time constants are given in Appendix Table A.1.

EXAMPLE 7.2 Three-phase short-circuit currents, unloaded synchronous generator

A 500-MVA 20-kV, 60-Hz synchronous generator with reactances $X''_d = 0.15$, $X'_d = 0.24, X_d = 1.1$ per unit and time constants $T''_d = 0.035$, $T'_d = 2.0$, $T_A = 0.20$ s is connected to a circuit breaker. The generator is operating at 5% above rated voltage and at no-load when a bolted three-phase short circuit occurs on the load side of the breaker. The breaker interrupts the fault

3 cycles after fault inception. Determine (a) the subtransient fault current in per-unit and kA rms; (b) maximum dc offset as a function of time; and (c) rms asymmetrical fault current, which the breaker interrupts, assuming maximum dc offset.

SOLUTION

a. The no-load voltage before the fault occurs is $E_g = 1.05$ per unit. From (7.2.2), the subtransient fault current that occurs in each of the three phases is

$$I'' = \frac{1.05}{0.15} = 7.0 \quad \text{per unit}$$

The generator base current is

$$I_{base} = \frac{S_{rated}}{\sqrt{3}V_{rated}} = \frac{500}{(\sqrt{3})(20)} = 14.43 \quad \text{kA}$$

The rms subtransient fault current in kA is the per-unit value multiplied by the base current:

$$I'' = (7.0)(14.43) = 101.0 \quad \text{kA}$$

b. From (7.2.5), the maximum dc offset that may occur in any one phase is

$$i_{dcmax}(t) = \sqrt{2}(101.0)e^{-t/0.20} = 142.9e^{-t/0.20} \quad \text{kA}$$

c. From (7.2.1), the rms ac fault current at $t = 3$ cycles $= 0.05$ s is

$$I_{ac}(0.05\ \text{s}) = 1.05\left[\left(\frac{1}{0.15} - \frac{1}{0.24}\right)e^{-0.05/0.035} + \left(\frac{1}{0.24} - \frac{1}{1.1}\right)e^{-0.05/2.0} + \frac{1}{1.1}\right]$$

$$= 4.920 \quad \text{per unit}$$

$$= (4.920)(14.43) = 71.01 \quad \text{kA}$$

Modifying (7.1.10) to account for the time-varying symmetrical component of fault current, we obtain

$$I_{rms}(0.05) = \sqrt{[I_{ac}(0.05)]^2 + [\sqrt{2}I''e^{-t/T_a}]^2}$$

$$= I_{ac}(0.05)\sqrt{1 + 2\left[\frac{I''}{I_{ac}(0.05)}\right]^2 e^{-2t/Ta}}$$

$$= (71.01)\sqrt{1 + 2\left[\frac{101}{71.01}\right]^2 e^{-2(0.05)/0.20}}$$

$$= (71.01)(1.8585)$$

$$= 132 \quad \text{kA}$$

■

7.3

POWER SYSTEM THREE-PHASE SHORT CIRCUITS

In order to calculate the subtransient fault current for a three-phase short circuit in a power system, we make the following assumptions:

1. Transformers are represented by their leakage reactances. Winding resistances, shunt admittances, and Δ–Y phase shifts are neglected.
2. Transmission lines are represented by their equivalent series reactances. Series resistances and shunt admittances are neglected.
3. Synchronous machines are represented by constant-voltage sources behind subtransient reactances. Armature resistance, saliency, and saturation are neglected.
4. All nonrotating impedance loads are neglected.
5. Induction motors are either neglected (especially for small motors rated less than 50 hp (40 kW)) or represented in the same manner as synchronous machines.

These assumptions are made for simplicity in this text, and in practice they should not be made for all cases. For example, in distribution systems, resistances of primary and secondary distribution lines may in some cases significantly reduce fault current magnitudes.

Figure 7.3 shows a single-line diagram consisting of a synchronous generator feeding a synchronous motor through two transformers and a transmission line. We shall consider a three-phase short circuit at bus 1. The positive-sequence equivalent circuit is shown in Figure 7.4(a), where the voltages E_g'' and E_m'' are the prefault internal voltages behind the subtransient reactances of the machines, and the closing of switch SW represents the fault. For purposes of calculating the subtransient fault current, E_g'' and E_m'' are assumed to be constant-voltage sources.

In Figure 7.4(b) the fault is represented by two opposing voltage sources with equal phasor values V_F. Using superposition, the fault current can then be calculated from the two circuits shown in Figure 7.4(c). However, if V_F equals the prefault voltage at the fault, then the second circuit in Figure 7.4(c) represents the system before the fault occurs. As such, $I_{F2}'' = 0$ and V_F,

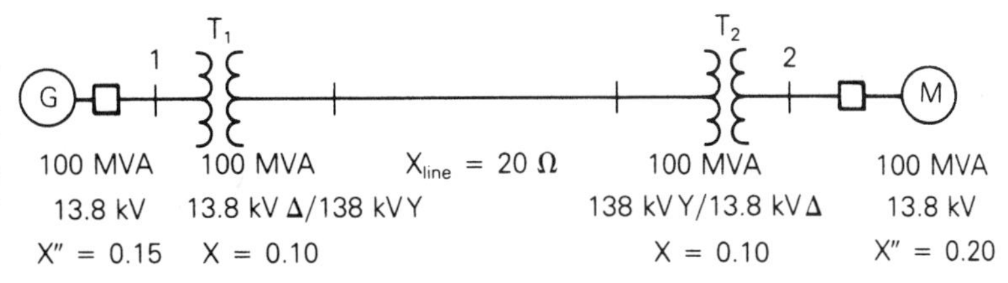

FIGURE 7.3 Single-line diagram of a synchronous generator feeding a synchronous motor

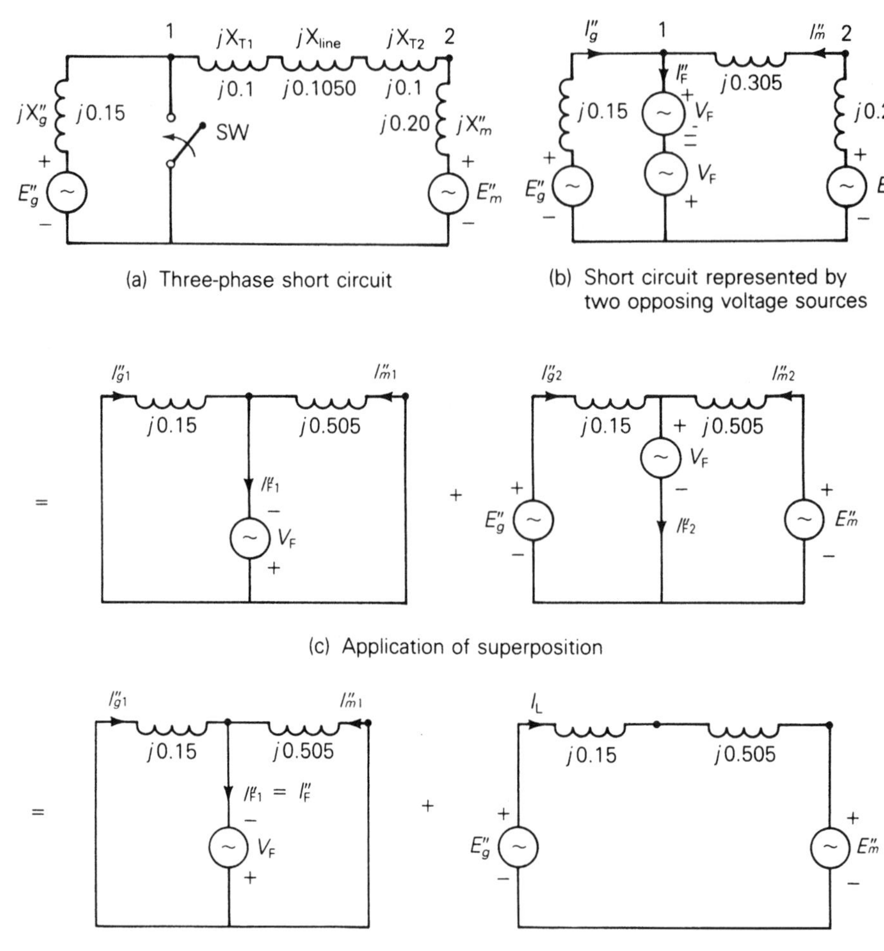

FIGURE 7.4 Application of superposition to a power system three-phase short circuit

which has no effect, can be removed from the second circuit, as shown in Figure 7.4(d). The subtransient fault current is then determined from the first circuit in Figure 7.4(d), $I''_{\mathrm{F}} = I''_{\mathrm{F1}}$. The contribution to the fault from the generator is $I''_g = I''_{g1} + I''_{g2} = I''_{g1} + I_{\mathrm{L}}$, where I_{L} is the prefault generator current. Similarly, $I''_m = I''_{m1} - I_{\mathrm{L}}$.

EXAMPLE 7.3 Three-phase short-circuit currents, power system

The synchronous generator in Figure 7.3 is operating at rated MVA, 0.95 p.f. lagging and at 5% above rated voltage when a bolted three-phase short circuit occurs at bus 1. Calculate the per-unit values of (a) subtransient fault current; (b) subtransient generator and motor currents, neglecting prefault current; and (c) subtransient generator and motor currents including prefault current.

SOLUTION

a. Using a 100-MVA base, the base impedance in the zone of the transmission line is

$$Z_{\text{base, line}} = \frac{(138)^2}{100} = 190.44 \quad \Omega$$

and

$$X_{\text{line}} = \frac{20}{190.44} = 0.1050 \quad \text{per unit}$$

The per-unit reactances are shown in Figure 7.4. From the first circuit in Figure 7.4(d), the Thévenin impedance as viewed from the fault is

$$Z_{\text{Th}} = jX_{\text{Th}} = j\frac{(0.15)(0.505)}{(0.15 + 0.505)} = j0.11565 \quad \text{per unit}$$

and the prefault voltage at the generator terminals is

$$V_{\text{F}} = 1.05\underline{/0^\circ} \quad \text{per unit}$$

The subtransient fault current is then

$$I''_{\text{F}} = \frac{V_{\text{F}}}{Z_{\text{Th}}} = \frac{1.05\underline{/0^\circ}}{j0.11565} = -j9.079 \quad \text{per unit}$$

b. Using current division in the first circuit of Figure 7.4(d),

$$I''_{g1} = \left(\frac{0.505}{0.505 + 0.15}\right) I''_{\text{F}} = (0.7710)(-j9.079) = -j7.000 \quad \text{per unit}$$

$$I''_{m1} = \left(\frac{0.15}{0.505 + 0.15}\right) I''_{\text{F}} = (0.2290)(-j9.079) = -j2.079 \quad \text{per unit}$$

c. The generator base current is

$$I_{\text{base, gen}} = \frac{100}{(\sqrt{3})(13.8)} = 4.1837 \quad \text{kA}$$

and the prefault generator current is

$$I_{\text{L}} = \frac{100}{(\sqrt{3})(1.05 \times 13.8)}\underline{/-\cos^{-1} 0.95} = 3.9845\underline{/-18.19^\circ} \quad \text{kA}$$

$$= \frac{3.9845\underline{/-18.19^\circ}}{4.1837} = 0.9524\underline{/-18.19^\circ}$$

$$= 0.9048 - j0.2974 \quad \text{per unit}$$

The subtransient generator and motor currents, including prefault current, are then

$$I_g'' = I_{g1}'' + I_L = -j7.000 + 0.9048 - j0.2974$$
$$= 0.9048 - j7.297 = 7.353\underline{/-82.9^\circ} \quad \text{per unit}$$
$$I_m'' = I_{m1}'' - I_L = -j2.079 - 0.9048 + j0.2974$$
$$= -0.9048 - j1.782 = 1.999\underline{/243.1^\circ} \quad \text{per unit}$$

An alternate method of solving Example 7.3 is to first calculate the internal voltages E_g'' and E_m'' using the prefault load current I_L. Then, instead of using superposition, the fault currents can be resolved directly from the circuit in Figure 7.4(a) (see Problem 7.11). However, in a system with many synchronous machines, the superposition method has the advantage that all machine voltage sources are shorted, and the prefault voltage is the only source required to calculate the fault current. Also, when calculating the contributions to fault current from each branch, prefault currents are usually small, and hence can be neglected. Otherwise, prefault load currents could be obtained from a power-flow program. ■

7.4 BUS IMPEDANCE MATRIX

We now extend the results of the previous section to calculate subtransient fault currents for three-phase faults in an N-bus power system. The system is modeled by its positive-sequence network, where lines and transformers are represented by series reactances and synchronous machines are represented by constant-voltage sources behind subtransient reactances. As before, all resistances, shunt admittances, and nonrotating impedance loads are neglected. For simplicity, we also neglect prefault load currents.

Consider a three-phase short circuit at any bus n. Using the superposition method described in Section 7.3, we analyze two separate circuits. (For example, see Figure 7.4d.) In the first circuit, all machine-voltage sources are short-circuited, and the only source is due to the prefault voltage at the fault. Writing nodal equations for the first circuit,

$$\boldsymbol{Y}_{\text{bus}}\boldsymbol{E}^{(1)} = \boldsymbol{I}^{(1)} \tag{7.4.1}$$

where $\boldsymbol{Y}_{\text{bus}}$ is the positive-sequence bus admittance matrix, $\boldsymbol{E}^{(1)}$ is the vector of bus voltages, and $\boldsymbol{I}^{(1)}$ is the vector of current sources. The superscript (1) denotes the first circuit. Solving (7.4.1),

$$\boldsymbol{Z}_{\text{bus}}\boldsymbol{I}^{(1)} = \boldsymbol{E}^{(1)} \tag{7.4.2}$$

where

$$\boldsymbol{Z}_{\text{bus}} = \boldsymbol{Y}_{\text{bus}}^{-1} \tag{7.4.3}$$

$\boldsymbol{Z}_{\text{bus}}$, the inverse of $\boldsymbol{Y}_{\text{bus}}$, is called the positive-sequence *bus impedance matrix*. Both $\boldsymbol{Z}_{\text{bus}}$ and $\boldsymbol{Y}_{\text{bus}}$ are symmetric matrices.

Since the first circuit contains only one source, located at faulted bus n, the current source vector contains only one nonzero component, $I_n^{(1)} = -I''_{\text{F}n}$. Also, the voltage at faulted bus n in the first circuit is $E_n^{(1)} = -V_{\text{F}}$. Rewriting (7.4.2),

$$\begin{bmatrix} Z_{11} & Z_{12} & \cdots & Z_{1n} & \cdots & Z_{1N} \\ Z_{21} & Z_{22} & \cdots & Z_{2n} & \cdots & Z_{2N} \\ \vdots & & & & & \\ Z_{n1} & Z_{n2} & \cdots & Z_{nn} & \cdots & Z_{nN} \\ \vdots & & & & & \\ Z_{N1} & Z_{N2} & \cdots & Z_{Nn} & \cdots & Z_{NN} \end{bmatrix} \begin{bmatrix} 0 \\ 0 \\ \vdots \\ -I''_{\text{F}n} \\ \vdots \\ 0 \end{bmatrix} = \begin{bmatrix} E_1^{(1)} \\ E_2^{(1)} \\ \vdots \\ -V_{\text{F}} \\ \vdots \\ E_N^{(1)} \end{bmatrix} \tag{7.4.4}$$

The minus sign associated with the current source in (7.4.4) indicates that the current injected into bus n is the negative of $I''_{\text{F}n}$, since $I''_{\text{F}n}$ flows away from bus n to the neutral. From (7.4.4), the subtransient fault current is

$$I''_{\text{F}n} = \frac{V_{\text{F}}}{Z_{nn}} \tag{7.4.5}$$

Also from (7.4.4) and (7.4.5), the voltage at any bus k in the first circuit is

$$E_k^{(1)} = Z_{kn}(-I''_{\text{F}n}) = \frac{-Z_{kn}}{Z_{nn}} V_{\text{F}} \tag{7.4.6}$$

The second circuit represents the prefault conditions. Neglecting prefault load current, all voltages throughout the second circuit are equal to the prefault voltage; that is, $E_k^{(2)} = V_{\text{F}}$ for each bus k. Applying superposition,

$$\begin{aligned} E_k = E_k^{(1)} + E_k^{(2)} &= \frac{-Z_{kn}}{Z_{nn}} V_{\text{F}} + V_{\text{F}} \\ &= \left(1 - \frac{Z_{kn}}{Z_{nn}}\right) V_{\text{F}} \qquad k = 1, 2, \ldots, N \end{aligned} \tag{7.4.7}$$

EXAMPLE 7.4 **Using Z_{bus} to compute three-phase short-circuit currents in a power system**

Faults at bus 1 and 2 in Figure 7.3 are of interest. The prefault voltage is 1.05 per unit and prefault load current is neglected. (a) Determine the 2 × 2 positive-sequence bus impedance matrix. (b) For a bolted three-phase short circuit at bus 1, use $\boldsymbol{Z}_{\text{bus}}$ to calculate the subtransient fault current and the contribution to the fault current from the transmission line. (c) Repeat part (b) for a bolted three-phase short circuit at bus 2.

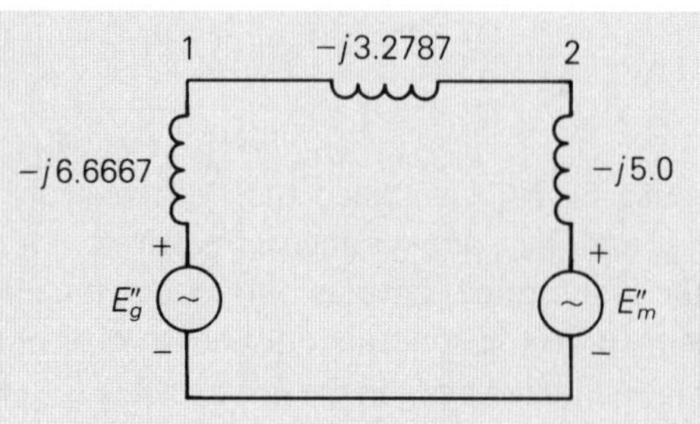

FIGURE 7.5 Circuit of Figure 7.4(a) showing per-unit admittance values

SOLUTION

a. The circuit of Figure 7.4(a) is redrawn in Figure 7.5 showing per-unit admittance rather than per-unit impedance values. Neglecting prefault load current, $E''_g = E''_m = V_F = 1.05\underline{/0^\circ}$ per unit. From Figure 7.5, the positive-sequence bus admittance matrix is

$$\boldsymbol{Y}_{\text{bus}} = -j\begin{bmatrix} 9.9454 & -3.2787 \\ -3.2787 & 8.2787 \end{bmatrix} \quad \text{per unit}$$

Inverting $\boldsymbol{Y}_{\text{bus}}$,

$$\boldsymbol{Z}_{\text{bus}} = \boldsymbol{Y}_{\text{bus}}^{-1} = +j\begin{bmatrix} 0.11565 & 0.04580 \\ 0.04580 & 0.13893 \end{bmatrix} \quad \text{per unit}$$

b. Using (7.4.5) the subtransient fault current at bus 1 is

$$I''_{F1} = \frac{V_F}{Z_{11}} = \frac{1.05\underline{/0^\circ}}{j0.11565} = -j9.079 \quad \text{per unit}$$

which agrees with the result in Example 7.3, part (a). The voltages at buses 1 and 2 during the fault are, from (7.4.7),

$$E_1 = \left(1 - \frac{Z_{11}}{Z_{11}}\right)V_F = 0$$

$$E_2 = \left(1 - \frac{Z_{21}}{Z_{11}}\right)V_F = \left(1 - \frac{j0.04580}{j0.11565}\right)1.05\underline{/0^\circ} = 0.6342\underline{/0^\circ}$$

The current to the fault from the transmission line is obtained from the voltage drop from bus 2 to 1 divided by the impedance of the line and transformers T_1 and T_2:

$$I_{21} = \frac{E_2 - E_1}{j(X_{\text{line}} + X_{T1} + X_{T2})} = \frac{0.6342 - 0}{j0.3050} = -j2.079 \quad \text{per unit}$$

which agrees with the motor current calculated in Example 7.3, part (b), where prefault load current is neglected.

c. Using (7.4.5), the subtransient fault current at bus 2 is

$$I''_{F2} = \frac{V_F}{Z_{22}} = \frac{1.05\underline{/0^\circ}}{j0.13893} = -j7.558 \quad \text{per unit}$$

and from (7.4.7),

$$E_1 = \left(1 - \frac{Z_{12}}{Z_{22}}\right) V_F = \left(1 - \frac{j0.04580}{j0.13893}\right) 1.05\underline{/0^\circ} = 0.7039\underline{/0^\circ}$$

$$E_2 = \left(1 - \frac{Z_{22}}{Z_{22}}\right) V_F = 0$$

The current to the fault from the transmission line is

$$I_{12} = \frac{E_1 - E_2}{j(X_{line} + X_{T1} + X_{T2})} = \frac{0.7039 - 0}{j0.3050} = -j2.308 \quad \text{per unit} \quad \blacksquare$$

Figure 7.6 shows a bus impedance equivalent circuit that illustrates the short-circuit currents in an N-bus system. This circuit is given the name *rake equivalent* in Neuenswander [5] due to its shape, which is similar to a garden rake.

The diagonal elements $Z_{11}, Z_{22}, \ldots, Z_{NN}$ of the bus impedance matrix, which are the *self-impedances*, are shown in Figure 7.6. The off-diagonal elements, or the *mutual impedances*, are indicated by the brackets in the figure.

Neglecting prefault load currents, the internal voltage sources of all synchronous machines are equal both in magnitude and phase. As such, they can be connected, as shown in Figure 7.7, and replaced by one equivalent source V_F from neutral bus 0 to a references bus, denoted r. This equivalent source is also shown in the rake equivalent of Figure 7.6.

FIGURE 7.6 Bus impedance equivalent circuit (*rake equivalent*)

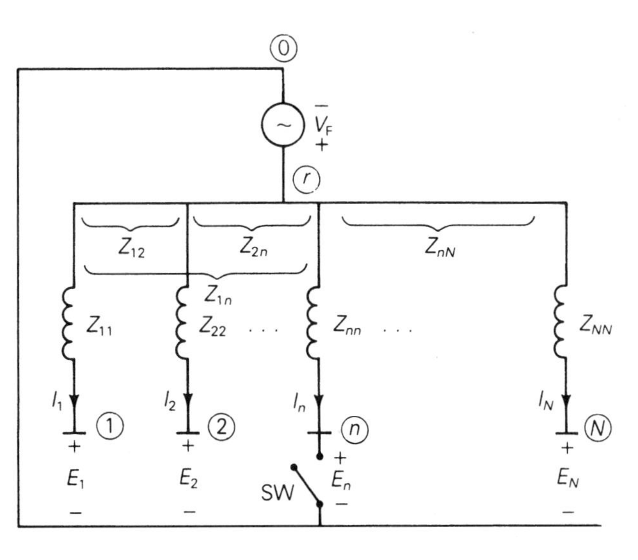

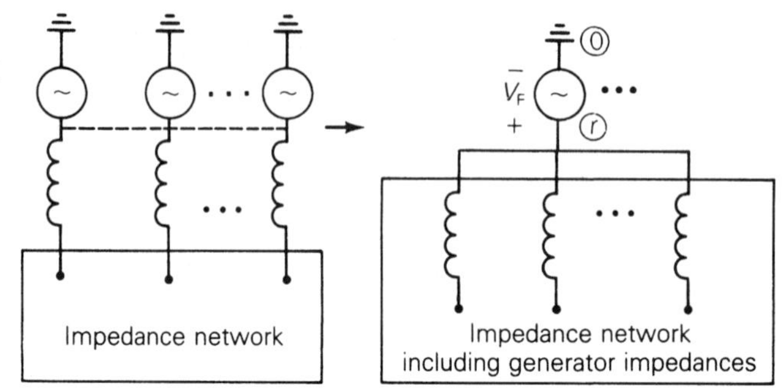

FIGURE 7.7 Parallel connection of unloaded synchronous machine internal-voltage sources

Using $\boldsymbol{Z}_{\text{bus}}$, the fault currents in Figure 7.6 are given by

$$\begin{bmatrix} Z_{11} & Z_{12} & \cdots & Z_{1n} & \cdots & Z_{1N} \\ Z_{21} & Z_{22} & \cdots & Z_{2n} & \cdots & Z_{2N} \\ \vdots & & & & & \vdots \\ Z_{n1} & Z_{n2} & \cdots & Z_{nn} & \cdots & Z_{nN} \\ \vdots & & & & & \\ Z_{N1} & Z_{N2} & \cdots & Z_{Nn} & \cdots & Z_{NN} \end{bmatrix} \begin{bmatrix} I_1 \\ I_2 \\ \vdots \\ I_n \\ \vdots \\ I_N \end{bmatrix} = \begin{bmatrix} V_F - E_1 \\ V_F - E_2 \\ \vdots \\ V_F - E_n \\ \vdots \\ V_F - E_N \end{bmatrix} \tag{7.4.8}$$

where $I_1, I_2, \ldots$ are the branch currents and $(V_F - E_1), (V_F - E_2), \ldots$ are the voltages across the branches.

If switch SW in Figure 7.6 is open, all currents are zero and the voltage at each bus with respect to the neutral equals V_F. This corresponds to prefault conditions, neglecting prefault load currents. If switch SW is closed, corresponding to a short circuit at bus n, $E_n = 0$ and all currents except I_n remain zero. The fault current is $I''_{Fn} = I_n = V_F/Z_{nn}$, which agrees with (7.4.5). This fault current also induces a voltage drop $Z_{kn}I_n = (Z_{kn}/Z_{nn})V_F$ across each branch k. The voltage at bus k with respect to the neutral then equals V_F minus this voltage drop, which agrees with (7.4.7).

As shown by Figure 7.6 as well as (7.4.5), subtransient fault currents throughout an N-bus system can be determined from the bus impedance matrix and the prefault voltage. $\boldsymbol{Z}_{\text{bus}}$ can be computed by first constructing $\boldsymbol{Y}_{\text{bus}}$, via nodal equations, and then inverting $\boldsymbol{Y}_{\text{bus}}$. Once $\boldsymbol{Z}_{\text{bus}}$ has been obtained, these fault currents are easily computed.

EXAMPLE 7.5

PowerWorld Simulator case Example 7_5 models the 5-bus power system whose one-line diagram is shown in Figure 6.2. Machine, line, and transformer data are given in Tables 7.3, 7.4, and 7.5. This system is initially unloaded. Prefault voltages at all the buses are 1.05 per unit. Use PowerWorld Simulator to determine the fault current for three-phase faults at each of the buses.

TABLE 7.3

Synchronous machine data for SYMMETRICAL SHORT CIRCUITS program*

Bus	Machine Subtransient Reactance—X''_d (per unit)
1	0.045
3	0.0225

*S_{base} = 100 MVA
V_{base} = 15 kV at buses 1, 3
= 345 kV at buses 2, 4, 5

TABLE 7.4

Line data for SYMMETRICAL SHORT CIRCUITS program

Bus-to-Bus	Equivalent Positive-Sequence Series Reactance (per unit)
2–4	0.1
2–5	0.05
4–5	0.025

TABLE 7.5

Transformer data for SYMMETRICAL SHORT CIRCUITS program

Bus-to-Bus	Leakage Reactance—X (per unit)
1–5	0.02
3–4	0.01

SOLUTION To fault a bus from the one-line, first right-click on the bus symbol to display the local menu, and then select "Fault." This displays the **Fault** dialog (see Figure 7.8). The selected bus will be automatically selected as the fault location. Verify that the Fault Location is "Bus Fault" and the Fault Type is "3 Phase Balanced" (unbalanced faults are covered in Chapter 9). Then select "**Calculate**," located in the bottom left corner of the dialog, to determine the fault currents and voltages. The results are shown in the tables at the bottom of the dialog. Additionally, the values can be animated on the one-line by changing the Oneline Display Field value. Since with a three-phase fault the system remains balanced, the magnitudes of the a phase, b phase and c phase values are identical. The 5 × 5 $\boldsymbol{Z}_{bus}$ matrix for this system is shown in Table 7.6, and the fault currents and bus voltages for faults at each of the buses are given in Table 7.7. Note that these fault currents are subtransient fault currents, since the machine reactance input data consist of direct axis subtransient reactances.

TABLE 7.6

$\boldsymbol{Z}_{bus}$ for Example 7.5

$$j\begin{bmatrix} 0.0279725 & 0.0177025 & 0.0085125 & 0.0122975 & 0.020405 \\ 0.0177025 & 0.0569525 & 0.0136475 & 0.019715 & 0.02557 \\ 0.0085125 & 0.0136475 & 0.0182425 & 0.016353 & 0.012298 \\ 0.0122975 & 0.019715 & 0.016353 & 0.0236 & 0.017763 \\ 0.020405 & 0.02557 & 0.012298 & 0.017763 & 0.029475 \end{bmatrix}$$

TABLE 7.7

Fault currents and bus voltages for Example 7.5

Fault Bus	Fault Current (per unit)	Contributions to Fault Current: Gen Line or TRSF	Bus-to-Bus	Current (per unit)
1	37.536			
		G 1	GRND–1	23.332
		T 1	5–1	14.204
2	18.436			
		L 1	4–2	6.864
		L 2	5–2	11.572
3	57.556			
		G 2	GRND–3	46.668
		T 2	4–3	10.888
4	44.456			
		L 1	2–4	1.736
		L 3	5–4	10.412
		T 2	3–4	32.308
5	35.624			
		L 2	2–5	2.78
		L 3	4–5	16.688
		T 1	1–5	16.152

$V_F = 1.05$ Fault Bus:	Per-Unit Bus Voltage Magnitudes during the Fault: Bus 1	Bus 2	Bus 3	Bus 4	Bus 5
1	0.0000	0.7236	0.5600	0.5033	0.3231
2	0.3855	0.0000	0.2644	0.1736	0.1391
3	0.7304	0.7984	0.0000	0.3231	0.6119
4	0.5884	0.6865	0.1089	0.0000	0.4172
5	0.2840	0.5786	0.3422	0.2603	0.0000

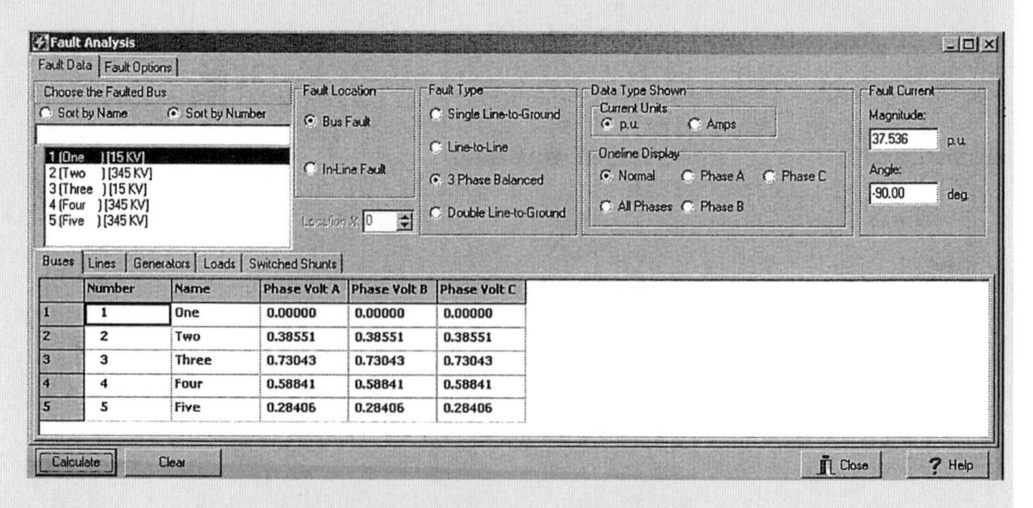

FIGURE 7.8 Fault Analysis Dialog for Example 7.5—fault at bus 1

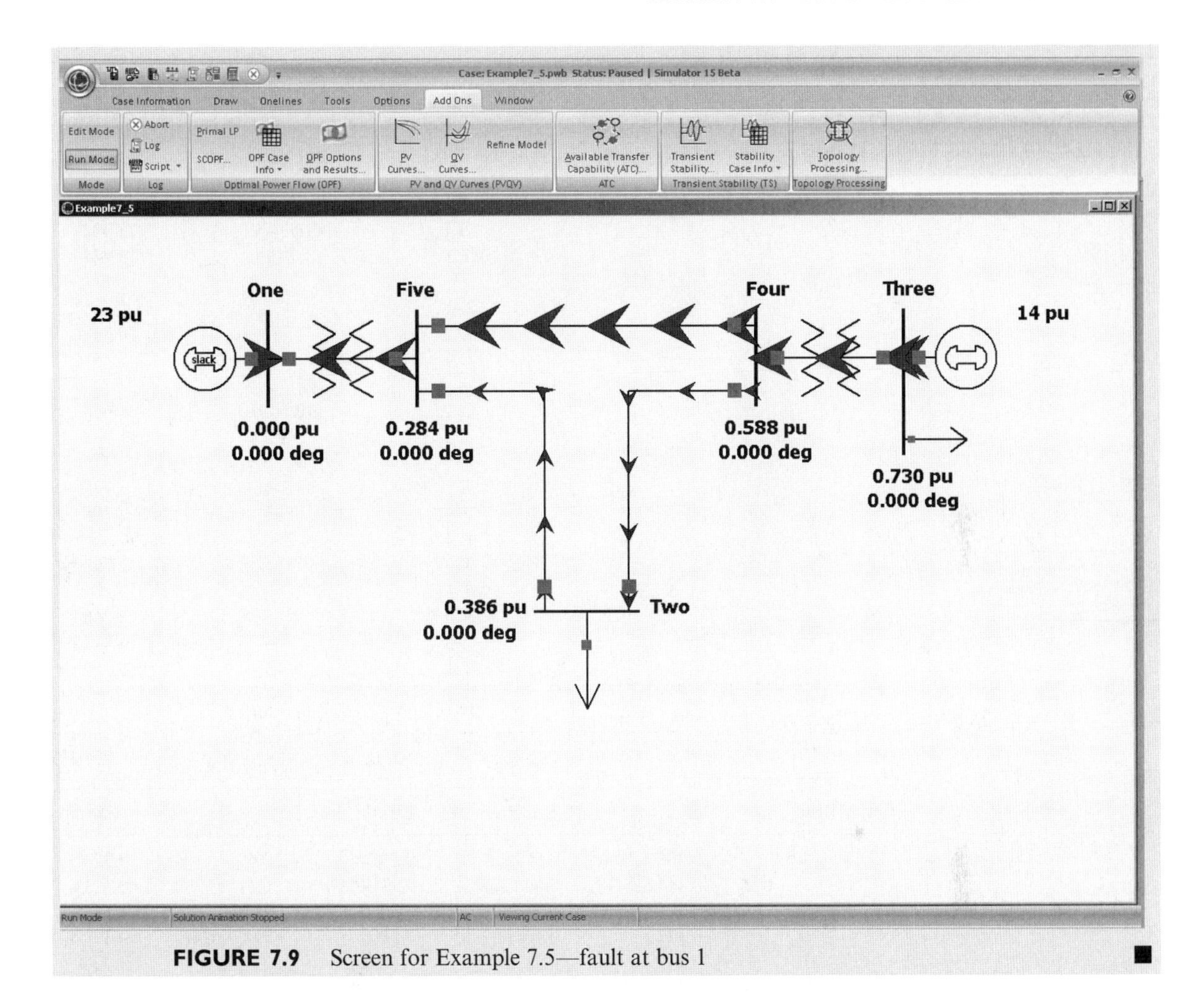

FIGURE 7.9 Screen for Example 7.5—fault at bus 1 ■

EXAMPLE 7.6

Redo Example 7.5 with an additional line installed between buses 2 and 4. This line, whose reactance is 0.075 per unit, is not mutually coupled to any other line.

SOLUTION The modified system is contained in PowerWorld Simulator case Example 7_6. $\boldsymbol{Z}_{\text{bus}}$ along with the fault currents and bus voltages are shown in Tables 7.8 and 7.9.

TABLE 7.8

$\boldsymbol{Z}_{\text{bus}}$ for Example 7.6

$$j\begin{bmatrix} 0.027723 & 0.01597 & 0.00864 & 0.01248 & 0.02004 \\ 0.01597 & 0.04501 & 0.01452 & 0.02097 & 0.02307 \\ 0.00864 & 0.01452 & 0.01818 & 0.01626 & 0.01248 \\ 0.01248 & 0.02097 & 0.01626 & 0.02349 & 0.01803 \\ 0.02004 & 0.02307 & 0.01248 & 0.01803 & 0.02895 \end{bmatrix}$$

TABLE 7.9

Fault currents and bus voltages for Example 7.6

Fault Bus	Fault Current (per unit)	Contributions to Fault Current		
		Gen Line or TRSF	Bus-to-Bus	Current (per unit)
1	37.872			
		G 1	GRND–1	23.332
		T 1	5–1	14.544
2	23.328			
		L 1	4–2	5.608
		L 2	5–2	10.24
		L 4	4–2	7.48
3	57.756			
		G 2	GRND–3	46.668
		T 2	4–3	11.088
4	44.704			
		L 1	2–4	1.128
		L 3	5–4	9.768
		L 4	2–4	1.504
		T 2	3–4	32.308
5	36.268			
		L 2	2–5	4.268
		L 3	4–5	15.848
		T 1	1–5	16.152

$V_F = 1.05$ Fault Bus:	Per-Unit Bus Voltage Magnitudes during the Fault				
	Bus 1	Bus 2	Bus 3	Bus 4	Bus 5
1	0.0000	0.6775	0.5510	0.4921	0.3231
2	0.4451	0.0000	0.2117	0.1127	0.2133
3	0.7228	0.7114	0.0000	0.3231	0.5974
4	0.5773	0.5609	0.1109	0.0000	0.3962
5	0.2909	0.5119	0.3293	0.2442	0.0000

■

7.5 CIRCUIT BREAKER AND FUSE SELECTION

A SHORT CIRCUITS computer program may be utilized in power system design to select, set, and coordinate protective equipment such as circuit breakers, fuses, relays, and instrument transformers. In this section we discuss basic principles of circuit breaker and fuse selection.

AC CIRCUIT BREAKERS

A *circuit breaker* is a mechanical switch capable of interrupting fault currents and of reclosing. When circuit-breaker contacts separate while carrying current, an arc forms. The breaker is designed to extinguish the arc by elongating and cooling it. The fact that ac arc current naturally passes through zero twice during its 60-Hz cycle aids the arc extinction process.

Circuit breakers are classified as *power* circuit breakers when they are intended for service in ac circuits above 1500 V, and as *low-voltage* circuit breakers in ac circuits up to 1500 V. There are different types of circuit breakers depending on the medium—air, oil, SF_6 gas, or vacuum—in which the arc is elongated. Also, the arc can be elongated either by a magnetic force or by a blast of air.

Some circuit breakers are equipped with a high-speed automatic reclosing capability. Since most faults are temporary and self-clearing, reclosing is based on the idea that if a circuit is deenergized for a short time, it is likely that whatever caused the fault has disintegrated and the ionized arc in the fault has dissipated.

When reclosing breakers are employed in EHV systems, standard practice is to reclose only once, approximately 15 to 50 cycles (depending on operating voltage) after the breaker interrupts the fault. If the fault persists and the EHV breaker recloses into it, the breaker reinterrupts the fault current and then "locks out," requiring operator resetting. Multiple-shot reclosing in EHV systems is not standard practice because transient stability (Chapter 11) may be compromised. However, for distribution systems (2.4–46 kV) where customer outages are of concern, standard reclosers are equipped for two or more reclosures.

For low-voltage applications, molded case circuit breakers with dual trip capability are available. There is a magnetic instantaneous trip for large fault currents above a specified threshold and a thermal trip with time delay for smaller fault currents.

Modern circuit-breaker standards are based on symmetrical interrupting current. It is usually necessary to calculate only symmetrical fault current at a system location, and then select a breaker with a symmetrical interrupting capability equal to or above the calculated current. The breaker has the additional capability to interrupt the asymmetrical (or total) fault current if the dc offset is not too large.

Recall from Section 7.1 that the maximum asymmetry factor K $(\tau = 0)$ is $\sqrt{3}$, which occurs at fault inception $(\tau = 0)$. After fault inception, the dc fault current decays exponentially with time constant $T = (L/R) = (X/\omega R)$, and the asymmetry factor decreases. Power circuit breakers with a 2-cycle rated interruption time are designed for an asymmetrical interrupting capability up to 1.4 times their symmetrical interrupting capability, whereas slower circuit breakers have a lower asymmetrical interrupting capability.

A simplified method for breaker selection is called the "E/X simplified method" [1, 7]. The maximum symmetrical short-circuit current at the system location in question is calculated from the prefault voltage and system reactance characteristics, using computer programs. Resistances, shunt admittances, nonrotating impedance loads, and prefault load currents are neglected. Then, if the X/R ratio at the system location is less than 15, a breaker with a symmetrical interrupting capability equal to or above the

TABLE 7.10

Preferred ratings for outdoor circuit breakers (symmetrical current basis of rating) [10] (Application Guide for AC High-Voltage Circuit Breakers Rated on a Symmetrical Current Basis, ANSI C37.010 (New York: American National Standards Institute, 1972). © 1972 IEEE)

Identification		Rated Values					
		Voltage		Insulation Level: Rated Withstand Test Voltage		Current	
Nominal Voltage Class (kV, rms)	Nominal 3-Phase MVA Class	Rated Max Voltage (kV, rms)	Rated Voltage Range Factor (K)	Low Frequency (kV, rms)	Impulse (kV, Crest)	Rated Continuous Current at 60 Hz (Amperes, rms)	Rated Short-Circuit Current (at Rated Max kV) (kA, rms)
Col 1	Col 2	Col 3	Col 4	Col 5	Col 6	Col 7	Col 8
14.4	250	15.5	2.67			600	8.9
14.4	500	15.5	1.29			1200	18
23	500	25.8	2.15			1200	11
34.5	1500	38	1.65			1200	22
46	1500	48.3	1.21			1200	17
69	2500	72.5	1.21			1200	19
115		121	1.0			1200	20
115		121	1.0			1600	40
115		121	1.0			2000	40
115		121	1.0			2000	63
115		121	1.0			3000	40
115		121	1.0			3000	63
138		145	1.0			1200	20
138	Not	145	1.0			1600	40
138		145	1.0			2000	40
138		145	1.0			2000	63
138		145	1.0			2000	80
138	Applicable	145	1.0			3000	40
138		145	1.0			3000	63
138		145	1.0			3000	80
161		169	1.0			1200	16
161		169	1.0			1600	31.5
161		169	1.0			2000	40
161		169	1.0			2000	50
230		242	1.0			1600	31.5
230		242	1.0			2000	31.5
230		242	1.0			3000	31.5
230		242	1.0			2000	40
230		242	1.0			3000	40
230		242	1.0			3000	63
345		362	1.0			2000	40
345		362	1.0			3000	40
500		550	1.0			2000	40
500		550	1.0			3000	40
700		765	1.0			2000	40
700		765	1.0			3000	40

TABLE 7.10

(continued)

Rated Values			Related Required Capabilities		
			Current Values		
			Max Symmetrical Interrupting Capability	3-Second Short-Time Current Carrying Capability	Closing and Latching Capability 1.6K Times Rated Short-Circuit Current
			K Times Rated Short-Circuit Current		
Rated Interrupting Time (Cycles)	Rated Permissible Tripping Delay (Seconds)	Rated Max Voltage Divided by K (kV, rms)	(kA, rms)	(kA, rms)	(kA, rms)
Col 9	Col 10	Col 11	Col 12	Col 13	Col 14
5	2	5.8	24	24	38
5	2	12	23	23	37
5	2	12	24	24	38
5	2	23	36	36	58
5	2	40	21	21	33
5	2	60	23	23	37
3	1	121	20	20	32
3	1	121	40	40	64
3	1	121	40	40	64
3	1	121	63	63	101
3	1	121	40	40	64
3	1	121	63	63	101
3	1	145	20	20	32
3	1	145	40	40	64
3	1	145	40	40	64
3	1	145	63	63	101
3	1	145	80	80	128
3	1	145	40	40	64
3	1	145	63	63	101
3	1	145	80	80	128
3	1	169	16	16	26
3	1	169	31.5	31.5	50
3	1	169	40	40	64
3	1	169	50	50	80
3	1	242	31.5	31.5	50
3	1	242	31.5	31.5	50
3	1	242	31.5	31.5	50
3	1	242	40	40	64
3	1	242	40	40	64
3	1	242	63	63	101
3	1	362	40	40	64
3	1	362	40	40	64
2	1	550	40	40	64
2	1	550	40	40	64
2	1	765	40	40	64
2	1	765	40	40	64

calculated current at the given operating voltage is satisfactory. However, if X/R is greater than 15, the dc offset may not have decayed to a sufficiently low value. In this case, a method for correcting the calculated fault current to account for dc and ac time constants as well as breaker speed can be used [10]. If X/R is unknown, the calculated fault current should not be greater than 80% of the breaker interrupting capability.

When selecting circuit breakers for generators, two cycle breakers are employed in practice, and the subtransient fault current is calculated; therefore subtransient machine reactances X_d'' are used in fault calculations. For synchronous motors, subtransient reactances X_d'' or transient reactances X_d' are used, depending on breaker speed. Also, induction motors can momentarily contribute to fault current. Large induction motors are usually modeled as sources in series with X_d'' or X_d', depending on breaker speed. Smaller induction motors (below 50 hp (40 kW)) are often neglected entirely.

Table 7.10 shows a schedule of preferred ratings for outdoor power circuit breakers. We describe some of the more important ratings shown next.

Voltage ratings

Rated maximum voltage: Designates the maximum rms line-to-line operating voltage. The breaker should be used in systems with an operating voltage less than or equal to this rating.

Rated low frequency withstand voltage: The maximum 60-Hz rms line-to-line voltage that the circuit breaker can withstand without insulation damage.

Rated impulse withstand voltage: The maximum crest voltage of a voltage pulse with standard rise and delay times that the breaker insulation can withstand.

Rated voltage range factor K: The range of voltage for which the symmetrical interrupting capability times the operating voltage is constant.

Current ratings

Rated continuous current: The maximum 60-Hz rms current that the breaker can carry continuously while it is in the closed position without overheating.

Rated short-circuit current: The maximum rms symmetrical current that the breaker can safely interrupt at rated maximum voltage.

Rated momentary current: The maximum rms asymmetrical current that the breaker can withstand while in the closed position without damage. Rated momentary current for standard breakers is 1.6 times the symmetrical interrupting capability.

Rated interrupting time: The time in cycles on a 60-Hz basis from the instant the trip coil is energized to the instant the fault current is cleared.

FIGURE 7.10

Symmetrical interrupting capability of a 69-kV class breaker

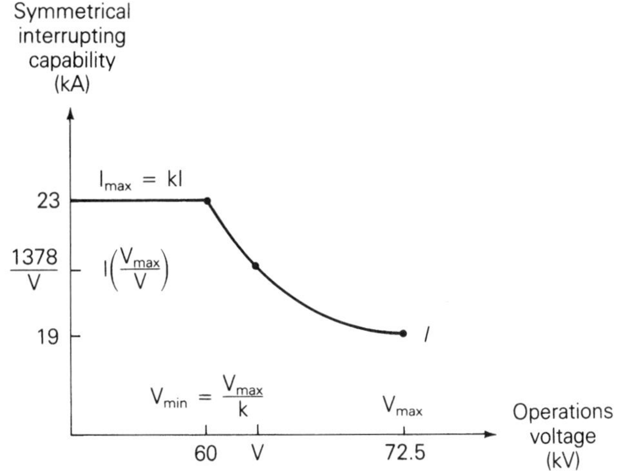

Rated interrupting MVA: For a three-phase circuit breaker, this is $\sqrt{3}$ times the rated maximum voltage in kV times the rated short-circuit current in kA. It is more common to work with current and voltage ratings than with MVA rating.

As an example, the symmetrical interrupting capability of the 69-kV class breaker listed in Table 7.10 is plotted versus operating voltage in Figure 7.10. As shown, the symmetrical interrupting capability increases from its rated short-circuit current $I = 19$ kA at rated maximum voltage $V_{max} = 72.5$ kV up to $I_{max} = KI = (1.21)(19) = 23$ kA at an operating voltage $V_{min} = V_{max}/K = 72.5/1.21 = 60$ kV. At operating voltages V between V_{min} and V_{max}, the symmetrical interrupting capability is $I \times V_{max}/V = 1378/V$ kA. At operating voltages below V_{min}, the symmetrical interrupting capability remains at $I_{max} = 23$ kA.

Breakers of the 115-kV class and higher have a voltage range factor $K = 1.0$; that is, their symmetrical interrupting current capability remains constant.

EXAMPLE 7.7 Circuit breaker selection

The calculated symmetrical fault current is 17 kA at a three-phase bus where the operating voltage is 64 kV. The X/R ratio at the bus is unknown. Select a circuit breaker from Table 7.10 for this bus.

SOLUTION The 69-kV-class breaker has a symmetrical interrupting capability $I(V_{max}/V) = 19(72.5/64) = 21.5$ kA at the operating voltage $V = 64$ kV. The calculated symmetrical fault current, 17 kA, is less than 80% of this capability (less than $0.80 \times 21.5 = 17.2$ kA), which is a requirement when X/R is unknown. Therefore, we select the 69-kV-class breaker from Table 7.10. ■

FUSES

Figure 7.11(a) shows a cutaway view of a fuse, which is one of the simplest overcurrent devices. The fuse consists of a metal "fusible" link or links encapsulated in a tube, packed in filler material, and connected to contact terminals. Silver is a typical link metal, and sand is a typical filler material.

During normal operation, when the fuse is operating below its continuous current rating, the electrical resistance of the link is so low that it simply acts as a conductor. If an overload current from one to about six times its continuous current rating occurs and persists for more than a short interval of time, the temperature of the link eventually reaches a level that causes a restricted segment of the link to melt. As shown in Figure 7.11(b), a gap is then formed and an electric arc is established. As the arc causes the link metal to burn back, the gap width increases. The resistance of the arc eventually reaches such a high level that the arc cannot be sustained and it is extinguished, as in Figure 7.11(c). The current flow within the fuse is then completely cut off.

FIGURE 7.11 Typical fuse

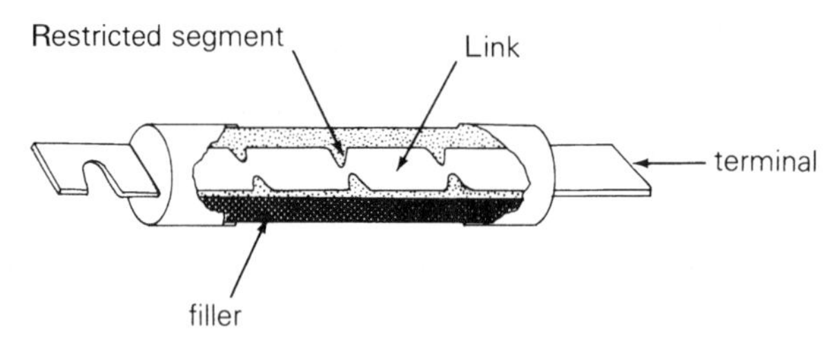

(a) Cutaway view

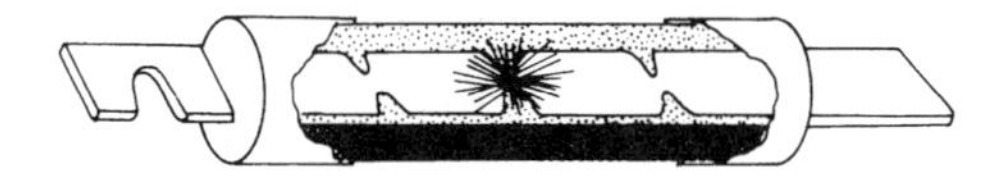

(b) The link melts and an arc is established under sustained overload current

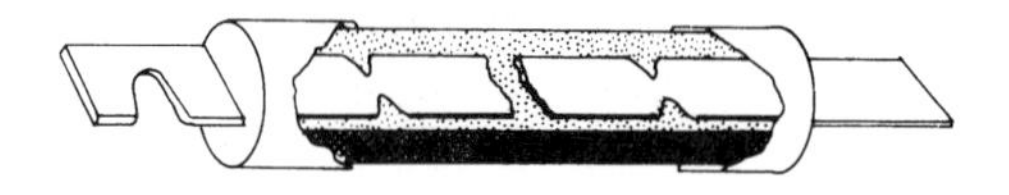

(c) The "open" link after clearing the overload current.

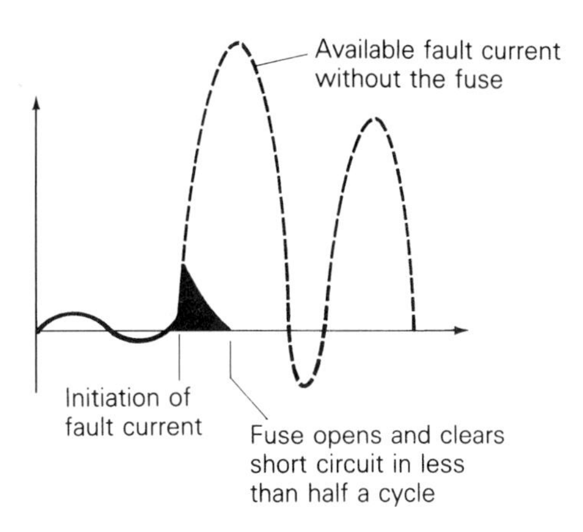

FIGURE 7.12

Operation of a current-limiting fuse

If the fuse is subjected to fault currents higher than about six times its continuous current rating, several restricted segments melt simultaneously, resulting in rapid arc suppression and fault clearing. Arc suppression is accelerated by the filler material in the fuse.

Many modern fuses are current limiting. As shown in Figure 7.12, a current-limiting fuse has such a high speed of response that it cuts off a high fault current in less than a half cycle—before it can build up to its full peak value. By limiting fault currents, these fuses permit the use of motors, transformers, conductors, and bus structures that could not otherwise withstand the destructive forces of high fault currents.

Fuse specification is normally based on the following four factors.

1. *Voltage rating*. This rms voltage determines the ability of a fuse to suppress the internal arc that occurs after the fuse link melts. A blown fuse should be able to withstand its voltage rating. Most low-voltage fuses have 250- or 600-V ratings. Ratings of medium-voltage fuses range from 2.4 to 34.5 kV.

2. *Continuous current rating*. The fuse should carry this rms current indefinitely, without melting and clearing.

3. *Interrupting current rating*. This is the largest rms asymmetrical current that the fuse can safely interrupt. Most modern, low-voltage current-limiting fuses have a 200-kA interrupting rating. Standard interrupting ratings for medium-voltage current-limiting fuses include 65, 80, and 100 kA.

4. *Time response*. The melting and clearing time of a fuse depends on the magnitude of the overcurrent or fault current and is usually specified by a "time–current" curve. Figure 7.13 shows the time–current curve of a 15.5-kV, 100-A (continuous) current-limiting fuse. As shown, the fuse link melts within 2 s and clears within 5 s for a 500-A current. For a 5-kA current, the fuse link melts in less than 0.01 s and clears within 0.015 s.

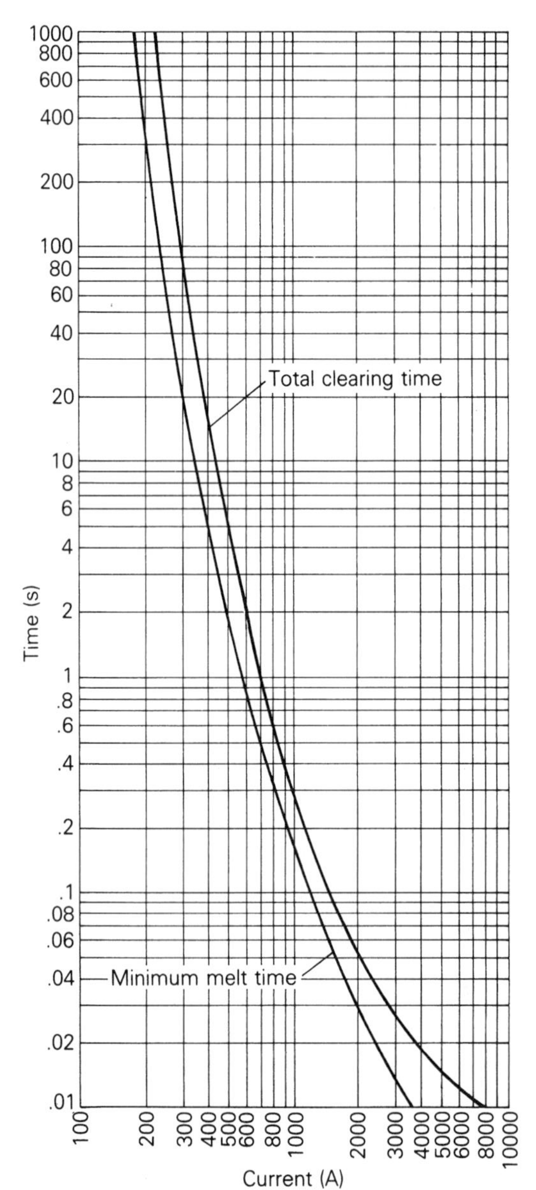

FIGURE 7.13 Time–current curves for a 15.5-kV, 100-A current-limiting fuse

It is usually a simple matter to coordinate fuses in a power circuit such that only the fuse closest to the fault opens the circuit. In a radial circuit, fuses with larger continuous current ratings are located closer to the source, such that the fuse closest to the fault clears before other, upstream fuses melt.

Fuses are inexpensive, fast operating, easily coordinated, and reliable, and they do not require protective relays or instrument transformers. Their chief disadvantage is that the fuse or the fuse link must be manually replaced after it melts. They are basically one-shot devices that are, for example, incapable of high-speed reclosing.

MULTIPLE CHOICE QUESTIONS

SECTION 7.1

7.1 The asymmetrical short-circuit current in series R–L circuit for a simulated solid or "bolted fault" can be considered as a combination of symmetrical (ac) component that is a __________, and dc-offset current that decays __________, and depends on __________. Fill in the Blanks.

7.2 Even though the fault current is not symmetrical and not strictly periodic, the rms asymmetrical fault current is computed as the rms ac fault current times an "asymmetry factor," which is a function of __________. Fill in the Blank.

SECTION 7.2

7.3 The amplitude of the sinusoidal symmetrical ac component of the three-phase short-circuit current of an unloaded synchronous machine decreases from a high initial value to a lower steady-state value, going through the stages of __________ and __________ periods. Fill in the Blanks.

7.4 The duration of subtransient fault current is dictated by __________ time constant, and that of transient fault current is dictated by __________ time constant. Fill in the Blanks.

7.5 The reactance that plays a role under steady-state operation of a synchronous machine is called __________. Fill in the Blank.

7.6 The dc-offset component of the three-phase short-circuit current of an unloaded synchronous machine is different in the three phases and its exponential decay is dictated by __________. Fill in the Blank.

SECTION 7.3

7.7 Generally, in power-system short-circuit studies, for calculating subtransient fault currents, transformers are represented by their __________, transmission lines by their equivalent __________, and synchronous machines by __________ behind their subtransient reactances. Fill in the Blanks.

7.8 In power-system fault studies, all nonrotating impedance loads are usually neglected.
(a) True (b) False

7.9 Can superposition be applied in power-system short-circuit studies for calculating fault currents?
(a) Yes (b) No

7.10 Before proceeding with per-unit fault current calculations, based on the single-line diagram of the power system, a positive-sequence equivalent circuit is set up on a chosen base system.
(a) True (b) False

SECTION 7.4

7.11 The inverse of the bus-admittance matrix is called ________ matrix. Fill in the Blank.

7.12 For a power system, modeled by its positive-sequence network, both bus-admittance matrix and bus-impedance matrix are symmetric.
(a) True (b) False

7.13 The bus-impedance equivalent circuit can be represented in the form of a "rake" with the diagonal elements, which are ______, and the non-diagonal (off-diagonal) elements, which are ________. Fill in the Blanks.

SECTION 7.5

7.14 A circuit breaker is designed to extinguish the arc by ________. Fill in the Blank.

7.15 Power-circuit breakers are intended for service in ac circuit above ________ V. Fill in the Blank.

7.16 In circuit breakers, besides air or vacuum, what gaseous medium, in which the arc is elongated, is used?

7.17 Oil can be used as a medium to extinguish the arc in circuit breakers.
(a) True (b) False

7.18 Besides a blast of air/gas, the arc in a circuit breaker can be elongated by ______. Fill in the Blank.

7.19 For distribution systems, standard reclosers are equipped for two or more reclosures, where as multiple-shot reclosing in EHV systems is not a standard practice.
(a) True (b) False

7.20 Breakers of the 115-kV class and higher have a voltage range factor K = _______, such that their symmetrical interrupting current capability remains constant. Fill in the Blank.

7.21 A typical fusible link metal in fuses is _______, and a typical filler material is _______. Fill in the Blanks.

7.22 The melting and clearing time of a current-limiting fuse is usually specified by a _______ curve.

PROBLEMS

SECTION 7.1

7.1 In the circuit of Figure 7.1, V = 277 volts, L = 2 mH, R = 0.4 Ω, and $\omega = 2\pi 60$ rad/s. Determine (a) the rms symmetrical fault current; (b) the rms asymmetrical fault current at the instant the switch closes, assuming maximum dc offset; (c) the rms asymmetrical fault current 5 cycles after the switch closes, assuming maximum dc offset; (d) the dc offset as a function of time if the switch closes when the instantaneous source voltage is 300 volts.

7.2 Repeat Example 7.1 with V = 4 kV, X = 2 Ω, and R = 1 Ω.

7.3 In the circuit of Figure 7.1, let R = 0.125 Ω, L = 10 mH, and the source voltage is e(t) = 151 sin(377t + α) V. Determine the current response after closing the switch for the following cases: (a) no dc offset; (b) maximum dc offset. Sketch the current waveform up to t = 0.10 s corresponding to case (a) and (b).

7.4 Consider the expression for $i(t)$ given by

$$i(t) = \sqrt{2}I_{\text{rms}}[\sin(\omega t - \theta_z) + \sin\theta_z . \bar{e}^{(\omega R/X)t}]$$

where $\theta_z = \tan^{-1}(\omega L/R)$.
(a) For (X/R) equal to zero and infinity, plot $i(t)$ as a function of (ωt).
(b) Comment on the dc offset of the fault current waveforms.
(c) Find the asymmetrical current factor and the time of peak, t_p, in milliseconds, for (X/R) ratios of zero and infinity.

7.5 If the source impedance at a 13.2 kV distribution substation bus is $(0.5 + j1.5)$ Ω per phase, compute the rms and maximum peak instantaneous value of the fault current, for a balanced three-phase fault. For the system (X/R) ratio of 3.0, the asymmetrical factor is 1.9495 and the time of peak is 7.1 ms (see Problem 7.4). Comment on the withstanding peak current capability to which all substation electrical equipment need to be designed.

SECTION 7.2

7.6 A 1000-MVA 20-kV, 60-Hz three-phase generator is connected through a 1000-MVA 20-kV Δ/345-kV Y transformer to a 345-kV circuit breaker and a 345-kV transmission line. The generator reactances are $X''_d = 0.17$, $X'_d = 0.30$, and $X_d = 1.5$ per unit, and its time constants are $T''_d = 0.05$, $T'_d = 1.0$, and $T_A = 0.10$ s. The transformer series reactance is 0.10 per unit; transformer losses and exciting current are neglected. A three-phase short-circuit occurs on the line side of the circuit breaker when the generator is operated at rated terminal voltage and at no-load. The breaker interrupts the fault 3 cycles after fault inception. Determine (a) the subtransient current through the breaker in per-unit and in kA rms; and (b) the rms asymmetrical fault current the breaker interrupts, assuming maximum dc offset. Neglect the effect of the transformer on the time constants.

7.7 For Problem 7.6, determine (a) the instantaneous symmetrical fault current in kA in phase *a* of the generator as a function of time, assuming maximum dc offset occurs in this generator phase; and (b) the maximum dc offset current in kA as a function of time that can occur in any one generator phase.

7.8 A 300-MVA, 13.8-kV, three-phase, 60-Hz, Y-connected synchronous generator is adjusted to produce rated voltage on open circuit. A balanced three-phase fault is applied to the terminals at $t = 0$. After analyzing the raw data, the symmetrical transient current is obtained as

$$i_{ac}(t) = 10^4(1 + e^{-t/\tau_1} + 6e^{-t/\tau_2}) \quad A$$

where $\tau_1 = 200$ ms and $\tau_2 = 15$ ms. (a) Sketch $i_{ac}(t)$ as a function of time for $0 \leqslant t \leqslant 500$ ms. (b) Determine X_d'' and X_d in per-unit based on the machine ratings.

7.9 Two identical synchronous machines, each rated 60 MVA, 15 kV, with a subtransient reactance of 0.1 pu, are connected through a line of reactance 0.1 pu on the base of the machine rating. One machine is acting as a synchronous generator, while the other is working as a motor drawing 40 MW at 0.8 pf leading with a terminal voltage of 14.5 kV, when a symmetrical three-phase fault occurs at the motor terminals. Determine the subtransient currents in the generator, the motor, and the fault by using the internal voltages of the machines. Choose a base of 60 MVA, 15 kV in the generator circuit.

SECTION 7.3

7.10 Recalculate the subtransient current through the breaker in Problem 7.6 if the generator is initially delivering rated MVA at 0.80 p.f. lagging and at rated terminal voltage.

7.11 Solve Example 7.4, parts (a) and (c) without using the superposition principle. First calculate the internal machine voltages E_g'' and E_m'', using the prefault load current. Then determine the subtransient fault, generator, and motor currents directly from Figure 7.4(a). Compare your answers with those of Example 7.3.

7.12 Equipment ratings for the four-bus power system shown in Figure 7.14 are as follows:

Generator G1:	500 MVA, 13.8 kV, $X'' = 0.20$ per unit
Generator G2:	750 MVA, 18 kV, $X'' = 0.18$ per unit
Generator G3:	1000 MVA, 20 kV, $X'' = 0.17$ per unit
Transformer T1:	500 MVA, 13.8 Δ/500 Y kV, $X = 0.12$ per unit
Transformer T2:	750 MVA, 18 Δ/500 Y kV, $X = 0.10$ per unit
Transformer T3:	1000 MVA, 20 Δ/500 Y kV, $X = 0.10$ per unit
Each 500-kV line:	$X_1 = 50\ \Omega$

A three-phase short circuit occurs at bus 1, where the prefault voltage is 525 kV. Prefault load current is neglected. Draw the positive-sequence reactance diagram in

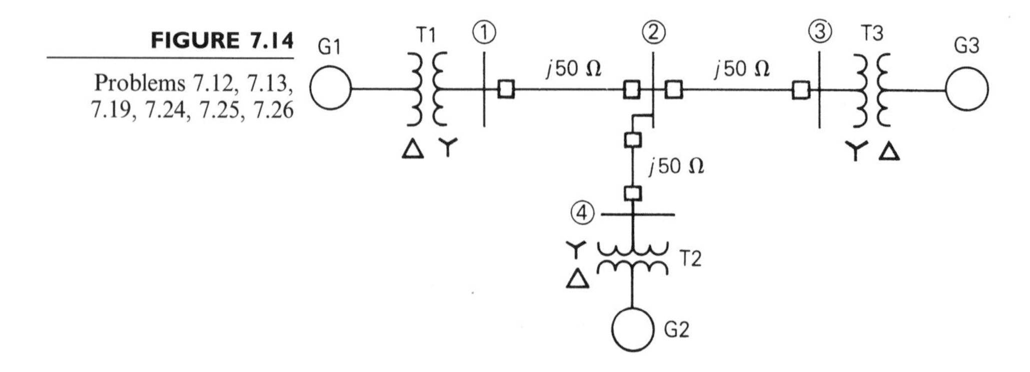

FIGURE 7.14 Problems 7.12, 7.13, 7.19, 7.24, 7.25, 7.26

per-unit on a 1000-MVA, 20-kV base in the zone of generator G3. Determine (a) the Thévenin reactance in per-unit at the fault, (b) the subtransient fault current in per-unit and in kA rms, and (c) contributions to the fault current from generator G1 and from line 1–2.

7.13 For the power system given in Problem 7.12, a three-phase short circuit occurs at bus 2, where the prefault voltage is 525 kV. Prefault load current is neglected. Determine the (a) Thévenin equivalent at the fault, (b) subtransient fault current in per-unit and in kA rms, and (c) contributions to the fault from lines 1–2, 2–3, and 2–4.

7.14 Equipment ratings for the five-bus power system shown in Figure 7.15 are as follows:

Generator G1: 50 MVA, 12 kV, $X'' = 0.2$ per unit

Generator G2: 100 MVA, 15 kV, $X'' = 0.2$ per unit

Transformer T1: 50 MVA, 10 kV Y/138 kV Y, $X = 0.10$ per unit

Transformer T2: 100 MVA, 15 kV Δ/138 kV Y, $X = 0.10$ per unit

Each 138-kV line: $X_1 = 40\ \Omega$

A three-phase short circuit occurs at bus 5, where the prefault voltage is 15 kV. Prefault load current is neglected. (a) Draw the positive-sequence reactance diagram in per-unit on a 100-MVA, 15-kV base in the zone of generator G2. Determine: (b) the Thévenin equivalent at the fault, (c) the subtransient fault current in per-unit and in kA rms, and (d) contributions to the fault from generator G2 and from transformer T2.

FIGURE 7.15

Problems 7.14, 7.15, 7.20

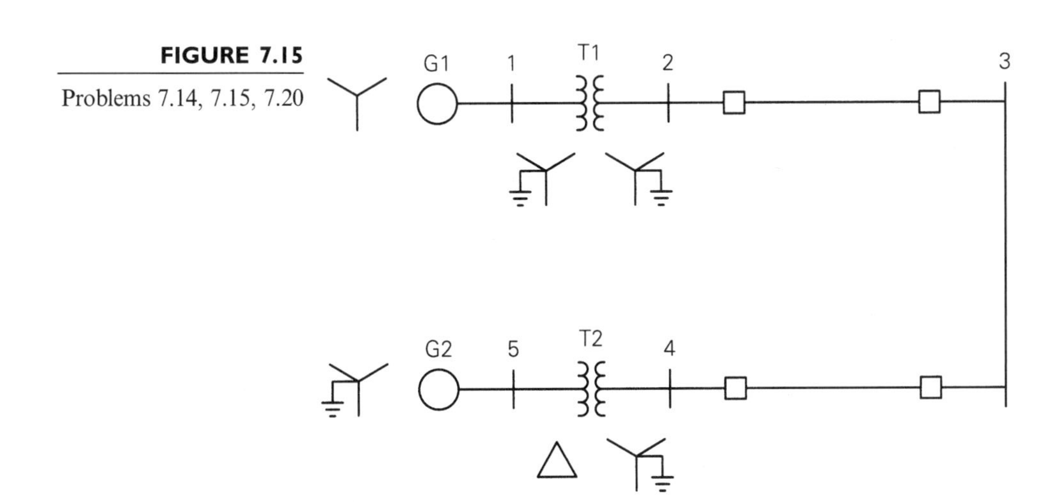

7.15 For the power system given in Problem 7.14, a three-phase short circuit occurs at bus 4, where the prefault voltage is 138 kV. Prefault load current is neglected. Determine (a) the Thévenin equivalent at the fault, (b) the subtransient fault current in per-unit and in kA rms, and (c) contributions to the fault from transformer T2 and from line 3–4.

7.16 In the system shown in Figure 7.16, a three-phase short circuit occurs at point F. Assume that prefault currents are zero and that the generators are operating at rated voltage. Determine the fault current.

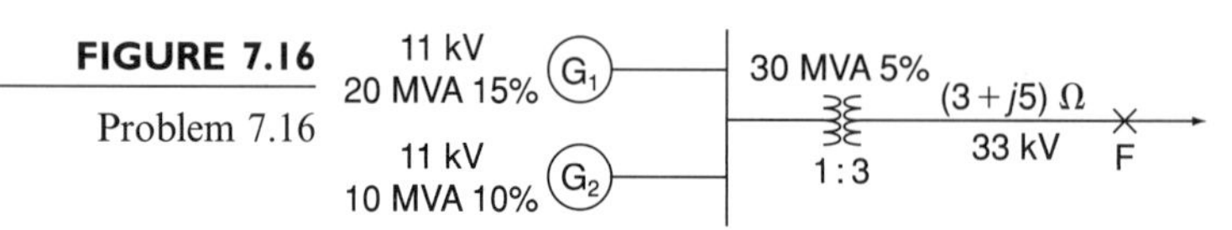

FIGURE 7.16 Problem 7.16

7.17 A three-phase short circuit occurs at the generator bus (bus 1) for the system shown in Figure 7.17. Neglecting prefault currents and assuming that the generator is operating at its rated voltage, determine the subtransient fault current using superposition.

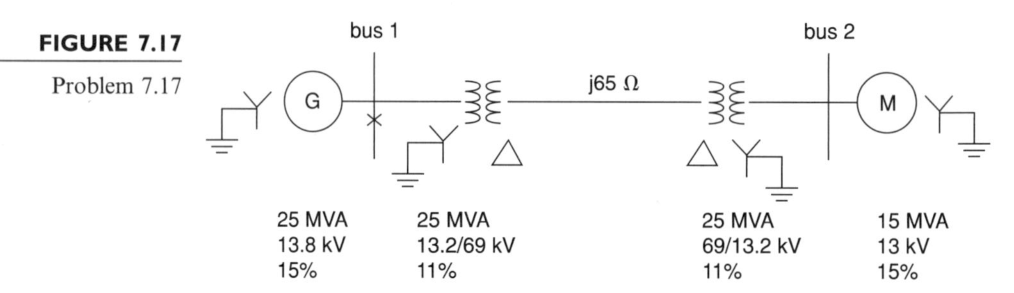

FIGURE 7.17 Problem 7.17

SECTION 7.4

7.18 (a) The bus impedance matrix for a three-bus power system is

$$\boldsymbol{Z}_{\text{bus}} = j\begin{bmatrix} 0.12 & 0.08 & 0.04 \\ 0.08 & 0.12 & 0.06 \\ 0.04 & 0.06 & 0.08 \end{bmatrix} \quad \text{per unit}$$

where subtransient reactances were used to compute $\boldsymbol{Z}_{\text{bus}}$. Prefault voltage is 1.0 per unit and prefault current is neglected. (a) Draw the bus impedance matrix equivalent circuit (rake equivalent). Identify the per-unit self- and mutual impedances as well as the prefault voltage in the circuit. (b) A three-phase short circuit occurs at bus 2. Determine the subtransient fault current and the voltages at buses 1, 2, and 3 during the fault.
(b) For 7.18 Repeat for the case of

$$\boldsymbol{Z}_{\text{bus}} = j\begin{bmatrix} 0.4 & 0.1 & 0.3 \\ 0.1 & 0.8 & 0.5 \\ 0.3 & 0.5 & 1.2 \end{bmatrix} \quad \text{per unit}$$

7.19 Determine $\boldsymbol{Y}_{\text{bus}}$ in per-unit for the circuit in Problem 7.12. Then invert $\boldsymbol{Y}_{\text{bus}}$ to obtain $\boldsymbol{Z}_{\text{bus}}$.

7.20 Determine $\boldsymbol{Y}_{\text{bus}}$ in per-unit for the circuit in Problem 7.14. Then invert $\boldsymbol{Y}_{\text{bus}}$ to obtain $\boldsymbol{Z}_{\text{bus}}$.

7.21 Figure 7.18 shows a system reactance diagram. (a) Draw the admittance diagram for the system by using source transformations. (b) Find the bus admittance matrix $\boldsymbol{Y}_{\text{bus}}$. (c) Find the bus impedance $\boldsymbol{Z}_{\text{bus}}$ matrix by inverting $\boldsymbol{Y}_{\text{bus}}$.

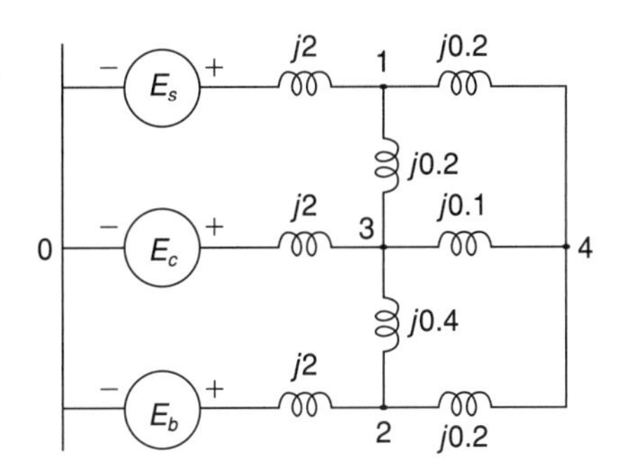

FIGURE 7.18 Problem 7.21

7.22 For the network shown in Figure 7.19, impedances labeled 1 through 6 are in per-unit. (a) Determine $\boldsymbol{Y}_{\text{bus}}$. Preserve all buses. (b) Using MATLAB or a similar computer program, invert $\boldsymbol{Y}_{\text{bus}}$ to obtain $\boldsymbol{Z}_{\text{bus}}$.

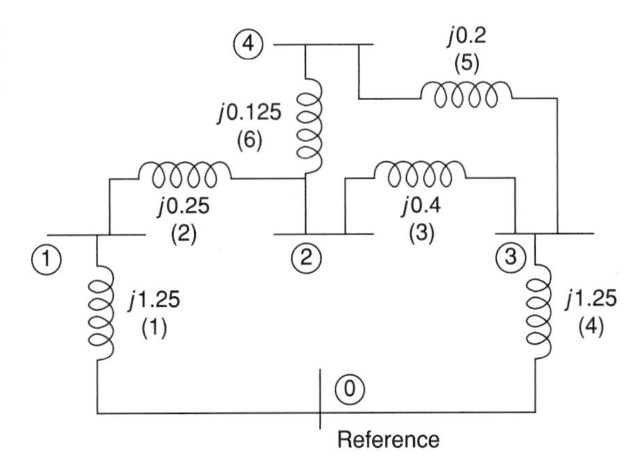

FIGURE 7.19 Problem 7.22

7.23 A single-line diagram of a four-bus system is shown in Figure 7.20, for which Z_{BUS} is given below:

$$\boldsymbol{Z}_{\text{BUS}} = j\begin{bmatrix} 0.25 & 0.2 & 0.16 & 0.14 \\ 0.2 & 0.23 & 0.15 & 0.151 \\ 0.16 & 0.15 & 0.196 & 0.1 \\ 0.14 & 0.151 & 0.1 & 0.195 \end{bmatrix} \text{ per unit}$$

Let a three-phase fault occur at bus 2 of the network.
(a) Calculate the initial symmetrical rms current in the fault.
(b) Determine the voltages during the fault at buses 1, 3, and 4.
(c) Compute the fault currents contributed to bus 2 by the adjacent unfaulted buses 1, 3, and 4.
(d) Find the current flow in the line from bus 3 to bus 1. Assume the prefault voltage V_f at bus 2 to be $1\underline{/0^\circ}$ pu, and neglect all prefault currents.

FIGURE 7.20

Single-line diagram for Problem 7.23

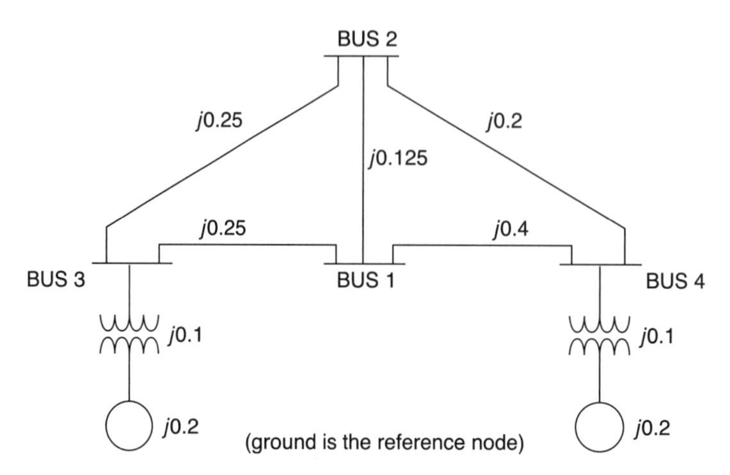

PW **7.24** PowerWorld Simulator case Problem 7_24 models the system shown in Figure 7.14 with all data on a 1000-MVA base. Using PowerWorld Simulator, determine the current supplied by each generator and the per-unit bus voltage magnitudes at each bus for a fault at bus 2.

PW **7.25** Repeat Problem 7.24, except place the fault at bus 1.

PW **7.26** Repeat Problem 7.24, except place the fault midway between buses 2 and 4. Determining the values for line faults requires that the line be split, with a fictitious bus added at the point of the fault. The original line's impedance is then allocated to the two new lines based on the fault location, 50% each for this problem. Fault calculations are then the same as for a bus fault. This is done automatically in PowerWorld Simulator by first right-clicking on a line, and then selecting "Fault." The Fault dialog appears as before, except now the fault type is changed to "In-Line Fault." Set the location percentage field to 50% to model a fault midway between buses 2 and 4.

PW **7.27** One technique for limiting fault current is to place reactance in series with the generators. Such reactance can be modeled in Simulator by increasing the value of the generator's positive sequence internal impedance. For the Problem 7.24 case, how much per-unit reactance must be added to G3 to limit its maximum fault current to 2.5 per unit for all 3 phase bus faults? Where is the location of the most severe bus fault?

PW **7.28** Using PowerWorld Simulator case Example 6.13, determine the per-unit current and actual current in amps supplied by each of the generators for a fault at the PETE69 bus. During the fault, what percentage of the system buses have voltage magnitudes below 0.75 per unit?

PW **7.29** Repeat Problem 7.28, except place the fault at the BOB69 bus.

PW **7.30** Redo Example 7.5, except first open the generator at bus 3.

SECTION 7.5

7.31 A three-phase circuit breaker has a 15.5-kV rated maximum voltage, 9.0-kA rated short-circuit current, and a 2.67-rated voltage range factor. (a) Determine the symmetrical interrupting capability at 10-kV and 5-kV operating voltages. (b) Can this breaker be safely installed at a three-phase bus where the symmetrical fault current is 10 kA, the operating voltage is 13.8 kV, and the (X/R) ratio is 12?

7.32 A 500-kV three-phase transmission line has a 2.2-kA continuous current rating and a 2.5-kA maximum short-time overload rating, with a 525-kV maximum operating voltage. Maximum symmetrical fault current on the line is 30 kA. Select a circuit breaker for this line from Table 7.10.

7.33 A 69-kV circuit breaker has a voltage range factor $K = 1.21$, a continuous current rating of 1200 A, and a rated short-circuit current of 19,000 A at the maximum rated voltage of 72.5 kV. Determine the maximum symmetrical interrupting capability of the breaker. Also, explain its significance at lower operating voltages.

7.34 As shown in Figure 7.21, a 25-MVA, 13.8-kV, 60-Hz synchronous generator with $X_d'' = 0.15$ per unit is connected through a transformer to a bus that supplies four identical motors. The rating of the three-phase transformer is 25 MVA, 13.8/6.9 kV, with a leakage reactance of 0.1 per unit. Each motor has a subtransient reactance $X_d'' = 0.2$ per unit on a base of 5 MVA and 6.9 kV. A three-phase fault occurs at point P, when the bus voltage at the motors is 6.9 kV. Determine: (a) the subtransient fault current, (b) the subtransient current through breaker A, (c) the symmetrical short-circuit interrupting current (as defined for circuit breaker applications) in the fault and in breaker A.

FIGURE 7.21

Problem 7.34

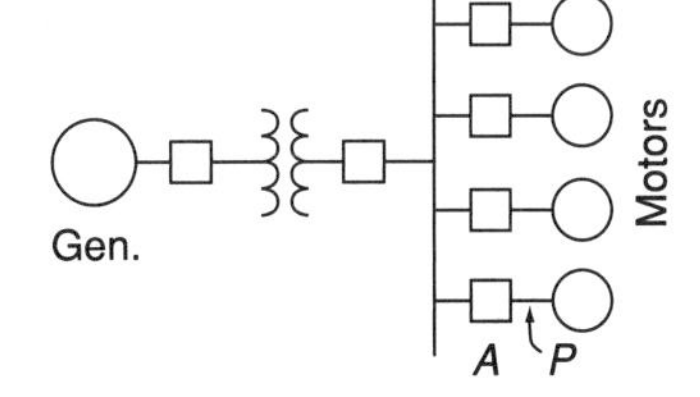

CASE STUDY QUESTIONS

A. Why are arcing (high-impedance) faults more difficult to detect than low-impedance faults?

B. What methods are available to prevent the destructive effects of arcing faults from occurring?

DESIGN PROJECT 4 (*CONTINUED*): POWER FLOW/SHORT CIRCUITS

Additional time given: 3 weeks
Additional time required: 10 hours

This is a continuation of Design Project 4. Assignments 1 and 2 are given in Chapter 6.

Assignment 3: Symmetrical Short Circuits

For the single-line diagram that you have been assigned (Figure 6.13 or 6.14), convert the positive-sequence reactance data to per-unit using the given base quantities. For synchronous machines, use subtransient reactance. Then using PowerWorld Simulator, create the machine, transmission line, and transformer input data files. Next, run the program to compute subtransient fault currents for a bolted three-phase-to-ground fault at bus 1, then at bus 2, then at bus 3, and so on. Also compute bus voltages during the faults and the positive-sequence bus impedance matrix. Assume 1.0 per-unit prefault voltage. Neglect prefault load currents and all losses.

Your output for this assignment consists of three input data files and three output data (fault currents, bus voltages, and the bus impedance matrix) files.

This project continues in Chapter 9.

REFERENCES

1. Westinghouse Electric Corporation, *Electrical Transmission and Distribution Reference Book*, 4th ed. (East Pittsburgh, PA: 1964).
2. E. W. Kimbark, *Power System Stability, Synchronous Machines*, vol. 3 (New York: Wiley, 1956).
3. A. E. Fitzgerald, C. Kingsley, and S. Umans, *Electric Machinery*, 5th ed. (New York: McGraw-Hill, 1990).
4. M. S. Sarma, *Electric Machines* 2nd ed. (Boston: PWS Publishing, 1994).
5. J. R. Neuenswander, *Modern Power Systems* (New York: Intext Educational Publishers, 1971).
6. H. E. Brown, *Solution of Large Networks by Matrix Methods* (New York: Wiley, 1975).
7. G. N. Lester, "High Voltage Circuit Breaker Standards in the USA—Past, Present and Future," *IEEE Transactions PAS*, vol. PAS–93 (1974): pp. 590–600.
8. W. D. Stevenson, Jr., *Elements of Power System Analysis*, 4th ed. (New York: McGraw-Hill, 1982).
9. C. A. Gross, *Power System Analysis* (New York: Wiley, 1979).
10. *Application Guide for AC High-Voltage Circuit Breakers Rated on a Symmetrical Current Basis*, ANSI C 37.010 (New York: American National Standards Institute, 1972).
11. F. Shields, "The Problem of Arcing Faults in Low-Voltage Power Distribution Systems," *IEEE Transactions on Industry and General Applications*, vol. IGA-3, no. 1, (January/February 1967), pp. 15–25.

Generator stator showing completed windings for a 757-MVA, 3600-RPM, 60-Hz synchronous generator (Courtesy of General Electric)

8

SYMMETRICAL COMPONENTS

The method of symmetrical components, first developed by C. L. Fortescue in 1918, is a powerful technique for analyzing unbalanced three-phase systems. Fortescue defined a linear transformation from phase components to a new set of components called *symmetrical components*. The advantage of this transformation is that for balanced three-phase networks the equivalent circuits obtained for the symmetrical components, called *sequence networks*, are separated into three uncoupled networks. Furthermore, for unbalanced three-phase systems, the three sequence networks are connected only at points of unbalance. As a result, sequence networks for many cases of unbalanced three-phase systems are relatively easy to analyze.

The symmetrical component method is basically a modeling technique that permits systematic analysis and design of three-phase systems. Decoupling a detailed three-phase network into three simpler sequence networks reveals complicated phenomena in more simplistic terms. Sequence network results can then be superposed to obtain three-phase network results. As an example, the application of symmetrical components to unsymmetrical short-circuit studies (see Chapter 9) is indispensable.

The objective of this chapter is to introduce the concept of symmetrical components in order to lay a foundation and provide a framework for later chapters covering both equipment models as well as power system analysis and design methods. In Section 8.1, we define symmetrical components. In Sections 8.2–8.7, we present sequence networks of loads, series impedances, transmission lines, rotating machines, and transformers. We discuss complex power in sequence networks in Section 8.8. Although Fortescue's original work is valid for polyphase systems with n phases, we will consider only three-phase systems here.

CASE STUDY The following article provides an overview of circuit breakers with high voltage ratings at or above 72.5 kV [4]. Circuit breakers are broadly classified by the medium used to extinguish the arc: bulk oil, minimum oil, air-blast, vacuum, and sulfur hexafluoride (SF_6). For high voltages, oil circuit breakers dominated in the early 1900s through the 1950s for applications up to 362 kV, with minimum oil circuit breakers developed up to 380 kV. The development of air-blast circuit breakers started in Europe in the 1920s and became prevalent in the 1950s. Air-blast circuit breakers, which use air under high pressure that is blown between the circuit breaker contacts to extinguish the arc, have been used at voltages up to 800 kV and many are still in operation today. Air-blast circuit breakers were manufactured until the 1980s when they were supplanted by lower cost and simpler SF_6 puffer-type circuit breakers. SF_6 gas possesses exceptional arc-interrupting properties that have led to a worldwide change to SF_6 high-voltage circuit breakers, which are more reliable, more efficient and more compact than other types of circuit breakers. Vacuum circuit breakers are commonly used at medium voltages between 1 and 72.5 kV.

Circuit Breakers Go High Voltage: The Low Operating Energy of SF_6 Circuit Breakers Improves Reliability and Reduces Wear and Tear

DENIS DUFOURNET

The first sulfur hexafluoride (SF_6) gas industrial developments were in the medium voltage range. This equipment confirmed the advantages of a technique that uses SF_6 at a low-pressure level concurrently with the auto-pneumatic blast system to interrupt the arc that was called later puffer.

High-voltage SF_6 circuit breakers with self-blast interrupters have found worldwide acceptance because their high current interrupting capability is obtained with a low operating energy that can be provided by low-cost, spring-operated mechanisms. The low-operating energy required reduces the stress and wear of the mechanical components and significantly improves the overall reliability of the circuit breaker. This switching principle was first introduced in the high-voltage area about 20 years ago, starting with the voltage level of 72.5 kV. Today this technique is available up to 800 kV. Furthermore it is used for generator circuit breaker applications with short circuit currents of 63 kA and above.

Service experience shows that when the SF_6 circuit breakers of the self-blast technology were first designed, the expectations of the designers had been fulfilled completely with respect to reliability and day-to-day operation.

A HISTORY OF CIRCUIT BREAKERS

Bulk oil circuit breakers dominated in the early 1900s and remained in use throughout the 1950s, for applications up to 362 kV for which they had eight breaks in series. They were replaced by minimum oil and air-blast circuit breakers for high-voltage applications.

Minimum oil circuit breakers, as shown in Figure 1, have arc control structures that improve the arc cooling process and significantly reduce the volume of oil. They were developed up to 380 kV, in particular for the first 380 kV network in the world (Harspränget–Halsberg line in Sweden in 1952). There were tentative extensions to 765 kV, 50 kA, but minimum oil circuit breakers were supplanted in the EHV range by air-blast circuit breakers that were the first to be applied in 525, 735, and 765 kV networks, respectively in Russia (1960), Canada (1965), and the United States (1969).

Air-blast circuit breakers, as shown in Figure 2, use air under high pressure that is blown through the arc space between the opening contacts to extinguish the arc. The development of air-blast circuit breakers started in Europe in the 1920s, with further development in 1930s and 1940s, and became prevalent in the 1950s.

Air-blast circuit breakers were very successful in North America and Europe. They had an interrupting capability of 63 kA, later increased to 90 kA in the 1970s. Many circuit breakers of this type are still in operation today, in particular in North America, at 550 and 800 kV.

Air-blast circuit breakers were manufactured until the 1980s when they were supplanted by the lower cost and less complex SF_6 puffer-type circuit breakers.

Figure 1
Minimum oil circuit breaker 145 kV type orthojector (Courtesy of Alstom Grid)

Figure 2
Air-blast circuit breaker type PK12 applied to 765 kV in North America (Courtesy of Alstom Grid)

The first industrial application of SF_6 dates from 1937 when it was used in the United States as an insulating medium for cables (patent by F.S. Cooper of General Electric). With the advent of the nuclear power industry in the 1950s, SF_6 was produced in large quantities and its use extended to circuit breakers as a quenching medium.

The first application of SF_6 for current interruption was done in 1953 when 15–161 kV switches were developed by Westinghouse. The first high-voltage SF_6 circuit breakers were built also by Westinghouse in 1956, the interrupting capability was then limited to 5 kA under 115 kV, with each pole having six interrupting units in series. In 1959, Westinghouse produced the first SF_6 circuit breakers with high current interrupting capabilities: 41.8 kA under 138 kV (10,000 MVA) and 37.8 kA under 230 kV (15,000 MVA). These circuit breakers were of the dual pressure type based on the axial blast principles used in air-blast circuit breakers. They were supplanted by the SF_6 puffer circuit breakers.

In 1967, the puffer-type technique was introduced for high-voltage circuit breakers where the relative movement of a piston and a cylinder linked to the moving contact produced the pressure build-up necessary to blast the arc. The puffer technique, shown in Figure 3, was applied in the first 245 kV metal-enclosed gas insulated circuit breaker installed in France in 1969.

The excellent properties of SF_6 lead to the fast extension of this technique in the 1970s and to its use for the development of circuit breakers with high current interrupting capability, up to 800 kV.

The achievement, around 1983, of the first single-break 245 kV and the corresponding 420 kV, 550 kV, and 800 kV, with, respectively, two, three, and four chambers per pole, lead to the dominance of SF_6 circuit breakers in the complete high-voltage range.

Several characteristics of SF_6 puffer circuit breakers can explain their success:

- simplicity of the interrupting chamber which does not need an auxiliary chamber for breaking
- autonomy provided by the puffer technique
- the possibility to obtain the highest performances, up to 63 kA, with a reduced number of interrupting chambers (Figure 4)

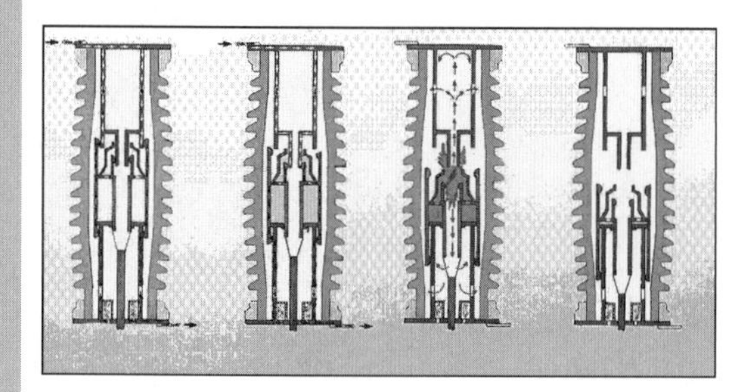

Figure 3
Puffer-type circuit breaker

Figure 4
800 kV 50 kA circuit breaker type FX with closing resistors (Courtesy of Alstom Grid)

- short interrupting time of 2-2. 5 cycles at 60 Hz
- high electrical endurance, allowing at least 25 years of operation without reconditioning
- possible compact solutions when used for gas-insulated switchgear (GIS) or hybrid switchgears
- integrated closing resistors or synchronized operations to reduce switching over voltages
- reliability and availability
- low noise level
- no compressor for SF_6 gas.

The reduction in the number of interrupting chambers per pole has led to a considerable simplification of circuit breakers as the number of parts as well as the number of seals was decreased. As a direct consequence, the reliability of circuit breakers was improved, as verified later by CIGRE surveys.

SELF-BLAST TECHNOLOGY

The last 20 years have seen the development of the self-blast technique for SF_6 interrupting chambers. This technique has proven to be very efficient and has been widely applied for high-voltage circuit breakers up to 800 kV. It has allowed the development of new ranges of circuit breakers operated by low energy spring-operated mechanisms.

Another aim of this evolution was to further increase the reliability by reducing dynamic forces in the pole and its mechanism.

These developments have been facilitated by the progress made in digital simulations that were widely used to optimize the geometry of the interrupting chamber and the mechanics between the poles and the mechanism.

The reduction of operating energy was achieved by lowering energy used for gas compression and by making a larger use of arc energy to produce the pressure necessary to quench the arc and obtain current interruption.

Low-current interruption, up to about 30% of rated short-circuit current, is obtained by a puffer blast where the overpressure necessary to quench the arc is produced by gas compression in a volume limited by a fixed piston and a moving cylinder.

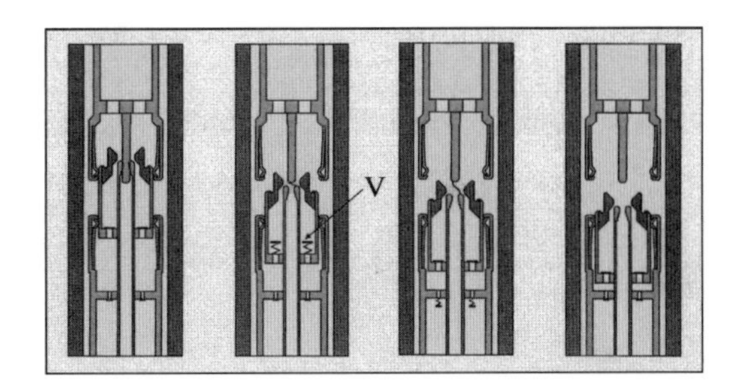

Figure 5
Self blast (or double volume) interrupting chamber

Figure 5 shows the self-blast interruption principle where a valve (V) was introduced between the expansion and the compression volume.

When interrupting low currents, the valve (V) opens under the effect of the overpressure generated in the compression volume. The interruption of the arc is made as in a puffer circuit breaker thanks to the compression of the gas obtained by the piston action.

In the case of high-current interruption, the arc energy produces a high overpressure in the expansion volume, which leads to the closure of the valve (V) and thus isolating the expansion volume from the compression volume. The overpressure necessary for breaking is obtained by the optimal use of the thermal effect and of the nozzle clogging effect produced whenever the cross-section of the arc significantly reduces the exhaust of gas in the nozzle.

This technique, known as self-blast, has been used extensively for more than 15 years for the development of many types of interrupting chambers and circuit breakers (Figure 6).

The better knowledge of arc interruption obtained by digital simulations and validation of performances by interrupting tests has contributed to a higher reliability of these self-blast circuit breakers. In addition, the reduction in

Figure 6
Dead tank circuit breaker 145 kV with spring-operating mechanism and double motion self blast interrupting chambers (Courtesy of Alstom Grid)

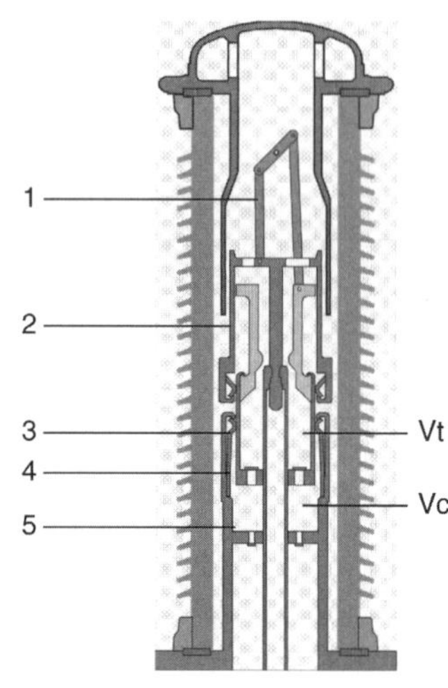

Figure 7
Double motion interrupting chamber

operating energy, allowed by the self-blast technique, leads to a higher mechanical endurance.

DOUBLE MOTION PRINCIPLE

The self-blast technology was further optimized by using the double-motion principle. This leads to further reduction of the operating energy by reducing the kinetic energy consumed during opening. The method consists of displacing the two arcing contacts in opposite directions. With such a system, it was possible to reduce the necessary opening energy for circuit breakers drastically.

Figure 7 shows the arcing chamber of a circuit breaker with the double motion principle. The pole columns are equipped with helical springs mounted in the crankcase.

These springs contain the necessary energy for an opening operation. The energy of the spring is transmitted to the arcing chamber via an insulating rod.

To interrupt an arc, the contact system must have sufficient velocity to avoid reignitions. Furthermore, a pressure rise must be generated to establish a gas flow in the chamber.

The movable upper contact system is connected to the nozzle of the arcing chamber via a linkage system. This allows the movement of both arcing contacts in opposite directions. Therefore the velocity of one contact can be reduced by 50% because the relative velocity of both contacts is still 100%. The necessary kinetic energy scales with the square of the velocity, allowing—theoretically—an energy reduction in the opening spring by a factor of 4. In reality, this value can't be achieved because the moving mass has to be increased. As in the self-blast technique described previously, the arc itself mostly generates the pressure rise.

Because the pressure generation depends on the level of the short-circuit current, an additional small piston is necessary to interrupt small currents (i.e., less than 30% of the rated short-circuit current). Smaller pistons mean less operating energy.

The combination of both double motion of contacts and self-blast technique allows for the significant reduction of opening energy.

GENERATOR CIRCUIT BREAKERS

Generator circuit breakers are connected between a generator and the step-up voltage transformer. They are generally used at the outlet of high-power generators (100–1,800 MVA) to protect them in a sure, quick, and economical manner. Such circuit breakers must be able to allow the passage of high permanent currents under continuous service (6,300–40,000 A), and have a high breaking capacity (63–275 kA).

They belong to the medium voltage range, but the transient recovery voltage (TRV) withstand capability is such that the interrupting principles developed for the high-voltage range has been used. Two particular embodiments of the thermal blast and self-blast techniques have been developed and applied to generator circuit breakers.

Thermal Blast Chamber with Arc-Assisted Opening

In this interruption principle arc energy is used, on the one hand to generate the blast by thermal expansion and, on the other hand, to accelerate the moving part of the circuit breaker when interrupting high currents (Figure 8).

The overpressure produced by the arc energy downstream of the interruption zone is applied on an auxiliary piston linked with the moving part. The resulting force accelerates the moving part, thus increasing the energy available for tripping.

It is possible with this interrupting principle to increase the tripping energy delivered by the operating mechanism by about 30% and to maintain the opening speed irrespective of the short circuit current.

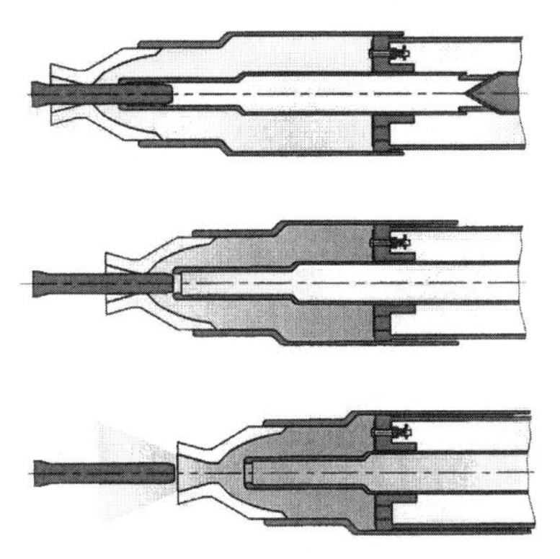

Figure 8
Thermal blast chamber with arc-assisted opening

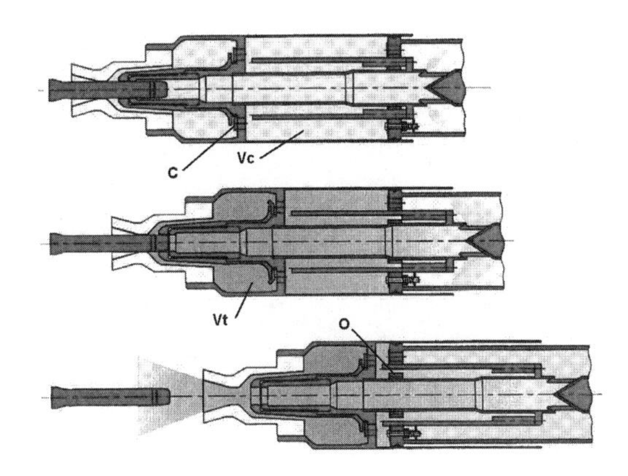

Figure 9
Self-blast chamber with rear exhaust

It is obviously better suited to circuit breakers with high breaking currents such as generator circuit breakers that are required to interrupt currents as high as 120 kA or even 160 kA.

Self-Blast Chamber with Rear Exhaust

This principle works as follows (Figure 9): In the first phase, the relative movement of the piston and the blast cylinder is used to compress the gas in the compression volume Vc. This overpressure opens the valve C and is then transmitted to expansion volume Vt.

In the second phase, gas in volume Vc is exhausted to the rear through openings (O).

The gas compression is sufficient for the interruption of low currents. During high short-circuit current interruption, volume Vt is pressurized by the thermal energy of the arc. This high pressure closes valve C. The pressure in volume Vc on the other hand is limited by an outflow of gas through the openings (O). The high overpressure generated in volume Vt produces the quenching blast necessary to extinguish the arc at current zero.

In this principle the energy that has to be delivered by the operating mechanism is limited and low energy spring operated mechanism can be used.

Figure 10 shows a generator circuit breaker with such type of interrupting chamber.

Figure 10
Generator circuit breaker SF_6 17, 5 kV 63 kA 60 Hz

EVOLUTION OF TRIPPING ENERGY

Figure 11 summarizes the evolution of tripping energy for 245 and 420 kV, from 1974 to 2003. It shows that the operating energy has been divided by a factor of five–seven during this period of nearly three decades. This illustrates the great

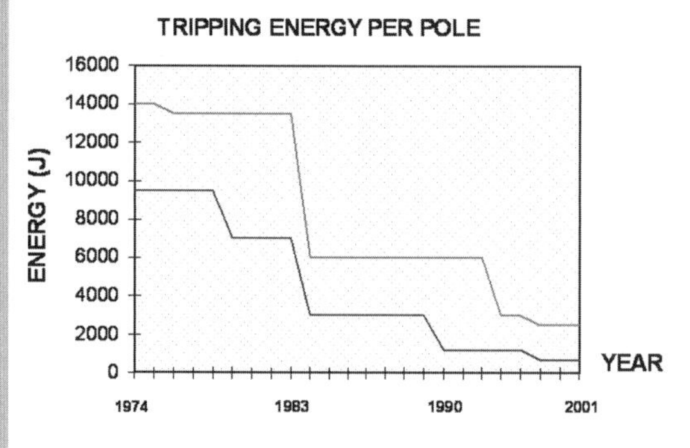

Figure 11
Evolution of tripping energy since 1974 of 245 and 420 kV circuit breakers

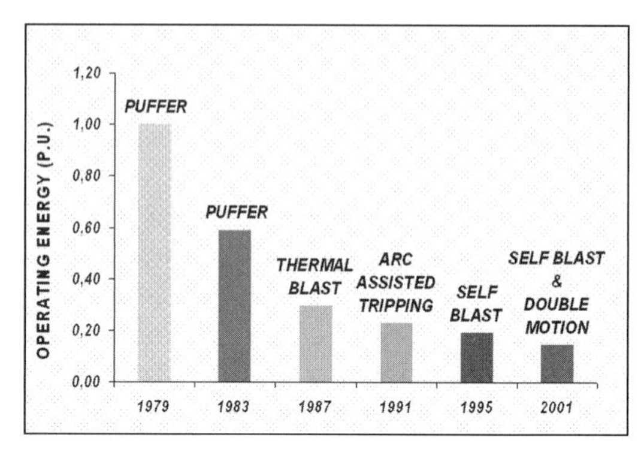

Figure 12
Operating energy as function of interrupting principle

progress that has been made in interrupting techniques for high-voltage circuit breakers during that period.

Figure 12 shows the continuous reduction of the necessary operating energy obtained through the technological progress.

OUTLOOK FOR THE FUTURE

Several interrupting techniques have been presented that all aim to reduce the operating energy of high-voltage circuit breakers. To date they have been widely applied, resulting in the lowering of drive energy, as shown in Figures 11 and 12.

Present interrupting technologies can be applied to circuit breakers with the higher rated interrupting currents (63–80 kA) required in some networks with increasing power generation (Figure 13).

Progress can still be made by the further industrialization of all components and by introducing new drive technologies. Following the remarkable evolution in chamber technology, the operating mechanism represents a not negligible contribution to the moving mass of circuit breakers, especially in the extra high-voltage range ($\geq$ 420 kV). Therefore progress in high-voltage circuit breakers can still be expected with the implementation of the same interrupting principles.

If one looks further in the future, other technology developments could possibly lead to a

Figure 13
GIS circuit breaker 550 kV 63 kA 50/60 Hz

further reduction in the SF_6 content of circuit breakers.

CONCLUSIONS

Over the last 50 years, high-voltage circuit breakers have become more reliable, more efficient, and more compact because the interrupting capability per break has been increased dramatically. These developments have not only produced major savings, but they have also had a massive impact on the layout of substations with respect to space requirements.

New types of SF_6 interrupting chambers, which implement innovative interrupting principles, have been developed during the last three decades with the objective of reducing the operating energy of the circuit breaker. This has led to reduced stress and wear of the mechanical components and consequently to an increased reliability of circuit breakers.

Service experience shows that the expectations of the designers, with respect to reliability and day-to-day operation, have been fulfilled.

FOR FURTHER READING

W.M. Leeds, R.E. Friedrich, C.L. Wagner, and T.E. Browne Jr, "Application of switching surge, arc and gas flow studies to the design of SF_6 breakers," presented at CIGRE Session 1970, paper 13-11.

E. Thuries, "Development of air-blast circuit-breakers," presented at CIGRE Session 1972, paper 13-09.

D. Dufournet and E. Thuries "Recent development of HV circuit-breakers," presented at 11th CEPSI Conference, Kuala Lumpur, Malaisia, Oct. 1996.

D. Dufournet, F. Sciullo, J. Ozil, and A. Ludwig, "New interrupting and drive techniques to increase high-voltage circuit breakers performance and reliability," presented at CIGRE session, 1998, paper 13-104.

A. Ludwig, D. Dufournet, and E. Mikes, "Improved performance and reliability of high-voltage circuit breakers with spring mechanisms through new breaking and operating elements," presented at 12th CEPSI Conference, Pattaya, Thailand, 1998.

D. Dufournet, J.M. Willieme, and G.F. Montillet, "Design and implementation of a SF_6 interrupting chamber applied to low range generator circuit breakers suitable for interruption of current having a non-zero passage," *IEEE Trans. Power Delivery*, vol. 17, no. 4, pp. 963–967, Oct. 2002.

D. Dufournet, "Generator circuit breakers: SF_6 Breaking chamber–interruption of current with non-zero passage. Influence of cable connection on TRV of system fed faults," presented at CIGRE 2002, Paris, France, Aug. 2002, paper 13-101.

D. Dufournet, C. Lindner, D. Johnson, and D. Vondereck, "Technical trends in circuit breaker switching technologies," presented at CIGRE SC A3 Colloquium, Sarajevo, 2003.

BIOGRAPHY

Denis Dufournet is with AREVA T&D.

8.1

DEFINITION OF SYMMETRICAL COMPONENTS

Assume that a set of three-phase voltages designated V_a, V_b, and V_c is given. In accordance with Fortescue, these phase voltages are resolved into the following three sets of sequence components:

1. *Zero-sequence* components, consisting of three phasors with equal magnitudes and with zero phase displacement, as shown in Figure 8.1(a)
2. *Positive-sequence* components, consisting of three phasors with equal magnitudes, $\pm 120°$ phase displacement, and positive sequence, as in Figure 8.1(b)
3. *Negative-sequence* components, consisting of three phasors with equal magnitudes, $\pm 120°$ phase displacement, and negative sequence, as in Figure 8.1(c)

In this text we will work only with the zero-, positive-, and negative-sequence components of phase a, which are V_{a0}, V_{a1}, and V_{a2}, respectively. For simplicity, we drop the subscript a and denote these sequence components as V_0, V_1, and V_2. They are defined by the following transformation:

$$\begin{bmatrix} V_a \\ V_b \\ V_c \end{bmatrix} = \begin{bmatrix} 1 & 1 & 1 \\ 1 & a^2 & a \\ 1 & a & a^2 \end{bmatrix} \begin{bmatrix} V_0 \\ V_1 \\ V_2 \end{bmatrix} \tag{8.1.1}$$

FIGURE 8.1 Resolving phase voltages into three sets of sequence components

$V_{a0}\ V_{b0}\ V_{c0} = V_0$

(a) Zero-sequence components

V_{c1}, $V_{a1} = V_1$, V_{b1}

(b) Positive-sequence components

V_{b2}, $V_{a2} = V_2$, V_{c2}

(c) Negative-sequence components

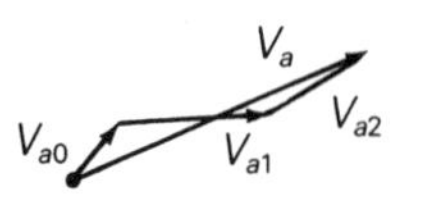

Phase *a*

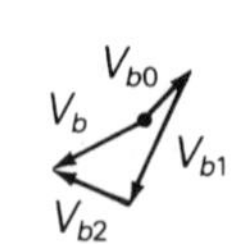

Phase *b*

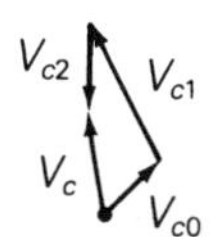

Phase *c*

where

$$a = 1\underline{/120^\circ} = \frac{-1}{2} + j\frac{\sqrt{3}}{2} \tag{8.1.2}$$

Writing (8.1.1) as three separate equations:

$$V_a = V_0 + V_1 + V_2 \tag{8.1.3}$$

$$V_b = V_0 + a^2V_1 + aV_2 \tag{8.1.4}$$

$$V_c = V_0 + aV_1 + a^2V_2 \tag{8.1.5}$$

In (8.1.2), a is a complex number with unit magnitude and a 120° phase angle. When any phasor is multiplied by a, that phasor rotates by 120° (counterclockwise). Similarly, when any phasor is multiplied by $a^2 = (1\underline{/120^\circ}) \cdot (1\underline{/120^\circ}) = 1\underline{/240^\circ}$, the phasor rotates by 240°. Table 8.1 lists some common identities involving a.

The complex number a is similar to the well-known complex number $j = \sqrt{-1} = 1\underline{/90^\circ}$. Thus the only difference between j and a is that the angle of j is 90°, and that of a is 120°.

Equation (8.1.1) can be rewritten more compactly using matrix notation. We define the following vectors $\boldsymbol{V}_p$ and $\boldsymbol{V}_s$, and matrix $\boldsymbol{A}$:

$$\boldsymbol{V}_p = \begin{bmatrix} V_a \\ V_b \\ V_c \end{bmatrix} \tag{8.1.6}$$

$$\boldsymbol{V}_s = \begin{bmatrix} V_0 \\ V_1 \\ V_2 \end{bmatrix} \tag{8.1.7}$$

$$\boldsymbol{A} = \begin{bmatrix} 1 & 1 & 1 \\ 1 & a^2 & a \\ 1 & a & a^2 \end{bmatrix} \tag{8.1.8}$$

TABLE 8.1

Common identities involving $a = 1\underline{/120^\circ}$

$a^4 = a = 1\underline{/120^\circ}$
$a^2 = 1\underline{/240^\circ}$
$a^3 = 1\underline{/0^\circ}$
$1 + a + a^2 = 0$
$1 - a = \sqrt{3}\underline{/-30^\circ}$
$1 - a^2 = \sqrt{3}\underline{/+30^\circ}$
$a^2 - a = \sqrt{3}\underline{/270^\circ}$
$ja = 1\underline{/210^\circ}$
$1 + a = -a^2 = 1\underline{/60^\circ}$
$1 + a^2 = -a = 1\underline{/-60^\circ}$
$a + a^2 = -1 = 1\underline{/180^\circ}$

$\boldsymbol{V}_p$ is the column vector of phase voltages, $\boldsymbol{V}_s$ is the column vector of sequence voltages, and $\boldsymbol{A}$ is a 3×3 transformation matrix. Using these definitions, (8.1.1) becomes

$$\boldsymbol{V}_p = \boldsymbol{A}\boldsymbol{V}_s \tag{8.1.9}$$

The inverse of the $\boldsymbol{A}$ matrix is

$$\boldsymbol{A}^{-1} = \frac{1}{3}\begin{bmatrix} 1 & 1 & 1 \\ 1 & a & a^2 \\ 1 & a^2 & a \end{bmatrix} \tag{8.1.10}$$

Equation (8.1.10) can be verified by showing that the product $\boldsymbol{A}\boldsymbol{A}^{-1}$ is the unit matrix. Also, premultiplying (8.1.9) by $\boldsymbol{A}^{-1}$ gives

$$V_s = \boldsymbol{A}^{-1} V_p \tag{8.1.11}$$

Using (8.1.6), (8.1.7), and (8.1.10), then (8.1.11) becomes

$$\begin{bmatrix} V_0 \\ V_1 \\ V_2 \end{bmatrix} = \frac{1}{3} \begin{bmatrix} 1 & 1 & 1 \\ 1 & a & a^2 \\ 1 & a^2 & a \end{bmatrix} \begin{bmatrix} V_a \\ V_b \\ V_c \end{bmatrix} \tag{8.1.12}$$

Writing (8.1.12) as three separate equations,

$$V_0 = \tfrac{1}{3}(V_a + V_b + V_c) \tag{8.1.13}$$

$$V_1 = \tfrac{1}{3}(V_a + aV_b + a^2 V_c) \tag{8.1.14}$$

$$V_2 = \tfrac{1}{3}(V_a + a^2 V_b + aV_c) \tag{8.1.15}$$

Equation (8.1.13) shows that there is no zero-sequence voltage in a *balanced* three-phase system because the sum of three balanced phasors is zero. In an unbalanced three-phase system, line-to-neutral voltages may have a zero-sequence component. However, line-to-line voltages never have a zero-sequence component, since by KVL their sum is always zero.

The symmetrical component transformation can also be applied to currents, as follows. Let

$$I_p = \boldsymbol{A} I_s \tag{8.1.16}$$

where I_p is a vector of phase currents,

$$I_p = \begin{bmatrix} I_a \\ I_b \\ I_c \end{bmatrix} \tag{8.1.17}$$

and I_s is a vector of sequence currents,

$$I_s = \begin{bmatrix} I_0 \\ I_1 \\ I_2 \end{bmatrix} \tag{8.1.18}$$

Also,

$$I_s = \boldsymbol{A}^{-1} I_p \tag{8.1.19}$$

Equations (8.1.16) and (8.1.19) can be written as separate equations as follows. The phase currents are

$$I_a = I_0 + I_1 + I_2 \tag{8.1.20}$$

$$I_b = I_0 + a^2 I_1 + aI_2 \tag{8.1.21}$$

$$I_c = I_0 + aI_1 + a^2 I_2 \tag{8.1.22}$$

and the sequence currents are

$$I_0 = \tfrac{1}{3}(I_a + I_b + I_c) \qquad (8.1.23)$$

$$I_1 = \tfrac{1}{3}(I_a + aI_b + a^2I_c) \qquad (8.1.24)$$

$$I_2 = \tfrac{1}{3}(I_a + a^2I_b + aI_c) \qquad (8.1.25)$$

In a three-phase Y-connected system, the neutral current I_n is the sum of the line currents:

$$I_n = I_a + I_b + I_c \qquad (8.1.26)$$

Comparing (8.1.26) and (8.1.23),

$$I_n = 3I_0 \qquad (8.1.27)$$

The neutral current equals three times the zero-sequence current. In a balanced Y-connected system, line currents have no zero-sequence component, since the neutral current is zero. Also, in any three-phase system with no neutral path, such as a Δ-connected system or a three-wire Y-connected system with an ungrounded neutral, line currents have no zero-sequence component.

The following three examples further illustrate symmetrical components.

EXAMPLE 8.1 Sequence components: balanced line-to-neutral voltages

Calculate the sequence components of the following balanced line-to-neutral voltages with *abc* sequence:

$$V_p = \begin{bmatrix} V_{an} \\ V_{bn} \\ V_{cn} \end{bmatrix} = \begin{bmatrix} 277\underline{/0^\circ} \\ 277\underline{/-120^\circ} \\ 277\underline{/+120^\circ} \end{bmatrix} \quad \text{volts}$$

SOLUTION Using (8.1.13)–(8.1.15):

$$V_0 = \tfrac{1}{3}[277\underline{/0^\circ} + 277\underline{/-120^\circ} + 277\underline{/+120^\circ}] = 0$$

$$V_1 = \tfrac{1}{3}[277\underline{/0^\circ} + 277\underline{/(-120^\circ + 120^\circ)} + 277\underline{/(120^\circ + 240^\circ)}]$$

$$= 277\underline{/0^\circ} \quad \text{volts} = V_{an}$$

$$V_2 = \tfrac{1}{3}[277\underline{/0^\circ} + 277\underline{/(-120^\circ + 240^\circ)} + 277\underline{/(120^\circ + 120^\circ)}]$$

$$= \tfrac{1}{3}[277\underline{/0^\circ} + 277\underline{/120^\circ} + 277\underline{/240^\circ}] = 0$$

This example illustrates the fact that balanced three-phase systems with *abc* sequence (or positive sequence) have no zero-sequence or negative-sequence components. For this example, the positive-sequence voltage V_1 equals V_{an}, and the zero-sequence and negative-sequence voltages are both zero. ■

EXAMPLE 8.2 Sequence components: balanced *acb* currents

A Y-connected load has balanced currents with *acb* sequence given by

$$\boldsymbol{I}_p = \begin{bmatrix} I_a \\ I_b \\ I_c \end{bmatrix} = \begin{bmatrix} 10\underline{/0^\circ} \\ 10\underline{/+120^\circ} \\ 10\underline{/-120^\circ} \end{bmatrix} \quad \text{A}$$

Calculate the sequence currents.

SOLUTION Using (8.1.23)–(8.1.25):

$$I_0 = \tfrac{1}{3}[10\underline{/0^\circ} + 10\underline{/120^\circ} + 10\underline{/-120^\circ}] = 0$$

$$I_1 = \tfrac{1}{3}[10\underline{/0^\circ} + 10\underline{/(120^\circ + 120^\circ)} + 10\underline{/(-120^\circ + 240^\circ)}]$$

$$= \tfrac{1}{3}[10\underline{/0^\circ} + 10\underline{/240^\circ} + 10\underline{/120^\circ}] = 0$$

$$I_2 = \tfrac{1}{3}[10\underline{/0^\circ} + 10\underline{/(120^\circ + 240^\circ)} + 10\underline{/(-120^\circ + 120^\circ)}]$$

$$= 10\underline{/0^\circ} \text{ A} = I_a$$

This example illustrates the fact that balanced three-phase systems with *acb* sequence (or negative sequence) have no zero-sequence or positive-sequence components. For this example the negative-sequence current I_2 equals I_a, and the zero-sequence and positive-sequence currents are both zero. ■

EXAMPLE 8.3 Sequence components: unbalanced currents

A three-phase line feeding a balanced-Y load has one of its phases (phase *b*) open. The load neutral is grounded, and the unbalanced line currents are

$$\boldsymbol{I}_p = \begin{bmatrix} I_a \\ I_b \\ I_c \end{bmatrix} = \begin{bmatrix} 10\underline{/0^\circ} \\ 0 \\ 10\underline{/120^\circ} \end{bmatrix} \quad \text{A}$$

Calculate the sequence currents and the neutral current.

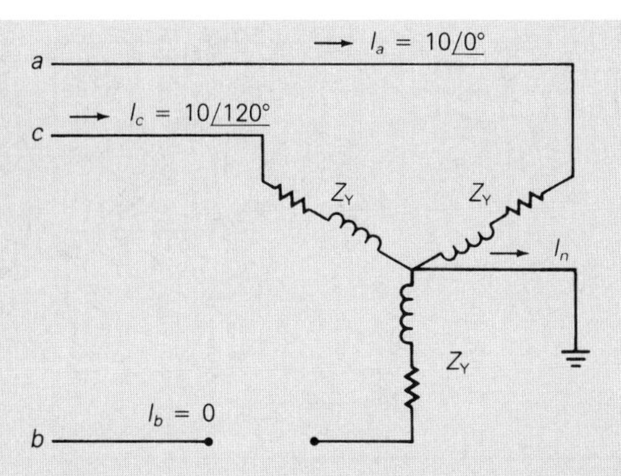

FIGURE 8.2
Circuit for Example 8.3

SOLUTION The circuit is shown in Figure 8.2. Using (8.1.23)–(8.1.25):

$$I_0 = \tfrac{1}{3}[10\underline{/0°} + 0 + 10\underline{/120°}]$$
$$= 3.333\underline{/60°} \quad \text{A}$$
$$I_1 = \tfrac{1}{3}[10\underline{/0°} + 0 + 10\underline{/(120° + 240°)}] = 6.667\underline{/0°} \quad \text{A}$$
$$I_2 = \tfrac{1}{3}[10\underline{/0°} + 0 + 10\underline{/(120° + 120°)}]$$
$$= 3.333\underline{/-60°} \quad \text{A}$$

Using (8.1.26) the neutral current is

$$I_n = (10\underline{/0°} + 0 + 10\underline{/120°})$$
$$= 10\underline{/60°} \text{ A} = 3I_0$$

This example illustrates the fact that *unbalanced* three-phase systems may have nonzero values for all sequence components. Also, the neutral current equals three times the zero-sequence current, as given by (8.1.27). ■

8.2

SEQUENCE NETWORKS OF IMPEDANCE LOADS

Figure 8.3 shows a balanced-Y impedance load. The impedance of each phase is designated Z_Y, and a neutral impedance Z_n is connected between the load neutral and ground. Note from Figure 8.3 that the line-to-ground voltage V_{ag} is

$$V_{ag} = Z_Y I_a + Z_n I_n$$
$$= Z_Y I_a + Z_n(I_a + I_b + I_c)$$
$$= (Z_Y + Z_n)I_a + Z_n I_b + Z_n I_c \tag{8.2.1}$$

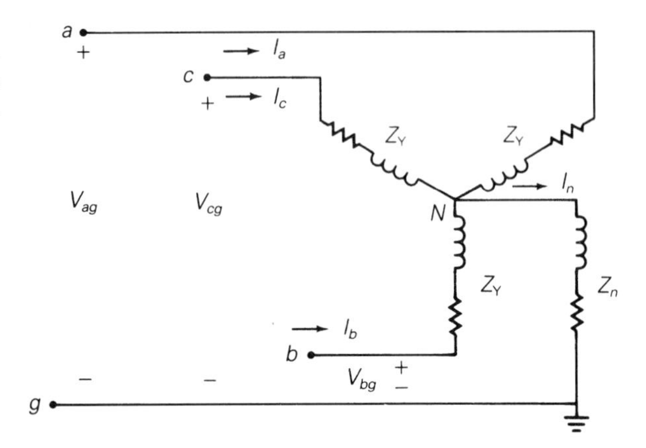

FIGURE 8.3 Balanced-Y impedance load

Similar equations can be written for V_{bg} and V_{cg}:

$$V_{bg} = Z_n I_a + (Z_Y + Z_n) I_b + Z_n I_c \tag{8.2.2}$$

$$V_{cg} = Z_n I_a + Z_n I_b + (Z_Y + Z_n) I_c \tag{8.2.3}$$

Equations (8.2.1)–(8.2.3) can be rewritten in matrix format:

$$\begin{bmatrix} V_{ag} \\ V_{bg} \\ V_{cg} \end{bmatrix} = \begin{bmatrix} (Z_Y + Z_n) & Z_n & Z_n \\ Z_n & (Z_Y + Z_n) & Z_n \\ Z_n & Z_n & (Z_Y + Z_n) \end{bmatrix} \begin{bmatrix} I_a \\ I_b \\ I_c \end{bmatrix} \tag{8.2.4}$$

Equation (8.2.4) is written more compactly as

$$\boldsymbol{V}_p = \boldsymbol{Z}_p \boldsymbol{I}_p \tag{8.2.5}$$

where $\boldsymbol{V}_p$ is the vector of line-to-ground voltages (or phase voltages), $\boldsymbol{I}_p$ is the vector of line currents (or phase currents), and $\boldsymbol{Z}_p$ is the 3×3 phase impedance matrix shown in (8.2.4). Equations (8.1.9) and (8.1.16) can now be used in (8.2.5) to determine the relationship between the sequence voltages and currents, as follows:

$$\boldsymbol{A}\boldsymbol{V}_s = \boldsymbol{Z}_p \boldsymbol{A} \boldsymbol{I}_s \tag{8.2.6}$$

Premultiplying both sides of (8.2.6) of $\boldsymbol{A}^{-1}$ gives

$$\boldsymbol{V}_s = (\boldsymbol{A}^{-1} \boldsymbol{Z}_p \boldsymbol{A}) \boldsymbol{I}_s \tag{8.2.7}$$

or

$$\boldsymbol{V}_s = \boldsymbol{Z}_s \boldsymbol{I}_s \tag{8.2.8}$$

where

$$\boldsymbol{Z}_s = \boldsymbol{A}^{-1} \boldsymbol{Z}_p \boldsymbol{A} \tag{8.2.9}$$

The impedance matrix $\boldsymbol{Z}_s$ defined by (8.2.9) is called the *sequence impedance matrix*. Using the definition of $\boldsymbol{A}$, its inverse $\boldsymbol{A}^{-1}$, and $\boldsymbol{Z}_p$ given by (8.1.8), (8.1.10), and (8.2.4), the sequence impedance matrix $\boldsymbol{Z}_s$ for the balanced-Y load is

$$\boldsymbol{Z}_s = \frac{1}{3}\begin{bmatrix} 1 & 1 & 1 \\ 1 & a & a^2 \\ 1 & a^2 & a \end{bmatrix}\begin{bmatrix} (Z_Y + Z_n) & Z_n & Z_n \\ Z_n & (Z_Y + Z_n) & Z_n \\ Z_n & Z_n & (Z_Y + Z_n) \end{bmatrix} \times \begin{bmatrix} 1 & 1 & 1 \\ 1 & a^2 & a \\ 1 & a & a^2 \end{bmatrix} \tag{8.2.10}$$

Performing the indicated matrix multiplications in (8.2.10), and using the identity $(1 + a + a^2) = 0$,

$$\boldsymbol{Z}_s = \frac{1}{3}\begin{bmatrix} 1 & 1 & 1 \\ 1 & a & a^2 \\ 1 & a^2 & a \end{bmatrix}\begin{bmatrix} (Z_Y + 3Z_n) & Z_Y & Z_Y \\ (Z_Y + 3Z_n) & a^2 Z_Y & aZ_Y \\ (Z_Y + 3Z_n) & aZ_Y & a^2 Z_Y \end{bmatrix}$$

$$= \begin{bmatrix} (Z_Y + 3Z_n) & 0 & 0 \\ 0 & Z_Y & 0 \\ 0 & 0 & Z_Y \end{bmatrix} \tag{8.2.11}$$

As shown in (8.2.11), the sequence impedance matrix $\boldsymbol{Z}_s$ for the balanced-Y load of Figure 8.3 is a diagonal matrix. Since $\boldsymbol{Z}_s$ is diagonal, (8.2.8) can be written as three *uncoupled* equations. Using (8.1.7), (8.1.18), and (8.2.11) in (8.2.8),

$$\begin{bmatrix} V_0 \\ V_1 \\ V_2 \end{bmatrix} = \begin{bmatrix} (Z_Y + 3Z_n) & 0 & 0 \\ 0 & Z_Y & 0 \\ 0 & 0 & Z_Y \end{bmatrix}\begin{bmatrix} I_0 \\ I_1 \\ I_2 \end{bmatrix} \tag{8.2.12}$$

Rewriting (8.2.12) as three separate equations,

$$V_0 = (Z_Y + 3Z_n)I_0 = Z_0 I_0 \tag{8.2.13}$$

$$V_1 = Z_Y I_1 = Z_1 I_1 \tag{8.2.14}$$

$$V_2 = Z_Y I_2 = Z_2 I_2 \tag{8.2.15}$$

As shown in (8.2.13), the zero-sequence voltage V_0 depends only on the zero-sequence current I_0 and the impedance $(Z_Y + 3Z_n)$. This impedance is called the *zero-sequence impedance* and is designated Z_0. Also, the positive-sequence voltage V_1 depends only on the positive-sequence current I_1 and an impedance $Z_1 = Z_Y$ called the *positive-sequence impedance*. Similarly, V_2 depends only on I_2 and the *negative-sequence impedance* $Z_2 = Z_Y$.

Equations (8.2.13)–(8.2.15) can be represented by the three networks shown in Figure 8.4. These networks are called the *zero-sequence*, *positive-sequence*, and *negative-sequence networks*. As shown, each sequence network

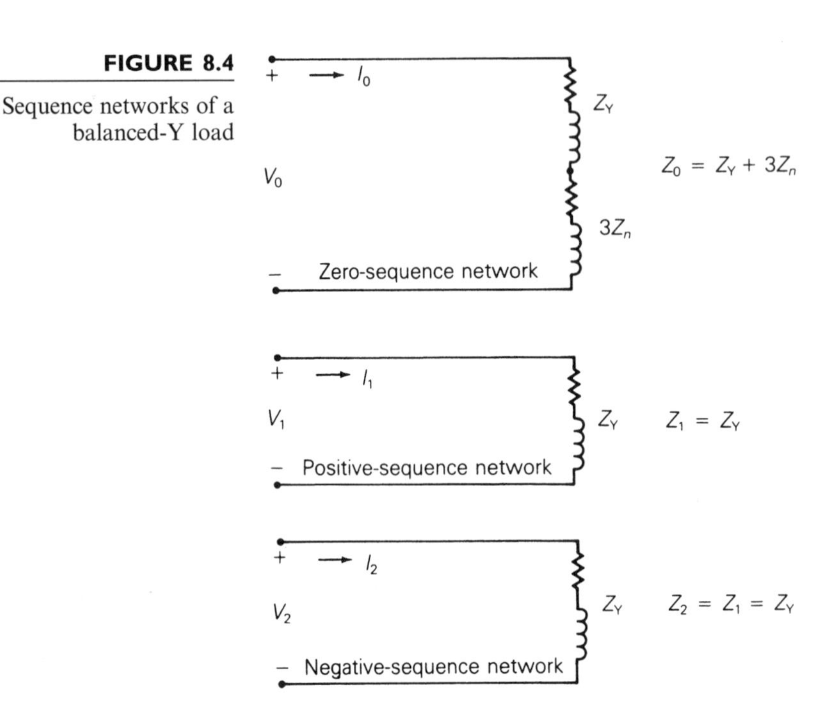

FIGURE 8.4 Sequence networks of a balanced-Y load

is separate, uncoupled from the other two. The separation of these sequence networks is a consequence of the fact that $\boldsymbol{Z}_s$ is a diagonal matrix for a balanced-Y load. This separation underlies the advantage of symmetrical components.

Note that the neutral impedance does not appear in the positive- and negative-sequence networks of Figure 8.4. This illustrates the fact that positive- and negative-sequence currents do not flow in neutral impedances. However, the neutral impedance is multiplied by 3 and placed in the zero-sequence network of the figure. The voltage $I_0(3Z_n)$ across the impedance $3Z_n$ is the voltage drop $(I_n Z_n)$ across the neutral impedance Z_n in Figure 8.3, since $I_n = 3I_0$.

When the neutral of the Y load in Figure 8.3 has no return path, then the neutral impedance Z_n is infinite and the term $3Z_n$ in the zero-sequence network of Figure 8.4 becomes an open circuit. Under this condition of an open neutral, no zero-sequence current exists. However, when the neutral of the Y load is solidly grounded with a zero-ohm conductor, then the neutral impedance is zero and the term $3Z_n$ in the zero-sequence network becomes a short circuit. Under this condition of a solidly grounded neutral, zero-sequence current I_0 can exist when there is a zero-sequence voltage caused by unbalanced voltages applied to the load.

Figure 2.15 shows a balanced-Δ load and its equivalent balanced-Y load. Since the Δ load has no neutral connection, the equivalent Y load in Figure 2.15 has an open neutral. The sequence networks of the equivalent Y load corresponding to a balanced-Δ load are shown in Figure 8.5. As shown,

FIGURE 8.5

Sequence networks for an equivalent Y representation of a balanced-Δ load

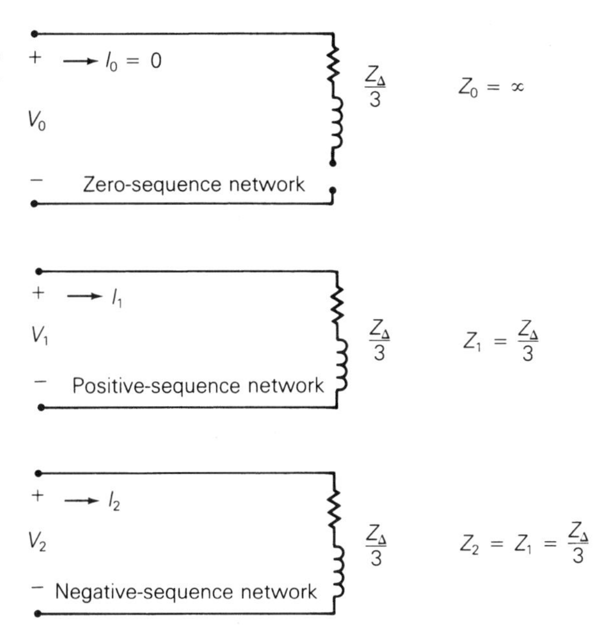

the equivalent Y impedance $Z_Y = Z_\Delta/3$ appears in each of the sequence networks. Also, the zero-sequence network has an open circuit, since $Z_n = \infty$ corresponds to an open neutral. No zero-sequence current occurs in the equivalent Y load.

The sequence networks of Figure 8.5 represent the balanced-Δ load as viewed from its terminals, but they do not represent the internal load characteristics. The currents I_0, I_1, and I_2 in Figure 8.5 are the sequence components of the line currents feeding the Δ load, not the load currents within the Δ. The Δ load currents, which are related to the line currents by (2.5.14), are not shown in Figure 8.5.

EXAMPLE 8.4 **Sequence networks: balanced-Y and balanced-Δ loads**

A balanced-Y load is in parallel with a balanced-Δ-connected capacitor bank. The Y load has an impedance $Z_Y = (3 + j4)\ \Omega$ per phase, and its neutral is grounded through an inductive reactance $X_n = 2\ \Omega$. The capacitor bank has a reactance $X_c = 30\ \Omega$ per phase. Draw the sequence networks for this load and calculate the load-sequence impedances.

SOLUTION The sequence networks are shown in Figure 8.6. As shown, the Y-load impedance in the zero-sequence network is in series with three times the neutral impedance. Also, the Δ-load branch in the zero-sequence network is open, since no zero-sequence current flows into the Δ load. In the positive- and negative-sequence circuits, the Δ-load impedance is divided by 3 and placed in parallel with the Y-load impedance. The equivalent sequence impedances are

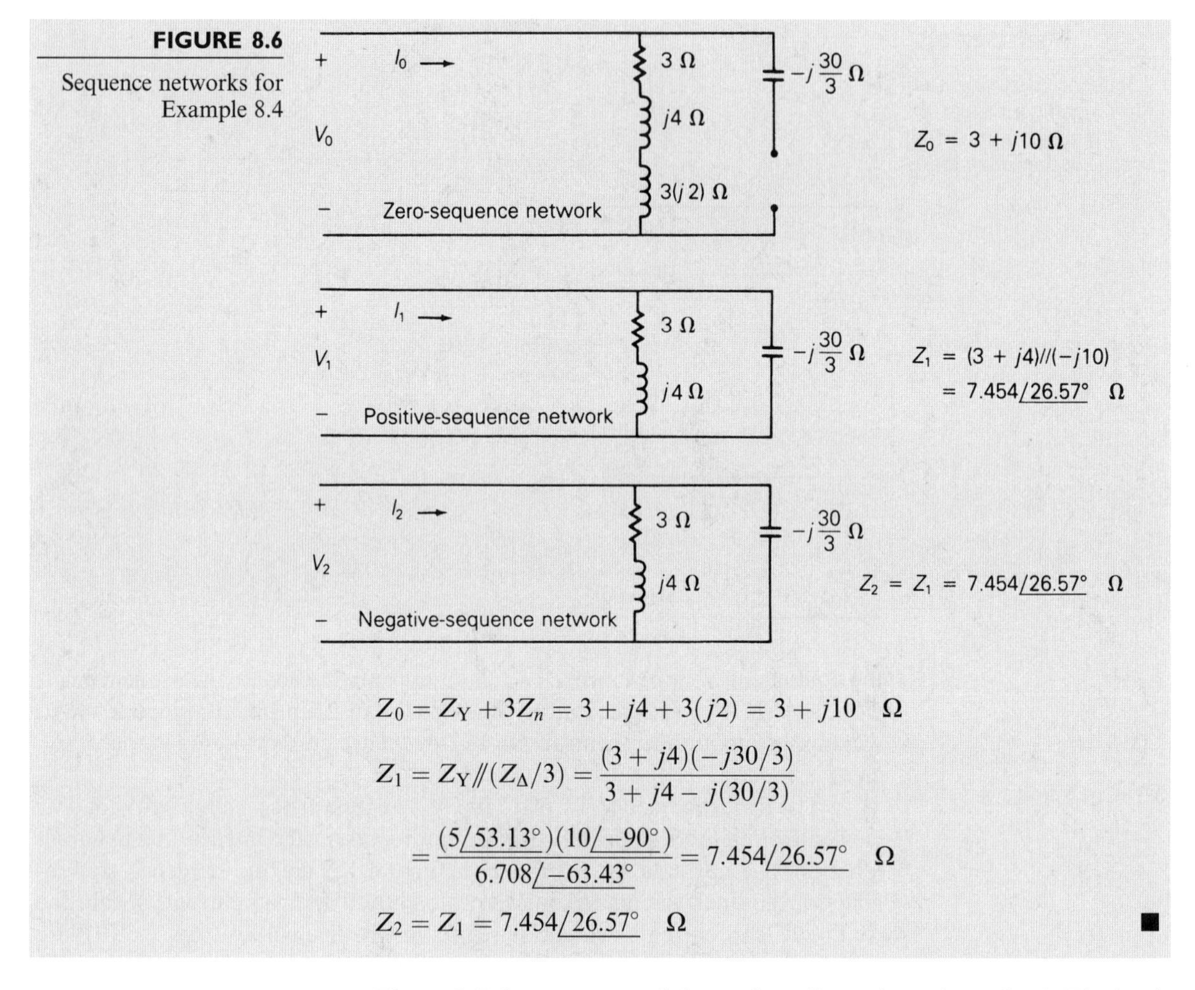

FIGURE 8.6 Sequence networks for Example 8.4

$$Z_0 = Z_Y + 3Z_n = 3 + j4 + 3(j2) = 3 + j10 \quad \Omega$$

$$Z_1 = Z_Y /\!/ (Z_\Delta/3) = \frac{(3+j4)(-j30/3)}{3 + j4 - j(30/3)}$$

$$= \frac{(5\underline{/53.13^\circ})(10\underline{/-90^\circ})}{6.708\underline{/-63.43^\circ}} = 7.454\underline{/26.57^\circ} \quad \Omega$$

$$Z_2 = Z_1 = 7.454\underline{/26.57^\circ} \quad \Omega$$

■

Figure 8.7 shows a general three-phase linear impedance load. The load could represent a balanced load such as the balanced-Y or balanced-Δ load, or an unbalanced impedance load. The general relationship between the line-to-ground voltages and line currents for this load can be written as

$$\begin{bmatrix} V_{ag} \\ V_{bg} \\ V_{cg} \end{bmatrix} = \begin{bmatrix} Z_{aa} & Z_{ab} & Z_{ac} \\ Z_{ab} & Z_{bb} & Z_{bc} \\ Z_{ac} & Z_{bc} & Z_{cc} \end{bmatrix} \begin{bmatrix} I_a \\ I_b \\ I_c \end{bmatrix} \tag{8.2.16}$$

or

$$\boldsymbol{V}_p = \boldsymbol{Z}_p \boldsymbol{I}_p \tag{8.2.17}$$

where $\boldsymbol{V}_p$ is the vector of line-to-neutral (or phase) voltages, $\boldsymbol{I}_p$ is the vector of line (or phase) currents, and $\boldsymbol{Z}_p$ is a 3×3 phase impedance matrix. It is assumed here that the load is nonrotating, and that $\boldsymbol{Z}_p$ is a symmetric matrix, which corresponds to a bilateral network.

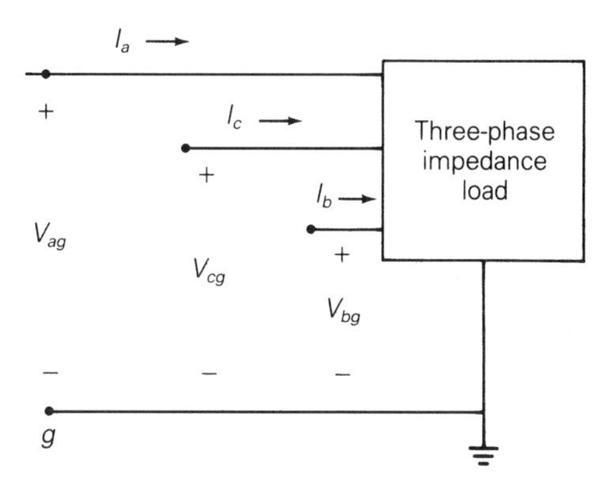

FIGURE 8.7 General three-phase impedance load (linear, bilateral network, nonrotating equipment)

Since (8.2.17) has the same form as (8.2.5), the relationship between the sequence voltages and currents for the general three-phase load of Figure 8.6 is the same as that of (8.2.8) and (8.2.9), which are rewritten here:

$$\boldsymbol{V}_s = \boldsymbol{Z}_s \boldsymbol{I}_s \tag{8.2.18}$$

$$\boldsymbol{Z}_s = \boldsymbol{A}^{-1} \boldsymbol{Z}_p \boldsymbol{A} \tag{8.2.19}$$

The sequence impedance matrix $\boldsymbol{Z}_s$ given by (8.2.19) is a 3×3 matrix with nine sequence impedances, defined as follows:

$$\boldsymbol{Z}_s = \begin{bmatrix} Z_0 & Z_{01} & Z_{02} \\ Z_{10} & Z_1 & Z_{12} \\ Z_{20} & Z_{21} & Z_2 \end{bmatrix} \tag{8.2.20}$$

The diagonal impedances Z_0, Z_1, and Z_2 in this matrix are the self-impedances of the zero-, positive-, and negative-sequence networks. The off-diagonal impedances are the mutual impedances between sequence networks. Using the definitions of $\boldsymbol{A}, \boldsymbol{A}^{-1}, \boldsymbol{Z}_p$, and $\boldsymbol{Z}_s$, (8.2.19) is

$$\begin{bmatrix} Z_0 & Z_{01} & Z_{02} \\ Z_{10} & Z_1 & Z_{12} \\ Z_{20} & Z_{21} & Z_2 \end{bmatrix} = \frac{1}{3}\begin{bmatrix} 1 & 1 & 1 \\ 1 & a & a^2 \\ 1 & a^2 & a \end{bmatrix}\begin{bmatrix} Z_{aa} & Z_{ab} & Z_{ac} \\ Z_{ab} & Z_{bb} & Z_{bc} \\ Z_{ac} & Z_{bc} & Z_{cc} \end{bmatrix}\begin{bmatrix} 1 & 1 & 1 \\ 1 & a^2 & a \\ 1 & a & a^2 \end{bmatrix} \tag{8.2.21}$$

Performing the indicated multiplications in (8.2.21), and using the identity $(1 + a + a^2) = 0$, the following separate equations can be obtained (see Problem 8.18):

Diagonal sequence impedances

$$Z_0 = \tfrac{1}{3}(Z_{aa} + Z_{bb} + Z_{cc} + 2Z_{ab} + 2Z_{ac} + 2Z_{bc}) \tag{8.2.22}$$

$$Z_1 = Z_2 = \tfrac{1}{3}(Z_{aa} + Z_{bb} + Z_{cc} - Z_{ab} - Z_{ac} - Z_{bc}) \tag{8.2.23}$$

Off-diagonal sequence impedances

$$Z_{01} = Z_{20} = \tfrac{1}{3}(Z_{aa} + a^2 Z_{bb} + aZ_{cc} - aZ_{ab} - a^2 Z_{ac} - Z_{bc}) \tag{8.2.24}$$

$$Z_{02} = Z_{10} = \tfrac{1}{3}(Z_{aa} + aZ_{bb} + a^2 Z_{cc} - a^2 Z_{ab} - aZ_{ac} - Z_{bc}) \tag{8.2.25}$$

$$Z_{12} = \tfrac{1}{3}(Z_{aa} + a^2 Z_{bb} + aZ_{cc} + 2aZ_{ab} + 2a^2 Z_{ac} + 2Z_{bc}) \tag{8.2.26}$$

$$Z_{21} = \tfrac{1}{3}(Z_{aa} + aZ_{bb} + a^2 Z_{cc} + 2a^2 Z_{ab} + 2aZ_{ac} + 2Z_{bc}) \tag{8.2.27}$$

A *symmetrical load* is defined as a load whose sequence impedance matrix is diagonal; that is, all the mutual impedances in (8.2.24)–(8.2.27) are zero. Equating these mutual impedances to zero and solving, the following conditions for a symmetrical load are determined. When both

$$Z_{aa} = Z_{bb} = Z_{cc} \tag{8.2.28}$$

and

$$Z_{ab} = Z_{ac} = Z_{bc} \tag{8.2.29}$$

(conditions for a symmetrical load)

then

$$Z_{01} = Z_{10} = Z_{02} = Z_{20} = Z_{12} = Z_{21} = 0 \tag{8.2.30}$$

$$Z_0 = Z_{aa} + 2Z_{ab} \tag{8.2.31}$$

$$Z_1 = Z_2 = Z_{aa} - Z_{ab} \tag{8.2.32}$$

The conditions for a symmetrical load are that the diagonal phase impedances be equal and that the off-diagonal phase impedances be equal. These conditions can be verified by using (8.2.28) and (8.2.29) with the

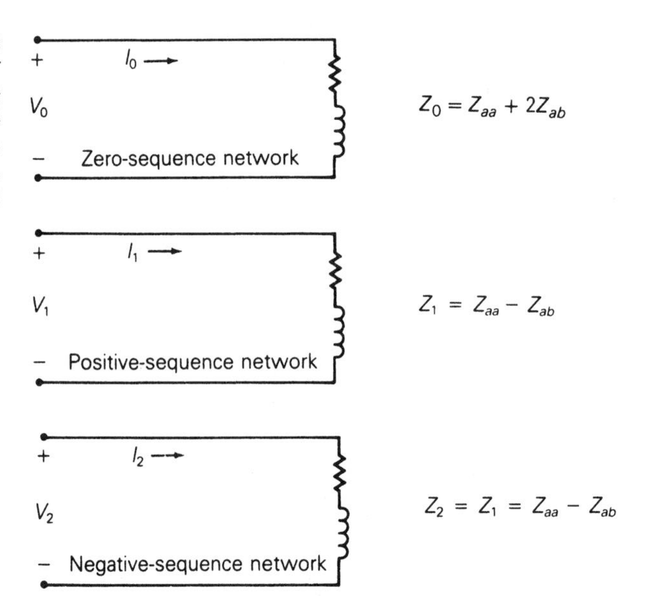

FIGURE 8.8 Sequence networks of a three-phase symmetrical impedance load (linear, bilateral network, nonrotating equipment)

identity $(1 + a + a^2) = 0$ in (8.2.24)–(8.2.27) to show that all the mutual sequence impedances are zero. Note that the positive- and negative-sequence impedances are equal for a symmetrical load, as shown by (8.2.32), and for a nonsymmetrical load, as shown by (8.2.23). This is always true for linear, symmetric impedances that represent nonrotating equipment such as transformers and transmission lines. However, the positive- and negative-sequence impedances of rotating equipment such as generators and motors are generally not equal. Note also that the zero-sequence impedance Z_0 is not equal to the positive- and negative-sequence impedances of a symmetrical load unless the mutual phase impedances $Z_{ab} = Z_{ac} = Z_{bc}$ are zero.

The sequence networks of a symmetrical impedance load are shown in Figure 8.8. Since the sequence impedance matrix $\boldsymbol{Z}_s$ is diagonal for a symmetrical load, the sequence networks are separate or uncoupled.

8.3 SEQUENCE NETWORKS OF SERIES IMPEDANCES

Figure 8.9 shows series impedances connected between two three-phase buses denoted abc and $a'b'c'$. Self-impedances of each phase are denoted Z_{aa}, Z_{bb}, and Z_{cc}. In general, the series network may also have mutual impedances between phases. The voltage drops across the series-phase impedances are given by

$$\begin{bmatrix} V_{an} - V_{a'n} \\ V_{bn} - V_{b'n} \\ V_{cn} - V_{c'n} \end{bmatrix} = \begin{bmatrix} V_{aa'} \\ V_{bb'} \\ V_{cc'} \end{bmatrix} = \begin{bmatrix} Z_{aa} & Z_{ab} & Z_{ac} \\ Z_{ab} & Z_{bb} & Z_{bc} \\ Z_{ac} & Z_{cb} & Z_{cc} \end{bmatrix} \begin{bmatrix} I_a \\ I_b \\ I_c \end{bmatrix} \tag{8.3.1}$$

Both self-impedances and mutual impedances are included in (8.3.1). It is assumed that the impedance matrix is symmetric, which corresponds to a bilateral network. It is also assumed that these impedances represent

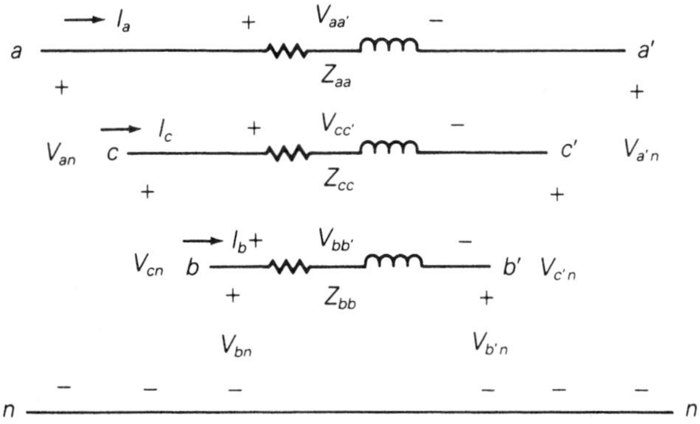

FIGURE 8.9 Three-phase series impedances (linear, bilateral network, nonrotating equipment)

nonrotating equipment. Typical examples are series impedances of transmission lines and of transformers. Equation (8.3.1) has the following form:

$$\boldsymbol{V}_p - \boldsymbol{V}_{p'} = \boldsymbol{Z}_p \boldsymbol{I}_p \tag{8.3.2}$$

where $\boldsymbol{V}_p$ is the vector of line-to-neutral voltages at bus abc, $\boldsymbol{V}_{p'}$ is the vector of line-to-neutral voltages at bus $a'b'c'$, $\boldsymbol{I}_p$ is the vector of line currents, and $\boldsymbol{Z}_p$ is the 3×3 phase impedance matrix for the series network. Equation (8.3.2) is now transformed to the sequence domain in the same manner that the load-phase impedances were transformed in Section 8.2. Thus,

$$\boldsymbol{V}_s - \boldsymbol{V}_{s'} = \boldsymbol{Z}_s \boldsymbol{I}_s \tag{8.3.3}$$

where

$$\boldsymbol{Z}_s = \boldsymbol{A}^{-1} \boldsymbol{Z}_p \boldsymbol{A} \tag{8.3.4}$$

From the results of Section 8.2, this sequence impedance $\boldsymbol{Z}_s$ matrix is diagonal under the following conditions:

$$\left.\begin{array}{l} Z_{aa} = Z_{bb} = Z_{cc} \\ \text{and} \\ Z_{ab} = Z_{ac} = Z_{bc} \end{array}\right\} \begin{array}{l}\text{conditions for} \\ \text{symmetrical} \\ \text{series impedances}\end{array} \tag{8.3.5}$$

When the phase impedance matrix $\boldsymbol{Z}_p$ of (8.3.1) has both equal self-impedances and equal mutual impedances, then (8.3.4) becomes

$$Z_s = \begin{bmatrix} Z_0 & 0 & 0 \\ 0 & Z_1 & 0 \\ 0 & 0 & Z_2 \end{bmatrix} \tag{8.3.6}$$

where

$$Z_0 = Z_{aa} + 2Z_{ab} \tag{8.3.7}$$

and

$$Z_1 = Z_2 = Z_{aa} - Z_{ab} \tag{8.3.8}$$

and (8.3.3) becomes three uncoupled equations, written as follows:

$$V_0 - V_{0'} = Z_0 I_0 \tag{8.3.9}$$

$$V_1 - V_{1'} = Z_1 I_1 \tag{8.3.10}$$

$$V_2 - V_{2'} = Z_2 I_2 \tag{8.3.11}$$

Equations (8.3.9)–(8.3.11) are represented by the three uncoupled sequence networks shown in Figure 8.10. From the figure it is apparent that for symmetrical series impedances, positive-sequence currents produce only positive-sequence voltage drops. Similarly, negative-sequence currents produce only negative-sequence voltage drops, and zero-sequence currents produce only zero-sequence voltage drops. However, if the series impedances

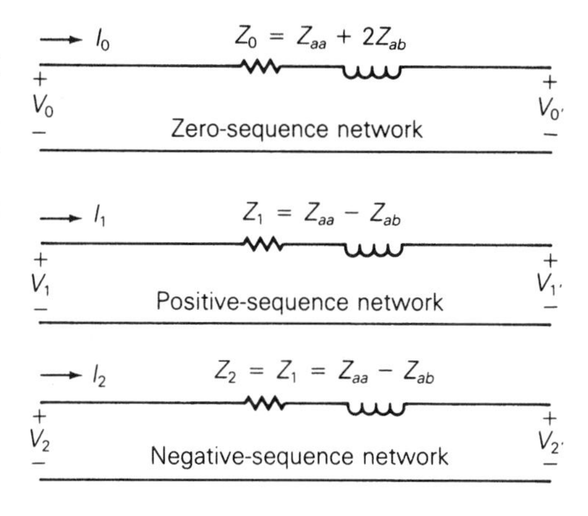

FIGURE 8.10 Sequence networks of three-phase symmetrical series impedances (linear, bilateral network, nonrotating equipment)

are not symmetrical, then $\boldsymbol{Z}_s$ is not diagonal, the sequence networks are coupled, and the voltage drop across any one sequence network depends on all three sequence currents.

8.4 SEQUENCE NETWORKS OF THREE-PHASE LINES

Section 4.7 develops equations suitable for computer calculation of the series phase impedances, including resistances and inductive reactances, of three-phase overhead transmission lines. The series phase impedance matrix $\boldsymbol{Z}_P$ for an untransposed line is given by Equation (4.7.19), and $\hat{\boldsymbol{Z}}_P$ for a completely transposed line is given by (4.7.21)–(4.7.23). Equation (4.7.19) can be transformed to the sequence domain to obtain

$$\boldsymbol{Z}_S = \boldsymbol{A}^{-1}\boldsymbol{Z}_P\boldsymbol{A} \tag{8.4.1}$$

$\boldsymbol{Z}_S$ is the 3×3 series sequence impedance matrix whose elements are

$$\boldsymbol{Z}_S = \begin{bmatrix} Z_0 & Z_{01} & Z_{02} \\ Z_{10} & Z_1 & Z_{12} \\ Z_{20} & Z_{21} & Z_2 \end{bmatrix} \quad \Omega/\text{m} \tag{8.4.2}$$

In general $\boldsymbol{Z}_S$ is not diagonal. However, if the line is completely transposed,

$$\hat{\boldsymbol{Z}}_S = \boldsymbol{A}^{-1}\hat{\boldsymbol{Z}}_P\boldsymbol{A} = \begin{bmatrix} \hat{Z}_0 & 0 & 0 \\ 0 & \hat{Z}_1 & 0 \\ 0 & 0 & \hat{Z}_2 \end{bmatrix} \tag{8.4.3}$$

where, from (8.3.7) and (8.3.8),

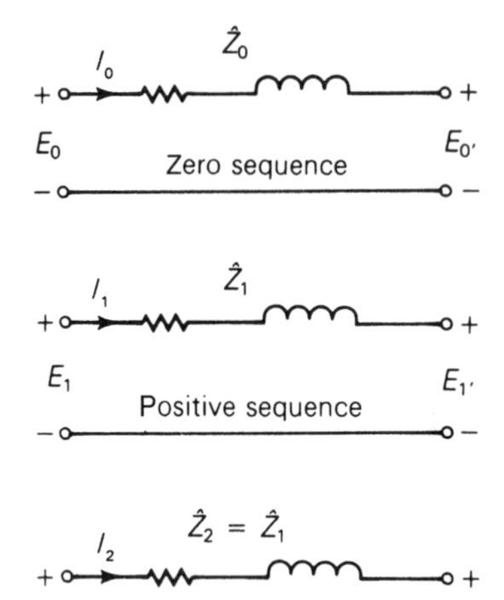

FIGURE 8.11 Circuit representation of the series sequence impedances of a completely transposed three-phase line

$$\hat{Z}_0 = \hat{Z}_{aaeq} + 2\hat{Z}_{abeq} \tag{8.4.4}$$

$$\hat{Z}_1 = \hat{Z}_2 = \hat{Z}_{aaeq} - \hat{Z}_{abeq} \tag{8.4.5}$$

A circuit representation of the series sequence impedances of a completely transposed three-phase line is shown in Figure 8.11.

Section 4.11 develops equations suitable for computer calculation of the shunt phase admittances of three-phase overhead transmission lines. The shunt admittance matrix $\boldsymbol{Y}_P$ for an untransposed line is given by Equation (4.11.16), and $\hat{\boldsymbol{Y}}_P$ for a completely transposed three-phase line is given by (4.11.17).

Equation (4.11.16) can be transformed to the sequence domain to obtain

$$\boldsymbol{Y}_S = \boldsymbol{A}^{-1}\boldsymbol{Y}_P\boldsymbol{A} \tag{8.4.6}$$

where

$$\boldsymbol{Y}_S = \mathbf{G}_S + j(2\pi f)\mathbf{C}_S \tag{8.4.7}$$

$$\mathbf{C}_S = \begin{bmatrix} C_0 & C_{01} & C_{02} \\ C_{10} & C_1 & C_{12} \\ C_{20} & C_{21} & C_2 \end{bmatrix} \quad \text{F/m} \tag{8.4.8}$$

In general, $\mathbf{C}_S$ is not diagonal. However, for the completely transposed line,

$$\hat{\boldsymbol{Y}}_S = \boldsymbol{A}^{-1}\hat{\boldsymbol{Y}}_P\boldsymbol{A} = \begin{bmatrix} \hat{y}_0 & 0 & 0 \\ 0 & \hat{y}_1 & 0 \\ 0 & 0 & \hat{y}_2 \end{bmatrix} = j(2\pi f)\begin{bmatrix} \hat{C}_0 & 0 & 0 \\ 0 & \hat{C}_1 & 0 \\ 0 & 0 & \hat{C}_2 \end{bmatrix} \tag{8.4.9}$$

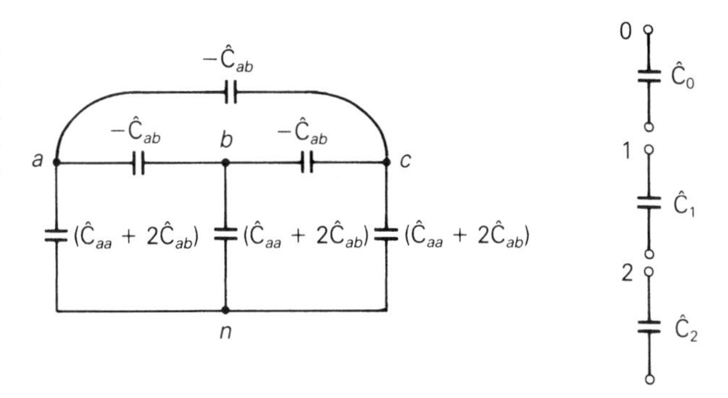

FIGURE 8.12 Circuit representations of the capacitances of a completely transposed three-phase line

where

$$\hat{C}_0 = \hat{C}_{aa} + 2\hat{C}_{ab} \quad \text{F/m} \tag{8.4.10}$$

$$\hat{C}_1 = \hat{C}_2 = \hat{C}_{aa} - \hat{C}_{ab} \quad \text{F/m} \tag{8.4.11}$$

Since $\hat{C}_{ab}$ is negative, the zero-sequence capacitance $\hat{C}_0$ is usually much less than the positive- or negative-sequence capacitance.

Circuit representations of the phase and sequence capacitances of a completely transposed three-phase line are shown in Figure 8.12.

8.5 SEQUENCE NETWORKS OF ROTATING MACHINES

A Y-connected synchronous generator grounded through a neutral impedance Z_n is shown in Figure 8.13. The internal generator voltages are designated E_a, E_b, and E_c, and the generator line currents are designated I_a, I_b, and I_c.

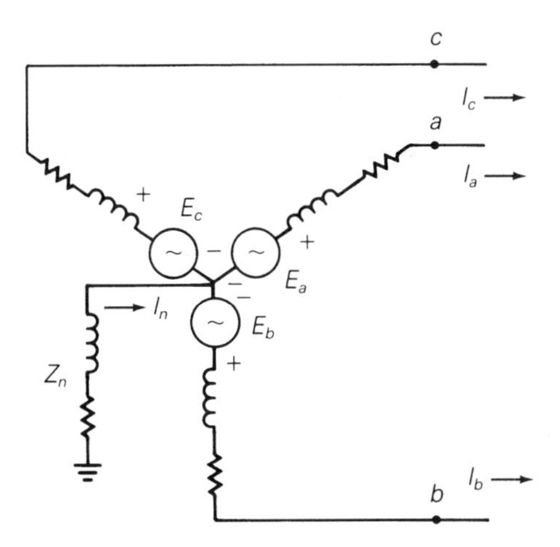

FIGURE 8.13 Y-connected synchronous generator

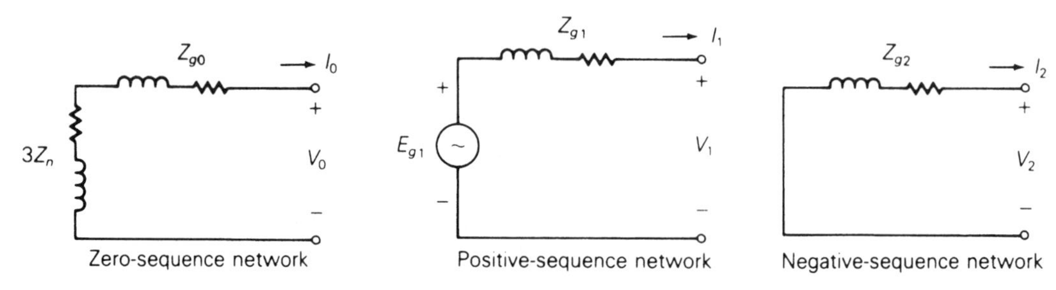

FIGURE 8.14 Sequence networks of a Y-connected synchronous generator

The sequence networks of the generator are shown in Figure 8.14. Since a three-phase synchronous generator is designed to produce balanced internal phase voltages E_a, E_b, E_c with only a positive-sequence component, a source voltage E_{g1} is included only in the positive-sequence network. The sequence components of the line-to-ground voltages at the generator terminals are denoted V_0, V_1, and V_2 in Figure 8.14.

The voltage drop in the generator neutral impedance is $Z_n I_n$, which can be written as $(3Z_n)I_0$, since, from (8.1.27), the neutral current is three times the zero-sequence current. Since this voltage drop is due only to zero-sequence current, an impedance $(3Z_n)$ is placed in the zero-sequence network of Figure 8.14 in series with the generator zero-sequence impedance Z_{g0}.

The sequence impedances of rotating machines are generally not equal. A detailed analysis of machine-sequence impedances is given in machine theory texts. We give only a brief explanation here.

When a synchronous generator stator has balanced three-phase positive-sequence currents under steady-state conditions, the net mmf produced by these positive-sequence currents rotates at the synchronous rotor speed in the same direction as that of the rotor. Under this condition, a high value of magnetic flux penetrates the rotor, and the positive-sequence impedance Z_{g1} has a high value. Under steady-state conditions, the positive-sequence generator impedance is called the *synchronous impedance*.

When a synchronous generator stator has balanced three-phase negative-sequence currents, the net mmf produced by these currents rotates at synchronous speed in the direction opposite to that of the rotor. With respect to the rotor, the net mmf is not stationary but rotates at twice synchronous speed. Under this condition, currents are induced in the rotor windings that prevent the magnetic flux from penetrating the rotor. As such, the negative-sequence impedance Z_{g2} is less than the positive-sequence synchronous impedance.

When a synchronous generator has only zero-sequence currents, which are line (or phase) currents with equal magnitude and phase, then the net mmf produced by these currents is theoretically zero. The generator zero-sequence impedance Z_{g0} is the smallest sequence impedance and is due to leakage flux, end turns, and harmonic flux from windings that do not produce a perfectly sinusoidal mmf.

Typical values of machine-sequence impedances are listed in Table A.1 in the Appendix. The positive-sequence machine impedance is synchronous, transient, or subtransient. *Synchronous* impedances are used for steady-state conditions, such as in power-flow studies, which are described in Chapter 6. *Transient* impedances are used for stability studies, which are described in Chapter 13, and *subtransient* impedances are used for short-circuit studies, which are described in Chapters 7 and 9. Unlike the positive-sequence impedances, a machine has only one negative-sequence impedance and only one zero-sequence impedance.

The sequence networks for three-phase synchronous motors and for three-phase induction motors are shown in Figure 8.15. Synchronous motors have the same sequence networks as synchronous generators, except that the sequence currents for synchronous motors are referenced *into* rather than out of the sequence networks. Also, induction motors have the same sequence networks as synchronous motors, except that the positive-sequence voltage

FIGURE 8.15

Sequence networks of three-phase motors

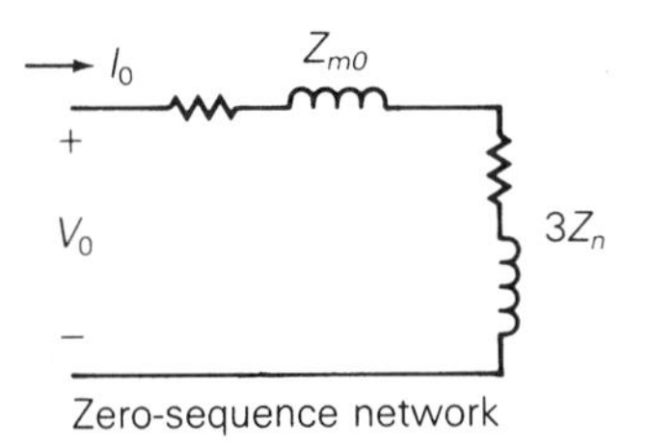

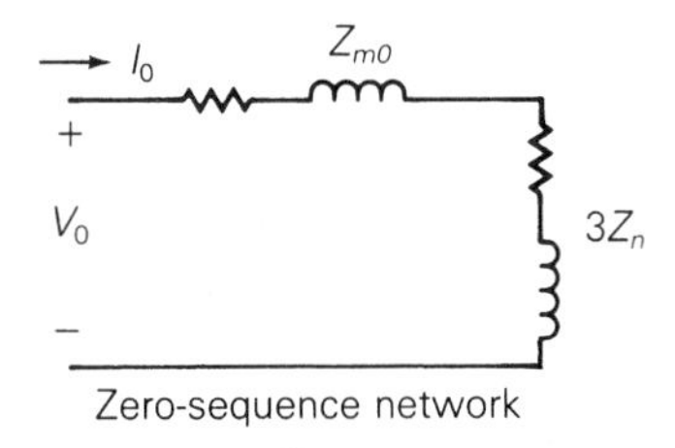

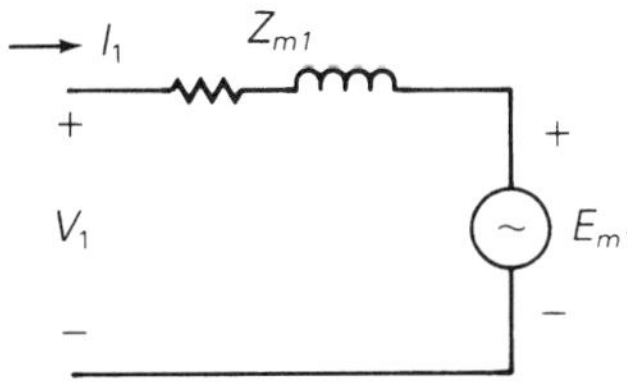

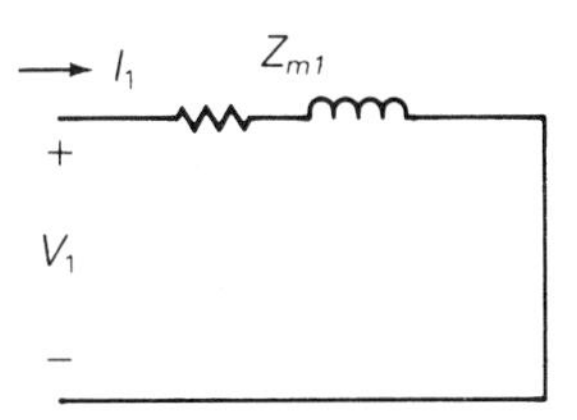

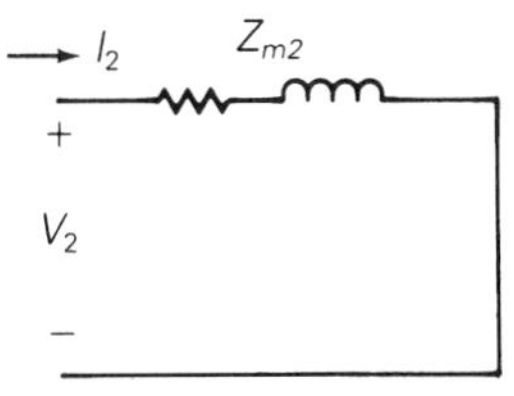

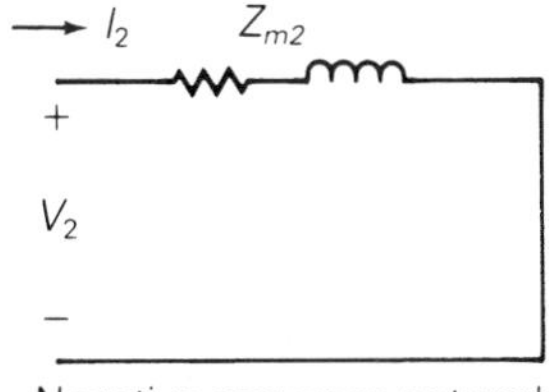

(a) Synchronous motor

(b) Induction motor

source E_{m1} is removed. Induction motors do not have a dc source of magnetic flux in their rotor circuits, and therefore E_{m1} is zero (or a short circuit).

The sequence networks shown in Figures 8.14 and 8.15 are simplified networks for rotating machines. The networks do not take into account such phenomena as machine saliency, saturation effects, and more complicated transient effects. These simplified networks, however, are in many cases accurate enough for power system studies.

EXAMPLE 8.5 Currents in sequence networks

Draw the sequence networks for the circuit of Example 2.5 and calculate the sequence components of the line current. Assume that the generator neutral is grounded through an impedance $Z_n = j10\ \Omega$, and that the generator sequence impedances are $Z_{g0} = j1\ \Omega$, $Z_{g1} = j15\ \Omega$, and $Z_{g2} = j3\ \Omega$.

SOLUTION The sequence networks are shown in Figure 8.16. They are obtained by interconnecting the sequence networks for a balanced-Δ load, for

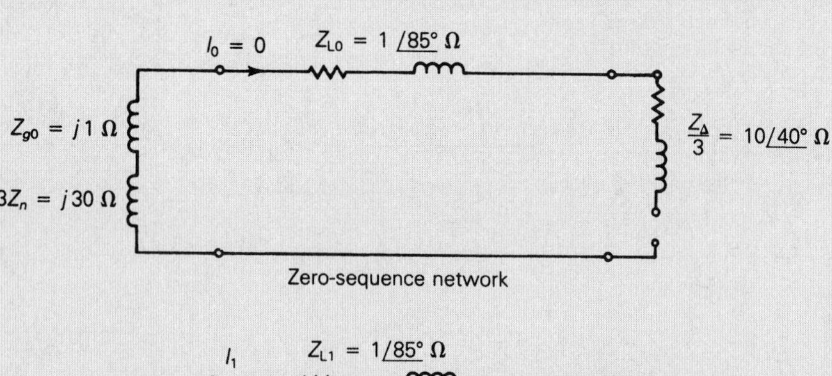

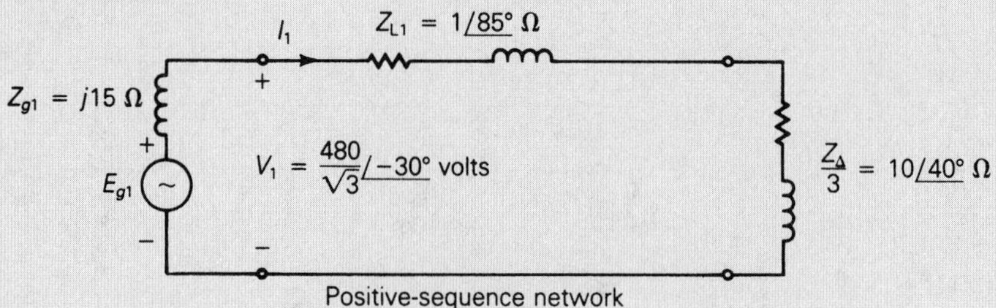

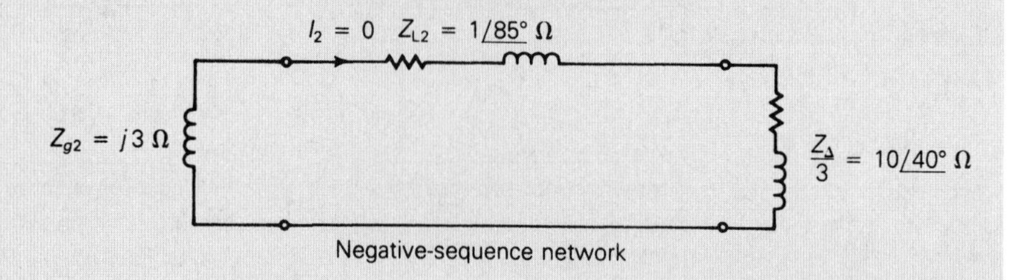

FIGURE 8.16 Sequence networks for Example 8.5

series-line impedances, and for a synchronous generator, which are given in Figures 8.5, 8.10, and 8.14.

It is clear from Figure 8.16 that $I_0 = I_2 = 0$ since there are no sources in the zero- and negative-sequence networks. Also, the positive-sequence generator terminal voltage V_1 equals the generator line-to-neutral terminal voltage. Therefore, from the positive-sequence network shown in the figure and from the results of Example 2.5,

$$I_1 = \frac{V_1}{(Z_{L1} + \frac{1}{3}Z_\Delta)} = 25.83\underline{/-73.78^\circ} \text{ A} = I_a$$

Note that from (8.1.20), I_1 equals the line current I_a, since $I_0 = I_2 = 0$. ■

The following example illustrates the superiority of using symmetrical components for analyzing unbalanced systems.

EXAMPLE 8.6 Solving unbalanced three-phase networks using sequence components

A Y-connected voltage source with the following unbalanced voltage is applied to the balanced line and load of Example 2.5.

$$\begin{bmatrix} V_{ag} \\ V_{bg} \\ V_{cg} \end{bmatrix} = \begin{bmatrix} 277\underline{/0^\circ} \\ 260\underline{/-120^\circ} \\ 295\underline{/+115^\circ} \end{bmatrix} \quad \text{volts}$$

The source neutral is solidly grounded. Using the method of symmetrical components, calculate the source currents I_a, I_b, and I_c.

SOLUTION Using (8.1.13)–(8.1.15), the sequence components of the source voltages are:

$$V_0 = \tfrac{1}{3}(277\underline{/0^\circ} + 260\underline{/-120^\circ} + 295\underline{/115^\circ})$$
$$= 7.4425 + j14.065 = 15.912\underline{/62.11^\circ} \quad \text{volts}$$

$$V_1 = \tfrac{1}{3}(227\underline{/0^\circ} + 260\underline{/-120^\circ + 120^\circ} + 295\underline{/115^\circ + 240^\circ})$$
$$= \tfrac{1}{3}(277\underline{/0^\circ} + 260\underline{/0^\circ} + 295\underline{/-5^\circ})$$
$$= 276.96 - j8.5703 = 277.1\underline{/-1.772^\circ} \quad \text{volts}$$

$$V_2 = \tfrac{1}{3}(277\underline{/0^\circ} + 260\underline{/-120^\circ + 240^\circ} + 295\underline{/115^\circ + 120^\circ})$$
$$= \tfrac{1}{3}(277\underline{/0^\circ} + 260\underline{/120^\circ} + 295\underline{/235^\circ})$$
$$= -7.4017 - j5.4944 = 9.218\underline{/216.59^\circ} \quad \text{volts}$$

These sequence voltages are applied to the sequence networks of the line and load, as shown in Figure 8.17. The sequence networks of this figure

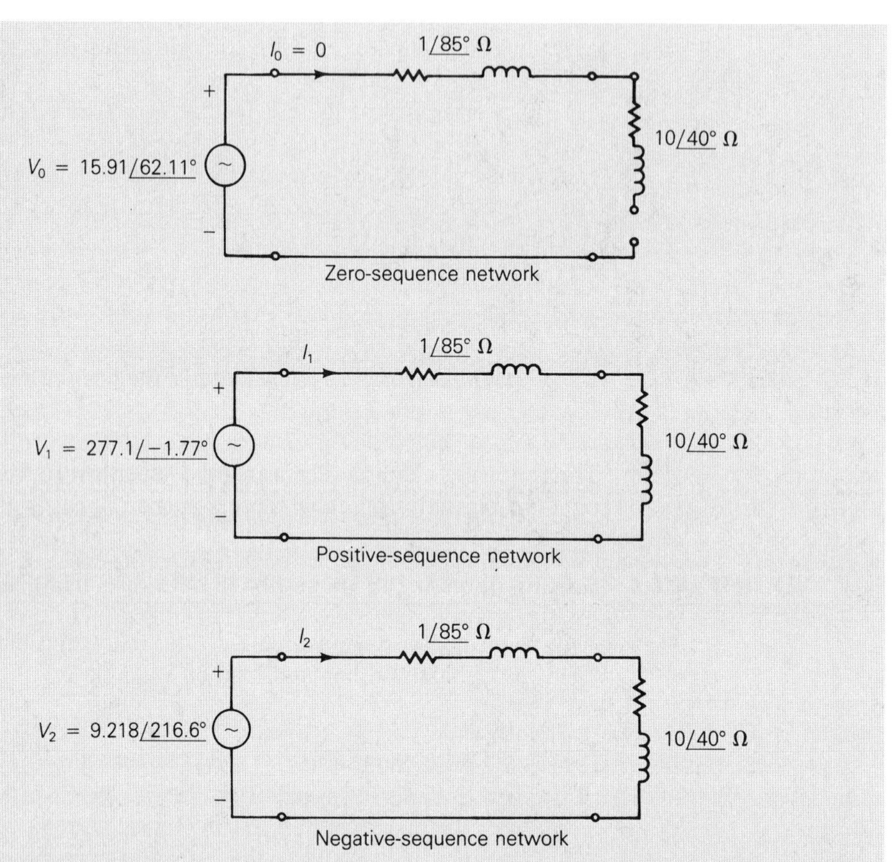

FIGURE 8.17 Sequence networks for Example 8.6

are uncoupled, and the sequence components of the source currents are easily calculated as follows:

$$I_0 = 0$$

$$I_1 = \frac{V_1}{Z_{L1} + \frac{Z_\Delta}{3}} = \frac{277.1/\underline{-1.772^\circ}}{10.73/\underline{43.78^\circ}} = 25.82/\underline{-45.55^\circ} \quad \text{A}$$

$$I_2 = \frac{V_2}{Z_{L2} + \frac{Z_\Delta}{3}} = \frac{9.218/\underline{216.59^\circ}}{10.73/\underline{43.78^\circ}} = 0.8591/\underline{172.81^\circ} \quad \text{A}$$

Using (8.1.20)–(8.1.22), the source currents are:

$$I_a = (0 + 25.82/\underline{-45.55^\circ} + 0.8591/\underline{172.81^\circ})$$
$$= 17.23 - j18.32 = 25.15/\underline{-46.76^\circ} \quad \text{A}$$

$$I_b = (0 + 25.82/\underline{-45.55^\circ + 240^\circ} + 0.8591/\underline{172.81^\circ + 120^\circ})$$
$$= (25.82/\underline{194.45^\circ} + 0.8591/\underline{292.81^\circ})$$
$$= -24.67 - j7.235 = 25.71/\underline{196.34^\circ} \quad \text{A}$$

$$I_c = (0 + 25.82\underline{/-45.55^\circ + 120^\circ} + 0.8591\underline{/172.81^\circ + 240^\circ})$$

$$= (25.82\underline{/74.45^\circ} + 0.8591\underline{/52.81^\circ})$$

$$= 7.441 + j25.56 = 26.62\underline{/73.77^\circ} \quad \text{A}$$

You should calculate the line currents for this example without using symmetrical components, in order to verify this result and to compare the two solution methods (see Problem 8.33). Without symmetrical components, coupled KVL equations must be solved. With symmetrical components, the conversion from phase to sequence components decouples the networks as well as the resulting KVL equations, as shown above. ■

8.6

PER-UNIT SEQUENCE MODELS OF THREE-PHASE TWO-WINDING TRANSFORMERS

Figure 8.18(a) is a schematic representation of an ideal Y–Y transformer grounded through neutral impedances Z_N and Z_n. Figures 8.18(b–d) show the per-unit sequence networks of this ideal transformer.

When balanced positive-sequence currents or balanced negative-sequence currents are applied to the transformer, the neutral currents are zero and there are no voltage drops across the neutral impedances. Therefore, the per-unit positive- and negative-sequence networks of the ideal Y–Y transformer, Figures 8.18(b) and (c), are the same as the per-unit single-phase ideal transformer, Figure 3.9(a).

Zero-sequence currents have equal magnitudes and equal phase angles. When per-unit sequence currents $I_{A0} = I_{B0} = I_{C0} = I_0$ are applied to the high-voltage windings of an ideal Y–Y transformer, the neutral current $I_N = 3I_0$ flows through the neutral impedance Z_N, with a voltage drop $(3Z_N)I_0$. Also, per-unit zero-sequence current I_0 flows in each low-voltage winding [from (3.3.9)], and therefore $3I_0$ flows through neutral impedance Z_n, with a voltage drop $(3I_0)Z_n$. The per-unit zero-sequence network, which includes the impedances $(3Z_N)$ and $(3Z_n)$, is shown in Figure 8.18(b).

Note that if either one of the neutrals of an ideal transformer is ungrounded, then no zero sequence can flow in either the high- or low-voltage windings. For example, if the high-voltage winding has an open neutral, then $I_N = 3I_0 = 0$, which in turn forces $I_0 = 0$ on the low-voltage side. This can be shown in the zero-sequence network of Figure 8.18(b) by making $Z_N = \infty$, which corresponds to an open circuit.

The per-unit sequence networks of a practical Y–Y transformer are shown in Figure 8.19(a). These networks are obtained by adding external impedances to the sequence networks of the ideal transformer, as follows. The leakage impedances of the high-voltage windings are series impedances like the series impedances shown in Figure 8.9, with no coupling between phases

FIGURE 8.18

Ideal Y–Y transformer

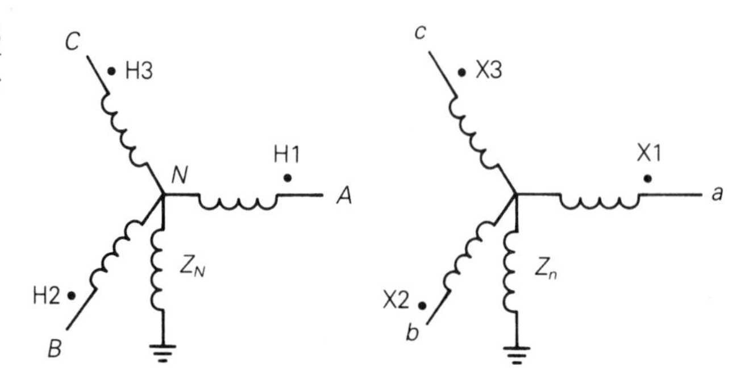

(a) Schematic representation

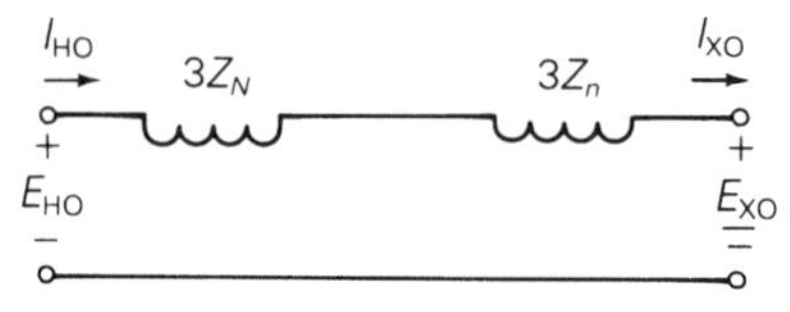

(b) Per-unit zero-sequence network

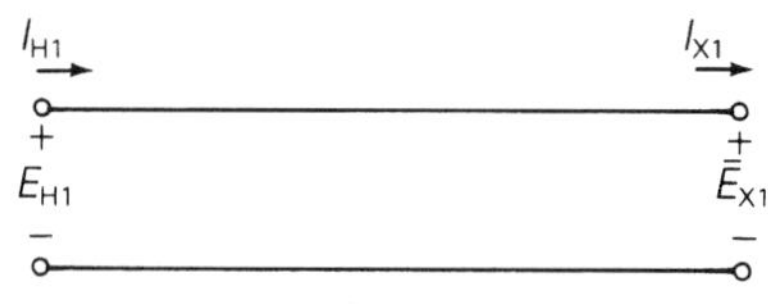

(c) Per-unit positive-sequence network

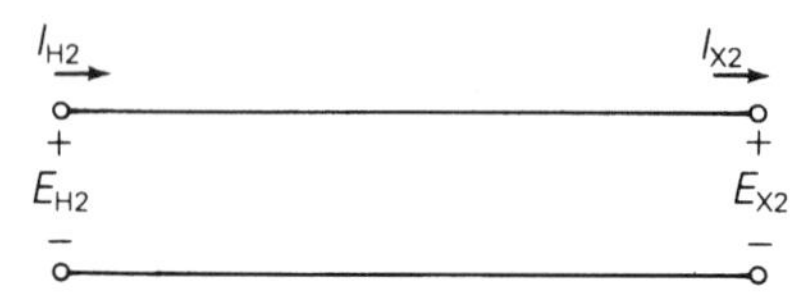

(d) Per-unit negative-sequence network

($Z_{ab} = 0$). If the phase a, b, and c windings have equal leakage impedances $Z_H = R_H + jX_H$, then the series impedances are *symmetrical* with sequence networks, as shown in Figure 8.10, where $Z_{H0} = Z_{H1} = Z_{H2} = Z_H$. Similarly, the leakage impedances of the low-voltage windings are symmetrical series impedances with $Z_{X0} = Z_{X1} = Z_{X2} = Z_X$. These series leakage impedances are shown in per-unit in the sequence networks of Figure 8.19(a).

The shunt branches of the practical Y–Y transformer, which represent exciting current, are equivalent to the Y load of Figure 8.3. Each phase in Figure 8.3 represents a core loss resistor in parallel with a magnetizing inductance. Assuming these are the same for each phase, then the Y load is *symmetrical*, and the sequence networks are shown in Figure 8.4. These shunt

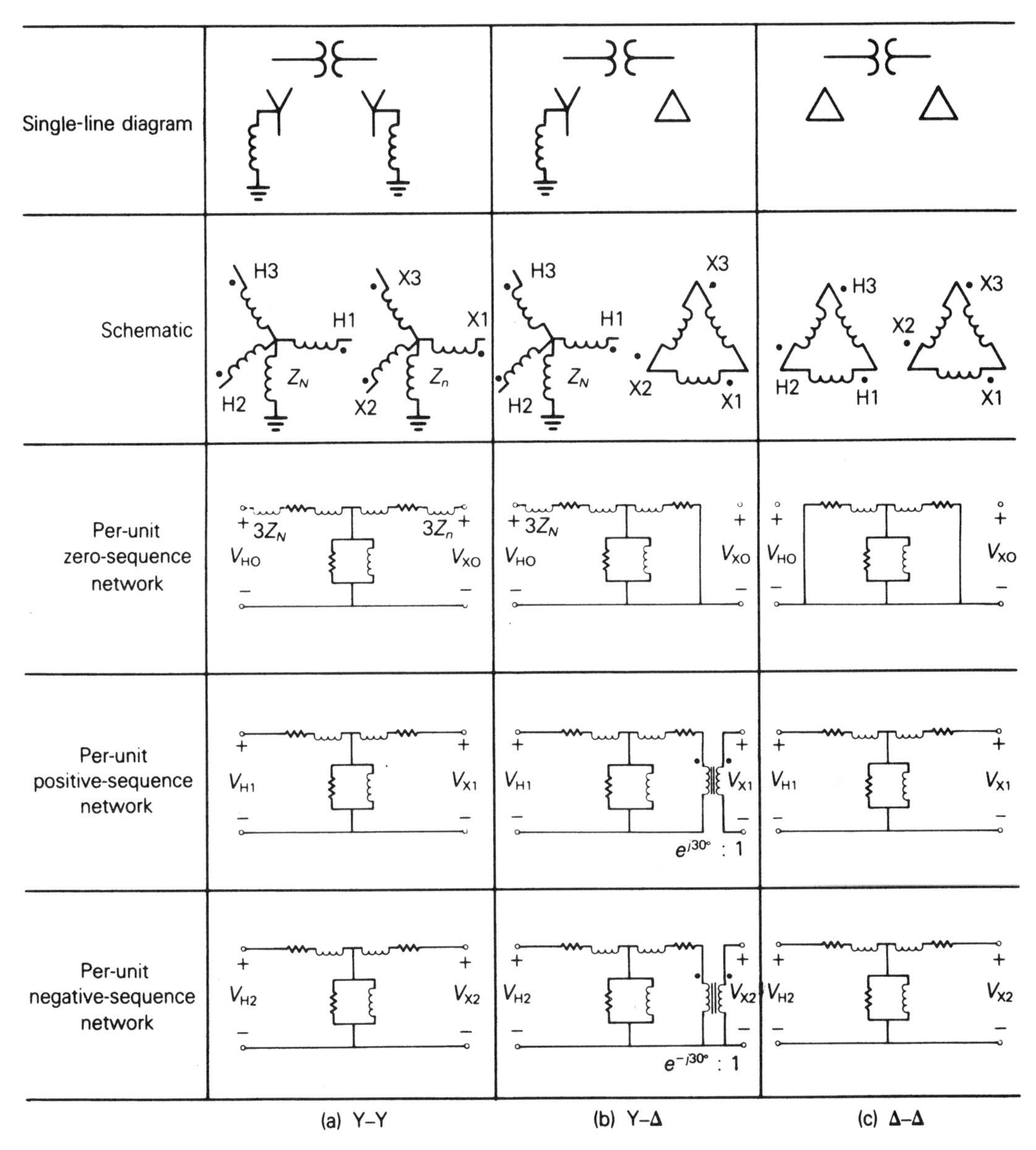

FIGURE 8.19 Per-unit sequence networks of practical Y–Y, Y–Δ, and Δ–Δ transformers

branches are also shown in Figure 8.19(a). Note that $(3Z_N)$ and $(3Z_n)$ have already been included in the zero-sequence network.

The per-unit positive- and negative-sequence transformer impedances of the practical Y–Y transformer in Figure 8.19(a) are identical, which is always true for nonrotating equipment. The per-unit zero-sequence network, however, depends on the neutral impedances Z_N and Z_n.

The per-unit sequence networks of the Y–Δ transformer, shown in Figure 8.19(b), have the following features:

1. The per-unit impedances do not depend on the winding connections. That is, the per-unit impedances of a transformer that is connected Y–Y, Y–Δ, Δ–Y, or Δ–Δ are the same. However, the base voltages do depend on the winding connections.

2. A phase shift is included in the per-unit positive- and negative-sequence networks. For the American standard, the positive-sequence voltages and currents on the high-voltage side of the Y–Δ transformer lead the corresponding quantities on the low-voltage side by 30°. For negative sequence, the high-voltage quantities lag by 30°.

3. Zero-sequence currents can flow in the Y winding if there is a neutral connection, and corresponding zero-sequence currents flow within the Δ winding. However, no zero-sequence current enters or leaves the Δ winding.

The phase shifts in the positive- and negative-sequence networks of Figure 8.19(b) are represented by the phase-shifting transformer of Figure 3.4. Also, the zero-sequence network of Figure 8.19(b) provides a path on the Y side for zero-sequence current to flow, but no zero-sequence current can enter or leave the Δ side.

The per-unit sequence networks of the Δ–Δ transformer, shown in Figure 8.19(c), have the following features:

1. The positive- and negative-sequence networks, which are identical, are the same as those for the Y–Y transformer. It is assumed that the windings are labeled so there is no phase shift. Also, the per-unit impedances do not depend on the winding connections, but the base voltages do.

2. Zero-sequence currents *cannot* enter or leave either Δ winding, although they can circulate within the Δ windings.

EXAMPLE 8.7 Solving unbalanced three-phase networks with transformers using per-unit sequence components

A 75-kVA, 480-volt Δ/208-volt Y transformer with a solidly grounded neutral is connected between the source and line of Example 8.6. The transformer leakage reactance is $X_{eq} = 0.10$ per unit; winding resistances and exciting current are neglected. Using the transformer ratings as base quantities, draw the per-unit sequence networks and calculate the phase a source current I_a.

SOLUTION The base quantities are $S_{base1\phi} = 75/3 = 25$ kVA, $V_{baseHLN} = 480/\sqrt{3} = 277.1$ volts, $V_{baseXLN} = 208/\sqrt{3} = 120.1$ volts, and $Z_{baseX} = (120.1)^2/25{,}000 = 0.5770\ \Omega$. The sequence components of the actual source voltages are given in Figure 8.17. In per-unit, these voltages are

$$V_0 = \frac{15.91\underline{/62.11^\circ}}{277.1} = 0.05742\underline{/62.11^\circ} \quad \text{per unit}$$

$$V_1 = \frac{277.1\underline{/-1.772^\circ}}{277.1} = 1.0\underline{/-1.772^\circ} \quad \text{per unit}$$

$$V_2 = \frac{9.218\underline{/216.59^\circ}}{277.1} = 0.03327\underline{/216.59^\circ} \quad \text{per unit}$$

The per-unit line and load impedances, which are located on the low-voltage side of the transformer, are

$$Z_{L0} = Z_{L1} = Z_{L2} = \frac{1\underline{/85^\circ}}{0.577} = 1.733\underline{/85^\circ} \quad \text{per unit}$$

$$Z_{\text{load1}} = Z_{\text{load2}} = \frac{Z_\Delta}{3(0.577)} = \frac{10\underline{/40^\circ}}{0.577} = 17.33\underline{/40^\circ} \quad \text{per unit}$$

FIGURE 8.20

Per-unit sequence networks for Example 8.7

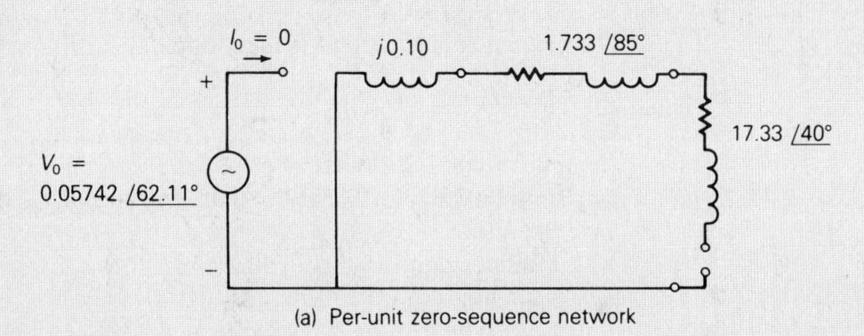

(a) Per-unit zero-sequence network

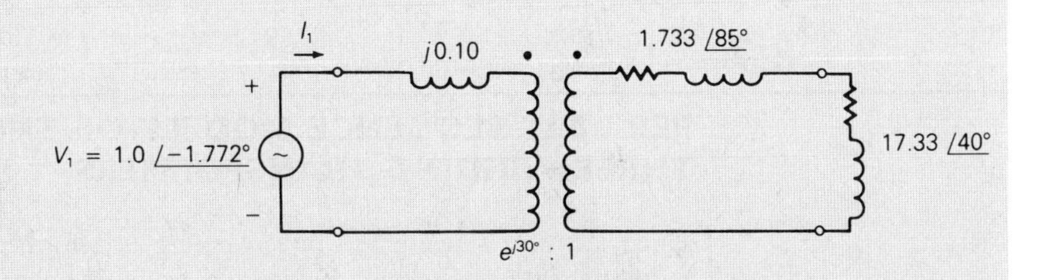

(b) Per-unit positive-sequence network

(c) Per-unit negative-sequence network

The per-unit sequence networks are shown in Figure 8.20. Note that the per-unit line and load impedances, when referred to the high-voltage side of the phase-shifting transformer, do not change [(see (3.1.26)]. Therefore, from Figure 8.20, the sequence components of the source currents are

$$I_0 = 0$$

$$I_1 = \frac{V_1}{jX_{eq} + Z_{L1} + Z_{load1}} = \frac{1.0\underline{/-1.772^\circ}}{j0.10 + 1.733\underline{/85^\circ} + 17.33\underline{/40^\circ}}$$

$$= \frac{1.0\underline{/-1.772^\circ}}{13.43 + j12.97} = \frac{1.0\underline{/-1.772^\circ}}{18.67\underline{/44.0^\circ}} = 0.05356\underline{/-45.77^\circ} \quad \text{per unit}$$

$$I_2 = \frac{V_2}{jX_{eq} + Z_{L2} + Z_{load2}} = \frac{0.03327\underline{/216.59^\circ}}{18.67\underline{/44.0^\circ}}$$

$$= 0.001782\underline{/172.59^\circ} \quad \text{per unit}$$

The phase a source current is then, using (8.1.20),

$$I_a = I_0 + I_1 + I_2$$

$$= 0 + 0.05356\underline{/-45.77^\circ} + 0.001782\underline{/172.59^\circ}$$

$$= 0.03511 - j0.03764 = 0.05216\underline{/-46.19^\circ} \quad \text{per unit}$$

Using $I_{baseH} = \dfrac{75{,}000}{480\sqrt{3}} = 90.21$ A,

$$I_a = (0.05216)(90.21)\underline{/-46.19^\circ} = 4.705\underline{/-46.19^\circ} \quad \text{A}$$

■

8.7 PER-UNIT SEQUENCE MODELS OF THREE-PHASE THREE-WINDING TRANSFORMERS

Three identical single-phase three-winding transformers can be connected to form a three-phase bank. Figure 8.21 shows the general per-unit sequence networks of a three-phase three-winding transformer. Instead of labeling the windings 1, 2, and 3, as was done for the single-phase transformer, the letters H, M, and X are used to denote the high-, medium-, and low-voltage windings, respectively. By convention, a common S_{base} is selected for the H, M, and X terminals, and voltage bases V_{baseH}, V_{baseM}, and V_{baseX} are selected in proportion to the rated line-to-line voltages of the transformer.

For the general zero-sequence network, Figure 8.21(a), the connection between terminals H and H′ depends on how the high-voltage windings are connected, as follows:

1. Solidly grounded Y—Short H to H′.
2. Grounded Y through Z_N—Connect $(3Z_N)$ from H to H′.

FIGURE 8.21

Per-unit sequence networks of a three-phase three-winding transformer

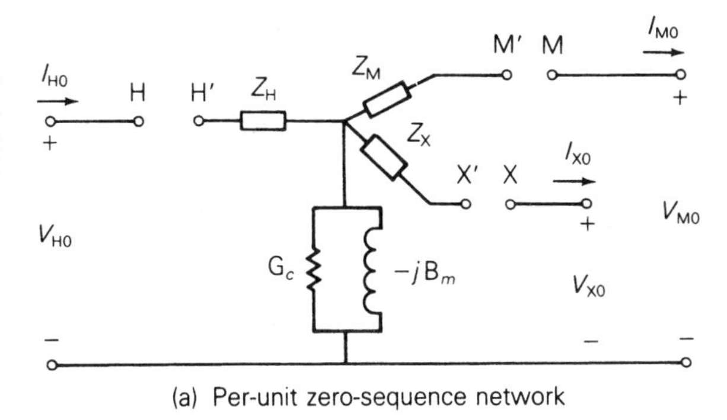

(a) Per-unit zero-sequence network

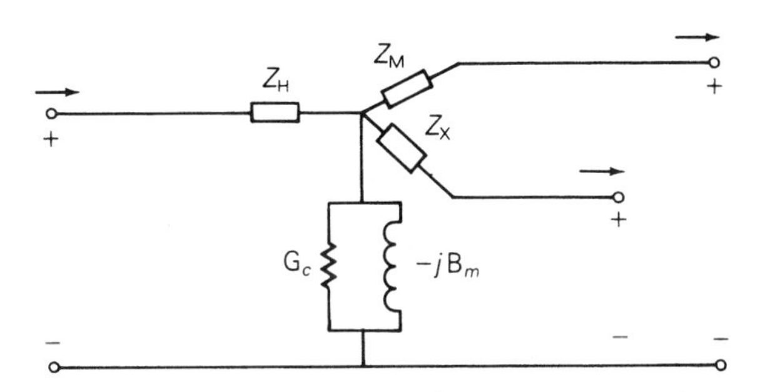

(b) Per-unit positive- or negative-sequence network (phase shift not shown)

3. Ungrounded Y—Leave H–H′ open as shown.

4. Δ—Short H′ to the reference bus.

Terminals X–X′ and M–M′ are connected in a similar manner.

The impedances of the per-unit negative-sequence network are the same as those of the per-unit positive-sequence network, which is always true for nonrotating equipment. Phase-shifting transformers, not shown in Figure 8.21(b), can be included to model phase shift between Δ and Y windings.

EXAMPLE 8.8 Three-winding three-phase transformer: per-unit sequence networks

Three transformers, each identical to that described in Example 3.9, are connected as a three-phase bank in order to feed power from a 900-MVA, 13.8-kV generator to a 345-kV transmission line and to a 34.5-kV distribution line. The transformer windings are connected as follows:

13.8-kV windings (X):	Δ, to generator
199.2-kV windings (H):	solidly grounded Y, to 345-kV line
19.92-kV windings (M):	grounded Y through $Z_n = j0.10\ \Omega$, to 34.5-kV line

The positive-sequence voltages and currents of the high- and medium-voltage Y windings lead the corresponding quantities of the low-voltage Δ winding by 30°. Draw the per-unit sequence networks, using a three-phase base of 900 MVA and 13.8 kV for terminal X.

SOLUTION The per-unit sequence networks are shown in Figure 8.22. Since $V_{baseX} = 13.8$ kV is the rated line-to-line voltage of terminal X, $V_{baseM} = \sqrt{3}(19.92) = 34.5$ kV, which is the rated line-to-line voltage of terminal M. The base impedance of the medium-voltage terminal is then

$$Z_{baseM} = \frac{(34.5)^2}{900} = 1.3225 \quad \Omega$$

Therefore, the per-unit neutral impedance is

$$Z_n = \frac{j0.10}{1.3225} = j0.07561 \quad \text{per unit}$$

FIGURE 8.22 Per-unit sequence networks for Example 8.8

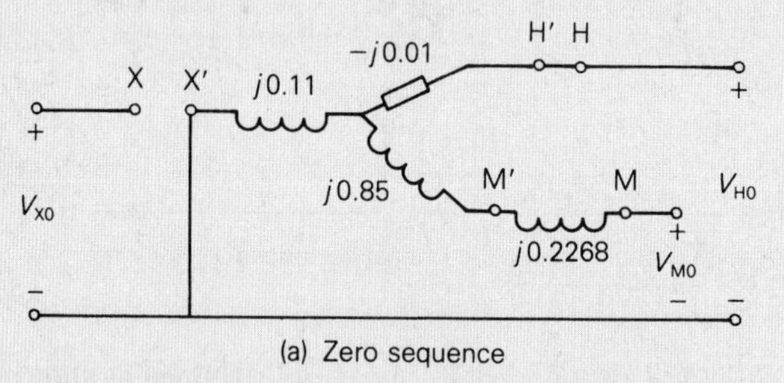

(a) Zero sequence

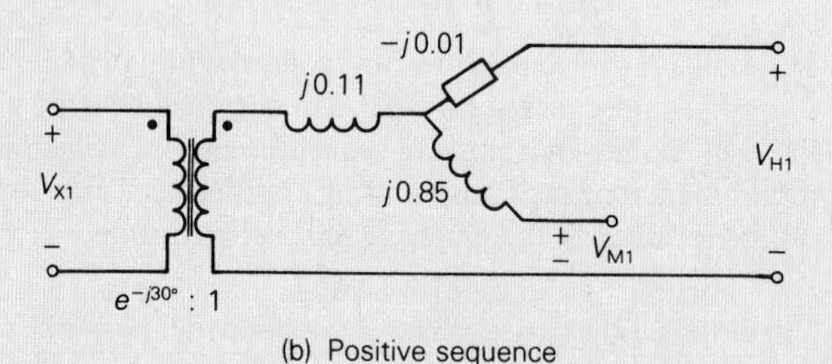

(b) Positive sequence

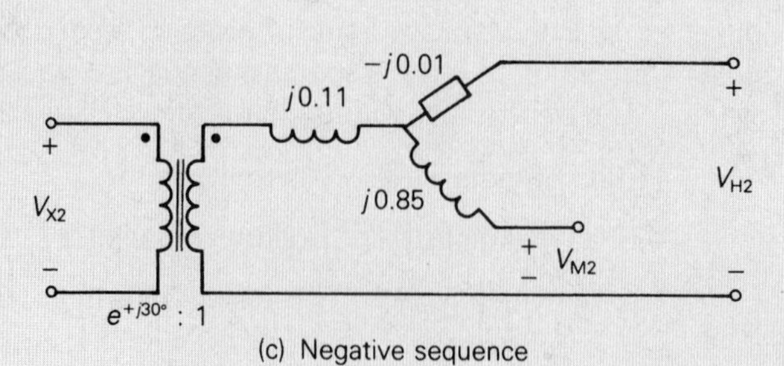

(c) Negative sequence

and $(3Z_n) = j0.2268$ is connected from terminal M to M′ in the per-unit zero-sequence network. Since the high-voltage windings have a solidly grounded neutral, H to H′ is shorted in the zero-sequence network. Also, phase-shifting transformers are included in the positive- and negative-sequence networks. ■

8.8 POWER IN SEQUENCE NETWORKS

The power delivered to a three-phase network can be determined from the power delivered to the sequence networks. Let S_p denote the total complex power delivered to the three-phase load of Figure 8.7, which can be calculated from

$$S_p = V_{ag}I_a^* + V_{bg}I_b^* + V_{cg}I_c^* \tag{8.8.1}$$

Equation (8.8.1) is also valid for the total complex power delivered by the three-phase generator of Figure 8.13, or for the complex power delivered to any three-phase bus. Rewriting (8.8.1) in matrix format,

$$S_p = [V_{ag}\, V_{bg}\, V_{cg}] \begin{bmatrix} I_a^* \\ I_b^* \\ I_c^* \end{bmatrix}$$

$$= \boldsymbol{V}_p^{\mathrm{T}} \boldsymbol{I}_p^* \tag{8.8.2}$$

where T denotes transpose and * denotes complex conjugate. Now, using (8.1.9) and (8.1.16),

$$S_p = (\boldsymbol{A}\boldsymbol{V}_s)^{\mathrm{T}}(\boldsymbol{A}\boldsymbol{I}_s)^*$$

$$= \boldsymbol{V}_s^{\mathrm{T}}[\boldsymbol{A}^{\mathrm{T}}\boldsymbol{A}^*]\boldsymbol{I}_s^* \tag{8.8.3}$$

Using the definition of $\boldsymbol{A}$, which is (8.1.8), to calculate the term within the brackets of (8.8.3), and noting that a and a^2 are conjugates,

$$\boldsymbol{A}^{\mathrm{T}}\boldsymbol{A}^* = \begin{bmatrix} 1 & 1 & 1 \\ 1 & a^2 & a \\ 1 & a & a^2 \end{bmatrix}^{\mathrm{T}} \begin{bmatrix} 1 & 1 & 1 \\ 1 & a^2 & a \\ 1 & a & a^2 \end{bmatrix}^*$$

$$= \begin{bmatrix} 1 & 1 & 1 \\ 1 & a^2 & a \\ 1 & a & a^2 \end{bmatrix} \begin{bmatrix} 1 & 1 & 1 \\ 1 & a & a^2 \\ 1 & a^2 & a \end{bmatrix}$$

$$= \begin{bmatrix} 3 & 0 & 0 \\ 0 & 3 & 0 \\ 0 & 0 & 3 \end{bmatrix} = 3\boldsymbol{U} \tag{8.8.4}$$

Equation (8.8.4) can now be used in (8.8.3) to obtain

$$S_p = 3V_s^T I_s^*$$

$$= 3[V_0 + V_1 + V_2]\begin{bmatrix} I_0^* \\ I_1^* \\ I_2^* \end{bmatrix} \tag{8.8.5}$$

$$S_p = 3(V_0 I_0^* + V_1 I_1^* + V_2 I_2^*)$$

$$= 3S_s \tag{8.8.6}$$

Thus, the total complex power S_p delivered to a three-phase network equals *three* times the total complex power S_s delivered to the sequence networks.

The factor of 3 occurs in (8.8.6) because $A^T A^* = 3U$, as shown by (8.8.4). It is possible to eliminate this factor of 3 by defining a new transformation matrix $A_1 = (1/\sqrt{3})A$ such that $A_1^T A_1^* = U$, which means that A_1 is a *unitary* matrix. Using A_1 instead of A, the total complex power delivered to three-phase networks would equal the total complex power delivered to the sequence networks. However, standard industry practice for symmetrical components is to use A, defined by (8.1.8).

EXAMPLE 8.9 Power in sequence networks

Calculate S_p and S_s delivered by the three-phase source in Example 8.6. Verify that $S_p = 3S_s$.

SOLUTION Using (8.5.1),

$$S_p = (277\underline{/0°})(25.15\underline{/+46.76°}) + (260\underline{/-120°})(25.71\underline{/-196.34°})$$

$$+ (295\underline{/115°})(26.62\underline{/-73.77°})$$

$$= 6967\underline{/46.76°} + 6685\underline{/43.66°} + 7853\underline{/41.23°}$$

$$= 15{,}520 + j14{,}870 = 21{,}490\underline{/43.78°} \quad \text{VA}$$

In the sequence domain,

$$S_s = V_0 I_0^* + V_1 I_1^* + V_2 I_2^*$$

$$= 0 + (277.1\underline{/-1.77°})(25.82\underline{/45.55°})$$

$$+ (9.218\underline{/216.59°})(0.8591\underline{/-172.81°})$$

$$= 7155\underline{/43.78°} + 7.919\underline{/43.78°}$$

$$= 5172 + j4958 = 7163\underline{/43.78°} \quad \text{VA}$$

Also,

$$3S_s = 3(7163\underline{/43.78°}) = 21{,}490\underline{/43.78°} = S_p$$

■

MULTIPLE CHOICE QUESTIONS

SECTION 8.1

8.1 Positive-sequence components consist of three phasors with _________ magnitudes, and _________ phase displacement in positive sequence; negative-sequence components consist of three phasors with _________ magnitudes, and _________ phase displacement in negative sequence; and zero-sequence components consist of three phasors with _________ magnitudes, and _________ phase displacement. Fill in the Blanks.

8.2 In symmetrical-component theory, express the complex-number operator $a = 1\underline{/120°}$ in exponential and rectangular forms.

8.3 In terms of sequence components of phase a given by $V_{a0} = V_0, V_{a1} = V_1$ and $V_{a2} = V_2$, give expressions for the phase voltages V_a, V_b, and V_c.
$V_a =$ _______________; $V_b =$ _______________; $V_c =$ _______________

8.4 The sequence components V_0, V_1, and V_2 can be expressed in terms of phase components V_a, V_b, and V_c.
$V_0 =$ _______________; $V_1 =$ _______________; $V_2 =$ _______________

8.5 In a balanced three-phase system, what is the zero-sequence voltage?
$V_0 =$ _______________

8.6 In an unblanced three-phase system, line-to-neutral voltage _________ have a zero-sequence component, whereas line-to-line voltages _________ have a zero-sequence component. Fill in the Blanks.

8.7 Can the symmetrical component transformation be applied to currents, just as applied to voltages?
(a) Yes (b) No

8.8 In a three-phase Wye-connected system with a neutral, express the neutral current in terms of phase currents and sequence-component terms.
$I_n =$ _______________ $=$ _______________

8.9 In a balanced Wye-connected system, what is the zero-sequence component of the line currents?

8.10 In a delta-connected three-phase system, line currents have no zero-sequence component.
(a) True (b) False

8.11 Balanced three-phase systems with positive sequence do not have zero-sequence and negative-sequence components.
(a) True (b) False

8.12 Unbalanced three-phase systems may have nonzero values for all sequence components.
(a) True (b) False

SECTION 8.2

8.13 For a balanced-Y impedance load with per-phase impedance of Z_Y and A neutral impedance Z_n connected between the load neutral and the ground, the 3×3 phase-impedance matrix will consist of equal diagonal elements given by _________, and equal nondiagonal elements given by _________. Fill in the Blanks.

8.14 Express the sequence impedance matrix $\boldsymbol{Z}_s$ in terms of the phase-impedance matrix $\boldsymbol{Z}_p$, and the transformation matrix $\boldsymbol{A}$ which relates $\boldsymbol{V}_p = \boldsymbol{A}\boldsymbol{V}_s$ and $\boldsymbol{I}_p = \boldsymbol{A}\boldsymbol{I}_s$.
$\boldsymbol{Z}_s =$ ________. Fill in the Blank.

8.15 The sequence impedance matrix Z_s for a balanced-Y load is a diagonal matrix and the sequence networks are uncoupled.
(a) True (b) False

8.16 For a balanced-Y impedance load with per-phase impedance of Z_Y and a neutral impedance Z_n, the zero-sequence voltage $V_0 = Z_0 I_0$, where $Z_0 =$ ________. Fill in the Blank.

8.17 For a balanced-Δ load with per-phase impedance of Z_Δ the equivalent Y-load will have an open neutral; for the corresponding uncoupled sequence networks, $Z_0 =$ ________, $Z_1 =$ ________, and $Z_2 =$ ________. Fill in the Blanks.

8.18 For a three-phase symmetrical impedance load, the sequence impedance matrix is ________ and hence the sequence networks are coupled/uncoupled.

SECTION 8.3

8.19 Sequence networks for three-phase symmetrical series impedances are coupled/uncoupled; positive-sequence currents produce only ________ voltage drops.

SECTION 8.4

8.20 The series sequence impedance matrix of a completely transposed three-phase line is ________, with its nondiagonal elements equal to ________. Fill in the Blanks.

SECTION 8.5

8.21 A Y-connected synchronous generator grounded through a neutral impedance Z_n, with a zero-sequence impedance Z_{g0}, will have zero-sequence impedance $Z_0 =$ ________ in its zero-sequence network. Fill in the Blank.

8.22 In sequence networks, a Y-connected synchronous generator is represented by its source per-unit voltage only in ________ network, while synchronous/transient/subtransient impedance is used in positive-sequence network for short-circuit studies.

8.23 In the positive-sequence network of a synchronous motor, a source voltage is represented, whereas in that of an induction motor, the source voltage does/does not come into picture.

8.24 With symmetrical components, the conversion from phase to sequence components decouples the networks and the resulting kVL equations.
(a) True (b) False

SECTION 8.6

8.25 Consider the per-unit sequence networks of Y-Y, Y-Δ, and Δ – Δ transformers, with neutral impedances of Z_N on the high-voltage Y-side, and Z_n on the low-voltage Y-side. Answer the following:

(i) Zero-sequence currents can/cannot flow in the Y winding with a neutral connection; corresponding zero-sequence currents do/do not flow within the delta winding;

however zero-sequence current does/does not enter or leave the Δ winding. In zero-sequence network, 1/2/3 times the neutral impedance comes into play in series.

(ii) In Y(HV)- Δ(LV) transformers, if a phase shift is included as per the American-standard notation, the ratio _________ is used in positive-sequence network, and the ratio _________ is used in the negative-sequence network.

(iii) The base voltages depend on the winding connections; the per-unit impedances do/do not depend on the winding connections.

SECTION 8.7

8.26 In per-unit sequence models of three-phase three-winding transformers, for the general zero-sequence network, the connection between terminals H and H′ depends on how the high-voltage windings are connected:

(i) For solidly grounded Y, ________ H to H′.
(ii) For grounded Y through Z_n, connect _________ from H to H′.
(iii) For ungrounded Y, leave H–H′ _________.
(iv) For Δ, ________ H′ to the reference bus.

SECTION 8.8

8.27 The total complex power delivered to a three-phase network equals 1/2/3 times the total complex power delivered to the sequence networks.

8.28 Express the complex power S_s Delivered to the sequence networks in terms of sequence voltages and sequence currents.
$S_s =$ ___________

PROBLEMS

SECTION 8.1

8.1 Using the operator $a = 1\angle 120°$, evaluate the following in polar form: (a) $(a-1)/(1+a-a^2)$, (b) $(a^2+a+j)/(ja+a^2)$, (c) $(1+a)(1+a^2)$, (d) $(a-a^2)(a^2-1)$.

8.2 Using $a = 1\angle 120°$, evaluate the following in rectangular form:

a. a^{10}

b. $(ja)^{10}$

c. $(1-a)^3$

d. e^a

Hint for (d): $e^{(x+jy)} = e^x e^{jy} = e^x\angle y$, where y is in radians.

8.3 Determine the symmetrical components of the following line currents: (a) $I_a = 5\angle 90°$, $I_b = 5\angle 320°$, $I_c = 5\angle 220°$ A; (b) $I_a = j50$, $I_b = 50$, $I_c = 0$ A.

8.4 Find the phase voltages V_{an}, V_{bn}, and V_{cn} whose sequence components are: $V_0 = 50\angle 80°$, $V_1 = 100\angle 0°$, $V_2 = 50\angle 90°$ V.

8.5 For the unbalanced three-phase system described by

$$I_a = 12\underline{/0^\circ}\,\text{A}, \quad I_b = 6\underline{/-90^\circ}\,\text{A}, \quad I_C = 8\underline{/150^\circ}\,\text{A}$$

compute the symmetrical components I_0, I_1, I_2.

8.6 (a) Given the symmetrical components to be

$$V_0 = 10\underline{/0^\circ}\,V, \quad V_1 = 80\underline{/30^\circ}\,V, \quad V_2 = 40\underline{/-30^\circ}\,V$$

determine the unbalanced phase voltages V_a, V_b, and V_c.
(b) Using the results of part (a), calculate the line-to-line voltages V_{ab}, V_{bc}, and V_{ca}. Then determine the symmetrical components of these ling-to-line voltages, the symmetrical components of the corresponding phase voltages, and the phase voltages. Compare them with the result of part (a). Comment on why they are different, even though either set will result in the same line-to-line voltages.

8.7 One line of a three-phase generator is open circuited, while the other two are short-circuited to ground. The line currents are $I_a = 0$, $I_b = 1000\underline{/150^\circ}$, and $I_c = 1000\underline{/+30^\circ}$ A. Find the symmetrical components of these currents. Also find the current into the ground.

8.8 Let an unbalanced, three-phase, Wye-connected load (with phase impedances of Z_a, Z_b, and Z_c) be connected to a balanced three-phase supply, resulting in phase voltages of V_a, V_b, and V_c across the corresponding phase impedances.
Choosing V_{ab} as the reference, show that

$$V_{ab,0} = 0; \quad V_{ab,1} = \sqrt{3}V_{a,1}e^{j30^\circ}; \quad V_{ab,2} = \sqrt{3}V_{a,2}e^{-j30^\circ}.$$

8.9 Reconsider Problem 8.8 and choosing V_{bc} as the reference, show that

$$V_{bc,0} = 0; \quad V_{bc,1} = -j\sqrt{3}V_{a,1}; \quad V_{bc,2} = j\sqrt{3}V_{a,2}.$$

8.10 Given the line-to-ground voltages $V_{ag} = 280\underline{/0^\circ}$, $V_{bg} = 250\underline{/-110^\circ}$, and $V_{cg} = 290\underline{/130^\circ}$ volts, calculate (a) the sequence components of the line-to-ground voltages, denoted $V_{\text{L}g0}$, $V_{\text{L}g1}$, and $V_{\text{L}g2}$; (b) line-to-line voltages V_{ab}, V_{bc}, and V_{ca}; and (c) sequence components of the line-to-line voltages $V_{\text{LL}0}$, $V_{\text{LL}1}$, and $V_{\text{LL}2}$. Also, verify the following general relation: $V_{\text{LL}0} = 0$, $V_{\text{LL}1} = \sqrt{3}V_{\text{L}g1}\underline{/+30^\circ}$, and $V_{\text{LL}2} = \sqrt{3}V_{\text{L}g2}\underline{/-30^\circ}$ volts.

8.11 A balanced Δ-connected load is fed by a three-phase supply for which phase C is open and phase A is carrying a current of $10\underline{/0^\circ}$ A. Find the symmetrical components of the line currents. (Note that zero-sequence currents are not present for any three-wire system.)

8.12 A Y-connected load bank with a three-phase rating of 500 kVA and 2300 V consists of three identical resistors of 10.58 Ω. The load bank has the following applied voltages: $V_{\text{ab}} = 1840\underline{/82.8^\circ}$, $V_{\text{bc}} = 2760\underline{/-41.4^\circ}$, and $V_{\text{ca}} = 2300\underline{/180^\circ}$ V. Determine the symmetrical components of (a) the line-to-line voltages $V_{\text{ab}0}$, $V_{\text{ab}1}$, and $V_{\text{ab}2}$; (b) the line-to-neutral voltages $V_{\text{an}0}$, $V_{\text{an}1}$, and $V_{\text{an}2}$; (c) and the line currents $I_{\text{a}0}$, $I_{\text{a}1}$, and $I_{\text{a}2}$. (Note that the absence of a neutral connection means that zero-sequence currents are not present.)

SECTION 8.2

8.13 The currents in a Δ load are $I_{ab} = 10\underline{/0^\circ}$, $I_{bc} = 15\underline{/-90^\circ}$, and $I_{ca} = 20\underline{/90^\circ}$ A. Calculate (a) the sequence components of the Δ-load currents, denoted $I_{\Delta 0}$, $I_{\Delta 1}$, $I_{\Delta 2}$; (b) the line currents I_a, I_b, and I_c, which feed the Δ load; and (c) sequence components of the line currents $I_{\text{L}0}$, $I_{\text{L}1}$, and $I_{\text{L}2}$. Also, verify the following general relation: $I_{\text{L}0} = 0$, $I_{\text{L}1} = \sqrt{3}I_{\Delta 1}\underline{/-30^\circ}$, and $I_{\text{L}2} = \sqrt{3}I_{\Delta 2}\underline{/+30^\circ}$ A.

8.14 The voltages given in Problem 8.10 are applied to a balanced-Y load consisting of $(12 + j16)$ ohms per phase. The load neutral is solidly grounded. Draw the sequence networks and calculate I_0, I_1, and I_2, the sequence components of the line currents. Then calculate the line currents I_a, I_b, and I_c.

8.15 Repeat Problem 8.14 with the load neutral open.

8.16 Repeat Problem 8.14 for a balanced-Δ load consisting of $(12 + j16)$ ohms per phase.

8.17 Repeat Problem 8.14 for the load shown in Example 8.4 (Figure 8.6).

8.18 Perform the indicated matrix multiplications in (8.2.21) and verify the sequence impedances given by (8.2.22)–(8.2.27).

8.19 The following unbalanced line-to-ground voltages are applied to the balanced-Y load shown in Figure 3.3: $V_{ag} = 100\angle 0°$, $V_{bg} = 75\angle 180°$, and $V_{cg} = 50\angle 90°$ volts. The Y load has $Z_Y = 3 + j4\ \Omega$ per phase with neutral impedance $Z_n = j1\ \Omega$. (a) Calculate the line currents I_a, I_b, and I_c without using symmetrical components. (b) Calculate the line currents I_a, I_b, and I_c using symmetrical components. Which method is easier?

8.20 (a) Consider three equal impedances of $(j27)\ \Omega$ connected in Δ. Obtain the sequence networks.
(b) Now, with a mutual impedance of $(j6)\ \Omega$ between each pair of adjacent branches in the Δ-connected load of part (a), how would the sequence networks change?

8.21 The three-phase impedance load shown in Figure 8.7 has the following phase impedance matrix:

$$Z_P = \begin{bmatrix} (6 + j10) & 0 & 0 \\ 0 & (6 + j10) & 0 \\ 0 & 0 & (6 + j10) \end{bmatrix} \Omega$$

Determine the sequence impedance matrix Z_S for this load. Is the load symmetrical?

8.22 The three-phase impedance load shown in Figure 8.7 has the following sequence impedance matrix:

$$Z_S = \begin{bmatrix} (8 + j12) & 0 & 0 \\ 0 & 4 & 0 \\ 0 & 0 & 4 \end{bmatrix} \Omega$$

Determine the phase impedance matrix Z_P for this load. Is the load symmetrical?

8.23 Consider a three-phase balanced Y-connected load with self and mutual impedances as shown in Figure 8.23. Let the load neutral be grounded through an impedance Z_n. Using Kirchhoff's laws, develop the equations for line-to-neutral voltages, and then determine the elements of the phase impedance matrix. Also find the elements of the corresponding sequence impedance matrix.

8.24 A three-phase balanced voltage source is applied to a balanced Y-connected load with ungrounded neutral. The Y-connected load consists of three mutually coupled reactances, where the reactance of each phase is $j12\ \Omega$ and the mutual coupling between any two phases is $j4\ \Omega$. The line-to-line source voltage is $100\sqrt{3}$ V. Determine the line currents (a) by mesh analysis without using symmetrical components, and (b) using symmetrical components.

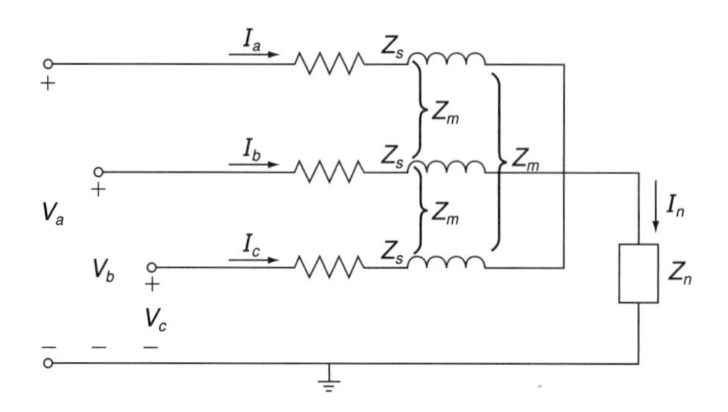

FIGURE 8.23 Problem 8.23

8.25 A three-phase balanced Y-connected load with series impedances of $(8 + j24)$ Ω per phase and mutual impedance between any two phases of $j4$ Ω is supplied by a three-phase unbalanced source with line-to-neutral voltages of $V_{an} = 200\underline{/25^\circ}$, $V_{bn} = 100\underline{/-155^\circ}$, $V_{cn} = 80\underline{/100^\circ}$ V. The load and source neutrals are both solidly grounded. Determine: (a) the load sequence impedance matrix, (b) the symmetrical components of the line-to-neutral voltages, (c) the symmetrical components of the load currents, and (d) the load currents.

SECTION 8.3

8.26 Repeat Problem 8.14 but include balanced three-phase line impedances of $(3 + j4)$ ohms per phase between the source and load.

8.27 Consider the flow of unbalanced currents in the symmetrical three-phase line section with neutral conductor as shown in Figure 8.24. (a) Express the voltage drops across the line conductors given by $V_{aa'}$, $V_{bb'}$, and $V_{cc'}$ in terms of line currents, self-impedances defined by $Z_s = Z_{aa} + Z_{nn} - 2Z_{an}$, and mutual impedances defined by $Z_m = Z_{ab} + Z_{nn} - 2Z_{an}$. (b) Show that the sequence components of the voltage drops between the ends of the line section can be written as $V_{aa'0} = Z_0 I_{a0}$, $V_{aa'1} = Z_1 I_{a1}$, and $V_{aa'2} = Z_2 I_{a2}$, where $Z_0 = Z_s + 2Z_m = Z_{aa} + 2Z_{ab} + 3Z_{nn} - 6Z_{an}$ and $Z_1 = Z_2 = Z_s - Z_m = Z_{aa} - Z_{ab}$.

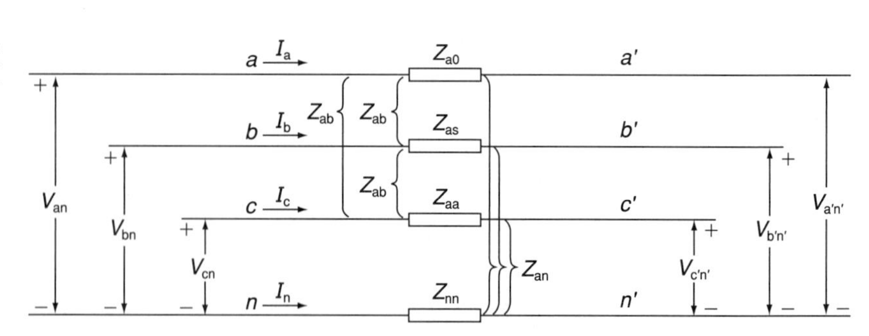

FIGURE 8.24 Problem 8.27

8.28 Let the terminal voltages at the two ends of the line section shown in Figure 8.24 be given by:

$$V_{an} = (182 + j70)\ \text{kV} \qquad V_{an'} = (154 + j28)\ \text{kV}$$
$$V_{bn} = (72.24 - j32.62)\ \text{kV} \qquad V_{bn'} = (44.24 + j74.62)\ \text{kV}$$
$$V_{cn} = (-170.24 + j88.62)\ \text{kV} \qquad V_{cn'} = (-198.24 + j46.62)\ \text{kV}$$

The line impedances are given by:

$$Z_{aa} = j60\ \Omega \qquad Z_{ab} = j20\ \Omega \qquad Z_{nn} = j80\ \Omega \qquad Z_{an} = 0$$

(a) Compute the line currents using symmetrical components. (*Hint:* See Problem 8.27.) (b) Compute the line currents without using symmetrical components.

8.29 A completely transposed three-phase transmission line of 200 km in length has the following symmetrical sequence impedances and sequence admittances:

$$Z_1 = Z_2 = j0.5\ \Omega/\text{km}; \quad Z_0 = j2\ \Omega/\text{km}$$
$$Y_1 = Y_2 = j3 \times 10^{-9}\ \text{s/m}; \quad Y_0 = j1 \times 10^{-9}\ \text{s/m}$$

Set up the nominal Π sequence circuits of this medium-length line.

SECTION 8.5

8.30 As shown in Figure 8.25, a balanced three-phase, positive-sequence source with $V_{AB} = 480\underline{/0^\circ}$ volts is applied to an unbalanced Δ load. Note that one leg of the Δ is open. Determine: (a) the load currents I_{AB} and I_{BC}; (b) the line currents I_A, I_B, and I_C, which feed the Δ load; and (c) the zero-, positive-, and negative-sequence components of the line currents.

FIGURE 8.25

Problem 8.30

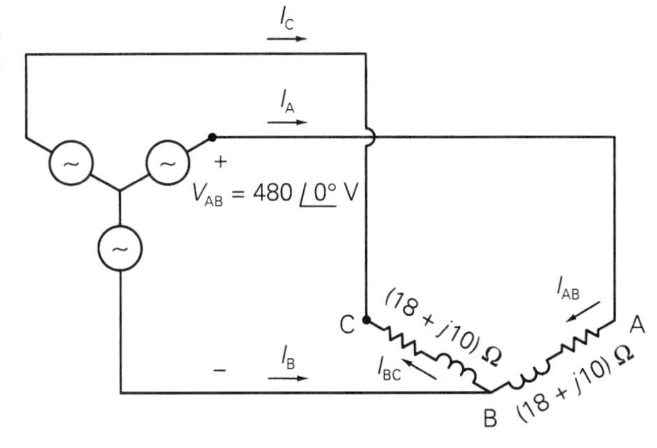

8.31 A balanced Y-connected generator with terminal voltage $V_{bc} = 200\underline{/0^\circ}$ volts is connected to a balanced-Δ load whose impedance is $10\underline{/40^\circ}$ ohms per phase. The line impedance between the source and load is $0.5\underline{/80^\circ}$ ohm for each phase. The generator neutral is grounded through an impedance of $j5$ ohms. The generator sequence impedances are given by $Z_{g0} = j7$, $Z_{g1} = j15$, and $Z_{g2} = j10$ ohms. Draw the sequence networks for this system and determine the sequence components of the line currents.

8.32 In a three-phase system, a synchronous generator supplies power to a 200-volt synchronous motor through a line having an impedance of $0.5\angle 80^\circ$ ohm per phase. The motor draws 5 kW at 0.8 p.f. leading and at rated voltage. The neutrals of both the generator and motor are grounded through impedances of $j5$ ohms. The sequence impedances of both machines are $Z_0 = j5$, $Z_1 = j15$, and $Z_2 = j10$ ohms. Draw the sequence networks for this system and find the line-to-line voltage at the generator terminals. Assume balanced three-phase operation.

8.33 Calculate the source currents in Example 8.6 without using symmetrical components. Compare your solution method with that of Example 8.6. Which method is easier?

8.34 A Y-connected synchronous generator rated 20 MVA at 13.8 kV has a positive-sequence reactance of $j2.38\ \Omega$, negative-sequence reactance of $j3.33\ \Omega$, and zero-sequence reactance of $j0.95\ \Omega$. The generator neutral is solidly grounded. With the generator operating unloaded at rated voltage, a so-called single line-to-ground fault occurs at the machine terminals. During this fault, the line-to-ground voltages at the generator terminals are $V_{ag} = 0$, $V_{bg} = 8.071\angle -102.25^\circ$, and $V_{cg} = 8.071\angle 102.25^\circ$ kV. Determine the sequence components of the generator fault currents and the generator fault currents. Draw a phasor diagram of the pre-fault and post-fault generator terminal voltages. (*Note:* For this fault, the sequence components of the generator fault currents are all equal to each other.)

8.35 Figure 8.26 shows a single-line diagram of a three-phase, interconnected generator-reactor system, in which the given per-unit reactances are based on the ratings of the individual pieces of equipment. If a three-phase short-circuit occurs at fault point F, obtain the fault MVA and fault current in kA, if the pre-fault busbar line-to-line voltage is 13.2 kV. Choose 100 MVA as the base MVA for the system.

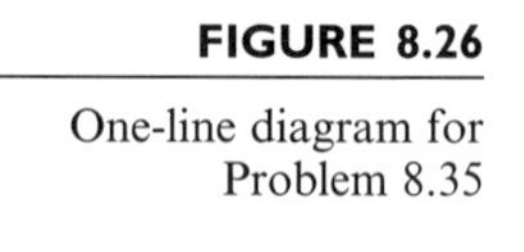

FIGURE 8.26 One-line diagram for Problem 8.35

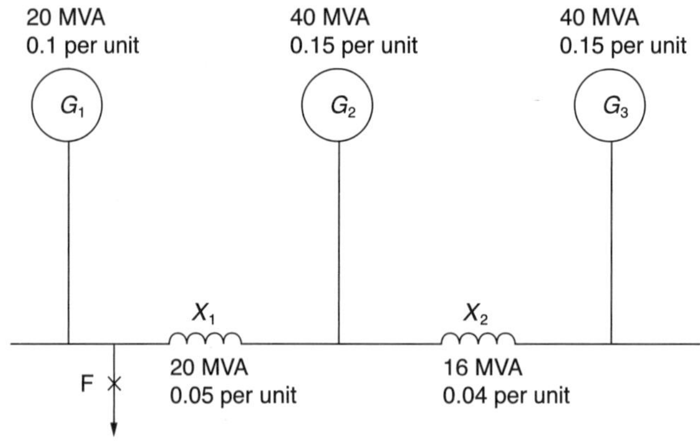

8.36 Consider Figures 8.13 and 8.14 of the text with reference to a Y-connected synchronous generator (grounded through a neutral impedance Z_n) operating at no load. For a line-to-ground fault occurring on phase a of the generator, list the constraints on the currents and voltages in the phase domain, transform those into the sequence domain, and then obtain a sequence-network representation. Also, find the expression for the fault current in phase a.

8.37 Reconsider the synchronous generator of Problem 8.36. Obtain sequence-network representations for the following fault conditions.
(a) A short-circuit between phases b and c.
(b) A double line-to-ground fault with phases b and c grounded.

SECTION 8.6

8.38 Three single-phase, two-winding transformers, each rated 450 MVA, 20 kV/288.7 kV, with leakage reactance $X_{eq} = 0.12$ per unit, are connected to form a three-phase bank. The high-voltage windings are connected in Y with a solidly grounded neutral. Draw the per-unit zero-, positive-, and negative-sequence networks if the low-voltage windings are connected: (a) in Δ with American standard phase shift, (b) in Y with an open neutral. Use the transformer ratings as base quantities. Winding resistances and exciting current are neglected.

8.39 The leakage reactance of a three-phase, 500-MVA, 345 Y/23 Δ-kV transformer is 0.09 per unit based on its own ratings. The Y winding has a solidly grounded neutral. Draw the sequence networks. Neglect the exciting admittance and assume American standard phase shift.

8.40 Choosing system bases to be 360/24 kV and 100 MVA, redraw the sequence networks for Problem 8.39.

8.41 Draw the zero-sequence reactance diagram for the power system shown in Figure 3.33. The zero-sequence reactance of each generator and of the synchronous motor is 0.05 per unit based on equipment ratings. Generator 2 is grounded through a neutral reactor of 0.06 per unit on a 100-MVA, 18-kV base. The zero-sequence reactance of each transmission line is assumed to be three times its positive-sequence reactance. Use the same base as in Problem 3.29.

8.42 Three identical Y-connected resistors of $1.0\angle 0^\circ$ per unit form a load bank, which is supplied from the low-voltage Y-side of a $Y - \Delta$ transformer. The neutral of the load is not connected to the neutral of the system. The positive- and negative-sequence currents flowing toward the resistive load are given by

$$I_{a,1} = 1\angle 4.5^\circ \text{ per unit;} \quad I_{a,2} = 0.25\angle 250^\circ \text{ per unit}$$

and the corresponding voltages on the low-voltage Y-side of the transformer are

$$V_{an,1} = 1\angle 45^\circ \text{ per unit (Line-to-neutral voltage base)}$$

$$V_{an,2} = 0.25\angle 250^\circ \text{ per unit (Line-to-neutral voltage base)}$$

Determine the line-to-line voltages and the line currents in per unit on the high-voltage side of the transformer. Account for the phase shift.

SECTION 8.7

8.43 Draw the positive-, negative-, and zero-sequence circuits for the transformers shown in Figure 3.34. Include ideal phase-shifting transformers showing phase shifts determined in Problem 3.32. Assume that all windings have the same kVA rating and that the equivalent leakage reactance of any two windings with the third winding open is 0.10 per unit. Neglect the exciting admittance.

8.44 A single-phase three-winding transformer has the following parameters: $Z_1 = Z_2 = Z_3 = 0 + j0.05$, $G_c = 0$, and $B_m = 0.2$ per unit. Three identical transformers, as described, are connected with their primaries in Y (solidly grounded neutral) and with their secondaries and tertiaries in Δ. Draw the per-unit sequence networks of this transformer bank.

SECTION 8.8

8.45 For Problem 8.14, calculate the real and reactive power delivered to the three-phase load.

8.46 A three-phase impedance load consists of a balanced-Δ load in parallel with a balanced-Y load. The impedance of each leg of the Δ load is $Z_\Delta = 6 + j6\ \Omega$, and the impedance of each leg of the Y load is $Z_Y = 2 + j2\ \Omega$. The Y load is grounded through a neutral impedance $Z_n = j1\ \Omega$. Unbalanced line-to-ground source voltages V_{ag}, V_{bg}, and V_{cg} with sequence components $V_0 = 10\angle 60°$, $V_1 = 100\angle 0°$, and $V_2 = 15\angle 200°$ volts are applied to the load. (a) Draw the zero-, positive-, and negative-sequence networks. (b) Determine the complex power delivered to each sequence network. (c) Determine the total complex power delivered to the three-phase load.

8.47 For Problem 8.12, compute the power absorbed by the load using symmetrical components. Then verify the answer by computing directly without using symmetrical components.

8.48 For Problem 8.25, determine the complex power delivered to the load in terms of symmetrical components. Verify the answer by adding up the complex power of each of the three phases.

8.49 Using the voltages of Problem 8.6(a) and the currents of Problem 8.5, compute the complex power dissipated based on (a) phase components, and (b) symmetrical components.

CASE STUDY QUESTIONS

A. What are the advantages of SF_6 circuit breakers for applications at or above 72.5 kV?

B. What are the properties of SF_6 that make it make it advantageous as a medium for interrupting an electric arc?

REFERENCES

1. Westinghouse Electric Corporation, *Applied Protective Relaying* (Newark, NJ: Westinghouse, 1976).

2. P. M. Anderson, *Analysis of Faulted Power Systems* (Ames, IA: Iowa State University Press, 1973).

3. W. D. Stevenson, Jr., *Elements of Power System Analysis*, 4th ed. (New York: McGraw-Hill, 1982).

4. D. Dufournet, "Circuit Breakers Go High Voltage," *IEEE Power & Energy Magazine*, 7, 1(January/February 2009), pp. 34–40.

The converter switch yard at Bonneville Power Administrations Celilo Converter Station in The Dallas, OR, USA. This station converts ac power to HVDC for transmission of up to 1,440 MW at ± 400 KV over an 856-m16 bipolar line between the Dallas, OR and Los Angeles, CA (AP Photo/Rick Bowmer/ CP Images)

9

UNSYMMETRICAL FAULTS

Short circuits occur in three-phase power systems as follows, in order of frequency of occurrence: single line-to-ground, line-to-line, double line-to-ground, and balanced three-phase faults. The path of the fault current may have either either zero impedance, which is called a *bolted* short circuit, or nonzero impedance. Other types of faults include one-conductor-open and two-conductors-open, which can occur when conductors break or when one or two phases of a circuit breaker inadvertently open.

Although the three-phase short circuit occurs the least, we considered it first, in Chapter 7, because of its simplicity. When a balanced three-phase fault occurs in a balanced three-phase system, there is only positive-sequence fault current; the zero-, positive-, and negative-sequence networks are completely uncoupled.

When an unsymmetrical fault occurs in an otherwise balanced system, the sequence networks are interconnected only at the fault location. As such, the computation of fault currents is greatly simplified by the use of sequence networks.

As in the case of balanced three-phase faults, unsymmetrical faults have two components of fault current: an ac or symmetrical component—including subtransient, transient, and steady-state currents—and a dc component. The simplified E/X method for breaker selection described in Section 7.5 is also applicable to unsymmetrical faults. The dc offset current need not be considered unless it is too large—for example, when the X/R ratio is too large.

We begin this chapter by using the per-unit zero-, positive-, and negative-sequence networks to represent a three-phase system. Also, we make certain assumptions to simplify fault-current calculations, and briefly review the balanced three-phase fault. We present single line-to-ground, line-to-line, and double line-to-ground faults in Sections 9.2, 9.3, and 9.4. The use of the positive-sequence bus impedance matrix for three-phase fault calculations in Section 7.4 is extended in Section 9.5 to unsymmetrical fault calculations by considering a bus impedance matrix for each sequence network. Examples using PowerWorld Simulator, which is based on the use of bus impedance matrices, are also included. The PowerWorld Simulator computes symmetrical fault currents for both three-phase and unsymmetrical faults. The Simulator may be used in power system design to select, set, and coordinate protective equipment.

CASE STUDY

When short circuits are not interrupted promptly, electrical fires and explosions can occur. To minimize the probability of electrical fire and explosion, the following are recommended:

Careful design of electric power system layouts
Quality equipment installation
Power system protection that provides rapid detection and isolation of faults (see Chapter 10)
Automatic fire-suppression systems
Formal maintenance programs and inspection intervals
Repair or retirement of damaged or decrepit equipment

The following article describes incidents at three U.S. utilities during the summer of 1990 [8].

Fires at U.S. Utilities

GLENN ZORPETTE

Electrical fires in substations were the cause of three major midsummer power outages in the United States, two on Chicago's West Side and one in New York City's downtown financial district. In Chicago, the trouble began Saturday night, July 28, with a fire in switch house No. 1 at the Commonwealth Edison Co.'s Crawford substation, according to spokesman Gary Wald.

Some 40,000 residents of Chicago's West Side lost electricity. About 25,000 had service restored within a day or so and the rest, within three days. However, as part of the restoration, Commonwealth Edison installed a temporary line configuration around the Crawford substation. But when a second fire broke out on Aug. 5 in a different, nearby substation, some of the protective systems that would have isolated that fire were inoperable because of that configuration. Thus, what would have been a minor mishap resulted in a one-day loss of power to 25,000 customers—the same 25,000 whose electricity was restored first after the Crawford fire.

The New York outage began around midday on Aug. 13, after an electrical fire broke out in switching equipment at Consolidated Edison's Seaport substation, a point of entry into Manhattan for five 138-kilovolt transmission lines. To interrupt the flow of energy to the fire, Edison had to disconnect the five lines, which cut power to four networks in downtown Manhattan, according to Con Ed spokeswoman Martha Liipfert.

Power was restored to three of the networks within about five hours, but the fourth network, Fulton—which carried electricity to about 2400 separate residences and 815 businesses—was out until Aug. 21. Liipfert said much of the equipment in the Seaport substation will have to be replaced, at an estimated cost of about $25 million.

Mounting concern about underground electrical vaults in some areas was tragically validated by an explosion in Pasadena, Calif., that killed three city workers in a vault. Partly in response to the explosion, the California Public Utilities Commission adopted new regulations last Nov. 21 requiring that utilities in the state set up formal maintenance programs, inspection intervals, and guidelines for rejecting decrepit or inferior equipment. "They have to maintain a paper trail, and we as a commission will do inspections of underground vaults and review their records to make sure they're maintaining their vaults and equipment in good order," said Russ Copeland, head of the commission's utility safety branch.

9.1 SYSTEM REPRESENTATION

A three-phase power system is represented by its sequence networks in this chapter. The zero-, positive-, and negative-sequence networks of system components—generators, motors, transformers, and transmission lines—as developed in Chapter 8 can be used to construct system zero-, positive-, and negative-sequence networks. We make the following assumptions:

1. The power system operates under balanced steady-state conditions before the fault occurs. Thus the zero-, positive-, and negative-sequence networks are uncoupled before the fault occurs. During unsymmetrical faults they are interconnected only at the fault location.

2. Prefault load current is neglected. Because of this, the positive-sequence internal voltages of all machines are equal to the prefault voltage V_F. Therefore, the prefault voltage at each bus in the positive-sequence network equals V_F.

3. Transformer winding resistances and shunt admittances are neglected.

4. Transmission-line series resistances and shunt admittances are neglected.

5. Synchronous machine armature resistance, saliency, and saturation are neglected.

6. All nonrotating impedance loads are neglected.

7. Induction motors are either neglected (especially for motors rated 50 hp (40 kW) or less) or represented in the same manner as synchronous machines.

Note that these assumptions are made for simplicity in this text, and in practice should not be made for all cases. For example, in primary and secondary distribution systems, prefault currents may be in some cases comparable to short-circuit currents, and in other cases line resistances may significantly reduce fault currents.

Although fault currents as well as contributions to fault currents on the fault side of Δ–Y transformers are not affected by Δ–Y phase shifts, contributions to the fault from the other side of such transformers are affected by Δ–Y phase shifts for unsymmetrical faults. Therefore, we include Δ–Y phase-shift effects in this chapter.

We consider faults at the general three-phase bus shown in Figure 9.1. Terminals *abc*, denoted the *fault terminals*, are brought out in order to make external connections that represent faults. Before a fault occurs, the currents I_a, I_b, and I_c are zero.

Figure 9.2(a) shows general sequence networks as viewed from the fault terminals. Since the prefault system is balanced, these zero-, positive-, and negative-sequence networks are uncoupled. Also, the sequence components of the fault currents, I_0, I_1, and I_2, are zero before a fault occurs. The general sequence networks in Figure 9.2(a) are reduced to their Thévenin equivalents as viewed from the fault terminals in Figure 9.2(b). Each sequence network has a Thévenin equivalent impedance. Also, the positive-sequence network has a Thévenin equivalent voltage source, which equals the prefault voltage V_F.

FIGURE 9.1

General three-phase bus

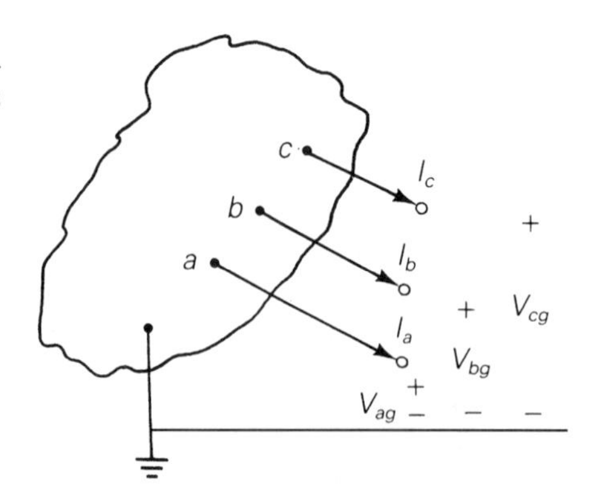

FIGURE 9.2

Sequence networks at a general three-phase bus in a balanced system

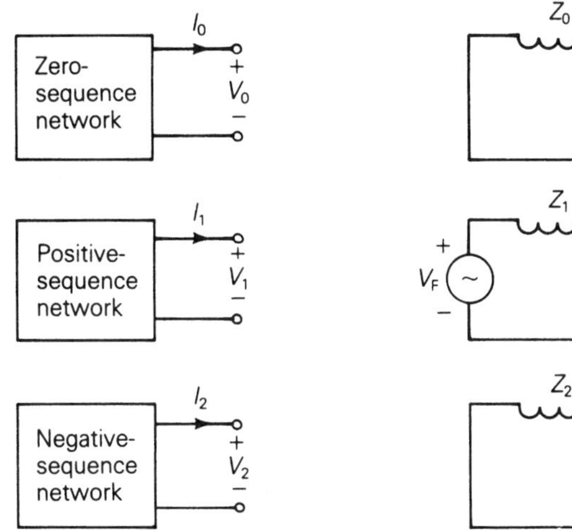

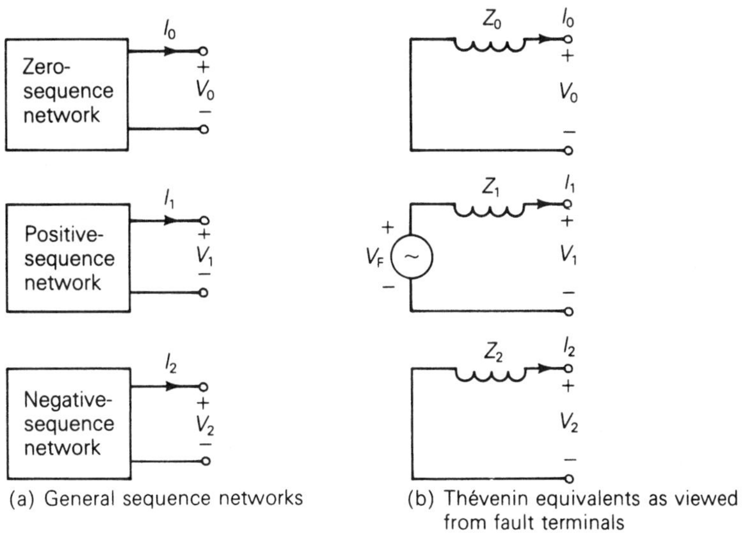

(a) General sequence networks

Z_0 I_0 + V_0 −

Z_1 I_1 + V_F − + V_1 −

Z_2 I_2 + V_2 −

(b) Thévenin equivalents as viewed from fault terminals

EXAMPLE 9.1 Power-system sequence networks and their Thévenin equivalents

A single-line diagram of the power system considered in Example 7.3 is shown in Figure 9.3, where negative- and zero-sequence reactances are also given. The neutrals of the generator and Δ–Y transformers are solidly grounded. The motor neutral is grounded through a reactance $X_n = 0.05$ per unit on the motor base. (a) Draw the per-unit zero-, positive-, and negative-sequence networks on a 100-MVA, 13.8-kV base in the zone of the generator. (b) Reduce the sequence networks to their Thévenin equivalents, as viewed from bus 2. Prefault voltage is $V_F = 1.05\underline{/0°}$ per unit. Prefault load current and Δ–Y transformer phase shift are neglected.

FIGURE 9.3

Single-line diagram for Example 9.1

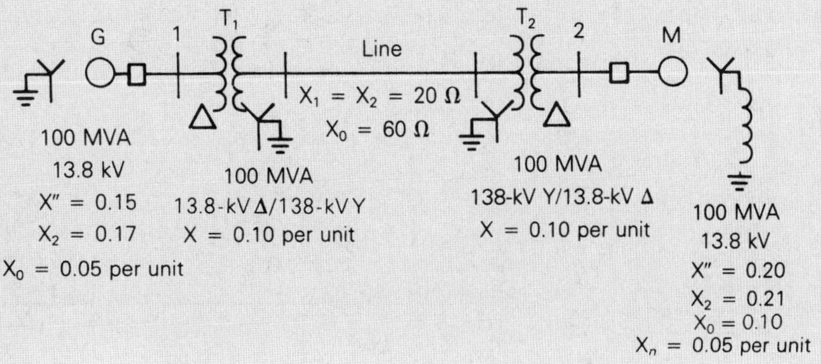

G 1 T_1 Line T_2 2 M

$X_1 = X_2 = 20\ \Omega$

$X_0 = 60\ \Omega$

100 MVA
13.8 kV
$X'' = 0.15$
$X_2 = 0.17$
$X_0 = 0.05$ per unit

100 MVA
13.8-kV Δ/138-kV Y
$X = 0.10$ per unit

100 MVA
138-kV Y/13.8-kV Δ
$X = 0.10$ per unit

100 MVA
13.8 kV
$X'' = 0.20$
$X_2 = 0.21$
$X_0 = 0.10$
$X_n = 0.05$ per unit

SOLUTION

a. The sequence networks are shown in Figure 9.4. The positive-sequence network is the same as that shown in Figure 7.4(a). The negative-sequence network is similar to the positive-sequence network, except that there are no sources, and negative-sequence machine reactances are shown. Δ–Y phase shifts are omitted from the positive- and negative-sequence networks for this example. In the zero-sequence network the zero-sequence generator, motor, and transmission-line reactances are shown. Since the motor neutral is grounded through a neutral reactance X_n, $3X_n$ is included in the zero-sequence motor circuit. Also, the zero-sequence Δ–Y transformer models are taken from Figure 8.19.

b. Figure 9.5 shows the sequence networks reduced to their Thévenin equivalents, as viewed from bus 2. For the positive-sequence equivalent, the Thévenin voltage source is the prefault voltage $V_F = 1.05\underline{/0°}$ per unit.

FIGURE 9.4

Sequence networks for Example 9.1

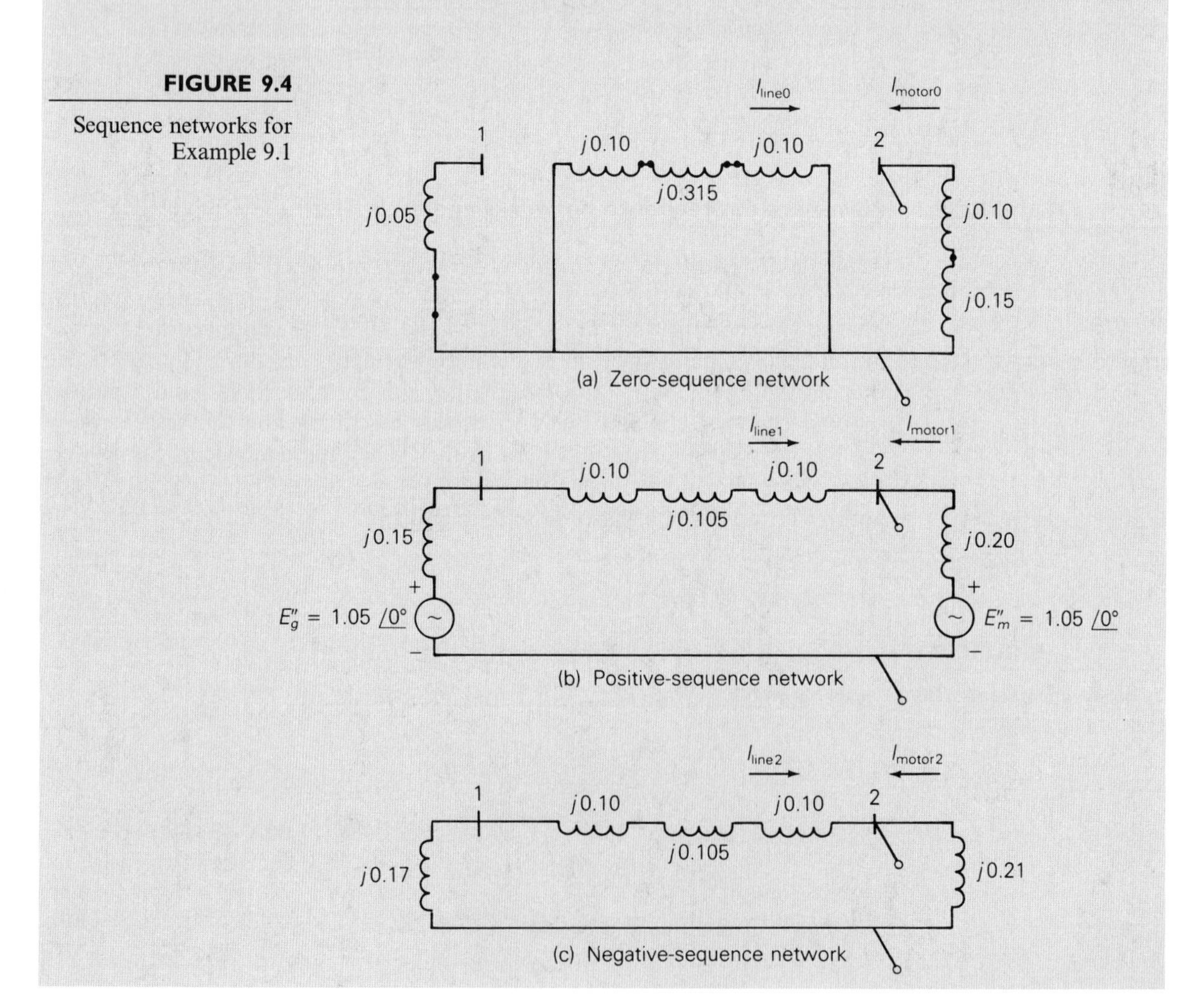

FIGURE 9.5

Thévenin equivalents of sequence networks for Example 9.1

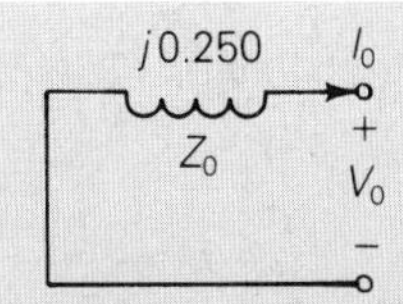

(a) Zero-sequence network

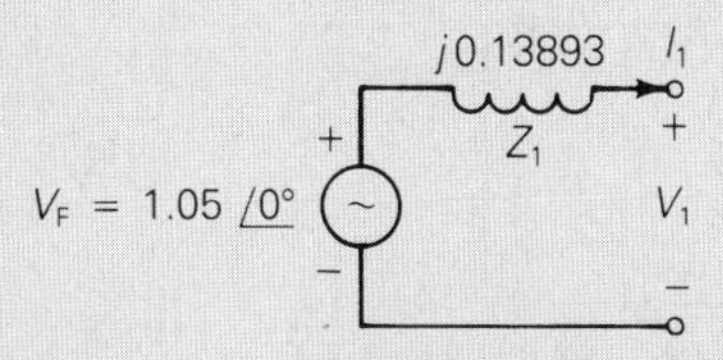

(b) Positive-sequence network

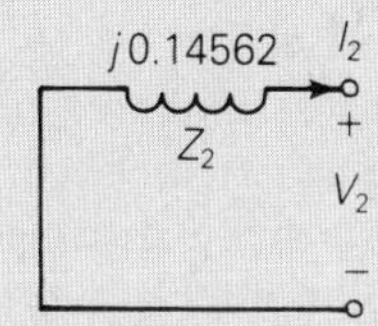

(c) Negative-sequence network

From Figure 9.4, the positive-sequence Thévenin impedance at bus 2 is the motor impedance $j0.20$, as seen to the right of bus 2, in parallel with $j(0.15 + 0.10 + 0.105 + 0.10) = j0.455$, as seen to the left; the parallel combination is $j0.20 /\!/ j0.455 = j0.13893$ per unit. Similarly, the negative-sequence Thévenin impedance is $j0.21 /\!/ j(0.17 + 0.10 + 0.105 + 0.10) = j0.21 /\!/ j0.475 = j0.14562$ per unit. In the zero-sequence network of Figure 9.4, the Thévenin impedance at bus 2 consists only of $j(0.10 + 0.15) = j0.25$ per unit, as seen to the right of bus 2; due to the Δ connection of transformer T_2, the zero-sequence network looking to the left of bus 2 is open. ■

Recall that for three-phase faults, as considered in Chapter 7, the fault currents are balanced and have only a positive-sequence component. Therefore we work only with the positive-sequence network when calculating three-phase fault currents.

EXAMPLE 9.2 Three-phase short-circuit calculations using sequence networks

Calculate the per-unit subtransient fault currents in phases a, b, and c for a bolted three-phase-to-ground short circuit at bus 2 in Example 9.1.

SOLUTION The terminals of the positive-sequence network in Figure 9.5(b) are shorted, as shown in Figure 9.6. The positive-sequence fault current is

FIGURE 9.6

Example 9.2: Bolted three-phase-to-ground fault at bus 2

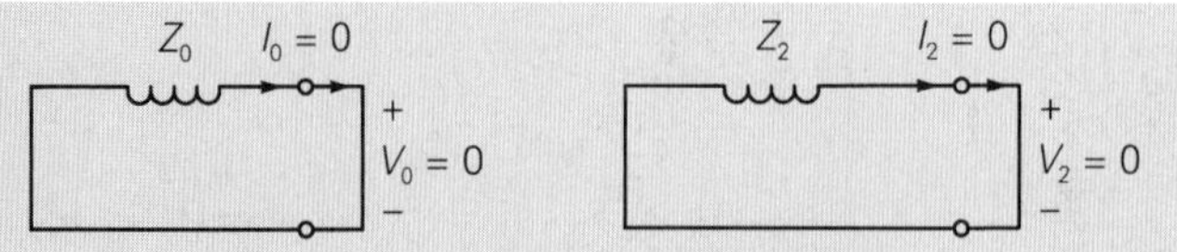

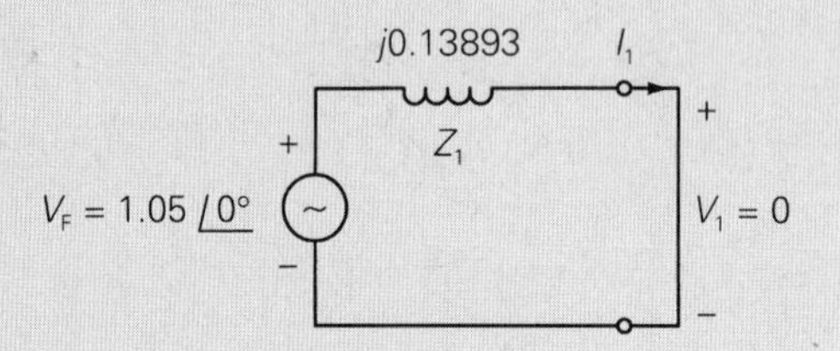

$$I_1 = \frac{V_F}{Z_1} = \frac{1.05\underline{/0°}}{j0.13893} = -j7.558 \quad \text{per unit}$$

which is the same result as obtained in part (c) of Example 7.4. Note that since subtransient machine reactances are used in Figures 9.4–9.6, the current calculated above is the positive-sequence subtransient fault current at bus 2. Also, the zero-sequence current I_0 and negative-sequence current I_2 are both zero. Therefore, the subtransient fault currents in each phase are, from (8.1.16),

$$\begin{bmatrix} I''_a \\ I''_b \\ I''_c \end{bmatrix} = \begin{bmatrix} 1 & 1 & 1 \\ 1 & a^2 & a \\ 1 & a & a^2 \end{bmatrix} \begin{bmatrix} 0 \\ -j7.558 \\ 0 \end{bmatrix} = \begin{bmatrix} 7.558\underline{/-90°} \\ 7.558\underline{/150°} \\ 7.558\underline{/30°} \end{bmatrix} \quad \text{per unit}$$

■

The sequence components of the line-to-ground voltages at the fault terminals are, from Figure 9.2(b),

$$\begin{bmatrix} V_0 \\ V_1 \\ V_2 \end{bmatrix} = \begin{bmatrix} 0 \\ V_F \\ 0 \end{bmatrix} - \begin{bmatrix} Z_0 & 0 & 0 \\ 0 & Z_1 & 0 \\ 0 & 0 & Z_2 \end{bmatrix} \begin{bmatrix} I_0 \\ I_1 \\ I_2 \end{bmatrix} \tag{9.1.1}$$

During a bolted three-phase fault, the sequence fault currents are $I_0 = I_2 = 0$ and $I_1 = V_F/Z_1$; therefore, from (9.1.1), the sequence fault voltages are $V_0 = V_1 = V_2 = 0$, which must be true since $V_{ag} = V_{bg} = V_{cg} = 0$. However, fault voltages need not be zero during unsymmetrical faults, which we consider next.

9.2 SINGLE LINE-TO-GROUND FAULT

Consider a single line-to-ground fault from phase a to ground at the general three-phase bus shown in Figure 9.7(a). For generality, we include a fault

FIGURE 9.7

Single line-to-ground fault

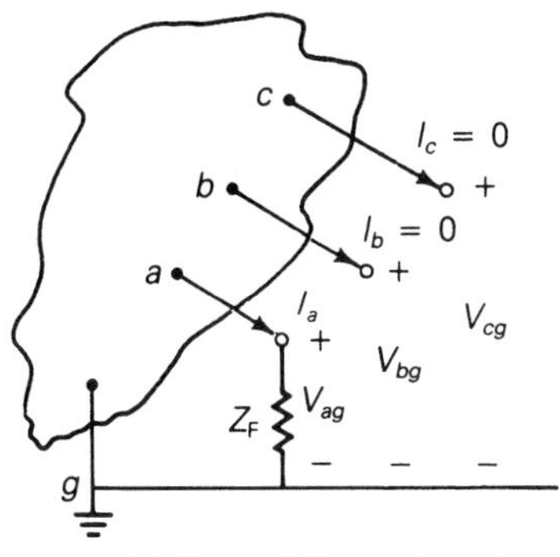

(a) General three-phase bus

Fault conditions in phase domain:

$V_{ag} = Z_F I_a$

$I_b = I_c = 0$

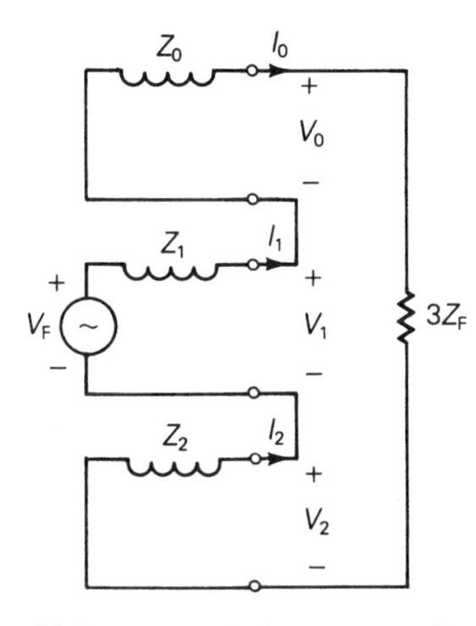

(b) Interconnected sequence networks

Fault conditions in sequence domain:

$I_0 = I_1 = I_2$

$(V_0 + V_1 + V_2) = 3Z_F I_1$

impedance Z_F. In the case of a bolted fault, $Z_F = 0$, whereas for an arcing fault, Z_F is the arc impedance. In the case of a transmission-line insulator flashover, Z_F includes the total fault impedance between the line and ground, including the impedances of the arc and the transmission tower, as well as the tower footing if there are no neutral wires.

The relations to be derived here apply only to a single line-to-ground fault on phase a. However, since any of the three phases can be arbitrarily labeled phase a, we do not consider single line-to-ground faults on other phases.

From Figure 9.7(a):

$$\left.\begin{array}{l}\text{Fault conditions in phase domain}\\ \text{Single line-to-ground fault}\end{array}\right\}\begin{array}{ll} I_b = I_c = 0 & (9.2.1)\\ V_{ag} = Z_F I_a & (9.2.2)\end{array}$$

We now transform (9.2.1) and (9.2.2) to the sequence domain. Using (9.2.1) in (8.1.19),

$$\begin{bmatrix} I_0 \\ I_1 \\ I_2 \end{bmatrix} = \frac{1}{3}\begin{bmatrix} 1 & 1 & 1 \\ 1 & a & a^2 \\ 1 & a^2 & a \end{bmatrix}\begin{bmatrix} I_a \\ 0 \\ 0 \end{bmatrix} = \frac{1}{3}\begin{bmatrix} I_a \\ I_a \\ I_a \end{bmatrix} \tag{9.2.3}$$

Also, using (8.1.3) and (8.1.20) in (9.2.2),

$$(V_0 + V_1 + V_2) = Z_F(I_0 + I_1 + I_2) \tag{9.2.4}$$

From (9.2.3) and (9.2.4):

Fault conditions in sequence domain
Single line-to-ground fault

$$\left.\begin{aligned} &I_0 = I_1 = I_2 &(9.2.5)\\ &(V_0 + V_1 + V_2) = (3Z_F)I_1 &(9.2.6)\end{aligned}\right\}$$

Equations (9.2.5) and (9.2.6) can be satisfied by interconnecting the sequence networks in series at the fault terminals through the impedance $(3Z_F)$, as shown in Figure 9.7(b). From this figure, the sequence components of the fault currents are:

$$I_0 = I_1 = I_2 = \frac{V_F}{Z_0 + Z_1 + Z_2 + (3Z_F)} \tag{9.2.7}$$

Transforming (9.2.7) to the phase domain via (8.1.20),

$$I_a = I_0 + I_1 + I_2 = 3I_1 = \frac{3V_F}{Z_0 + Z_1 + Z_2 + (3Z_F)} \tag{9.2.8}$$

Note also from (8.1.21) and (8.1.22),

$$I_b = (I_0 + a^2 I_1 + aI_2) = (1 + a^2 + a)I_1 = 0 \tag{9.2.9}$$

$$I_c = (I_0 + aI_1 + a^2 I_2) = (1 + a + a^2)I_1 = 0 \tag{9.2.10}$$

These are obvious, since the single line-to-ground fault is on phase a, not phase b or c.

The sequence components of the line-to-ground voltages at the fault are determined from (9.1.1). The line-to-ground voltages at the fault can then be obtained by transforming the sequence voltages to the phase domain.

EXAMPLE 9.3 Single line-to-ground short-circuit calculations using sequence networks

Calculate the subtransient fault current in per-unit and in kA for a bolted single line-to-ground short circuit from phase a to ground at bus 2 in Example 9.1. Also calculate the per-unit line-to-ground voltages at faulted bus 2.

SOLUTION The zero-, positive-, and negative-sequence networks in Figure 9.5 are connected in series at the fault terminals, as shown in Figure 9.8.

FIGURE 9.8

Example 9.3: Single line-to-ground fault at bus 2

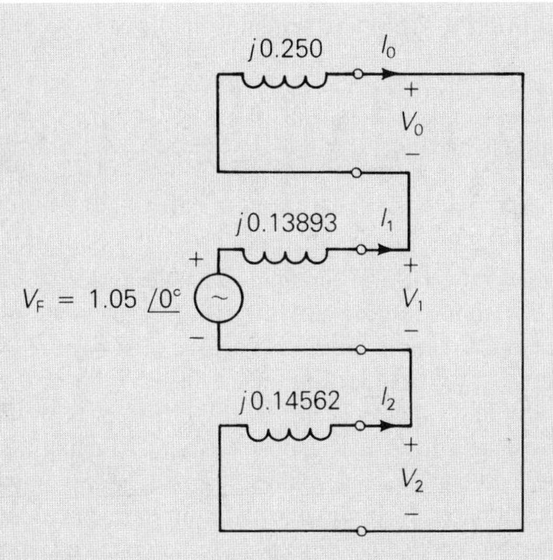

Since the short circuit is bolted, $Z_F = 0$. From (9.2.7), the sequence currents are:

$$I_0 = I_1 = I_2 = \frac{1.05\underline{/0°}}{j(0.25 + 0.13893 + 0.14562)}$$

$$= \frac{1.05}{j0.53455} = -j1.96427 \quad \text{per unit}$$

From (9.2.8), the subtransient fault current is

$$I_a'' = 3(-j1.96427) = -j5.8928 \quad \text{per unit}$$

The base current at bus 2 is $100/(13.8\sqrt{3}) = 4.1837$ kA. Therefore,

$$I_a'' = (-j5.8928)(4.1837) = 24.65\underline{/-90°} \quad \text{kA}$$

From (9.1.1), the sequence components of the voltages at the fault are

$$\begin{bmatrix} V_0 \\ V_1 \\ V_2 \end{bmatrix} = \begin{bmatrix} 0 \\ 1.05\underline{/0°} \\ 0 \end{bmatrix} - \begin{bmatrix} j0.25 & 0 & 0 \\ 0 & j0.13893 & 0 \\ 0 & 0 & j0.14562 \end{bmatrix} \begin{bmatrix} -j1.96427 \\ -j1.96427 \\ -j1.96427 \end{bmatrix}$$

$$= \begin{bmatrix} -0.49107 \\ 0.77710 \\ -0.28604 \end{bmatrix} \quad \text{per unit}$$

Transforming to the phase domain, the line-to-ground voltages at faulted bus 2 are

$$\begin{bmatrix} V_{ag} \\ V_{bg} \\ V_{cg} \end{bmatrix} = \begin{bmatrix} 1 & 1 & 1 \\ 1 & a^2 & a \\ 1 & a & a^2 \end{bmatrix} \begin{bmatrix} -0.49107 \\ 0.77710 \\ -0.28604 \end{bmatrix} = \begin{bmatrix} 0 \\ 1.179\underline{/231.3^\circ} \\ 1.179\underline{/128.7^\circ} \end{bmatrix} \quad \text{per unit}$$

Note that $V_{ag} = 0$, as specified by the fault conditions. Also $I_b'' = I_c'' = 0$.

Open PowerWorld Simulator case Example 9_3 to see this example. The process for simulating an unsymmetrical fault is almost identical to that for a balanced fault. That is, from the one-line, first right-click on the bus symbol corresponding to the fault location. This displays the local menu. Select **"Fault.."** to display the Fault dialog. Verify that the correct bus is selected, and then set the Fault Type field to "Single Line-to-Ground." Finally, click on **Calculate** to determine the fault currents and voltages. The results are shown in the tables at the bottom of the dialog. Notice that with an unsymmetrical fault the phase magnitudes are no longer identical. The values can be animated on the one line by changing the Oneline Display field value, which is shown on the Fault Options page.

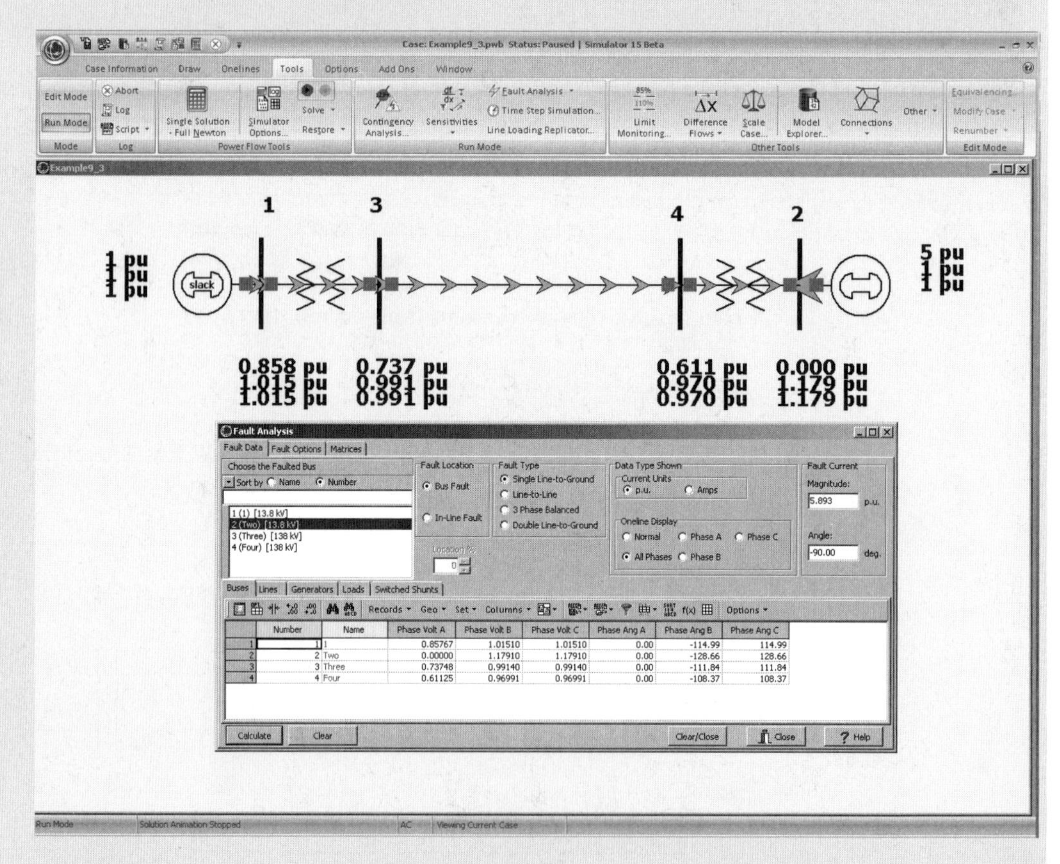

FIGURE 9.9 Screen for Example 9.3–fault at bus 2 ■

9.3

LINE-TO-LINE FAULT

Consider a line-to-line fault from phase b to c, shown in Figure 9.10(a). Again, we include a fault impedance Z_F for generality. From Figure 9.10(a):

$$\left.\begin{array}{l}\text{Fault conditions in phase domain}\\ \text{Line-to-line fault}\end{array}\right\} \begin{array}{l} I_a = 0 \quad (9.3.1)\\ I_c = -I_b \quad (9.3.2)\end{array}$$

$$V_{bg} - V_{cg} = Z_F I_b \tag{9.3.3}$$

We transform (9.3.1)–(9.3.3) to the sequence domain. Using (9.3.1) and (9.3.2) in (8.1.19),

FIGURE 9.10

Line-to-line fault

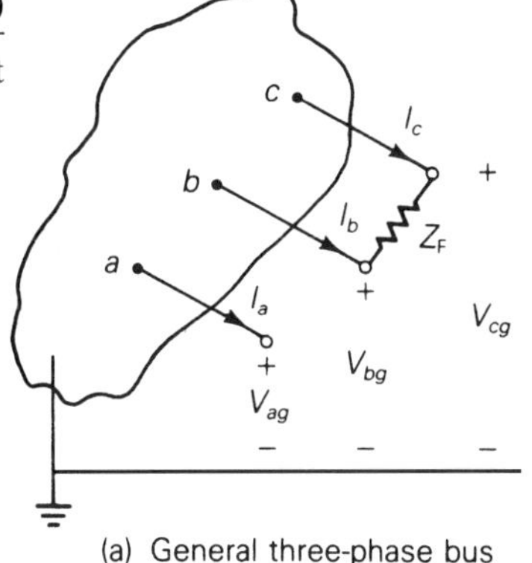

Fault conditions in phase domain:

$I_a = 0$

$I_c = -I_b$

$(V_{bg} - V_{cg}) = Z_F I_b$

(a) General three-phase bus

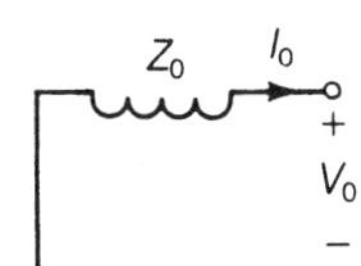

Fault conditions in sequence domain:

$I_0 = 0$

$I_2 = -I_1$

$(V_1 - V_2) = Z_F I_1$

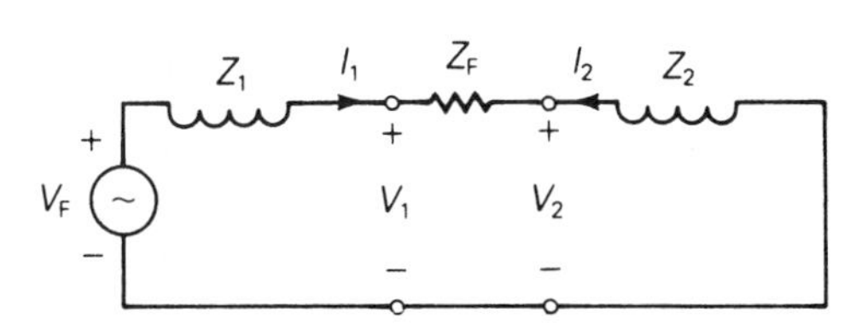

(b) Interconnected sequence networks

$$\begin{bmatrix} I_0 \\ I_1 \\ I_2 \end{bmatrix} = \frac{1}{3}\begin{bmatrix} 1 & 1 & 1 \\ 1 & a & a^2 \\ 1 & a^2 & a \end{bmatrix}\begin{bmatrix} 0 \\ I_b \\ -I_b \end{bmatrix} = \begin{bmatrix} 0 \\ \frac{1}{3}(a-a^2)I_b \\ \frac{1}{3}(a^2-a)I_b \end{bmatrix} \tag{9.3.4}$$

Using (8.1.4), (8.1.5), and (8.1.21) in (9.3.3),

$$(V_0 + a^2V_1 + aV_2) - (V_0 + aV_1 + a^2V_2) = Z_F(I_0 + a^2I_1 + aI_2) \tag{9.3.5}$$

Noting from (9.3.4) that $I_0 = 0$ and $I_2 = -I_1$, (9.3.5) simplifies to

$$(a^2 - a)V_1 - (a^2 - a)V_2 = Z_F(a^2 - a)I_1$$

or

$$V_1 - V_2 = Z_F I_1 \tag{9.3.6}$$

Therefore, from (9.3.4) and (9.3.6):

Fault conditions in sequence domain / Line-to-line fault

$$I_0 = 0 \tag{9.3.7}$$

$$I_2 = -I_1 \tag{9.3.8}$$

$$V_1 - V_2 = Z_F I_1 \tag{9.3.9}$$

Equations (9.3.7)–(9.3.9) are satisfied by connecting the positive- and negative-sequence networks in parallel at the fault terminals through the fault impedance Z_F, as shown in Figure 9.10(b). From this figure, the fault currents are:

$$I_1 = -I_2 = \frac{V_F}{(Z_1 + Z_2 + Z_F)} \qquad I_0 = 0 \tag{9.3.10}$$

Transforming (9.3.10) to the phase domain and using the identity $(a^2 - a) = -j\sqrt{3}$, the fault current in phase b is

$$\begin{aligned} I_b &= I_0 + a^2I_1 + aI_2 = (a^2 - a)I_1 \\ &= -j\sqrt{3}I_1 = \frac{-j\sqrt{3}V_F}{(Z_1 + Z_2 + Z_F)} \end{aligned} \tag{9.3.11}$$

Note also from (8.1.20) and (8.1.22) that

$$I_a = I_0 + I_1 + I_2 = 0 \tag{9.3.12}$$

and

$$I_c = I_0 + aI_1 + a^2I_2 = (a - a^2)I_1 = -I_b \tag{9.3.13}$$

which verify the fault conditions given by (9.3.1) and (9.3.2). The sequence components of the line-to-ground voltages at the fault are given by (9.1.1).

EXAMPLE 9.4 Line-to-line short-circuit calculations using sequence networks

Calculate the subtransient fault current in per-unit and in kA for a bolted line-to-line fault from phase b to c at bus 2 in Example 9.1.

FIGURE 9.11

Example 9.4: Line-to-line fault at bus 2

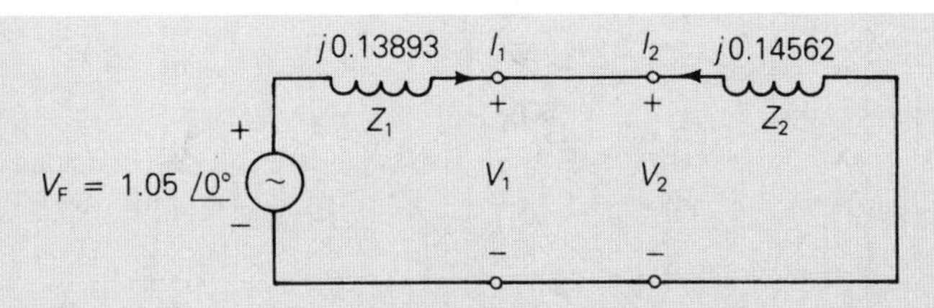

SOLUTION The positive- and negative-sequence networks in Figure 9.5 are connected in parallel at the fault terminals, as shown in Figure 9.11. From (9.3.10) with $Z_F = 0$, the sequence fault currents are

$$I_1 = -I_2 = \frac{1.05\underline{/0^\circ}}{j(0.13893 + 0.14562)} = 3.690\underline{/-90^\circ}$$

$$I_0 = 0$$

From (9.3.11), the subtransient fault current in phase b is

$$I_b'' = (-j\sqrt{3})(3.690\underline{/-90^\circ}) = -6.391 = 6.391\underline{/180^\circ} \quad \text{per unit}$$

Using 4.1837 kA as the base current at bus 2,

$$I_b'' = (6.391\underline{/180^\circ})(4.1837) = 26.74\underline{/180^\circ} \quad \text{kA}$$

Also, from (9.3.12) and (9.3.13),

$$I_a'' = 0 \qquad I_c'' = 26.74\underline{/0^\circ} \quad \text{kA}$$

The line-to-line fault results for this example can be shown in PowerWorld Simulator by repeating the Example 9.3 procedure, with the exception that the Fault Type field value should be "Line-to-Line." ■

9.4 DOUBLE LINE-TO-GROUND FAULT

A double line-to-ground fault from phase b to phase c through fault impedance Z_F to ground is shown in Figure 9.12(a). From this figure:

Fault conditions in the phase domain, Double line-to-ground fault:

$$I_a = 0 \tag{9.4.1}$$

$$V_{cg} = V_{bg} \tag{9.4.2}$$

$$V_{bg} = Z_F(I_b + I_c) \tag{9.4.3}$$

Transforming (9.4.1) to the sequence domain via (8.1.20),

$$I_0 + I_1 + I_2 = 0 \tag{9.4.4}$$

Also, using (8.1.4) and (8.1.5) in (9.4.2),

$$(V_0 + aV_1 + a^2V_2) = (V_0 + a^2V_1 + aV_2)$$

FIGURE 9.12

Double line-to-ground fault

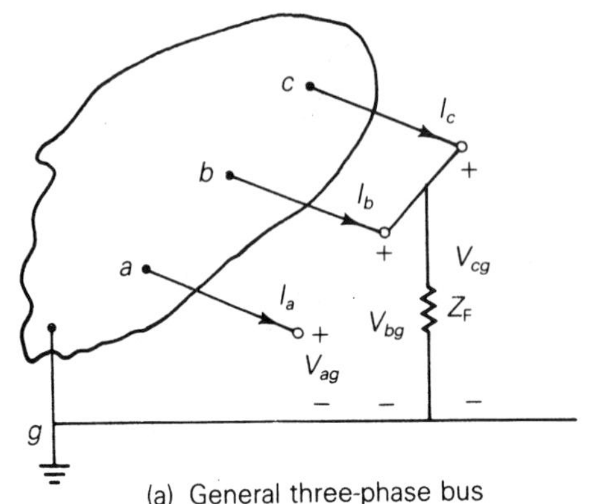

(a) General three-phase bus

Fault conditions in phase domain:

$I_a = 0$

$V_{bg} = V_{cg} = Z_F(I_b + I_c)$

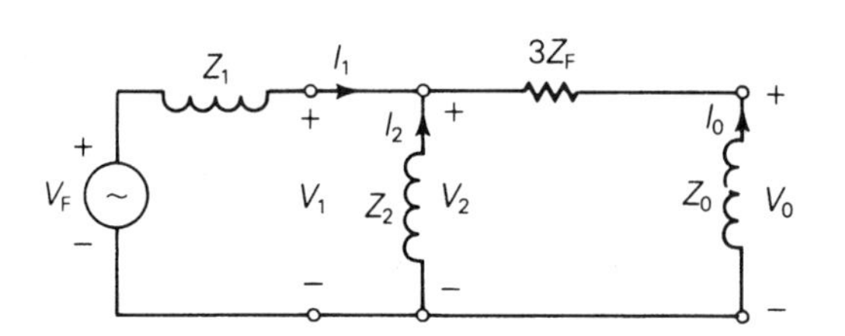

(b) Interconnected sequence networks

Fault conditions in sequence domain:

$I_0 + I_1 + I_2 = 0$

$V_0 - V_1 = (3Z_F)I_0$

$V_1 = V_2$

Simplifying:

$$(a^2 - a)V_2 = (a^2 - a)V_1$$

or

$$V_2 = V_1 \tag{9.4.5}$$

Now, using (8.1.4), (8.1.21), and (8.1.22) in (9.4.3),

$$(V_0 + a^2V_1 + aV_2) = Z_F(I_0 + a^2I_1 + aI_2 + I_0 + aI_1 + a^2I_2) \tag{9.4.6}$$

Using (9.4.5) and the identity $a^2 + a = -1$ in (9.4.6),

$$(V_0 - V_1) = Z_F(2I_0 - I_1 - I_2) \tag{9.4.7}$$

From (9.4.4), $I_0 = -(I_1 + I_2)$; therefore, (9.4.7) becomes

$$V_0 - V_1 = (3Z_F)I_0 \tag{9.4.8}$$

From (9.4.4), (9.4.5), and (9.4.8), we summarize:

Fault conditions in the sequence domain
Double line-to-ground fault

$$I_0 + I_1 + I_2 = 0 \tag{9.4.9}$$

$$V_2 = V_1 \tag{9.4.10}$$

$$V_0 - V_1 = (3Z_F)I_0 \tag{9.4.11}$$

Equations (9.4.9)–(9.4.11) are satisfied by connecting the zero-, positive-, and negative-sequence networks in parallel at the fault terminal; additionally, $(3Z_F)$ is included in series with the zero-sequence network. This connection is shown in Figure 9.12(b). From this figure the positive-sequence fault current is

$$I_1 = \frac{V_F}{Z_1 + [Z_2 /\!/ (Z_0 + 3Z_F)]} = \frac{V_F}{Z_1 + \left[\dfrac{Z_2(Z_0 + 3Z_F)}{Z_2 + Z_0 + 3Z_F}\right]} \tag{9.4.12}$$

Using current division in Figure 9.12(b), the negative- and zero-sequence fault currents are

$$I_2 = (-I_1)\left(\frac{Z_0 + 3Z_F}{Z_0 + 3Z_F + Z_2}\right) \tag{9.4.13}$$

$$I_0 = (-I_1)\left(\frac{Z_2}{Z_0 + 3Z_F + Z_2}\right) \tag{9.4.14}$$

These sequence fault currents can be transformed to the phase domain via (8.1.16). Also, the sequence components of the line-to-ground voltages at the fault are given by (9.1.1).

EXAMPLE 9.5 Double line-to-ground short-circuit calculations using sequence networks

Calculate (a) the subtransient fault current in each phase, (b) neutral fault current, and (c) contributions to the fault current from the motor and from the transmission line, for a bolted double line-to-ground fault from phase b to c to ground at bus 2 in Example 9.1. Neglect the Δ–Y transformer phase shifts.

SOLUTION

a. The zero-, positive-, and negative-sequence networks in Figure 9.5 are connected in parallel at the fault terminals in Figure 9.13. From (9.4.12) with $Z_F = 0$,

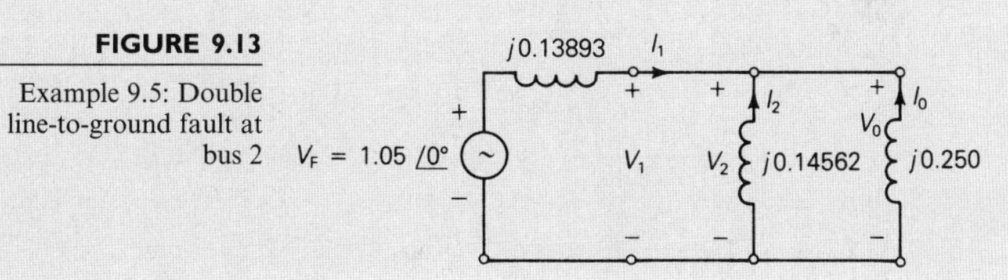

FIGURE 9.13 Example 9.5: Double line-to-ground fault at bus 2

$$I_1 = \frac{1.05\underline{/0^\circ}}{j\left[0.13893 + \dfrac{(0.14562)(0.25)}{0.14562 + 0.25}\right]} = \frac{1.05\underline{/0^\circ}}{j0.23095}$$

$$= -j4.5464 \quad \text{per unit}$$

From (9.4.13) and (9.4.14),

$$I_2 = (+j4.5464)\left(\frac{0.25}{0.25 + 0.14562}\right) = j2.8730 \quad \text{per unit}$$

$$I_0 = (+j4.5464)\left(\frac{0.14562}{0.25 + 0.14562}\right) = j1.6734 \quad \text{per unit}$$

Transforming to the phase domain, the subtransient fault currents are:

$$\begin{bmatrix} I_a'' \\ I_b'' \\ I_c'' \end{bmatrix} = \begin{bmatrix} 1 & 1 & 1 \\ 1 & a^2 & a \\ 1 & a & a^2 \end{bmatrix} \begin{bmatrix} +j1.6734 \\ -j4.5464 \\ +j2.8730 \end{bmatrix} = \begin{bmatrix} 0 \\ 6.8983\underline{/158.66^\circ} \\ 6.8983\underline{/21.34^\circ} \end{bmatrix} \quad \text{per unit}$$

Using the base current of 4.1837 kA at bus 2,

$$\begin{bmatrix} I_a'' \\ I_b'' \\ I_c'' \end{bmatrix} = \begin{bmatrix} 0 \\ 6.8983\underline{/158.66^\circ} \\ 6.8983\underline{/21.34^\circ} \end{bmatrix} (4.1837) = \begin{bmatrix} 0 \\ 28.86\underline{/158.66^\circ} \\ 28.86\underline{/21.34^\circ} \end{bmatrix} \quad \text{kA}$$

b. The neutral fault current is

$$I_n = (I_b'' + I_c'') = 3I_0 = j5.0202 \quad \text{per unit}$$

$$= (j5.0202)(4.1837) = 21.00\underline{/90^\circ} \quad \text{kA}$$

c. Neglecting Δ–Y transformer phase shifts, the contributions to the fault current from the motor and transmission line can be obtained from Figure 9.4. From the zero-sequence network, Figure 9.4(a), the contribution to the zero-sequence fault current from the line is zero, due to the transformer connection. That is,

$$I_{\text{line } 0} = 0$$

$$I_{\text{motor } 0} = I_0 = j1.6734 \quad \text{per unit}$$

From the positive-sequence network, Figure 9.4(b), the positive terminals of the internal machine voltages can be connected, since $E_g'' = E_m''$. Then, by current division,

$$I_{\text{line } 1} = \frac{X_m''}{X_m'' + (X_g'' + X_{T1} + X_{\text{line } 1} + X_{T2})} I_1$$

$$= \frac{0.20}{0.20 + (0.455)}(-j4.5464) = -j1.3882 \quad \text{per unit}$$

$$I_{\text{motor } 1} = \frac{0.455}{0.20 + 0.455}(-j4.5464) = -j3.1582 \quad \text{per unit}$$

From the negative-sequence network, Figure 9.4(c), using current division,

$$I_{\text{line }2} = \frac{0.21}{0.21 + 0.475}(j2.8730) = j0.8808 \quad \text{per unit}$$

$$I_{\text{motor }2} = \frac{0.475}{0.21 + 0.475}(j2.8730) = j1.9922 \quad \text{per unit}$$

Transforming to the phase domain with base currents of 0.41837 kA for the line and 4.1837 kA for the motor,

$$\begin{bmatrix} I''_{\text{line }a} \\ I''_{\text{line }b} \\ I''_{\text{line }c} \end{bmatrix} = \begin{bmatrix} 1 & 1 & 1 \\ 1 & a^2 & a \\ 1 & a & a^2 \end{bmatrix} \begin{bmatrix} 0 \\ -j1.3882 \\ j0.8808 \end{bmatrix}$$

$$= \begin{bmatrix} 0.5074\underline{/-90^\circ} \\ 1.9813\underline{/172.643^\circ} \\ 1.9813\underline{/7.357^\circ} \end{bmatrix} \quad \text{per unit}$$

$$= \begin{bmatrix} 0.2123\underline{/-90^\circ} \\ 0.8289\underline{/172.643^\circ} \\ 0.8289\underline{/7.357^\circ} \end{bmatrix} \quad \text{kA}$$

$$\begin{bmatrix} I''_{\text{motor }a} \\ I''_{\text{motor }b} \\ I''_{\text{motor }c} \end{bmatrix} = \begin{bmatrix} 1 & 1 & 1 \\ 1 & a^2 & a \\ 1 & a & a^2 \end{bmatrix} \begin{bmatrix} j1.6734 \\ -j3.1582 \\ j1.9922 \end{bmatrix}$$

$$= \begin{bmatrix} 0.5074\underline{/90^\circ} \\ 4.9986\underline{/153.17^\circ} \\ 4.9986\underline{/26.83^\circ} \end{bmatrix} \quad \text{per unit}$$

$$= \begin{bmatrix} 2.123\underline{/90^\circ} \\ 20.91\underline{/153.17^\circ} \\ 20.91\underline{/26.83^\circ} \end{bmatrix} \quad \text{kA}$$

The double line-to-line fault results for this example can be shown in PowerWorld Simulator by repeating the Example 9.3 procedure, with the exception that the Fault Type field value should be "Double Line-to-Ground." ■

EXAMPLE 9.6 Effect of Δ–Y transformer phase shift on fault currents

Rework Example 9.5, with the Δ–Y transformer phase shifts included. Assume American standard phase shift.

SOLUTION The sequence networks of Figure 9.4 are redrawn in Figure 9.14 with ideal phase-shifting transformers representing Δ–Y phase shifts. In

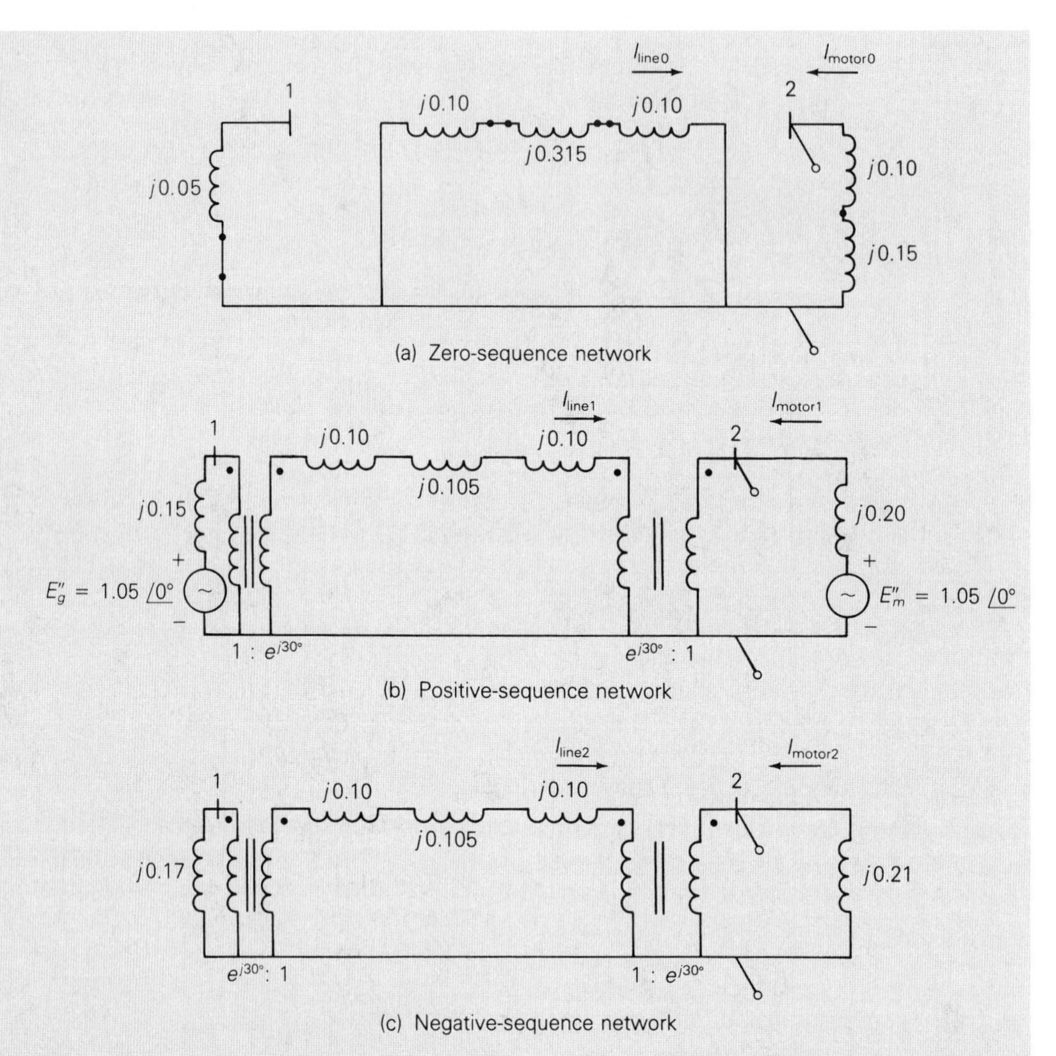

FIGURE 9.14 Sequence networks for Example 9.6

accordance with the American standard, positive-sequence quantities on the high-voltage side of the transformers lead their corresponding quantities on the low-voltage side by 30°. Also, the negative-sequence phase shifts are the reverse of the positive-sequence phase shifts.

a. Recall from Section 3.1 and (3.1.26) that per-unit impedance is unchanged when it is referred from one side of an ideal phase-shifting transformer to the other. Accordingly, the Thévenin equivalents of the sequence networks in Figure 9.14, as viewed from fault bus 2, are the same as those given in Figure 9.5. Therefore, the sequence components as well as the phase components of the fault currents are the same as those given in Example 9.5(a).

b. The neutral fault current is the same as that given in Example 9.5(b).

c. The zero-sequence network, Figure 9.14(a), is the same as that given in Figure 9.4(a). Therefore, the contributions to the zero-sequence fault current from the line and motor are the same as those given in Example 9.5(c).

$$I_{\text{line }0} = 0 \qquad I_{\text{motor }0} = I_0 = j1.6734 \quad \text{per unit}$$

The contribution to the positive-sequence fault current from the line in Figure 9.13(b) leads that in Figure 9.4(b) by 30°. That is,

$$I_{\text{line }1} = (-j1.3882)(1\underline{/30^\circ}) = 1.3882\underline{/-60^\circ} \quad \text{per unit}$$

$$I_{\text{motor }1} = -j3.1582 \quad \text{per unit}$$

Similarly, the contribution to the negative-sequence fault current from the line in Figure 9.14(c) lags that in Figure 9.4(c) by 30°. That is,

$$I_{\text{line }2} = (j0.8808)(1\underline{/-30^\circ}) = 0.8808\underline{/60^\circ} \quad \text{per unit}$$

$$I_{\text{motor }2} = j1.9922 \quad \text{per unit}$$

Thus, the sequence currents as well as the phase currents from the motor are the same as those given in Example 9.5(c). Also, the sequence currents from the line have the same magnitudes as those given in Example 9.5(c), but the positive- and negative-sequence line currents are shifted by +30° and −30°, respectively. Transforming the line currents to the phase domain:

$$\begin{bmatrix} I''_{\text{line }a} \\ I''_{\text{line }b} \\ I''_{\text{line }c} \end{bmatrix} = \begin{bmatrix} 1 & 1 & 1 \\ 1 & a^2 & a \\ 1 & a & a^2 \end{bmatrix} \begin{bmatrix} 0 \\ 1.3882\underline{/-60^\circ} \\ 0.8808\underline{/60^\circ} \end{bmatrix}$$

$$= \begin{bmatrix} 1.2166/-21.17^\circ \\ 2.2690\underline{/180^\circ} \\ 1.2166\underline{/21.17^\circ} \end{bmatrix} \quad \text{per unit}$$

$$= \begin{bmatrix} 0.5090\underline{/-21.17^\circ} \\ 0.9492\underline{/180^\circ} \\ 0.5090\underline{/21.17^\circ} \end{bmatrix} \quad \text{kA}$$

In conclusion, Δ–Y transformer phase shifts have no effect on the fault currents and no effect on the contribution to the fault currents on the fault side of the Δ–Y transformers. However, on the other side of the Δ–Y transformers, the positive- and negative-sequence components of the contributions to the fault currents are shifted by ±30°, which affects both the magnitude as well as the angle of the phase components of these fault contributions for unsymmetrical faults. ■

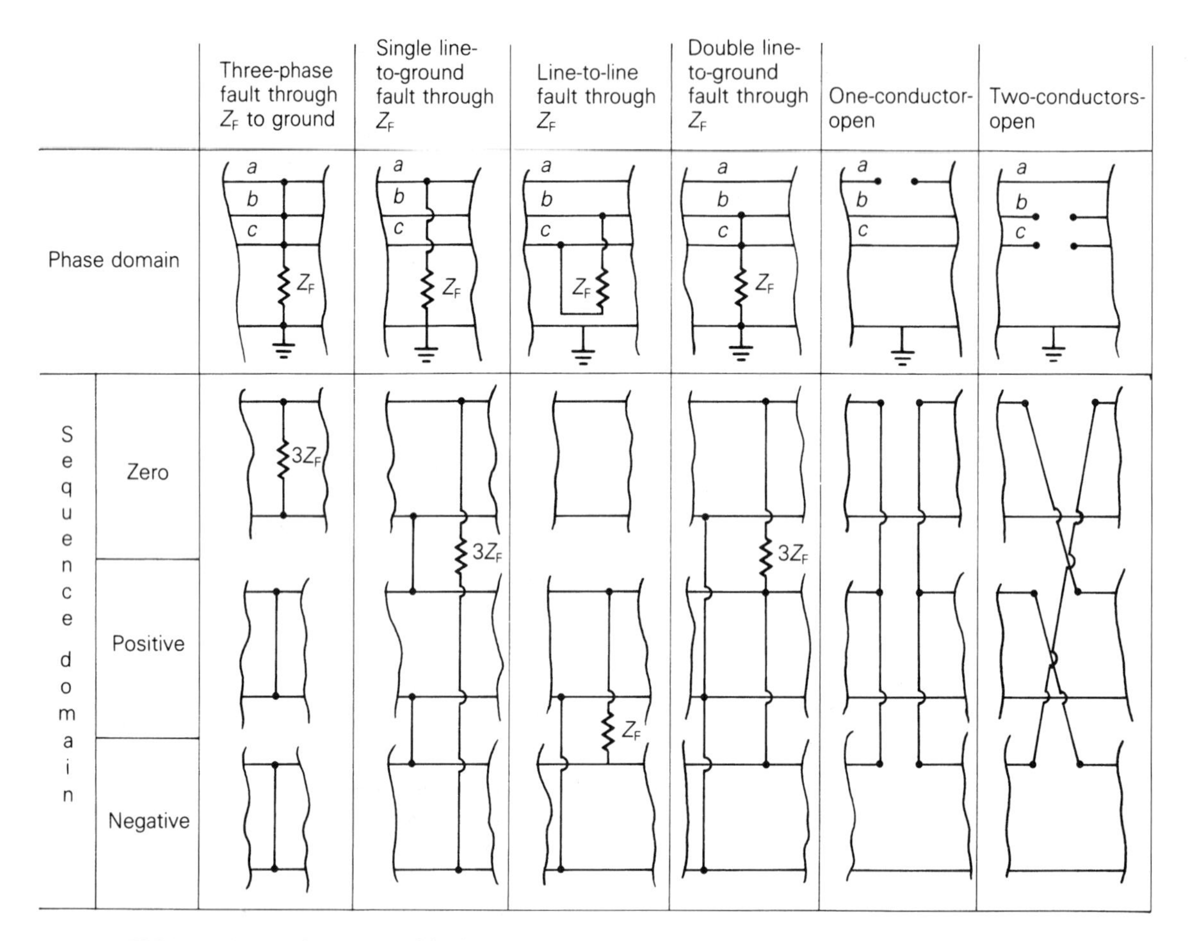

FIGURE 9.15 Summary of faults

Figure 9.15 summarizes the sequence network connections for both the balanced three-phase fault and the unsymmetrical faults that we have considered. Sequence network connections for two additional faults, one-conductor-open and two-conductors-open, are also shown in Figure 9.15 and are left as an exercise for you to verify (see Problems 9.26 and 9.27).

9.5 SEQUENCE BUS IMPEDANCE MATRICES

We use the positive-sequence bus impedance matrix in Section 7.4 for calculating currents and voltages during balanced three-phase faults. This method is extended here to unsymmetrical faults by representing each sequence network as a bus impedance equivalent circuit (or as a rake equivalent). A bus impedance matrix can be computed for each sequence network by inverting the corresponding bus admittance network. For simplicity, resistances, shunt admittances, nonrotating impedance loads, and prefault load currents are neglected.

Figure 9.16 shows the connection of sequence rake equivalents for both symmetrical and unsymmetrical faults at bus n of an N-bus three-phase power system. Each bus impedance element has an additional subscript, 0, 1, or 2, that identifies the sequence rake equivalent in which it is located. Mutual impedances are not shown in the figure. The prefault voltage V_F is

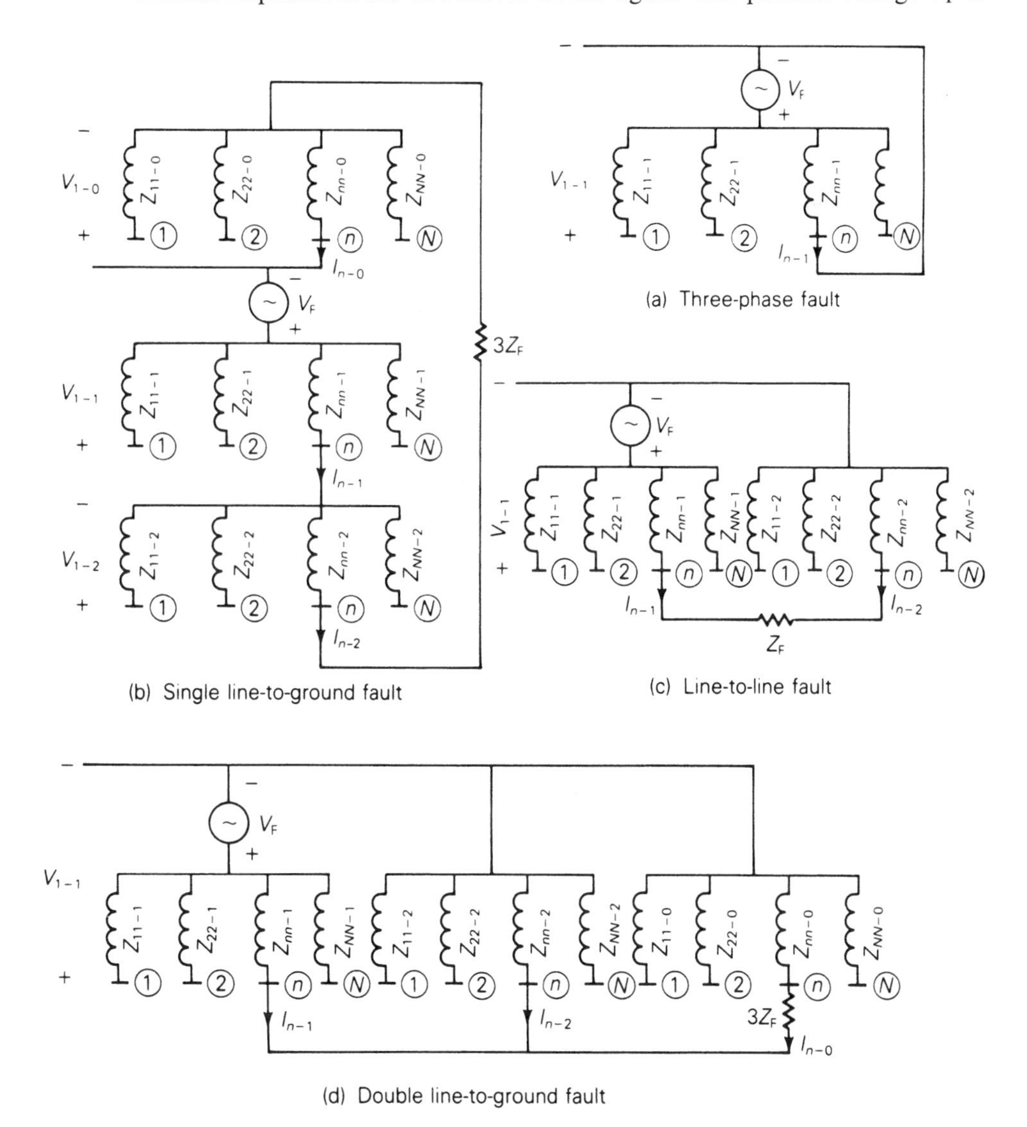

FIGURE 9.16 Connection of rake equivalent sequence networks for three-phase system faults (mutual impedances not shown)

included in the positive-sequence rake equivalent. From the figure the sequence components of the fault current for each type of fault at bus n are as follows:

Balanced three-phase fault:

$$I_{n-1} = \frac{V_F}{Z_{nn-1}} \tag{9.5.1}$$

$$I_{n-0} = I_{n-2} = 0 \tag{9.5.2}$$

Single line-to-ground fault (phase *a* to ground):

$$I_{n-0} = I_{n-1} = I_{n-2} = \frac{V_F}{Z_{nn-0} + Z_{nn-1} + Z_{nn-2} + 3Z_F} \tag{9.5.3}$$

Line-to-line fault (phase *b* to *c*):

$$I_{n-1} = -I_{n-2} = \frac{V_F}{Z_{nn-1} + Z_{nn-2} + Z_F} \tag{9.5.4}$$

$$I_{n-0} = 0 \tag{9.5.5}$$

Double line-to-ground fault (phase *b* to *c* to ground):

$$I_{n-1} = \frac{V_F}{Z_{nn-1} + \left[\dfrac{Z_{nn-2}(Z_{nn-0} + 3Z_F)}{Z_{nn-2} + Z_{nn-0} + 3Z_F}\right]} \tag{9.5.6}$$

$$I_{n-2} = (-I_{n-1})\left(\frac{Z_{nn-0} + 3Z_F}{Z_{nn-0} + 3Z_F + Z_{nn-2}}\right) \tag{9.5.7}$$

$$I_{n-0} = (-I_{n-1})\left(\frac{Z_{nn-2}}{Z_{nn-0} + 3Z_F + Z_{nn-2}}\right) \tag{9.5.8}$$

Also from Figure 9.16, the sequence components of the line-to-ground voltages at any bus k during a fault at bus n are:

$$\begin{bmatrix} V_{k-0} \\ V_{k-1} \\ V_{k-2} \end{bmatrix} = \begin{bmatrix} 0 \\ V_F \\ 0 \end{bmatrix} - \begin{bmatrix} Z_{kn-0} & 0 & 0 \\ 0 & Z_{kn-1} & 0 \\ 0 & 0 & Z_{kn-2} \end{bmatrix} \begin{bmatrix} I_{n-0} \\ I_{n-1} \\ I_{n-2} \end{bmatrix} \tag{9.5.9}$$

If bus k is on the unfaulted side of a Δ–Y transformer, then the phase angles of V_{k-1} and V_{k-2} in (9.5.9) are modified to account for Δ–Y phase shifts. Also, the above sequence fault currents and sequence voltages can be transformed to the phase domain via (8.1.16) and (8.1.9).

EXAMPLE 9.7 Single line-to-ground short-circuit calculations using $Z_{bus\,0}$, $Z_{bus\,1}$, and $Z_{bus\,2}$

Faults at buses 1 and 2 for the three-phase power system given in Example 9.1 are of interest. The prefault voltage is 1.05 per unit. Prefault load current is neglected. (a) Determine the per-unit zero-, positive-, and negative-sequence bus impedance matrices. Find the subtransient fault current in per-unit for a bolted single line-to-ground fault current from phase a to ground (b) at bus 1 and (c) at bus 2. Find the per-unit line-to-ground voltages at (d) bus 1 and (e) bus 2 during the single line-to-ground fault at bus 1.

SOLUTION

a. Referring to Figure 9.4(a), the zero-sequence bus admittance matrix is

$$\boldsymbol{Y}_{\text{bus }0} = -j\left[\begin{array}{c|c} 20 & 0 \\ \hline 0 & 4 \end{array}\right] \quad \text{per unit}$$

Inverting $\boldsymbol{Y}_{\text{bus }0}$,

$$\boldsymbol{Z}_{\text{bus }0} = j\left[\begin{array}{c|c} 0.05 & 0 \\ \hline 0 & 0.25 \end{array}\right] \quad \text{per unit}$$

Note that the transformer leakage reactances and the zero-sequence transmission-line reactance in Figure 9.4(a) have no effect on $\boldsymbol{Z}_{\text{bus }0}$. The transformer Δ connections block the flow of zero-sequence current from the transformers to bus 1 and 2.

The positive-sequence bus admittance matrix, from Figure 9.4(b), is

$$\boldsymbol{Y}_{\text{bus }1} = -j\left[\begin{array}{c|c} 9.9454 & -3.2787 \\ \hline -3.2787 & 8.2787 \end{array}\right] \quad \text{per unit}$$

Inverting $\boldsymbol{Y}_{\text{bus }1}$,

$$\boldsymbol{Z}_{\text{bus }1} = j\left[\begin{array}{c|c} 0.11565 & 0.04580 \\ \hline 0.04580 & 0.13893 \end{array}\right] \quad \text{per unit}$$

Similarly, from Figure 9.4(c)

$$\boldsymbol{Y}_{\text{bus }2} = -j\left[\begin{array}{c|c} 9.1611 & -3.2787 \\ \hline -3.2787 & 8.0406 \end{array}\right]$$

Inverting $\boldsymbol{Y}_{\text{bus }2}$,

$$\boldsymbol{Z}_{\text{bus }2} = j\left[\begin{array}{c|c} 0.12781 & 0.05212 \\ \hline 0.05212 & 0.14562 \end{array}\right] \quad \text{per unit}$$

b. From (9.5.3), with $n = 1$ and $Z_{\text{F}} = 0$, the sequence fault currents are

$$I_{1-0} = I_{1-1} = I_{1-2} = \frac{V_{\text{F}}}{Z_{11-0} + Z_{11-1} + Z_{11-2}}$$

$$= \frac{1.05\underline{/0^\circ}}{j(0.05 + 0.11565 + 0.12781)} = \frac{1.05}{j0.29346} = -j3.578 \quad \text{per unit}$$

The subtransient fault currents at bus 1 are, from (8.1.16),

$$\begin{bmatrix} I''_{1a} \\ I''_{1b} \\ I''_{1c} \end{bmatrix} = \begin{bmatrix} 1 & 1 & 1 \\ 1 & a^2 & a \\ 1 & a & a^2 \end{bmatrix} \begin{bmatrix} -j3.578 \\ -j3.578 \\ -j3.578 \end{bmatrix} = \begin{bmatrix} -j10.73 \\ 0 \\ 0 \end{bmatrix} \quad \text{per unit}$$

c. Again from (9.5.3), with $n = 2$ and $Z_F = 0$,

$$I_{2-0} = I_{2-1} = I_{2-2} = \frac{V_F}{Z_{22-0} + Z_{22-1} + Z_{22-2}}$$

$$= \frac{1.05\underline{/0^\circ}}{j(0.25 + 0.13893 + 0.14562)} = \frac{1.05}{j0.53455}$$

$$= -j1.96427 \quad \text{per unit}$$

and

$$\begin{bmatrix} I''_{2a} \\ I''_{2b} \\ I''_{2c} \end{bmatrix} = \begin{bmatrix} 1 & 1 & 1 \\ 1 & a^2 & a \\ 1 & a & a^2 \end{bmatrix} \begin{bmatrix} -j1.96427 \\ -j1.96427 \\ -j1.96427 \end{bmatrix} = \begin{bmatrix} -j5.8928 \\ 0 \\ 0 \end{bmatrix} \quad \text{per unit}$$

This is the same result as obtained in Example 9.3.

d. The sequence components of the line-to-ground voltages at bus 1 during the fault at bus 1 are, from (9.5.9), with $k = 1$ and $n = 1$,

$$\begin{bmatrix} V_{1-0} \\ V_{1-1} \\ V_{1-2} \end{bmatrix} = \begin{bmatrix} 0 \\ 1.05\underline{/0^\circ} \\ 0 \end{bmatrix} - \begin{bmatrix} j0.05 & 0 & 0 \\ 0 & j0.11565 & 0 \\ 0 & 0 & j0.12781 \end{bmatrix} \begin{bmatrix} -j3.578 \\ -j3.578 \\ -j3.578 \end{bmatrix}$$

$$= \begin{bmatrix} -0.1789 \\ 0.6362 \\ -0.4573 \end{bmatrix} \quad \text{per unit}$$

and the line-to-ground voltages at bus 1 during the fault at bus 1 are

$$\begin{bmatrix} V_{1-ag} \\ V_{1-bg} \\ V_{1-cg} \end{bmatrix} = \begin{bmatrix} 1 & 1 & 1 \\ 1 & a^2 & a \\ 1 & a & a^2 \end{bmatrix} \begin{bmatrix} -0.1789 \\ +0.6362 \\ -0.4573 \end{bmatrix}$$

$$= \begin{bmatrix} 0 \\ 0.9843\underline{/254.2^\circ} \\ 0.9843\underline{/105.8^\circ} \end{bmatrix} \quad \text{per unit}$$

e. The sequence components of the line-to-ground voltages at bus 2 during the fault at bus 1 are, from (9.5.9), with $k = 2$ and $n = 1$,

$$\begin{bmatrix} V_{2-0} \\ V_{2-1} \\ V_{2-2} \end{bmatrix} = \begin{bmatrix} 0 \\ 1.05\underline{/0^\circ} \\ 0 \end{bmatrix} - \begin{bmatrix} 0 & 0 & 0 \\ 0 & j0.04580 & 0 \\ 0 & 0 & j0.05212 \end{bmatrix} \begin{bmatrix} -j3.578 \\ -j3.578 \\ -j3.578 \end{bmatrix}$$

$$= \begin{bmatrix} 0 \\ 0.8861 \\ -0.18649 \end{bmatrix} \quad \text{per unit}$$

Note that since both bus 1 and 2 are on the low-voltage side of the Δ–Y transformers in Figure 9.3, there is no shift in the phase angles of these sequence voltages. From the above, the line-to-ground voltages at bus 2 during the fault at bus 1 are

$$\begin{bmatrix} V_{2-ag} \\ V_{2-bg} \\ V_{2-cg} \end{bmatrix} = \begin{bmatrix} 1 & 1 & 1 \\ 1 & a^2 & a \\ 1 & a & a^2 \end{bmatrix} \begin{bmatrix} 0 \\ 0.8861 \\ -0.18649 \end{bmatrix}$$

$$= \begin{bmatrix} 0.70 \\ 0.9926\underline{/249.4^\circ} \\ 0.9926\underline{/110.6^\circ} \end{bmatrix} \quad \text{per unit}$$

■

PowerWorld Simulator computes the symmetrical fault current for each of the following faults at any bus in an N-bus power system: balanced three-phase fault, single line-to-ground fault, line-to-line fault, or double line-to-ground fault. For each fault, the Simulator also computes bus voltages and contributions to the fault current from transmission lines and transformers connected to the fault bus.

Input data for the Simulator include machine, transmission-line, and transformer data, as illustrated in Tables 9.1, 9.2, and 9.3 as well as the prefault voltage V_F and fault impedance Z_F. When the machine positive-sequence reactance input data consist of direct axis subtransient reactances, the computed symmetrical fault currents are subtransient fault currents. Alternatively, transient or steady-state fault currents are computed when

TABLE 9.1
Synchronous machine data for Example 9.8

Bus	X_0 per unit	$X_1 = X_d''$ per unit	X_2 per unit	Neutral Reactance X_n per unit
1	0.0125	0.045	0.045	0
3	0.005	0.0225	0.0225	0.0025

TABLE 9.2

Line data for Example 9.8

Bus-to-Bus	X_0 per unit	X_1 per unit
2–4	0.3	0.1
2–5	0.15	0.05
4–5	0.075	0.025

TABLE 9.3

Transformer data for Example 9.8

Low-Voltage (connection) bus	High-Voltage (connection) bus	Leakage Reactance per unit	Neutral Reactance per unit
1 (Δ)	5 (Y)	0.02	0
3 (Δ)	4 (Y)	0.01	0

$S_{base} = 100$ MVA

$$V_{base} = \begin{cases} 15 \text{ kV at buses } 1,\ 3 \\ 345 \text{ kV at buses } 2,\ 4,\ 5 \end{cases}$$

these input data consist of direct axis transient or synchronous reactances. Transmission-line positive- and zero-sequence series reactances are those of the equivalent π circuits for long lines or of the nominal π circuit for medium or short lines. Also, recall that the negative-sequence transmission-line reactance equals the positive-sequence transmission-line reactance. All machine, line, and transformer reactances are given in per-unit on a common MVA base. Prefault load currents are neglected.

The Simulator computes (but does not show) the zero-, positive-, and negative-sequence bus impedance matrices $\boldsymbol{Z}_{\text{bus }0}, \boldsymbol{Z}_{\text{bus }1}$, and $\boldsymbol{Z}_{\text{bus }2}$, by inverting the corresponding bus admittance matrices.

After $\boldsymbol{Z}_{\text{bus }0}, \boldsymbol{Z}_{\text{bus }1}$, and $\boldsymbol{Z}_{\text{bus }2}$ are computed, (9.5.1)–(9.5.9) are used to compute the sequence fault currents and the sequence voltages at each bus during a fault at bus 1 for the fault type selected by the program user (for example, three-phase fault, or single line-to-ground fault, and so on). Contributions to the sequence fault currents from each line or transformer branch connected to the fault bus are computed by dividing the sequence voltage across the branch by the branch sequence impedance. The phase angles of positive- and negative-sequence voltages are also modified to account for Δ–Y transformer phase shifts. The sequence currents and sequence voltages are then transformed to the phase domain via (8.1.16) and (8.1.9). All these computations are then repeated for a fault at bus 2, then bus 3, and so on to bus N.

Output data for the fault type and fault impedance selected by the user consist of the fault current in each phase, contributions to the fault current from each branch connected to the fault bus for each phase, and the line-to-ground voltages at each bus—for a fault at bus 1, then bus 2, and so on to bus N.

EXAMPLE 9.8 PowerWorld Simulator

Consider the five-bus power system whose single-line diagram is shown in Figure 6.2. Machine, line, and transformer data are given in Tables 9.1, 9.2, and 9.3. Note that the neutrals of both transformers and generator 1 are solidly grounded, as indicated by a neutral reactance of zero for these equipments. However, a neutral reactance = 0.0025 per unit is connected to the generator 2 neutral. The prefault voltage is 1.05 per unit. Using PowerWorld Simulator, determine the fault currents and voltages for a bolted single line-to-ground fault at bus 1, then bus 2, and so on to bus 5.

SOLUTION Open PowerWorld Simulator case Example 9.8 to see this example. Tables 9.4 and 9.5 summarize the PowerWorld Simulator results for each of the faults. Note that these fault currents are subtransient currents, since the machine positive-sequence reactance input consists of direct axis subtransient reactances.

TABLE 9.4

Fault currents for Example 9.8

Fault Bus	Single Line-to-Ground Fault Current (Phase A) per unit/degrees	GEN LINE OR TRSF	Bus-to-Bus	Contributions to Fault Current: Phase A	Current Phase B	Phase C
				per unit/degrees		
1	46.02/−90.00	G1	GRND–1	34.41/ −90.00	5.804/ −90.00	5.804/ −90.00
		T1	5–1	11.61/ −90.00	5.804/ 90.00	5.804/ 90.00
2	14.14/−90.00	L1	4–2	5.151/ −90.00	0.1124/ 90.00	0.1124/ 90.00
		L2	5–2	8.984/ −90.00	0.1124/ −90.00	0.1124/ −90.00
3	64.30/−90.00	G2	GRND–3	56.19/ −90.00	4.055/ −90.00	4.055/ −90.00
		T2	4–3	8.110/ −90.00	4.055/ 90.00	4.055/ 90.00
4	56.07/−90.00	L1	2–4	1.742/ −90.00	0.4464/ 90.00	0.4464/ 90.00
		L3	5–4	10.46/ −90.00	2.679/ 90.00	2.679/ 90.00
		T2	3–4	43.88/ −90.00	3.125/ −90.00	3.125/ −90.00
5	42.16/−90.00	L2	2–5	2.621/ −90.00	0.6716/ 90.00	0.6716/ 90.00
		L3	4–5	15.72/ −90.00	4.029/ 90.00	4.029/ 90.00
		T1	1–5	23.82/ −90.00	4.700/ −90.00	4.700/ −90.00

TABLE 9.5

Bus voltages for Example 9.8

$V_{prefault} = 1.05\angle 0$		Bus Voltages during Fault		
Fault Bus	Bus	Phase A	Phase B	Phase C
1	1	0.0000∠0.00	0.9537∠−107.55	0.9537∠107.55
	2	0.5069∠0.00	0.9440∠−105.57	0.9440∠105.57
	3	0.7888∠0.00	0.9912∠−113.45	0.9912∠113.45
	4	0.6727∠0.00	0.9695∠−110.30	0.9695∠110.30
	5	0.4239∠0.00	0.9337∠−103.12	0.9337∠103.12
2	1	0.8832∠0.00	1.0109∠--115.90	1.0109∠115.90
	2	0.0000∠0.00	1.1915∠−130.26	1.1915∠130.26
	3	0.9214∠0.00	1.0194∠−116.87	1.0194∠116.87
	4	0.8435∠0.00	1.0158∠−116.47	1.0158∠116.47
	5	0.7562∠0.00	1.0179∠−116.70	1.0179∠116.70
3	1	0.6851∠0.00	0.9717∠−110.64	0.9717∠110.64
	2	0.4649∠0.00	0.9386∠−104.34	0.9386∠104.34
	3	0.0000∠0.00	0.9942∠−113.84	0.9942∠113.84
	4	0.3490∠0.00	0.9259∠−100.86	0.9259∠100.86
	5	0.5228∠0.00	0.9462∠−106.04	0.9462∠106.04
4	1	0.5903∠0.00	0.9560∠−107.98	0.9560∠107.98
	2	0.2309∠0.00	0.9401∠−104.70	0.9401∠104.70
	3	0.4387∠0.00	0.9354∠−103.56	0.9354∠103.56
	4	0.0000∠0.00	0.9432∠−105.41	0.9432∠105.41
	5	0.3463∠0.00	0.9386∠−104.35	0.9386∠104.35
5	1	0.4764∠0.00	0.9400∠−104.68	0.9400∠104.68
	2	0.1736∠0.00	0.9651∠−109.57	0.9651∠109.57
	3	0.7043∠0.00	0.9751∠−111.17	0.9751∠111.17
	4	0.5209∠0.00	0.9592∠−108.55	0.9592∠108.55
	5	0.0000∠0.00	0.9681∠−110.07	0.9681∠110.07

■

MULTIPLE CHOICE QUESTIONS

SECTION 9.1

9.1 For power-system fault studies, it is assumed that the system is operating under balanced steady-state conditions prior to the fault, and sequence networks are uncoupled before the fault occurs.
(a) True (b) False

9.2 The first step in power-system fault calculations is to develop sequence networks based on the single-line diagram of the system, and then reduce them to their Thévenin equivalents, as viewed from the fault location.
(a) True (b) False

9.3 When calculating symmetrical three-phase fault currents, only _______ sequence network needs to be considered. Fill in the Blank.

9.4 In order of frequency of occurance of short-circuit faults in three-phase power systems, list those: ________, ________, ________, ________. Fill in the Blanks.

9.5 For a bolted three-phase-to-ground fault, sequence-fault currents ________ are zero, sequence fault voltages are ________, and line-to-ground voltages are ________. Fill in the Blanks.

SECTION 9.2

9.6 For a single-line-to-ground fault with a fault-impedance Z_F, the sequence networks are to be connected ________ at the fault terminals through the impedance ________; the sequence components of the fault currents are ___________. Fill in the Blanks.

SECTION 9.3

9.7 For a line-to-line fault with a fault impedance Z_F, the positive-and negative-sequence networks are to be connected _____________ at the fault terminals through the impedance of 1/2/3 times Z_F; the zero-sequence current is ________. Fill in the Blanks.

SECTION 9.4

9.8 For a double line-to-ground fault through a fault impedance Z_F, the sequence networks are to be connected _____________, at the fault terminal; additionally, _________ is to be included in series with the zero-sequence network. Fill in the Blanks.

SECTION 9.5

9.9 The sequence bus-impedance matrices can also be used to calculate fault currents and voltages for symmetrical as well as unsymmetrical faults by representing each sequence network as a bus-impedance rake-equivalent circuit.
(a) True (b) False

PROBLEMS

SECTION 9.1

9.1 The single-line diagram of a three-phase power system is shown in Figure 9.17. Equipment ratings are given as follows:

Synchronous generators:

G1	1000 MVA	15 kV	$X_d'' = X_2 = 0.18, X_0 = 0.07$ per unit
G2	1000 MVA	15 kV	$X_d'' = X_2 = 0.20, X_0 = 0.10$ per unit
G3	500 MVA	13.8 kV	$X_d'' = X_2 = 0.15, X_0 = 0.05$ per unit
G4	750 MVA	13.8 kV	$X_d'' = 0.30, X_2 = 0.40, X_0 = 0.10$ per unit

Transformers:

T1	1000 MVA	15 kV Δ/765 kV Y	X = 0.10 per unit
T2	1000 MVA	15 kV Δ/765 kV Y	X = 0.10 per unit
T3	500 MVA	15 kV Y/765 kV Y	X = 0.12 per unit
T4	750 MVA	15 kV Y/765 kV Y	X = 0.11 per unit

FIGURE 9.17

Problem 9.1

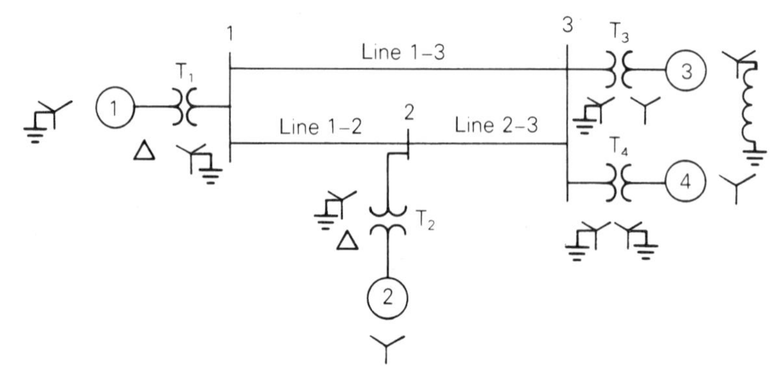

Transmission lines:

1–2	765 kV	$X_1 = 50\ \Omega,\ X_0 = 150\ \Omega$
1–3	765 kV	$X_1 = 40\ \Omega,\ X_0 = 100\ \Omega$
2–3	765 kV	$X_1 = 40\ \Omega,\ X_0 = 100\ \Omega$

The inductor connected to Generator 3 neutral has a reactance of 0.05 per unit using generator 3 ratings as a base. Draw the zero-, positive-, and negative-sequence reactance diagrams using a 1000-MVA, 765-kV base in the zone of line 1–2. Neglect the Δ–Y transformer phase shifts.

9.2 Faults at bus n in Problem 9.1 are of interest (the instructor selects $n = 1, 2$, or 3). Determine the Thévenin equivalent of each sequence network as viewed from the fault bus. Prefault voltage is 1.0 per unit. Prefault load currents and Δ–Y transformer phase shifts are neglected. (*Hint*: Use the Y–Δ conversion in Figure 2.27.)

9.3 Determine the subtransient fault current in per-unit and in kA during a bolted three-phase fault at the fault bus selected in Problem 9.2.

9.4 In Problem 9.1 and Figure 9.17, let 765 kV be replaced by 500 kV, keeping the rest of the data to be the same. Repeat (a) Problems 9.1, (b) 9.2, and (c) 9.3.

9.5 Equipment ratings for the four-bus power system shown in Figure 7.14 are given as follows:

Generator G1:	500 MVA, 13.8 kV, $X''_d = X_2 = 0.20$, $X_0 = 0.10$ per unit
Generator G2:	750 MVA, 18 kV, $X''_d = X_2 = 0.18$, $X_0 = 0.09$ per unit
Generator G3:	1000 MVA, 20 kV, $X''_d = 0.17$, $X_2 = 0.20$, $X_0 = 0.09$ per unit
Transformer T1:	500 MVA, 13.8 kV Δ/500 kV Y, $X = 0.12$ per unit
Transformer T2:	750 MVA, 18 kV Δ/500 kV Y, $X = 0.10$ per unit
Transformer T3:	1000 MVA, 20 kV Δ/500 kV Y, $X = 0.10$ per unit
Each line:	$X_1 = 50$ ohms, $X_0 = 150$ ohms

The inductor connected to generator G3 neutral has a reactance of 0.028 Ω. Draw the zero-, positive-, and negative-sequence reactance diagrams using a 1000-MVA, 20-kV base in the zone of generator G3. Neglect Δ–Y transformer phase shifts.

9.6 Faults at bus n in Problem 9.5 are of interest (the instructor selects $n = 1, 2, 3$, or 4). Determine the Thévenin equivalent of each sequence network as viewed from the fault bus. Prefault voltage is 1.0 per unit. Prefault load currents and Δ–Y phase shifts are neglected.

9.7 Determine the subtransient fault current in per-unit and in kA during a bolted three-phase fault at the fault bus selected in Problem 9.6.

9.8 Equipment ratings for the five-bus power system shown in Figure 7.15 are given as follows:

Generator G1:	50 MVA, 12 kV, $X''_d = X_2 = 0.20$, $X_0 = 0.10$ per unit
Generator G2:	100 MVA, 15 kV, $X''_d = 0.2$, $X_2 = 0.23$, $X_0 = 0.1$ per unit
Transformer T1:	50 MVA, 10 kV Y/138 kV Y, X = 0.10 per unit
Transformer T2:	100 MVA, 15 kV Δ/138 kV Y, X = 0.10 per unit
Each 138-kV line:	$X_1 = 40$ ohms, $X_0 = 100$ ohms

Draw the zero-, positive-, and negative-sequence reactance diagrams using a 100-MVA, 15-kV base in the zone of generator G2. Neglect Δ–Y transformer phase shifts.

9.9 Faults at bus n in Problem 9.8 are of interest (the instructor selects $n = 1, 2, 3, 4,$ or 5). Determine the Thévenin equivalent of each sequence network as viewed from the fault bus. Prefault voltage is 1.0 per unit. Prefault load currents and Δ–Y phase shifts are neglected.

9.10 Determine the subtransient fault current in per-unit and in kA during a bolted three-phase fault at the fault bus selected in Problem 9.9.

9.11 Consider the system shown in Figure 9.18. (a) As viewed from the fault at F, determine the Thévenin equivalent of each sequence network. Neglect Δ–Y phase shifts. (b) Compute the fault currents for a balanced three-phase fault at fault point F through three fault impedances $Z_{FA} = Z_{FB} = Z_{FC} = j0.5$ per unit. Equipment data in per-unit on the same base are given as follows:

Synchronous generators:

G1	$X_1 = 0.2$	$X_2 = 0.12$	$X_0 = 0.06$
G2	$X_1 = 0.33$	$X_2 = 0.22$	$X_0 = 0.066$

Transformers:

T1	$X_1 = X_2 = X_0 = 0.2$
T2	$X_1 = X_2 = X_0 = 0.225$

FIGURE 9.18

Problem 9.11

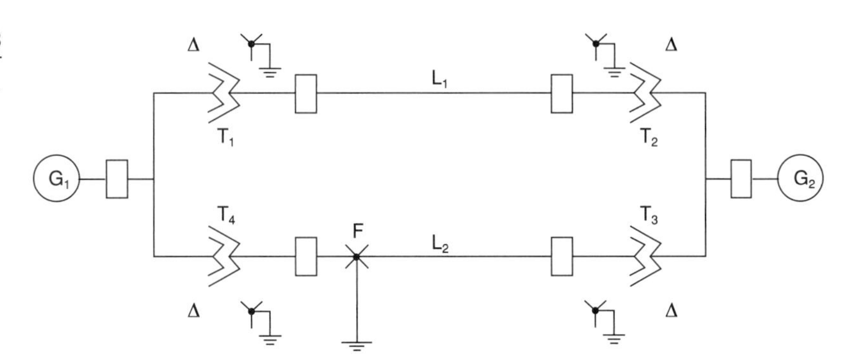

T3 $X_1 = X_2 = X_0 = 0.27$

T4 $X_1 = X_2 = X_0 = 0.16$

Transmission lines:

L1 $X_1 = X_2 = 0.14$ $X_0 = 0.3$

L1 $X_1 = X_2 = 0.35$ $X_0 = 0.6$

9.12 Equipment ratings and per-unit reactances for the system shown in Figure 9.19 are given as follows:

Synchronous generators:

G1 100 MVA 25 kV $X_1 = X_2 = 0.2$ $X_0 = 0.05$

G2 100 MVA 13.8 kV $X_1 = X_2 = 0.2$ $X_0 = 0.05$

Transformers:

T1 100 MVA 25/230 kV $X_1 = X_2 = X_0 = 0.05$

T2 100 MVA 13.8/230 kV $X_1 = X_2 = X_0 = 0.05$

Transmission lines:

TL12 100 MVA 230 kV $X_1 = X_2 = 0.1$ $X_0 = 0.3$

TL13 100 MVA 230 kV $X_1 = X_2 = 0.1$ $X_0 = 0.3$

TL23 100 MVA 230 kV $X_1 = X_2 = 0.1$ $X_0 = 0.3$

Using a 100-MVA, 230-kV base for the transmission lines, draw the per-unit sequence networks and reduce them to their Thévenin equivalents, "looking in" at bus 3. Neglect Δ–Y phase shifts. Compute the fault currents for a bolted three-phase fault at bus 3.

FIGURE 9.19

Problem 9.12

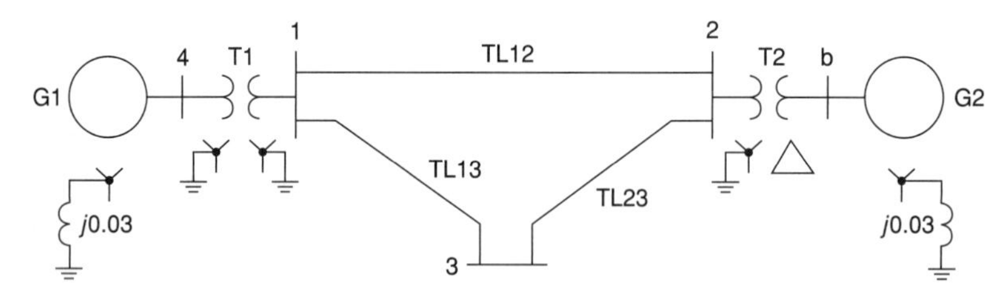

9.13 Consider the one-line diagram of a simple power system shown in Figure 9.20. System data in per-unit on a 100-MVA base are given as follows:

Synchronous generators:

G1 100 MVA 20 kV $X_1 = X_2 = 0.15$ $X_0 = 0.05$

G2 100 MVA 20 kV $X_1 = X_2 = 0.15$ $X_0 = 0.05$

Transformers:

T1 100 MVA 20/220 kV $X_1 = X_2 = X_0 = 0.1$

T2 100 MVA 20/220 kV $X_1 = X_2 = X_0 = 0.1$

Transmission lines:

L12	100 MVA	220 kV	$X_1 = X_2 = 0.125$	$X_0 = 0.3$
L13	100 MVA	220 kV	$X_1 = X_2 = 0.15$	$X_0 = 0.35$
L23	100 MVA	220 kV	$X_1 = X_2 = 0.25$	$X_0 = 0.7125$

The neutral of each generator is grounded through a current-limiting reactor of 0.08333 per unit on a 100-MVA base. All transformer neutrals are solidly grounded. The generators are operating no-load at their rated voltages and rated frequency with their EMFs in phase. Determine the fault current for a balanced three-phase fault at bus 3 through a fault impedance $Z_F = 0.1$ per unit on a 100-MVA base. Neglect Δ–Y phase shifts.

FIGURE 9.20
Problem 9.13

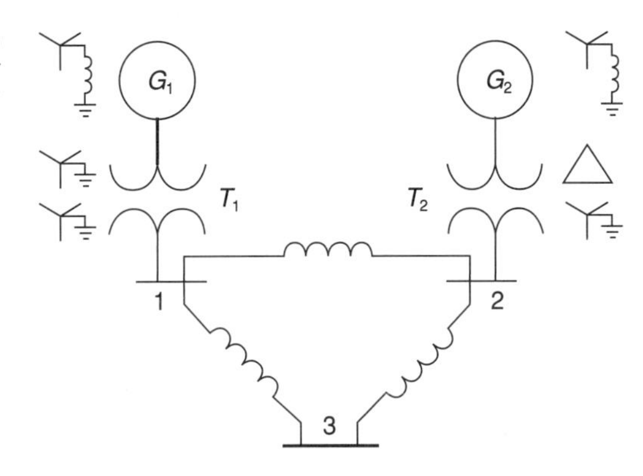

SECTIONS 9.2–9.4

9.14 Determine the subtransient fault current in per-unit and in kA, as well as the per-unit line-to-ground voltages at the fault bus for a bolted single line-to-ground fault at the fault bus selected in Problem 9.2.

9.15 Repeat Problem 9.14 for a single line-to-ground arcing fault with arc impedance $Z_F = 30 + j0\ \Omega$.

9.16 Repeat Problem 9.14 for a bolted line-to-line fault.

9.17 Repeat Problem 9.14 for a bolted double line-to-ground fault.

9.18 Repeat Problems 9.1 and 9.14 including Δ–Y transformer phase shifts. Assume American standard phase shift. Also calculate the sequence components and phase components of the contribution to the fault current from generator n ($n = 1, 2$, or 3 as specified by the instructor in Problem 9.2).

9.19 (a) Repeat Problem 9.14 for the case of Problem 9.4 (b).
(b) Repeat Problem 9.19(a) for a single line-to-ground arcing fault with arc impedance $Z_F = (15 + j0)\ \Omega$.
(c) Repeat Problem 9.19(a) for a bolted line-to-line fault.
(d) Repeat Problem 9.19(a) for a bolted double line-to-ground fault.
(e) Repeat Problems 9.4(a) and 9.19(a) including Δ–Y transformer phase shifts. Assume American standard phase shift. Also calculate the sequence components and phase components of the contribution to the fault current from generator n ($n = 1, 2$, or 3) as specified by the instructor in Problem 9.4(b).

9.20 A 500-MVA, 13.8-kV synchronous generator with $X_d'' = X_2 = 0.20$ and $X_0 = 0.05$ per unit is connected to a 500-MVA, 13.8-kV Δ/500-kV Y transformer with 0.10 per-unit leakage reactance. The generator and transformer neutrals are solidly grounded. The generator is operated at no-load and rated voltage, and the high-voltage side of the transformer is disconnected from the power system. Compare the subtransient fault currents for the following bolted faults at the transformer high-voltage terminals: three-phase fault, single line-to-ground fault, line-to-line fault, and double line-to-ground fault.

9.21 Determine the subtransient fault current in per-unit and in kA, as well as contributions to the fault current from each line and transformer connected to the fault bus for a bolted single line-to-ground fault at the fault bus selected in Problem 9.6.

9.22 Repeat Problem 9.21 for a bolted line-to-line fault.

9.23 Repeat Problem 9.21 for a bolted double line-to-ground fault.

9.24 Determine the subtransient fault current in per-unit and in kA, as well as contributions to the fault current from each line, transformer, and generator connected to the fault bus for a bolted single line-to-ground fault at the fault bus selected in Problem 9.9.

9.25 Repeat Problem 9.24 for a single line-to-ground arcing fault with arc impedance $Z_F = 0.05 + j0$ per unit.

9.26 Repeat Problem 9.24 for a bolted line-to-line fault.

9.27 Repeat Problem 9.24 for a bolted double line-to-ground fault.

9.28 As shown in Figure 9.21(a), two three-phase buses abc and $a'b'c'$ are interconnected by short circuits between phases b and b' and between c and c', with an open circuit between phases a and a'. The fault conditions in the phase domain are $I_a = I_{a'} = 0$ and $V_{bb'} = V_{cc'} = 0$. Determine the fault conditions in the sequence domain and verify the interconnection of the sequence networks as shown in Figure 9.15 for this one-conductor-open fault.

9.29 Repeat Problem 9.28 for the two-conductors-open fault shown in Figure 9.21(b). The fault conditions in the phase domain are

$$I_b = I_{b'} = I_c = I_{c'} = 0 \quad \text{and} \quad V_{aa'} = 0$$

FIGURE 9.21

Problems 9.28 and 9.29: open conductor faults

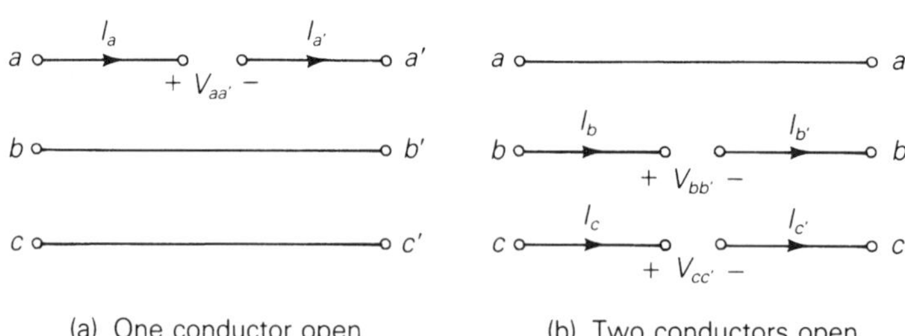

9.30 For the system of Problem 9.11, compute the fault current and voltages at the fault for the following faults at point F: (a) a bolted single line-to-ground fault; (b) a line-to-line fault through a fault impedance $Z_F = j0.05$ per unit; (c) a double line-to-ground fault from phase B to C to ground, where phase B has a fault impedance $Z_F = j0.05$ per unit, phase C also has a fault impedance $Z_F = j0.05$ per unit, and the common line-to-ground fault impedance is $Z_G = j0.033$ per unit.

9.31 For the system of Problem 9.12, compute the fault current and voltages at the fault for the following faults at bus 3: (a) a bolted single line-to-ground fault, (b) a bolted line-to-line fault, (c) a bolted double line-to-ground fault. Also, for the single line-to-ground fault at bus 3, determine the currents and voltages at the terminals of generators G1 and G2.

9.32 For the system of Problem 9.13, compute the fault current for the following faults at bus 3: (a) a single line-to-ground fault through a fault impedance $Z_F = j0.1$ per unit, (b) a line-to-line fault through a fault impedance $Z_F = j0.1$ per unit, (c) a double line-to-ground fault through a common fault impedance to ground $Z_F = j0.1$ per unit.

9.33 For the three-phase power system with single-line diagram shown in Figure 9.22, equipment ratings and per-unit reactances are given as follows:

Machines 1 and 2:	100 MVA	20 kV	$X_1 = X_2 = 0.2$
	$X_0 = 0.04$	$X_n = 0.04$	
Transformers 1 and 2:	100 MVA	$20\Delta/345Y$ kV	
	$X_1 = X_2 = X_0 = 0.08$		

Select a base of 100 MVA, 345 kV for the transmission line. On that base, the series reactances of the line are $X_1 = X_2 = 0.15$ and $X_0 = 0.5$ per unit. With a nominal system voltage of 345 kV at bus 3, machine 2 is operating as a motor drawing 50 MVA at 0.8 power factor lagging. Compute the change in voltage at bus 3 when the transmission line undergoes (a) a one-conductor-open fault, (b) a two-conductor-open fault along its span between buses 2 and 3.

FIGURE 9.22
Problem 9.33

9.34 At the general three-phase bus shown in Figure 9.7(a) of the text, consider a simultaneous single line-to-ground fault on phase a and line-to-line fault between phases b and c, with no fault impedances. Obtain the sequence-network interconnection satisfying the current and voltage constraints.

9.35 Thévenin equivalent sequence networks looking into the faulted bus of a power system are given with $Z_1 = j0.15$, $Z_2 = j0.15$, $Z_0 = j0.2$, and $E_1 = 1\underline{/0^\circ}$ per unit. Compute the fault currents and voltages for the following faults occurring at the faulted bus:
(a) Balanced three-phase fault
(b) Single line-to-ground fault
(c) Line-line fault
(d) Double line-to-ground fault
Which is the worst fault from the viewpoint of the fault current?

9.36 The single-line diagram of a simple power system is shown in Figure 9.23 with per unit values. Determine the fault current at bus 2 for a three-phase fault. Ignore the effect of phase shift.

FIGURE 9.23

For Problem 9.36

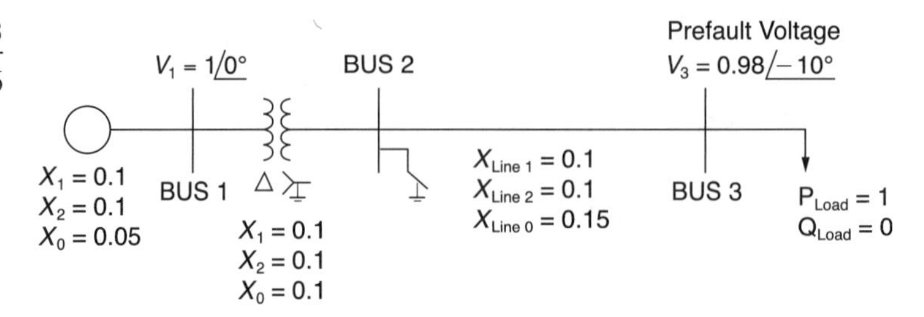

9.37 Consider a simple circuit configuration shown in Figure 9.24 to calculate the fault currents I_1, I_2, and I with the switch closed.
(a) Compute E_1 and E_2 prior to the fault based on the prefault voltage $V = 1\underline{/0°}$, and then, with the switch closed, determine I_1, I_2, and I.
(b) Start by ignoring prefault currents, with $E_1 = E_2 = 1\underline{/0°}$. Then superimpose the load currents, which are the prefault currents, $I_1 = -I_2 = 1\underline{/0°}$. Compare the results with those of part (a).

FIGURE 9.24

For Problem 9.37

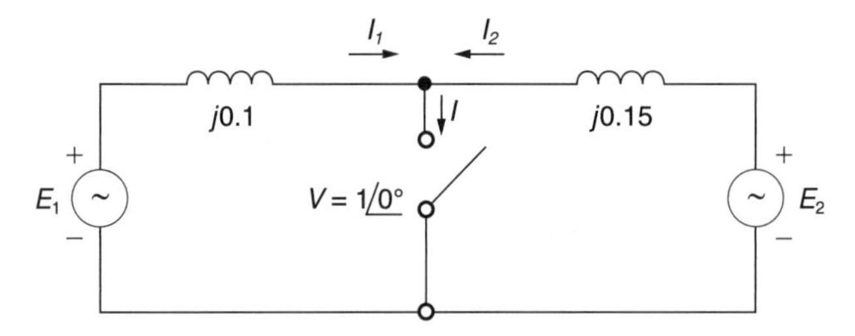

SECTION 9.5

9.38 The zero-, positive-, and negative-sequence bus impedance matrices for a three-bus three-phase power system are

$$\boldsymbol{Z}_{\text{bus }0} = j\begin{bmatrix} 0.10 & 0 & 0 \\ 0 & 0.20 & 0 \\ 0 & 0 & 0.10 \end{bmatrix} \text{ per unit}$$

$$\boldsymbol{Z}_{\text{bus }1} = \boldsymbol{Z}_{\text{bus }2} = j\begin{bmatrix} 0.12 & 0.08 & 0.04 \\ 0.08 & 0.12 & 0.06 \\ 0.04 & 0.06 & 0.08 \end{bmatrix}$$

Determine the per-unit fault current and per-unit voltage at bus 2 for a bolted three-phase fault at bus 1. The prefault voltage is 1.0 per unit.

9.39 Repeat Problem 9.38 for a bolted single line-to-ground fault at bus 1.

9.40 Repeat Problem 9.38 for a bolted line-to-line fault at bus 1.

9.41 Repeat Problem 9.38 for a bolted double line-to-ground fault at bus 1.

9.42 (a) Compute the 3×3 per-unit zero-, positive-, and negative-sequence bus impedance matrices for the power system given in Problem 9.1. Use a base of 1000 MVA and 765 kV in the zone of line 1–2.

9.42 (b) Using the bus impedance matrices determined in Problem 9.42, verify the fault currents for the faults given in Problems 9.3, 9.14, 9.15, 9.16, and 9.17.

9.43 The zero-, positive-, and negative-sequence bus impedance matrices for a two-bus three-phase power system are

$$Z_{\text{bus }0} = j\left[\begin{array}{c|c} 0.10 & 0 \\ \hline 0 & 0.10 \end{array}\right] \quad \text{per unit}$$

$$Z_{\text{bus }1} = Z_{\text{bus }2} = j\left[\begin{array}{c|c} 0.20 & 0.10 \\ \hline 0.10 & 0.30 \end{array}\right] \quad \text{per unit}$$

Determine the per-unit fault current and per-unit voltage at bus 2 for a bolted three-phase fault at bus 1. The prefault voltage is 1.0 per unit.

9.44 Repeat Problem 9.43 for a bolted single line-to-ground fault at bus 1.

9.45 Repeat Problem 9.43 for a bolted line-to-line fault at bus 1.

9.46 Repeat Problem 9.43 for a bolted double line-to-ground fault at bus 1.

9.47 Compute the 3×3 per-unit zero-, positive-, and negative-sequence bus impedance matrices for the power system given in Problem 4(a). Use a base of 1000 MVA and 500 kV in the zone of line 1–2.

9.48 Using the bus impedance matrices determined in Problem 9.47, verify the fault currents for the faults given in Problems 9.4(b), 9.4(c), 9.19 (a through d).

9.49 Compute the 4×4 per-unit zero-, positive-, and negative-sequence bus impedance matrices for the power system given in Problem 9.5. Use a base of 1000 MVA and 20 kV in the zone of generator G3.

9.50 Using the bus impedance matrices determined in Problem 9.42, verify the fault currents for the faults given in Problems 9.7, 9.21, 9.22, and 9.23.

9.51 Compute the 5×5 per-unit zero-, positive-, and negative-sequence bus impedance matrices for the power system given in Problem 9.8. Use a base of 100 MVA and 15 kV in the zone of generator G2.

9.52 Using the bus impedance matrices determined in Problem 9.51, verify the fault currents for the faults given in Problems 9.10, 9.24, 9.25, 9.26, and 9.27.

9.53 The positive-sequence impedance diagram of a five-bus network with all values in per-unit on a 100-MVA base is shown in Figure 9.25. The generators at buses 1 and 3 are rated 270 and 225 MVA, respectively. Generator reactances include subtransient values plus reactances of the transformers connecting them to the buses. The turns ratios of the transformers are such that the voltage base in each generator circuit is equal to the voltage rating of the generator. (a) Develop the positive-sequence bus admittance matrix $Y_{\text{bus }1}$. (b) Using MATLAB or another computer program, invert $Y_{\text{bus }1}$ to obtain $Z_{\text{bus }1}$. (c) Determine the subtransient current for a three-phase fault at bus 4 and the contributions to the fault current from each line. Neglect prefault currents and assume a prefault voltage of 1.0 per unit.

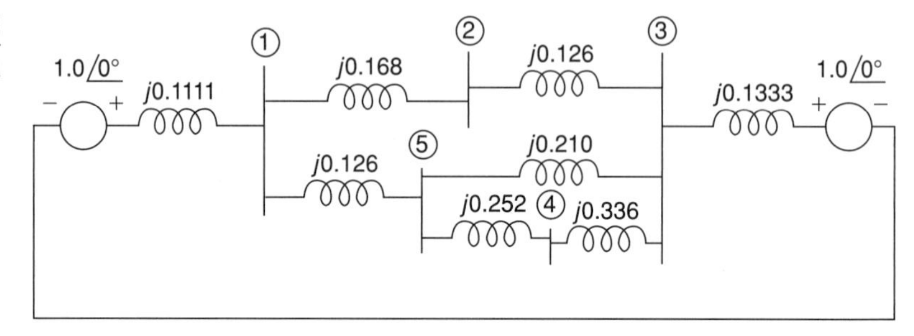

FIGURE 9.25

Problems 9.53 and 9.54

9.54 For the five-bus network shown in Figure 9.25, a bolted single-line-to-ground fault occurs at the bus 2 end of the transmission line between buses 1 and 2. The fault causes the circuit breaker at the bus 2 end of the line to open, but all other breakers remain closed. The fault is shown in Figure 9.26. Compute the subtransient fault current with the circuit breaker at the bus-2 end of the faulted line open. Neglect prefault current and assume a prefault voltage of 1.0 per unit.

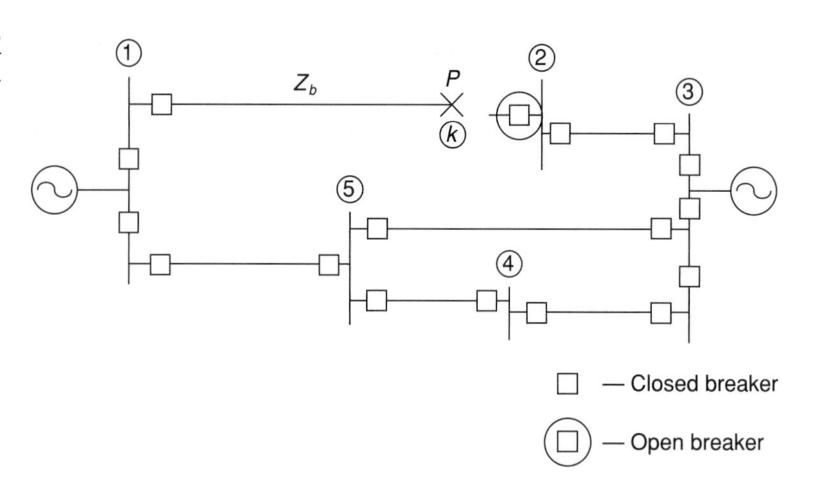

FIGURE 9.26

Problem 9.54

9.55 A single-line diagram of a four-bus system is shown in Figure 9.27. Equipment ratings and per-unit reactances are given as follows.

Machines 1 and 2:	100 MVA	20 kV	$X_1 = X_2 = 0.2$
	$X_0 = 0.04$	$X_n = 0.05$	
Transformers T_1 and T_2:	100 MVA	$20\Delta/345Y$ kV	
	$X_1 = X_2 = X_0 = 0.08$		

On a base of 100 MVA and 345 kV in the zone of the transmission line, the series reactances of the transmission line are $X_1 = X_2 = 0.15$ and $X_0 = 0.5$ per unit. (a) Draw

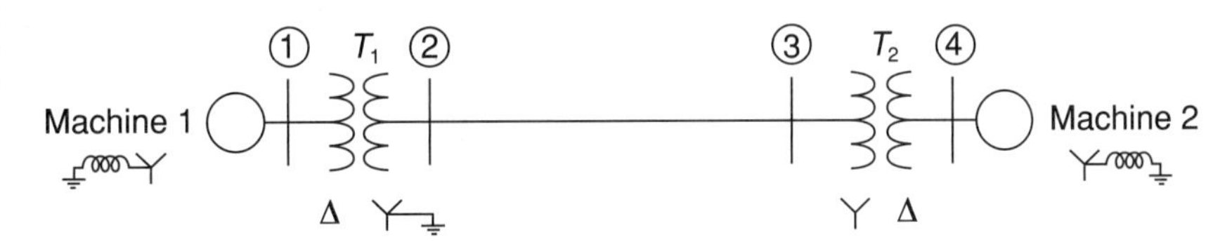

FIGURE 9.27 Problem 9.55

each of the sequence networks and determine the bus impedance matrix for each of them. (b) Assume the system to be operating at nominal system voltage without prefault currents, when a bolted line-to-line fault occurs at bus 3. Compute the fault current, the line-to-line voltages at the faulted bus, and the line-to-line voltages at the terminals of machine 2. (c) Assume the system to be operating at nominal system voltage without prefault currents, when a bolted double-line-to-ground fault occurs at the terminals of machine 2. Compute the fault current and the line-to-line voltages at the faulted bus.

9.56 The system shown in Figure 9.28 is the same as in Problem 9.48 except that the transformers are now Y–Y connected and solidly grounded on both sides. (a) Determine the bus impedance matrix for each of the three sequence networks. (b) Assume the system to be operating at nominal system voltage without prefault currents, when a bolted single-line-to-ground fault occurs on phase A at bus 3. Compute the fault current, the current out of phase C of machine 2 during the fault, and the line-to-ground voltages at the terminals of machine 2 during the fault.

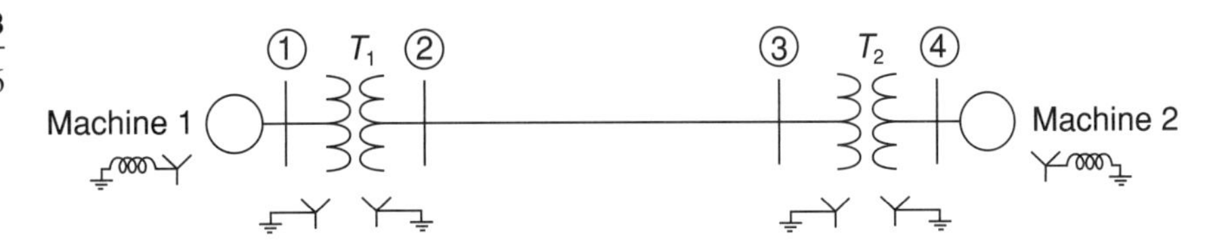

FIGURE 9.28 Problem 9.56

9.57 The results in Table 9.5 show that during a phase "a" single line-to-ground fault the phase angle on phase "a" voltages is always zero. Explain why we would expect this result.

PW **9.58** The results in Table 9.5 show that during the single line-to-ground fault at bus 2 the "b" and "c" phase voltage magnitudes at bus 2 actually rise above the pre-fault voltage of 1.05 per unit. Use PowerWorld Simulator with case Example 9_8 to determine the type of bus 2 fault that gives the highest per-unit voltage magnitude.

PW **9.59** Using PowerWorld Simulator case Example 9_8, plot the variation in the bus 2 phase "a," "b," and "c" voltage magnitudes during a single line-to-ground fault at bus 2 as the fault reactance is varied from 0 to 0.30 per unit in 0.05 per-unit steps (the fault impedance is specified on the Fault Options page of the Fault Analysis dialog).

PW **9.60** Using the Example 9.8 case determine the fault current, except with a line-to-line fault at each of the buses. Compare the fault currents with the values given in Table 9.4.

PW **9.61** Using the Example 9.8 case determine the fault current, except with a bolted double line-to-ground fault at each of the buses. Compare the fault currents with the values given in Table 9.4.

PW **9.62** Re-determine the Example 9.8 fault currents, except with a new line installed between buses 2 and 4. The parameters for this new line should be identical to those of the existing line between buses 2 and 4. The new line is not mutually coupled to any other line. Are the fault currents larger or smaller than the Example 9.8 values?

PW **9.63** Re-determine the Example 9.8 fault currents, except with a second generator added at bus 3. The parameters for the new generator should be identical to those of the existing generator at bus 3. Are the fault currents larger or smaller than the Example 9.8 values?

PW **9.64** Using PowerWorld Simulator case Chapter 9_Design, calculate the per-unit fault current and the current supplied by each of the generators for a single line-to-ground fault at the PETE69 bus. During the fault, what percentage of buses have voltage magnitude below 0.75 per unit?

PW **9.65** Repeat Problem 9.64, except place the fault at the TIM69 bus.

DESIGN PROJECT 4 (*CONTINUED*): POWER FLOW/SHORT CIRCUITS

Additional time given: 3 weeks
Additional time required: 10 hours

This is a continuation of Design Project 4. Assignments 1 and 2 are given in Chapter 6. Assignment 3 is given in Chapter 7.

Assignment 4: Short Circuits—Breaker/Fuse Selection

For the single-line diagram that you have been assigned (Figure 6.22 or 6.23), convert the zero-, positive-, and negative-sequence reactance data to per-unit using the given system base quantities. Use subtransient machine reactances. Then using PowerWorld Simulator, create the generator, transmission line, and transformer input data files. Next run the Simulator to compute subtransient fault currents for (1) single-line-to-ground, (2) line-to-line, and (3) double-line-to-ground bolted faults at each bus. Also compute the zero-, positive-, and negative-sequence bus impedance matrices. Assume 1.0 per-unit prefault voltage. Also, neglect prefault load currents and all losses.

For students assigned to Figure 6.22: Select a suitable circuit breaker from Table 7.10 for *each* location shown on your single-line diagram. Each breaker that you select should: (1) have a rated voltage larger than the maximum system operating voltage, (2) have a rated continuous current at least 30% larger than normal load current (normal load currents are computed in Assignment 2), and (3) have a rated short-circuit current larger than the maximum fault current for any type of fault at the bus where the breaker is located (fault currents are computed in Assignments 3 and 4). This conservative practice of selecting a breaker to interrupt the entire fault current, not just the contribution to the fault through the breaker, allows for future increases in fault currents. *Note:* Assume that the (X/R) ratio at each bus is less than 15, such that the breakers are capable of interrupting the dc-offset in addition to the subtransient fault current. Circuit breaker cost should also be a factor in your selection. Do *not* select a breaker that interrupts 63 kA if a 40-kA or a 31.5-kA breaker will do the job.

For students assigned to Figure 6.23: Enclosed [9, 10] are "melting time" and "total clearing time" curves for K rated fuses with continuous current ratings from 15 to 200 A. Select suitable branch and tap fuses from these curves for each of the following three locations on your single-line diagram: bus 2, bus 4, and bus 7. Each fuse you select should have a continuous current rating that is at least 15% higher but not more than 50% higher than the normal load current at that bus (normal load currents are computed in Assignment 2). Assume that cables to the load can withstand 50% continuous overload currents. Also, branch fuses should be coordinated with tap fuses; that is, for every fault current, the tap fuse should clear before the branch fuse melts. For each of the three buses, assume a reasonable X/R ratio and determine the asymmetrical fault current for a three-phase bolted fault (subtransient current is computed in Assignment 3). Then for the fuses that you select from [9, 10], determine the clearing time CT of tap fuses and the melting time MT of branch fuses. The ratio MT/CT should be less than 0.75 for good coordination.

DESIGN PROJECT 6

Time given: 3 weeks
Approximate time required: 10 hours

As a protection engineer for Metropolis Light and Power (MLP) your job is to ensure that the transmission line and transformer circuit breaker ratings are sufficient to interrupt the fault current associated with any type of fault (balanced three phase, single line-to-ground, line-to-line, and double line-to-ground). The MLP power system is modeled in case Chapter9_Design. This case models the positive, negative and zero sequence values for each system device. Note that the 69/138 kV transformers are grounded wye on the low side and delta on the high side; the 138 kV/345 kV transformers grounded wye on both sides. In this design problem your job is to evaluate the circuit breaker ratings for the three 345 kV transmission lines and the six 345/138 kV transformers. You need not consider the 138 or 69 kV transmission lines, or the 138/69 kV transformers.

Design Procedure

1. Load Chapter9_Design into PowerWorld Simulator. Perform an initial power flow solution to get the base case system operating point.
2. Apply each of the four fault types to each of the 345 kV buses and to the 138 kV buses attached to 345/138 kV transformers to determine

the maximum fault current that each of the 345 kV lines and 345/138 kV transformers will experience.

3. For each device select a suitable circuit breaker from Table 7.10. Each breaker that you select should a) have a rated voltage larger than the maximum system operating voltage, b) have a rated continuous current at least 30% larger than the normal rated current for the line, c) have a rated short circuit current larger than the maximum fault current for any type of fault at the bus where the breaker is located. This conservative practice of selecting a breaker to interrupt the entire fault current, not just the contribution to the fault current through the breaker allows for future increases in fault currents. Since higher rated circuit breakers cost more, you should select the circuit breaker with the lowest rating that satisfies the design constraints.

Simplifying Assumptions

1. You need only consider the base case conditions given in the Chapter9_Design case.
2. You may assume that the X/R ratios at each bus is sufficiently small (less than 15) so that the dc offset has decayed to a sufficiently low value (see Section 7.7 for details).
3. As is common with commercial software, including PowerWorld Simulator, the Δ-Y transformer phase shifts are neglected.

CASE STUDY QUESTIONS

A. Are safety hazards associated with generation, transmission, and distribution of electric power by the electric utility industry greater than or less than safety hazards associated with the transportation industry? The chemical products industry? The medical services industry? The agriculture industry?

B. What is the public's perception of the electric utility industry's safety record?

REFERENCES

1. Westinghouse Electric Corporation, *Electrical Transmission and Distribution Reference Book*, 4th ed. (East Pittsburgh, PA, 1964).
2. Westinghouse Electric Corporation, *Applied Protective Relaying* (Newark, NJ, 1976).
3. P. M. Anderson, *Analysis of Faulted Power Systems* (Ames: Iowa State University Press, 1973).

4. J. R. Neuenswander, *Modern Power Systems* (New York: Intext Educational Publishers, 1971).

5. H. E. Brown, *Solution of Large Networks by Matrix Methods* (New York: Wiley, 1975).

6. W. D. Stevenson, Jr., *Elements of Power System Analysis*, 4th ed. (New York: McGraw-Hill, 1982).

7. C. A. Gross, *Power System Analysis* (New York: Wiley, 1979).

8. Glenn Zorpette, "Fires at U.S. Utilities," *IEEE Spectrum*, *28*, 1 (January 1991), p. 64.

9. McGraw Edison Company, *Fuse Catalog*, R240-91-1 (Canonsburg, PA: Mcgraw Edison, April 1985).

10. Westinghouse Electric Corporation, *Electric Utility Engineering Reference Book: Distribution Systems* (Pittsburgh, PA: Westinghouse, 1959).

Lightning slices through rainy skies above a city, in a time-exposure view (Jhaz Photography/Shutterstock)

10 SYSTEM PROTECTION

Short circuits occur in power systems when equipment insulation fails, due to system overvoltages caused by lightning or switching surges, to insulation contamination, or to other mechanical and natural causes. Careful design, operation, and maintenance can minimize the occurrence of short circuits but cannot eliminate them. We discussed methods for calculating short-circuit currents for balanced and unbalanced faults in Chapters 7 and 9. Such currents can be several orders of magnitude larger than normal operating currents and, if allowed to persist, may cause insulation damage, conductor melting, fire, and explosion. Windings and busbars may also suffer mechanical damage due to high magnetic forces during faults. Clearly, faults must be quickly removed from a power system. Standard EHV protective equipment is designed to clear faults within 3 cycles, whereas lower-voltage protective equipment typically operates within 5–20 cycles.

This chapter provides an introduction to power system protection. Blackburn defines protection as "the science, skill, and art of applying and setting relays and/or fuses to provide maximum sensitivity to faults and undesirable conditions, but to avoid their operation on all permissible or tolerable conditions" [1]. The basic idea is to define the undesirable conditions and look for differences between the undesirable and permissible conditions that relays or fuses can sense. It is also important to remove only the faulted equipment from the system while maintaining as much of the unfaulted system as possible in service, in order to continue to supply as much of the load as possible.

Although fuses and reclosers (circuit breakers with built-in instrument transformers and relays) are widely used to protect primary distribution systems (with voltages in the 2.4–46 kV range), we focus primarily in this chapter on circuit breakers and relays, which are used to protect HV (115–230 kV) and EHV (345–765 kV) power systems. The IEEE defines a relay as "a device whose function is to detect defective lines or apparatus or other power system conditions of an abnormal or dangerous nature and to initiate appropriate control action" [1]. In practice, a relay is a device that closes or opens a contact when energized. Relays are also used in low-voltage (600-V and below) power systems and almost anywhere that electricity is used. They are used in heating, air conditioning, stoves, clothes washers and dryers, refrigerators, dishwashers, telephone networks, traffic controls, airplane and other transportation systems, and robotics, as well as many other applications.

Problems with the protection equipment itself can occur. A second line of defense, called *backup* relays, may be used to protect the first line of defense, called *primary* relays. In HV and EHV systems, separate current- or voltage-measuring devices, separate trip coils on the circuit breakers, and separate batteries for the trip coils may be used. Also, the various protective devices must be properly coordinated such that primary relays assigned to protect equipment in a particular zone operate first. If the primary relays fail, then backup relays should operate after a specified time delay.

This chapter begins with a discussion of the basic system-protection components.

CASE STUDY

The following article describes key technology areas that have been identified to modernize the transmission and distribution of electric power [14]. Included in this article are discussions of the following: (1) advanced control features including flexible ac transmission systems (FACTS); (2) advanced protection including digital technology for relays, communication and operation; (3) synchronized phasor measurement units (PMUs) achieved through a Global Positioning System (GPS); (4) automatic calibration of instrument transformers using PMUs; (5) precise state measurements and estimates; and (6) intelligent visualization techniques.

The Future of Power Transmission: Technological Advances for Improved Performance

BY STANLEY H. HOROWITZ,
ARUN G. PHADKE, AND BRUCE A. RENZ

The electric power system is on the verge of significant transformation. For the past five years or so, work has been under way to conceptualize the shape of a 21st-century grid that exploits the huge progress that has been made in digital technology and advanced materials. The National Energy Technology Laboratory (NETL) has identified five foundational key technology areas (KTAs), as shown in Figure 1.

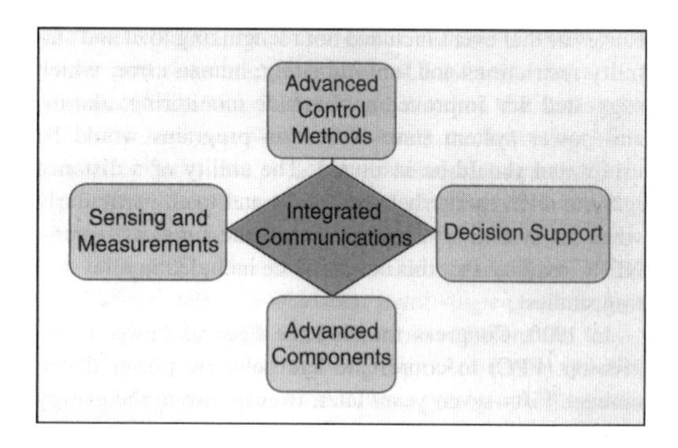

Figure 1
NETL's five key technology areas

Foremost among these KTAs will be integrated communications. The communications requirements for transmission enhancement are clear. Broadband, secure, low-latency channels connecting transmission stations to each other and to control centers will enable advances in each of the other KTAs.

- Sensing and measurements will include phasor measurement data streaming over high-speed channels.
- Advanced components, such as all forms of flexible ac transmission system (FACTS) devices, HVDC, and new storage technologies will respond to control signals sent to address perturbations occurring in milliseconds.
- Advanced control (and protection) methods will include differential line relaying, adaptive settings, and various system integrity protection schemes that rely on low-latency communications.
- Improved interfaces and decision support will utilize instantaneous measurements from phasor measurement units (PMUs) and other sources to drive fast simulations and advanced visualization tools that can help system operators assess dynamic challenges to system stability.

Each of these elements will be applied to the modernization of the grid, at both the distribution level and the transmission level. Because it is clearly less advanced, distribution is receiving most of the initial focus. This is dramatically illustrated by the American Recovery and Reinvestment Act's Smart Grid Investment Grants (SGIGs), announced in October. Of the $3.4 billion awarded to 100 proposers (of the more than 400 that applied), only $148 million went to transmission applications; most of the rest was for distribution projects.

While the changes to distribution will be revolutionary, transmission will change in an evolutionary manner. Distributed generation and storage, demand response, advanced metering infrastructure (SGIGs will fund the deployment of 18 million smart meters), distribution automation, two-way power flow, and differentiated power quality together represent a sea change in distribution design that will require enormous financial and intellectual capital.

The role of transmission will not be diminished, however, by this new distribution paradigm. Large central power plants will continue to serve as our bulk power source, and many new ones will be fueled by renewable resources that would today be out of reach of the transmission grid. New lines will be built to connect these new plants, and new methods will be employed to accommodate their very different performance characteristics. Addressing the resulting greater variability of supply will be the job of the five KTAs listed above. As KTA technology speeds increase, control of transmission will advance from quasi-steady-state to dynamic.

The traditional communications technologies capable of supporting these strict requirements are fiber optics (e.g., optical ground wire) and microwave. Recently a third candidate has appeared on the scene. Research funded by the U.S. Department of Energy (DOE) and American Electrical Power in conjunction with a small Massachusetts smart-grid communications company, Amperion, has demonstrated the viability of broadband over power line (BPL) for application on transmission lines. Currently, a five-mile (or 8-km), 69-kV line is operating at megabit-per-second data rates with latency of less than 10 ms. The next step will be to extend this high-voltage BPL technology to 138 kV.

HOW WE GOT HERE

In 551 B.C., Confucius wrote, "Study the past if you would know the future." The future of the electrical power transmission system must be based on a study of the past considering its successes and failures, on knowledge of the existing system and all of its component disciplines, and on a thorough understanding of the latest technologies and their possible applications. The electrical power system, and in particular its transmission and distribution network, is a vital and integral part of today's society. Because it is essential to all our endeavors, we must be prepared to integrate new, exciting, and highly innovative concepts to guarantee that it performs reliably, safely, economically, and cleanly.

Although not unique in world events involving power systems, two widely known outages in the United States and Canada serve as examples of the history, analysis, and remedies for blackouts and can provide a basis for future actions. Widely publicized, the blackouts of 9 November 1965 in the northeastern United States and 23 August 2003 in the northeastern United States and Canada are typical events that can help shape our planning and operating efforts for the future.

In 1965 we learned that cooperation and interaction between utilities were essential. In response to the blackout, utilities established the National Electric Reliability Corporation (NERC) in 1968, which began distributing recommendations and information. These communications formed the basis for more reliable and secure planning, operating, and protective activities. The decisions of the newly formed NERC were, however, only recommendations. Deficiencies due to limitations in transmission planning, operations, and protection were recognized, and steps were taken to correct them. Transmission systems were strengthened considerably by the construction of 345-kV, 500-kV, and 765-kV lines. System planning studies were made cooperatively; operating parameters and system problems were studied jointly. Underfrequency load shedding became universal, with specific settings arrived at by agreement between utilities, and loss-of-field relaying was recognized as a system phenomenon and studied accordingly.

In 2003 another blackout of similar proportions affected the northeastern United States and parts of Canada. The causes of that event included not recognizing load and stability restrictions and, unfortunately, human error, which suggested that improved systemwide monitoring, alarms, and power system state estimation programs would be useful and should be instituted. The ability of a distance relay to differentiate between faults and load, particularly when the system is stressed, has become a major concern. NERC requires that this condition be included in relay setting studies.

In 1920, Congress founded the Federal Power Commission (FPC) to coordinate hydroelectric power development. Fifty-seven years later, in response to the energy crisis, the DOE was formed. The DOE included the FPC, renamed the Federal Energy Regulatory Commission (FERC), whose mandate was primarily to conduct hearings and approve price control and related topics, including electric practices on bulk transmission systems. After the 2003 event, FERC also became a regulatory instrument, reviewing transmission line improvements and rights-of-way. FERC review and approval, as with NERC, has now become mandatory. The actions of FERC and NERC will, in the future, be major components of system decisions and practices.

TECHNOLOGY'S ROLE GOING FORWARD

With the preceding as background, we can now review in greater detail some of the transmission enhancements that will be part of the 21st-century transmission system.

ADVANCED CONTROL

It is axiomatic that the fundamental basis for the reliable performance of the transmission system has to be the system itself. The primary components, system configuration, line specifications, and design of high-voltage equipment must be consistent with the mission of the power system, i.e., to deliver electric energy safely, reliably, economically, and in a timely fashion. Furthermore, high-voltage, electronic-based power equipment such as bulk storage systems (e.g., flow batteries), FACTS devices (including unified power flow controllers, static var compensators, and static synchronous compensators), and current-limiting devices (CLDs), which are based on high-temperature superconductivity, are now or will soon be available. Coupled with sophisticated communications and computing tools, these devices make the transmission system much more accommodating of variations in load and/or voltage.

Of these advanced control devices, FACTS represents the most mature technology. It is in somewhat limited use at present but has the potential to be an increasingly important element in the future. FACTS can provide control of ac transmission system parameters and thus increase power transfer capability and improve voltage regulation. Changes in generation and load patterns may make such flexibility extremely desirable. With the increased penetration of central renewable sources and with the continued variability of electricity markets, the value of these various electronics-based power devices will only grow.

In addition to FACTS, bulk storage, and CLDs, various new aspects of the distribution system such as demand response, distributed generation, plug-in hybrid electric vehicles (PHEVs), and other forms of distributed storage can be centrally coordinated and integrated to function as a "virtual power system" that supports the transmission system in times of stress.

ADVANCED PROTECTION

Recognizing that protection of specific equipment and localized systems is inadequate in the face of systemwide stress, in 1966 a joint IEEE/CIGRE questionnaire was circulated. The results indicated that protection schemes had to encompass wider areas of the transmission system. This effort required communication and control center involvement. The effort was termed special protection systems (SPS). The primary application of SPS at that time was for limited system events such as underfrequency and undervoltage, with some advanced generation controls. As system stress becomes a more common concern, the application of SPS takes on added importance and in fact becomes an important tool for protecting the grid against wide-area contingencies.

The SPS concept is no longer considered "special" and is now commonly referred to as system integrity protection systems (SIPS), remedial action systems (RAS), or wide-area protection and control (WAPC). These schemes are intended to address widespread power system constraints or to be invoked when such constraints could occur as a result of increased transfer limits. The Power System Relaying Committee of the IEEE initiated a recent survey on power system integrity protective schemes that was distributed worldwide with cooperation from CIGRE, NERC, IEE, and other utility organizations. The survey revealed very widespread application, with more than 100 schemes of various complexity and purpose. Emerging technologies in highspeed communication, wide-area measurement, and phasor measurement are all employed and will be vital components of the transmission system in the future.

One of the most exciting features of the transmission system of the future involves power system protection. This is due in large measure to the advantages of digital technology for relays, communication, and operation. Relays now have the ability to perform previously unimaginable functions, made feasible by evaluating operating and fault parameters and coupling this data with high-speed communication and computer-driven applications within the power system control center. With the ever-increasing restrictions on transmission line and generator construction and siting and the decreasing difference between normal and abnormal operation, loading, and stability, the margins between the relaying reliability concepts of dependability and security are becoming blurred. Consequently, the criteria of traditional protection and control are being challenged. The hallmark of relays is the tradeoff between dependability (the ability to always trip when required) and security (the characteristic of never tripping when not required).

Traditionally, relays and relay schemes have been designed to be dependable. Losing a transmission line element must be tolerated, whether the loss is for an actual fault or for an inadvertent or incorrect trip. When the system is stressed, however, an incorrect trip is not allowable. With the system stressed, losing another element could be the final step in bringing down the entire network. With digital logic and operations, it is possible to reorder protection priorities and require additional inputs before allowing a trip. This can be done with appropriate communication from a central center advising the relays.

Probably one of the most difficult decisions for a relay is to distinguish between heavy loads and faults. Heretofore, relays simply relied on the impedance measurement, with settings determined by off-line load studies using conditions based on experience. As in the two blackouts mentioned above, this criterion was not adequate for unusual system conditions that were not previously considered probable. Digital relays can now establish such parameters as power factor or voltage and remove the measured impedance from the tripping logic.

The bête noir of protection has traditionally been the multiterminal line. The current to the fault and the voltage at the fault defines the fault location. A relay designed to protect the system for this fault, however, sees only the current and voltage at that relay's specific location. The advent of high-speed communication and digital logic remedies this condition and allows all involved relays to receive the appropriate fault currents and voltages.

The increasing popularity of transmission line differential relaying also provides both dependability and security for faults in a multiterminal configuration. Although primarily a current-measuring relay, the digital construction allows far more protection, monitoring, and recording functions. Future applications will be available to accomplish the features mentioned above and in ways not yet implemented or even thought of.

One of the earliest advantages of the computer relay is its ability to monitor itself and either repair, replace, or report the problem. This feature is sure to be a major feature in future transmission line protection. In addition, the information stored in each relay during both normal and abnormal conditions and the ability to analyze and transfer this information to analyzers have made previously used oscillography and sequence-of-events recorders obsolete. Replacing these devices will result in very significant savings in both hardware and installation costs. AEP, in conjunction with Schweitzer Engineering and Tarigma Corp., has embarked on a revolutionary program that lets selected centers receive data from critical substations that will combine, display, and analyze fault data to a degree and in a time frame heretofore not possible. Combining the current, voltage, communication signals, and breaker performance from several stations on one record that can be analyzed at several control and engineering centers permits operations to be verified and personnel to be alerted to potential problems. A vital by-product of this advanced monitoring is the fact that it allows NERC requirements for monitoring and analysis to be met.

Perhaps even more exciting is the possibility of predicting the instability of a power swing. Modern protection theory knows how to detect the swing using zones of stability and instability. The problem is how to set the zones. With accurate synchronized phasor measurements from several buses, the goal of real-time instability protection seems achievable. Out-of-step relays could then establish blocking or tripping functions at the appropriate stations.

The role of underfrequency load shedding has already been discussed. Future schemes, however, could use real-time measurements at system interconnection boundaries, compute a dynamic area control error, and limit any potential widespread underfrequency by splitting the system.

Computer relays, if not already in universal use, will be in the near future. This will let utilities protect, monitor, and analyze system and equipment performance in ways and to a degree not possible before.

SYNCHRONIZED PHASOR MEASUREMENTS

It has been recognized in recent years that synchronized phasor measurements are exceedingly versatile tools of modern power system protection, monitoring, and control. Future power systems are going to depend on making use of these measurements to an ever-increasing extent. The principal function of these systems is to measure positive sequence voltages and currents with a precise time stamp (to within a microsecond) of the instant when the measurement was made. The time stamps are directly traceable to the Coordinated Universal Time (UTC) standard and are achieved by using Global Positioning System (GPS) transmissions for synchronization. Many PMUs also provide other measurements, such as individual phase voltages and currents, harmonics, local frequency, and rate of change of frequency. These measurements can be obtained as often as once per power frequency cycle, although for a number of applications a slower measurement rate may be preferable. In well-designed systems, measurement latency (i.e., the delay between when the measurement is made and when it becomes available for use) can be limited to fewer than 50 ms. The performance requirements of the PMUs are embodied in the IEEE synchrophasor standard (C37.118). A measurement system that incorporates PMUs deployed over large portions of the power system has come to be known as a wide-area measurement system (WAMS), and a power system protection, monitoring, and control application that utilizes these measurements is often referred to as a wide-area measurement protection and control system (WAMPACS).

AUTOMATIC CALIBRATION OF INSTRUMENT TRANSFORMERS

It is well known that current and voltage transformers used on high-voltage networks have ratio and phase-angle errors that affect the accuracy of the measurements made on the secondary of these transformers. Capacitive voltage transformers are known to have errors that change with ambient conditions as well as with the age of the capacitor elements. Inductive instrument transformers have errors that change when their secondary loading (burden) is manually changed. The PMU offers a unique opportunity for calibration of the instrument transformers in real time and as often as necessary. In simple terms, the technique is based on having some buses where a precise voltage transformer (with known calibration) is available and where a PMU is placed. Potential transformers used for revenue metering are an example of such a voltage source. Using measurements by the PMU at this location, the calibration at the remote end of a feeder connected to this bus can be found. This calibration is not affected by current transformer (CT) errors when the system loading is light. It is thus possible to calibrate all voltage transformers using current measurements at light system load. Using the voltage transformer calibration thus obtained and additional measurements during heavy system load, the current transformers can be calibrated. In practice, it has been found (in simulated case studies) that by combining several light and heavy load measurement sets a very accurate estimate of all the current and voltage transformers can be obtained. Although a single accurate voltage source is sufficient in principle, having a number of them scattered throughout the network provides a more secure calibration.

PRECISE STATE MEASUREMENTS AND ESTIMATES

State estimation of power systems using real-time measurements of active and reactive power flows in the network (supplemented with a few other measurements) was introduced in the late 1960s to improve the awareness among power system operators of the prevailing state of the power grid and its ability to handle contingency conditions that may occur in the immediate future. This was a big step forward in intelligent operation of the power grid. The limitations of this technology (such as nonsimultaneity of system measurements across the network) were rooted in the technology of that day. The fact that the data from a dynamically changing power system was not obtained simultaneously over a significant time span meant that the estimated state was an approximation of the actual system state. Consequently, the system state and its response to contingencies could only be reasonably accurate when the power system was in a quasi-steady state. Indeed, when the power system was undergoing significant changes due to evolving events, the state estimator could not always be counted on to converge to a usable solution.

The advent of wide-area measurements using GPS-synchronized PMUs led to a paradigm shift in the state estimation process. With this technology, the capability of directly measuring the state of the power system has become a reality. PMUs measure positive sequence voltages at network buses and positive sequence currents in transmission lines and transformers. Since the state of the power system is defined as a collection of positive sequence voltages at all network buses, it is clear that with sufficient numbers of PMU installations in the system one can measure the system state directly: no estimation is necessary. In fact, the transmission line currents provide a direct estimate of voltage at a remote bus in terms of the voltage at one end. It is therefore not necessary to install PMUs at all system buses. It has been found that by installing PMUs at about one-third of system buses with voltage and current measurements, it is possible to determine the complete system state vector. Feeding this information into the appropriate computers provides the information necessary for the adaptive protective function described above. Of course, a larger number of PMUs provides redundancy of measurements, which is always a desirable feature of estimation processes.

COMPLETE AND INCOMPLETE OBSERVABILITY

In order to achieve a state estimate in the traditional way, i.e., by using unsynchronized supervisory control and data acquisition (SCADA) measurements, a complete network tree must be measured. With PMUs, however, it is sufficient to measure isolated parts of the network, which provides islands of observable networks. This is possible since all phasors are synchronized to the same instant in time. The process has been described as *PMU placement for incomplete observability.* The remaining network buses can be estimated from the observed islands using approximation techniques. This is, of course, not as accurate as providing a sufficient number of PMUs in the first place. But it has been shown that combining incomplete observations with such an approximation technique to estimate the unobserved parts provides surprisingly useful results. Incomplete observability estimators are a natural step in the progression towards complete observability and will be a feature in future transmission systems.

Figure 2 illustrates the principle of complete and incomplete observability. In Figure 2(a), PMUs are placed at buses identified by dark circles. By making use of the current measurements and the network impedance data, it is possible to calculate the voltages at the buses identified by light blue circles. In this case, complete observability is achieved with two PMUs. Figure 2(b) illustrates the use of fewer PMUs than would be necessary for complete observability.

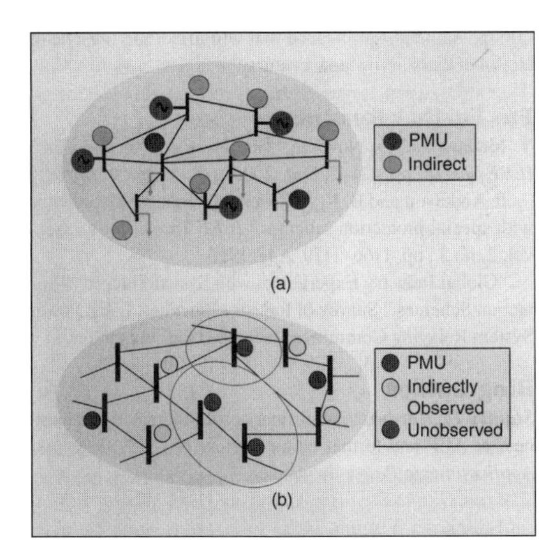

Figure 2
(a) Complete observability, (b) Incomplete observability

Even with current measurements, it is not possible to determine the voltages at the buses identified by the red circles. These buses form islands of incomplete observability. As mentioned earlier, these bus voltages can be estimated fairly accurately using voltages at surrounding buses.

STATE ESTIMATES OF INTERCONNECTED SYSTEMS

A common problem faced by interconnected power systems is that various parts of the system may be under different control centers, with each part having its own state estimator. This, of course, implies that each partial state estimate has its own reference bus. To perform studies such as contingency analysis on the interconnected power system, it is necessary to have a single state estimate for the entire network. This requires either that 1) a new system state using data from all partial control centers be determined or 2) an alternative must be found to modify the results of individual state estimates to put all states on a common reference. Option 1 is cumbersome and wasteful of computational effort. Option 2 becomes exceedingly simple with PMUs. At the simplest level, one can visualize putting a PMU at each of the reference buses, thus obtaining the phase-angle relationships between all partial estimates. These phase-angle corrections may then be used to form a combined state estimate for the entire interconnected network on a single reference. It has been found in practice that the

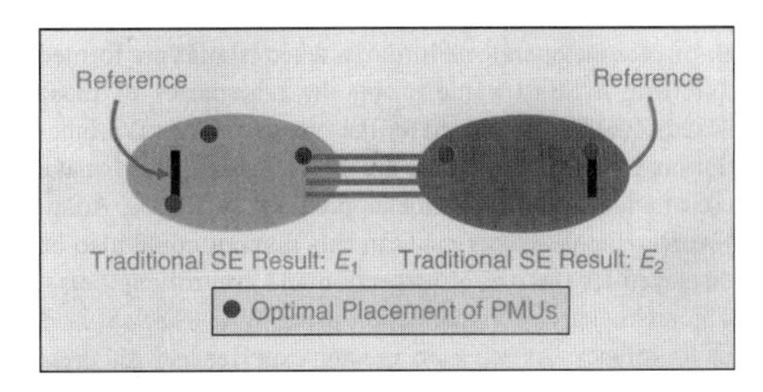

Figure 3
Connecting adjacent state estimates with phasor data

placement of a few PMUs in each partial system (rather than just one at the reference bus) leads to greater security and optimal performance. This principle is illustrated in Figure 3. Systems 1 and 2 are connected by tie lines and have state estimates S1 and S2 that are obtained independently, each with its own reference bus. With the use of PMU data from optimally selected buses (shown in red), it is possible to determine the angle difference between the two references and obtain a single state estimate for the interconnected system.

INTELLIGENT VISUALIZATION TECHNIQUES

The traditional visualization techniques used in energy management system (EMS) centers focus on showing network bus voltages and line flows, along with any constraint violations that may exist. It is, of course, possible to reproduce such displays using WAMS technology. Dynamic loading limits of transmission lines have been estimated with WAMS, and it would be relatively simple to show prevailing loading conditions and their proximity to the dynamic loading limits. Many PMUs offer the possibility of measuring system unbalances. It would then be possible to display unbalance currents to determine their sources and mitigation techniques to correct the unbalance.

With direct measurement of synchronized phasors, many more display options become possible. For example, a geographical display with phase angles at all network buses shown at the physical location of buses—and perhaps fitted with a surface in order to provide a hilly contour—would immediately show the distribution of positive sequence voltage phase angles.

Figure 4 shows such a visualization of a hypothetical network state for the entire United States. The map colors identify the magnitude and sign of the positive sequence voltage phase angle with respect to a center of angle reference. The lower plot is a footprint of equiangle loci from the map. Since the positive sequence voltage

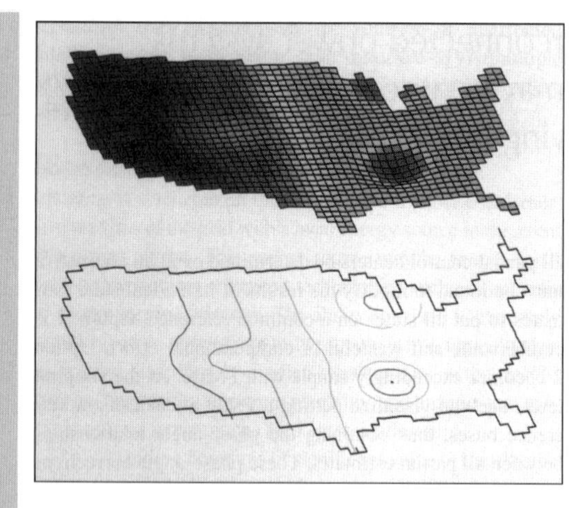

Figure 4
U.S. Phasor contour map

phase-angle profile of a network conveys a great deal of information regarding its power flow and loading conditions, such visualizations can instantly show the quality of the prevailing system state and its distance from a normal state. High-speed dynamic phenomena can be represented by animations of such visualizations.

Such a display would instantly show the general disposition of generation surplus and load surplus areas. Such a picture can be updated at scan rates of a few cycles, leading to visualization of dynamic conditions on the network. If thresholds for phase-angle differences between key buses have been established for secure operation of the network, then violation of those thresholds could lead to important alarms for the operator. Similarly, when islands are formed following a catastrophic event, the boundaries of those islands could be displayed for the operator. Several protection and control principles are being developed to make use of wide-area measurements provided by PMUs. Adaptive relaying decisions made in this manner could also be displayed for the use of protection and control engineers. The technology of visualization using WAMS schemes is in its infancy. As we gain greater experience with these systems, more interesting display ideas will undoubtedly be forthcoming.

CONCLUSION

Modernizing the U.S. power grid has become a national priority. Unprecedented levels of governmental funding have been committed in order to achieve this goal. The initial focus has been on the fundamental transformation of the distribution system. This is in itself a huge technical challenge that will be measured not in years but in decades. The end result is expected to be higher efficiency, reduced environmental impact, improved reliability, and lower exposure to terrorism.

The revolution in distribution must be accompanied by the continued evolution of the transmission system. Events like the 2003 blackout—more the result of human shortcomings than technological breakdowns—can be eliminated by exploiting the huge progress made in recent years in the digital and material sciences. Other industries have already harvested these opportunities; now it is our turn.

Technological development is an engineering challenge. This nation has time and again demonstrated its ability to meet such challenges whenever they have been clearly focused. But there is another challenge that may actually be more difficult. It is to find the political alignment that is needed to accept the vision and move forward aggressively. For transmission, that means recognizing that new lines, not just better lines, will be needed. It is simply not acceptable to wait ten or more years for a new line to move from concept to reality. Unlike many other parts of the world, the United States has allowed fragmented responsibility for transmission additions to slow the process to an unacceptable extent.

With the intense focus now on energy in general and electricity in particular, it should be possible to overcome both the technical and the political obstacles and to reestablish U. S. leadership in this vital arena. Doing so is a matter of huge national significance that will affect the lifestyle of all Americans in this new century.

FOR FURTHER READING

V. Madani and D. Novosel, "Getting a Grip on the Grid," *IEEE Spectr.*, pp. 42–47, Dec. 2005.

P. Anderson and B. K. LeReverend, "Industry Experience with Special Protection Schemes," *IEEE Trans. Power Syst.*, vol. 2, no. 3, pp. 1166–1179, Aug. 1996.

"Global Industry Experiences with System Integrity Protection Schemes," Survey of Industry Practices, IEEE Power System Relaying Committee, submitted for publication.

BIOGRAPHIES

Stanley H. Horowitz is a former consulting electrical engineer at AEP and former editor-in-chief of *IEEE Computer Applications in Power* magazine.

Arun G. Phadke is the University Distinguished Professor Emeritus at Virginia Tech.

Bruce A. Renz is president of Renz Consulting, LLC.

10.1

SYSTEM PROTECTION COMPONENTS

Protection systems have three basic components:

1. Instrument transformers
2. Relays
3. Circuit breakers

Figure 10.1 shows a simple overcurrent protection schematic with: (1) one type of instrument transformer—the current transformer (CT), (2) an overcurrent relay (OC), and (3) a circuit breaker (CB) for a single-phase line. The function of the CT is to reproduce in its secondary winding a current I' that is proportional to the primary current I. The CT converts primary currents in the kiloamp range to secondary currents in the 0–5 ampere range for convenience of measurement, with the following advantages.

Safety: Instrument transformers provide electrical isolation from the power system so that personnel working with relays will work in a safer environment.

Economy: Lower-level relay inputs enable relays to be smaller, simpler, and less expensive.

Accuracy: Instrument transformers accurately reproduce power system currents and voltages over wide operating ranges.

The function of the relay is to discriminate between normal operation and fault conditions. The OC relay in Figure 10.1 has an operating coil, which is connected to the CT secondary winding, and a set of contacts.

FIGURE 10.1

Overcurrent protection schematic

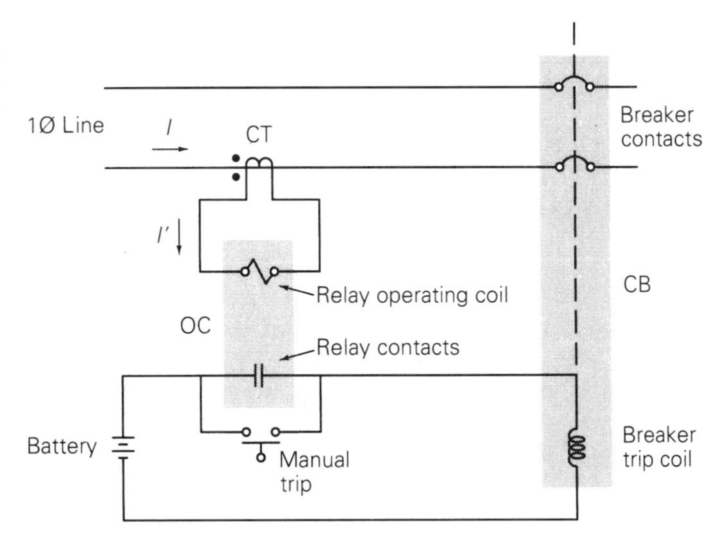

When $|I'|$ exceeds a specified "pickup" value, the operating coil causes the normally open contacts to close. When the relay contacts close, the trip coil of the circuit breaker is energized, which then causes the circuit breaker to open.

Note that the circuit breaker does not open until its operating coil is energized, either manually or by relay operation. Based on information from instrument transformers, a decision is made and "relayed" to the trip coil of the breaker, which actually opens the power circuit—hence the name *relay*.

System-protection components have the following design criteria [2]:

Reliability: Operate dependably when fault conditions occur, even after remaining idle for months or years. Failure to do so may result in costly damages.

Selectivity: Avoid unnecessary, false trips.

Speed: Operate rapidly to minimize fault duration and equipment damage. Any intentional time delays should be precise.

Economy: Provide maximum protection at minimum cost.

Simplicity: Minimize protection equipment and circuitry.

Since it is impossible to satisfy all these criteria simultaneously, compromises must be made in system protection.

10.2 INSTRUMENT TRANSFORMERS

There are two basic types of instrument transformers: voltage transformers (VTs), formerly called potential transformers (PTs), and current transformers (CTs). Figure 10.2 shows a schematic representation for the VT and CT.

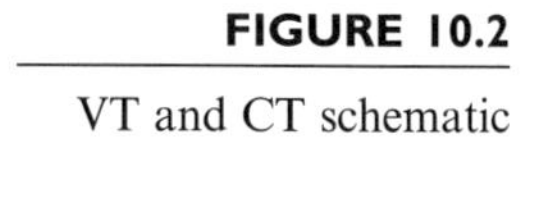

FIGURE 10.2

VT and CT schematic

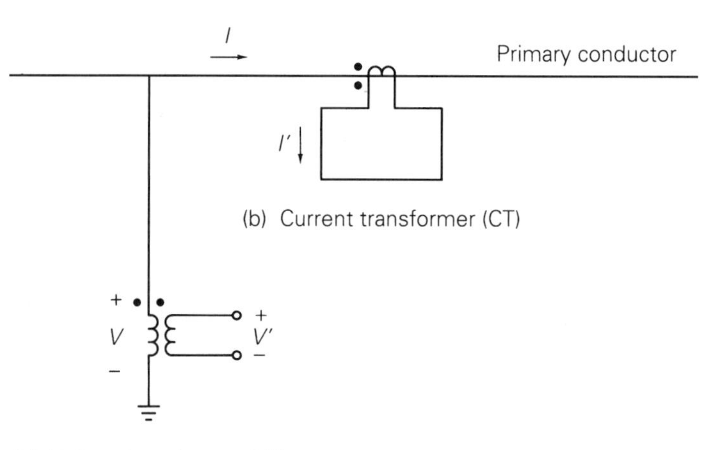

FIGURE 10.3

Three 34.5-kV voltage transformers with 34.5 kV : 115/67 volt VT ratios, at Lisle substation, Lisle, Illinois (Courtesy of Commonwealth Edison, an Exelon Company)

The transformer primary is connected to or into the power system and is insulated for the power system voltage. The VT reduces the primary voltage and the CT reduces the primary current to much lower, standardized levels suitable for operation of relays. Photos of VTs and CTs are shown in Figures 10.3–10.6 on pages 527–529.

For system-protection purposes, VTs are generally considered to be sufficiently accurate. Therefore, the VT is usually modeled as an ideal transformer, where

$$V' = (1/\mathrm{n})V \tag{10.2.1}$$

V' is a scaled-down representation of V and is in phase with V. A standard VT secondary voltage rating is 115 V (line-to-line). Standard VT ratios are given in Table 10.1 on page 530.

Ideally, the VT secondary is connected to a voltage-sensing device with infinite impedance, such that the entire VT secondary voltage is across the sensing device. In practice, the secondary voltage divides across the high-impedance sensing device and the VT series leakage impedances. VT leakage impedances are kept low in order to minimize voltage drops and phase-angle differences from primary to secondary.

The primary winding of a current transformer usually consists of a single turn, obtained by running the power system's primary conductor through the CT core. The normal current rating of CT secondaries is standardized at 5 A in the United States, whereas 1 A is standard in Europe and some other regions. Currents of 10 to 20 times (or greater) normal rating often occur in CT windings for a few cycles during short circuits. Standard CT ratios are given in Table 10.2 on page 530.

FIGURE 10.4 Three 500-kV coupling capacitor voltage transformers with 303.1 kV : 115/67 volt VT ratios, Westwing 500-kV Switching Substation (Courtesy of Arizona Public Service)

Ideally, the CT secondary is connected to a current-sensing device with zero impedance, such that the entire CT secondary current flows through the sensing device. In practice, the secondary current divides, with most flowing through the low-impedance sensing device and some flowing through the CT shunt excitation impedance. CT excitation impedance is kept high in order to minimize excitation current.

An approximate equivalent circuit of a CT is shown in Figure 10.7, where

Z' = CT secondary leakage impedance

X_e = (Saturable) CT excitation reactance

Z_B = Impedance of terminating device (relay, including leads)

The total impedance Z_B of the terminating device is called the *burden* and is typically expressed in values of less than an ohm. The burden on a CT may also be expressed as volt-amperes at a specified current.

Associated with the CT equivalent circuit is an excitation curve that determines the relationship between the CT secondary voltage E' and excitation

FIGURE 10.5

Three 25 kV class current transformers–window design (Courtesy of Kuhlman Electric Corporation)

FIGURE 10.6

500-kV class current transformers with 2000:5 CT ratios in front of 500-kV SF6 circuit breakers, Westwing 500-kV Switching Substation (Courtesy of Arizona Public Service)

TABLE 10.1

Standard VT ratios

Voltage Ratios						
1:1	2:1	2.5:1	4:1	5:1	20:1	40:1
60:1	100:1	200:1	300:1	400:1	600:1	800:1
1000:1	2000:1	3000:1	4500:1			

TABLE 10.2

Standard CT ratios

Current Ratios						
50:5	100:5	150:5	200:5	250:5	300:5	400:5
450:5	500:5	600:5	800:5	900:5	1000:5	1200:5
1500:5	1600:5	2000:5	2400:5	2500:5	3000:5	3200:5
4000:5	5000:5	6000:5				

current I_e. Excitation curves for a multiratio bushing CT with ANSI classification C100 are shown in Figure 10.8.

Current transformer performance is based on the ability to deliver a secondary output current I' that accurately reproduces the primary current I. Performance is determined by the highest current that can be reproduced without saturation to cause large errors. Using the CT equivalent circuit and excitation curves, the following procedure can be used to determine CT performance.

STEP 1 Assume a CT secondary output current I'.

STEP 2 Compute $E' = (Z' + Z_B)I'$.

STEP 3 Using E', find I_e from the excitation curve.

STEP 4 Compute $I = n(I' + I_e)$.

STEP 5 Repeat Steps 1–4 for different values of I', then plot I' versus I.

For simplicity, approximate computations are made with magnitudes rather than with phasors. Also, the CT error is the percentage difference between $(I' + I_e)$ and I', given by:

$$\text{CT error} = \frac{I_e}{I' + I_e} \times 100\% \tag{10.2.2}$$

The following examples illustrate the procedure.

FIGURE 10.7

CT equivalent circuit

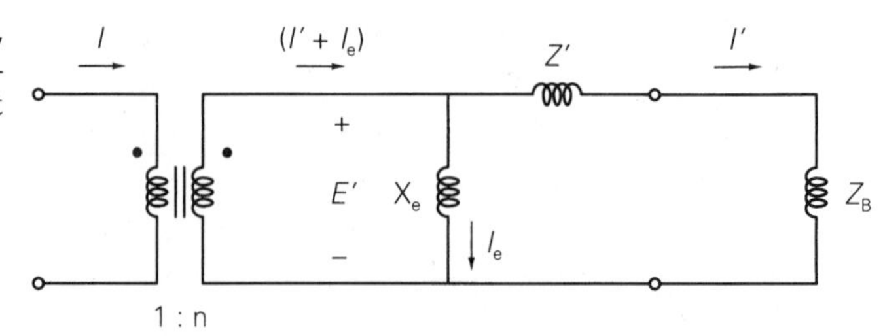

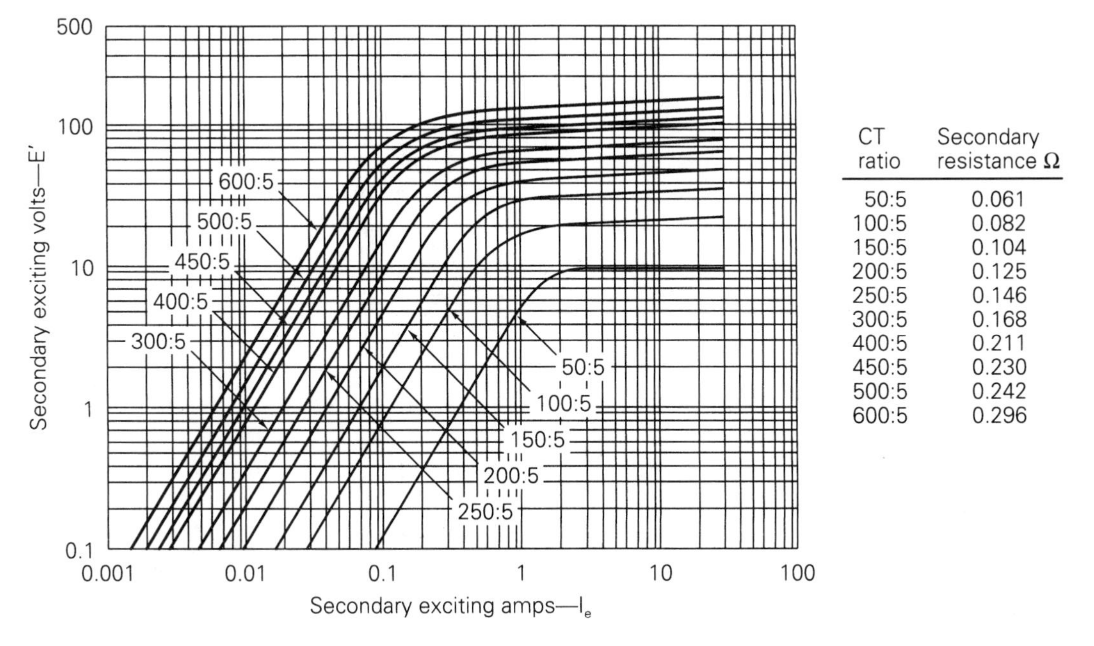

CT ratio	Secondary resistance Ω
50:5	0.061
100:5	0.082
150:5	0.104
200:5	0.125
250:5	0.146
300:5	0.168
400:5	0.211
450:5	0.230
500:5	0.242
600:5	0.296

FIGURE 10.8 Excitation curves for a multiratio bushing CT with a C100 ANSI accuracy classification [3] (Westinghouse Relay manual, A New Silent Sentinels Publication (Newark, NJ: Westinghouse Electric corporation, 1972))

EXAMPLE 10.1 Current transformer (CT) performance

Evaluate the performance of the multiratio CT in Figure 10.8 with a 100:5 CT ratio, for the following secondary output currents and burdens: (a) $I' = 5$ A and $Z_B = 0.5\ \Omega$; (b) $I' = 8$ A and $Z_B = 0.8\ \Omega$; and (c) $I' = 15$ A and $Z_B = 1.5\ \Omega$. Also, compute the CT error for each output current.

SOLUTION From Figure 10.8, the CT with a 100:5 CT ratio has a secondary resistance $Z' = 0.082\ \Omega$. Completing the above steps:

a. **STEP 1** $I' = 5$ A

STEP 2 From Figure 10.7,

$$E' = (Z' + Z_B)I' = (0.082 + 0.5)(5) = 2.91 \text{ V}$$

STEP 3 From Figure 10.8, $I_e = 0.25$ A

STEP 4 From Figure 10.7, $I = (100/5)(5 + 0.25) = 105$ A

$$\text{CT error} = \frac{0.25}{5.25} \times 100 = 4.8\%$$

b. **STEP 1** $I' = 8$ A

STEP 2 From Figure 10.7,

$$E' = (Z' + Z_B)I' = (0.082 + 0.8)(8) = 7.06 \text{ V}$$

STEP 3 From Figure 10.8, $I_e = 0.4$ A

STEP 4 From Figure 10.7, $I = (100/5)(8 + 0.4) = 168$ A

$$\text{CT error} = \frac{0.4}{8.4} \times 100 = 4.8\%$$

c. **STEP 1** $I' = 15$ A

STEP 2 From Figure 10.7,

$$E' = (Z' + Z_B)I' = (0.082 + 1.5)(15) = 23.73 \text{ V}$$

STEP 3 From Figure 10.8, $I_e = 20$ A

STEP 4 From Figure 10.7, $I = (100/5)(15 + 20) = 700$ A

$$\text{CT error} = \frac{20}{35} \times 100 = 57.1\%$$

Note that for the 15-A secondary current in (c), high CT saturation causes a large CT error of 57.1%. Standard practice is to select a CT ratio to give a little less than 5-A secondary output current at maximum normal load. From (a), the 100:5 CT ratio and 0.5 Ω burden are suitable for a maximum primary load current of about 100 A. This example is extended in Problem 10.2 to obtain a plot of I' versus I. ■

EXAMPLE 10.2 Relay operation versus fault current and CT burden

An overcurrent relay set to operate at 8 A is connected to the multiratio CT in Figure 10.8 with a 100:5 CT ratio. Will the relay detect a 200-A primary fault current if the burden Z_B is (a) 0.8 Ω, (b) 3.0 Ω?

SOLUTION Note that if an ideal CT is assumed, $(100/5) \times 8 = 160$-A primary current would cause the relay to operate.

a. From Example 10.1(b), a 168-A primary current with $Z_B = 0.8$ Ω produces a secondary output current of 8 A, which would cause the relay to operate. Therefore, the higher 200-A fault current will also cause the relay to operate.

b. **STEP 1** $I' = 8$ A

STEP 2 From Figure 10.7,

$$E' = (Z' + Z_B)I' = (0.05 + 3.0)(8) = 24.4 \text{ V}$$

STEP 3 From Figure 10.8, $I_e = 30$ A

STEP 4 From Figure 10.7, $I = (100/5)(8 + 30) = 760$ A

With a 3.0-Ω burden, 760 A is the lowest primary current that causes the relay to operate. Therefore, the relay will not operate for the 200-A fault current. ■

10.3 OVERCURRENT RELAYS

As shown in Figure 10.1, the CT secondary current I' is the input to the overcurrent relay operating coil. Instantaneous overcurrent relays respond to the magnitude of their input current, as shown by the trip and block regions in Figure 10.9. If the current magnitude $\mathrm{I}' = |I'|$ exceeds a specified adjustable current magnitude I_p, called the *pickup* current, then the relay contacts close "instantaneously" to energize the circuit breaker trip coil. If I' is less than the pickup current I_p, then the relay contacts remain open, blocking the trip coil.

Time-delay overcurrent relays also respond to the magnitude of their input current, but with an intentional time delay. As shown in Figure 10.10, the time delay depends on the magnitude of the relay input current. If I' is a large multiple of the pickup current I_p, then the relay operates (or trips) after a small time delay. For smaller multiples of pickup, the relay trips after a longer time delay. And if $\mathrm{I}' < \mathrm{I}_\mathrm{p}$, the relay remains in the blocking position.

Figure 10.11 shows two examples of a time-delay overcurrent relay: (a) Westinghouse electromechanical CO relay; and (b) Basler Electric digital relay. Characteristic curves of the Westinghouse CO-8 relay are shown in Figure 10.12. These relays have two settings:

Current tap setting: The pickup current in amperes.

Time-dial setting: The adjustable amount of time delay.

FIGURE 10.9

Instantaneous overcurrent relay block and trip regions

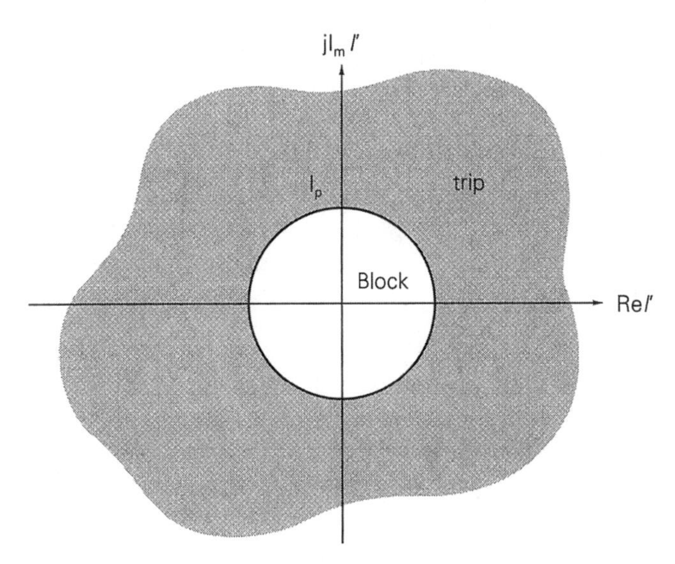

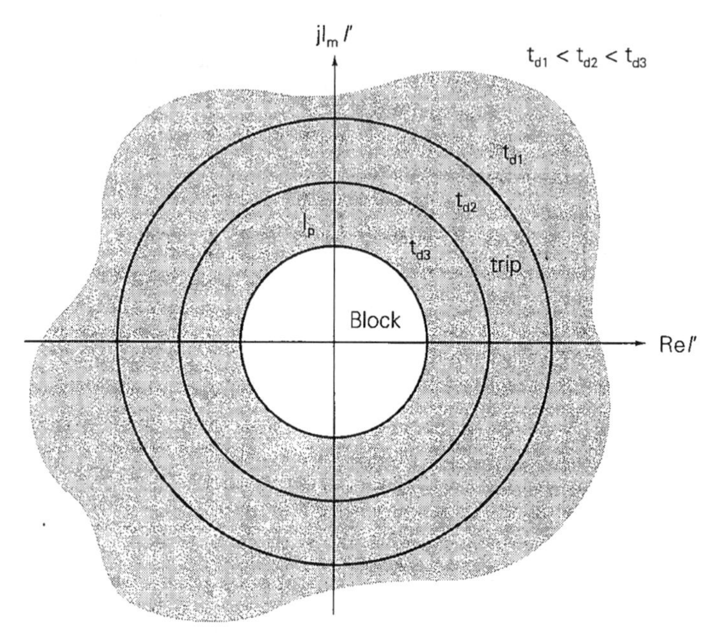

FIGURE 10.10 Time-delay overcurrent relay block and trip regions

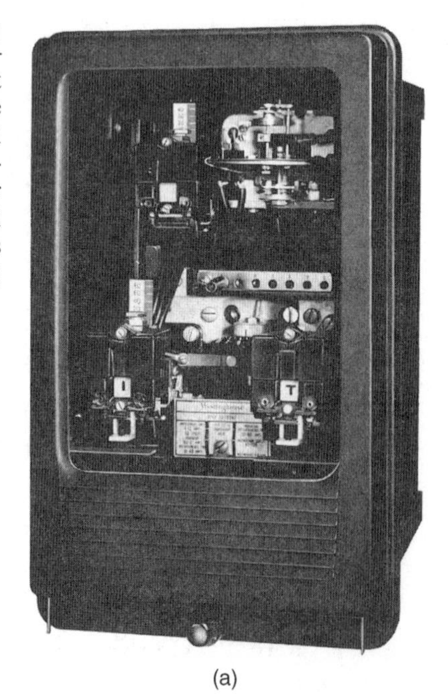

(a)

(b)

FIGURE 10.11 Time-delay overcurrent relays: (a) Westinghouse Electromechanical (Courtesy of ABB-Westinghouse) (b) Basler Electric Digital (Courtesy Danvers Electric)

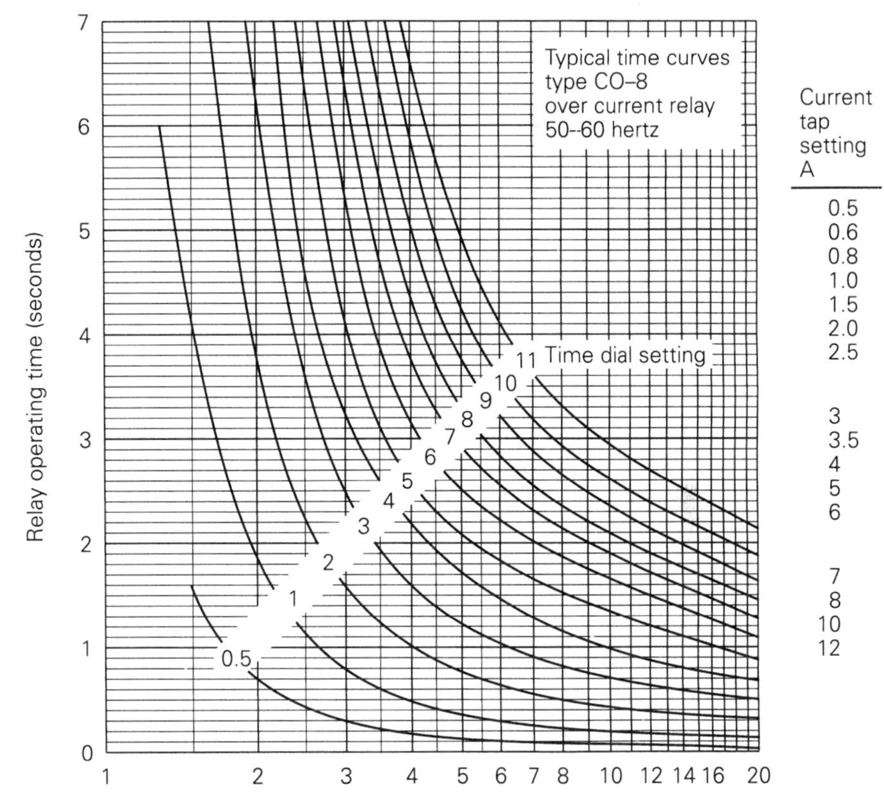

FIGURE 10.12 CO-8 time-delay overcurrent relay characteristics (Courtesy of Westinghouse Electric Corporation)

The characteristic curves are usually shown with operating time in seconds versus relay input current as a multiple of the pickup current. The curves are asymptotic to the vertical axis and decrease with some inverse power of current magnitude for values exceeding the pickup current. This inverse time characteristic can be shifted up or down by adjustment of the time-dial setting. Although discrete time-dial settings are shown in Figure 10.12, intermediate values can be obtained by interpolating between the discrete curves.

EXAMPLE 10.3 Operating time for a CO-8 time-delay overcurrent relay

The CO-8 relay with a current tap setting of 6 amperes and a time-dial setting of 1 is used with the 100:5 CT in Example 10.1. Determine the relay operating time for each case.

SOLUTION

a. From Example 10.1(a)

$$I' = 5\ \text{A} \qquad \frac{I'}{I_p} = \frac{5}{6} = 0.83$$

The relay does not operate. It remains in the blocking position.

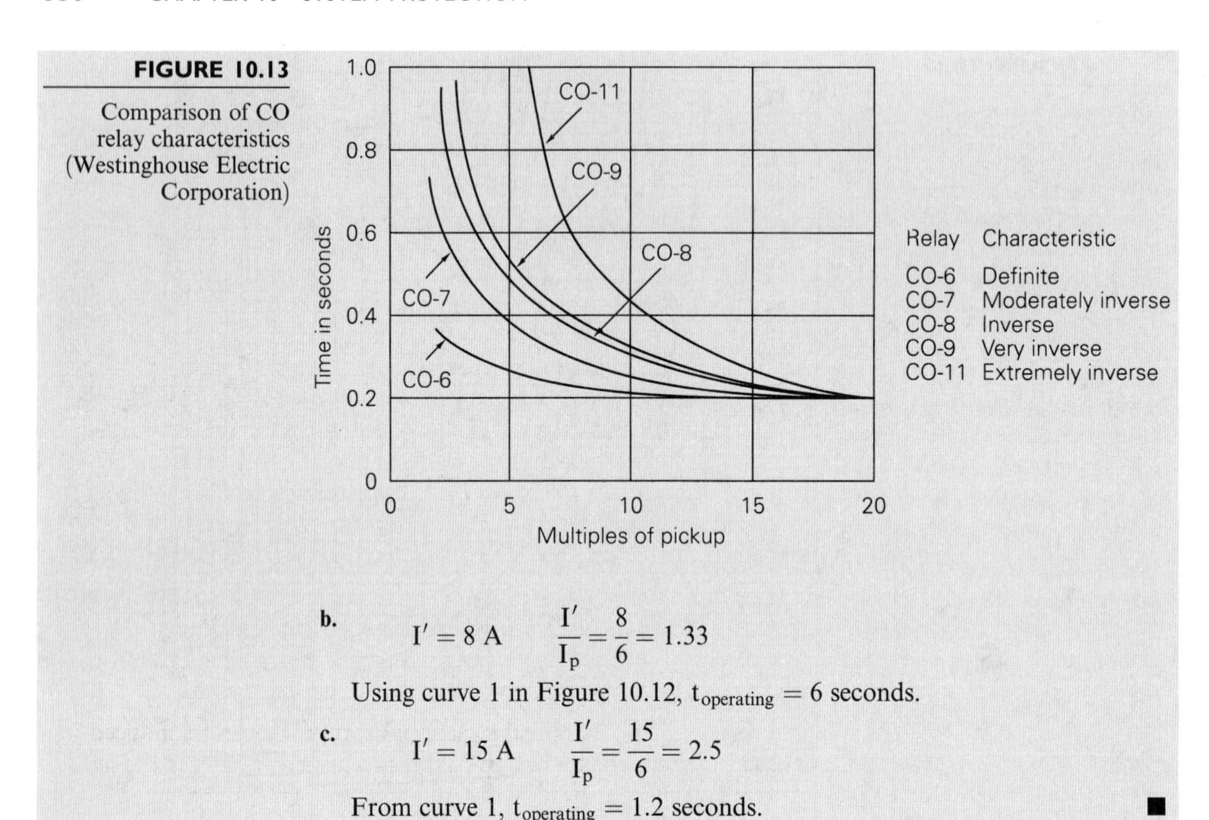

FIGURE 10.13

Comparison of CO relay characteristics (Westinghouse Electric Corporation)

b.
$$I' = 8\ \text{A} \qquad \frac{I'}{I_p} = \frac{8}{6} = 1.33$$

Using curve 1 in Figure 10.12, $t_{\text{operating}} = 6$ seconds.

c.
$$I' = 15\ \text{A} \qquad \frac{I'}{I_p} = \frac{15}{6} = 2.5$$

From curve 1, $t_{\text{operating}} = 1.2$ seconds. ■

Figure 10.13 shows the time-current characteristics of five CO time-delay overcurrent relays used in transmission and distribution lines. The time-dial settings are selected in the figure so that all relays operate in 0.2 seconds at 20 times the pickup current. The choice of relay time-current characteristic depends on the sources, lines, and loads. The definite (CO-6) and moderately inverse (CO-7) relays maintain a relatively constant operating time above 10 times pickup. The inverse (CO-8), very inverse (CO-9), and extremely inverse (CO-11) relays operate respectively faster on higher fault currents.

Figure 10.14 illustrates the operating principle of an electromechanical time-delay overcurrent relay. The ac input current to the relay operating coil sets up a magnetic field that is perpendicular to a conducting aluminum disc.

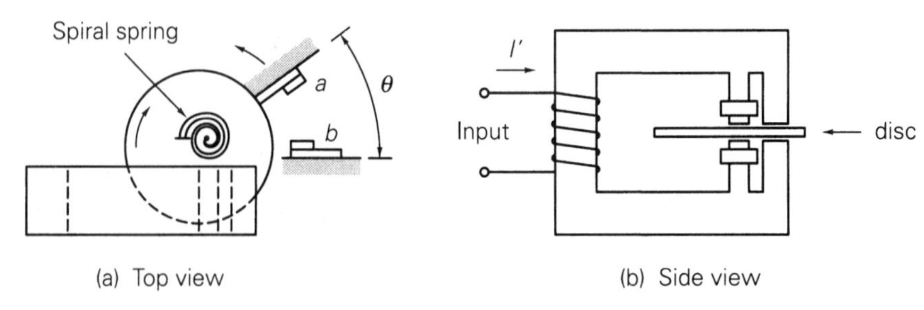

FIGURE 10.14

Electromechanical time-delay overcurrent relay—induction disc type

FIGURE 10.15 Solid-state relay panel (center) for a 345-kV transmission line, with electromechanical relays on each side, at Electric Junction Substation, Naperville, Illinois (Courtesy of Commonwealth Edison, an Exelon Company)

The disc can rotate and is restrained by a spiral spring. Current is induced in the disc, interacts with the magnetic field, and produces a torque. If the input current exceeds the pickup current, the disc rotates through an angle θ to close the relay contacts. The larger the input current, the larger is the torque and the faster the contact closing. After the current is removed or reduced below the pickup, the spring provides reset of the contacts.

A solid state relay panel between older-style electromechanical relays is shown in Figure 10.15.

10.4 RADIAL SYSTEM PROTECTION

Many radial systems are protected by time-delay overcurrent relays. Adjustable time delays can be selected such that the breaker closest to the fault opens, while other upstream breakers with larger time delays remain closed. That is, the relays can be coordinated to operate in sequence so as to interrupt minimum load during faults. Successful relay coordination is obtained when fault currents are much larger than normal load currents. Also, coordination of overcurrent relays usually limits the maximum number of breakers in a radial system to five or less, otherwise the relay closest to the source may have an excessive time delay.

Consider a fault at P_1 to the right of breaker B3 for the radial system of Figure 10.16. For this fault we want breaker B3 to open while B2 (and B1) remains closed. Under these conditions, only load L3 is interrupted. We could select a longer time delay for the relay at B2, so that B3 operates first. Thus,

FIGURE 10.16

Single-line diagram of a 34.5-kV radial system

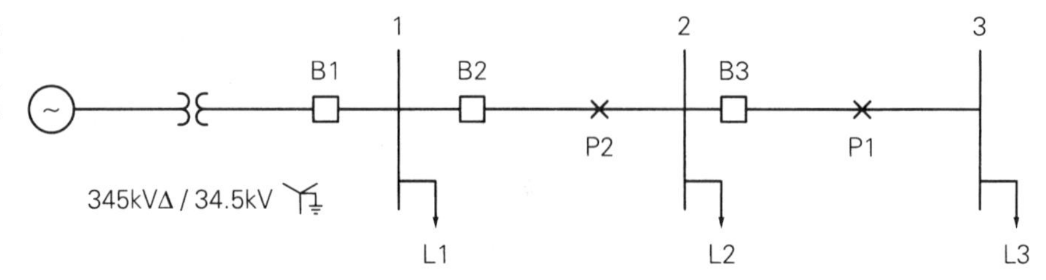

for any fault to the right of B3, B3 provides primary protection. Only if B3 fails to open will B2 open, after time delay, thus providing backup protection.

Similarly, consider a fault at P_2 between B2 and B3. We want B2 to open while B1 remains closed. Under these conditions, loads L2 and L3 are interrupted. Since the fault is closer to the source, the fault current will be larger than for the previous fault considered. B2, set to open for the previous, smaller fault current after time delay, will open more rapidly for this fault. We also select the B1 relay with a longer time delay than B2, so that B2 opens first. Thus, B2 provides primary protection for faults between B2 and B3, as well as backup protection for faults to the right of B3. Similarly, B1 provides primary protection for faults between B1 and B2, as well as backup protection for further downstream faults.

The *coordination time interval* is the time interval between the primary and remote backup protective devices. It is the difference between the time that the backup relaying operates and the time that circuit breakers clear the fault under primary relaying. Precise determination of relay operating times is complicated by several factors, including CT error, dc offset component of fault current, and relay overtravel. Therefore, typical coordination time intervals from 0.2 to 0.5 seconds are selected to account for these factors in most practical applications.

EXAMPLE 10.4 Coordinating time-delay overcurrent relays in a radial system

Data for the 60-Hz radial system of Figure 10.16 are given in Tables 10.3, 10.4, and 10.5. Select current tap settings (TSs) and time-dial settings (TDSs)

TABLE 10.3

Maximum loads—Example 10.4

Bus	S MVA	Lagging p.f.
1	11.0	0.95
2	4.0	0.95
3	6.0	0.95

TABLE 10.4

Symmetrical fault currents—Example 10.4

Bus	Maximum Fault Current (Bolted Three-Phase) A	Minimum Fault Current (L–G or L–L) A
1	3000	2200
2	2000	1500
3	1000	700

TABLE 10.5

Breaker, CT, and relay data—Example 10.4

Breaker	Breaker Operating Time	CT Ratio	Relay
B1	5 cycles	400:5	CO-8
B2	5 cycles	200:5	CO-8
B3	5 cycles	200:5	CO-8

FIGURE 10.17

Relay connections to trip all three phases

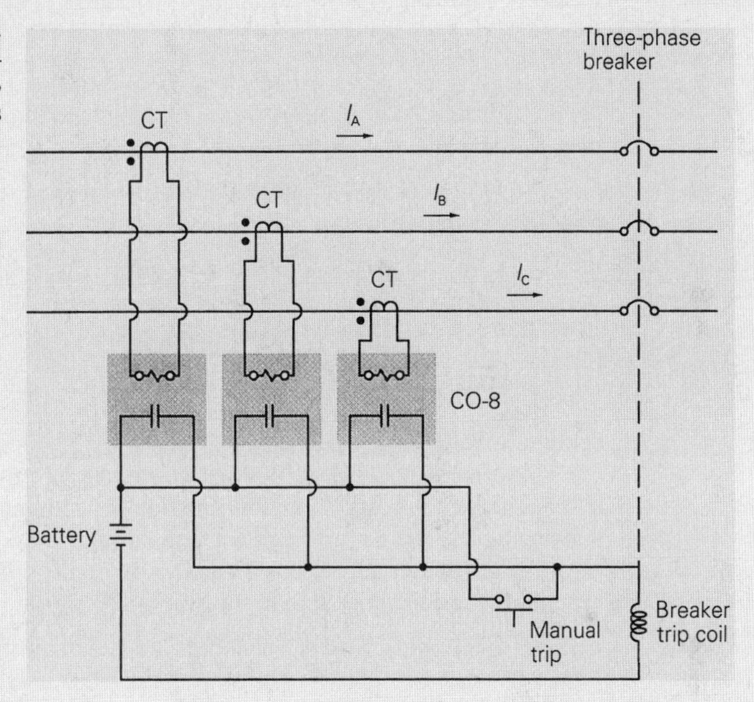

to protect the system from faults. Assume three CO-8 relays for each breaker, one for each phase, with a 0.3-second coordination time interval. The relays for each breaker are connected as shown in Figure 10.17, so that all three phases of the breaker open when a fault is detected on any one phase. Assume a 34.5-kV (line-to-line) voltage at all buses during normal operation. Also, future load growth is included in Table 10.3, such that maximum loads over the operating life of the radial system are given in this table.

SOLUTION First, select TSs such that the relays do not operate for maximum load currents. Starting at B3, the primary and secondary CT currents for maximum load L3 are

$$I_{L3} = \frac{S_{L3}}{V_3\sqrt{3}} = \frac{6 \times 10^6}{(34.5 \times 10^3)\sqrt{3}} = 100.4 \text{ A}$$

$$I'_{L3} = \frac{100.4}{(200/5)} = 2.51 \text{ A}$$

From Figure 10.12, we select for the B3 relay a 3-A TS, which is the lowest TS above 2.51 A.

Note that $|S_{L2} + S_{L3}| = |S_{L2}| + |S_{L3}|$ because the load power factors are identical. Thus, at B2, the primary and secondary CT currents for maximum load are

$$I_{L2} = \frac{S_{L2} + S_{L3}}{V_2\sqrt{3}} = \frac{(4+6) \times 10^6}{(34.5 \times 10^3)\sqrt{3}} = 167.3 \text{ A}$$

$$I'_{L2} = \frac{167.3}{(200/5)} = 4.18 \text{ A}$$

From Figure 10.12, select for the B2 relay a 5-A TS, the lowest TS above 4.18 A. At B1,

$$I_{L1} = \frac{S_{L1} + S_{L2} + S_{L3}}{V_1\sqrt{3}} = \frac{(11+4+6) \times 10^6}{(34.5 \times 10^3)\sqrt{3}} = 351.4 \text{ A}$$

$$I'_{L1} = \frac{351.4}{(400/5)} = 4.39 \text{ A}$$

Select a 5-A TS for the B1 relay.

Next select the TDSs. We first coordinate for the maximum fault currents in Table 10.4, checking coordination for minimum fault currents later. Starting at B3, the largest fault current through B3 is 2000 A, which occurs for the three-phase fault at bus 2 (just to the right of B3). Neglecting CT saturation, the fault-to-pickup current ratio at B3 for this fault is

$$\frac{I'_{3\text{Fault}}}{TS3} = \frac{2000/(200/5)}{3} = 16.7$$

Since we want to clear faults as rapidly as possible, select a 1/2 TDS for the B3 relay. Then, from the 1/2 TDS curve in Figure 10.12, the relay operating time is $T3 = 0.05$ seconds. Adding the breaker operating time (5 cycles = 0.083 s), primary protection clears this fault in $T3 + T_{\text{breaker}} = 0.05 + 0.083 = 0.133$ seconds.

For this same fault, the fault-to-pickup current ratio at B2 is

$$\frac{I'_{2\text{Fault}}}{TS2} = \frac{2000/(200/5)}{5} = 10.0$$

Adding the B3 relay operating time ($T3 = 0.05$ s), breaker operating time (0.083 s), and 0.3 s coordination time interval, we want a B2 relay operating time

$$T2 = T3 + T_{\text{breaker}} + T_{\text{coordination}} = 0.05 + 0.083 + 0.3 \approx 0.43 \text{ s}$$

From Figure 10.12, select TDS2 = 2.

Next select the TDS at B1. The largest fault current through B2 is 3000 A, for a three-phase fault at bus 1 (just to the right of B2). The fault-to-pickup current ratio at B2 for this fault is

TABLE 10.6 Solution—Example 10.4

Breaker	Relay	TS	TDS
B1	CO-8	5	3
B2	CO-8	5	2
B3	CO-8	3	1/2

$$\frac{I'_{2\text{Fault}}}{\text{TS2}} = \frac{3000/(200/5)}{5} = 15.0$$

From the 2 TDS curve in Figure 10.12, T2 = 0.38 s. For this same fault,

$$\frac{I'_{1\text{Fault}}}{\text{TS1}} = \frac{3000/(400/5)}{5} = 7.5$$

$$\text{T1} = \text{T2} + \text{T}_{\text{breaker}} + \text{T}_{\text{coordination}} = 0.38 + 0.083 + 0.3 \approx 0.76 \text{ s}$$

From Figure 10.12, select TDS1 = 3. The relay settings are shown in Table 10.6. Note that for reliable relay operation the fault-to-pickup current ratios with minimum fault currents should be greater than 2. Coordination for minimum fault currents listed in Table 10.4 is evaluated in Problem 10.11. ■

Note that separate relays are used for each phase in Example 10.4, and therefore these relays will operate for three-phase as well as line-to-line, single line-to-ground, and double line-to-ground faults. However, in many cases single line-to-ground fault currents are much lower than three-phase fault currents, especially for distribution feeders with high zero-sequence impedances. In these cases a separate ground relay with a lower current tap setting than the phase relays is used. The ground relay is connected to operate on zero-sequence current from three of the phase CTs connected in parallel or from a CT in the grounded neutral.

10.5 RECLOSERS AND FUSES

Automatic circuit reclosers are commonly used for distribution circuit protection. A *recloser* is a self-controlled device for automatically interrupting and reclosing an ac circuit with a preset sequence of openings and reclosures. Unlike circuit breakers, which have separate relays to control breaker opening and reclosing, reclosers have built-in controls. More than 80% of faults on overhead distribution circuits are temporary, caused by tree limb contact, by animal interference, by wind bringing bare conductors in contact, or by lightning. The automatic tripping-reclosing sequence of reclosers clears these temporary faults and restores service with only momentary outages, thereby significantly improving customer service. A disadvantage of reclosers is the increased hazard when a circuit is physically contacted by people—for

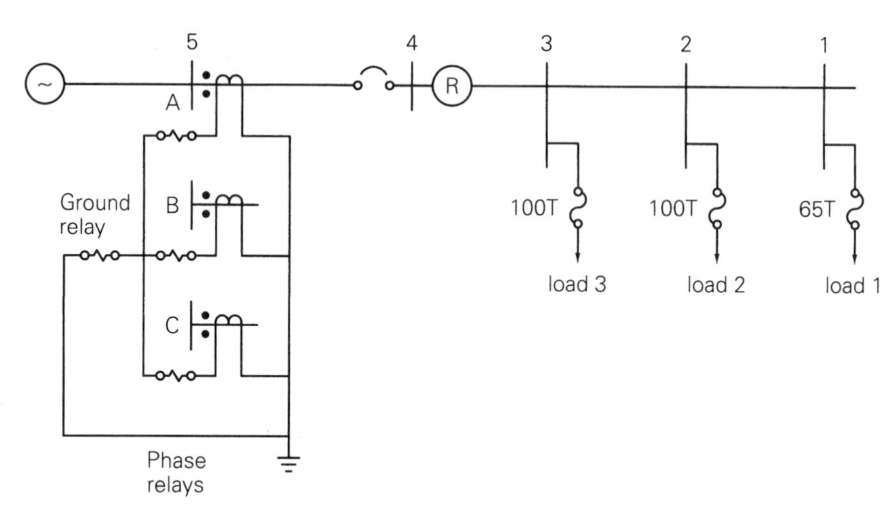

FIGURE 10.18 Single-line diagram of a 13.8-kV radial distribution feeder with fuse/recloser/relay protection

example, in the case of a broken conductor at ground level that remains energized. Also, reclosing should be locked out during live-line maintenance by utility personnel.

Figure 10.18 shows a common protection scheme for radial distribution circuits utilizing fuses, reclosers, and time-delay overcurrent relays. Data for the 13.8-kV feeder in this figure is given in Table 10.7. There are three load taps protected by fuses. The recloser ahead of the fuses is set to open and reclose for faults up to and beyond the fuses. For temporary faults the recloser can be set for one or more instantaneous or time-delayed trips and reclosures in order to clear the faults and restore service. If faults persist, the fuses operate for faults to their right (downstream), or the recloser opens after time delay and locks out for faults between the recloser and fuses. Separate time-delay overcurrent phase and ground relays open the substation breaker after multiple reclosures of the recloser.

Coordination of the fuses, recloser, and time-delay overcurrent relays is shown via the time-current curves in Figure 10.19. Type T (slow) fuses are selected because their time-current characteristics coordinate well with reclosers. The fuses are selected on the basis of maximum loads served from the taps. A 65 T fuse is selected for the bus 1 tap, which has a 60-A maximum

TABLE 10.7 Data for Figure 10.18

Bus	Maximum Load Current A	3φ Fault Current A	IL-G Fault Current A
1	60	1000	850
2	95	1500	1300
3	95	2000	1700
4	250	3000	2600
5	250	4000	4050

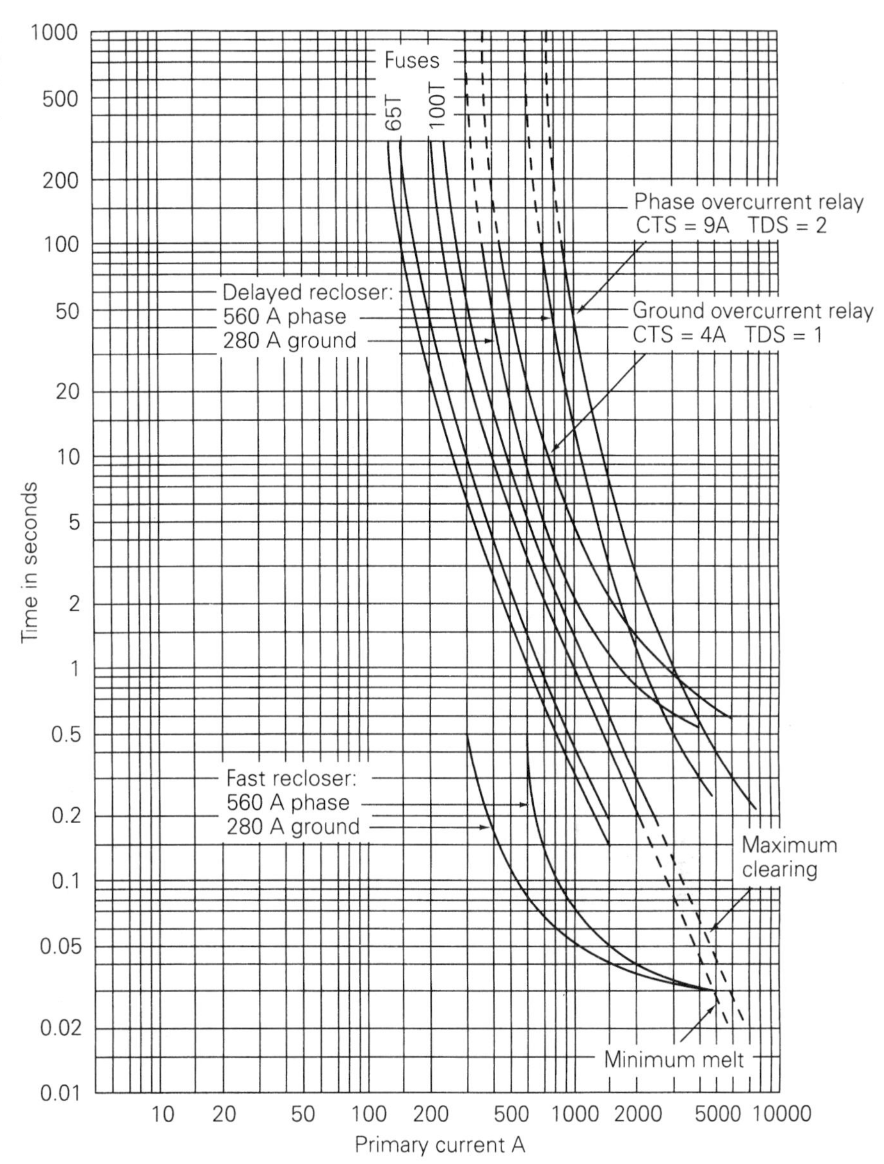

FIGURE 10.19 Time-current curves for the radial distribution circuit of Figure 10.18

load current, and 100 T is selected for the bus 2 and 3 taps, which have 95-A maximum load currents. The fuses should also have a rated voltage larger than the maximum bus voltage and an interrupting current rating larger than the maximum asymmetrical fault current at the fuse location. Type T fuses with voltage ratings of 15 kV and interrupting current ratings of 10 kA and higher are standard.

Standard reclosers have minimum trip ratings of 50, 70, 100, 140, 200, 280, 400, 560, 800, 1120, and 1600 A, with voltage ratings up to 38 kV and maximum interrupting currents up to 16 kA. A minimum trip rating of 200–250% of maximum load current is typically selected for the phases, in order to override cold load pickup with a safety factor. The minimum trip rating of the ground unit is typically set at maximum load and should be higher than the maximum allowable load unbalance. For the recloser in Figure 10.18, which carries a 250-A maximum load, minimum trip ratings of 560 A for each phase and 280 A for the ground unit are selected.

A popular operation sequence for reclosers is two fast operations, without intentional time delay, followed by two delayed operations. The fast operations allow temporary faults to self-clear, whereas the delayed operations allow downstream fuses to clear permanent faults. Note that the time-current curves of the fast recloser lie below the fuse curves in Figure 10.19, such that the recloser opens before the fuses melt. The fuse curves lie below the delayed recloser curves, such that the fuses clear before the recloser opens. The recloser is typically programmed to reclose $\frac{1}{2}$ s after the first fast trip, 2 s after the second fast trip, and 5–10 s after a delayed trip.

Time-delay overcurrent relays with an extremely inverse characteristic coordinate with both reclosers and type T fuses. A 300:5 CT ratio is selected to give a secondary current of $250 \times (5/300) = 4.17$ A at maximum load. Relay settings are selected to allow the recloser to operate effectively to clear faults before relay operation. A current tap setting of 9 A is selected for the CO-11 phase relays so that minimum pickup exceeds twice the maximum load. A time-dial setting of 2 is selected so that the delayed recloser trips at least 0.2 s before the relay. The ground relay is set with a current tap setting of 4 A and a time-dial setting of 1.

EXAMPLE 10.5 Fuse/recloser coordination

For the system of Figure 10.18, describe the operating sequence of the protective devices for the following faults: (a) a self-clearing, temporary, three-phase fault on the load side of tap 2; and (b) a permanent three-phase fault on the load side of tap 2.

SOLUTION

a. From Table 10.7, the three-phase fault current at bus 2 is 1500 A. From Figure 10.19, the 560-A fast recloser opens 0.05 s after the 1500-A fault current occurs, and then recloses $\frac{1}{2}$ s later. Assuming the fault has self-cleared, normal service is restored. During the 0.05-s fault duration, the 100 T fuse does not melt.

b. For a permanent fault the fast recloser opens after 0.05 s, recloses $\frac{1}{2}$ s later into the permanent fault, opens again after $\frac{1}{2}$ s, and recloses into the fault a second time after a 2-s delay. Then the 560-A delayed recloser opens 3 seconds later. During this interval the 100 T fuse clears the fault. The delayed recloser then recloses 5 to 10 s later, restoring service to loads 1 and 3. ■

10.6 DIRECTIONAL RELAYS

Directional relays are designed to operate for fault currents in only one direction. Consider the directional relay D in Figure 10.20, which is required to operate only for faults to the right of the CT. Since the line impedance is mostly reactive, a fault at P_1 to the right of the CT will have a fault current I from bus 1 to bus 2 that lags the bus voltage V by an angle of almost 90°. This fault current is said to be in the forward direction. On the other hand, a fault at P_2, to the left of the CT, will have a fault current I that leads V by almost 90°. This fault current is said to be in the reverse direction.

The directional relay has two inputs: the reference voltage $V = \mathrm{V}\underline{/0^\circ}$, and current $I = \mathrm{I}\underline{/\phi}$. The relay trip and block regions, shown in Figure 10.21, can be described by

$$
\begin{aligned}
-180^\circ < (\phi - \phi_1) < 0^\circ \quad &(\text{Trip}) \\
\text{Otherwise} \quad &(\text{Block})
\end{aligned}
\tag{10.6.1}
$$

where ϕ is the angle of the current with respect to the voltage and ϕ_1, typically 2° to 8°, defines the boundary between the trip and block regions.

The contacts of the overcurrent relay OC and the directional relay D are connected in series in Figure 10.20, so that the breaker trip coil is

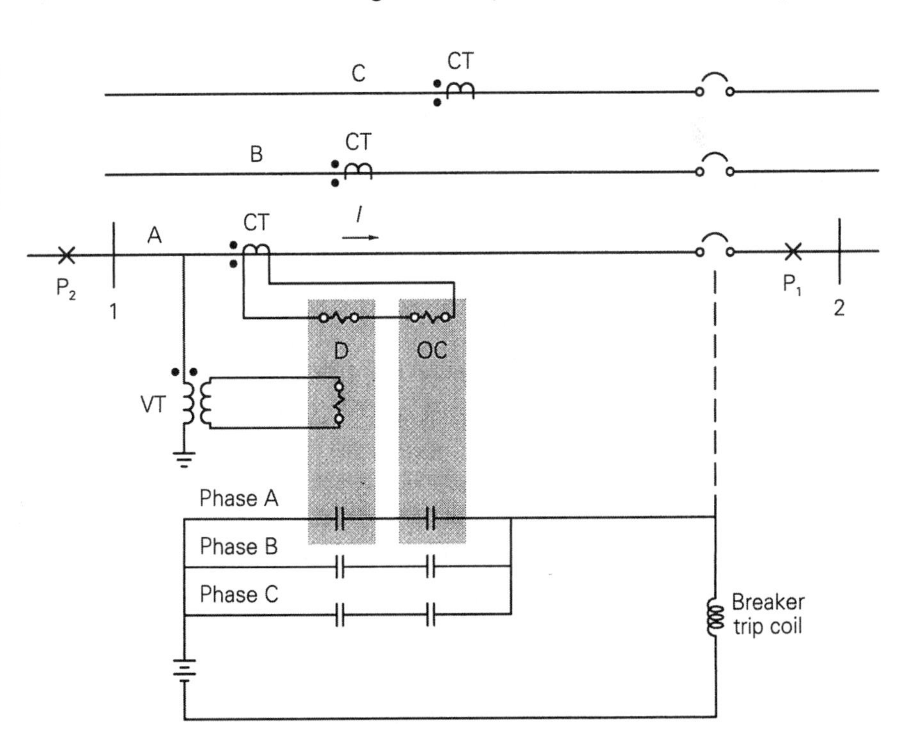

FIGURE 10.20 Directional relay in series with overcurrent relay (only phase A is shown)

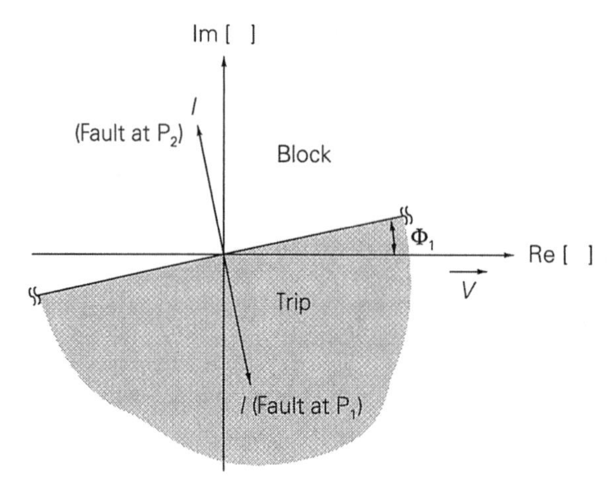

FIGURE 10.21 Directional relay block and trip regions in the complex plane

energized only when the CT secondary current (1) exceeds the OC relay pickup value, and (2) is in the forward tripping direction.

Although construction details differ, the operating principle of an electromechanical directional relay is similar to that of a watt-hour meter. There are two input coils, a voltage coil and a current coil, both located on a stator, and there is a rotating disc element. Suppose that the reference voltage is passed through a phase-shifting element to obtain $V_1 = \mathrm{V}\underline{/\phi_1 - 90^\circ}$. If V_1 and $I = \mathrm{I}\underline{/\phi}$ are applied to a watt-hour meter, the torque on the rotating element is

$$\mathrm{T} = \mathrm{kVI}\cos(\phi_1 - \phi - 90^\circ) = \mathrm{kVI}\sin(\phi_1 - \phi) \tag{10.6.2}$$

Note that for faults in the forward direction the current lags the voltage, and the angle $(\phi_1 - \phi)$ in (10.6.2) is close to 90°. This results in maximum positive torque on the rotating disc, which would cause the relay contacts to close. On the other hand, for faults in the reverse direction the current leads the voltage, and $(\phi_1 - \phi)$ is close to −90°. This results in maximum negative torque tending to rotate the disc element in the backward direction. Backward motion can be restrained by mechanical stops.

10.7 PROTECTION OF TWO-SOURCE SYSTEM WITH DIRECTIONAL RELAYS

It becomes difficult and in some cases impossible to coordinate overcurrent relays when there are two or more sources at different locations. Consider the system with two sources shown in Figure 10.22. Suppose there is a fault at P_1. We want B23 and B32 to clear the fault so that service to the three loads

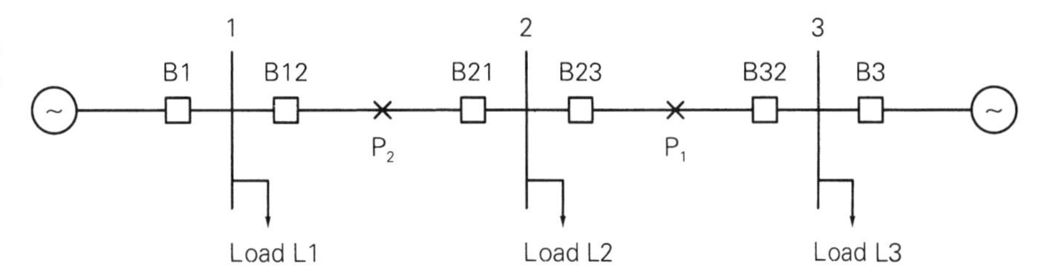

FIGURE 10.22

System with two sources

continues without interruption. Using time-delay overcurrent relays, we could set B23 faster than B21. Now consider a fault at P_2 instead. Breaker B23 will open faster than B21, and load L2 will be disconnected. When a fault can be fed from both the left and right, overcurrent relays cannot be coordinated. However, directional relays can be used to overcome this problem.

EXAMPLE 10.6 Two-source system protection with directional and time-delay overcurrent relays

Explain how directional and time-delay overcurrent relays can be used to protect the system in Figure 10.22. Which relays should be coordinated for a fault (a) at P_1, (b) at P_2? (c) Is the system also protected against bus faults?

SOLUTION Breakers B12, B21, B23, and B32 should respond only to faults on their "forward" or "line" sides. Directional overcurrent relays connected as shown in Figure 10.20 can be used for these breakers. Overcurrent relays alone can be used for breakers B1 and B3, which do not need to be directional.

a. For a fault at P_1, the B21 relay would not operate; B12 should coordinate with B23 so that B23 trips before B12 (and B1). Also, B3 should coordinate with B32.

b. For a fault at P_2, B23 would not operate; B32 should coordinate with B21 so that B21 trips before B32 (and B3). Also, B1 should coordinate with B12.

c. Yes, the directional overcurrent relays also protect the system against bus faults. If the fault is at bus 2, relays at B21 and B23 will not operate, but B12 and B32 will operate to clear the fault. B1 and B21 will operate to clear a fault at bus 1. B3 and B23 will clear a fault at bus 3. ■

10.8

ZONES OF PROTECTION

Protection of simple systems has been discussed so far. For more general power system configurations, a fundamental concept is the division of a

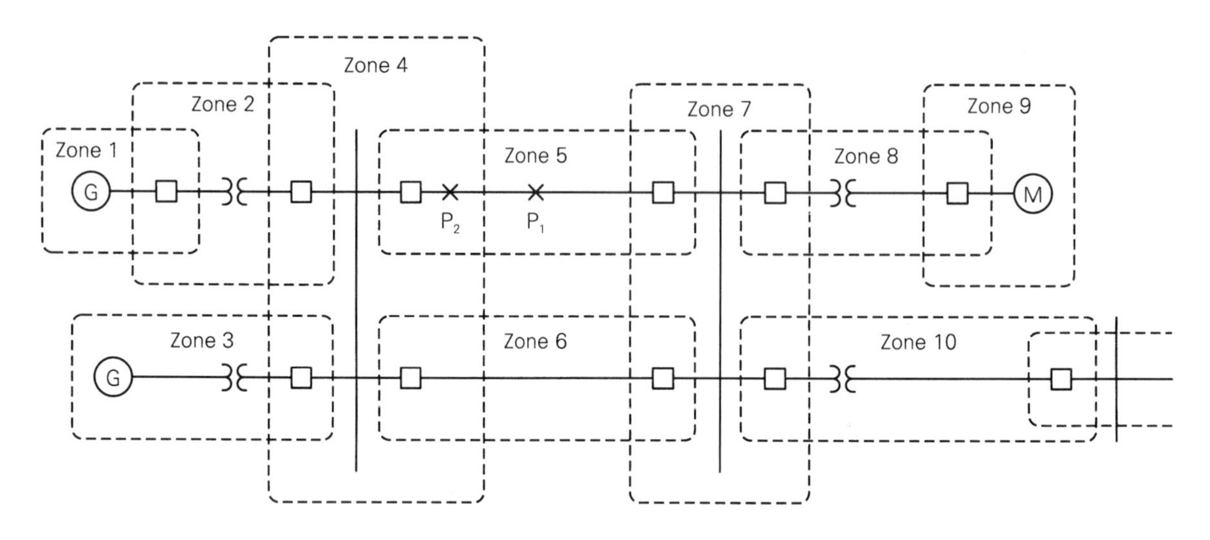

FIGURE 10.23 Power system protective zones

system into protective zones [1]. If a fault occurs anywhere within a zone, action will be taken to isolate that zone from the rest of the system. Zones are defined for:

generators,

transformers,

buses,

transmission and distribution lines, and

motors.

Figure 10.23 illustrates the protective zone concept. Each zone is defined by a closed, dashed line. Zone 1, for example, contains a generator and connecting leads to a transformer. In some cases a zone may contain more than one component. For example, zone 3 contains a generator-transformer unit and connecting leads to a bus, and zone 10 contains a transformer and a line. Protective zones have the following characteristics:

Zones are overlapped.

Circuit breakers are located in the overlap regions.

For a fault anywhere in a zone, all circuit breakers in that zone open to isolate the fault.

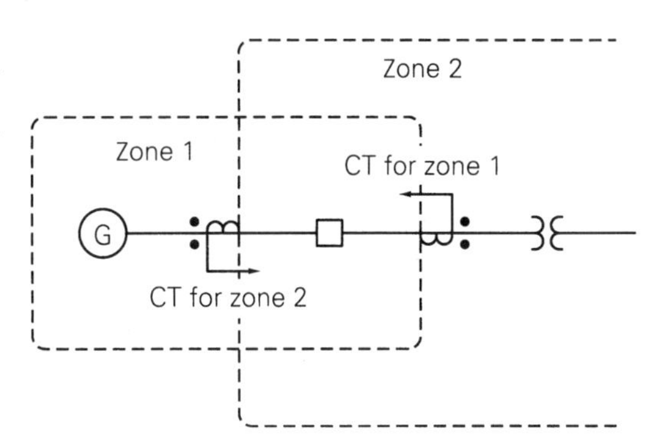

FIGURE 10.24

Overlapping protection around a circuit breaker

Neighboring zones are overlapped to avoid the possibility of unprotected areas. Without overlap the small area between two neighboring zones would not be located in any zone and thus would not be protected.

Since isolation during faults is done by circuit breakers, they should be inserted between equipment in a zone and each connection to the system. That is, breakers should be inserted in each overlap region. As such, they identify the boundaries of protective zones. For example, zone 5 in Figure 10.23 is connected to zones 4 and 7. Therefore, a circuit breaker is located in the overlap region between zones 5 and 4, as well as between zones 5 and 7.

If a fault occurs anywhere within a zone, action is taken to open all breakers in that zone. For example, if a fault occurs at P_1 on the line in zone 5, then the two breakers in zone 5 should open. If a fault occurs at P_2 within the overlap region of zones 4 and 5, then all five breakers in zones 4 and 5 should open. Clearly, if a fault occurs within an overlap region, two zones will be isolated and a larger part of the system will be lost from service. To minimize this possibility, overlap regions are kept as small as possible.

Overlap is accomplished by having two sets of instrument transformers and relays for each circuit breaker. For example, the breaker in Figure 10.24 shows two CTs, one for zone 1 and one for zone 2. Overlap is achieved by the order of the arrangement: first the equipment in the zone, second the breaker, and then the CT for that zone.

EXAMPLE 10.7 Zones of protection

Draw the protective zones for the power system shown in Figure 10.25. Which circuit breakers should open for a fault at P_1? at P_2?

SOLUTION Noting that circuit breakers identify zone boundaries, protective zones are drawn with dashed lines as shown in Figure 10.26. For a fault at P_1, located in zone 5, breakers B24 and B42 should open. For a fault at P_2, located in the overlap region of zones 4 and 5, breakers B24, B42, B21, and B23 should open.

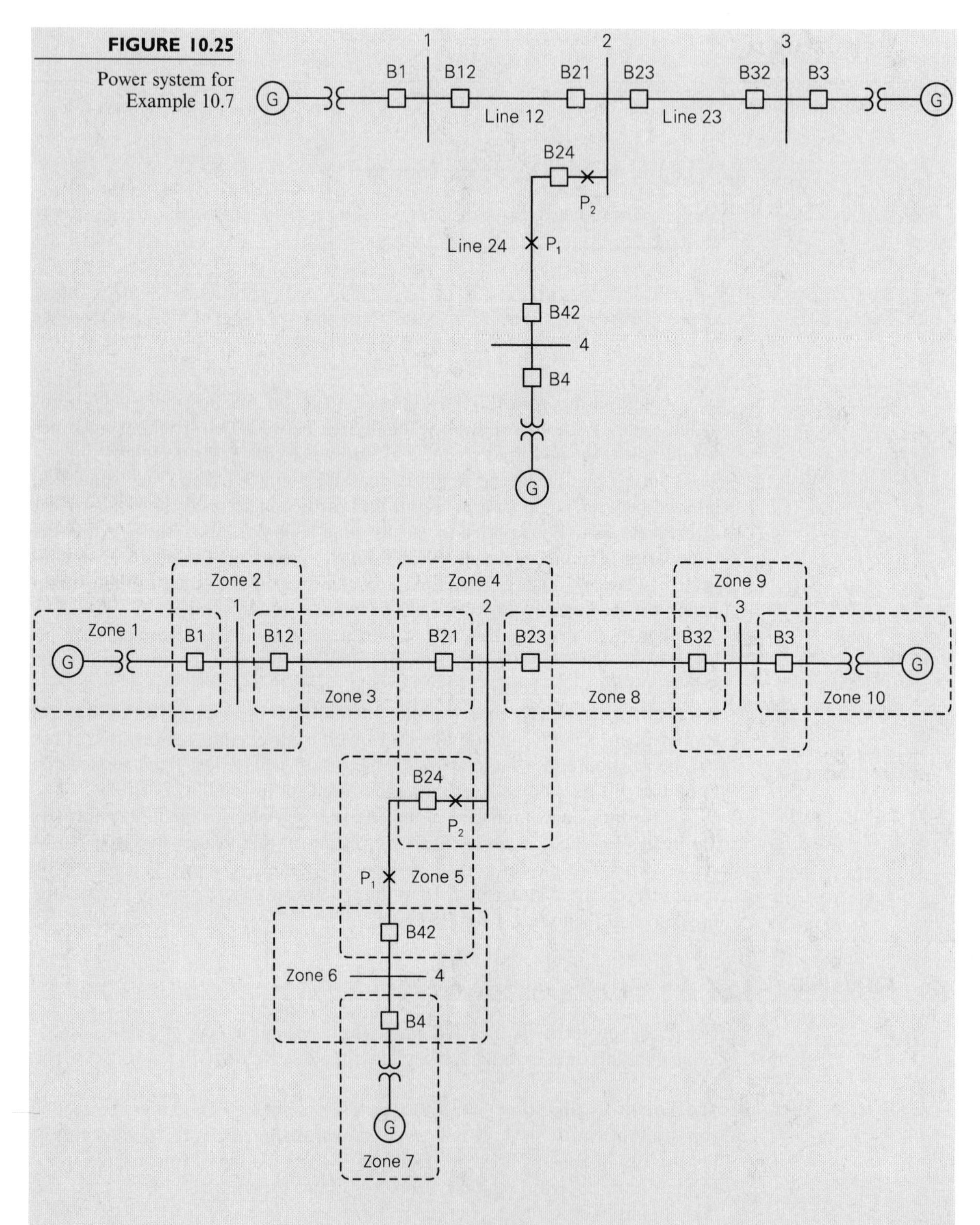

FIGURE 10.25 Power system for Example 10.7

FIGURE 10.26 Protective zones for Example 10.7 ■

10.9

LINE PROTECTION WITH IMPEDANCE (DISTANCE) RELAYS

Coordinating time-delay overcurrent relays can also be difficult for some radial systems. If there are too many radial lines and buses, the time delay for the breaker closest to the source becomes excessive.

Also, directional overcurrent relays are difficult to coordinate in transmission loops with multiple sources. Consider the use of these relays for the transmission loop shown in Figure 10.27. For a fault at P_1, we want the B21 relay to operate faster than the B32 relay. For a fault at P_2, we want B32 faster than B13. And for a fault at P_3, we want B13 faster than B21. Proper coordination, which depends on the magnitudes of the fault currents, becomes a tedious process. Furthermore, when consideration is given to various lines or sources out of service, coordination becomes extremely difficult.

To overcome these problems, relays that respond to a voltage-to-current ratio can be used. Note that during a three-phase fault, current increases while bus voltages close to the fault decrease. If, for example, current increases by a factor of 5 while voltage decreases by a factor of 2, then the voltage-to-current ratio decreases by a factor of 10. That is, the voltage-to-current ratio is

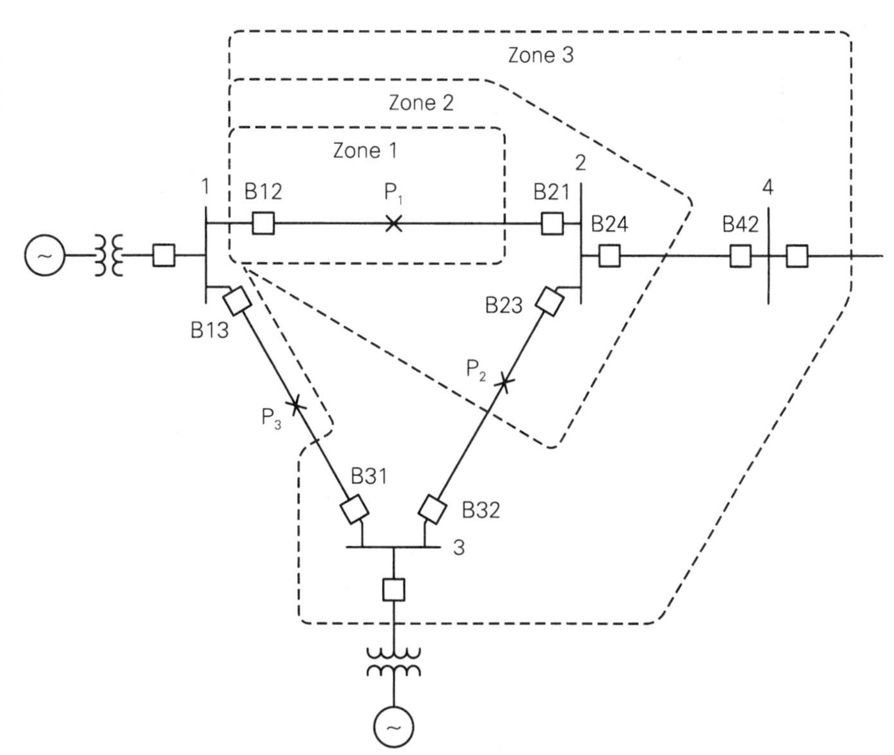

FIGURE 10.27 345-kV transmission loop

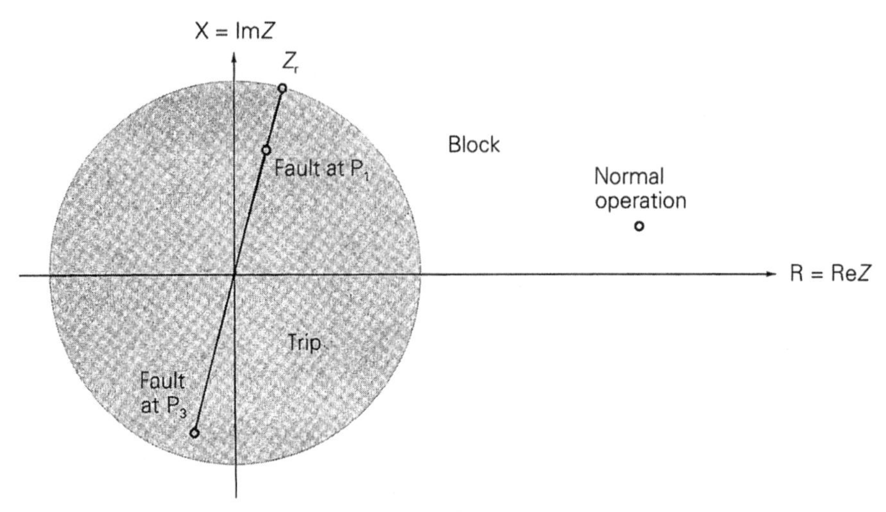

FIGURE 10.28 Impedance relay block and trip regions

more sensitive to faults than current alone. A relay that operates on the basis of voltage-to-current ratio is called an *impedance* relay. It is also called a *distance* relay or a *ratio* relay.

Impedance relay block and trip regions are shown in Figure 10.28, where the impedance Z is defined as the voltage-to-current ratio at the relay location. The relay trips for $|Z| < |Z_r|$, where Z_r is an adjustable relay setting. The impedance circle that defines the border between the block and trip regions passes through Z_r.

A straight line called the *line impedance locus* is shown for the impedance relay in Figure 10.28. This locus is a plot of positive sequence line impedances, predominantly reactive, as viewed between the relay location and various points along the line. The relay setting Z_r is a point in the R-X plane through which the impedance circle that defines the trip-block boundary must pass.

Consider an impedance relay for breaker B12 in Figure 10.27, for which $Z = V_1/I_{12}$. During normal operation, load currents are usually much smaller than fault currents, and the ratio Z has a large magnitude (and some arbitrary phase angle). Therefore, Z will lie outside the circle of Figure 10.28, and the relay will not trip during normal operation.

During a three-phase fault at P_1, however, Z appears to relay B12 to be the line impedance from the B12 relay to the fault. If $|Z_r|$ in Figure 10.28 is set to be larger than the magnitude of this impedance, then the B12 relay will trip. Also, during a three-phase fault at P_3, Z appears to relay B12 to be the negative of the line impedance from the relay to the fault. If $|Z_r|$ is larger than the magnitude of this impedance, the B12 relay will trip. Thus, the impedance relay of Figure 10.28 is not directional; a fault to the left or right of the relay can cause a trip.

Two ways to include directional capability with an impedance relay are shown in Figure 10.29. In Figure 10.29(a), an impedance relay with directional restraint is obtained by including a directional relay in series with an

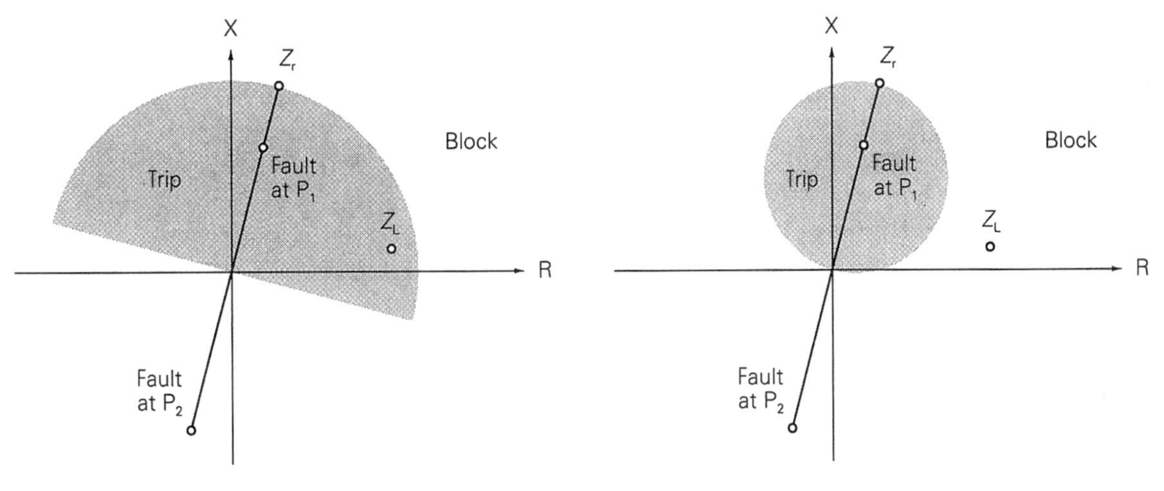

FIGURE 10.29 Impedance relays with directional capability

impedance relay, just as was done previously with an overcurrent relay. In Figure 10.29(b), a modified impedance relay is obtained by offsetting the center of the impedance circle from the origin. This modified impedance relay is sometimes called an *mho* relay. If either of these relays is used at B12 in Figure 10.27, a fault at P_1 will result in a trip decision, but a fault at P_3 will result in a block decision.

Note that the radius of the impedance circle for the modified impedance relay is half of the corresponding radius for the impedance relay with directional restraint. The modified impedance relay has the advantage of better selectivity for high power factor loads. For example, the high power factor load Z_L lies outside the trip region of Figure 10.29(b) but inside the trip region of Figure 10.29(a).

The *reach* of an impedance relay denotes how far down the line the relay detects faults. For example, an 80% reach means that the relay will detect any (solid three-phase) fault between the relay and 80% of the line length. This explains the term *distance* relay.

It is common practice to use three directional impedance relays per phase, with increasing reaches and longer time delays. For example, Figure 10.27 shows three protection zones for B12. The zone 1 relay is typically set for an 80% reach and instantaneous operation, in order to provide primary protection for line 1–2. The zone 2 relay is set for about 120% reach, extending beyond bus 2, with a typical time delay of 0.2 to 0.3 seconds. The zone 2 relay provides backup protection for faults on line 1–2 as well as remote backup for faults on line 2–3 or 2–4 in zone 2.

Note that in the case of a fault on line 2–3 we want the B23 relay to trip, not the B12 relay. Since the impedance seen by B12 for faults near bus 2, either on line 1–2 or line 2–3, is essentially the same, we cannot set the B12 zone 1 relay for 100% reach. Instead, an 80% reach is selected to avoid

FIGURE 10.30

Three-zone, directional impedance relay

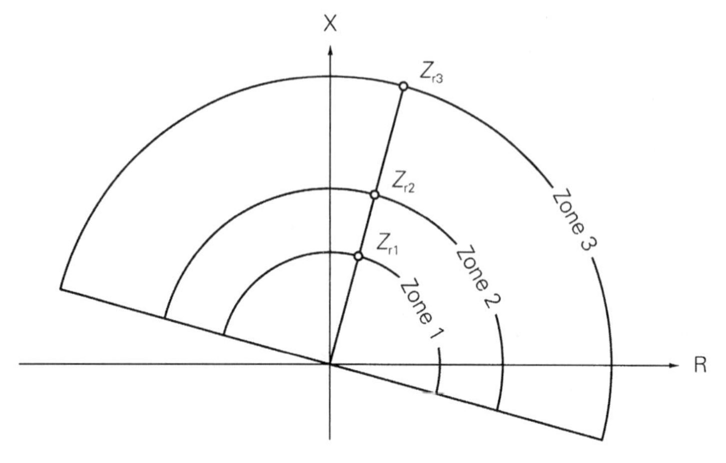

(a) Impedance relay with directional restraint

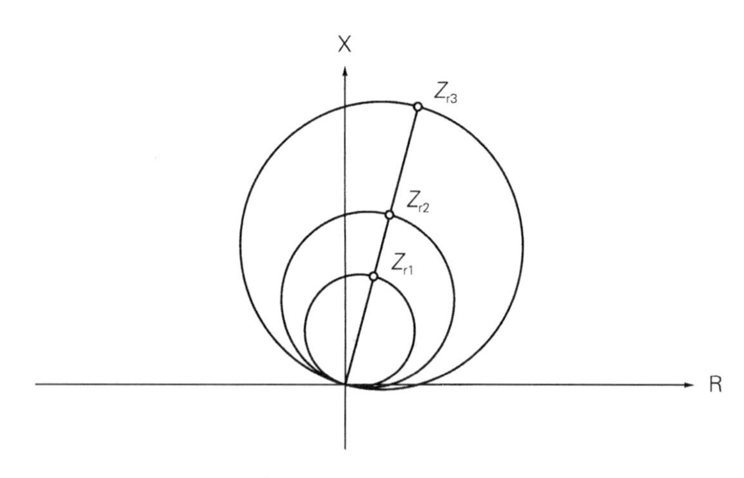

(b) Modified impedance relay (mho relay)

instantaneous operation of B12 for a fault on line 2–3 near bus 2. For example, if there is a fault at P_2 on line 2–3, B23 should trip instantaneously; if it fails, B12 will trip after time delay. Other faults at or near bus 2 also cause tripping of the B12 zone 2 relay after time delay.

Reach for the zone 3 B12 relay is typically set to extend beyond buses 3 and 4 in Figure 10.27, in order to provide remote backup for neighboring lines. As such, the zone 3 reach is set for 100% of line 1–2 plus 120% of either line 2–3 or 2–4, whichever is longer, with an even larger time delay, typically one second.

Typical block and trip regions are shown in Figure 10.30 for both types of three-zone, directional impedance relays. Relay connections for a three-zone impedance relay with directional restraint are shown in Figure 10.31.

FIGURE 10.31

Relay connections for a three-zone directional impedance relay (only phase A is shown)

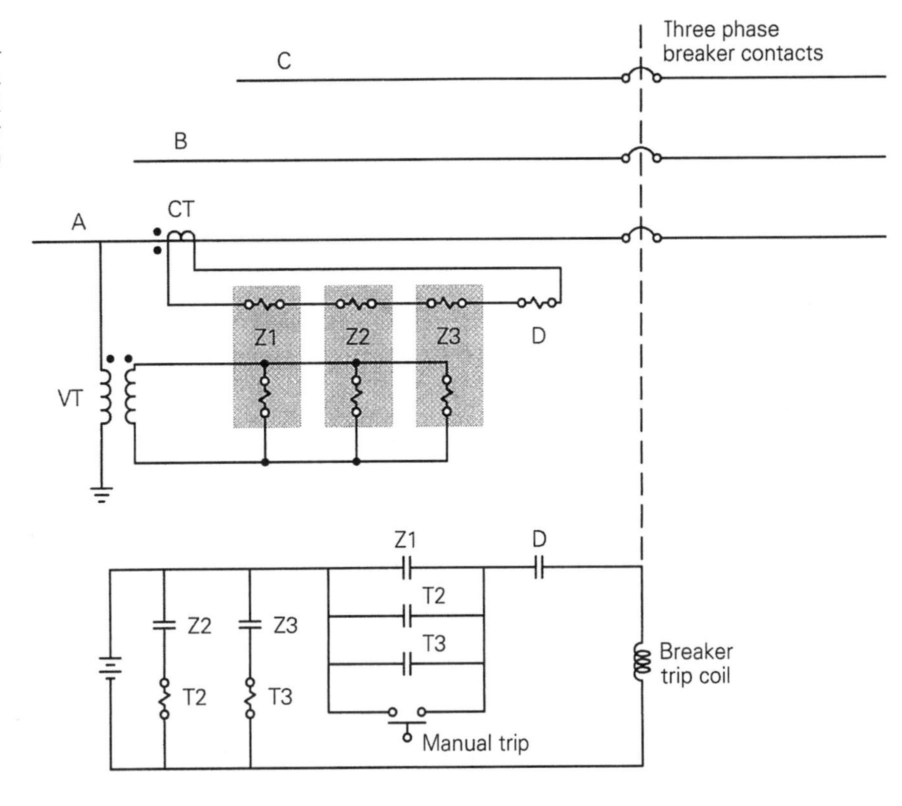

EXAMPLE 10.8 Three-zone impedance relay settings

Table 10.8 gives positive-sequence line impedances as well as CT and VT ratios at B12 for the 345-kV system shown in Figure 10.27. (a) Determine the settings Z_{r1}, Z_{r2}, and Z_{r3} for the B12 three-zone, directional impedance relays connected as shown in Figure 10.31. Consider only solid, three-phase faults.

TABLE 10.8

Data for Example 10.8

Line	Positive-Sequence Impedance Ω	
1–2	8 + j50	
2–3	8 + j50	
2–4	5.3 + j33	
1–3	4.3 + j27	
Breaker	**CT Ratio**	**VT Ratio**
B12	1500:5	3000:1

(b) Maximum current for line 1–2 during emergency loading conditions is 1500 A at a power factor of 0.95 lagging. Verify that B12 does not trip during normal and emergency loadings.

SOLUTION

a. Denoting V_{LN} as the line-to-neutral voltage at bus 1 and I_L as the line current through B12, the primary impedance Z viewed at B12 is

$$Z = \frac{V_{LN}}{I_L}\ \Omega$$

Using the CT and VT ratios given in Table 10.8, the secondary impedance viewed by the B12 impedance relays is

$$Z' = \frac{V_{LN}\Big/\left(\frac{3000}{1}\right)}{I_L\Big/\left(\frac{1500}{5}\right)} = \frac{Z}{10}$$

We set the B12 zone 1 relay for 80% reach, that is, 80% of the line 1–2 (secondary) impedance:

$$Z_{r1} = 0.80(8 + j50)/10 = 0.64 + j4 = 4.05\underline{/80.9^\circ}\ \Omega \quad \text{secondary}$$

Setting the B12 zone 2 relay for 120% reach:

$$Z_{r2} = 1.2(8 + j50)/10 = 0.96 + j6 = 6.08\underline{/80.9^\circ}\ \Omega \quad \text{secondary}$$

From Table 10.8, line 2–4 has a larger impedance than line 2–3. Therefore, we set the B12 zone 3 relay for 100% reach of line 1–2 plus 120% reach of line 2–4.

$$Z_{r3} = 1.0(8 + j50)/10 + 1.2(5.3 + j33)/10$$
$$= 1.44 + j8.96 = 9.07\underline{/80.9^\circ}\ \Omega \quad \text{secondary}$$

b. The secondary impedance viewed by B12 during emergency loading, using $V_{LN} = 345/\sqrt{3}\underline{/0^\circ} = 199.2\underline{/0^\circ}$ kV and $I_L = 1500\underline{/-\cos^{-1}(0.95)} = 1500\underline{/-18.19^\circ}$ A, is

$$Z' = Z/10 = \left(\frac{199.2 \times 10^3}{1500\underline{/-18.19^\circ}}\right)\Big/10 = 13.28\underline{/18.19^\circ}\ \Omega \quad \text{secondary}$$

Since this impedance exceeds the zone 3 setting of $9.07\underline{/80.9^\circ}\ \Omega$, the impedance during emergency loading lies outside the trip regions of the three-zone, directional impedance relay. Also, lower line loadings during normal operation will result in even larger impedances farther away from the trip regions. B12 will trip during faults but not during normal and emergency loadings. ■

Remote backup protection of adjacent lines using zone 3 of an impedance relay may be ineffective. In practice, buses have multiple lines of different lengths with sources at their remote ends. Contributions to fault currents from the multiple lines may cause the zone 3 relay to underreach. This "infeed effect" is illustrated in Problem 10.21.

The impedance relays considered so far use line-to-neutral voltages and line currents and are called *ground fault relays*. They respond to three-phase, single line-to-ground, and double line-to-ground faults very effectively. The impedance seen by the relay during unbalanced faults will generally not be the same as seen during three-phase faults and will not be truly proportional to the distance to the fault location. However, the relay can be accurately set for any fault location after computing impedance to the fault using fault currents and voltages. For other fault locations farther away (or closer), the impedance to the fault will increase (or decrease).

Ground fault relays are relatively insensitive to line-to-line faults. Impedance relays that use line-to-line voltages V_{ab}, V_{bc}, V_{ca} and line-current differences $I_a - I_b$, $I_b - I_c$, $I_c - I_a$ are called *phase relays*. Phase relays respond effectively to line-to-line faults and double line-to-ground faults but are relatively insensitive to single line-to-ground faults. Therefore, both phase and ground fault relays need to be used.

10.10 DIFFERENTIAL RELAYS

Differential relays are commonly used to protect generators, buses, and transformers. Figure 10.32 illustrates the basic method of differential relaying

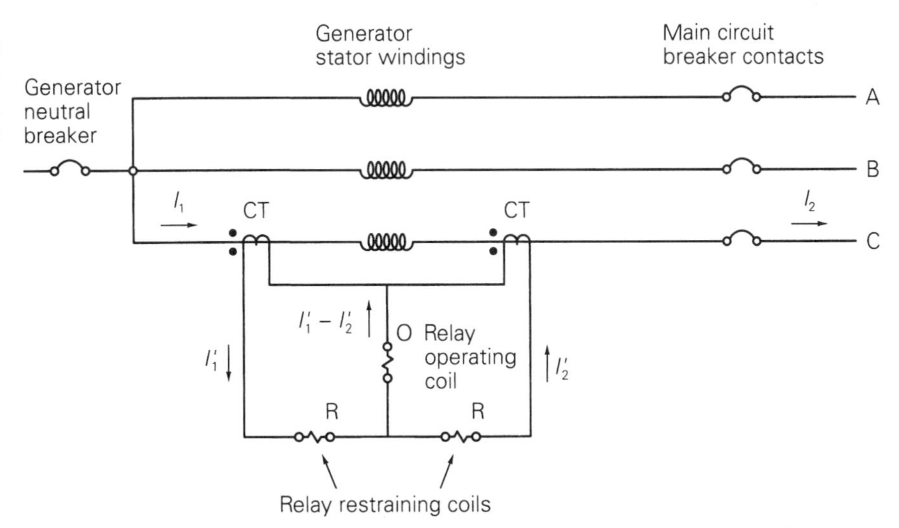

FIGURE 10.32 Differential relaying for generator protection (protection for one phase shown)

for generator protection. The protection of only one phase is shown. The method is repeated for the other two phases. When the relay in any one phase operates, all three phases of the main circuit breaker will open, as well as the generator neutral and field breakers (not shown).

For the case of no internal fault within the generator windings, $I_1 = I_2$, and, assuming identical CTs, $I_1' = I_2'$. For this case the current in the relay operating coil is zero, and the relay does not operate. On the other hand, for an internal fault such as a phase-to-ground or phase-to-phase short within the generator winding, $I_1 \neq I_2$, and $I_1' \neq I_2'$. Therefore, a difference current $I_1' - I_2'$ flows in the relay operating coil, which may cause the relay to operate. Since this relay operation depends on a *difference* current, it is called a *differential* relay.

An electromechanical differential relay called a *balance beam* relay is shown in Figure 10.33. The relay contacts close if the downward force on the right side exceeds the downward force on the left side. The electromagnetic force on the right, operating coil is proportional to the square of the operating coil mmf—that is, to $[\mathrm{N}_0(I_1' - I_2')]^2$. Similarly, the electromagnetic force on the left, restraining coil is proportional to $[\mathrm{N}_\mathrm{r}(I_1' + I_2')/2]^2$. The condition for relay operation is then

$$[\mathrm{N}_0(I_1' - I_2')]^2 > [\mathrm{N}_\mathrm{r}(I_1' + I_2')/2]^2 \tag{10.10.1}$$

Taking the square root:

$$|I_1' - I_2'| > \mathrm{k}|(I_1' + I_2')/2| \tag{10.10.2}$$

where

$$\mathrm{k} = \mathrm{N}_\mathrm{r}/\mathrm{N}_0 \tag{10.10.3}$$

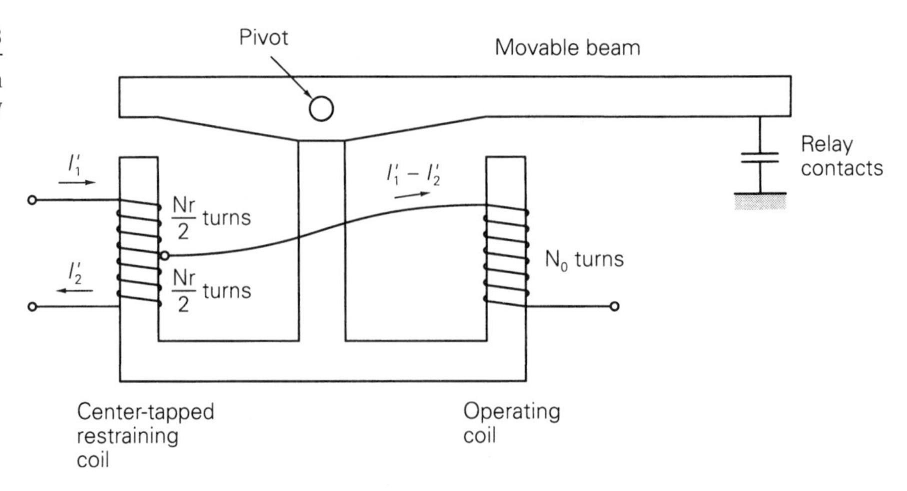

FIGURE 10.33 Balance beam differential relay

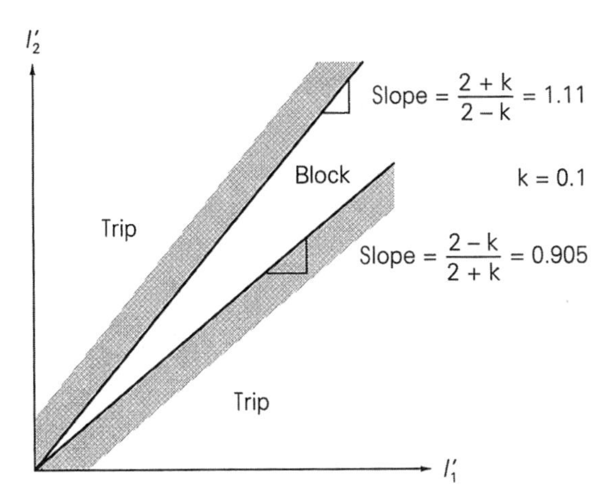

FIGURE 10.34 Differential relay block and trip regions

Assuming I_1' and I_2' are in phase, (10.10.2) is solved to obtain

$$\mathrm{I}_2' > \frac{2+\mathrm{k}}{2-\mathrm{k}}\mathrm{I}_1' \quad \text{for } \mathrm{I}_2' > \mathrm{I}_1'$$

$$\mathrm{I}_2' < \frac{2-\mathrm{k}}{2+\mathrm{k}}\mathrm{I}_1' \quad \text{for } I_2' < I_1' \tag{10.10.4}$$

Equation (10.10.4) is plotted in Figure 10.34 to obtain the block and trip regions of the differential relay for k = 0.1. Note that as k increases, the block region becomes larger; that is, the relay becomes less sensitive. In practice, no two CTs are identical, and the differential relay current $I_1' - I_2'$ can become appreciable during external faults, even though $I_1 = I_2$. The balanced beam relay solves this problem without sacrificing sensitivity during normal currents, since the block region increases as the currents increase, as shown in Figure 10.34. Also, the relay can be easily modified to enlarge the block region for very small currents near the origin, in order to avoid false trips at low currents.

Note that differential relaying provides primary zone protection without backup. Coordination with protection in adjacent zones is eliminated, which permits high speed tripping. Precise relay settings are unnecessary. Also, the need to calculate system fault currents and voltages is avoided.

10.11 BUS PROTECTION WITH DIFFERENTIAL RELAYS

Differential bus protection is illustrated by the single-line diagram of Figure 10.35. In practice, three differential relays are required, one for each phase. Operation of any one relay would cause all of the three-phase circuit breakers connected to the bus to open, thereby isolating the three-phase bus from service.

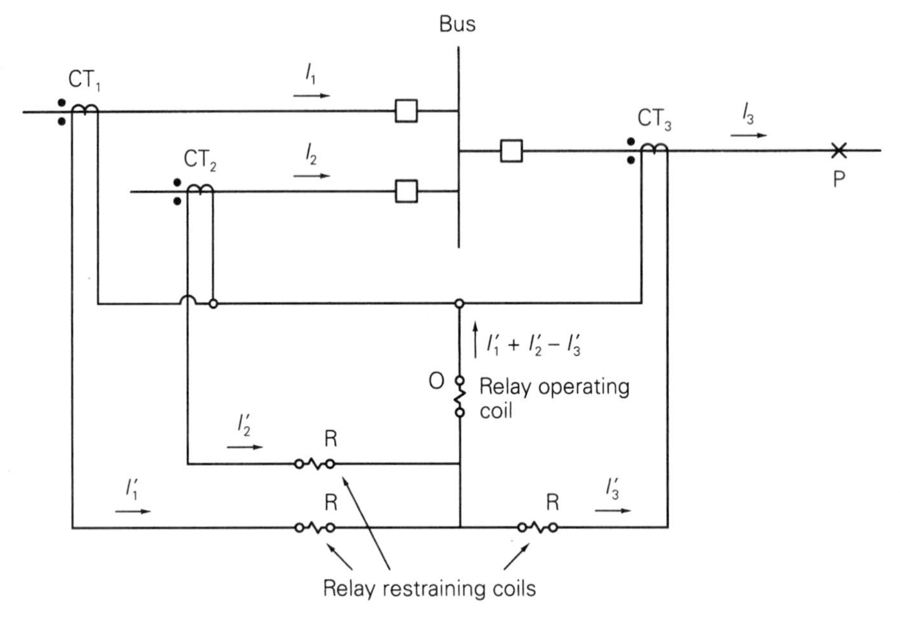

FIGURE 10.35 Single-line diagram of differential bus protection

For the case of no internal fault between the CTs—that is, no bus fault—$I_1 + I_2 = I_3$. Assuming identical CTs, the differential relay current $I_1' + I_2' - I_3'$ equals zero, and the relay does not operate. However, if there is a bus fault, the differential current $I_1' + I_2' - I_3'$, which is not zero, flows in the operating coil to operate the relay. Use of the restraining coils overcomes the problem of nonidentical CTs.

A problem with differential bus protection can result from different levels of fault currents and varying amounts of CT saturation. For example, consider an external fault at point P in Figure 10.35. Each of the CT_1 and CT_2 primaries carries part of the fault current, but the CT_3 primary carries the sum $I_3 = I_1 + I_2$. CT_3, energized at a higher level, will have more saturation, such that $I_3' \neq I_1' + I_2'$. If the saturation is too high, the differential current in the relay operating coil could result in a false trip. This problem becomes more difficult when there are large numbers of circuits connected to the bus. Various schemes have been developed to overcome this problem [1].

10.12 TRANSFORMER PROTECTION WITH DIFFERENTIAL RELAYS

The protection method used for power transformers depends on the transformer MVA rating. Fuses are often used to protect transformers with small MVA ratings, whereas differential relays are commonly used to protect transformers with ratings larger than 10 MVA.

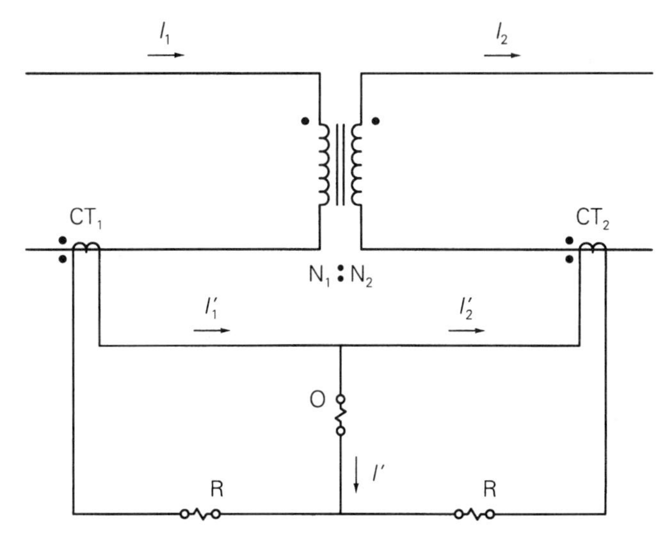

FIGURE 10.36 Differential protection of a single-phase, two-winding transformer

The differential protection method is illustrated in Figure 10.36 for a single-phase, two-winding transformer. Denoting the turns ratio of the primary and secondary CTs by $1/n_1$ and $1/n_2$, respectively (a CT with 1 primary turn and n secondary turns has a turns ratio $a = 1/n$), the CT secondary currents are

$$I_1' = \frac{I_1}{n_1} \qquad I_2' = \frac{I_2}{n_2} \tag{10.12.1}$$

and the current in the relay operating coil is

$$I' = I_1' - I_2' = \frac{I_1}{n_1} - \frac{I_2}{n_2} \tag{10.12.2}$$

For the case of no fault between the CTs—that is, no internal transformer fault—the primary and secondary currents for an ideal transformer are related by

$$I_2 = \frac{N_1 I_1}{N_2} \tag{10.12.3}$$

Using (10.12.3) in (10.12.2),

$$I' = \frac{I_1}{n_1}\left(1 - \frac{N_1/N_2}{n_2/n_1}\right) \tag{10.12.4}$$

To prevent the relay from tripping for the case of no internal transformer fault, where (10.12.3) and (10.12.4) are satisfied, the differential relay current I' must be zero. Therefore, from (10.12.4), we select

$$\frac{n_2}{n_1} = \frac{N_1}{N_2} \tag{10.12.5}$$

If an internal transformer fault between the CTs does occur, (10.12.3) is not satisfied and the differential relay current $I' = I_1' - I_2'$ is not zero. The relay will trip if the operating condition given by (10.10.4) is satisfied. Also, the value of k in (10.10.4) can be selected to control the size of the block region shown in Figure 10.34, thereby controlling relay sensitivity.

EXAMPLE 10.9 Differential relay protection for a single-phase transformer

A single-phase two-winding, 10-MVA, 80 kV/20 kV transformer has differential relay protection. Select suitable CT ratios. Also, select k such that the relay blocks for up to 25% mismatch between I_1' and I_2'.

SOLUTION The transformer-rated primary current is

$$I_{1\text{rated}} = \frac{10 \times 10^6}{80 \times 10^3} = 125 \text{ A}$$

From Table 10.2, select a 150:5 primary CT ratio to give $I_1' = 125(5/150) = 4.17$ A at rated conditions. Similarly, $I_{2\text{rated}} = 500$ A. Select a 600:5 secondary CT ratio to give $I_2' = 500(5/600) = 4.17$ A and a differential current $I' = I_1' - I_2' = 0$ (neglecting magnetizing current) at rated conditions. Also, for a 25% mismatch between I_1' and I_2', select a 1.25 upper slope in Figure 10.34. That is,

$$\frac{2+k}{2-k} = 1.25 \qquad k = 0.2222$$

■

A common problem in differential transformer protection is the mismatch of relay currents that occurs when standard CT ratios are used. If the primary winding in Example 10.9 has a 138-kV instead of 80-kV rating, then $I_{1\text{rated}} = 10 \times 10^6/138 \times 10^3 = 72.46$ A, and a 100:5 primary CT would give $I_1' = 72.46(5/100) = 3.62$ A at rated conditions. This current does not balance $I_2' = 4.17$ A using a 5:600 secondary CT, nor $I_2' = 3.13$ A using a 5:800 secondary CT. The mismatch is about 15%.

One solution to this problem is to use auxiliary CTs, which provide a wide range of turns ratios. A 5:5.76 auxiliary CT connected to the 5:600 secondary CT in the above example would reduce I_2' to $4.17(5/5.76) = 3.62$ A, which does balance I_1'. Unfortunately, auxiliary CTs add their own burden to the main CTs and also increase transformation errors. A better solution is to use tap settings on the relays themselves, which have the same effect as auxiliary CTs. Most transformer differential relays have taps that provide for differences in restraining windings in the order of 2 or 3 to 1.

When a transformer is initially energized, it can draw a large "inrush" current, a transient current that flows in the shunt magnetizing branch and decays after a few cycles to a small steady-state value. Inrush current appears as a differential current since it flows only in the primary winding. If a large inrush current does occur upon transformer energization, a differential relay will see a large differential current and trip out the transformer unless the protection method is modified to detect inrush current.

One method to prevent tripping during transformer inrush is based on the fact that inrush current is nonsinusoidal with a large second-harmonic component. A filter can be used to pass fundamental and block harmonic components of the differential current I' to the relay operating coil. Another method is based on the fact that inrush current has a large dc component, which can be used to desensitize the relay. Time-delay relays may also be used to temporarily desensitize the differential relay until the inrush current has decayed to a low value.

Figure 10.37 illustrates differential protection of a three-phase Y–Δ two-winding transformer. Note that a Y–Δ transformer produces 30° phase

FIGURE 10.37

Differential protection of a three-phase, Y–Δ, two-winding transformer

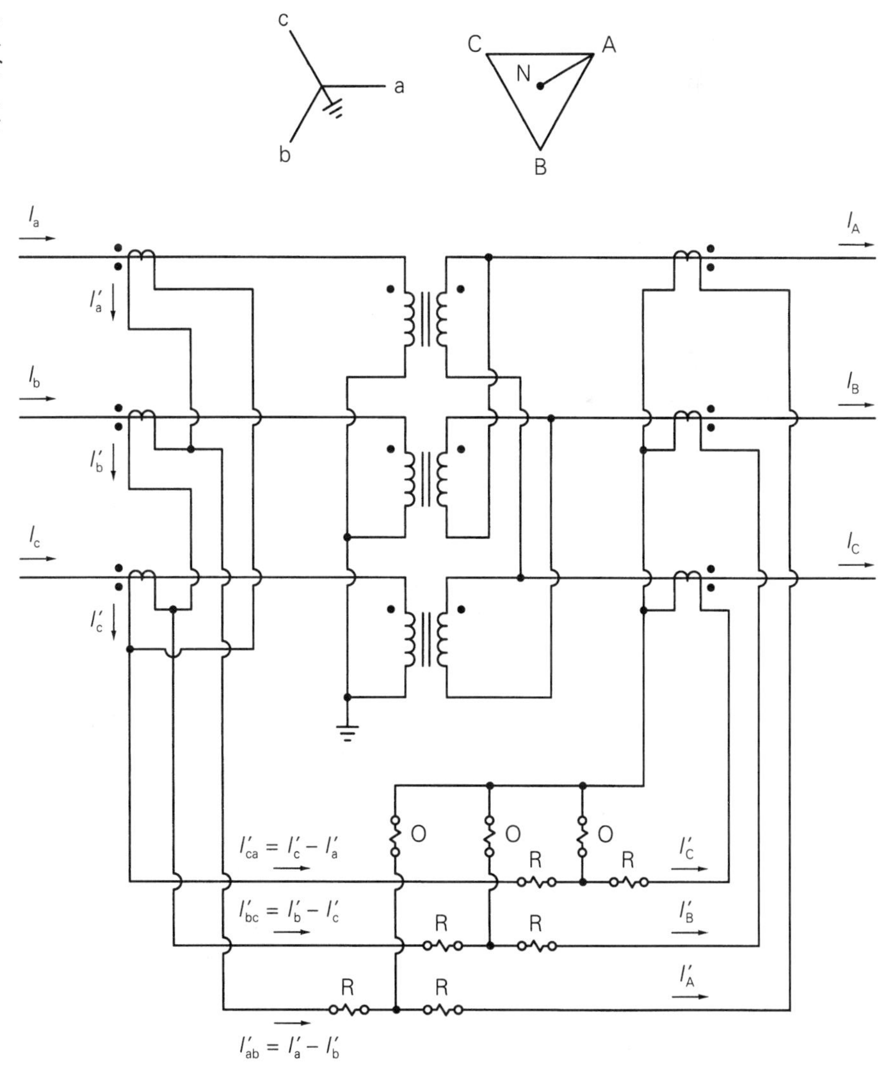

shifts in the line currents. The CTs must be connected to compensate for the 30° phase shifts, such that the CT secondary currents as seen by the relays are in phase. The correct phase-angle relationship is obtained by connecting CTs on the Y side of the transformer in Δ, and CTs on the Δ side in Y.

EXAMPLE 10.10 Differential relay protection for a three-phase transformer

A 30-MVA, 34.5 kV Y/138 kV Δ transformer is protected by differential relays with taps. Select CT ratios, CT connections, and relay tap settings. Also determine currents in the transformer and in the CTs at rated conditions. Assume that the available relay tap settings are 5:5, 5:5.5, 5:6.6, 5:7.3, 5:8, 5:9, and 5:10, giving relay tap ratios of 1.00, 1.10, 1.32, 1.46, 1.60, 1.80, and 2.00.

SOLUTION As shown in Figure 10.37, CTs are connected in Δ on the (34.5-kV) Y side of the transformer, and CTs are connected in Y on the (138-kV) Δ side, in order to obtain the correct phasing of the relay currents.

Rated current on the 138-kV side of the transformer is

$$I_{A\,\text{rated}} = \frac{30 \times 10^6}{\sqrt{3}(138 \times 10^3)} = 125.51 \text{ A}$$

Select a 150:5 CT on the 138-kV side to give $I'_A = 125.51(5/150) = 4.184$ A in the 138-kV CT secondaries and in the righthand restraining windings of Figure 10.37.

Next, rated current on the 34.5-kV side of the transformer is

$$I_{a\,\text{rated}} = \frac{30 \times 10^6}{\sqrt{3}(34.5 \times 10^3)} = 502.04 \text{ A}$$

Select a 500:5 CT on the 34.5-kV side to give $I'_a = 502.0(5/500) = 5.02$ A in the 34.5-kV CT secondaries and $I'_{ab} = 5.02\sqrt{3} = 8.696$ A in the lefthand restraining windings of Figure 10.37.

Finally, select relay taps to balance the currents in the restraining windings. The ratio of the currents in the left- to righthand restraining windings is

$$\frac{I'_{ab}}{I'_A} = \frac{8.696}{4.184} = 2.078$$

The closest relay tap ratio is $T'_{AB}/T'_A = 2.0$, corresponding to a relay tap setting of $T'_A : T'_{ab} = 5 : 10$. The percentage mismatch for this tap setting is

$$\left| \frac{(I'_A/T'_A) - (I'_{ab}/T'_{ab})}{(I'_{ab}/T'_{ab})} \right| \times 100 = \left| \frac{(4.184/5) - (8.696/10)}{(8.696/10)} \right| \times 100 = 3.77\%$$

This is a good mismatch; since transformer differential relays typically have their block regions adjusted between 20% and 60% (by adjusting k in Figure 10.34), a 3.77% mismatch gives an ample safety margin in the event of CT and relay differences. ■

For three-phase transformers (Y–Y, Y–Δ, Δ–Y, Δ–Δ), the general rule is to connect CTs on the Y side in Δ and CTs on the Δ side in Y. This arrangement compensates for the 30° phase shifts in Y–Δ or Δ–Y banks. Note also that zero-sequence current cannot enter a Δ side of a transformer or the CTs on that side, and zero-sequence current on a grounded Y side cannot enter the Δ-connected CTs on that side. Therefore, this arrangement also blocks zero-sequence currents in the differential relays during external ground faults. For internal ground faults, however, the relays can operate from the positive- and negative-sequence currents involved in these faults.

Differential protection methods have been modified to handle multi-winding transformers, voltage-regulating transformers, phase-angle regulating transformers, power-rectifier transformers, transformers with special connections (such as zig-zag), and other, special-purpose transformers. Also, other types of relays such as gas-pressure detectors for liquid-filled transformers are used.

10.13 PILOT RELAYING

Pilot relaying refers to a type of differential protection that compares the quantities at the terminals via a communication channel rather than by a direct wire interconnection of the relays. Differential protection of generators, buses, and transformers considered in previous sections does not require pilot relaying because each of these devices is at one geographical location where CTs and relays can be directly interconnected. However, differential relaying of transmission lines requires pilot relaying because the terminals are widely separated (often by many kilometers). In actual practice, pilot relaying is typically applied to short transmission lines (up to 80 km) with 69 to 115 kV ratings.

Four types of communication channels are used for pilot relaying:

1. *Pilot wires:* Separate electrical circuits operating at dc, 50 to 60 Hz, or audio frequencies. These could be owned by the power company or leased from the telephone company.

2. *Power-line carrier:* The transmission line itself is used as the communication circuit, with frequencies between 30 and 300 kHz being transmitted. The communication signals are applied to all three phases using an L–C voltage divider and are confined to the line under protection by blocking filters called *line traps* at each end.

3. *Microwave:* A 2 to 12 GHz signal transmitted by line-of-sight paths between terminals using dish antennas.
4. *Fiber optic cable:* Signals transmitted by light modulation through electrically nonconducting cable. This cable eliminates problems due to electrical insulation, inductive coupling from other circuits, and atmospheric disturbances.

Two common fault detection methods are *directional comparison*, where the power flows at the line terminals are compared, and *phase comparison*, where the relative phase angles of the currents at the terminals are compared. Also, the communication channel can either be required for trip operations, which is known as a *transfer trip system*, or not be required for trip operations, known as a *blocking system*. A particular pilot-relaying method is usually identified by specifying the fault-detection method and the channel use. The four basic combinations are directional comparison blocking, directional comparison transfer trip, phase comparison blocking, and phase comparison transfer trip.

Like differential relays, pilot relays provide primary zone protection without backup. Thus, coordination with protection in adjacent zones is eliminated, resulting in high-speed tripping. Precise relay settings are unnecessary. Also, the need to calculate system fault currents and voltages is eliminated.

10.14 DIGITAL RELAYING

In previous sections we described the operating principle of relays built with electromechanical components, including the induction disc time-delay overcurrent relay, Figure 10.14; the directional relay, similar in operation to a watt-hour meter; and the balance-beam differential relay, Figure 10.33. These electromechanical relays, introduced in the early 1900s, have performed well over the years and continue in relatively maintenance-free operation today. Solid-state relays using analog circuits and logic gates, with block-trip regions similar to those of electromechanical relays and with newer types of block/trip regions, have been available since the late 1950s. Such relays, widely used in HV and EHV systems, offer the reliability and ruggedness of their electromechanical counterparts at a competitive price. Beyond solid-state analog relays, a new generation of relays based on digital computer technology has been under development since the 1980s.

Benefits of digital relays include accuracy, improved sensitivity to faults, better selectivity, flexibility, user-friendliness, easy testing, and relay event monitoring/recording capabilities. Digital relaying also has the advantage that modifications to tripping characteristics, either changes in conventional settings or shaping of entirely new block/trip regions, could be made by updating software from a remote computer terminal. For example, the relay engineer could reprogram tripping characteristics of field-installed, in-service relays without leaving the engineering office. Alternatively, relay software could be updated in real time, based on operating conditions, from a central computer.

An important feature of power system protection is the decentralized, local nature of relays. Except for pilot relaying, each relay receives information from nearby local CTs and VTs and trips only local breakers. Interest in digital relaying is not directed at replacing local relays by a central computer. Instead, each electromechanical or solid-state analog relay would be replaced by a dedicated, local digital relay with a similar operating principle, such as time-delay overcurrent, impedance, or differential relaying. The central computer would interact with local digital relays in a supervisory role.

PROBLEMS

SECTION 10.2

10.1 The primary conductor in Figure 10.2 is one phase of a three-phase transmission line operating at 345 kV, 600 MVA, 0.95 power factor lagging. The CT ratio is 1200:5 and the VT ratio is 3000:1. Determine the CT secondary current I′ and the VT secondary voltage V′. Assume zero CT error.

10.2 A CO-8 relay with a current tap setting of 5 amperes is used with the 100:5 CT in Example 10.1. The CT secondary current I′ is the input to the relay operating coil. The CO-8 relay burden is shown in the following table for various relay input currents.

CO-8 relay input current I′, A	5	8	10	13	15
CO-8 relay burden Z_B, Ω	0.5	0.8	1.0	1.3	1.5

Primary current and CT error are computed in Example 10.1 for the 5-, 8-, and 15-A relay input currents. Compute the primary current and CT error for (a) I′ = 10 A and Z_B = 1.0 Ω, and for (b) I′ = 13 A and Z_B = 1.3 Ω. (c) Plot I′ versus I for the above five values of I′. (d) For reliable relay operation, the fault-to-pickup current ratio with minimum fault current should be greater than two. Determine the minimum fault current for application of this CT and relay with 5-A tap setting.

10.3 An overcurrent relay set to operate at 10 A is connected to the CT in Figure 10.8 with a 200:5 CT ratio. Determine the minimum primary fault current that the relay will detect if the burden Z_B is (a) 1.0 Ω, (b) 4.0 Ω, and (c) 5.0 Ω.

10.4 Given the open-delta VT connection shown in Figure 10.38, both VTs having a voltage rating of 240 kV : 120 V, the voltages are specified as $V_{AB} = 230\angle 0°$, $V_{BC} = 230\angle -120°$, and $V_{CA} = 230\angle 120°$ kV. Determine V_{ab}, V_{bc}, and V_{ca} for the following cases: (a) The dots are shown in Figure 10.38. (b) The dot near c is moved to b in Figure 10.38.

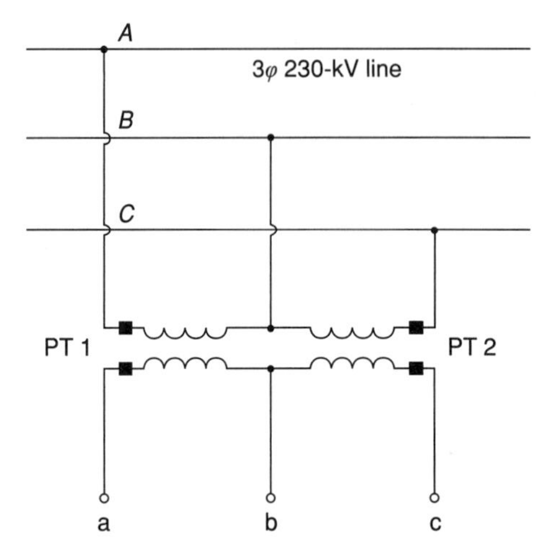

FIGURE 10.38 Problem 10.4

10.5 A CT with an excitation curve given in Figure 10.39 has a rated current ratio of 500:5 A and a secondary leakage impedance of $0.1 + j0.5\ \Omega$. Calculate the CT secondary output current and the CT error for the following cases: (a) The impedance of the terminating device is $4.9 + j0.5\ \Omega$ and the primary CT load current is 400 A. (b) The impedance of the terminating device is $4.9 + j0.5\ \Omega$ and the primary CT fault current is 1200 A. (c) The impedance of the terminating device is $14.9 + j1.5\ \Omega$ and the primary CT load current is 400 A. (d) The impedance of the terminating device is $14.9 + j1.5\ \Omega$ and the primary CT fault current is 1200 A.

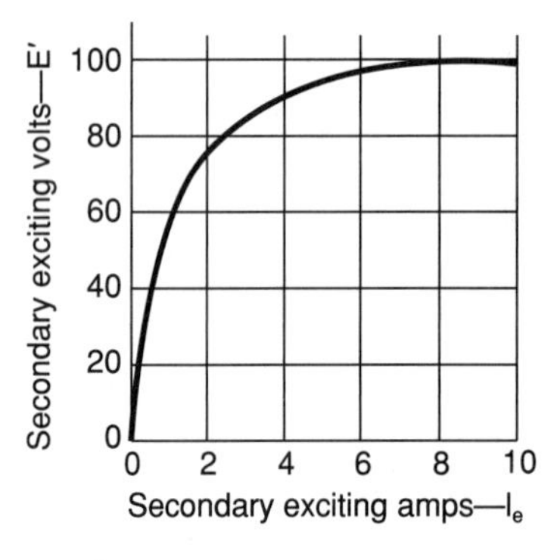

FIGURE 10.39 Problem 10.5

10.6 The CT of Problem 10.5 is utilized in conjunction with a current-sensitive device that will operate at current levels of 8 A or above. Check whether the device will detect the 1200-A fault current for cases (b) and (d) in Problem 10.5.

SECTION 10.3

10.7 The input current to a CO-8 relay is 10 A. Determine the relay operating time for the following current tap settings (TS) and time dial settings (TDS): (a) TS = 1.0, TDS = 1/2; (b) TS = 2.0, TDS = 1.5; (c) TS = 2.0, TDS = 7; (d) TS = 3.0, TDS = 7; and (e) TS = 12.0, TDS = 1.

10.8 The relay in Problem 10.2 has a time-dial setting of 4. Determine the relay operating time if the primary fault current is 500 A.

10.9 An RC circuit used to produce time delay is shown in Figure 10.40. For a step input voltage $v_i(t) = 2u(t)$ and $C = 10\ \mu F$, determine T_{delay} for the following cases: (a) $R = 100\ k\Omega$; and (b) $R = 1\ M\Omega$. Sketch the output $v_o(t)$ versus time for cases (a) and (b).

FIGURE 10.40

Problem 10.9

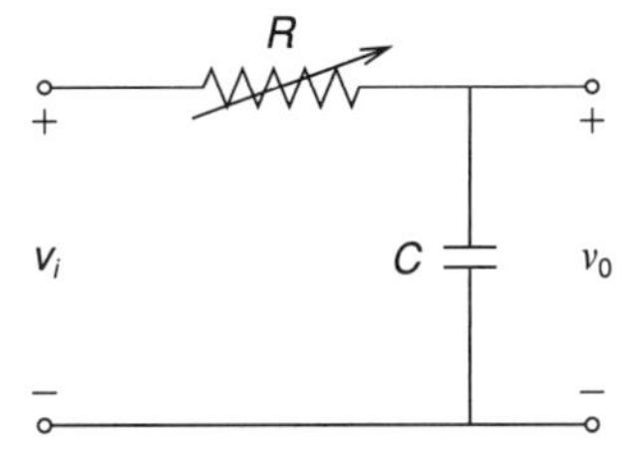

10.10 Reconsider case (b) of Problem 10.5. Let the load impedance $4.9 + j0.5\ \Omega$ be the input impedance to a CO-7 induction disc time-delay overcurrent relay. The CO-7 relay characteristic is shown in Figure 10.41. For a tap setting of 5 A and a time dial setting of 2, determine the relay operating time.

FIGURE 10.41

Problems 10.10 and 10.14

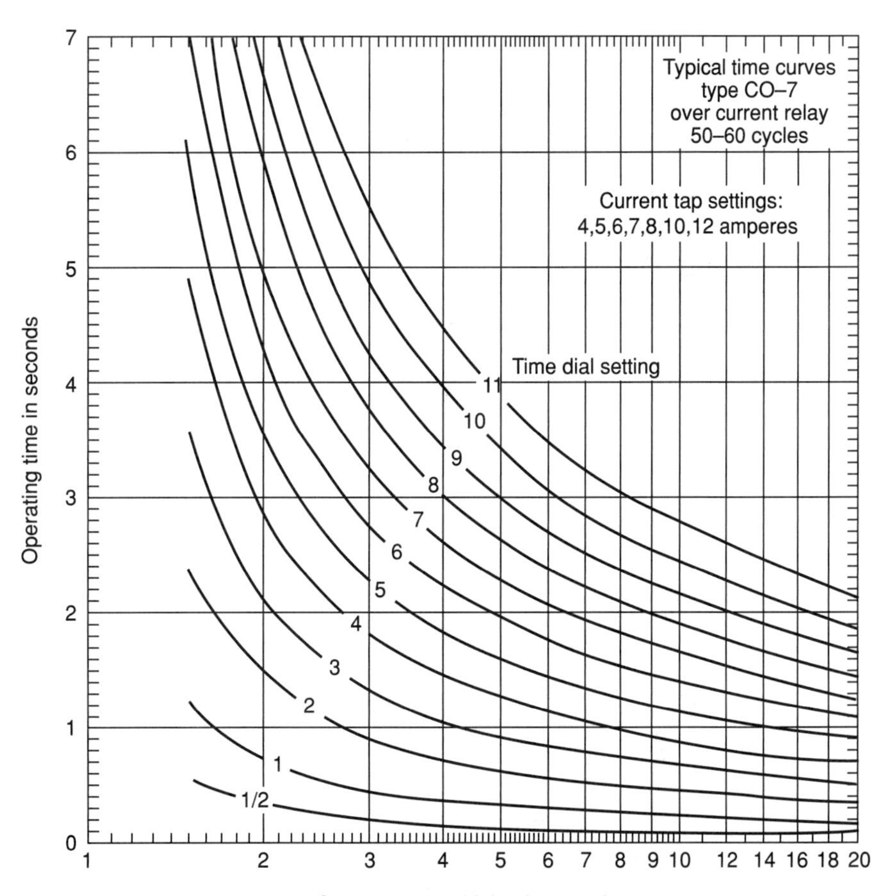

SECTION 10.4

10.11 Evaluate relay coordination for the minimum fault currents in Example 10.4. For the selected current tap settings and time dial settings, (a) determine the operating time of relays at B2 and B3 for the 700-A fault current. (b) Determine the operating time of relays at B1 and B2 for the 1500-A fault current. Are the fault-to-pickup current ratios $\geqslant 2.0$ (a requirement for reliable relay operation) in all cases? Are the coordination time intervals $\geqslant 0.3$ seconds in all cases?

10.12 Repeat Example 10.4 for the following system data. Coordinate the relays for the maximum fault currents.

	Maximum Load		Symmetrical Fault Current	
Bus	MVA	Lagging p.f.	Maximum A	Minimum A
1	9.0	0.95	5000	3750
2	9.0	0.95	3000	2250
3	9.0	0.95	2000	1500

Breaker	Breaker Operating Time	CT Ratio	Relay
B1	5 cycles	600:5	CO-8
B2	5 cycles	400:5	CO-8
B3	5 cycles	200:5	CO-8

10.13 Using the current tap settings and time dial settings that you have selected in Problem 10.12, evaluate relay coordination for the minimum fault currents. Are the fault-to-pickup current ratios $\geqslant 2.0$, and are the coordination time delays $\geqslant 0.3$ seconds in all cases?

10.14 An 11-kV radial system is shown in Figure 10.42. Assuming a CO-7 relay with relay characteristic given in Figure 10.41 and the same power factor for all loads, select relay settings to protect the system.

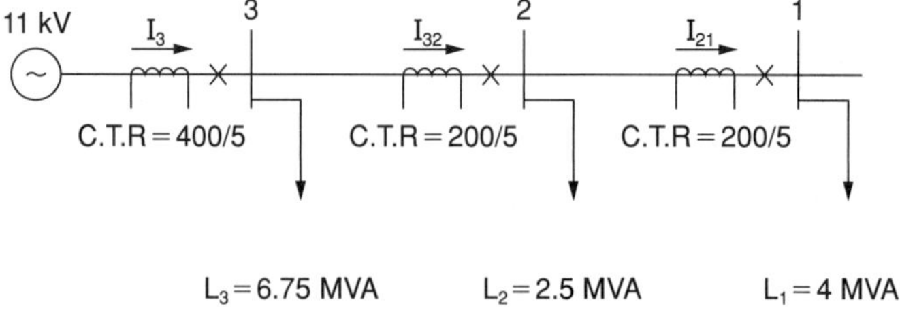

FIGURE 10.42 Problem 10.14

SECTION 10.5

10.15 Rework Example 10.5 for the following faults: (a) a three-phase, permanent fault on the load side of tap 3; (b) a single line-to-ground, permanent fault at bus 4 on the load side of the recloser; and (c) a three-phase, permanent fault at bus 4 on the source side of the recloser.

10.16 A three-phase 34.5-kV feeder supplying a 4-MVA load is protected by 80E power fuses in each phase, in series with a recloser. The time-current characteristic of the 80E fuse is shown in Figure 10.43. Analysis yields maximum and minimum fault currents of 1000 and 500 A, respectively. (a) To have the recloser clear the fault, find the maximum clearing time necessary for recloser operation. (b) To have the fuses clear the fault, find the minimum recloser clearing time. Assume that the recloser operating time is independent of fault current magnitude.

FIGURE 10.43

Problem 10.16

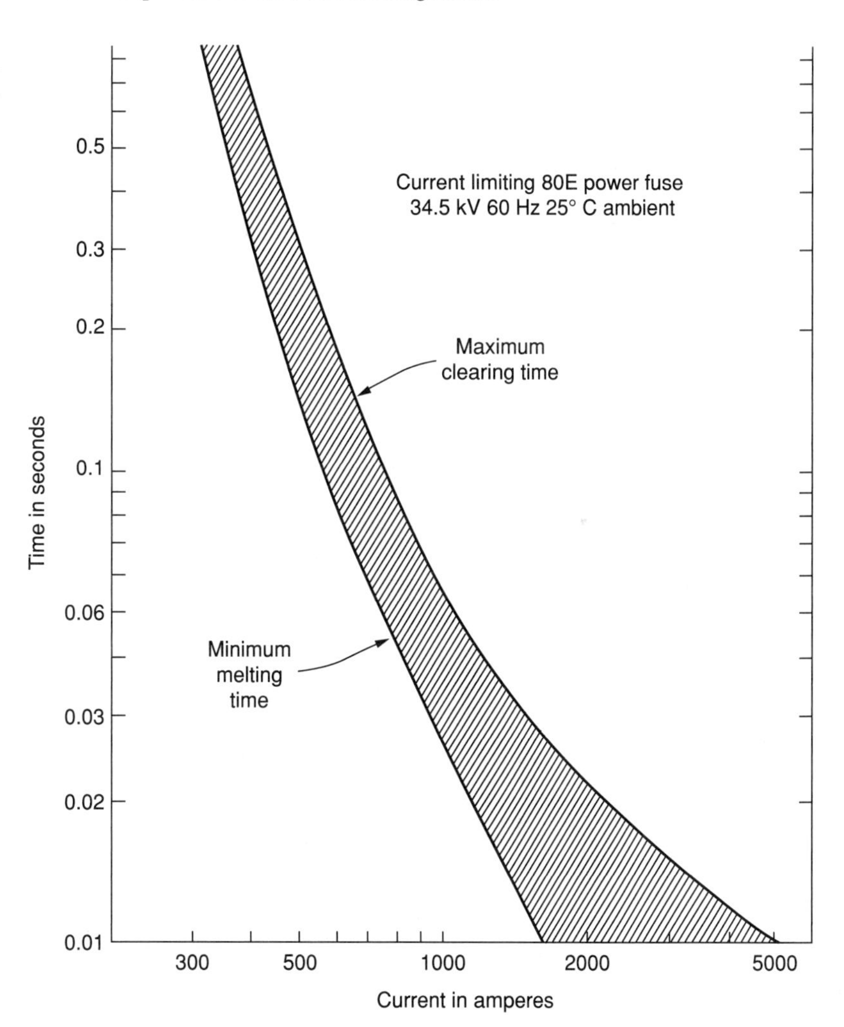

SECTION 10.7

10.17 For the system shown in Figure 10.44, directional overcurrent relays are used at breakers B12, B21, B23, B32, B34, and B43. Overcurrent relays alone are used at B1 and B4. (a) For a fault at P_1, which breakers do not operate? Which breakers should be coordinated? Repeat (a) for a fault at (b) P_2, (c) P_3. (d) Explain how the system is protected against bus faults.

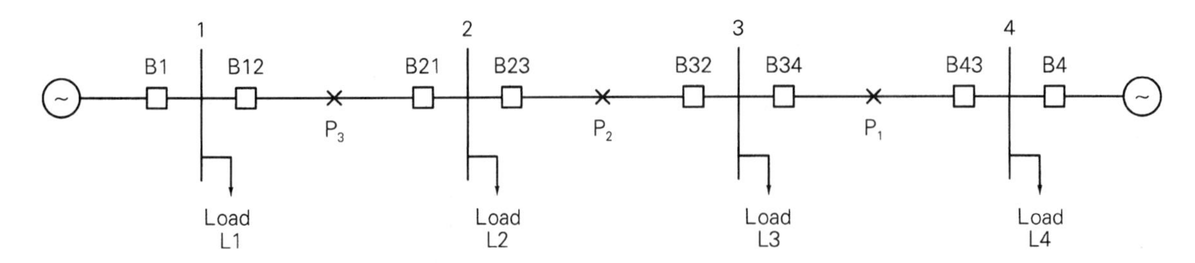

FIGURE 10.44 Problem 10.17

SECTION 10.8

10.18 (a) Draw the protective zones for the power system shown in Figure 10.45. Which circuit breakers should open for a fault at (a) P_1, (b) P_2, and (c) P_3?

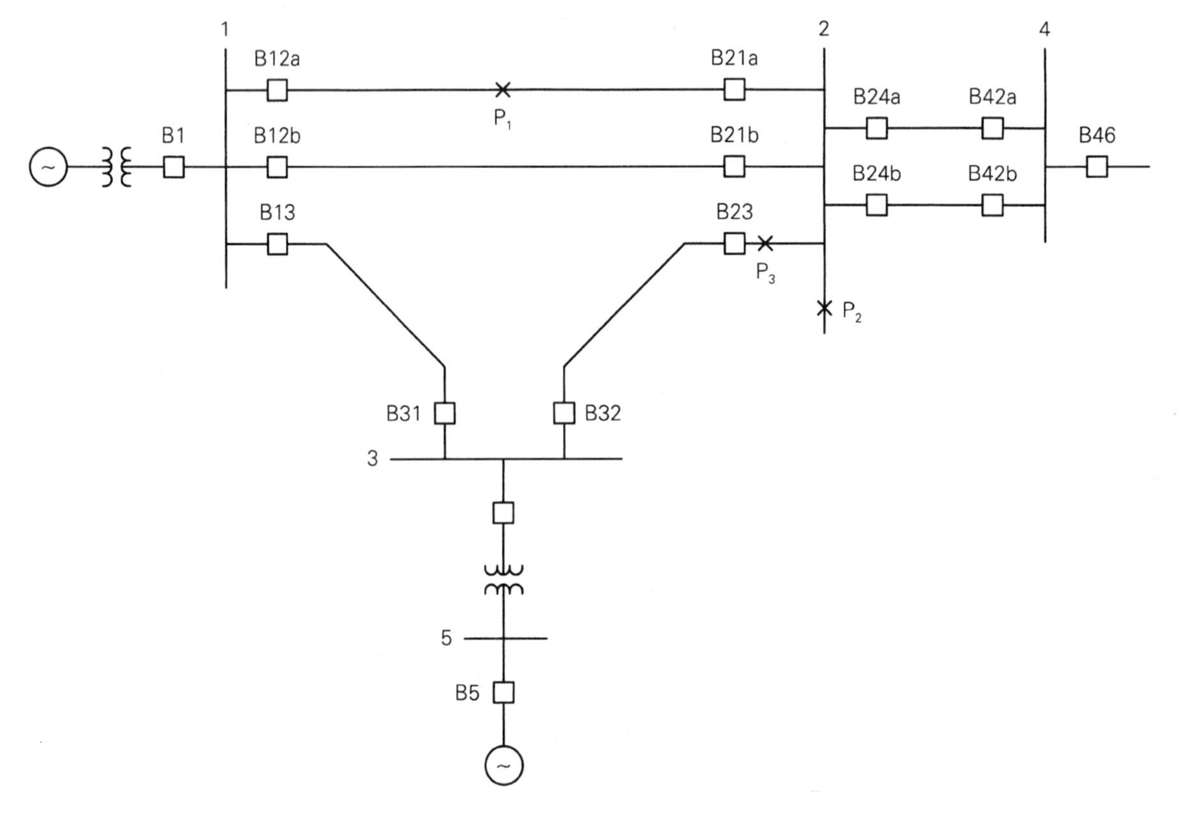

FIGURE 10.45 Problem 10.18

10.19 Figure 10.46 shows three typical bus arrangements. Although the number of lines connected to each arrangement varies widely in practice, four lines are shown for convenience and comparison. Note that the required number of circuit breakers per line is 1 for the ring bus, $1\frac{1}{2}$ for the breaker-and-a-half double-bus, and 2 for the double-breaker double-bus arrangement. For each arrangement: (a) Draw the protective zones. (b) Identify the breakers that open under primary protection for a fault on line 1.

FIGURE 10.46

Problem 10.19—typical bus arrangements

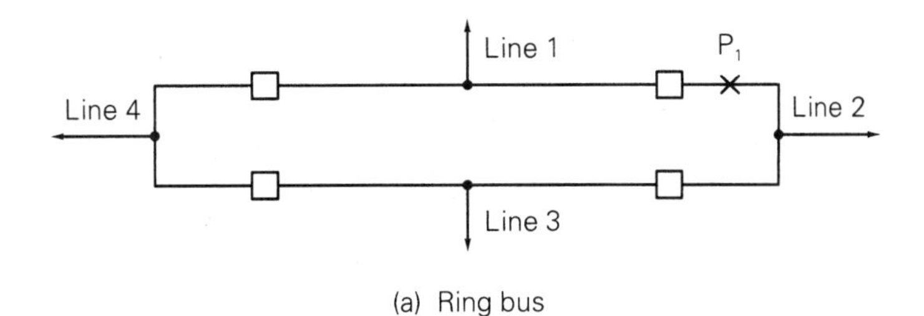

(a) Ring bus

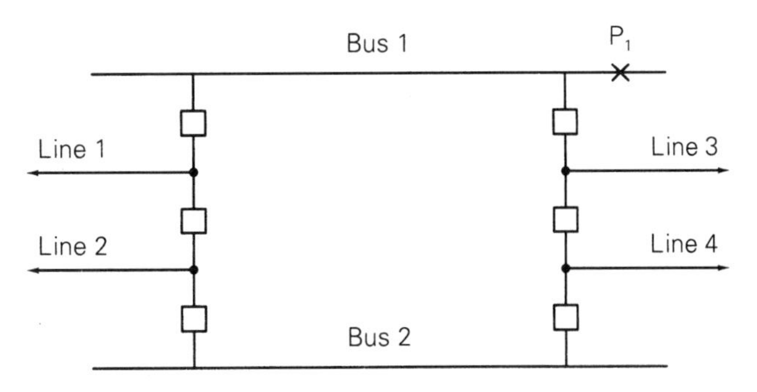

(b) Breaker-and-a-half double bus

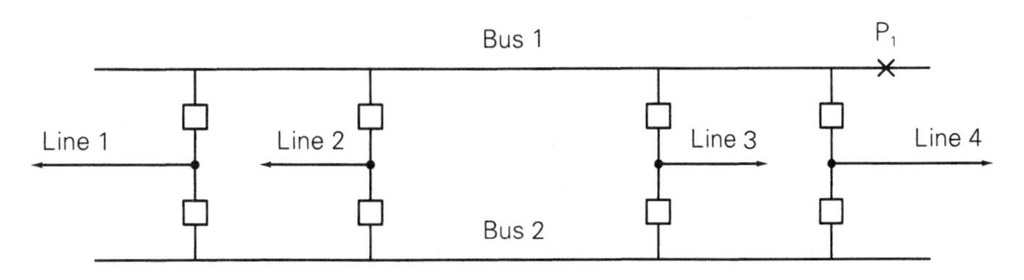

(c) Double-breaker double bus

(c) Identify the lines that are removed from service under primary protection during a bus fault at P_1. (d) Identify the breakers that open under backup protection in the event a breaker fails to clear a fault on line 1 (that is, a stuck breaker during a fault on line 1).

SECTION 10.9

10.20 Three-zone mho relays are used for transmission line protection of the power system shown in Figure 10.25. Positive-sequence line impedances are given as follows.

Line	Positive-Sequence Impedance, Ω
1–2	$6 + j60$
2–3	$4 + j40$
2–4	$5 + j50$

Rated voltage for the high-voltage buses is 500 kV. Assume a 1500:5 CT ratio and a 4500:1 VT ratio at B12. (a) Determine the settings Z_{r1}, Z_{r2}, and Z_{r3} for the mho relay at B12. (b) Maximum current for line 1–2 under emergency loading conditions is 1400 A at 0.90 power factor lagging. Verify that B12 does not trip during emergency loading conditions.

10.21 Line impedances for the power system shown in Figure 10.47 are $Z_{12} = Z_{23} = 3.0 + j40.0\ \Omega$, and $Z_{24} = 6.0 + j80.0\ \Omega$. Reach for the zone 3 B12 impedance relays is set for 100% of line 1–2 plus 120% of line 2–4. (a) For a bolted three-phase fault at bus 4, show that the apparent primary impedance "seen" by the B12 relays is

$$Z_{\text{apparent}} = Z_{12} + Z_{24} + (I_{32}/I_{12})Z_{24}$$

where (I_{32}/I_{12}) is the line 2–3 to line 1–2 fault current ratio. (b) If $|I_{32}/I_{12}| > 0.20$, does the B12 relay see the fault at bus 4?

Note: This problem illustrates the "infeed effect." Fault currents from line 2–3 can cause the zone 3 B12 relay to underreach. As such, remote backup of line 2–4 at B12 is ineffective.

FIGURE 10.47

Problem 10.21

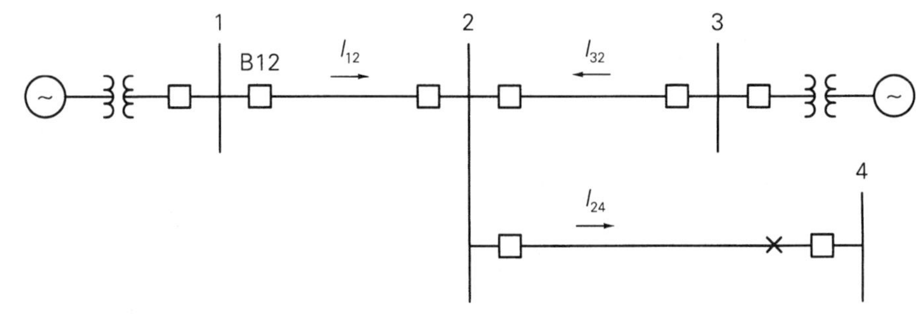

10.22 Consider the transmission line shown in Figure 10.48 with series impedance Z_L, negligible shunt admittance, and a load impedance Z_R at the receiving end. (a) Determine Z_R for the given conditions of $V_R = 1.0$ per unit and $S_R = 2 + j0.8$ per unit. (b) Construct the impedance diagram in the R-X plane for $Z_L = 0.1 + j0.3$ per unit. (c) Find Z_S for this condition and the angle δ between Z_S and Z_R.

FIGURE 10.48

Problem 10.22

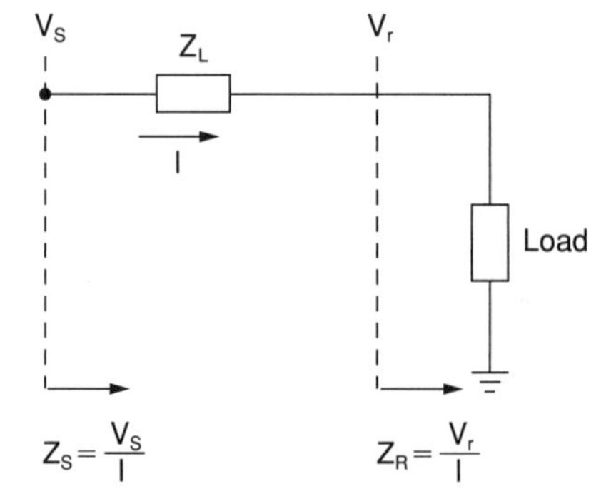

10.23 A simple system with circuit breaker-relay locations is shown in Figure 10.49. The six transmission-line circuit breakers are controlled by zone distance and directional relays, as shown in Figure 10.50. The three transmission lines have the same

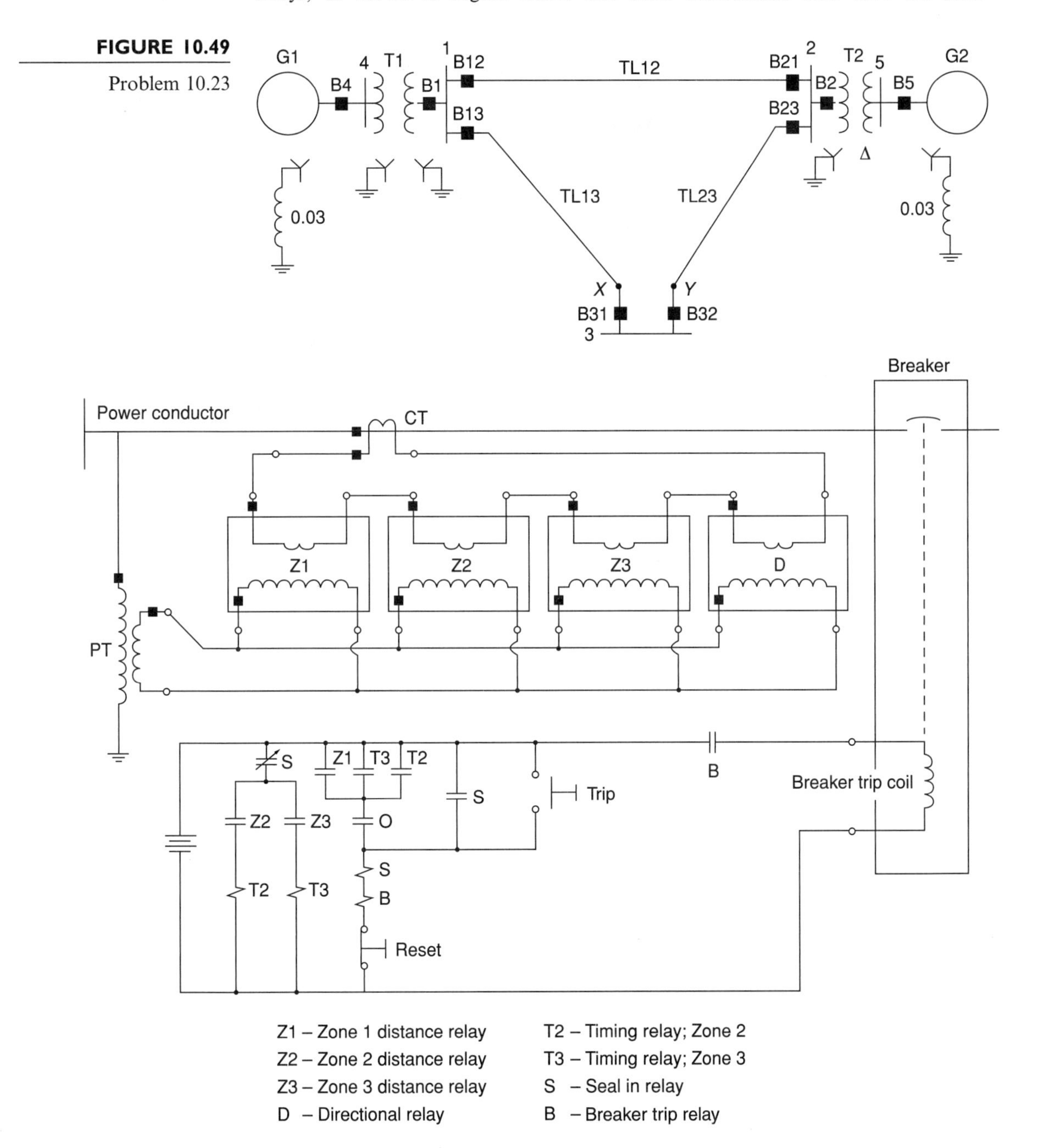

FIGURE 10.49 Problem 10.23

FIGURE 10.50 Three-zone distance-relay scheme (shown for one phase only) for Problem 10.23

positive-sequence impedance of j0.1 per unit. The reaches for zones 1, 2, and 3 are 80, 120, and 250%, respectively. Consider only three-phase faults. (a) Find the settings Z_r in per unit for all distance relays. (b) Convert the settings in Ω if the VTs are rated 133 kV:115 V and the CTs are rated 400:5 A. (c) For a fault at location X, which is 10% down line TL31 from bus 3, discuss relay operations.

SECTION 10.10

10.24 Select k such that the differential relay characteristic shown in Figure 10.34 blocks for up to 20% mismatch between I_1' and I_2'.

SECTION 10.11

10.25 Consider a protected bus that terminates four lines, as shown in Figure 10.51. Assume that the linear couplers have the standard $X_m = 5$ mΩ and a three-phase fault externally located on line 3 causes the fault currents shown in Figure 10.51. Note that the infeed current on line 3 to the fault is –j10 kA. (a) Determine V_o. (b) Let the fault be moved to an internal location on the protected bus between lines 3 and 4. Find V_o and discuss what happens. (c) By moving the external fault from line 3 to a corresponding point on (i) line 2 and (ii) line 4, determine V_o in each case.

FIGURE 10.51

Problem 10.25—Bus differential protection using linear couplers

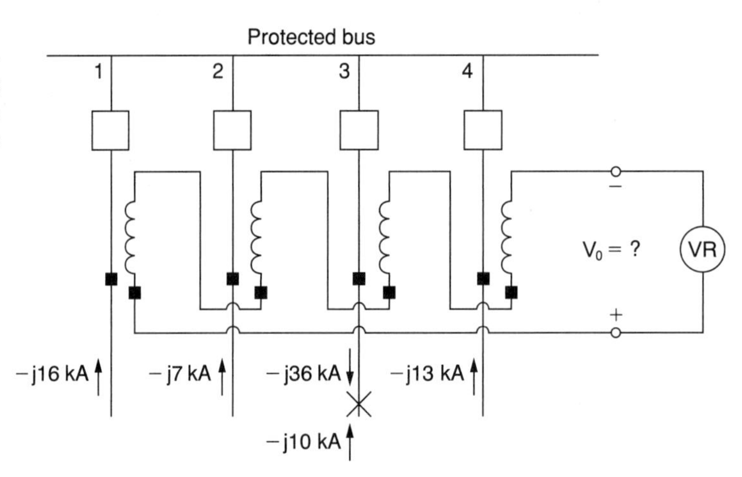

SECTION 10.12

10.26 A single-phase, 5-MVA, 20/8.66-kV transformer is protected by a differential relay with taps. Available relay tap settings are 5:5, 5:5.5, 5:6.6, 5:7.3, 5:8, 5:9, and 5:10, giving tap ratios of 1.00, 1.10, 1.32, 1.46, 1.60, 1.80, and 2.00. Select CT ratios and relay tap settings. Also, determine the percentage mismatch for the selected tap setting.

10.27 A three-phase, 500-MVA, 345 kV Δ/500 kV Y transformer is protected by differential relays with taps. Select CT ratios, CT connections, and relay tap settings. Determine the currents in the transformer and in the CTs at rated conditions. Also determine the percentage mismatch for the selected relay tap settings. Available relay tap settings are given in Problem 10.26.

10.28 For a Δ–Y connected, 15-MVA, 33:11 kV transformer with differential relay protection and CT ratios shown in Figure 10.52, determine the relay currents at full load and calculate the minimum relay current setting to allow 125% overload.

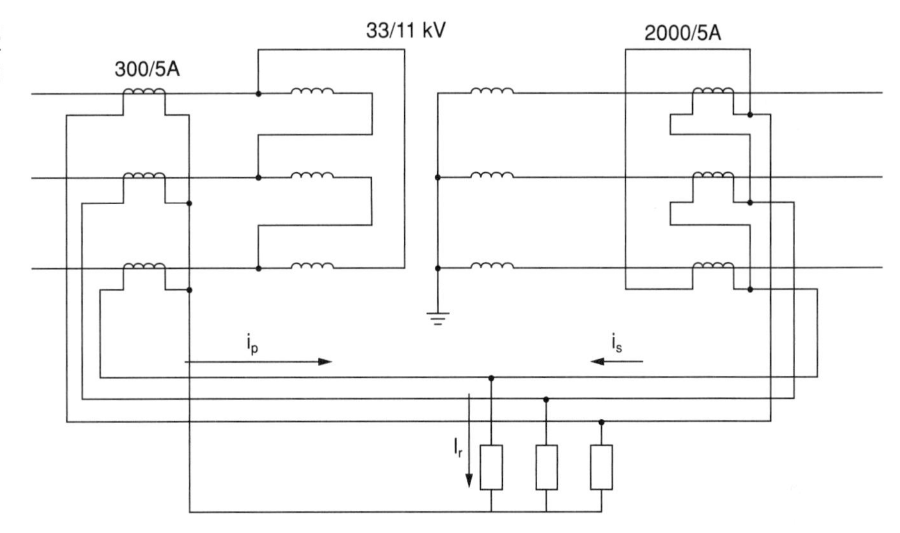

FIGURE 10.52
Problem 10.28

10.29 Consider a three-phase Δ–Y connected, 30-MVA, 33:11 kV transformer with differential relay protection. If the CT ratios are 500:5 A on the primary side and 2000:5 A on the secondary side, compute the relay current setting for faults drawing up to 200% of rated transformer current.

10.30 Determine the CT ratios for differential protection of a three-phase, Δ–Y connected, 15-MVA, 33:11 kV transformer, such that the circulating current in the transformer Δ does not exceed 5 A.

CASE STUDY QUESTIONS

A. What is a flexible ac transmission systems (FACTS) device? One example of a FACTS device is a static synchronous compensator ("STATCON"). Briefly describe what a STATCOM is and find the location of one STATCOM device currently that has been installed in the United States (use the Internet).

B. What is a phasor measurement unit (PMU) in electric power systems? What is a synchronized PMU? What is measurement system called that incorporates PMUs deployed over a large portion of a power system?

C. How could a PMU be used to improve the performance of an instrument transformer?

REFERENCES

1. J. L. Blackburn, *Protective Relaying* (New York: Dekker, 1997).
2. J. L. Blackburn et al., *Applied Protective Relaying* (Newark, NJ: Westinghouse Electric Corporation, 1976).

3. *Westinghouse Relay Manual, A New Silent Sentinels Publication* (Newark, NJ: Westinghouse Electric Corporation, 1972).
4. J. W. Ingleson et al., "Bibliography of Relay Literature. 1986–1987. IEEE Committee Report," *IEEE Transactions on Power Delivery, 4*, 3, pp. 1649–1658 (July 1989).
5. *IEEE Recommended Practice for Protection and Coordination of Industrial and Commercial Power Systems*—IEEE Buff Book, IEEE Standard 242-2001 (www.ieee.org, January 2001).
6. *Distribution Manual* (New York: Ebasco/Electrical World, 1990).
7. C. Russel Mason, *The Art and Science of Protective Relaying* (New York: Wiley, 1956).
8. C. A. Gross, *Power System Analysis* (New York: Wiley, 1979).
9. W. D. Stevenson, Jr., *Elements of Power System Analysis*, 4th ed. (New York: McGraw-Hill, 1982).
10. A. R. Bergen, *Power System Analysis* (Englewood Cliffs, NJ: Prentice-Hall, 1986).
11. S. H. Horowitz and A. G. Phadke, *Power System Relaying* (New York: Research Studies Press, 1992).
12. A. G. Phadke and J. S. Thorpe, *Computer Relaying for Power Systems* (New York: Wiley, 1988).
13. C. F. Henville, "Digital Relay Reports Verify Power System Models," *IEEE Computer Applications in Power*, 13, 1 (January 2000), pp. 26–32.
14. S. Horowitz, A. Phadke, and B. Renz, "The Future of Power Transmission," *IEEE Power & Energy Magazine,* 8, 2 (March/April 2010), pp. 34–40.
15. D. Reimert, *Protective Relaying for Power Generation Systems* (Boca Raton, FL: CRC Press, 2005).

1300-MW generating unit consisting of a cross-compound steam turbine and two 722-MVA synchronous generators (Courtesy of American Electric Power)

11

TRANSIENT STABILITY

Power system stability refers to the ability of synchronous machines to move from one steady-state operating point following a disturbance to another steady-state operating point, without losing synchronism [1]. There are three types of power system stability: steady-state, transient, and dynamic.

Steady-state stability, discussed in Chapter 5, involves slow or gradual changes in operating points. Steady-state stability studies, which are usually performed with a power-flow computer program (Chapter 6), ensure that phase angles across transmission lines are not too large, that bus voltages are close to nominal values, and that generators, transmission lines, transformers, and other equipment are not overloaded.

Transient stability, the main focus of this chapter, involves major disturbances such as loss of generation, line-switching operations, faults, and sudden load changes. Following a disturbance, synchronous machine frequencies undergo transient deviations from synchronous frequency (60 Hz), and machine power angles change. The objective of a transient stability study is to determine whether or not the machines will return to synchronous frequency with new steady-state power angles. Changes in power flows and bus voltages are also of concern.

Elgerd [2] gives an interesting mechanical analogy to the power system transient stability program. As shown in Figure 11.1, a number of masses representing synchronous machines are interconnected by a network of elastic strings representing transmission lines. Assume that this network is initially at rest in steady-state, with the net force on each string below its break point, when one of the strings is cut, representing the loss of a transmission line. As a result, the masses undergo transient oscillations and the forces on the strings fluctuate. The system will then either settle down to a new steady-state operating point with a new set of string forces, or additional strings will break, resulting in an even weaker network and eventual system collapse. That is, for a given disturbance, the system is either transiently stable or unstable.

In today's large-scale power systems with many synchronous machines interconnected by complicated transmission networks, transient stability studies are best performed with a digital computer program. For a specified disturbance, the program alternately solves, step by step, algebraic power-flow equations representing a network and nonlinear differential equations representing synchronous machines. Both predisturbance, disturbance, and postdisturbance computations are performed. The program output includes power angles and frequencies of synchronous machines, bus voltages, and power flows versus time.

In many cases, transient stability is determined during the first swing of machine power angles following a disturbance. During the first swing, which typically lasts about 1 second, the mechanical output power and the internal voltage of a generating unit are often assumed constant. However, where multiswings lasting several seconds are of concern, models of turbine-governors and excitation systems (for example, see Figures 12.3 and 12.5) as well as more detailed machine models can be employed to obtain accurate transient stability results over the longer time period.

Dynamic stability involves an even longer time period, typically several minutes. It is possible for controls to affect dynamic stability even though transient stability is maintained. The action of turbine-governors, excitation systems, tap-changing transformers, and controls from a power system dispatch center can interact to stabilize or destabilize a power system several minutes after a disturbance has occurred.

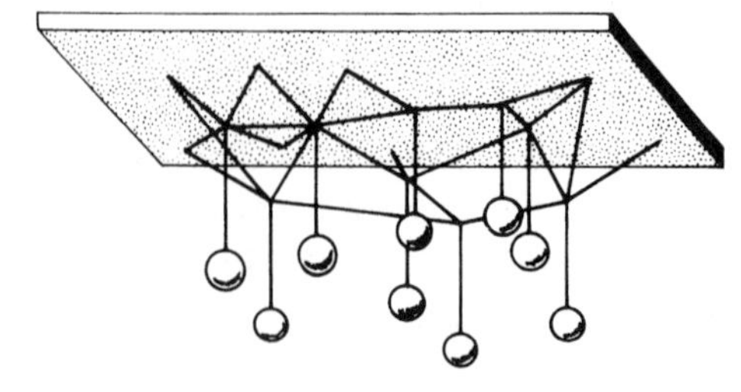

FIGURE 11.1 Mechanical analog of power system transient stability [2] (Electric energy systems theory: an introduction by Elgerd. Copyright 1982 by McGraw-Hill Companies, Inc.–Books. Reproduced with permission of McGraw-Hill Companies, Inc.–Books in the format Textbook via Copyright Clearance Center)

To simplify transient stability studies, the following assumptions are made:

1. Only balanced three-phase systems and balanced disturbances are considered. Therefore, only positive-sequence networks are employed.

2. Deviations of machine frequencies from synchronous frequency (60 Hz) are small, and dc offset currents and harmonics are neglected. Therefore, the network of transmission lines, transformers, and impedance loads is essentially in steady-state; and voltages, currents, and powers can be computed from algebraic power-flow equations.

In Section 11.1 we introduce the swing equation, which determines synchronous machine rotor dynamics. In Section 11.2 we give a simplified model of a synchronous machine and a Thévenin equivalent of a system consisting of lines, transformers, loads, and other machines. Then in Section 11.3 we present the equal-area criterion; this gives a direct method for determining the transient stability of one machine connected to a system equivalent. We discuss numerical integration techniques for solving swing equations step by step in Section 11.4 and use them in Section 11.5 to determine multimachine stability. Section 11.6 introduces a more detailed synchronous generator model, while Section 11.7 discusses how wind turbines are modeled in transient stability studies. Finally, Section 11.8 discusses design methods for improving power system transient stability.

CASE STUDY

One of the challenges to the operation of a large, interconnected power system is to ensure that the generators will remain in synchronism with one another following a large system disturbance such as the loss of a large generator or transmission line. Traditionally, such a stability assessment has been done by engineers performing lots of off-line studies using a variety of assumed system operating conditions. But the actual system operating point never exactly matches the assumed system conditions. The following article discusses a newer method for doing such an assessment in near real-time using the actual system operating conditions [13].

Real-Time Dynamic Security Assessment

ROBERT SCHAINKER, PETER MILLER, WADAD DUBBELDAY, PETER HIRSCH, AND GUORUI ZHANG

The electrical grid changes constantly with generation plants coming online or off-line as required to meet diurnal electrical demand, and with transmission lines coming online or off-line due to transmission outage events or according to maintenance schedules. In state-of-the-art electric utility control centers (illustrative example shown in Figure 1), grid operators use energy management systems (EMSs) to perform network and load monitoring, perform

Figure 1
Illustrative example of a state-of-the-art electric grid energy management system

necessary grid control actions, and manage to grid power flows within its terroritory or region of responsibility.

Limits to flows and voltages on the transmission system are assigned on the basis of transmission line thermal limits and/or off-line studies of voltage and transient dynamic stability. Power flow limits for each transmission line determined in these off-line studies are, by design, conservative, since system operators must always maintain the security and economic operation of their power system over a wide range of operating conditions. Also, the assumption that the grid power flows settle down to steady-state condition is reexamined in real time as the transmission grid conditions change in real time.

Dynamic security assessment (DSA) software analyses allow for the study of the transient and dynamic responses to a large number of potential system disturbances (contingencies) in a transient time frame, which is normally up to about 10 s after a disturbance/outage. Currently, these analyses are performed off line since the simulation process takes hours of computer time to complete for a typically large grid network, which must be simulated for each condition of a large set of all possible outage conditions that could occur. The current, long simulation time makes DSA calculations impractical for use in a real-time application, wherein an operator would need to perform real-world control actions within tens of minutes after a real-world outage to be sure that the grid will not go into an unstable voltage instability and/or a cascading blackout condition.

Therefore, if the DSA calculation could be completed in less than about 10 min, operators who control the grid during emergency conditions (terrorist induced or "nature" induced) can indeed have sufficient time to take appropriate corrective or preventive control actions to handle the identified critial events, which may cause grid instability, or cause cascading outages that would severely impact their utility grid region or their neighboring utility regions, which would potentially avert billion dollar expenses associated with regional blackouts.

The work that lead up to this article was motivated by the attempt to dramatically reduce the time for DSA calculations so that DSA analyses can be converted from off-line studies to routine, online use in order to aid grid operators in their real-time controller analyses. The large amount of time for DSA calculations occurs because grid transients for a large grid network must be calculated over about a 10 s time interval and properly represent a large interconnected power system network system, which must properly represent detailed static and/or dynamic models of power system components, such as transmission network solid-state flexible ac transmission system devices, all types of generators, power system stabilizers, various types of relays/protection systems, load models, and various types of faults or disturbances.

This article describes the methods and successful results obtained in developing a real-time version of the DSA tool. The material below is organized by first providing a description of the DSA software package generally used by the U.S. electric utility industry. Then discussed is a way to dramatically reduce the computation time to perform DSA calculations, which, among other useful techniques, uses a new distributed computational architecture. The results from applying this new version of DSA are then presented using a large utility system as an example. The results clearly show that, indeed, using the new DSA approach, calculations for a large power system can be performed fast enough for the real-time application to EMSs that operate today's grid systems. The article then ends with some insights and concluding remarks.

DYNAMIC SECURITY ASSESSMENT (DSA)

DSA software performs simulations of the impact of potential electric grid fault conditions for a preset time frame after a potential grid disturbance, usually over a time interval of 5–10 s after an outage contingency condition occurs. Contingency conditions studied include "normal" transmission line and/or power plant outages caused by acts of nature or equipment (e.g., outages due to lightning and/or generator "trips"), wear and tear (for example, equipment age failures), and outage conditions caused by human error and/or potential terrorist-induced equipment failures.

Recent efforts by the authors of this article have focused on improving the performance of the DSA calculation process with the eventual goal of implementing the DSA evaluation process in an online utility energy management system (EMS). Past DSA research projects have resulted in significant achievements in determining which outage contingency conditions are significant and not significant by rapidly separating the outage contingencies into "definitely safe" and "potentially harmful" groups. The "potentially harmful" group must be studied in more detail to accurately determine whether a "potentially harmful" contingency is in fact harmful.

DYNAMIC SECURITY ASSESSMENT MODELS

The DSA program uses a complete representation of all the generators (for example, fossil, nuclear, gas, oil, hydro, and wind generators) including their exciters, governors and stabilizers, transmission lines and many other linear and nonlinear components. For example, nonlinear devices embedded into DSA software include such items as:

1) synchronous machines
2) induction motors
3) static VAR compensators
4) thyristor-controlled series compensations
5) thyristor-controlled tap changers and/or phase regulators
6) thyristor-controlled braking resistors or braking capacitors
7) static load models (nonlinear loads)
8) high-voltage dc link
9) user-defined models, as appropriate.

In addition, DSA software models different types of electric grid protection relays:

1) load shedding relay
2) underfrequency load shedding relay
3) voltage difference load dropping relay
4) underfrequency generation rejection relay
5) underfrequency line tripping relay
6) impedance/default distance relay
7) series capacitor gap relay
8) rate of change of power relay.

Also modeled within the DSA software are static nonlinear load models, which are different from constant impedance load models.

DSA also models the following four types of static nonlinear loads:

1) constant current load
2) constant mega-voltage-ampere load
3) general exponential voltage and frequency-dependent load
4) thermostatically controlled load.

Additionally, DSA models each transmission line as a network impedance model with capacitance, inductance, and resistance. Each line also has thermal line rating limits. In addition, tap- and phase-shifting transformers are modeled.

Contingencies for DSA are defined in terms of the fault type, location, duration, and sequence of events making up a contingency scenario. Typical short-circuit faults are three-phase faults, single-line-to-ground and/or double-line-to-ground short-circuit faults. Automatic switching actions taken into account in the computation simulation are line removal or line closure into the grid network. The location of the short-circuit fault can be at the electrical bus, line end, or line section.

DSA ALGORITHM

The solution to all these devices operating in an electric grid requires solving a large set of differential equations. For a 5,000-node network with 300 generators, over 14,000 nonlinear differential equations must be simultaneously solved. DSA uses a numerical analysis method to solve these nonlinear differential equations. The numerical method uses a small time step of about 0.01 s, and at each time step, the method linearizes the equations to calculate the future time response. A classic Newton-Rahpson iteration approach is incorporated into the numerical method, and for a 10-s simulation, 1,000 such time steps are used.

The solution for a conventional transient stability program can take considerable time to solve for one contingency and even longer for multiple contingencies. Typically, for a 5,000-node network in which 300 contingencies are investigated, a 30-s transient stability simulation may take over two hours of calculation time, dependent on the type of computer used in the calculation.

One would need over 100-fold improvement in DSA simulation time performance to be able to do this calculation in about 10 min or less.

Based on these requirements, some of the significant ways for improving the DSA performance deployed by the authors herein are described below.

1) An improved stopping (called *early termination*) criteria when evaluating each contingency is used to reduce the overall time each contingency is simulated. That is, if the program simulation is for 10 s and after a short time duration, say less than 2 s of simulation time, it can be determined that the contingency case being investigated is unstable or stable, then the DSA program evaluating that contingency is stopped, and a flag is set to unstable or stable for the contingency case being investigated. If no stable/unstable determination can be made, then the DSA program for that contingency case runs the full 10-s simulation time period specified. Using this technique, the DSA program does not have to be run to completion for every contingency. It is run to completion only for those contingencies that are moderately stable or moderately unstable.

2) A novel distributed computing architecture (see Figure 2) was also used to improve the time it takes to perform the numerous contingency cases investigated. In general, there are two ways of performing distributed computation, and both were investigated. One is to parallelize the DSA algorithm and its calculation approach, using central processing units (CPUs) in parallel to perform the calculations. This will improve the performance somewhat, but due to the sparse nature of the differential equation matrices involved, this improvement has been found to be not very useful. A better technique is to run the full DSA software application on each of *n* computers (set up to communicate with each other) and distribute the contingencies (so each computer runs a different set of contingencies). The master computer distributes the contingencies to each of the slave/server computers as needed. Of course, this will work as long as the number of contingencies is equal, or exceeds the number of computers, which is certainly the case. Full distributed computation is thus achieved and the only slow down is due to the use of one master computer to orchestrate/distribute the contingencies to the other computers and receive/catalogue the solution results from the other computers as the results become available.

Using the above methods, and others, the authors developed a new DSA computation architecture and approach, which did improve the computation time by a factor of about 100+, based on the following improvement components: an improvement factor of about 2, due to not having to move data among computers and hard disk storage locations, an improvement factor of about 3 by using the "stopping" criteria discussed above, an improvement factor of about 4 by using five computers in the distributed computer architecture discussed above, and an

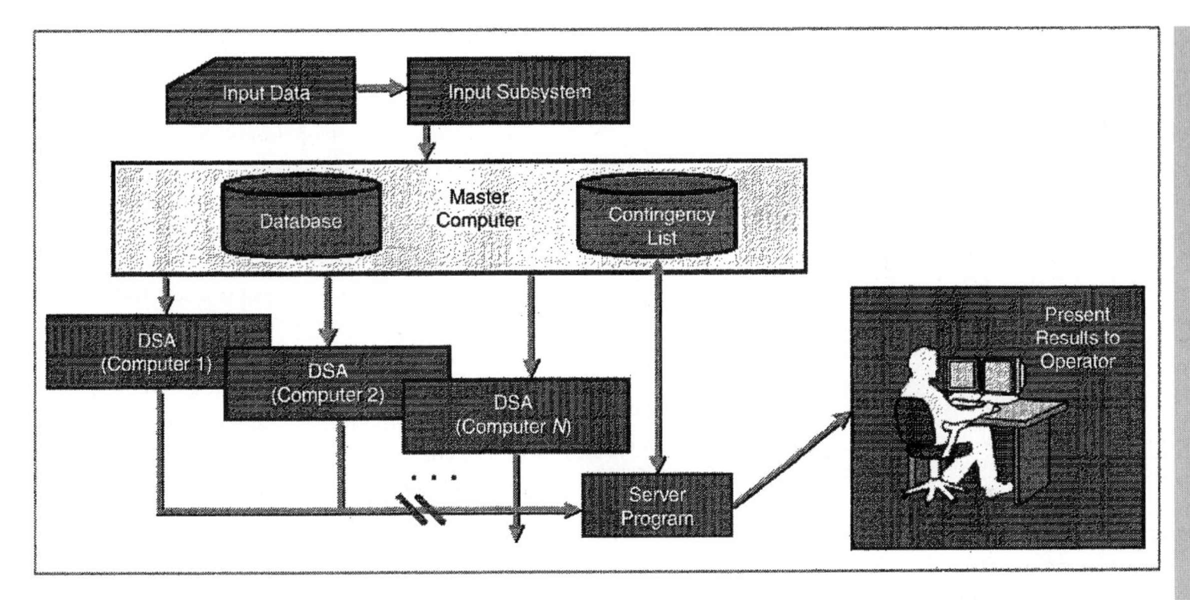

Figure 2
A schematic of the distributed computer architecture used to improve the DSA computation time

improvement factor of about 6 due to faster CPUs used to perform the calculations, as compared to those used circa 2000.

DSA INPUT DATA

The DSA input data consist of three sets of data:

- the power flow data, which contain all the transmission line configurations, tap-changing transformers, phaseshifting transformers, load representation, electric breaker information, relay information, and the type and location of the generation plants
- The dynamic data, which contain various types of generator models, including the generator exciter models, governor models, power system stabilizers, the exciter models, governor models, power system stabilizers, basic generator parameters (along with their limits and time constants), load models, and protection relay models
- The contingency data, which include the various types of faults, including the type, location, duration, and clearance of the faults and the switching actions after faults are cleared.

DSA OUTPUT DATA

The DSA program produces output results for each contingency and for each generator. The results are data and information on items such as the relative generator angle, the speed of the generator, and the voltage at each generator. This output is temporarily saved on the computer running the contingency and is then transferred to the master computer at the end of the contingency run. On the master computer, time-dependent plots for each contingency of the top three worst grid node response cases are made available to the user.

Figure 3 was produced using the DSA improvement methods described above. On the vertical axis of this figure is the computer run time needed to perform a DSA calculation for a utility grid system that has 5,839 electric buses, 11,680 transmission lines, and 779 generators. The computation time needed to run this large, representative utility test case with only the master computer and then,

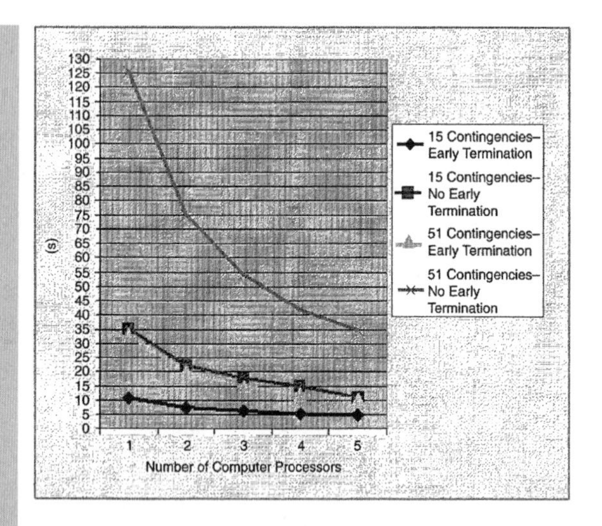

Figure 3
The DSA computation time performance, with and without early termination method

sequentially, with one, two, three, four, and the available portion of the master computer used as a "fifth" slave computer. Each point on the plots in the figure show the time required to do all the DSA calculations. Comparison data were plotted for cases where the number of contingencies was 15 and 51. Also, for comparison purposes, data were plotted for cases where the early termination logic was used for each contingency case computed and for when no early termination logic was used for each contingency case computed. The results were impressive showing a significant improvement in computing time. For the test case with 51 contingencies, the computing time ranged from 125 s (using only one computer) down to 35 s using all five computers (i.e., the master and the four slave computers). This set of runs showed an improvement factor of about 3.6 in computing time. For the test case with 15 contingencies, the computer run time ranged from 35 s (using only one computer) down to 10.3 s using all five computers (i.e., the master and the four slave computers). This set of runs showed an improvement factor of about 3.4 in computer run time. These results clearly show the power of the master-slave computer architecture developed and successfully investigated and tested.

A number of DSA algorithm improvements were also investigated. The most effective one investigated was the "early termination" method. Sample results are also shown in Figure 3. Using the early termination method and comparing it to the "no early termination" method, for the test case with 51 contingencies, the computer run time improved from 125 s to 33 s (using only one computer) and from 35 s to 10 s (using all five computers, i.e., the master and the four slave computers). This set of runs showed about an improvement factor of 3.8 to 3.5 in computer run time. For the test case with 15 contingencies, the computer run time improved from 35 s to 11 s (using only one computer) and from 12 s to 5 s (using all five computers—i.e., the master and the four slave computers). This set of runs showed an improvement factor of about 3.2 to 2.4 in computer run time. These results also clearly show the power of the master-slave computer architecture system developed and successfully tested.

DSA GRAPHICAL OUTPUT DISPLAYS

The DSA program provides several graphical output displays to show the following types of output results (some of which are illustrated in Figures 4–7):

- largest generator speed angles, for both stable and unstable contingency cases
- highest frequencies, for both stable and unstable contingency cases.

CONCLUSIONS

Using the distributed computer architecture for DSA calculations, grid operators can now quickly analyze a large number of system contingency outage events. Thus, they can evaluate the appropriate preventive or corrective control actions to effectively handle various severe system disturbances or even mitigate costly cascading blackouts, events that are either initiated by nature or terrorist induced.

Online dynamic security analysis (DSA) requires extensive computer resources, particularly for large electric power systems. With the recent advances in computer technology and the intra- and interenterprise communication networking, it now becomes

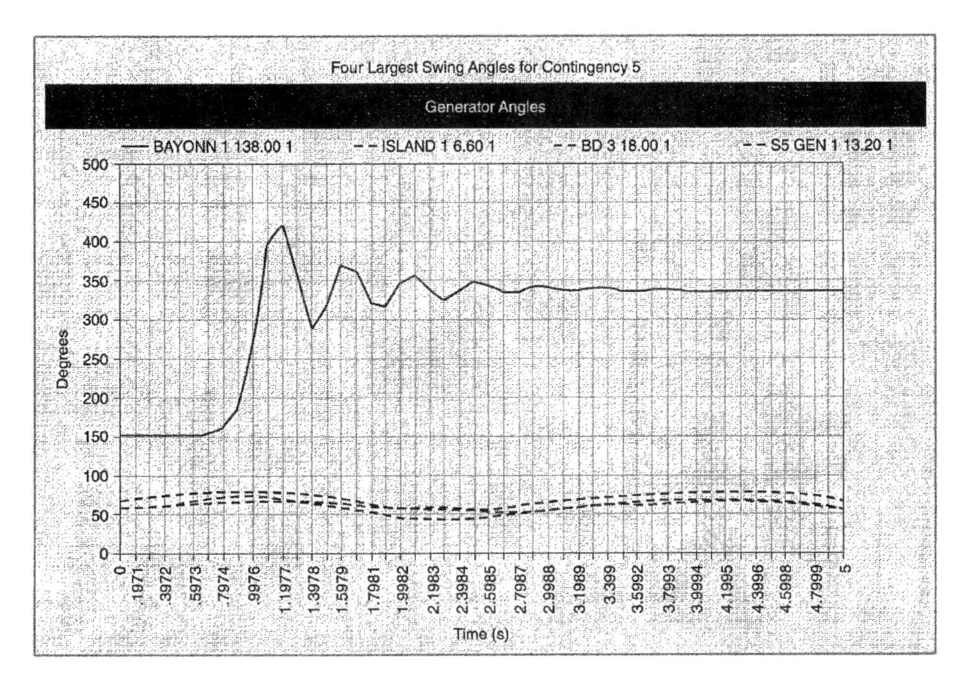

Figure 4
The DSA output plot for the largest generator swing angle for a stable contingency case

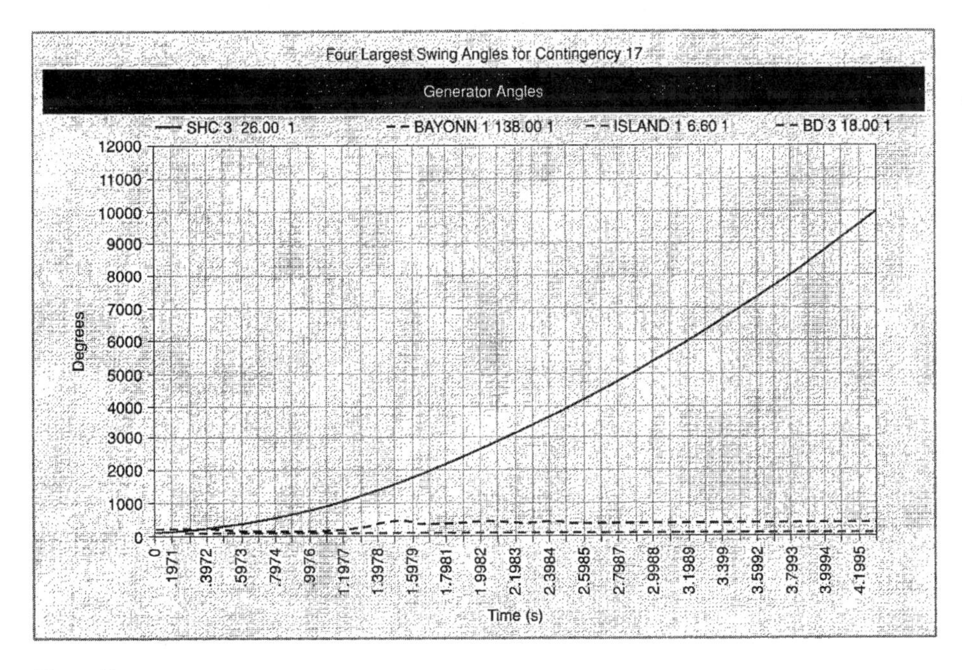

Figure 5
The DSA output plot for the largest generator swing angle for an unstable contingency case

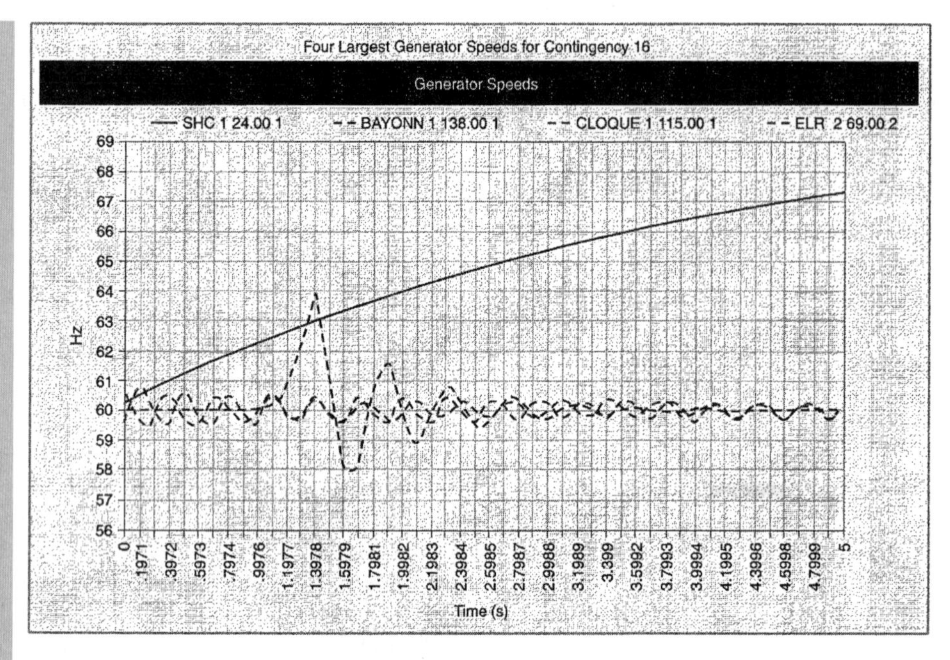

Figure 6
DSA output plot for largest generator speeds for an unstable case

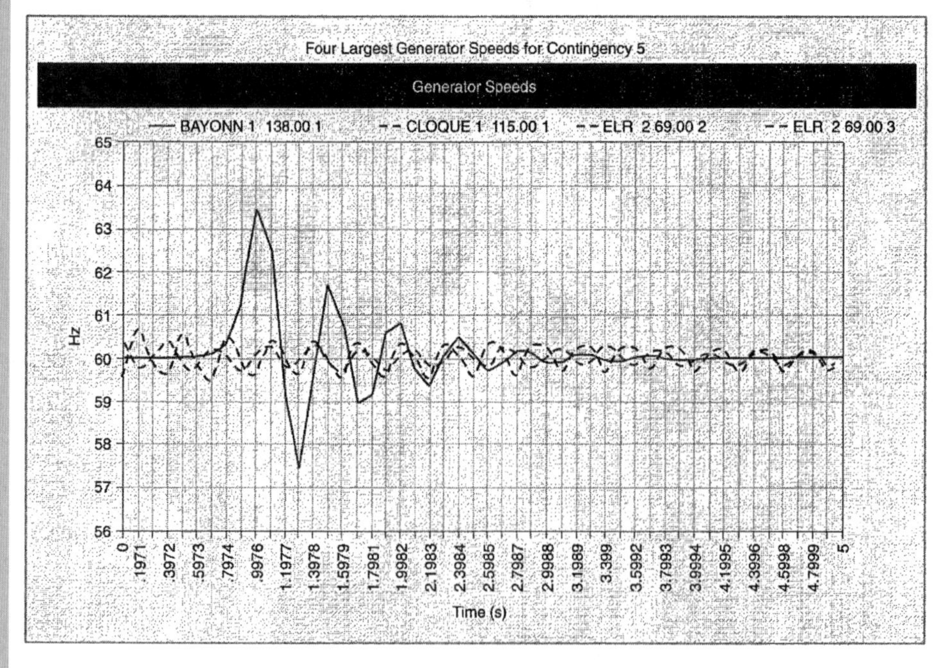

Figure 7
DSA output plot for largest generator speed display for stable case

cost-effective and possible to apply distributed computing to online DSA in order to meet real-time performance requirements needed in the electric utility industry.

Thus, the major conclusions of the work presented herein are:

- The distributed computing architecture to perform the dynamic security assessment (DSA) analysis of a large interconnected power system with a large number of contingencies has been demonstrated to be extremely fast. As such, this computerized approach should be implemented for real-time decision-making conditions, which are faced by utility and grid operators when any unplanned outage condition occurs that might lead to system instability or even cascading blackout conditions.
- The distributed computer approach developed was tested successfully with five computers in a master-slave arrangement that is scalable to any number of extra slave computers.
- The dynamic security analysis (DSA) using distributed computing can be fully integrated with utility operator EMSs using real-time operating conditions and grid State Estimation estimators.
- The dynamic security analysis (DSA) using distributed computing can also be used for performing operational planning studies for large power systems.
- The dynamic security analysis (DSA) using the distributed computing technology presented herein used the Oracle 9i relational database and its related software. This enables flexible software integration with a wide variety of IT infrastructure systems currently used by many electric utilities and/or grid operators.
- The proposed approach can be used to better utilize existing computer resources and communication networks of electric utilities. This will significantly improve the performance of DSA computations electric utilities perform routinely.
- The performance of the DSA approach presented herein is also fast enough for the real-time calculation of the interface transfer limits using real-time operating conditions for large interconnected power systems.

ACKNOWLEDGMENTS

The material presented in this paper was sponsored by the Department of Homeland through a Space and Naval Warfare Systems Center, San Diego, Contract N66001-04-C-0076.

FOR FURTHER READING

"Analytical methods for contingency selection and ranking for dynamic security analysis," EPRI, Palo Alto, CA, TR-104352, Project 3103-03 Final Rep., Sep. 1994.

"Simulation program for on-line dynamic security assessment," EPRI, Palo Alto, CA, TR-109751, Jan. 1998.

"Standard test cases for dynamic security assessment," EPRI, Palo Alto, CA, TR-105885, Dec. 1995.

A. A. Fouad and V. Vittal, *Power System Transient Stability Analysis Using the Transient Energy Function Method*. Englewood Cliffs, NJ: Prentice Hall 1992.

C. K. Tang, C. E. Graham, M. El-Kady, and R. T. H. Alden, "Transient stability index from conventional time domain simulation," *IEEE Trans. Power Syst.*, vol. 9, no. 3, Aug. 1993.

G. D. Irisarri, G. C. Ejebe, W. F. Tinney, and J. G. Waight, "Efficient computation of equilibrium points for transient energy analysis," *IEEE Trans. Power Syst.*, vol. 9, no. 2, May 1994.

BIOGRAPHIES

Robert Schainker is a technical executive and manager of the Security Program Department at the Electric Power Research Institute. He received his D.Sc. in applied mathematics and control systems, his M.S. in electrical engineering, and his B.S. in mechanical engineering at Washington University in St. Louis, Missouri.

Peter Miller is program manager, Homeland Security Advance Research Project Agency, Mission Support Office, Science and Technology, U.S. Department of Homeland Security. He holds a B.S. in mathematics from the City University of

New York, where he received the Borden Prize, and he holds an S.M. in computer science and electrical engineering from the Massachusetts Institute of Technology.

Wadad Dubbelday is an engineer at the Navy at the Space and Naval Warfare Systems Command (SPAWAR) office in San Diego, California. She holds a B.S. in physics from the Florida Institute of Technology and holds M.S. and Ph.D. degrees in electrical engineering-applied physics from the University of California at San Diego.

Peter Hirsch is a project manager in the Power Systems Assets Planning and Operations Department of the Electric Power Research Institute. He also is the manager of the software quality group within EPRI's Power Delivery and Markets Sector. He holds a B.S. in applied mathematics and engineering physics and M.S. and Ph.D. degrees in mathematics from the University of Wisconsin. He is a Senior Member of the IEEE.

Guorui Zhang is principal engineer at EPRI-Solutions, a subsidiary of the Electric Power Research Institute. He received his B.S. in computer software engineering at Singh University, China; he received his Ph.D. in electrical engineering at the University of Manchester Institute of Science and Technology, Manchester, England. He is a Senior Member of the IEEE.

11.1 THE SWING EQUATION

Consider a generating unit consisting of a three-phase synchronous generator and its prime mover. The rotor motion is determined by Newton's second law, given by

$$\mathrm{J}\alpha_m(t) = \mathrm{T}_m(t) - \mathrm{T}_e(t) = \mathrm{T}_a(t) \tag{11.1.1}$$

where J = total moment of inertia of the rotating masses, kg-m^2

α_m = rotor angular acceleration, rad/s^2

T_m = mechanical torque supplied by the prime mover minus the retarding torque due to mechanical losses, N-m

T_e = electrical torque that accounts for the total three-phase electrical power output of the generator, plus electrical losses, N-m

T_a = net accelerating torque, N-m

Also, the rotor angular acceleration is given by

$$\alpha_m(t) = \frac{d\omega_m(t)}{dt} = \frac{d^2\theta_m(t)}{dt^2} \tag{11.1.2}$$

$$\omega_m(t) = \frac{d\theta_m(t)}{dt} \tag{11.1.3}$$

where ω_m = rotor angular velocity, rad/s

θ_m = rotor angular position with respect to a stationary axis, rad

T_m and T_e are positive for generator operation. In steady-state T_m equals T_e, the accelerating torque T_a is zero, and, from (11.1.1), the rotor acceleration α_m is zero, resulting in a constant rotor velocity called *synchronous speed.* When T_m is greater than T_e, T_a is positive and α_m is therefore positive, resulting in increasing rotor speed. Similarly, when T_m is less than T_e, the rotor speed is decreasing.

It is convenient to measure the rotor angular position with respect to a synchronously rotating reference axis instead of a stationary axis. Accordingly, we define

$$\theta_m(t) = \omega_{msyn} t + \delta_m(t) \tag{11.1.4}$$

where ω_{msyn} = synchronous angular velocity of the rotor, rad/s

δ_m = rotor angular position with respect to a synchronously rotating reference, rad

Using (11.1.2) and (11.1.4), (11.1.1) becomes

$$J\frac{d^2\theta_m(t)}{dt^2} = J\frac{d^2\delta_m(t)}{dt^2} = T_m(t) - T_e(t) = T_a(t) \tag{11.1.5}$$

It is also convenient to work with power rather than torque, and to work in per-unit rather than in actual units. Accordingly, we multiply (11.1.5) by $\omega_m(t)$ and divide by S_{rated}, the three-phase voltampere rating of the generator:

$$\frac{J\omega_m(t)}{S_{rated}}\frac{d^2\delta_m(t)}{dt^2} = \frac{\omega_m(t)T_m(t) - \omega_m(t)T_e(t)}{S_{rated}}$$

$$= \frac{p_m(t) - p_e(t)}{S_{rated}} = p_{mp.u.}(t) - p_{ep.u.}(t) = p_{ap.u.}(t) \tag{11.1.6}$$

where $p_{mp.u.}$ = mechanical power supplied by the prime mover minus mechanical losses, per unit

$p_{ep.u.}$ = electrical power output of the generator plus electrical losses, per unit

Finally, it is convenient to work with a normalized inertia constant, called the H constant, which is defined as

$$H = \frac{\text{stored kinetic energy at synchronous speed}}{\text{generator voltampere rating}}$$

$$= \frac{\frac{1}{2}J\omega_{msyn}^2}{S_{rated}} \quad \text{joules/VA or per unit-seconds} \tag{11.1.7}$$

The H constant has the advantage that it falls within a fairly narrow range, normally between 1 and 10 p.u.-s, whereas J varies widely, depending on generating unit size and type. Solving (11.1.7) for J and using in (11.1.6),

$$2H\frac{\omega_m(t)}{\omega_{msyn}^2}\frac{d^2\delta_m(t)}{dt^2} = p_{mp.u.}(t) - p_{ep.u.}(t) = p_{ap.u.}(t) \tag{11.1.8}$$

Defining per-unit rotor angular velocity,

$$\omega_{\text{p.u.}}(t) = \frac{\omega_m(t)}{\omega_{m\text{syn}}} \tag{11.1.9}$$

Equation (11.1.8) becomes

$$\frac{2\text{H}}{\omega_{m\text{syn}}}\omega_{\text{p.u.}}(t)\frac{d^2\delta_m(t)}{dt^2} = p_{m\text{p.u.}}(t) - p_{e\text{p.u.}}(t) = p_{a\text{p.u.}}(t) \tag{11.1.10}$$

For a synchronous generator with P poles, the electrical angular acceleration α, electrical radian frequency ω, and power angle δ are

$$\alpha(t) = \frac{\text{P}}{2}\alpha_m(t) \tag{11.1.11}$$

$$\omega(t) = \frac{\text{P}}{2}\omega_m(t) \tag{11.1.12}$$

$$\delta(t) = \frac{\text{P}}{2}\delta_m(t) \tag{11.1.13}$$

Similarly, the synchronous electrical radian frequency is

$$\omega_{\text{syn}} = \frac{\text{P}}{2}\omega_{m\text{syn}} \tag{11.1.14}$$

The per-unit electrical frequency is

$$\omega_{\text{p.u.}}(t) = \frac{\omega(t)}{\omega_{\text{syn}}} = \frac{\frac{2}{\text{P}}\omega(t)}{\frac{2}{\text{P}}\omega_{\text{syn}}} = \frac{\omega_m(t)}{\omega_{m\text{syn}}} \tag{11.1.15}$$

Therefore, using (11.1.13–11.1.15), (11.1.10) can be written as

$$\frac{2\text{H}}{\omega_{syn}}\omega_{\text{p.u.}}(t)\frac{d^2\delta(t)}{dt^2} = p_{m\text{p.u.}}(t) - p_{e\text{p.u.}}(t) = p_{a\text{p.u.}}(t) \tag{11.1.16}$$

Frequently (11.1.16) is modified to also include a term that represents a damping torque anytime the generator deviates from its synchronous speed, with its value proportional to the speed deviation

$$\begin{aligned} &2H/\omega_{\text{syn}}w_{\text{p.u.}}(t)(d^2\delta(t)/(dt^2)) \\ &\quad = p_{m\text{p.u.}}(t) - p_{e\text{p.u.}}(t) - D/\omega_{\text{syn}}(d\delta(t)/(dt)) \\ &\quad = p_{a\text{p.u.}}(t) \end{aligned} \tag{11.1.17}$$

where D is either zero or a relatively small positive number with typical values between 0 and 2. The units of D are per unit power divided by per unit speed deviation.

Equation (11.1.17), called the per-unit *swing equation*, is the fundamental equation that determines rotor dynamics in transient stability studies. Note that it is nonlinear due to $p_{ep.u.}(t)$, which is shown in Section 11.2 to be a nonlinear function of δ. Equation (11.1.17) is also nonlinear due to the $\omega_{p.u.}(t)$ term. However, in practice the rotor speed does not vary significantly from synchronous speed during transients. That is, $\omega_{p.u.}(t) \simeq 1.0$, which is often assumed in (11.1.17) for hand calculations.

Equation (11.1.17) is a second-order differential equation that can be rewritten as two first-order differential equations. Differentiating (11.1.4), and then using (11.1.3) and (11.1.12)–(11.1.14), we obtain

$$\frac{d\delta(t)}{dt} = \omega(t) - \omega_{syn} \tag{11.1.18}$$

Using (11.1.18) in (11.1.17),

$$\frac{2\text{H}}{\omega_{syn}}\omega_{p.u.}(t)\frac{d\omega(t)}{dt} = p_{mp.u.}(t) - p_{ep.u.}(t) - D/\omega_{syn}\frac{d\delta(t)}{dt} = p_{ap.u.}(t) \tag{11.1.19}$$

Equations (11.1.18) and (11.1.19) are two first-order differential equations.

EXAMPLE 11.1 Generator per-unit swing equation and power angle during a short circuit

A three-phase, 60-Hz, 500-MVA, 15-kV, 32-pole hydroelectric generating unit has an H constant of 2.0 p.u.-s and $D = 0$. (a) Determine ω_{syn} and ω_{msyn}. (b) Give the per-unit swing equation for this unit. (c) The unit is initially operating at $p_{mp.u.} = p_{ep.u.} = 1.0$, $\omega = \omega_{syn}$, and $\delta = 10°$ when a three-phase-to-ground bolted short circuit at the generator terminals causes $p_{ep.u.}$ to drop to zero for $t \geqslant 0$. Determine the power angle 3 cycles after the short circuit commences. Assume $p_{mp.u.}$ remains constant at 1.0 per unit. Also assume $\omega_{p.u.}(t) = 1.0$ in the swing equation.

SOLUTION

a. For a 60-Hz generator,

$$\omega_{syn} = 2\pi 60 = 377 \text{ rad/s}$$

and, from (11.1.14), with P = 32 poles,

$$\omega_{msyn} = \frac{2}{\text{P}}\omega_{syn} = \left(\frac{2}{32}\right)377 = 23.56 \quad \text{rad/s}$$

b. From (11.1.16), with H = 2.0 p.u.-s,

$$\frac{4}{2\pi 60}\omega_{p.u.}(t)\frac{d^2\delta(t)}{dt^2} = p_{mp.u.}(t) - p_{ep.u.}(t)$$

c. The initial power angle is

$$\delta(0) = 10^\circ = 0.1745 \quad \text{radian}$$

Also, from (11.1.17), at $t = 0$,

$$\frac{d\delta(0)}{dt} = 0$$

Using $p_{mp.u.}(t) = 1.0$, $p_{ep.u.} = 0$, and $\omega_{p.u.}(t) = 1.0$, the swing equation from (**b**) is

$$\left(\frac{4}{2\pi 60}\right)\frac{d^2\delta(t)}{dt^2} = 1.0 \quad t \geqslant 0$$

Integrating twice and using the above initial conditions,

$$\frac{d\delta(t)}{dt} = \left(\frac{2\pi 60}{4}\right)t + 0$$

$$\delta(t) = \left(\frac{2\pi 60}{8}\right)t^2 + 0.1745$$

At $t = 3$ cycles $= \dfrac{3 \text{ cycles}}{60 \text{ cycles/second}} = 0.05$ second,

$$\delta(0.05) = \left(\frac{2\pi 60}{8}\right)(0.05)^2 + 0.1745$$

$$= 0.2923 \text{ radian} = 16.75^\circ$$

■

EXAMPLE 11.2 Equivalent swing equation: two generating units

A power plant has two three-phase, 60-Hz generating units with the following ratings:

Unit 1: 500 MVA, 15 kV, 0.85 power factor, 32 poles, $H_1 = 2.0$ p.u.-s, $D = 0$

Unit 2: 300 MVA, 15 kV, 0.90 power factor, 16 poles, $H_2 = 2.5$ p.u.-s, $D = 0$

(a) Give the per-unit swing equation of each unit on a 100-MVA system base. (b) If the units are assumed to "swing together," that is, $\delta_1(t) = \delta_2(t)$, combine the two swing equations into one equivalent swing equation.

SOLUTION

a. If the per-unit powers on the right-hand side of the swing equation are converted to the system base, then the H constant on the left-hand side must also be converted. That is,

$$H_{new} = H_{old}\frac{S_{old}}{S_{new}} \quad \text{per unit}$$

Converting H_1 from its 500-MVA rating to the 100-MVA system base,

$$H_{1new} = H_{1old}\frac{S_{old}}{S_{new}} = (2.0)\left(\frac{500}{100}\right) = 10 \quad \text{p.u.-s}$$

Similarly, converting H_2,

$$H_{2new} = (2.5)\left(\frac{300}{100}\right) = 7.5 \quad \text{p.u.-s}$$

The per-unit swing equations on the system base are then

$$\frac{2H_{1new}}{\omega_{syn}}\omega_{1p.u.}(t)\frac{d^2\delta_1(t)}{dt^2} = \frac{20.0}{2\pi 60}\omega_{1p.u.}(t)\frac{d^2\delta_1(t)}{dt^2}$$
$$= p_{m1p.u.}(t) - p_{e1p.u.}(t)$$

$$\frac{2H_{2new}}{\omega_{syn}}\omega_{2p.u.}(t)\frac{d^2\delta_2(t)}{dt^2} = \frac{15.0}{2\pi 60}\omega_{2p.u.}(t)\frac{d^2\delta_2(t)}{dt^2} = p_{m2p.u.}(t) - p_{e2p.u.}$$

b. Letting:

$$\delta(t) = \delta_1(t) = \delta_2(t)$$
$$\omega_{p.u.}(t) = \omega_{1p.u.}(t) = \omega_{2p.u.}(t)$$
$$p_{mp.u.}(t) = p_{m1p.u.}(t) + p_{m2p.u.}(t)$$
$$p_{ep.u.}(t) = p_{e1p.u.}(t) + p_{e2p.u.}(t)$$

and adding the above swing equations

$$\frac{2(H_{1new} + H_{2new})}{\omega_{syn}}\omega_{p.u.}(t)\frac{d^2\delta(t)}{dt^2}$$
$$= \frac{35.0}{2\pi 60}\omega_{p.u.}(t)\frac{d^2\delta(t)}{dt^2} = p_{mp.u.}(t) - p_{ep.u.}(t)$$

When transient stability studies involving large-scale power systems with many generating units are performed with a digital computer, computation time can be reduce by combining the swing equations of those units that swing together. Such units, which are called *coherent machines*, usually are connected to the same bus or are electrically close, and they are usually remote from network disturbances under study. ■

11.2

SIMPLIFIED SYNCHRONOUS MACHINE MODEL AND SYSTEM EQUIVALENTS

Figure 11.2 shows a simplified model of a synchronous machine, called the classical model, that can be used in transient stability studies. As shown, the synchronous machine is represented by a constant internal voltage E' behind its direct axis transient reactance X'_d. This model is based on the following assumptions:

1. The machine is operating under balanced three-phase positive-sequence conditions.
2. Machine excitation is constant.
3. Machine losses, saturation, and saliency are neglected.

In transient stability programs, more detailed models can be used to represent exciters, losses, saturation, and saliency. However, the simplified model reduces model complexity while maintaining reasonable accuracy in stability calculations.

Each generator in the model is connected to a system consisting of transmission lines, transformers, loads, and other machines. To a first approximation the system can be represented by an "infinite bus" behind a system reactance. An infinite bus is an ideal voltage source that maintains constant voltage magnitude, constant phase, and constant frequency.

Figure 11.3 shows a synchronous generator connected to a system equivalent. The voltage magnitude V_{bus} and 0° phase of the infinite bus are

FIGURE 11.2

Simplified synchronous machine model for transient stability studies

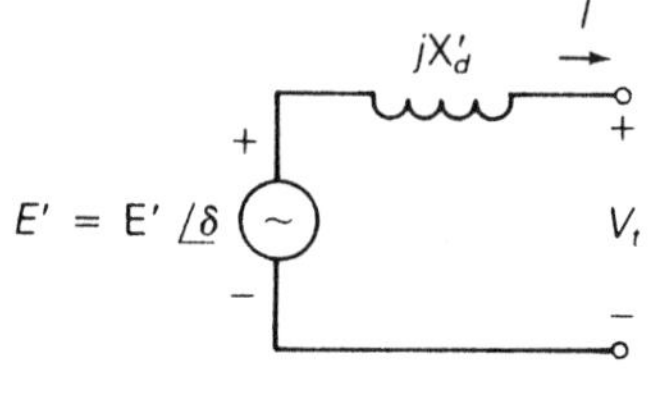

(a) Circuit diagram

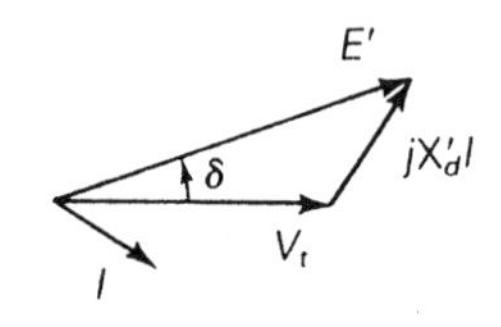

(b) Phasor diagram

FIGURE 11.3

Synchronous generator connected to a system equivalent

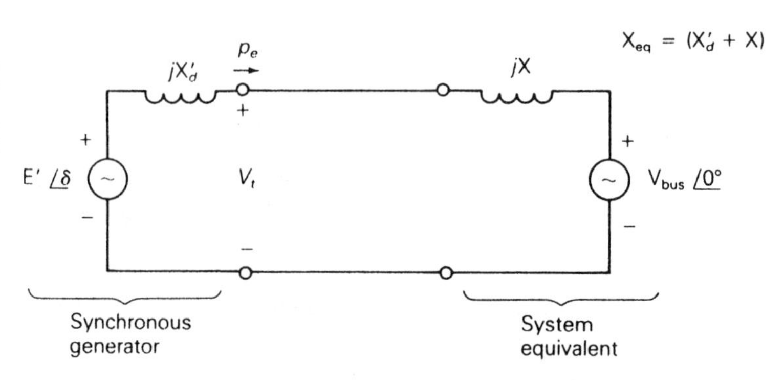

constant. The phase angle δ of the internal machine voltage is the machine power angle with respect to the infinite bus.

The equivalent reactance between the machine internal voltage and the infinite bus is $X_{eq} = (X'_d + X)$. From (6.7.3), the real power delivered by the synchronous generator to the infinite bus is

$$p_e = \frac{E' V_{bus}}{X_{eq}} \sin\delta \tag{11.2.1}$$

During transient disturbances both E' and V_{bus} are considered constant in (11.2.1). Thus p_e is a sinusoidal function of the machine power angle δ.

EXAMPLE 11.3 Generator internal voltage and real power output versus power angle

Figure 11.4 shows a single-line diagram of a three-phase, 60-Hz synchronous generator, connected through a transformer and parallel transmission lines to an infinite bus. All reactances are given in per-unit on a common system base. If the infinite bus receives 1.0 per unit real power at 0.95 p.f. lagging, determine (a) the internal voltage of the generator and (b) the equation for the electrical power delivered by the generator versus its power angle δ.

SOLUTION

a. The equivalent circuit is shown in Figure 11.5, from which the equivalent reactance between the machine internal voltage and infinite bus is

$$\begin{aligned} X_{eq} &= X'_d + X_{TR} + X_{12} \| (X_{13} + X_{23}) \\ &= 0.30 + 0.10 + 0.20 \| (0.10 + 0.20) \\ &= 0.520 \quad \text{per unit} \end{aligned}$$

The current into the infinite bus is

$$\begin{aligned} I &= \frac{P}{V_{bus}(\text{p.f.})} \underline{/ -\cos^{-1}(\text{p.f.})} = \frac{(1.0)}{(1.0)(0.95)} \underline{/ -\cos^{-1} 0.95} \\ &= 1.05263 \underline{/ -18.195^\circ} \quad \text{per unit} \end{aligned}$$

FIGURE 11.4 Single-line diagram for Example 11.3

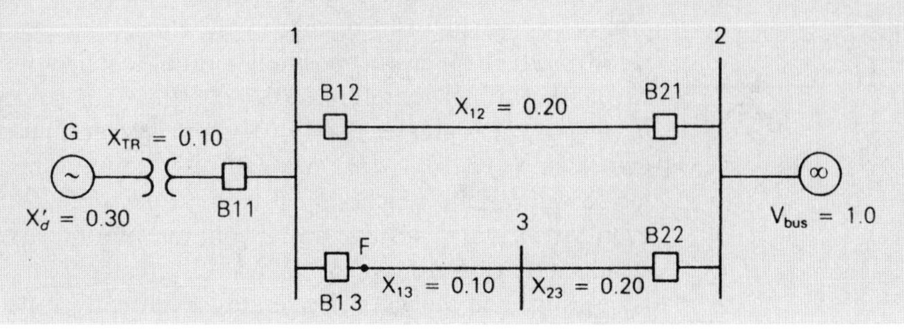

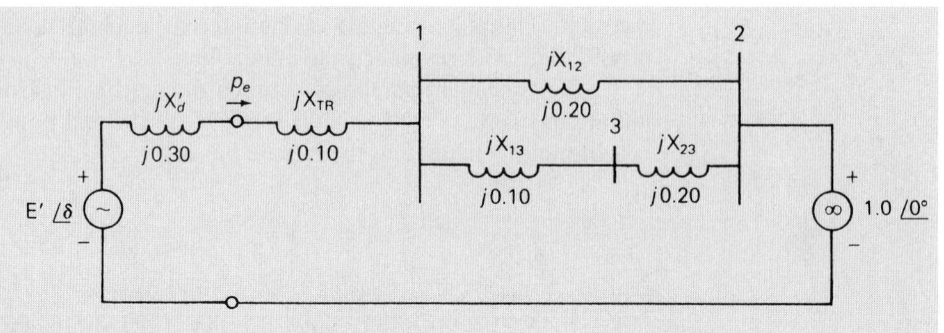

FIGURE 11.5 Equivalent circuit for Example 11.3

and the machine internal voltage is

$$
\begin{aligned}
E' = \mathrm{E}'\underline{/\delta} &= V_{\text{bus}} + j\mathrm{X}_{\text{eq}} I \\
&= 1.0\underline{/0^\circ} + (j0.520)(1.05263\underline{/-18.195^\circ}) \\
&= 1.0\underline{/0^\circ} + 0.54737\underline{/71.805^\circ} \\
&= 1.1709 + j0.5200 \\
&= 1.2812\underline{/23.946^\circ} \quad \text{per unit}
\end{aligned}
$$

b. From (11.2.1),

$$
p_e = \frac{(1.2812)(1.0)}{0.520} \sin\delta = 2.4638 \sin\delta \quad \text{per unit}
$$

■

11.3

THE EQUAL-AREA CRITERION

Consider a synchronous generating unit connected through a reactance to an infinite bus. Plots of electrical power p_e and mechanical power p_m versus power angle δ are shown in Figure 11.6. p_e is a sinusoidal function of δ, as given by (11.2.1).

Suppose the unit is initially operating in steady-state at $p_e = p_m = p_{m0}$ and $\delta = \delta_0$, when a step change in p_m from p_{m0} to p_{m1} occurs at $t = 0$. Due to rotor inertia, the rotor position cannot change instantaneously. That is, $\delta_m(0^+) = \delta_m(0^-)$; therefore, $\delta(0^+) = \delta(0^-) = \delta_0$ and $p_e(0^+) = p_e(0^-)$. Since $p_m(0^+) = p_{m1}$ is greater than $p_e(0^+)$, the acceleration power $p_a(0^+)$ is positive and, from (11.1.16), $(d^2\delta)/(dt^2)(0^+)$ is positive. The rotor accelerates and δ increases. When δ reaches δ_1, $p_e = p_{m1}$ and $(d^2\delta)/(dt^2)$ becomes zero. However, $d\delta/dt$ is still positive and δ continues to increase, overshooting its final steady-state operating point. When δ is greater than δ_1, p_m is less than p_e, p_a is negative, and the rotor decelerates. Eventually, δ reaches a maximum value

FIGURE 11.6

p_e and p_m versus δ

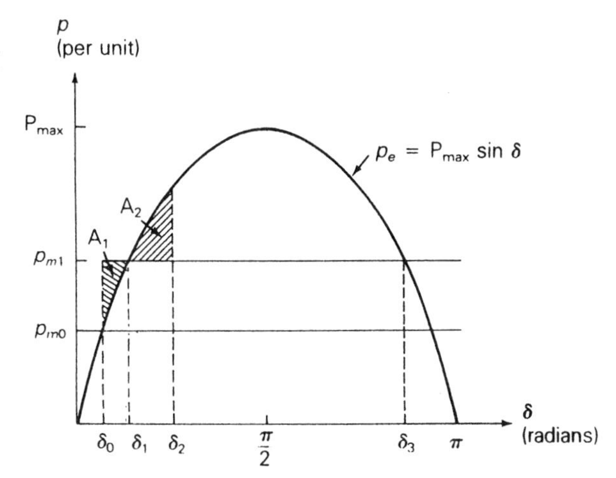

δ_2 and then swings back toward δ_1. Using (11.1.16), which has no damping, δ would continually oscillate around δ_1. However, damping due to mechanical and electrical losses causes δ to stabilize at its final steady-state operating point δ_1. Note that if the power angle exceeded δ_3, then p_m would exceed p_e and the rotor would accelerate again, causing a further increase in δ and loss of stability.

One method for determining stability and maximum power angle is to solve the nonlinear swing equation via numerical integration techniques using a digital computer. We describe this method, which is applicable to multimachine systems, in Section 11.4. However, there is also a direct method for determining stability that does not involve solving the swing equation; this method is applicable for one machine connected to an infinite bus or for two machines. We describe the method, called the *equal-area criterion*, in this section.

In Figure 11.6, p_m is greater than p_e during the interval $\delta_0 < \delta < \delta_1$, and the rotor is accelerating. The shaded area A_1 between the p_m and p_e curves is called the accelerating area. During the interval $\delta_1 < \delta < \delta_2$, p_m is less than p_e, the rotor is decelerating, and the shaded area A_2 is the decelerating area. At both the initial value $\delta = \delta_0$ and the maximum value $\delta = \delta_2$, $d\delta/dt = 0$. The equal-area criterion states that $A_1 = A_2$.

To derive the equal-area criterion for one machine connected to an infinite bus, assume $\omega_{\text{p.u.}}(t) = 1$ in (11.1.16), giving

$$\frac{2H}{\omega_{\text{syn}}} \frac{d^2\delta}{dt^2} = p_{m\text{p.u.}} - p_{e\text{p.u.}} \tag{11.3.1}$$

Multiplying by $d\delta/dt$ and using

$$\frac{d}{dt}\left[\frac{d\delta}{dt}\right]^2 = 2\left(\frac{d\delta}{dt}\right)\left(\frac{d^2\delta}{dt^2}\right)$$

(11.3.1) becomes

$$\frac{2\mathrm{H}}{\omega_{\mathrm{syn}}}\left(\frac{d^2\delta}{dt^2}\right)\left(\frac{d\delta}{dt}\right) = \frac{\mathrm{H}}{\omega_{\mathrm{syn}}}\frac{d}{dt}\left[\frac{d\delta}{dt}\right]^2 = (p_{m\mathrm{p.u.}} - p_{e\mathrm{p.u.}})\frac{d\delta}{dt} \tag{11.3.2}$$

Multiplying (11.3.2) by dt and integrating from δ_0 to δ,

$$\frac{\mathrm{H}}{\omega_{\mathrm{syn}}}\int_{\delta_0}^{\delta} d\left[\frac{d\delta}{dt}\right]^2 = \int_{\delta_0}^{\delta}(p_{m\mathrm{p.u.}} - p_{e\mathrm{p.u.}})\,d\delta$$

or

$$\frac{\mathrm{H}}{\omega_{\mathrm{syn}}}\left[\frac{d\delta}{dt}\right]^2\Bigg|_{\delta_0}^{\delta} = \int_{\delta_0}^{\delta}(p_{m\mathrm{p.u.}} - p_{e\mathrm{p.u.}})\,d\delta \tag{11.3.3}$$

The above integration begins at δ_0 where $d\delta/dt = 0$, and continues to an arbitrary δ. When δ reaches its maximum value, denoted δ_2, $d\delta/dt$ again equals zero. Therefore, the left-hand side of (11.3.3) equals zero for $\delta = \delta_2$ and

$$\int_{\delta_0}^{\delta_2}(p_{m\mathrm{p.u.}} - p_{e\mathrm{p.u.}})\,d\delta = 0 \tag{11.3.4}$$

Separating this integral into positive (accelerating) and negative (decelerating) areas, we arrive at the equal-area criterion

$$\int_{\delta_0}^{\delta_1}(p_{m\mathrm{p.u.}} - p_{e\mathrm{p.u.}})\,d\delta + \int_{\delta_1}^{\delta_2}(p_{m\mathrm{p.u.}} - p_{e\mathrm{p.u.}})\,d\delta = 0$$

or

$$\int_{\delta_0}^{\delta_1}\underbrace{(p_{m\mathrm{p.u.}} - p_{e\mathrm{p.u.}})\,d\delta}_{\mathrm{A}_1} = \int_{\delta_1}^{\delta_2}\underbrace{(p_{e\mathrm{p.u.}} - p_{m\mathrm{p.u.}})\,d\delta}_{\mathrm{A}_2} \tag{11.3.5}$$

In practice, sudden changes in mechanical power usually do not occur, since the time constants associated with prime mover dynamics are on the order of seconds. However, stability phenomena similar to that described above can also occur from sudden changes in electrical power, due to system faults and line switching. The following three examples are illustrative.

EXAMPLE 11.4 Equal-area criterion: transient stability during a three-phase fault

The synchronous generator shown in Figure 11.4 is initially operating in the steady-state condition given in Example 11.3, when a temporary three-phase-to-ground bolted short circuit occurs on line 1–3 at bus 1, shown as point F in Figure 11.4. Three cycles later the fault extinguishes by itself. Due to a

FIGURE 11.7

p–δ plot for Example 11.4

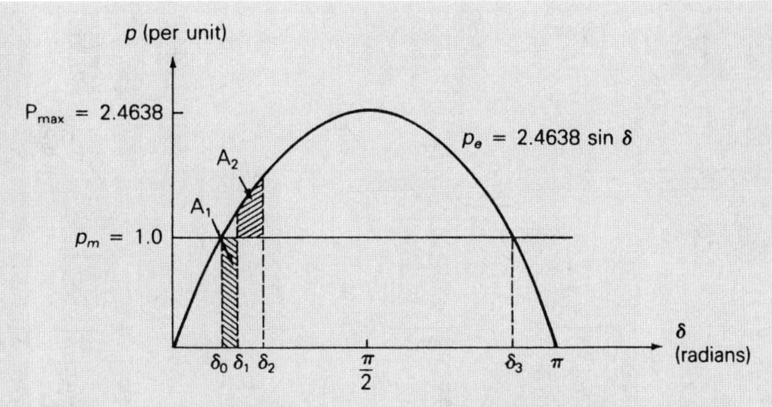

relay misoperation, all circuit breakers remain closed. Determine whether stability is or is not maintained and determine the maximum power angle. The inertia constant of the generating unit is 3.0 per unit-seconds on the system base. Assume p_m remains constant throughout the disturbance. Also assume $\omega_{\text{p.u.}}(t) = 1.0$ in the swing equation.

SOLUTION Plots of p_e and p_m versus δ are shown in Figure 11.7. From Example 11.3 the initial operating point is $p_e(0^-) = p_m = 1.0$ per unit and $\delta(0^+) = \delta(0^-) = \delta_0 = 23.95° = 0.4179$ radian. At $t = 0$, when the short circuit occurs, p_e instantaneously drops to zero and remains at zero during the fault since power cannot be transferred past faulted bus 1. From (11.1.16), with $\omega_{\text{p.u.}}(t) = 1.0$,

$$\frac{2\text{H}}{\omega_{\text{syn}}}\frac{d^2\delta(t)}{dt^2} = p_{m\text{p.u.}} \qquad 0 \leqslant t \leqslant 0.05 \quad \text{s}$$

Integrating twice with initial condition $\delta(0) = \delta_0$ and $\dfrac{d\delta(0)}{dt} = 0$,

$$\frac{d\delta(t)}{dt} = \frac{\omega_{\text{syn}}\, p_{m\text{p.u.}}}{2\text{H}} t + 0$$

$$\delta(t) = \frac{\omega_{\text{syn}}\, p_{m\text{p.u.}}}{4\text{H}} t^2 + \delta_0$$

At $t = 3$ cycles $= 0.05$ second,

$$\delta_1 = \delta(0.05 \text{ s}) = \frac{2\pi 60}{12}(0.05)^2 + 0.4179$$

$$= 0.4964 \text{ radian} = 28.44°$$

The accelerating area A_1, shaded in Figure 11.7, is

$$\text{A}_1 = \int_{\delta_0}^{\delta_1} p_m\, d\delta = \int_{\delta_0}^{\delta_1} 1.0\, d\delta = (\delta_1 - \delta_0) = 0.4964 - 0.4179 = 0.0785$$

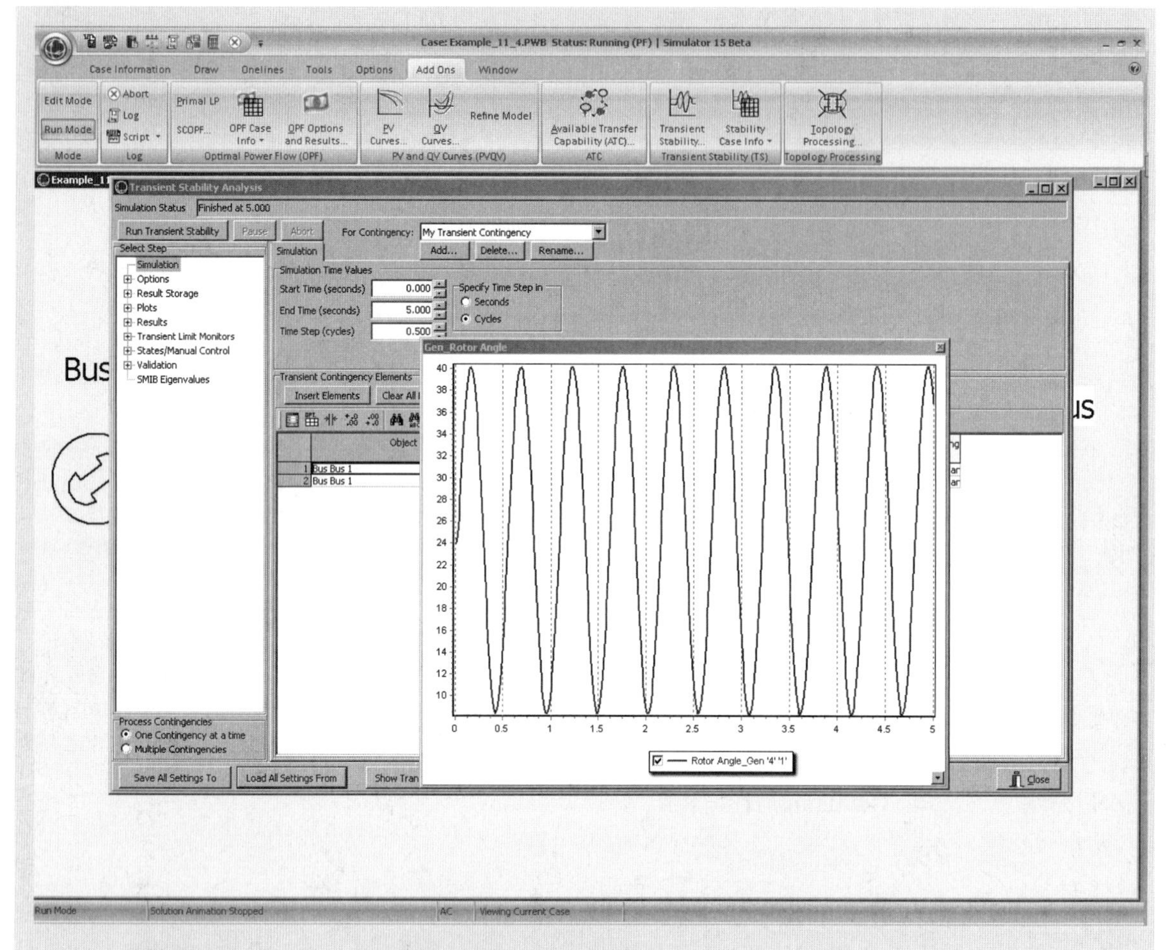

FIGURE 11.8 Variation in $\delta(t)$ without Damping

At $t = 0.05$ s the fault extinguishes and p_e instantaneously increases from zero to the sinusoidal curve in Figure 11.7. δ continues to increase until the decelerating area A_2 equals A_1. That is,

$$A_2 = \int_{\delta_1}^{\delta_2} (p_{\max} \sin \delta - p_m)\, d\delta$$

$$= \int_{0.4964}^{\delta_2} (2.4638 \sin \delta - 1.0)\, d\delta = A_1 = 0.0785$$

Integrating,

$$2.4638[\cos(0.4964) - \cos \delta_2] - (\delta_2 - 0.4964) = 0.0785$$

$$2.4638 \cos \delta_2 + \delta_2 = 2.5843$$

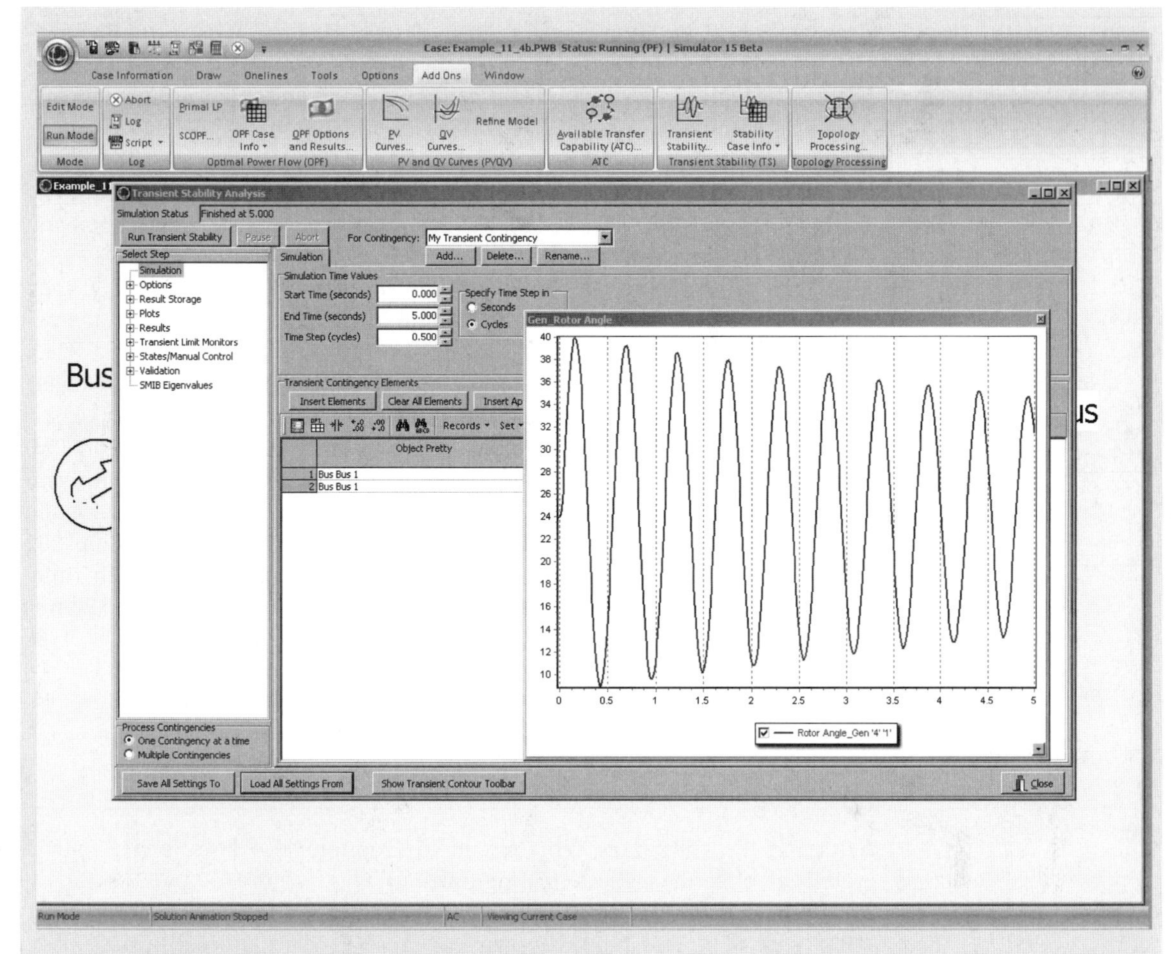

FIGURE 11.9 Variation in $\delta(t)$ with Damping

The above nonlinear algebraic equation can be solved iteratively to obtain

$$\delta_2 = 0.7003 \text{ radian} = 40.12^\circ$$

Since the maximum angle δ_2 does not exceed $\delta_3 = (180^\circ - \delta_0) = 156.05^\circ$, stability is maintained. In steady-state, the generator returns to its initial operating point $p_{ess} = p_m = 1.0$ per unit and $\delta_{ss} = \delta_0 = 23.95^\circ$.

Note that as the fault duration increases, the risk of instability also increases. The *critical clearing time*, denoted t_{cr}, is the longest fault duration allowable for stability.

To see this case modeled in PowerWorld Simulator, open case Example 11_4 (see Figure 11.8). Then select **Add-Ons, Transient Stability**, which displays the Transient Stability Analysis Form. Notice that in the Transient Stability Contingency Elements list, a fault is applied to bus 1 at $t = 0$ s and cleared at $t = 0.05$ s (three cycles later). To see the time variation in the generator angle (modeled at bus 4 in PowerWorld Simulator), click the **Run Transient Stability** button. When the simulation is finished, a graph showing this angle automatically appears, as shown in the figure. More detailed results are also available by clicking on **Results** in the list on the left side of the form. To rerun the example with a different fault duration, overwrite the Time (second) field in the Transient Contingency Elements list, and then again click the **Run Transient Stability** button.

Notice that because this system is modeled without damping (i.e., $D = 0$), the angle oscillations do not damp out with time. To extend the example, right-click on the Bus 4 generator on the one-line diagram and select **Generator Information Dialog**. Then click on the **Stability, Machine Models** tab to see the parameters associated with the GENCLS model (i.e., a classical model–a more detailed machine model is introduced in Section 11.6). Change the "D" field to 1.0, select **OK** to close the dialog, and then rerun the transient stability case. The results are as shown in Figure 11.9. While the inclusion of damping did not significantly alter the maximum for $\delta(t)$, the magnitude of the angle oscillations is now decreasing with time. For convenience this modified example is contained in PowerWorld Simulator case Example 11.4b. ■

EXAMPLE 11.5 Equal-area criterion: critical clearing time for a temporary three-phase fault

Assuming the temporary short circuit in Example 11.4 lasts longer than 3 cycles, calculate the critical clearing time.

SOLUTION The p–δ plot is shown in Figure 11.10. At the critical clearing angle, denoted δ_{cr}, the fault is extinguished. The power angle then increases to a maximum value $\delta_3 = 180° - \delta_0 = 156.05° = 2.7236$ radians, which gives

FIGURE 11.10 p–δ plot for Example 11.5

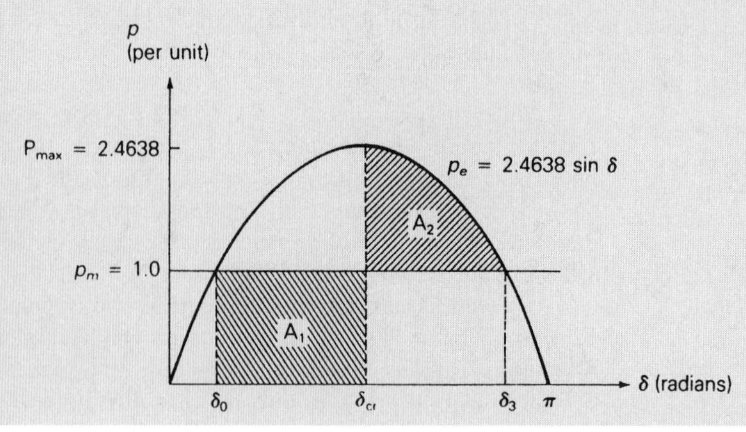

the maximum decelerating area. Equating the accelerating and decelerating areas,

$$A_1 = \int_{\delta_0}^{\delta_{cr}} p_m \, d\delta = A_2 = \int_{\delta_{cr}}^{\delta_3} (P_{max} \sin\delta - p_m) \, d\delta$$

$$\int_{0.4179}^{\delta_{cr}} 1.0 \, d\delta = \int_{\delta_{cr}}^{2.7236} (2.4638 \sin\delta - 1.0) \, d\delta$$

Solving for δ_{cr},

$$(\delta_{cr} - 0.4179) = 2.4638[\cos\delta_{cr} - \cos(2.7236)] - (2.7236 - \delta_{cr})$$

$$2.4638 \cos\delta_{cr} = +0.05402$$

$$\delta_{cr} = 1.5489 \text{ radians} = 88.74^\circ$$

From the solution to the swing equation given in Example 11.4,

$$\delta(t) = \frac{\omega_{syn} p_{mp.u.}}{4H} t^2 + \delta_0$$

Solving

$$t = \sqrt{\frac{4H}{\omega_{syn} p_{mp.u.}} (\delta(t) - \delta_0)}$$

Using $\delta(t_{cr}) = \delta_{cr} = 1.5489$ and $\delta_0 = 0.4179$ radian,

$$t_{cr} = \sqrt{\frac{12}{(2\pi 60)(1.0)} (1.5489 - 0.4179)}$$

$$= 0.1897 \text{ s} = 11.38 \text{ cycles}$$

If the fault is cleared before $t = t_{cr} = 11.38$ cycles, stability is maintained. Otherwise, the generator goes out of synchronism with the infinite bus; that is, stability is lost.

To see a time-domain simulation of this case, open Example 11.5 in PowerWorld Simulator (see Figure 11.11). Again select **Add-Ons, Transient Stability** to view the Transient Stability Analysis Form. In order to better visualize the results on a PowerWorld one-line diagram, there is an option to transfer the transient stability results to the one-line every n timesteps. To access this option, select the **Options** page from the list on the left side of the display, then the **General** tab, then check the **Transfer Results to Power Flow after Interval Check** field. With this option checked, click on the **Run Transient Stability**, which will run the case with a critical clearing time of 0.1895 seconds. The plot is set to dynamically update as well. The final results are shown in Figure 11.11. Because the one-line is reanimated every n timesteps (4 in this case), a potential downside to this option is it takes longer to run. Uncheck the option to restore full solution speed.

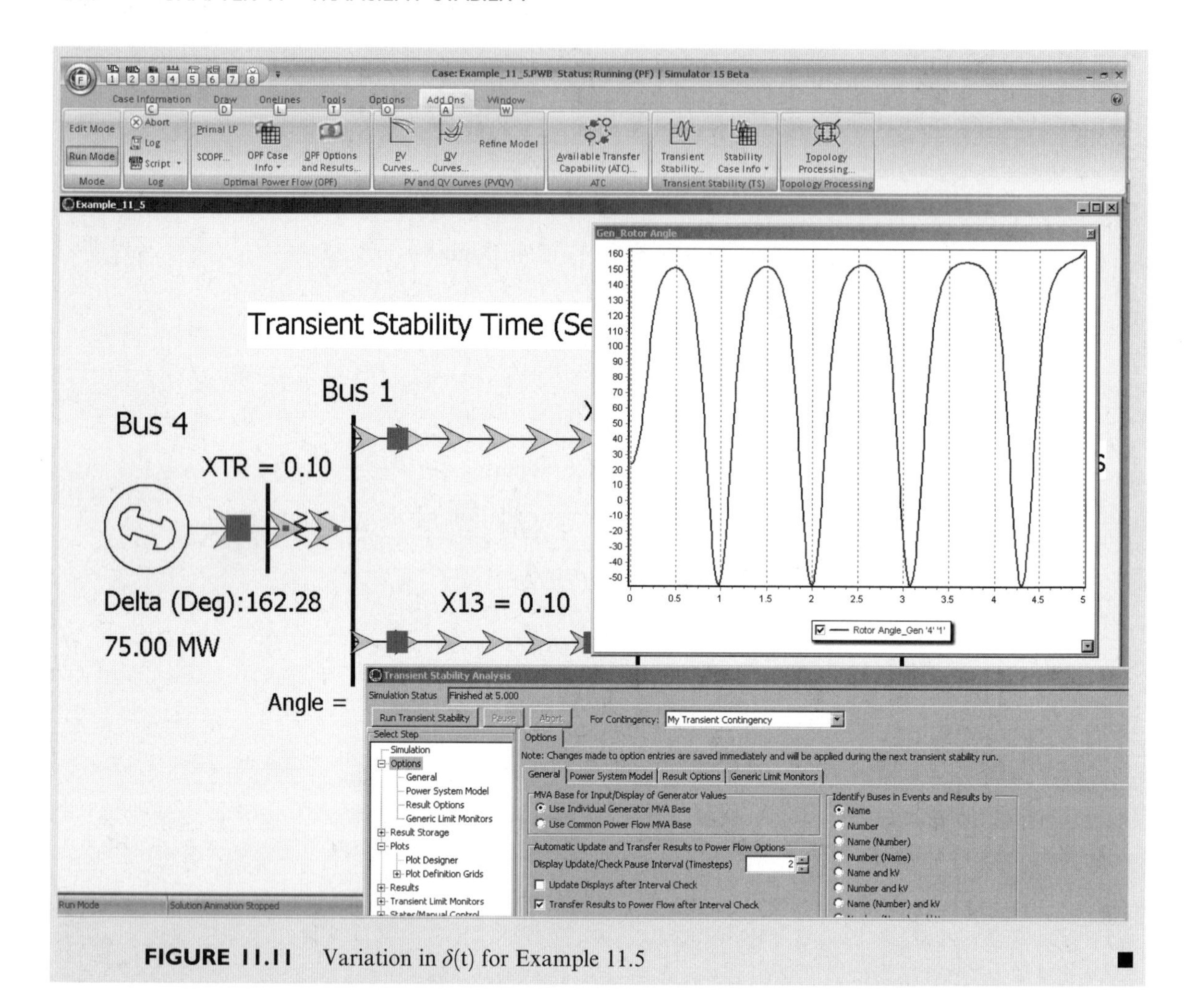

FIGURE 11.11 Variation in $\delta(t)$ for Example 11.5 ■

EXAMPLE 11.6 Equal-area criterion: critical clearing angle for a cleared three-phase fault

The synchronous generator in Figure 11.4 is initially operating in the steady-state condition given in Example 11.3 when a permanent three-phase-to-ground bolted short circuit occurs on line 1–3 at bus 3. The fault is cleared by opening the circuit breakers at the ends of line 1–3 and line 2–3. These circuit breakers then remain open. Calculate the critical clearing angle. As in previous examples, H = 3.0 p.u.-s, $p_m = 1.0$ per unit and $\omega_{\text{p.u.}} = 1.0$ in the swing equation.

SOLUTION From Example 11.3, the equation for the prefault electrical power, denoted p_{e1} here, is $p_{e1} = 2.4638 \sin\delta$ per unit. The faulted network is

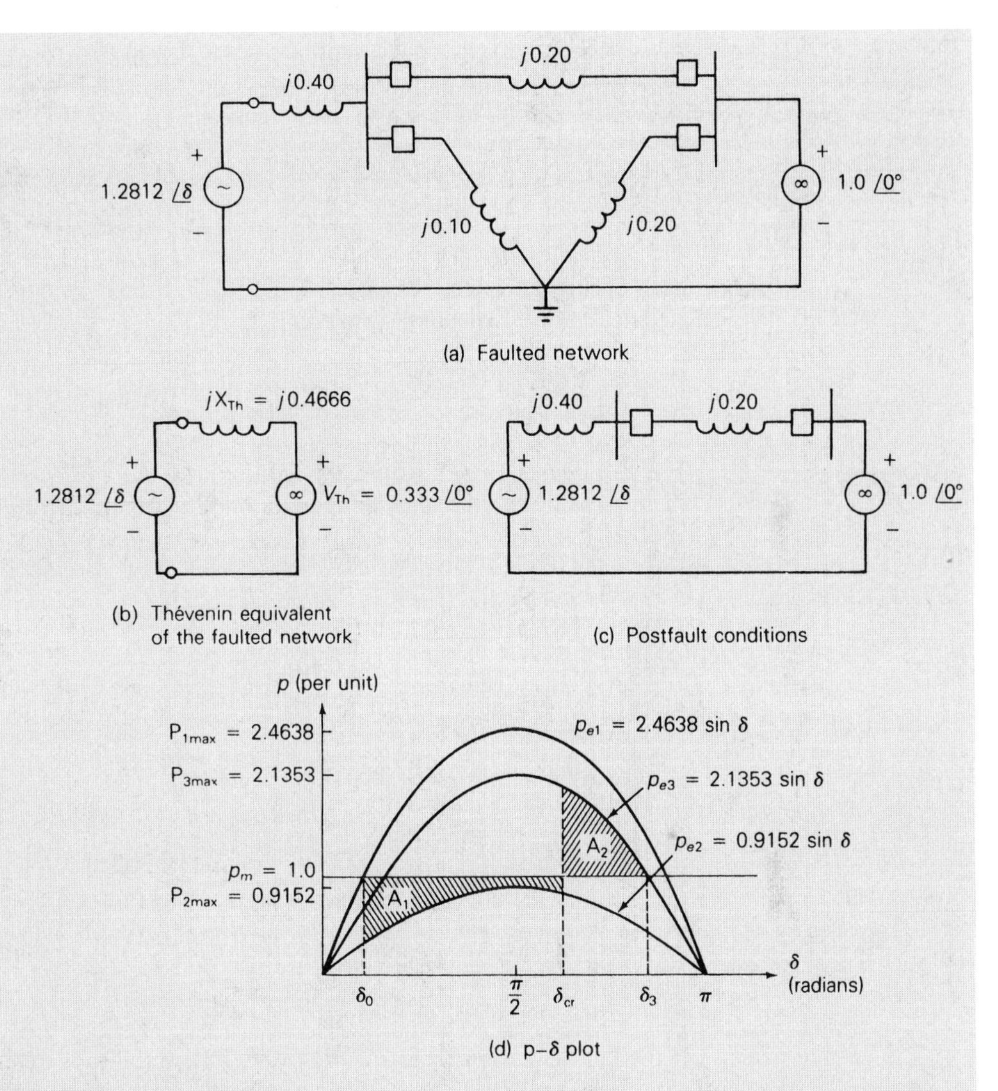

FIGURE 11.12 Example 11.6

shown in Figure 11.12(a), and the Thévenin equivalent of the faulted network, as viewed from the generator internal voltage source, is shown in Figure 11.12(b). The Thévenin reactance is

$$X_{Th} = 0.40 + 0.20 \| 0.10 = 0.46666 \quad \text{per unit}$$

and the Thévenin voltage source is

$$V_{Th} = 1.0\underline{/0°}\left[\frac{X_{13}}{X_{13} + X_{12}}\right] = 1.0\underline{/0°}\,\frac{0.10}{0.30}$$

$$= 0.33333\underline{/0°} \quad \text{per unit}$$

From Figure 11.12(b), the equation for the electrical power delivered by the generator to the infinite bus during the fault, denoted p_{e2}, is

$$p_{e2} = \frac{E'V_{Th}}{X_{Th}} \sin\delta = \frac{(1.2812)(0.3333)}{0.46666} \sin\delta = 0.9152 \sin\delta \quad \text{per unit}$$

The postfault network is shown in Figure 11.12(c), where circuit breakers have opened and removed lines 1–3 and 2–3. From this figure, the postfault electrical power delivered, denoted p_{e3}, is

$$p_{e3} = \frac{(1.2812)(1.0)}{0.60} \sin\delta = 2.1353 \sin\delta \quad \text{per unit}$$

The p–δ curves as well as the accelerating area A_1 and decelerating area A_2 corresponding to critical clearing are shown in Figure 11.12(d). Equating A_1 and A_2,

$$A_1 = \int_{\delta_0}^{\delta_{cr}} (p_m - P_{2max} \sin\delta)\, d\delta = A_2 = \int_{\delta_{cr}}^{\delta_3} (P_{3max} \sin\delta - p_m)\, d\delta$$

$$\int_{0.4179}^{\delta_{cr}} (1.0 - 0.9152 \sin\delta)\, d\delta = \int_{\delta_{cr}}^{2.6542} (2.1353 \sin\delta - 1.0)\, d\delta$$

Solving for δ_{cr},

$$(\delta_{cr} - 0.4179) + 0.9152(\cos\delta_{cr} - \cos 0.4179)$$
$$= 2.1353(\cos\delta_{cr} - \cos 2.6542) - (2.6542 - \delta_{cr})$$

$$-1.2201 \cos\delta_{cr} = 0.4868$$

$$\delta_{cr} = 1.9812 \text{ radians} = 111.5°$$

If the fault is cleared before $\delta = \delta_{cr} = 111.5°$, stability is maintained. Otherwise, stability is lost. To see this case in PowerWorld Simulator open case Example 11_6. ■

11.4 NUMERICAL INTEGRATION OF THE SWING EQUATION

The equal-area criterion is applicable to one machine and an infinite bus or to two machines. For multimachine stability problems, however, numerical integration techniques can be employed to solve the swing equation for each machine.

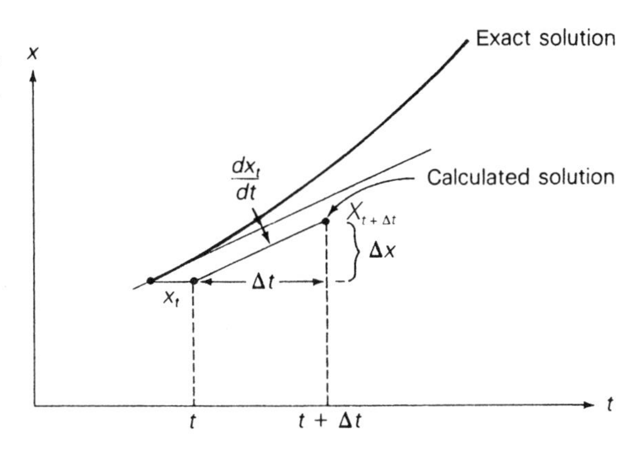

FIGURE 11.13 Euler's method

Given a first-order differential equation

$$\frac{dx}{dt} = f(x) \tag{11.4.1}$$

one relatively simple integration technique is Euler's method [1], illustrated in Figure 11.13. The integration step size is denoted Δt. Calculating the slope at the beginning of the integration interval, from (11.4.1),

$$\frac{dx_t}{dt} = f(x_t) \tag{11.4.2}$$

The new value $x_{t+\Delta t}$ is calculated from the old value x_t by adding the increment Δx,

$$x_{t+\Delta t} = x_t + \Delta x = x_t + \left(\frac{dx_t}{dt}\right)\Delta t \tag{11.4.3}$$

As shown in the figure, Euler's method assumes that the slope is constant over the entire interval Δt. An improvement can be obtained by calculating the slope at both the beginning and end of the interval, and then averaging these slopes. The modified Euler's method is illustrated in Figure 11.14. First, the

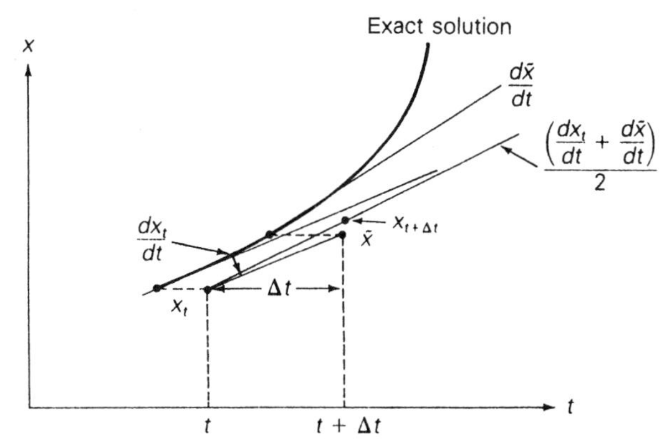

FIGURE 11.14 Modified Euler's method

slope at the beginning of the interval is calculated from (11.4.1) and used to calculate a preliminary value $\tilde{x}$ given by

$$\tilde{x} = x_t + \left(\frac{dx_t}{dt}\right)\Delta t \tag{11.4.4}$$

Next, the slope at $\tilde{x}$ is calculated:

$$\frac{d\tilde{x}}{dt} = f(\tilde{x}) \tag{11.4.5}$$

Then, the new value is calculated using the average slope:

$$x_{t+\Delta t} = x_t + \frac{\left(\frac{dx_t}{dt} + \frac{d\tilde{x}}{dt}\right)}{2}\Delta t \tag{11.4.6}$$

We now apply the modified Euler's method to calculate machine frequency ω and power angle δ. Letting x be either δ or ω, the old values at the beginning of the interval are denoted δ_t and ω_t. From (11.1.17) and (11.1.18), the slopes at the beginning of the interval are

$$\frac{d\delta_t}{dt} = \omega_t - \omega_{\text{syn}} \tag{11.4.7}$$

$$\frac{d\omega_t}{dt} = \frac{p_{ap.u.t}\,\omega_{\text{syn}}}{2H\omega_{\text{p.u.}t}} \tag{11.4.8}$$

where $p_{ap.u.t}$ is the per-unit accelerating power calculated at $\delta = \delta_t$, and $\omega_{\text{p.u.}t} = \omega_t/\omega_{\text{syn}}$. Applying (11.4.4), preliminary values are

$$\tilde{\delta} = \delta_t + \left(\frac{d\delta_t}{dt}\right)\Delta t \tag{11.4.9}$$

$$\tilde{\omega} = \omega_t + \left(\frac{d\omega_t}{dt}\right)\Delta t \tag{11.4.10}$$

Next, the slopes at $\tilde{\delta}$ and $\tilde{\omega}$ are calculated, again using (11.1.17) and (11.1.18):

$$\frac{d\tilde{\delta}}{dt} = \tilde{\omega} - \omega_{\text{syn}} \tag{11.4.11}$$

$$\frac{d\tilde{\omega}}{dt} = \frac{\tilde{p}_{ap.u.}\,\omega_{\text{syn}}}{2H\tilde{\omega}_{\text{p.u.}}} \tag{11.4.12}$$

where $\tilde{p}_{ap.u.}$ is the per-unit accelerating power calculated at $\delta = \tilde{\delta}$, and $\tilde{\omega}_{\text{p.u.}} = \tilde{\omega}/\omega_{\text{syn}}$. Applying (11.4.6), the new values at the end of the interval are

$$\delta_{t+\Delta t} = \delta_t + \frac{\left(\frac{d\delta_t}{dt} + \frac{d\tilde{\delta}}{dt}\right)}{2}\Delta t \qquad (11.4.13)$$

$$\omega_{t+\Delta t} = \omega_t + \frac{\left(\frac{d\omega_t}{dt} + \frac{d\tilde{\omega}}{dt}\right)}{2}\Delta t \qquad (11.4.14)$$

This procedure, given by (11.4.7)–(11.4.13), begins at $t = 0$ with specified initial values δ_0 and ω_0, and continues iteratively until $t = T$, a specified final time. Calculations are best performed using a digital computer.

EXAMPLE 11.7 Euler's method: computer solution to swing equation and critical clearing time

Verify the critical clearing angle determined in Example 11.6, and calculate the critical clearing time by applying the modified Euler's method to solve the swing equation for the following two cases:

Case 1 The fault is cleared at $\delta = 1.95$ radians $= 112°$ (which is less than δ_{cr})

Case 2 The fault is cleared at $\delta = 2.09$ radians $= 120°$ (which is greater than δ_{cr})

For calculations, use a step size $\Delta t = 0.01$ s, and solve the swing equation from $t = 0$ to $t = T = 0.85$ s.

SOLUTION Equations (11.4.7)–(11.4.14) are solved by a digital computer program written in BASIC. From Example 11.6, the initial conditions at $t = 0$ are

$$\delta_0 = 0.4179 \quad \text{rad}$$

$$\omega_0 = \omega_{syn} = 2\pi 60 \quad \text{rad/s}$$

Also, the H constant is 3.0 p.u.-s, and the faulted accelerating power is

$$p_{a\text{p.u.}} = 1.0 - 0.9152 \sin\delta$$

The postfault accelerating power is

$$p_{a\text{p.u.}} = 1.0 - 2.1353 \sin\delta \quad \text{per unit}$$

The computer program and results at 0.02 s printout intervals are listed in Table 11.1. As shown, these results agree with Example 11.6, since the system is stable for Case 1 and unstable for Case 2. Also from Table 11.1, the critical clearing time is between 0.34 and 0.36 s.

TABLE 11.1 Computer calculation of swing curves for Example 11.7

Case 1 Stable			Case 2 Unstable		
Time s	Delta rad	Omega rad/s	Time s	Delta rad	Omega rad/s
0.000	0.418	376.991	0.000	0.418	376.991
0.020	0.426	377.778	0.020	0.426	377.778
0.040	0.449	378.547	0.040	0.449	378.547
0.060	0.488	379.283	0.060	0.488	379.283
0.080	0.541	379.970	0.080	0.541	379.970
0.100	0.607	380.599	0.100	0.607	380.599
0.120	0.685	381.159	0.120	0.685	381.159
0.140	0.773	381.646	0.140	0.773	381.646
0.160	0.870	382.056	0.160	0.870	382.056
0.180	0.975	382.392	0.180	0.975	382.392
0.200	1.086	382.660	0.200	1.086	382.660
0.220	1.202	382.868	0.220	1.202	382.868
0.240	1.321	383.027	0.240	1.321	383.027
0.260	1.443	383.153	0.260	1.443	383.153
0.280	1.567	383.262	0.280	1.567	383.262
0.300	1.694	383.370	0.300	1.694	383.370
0.320	1.823	383.495	0.320	1.823	383.495
0.340	1.954	383.658	0.340	1.954	383.658
	Fault Cleared		0.360	2.090	383.876
0.360	2.076	382.516		Fault Cleared	
0.380	2.176	381.510	0.380	2.217	382.915
0.400	2.257	380.638	0.400	2.327	382.138
0.420	2.322	379.886	0.420	2.424	381.546
0.440	2.373	379.237	0.440	2.511	381.135
0.460	2.413	378.674	0.460	2.591	380.902
0.480	2.441	378.176	0.480	2.668	380.844
0.500	2.460	377.726	0.500	2.746	380.969
0.520	2.471	377.307	0.520	2.828	381.288
0.540	2.473	376.900	0.540	2.919	381.824
0.560	2.467	376.488	0.560	3.022	382.609
0.580	2.453	376.056	0.580	3.145	383.686
0.600	2.429	375.583	0.600	3.292	385.111
0.620	2.396	375.053	0.620	3.472	386.949
0.640	2.351	374.446	0.640	3.693	389.265
0.660	2.294	373.740	0.660	3.965	392.099
0.680	2.221	372.917	0.680	4.300	395.426
0.700	2.130	371.960	0.700	4.704	399.079
0.720	2.019	370.855	0.720	5.183	402.689
0.740	1.884	369.604	0.740	5.729	405.683
0.760	1.723	368.226	0.760	6.325	407.477
0.780	1.533	366.773	0.780	6.941	407.812
0.800	1.314	365.341	0.800	7.551	406.981
0.820	1.068	364.070	0.820	8.139	405.711
0.840	0.799	363.143	0.840	8.702	404.819
0.860	0.516	362.750	0.860	9.257	404.934

Program Listing

```
10 REM EXAMPLE 13.7
20 REM SOLUTION TO SWING EQUATION
30 REM THE STEP SIZE IS DELTA
40 REM THE CLEARING ANGLE IS DLTCLR
50 DELTA + .01
60 DLTCLR = 1.95
70 J = 1
80 PMAX = .9152
90 PI = 3.1415927 #
100 T = 0
110 X1 = .4179
120 X2 = 2 * PI * 60
130 LPRINT "TIME DELTA OMEGA"
140 LPRINT "s rad rad/s"
150 LPRINT USING "#####.###"; T;X1;X2
160 FOR K = 1 TO 86
170 REM LINE 180 IS EQ(13.4.7)
180 X3 = X2 - (2 * PI * 60)
190 IF J = 2 THEN GOTO 240
200 IF X1 > DLTCLR OR X1 = DLTCLR THEN
   PMAX = 2.1353
210 IF X1 > DLTCLR OR X1 = DLTCLR THEN
   LPRINT "FAULT CLEARED"
220 IF X1 > DLTCLR OR X1 = DLTCLR THEN
   J = 2
230 REM LINES 240 AND 250 ARE EQ(13.4.8)
240 X4 = 1 - PMAX * SIN(X1)
250 X5 = X4 * (2 * PI * 60) * (2 * PI * 60)/(6 * X2)
260 REM LINE 270 IS EQ(13.4.9)
270 X6 = X1 + X3 * DELTA
280 REM LINE 290 IS EQ(13.4.10)
290 X7 = X2 + X5 * DELTA
300 REM LINE 310 IS EQ(13.4.11)
310 X8 = X7 - 2 * PI * 60
320 REM LINES 330 AND 340 ARE EQ(13.4.12)
330 X9 = 1 - PMAX * SIN(X6)
340 X10 = X9 * (2 * PI * 60) * (2 * PI * 60)/(6 * X7)
350 REM LINE 360 IS EQ(13.4.13)
360 X1 = X1 + (X3 + X8) * (DELTA/2)
370 REM LINE 380 IS EQ(13.4.14)
380 X2 = X2 + (X5 + X10) * (DELTA/2)
390 T = K * DELTA
400 Z = K/2
410 M = INT(Z)
420 IF M = Z THEN LPRINT USING
   "#####.###"; T;X1;X2
430 NEXT K
440 END
```

■

In addition to Euler's method, there are many other numerical integration techniques, such as Runge–Kutta, Picard's method, and Milne's predictor-corrector method [1]. Comparison of the methods shows a trade-off of accuracy versus computation complexity. The Euler method is a relatively simple method to compute, but requires a small step size Δt for accuracy. Some of the other methods can use a larger step size for comparable accuracy, but the computations are more complex.

To see this case in PowerWorld Simulator open case Example 11_7, which plots both the generator angle and speed. However, rather than showing the speed in radians per second, Hertz is used. Also, in addition to plotting the angle and speed versus time, the case includes a "phase portrait" in which the speed is plotted as a function of the angle. Numeric results are available by clicking on the **Results** page. The results shown in PowerWorld differ slightly from those in the table because PowerWorld uses a more exact second order integration method.

11.5 MULTIMACHINE STABILITY

The numerical integration methods discussed in Section 11.4 can be used to solve the swing equations for a multimachine stability problem. However, a method is required for computing machine output powers for a general network. Figure 11.15 shows a general N-bus power system with M synchronous machines. Each machine is the same as that represented by the simplified model of Figure 11.2, and the internal machine voltages are denoted $E_1', E_2', \ldots, E_M'$. The M machine terminals are connected to system buses denoted $G1, G2, \ldots, GM$ in Figure 11.15. All loads are modeled here as constant admittances. Writing nodal equations for this network,

$$\begin{bmatrix} \boldsymbol{Y}_{11} & \boldsymbol{Y}_{12} \\ \boldsymbol{Y}_{12}^{\mathrm{T}} & \boldsymbol{Y}_{22} \end{bmatrix} \begin{bmatrix} \boldsymbol{V} \\ \boldsymbol{E} \end{bmatrix} = \begin{bmatrix} \boldsymbol{0} \\ \boldsymbol{I} \end{bmatrix} \tag{11.5.1}$$

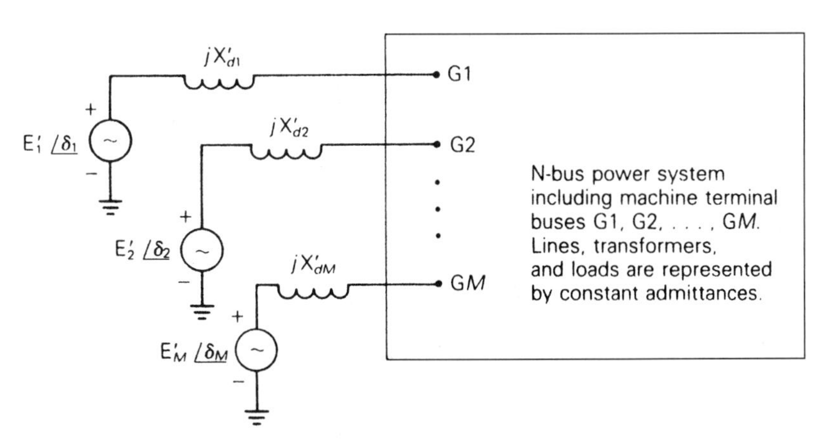

jX'_{d1}
G1
$E_1'\,\underline{/\delta_1}$
jX'_{d2}
G2
$E_2'\,\underline{/\delta_2}$
jX'_{dM}
GM
$E_M'\,\underline{/\delta_M}$
N-bus power system including machine terminal buses G1, G2, . . . , GM. Lines, transformers, and loads are represented by constant admittances.

FIGURE 11.15 N-bus power-system representation for transient stability studies

where

$$\boldsymbol{V} = \begin{bmatrix} V_1 \\ V_2 \\ \vdots \\ V_N \end{bmatrix} \quad \text{is the } N \text{ vector of bus voltages} \tag{11.5.2}$$

$$\boldsymbol{E} = \begin{bmatrix} E'_1 \\ E'_2 \\ \vdots \\ E'_M \end{bmatrix} \quad \text{is the } M \text{ vector of machine voltages} \tag{11.5.3}$$

$$\boldsymbol{I} = \begin{bmatrix} I_1 \\ I_2 \\ \vdots \\ I_M \end{bmatrix} \quad \begin{matrix}\text{is the } M \text{ vector of machine currents} \\ \text{(these are current sources)}\end{matrix} \tag{11.5.4}$$

$$\left[\begin{array}{c|c} \boldsymbol{Y}_{11} & \boldsymbol{Y}_{12} \\ \hline \boldsymbol{Y}_{12}^{\mathrm{T}} & \boldsymbol{Y}_{22} \end{array}\right] \quad \text{is an } (N+M) \times (N+M) \text{ admittance matrix} \tag{11.5.5}$$

The admittance matrix in (11.5.5) is partitioned in accordance with the N system buses and M internal machine buses, as follows:

$\boldsymbol{Y}_{11}$ is $N \times N$

$\boldsymbol{Y}_{12}$ is $N \times M$

$\boldsymbol{Y}_{22}$ is $M \times M$

$\boldsymbol{Y}_{11}$ is similar to the bus admittance matrix used for power flows in Chapter 7, except that load admittances and inverted generator impedances are included. That is, if a load is connected to bus n, then that load admittance is added to the diagonal element Y_{11nn}. Also, $(1/j\mathrm{X}'_{dn})$ is added to the diagonal element Y_{11GnGn}.

$\boldsymbol{Y}_{22}$ is a diagonal matrix of inverted generator impedances; that is,

$$\boldsymbol{Y}_{22} = \begin{bmatrix} \dfrac{1}{j\mathrm{X}'_{d1}} & & & 0 \\ & \dfrac{1}{j\mathrm{X}'_{d2}} & & \\ & & \ddots & \\ 0 & & & \dfrac{1}{j\mathrm{X}'_{dM}} \end{bmatrix} \tag{11.5.6}$$

Also, the kmth element of $\boldsymbol{Y}_{12}$ is

$$Y_{12km} = \begin{cases} \dfrac{-1}{jX'_{dn}} & \text{if } k = Gn \text{ and } m = n \\ 0 & \text{otherwise} \end{cases} \tag{11.5.7}$$

Writing (11.5.1) as two separate equations,

$$\boldsymbol{Y}_{11}\boldsymbol{V} + \boldsymbol{Y}_{12}\boldsymbol{E} = \boldsymbol{0} \tag{11.5.8}$$

$$\boldsymbol{Y}_{12}^{\mathrm{T}}\boldsymbol{V} + \boldsymbol{Y}_{22}\boldsymbol{E} = \boldsymbol{I} \tag{11.5.9}$$

Assuming $\boldsymbol{E}$ is known, (11.5.8) is a linear equation in $\boldsymbol{V}$ that can be solved either iteratively or by Gauss elimination. Using the Gauss-Seidel iterative method given by (7.2.9), the kth component of $\boldsymbol{V}$ is

$$V_k(i+1) = \frac{1}{Y_{11kk}}\left[-\sum_{n=1}^{M} Y_{12kn}E_n - \sum_{n=1}^{k-1} Y_{11kn}V_n(i+1) - \sum_{n=k+1}^{N} Y_{11kn}V_n(i)\right] \tag{11.5.10}$$

After $\boldsymbol{V}$ is computed, the machine currents can be obtained from (11.5.9). That is,

$$\boldsymbol{I} = \begin{bmatrix} I_1 \\ I_2 \\ \vdots \\ I_M \end{bmatrix} = \boldsymbol{Y}_{12}^{\mathrm{T}}\boldsymbol{V} + \boldsymbol{Y}_{22}\boldsymbol{E} \tag{11.5.11}$$

The (real) electrical power output of machine n is then

$$p_{en} = \mathrm{Re}[E_n I_n^*] \qquad n = 1, 2, \ldots, M \tag{11.5.12}$$

We are now ready to outline a computation procedure for solving a transient stability problem. The procedure alternately solves the swing equations representing the machines and the above algebraic power-flow equations representing the network. We use the modified Euler method of Section 11.4 to solve the swing equations and the Gauss–Seidel iterative method to solve the power-flow equations. We now give the procedure in the following 11 steps.

TRANSIENT STABILITY COMPUTATION PROCEDURE

STEP 1 Run a prefault power-flow program to compute initial bus voltages V_k, $k = 1, 2, \ldots, N$, initial machine currents I_n, and initial machine electrical power outputs p_{en}, $n = 1, 2, \ldots, M$. Set machine mechanical power outputs, $p_{mn} = p_{en}$. Set initial machine frequencies, $\omega_n = \omega_{\mathrm{syn}}$. Compute the load admittances.

STEP 2 Compute the internal machine voltages:

$$E_n = E_n\underline{/\delta_n} = V_{Gn} + (jX'_{dn})I_n \quad n = 1, 2, \ldots, M$$

where V_{Gn} and I_n are computed in Step 1. The magnitudes E_n will remain constant throughout the study. The angles δ_n are the initial power angles.

STEP 3 Compute Y_{11}. Modify the $(N \times N)$ power-flow bus admittance matrix by including the load admittances and inverted generator impedances.

STEP 4 Compute Y_{22} from (11.5.6) and Y_{12} from (11.5.7).

STEP 5 Set time $t = 0$.

STEP 6 Is there a switching operation, change in load, short circuit, or change in data? For a switching operation or change in load, modify the bus admittance matrix. For a short circuit, set the faulted bus voltage [in (11.5.10)] to zero.

STEP 7 Using the internal machine voltages $E_n = E_n/\!\underline{\delta_n}$, $n = 1, 2, \ldots, M$, with the values of δ_n at time t, compute the machine electrical powers p_{en} at time t from (11.5.10) to (11.5.12).

STEP 8 Using p_{en} computed in Step 7 and the values of δ_n and ω_n at time t, compute the preliminary estimates of power angles $\tilde{\delta}_n$ and machine speeds $\tilde{\omega}_n$ at time $(t + \Delta t)$ from (11.4.7) to (11.4.10).

STEP 9 Using $E_n = E_n/\!\underline{\tilde{\delta}_n}$, $n = 1, 2, \ldots, M$, compute the preliminary estimates of the machine electrical powers $\tilde{p}_{en}$ at time $(t + \Delta t)$ from (11.5.10) to (11.5.12).

STEP 10 Using $\tilde{p}_{en}$ computed in Step 9, as well as $\tilde{\delta}_n$ and $\tilde{\omega}_n$ computed in Step 8, compute the final estimates of power angles δ_n and machine speeds ω_n at time $(t + \Delta t)$ from (11.4.11) to (11.4.14).

STEP 11 Set time $t = t + \Delta t$. Stop if $t \geq T$. Otherwise, return to Step 6.

An important transient stability parameter is the step size (time step), Δt, used in the numerical integration. Because the time required to solve a transient stability problem varies inversely with the time step, a larger value would be preferred. However, if too large a value is chosen, then the solution accuracy may suffer, and for some integration methods, such as Euler's, the solution can experience numeric instability. To see an example of numeric instability, re-do the PowerWorld Simulator Example 11.7, except change the time step to 0.02 seconds. A typical time step for commercial transient stability simulations is 1/2 cycle (0.00833 seconds for a 60 Hz system).

EXAMPLE 11.8 Modifying power-flow Y_{bus} for application to multimachine stability

Consider a transient stability study for the power system given in Example 6.9, with the 184-Mvar shunt capacitor of Example 6.14 installed at bus 2. Machine transient reactances are $X'_{d1} = 0.05$ and $X'_{d2} = 0.025$ per unit on the system base. Determine the admittance matrices Y_{11}, Y_{22}, and Y_{12}.

SOLUTION From Example 6.9, the power system has $N = 5$ buses and $M = 2$ machines. The second row of the 5×5 bus admittance matrix used for power flows is calculated in Example 6.9. Calculating the other rows in the same manner, we obtain

$$Y_{bus} = \begin{bmatrix} (3.728-j49.72) & 0 & 0 & 0 & (-3.728+j49.72) \\ 0 & (2.68-j26.46) & 0 & (-0.892+j9.92) & (-1.784+j19.84) \\ 0 & 0 & (7.46-j99.44) & (-7.46+j99.44) & 0 \\ 0 & (-0.892+j9.92) & (-7.46+j99.44) & (11.92-j148.) & (-3.572+j39.68) \\ (-3.728+j49.72) & (-1.784+j19.84) & 0 & (-3.572+j39.68) & (9.084-j108.6) \end{bmatrix} \text{ per unit}$$

To obtain Y_{11}, Y_{bus} is modified by including load admittances and inverted generator impedances. From Table 6.1, the load at bus 3 is $P_{L3} + jQ_{L3} = 0.8 + j0.4$ per unit and the voltage at bus 3 is $V_3 = 1.05$ per unit. Representing this load as a constant admittance,

$$Y_{\text{load }3} = \frac{P_{L3} - jQ_{L3}}{V_3^2} = \frac{0.8 - j0.4}{(1.05)^2} = 0.7256 - j0.3628 \quad \text{per unit}$$

Similarly, the load admittance at bus 2 is

$$Y_{\text{load }2} = \frac{P_{L2} - jQ_{L2}}{V_2^2} = \frac{8 - j2.8 + j1.84}{(0.959)^2} = 8.699 - j1.044$$

where V_2 is obtained from Example 6.14 and the 184-Mvar (1.84 per unit) shunt capacitor bank is included in the bus 2 load.

The inverted generator impedances are: for machine 1 connected to bus 1,

$$\frac{1}{jX'_{d1}} = \frac{1}{j0.05} = -j20.0 \text{ per unit}$$

and for machine 2 connected to bus 3,

$$\frac{1}{jX'_{d2}} = \frac{1}{j0.025} = -j40.0 \text{ per unit}$$

To obtain Y_{11}, add $(1/jX'_{d1})$ to the first diagonal element of Y_{bus}, add $Y_{\text{load }2}$ to the second diagonal element, and add $Y_{\text{load }3} + (1/jX'_{d2})$ to the third diagonal element. The 5×5 matrix Y_{11} is then

$$Y_{11} = \begin{bmatrix} (3.728-j69.72) & 0 & 0 & 0 & (-3.728+j49.72) \\ 0 & (11.38-j29.50) & 0 & (-0.892+j9.92) & (-1.784+j19.84) \\ 0 & 0 & (8.186-j139.80) & (-7.46+j99.44) & 0 \\ 0 & (-0.892+j9.92) & (-7.46+j99.44) & (11.92-j148.) & (-3.572-j39.68) \\ (-3.728+j49.72) & (-1.784+j19.84) & 0 & (-3.572+j39.68) & (9.084-j108.6) \end{bmatrix} \text{ per unit}$$

From (11.5.6), the 2×2 matrix Y_{22} is

$$Y_{22} = \begin{bmatrix} \dfrac{1}{jX'_{d1}} & 0 \\ 0 & \dfrac{1}{jX'_{d2}} \end{bmatrix} = \begin{bmatrix} -j20.0 & 0 \\ 0 & -j40.0 \end{bmatrix} \text{ per unit}$$

From Figure 6.2, generator 1 is connected to bus 1 (therefore, bus G1 = 1 and generator 2 is connected to bus 3 (therefore G2 = 3). From (11.5.7), the 5×2 matrix Y_{12} is

$$Y_{12} = \begin{bmatrix} j20.0 & 0 \\ 0 & 0 \\ 0 & j40.0 \\ 0 & 0 \\ 0 & 0 \end{bmatrix} \text{ per unit}$$

■

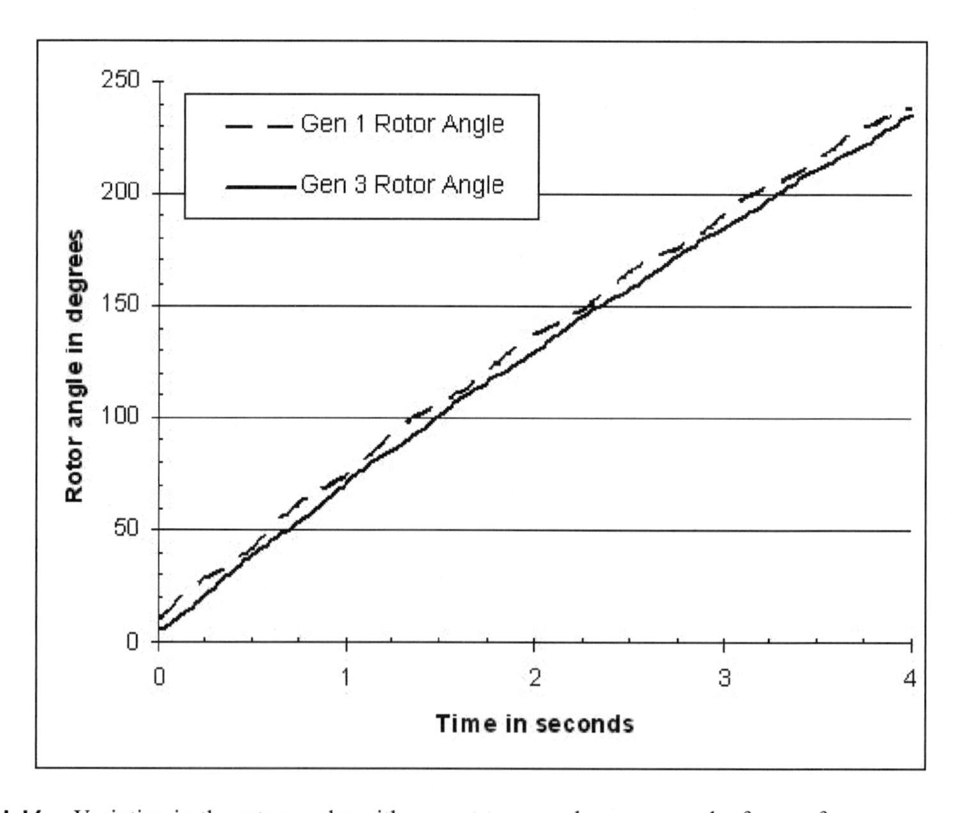

FIGURE 11.16 Variation in the rotor angles with respect to a synchronous speed reference frame

To see this case in PowerWorld Simulator open case Example 11_8. To see the $\mathbf{Y}_{11}$ matrix entries, first display the Transient Stability Analysis Form, and then select **States/Manual Control, Transient Stability Ybus**. By default, this case is set to solve a self-clearing fault at bus 4 that extinguishes itself after three cycles (0.05 s). Both generators are modeled with $H = 5.0$ p.u.-s and $D = 1.0$ p.u.

For the bus 4 fault, Figure 11.16 shows the variation in the rotor angles for the two generators with respect to a 60 Hz synchronous reference frame. The angles are increasing with time because neither of the generators is modeled with a governor, and there is no infinite bus. While it is clear that the generator angles remain together, it is very difficult to tell from Figure 11.16 the exact variation in the angle differences. Therefore transient stability programs usually report angle differences, either with respect to the angle at a specified bus or with respect to the average of all the generator angles. The latter is shown in Figure 11.17 which displays the results from the PowerWorld Simulator Example 13_8 case.

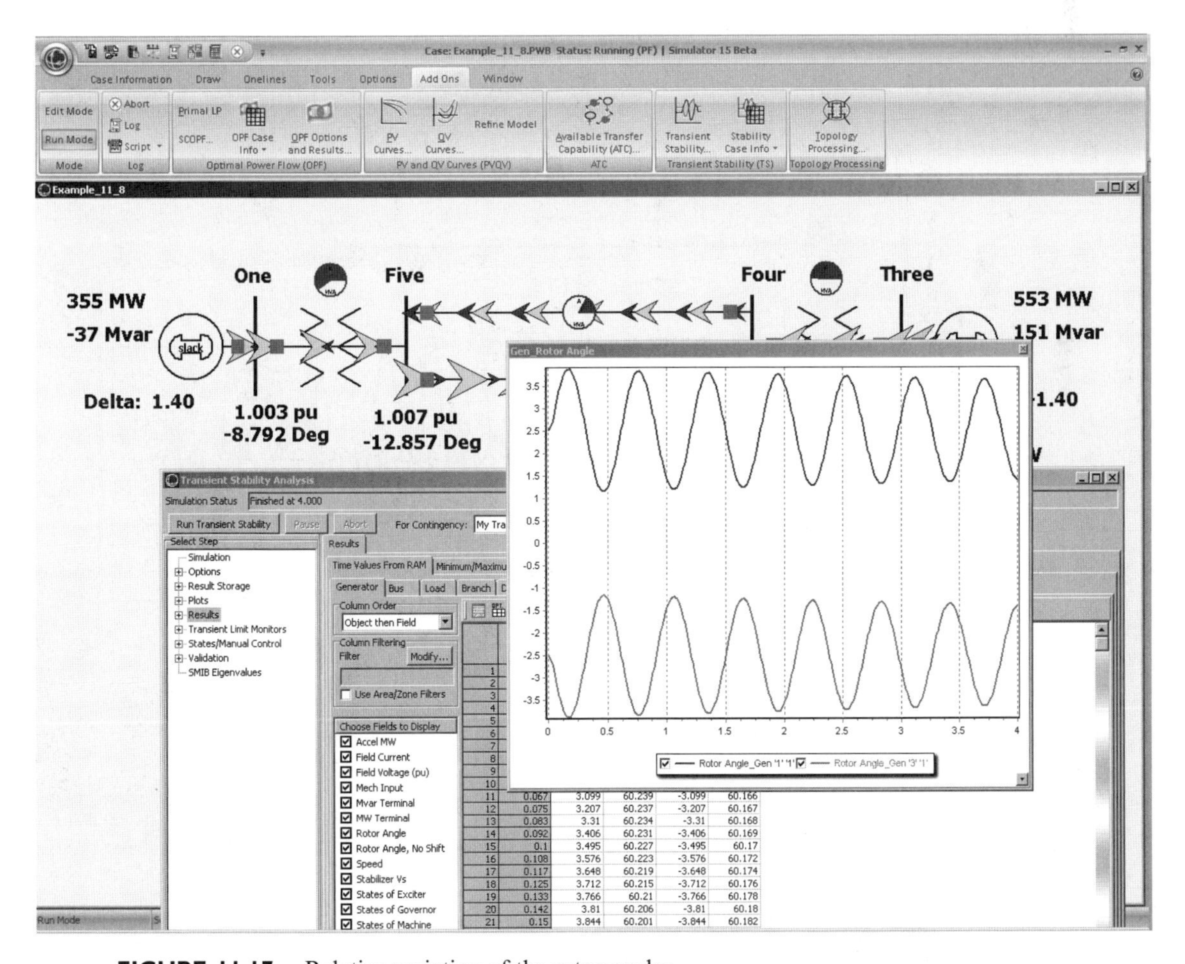

FIGURE 11.17 Relative variation of the rotor angles

EXAMPLE 11.9 Stability results for 37 bus, 9 generator system

PowerWorld Simulator case Example 13_9 demonstrates a transient stability solution using the 37 bus system introduced in Chapter 6 with the system augmented to include classical models for each of the generators. By default, the case models a transmission line fault on the 69 kV line from bus 44 (LAUF69) to bus 14 (WEBER69) with the fault at the LAUF69 end of the line. The fault is cleared after 0.1 seconds by opening this transmission line. The results from this simulation are shown in Figure 11.18, with the largest generator angle variation occurring (not surprisingly) at the bus 44 generator. Notice that during and initially after the fault, the bus 44 generator's angle increases relative to all the other angles in the system. The critical clearing time for this fault is about 0.262 seconds.

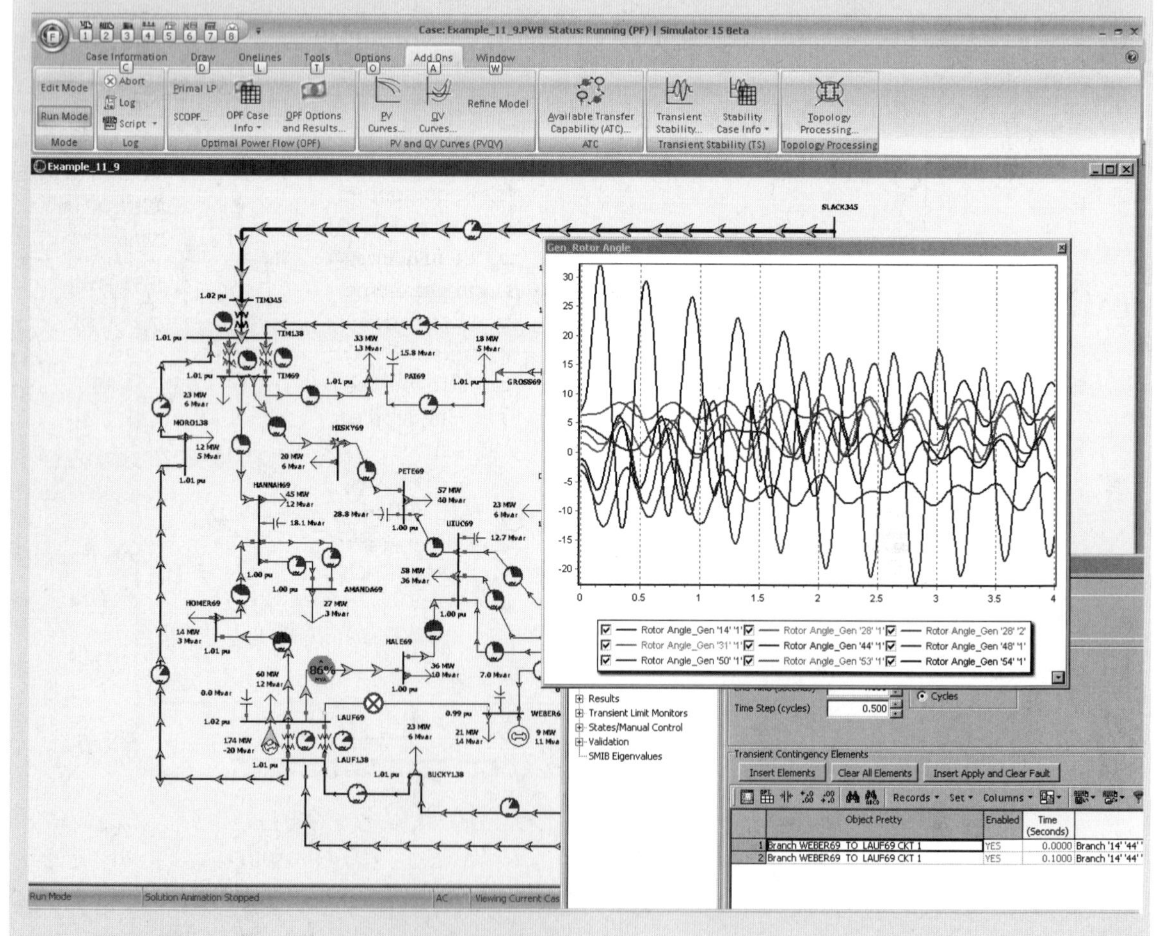

FIGURE 11.18 Rotor Angles for Example 11.9 case ■

11.6

A TWO-AXIS SYNCHRONOUS MACHINE MODEL

While the classical model for a synchronous machine provides a useful mechanism for introducing transient stability concepts, it is only appropriate for the most basic of system studies. Also, it cannot be coupled with the exciter and governor models that will be introduced in the next chapter. In this section a more realistic synchronous machine model is introduced.

The analysis of more detailed synchronous machine models requires that each machine model be expressed in a frame of reference that rotates at the same speed as its rotor. The standard approach is to use a d-q reference frame in which the major "direct" (d) axis is aligned with the rotor poles, and the quadrature (q) axis leads the direct axis by 90°. The rotor angle δ is then defined as the angle by which the q-axis leads the network reference frame (see Figure 11.19). The equation for transforming the network quantities to the d-q reference frame is given by (11.6.1) and from the d-q reference frame by (11.6.2),

$$\begin{bmatrix} V_r \\ V_i \end{bmatrix} = \begin{bmatrix} \sin\delta & \cos\delta \\ -\cos\delta & \sin\delta \end{bmatrix} \begin{bmatrix} V_d \\ V_q \end{bmatrix} \tag{11.6.1}$$

$$\begin{bmatrix} V_d \\ V_q \end{bmatrix} = \begin{bmatrix} \sin\delta & -\cos\delta \\ \cos\delta & \sin\delta \end{bmatrix} \begin{bmatrix} V_{real} \\ V_{imag} \end{bmatrix} \tag{11.6.2}$$

where the terminal voltage in the network reference frame is $V_T = V_r + jV_i$. A similar conversion is done for the currents.

Numerous different transient stability models exist for synchronous machines, most of which are beyond the scope of this text. The two-axis model,

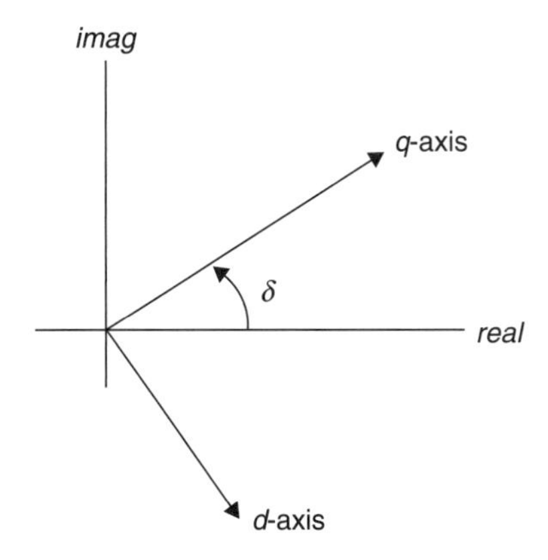

FIGURE 11.19

Reference frame transformations

which models the dynamics associated with the synchronous generator field winding and one damper winding, while neglecting the faster subtransient damper dynamics and stator transients, provides a nice compromise. For accessibility, machine saturation is not considered. With the two-axis model, the electrical behavior of the generator is represented by two algebraic equations and two differential equations

$$E_q' = V_q + R_a I_q + X_d' I_d \tag{11.6.3}$$

$$E_d' = V_d + R_a I_d - X_q' I_q \tag{11.6.4}$$

$$\frac{dE_q'}{dt} = \frac{1}{T_{do}'}\left(-E_q' - (X_d - X_d')I_d + E_{fd}\right) \tag{11.6.5}$$

$$\frac{dE_d'}{dt} = \frac{1}{T_{qo}'}\left(-E_d' + (X_q - X_q')I_q\right) \tag{11.6.6}$$

where $V_d + jV_q$ and $I_d + jI_q$ are the generator's terminal voltage and current shifted into the generator's reference frame, and E_{fd} is proportional to the field voltage. The per unit electrical torque, $T_{elec.}$ is then

$$T_e = V_d I_d + V_q I_q + R_a(I_d^2 + I_q^2) \tag{11.6.7}$$

While $p_e = T_e \omega_{p.u.}$, it is often assumed that $\omega_{p.u.} = 1.0$ [12, pp. 175] with the result being an assumption that $p_e = T_e$. When (11.6.5) and (11.6.6) are combined with generator mechanical equations presented in (11.1.18) and (11.1.19), substituting (11.6.7) for $p_{ep.u}$, the result is a synchronous generator model containing four first-order differential equations.

The initial value for δ can be determined by noting that in steady-state the angle of the internal voltage,

$$E = V_T + jX_q I \tag{11.6.8}$$

Hence the initial value of δ is the angle on E. Once δ has been determined, (11.6.2) is used to transfer the generator terminal voltage and current into the generator's reference frame, and then (11.6.3), (11.6.4) and (11.6.5) (assuming the left-hand side is zero) are used to determine the initial values of E_q', E_d', and E_{fd}. In this chapter the field voltage, E_{fd}, will be assumed constant (the use of the generator exciters to control the field voltage is a topic for the next chapter).

EXAMPLE 11.10 Two-Axis Model Example

For the system from Example 11.3, with the synchronous generator modeled using a two-axis model, determine a) the initial conditions, and then b) use PowerWorld Simulator to determine the critical clearing time for the Example 11.6

fault (three phase fault at bus 3, cleared by opening lines 1–3 and 2–3). Assume $H = 3.0$ per unit-seconds, $R_a = 0, X_d = 2.1, X_q = 2.0, X_d = 0.3, X_q = 0.5$, all per unit using the 100 MVA system base.

SOLUTION

a. From Example 11.3, the current out of the generator is

$$I = 1.0526\underline{/-18.20^\circ} = 1 - j0.3288$$

which gives a generator terminal voltage of

$$V_T = 1.0\underline{/0^\circ} + (j0.22)(1.0526\underline{/-18.20^\circ}) = 1.0946\underline{/11.59^\circ} = 1.0723 + j0.220$$

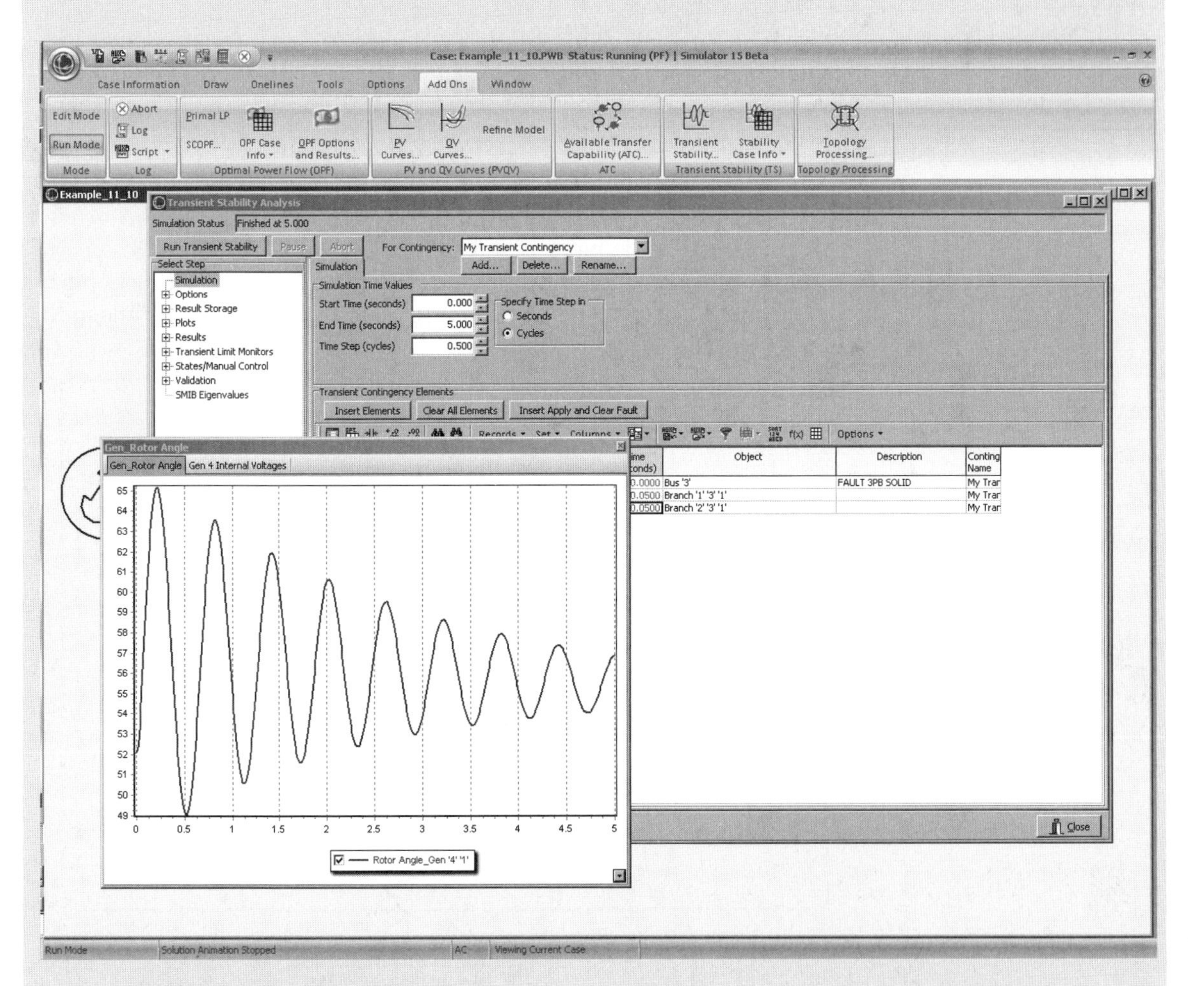

FIGURE 11.20 Variation in generator 4 rotor angle with a fault clearing time of 0.05 seconds

From (11.6.8),

$$\bar{E} = 1.0946\underline{/11.59^\circ} + (j2.0)(1.052\underline{/-18.2^\circ}) = 2.814\angle 52.1^\circ$$
$$\rightarrow \delta = 52.1^\circ$$

Using (11.6.2) gives

$$\begin{bmatrix} V_d \\ V_q \end{bmatrix} = \begin{bmatrix} 0.7889 & -0.6146 \\ 0.6146 & 0.7889 \end{bmatrix} \begin{bmatrix} 1.0723 \\ 0.220 \end{bmatrix} = \begin{bmatrix} 0.7107 \\ 0.8326 \end{bmatrix}$$

and

$$\begin{bmatrix} I_d \\ I_q \end{bmatrix} = \begin{bmatrix} 0.7889 & -0.6146 \\ 0.6146 & 0.7889 \end{bmatrix} \begin{bmatrix} 1.000 \\ -0.3287 \end{bmatrix} = \begin{bmatrix} 0.9909 \\ 0.3553 \end{bmatrix}$$

Then, solving (11.6.3), (11.6.4) and (11.6.5) gives

$$E_q' = 0.8326 + (0.3)(0.9909) = 1.1299$$
$$E_d' = 0.7107 - (0.5)(0.3553) = 0.5330$$
$$E_{fd} = 1.1299 + (2.1 - 0.3)(0.9909) = 2.9135$$

b. Open PowerWorld Simulator case Example 11_10 (see Figure 11.20). **Select Add-Ons, Transient Stability** to view the Transient Stability Analysis Form. Initially the bus 3 fault is set to clear at 0.05 seconds. Select **Run Transient Stability** to create the results shown in Figure 11.20. In comparing these results with those from Example 11.4, notice that while the initial

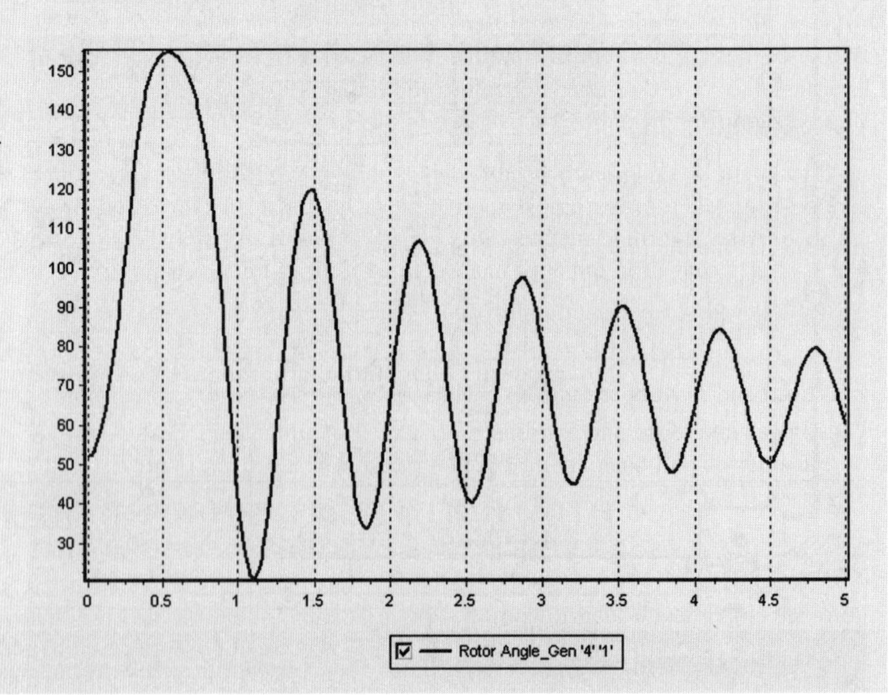

FIGURE 11.21 Variation in generator 4 rotor angle with a fault clearing time of 0.30 seconds

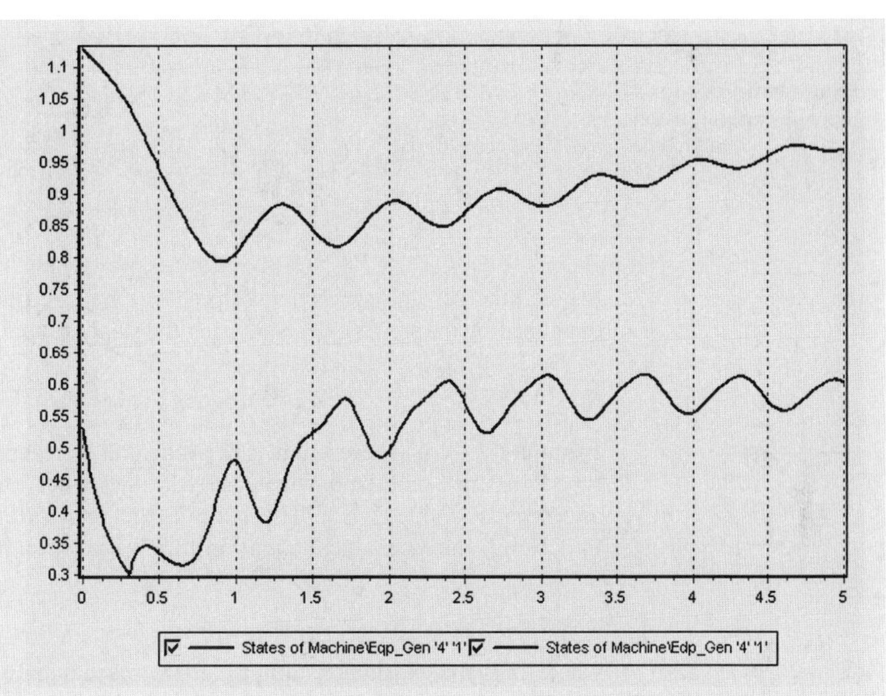

FIGURE 11.22 Variation in generator 4 E_q' and E_d' with a fault clearing time of 0.30 seconds

value of δ is different, the initial angle increase of about 13° is similar to the increase of 16° in Example 11.4. A key difference between the two is the substantial amount of damping in this case. This damping arises because of the explicit modeling of the field and damper windings with the two-axis model. The critical clearing time can be determined by gradually increasing the clearing time until the generator loses synchronism. This occurs at about 0.30 seconds, with the almost critically cleared angle shown in Figure 11.21. Since there are now two additional state variables for generator 4, E_q' and E_d', their values can also be shown. This is done in Figure 11.22, again for the 0.30 second clearing time. ■

11.7 WIND TURBINE MACHINE MODELS

As wind energy continues its rapid growth, wind turbine models need to be included in transient stability analysis. As was introduced in Chapter 6, there are four main types of wind turbines that must be considered. Model types 1 and 2 are based on an induction machine models. As is the case with a synchronous machine, the stator windings of the induction machine are connected to the rest of the electric network. However, rather than having a dc field winding on the rotor, the ac rotor currents are induced by the relative motion between the rotating magnetic field setup by the stator currents, and the rotor. Usually the difference

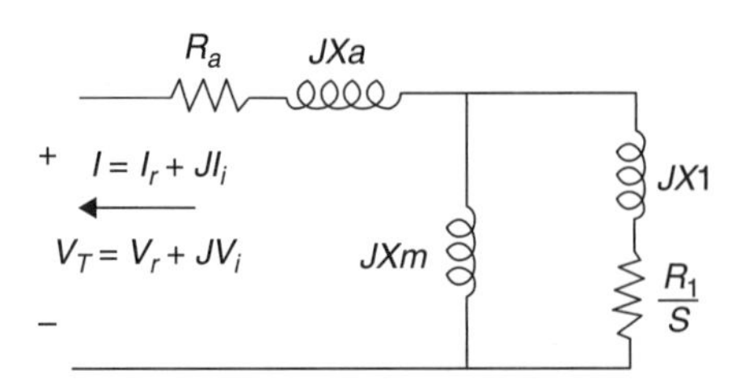

FIGURE 11.23 Equivalent circuit for a single cage induction machine

between the per unit synchronous speed, n_s, and the per unit rotor speed, n_r, is quantified by the slip (S), defined (using the standard motor convention), as

$$S = \frac{n_s - n_r}{n_s} \tag{11.7.1}$$

From (11.7.1), it is clear that if the machine were operating at synchronous speed, its slip would be 0, with positive values when it is operating as a motor and negative values when it is operating as a generator. Expressing all values in per unit, the mechanical equation for an induction machine is

$$\frac{dS}{dt} = \frac{1}{2H}(T_m - T_e) \tag{11.7.2}$$

where H is the inertia constant, T_m is the mechanical torque, and T_e the electrical torque, defined in (11.7.10).

The simplified electric circuit for a single-cage induction machine is shown in Figure 11.23, using the generator convention in which current out of the machine is assumed to be positive. Similar to what is done for synchronous machines, an induction machine can be modeled as an equivalent voltage behind the stator resistance and a transient reactance X'. Referring to Figure 11.23, the values used in this representation are

$$X' = X_a + \frac{X_1 X_m}{X_1 + X_m} \tag{11.7.3}$$

where X' is the apparent reactance seen when the rotor is locked (i.e., slip is 1),

$$X = X_a + X_m \tag{11.7.4}$$

X is the synchronous reactance, and

$$T'_o = \frac{(X_1 + X_m)}{\omega_o R_1} \tag{11.7.5}$$

is the open-circuit time constant for the rotor. Also, X_a is commonly called the leakage reactance.

Electrically the induction machine is modeled using two algebraic and two differential equations. However, in contrast to synchronous machines, because the reactances of induction machines do not depend upon the rotor position values, they are specified in the network reference frame. The equations are

$$V_r = E'_r - R_a I_r + X' I_i \tag{11.7.6}$$

$$V_i = E_i' - R_a I_i - X' I_r \tag{11.7.7}$$

$$\frac{dE_r'}{dt} = \omega_o S E_i' - \frac{1}{T_o'}\left((E_r' - (X - X')I_i\right) \tag{11.7.8}$$

$$\frac{dE_i'}{dt} = -\omega_o S E_r' - \frac{1}{T_o'}\left((E_i' + (X - X')I_r\right) \tag{11.7.9}$$

The induction machine electric torque is then given by

$$T_e = (E_r' I_r + E_i' I_i)/\omega_o \tag{11.7.10}$$

and the terminal real power injection by

$$P_e = (V_r I_r + V_i I_i) \tag{11.7.11}$$

The transient stability initial conditions are determined by setting (11.7.8) and (11.7.9) to zero, and then using the power flow real power injection and terminal voltage as inputs to solve (11.7.6), (11.7.7), (11.7.8), (11.7.9) and (11.7.11) for the other variables. The Newton-Raphson approach (Section 6.3) is commonly used. Since the induction machine reactive power injection will not normally match the power flow value, the difference is modeled by including a shunt capacitor whose susceptance is determined to match the initial power flow conditions. The reactive power produced by the machine is given by

$$Q_e = (-V_r I_i + V_i I_r) \tag{11.7.12}$$

with the value negative since induction machines consume reactive power.

EXAMPLE 11.11 Induction Generator Example

For the system from Example 11.3, assume the synchronous generator is replaced with an induction generator and shunt capacitor in order to represent a wind farm with the same initial real and reactive power output as in Example 11.3. The induction generator parameters are $H = 0.9$ per unit-seconds, $R_a = 0.013, X_a = 0.067, X_m = 3.8, R_1 = 0.0124, X_1 = 0.17$ (all per unit using the 100 MVA system base). This system is modeled in PowerWorld Simulator case Example 11_11 (see Figure 11.24). (a) Use the previous equations to verify the initial conditions of $S = -0.0111, E_r' = 0.9314, E_i' = 0.4117, I_r = 0.7974, I_i = 0.6586$. (b) Plot the terminal voltage for the fault sequence from Example 11.6.

SOLUTION

a. Using (11.7.3) to (11.7.5) the values of X', X, and T_o' are determined to be 0.2297, 3.867, per unit and 0.85 seconds respectively. With

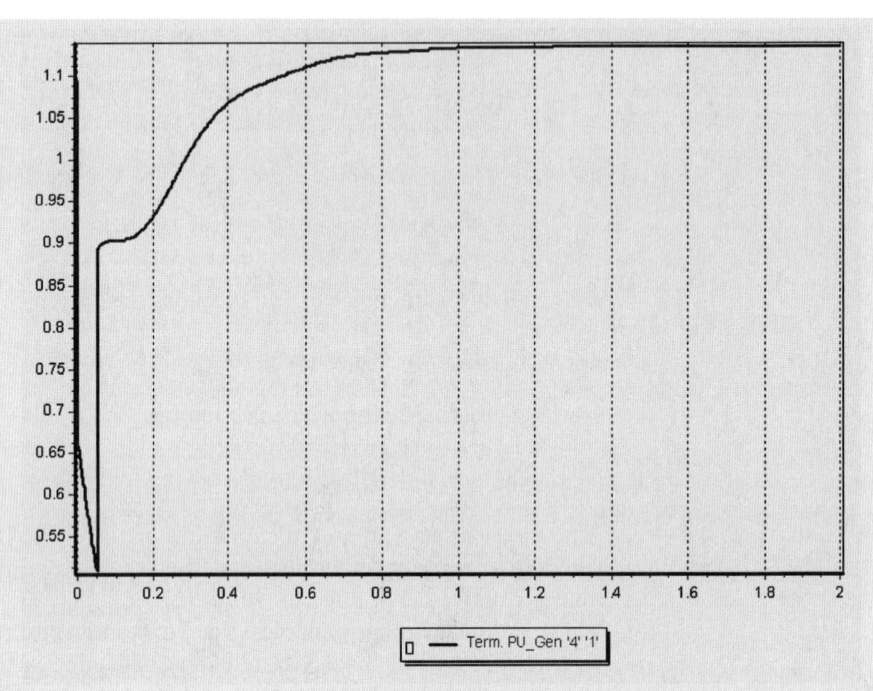

FIGURE 11.24 Example 11.11 Generator 4 voltage magnitude for a fault clearing time of 0.05 seconds

$V_T = 1.0723 + j0.220$ and $P_e = 1.0$ per unit, we can verify (11.7.6) and (11.7.7) as

$$V_r = 0.9314 - (0.013)(0.7974) + (0.2297(0.6586) = 1.0723$$
$$V_i = 0.4117 - (0.013)(0.6586) - (0.2297(0.7974) = 0.2200$$

And (11.7.8), (11.7.9) as

$$\frac{dE'_r}{dt} = 2\pi 60(-0.0111)(0.4117) - \frac{1}{0.85}(0.9314 - (3.637)(0.6586)) = 0.0$$
$$\frac{dE'_i}{dt} = -2\pi 60(-0.0111)(0.9314) - \frac{1}{0.85}(0.4117 + (3.637)(0.7974)) = 0.0$$
$$P_e = (1.0723)(0.7974) + (0.220)(0.6586) = 1.000$$
$$Q_e = -(1.0723)(0.6586) + (0.220)(0.7974) = -0.531$$

You can see the initial values in PowerWorld Simulator by first displaying the **States/Manual Control** page of the Transient Stability Analysis form, which initializes the transient stability. Then from the one-line diagram view the Generator Information Dialog for the generator at bus 4, and select **Stability, Terminal and State, Terminal Values**. Because the generator in the power flow is producing 57.2 Mvar, and the induction machine is consuming 53.1 Mvar, a shunt capacitor that produces 110.3 Mvar with a 1.0946 terminal voltage must be modeled.

b. The following figure plots the terminal voltage for the three cycle fault. ■

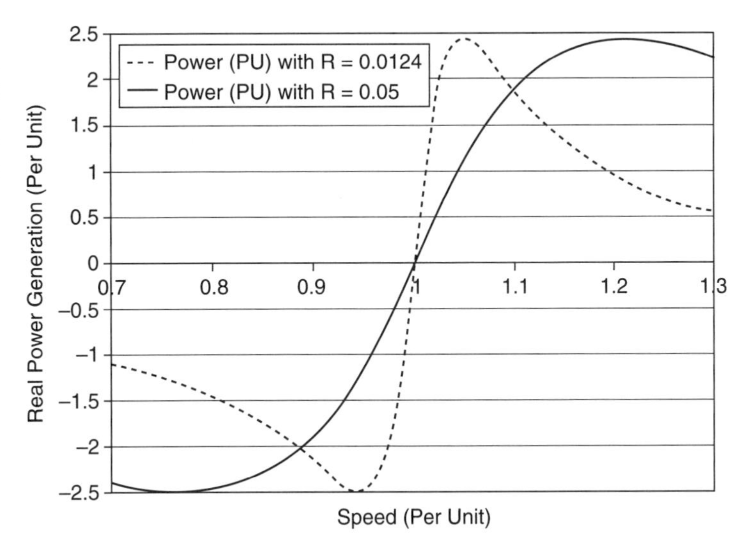

FIGURE 11.25

Affect of varying external resistance on an induction machine torque-speed curve

Both the Type 1 and 2 wind turbine models utilize induction generators, but whereas the Type 1 models have a conventional squirrel cage rotor with fixed rotor resistance, the Type 2 models are wound rotor induction machines that utilize a control system to vary the rotor resistance. The reason for this is to provide a more steady power output from the wind turbine during wind variation. From (11.7.5) it is clear that increasing this external resistance has the affect of decreasing the open circuit time constant. The inputs to the rotor resistance control system are turbine speed and electrical power output, while the output is the external resistance that is in series with R_1 from Figure 11.22. Figure 11.25 plots the variation in the real power output for the Example 11.11 generator as a function of speed for the original rotor resistance of 0.0124 and for a total rotor resistance of 0.05 per unit. With a total resistance of 0.05, the operating point slip changes to about −0.045, which corresponds to a per unit speed of 1.045.

Most new wind turbines are either Type 3 or Type 4. Type 3 wind turbines are used to represent doubly-fed asynchronous generators (DFAGs), also sometimes referred to as doubly-fed induction generators (DFIGs). A DFAG consists of a traditional wound rotor induction machine, but with the rotor windings connected to the ac network through an ac-dc-ac converter—the machine is "doubly-fed" through both the stator and rotor windings (see Figure 11.26). The advantages of this arrangement are that it allows for separate control of both the real and reactive power (like a synchronous machine), and the ability to transfer power both ways through the rotor converter allows for a much wider speed range. Because the stator is directly connected to the ac grid, the rotor circuit converter need only be sized to about 30% of the machines rated capacity. Another consequence of this design is the absence of an electrical coupling with the mechanical equation such as was seen with (11.1.10) for the synchronous machines and in (11.7.2) for the Type 1 and 2 induction machine models.

From a transient stability perspective the DFAG dynamics are driven by the converter, with the result that the machine can be well approximated as a

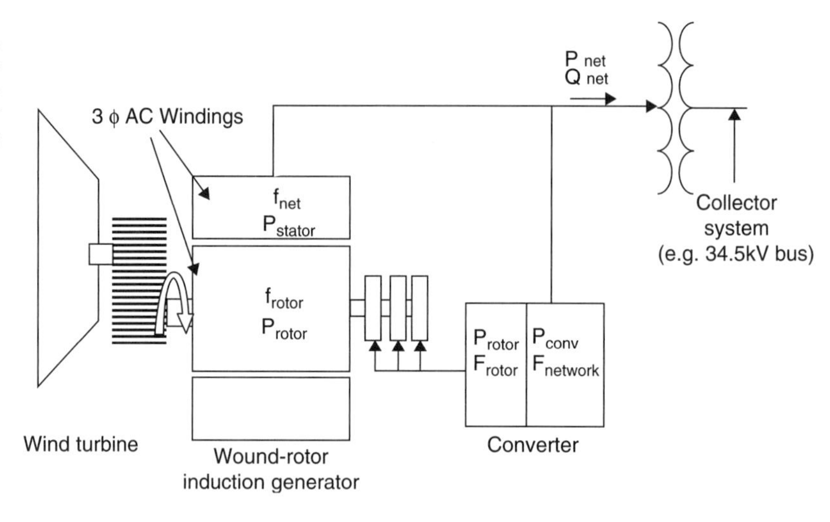

FIGURE 11.26 Doubly-fed asynchronous generator components

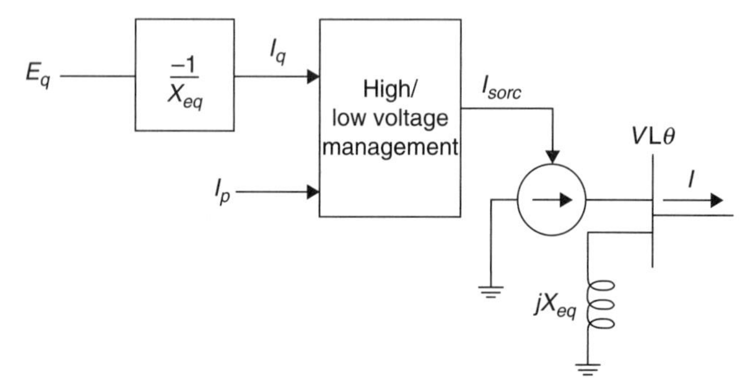

FIGURE 11.27 Type 3 DFAG model circuit diagram

voltage-source converter (VSC). A VSC can be modeled as a synthesized current injection in parallel with an effective reactance, X_{eq}, (Figure 11.27) in which the current in phase with the terminal voltage, I_p, and the reactive power current, I_q, can be controlled independently. Low and high voltage current management is used to limit these values during system disturbances. With a terminal voltage angle of θ, the current injection on the network reference is

$$I_{sorc} = \left(I_p + jI_q\right)(1\angle\theta) \tag{11.7.13}$$

And the reactive voltage is

$$E_q = -I_q X_{eq} \tag{11.7.14}$$

Type 4 wind turbines utilize a completely asynchronous design in which the full output of the machine is connected to the ac network through an ac-dc-ac converter (see Figure 11.28). Because the converter completely decouples the electric generator from the rest of the network, there is considerable freedom in selecting the electric machine type. For example, a conventional synchronous generator, a permanent magnet synchronous generator, or even a squirrel cage induction machine.

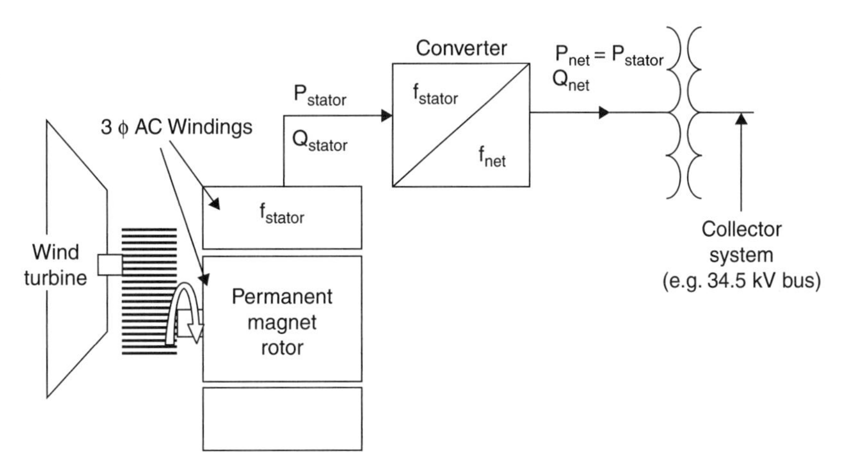

FIGURE 11.28 Type 4 full converter components

From a transient stability perspective, electrically, the Type 4 model is similar to the Type 3 in that it can also be represented as a VSC. The key difference is lack of the effective reactance, with Ip and Iq being the direct control variables for the Type 4 model. As is the case with a DFAG, there is no electrical coupling with the turbine dynamics.

EXAMPLE 11.12 Doubly-Fed Asynchronous Generator Example

For the system from Example 11.3, assume the synchronous generator is replaced with a Type 3 DFAG generator in order to represent a wind farm with the initial current into the infinite bus set to 1.0 (unity power factor). The DFAG reactance $X_{eq} = 0.8$ per unit using a 100 MVA system base. Determine the initial values for I_p, I_q, and E_q.

SOLUTION With $I = 1.0$ and an impedance of $j0.12$ between the machine's terminal and the infinite bus, the terminal voltage is

$$V_T = 1.0 + (1.0)(j0.22) = 1.0 + j0.22 = 1.0239\underline{/12.41^\circ}$$

The amount supplied by I_{sorc} is I plus the amount modeled as going into X_{eq}

$$I_{sorc} = I - \frac{V_T}{jX_{eq}} = 1.00 + \frac{1.0 + j0.220}{j0.8}$$

$$I_{sorc} = 1.275 - j1.25$$

The values of I_p and I_q are then calculated by shifting these values backwards by the angle of the terminal voltage

$$I_p + jI_q = (1.275 - j1.25) * (1\underline{/12.41^\circ}) = 0.977 - j1.495$$

And then

$$E_q = -(-1.495)(0.8) = 1.196$$

You can see the initial values in PowerWorld Simulator by first displaying the **States/Manual Control** page of the Transient Stability Analysis form, which initializes the transient stability. Then from the one-line diagram view the Generator Information Dialog for the generator at bus 4, and select **Stability, Terminal and State, Terminal Values.** ■

11.8 DESIGN METHODS FOR IMPROVING TRANSIENT STABILITY

Design methods for improving power system transient stability include the following:

1. Improved steady-state stability
   a. Higher system voltage levels
   b. Additional transmission lines
   c. Smaller transmission-line series reactances
   d. Smaller transformer leakage reactances
   e. Series capacitive transmission-line compensation
   f. Static var compensators and flexible ac transmission systems (FACTS)
2. High-speed fault clearing
3. High-speed reclosure of circuit breakers
4. Single-pole switching
5. Larger machine inertia, lower transient rectance
6. Fast responding, high-gain exciters
7. Fast valving
8. Braking resistors

We discuss these design methods in the following paragraphs.

1. Increasing the maximum power transfer in steady-state can also improve transient stability, allowing for increased power transfer through the unfaulted portion of a network during disturbances. Upgrading voltage on existing transmission or opting for higher voltages on new transmission increases line loadability (5.5.6). Additional parallel lines increase power-transfer capability. Reducing system reactances also increases power-transfer capability. Lines with bundled phase conductors have lower series reactances than lines that are not bundled. Oversized transformers with lower leakage reactances also help. Series capacitors reduce the total series reactances of a line by compensating for the series line inductance.

The case study for Chapter 5 discusses FACTS technologies to improve line loadability and maintain stability.

2. High-speed fault clearing is fundamental to transient stability. Standard practice for EHV systems is 1-cycle relaying and 2-cycle circuit breakers, allowing for fault clearing within 3 cycles (0.05 s). Ongoing research is presently aimed at reducing these to one-half cycle relaying and 1-cycle circuit breakers.

3. The majority of transmission-line short circuits are temporary, with the fault arc self-extinguishing within 5–40 cycles (depending on system voltage) after the line is deenergized. High-speed reclosure of circuit breakers can increase postfault transfer power, thereby improving transient stability. Conservative practice for EHV systems is to employ high-speed reclosure only if stability is maintained when reclosing into a permanent fault with subsequent reopening and lockout of breakers.

4. Since the majority of short circuits are single line-to-ground, relaying schemes and independent-pole circuit breakers can be used to clear a faulted phase while keeping the unfaulted phases of a line operating, thereby maintaining some power transfer across the faulted line. Studies have shown that single line-to-ground faults are self-clearing even when only the faulted phase is deenergized. Capacitive coupling between the energized unfaulted phases and the deenergized faulted phase is, in most cases, not strong enough to maintain an arcing short circuit [5].

5. Inspection of the swing equation, (11.1.16), shows that increasing the per-unit inertia constant H of a synchronous machine reduces angular acceleration, thereby slowing down angular swings and increasing critical clearing times. Stability is also improved by reducing machine transient reactances, which increases power-transfer capability during fault or postfault periods [see (11.2.1)]. Unfortunately, present-day generator manufacturing trends are toward lower H constants and higher machine reactances, which are a detriment to stability.

6. Modern machine excitation systems with fast thyristor controls and high amplifier gains (to overcome generator saturation) can rapidly increase generator field excitation after sensing low terminal voltage during faults. The effect is to rapidly increase internal machine voltages during faults, thereby increasing generator output power during fault and postfault periods. Critical clearing times are also increased [6].

7. Some steam turbines are equipped with fast valving to divert steam flows and rapidly reduce turbine mechanical power outputs. During faults near the generator, when electrical power output is reduced, fast valving action acts to balance mechanical and electrical power, providing reduced acceleration and longer critical clearing times. The turbines are designed to withstand thermal stresses due to fast valving [7].

8. In power systems with generation areas that can be temporarily separated from load areas, braking resistors can improve stability.

When separation occurs, the braking resistor is inserted into the generation area for a second or two, preventing or slowing acceleration in the generation area. Shelton et al. [8] describe a 3-GW-s braking resistor.

PROBLEMS

SECTION 11.1

11.1 A three-phase, 60-Hz, 500-MVA, 11.8-kV, 4-pole steam turbine-generating unit has an H constant of 6 p.u.-s. Determine: (a) ω_{syn} and $\omega_{m\text{syn}}$; (b) the kinetic energy in joules stored in the rotating masses at synchronous speed; (c) the mechanical angular acceleration α_m and electrical angular acceleration α if the unit is operating at synchronous speed with an accelerating power of 500 MW.

11.2 Calculate J in kg-m^2 for the generating unit given in Problem 11.1.

11.3 Generator manufacturers often use the term WR^2, which is the weight in newtons of all the rotating parts of a generating unit (including the prime mover) multiplied by the square of the radius of gyration in meters. WR^2/9.81 is then the total moment of inertia of the rotating parts in kg-m^2. (a) Determine a formula for the stored kinetic energy in joules of a generating unit in terms of WR^2 and rotor angular velocity ω_m. (b) Show that

$$\text{H} = \frac{5.59 \times 10^{-4}\text{WR}^2(\text{rpm})^2}{\text{S}_{\text{rated}}} \quad \text{per unit-seconds}$$

where S_{rated} is the voltampere rating of the generator, and rpm is the synchronous speed in r/min. (c) Evaluate H for a three-phase generating unit rated 800 MVA, 3600 r/min, with WR^2 = 1,650,000 N-m^2. For conversion factors see the inside front and back covers.

11.4 The generating unit in Problem 11.1 is initially operating at $p_{m\text{p.u.}} = p_{e\text{p.u.}} = 0.7$ per unit, $\omega = \omega_{\text{syn}}$, and $\delta = 12°$ when a fault reduces the generator electrical power output by 60%. Determine the power angle δ five cycles after the fault commences. Assume that the accelerating power remains constant during the fault. Also assume that $\omega_{\text{p.u.}}(t) = 1.0$ in the swing equation.

11.5 How would the value of H change if a generator's assumed operating frequency is changed from 60 Hz to 50 Hz?

11.6 Repeat Example 11.1 except assume the number of poles is changed from 32 to 16, H is changed from 2.0 p.u.-s to 1.5 p.u.-s, and the unit is initially operating with an electrical and mechanical power of 0.5 p.u.

SECTION 11.2

11.7 Given that for a moving mass $W_{\text{kinetic}} = 1/2\ \text{Mv}^2$, how fast would a 80,000 kg diesel locomotive need to go to equal the energy stored in a 60-Hz, 100-MVA, 60 Hz, 2-pole generator spinning at synchronous speed with an H of 3.0 p.u.-s?

11.8 The synchronous generator in Figure 11.4 delivers 0.8 per-unit real power at 1.05 per-unit terminal voltage. Determine: (a) the reactive power output of the generator; (b) the generator internal voltage; and (c) an equation for the electrical power delivered by the generator versus power angle δ.

11.9 The generator in Figure 11.4 is initially operating in the steady-state condition given in Problem 11.8 when a three-phase-to-ground bolted short circuit occurs at bus 3.

Determine an equation for the electrical power delivered by the generator versus power angle δ during the fault.

11.10 For the five bus system from Example 6.9, assume the transmission lines and transformers are modeled with just their per unit reactance (e.g., neglect their resistance and B shunt values). If bus one is assumed to be an infinite bus, what is the equivalent (Thevenin) reactance looking into the system from the bus three terminal? Neglect any impedances associated with the loads.

11.11 Repeat Problem 11.10, except assume there is a three-phase-to-ground bolted short circuit at bus five.

SECTION 11.3

11.12 The generator in Figure 11.4 is initially operating in the steady-state condition given in Example 11.3 when circuit breaker B12 inadvertently opens. Use the equal-area criterion to calculate the maximum value of the generator power angle δ. Assume $\omega_{\text{p.u.}}(t) = 1.0$ in the swing equation.

11.13 The generator in Figure 11.4 is initially operating in the steady-state condition given in Example 11.3 when a temporary three-phase-to-ground short circuit occurs at point F. Three cycles later, circuit breakers B13 and B22 permanently open to clear the fault. Use the equal-area criterion to determine the maximum value of the power angle δ.

11.14 If breakers B13 and B22 in Problem 11.13 open later than 3 cycles after the fault commences, determine the critical clearing time.

11.15 Building upon Problem 11.11, assume a 60 Hz nominal system frequency, that the bus fault actually occurs on the line between buses five and two but at the bus two end, and that the fault is cleared by opening breakers B21 and B52. Again, neglecting the loads, assume that the generator at bus three is modeled with the classical generator model having a per unit value (on its 800 MVA base) of X_d' 0.24, and H = 3 p.u.-s. Before the fault occurs the generator is delivering 300 MW into the infinite bus at unity power factor (hence its terminal voltage is not 1.05 as was assumed in Example 6.9). Further, assume the fault is cleared after 3 cycles. Determine: (a) the initial generator one power angle, (b) the power angle when the fault is cleared, and (c) the maximum value of the power angle using the equal area criteria.

11.16 Analytically determine whether there is a critical clearing time for Problem 11.15.

SECTION 11.4

11.17 Consider the first order differential equation, $\dfrac{dx_1}{dt} = -x_2$, with an initial value $x(0) = 10$. With an integration step size of 0.1 seconds, determine the value of $x(0.5)$ using (a) Euler's method, (b) the modified Euler's method.

11.18 The following set of differential equations can be used to represent that behavior of a simple spring-mass system, with x_1(t) the mass's position and x_2(t) its velocity:

$$\frac{dx_1}{dt} = x_2$$

$$\frac{dx_2}{dt} = -x_1$$

For the initial condition of $x_1(0) = 1.0$, $x_2(0) = 0$, and a step size 0.1 seconds, determine the values $x_1(0.3)$ and $x_2(0.3)$ using (a) Euler's method, (b) the modified Euler's method.

11.19 A 60 Hz generator is supplying 400 MW (and 0 Mvar) to an infinite bus (with 1.0 per unit voltage) through two parallel transmission lines. Each transmission line has a per unit impedance (100 MVA base) of $0.09j$. The per unit transient reactance for the generator is $0.0375j$, the per unit inertia constant for the generator (H) is 20 seconds, and damping is 0.1 per unit (all with a 100 MVA base). At time = 0, one of the transmission lines experiences a balanced three phase short to ground one third (1/3) of the way down the line from the generator to the infinite bus. (a) Using the classical generator model, determine the prefault internal voltage magnitude and angle of the generator. (b) Express the system dynamics during the fault as a set of first order differential equations. (c) Using Euler's method, determine the generator internal angle at the end of the second timestep. Use an integration step size of one cycle.

PW **11.20** Open PowerWorld Simulator case Problem 11_20. This case models the Example 11.4 system with damping at the bus 1 generator, and with a line fault midway between buses 1 and 3. The fault is cleared by opening the line. Determine the critical clearing time for this fault.

PW **11.21** Open PowerWorld Simulator case Problem 11_21. This case models the Example 11.4 system with damping at the bus 1 generator, and with a line fault midway between buses 1 and 2. The fault is cleared by opening the line. Determine the critical clearing time (to the nearest 0.01 second) for this fault.

SECTION 11.5

11.22 Consider the six-bus power system shown in Figure 11.29, where all data are given in per-unit on a common system base. All resistances as well as transmission-line capacitances are neglected. (a) Determine the 6×6 per-unit bus admittance matrix $\boldsymbol{Y}_{\text{bus}}$ suitable for a power-flow computer program. (b) Determine the per-unit admittance matrices $\boldsymbol{Y}_{11}$, $\boldsymbol{Y}_{12}$, and $\boldsymbol{Y}_{22}$ given in (11.5.5), which are suitable for a transient stability study.

11.23 Modify the matrices $\boldsymbol{Y}_{11}$, $\boldsymbol{Y}_{12}$, and $\boldsymbol{Y}_{22}$ determined in Problem 11.22 for (a) the case when circuit breakers B32 and B51 open to remove line 3–5; and (b) the case when the load $P_{L3} + jQ_{L3}$ is removed.

PW **11.24** Open PowerWorld Simulator case Problem 11_24, which models the Example 6.9 with transient stability data added for the generators. Determine the critical clearing time (to the nearest 0.01 second) for a fault on the line between buses 2 and 5 at the bus 5 end which is cleared by opening the line.

PW **11.25** With PowerWorld Simulator using the Example 11.9 case determine the critical clearing time (to the closest 0.01 second) for a transmission line fault on the transmission line between bus 44 (LAUF69) and bus 4 (WEBER69), with the fault occurring near bus 44.

SECTION 11.6

PW **11.26** PowerWorld Simulator case Problem 11_26 duplicates Example 11.10, except with the synchronous generator initially supplying 100 MW at unity power factor to the infinite bus. (a) Derive the initial values for δ, E_q', E_d', and E_{fd}. (b) Determine the critical clearing time for the Example 11.10 fault.

PW **11.27** PowerWorld Simulator case Problem 11_27 duplicates the system from Problem 11.24, except the generators are modeled using a two-axis model, with the same X_d'

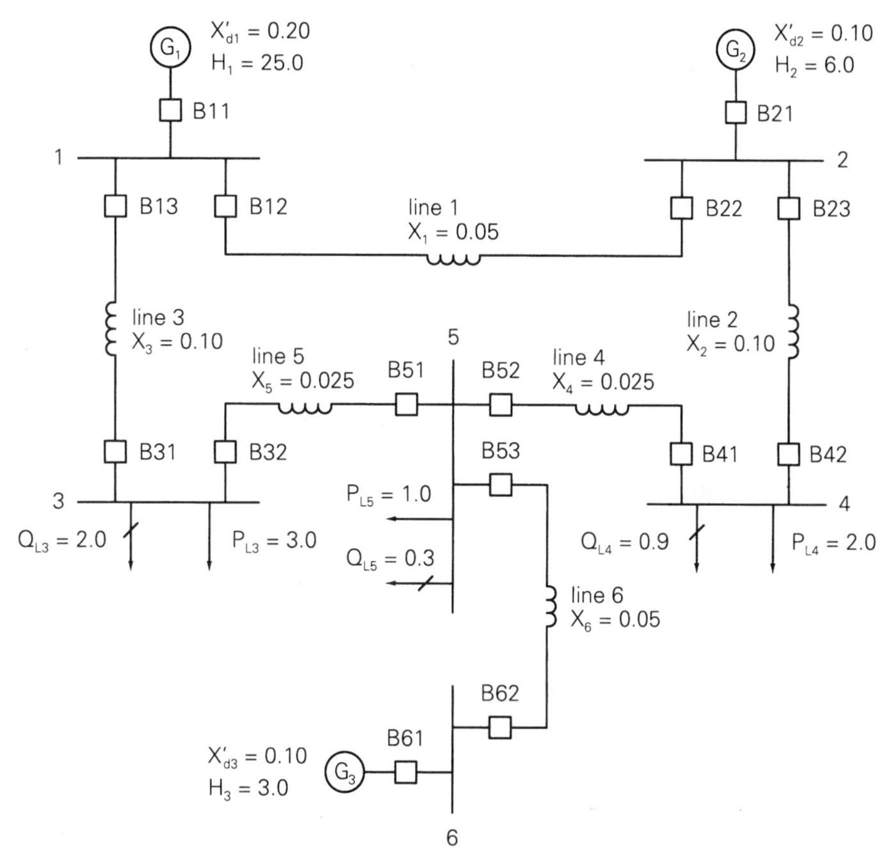

FIGURE 11.29 Single-line diagram of a six-bus power system (per-unit values are shown)

and H parameters are in Problem 11.24. Compare the critical clearing time between this case and the problem 11.24 case.

SECTION 11.7

PW **11.28** PowerWorld Simulator case Problem 11_28 duplicates Example 11.11 except the wind turbine generator is set so it is initially supplying 100 MW to the infinite bus at unity power factor. (a) Use the induction machine equations to verify the initial conditions of $S = -0.0129, E_r' = 0.8475, E_i' = 0.4230, I_r = 0.8433, I_i = 0.7119$. (b) Plot the terminal voltage for the fault sequence from Example 11.6.

11.29 Redo Example 11.12 with the assumption the generator is supplying $100 + j22$ MVA to the infinite bus.

CASE STUDY QUESTIONS

A. How is dynamic security assessment (DSA) software used in actual power system operations?

B. What techniques are used to decrease the time required to solve the DSA problem?

REFERENCES

1. G. W. Stagg and A. H. El-Abiad, *Computer Methods in Power Systems* (New York: McGraw-Hill, 1968).
2. O. I. Elgerd, *Electric Energy Systems Theory*, 2d ed. (New York: McGraw-Hill, 1982).
3. C. A. Gross, *Power System Analysis* (New York: Wiley, 1979).
4. W. D. Stevenson, Jr., *Elements of Power System Analysis*, 4th ed. (New York: McGraw-Hill, 1982).
5. E. W. Kimbark, "Suppression of Ground-Fault Arcs on Single-Pole Switched EHV Lines by Shunt Reactors," *IEEE Trans PAS, 83* (March 1964), pp. 285–290.
6. K. R. McClymont et al., "Experience with High-Speed Rectifier Excitation Systems," *IEEE Trans PAS*, vol. PAS-87 (June 1986), pp. 1464–1470.
7. E. W. Cushing et al., "Fast Valving as an Aid to Power System Transient Stability and Prompt Resynchronization and Rapid Reload after Full Load Rejection," *IEEE Trans PAS*, vol. PAS-90 (November/December 1971), pp. 2517–2527.
8. M. L. Shelton et al., "Bonneville Power Administration 1400 MW Braking Resistor," *IEEE Trans PAS*, vol. PAS-94 (March/April 1975), pp. 602–611.
9. P. W. Saver and M. A. Pai, *Power System Dynamics and Stability* (Prentice Hall, 1997).
10. P. Kundar, *Power System Stability and Control* (McGraw-Hill, 1994).
11. R. Schainker et al., "Real-Time Dynamic Security Assessment", *IEEE Power and Energy Magazine*, 4, 2 (March/April, 2006), pp. 51–58.
12. J. Arrillaga, C.P. Arnold, *Computer Analysis of Power Systems*, John Wiley & Sons Ltd, 1990.
13. K. Clark, N. W. Miller, J. J. Sanchez-Gasca, "Modeling of GE Wind Turbine-Generators for Grid Studies," Version 4.4, GE Energy, Schenectady, NY, September 2009.
14. E.H. Camm, et. al., "Characteristics of Wind Turbine Generators for Wind Power Plants," Proc. IEEE 2009 General Meeting, Calgary, AB, July 2009.

ISO New England's state-of-the-art control center helps to insure the reliable operation of New England's bulk power generation and transmission system (Photograph © Adam Laipson)

12

POWER SYSTEM CONTROLS

Automatic control systems are used extensively in power systems. Local controls are employed at turbine-generator units and at selected voltage-controlled buses. Central controls are employed at area control centers.

Figure 12.1 shows two basic controls of a steam turbine-generator: the voltage regulator and turbine-governor. The voltage regulator adjusts the power output of the generator exciter in order to control the magnitude of generator terminal voltage V_t. When a reference voltage V_{ref} is raised (or lowered), the output voltage V_r of the regulator increases (or decreases) the exciter voltage E_{fd} applied to the generator field winding, which in turn acts to increase (or decrease) V_t. Also a voltage transformer and rectifier monitor V_t, which is used as a feedback signal in the voltage regulator. If V_t decreases,

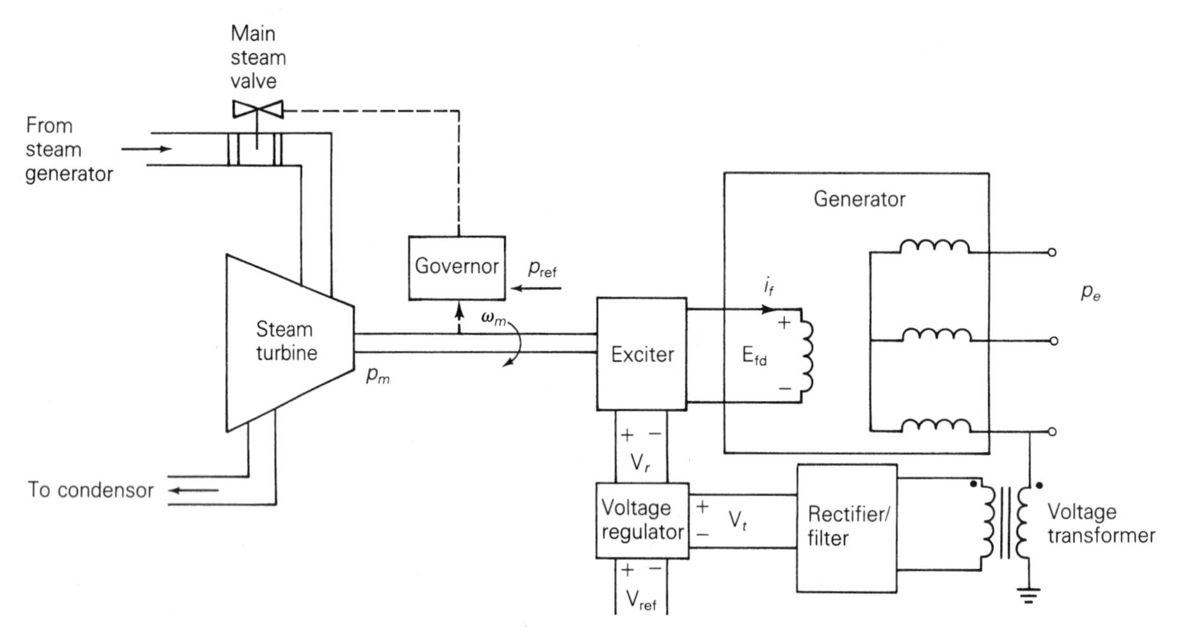

FIGURE 12.1 Voltage regulator and turbine-governor controls for a steam-turbine generator

the voltage regulator increases V_r to increase E_{fd}, which in turn acts to increase V_t.

The turbine-governor shown in Figure 12.1 adjusts the steam valve position to control the mechanical power output p_m of the turbine. When a reference power level p_{ref} is raised (or lowered), the governor moves the steam valve in the open (or close) direction to increase (or decrease) p_m. The governor also monitors rotor speed ω_m, which is used as a feedback signal to control the balance between p_m and the electrical power output p_e of the generator. Neglecting losses, if p_m is greater than p_e, ω_m increases, the governor moves the steam valve in the close direction to reduce p_m. Similarly, if p_m is less than p_e, ω_m decreases, the governor moves the valve in the open direction.

In addition to voltage regulators at generator buses, equipment is used to control voltage magnitudes at other selected buses. Tap-changing transformers, switched capacitor banks, and static var systems can be automatically regulated for rapid voltage control.

Central controls also play an important role in modern power systems. Today's systems are composed of interconnected areas, where each area has its own control center. There are many advantages to interconnections. For example, interconnected areas can share their reserve power to handle anticipated load peaks and unanticipated generator outages. Interconnected areas can also tolerate larger load changes with smaller frequency deviations than an isolated area.

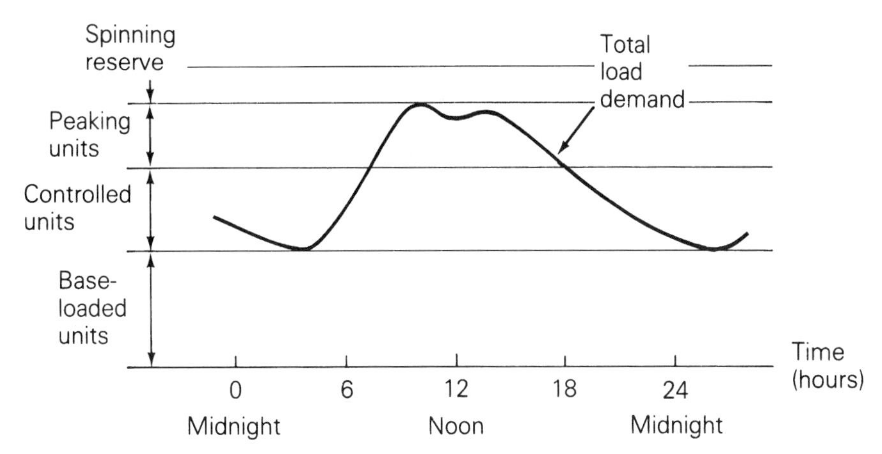

FIGURE 12.2
Daily load cycle

Figure 12.2 shows how a typical area meets its daily load cycle. The base load is carried by base-loaded generators running at 100% of their rating for 24 hours. Nuclear units and large fossil-fuel units are typically base-loaded. The variable part of the load is carried by units that are controlled from the central control center. Medium-sized fossil-fuel units and hydro units are used for control. During peak load hours, smaller, less efficient units such as gas-turbine or diesel-generating units are employed. In addition, generators operating at partial output (with *spinning reserve*) and standby generators provide a reserve margin.

The central control center monitors information including area frequency, generating unit outputs, and tie-line power flows to interconnected areas. This information is used by automatic *load-frequency control* (LFC) in order to maintain area frequency at its scheduled value (60 Hz) and net tie-line power flow out of the area at its scheduled value. Raise and lower reference power signals are dispatched to the turbine-governors of controlled units.

Operating costs vary widely among controlled units. Larger units tend to be more efficient, but the varying cost of different fuels such as coal, oil, and gas is an important factor. *Economic dispatch* determines the megawatt outputs of the controlled units that minimize the total operating cost for a given load demand. Economic dispatch is coordinated with LFC such that reference power signals dispatched to controlled units move the units toward their economic loadings and satisfy LFC objectives. *Optimal power flow* combines economic dispatch with power flow so as to optimize generation without exceeding limits on transmission line loadability.

In this chapter, we investigate automatic controls employed in power systems under normal operation. Sections 12.1 and 12.2 describe the operation of the two generator controls: voltage regulator and turbine-governor. We discuss load-frequency control in Section 12.3, economic dispatch in Section 12.4, and optimal power flow in Section 12.5.

CASE STUDY An important but often overlooked aspect of power system operations is the restoration of the system following a large blackout. During restoration the operating condition of the power system is usually quite different from that seen during normal operation. The following article presents an overview of some of the unique issues that need to be considered during system restoration [15].

Overcoming Restoration Challenges Associated with Major Power System Disturbances

M. M. ADIBI AND L. H. FINK

Recognizing that power system blackouts are likely to occur, it is prudent to consider the necessary measures that reduce their extent, intensity, and duration. Immediately after a major disturbance, the power system's frequency rise and decay are arrested automatically by load rejection, load shedding, controlled separation, and isolation mechanisms. The success rate of these automatic restoration mechanisms has been about 50%! The challenge is to coordinate the control and protective mechanisms with the operation of the generating plants and the electrical system. During the subsequent restoration, plant operators, in coordination with system operators, attempt manually to maintain a balance between load and generation. The duration of these manual procedures has invariably been much longer than equipment limitations can accommodate. Especially in light of the industry's reconfiguration, there is a danger that the operation of power plants and the power system may not maintain the necessary coordination resulting in greater impacts.

Records of major disturbances indicate that the initial system faults have been cleared in milliseconds, and systems have separated into unbalanced load and generation subsystems several seconds later. Blackouts have taken place several minutes after the separations, and the power systems have been restored several hours after the blackouts. Most of these power outages have been of extended duration. For instance, as found in a review covering 24 recent power failures, seven have lasted more than 6 h. The U.S.-Canada report on the 14 August 2003 blackout cites seven major power disturbances with the greatest impact, lasting between 10–50 h, as shown in Figure 1. These failures clearly indicate a need for renewed emphasis on developing restoration methodologies and implementation plans.

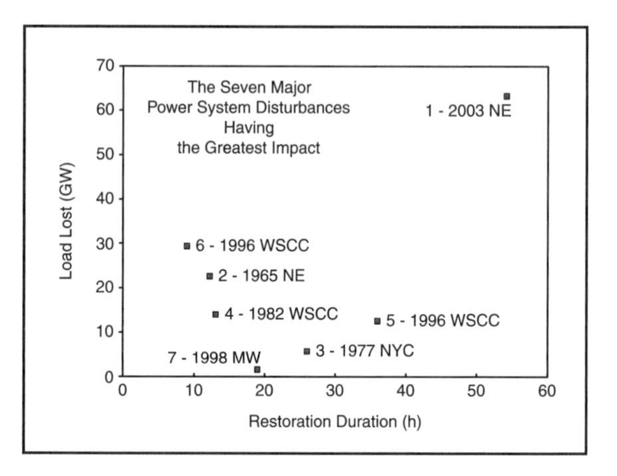

Figure 1
The seven disturbances with the greatest impact

RESTORATION ISSUES

A study of annual system disturbances reported by the North American Electric Reliability Council (NERC) over a ten-year period shows 117 power system disturbances that have had one or more restoration problem(s) belonging to a number of functional groups.

In 23 cases, problems were due to reactive power unbalance, involving generator underexcitations, sustained overvoltage, and switched capacitors/reactors. In 11 cases, problems were due to load and generation unbalance, including responses to sudden increases in load, and underfrequency load shedding. In 29 cases, problems were due to inadequate load and generation coordination, including lack of black-start capability, problems with switching operations, line overloads, and control center coordination. In 56 cases, problems were due to monitoring and control inadequacies, including communication, supervisory control and data acquisition (SCADA) system capabilities, computer overloading, display capabilities, simulation tools, and system status determination. In 15 cases, problems were due to protective systems, including interlocking schemes, synchronization and synchrocheck, standing phase angles, and problems with other types of relays as described later. In 20 cases, problems were due to depletion of energy storage, including low-pressure compressed air/gas and discharged batteries. In 41 cases, problems were due to system restoration plan inadequacies, including lack of planned procedure, outdated procedure, procedure not being followed, inadequate training, and, incredibly, lack of standard communication vocabulary.

Certainly, this summary does not include all the restoration problems encountered. The more common and significant problems are briefly described in this article.

RESTORATION PLANNING

Most operating companies maintain restoration plans based on their restoration objectives, operating philosophies and practices, and familiarity with the characteristics of their power plant restart capabilities and power system reintegration peculiarities. While these plans have successfully restored power systems in the past, they can be improved significantly by simulating steady-state, transient, and dynamic behavior of the power system under various restoration operating conditions and by employing engineering and operating judgment reflecting many factors not readily modeled.

Most power systems have certain characteristics in common and behave in a similar manner during the restoration process. It is therefore possible to establish a general procedure and guidelines to enhance rapid restoration. A detailed plan, however, must be developed specifically to meet the particular requirements of an individual power system. Once a plan has been developed and tested (by simulation and training drills), an online restoration guidance program capable of guiding the operator in making decisions on what steps to take and when to take them goes a long way toward minimizing the duration of blackout and, consequently, the impact of the blackout.

Figure 2 shows a general procedure comprising three temporal restoration stages. The basic distinction between the first stage and the succeeding stages is that, during the first stage, time is critical and many urgent actions must be taken quickly. The basic distinction between the two initial stages and the third stage is that, in the initial stages, blocks of load are control means to maintain stability, whereas in the third stage, the restoration of load is the primary objective.

In the first stage, the postdisturbance system status is evaluated, a "target system" for restoration

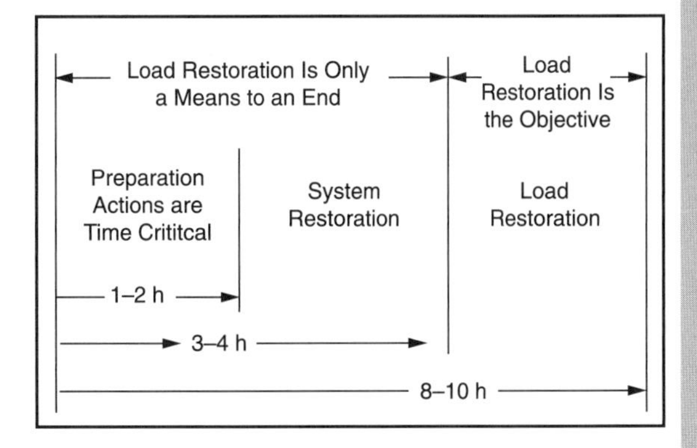

Figure 2
Typical restoration stages

TABLE 1 Initial sources of power and critical loads

	Minutes	Success Probability
Availability of Initial Sources		
Run-of-the-river hydro	5–10	High
Pump-storage hydro	5–10	High
Combustion turbine (CT)	5–15	1 in 2 or 3 CTs
Full or partial load rejection	Short	G T 50%
Low-frequency isolation scheme	Short	G T 50%
Controlled islanding	Short	Special cases
Tie-line with adjacent systems	Short	Not relied on*
Critical Loads		**Priorities**
Cranking drum-type units		High
Pipe-type cables pumping system		High
Transmission stations		Medium
Distribution stations		Medium
Industrial loads		Low**

*Policy: Provide remote cranking power
**Used in the initial stage to an advantage

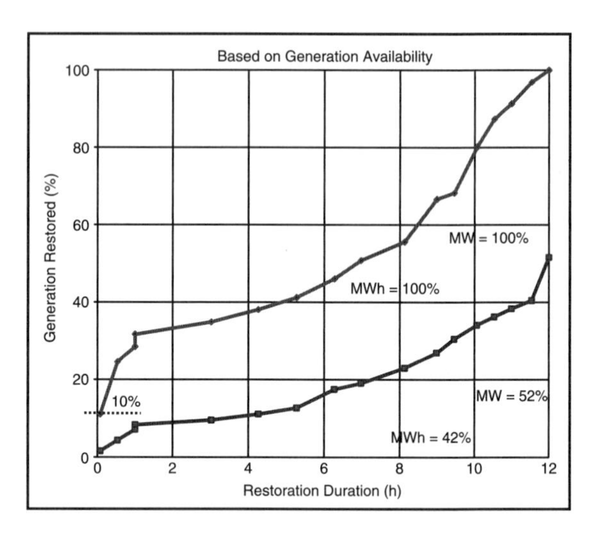

Figure 3
Significance of initial source

is defined, a strategy for rebuilding the transmission network is selected, the system is sectionalized into a few subsystems, and steps are taken to supply the critical loads with the initial sources of power that are available in each subsystem. The postrestoration "target system" will be more or less like the predisturbance system, depending on the severity of the disturbance, but it is important that it be clearly defined in advance to avoid missteps causing the system to go "back to square one." Table 1 lists the types of initial sources of power that may be available and the critical loads. The effect of having an initial source of power, both on the duration of the restoration and on the minimization of the unserved load, is illustrated in Figure 3. In this particular case, the choices for the initial source of power were installing combustion turbines, providing a low-frequency isolation mechanism, or equipping the base-loaded unit with full-load rejection capability. The full-load rejection alternative was selected as providing the best balance between cost and reliability.

In the second stage, the overall goal is reintegration of the bulk power network, as a means of achieving the goals of load restoration in the third stage. To this end, skeleton transmission paths are energized, subsystems defined in the first stage are resynchronized, and sufficient load is restored to stabilize generation and voltage. Larger, base-load units are prepared for restart. Such tasks as energizing higher voltage lines and synchronizing subsystems require either reliable guidelines that have been prepared in advance or tools for analysis prior to critical switching actions.

In the third stage, the primary objective of restoration is to minimize the unserved load, and the scheduling of load pickup will be based on response rate capabilities of available generators. The effective system response rate and the responsive reserve increase with the increase in the number of generators and load restoration can be accomplished in increasingly larger steps.

FREQUENCY CONTROL

Sustained operation of power systems is impossible unless generator frequencies and bus voltages are kept within strict limits. During normal operation, these requirements are met by automatic control loops under operator supervision. During restoration, when individual generators are being brought up to speed and large blocks of load are being reconnected, perturbations outside the range of automatic controls are inevitable;

hence, hands-on control by system operators is necessary.

"System" frequency is the mean frequency of all the machines that are online, and deviations by individual machines must be strictly minimized to avoid mechanical damage to the generator and disruption of the entire system. This is generally accomplished by picking up loads in increments that can be accommodated by the inertia and response of the restored and synchronized generators.

Smaller radial loads should be restored prior to larger loads while maintaining a reasonably constant real-to-reactive power ratio. Feeders equipped with underfrequency relays are picked up at the subsequent phases of restoration when system frequency has stabilized. Common practice in the initial stages of restart and reintegration is to rely on black-start combustion turbines (CT units), low-head short-conduit hydro units (hydro units), and drum-type boiler-turbine units (steam units). Figure 4 shows typical frequency response of these units to a 10% sudden load increase.

During restoration, operators must consider a prime mover's frequency response to a sudden increase in load. Such sudden load increases occur when picking up large network loads or when one of the online generators trips off. Load pickup in small increments tends to prolong the restoration duration, but in picking up large increments, there is always the risk of triggering a frequency decline and causing a recurrence of the system outage. The allowable size of load pickup depends on the rate of response of prime movers already online, which are likely to be under manual control at this point. Typical response rates are 5, 10, and 15% load for a frequency dip of about 0.5 Hz for steam, CT, and hydro units, respectively, as shown in Figure 5.

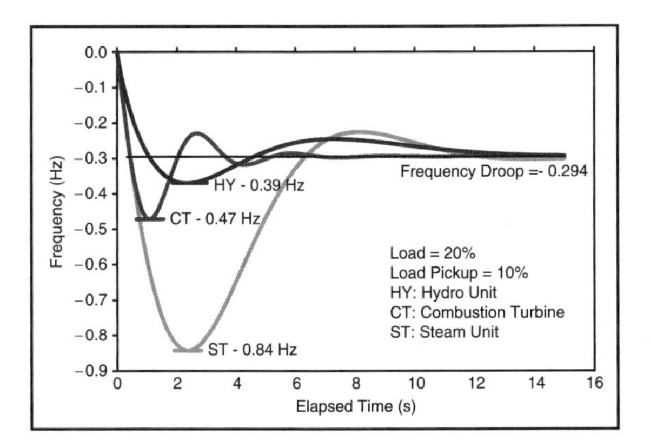

Figure 4
Prime movers' frequency responses

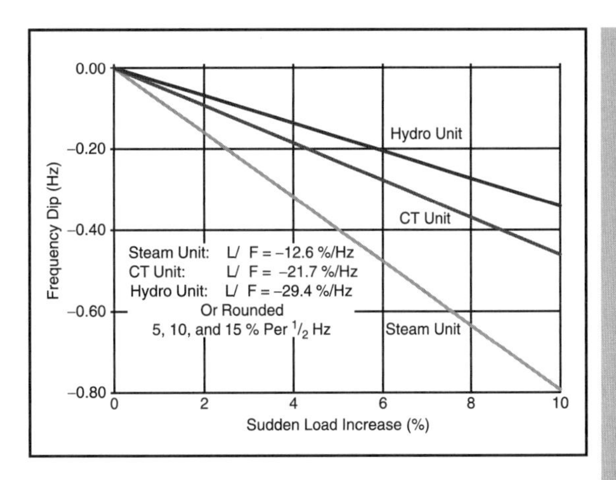

Figure 5
Prime movers' response rates

A set of guidelines for controlling system frequency has been established. These include: 1) black-starting combustion turbines in the automatic mode and at their maximum ramp rate; 2) placing these units under manual mode soon after they are paralleled (for black-start CT units with no automatic mode, these steps become 1) adjusting the governor speed droop to about 2%, and 2) returning the governor speed droop back to 5% soon after it is paralleled); 3) firming generation to meet the largest contingency, i.e., loss of the largest unit; 4) distributing reserve according to the online generators' dynamic response rates; and 5) ensuring that the size of load to be picked up is less than online generators' response rates.

The above guidelines permit the largest load increment that would keep the frequency within acceptable limits, the effective generation reserve that would meet the largest contingency, and the governor speed droop that would improve prime movers' frequency response.

VOLTAGE CONTROL

During the early stages of restoring high-voltage overhead and underground transmission lines, there are concerns with three related overvoltage areas: sustained power frequency overvoltages, switching transients, and harmonic resonance.

Overhead transmission system

Sustained power frequency overvoltages are caused by charging currents of lightly loaded transmission lines. If excessive, these currents may cause generator underexcitation or even self-excitation and instability. Sustained overvoltages also overexcite transformers, generate harmonic distortions, and cause transformer overheating.

Switching transients are caused by energizing large segments of the transmission system or by switching capacitive elements. Such transients are usually highly damped and of short duration. However, in conjunction with sustained overvoltages, they may result in arrester failures. They are not usually a significant factor at transmission voltages below 100 kV. At higher voltages, however, they may become significant because arrester operating voltages are relatively close to normal system voltage, and high-voltage lines are usually long so that energy stored on the lines may be large. In most cases, though, with no sustained traveling wave transients, surge arresters have sufficient energy-absorbing capability to clamp harmful overvoltages to safe levels without sustaining damage. The likely effects of transient overvoltages are determined by the study of special system conditions. Computer-aided analysis has proven to be a valuable tool in understanding switching surge overvoltages.

Harmonic resonance voltages are oscillatory undamped or only weakly damped temporary overvoltages of long duration. They originate from switching operations and nonlinear equipment, reflecting several factors that are characteristic of the networks during restoration. First, the natural frequency of the series-resonant circuit formed by the source inductance and line-charging capacitance may, under normal operating conditions, be a low multiple of 60 Hz. Next, "magnetizing in-rush" caused by energizing a transformer produces many harmonics. Finally, during early stages of restoration, the lines are lightly loaded; resonance therefore is lightly damped, which in turn means the resulting resonance voltages may be very high. If transformers become overexcited due to power frequency overvoltage, harmonic resonance voltages will be sustained or even escalate.

Power transformers, surge arresters, and circuit breakers are the equipment first affected by overvoltages. For transformers, concerns and constraints are with exceeding basic insulation levels (BILs), overexcitation, harmonic generation, and excessive heating. For circuit breakers, concerns are with higher transient recovery voltages, restriking, flashover, and lowering of the interruption capability. For surge arresters, overvoltages cause operation, prevent resealing, and damage the arresters.

Thus, for control of voltages during system restoration, factors to be considered are the length of line to be energized, the size of underlying loads, and the adequacy of online generation (minimum source impedance). In general, it is desirable to energize as large a section of a line as the resulting sustained and transient overvoltages will allow. Energizing small sections tends to prolong the restoration process, but energizing a large section involves a risk of damaging equipment insulation. Energizing lines with inadequate source impedance could result in higher sustained and transient voltages than equipment can withstand. The startup of more out-of-sequence generators, however, would use critical time and delay the overall restoration process. Underlying loads at the receiving or sending end of lines tend to reduce the sustained and transient voltages. In the case of switching transients, operators need to know the minimum load that would avoid transient overvoltages.

In developing restoration guidelines, the above concerns can be addressed by simple analysis and simulation, as shown in Figures 6 and 7. Sending- and receiving-end sustained and transient voltages

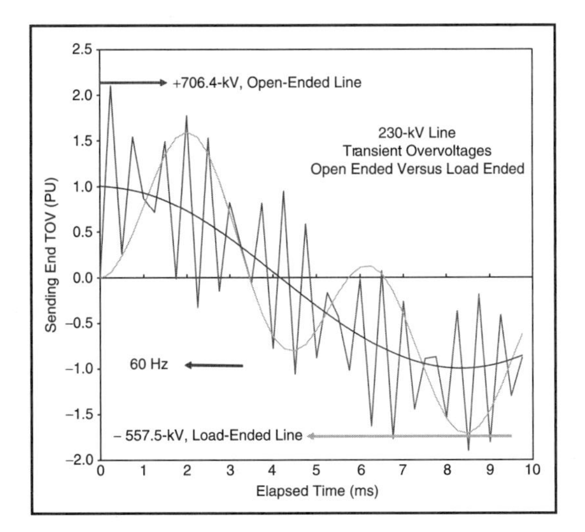

Figure 6
Sending-end transient overvoltages (TOV) of a 230-kV line

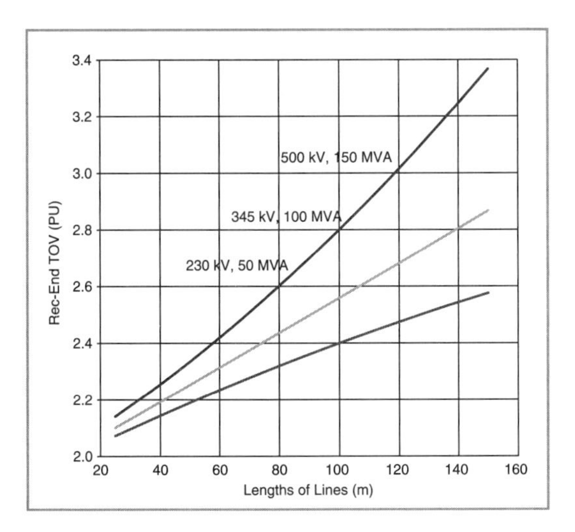

Figure 7
Sending-end transient overvoltages (TOV) of open-ended lines

can be determined for energizing lines of different voltage levels and lengths, energized by different sizes of generators, and with different sizes of line-end cold loads. These results can be used to provide qualitative guidelines to assist operators in energizing high (230 kV) and extrahigh (345 and 500 kV) voltage lines, as shown in Figure 7.

Underground transmission systems

Over 90% of underground transmission lines are of high-pressure oil-filled (HPOF) pipe-type cable, with voltages ranging from 69–500 kV. The primary concern with such cables is the integrity of the insulation. After a blackout, power supply to the pumping plants that maintain the oil pressure is lost. As the cable cools, dissolved gases are liberated, forming gas pockets in the insulation. Reenergization of the cable could then result in immediate failure of pothead terminators. Hence, the pumping plant is a critical load of very high priority.

Another concern during cable reenergization is the ability of the energizing system to absorb the cable's charging current. It should be noted that: 1) cables are loaded at well below their surge impedance loading, 2) the Mvar charging currents per mile (or per km) of a cable is about ten times that of an overhead line of the same voltage class, and 3) about 2 MW of load is picked up per one MVAr of charging.

GENERATOR REACTIVE CAPABILITY

Transformer tap selection

The generator reactive capability (GRC) curves furnished by manufacturers and used in operation planning typically have a greater range than can be realized during actual operation. Generally, these manufacturers' GRC curves are strictly a function of the synchronous machine design parameters and do not consider plant and system operating conditions as limiting factors. Concern over GRC is warranted by the need for reactive power to provide voltage support for large blocks of power transfer.

Figure 8 shows the rated and actual reactive power capability limits for a 460-MVA generator at 237-kV system bus voltage. The rated limits represent, respectively, the overexcitation limit due to rotor overheating, the underexcitation limit due to the stator core-end overheating (and the minimum excitation limiter relay settings), and the overload limit due to the stator overheating. However, more restrictive operating limits are imposed by the plant

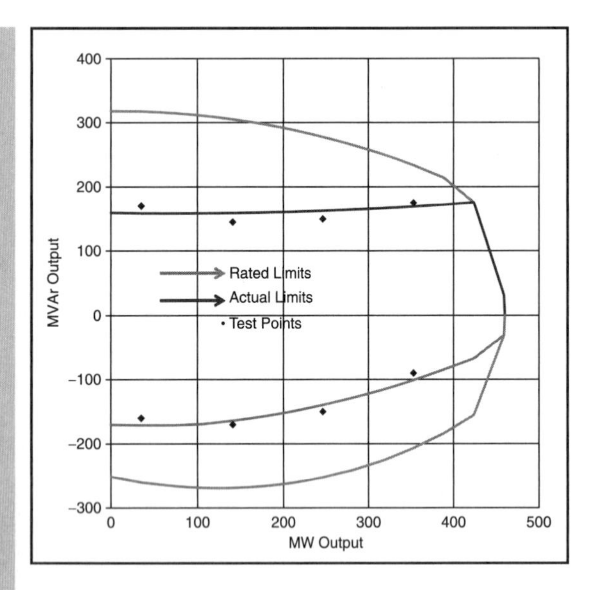

Figure 8
Rated and actual generator Mvar capability

auxiliary bus voltage limits (typically $\pm 5\%$), the generator terminal voltage limits ($\pm 5\%$), and the system bus minimum and maximum voltages during peak and light load conditions.

The high- and low-voltage limits for the auxiliary bus, generator terminal, and system bus are interrelated by the tap positions on the generator step-up (GSU) and auxiliary (AUX) transformers. It should be noted that, in general (particularly in the United States), the GSU and AUX transformers are not equipped with underload tap changers, and therefore these tap positions are very infrequently changed after installation. Consequently, as power system operating conditions change, it is necessary to check these transformer tap positions and ascertain that adequate over- and underexcitation reactive power is available to meet the needs of the power system under both peak and light load conditions.

Remote black-start

Black-start combustion turbines are often considered for remote cranking of steam electric stations under partial or complete power system collapse. Typically, this type of combustion turbine can be started within 5–15 min, which is well within the 30–45 min critical time interval allowed for the hot restart of drum-type boiler-turbine-generators (B-T-Gs).

In planning for black-start of a steam plant, a number of constraints must be considered. Among the more important ones are the "sustained" voltage drop (up to 20%, lasting over 10 s) at the steam plant's large auxiliary motor terminals, the over- and underexcitation limits of the combustion turbine, and the settings of the protective relays installed in the system between the black-start source and the steam plant.

In black-start operation, there are many limiting factors that impose severe demands on the reactive capability of the black-start source. One extreme condition is the initial absorption of the charging currents of the high-voltage (HV) and extra-high-voltage (EHV) lines to the steam plant. Another is supplying the reactive power demand for starting the largest auxiliary motor in the steam plant.

As shown in Figure 9, the 42-MVA CT must be capable of absorbing about 15 Mvar when energizing

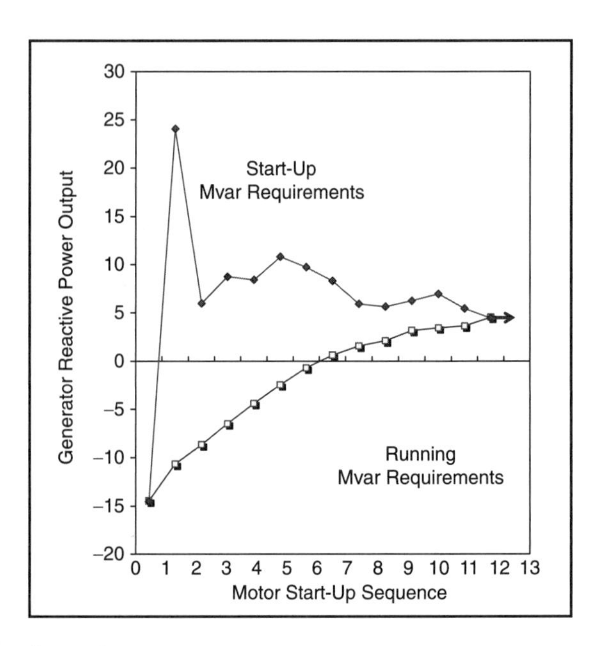

Figure 9
Motor start-up sequence

TABLE 2 Cumulative starting and running reactive loads of a 75-MW steam plant

Sequence of Starting	Horsepower	Starting Mvar*
0–Local load		4.5
1–Induced draft fan	6,000	34.9
2–Forced draft fan	3,000	23.5
3–Circulating water pump	3,000	25.3
4–Primary air fan	2,500	24.6
5–Startup BF pump	2,500	26.2
6–Boiler circulating pump	2,000	25.2
7–Condensate pump	1,500	23.9
8–River water pump	900	21.8
9–Auxilary cooling water pump	700	21.4
10–Coal mill	700	21.8
11–Air compressor	700	22.2
12–Closed cooling water	350	20.9

*Motor's starting reactive power plus previous motors' running reactive powers.

the 230- and 345-kV path between the CT and the steam plant and of supplying about 25 MVAr when stating the 6,000 hp induced draft fan in the 75-MW drum-type boiler-turbine-generator (BTG). These limiting conditions can be met by optimum selections of the GSU and AUX transformer taps and by adjusting the voltage set-point at the cranking source. Table 2 lists, and Figure 9 shows, the cumulative starting and running reactive power requirements of the auxiliary motors in the steam plant.

To determine the set-point and the optimum tap positions for all the transformers in the path between the cranking source and the steam unit's auxiliary bus, a number of analytical tools, including optimal power flows, are used to arrive at an approximate solution.

Nuclear plant requirements

A high restoration priority for utilities having nuclear plants is to provide two independent off-site sources of power within 4 h after an outage to enable controlled reactor shutdown and subsequent restoration to service within 24 h after the scram. Otherwise, the reactor must go through a cooling-down cycle which renders it unavailable for two or more days. Therefore, in areas with significant nuclear power, full power restoration may not be achievable for several days after a blackout.

Nuclear units are typically large—over 600 MW—and usually remotely located. They cannot be provided with the required off-site power sources until the EHV transmission lines are restored. There are a variety of factors, such as adequate reactive absorbing capability, the minimum source requirement, and negative sequence currents, that must be considered before EHV transmission lines are energized. In general, these requirements cannot be met until the third stage of restoration.

PROTECTION SYSTEM ISSUES

Distance, differential, and excitation relays

The performance of protective systems may be measured by the relative percentages of 1) correct and desired relay operations, 2) correct and undesired operations, 3) wrong tripping, and 4) failure to trip. The primary reason for the second and fourth categories is change in the power system topology. During restoration, the power system undergoes continual changes and, therefore, is subject to correct and undesired operations and failure to trip.

It is important that the performance of relays and relay schemes be evaluated under restoration conditions. The foreknowledge that certain system operating conditions could cause correct and undesired operations or failure to trip makes it possible to avoid such operating conditions during development and execution of the restoration plan.

The protective relays that could affect the restoration procedure include:

- distance relays without potential restraints
- out-of-step relays
- synchro-check relays
- negative sequence voltage relays
- differential relays lacking harmonic restraints
- V/Hz relays
- generator underexcitation relays
- loss-of-field relays
- underfrequency switched reactor/capacitor relays.

Standing phase angle reduction

The presence of excessive standing phase angle (SPA) differences across open circuit breakers causes significant delays in restoration. The SPA may occur across a tie-line between two connected systems or between two connected subsystems. It must be brought to a safe limit before an attempt is made to close the breaker.

To determine a safe SPA value, the impact on the T-G shaft torque of closing a breaker should be evaluated. The T-G shaft torque is the sum of its constant-load torque that exists before and the transient mechanical torque immediately after closing the breaker. The constant-load torque can readily be reduced by lowering the generator's real power output, but the transient torque can be reduced only by reducing the SPA, a feat not readily accomplished.

There has been a need for an efficient methodology to serve as a guideline for reducing excessive SPA without resorting to the raising and lowering of various generation levels on a trial and error basis.

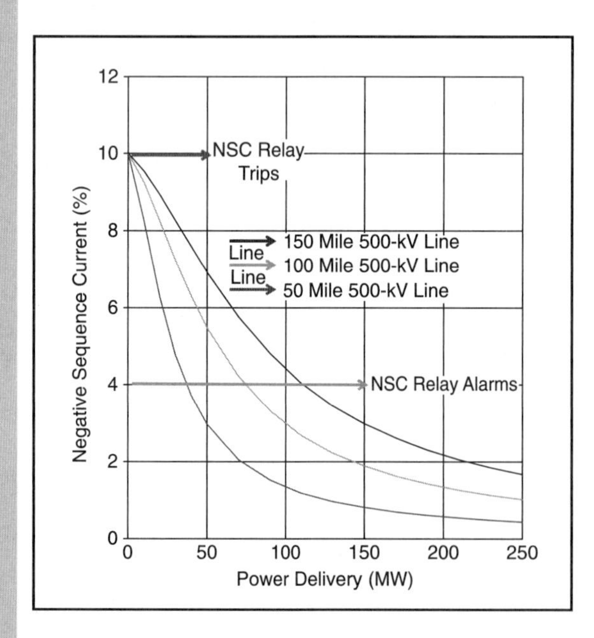

Figure 10
Current imbalance in 500-kV lines
*1 mile = 1.6 km

Asymmetry issues during restoration

Many existing EHV lines have asymmetrical (horizontal) conductor spacing without being transposed. These characteristics generate unacceptable negative sequence currents (NSCs). Under light load conditions and during restoration, NSC has caused cascade tripping of a number of generators (resulting in wide-area blackouts), has prevented synchronization of incoming generators, and has blocked remote black-start of large thermal units.

As shown in Figure 10, typically the generator NSC relays are set at 4% for alarms and 10% for tripping. It is important that when attempting to provide an off-site source to a nuclear or other large thermal power plant (e.g., in a remote black-start), the extent of NSC be determined and, if unacceptable, either the operation be deferred to after the initial restoration phase or the appropriate "underlying load" be determined for connection to the receiving end of the EHV lines.

ESTIMATING RESTORATION DURATION

Typically, restoration duration is estimated based on the availability of various prime movers. In these estimates, it is assumed that load can be picked up as soon as generation becomes available and that the time required for switching operations to energize transformers, lines, start-up of large motors, etc. is much less than the time required for generation availability. Case studies supported by field tests have shown that the above assumptions are not necessarily correct, and the restoration duration should be estimated using both the generation availability and the switching operations.

For example, in using the 20-MW combustion turbine of Table 3 for remote cranking of the 275-MW drum-type B-T-G, if the time estimate for cranking operation is well within 30–45 min, the hot restart of the B-T-G could be planned; if not, the B-T-G should be scheduled for a cold start-up, which requires an elapsed time of 3–4 h after the system collapse.

The critical path method (CPM) is a technique that can be used in restoration planning, scheduling, and evaluation. Its strength lies in the fact that

TABLE 3 Typical prime movers start-up timings

General Data			
Type	CT	Drum	SCOT*
Unit size, MW	20	275	800
Fuel	Gas	Oil	Coal
Duty	Peaking	Cycle	Base
Min. load, MW	5	30	420**
Hot Restart (h)			
Max. elapsed time	0.0	$\frac{1}{2}$–$\frac{3}{4}$	3
Light-off to synch.	0.1	$1\frac{1}{2}$	**4**
Synch to min. load	0.1	$\frac{3}{4}$	$1\frac{1}{2}$
Min. load to full load	0.1	1	$1\frac{1}{2}$
Cold Start-Up (h)			
Min. elapsed time	0.0	**3–4**	8
Light-off to synch.	0.1	7	16
Synch to min. load	0.0	$1\frac{3}{4}$	**3**
Min. load to full load	0.1	1	$1\frac{1}{2}$

*SCOT: Super-critical once-through.
**Gen. is manually loaded to this load level.

precise time estimates do not have to be made for each action. Using CPM, the restoration plan is broken down to levels, tasks, and basic operating actions. Operating actions include opening/closing breakers, raising/lowering transformer taps, adjusting voltage and frequency set-points, and starting auxiliary motors. Then, based on operator experience, the optimistic time, the pessimistic time, and the most likely time for each action can be determined. Estimation of duration of various tasks may dictate revision of the overall restoration plan. In any case, one can estimate the duration of the restoration with some degree of probability and not base the duration estimate merely on the timing of the prime movers.

RESTORATION TRAINING

During restoration, system operators are faced with a state of their system that is quite different from that to which they are accustomed in day-to-day operation and for which the EMS application programs at their disposal were not designed and are not well adopted.

There are distinct differences between the normal state and the restorative state in the type of models relevant to each, the objective pursued, and the information available at the control center to support the models. In the normal state, the primary objective is to minimize the cost of producing power, subject to observing certain security constraints. EMS application programs that help attain this goal incorporate models that represent the simplified, primarily steady-state behavior of power systems and incorporate data that is primarily obtained automatically from the system.

During the restoration process, by contrast, the objective is quite different. The objective here is to minimize the restoration time and the amount of unserved kilowatthours of energy, with security as a subsidiary objective, and without regard to production cost. At the same time, many dynamic phenomena neglected (or unimportant) during normal operation play a critical role during restoration and must be taken into account.

Available generic simulators can provide procedural training in the early stages of operator training. However, for exercising and preparing operators to cope with system-specific and time-critical emergency operations such as restoration, high-fidelity system-referenced simulators are needed. It is important to note that during the restart and reintegration phases of restoration, a power system often consists of one or more islands, most of the automatic controls have tripped or are deactivated, and the system is primarily under manual control. During these early phases of restoration, wider voltage and frequency ranges are tolerated. Under these large perturbations of long duration, the models and simulations that have been developed for small perturbations will not accurately represent the behavior of the power system and its components.

As shown in Figure 11, drills provide an excellent testing ground for the restoration plan and training of personnel. If conducted realistically, they will uncover potential problems with the existing plan. A key to good training and problem solving lies in the extent of the exercise. It should involve as many of the people and events that would be involved in an actual bulk power system restoration as possible. The exercises must be run frequently, and conditions must be varied so that operators will be trained in the handling of unpredictable events.

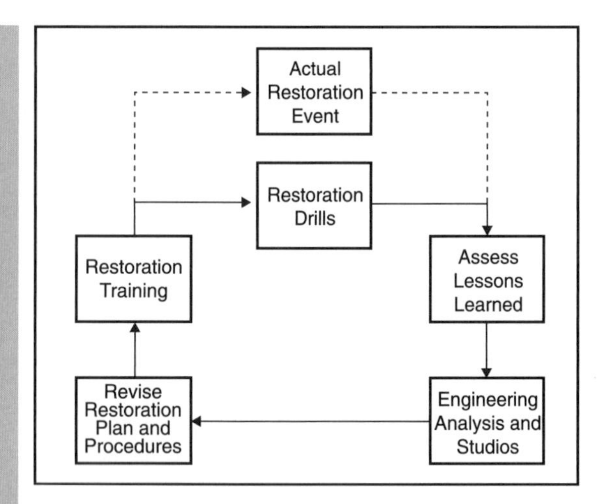

Figure 11
System restoration cycle

ACKNOWLEDGMENTS

This article has been compiled from the IEEE PES Power System Restoration Working Group papers. These papers are included in the book referenced under "For Further Reading." The contributing members included M. M. Adibi (chair); R. W. Alexander, PP&L; J. N. Borkoski, BG&E; L. H. Fink, consultant; R. J. Kafka, PEPCO; D. P. Milanicz, BG&E; T. L. Volkmann, NSP; and J. N. Wrubel, PSE&G.

FOR FURTHER READING

M. M. Adibi, *Power System Restoration—Methodologies and Implementation Strategies*. New York: Wiley, 2000.

BIOGRAPHIES

M. M. Adibi, as a program manager at IBM, conducted a 1967 investigation of the 1965 Northeast blackout for the Department of Public Service of the State of New York. In 1969, following the PJM blackout of 1967, he investigated bulk power security assessment for Edison Electric Institute. Since 1979 and in the aftermath of the 1977 New York blackout, he has developed restoration plans for over a dozen international utilities and has chaired the IEEE Power System Restoration Working Group. He is a Fellow of the IEEE.

L. H. Fink has had nearly 50 years of experience in electric power utility systems engineering and research with the Philadelphia Electric Company, the U.S. Department of Energy (where he developed and for five years managed the national research program Systems Engineering for Power), and subsequent consulting. His numerous publications deal with high-voltage underground cable systems, power plant and power system control, voltage dynamic phenomena, and system security analysis. He is a fellow of the IEEE.

12.1 GENERATOR-VOLTAGE CONTROL

The *exciter* delivers dc power to the field winding on the rotor of a synchronous generator. For older generators, the exciter consists of a dc generator driven by the rotor. The dc power is transferred to the rotor via slip rings and brushes. For newer generators, *static* or *brushless* exciters are often employed.

For static exciters, ac power is obtained directly from the generator terminals or a nearby station service bus. The ac power is then rectified via thyristors and transferred to the rotor of the synchronous generator via slip rings and brushes.

For brushless exciters, ac power is obtained from an "inverted" synchronous generator whose three-phase armature windings are located on the main generator rotor and whose field winding is located on the stator.

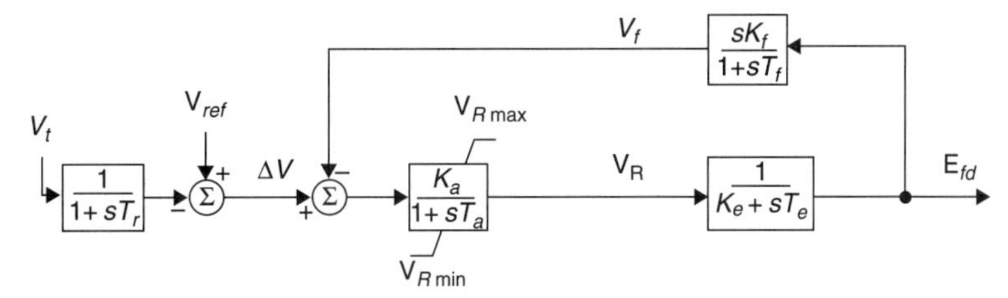

FIGURE 12.3

Block diagram for the IEEE Type 1 Exciter (neglecting saturation)

The ac power from the armature windings is rectified via diodes mounted on the rotor and is transferred directly to the field winding. For this design, slip rings and brushes are eliminated.

Block diagrams of several standard types of generator-voltage control systems have been developed by the IEEE Power and Energy Society, beginning in 1968 with [1] and most recently in 2005 with IEEE Std 421.5-2005. A block diagram for what is commonly known as the IEEE Type 1 exciter, which uses a shaft-driven dc generator to create the field current, is shown in Figure 12.3 (neglecting saturation).

In Figure 12.3, the leftmost block, $1/(1+s\mathrm{T}_r)$, represents the delay associated with measuring the terminal voltage V_t. where s is the Laplace operator and T_r is the measurement time constant. Note that if a unit step is applied to a $1/(1+s\mathrm{T}_r)$ block, the output rises exponentially to unity with time constant T_r. The measured generator terminal voltage V_t is compared with a voltage reference $\mathrm{V}_{\mathrm{ref}}$ to obtain a voltage error, $\Delta\mathrm{V}$, which in turn is applied to the voltage regulator. The voltage regulator is modeled as an amplifier with gain K_a and a time constant T_a, while the last forward block represents the dynamics of the exciter's dc generator. The output is the field voltage E_{fd}, which is applied to the generator field winding and acts to adjust the generator terminal voltage, as in (11.6.5). The feedback block in Figure 12.3 is used to improve the dynamic response of the exciter by reducing excessive overshoot. This feedback is represented by $(s\mathrm{K}_f)/(1+s\mathrm{T}_f)$, which provides a filtered first derivative negative feedback.

For any transient stability study, the initial values for the state variables need to be determined. This is done by assuming that the system is initially operating in steady-state, and recognizing that in steady-state all the derivatives will be zero. Then, by knowing the initial field voltage (found as in Example 11.10) and terminal voltage, all the other variables can be determined.

For wind turbines, how their voltage is controlled depends upon the type of the wind turbines. Type 1 wind turbines, squirrel cage induction machines, have no direct voltage control. Type 2 wind turbines are wound rotor induction machines with variable external resistance. While they do not have direct voltage control, the external resistance control system is usually modeled as a type of exciter. The block diagram for such a model is shown in Figure 12.4. The purpose for this control is to allow for a more constant power output from the wind turbine. For example, if a wind gust were to cause the turbine blades to accelerate, this controller would quickly increase the external resistance, flatting the torque-speed curve as shown in Figure 11.25.

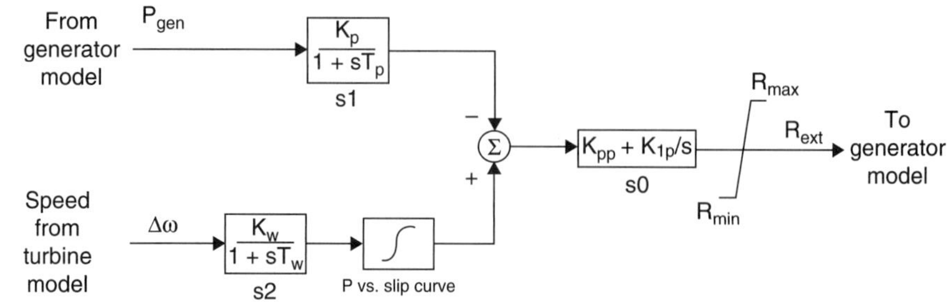

FIGURE 12.4 Simplified block diagram for a Type 2 wind turbine R_{ext} control system

Similar to synchronous machines, the Type 3 and 4 wind turbines have the ability to perform voltage or reactive power control. Common control modes include constant power factor control, coordinated control across a wind farm to maintain a constant voltage at the interconnection point, and constant reactive power control. Figure 12.5 shows a simplified version of a Type 3 wind turbine exciter, in which Q_{cmd} is the commanded reactive power, V_t is the terminal voltage, and the output, E_q is the input to the DFAG model shown in Figure 11.27. For fixed reactive power Q_{cmd} is a constant, while for power factor control, Q_{cmd} varies linearly with the real power output.

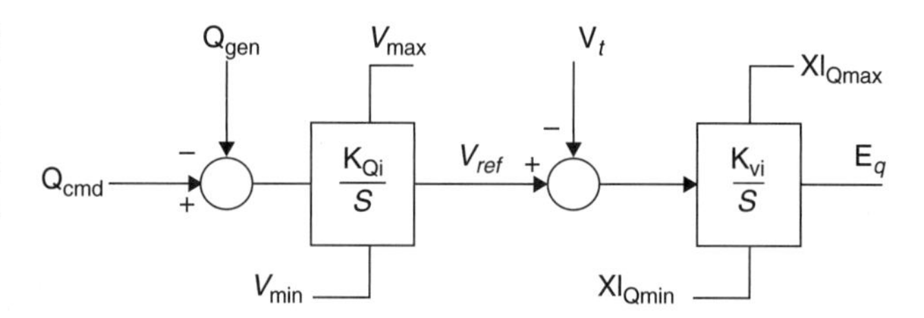

FIGURE 12.5 Simplified block diagram for a Type 3 wind turbine reactive power control system

EXAMPLE 12.1 Synchronous Generator Exciter Response

Using the system from Example 11.10, assume the two-axis generator is augmented with an IEEE Type 1 exciter with $T_r = 0$, $K_a = 100$, $T_a = 0.05$, $V_{rmax} = 5$, $V_{rmin} = -5$, $K_e = 1$, $T_e = 0.26$, $K_f = 0.01$ and $T_f = 1.0$. (a) Determine the initial values of V_r, V_f, and V_{ref}, (b) Using the fault sequence from Example 11.10, determine the bus 4 terminal voltage after 1 second and then after 5 seconds.

SOLUTION

a. The initial field voltage and terminal voltage, E_{fd} and V_t, do not depend on the exciter, so their values are equal to those found in Example 11.10, that is, 2.9135 and 1.0946 respectively. Since the system is initially in steady-state,

$$V_r = (K_e)(E_{fd}) = (1.0)(2.9135) = 2.9135$$

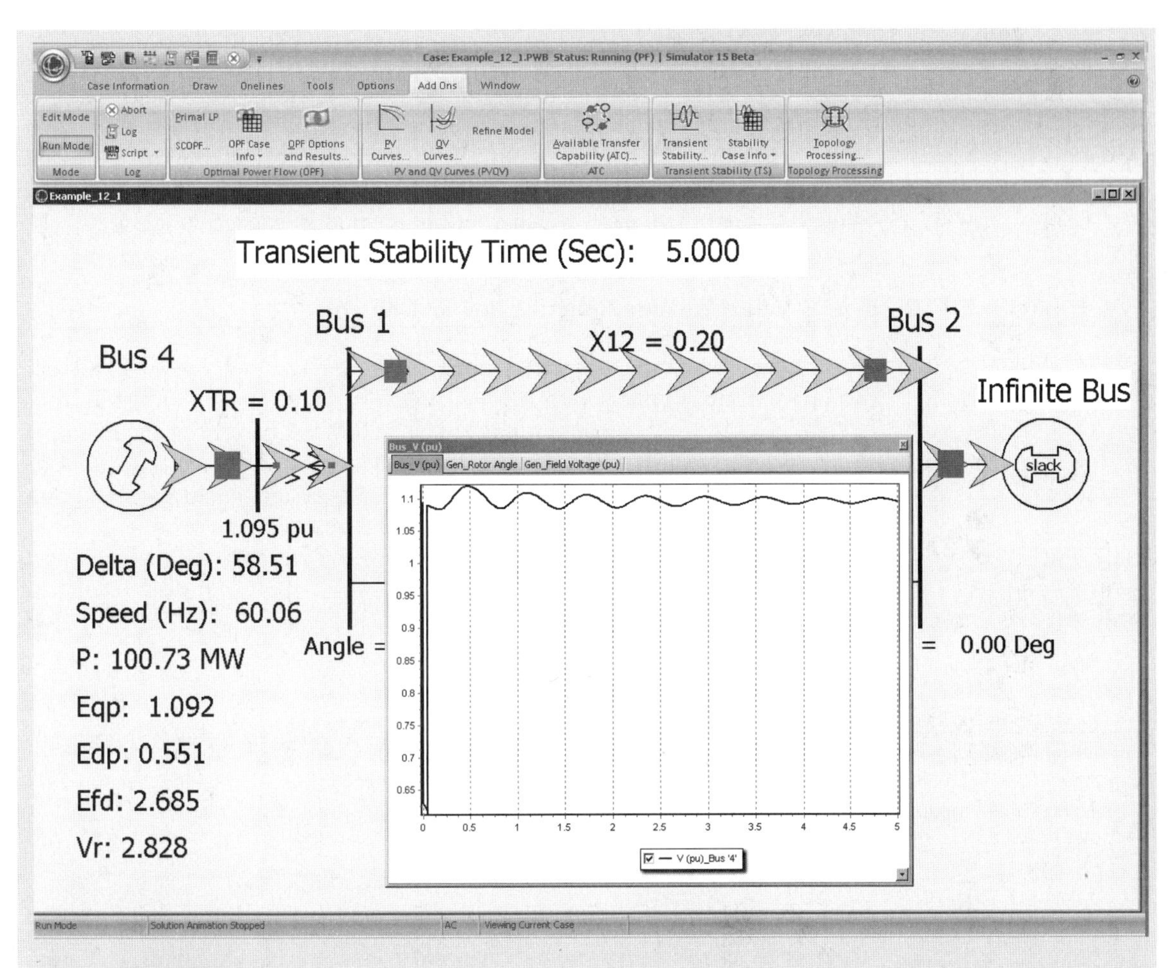

FIGURE 12.6 Example 12.1 results

Because V_f is the output of the filtered derivative feedback, its initial value is zero. Finally, writing the equation for the second summation block in Figure 12.3

$$\left(V_{ref} - V_t - V_f\right)(K_a) = V_r$$

$$V_{ref} = \frac{V_r}{K_a} + V_t + V_f = \frac{2.9135}{100} + 1.0946 = 1.1237$$

b. Open PowerWorld Simulator case Example 12_1 and display the Transient Stability Analysis Form (see Figure 12.6). To see the initial conditions, select the **States/Manual Control** page, and then select the **Transfer Present State to Power Flow** button to update the one-line display. From this page, it is also possible to just do a specified number of timesteps by selecting the **Do Specified Number of Timesteps(s)** button or to run to a specified simulation time using the **Run Until Specified Time** button. To determine the terminal voltage after one second, select the **Run Until Specified Time** button. The value is 1.104 pu. To finish the simulation, select the **Continue** button. The terminal voltage at five seconds is 1.095 pu, which is close to the pre-fault voltage, indicating the exciter is restoring the voltage to its setpoint value. In contrast, the bus 4 terminal voltage after five seconds in the Example 11.10 case, which does not include an exciter, is 1.115 pu. ■

EXAMPLE 12.2 Type 3 Wind Turbine Reactive Power Control

Assume the Type 3 wind turbine from Example 11.12 has a Figure 12.5 reactive power control system with $K_{Qi} = 0.4$, $K_{Vi} = 40$, $XI_{Qmax} = 1.45$, $XI_{Qmin} = 0.5$, $V_{max} = 1.1$, $V_{min} = 0.9$ (per unit using a 100 MVA base). For the Example 11.12 system conditions, determine the initial values for V_{ref}, Q_{cmd}, and estimate the maximum amount of reactive power this system could supply during a fault that depresses the terminal voltage to 0.5 pu.

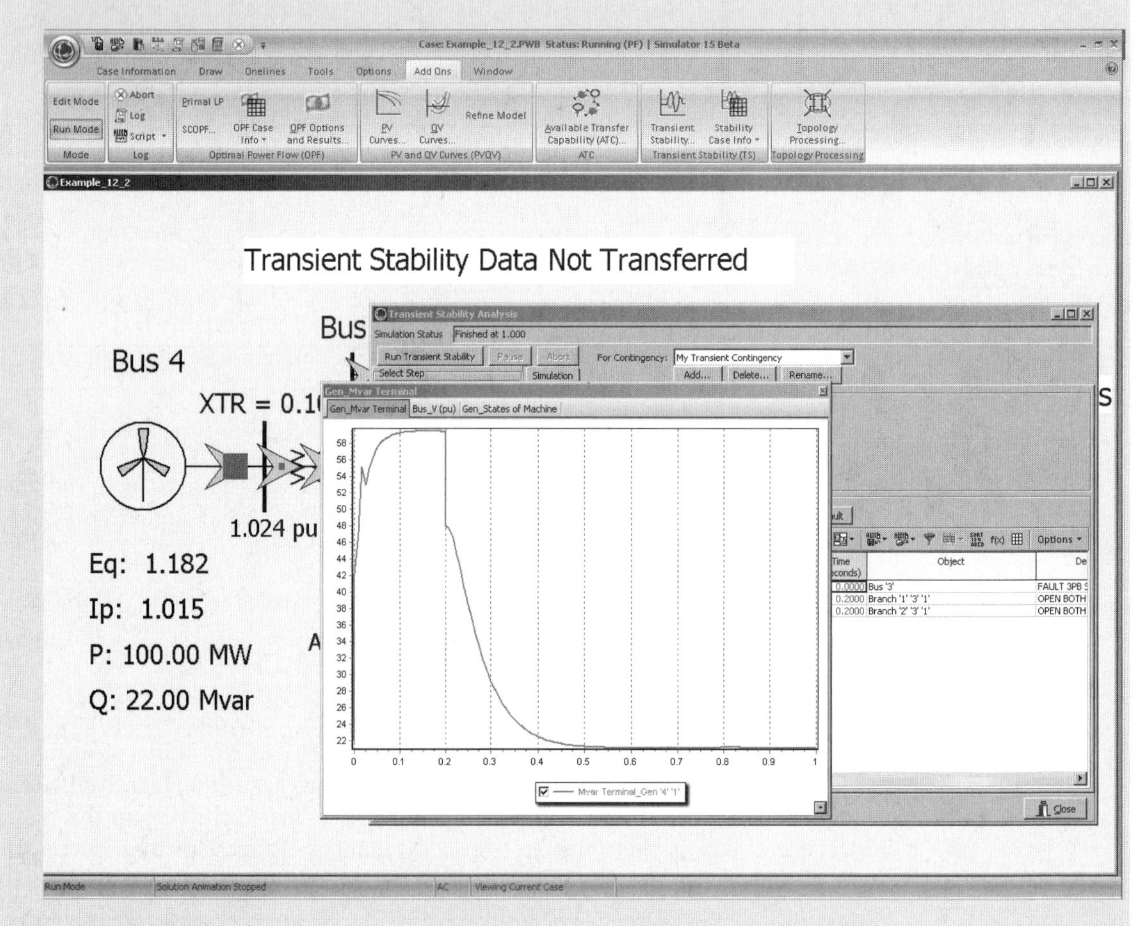

FIGURE 12.7 Example 12.2 variation in generator reactive power output

SOLUTION Since in steady-state the inputs to each of the two integrator blocks in Figure 12.5 must be zero, V_{ref} is just equal to the initial terminal voltage magnitude from Example 11.12, that is, 1.0239 pu, and Q_{cmd} is the initial reactive power output, which is 0.22 per unit (22 Mvar), found from the imaginary part of the product of the terminal voltage and the conjugate of the terminal current. During the fault with its low terminal voltage, the positive input into the K_{Vi} integration block will cause Eq to rapidly rise to its limit $XI_{\text{Qmax}} = 1.45$. The reactive component of I_{sorc} will then be $-1.45/0.8 = -1.8125$ pu. The total per unit reactive power injection with $V_t = 0.5$ during the fault is then

$$Q_{net} = (V_t)(1.8125) - \frac{V_t^2}{0.8} = 0.593 \text{ pu} = 59.3 \text{ Mvar}$$

This result can be confirmed by opening PowerWorld Simulator case Example 12_2 which models such a fault condition (see Figure 12.7). After the fault is cleared, the reactive power controller restores the machine's reactive power output to its pre-fault value. ■

Block diagrams such as those shown in Figure 12.3 are used for computer representation of generator-voltage control in transient stability computer programs (see Chapter 11). In practice, high-gain, fast-responding exciters provide large, rapid increases in field voltage E_{fd} during short circuits at the generator terminals in order to improve transient stability after fault clearing. Equations represented in the block diagram can be used to compute the transient response of generator-voltage control.

12.2 TURBINE-GOVERNOR CONTROL

Turbine-generator units operating in a power system contain stored kinetic energy due to their rotating masses. If the system load suddenly increases, stored kinetic energy is released to initially supply the load increase. Also, the electrical torque T_e of each turbine-generating unit increases to supply the load increase, while the mechanical torque T_m of the turbine initially remains constant. From Newton's second law, $J\alpha = T_m - T_e$, the acceleration α is therefore negative. That is, each turbine-generator decelerates and the rotor speed drops as kinetic energy is released to supply the load increase. The electrical frequency of each generator, which is proportional to rotor speed for synchronous machines, also drops.

From this, we conclude that either rotor speed or generator frequency indicates a balance or imbalance of generator electrical torque T_e and turbine mechanical torque T_m. If speed or frequency is decreasing, then T_e is greater than T_m (neglecting generator losses). Similarly, if speed or frequency is increasing, T_e is less than T_m. Accordingly, generator frequency is an appropriate control signal for governing the mechanical output power of the turbine.

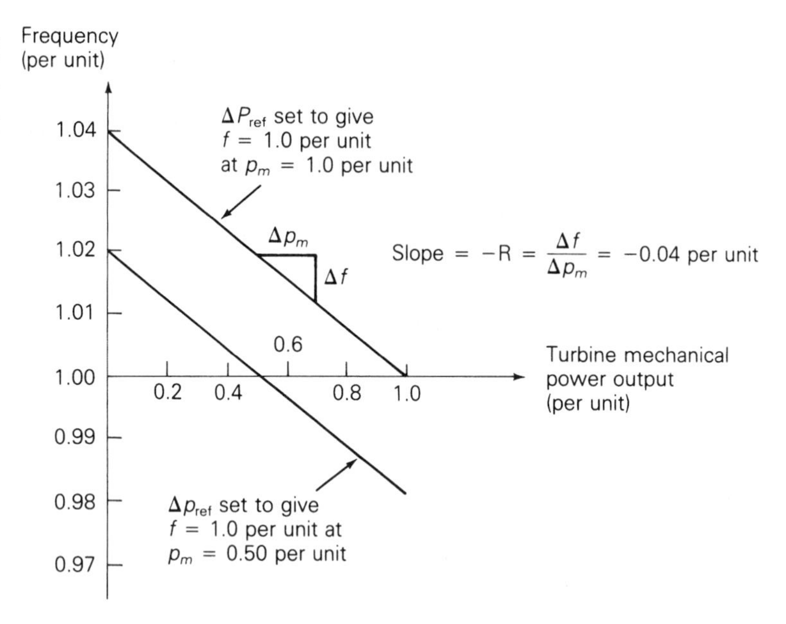

FIGURE 12.8 Steady-state frequency–power relation for a turbine-governor

The steady-state frequency–power relation for turbine-governor control is

$$\Delta p_m = \Delta p_{\text{ref}} - \frac{1}{\text{R}}\Delta f \tag{12.2.1}$$

where Δf is the change in frequency, Δp_m is the change in turbine mechanical power output, and Δp_{ref} is the change in a reference power setting. R is called the *regulation constant*. The equation is plotted in Figure 12.8 as a family of curves, with Δp_{ref} as a parameter. Note that when Δp_{ref} is fixed, Δp_m is directly proportional to the drop in frequency.

Figure 12.8 illustrates a steady-state frequency–power relation. When an electrical load change occurs, the turbine-generator rotor accelerates or decelerates, and frequency undergoes a transient disturbance. Under normal operating conditions, the rotor acceleration eventually becomes zero, and the frequency reaches a new steady-state, shown in the figure.

The regulation constant R in (12.2.1) is the negative of the slope of the Δf versus Δp_m curves shown in Figure 12.8. The units of R are Hz/MW when Δf is in Hz and Δp_m is in MW. When Δf and Δp_m are given in per-unit, however, R is also in per-unit.

EXAMPLE 12.3 Turbine-governor response to frequency change at a generating unit

A 500-MVA, 60-Hz turbine-generator has a regulation constant R = 0.05 per unit based on its own rating. If the generator frequency increases by 0.01 Hz in steady-state, what is the decrease in turbine mechanical power output? Assume a fixed reference power setting.

SOLUTION The per-unit change in frequency is

$$\Delta f_{\text{p.u.}} = \frac{\Delta f}{f_{\text{base}}} = \frac{0.01}{60} = 1.6667 \times 10^{-4} \quad \text{per unit}$$

Then, from (12.2.1), with $\Delta p_{\text{ref}} = 0$,

$$\Delta p_{m\text{p.u.}} = \left(\frac{-1}{0.05}\right)(1.6667 \times 10^{-4}) = -3.3333 \times 10^{-4} \quad \text{per unit}$$

$$\Delta p_m = (\Delta p_{m\text{p.u.}})\text{S}_{\text{base}} = (-3.3333 \times 10^{-4})(500) = -1.6667 \quad \text{MW}$$

The turbine mechanical power output decreases by 1.67 MW. ■

The steady-state frequency–power relation for one area of an interconnected power system can be determined by summing (12.2.1) for each turbine-generating unit in the area. Noting that Δf is the same for each unit,

$$\begin{aligned}\Delta p_m &= \Delta p_{m1} + \Delta p_{m2} + \Delta p_{m3} + \cdots \\ &= (\Delta p_{\text{ref}1} + \Delta p_{\text{ref}2} + \cdots) - \left(\frac{1}{\text{R}_1} + \frac{1}{\text{R}_2} + \cdots\right)\Delta f \\ &= \Delta p_{\text{ref}} - \left(\frac{1}{\text{R}_1} + \frac{1}{\text{R}_2} + \cdots\right)\Delta f \end{aligned} \tag{12.2.2}$$

where Δp_m is the total change in turbine mechanical powers and Δp_{ref} is the total change in reference power settings within the area. We define the *area frequency response characteristic* β as

$$\beta = \left(\frac{1}{\text{R}_1} + \frac{1}{\text{R}_2} \cdots\right) \tag{12.2.3}$$

Using (12.2.3) in (12.2.2),

$$\Delta p_m = \Delta p_{\text{ref}} - \beta \Delta f \tag{12.2.4}$$

Equation (12.2.4) is the area steady-state frequency–power relation. The units of β are MW/Hz when Δf is in Hz and Δp_m is in MW. β can also be given in per-unit. In practice, β is somewhat higher than that given by (12.2.3) due to system losses and the frequency dependence of loads.

A standard value for the regulation constant is $\text{R} = 0.05$ per unit. When all turbine-generating units have the same per-unit value of R based on their own ratings, then each unit shares total power changes in proportion to its own rating. Figure 12.9 shows a block diagram for a simple steam turbine governor commonly known as the TGOV1 model. The $1/(1 + s\text{T}_1)$ models the time delays associated with the governor, where s is again the Laplace

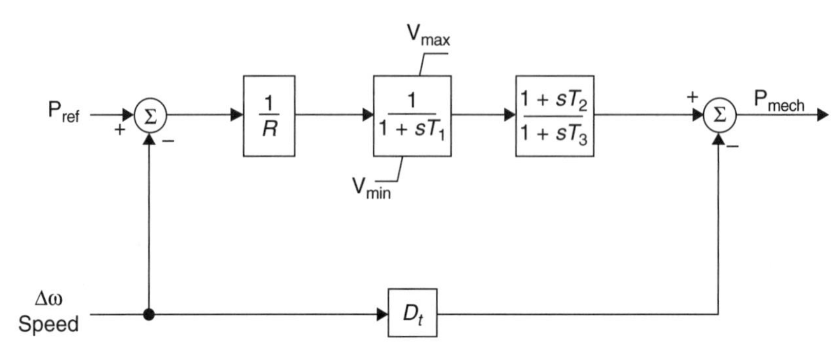

FIGURE 12.9 Turbine-governor block diagram

operator and T_1 is the time constant. The limits on the output of this block account for the fact that turbines have minimum and maximum outputs. The second block diagram models the delays associated with the turbine; for non-reheat turbines T_2 should be zero. Typical values are $R = 0.05$ pu, $T_1 = 0.5$ seconds, $T_3 = 0.5$ for a non-reheat turbine or $T_2 = 2.5$, and $T_3 = 7.5$ seconds otherwise. D_t is a turbine damping coefficient that is usually 0.02 or less (often zero). Additional turbine block diagrams are available in [3].

EXAMPLE 12.4 Response of turbine-governors to a load change in an interconnected power system

An interconnected 60-Hz power system consists of one area with three turbine-generator units rated 1000, 750, and 500 MVA, respectively. The regulation constant of each unit is R = 0.05 per unit based on its own rating. Each unit is initially operating at one-half of its own rating, when the system load suddenly increases by 200 MW. Determine (a) the per-unit area frequency response characteristic β on a 1000 MVA system base, (b) the steady-state drop in area frequency, and (c) the increase in turbine mechanical power output of each unit. Assume that the reference power setting of each turbine-generator remains constant. Neglect losses and the dependence of load on frequency.

SOLUTION

a. The regulation constants are converted to per-unit on the system base using

$$R_{\text{p.u.new}} = R_{\text{p.u.old}} \frac{S_{\text{base(new)}}}{S_{\text{base(old)}}}$$

We obtain

$$R_{1\text{p.u.new}} = R_{1\text{p.u.old}} = 0.05$$

$$R_{2\text{p.u.new}} = (0.05)\left(\frac{1000}{750}\right) = 0.06667$$

$$R_{3\text{p.u.new}} = (0.05)\left(\frac{1000}{550}\right) = 0.10 \quad \text{per unit}$$

Using (12.2.3),

$$\beta = \frac{1}{R_1} + \frac{1}{R_2} + \frac{1}{R_3} = \frac{1}{0.05} + \frac{1}{0.06667} + \frac{1}{0.10} = 45.0 \quad \text{per unit}$$

b. Neglecting losses and dependence of load on frequency, the steady-state increase in total turbine mechanical power equals the load increase, 200 MW or 0.20 per unit. Using (12.2.4) with $\Delta p_{\text{ref}} = 0$,

$$\Delta f = \left(\frac{-1}{\beta}\right)\Delta p_m = \left(\frac{-1}{45}\right)(0.20) = -4.444 \times 10^{-3} \quad \text{per unit}$$

$$= (-4.444 \times 10^{-3})(60) = -0.2667 \quad \text{Hz}$$

The steady-state frequency drop is 0.2667 Hz.

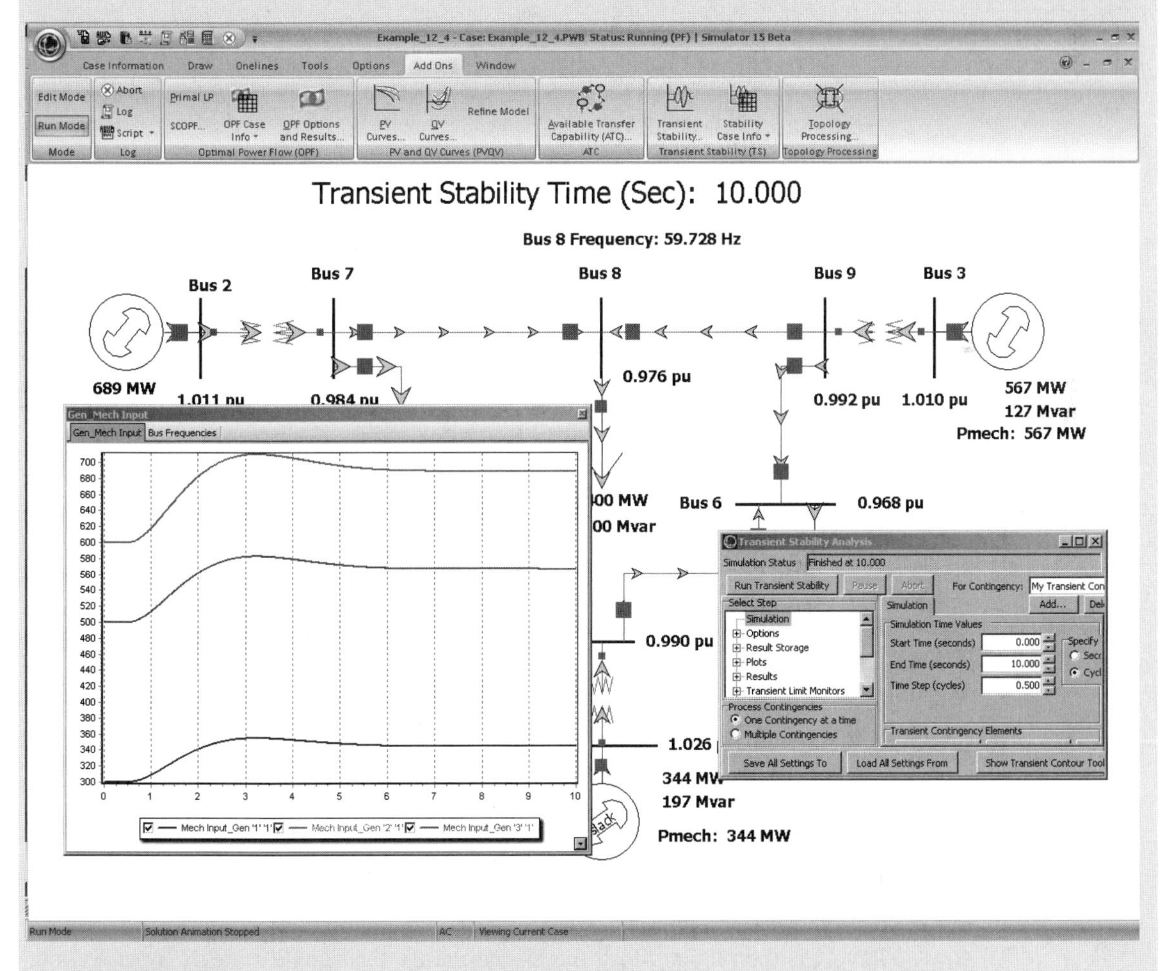

FIGURE 12.10 System one-line with generator mechanical power variation for Example 12.4

c. From (12.2.1), using $\Delta f = -4.444 \times 10^{-3}$ per unit,

$$\Delta p_{m1} = \left(\frac{-1}{0.05}\right)(-4.444 \times 10^{-3}) = 0.08888 \quad \text{per unit}$$
$$= 88.88 \quad \text{MW}$$

$$\Delta p_{m2} = \left(\frac{-1}{0.06667}\right)(-4.444 \times 10^{-3}) = 0.06666 \quad \text{per unit}$$
$$= 66.66 \quad \text{MW}$$

$$\Delta p_{m3} = \left(\frac{-1}{0.10}\right)(-4.444 \times 10^{-3}) = 0.04444 \quad \text{per unit}$$
$$= 44.44 \quad \text{MW}$$

Note that unit 1, whose MVA rating is $33\frac{1}{3}\%$ larger than that of unit 2 and 100% larger than that of unit 3, picks up $33\frac{1}{3}\%$ more load than unit 2 and 100% more load than unit 3. That is, each unit shares the total load change in proportion to its own rating.

PowerWorld Simulator case Example 12_4 contains a lossless nine bus, three generator system that duplicates the conditions from this example (see Figure 12.10). The generators at buses 1, 2 and 3 have ratings of 500, 1000, and 750 MVA respectively, with initial outputs of 300, 600, 500 MWs. Each is modeled with a two-axis synchronous machine model (see Section 11.6), an IEEE T_1 exciter and a TGOV1 governor model with the parameters equal to the defaults given earlier. At time $t = 0.5$ seconds, the load at bus 8 is increased from 200 to 400 MW. Figure 12.10 shows the results of a 10-second transient stability simulation. The final generator outputs are 344.5, 589.0, and 466.7 MWs, while the final frequency decline is 0.272 Hz, closely matching the results predicted in the example (the frequency decline exactly matches the 0.266 Hz prediction if the simulation is extended to 20 seconds). ■

The power output from wind turbines can be controlled by changing the pitch angle of the blades. When the available power in the wind is above the rating for the turbine, its blades are pitched to limit the mechanical power delivered to the electric machine. When the available power is less than the machine's rating, the pitch angle is set to its minimum. Figure 12.11 shows the generic pitch control model for Type 3 and 4 wind turbines, with the inputs being the per unit speed of the turbine, ω_r, the desired speed (normally 1.2 pu), the ordered per unit electrical output, and a setpoint power. These signals are combined as shown on the figure to produce a commanded angle for the blades, θ_{cmd}, expressed in degrees. The right side of the block diagram models the dynamics and limits associated with changing the pitch angle of the blades; R_{Theta} is the rate at which the blades change their angle in degrees per second. Typical values are $T_p = 0.3$ seconds, $\text{Theta}_{\text{min/max}}$ between 0° and 27°, rate limits of ±10°/second, $K_{pp} = 150$, $K_{ip} = 25$, $K_{pc} = 3$, $K_{ic} = 30$.

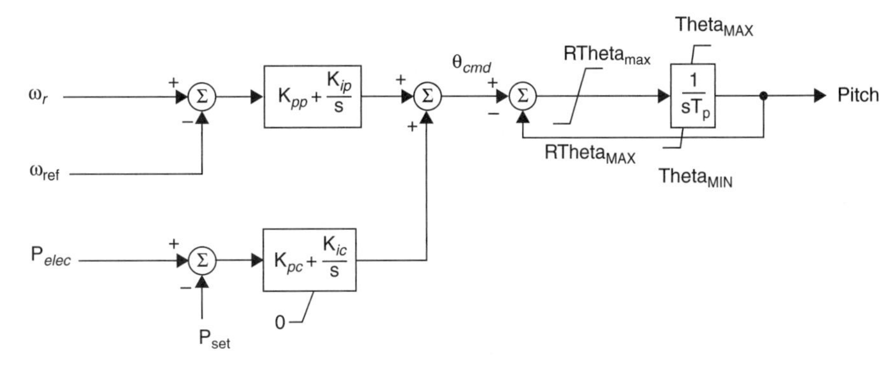

FIGURE 12.11 Pitch control for a Type 3 or 4 wind turbine model

In general, the larger the size of the interconnected system, the better the frequency response since there are more generators to share the task. However, as stated in [16], "Owners/operators of generator units have strong economic reasons to operate generator units in many ways that prevent effective governing response." For example, operating the unit at its full capacity, which maximizes the income that can be derived from the unit, but prevents the unit from increasing its output. This is certainly an issue with wind turbines since their "fuel" is essentially free. Also, the Type 3 and 4 units do not contribute inertial response.

12.3 LOAD-FREQUENCY CONTROL

As shown in Section 12.2, turbine-governor control eliminates rotor accelerations and decelerations following load changes during normal operation. However, there is a steady-state frequency error Δf when the change in turbine-governor reference setting Δp_{ref} is zero. One of the objectives of load-frequency control (LFC), therefore, is to return Δf to zero.

In a power system consisting of interconnected areas, each area agrees to export or import a scheduled amount of power through transmission-line interconnections, or tie-lines, to its neighboring areas. Thus, a second LFC objective is to have each area absorb its own load changes during normal operation. This objective is achieved by maintaining the net tie-line power flow out of each area at its scheduled value.

The following summarizes the two basic LFC objectives for an interconnected power system:

1. Following a load change, each area should assist in returning the steady-state frequency error Δf to zero.

2. Each area should maintain the net tie-line power flow out of the area at its scheduled value, in order for the area to absorb its own load changes.

The following control strategy developed by N. Cohn [4] meets these LFC objectives. We first define the *area control error* (ACE) as follows:

$$\begin{aligned} \text{ACE} &= (p_{\text{tie}} - p_{\text{tie, sched}}) + \text{B}_f(f - 60) \\ &= \Delta p_{\text{tie}} + \text{B}_f \Delta f \end{aligned} \tag{12.3.1}$$

where Δp_{tie} is the deviation in net tie-line power flow out of the area from its scheduled value $p_{\text{tie, sched}}$, and Δf is the deviation of area frequency from its scheduled value (60 Hz). Thus, the ACE for each area consists of a linear combination of tie-line error Δp_{tie} and frequency error Δf. The constant B_f is called a *frequency bias constant*.

The change in reference power setting $\Delta p_{\text{ref}i}$ of each turbine-governor operating under LFC is proportional to the integral of the area control error. That is,

$$\Delta p_{\text{ref}i} = -\text{K}_i \int \text{ACE}\, dt \tag{12.3.2}$$

Each area monitors its own tie-line power flows and frequency at the area control center. The ACE given by (12.3.1) is computed and a percentage of the ACE is allocated to each controlled turbine-generator unit. Raise or lower commands are dispatched to the turbine-governors at discrete time intervals of two or more seconds in order to adjust the reference power settings. As the commands accumulate, the integral action in (12.3.2) is achieved.

The constant K_i in (12.3.2) is an integrator gain. The minus sign in (12.3.2) indicates that if either the net tie-line power flow out of the area or the area frequency is low—that is, if the ACE is negative—then the area should increase its generation.

When a load change occurs in any area, a new steady-state operation can be obtained only after the power output of every turbine-generating unit in the interconnected system reaches a constant value. This occurs only when all reference power settings are zero, which in turn occurs only when the ACE of every area is zero. Furthermore, the ACE is zero in every area only when both Δp_{tie} and Δf are zero. Therefore, in steady-state, both LFC objectives are satisfied.

EXAMPLE 12.5 Response of LFC to a load change in an interconnected power system

As shown in Figure 12.12, a 60-Hz power system consists of two interconnected areas. Area 1 has 2000 MW of total generation and an area frequency response characteristic $\beta_1 = 700$ MW/Hz. Area 2 has 4000 MW of total generation and $\beta_2 = 1400$ MW/Hz. Each area is initially generating one-half of its total generation, at $\Delta p_{\text{tie1}} = \Delta p_{\text{tie2}} = 0$ and at 60 Hz when the load in area 1 suddenly increases by 100 MW. Determine the steady-state frequency error Δf and the steady-state tie-line error Δp_{tie} of each area for the following two

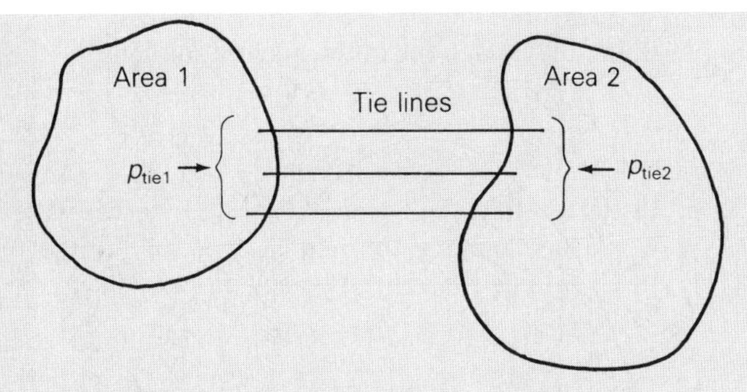

FIGURE 12.12 Example 12.5

cases: (a) without LFC, and (b) with LFC given by (12.3.1) and (12.3.2). Neglect losses and the dependence of load on frequency.

SOLUTION

a. Since the two areas are interconnected, the steady-state frequency error Δf is the same for both areas. Adding (12.2.4) for each area,

$$(\Delta p_{m1} + \Delta p_{m2}) = (\Delta p_{ref1} + \Delta p_{ref2}) - (\beta_1 + \beta_2)\Delta f$$

Neglecting losses and the dependence of load on frequency, the steady-state increase in total mechanical power of both areas equals the load increase, 100 MW. Also, without LFC, Δp_{ref1} and Δp_{ref2} are both zero. The above equation then becomes

$$100 = -(\beta_1 + \beta_2)\Delta f = -(700 + 1400)\Delta f$$

$$\Delta f = -100/2100 = -0.0476 \quad \text{Hz}$$

Next, using (12.2.4) for each area, with $\Delta p_{ref} = 0$,

$$\Delta p_{m1} = -\beta_1 \Delta f = -(700)(-0.0476) = 33.33 \quad \text{MW}$$

$$\Delta p_{m2} = -\beta_2 \Delta f = -(1400)(-0.0476) = 66.67 \quad \text{MW}$$

In response to the 100-MW load increase in area 1, area 1 picks up 33.33 MW and area 2 picks up 66.67 MW of generation. The 66.67-MW increase in area 2 generation is transferred to area 1 through the tie-lines. Therefore, the change in net tie-line power flow out of each area is

$$\Delta p_{tie2} = +66.67 \quad \text{MW}$$

$$\Delta p_{tie1} = -66.67 \quad \text{MW}$$

b. From (12.3.1), the area control error for each area is

$$\text{ACE}_1 = \Delta p_{tie1} + \text{B}_1 \Delta f_1$$

$$\text{ACE}_2 = \Delta p_{tie2} + \text{B}_2 \Delta f_2$$

Neglecting losses, the sum of the net tie-line flows must be zero; that is, $\Delta p_{tie1} + \Delta p_{tie2} = 0$ or $\Delta p_{tie2} = -\Delta p_{tie1}$. Also, in steady-state $\Delta f_1 = \Delta f_2 = \Delta f$.

Using these relations in the above equations,

$$\mathrm{ACE}_1 = \Delta p_{\mathrm{tie1}} + \mathrm{B}_1 \Delta f$$

$$\mathrm{ACE}_2 = -\Delta p_{\mathrm{tie1}} + \mathrm{B}_2 \Delta f$$

In steady-state, $\mathrm{ACE}_1 = \mathrm{ACE}_2 = 0$; otherwise, the LFC given by (12.3.2) would be changing the reference power settings of turbine-governors on LFC. Adding the above two equations,

$$\mathrm{ACE}_1 + \mathrm{ACE}_2 = 0 = (\mathrm{B}_1 + \mathrm{B}_2)\Delta f$$

Therefore, $\Delta f = 0$ and $\Delta p_{\mathrm{tie1}} = \Delta p_{\mathrm{tie2}} = 0$. That is, in steady-state the frequency error is returned to zero, area 1 picks up its own 100-MW load increase, and area 2 returns to its original operating condition—that is, the condition before the load increase occurred.

We note that turbine-governor controls act almost instantaneously, subject only to the time delays shown in Figure 12.9. However, LFC acts more slowly. LFC raise and lower signals are dispatched from the area control center to turbine-governors at discrete-time intervals of 2 or more seconds. Also, it takes time for the raise or lower signals to accumulate. Thus, case (a) represents the first action. Turbine-governors in both areas rapidly respond to the load increase in area 1 in order to stabilize the frequency drop. Case (b) represents the second action. As LFC signals are dispatched to turbine-governors, Δf and Δp_{tie} are slowly returned to zero. ■

The choice of the B_f and K_i constants in (12.3.1) and (12.3.2) affects the transient response to load changes—for example, the speed and stability of the response. The frequency bias B_f should be high enough such that each area adequately contributes to frequency control. Cohn [4] has shown that choosing B_f equal to the area frequency response characteristic, $\mathrm{B}_f = \beta$, gives satisfactory performance of the interconnected system. The integrator gain K_i should not be too high; otherwise, instability may result. Also, the time interval at which LFC signals are dispatched, 2 or more seconds, should be long enough so that LFC does not attempt to follow random or spurious load changes. A detailed investigation of the effect of B_f, K_i, and LFC time interval on the transient response of LFC and turbine-governor controls is beyond the scope of this text.

Two additional LFC objectives are to return the integral of frequency error and the integral of net tie-line error to zero in steady-state. By meeting these objectives, LFC controls both the time of clocks that are driven by 60-Hz motors and energy transfers out of each area. These two objectives are achieved by making temporary changes in the frequency schedule and tie-line schedule in (12.3.1).

Finally, note that LFC maintains control during normal changes in load and frequency—that is, changes that are not too large. During emergencies, when large imbalances between generation and load occur, LFC is bypassed and other, emergency controls are applied.

12.4

ECONOMIC DISPATCH

Section 12.3 describes how LFC adjusts the reference power settings of turbine-governors in an area to control frequency and net tie-line power flow out of the area. This section describes how the real power output of each controlled generating unit in an area is selected to meet a given load and to minimize the total operating costs in the area. This is the *economic dispatch* problem [5].

We begin this section by considering an area with only fossil-fuel generating units, with no constraints on maximum and minimum generator outputs, and with no transmission losses. The economic dispatch problem is first solved for this idealized case. Then we include inequality constraints on generator outputs; then we include transmission losses. Next, we discuss the coordination of economic dispatch with LFC. Finally, we briefly discuss the dispatch of other types of units including nuclear, pumped-storage hydro, and hydro units.

FOSSIL-FUEL UNITS, NO INEQUALITY CONSTRAINTS, NO TRANSMISSION LOSSES

Figure 12.11 shows the operating cost C_i of a fossil-fuel generating unit versus real power output P_i. Fuel cost is the major portion of the variable cost of operation, although other variable costs, such as maintenance, could have been included in the figure. Fixed costs, such as the capital cost of installing the unit, are not included. Only those costs that are a function of unit power output—that is, those costs that can be controlled by operating strategy—enter into the economic dispatch formulation.

In practice, C_i is constructed of piecewise continuous functions valid for ranges of output P_i, based on empirical data. The discontinuities in Figure 12.13 may be due to the firing of equipment such as additional boilers or condensers as power output is increased. It is often convenient to express C_i in terms of kJ/h, which is relatively constant over the lifetime of the unit, rather than $/hr, which can change monthly or daily. C_i can be converted to $/hr by multiplying the fuel input in kJ/h by the cost of fuel in $/kJ.

Figure 12.14 shows the unit incremental operating cost dC_i/dP_i versus unit output P_i, which is the slope or derivative of the C_i versus P_i curve in Figure 12.13. When C_i consists of only fuel costs, dC_i/dP_i is the ratio of the incremental fuel energy input in kilojoules to incremental energy output in kWh, which is called incremental *heat rate*. Note that the reciprocal of the heat rate, which is the ratio of output energy to input energy, gives a measure of fuel efficiency for the unit. For the unit shown in Figure 12.13, maximum

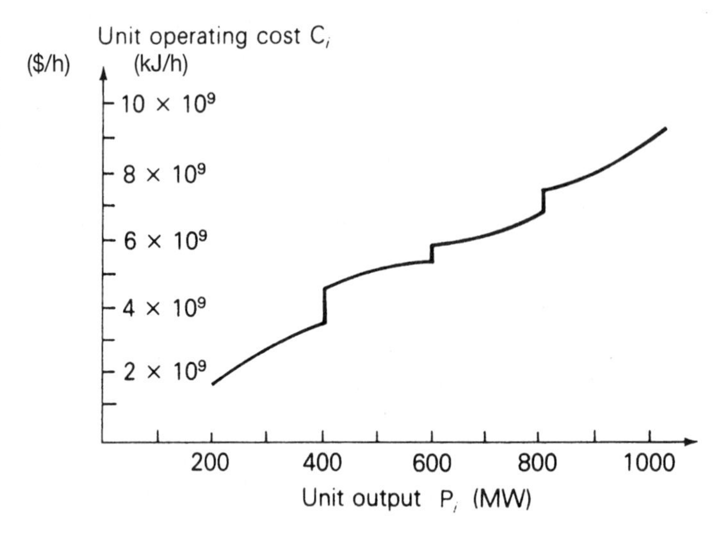

FIGURE 12.13 Unit operating cost versus real power output—fossil-fuel generating unit

efficiency occurs at $P_i = 600$ MW, where the heat rate is $\mathbf{C/P} = 5.4 \times 10^9/600 \times 10^3 = 9000$ kJ/kWh. The efficiency at this output is

$$\text{percentage efficiency} = \left(\frac{1}{9000}\frac{\text{kWh}}{\text{kJ}}\right)\left(3413\frac{\text{kJ}}{\text{kWh}}\right) \times 100 = 37.92\%$$

The dC_i/dP_i curve in Figure 12.14 is also represented by piecewise continuous functions valid for ranges of output P_i. For analytical work, the actual curves are often approximated by straight lines. The ratio dC_i/dP_i can also be converted to \$/kWh by multiplying the incremental heat rate in kJ/kWh by the cost of fuel in \$/kJ.

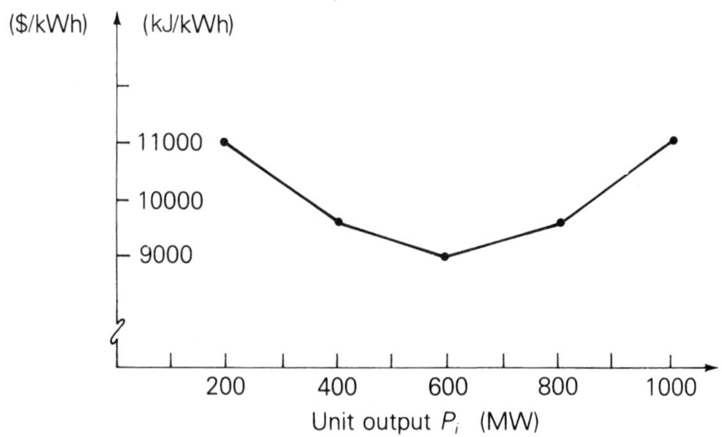

FIGURE 12.14 Unit incremental operating cost versus real power output—fossil-fuel generating unit

For the area of an interconnected power system consisting of N units operating on economic dispatch, the total variable cost C_T of operating these units is

$$\begin{aligned} C_T &= \sum_{i=i}^{N} C_i \\ &= C_1(P_1) + C_2(P_2) + \cdots + C_N(P_N) \quad \$/\text{h} \end{aligned} \tag{12.4.1}$$

where C_i, expressed in \$/h, includes fuel cost as well as any other variable costs of unit i. Let P_T equal the total load demand in the area. Neglecting transmission losses,

$$P_1 + P_2 + \cdots + P_N = P_T \tag{12.4.2}$$

Due to relatively slow changes in load demand, P_T may be considered constant for periods of 2 to 10 minutes. The economic dispatch problem can be stated as follows:

> Find the values of unit outputs $P_1, P_2, \ldots, P_N$ that minimize C_T given by (12.4.1), subject to the equality constraint given by (12.4.2).

A criterion for the solution to this problem is: All units on economic dispatch should operate at equal incremental operating cost. That is,

$$\frac{dC_1}{dP_1} = \frac{dC_2}{dP_2} = \cdots = \frac{dC_N}{dP_N} \tag{12.4.3}$$

An intuitive explanation of this criterion is the following. Suppose one unit is operating at a higher incremental operating cost than the other units. If the output power of that unit is reduced and transferred to units with lower incremental operating costs, then the total operating cost C_T decreases. That is, reducing the output of the unit with the *higher* incremental cost results in a *greater cost decrease* than the cost increase of adding that same output reduction to units with lower incremental costs. Therefore, all units must operate at the same incremental operating cost (the economic dispatch criterion).

A mathematical solution to the economic dispatch problem can also be given. The minimum value of C_T occurs when the total differential dC_T is zero. That is,

$$dC_T = \frac{\partial C_T}{\partial P_1} dP_1 + \frac{\partial C_T}{\partial P_2} dP_2 + \cdots + \frac{\partial C_T}{\partial P_N} dP_N = 0 \tag{12.4.4}$$

Using (12.4.1), (12.4.4) becomes

$$dC_T = \frac{dC_1}{dP_1} dP_1 + \frac{dC_2}{dP_2} dP_2 + \cdots + \frac{dC_N}{dP_N} dP_N = 0 \tag{12.4.5}$$

Also, assuming P_T is constant, the differential of (12.4.2) is

$$dP_1 + dP_2 + \cdots + dP_N = 0 \tag{12.4.6}$$

Multiplying (12.4.6) by λ and subtracting the resulting equation from (12.4.5),

$$\left(\frac{dC_1}{dP_1} - \lambda\right) dP_1 + \left(\frac{dC_2}{dP_2} - \lambda\right) dP_2 + \cdots + \left(\frac{dC_N}{dP_N} - \lambda\right) dP_N = 0 \qquad (12.4.7)$$

Equation (12.4.7) is satisfied when each term in parentheses equals zero. That is,

$$\frac{dC_1}{dP_1} = \frac{dC_2}{dP_2} = \cdots = \frac{dC_N}{dP_N} = \lambda \qquad (12.4.8)$$

Therefore, all units have the same incremental operating cost, denoted here by λ, in order to minimize the total operating cost C_T.

EXAMPLE 12.6 Economic dispatch solution neglecting generator limits and line losses

An area of an interconnected power system has two fossil-fuel units operating on economic dispatch. The variable operating costs of these units are given by

$$C_1 = 10P_1 + 8 \times 10^{-3}P_1^2 \quad \$/h$$

$$C_2 = 8P_2 + 9 \times 10^{-3}P_2^2 \quad \$/h$$

where P_1 and P_2 are in megawatts. Determine the power output of each unit, the incremental operating cost, and the total operating cost C_T that minimizes C_T as the total load demand P_T varies from 500 to 1500 MW. Generating unit inequality constraints and transmission losses are neglected.

SOLUTION The incremental operating costs of the units are

$$\frac{dC_1}{dP_1} = 10 + 16 \times 10^{-3}P_1 \quad \$/MWh$$

$$\frac{dC_2}{dP_2} = 8 + 18 \times 10^{-3}P_2 \quad \$/MWh$$

Using (12.4.8), the minimum total operating cost occurs when

$$\frac{dC_1}{dP_1} = 10 + 16 \times 10^{-3}P_1 = \frac{dC_2}{dP_2} = 8 + 18 \times 10^{-3}P_2$$

Using $P_2 = P_T - P_1$, the preceding equation becomes

$$10 + 16 \times 10^{-3}P_1 = 8 + 18 \times 10^{-3}(P_T - P_1)$$

Solving for P_1,

$$P_1 = \frac{18 \times 10^{-3}P_T - 2}{34 \times 10^{-3}} = 0.5294P_T - 58.82 \quad MW$$

TABLE 12.1

Economic dispatch solution for Example 12.6

P_T MW	P_1 MW	P_2 MW	dC_1/dP_1 \$/MWh	C_T \$/h
500	206	294	13.29	5529
600	259	341	14.14	6901
700	312	388	14.99	8358
800	365	435	15.84	9899
900	418	482	16.68	11525
1000	471	529	17.53	13235
1100	524	576	18.38	15030
1200	576	624	19.22	16910
1300	629	671	20.07	18875
1400	682	718	20.92	20924
1500	735	765	21.76	23058

Also, the incremental operating cost when C_T is minimized is

$$\frac{dC_2}{dP_2} = \frac{dC_1}{dP_1} = 10 + 16 \times 10^{-3} P_1 = 10 + 16 \times 10^{-3}(0.5294 P_T - 58.82)$$

$$= 9.0589 + 8.4704 \times 10^{-3} P_T \quad \$/\text{MWh}$$

and the minimum total operating cost is

$$C_T = C_1 + C_2 = (10P_1 + 8 \times 10^{-3} P_1^2) + (8P_2 + 9 \times 10^{-3} P_2^2) \quad \$/\text{h}$$

The economic dispatch solution is shown in Table 12.1 for values of P_T from 500 to 1500 MW. ■

EFFECT OF INEQUALITY CONSTRAINTS

Each generating unit must not operate above its rating or below some minimum value. That is,

$$P_{i\min} < P_i < P_{i\max} \qquad i = 1, 2, \ldots, N \tag{12.4.9}$$

Other inequality constraints may also be included in the economic dispatch problem. For example, some unit outputs may be restricted so that certain transmission lines or other equipments are not overloaded. Also, under adverse weather conditions, generation at some units may be limited to reduce emissions.

When inequality constraints are included, we modify the economic dispatch solution as follows. If one or more units reach their limit values, then these units are held at their limits, and the remaining units operate at equal incremental operating cost λ. The incremental operating cost of the area equals the common λ for the units that are not at their limits.

EXAMPLE 12.7 Economic dispatch solution including generator limits

Rework Example 12.6 if the units are subject to the following inequality constraints:

$$100 \leqq P_1 \leqq 600 \quad \text{MW}$$

$$400 \leqq P_2 \leqq 1000 \quad \text{MW}$$

SOLUTION At light loads, unit 2 operates at its lower limit of 400 MW, where its incremental operating cost is $dC_2/dP_2 = 15.2$ \$/MWh. Additional load comes from unit 1 until $dC_1/dP_1 = 15.2$ \$/MWh, or

$$\frac{dC_1}{dP_1} = 10 + 16 \times 10^{-3} P_1 = 15.2$$

$$P_1 = 325 \quad \text{MW}$$

For P_T less than 725 MW, where P_1 is less than 325 MW, the incremental operating cost of the area is determined by unit 1 alone.

At heavy loads, unit 1 operates at its upper limit of 600 MW, where its incremental operating cost is $dC_1/dP_1 = 19.60$ \$/MWh. Additional load comes from unit 2 for all values of dC_2/dP_2 greater than 19.60 \$/MWh. At $dC_2/dP_2 = 19.60$ \$/MWh,

$$\frac{dC_2}{dP_2} = 8 + 18 \times 10^{-3} P_2 = 19.60$$

$$P_2 = 644 \quad \text{MW}$$

For P_T greater than 1244 MW, where P_2 is greater than 644 MW, the incremental operating cost of the area is determined by unit 2 alone.

For $725 < P_T < 1244$ MW, neither unit has reached a limit value, and the economic dispatch solution is the same as that given in Table 12.1.

The solution to this example is summarized in Table 12.2 for values of P_T from 500 to 1500 MW.

TABLE 12.2 Economic dispatch solution for Example 12.7

P_T MW	P_1 MW	P_2 MW		dC/dP \$/MWh	C_T \$/h
500	100	400	$\frac{dC_1}{dP_1}$	11.60	5720
600	200	400		13.20	6960
700	300	400		14.80	8360
725	325	400		15.20	8735
800	365	435		15.84	9899
900	418	482		16.68	11525
1000	471	529		17.53	13235
1100	524	576		18.38	15030
1200	576	624		19.22	16910
1244	600	644	$\frac{dC_2}{dP_2}$	19.60	17765
1300	600	700		20.60	18890
1400	600	800		22.40	21040
1500	600	900		24.20	23370

■

EXAMPLE 12.8 PowerWorld Simulator—economic dispatch, including generator limits

PowerWorld Simulator case Example 12_8 uses a five-bus, three-generator lossless case to show the interaction between economic dispatch and the transmission system (see Figure 12.15). The variable operating costs for each of the units are given by

$$C_1 = 10P_1 + 0.016P_1^2 \ \$/\mathrm{h}$$

$$C_2 = 8P_2 + 0.018P_2^2 \ \$/\mathrm{h}$$

$$C_4 = 12P_4 + 0.018P_4^2 \ \$/\mathrm{h}$$

where P_1, P_2, and P_4 are the generator outputs in megawatts. Each generator has minimum/maximum limits of

$$100 \leq P_1 \leq 400 \text{ MW}$$

$$150 \leq P_2 \leq 500 \text{ MW}$$

$$50 \leq P_4 \leq 300 \text{ MW}$$

In addition to solving the power-flow equations, PowerWorld Simulator can simultaneously solve the economic dispatch problem to optimally

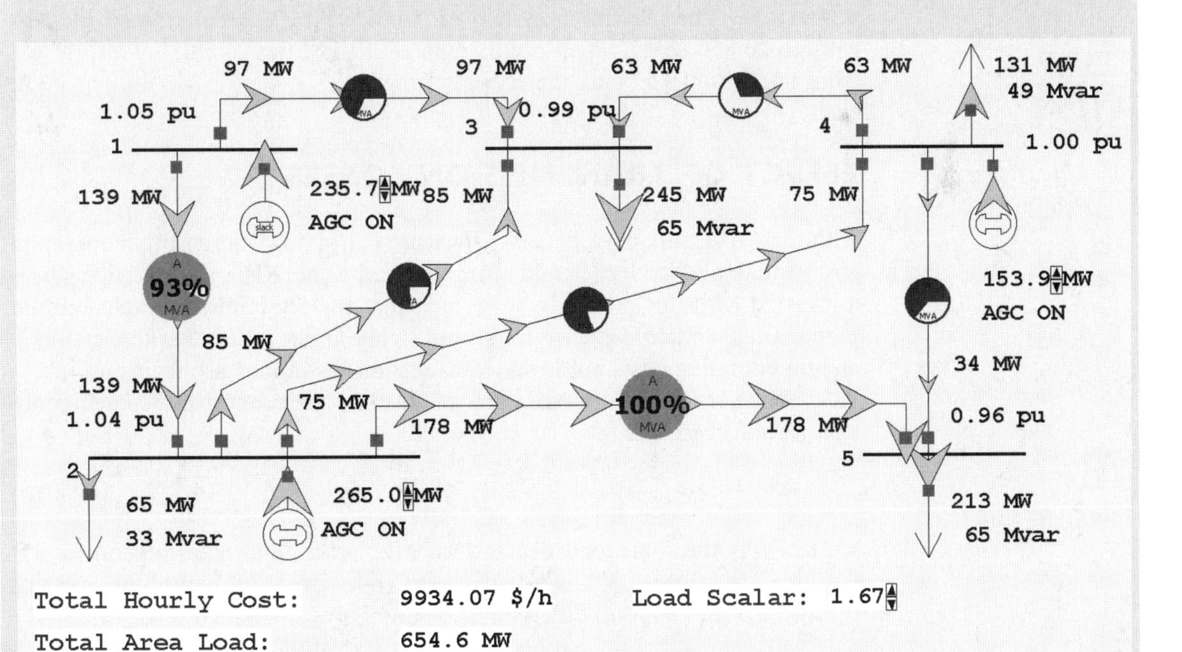

FIGURE 12.15 Example 12.8 with maximum economic loading

allocate the generation in an area. To turn on this option, select **Case Information, Aggregation, Areas...** to view a list of each of the control areas in a case (just one in this example). Then toggle the AGC Status field to ED. Now anytime the power-flow equations are solved, the generator outputs are also changed using the economic dispatch.

Initially the case has a total load of 392 MW, with an economic dispatch of $P_1 = 141$ MW, $P_2 = 181$, and $P_4 = 70$, with an incremental operating cost, λ, of 14.52 \$/MWh. To view a graph showing the incremental cost curves for all of the area generators, right-click on any generator to display the generator's local menu, and then select "All Area Gen IC Curves" (right-click on the graph's axes to change their scaling).

To see how changing load impacts the economic dispatch and power-flow solutions, first select **Tools, Play** to begin the simulation. Then, on the one-line, click on the up/down arrows next to the Load Scalar field. This field is used to scale the load at each bus in the system. Notice that the change in the Total Hourly Cost field is well approximated by the change in the load multiplied by the incremental operating cost.

Determine the maximum amount of load this system can supply without overloading any transmission line with the generators dispatched using economic dispatch.

SOLUTION The maximum system economic loading is determined numerically to be 655 MW (which occurs with a Load Scalar of 1.67), with the line from bus 2 to bus 5 being the critical element. ■

EFFECT OF TRANSMISSION LOSSES

Although one unit may be very efficient with a low incremental operating cost, it may also be located far from the load center. The transmission losses associated with this unit may be so high that the economic dispatch solution requires the unit to decrease its output, while other units with higher incremental operating costs but lower transmission losses increase their outputs.

When transmission losses are included in the economic dispatch problem, (12.4.2) becomes

$$P_1 + P_2 + \cdots + P_N - P_L = P_T \tag{12.4.10}$$

where P_T is the total load demand and P_L is the total transmission loss in the area. In general, P_L is not constant, but depends on the unit outputs $P_1, P_2, \ldots, P_N$. The total differential of (12.4.10) is

$$(dP_1 + dP_2 + \cdots + dP_N) - \left(\frac{\partial P_L}{\partial P_1}\,dP_1 + \frac{\partial P_L}{\partial P_2}\,dP_2 + \cdots + \frac{\partial P_L}{\partial P_N}\,dP_N\right) = 0 \tag{12.4.11}$$

Multiplying (12.4.11) by λ and subtracting the resulting equation from (12.4.5),

$$\left(\frac{dC_1}{dP_1} + \lambda\frac{\partial P_L}{\partial P_1} - \lambda\right) dP_1 + \left(\frac{dC_2}{dP_2} + \lambda\frac{\partial P_L}{\partial P_2} - \lambda\right) dP_2 + \cdots + \left(\frac{dC_N}{dP_N} + \lambda\frac{\partial P_L}{\partial P_N} - \lambda\right) dP_N = 0 \qquad (12.4.12)$$

Equation (12.4.12) is satisfied when each term in parentheses equals zero. That is,

$$\frac{dC_i}{dP_i} + \lambda\frac{\partial P_L}{\partial P_i} - \lambda = 0$$

or

$$\lambda = \frac{dC_i}{dP_i}(L_i) = \frac{dC_i}{dP_i}\left(\frac{1}{1 - \dfrac{\partial P_L}{\partial P_i}}\right) \qquad i = 1, 2, \ldots, N \qquad (12.4.13)$$

Equation (12.4.13) gives the economic dispatch criterion, including transmission losses. Each unit that is not at a limit value operates such that its incremental operating cost dC_i/dP_i multiplied by the *penalty factor* L_i is the same. Note that when transmission losses are negligible, $\partial P_L/\partial P_i = 0$, $L_i = 1$, and (12.4.13) reduces to (12.4.8).

EXAMPLE 12.9 Economic dispatch solution including generator limits and line losses

Total transmission losses for the power system area given in Example 12.7 are given by

$$P_L = 1.5 \times 10^{-4} P_1^2 + 2 \times 10^{-5} P_1 P_2 + 3 \times 10^{-5} P_2^2 \quad \text{MW}$$

where P_1 and P_2 are given in megawatts. Determine the output of each unit, total transmission losses, total load demand, and total operating cost C_T when the area $\lambda = 16.00$ \$/MWh.

SOLUTION Using the incremental operating costs from Example 12.6 in (12.4.13),

$$\frac{dC_1}{dP_1}\left(\frac{1}{1 - \dfrac{\partial P_L}{\partial P_1}}\right) = \frac{10 + 16 \times 10^{-3} P_1}{1 - (3 \times 10^{-4} P_1 + 2 \times 10^{-5} P_2)} = 16.00$$

$$\frac{dC_2}{dP_2}\left(\frac{1}{1 - \dfrac{\partial P_L}{\partial P_2}}\right) = \frac{8 + 18 \times 10^{-3} P_2}{1 - (6 \times 10^{-5} P_2 + 2 \times 10^{-5} P_1)} = 16.00$$

Rearranging the above two equations,

$$20.8 \times 10^{-3} P_1 + 32 \times 10^{-5} P_2 = 6.00$$

$$32 \times 10^{-5} P_1 + 18.96 \times 10^{-3} P_2 = 8.00$$

Solving,

$$P_1 = 282 \quad \text{MW} \qquad P_2 = 417 \quad \text{MW}$$

Using the equation for total transmission losses,

$$P_L = 1.5 \times 10^{-4}(282)^2 + 2 \times 10^{-5}(282)(417) + 3 \times 10^{-5}(417)^2$$

$$= 19.5 \quad \text{MW}$$

From (12.4.10), the total load demand is

$$P_T = P_1 + P_2 - P_L = 282 + 417 - 19.5 = 679.5 \quad \text{MW}$$

Also, using the cost formulas given in Example 12.6, the total operating cost is

$$C_T = C_1 + C_2 = 10(282) + 8 \times 10^{-3}(282)^2 + 8(417) + 9 \times 10^{-3}(417)^2$$

$$= 8357 \quad \$/\text{h}$$

Note that when transmission losses are included, λ given by (12.4.13) is no longer the incremental operating cost of the area. Instead, λ is the unit incremental operating cost dC_i/dP_i multiplied by the unit penalty factor L_i. ■

EXAMPLE 12.10 PowerWorld Simulator—economic dispatch, including generator limits and line losses

Example 12.10 repeats the Example 12.8 power system, except that now losses are included, with each transmission line modeled with an R/X ratio of 1/3 (see Figure 12.16). The current value of each generator's loss sensitivity, $\partial P_L/\partial P_G$, is shown immediately below the generator's MW output field. Calculate the penalty factors L_i and verify that the economic dispatch shown in the figure is optimal. Assume a Load Scalar of 1.0.

SOLUTION From (12.4.13) the condition for optimal dispatch is

$$\lambda = dC_i/dP_i(1/(1 - \partial P_L/\partial P_i) = dC_i/dP_i L_i \qquad i = 1, 2, \ldots, N$$

with

$$L_i = 1/(1 - \partial P_L/\partial P_i)$$

Therefore, $L_1 = 1.0$, $L_2 = 0.9733$, and $L_4 = 0.9238$.

$$\text{With } P_1 = 130.1 \text{ MW}, \quad dC_1/dP_1 * L_1 = (10 + 0.032 * 130.1) * 1.0$$
$$= 14.16 \quad \$/\text{MWh}$$

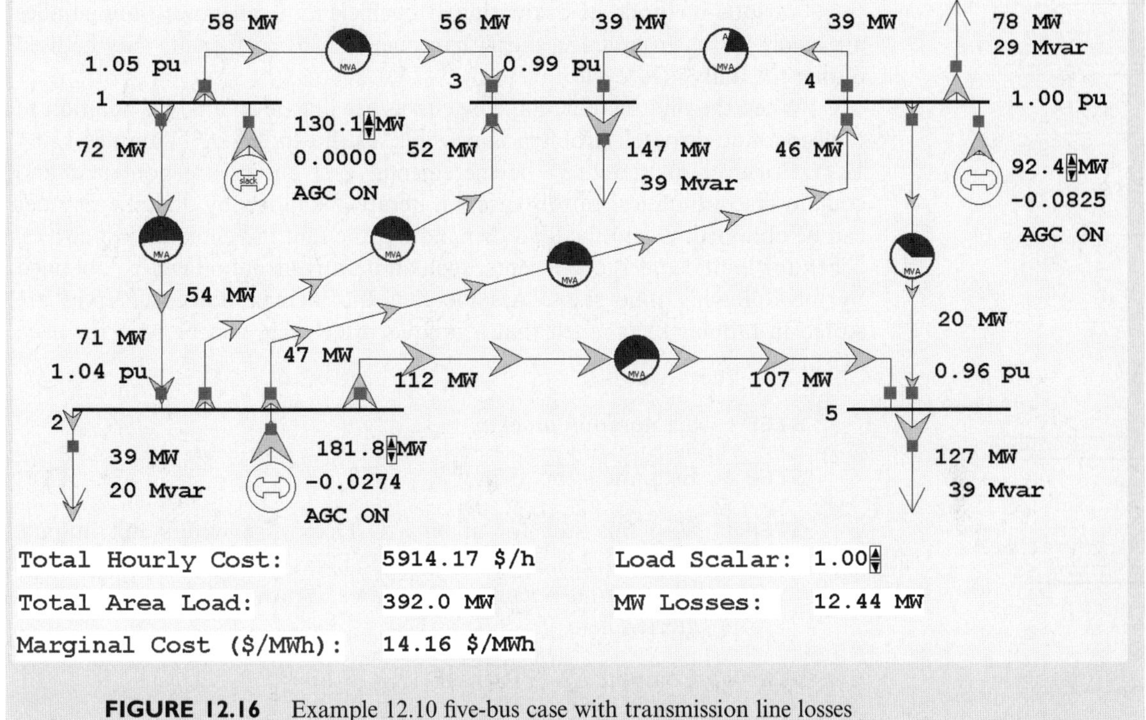

FIGURE 12.16 Example 12.10 five-bus case with transmission line losses

$$\text{With } P_2 = 181.8 \text{ MW}, \quad dC_2/dP_2 * L_2 = (8 + 0.036 * 181.8) * 0.9733 = 14.16 \quad \$/\text{MWh}$$

$$\text{With } P_4 = 92.4 \text{ MW}, \quad dC_4/dP_4 * L_4 = (12 + 0.036 * 92.4) * 0.9238 = 14.16 \quad \$/\text{MWh}$$

■

In Example 12.9, total transmission losses are expressed as a quadratic function of unit output powers. For an area with N units, this formula generalizes to

$$P_L = \sum_{i=1}^{N} \sum_{j=1}^{N} P_i B_{ij} P_j \tag{12.4.14}$$

where the B_{ij} terms are called *loss coefficients* or B *coefficients*. The B coefficients are not truly constant, but vary with unit loadings. However, the B coefficients are often assumed constant in practice since the calculation of $\partial P_L / \partial P_i$ is thereby simplified. Using (12.4.14),

$$\frac{\partial P_L}{\partial P_i} = 2 \sum_{j=1}^{N} B_{ij} P_j \tag{12.4.15}$$

This equation can be used to compute the penalty factor L_i in (12.4.13).

Various methods of evaluating B coefficients from power-flow studies are available [6]. In practice, more than one set of B coefficients may be used during the daily load cycle.

When the unit incremental cost curves are linear, an analytic solution to the economic dispatch problem is possible, as illustrated by Examples 12.6–12.8. However, in practice, the incremental cost curves are nonlinear and contain discontinuities. In this case, an iterative solution by digital computer can be obtained. Given the load demand P_T, the unit incremental cost curves, generator limits, and B coefficients, such an iterative solution can be obtained by the following nine steps. Assume that the incremental cost curves are stored in tabular form, such that a unique value of P_i can be read for each dC_i/dP_i.

STEP 1 Set iteration index $m = 1$.

STEP 2 Estimate mth value of λ.

STEP 3 Skip this step for all $m > 1$. Determine initial unit outputs P_i $(i = 1, 2, \ldots, N)$. Use $dC_i/dP_i = \lambda$ and read P_i from each incremental operating cost table. Transmission losses are neglected here.

STEP 4 Compute $\partial P_L/\partial P_i$ from (12.4.15) $(i = 1, 2, \ldots, N)$.

STEP 5 Compute dC_i/dP_i from (12.4.13) $(i = 1, 2, \ldots, N)$.

STEP 6 Determine updated values of unit output P_i $(i = 1, 2, \ldots, N)$. Read P_i from each incremental operating cost table. If P_i exceeds a limit value, set P_i to the limit value.

STEP 7 Compare P_i determined in Step 6 with the previous value $(i = 1, 2, \ldots, N)$. If the change in each unit output is less than a specified tolerance ε_1, go to Step 8. Otherwise, return to Step 4.

STEP 8 Compute P_L from (12.4.14).

STEP 9 If $\left| \left(\sum_{i=1}^{N} P_i \right) - P_L - P_T \right|$ is less than a specified tolerance ε_2, stop. Otherwise, set $m = m + 1$ and return to Step 2.

Instead of having their values stored in tabular form for this procedure, the incremental cost curves could instead be represented by nonlinear functions such as polynomials. Then, in Step 3 and Step 5, each unit output P_i would be computed from the nonlinear functions instead of being read from a table. Note that this procedure assumes that the total load demand P_T is constant. In practice, this economic dispatch program is executed every few minutes with updated values of P_T.

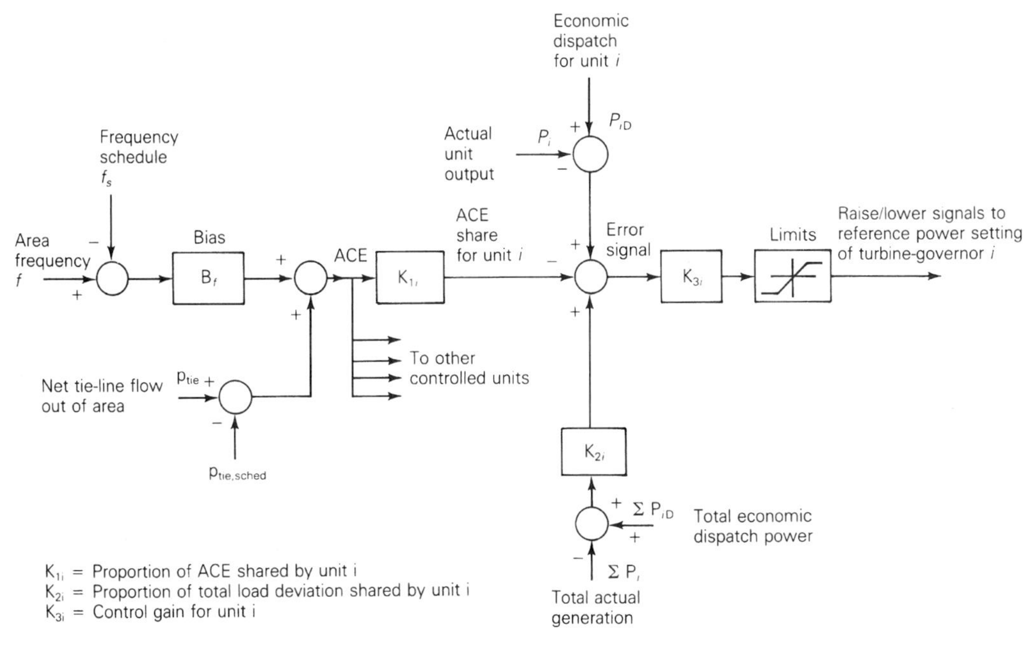

FIGURE 12.17 Automatic generation control [11] (A. J. Wood and B. F. Wollenberg, Power Generation, Operation, and Control (New York: Wiley, 1989))

COORDINATION OF ECONOMIC DISPATCH WITH LFC

Both the load-frequency control (LFC) and economic dispatch objectives are achieved by adjusting the reference power settings of turbine-governors on control. Figure 12.17 shows an *automatic generation control* strategy for achieving both objectives in a coordinated manner. As shown, the area control error (ACE) is first computed, and a share K_{1i} ACE is allocated to each unit. Second, the deviation of total actual generation from total desired generation is computed, and a share $K_{2i}\sum(P_{iD} - P_i)$ is allocated to unit i. Third, the deviation of actual generation from desired generation of unit i is computed, and $(P_{iD} - P_i)$ is allocated to unit i. An error signal formed from these three components and multiplied by a control gain K_{3i} determines the raise or lower signals that are sent to the turbine-governor of each unit i on control.

In practice, raise or lower signals are dispatched to the units at discrete time intervals of 2 to 10 seconds. The desired outputs P_{iD} of units on control, determined from the economic dispatch program, are updated at slower intervals, typically every 2 to 10 minutes.

OTHER TYPES OF UNITS

The economic dispatch criterion has been derived for a power system area consisting of fossil-fuel generating units. In practice, however, an area has a mix of different types of units including fossil-fuel, nuclear, pumped-storage hydro, hydro, wind, and other types.

Although the fixed costs of a nuclear unit may be high, their operating costs are low due to inexpensive nuclear fuel. As such, nuclear units are normally base-loaded at their rated outputs. That is, the reference power settings of turbine-governors for nuclear units are held constant at rated output; therefore, these units do not participate in LFC or economic dispatch.

Pumped-storage hydro is a form of energy storage. During off-peak hours these units are operated as synchronous motors to pump water to a higher elevation. Then during peak-load hours the water is released and the units are operated as synchronous generators to supply power. As such, pumped-storage hydro units are used for light-load build-up and peak-load shaving. Economic operation of the area is improved by pumping during off-peak hours when the area λ is low, and by generating during peak-load hours when λ is high. Techniques are available for incorporating pumped-storage hydro units into economic dispatch of fossil-fuel units [7].

In an area consisting of hydro plants located along a river, the objective is to maximize the energy generated over the yearly water cycle rather than to minimize total operating costs. Reservoirs are used to store water during high-water or light-load periods, although some water may have to be released through spillways. Also, there are constraints on water levels due to river transportation, irrigation, or fishing requirements. Optimal strategies are available for coordinating outputs of plants along a river [8]. Economic dispatch strategies for mixed fossil-fuel/hydro systems are also available [9, 10, 11].

Techniques are also available for including reactive power flows in the economic dispatch formulation, whereby both active and reactive powers are selected to minimize total operating costs. In particular, reactive injections from generators, switched capacitor banks, and static var systems, along with transformer tap settings, can be selected to minimize transmission-line losses [11]. However, electric utility companies usually control reactive power locally. That is, the reactive power output of each generator is selected to control the generator terminal voltage, and the reactive power output of each capacitor bank or static var system located at a power system bus is selected to control the voltage magnitude at that bus. In this way, the reactive power flows on transmission lines are low, and the need for central dispatch of reactive power is eliminated.

12.5 OPTIMAL POWER FLOW

Economic dispatch has one significant shortcoming—it ignores the limits imposed by the devices in the transmission system. Each transmission line and transformer has a limit on the amount of power that can be transmitted through it, with the limits arising because of thermal, voltage, or stability considerations (Section 5.6). Traditionally, the transmission system was designed so that when the generation was dispatched economically there would be no limit violations. Hence, just solving economic dispatch was usually sufficient. However, with the worldwide trend toward deregulation of the electric utility industry, the transmission system is becoming increasingly constrained. For example, in the PJM power market in the eastern United States the costs associated with active transmission line and transformer limit violations increased from $65 million in 1999 to almost $2.1 billion in 2005 [14].

The solution to the problem of optimizing the generation while enforcing the transmission lines is to combine economic dispatch with the power flow. The result is known as the optimal power flow (OPF). There are several methods for solving the OPF, with the linear programming (LP) approach the most common [13] (this is the technique used with PowerWorld Simulator). The LP OPF solution algorithm iterates between solving the power flow to determine the flow of power in the system devices and solving an LP to economically dispatch the generation (and possibility other controls) subject to the transmission system limits. In the absence of system elements loaded to their limits, the OPF generation dispatch will be identical to the economic dispatch solution, and the marginal cost of energy at each bus will be identical to the system λ. However, when one or more elements are loaded to their limits the economic dispatch becomes constrained, and the bus marginal energy prices are no longer identical. In some electricity markets these marginal prices are known as the Locational Marginal Prices (LMPs) and are used to determine the wholesale price of electricity at various locations in the system. For example, the real-time LMPs for the Midwest ISO are available online at www.midwestmarket.org.

EXAMPLE 12.11 PowerWorld Simulator—optimal power flow

PowerWorld Simulator case Example 12_11 duplicates the five-bus case from Example 12.8, except that the case will be solved using PowerWorld Simulator's LP OPF algorithm (see Figure 12.18). To turn on the OPF option, first select **Case Information, Aggregution, Areas...**, and toggle the AGC Status field to OPF. Finally, rather than solving the case with the "Single Solution" button, select **Add-ons, Primal LP** to solve using the LP OPF. Initially the OPF solution matches the ED solution from Example 12.8 since there are no overloaded lines. The green-colored fields immediately to the right of the buses show the marginal cost of supplying electricity to each bus in the system (i.e., the bus LMPs). With the system initially unconstrained, the bus marginal prices are all identical at $14.53/MWh, with a Load Scalar of 1.0.

Now increase the Load Scalar field from 1.00 to the maximum economic loading value, determined to be 1.67 in Example 12.8, and again select **Add-ons, Primal LP**. The bus marginal prices are still all identical, now at a value of $17.52/MWh, with the line from bus 2 to 5 just reaching its maximum

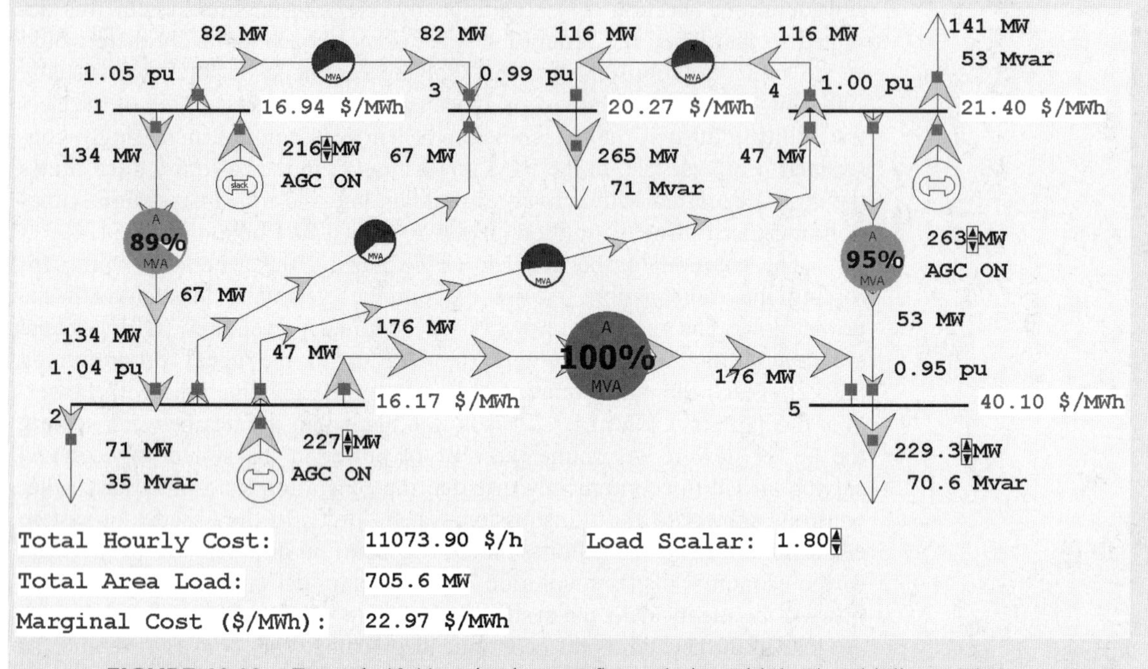

FIGURE 12.18 Example 12.11 optimal power flow solution with load multiplier = 1.80

value. For load scalar values above 1.67, the line from bus 2 to bus 5 becomes constrained, with a result that the bus marginal prices on the constrained side of the line become higher than those on the unconstrained side.

With the load scalar equal to 1.80, numerically verify that the price of power at bus 5 is approximately $40.08/MWh.

SOLUTION The easiest way to numerically verify the bus 5 price is to increase the load at bus 5 by a small amount and compare the change in total system operating cost. With a load scalar of 1.80, the bus 5 MW load is 229.3 MW with a case hourly cost of $11,074. Increasing the bus 5 load by 1.8 MW and resolving the LP OPF gives a new cost of $11,147, a change of about $40.5/MWh (note that this increase in load also increases the bus 5 price to over $42/MWh). Because of the constraint, the price of power at bus 5 is actually more than double the incremental cost of the most expensive generator! ■

PROBLEMS

SECTION 12.1

12.1 The block-diagram representation of a closed-loop automatic regulating system, in which generator voltage control is accomplished by controlling the exciter voltage, is shown in Figure 12.19. T_a, T_e, and T_f are the time constants associated with the amplifier, exciter, and generator field circuit, respectively. (a) Find the open-loop transfer function G(s). (b) Evaluate the minimum open-loop gain such that the steady-state error Δe_{ss} does not exceed 1%. (c) Discuss the nature of the dynamic response of the system to a step change in the reference input voltage.

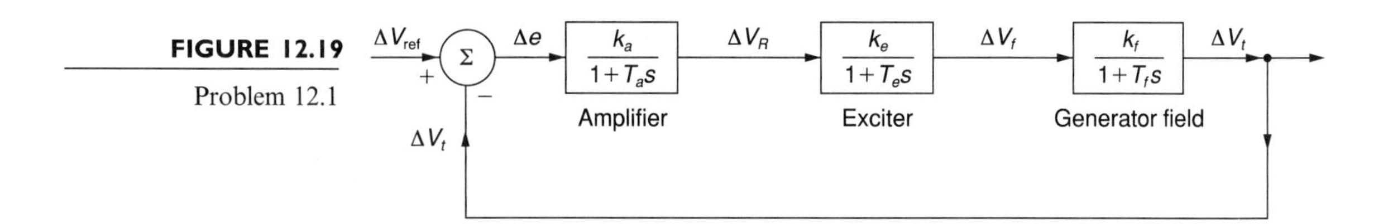

FIGURE 12.19 Problem 12.1

12.2 The Automatic Voltage Regulator (AVR) system of a generator is represented by the simplified block diagram shown in Figure 12.20, in which the sensor is modeled by a simple first-order transfer function. The voltage is sensed through a voltage transformer and then rectified through a bridge rectifier. Parameters of the AVR system are given as follows.

	Gain	Time Constant (seconds)
Amplifier	K_A	$\tau_A = 0.1$
Exciter	$K_E = 1$	$\tau_E = 0.4$
Generator	$K_G = 1$	$\tau_G = 1.0$
Sensor	$K_R = 1$	$\tau_R = 0.05$

(a) Determine the open-loop transfer function of the block diagram and the closed-loop transfer function relating the generator terminal voltage $V_t(s)$ to the reference voltage $V_{ref}(s)$. (b) For the range of K_A from 0 to 12.16, comment on the stability of the system. (c) For $K_A = 10$, evaluate the steady-state step response and steady-state error.

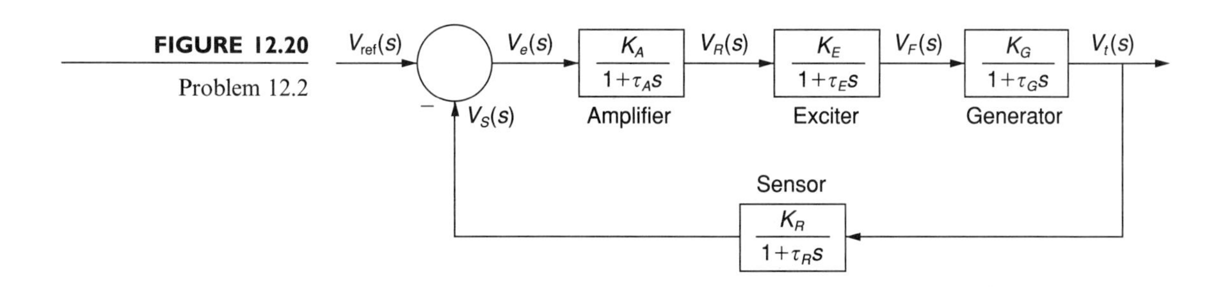

FIGURE 12.20 Problem 12.2

PW **12.3** Open PowerWorld Simulator case Problem 12_3. This case models the system from Example 12.1 except with the rate feedback gain constant, K_f, has been set to zero and the simulation end time was increased to 30 seconds. Without rate feedback the system voltage response will become unstable if the amplifier gain, K_a, becomes too large. In the simulation this instability will be indicated by undamped oscillations in the terminal voltage (because of the limits on V_r the response does not grow to infinity but rather bounces between the limits). Using transient stability simulations, iteratively determine the approximate value of K_a at which the system becomes unstable. The value of K_a can be on the **Generator Information Dialog, Stability, Exciters** page.

PW

12.4 One of the disadvantages of the IEEET1 exciter is following a fault the terminal voltage does not necessarily return to its pre-fault value. Using PowerWorld Simulator case Problem 12_3 determine the pre-fault bus 4 terminal voltage and field voltage. Then use the simulation to determine the final, post-fault values for these fields for $K_a = 100$, 200, 50, and 10. Referring to Figure 12.3, what is the relationship between the reference voltage, and the steady-state terminal voltage and the field voltage?

SECTION 12.2

12.5 An area of an interconnected 60-Hz power system has three turbine-generator units rated 200, 300, and 500 MVA. The regulation constants of the units are 0.03, 0.04, and 0.06 per unit, respectively, based on their ratings. Each unit is initially operating at one-half its own rating when the load suddenly decreases by 150 MW. Determine (a) the unit area frequency response characteristic β on a 100-MVA base, (b) the steady-state increase in area frequency, and (c) the MW decrease in mechanical power output of each turbine. Assume that the reference power setting of each turbine-governor remains constant. Neglect losses and the dependence of load on frequency.

12.6 Each unit in Problem 12.5 is initially operating at one-half its own rating when the load suddenly increases by 100 MW. Determine (a) the steady-state decrease in area frequency, and (b) the MW increase in mechanical power output of each turbine. Assume that the reference power setting of each turbine-generator remains constant. Neglect losses and the dependence of load on frequency.

12.7 Each unit in Problem 12.5 is initially operating at one-half its own rating when the frequency increases by 0.005 per unit. Determine the MW decrease of each unit. The reference power setting of each turbine-governor is fixed. Neglect losses and the dependence of load on frequency.

12.8 Repeat Problem 12.7 if the frequency decreases by 0.005 per unit. Determine the MW increase of each unit.

12.9 An interconnected 60-Hz power system consisting of one area has two turbine-generator units, rated 500 and 750 MVA, with regulation constants of 0.04 and 0.05 per unit, respectively, based on their respective ratings. When each unit carries a 300-MVA steady-state load, let the area load suddenly increase by 250 MVA. (a) Compute the area frequency response characteristic β on a 1000-MVA base. (b) Calculate Δf in per-unit on a 60-Hz base and in Hz.

PW **12.10** Open PowerWorld Simulator case Problem 12_10. The case models the system from Example 12.4 except 1) the load increases is a 50% rise at bus 6 for a total increase of 250 MW (from 500 MW to 750 MW), 2) the value of R for generator 1 is changed from 0.05 to 0.04 per unit. Repeat Example 12.4 using these modified values.

PW **12.11** Open PowerWorld Simulator case Problem 12_11, which includes a transient stability representation of the system from Example 6.13. Each generator is modeled using a two-axis machine model, an IEEE Type 1 exciter and a TGOV1 governor with R = 0.05 per unit (a summary of the generator models is available by selecting either **Stability Case Info, Transient Stability Generator Summary** which includes the generator MVA base, or **Stability Case Info, Transient Stability Case Summary**). The contingency is the loss of the generator at bus 50, which initially has 42.1 MW of generation. Analytically determine the steady-state frequency error in Hz following this contingency. Use PowerWorld Simulator to confirm this result; also determine the magnitude and time of the largest bus frequency deviation.

PW **12.12** Repeat Problem 12.12 except first double the H value for each of the machines. This can be most easily accomplished by selecting **Stability Case Info, Transient Stability Case Summary** to view the summary form. Right click on the line corresponding to the Machine Model class, and then select Show Dialog to view an editable form of the model parameters. Compare the magnitude and time of the largest bus frequency deviations between Problem 12.12 and 12.11.

12.13 For a large, 60 Hz, interconnected electrical system assume that following the loss of two 1400 MW generators (for a total generation loss of 2800 MW) the change in frequency is −0.12 Hz. If all the on-line generators that are available to participate in frequency regulation have an R of 0.05 per unit (on their own MVA base), estimate the total MVA rating of these units.

SECTION 12.3

12.14 A 60-Hz power system consists of two interconnected areas. Area 1 has 1200 MW of generation and an area frequency response characteristic $\beta_1 = 600$ MW/Hz. Area 2 has 1800 MW of generation and $\beta_2 = 800$ MW/Hz. Each area is initially operating at one-half its total generation, at $\Delta p_{tie1} = \Delta p_{tie2} = 0$ and at 60 Hz, when the load in area 1 suddenly increases by 400 MW. Determine the steady-state frequency error and the steady-state tie-line error Δp_{tie} of each area. Assume that the reference power settings of all turbine-governors are fixed. That is, LFC is not employed in any area. Neglect losses and the dependence of load on frequency.

12.15 Repeat Problem 12.14 if LFC is employed in area 2 alone. The area 2 frequency bias coefficient is set at $B_{f2} = \beta_2 = 800$ MW/Hz. Assume that LFC in area 1 is inoperative due to a computer failure.

12.16 Repeat Problem 12.14 if LFC is employed in both areas. The frequency bias coefficients are $B_{f1} = \beta_1 = 600$ MW/Hz and $B_{f2} = \beta_2 = 800$ MW/Hz.

12.17 Rework Problems 12.15 through 12.16 when the load in area 2 suddenly decreases by 300 MW. The load in area 1 does not change.

12.18 On a 1000-MVA common base, a two-area system interconnected by a tie line has the following parameters:

Area	1	2
Area Frequency Response Characteristic	$\beta_1 = 0.05$ per unit	$\beta_2 = 0.0625$ per unit
Frequency-Dependent Load Coefficient	$D_1 = 0.6$ per unit	$D_2 = 0.9$ per unit
Base Power	1000 MVA	1000 MVA
Governor Time Constant	$\tau_{g1} = 0.25$ s	$\tau_{g2} = 0.3$ s
Turbine Time Constant	$\tau_{t1} = 0.5$ s	$\tau_{t2} = 0.6$ s

The two areas are operating in parallel at the nominal frequency of 60 Hz. The areas are initially operating in steady state with each area supplying 1000 MW when a sudden load change of 187.5 MW occurs in area 1. Compute the new steady-state frequency and change in tie-line power flow (a) without LFC, and (b) with LFC.

SECTION 12.4

12.19 The fuel-cost curves for two generators are given as follows:

$$C_1(P_1) = 600 + 15 \cdot P_1 + 0.05 \cdot (P_1)^2$$
$$C_2(P_2) = 700 + 20 \cdot P_2 + 0.04 \cdot (P_2)^2$$

Assuming the system is lossless, calculate the optimal dispatch values of P_1 and P_2 for a total load of 1000 MW, the incremental operating cost, and the total operating cost.

12.20 Rework problem 12.19 except assume that the limit outputs are subject to the following inequality constraints:

$$200 \le P_1 \le 800 \text{ MW}$$
$$100 \le P_2 \le 500 \text{ MW}$$

12.21 Rework problem 12.19 except assume the 1000 MW value also includes losses, and that the penalty factor for the first unit is 1.0, and for the second unit 0.95.

12.22 The fuel-cost curves for a two generators power system are given as follows:

$$C_1(P_1) = 600 + 15 \cdot P_1 + 0.05 \cdot (P_1)^2$$
$$C_2(P_2) = 700 + 20 \cdot P_2 + 0.04 \cdot (P_2)^2$$

While the system losses can be approximated as

$$P_L = 2 \times 10^{-4}(P_1)^2 + 3 \times 10^{-4}(P_2)^2 - 4 \times 10^{-4} P_1 P_2 \text{ MW}$$

If the system is operating with a marginal cost (λ) of \$60/h, determine the output of each unit, the total transmission losses, the total load demand, and the total operating cost.

12.23 Expand the summations in (12.4.14) for $N = 2$, and verify the formula for $\partial P_L/\partial P_i$ given by (12.4.15). Assume $B_{ij} = B_{ji}$.

12.24 Given two generating units with their respective variable operating costs as:

$$C_1 = 0.01P^2_{G1} + 2P_{G1} + 100 \quad \$/hr \qquad \text{for } 25 \leq P_{G1} \leq 150 \text{ MW}$$

$$C_2 = 0.004P^2_{G2} + 2.6P_{G2} + 80 \quad \$/hr \qquad \text{for } 30 \leq P_{G2} \leq 200 \text{ MW}$$

Determine the economically optimum division of generation for $55 \leq P_L \leq 350$ MW. In particular, for $P_L = 282$ MW, compute P_{G1} and P_{G2}. Neglect transmission losses.

PW **12.25** Resolve Example 12.8, except with the generation at bus 2 set to a fixed value (i.e., modeled as off of AGC). Plot the variation in the total hourly cost as the generation at bus 2 is varied between 1000 and 200 MW in 5-MW steps, resolving the economic dispatch at each step. What is the relationship between bus 2 generation at the minimum point on this plot and the value from economic dispatch in Example 12.8? Assume a Load Scalar of 1.0.

PW **12.26** Using PowerWorld case Example 12_10, with the Load Scalar equal to 1.0, determine the generation dispatch that minimizes system losses (*Hint:* Manually vary the generation at buses 2 and 4 until their loss sensitivity values are zero). Compare the operating cost between this solution and the Example 12.10 economic dispatch result. Which is better?

PW **12.27** Repeat Problem 12.26, except with the Load Scalar equal to 1.4.

SECTION 12.5

PW **12.28** Using LP OPF with PowerWorld Simulator case Example 12_11, plot the variation in the bus 5 marginal price as the Load Scalar is increased from 1.0 in steps of 0.02. What is the maximum possible load scalar without overloading any transmission line? Why is it impossible to operate without violations above this value?

PW **12.29** Load PowerWorld Simulator case Problem 12_30. This case models a slightly modified version of the 37 bus case from Example 6.13 with generator cost information, but also with two of the three lines between buses BLT69 and UIUC69 open. When the case is loaded the "Total Cost" field shows the economic dispatch solution, which results in an overload on the remaining line between buses BLT69 and UIUC69. Before solving the case, select **LP OPF, OFP Buses** to view the bus LMPs, noting that they are all identical. Then Select **LP OPF, Primal LP** to solve the case using the OPF, and again view the bus LMPs. Verify the LMP at the UIUC69 bus by manually changing the load at the bus by one MW, and then noting the change in the Total Cost field. Repeat for the DEMAR69 bus. Note, because of solution convergence tolerances the manually calculated results may not exactly match the OFP calculated bus LMPs.

CASE STUDY QUESTIONS

1. What is meant by generator black-start capability, why is it needed, and what types of generators are best at providing black-start capability?
2. Why are overvoltages a concern during the restoration of the transmission system?
3. Research a recent large power system outage, describing some of the unique aspects associated with the restoration of the power system.

REFERENCES

1. IEEE Committee Report, "Computer Representation of Excitation Systems," *IEEE Transactions PAS*, vol. PAS-87 (June 1968), pp. 1460–1464.
2. M. S. Sarma, *Electric Machines* 2d ed. (Boston, PWS Publishing 1994).
3. IEEE Committee Report, "Dynamic Models for Steam and Hydro Turbines in Power System Studies," *IEEE Transactions PAS*, vol. PAS-92, no. 6 (November/December 1973), pp. 1904–1915.
4. N. Cohn, *Control of Generation and Power Flow on Interconnected Systems* (New York: Wiley, 1971).
5. L. K. Kirchmayer, *Economic Operation of Power Systems* (New York: Wiley, 1958).
6. L. K. Kirchmayer and G. W. Stagg, "Evaluation of Methods of Coordinating Incremental Fuel Costs and Incremental Transmission Losses," *Transactions AIEE*, vol. 71, part III (1952), pp. 513–520.
7. G. H. McDaniel and A. F. Gabrielle, "Dispatching Pumped Storage Hydro," *IEEE Transmission PAS*, vol. PAS-85 (May 1966), pp. 465–471.
8. E. B. Dahlin and E. Kindingstad, "Adaptive Digital River Flow Predictor for Power Dispatch," *IEEE Transactions PAS*, vol. PAS-83 (April 1964), pp. 320–327.
9. L. K. Kirchmayer, *Economic Control of Interconnected Systems* (New York: Wiley, 1959).
10. J. H. Drake et al., "Optimum Operation of a Hydrothermal System," *Transactions AIEE* (*Power Apparatus and Systems*), vol. 62 (August 1962), pp. 242–250.

12. A. J. Wood and B. F. Wollenberg, *Power Generation, Operation, and Control* (New York: Wiley, 1989).

12. R. J. Thomas et al., "Transmission System Planning—The Old World Meets The New," *Proceedings of The IEEE*, 93, 11 (November 2005), pp. 2026–2035.

13. B. Stott and J. L. Marinho, "Linear Programming for Power System Network Security Applications," *IEEE Trans. on Power Apparatus and Systems*, vol. PAS-98, (May/June 1979), pp. 837–848.
14. 2005 PJM State of the Market Report, available online at http://www.pjm.com/markets/market-monitor/som.html.

15. M. M. Adibi and L. H. Fink, "Overcoming Restoration Challenges Associated with Major Power System Disturbances", *IEEE Power and Energy Magazine*, 4, 5 (September/October 2006), pp. 68–77.

16. IEEE PES Task Force Report, "Interconnected Power System Response to Generation Governing: Present Practice and Outstanding Concerns," IEEE 07TP180, May 2007.

17. K. Clark, N. W. Miller, J. J. Sanchez-Gasca, "Modeling of GE Wind Turbine-Generators for Grid Studies," Version 4.4, GE Energy, Schenectady, NY, September 2009.

18. E. H. Camm, et. al., "Characteristics of Wind Turbine Generators for Wind Power Plants," Proc. IEEE 2009 General Meeting, Calgary, AB, July 2009.

Three-phase 600 MVA, 22-kV Generator Step-up Transformer at the Jim Bridger Power Plant, Rock Springs, WY, USA (Courtesy of PacifiCorp)

13

TRANSMISSION LINES: TRANSIENT OPERATION

Transient overvoltages caused by lightning strikes to transmission lines and by switching operations are of fundamental importance in selecting equipment insulation levels and surge-protection devices. We must, therefore, understand the nature of transmission-line transients.

During our study of the steady-state performance of transmission lines in Chapter 5, the line constants R, L, G, and C were recognized as distributed rather than lumped constants. When a line with distributed constants is subjected to a disturbance such as a lightning strike or a switching operation, voltage and current waves arise and travel along the line at a velocity near the speed of light. When these waves arrive at the line terminals, reflected voltage and current waves arise and travel back down the line, superimposed on the initial waves.

Because of line losses, traveling waves are attenuated and essentially die out after a few reflections. Also, the series inductances of transformer windings effectively block the disturbances, thereby preventing them from entering generator windings. However, due to the reinforcing action of several reflected waves, it is possible for voltage to build up to a level that could cause transformer insulation or line insulation to arc over and suffer damage.

Circuit breakers, which can operate within 50 ms, are too slow to protect against lightning or switching surges. Lightning surges can rise to peak levels within a few microseconds and switching surges within a few hundred microseconds—fast enough to destroy insulation before a circuit breaker could open. However, protective devices are available. Called surge arresters, these can be used to protect equipment insulation against transient overvoltages. These devices limit voltage to a ceiling level and absorb the energy from lightning and switching surges.

We begin this chapter with a discussion of traveling waves on single-phase lossless lines (Section 13.1). We present boundary conditions in Section 13.2 and the Bewley lattice diagram for organizing reflections in Section 13.3. We derive discrete-time models of single-phase lines and of lumped RLC elements in Section 13.4, and discuss the effects of line losses and multiconductor lines in Sections 13.5 and 13.6. In Section 13.7 we discuss power system overvoltages including lightning surges, switching surges, and power-frequency overvoltages, followed by an introduction to insulation coordination in Section 13.8.

CASE STUDY

Two case-study reports are presented here. The first describes metal oxide varistor (MOV) arresters used by electric utilities to protect transmission and distribution equipment against transient overvoltages in power systems with rated voltages through 345 kV [22]. The second describes the presently installed capacity of wind generation in North America and the impacts of wind generation on operations. Some Independent System Operators (ISOs) have developed wind-forecasting tools for real-time and day-ahead forecasts to help in determining the impact of variable wind-generation resources on dispatch requirements and in establishing market prices for reliable and economic system operation. As the amount of wind generation installed in power grids increases, the occurrence of large and rapid changes in wind-power production is becoming a significant grid management issue [23].

*VariSTAR® Type AZE Surge Arresters for Systems through 345 kV ANSI/IEEE C62.11 Certified Station Class Arresters**

GENERAL

The VariSTAR AZE Surge Arrester offers the latest in metal oxide varistor (MOV) technology for the economical protection of power and substation equipment. This arrester is gapless and constructed of a single series column of MOV disks. The arrester is designed and tested to the requirements of ANSI/IEEE Standard C62.11, and is available in ratings suitable for the transient overvoltage protection of electrical equipment on systems through 345 kV.

Cooper Power Systems assures the design integrity of the AZE arrester through a rigorous testing program conducted at our Thomas A. Edison Technical Center and at the factory in Olean, NY. The availability of complete "in-house" testing facilities

TABLE 1 AZE Series S (AZES) Ratings and Characteristics

Arrester Characteristic	Rating
System Application Voltages	3–345 kV
Arrester Voltage Ratings	3–360 kV
Rated Discharge Energy, (kJ/kV of MCOV)	
Arrester Ratings: 3–108 kV	3.4
120–240 kV	5.6
258–360 kV	8.9
System Frequency	50/60 Hz
Impulse Classifying Current	10 kA
High Current Withstand	100 kA
Pressure Relief Rating, kA rms sym	
Metal-Top Designs	65 kA
Cubicle-Mount Designs	40 kA
Cantilever Strength (in-lbs)*	
Metal-Top Designs:	
3–240 kV	90,000†
258–360 kV	120,000†

*Maximum working load should not exceed 40% of this value.
(August 1997. New Issue. © Cooper Industries, Inc.)
†90,000 in-lb = 10,000 N-m
120,000 in-lb = 13,500 N-m

assures that as continuous process improvements are made, they are professionally validated to high technical standards.

Table 1, shown above, contains information on some of the specific ratings and characteristics of AZE Series S (AZES) surge arresters.

CONSTRUCTION

External

The Type AZE station class arrester is available in two design configurations—a metal-top design in ratings 3–360 kV and a cubicle-mount design in ratings 3–48 kV. Cubicle-mount designs are ideally suited for confined spaces where clearances between live parts are limited.

The wet-process porcelain housing features an alternating shed design (ratings > 48 kV) that provides excellent resistance to the effects of atmospheric housing contamination. AZE arresters are available with optional extra creepage porcelains for use in areas with extreme natural atmospheric and man-made pollution.

The dielectric properties of the porcelain are coordinated with the electrical protective characteristics of the arrester. The unit end castings are of a corrosion-resistant aluminum alloy configured for interchangeable mounting with other manufacturers' arresters for ease in upgrading to the VariSTAR arrester technology. This three-footed mounting is provided on a 8.75 to 10 inch (22 to 25 cm) diameter pattern for customer supplied 0.5 inch (1.3 cm) diameter hardware.

High cantilever strength assures mechanical integrity (Table 1 lists the cantilever strength of metal-top AZES

Figure 1
120 kV rated VariSTAR Type AZE surge arrester
(August 1997. New Issue. © Cooper Industries, Inc.)

TABLE 2 Discharge Voltages—Maximum Guaranteed Protective Characteristics for AZES Surge Arresters

Arrester Rating (kV, rms)	Arrester MCOV (kV, rms)	Front-of-Wave Protective Level (kV)* 10 kA	Lightning Impulse Discharge Voltages (8/20 μsec, kV)						Switching Impulse Discharge Voltages (kV)**		
			1.5 kA	3 kA	5 kA	10 kA	20 kA	40 kA	500 A	1000 A	2000 A
3	2.55	9.7	7.4	7.8	8.1	8.6	9.8	12.2	6.8		
6	5.10	19.2	14.8	15.5	16.1	17.0	19.1	23.2	13.5		
9	7.65	28.8	22.1	23.3	24.1	25.5	28.5	34.1	20.2		
10	8.40	31.5	24.3	25.6	26.5	27.9	31.2	37.3	22.2		
12	10.2	38.3	29.5	31.0	32.1	33.9	37.8	45.0	27.0		
15	12.7	47.6	36.7	38.6	39.9	42.1	47.0	55.8	33.6		
18	15.3	57.3	44.2	46.5	48.1	50.7	56.5	66.9	40.4		
21	17.0	63.6	49.1	51.7	53.4	56.3	62.7	74.2	44.9		
24	19.5	73.0	56.3	59.3	61.3	64.6	71.9	84.9	51.5		
27	22.0	81.4	63.6	66.9	69.1	72.8	81.0	95.7	58.1		
30	24.4	91.2	70.5	74.1	76.6	80.7	89.8	106	64.4		
33	27.5	103	79.4	83.6	86.3	91.0	101	119	72.6		
36	29.0	108	83.8	88.1	91.0	95.9	107	126	76.6		
39	31.5	118	91.0	95.7	98.9	104	116	136	83.2		
42	34.0	127	98.2	103	107	112	125	147	89.8		
45	36.5	136	105	111	115	120	134	158	96.4		
48	39.0	146	113	118	122	129	143	169	103		
54	42.0	157	121	128	132	139	154	181	111		
60	48.0	179	139	146	151	158	176	207	127		
66	53.0	198	153	161	166	175	195	229	140		
72	57.0	212	165	173	179	188	209	246	151		
78	62.0	232	179	188	194	205	228	267	164		
84	68.0	253	196	207	213	224	250	293	180		
90	70.0	261	202	213	220	231	257	302	185		
96	76.0	284	219	231	238	251	279	327	201		
108	84.0	313	243	255	263	277	308	362	222		
120	98.0	337	267	277	283	298	326	379	241	250	
132	106	365	288	300	306	323	354	411	261	271	
138	111	382	302	314	321	338	370	430	273	284	
144	115	396	313	325	332	350	383	446	283	294	
162	130	447	354	368	376	396	433	504	320	332	
168	131	451	356	371	379	399	437	508	323	335	
172	140	481	381	396	405	426	467	542	345	358	
180	144	495	392	407	416	438	480	558	355	368	
192	152	523	414	430	439	463	506	589	374	388	
198	160	550	435	453	462	487	533	620	394	409	
204	165	567	449	467	477	502	550	639	406	421	
216	174	598	473	492	503	529	580	674	428	444	
228	182	626	495	515	526	554	606	705	448	465	
240	190	653	517	537	549	578	633	736	468	485	
258	209	684	547	568	580	605	666	771	502	526	535
264	212	693	555	576	588	613	675	782	509	533	543
276	220	720	575	598	611	637	701	811	528	553	563
288	230	751	602	625	639	665	732	848	552	578	589
294	235	767	615	639	652	679	748	866	564	591	602
300	239	781	625	650	663	691	761	881	574	601	612
312	245	801	630	655	669	709	780	903	578	606	617
330	267	872	698	726	741	772	850	985	641	671	683
336	269	879	704	731	747	778	856	991	645	676	689
360	289	945	756	785	802	836	920	1064	693	727	740

*Based on a current impulse that results in a discharge voltage cresting in 0.5 μs.
**45–60 μs rise time current surge.
(August 1997. New Issue.)

arresters). Cooper Power Systems recommends that a load limit of 445 N not be exceeded on the line terminal of cubicle mount designs. Loads exceeding this limit could cause a shortening of arrester life. Housings are available in standard grey or optional brown glaze color.

Standard line and ground terminal connectors accommodate up to a 0.75 inch (2 cm) diameter conductor. Insulating bases and discharge counters are optionally available for in-service monitoring of arrester discharge activity.

The end fittings and porcelain housing of each arrester unit are sealed and tested by means of a sensitive helium mass spectrometer; this assures that the quality and insulation protection provided by the arrester is never compromised over its lifetime by the entrance of moisture. A corrosion-resistant nameplate is provided and contains all information required by Standards. In addition, stacking arrangement information is provided for multi-unit arresters. Voltage grading rings are included for arresters rated 172 kV and above.

Change in the Air: Operational Challenges in Wind-Power Production and Prediction

BY WILLIAM GRANT, DAVE EDELSON, JOHN DUMAS, JOHN ZACK, MARK AHLSTROM, JOHN KEHLER, PASCAL STORCK, JEFF LERNER, KEITH PARKS, AND CATHY FINLEY

Wind generation continues to develop and has rapidly become a major player in the operation of the electrical grids in North America. As wind grows to represent a larger percentage of total generation resources and continues to generate a larger share of the energy consumed by end users, more effort is being directed at developing the tools and information that grid operators need to operate the system reliably. Some of the issues are common throughout the various regions, while some areas have unique problems due to system limitations. Operational differences vary, from the size of the balancing authority (BA) to the amount and type of ramp available to follow the load and wind output, the availability of units to cycle during low-load periods, and the size and type of reserves, including demand-side management, available for unplanned events. One tool that has been identified as necessary regardless of the system being operated is the ability of the grid operator to forecast wind plant output, including wind events on the system. This article will highlight how different grid and market operators are addressing this issue and why forecasting is important. The article will also address current forecasting practices and future challenges in improving forecasting capabilities.

ELECTRICAL GRID OPERATION

The electrical grid in North America is divided into three major, independently operated interconnections that cover the continental United States, parts of Canada, and a small part of Mexico. Two smaller interconnections cover Quebec and Alaska. Within each major interconnection, a number of organized multilateral markets and less formal bilateral markets operate. The major organized market areas in North America are shown in Figure 1.

WESTERN INTERCONNECTION

The Western Interconnection has an installed capacity of approximately 200,000 MW available to meet a forecast peak load of 160,864 MW. One reliability coordinator operating from two offices in the region monitors the Western Electricity Coordinating Council (WECC) grid operations, which cover the western part of North America, from New Mexico, Colorado, Wyoming, Montana, and Alberta to the West Coast. Two market operators perform market functions within WECC, the Alberta Electric System

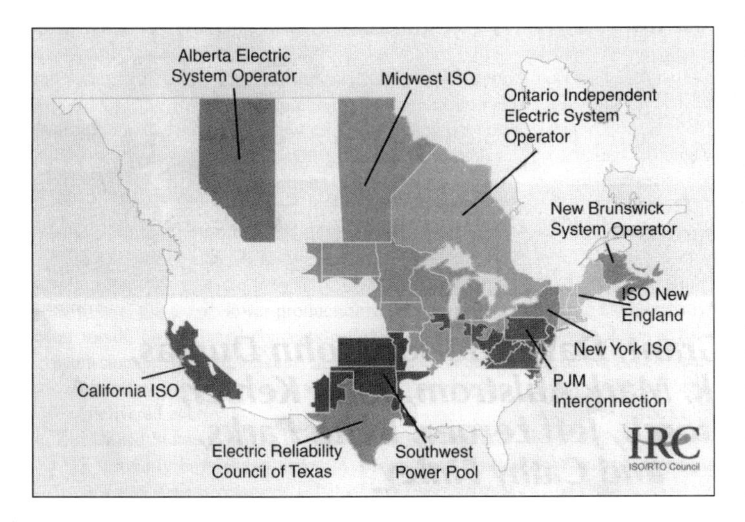

Figure 1
Major organized markets in North America

Operator (AESO) and the California Independent System Operator (CAISO).

ELECTRIC RELIABILITY COUNCIL OF TEXAS (ERCOT)

ERCOT has an installed capacity of 72,712 MW available to meet a forecast peak load of 63,491 MW. One reliability coordinator in the region covers most of Texas except for the Panhandle, the area surrounding El Paso, and the eastern portions next to the Louisiana and Arkansas borders. One market operator operates in the ERCOT region.

EASTERN INTERCONNECTION

The Eastern Interconnection has an installed capacity of approximately 755,000 MW to meet a forecast peak load of 630,000 MW (this includes the Quebec forecast of 20,988 MW peak load, with approximately 31,000 MW of capacity). This is the largest interconnection in North America, covering the area from Texas (except the ERCOT region), Kansas, Nebraska, and North and South Dakota all the way to the eastern seaboard. The Eastern Interconnection is operated by 19 registered reliability coordinators throughout the region and has several market operators. These include the 13-state regional transmission organization (RTO) of the previous Pennsylvania–New Jersey–Maryland pool (PJM); the NY Independent System Operator (NYISO); Independent System Operator–New England (ISO-NE); Independent Electricity System Operator, Ontario (IESO); the Midwest Independent System Operator (MISO); and the Southwest Power Pool (SPP).

MARKET STRUCTURE

In the operational markets in North America, the rules for variable resources vary from market to market. Almost all of the markets require a day-ahead forecast. The rules vary on how the imbalance between the forecast schedule and real-time output is treated financially. A few of the markets have developed requirements for the wind generators or their scheduling agents to provide meteorological data from the specific wind facilities. These data include wind speed, wind direction, barometric pressure, and temperature. A few of the market operators have developed wind-forecasting tools to help them forecast wind plant output utilizing the data provided. This is then used to help them determine the impact of the variable resources on the dispatch requirements of the market footprint. In the SPP imbalance market, the BAs are responsible for providing the ancillary services. In MISO, an ancillary services market offers this function. The following paragraphs give an example of specific market rules operating in the NYISO market.

Wind generators provide real-time offers indicating their economic willingness to produce power. When a wind resource is economical, it is compensated for all energy produced with no charges for schedule deviations. When a wind resource is not economical, it may be dispatched down and compensated at the lesser of its actual production or its schedule, and it may be liable for overgeneration charges.

If the wind resource is scheduling day-ahead, energy imbalances are settled at the balancing market prices.

Since June 2008, the NYISO has operated a centralized wind-forecasting program in which it has contracted with a third-party wind-forecast vendor to produce both a day-ahead and a real-time forecast for nearly all wind resources in the balancing area.

The day-ahead forecast is produced twice a day. The real-time forecast is produced every 15 minutes throughout the day. In support of the production of forecasts, the wind resources are required to supply site-level meteorological data to NYISO at least every 15 minutes. The wind resources must also supply the availability of their turbines (in aggregate) in support of the NYSIO centralized wind-forecasting program.

FORECASTS

Day Ahead

The day-ahead forecast is used for reliability and allows NYISO to consider the anticipated levels of wind power for the next operating day when making day-ahead unit commitment decisions.

Real Time

The real-time forecasts are used in NYISO's real-time security-constrained dispatch. The real-time forecasts are blended with persistence schedules, placing greater weight on persistence schedules in the nearer commitment/dispatch intervals and gradually shifting weight to the forecasts as the commitment intervals move farther out in time. It is important to note that NYISO re-dispatches the entire system every five minutes, which lessens the variability of the wind resources from one dispatch interval to the next. The variability of wind output from one dispatch interval to the next would be far greater if the system were only redispatched once per hour. Wind forecasts assist in making sure there is enough flexibility in the system beyond the five-minute dispatch time horizon, since persistence forecasting works very well for the five-minute time horizon.

Future forecasting efforts at NYISO aim to predict significant ramp events such as:

- fast drop off of output due to high wind-speed cutouts
- sudden increases or decreases in output due to fast-moving weather phenomena.

WIND GROWTH IN REGIONS

Wind generation continues to grow in North America. According to the latest numbers published by the American Wind Energy Association (AWEA), an additional 8,500 MW of new wind capacity (42% of all new capacity) was added during 2008 in North America, bringing total installed wind capacity to more than 26,200 MW. Estimates for 2009 are coming in around 5,000 MW of new installed capacity. (The decline in new wind generation from 2008 is attributed to the economic slowdown.) Development of national policy and transmission planning will have a large influence on future wind development. Figure 2 shows the growth of wind power in the United States over the last decade, while Figure 3 shows the location of wind development.

Installed wind capacity is becoming a larger percentage of the total capacity on the electrical grids. In ERCOT, the installed wind capacity of more than 8 GW is approximately 11% of the installed generation capacity of approximately 72 GW. The total energy produced for load for ERCOT in 2008 was reported at 308,959,455 MWh, of which wind produced approximately 5% (15,237,876 MWh). The peak load for 2008 was 62,174 MW.

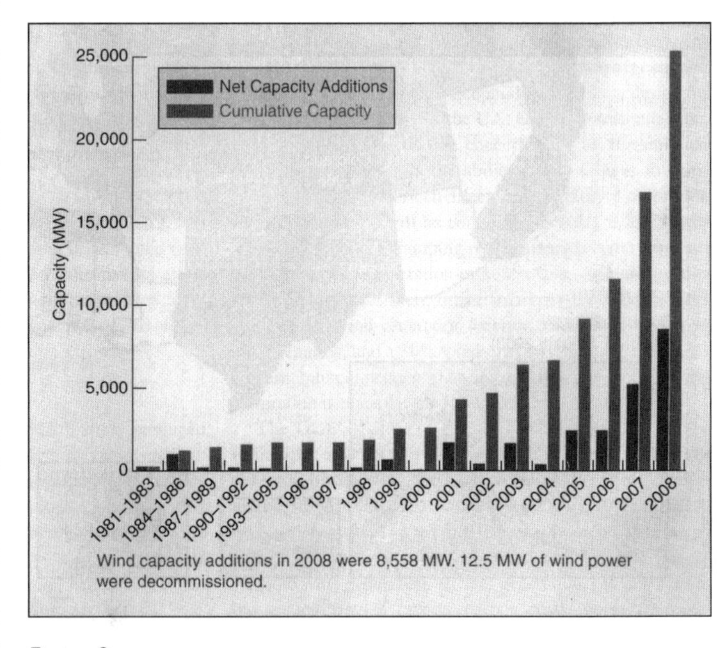

Figure 2
U.S. annual and cumulative wind-capacity data (from AWEA)

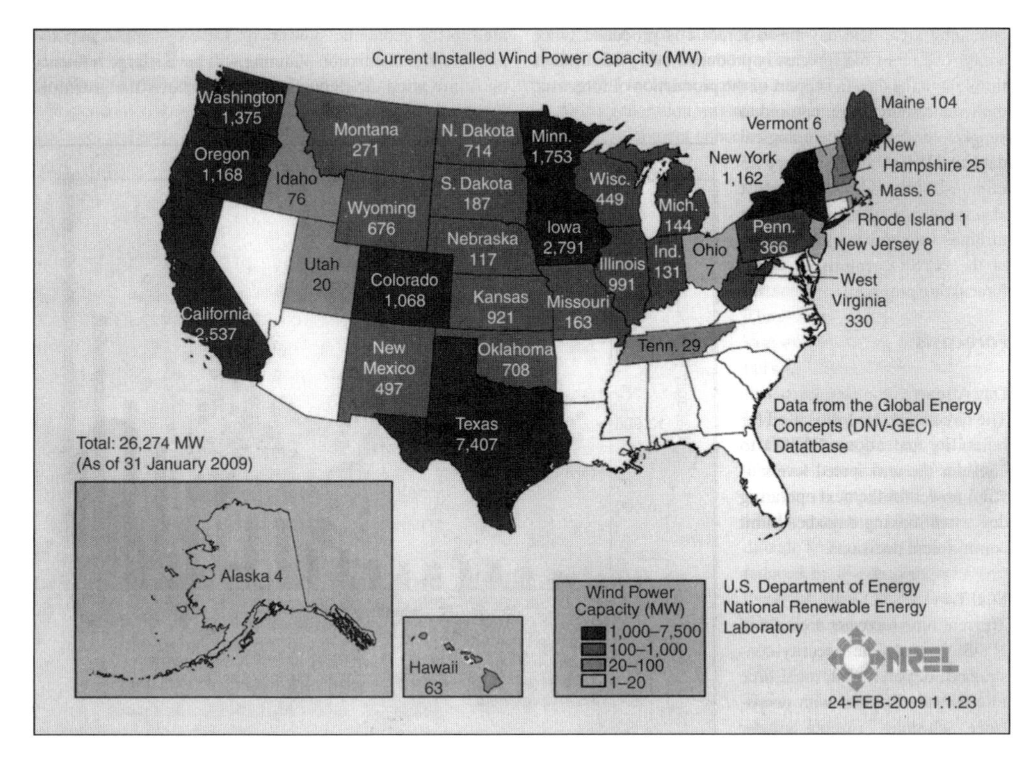

Figure 3
Current installed wind capacity by state

There are similar patterns in other balancing authorities. The Southwestern Public Service (SPS) balancing authority in SPP has 840 MW of installed wind, with a peak load reported in 2008 of 5,503 MW. The installed wind name-plate is approximately 13% of the acquired resources for the BA. Although the penetration of wind on the SPS system in terms of nameplate capacity as a fraction of the peak load is modest at the present time, there is a large backlog of wind generation in the interconnection queue. A recent study noted that at very high levels of wind penetration in the SPS BA (approximately 8,300 MW), the wind variations in SPS would be erratic and extreme in magnitude. Maximum ten-minute and hourly changes in wind output from the wind plants modeled represented levels of magnitude in the range of +26% to −23%, while the hourly changes were in the range of +57% to −52%. Although large wind-power output swings occurred in both the upward and downward directions, virtually all of the identified system issues were associated with increases in wind output, particularly during off-peak periods as load decreased and fewer traditional resources were online. This illustrates the point that many of the system issues caused by variability in wind generation are occurring on systems with relatively high wind penetration during times of high wind-power generation and low and/or decreasing load. It also illustrates the challenges of operating a small BA with a high wind penetration, and the value of aggregation across large geographical regions. Figure 4 shows the current ramping volumes and frequency of occurrence at the current installed capacity of 840 MW.

IMPACT ON OPERATIONS

With a larger percentage of the generation portfolio involving variable resources, wind-generation patterns are taking on a greater importance in the unit commitment process. The size of the BA and the type of resources available to it are also major considerations in the unit commitment process.

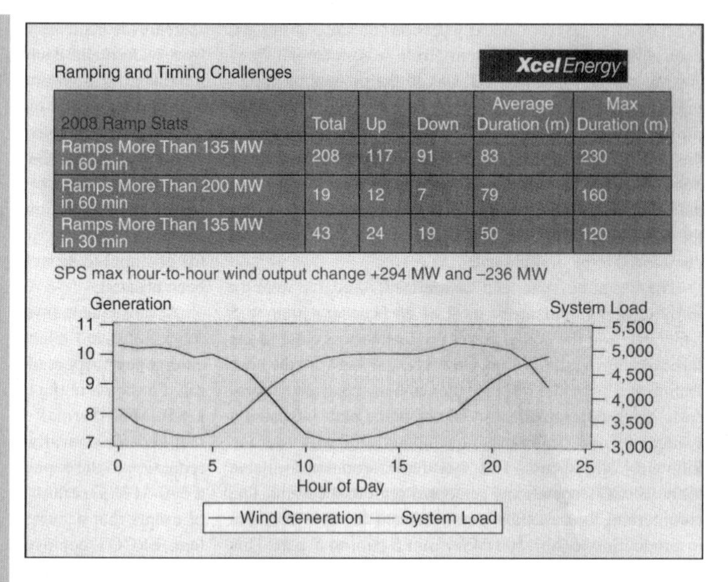

2008 Ramp Stats	Total	Up	Down	Average Duration (m)	Max Duration (m)
Ramps More Than 135 MW in 60 min	208	117	91	83	230
Ramps More Than 200 MW in 60 min	19	12	7	79	160
Ramps More Than 135 MW in 30 min	43	24	19	50	120

Figure 4
Ramp events on the SPS system at current levels of wind penetration

This process is still being performed many different ways throughout the interconnections. More market operators are starting to require specific site-location information from the variable-generation facilities. These data are being utilized in centralized wind forecasting software at the market level to determine unit commitment, ramp-rate requirements, and reserve requirements and to determine transmission constraints. Market participants are also using wind forecasts to minimize their exposure to increased costs due to scheduling errors and imbalance charges (where applicable), and, especially in nonmarket areas, to minimize errors on fuel nomination and unscheduled unit starts. Some BAs are limited by the lack of quick-start units and the absence of interruptible loads. Larger BAs, especially those with a market footprint such as ERCOT, have incorporated wind deviations and wind-forecast uncertainty into their market products like regulation and 30-minute reserves. Rules in ERCOT, for example, have been implemented to have the wind plants limit or regulate their ramp rates during periods of instructed curtailments. Wind generation installed after 2004 is required to bid into the ramp-down market. The smaller the BA, the fewer options the operators have to balance real-time load to resources and the higher the importance of the accuracy of the forecast.

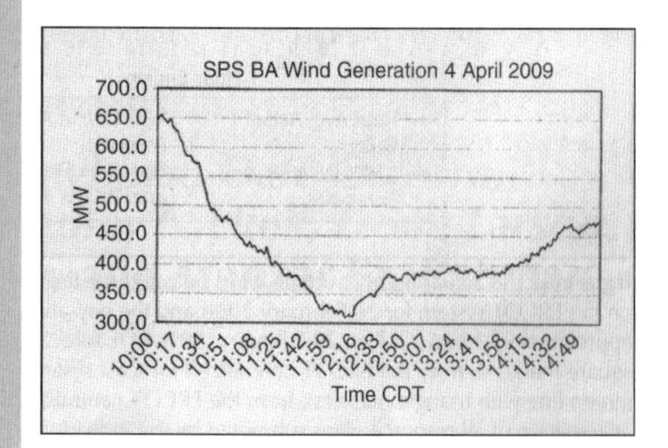

Figure 5
SPS 4 April 2009 high wind cutout event

WIND EVENTS EXPERIENCED

The level of wind penetration within the SPS BA highlights the importance of forecasting significant wind events so that system operators (SOs) can prepare resources for them. For example, on 4 April 2009, the SPS BA experienced a high wind cutout event across the peak hours that caused wind output to drop from 650 MW to 450 MW within one hour; wind output continued to drop, to approximately 310 MW by the end of the second hour (Figure 5). On this day, the peak load for the SPS BA was 3,214 MW, and the decrease in wind output was more than 10% of the actual generation at the start of the event. This event was well forecasted, and resources were available to meet the ramping requirements.

As the amount of wind penetration increases on grid systems, the occurrence of large and rapid changes in wind-power production (ramps) is becoming a significant grid management issue. The issues caused by these ramps include the following:

- Operators must ensure that there is a sufficient ramping capability from conventional generators to compensate for movement in wind output.
- Unexpected movements in wind generation can place added stress on ancillary services.

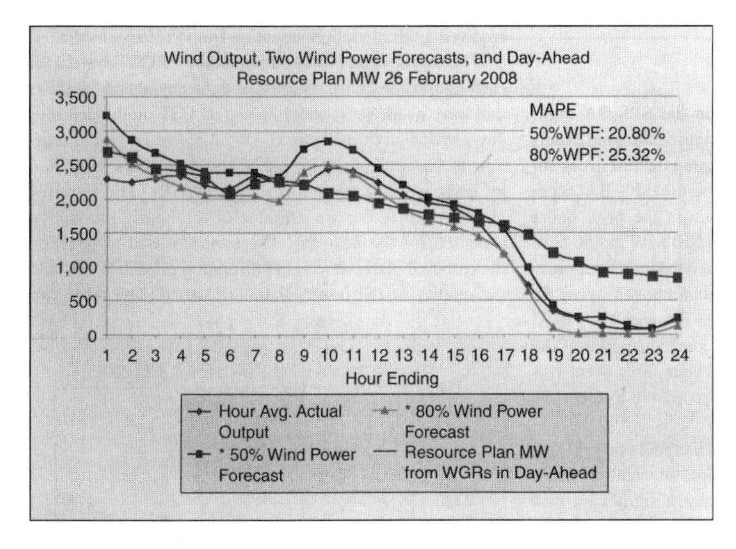

Figure 6
The actual hourly average wind output (blue line with diamond markers) on the ERCOT system for 26 February 2008 and the day-ahead "forecasts" made at approximately 3 p.m. CST on 25 February. 50% POE forecast (red line with small square markers), from the ERCOT centralized forecast system. 80% POE forecast (green line with triangle markers), from the ERCOT centralized forecast system. The aggregation of all resource plans submitted by the individual WGRs (blue-gray line with large square markers)

- Changes in wind output can cause significant changes in transmission congestion.

Some of these issues contributed to an incident on the ERCOT system on 26 February 2008. On that day, an unexpected downward ramp in wind-power production contributed to the need to declare a system emergency. This event received a fair amount of attention from the news media.

The actual wind power output on the ERCOT system for each hour of 26 February 2008 and the day-ahead forecasts that were available at about 3:00 p.m. CST on the previous day are depicted in Figure 6. One forecast is the aggregation of the individual day-ahead resource plans submitted by each wind-generation resource (WGR). The resource plan is each resource's estimate of the hourly production expected for the next day. The methods used to generate the resource plans were at the discretion of each WGR, and the quality of the results therefore varied. The other two predictions come from an early version of a state-of-the-art, centralized forecasting system (from a commercial forecasting service) that ERCOT was in the process of implementing at the time of the event. One of these is the 50% probability of exceedance (POE) forecast, and the other is the 80% POE forecast (i.e., a more conservative estimate of how much production can be expected). This system was delivering forecasts in test mode at the time of the event, but these were not available to the SO since they were not yet considered to be operational.

The resource plan and centralized forecasts were in fairly close agreement for most of 26 February prior to 5 p.m. CST, and the actual power production was close to the forecasts during this period. However, starting with the hour ending at 5 p.m., the resource plan and the centralized forecasts diverged dramatically. The resource plan indicated a gradual decrease in production to about 1,000 MW over the following three hours. This was the information available to the ERCOT operational personnel prior to the event. The commercial forecast, however, predicted a rapid decrease in production to 200–500 MW from 5 p.m. to 7 p.m. This information was not available to ERCOT operational personnel, as explained above. Actual production decreased quite rapidly during the 5 p.m. to 7 p.m. period, in close agreement with the commercial forecast. This event was well forecast by the centralized system prior to its occurrence.

A meteorological analysis of this case revealed that it was not associated with a noteworthy weather event. The systemwide decrease in wind-power production was caused by a widespread decrease in wind speeds associated with a weakening atmospheric pressure gradient as a high-pressure system moved into northern and central Texas and by a simultaneous stabilization of the atmospheric boundary at sunset that cut off the vertical turbulent transport of winds from higher levels of the atmosphere. Both of these processes were well simulated by the weather-forecast models more than a day prior to the event. The event thus had the potential to be well forecast if state-of-the-art tools had been utilized.

A subsequent investigation indicated that there were other, non-wind-related contributing factors to the "system emergency" aspect of this incident. Some of these were a rapid increase in the load while the wind was dying down; a 650-MW thermal unit that was in the day-ahead plan that became unavailable in the current day, resulting in a reduction to the expected excess capability; and the trip of a 370-MW generator prior to the event. Given the series of events that occurred and the resulting state of the system, ERCOT deployed load acting as a resource (LaaR) that was providing responsive reserve service to restore the system to normal operation. This event, however, underscores two points: 1) even moderately large ramps in wind production can contribute to significant grid management issues, and 2) even fairly routine weather events can produce ramps that can have a significant impact on grid operations.

CURRENT PRACTICES

CURRENT WIND-FORECASTING PRACTICES

Today's state-of-the-art wind-power production forecasts typically use a combination of physics-based and statistical models. Physics-based atmospheric models that are used for weather forecasting are typically referred to as numerical weather prediction (NWP) models.

NWP models have important advantages. Because they consist of a set of equations based on the fundamental principles of physics, no training sample is needed to make forecasts and they are not constrained by history. For an unusual but realistic set of conditions, an NWP model can predict an event that has never previously happened in quite the same way. But because of the complexity of doing such a simulation, they have a large computational cost. And even with the most detailed models, their representation of the atmosphere is limited by the spatial resolution of the model grid, the fidelity of the simulation, and the unavoidably incomplete knowledge of the initial state of the atmosphere.

Statistical models, on the other hand, are based on empirical relationships between a set of predictor (input) and forecast (output) variables. Because these relationships are derived from a training sample of historical data that includes values of both the predictor and forecast variables, statistical models have the advantage of "learning from experience" without needing explicit knowledge of the underlying physical relationships. Some of these statistical models can become quite sophisticated, finding complex multivariable and nonlinear relationships between many predictor variables and the desired forecast variable. Such systems include computational learning systems such as artificial neural networks, support vector machines, and similar technologies.

Statistical models are used in a number of ways in the wind-power production forecasting process. The basic approach is to use values from NWP models and measured data from the wind plant to predict the desired variables (e.g., hub height wind speed, wind-power output, and so on) at the wind plant location. Because they can essentially learn from experience, the statistical models add value to NWP forecasts by accounting for subtle effects due to the local terrain and other details that can't realistically be represented in the NWP models themselves. But because they need to learn from historical examples, statistical models tend to predict typical events better than rare events (unless they are specifically formulated for extreme event prediction and are trained on a sample that has a good representation of rare events).

Many forecasting systems also use an ensemble of individual forecasts rather than a single forecast. The basic concept behind the use of a forecast ensemble is that there is uncertainty in any forecasting procedure due to uncertainty in the input data and model configuration. The ensemble approach attempts to account for this uncertainty by generating a set of forecasts through perturbing the input data and/or model parameters within their reasonable ranges of uncertainty. This requires considerable computational resources and prudent choices, but if done well, the spread of the ensemble members can be a useful representation of the uncertainty in the forecast.

The relative value of different data sources and forecasting techniques varies significantly with the forecast look-ahead period. Very short-term forecasts (from zero to six hours out) typically rely heavily on statistical models that exploit recent data from the wind plant or nearby locations more than NWP values. For longer-term forecasts, the statistical model will depend much more heavily on the NWP forecast values. After about six to 10 days, the skill of NWP models is typically less than that of a climatology forecast (e.g., the long-term average by season and time of day).

Wind-forecasting services are available from several professional forecast providers. While essentially all state-of-the-art forecasting systems use similar input data, the details in terms of models and techniques vary substantially from one forecast provider to another. Recent comparisons have generally indicated there is no single approach that works best for all times, conditions, and locations. This suggests that there may be a benefit in using multiple forecast providers—essentially, obtaining an ensemble of forecast providers—especially if one can develop some skill in identifying which forecast algorithm is likely to perform better under various conditions or for different types of decisions.

To build on this further, there are also different types of forecast products for different purposes, including those that are tuned to provide a specific wind-power output value, predict events such as ramps, or provide a probability distribution around such values or events. There are also situational awareness products that display information (such as animated geographical displays of the wind and weather) that provides a higher level of understanding and confidence in the forecast. The value of tuning forecasting products to specific business or operating problems is very significant.

WIND PLANT DATA ISSUES

Meteorological and operational data from the wind plant play an important role in determining the forecast performance that can be achieved for a specific facility. Data from the wind-generation facility are used to optimize the relationship between meteorological variables and facility-specific power output and to correct weather forecast model errors.

Turbine availability information is the first significant data issue. (*Turbine availability* means knowing which turbines in the project are available to run and which are shut down for maintenance or other reasons.) Misreporting of turbine availability confuses the procedures that construct relationships between meteorological variables and power production.

The second issue relates to the spatial representation of conditions within the wind plant. Small wind plants with relatively homogeneous characteristics may be well represented by one meteorological tower or the total output from all turbines. Large plants in complex terrain may need multiple meteorological towers or turbine-level data to adequately represent the spatial variability within the wind plant. An inadequate representation of the spatial variability in the wind plant data set can lead to lower forecasting quality.

An example of the impact of the quality of wind plant data on forecast performance is shown in Figure 7. This figure shows the anemometer-measured wind speed versus the reported power production for two adjacent wind plants of similar

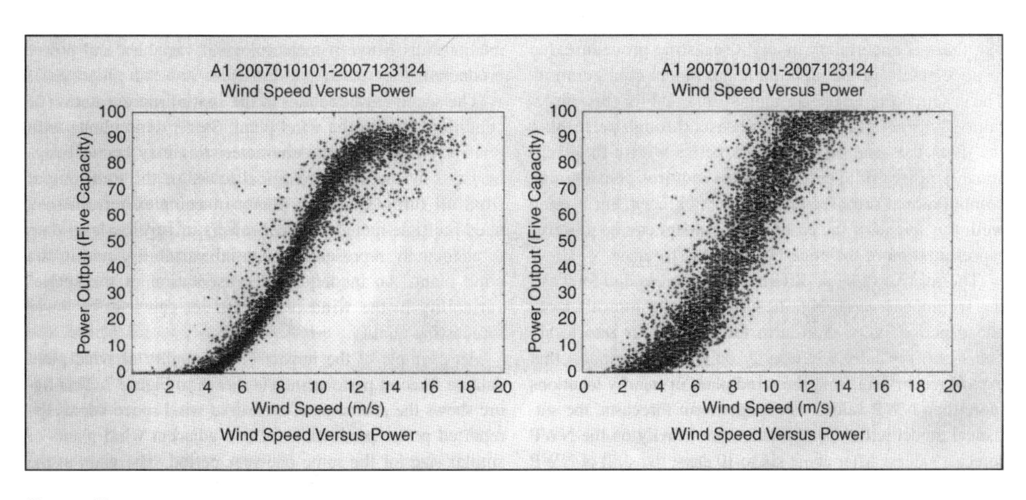

Figure 7
Measured hourly average wind speed versus measured hourly average power production over the same one-year period for two adjacent wind plants of similar size

size for the same one-year period. The plant at the left supplied data from six meteorological towers distributed throughout the plant and also consistently provided accurate turbine availability data. The plant at the right provided data from only one meteorological tower and less reliable turbine availability information. The relationship of wind speed to power production has much less scatter for the plant on the left, resulting in 23% lower mean absolute error (MAE) for a four-hour forecast for the one-year period depicted in the scatter plots.

RAMP FORECASTING

There are large variations in the sensitivity to prediction error among forecast users. For example, in many market-related forecast applications the user is sensitive to the total accumulation of the absolute error over a period of time. The accumulation of error is linked to the economic impact of imbalance charges or the cost of selling and/or buying energy in the spot market to cover schedule deviations. The sensitivity of a particular generator will depend on the market structure in which it operates, but in many cases it does not matter if the error accumulation occurs because of a modest error over all hours or as a result of a few hours with big errors and many hours with small errors. However, SOs typically have a lower sensitivity to the small errors associated with routine changes in the wind and a much higher sensitivity to the accuracy of predictions of the relatively infrequent large changes over short time periods (wind ramps). Unexpected wind ramps can have a large impact on the SOs' ability to keep power systems within their operating specifications and avoid catastrophic events.

Wind ramps result from many different weather events. Events that seem similar to the SO (resulting in changes in delivered power) may appear to be very different to a meteorologist, and the ability to predict ramp events greatly depends on what meteorological feature is causing the ramp.

Generally speaking, the larger and longer-lived the feature, the better it can be predicted. Meteorological features that are highly localized can be difficult to predict with much certainty. A large range of such features can affect the area of a wind plant and cause ramps in the power delivered from the plant.

The general public also tends to underestimate the complexity of common atmospheric events. For example, most people visualize weather events as predominantly horizontal phenomena, with weather and winds traveling along from one location to another and causing similar effects as they go. In this view, if we just have "upstream" measurements, we should be able to "see the changes coming" and better estimate their timing. In reality, this is only true to a very limited extent.

As will be further described below, some events that cause significant ramps in wind-power output are more vertical in nature and can't be detected "upstream" at all. For example, the typical diurnal pattern of wind is caused by changes in vertical turbulent mixing induced by variations in the vertical profile of temperature. Depending on how solar heating interacts with the surface and causes convective mixing of the lower atmosphere, the rate at which hub-height winds slow down in the afternoon (and winds at six feet speed up due to mixing of the faster winds aloft to our level) can be highly variable. The timing of these wind changes during the day can be difficult to predict precisely, and upstream measurements provide little help.

The nature of up-ramp and down-ramp events may also differ. For example, situations that can cause up-ramps include the following:

- **Cold frontal passage:** The strongest winds tend to be behind the front and can persist for many hours following frontal passage. As a large feature, these events are usually predicted quite well in a general sense, though the exact timing of the passage may vary since weather systems speed up or slow down in complex ways. This results in uncertainty around the timing of the ramp.
- **Thunderstorm outflow:** These events can be very localized, abrupt, and difficult to predict. The extent to which this will create a significant systemwide ramp will depend on the size of the thunderstorm complex and the geographical dispersion of the wind plants.
- **Rapid intensification of an area of low pressure:** These are larger-scale features, which are usually forecast pretty well within 12–24 hours of occurrence. The longer the forecast lead time, the more error there is in the forecast of these events.

- **Onset of mountain wave events (in the lee of mountain ranges):** Large-amplitude mountain waves can develop when the midlevel winds are sufficiently strong and blowing nearly perpendicular to the mountain ridge line and a layer of very stable air exists at or just above mountaintop level. The net result of these mountain waves is strong, extremely gusty downslope winds. It is difficult to forecast the onset and intensity of mountain wave events because small differences in topographic shape and orientation and small differences in atmospheric conditions can mean the difference between an event and a non-event. Mountain waves can also be highly localized, with one area experiencing extremely strong winds while areas just a few miles away are calm. The type of surface cover (snow versus no snow) can also effect whether or not these strong winds actually reach the surface at hub height.
- **Flow channeling:** Relatively subtle changes in wind direction in the area of a mountain valley or gorge such that wind begins to move parallel to the direction of the valley can quickly create a local "wind tunnel" effect in which the strongest winds occur inside the valley.
- **Sea breeze:** Localized winds caused by the temperature differences between water and land are well known near coastal areas, but it can be difficult to predict the timing, duration, and in particular the distance to which these winds will propagate inland from the coast before dissipating.
- **Thermal stability/vertical mixing:** As noted earlier, this is the erosion of the stable, near-surface boundary layer in the morning (often in the few hours after sunrise but sometimes later in the day). The extent to which this occurs depends on what type of land use or cover (snow, etc.) is currently on the surface, the amount of clouds present, and the strength of the winds in the lowest levels of the atmosphere.

Similarly, a wide range of different events can cause wind down-ramp events. Turbines reaching their cut-out speed are often cited as a major cause of down ramps (wind turbines are designed to shut themselves down at 25 m/s or about 55 miles per hour to protect the equipment), but these events are not as common as many believe. More often, the down ramp is caused by decreasing winds from meteorological causes rather than turbine cutouts from increasing winds.

Examples of meteorological events that can cause down ramps include the following:

- **Near-surface boundary-layer stabilization at sunset/nightfall:** The complexity of this forecast problem cannot be overstated, as boundary-layer heating and cooling rates that affect the timing and intensity of ramp events depend on a number of variables, such as what type of land use or snow is covering the surface, the amount of clouds, and the strength of the winds in the lowest levels of the atmosphere, as well as the underlying soil moisture level.
- **Relaxation of the pressure gradient as high pressure moves in following a cold front passage:** As noted above, the strongest winds tend to be behind the cold front, and the speed at which these winds fall off once the front has moved through can be challenging to predict.
- **Pressure changes following the passage of thunderstorm complexes:** Given the localized nature of thunderstorms and the dramatic pressure changes that can result, these events are not easily forecast with NWP models.
- **A decrease in wind speed as a warm front passes through:** Warm fronts tend to be very slow-moving, and the winds immediately along the front tend to be weaker than the winds both north and south of the front. This can create a down-ramp/up-ramp event as the front passes. This occurs in the central plains and the east of the United States, where well-developed warm fronts are observed. The complex terrain of the western states makes it difficult for consistent warm air masses to develop.

Down ramps may be more difficult to forecast because they are usually not directly associated with sharply defined meteorological features—thunderstorm complexes being the exception. Areas of complex terrain are also especially challenging, as there can be many terrain-induced local flows that aren't captured by typical forecast models.

Thus it is difficult to make general, sweeping statements about the "forecastability" of ramp events

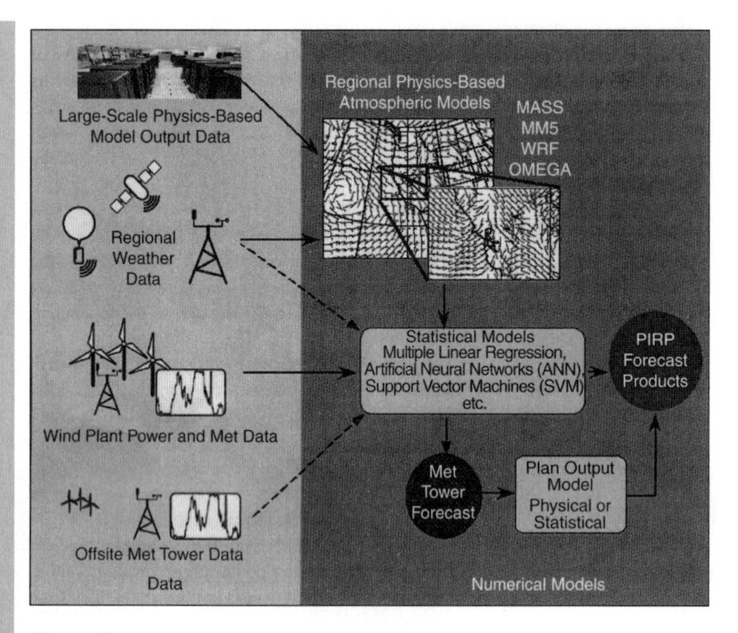

Figure 8
A schematic of the components of a typical state-of-the-art wind-power production forecasting system

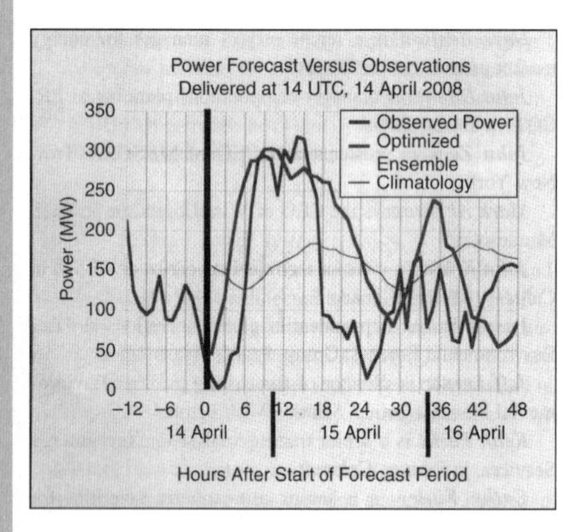

Figure 9
State-of-the-art and climatology forecasts for two observed ramp events in the power production from existing facilities on the AESO system

because they can be caused by many things, some of which can be predicted fairly well while others are difficult (if not impossible) to predict with current forecasting models. To really say something about forecasting ramp events, we need to know what is causing the ramp events, and that will depend on where the wind plants are located and many complex conditions and weather events.

A close examination of the detailed data from state-of-the-art forecasting systems (see Figure 8) reveals that they often contain information about the likelihood and characteristics of ramp events during the forecast period. The challenge is to communicate this information to SOs in an actionable way.

This potentially useful information is frequently suppressed in the most common type of wind-power production forecast: the deterministic interval forecast that provides a single estimate of hourly power production. Deterministic interval forecasts are typically produced by a statistical model that accepts several inputs (results from NWP models, recent observational data, and so on) and produces an estimate of power production for a specific time interval. These models are trained with a historical sample of input and output data and a specified performance criterion, such as minimizing the least-squared error.

Because small errors in forecasting the timing of a ramp produce large power errors, approaches that focus only on minimizing power error are not appropriate for ramp forecasting. The optimization algorithm tends to "hedge" during the ramp periods by lowering the ramp amplitude and stretching out ramp duration so that the possibility of very large errors is reduced. This is the best approach if one wants to achieve the lowest root mean square error (RMSE) for all the forecast intervals, but it is not a good approach if one wants to provide the most useful information about ramp events.

An example of the effectiveness of hedging is given in Figure 9. Two forecasts of power production and the actual production are depicted in this chart. One forecast (red) is a climatology forecast, which is simply the average production by time of day for this month of the year, with no information about the current meteorological situation. The second forecast (blue) is produced by a forecast system using an ensemble of NWP and statistical models. A large upward ramp occurs six hours into the forecast period, and a large downward ramp occurs 18 hours into the forecast period. In this case, the red forecast has a lower RMSE than the blue forecast for the two ramp periods due to the phase errors in the ramp forecasts, even though the blue forecast obviously provides more information about the ramp.

This is typically the case when judging forecast accuracy based on interval-by-interval power error values. A state-of-the-art forecast will have much lower error than a climatology forecast for nonramp periods, but the climatology forecast will be slightly better for ramp periods (because it is strongly hedged against ramp-timing errors). However, the state-of-the-art forecast contains much more information about the *possibility* of a ramp event. The challenge is to extract this imperfect ramp event information from the forecasting system and present it in a way that provides effective decision-making guidance for the SO. Displaying forecast information in the form of up- and down-ramp event probabilities as a function of time and providing situational-awareness graphics that let an operator "follow" ramp-causing weather events may be useful. The bottom line is that one type of forecast cannot meet all objectives, and users with multiple forecast objectives need different types of forecasts.

There also comes a point where a fully automated forecast based on numerical weather modeling alone will see diminishing returns. A human forecaster can add a great deal of value in forecasting ramp events, especially when looking from one to four hours out. There are patterns and features (such as rapidly evolving thunderstorm complexes on a radar display) that humans can still detect and interpret far better than numerical models or computational learning systems. The challenge becomes how to use the human input to best deliver forecasts and information to the SOs.

There are several baseline metrics that are used to determine the accuracy and value of a wind-energy forecast. These evolved by asking if there were no custom forecasts available, what would one use to predict wind power at a particular site, and how good will that forecast be? Figure 10 illustrates the concept of increased forecast error with lead time and shows how an unskilled forecast compares with that produced by an advanced statistically or physically based forecast system. The figure shows that in the forecast range from zero to six hours out, the persistence

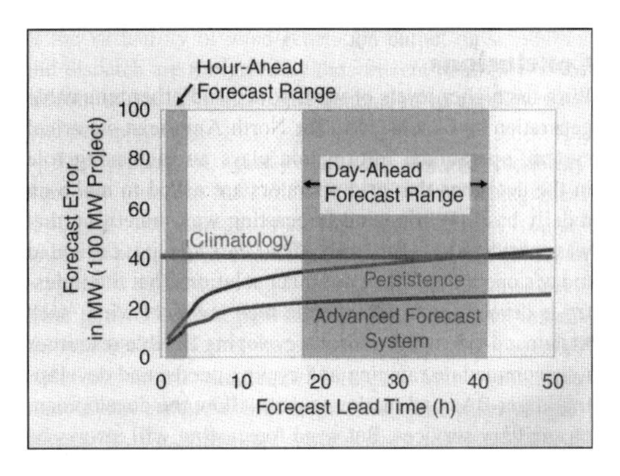

Figure 10
Forecast error (MW) as a function of the forecast lead time (hours) for a wind-power plant with 100 MW of hypothetical capacity

method provides a good forecast. For this reason, it is still used in the forecast process in several markets. Most forecast service providers are judged on their ability to beat the persistence model using advanced statistical methods as explained above. The persistence model rapidly degrades beyond the hours-ahead forecast, when errors can quickly exceed 25% of project capacity.

For day-ahead power scheduling, wind-power forecast performance is often judged against a long-term average climatology often derived from historical observations, met tower data, or a calibrated retrospective NWP model simulation. An advanced forecasting system will typically improve on errors due to the use of climatology by 40% to 60%. As the forecast length increases, the accuracy of the NWP model forecasting solution decreases, approaching that of climatology.

Depending on the weather pattern, geographic location, and NWP model or ensemble system used, the crossover point whereby a forecast no longer adds value over climatology varies between 10 and 15 days. After this time, only broad statements can be made about wind-power forecasts, as the wind uncertainty at such long lead times increases greatly. A hypothetical example: Southern California wind project A can expect a 50% capacity factor for the month of March, which deviates from the ten-year average by +5%, due to a positive ENSO index forecast.

One of the shortcomings of the current practice of measuring forecast performance is the period over which forecasts are averaged and the inclusion of all events despite the importance and attention that should be given to specific events (e.g., ramps, wind speeds below a prescribed threshold for O&M execution, and production during peak load periods). On the one hand, a meaningful statistical sample is required in order to draw valid conclusions from the observations. On the other hand, discriminating those events that have the biggest financial impact on the energy market (even though these events are relatively infrequent) would be more meaningful to operators, BAs, and independent SOs (ISOs). Verification statistics from rare events oftentimes are not consistent with those of monthly or seasonal averages. With the maturation of operational wind-power projects in North America, the data-mining potential lends itself to verifying most events with a greater level of statistical significance.

CONCLUSIONS

With the higher levels of wind power and other renewable generation now included in the North American electrical system, operational information plays an increasing role in the decisions that grid operators are asked to make on a daily basis. While wind forecasting was something that was considered a luxury only a few years ago, it is critical to today's operation. There are other solutions that the industry is developing to incorporate high levels of wind, such as demand-side management, developing flexible resources to accommodate ramping and cycling needs, and developing larger BAs and market tools to allow the development of ancillary services. But wind forecasting will always be a crucial part of the solution and the primary tool that grid operators will rely on to implement policies and procedures to reliably and economically operate the system.

FOR FURTHER READING

J. Dumas, "ERCOT Feb 26, 2008 EECP event," *UWIG*, Texas, Apr. 2008 [Online]. Available: http://www.uwig.org/FortWorth/workshop/Dumas.pdf

GE Energy, "Analysis of wind generation impact on ERCOT ancillary services requirements," Mar. 2008 [Online]. Available: http://www.uwig.org/AttchA-ERCOT_A-S_Study_Exec_Sum. pdf

N. Miller and G. Jordon, "Impact of control area size on viability of wind generation: A case study for New York," in *Proc. American Wind Energy Association Windpower*, 2006, Pittsburgh, PA [Online]. Available: http://www.nrel.gov/wind/systemsintegration/pdfs/2007/milligan_wind_integration_impacts.pdf

M. Ahlstrom, L. Jones, R. Zavadil, and W. Grant, "The future of wind forecasting and utility operations," *IEEE Power Energy Mag. (Special Issue on Working with Wind; Integrating Wind into the Power System)*, vol. 5. no. 6, pp. 57–64, Nov./Dec. 2005.

J. C. Smith, B. Oakleaf, M. Ahlstrom, D. Savage, C. Fin-ley, R. Zavadil, and J. Reboul, "The role of wind forecasting in utility system operation," *CIGRE*, Paper C2-301, Aug. 2008.

NERC, "Special report: Accommodating high levels of variable generation," Apr. 2009.

AMEC, "Results of the wind penetration study for the Southwestern Public Service Company portion of Southwest Power Pool," Mar. 2009.

BIOGRAPHIES

William Grant is manager of the Transmission Control Center and Wind Integration at Southwestern Public Service.

Dave Edelson is a senior project manager for energy market products at the NYISO.

John Dumas is manager of operations planning at ERCOT, in Taylor, Texas.

John Zack is president and CEO of MESO, in Troy, New York.

Mark Ahlstrom is the CEO of WindLogics, in St. Paul, Minnesota.

John Kehler is a senior technical specialist at AESO, in Calgary, Alberta, Canada.

Pascal Storck is president of global operations at 3Tier Environmental Forecast Group, Seattle, Washington.

Jeff Lerner is director of forecasting at 3Tier Environmental Forecast Group, Seattle, Washington.

Keith Parks is a senior trading analyst for Xcel Energy Services, in Denver, Colorado.

Cathy Finley is a senior atmospheric scientist for WindLogics, in Grand Rapids, Minnesota.

13.1

TRAVELING WAVES ON SINGLE-PHASE LOSSLESS LINES

We first consider a single-phase two-wire lossless transmission line. Figure 13.1 shows a line section of length Δx meters. If the line has a loop inductance L H/m and a line-to-line capacitance C F/m, then the line section has a series inductance L Δx H and shunt capacitance C Δx F, as shown. In Chapter 5, the direction of line position x was selected to be from the receiving end $(x = 0)$ to the sending end $(x = l)$; this selection was unimportant, since the variable x was subsequently eliminated when relating the steady-state sending-end quantities V_s and I_s to the receiving-end quantities V_R and I_R. Here, however, we are interested in voltages and current waveforms traveling along the line. Therefore, we select the direction of increasing x as being from the sending end $(x = 0)$ toward the receiving end $(x = l)$.

Writing a KVL and KCL equation for the circuit in Figure 13.1,

$$v(x + \Delta x, t) - v(x, t) = -\mathrm{L}\Delta x \frac{\partial i(x, t)}{\partial t} \tag{13.1.1}$$

$$i(x + \Delta x, t) - i(x, t) = -\mathrm{C}\Delta x \frac{\partial v(x, t)}{\partial t} \tag{13.1.2}$$

Dividing (13.1.1) and (13.1.2) by Δx and taking the limit as $\Delta x \to 0$, we obtain

$$\frac{\partial v(x, t)}{\partial x} = -\mathrm{L} \frac{\partial i(x, t)}{\partial t} \tag{13.1.3}$$

$$\frac{\partial i(x, t)}{\partial x} = -\mathrm{C} \frac{\partial v(x, t)}{\partial t} \tag{13.1.4}$$

We use partial derivatives here because $v(x, t)$ and $i(x, t)$ are differentiated with respect to both position x and time t. Also, the negative signs in (13.1.3) and (13.1.4) are due to the reference direction for x. For example, with a positive value of $\partial i/\partial t$ in Figure 13.1, $v(x, t)$ decreases as x increases.

Taking the Laplace transform of (13.1.3) and (13.1.4),

$$\frac{d\mathrm{V}(x, s)}{dx} = -s\mathrm{L}\mathrm{I}(x, s) \tag{13.1.5}$$

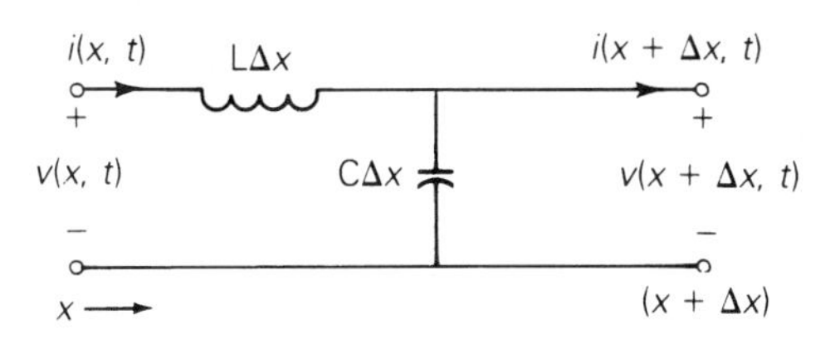

FIGURE 13.1 Single-phase two-wire lossless line section of length Δx

$$\frac{d\mathrm{I}(x,s)}{dx} = -s\mathrm{C}\mathrm{V}(x,s) \tag{13.1.6}$$

where zero initial conditions are assumed. $\mathrm{V}(x,s)$ and $\mathrm{I}(x,s)$ are the Laplace transforms of $v(x,t)$ and $i(x,t)$. Also, ordinary rather than partial derivatives are used since the derivatives are now with respect to only one variable, x.

Next we differentiate (13.1.5) with respect to x and use (13.1.6), in order to eliminate $\mathrm{I}(x,s)$:

$$\frac{d^2\mathrm{V}(x,s)}{dx^2} = -s\mathrm{L}\frac{d\mathrm{I}(x,s)}{dx} = s^2\mathrm{LCV}(x,s)$$

or

$$\frac{d^2\mathrm{V}(x,s)}{dx^2} - s^2\mathrm{LCV}(x,s) = 0 \tag{13.1.7}$$

Similarly, (13.1.6) can be differentiated in order to obtain

$$\frac{d^2\mathrm{I}(x,s)}{dx^2} - s^2\mathrm{LCI}(x,s) = 0 \tag{13.1.8}$$

Equation (13.1.7) is a linear, second-order homogeneous differential equation. By inspection, its solution is

$$\mathrm{V}(x,s) = \mathrm{V}^+(s)e^{-sx/v} + \mathrm{V}^-(s)e^{+sx/v} \tag{13.1.9}$$

where

$$v = \frac{1}{\sqrt{\mathrm{LC}}} \quad \mathrm{m/s} \tag{13.1.10}$$

Similarly, the solution to (13.1.8) is

$$\mathrm{I}(x,s) = \mathrm{I}^+(s)e^{-sx/v} + \mathrm{I}^-(s)e^{+sx/v} \tag{13.1.11}$$

You can quickly verify that these solutions satisfy (13.1.7) and (13.1.8). The "constants" $\mathrm{V}^+(s), \mathrm{V}^-(s), \mathrm{I}^+(s)$, and $\mathrm{I}^-(s)$, which in general are functions of s but are independent of x, can be evaluated from the boundary conditions at the sending and receiving ends of the line. The superscripts $+$ and $-$ refer to waves traveling in the positive x and negative x directions, soon to be explained.

Taking the inverse Laplace transform of (13.1.9) and (13.1.11), and recalling the time shift properly, $\mathscr{L}[f(t-\tau)] = \mathrm{F}(s)e^{-s\tau}$, we obtain

$$v(x,t) = v^+\left(t - \frac{x}{v}\right) + v^-\left(t + \frac{x}{v}\right) \tag{13.1.12}$$

$$i(x,t) = i^+\left(t - \frac{x}{v}\right) + i^-\left(t + \frac{x}{v}\right) \tag{13.1.13}$$

where the functions $v^+(\ ), v^-(\ ), i^+(\ )$, and $i^-(\ )$, can be evaluated from the boundary conditions.

FIGURE 13.2 The function $f^+(u)$, where $u = \left(t - \frac{x}{v}\right)$

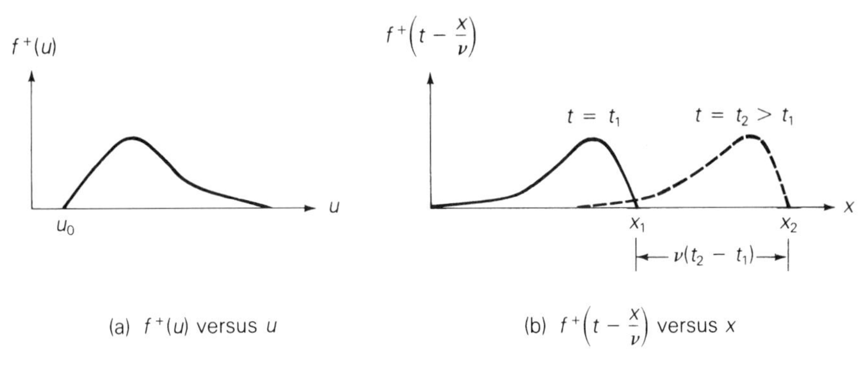

(a) $f^+(u)$ versus u

(b) $f^+\left(t - \frac{x}{\nu}\right)$ versus x

We now show that $v^+(t - x/v)$ represents a voltage wave traveling in the positive x direction with velocity $v = 1/\sqrt{\mathrm{LC}}$ m/s. Consider any wave $f^+(u)$, where $u = t - x/v$. Suppose that this wave begins at $u = u_0$, as shown in Figure 13.2(a). At time $t = t_1$, the wavefront is at $u_0 = (t_1 - x_1/v)$, or at $x_1 = v(t_1 - u_0)$. At a later time, t_2, the wavefront is at $u_0 = (t_2 - x_2/v)$ or at $x_2 = v(t_2 - u_0)$. As shown in Figure 13.2(b), the wavefront has moved in the positive x direction a distance $(x_2 - x_1) = v(t_2 - t_1)$ during time $(t_2 - t_1)$. The velocity is therefore $(x_2 - x_1)/(t_2 - t_1) = v$.

Similarly, $i^+(t - x/v)$ represents a current wave traveling in the positive x direction with velocity v. We call $v^+(t - x/v)$ and $i^+(t - x/v)$ the *forward* traveling voltage and current waves. It can be shown analogously that $v^-(t + x/v)$ and $i^-(t + x/v)$ travel in the negative x direction with velocity v. We call $v^-(t + x/v)$ and $i^-(t + x/v)$ the *backward* traveling voltage and current waves.

Recall from (5.4.16) that for a lossless line $f\lambda = 1/\sqrt{\mathrm{LC}}$. It is now evident that the term $1/\sqrt{\mathrm{LC}}$ in this equation is v, the velocity of propagation of voltage and current waves along the lossless line. Also, recall from Chapter 4 that L is proportional to μ and C is proportional to ε. For overhead lines, $v = 1/\sqrt{\mathrm{LC}}$ is approximately equal to $1/\sqrt{\mu\varepsilon} = 1/\sqrt{\mu_0\varepsilon_0} = 3 \times 10^8$ m/s, the speed of light in free space. For cables, the relative permitivity $\varepsilon/\varepsilon_0$ may be 3 to 5 or even higher, resulting in a value of v lower than that for overhead lines.

We next evaluate the terms $\mathrm{I}^+(s)$ and $\mathrm{I}^-(s)$. Using (13.1.9) and (13.1.10) in (13.1.6),

$$\frac{s}{v}[-\mathrm{I}^+(s)e^{-sx/v} + \mathrm{I}^-(s)e^{+sx/v}] = -s\mathrm{C}[\mathrm{V}^+(s)e^{-sx/v} + \mathrm{V}^-(s)e^{+sx/v}]$$

Equating the coefficients of $e^{-sx/v}$ on both sides of this equation,

$$\mathrm{I}^+(s) = (v\mathrm{C})\mathrm{V}^+(s) = \frac{\mathrm{V}^+(s)}{\sqrt{\frac{\mathrm{L}}{\mathrm{C}}}} = \frac{\mathrm{V}^+(s)}{\mathrm{Z}_c} \tag{13.1.14}$$

where

$$Z_c = \sqrt{\frac{L}{C}} \quad \Omega \tag{13.1.15}$$

Similarly, equating the coefficients of $e^{+sx/v}$,

$$I^-(s) = \frac{-V^-(s)}{Z_c} \tag{13.1.16}$$

Thus, we can rewrite (13.1.11) and (13.1.13) as

$$I(x,s) = \frac{1}{Z_c}[V^+(s)e^{-sx/v} - V^-(s)e^{+sx/v}] \tag{13.1.17}$$

$$i(x,t) = \frac{1}{Z_c}\left[v^+\left(t - \frac{x}{v}\right) - v^-\left(t + \frac{x}{v}\right)\right] \tag{13.1.18}$$

Recall from (5.4.3) that $Z_c = \sqrt{L/C}$ is the characteristic impedance (also called surge impedance) of a lossless line.

13.2 BOUNDARY CONDITIONS FOR SINGLE-PHASE LOSSLESS LINES

Figure 13.3 shows a single-phase two-wire lossless line terminated by an impedance $Z_R(s)$ at the receiving end and a source with Thévenin voltage $E_G(s)$ and with Thévenin impedance $Z_G(s)$ at the sending end. $V(x,s)$ and $I(x,s)$ are the Laplace transforms of the voltage and current at position x. The line has length l, surge impedance $Z_c = \sqrt{L/C}$, and velocity $v = 1/\sqrt{LC}$. We assume that the line is initially unenergized.

From Figure 13.3, the boundary condition at the receiving end is

$$V(l,s) = Z_R(s)I(l,s) \tag{13.2.1}$$

Using (13.1.9) and (13.1.17) in (13.2.1),

$$V^+(s)e^{-sl/v} + V^-(s)e^{+sl/v} = \frac{Z_R(s)}{Z_c}[V^+(s)e^{-sl/v} - V^-(s)e^{+sl/v}]$$

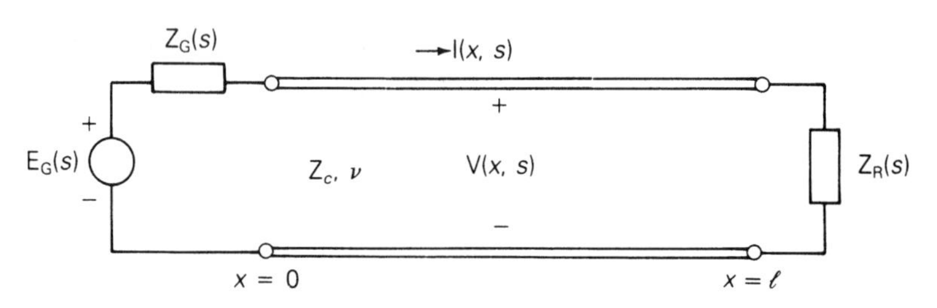

FIGURE 13.3 Single-phase two-wire lossless line with source and load terminations

Solving for $V^-(l,s)$

$$V^-(l,s) = \Gamma_R(s)V^+(s)e^{-2s\tau} \tag{13.2.2}$$

where

$$\Gamma_R(s) = \frac{\dfrac{Z_R(s)}{Z_c} - 1}{\dfrac{Z_R(s)}{Z_c} + 1} \quad \text{per unit} \tag{13.2.3}$$

$$\tau = \frac{l}{v} \quad \text{seconds} \tag{13.2.4}$$

$\Gamma_R(s)$ is called the *receiving-end voltage reflection coefficient*. Also, τ, called the *transit time* of the line, is the time it takes a wave to travel the length of the line.

Using (13.2.2) in (13.1.9) and (13.1.17),

$$V(x,s) = V^+(s)[e^{-sx/v} + \Gamma_R(s)e^{s[(x/v)-2\tau]}] \tag{13.2.5}$$

$$I(x,s) = \frac{V^+(s)}{Z_c}[e^{-sx/v} - \Gamma_R(s)e^{s[(x/v)-2\tau]}] \tag{13.2.6}$$

From Figure 13.3 the boundary condition at the sending end is

$$V(0,s) = E_G(s) - Z_G(s)I(0,s) \tag{13.2.7}$$

Using (13.2.5) and (13.2.6) in (13.2.7),

$$V^+(s)[1 + \Gamma_R(s)e^{-2s\tau}] = E_G(s) - \left[\frac{Z_G(s)}{Z_c}\right]V^+(s)[1 - \Gamma_R(s)e^{-2s\tau}]$$

Solving for $V^+(s)$,

$$V^+(s)\left\{\left[\frac{Z_G(s)}{Z_c} + 1\right] - \Gamma_R(s)e^{-2s\tau}\left[\frac{Z_G(s)}{Z_c} - 1\right]\right\} = E_G(s)$$

$$V^+(s)\left[\frac{Z_G(s)}{Z_c} + 1\right]\{1 - \Gamma_R(s)\Gamma_S(s)e^{-2s\tau}\} = E_G(s)$$

or

$$V^+(s) = E_G(s)\left[\frac{Z_c}{Z_G(s) + Z_c}\right]\left[\frac{1}{1 - \Gamma_R(s)\Gamma_S(s)e^{-2s\tau}}\right] \tag{13.2.8}$$

where

$$\Gamma_S(s) = \frac{\dfrac{Z_G(s)}{Z_c} - 1}{\dfrac{Z_G(s)}{Z_c} + 1} \tag{13.2.9}$$

$\Gamma_S(s)$ is called the *sending-end voltage reflection coefficient*. Using (13.2.9) in (13.2.5) and (13.2.6), the complete solution is

$$V(x,s) = E_G(s)\left[\frac{Z_c}{Z_G(s)+Z_c}\right]\left[\frac{e^{-sx/v}+\Gamma_R(s)e^{s[(x/v)-2\tau]}}{1-\Gamma_R(s)\Gamma_S(s)e^{-2s\tau}}\right] \quad (13.2.10)$$

$$I(x,s) = \left[\frac{E_G(s)}{Z_G(s)+Z_c}\right]\left[\frac{e^{-sx/v}-\Gamma_R(s)e^{s[(x/v)-2\tau]}}{1-\Gamma_R(s)\Gamma_S(s)e^{-2s\tau}}\right] \quad (13.2.11)$$

where

$$\Gamma_R(s) = \frac{\dfrac{Z_R(s)}{Z_c}-1}{\dfrac{Z_R(s)}{Z_c}+1} \quad \text{per unit}$$

$$\Gamma_S(s) = \frac{\dfrac{Z_G(s)}{Z_c}-1}{\dfrac{Z_G(s)}{Z_c}+1} \quad \text{per unit} \quad (13.2.12)$$

$$Z_c = \sqrt{\frac{L}{C}} \ \Omega \qquad v = \frac{1}{\sqrt{LC}} \ \text{m/s} \qquad \tau = \frac{l}{v} \ \text{s} \quad (13.2.13)$$

The following four examples illustrate this general solution. All four examples refer to the line shown in Figure 13.3, which has length l, velocity v, characteristic impedance Z_c, and is initially unenergized.

EXAMPLE 13.1 Single-phase lossless-line transients: step-voltage source at sending end, matched load at receiving end

Let $Z_R = Z_c$ and $Z_G = 0$. The source voltage is a step, $e_G(t) = Eu_{-1}(t)$. (a) Determine $v(x,t)$ and $i(x,t)$. Plot the voltage and current versus time t at the center of the line and at the receiving end.

SOLUTION

a. From (13.2.12) with $Z_R = Z_c$ and $Z_G = 0$,

$$\Gamma_R(s) = \frac{1-1}{1+1} = 0 \qquad \Gamma_S(s) = \frac{0-1}{0+1} = -1$$

The Laplace transform of the source voltage is $E_G(s) = E/s$. Then, from (13.2.10) and (13.2.11),

$$V(x,s) = \left(\frac{E}{s}\right)(1)(e^{-sx/v}) = \frac{Ee^{-sx/v}}{s}$$

$$I(x,s) = \frac{(E/Z_c)}{s}e^{-sx/v}$$

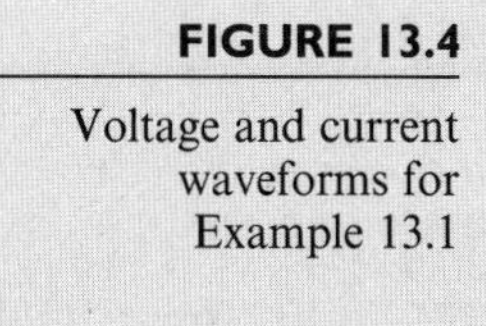

FIGURE 13.4

Voltage and current waveforms for Example 13.1

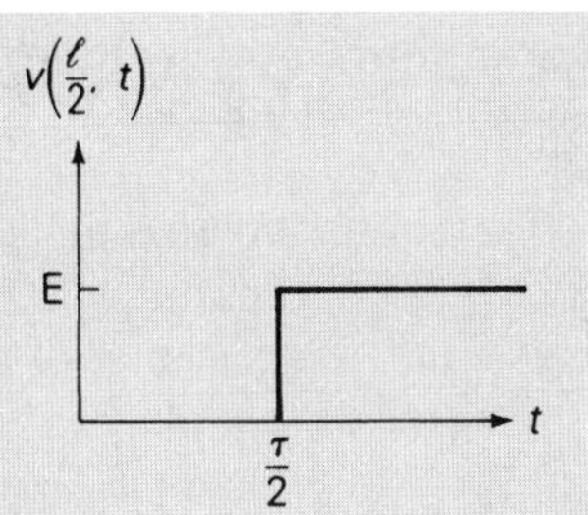

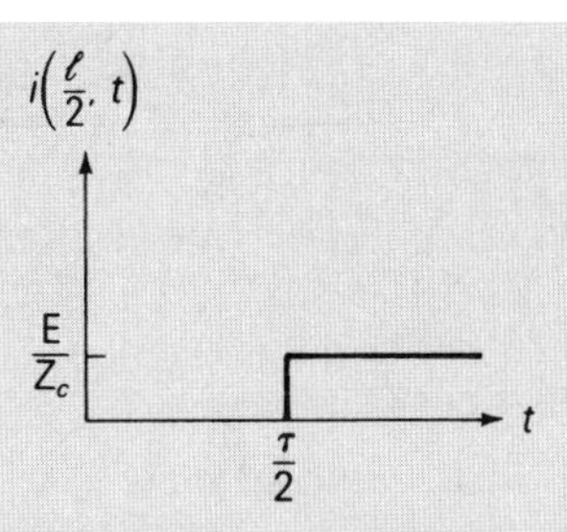

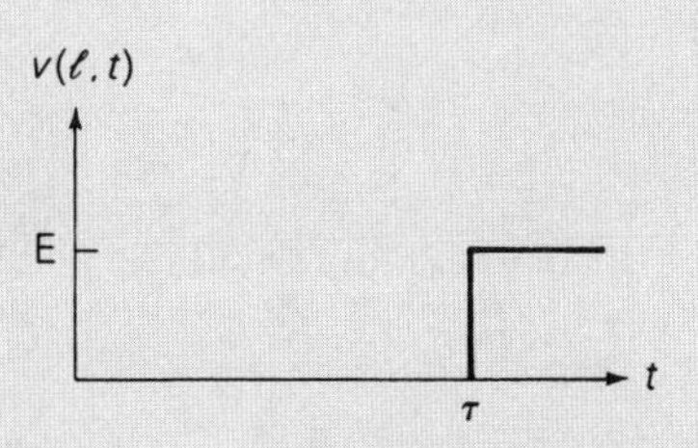

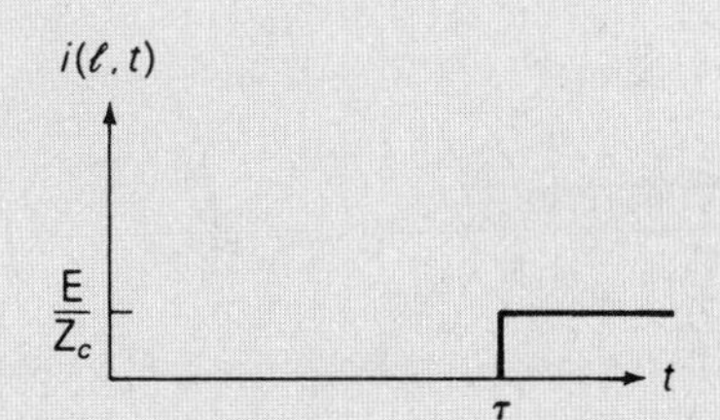

Taking the inverse Laplace transform,

$$v(x,t) = \mathrm{E}u_{-1}\left(t - \frac{x}{\nu}\right)$$

$$i(x,t) = \frac{\mathrm{E}}{Z_c}u_{-1}\left(t - \frac{x}{\nu}\right)$$

b. At the center of the line, where $x = l/2$,

$$v\left(\frac{l}{2}, t\right) = \mathrm{E}u_{-1}\left(t - \frac{\tau}{2}\right) \qquad i\left(\frac{l}{2}, t\right) = \frac{\mathrm{E}}{Z_c}u_{-1}\left(t - \frac{\tau}{2}\right)$$

At the receiving end, where $x = l$,

$$v(l,t) = \mathrm{E}u_{-1}(t - \tau) \qquad i(l,t) = \frac{\mathrm{E}}{Z_c}u_{-1}(t - \tau)$$

These waves, plotted in Figure 13.4, can be explained as follows. At $t = 0$ the ideal step voltage of E volts, applied to the sending end, encounters Z_c, the characteristic impedance of the line. Therefore, a forward traveling step voltage wave of E volts is initiated at the sending end. Also, since the ratio of the forward traveling voltage to current is Z_c, a forward traveling step current wave of (E/Z_c) amperes is initiated. These waves travel in the positive x direction, arriving at the center of the line at $t = \tau/2$, and at the end of the line at $t = \tau$. The receiving-end load is *matched* to the line; that is, $Z_R = Z_c$. For a matched load, $\Gamma_R = 0$, and therefore no backward traveling waves are initiated. In steady-state, the line with matched load is energized at E volts with current E/Z_c amperes. ■

EXAMPLE 13.2 Single-phase lossless-line transients: step-voltage source matched at sending end, open receiving end

The receiving end is open. The source voltage at the sending end is a step $e_G(t) = Eu_{-1}(t)$, with $Z_G(s) = Z_c$. (a) Determine $v(x,t)$ and $i(x,t)$. (b) Plot the voltage and current versus time t at the center of the line.

SOLUTION

a. From (13.2.12),

$$\Gamma_R(s) = \lim_{Z_R \to \infty} \frac{\frac{Z_R}{Z_c} - 1}{\frac{Z_R}{Z_c} + 1} = 1 \qquad \Gamma_S(s) = \frac{1-1}{1+1} = 0$$

The Laplace transform of the source voltage is $E_G(s) = E/s$. Then, from (13.2.10) and (13.2.11),

$$V(x,s) = \frac{E}{s}\left(\frac{1}{2}\right)[e^{-sx/v} + e^{s[(x/v)-2\tau]}]$$

$$I(x,s) = \frac{E}{s}\left(\frac{1}{2Z_c}\right)[e^{-sx/v} - e^{s[(x/v)-2\tau]}]$$

Taking the inverse Laplace transform,

$$v(x,t) = \frac{E}{2}u_{-1}\left(t - \frac{x}{v}\right) + \frac{E}{2}u_{-1}\left(t + \frac{x}{v} - 2\tau\right)$$

$$i(x,t) = \frac{E}{2Z_c}u_{-1}\left(t - \frac{x}{v}\right) - \frac{E}{2Z_c}u_{-1}\left(t + \frac{x}{v} - 2\tau\right)$$

b. At the center of the line, where $x = l/2$,

$$v\left(\frac{l}{2}, t\right) = \frac{E}{2}u_{-1}\left(t - \frac{\tau}{2}\right) + \frac{E}{2}u_{-1}\left(t - \frac{3\tau}{2}\right)$$

$$i\left(\frac{l}{2}, t\right) = \frac{E}{2Z_c}u_{-1}\left(t - \frac{\tau}{2}\right) - \frac{E}{2Z_c}u_{-1}\left(t - \frac{3\tau}{2}\right)$$

These waves are plotted in Figure 13.5. At $t = 0$, the step voltage source of E volts encounters the source impedance $Z_G = Z_c$ in series with the characteristic impedance of the line, Z_c. Using voltage division, the sending-end voltage at $t = 0$ is $E/2$. Therefore, a forward traveling step voltage wave of $E/2$ volts and a forward traveling step current wave of $E/(2Z_c)$ amperes are initiated at the sending end. These waves arrive at the center of the line at $t = \tau/2$. Also, with $\Gamma_R = 1$, the backward traveling voltage wave equals the forward traveling voltage wave, and the backward traveling current wave is the negative of the forward traveling current wave. These backward traveling waves, which are initiated at the receiving end at $t = \tau$ when the forward traveling waves

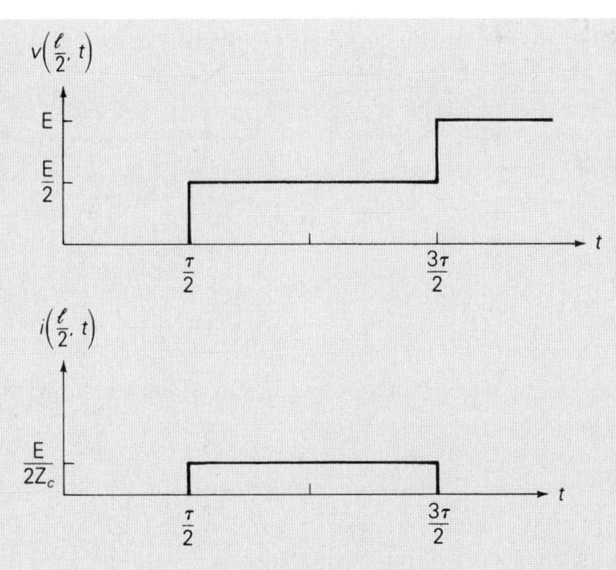

FIGURE 13.5 Voltage and current waveforms for Example 13.2

arrive there, arrive at the center of the line at $t = 3\tau/2$ and are superimposed on the forward traveling waves. No additional forward or backward traveling waves are initiated because the source impedance is matched to the line; that is, $\Gamma_S(s) = 0$. In steady-state, the line, which is open at the receiving end, is energized at E volts with zero current. ■

EXAMPLE 13.3 Single-phase lossless-line transients: step-voltage source matched at sending end, capacitive load at receiving end

The receiving end is terminated by a capacitor with C_R farads, which is initially unenergized. The source voltage at the sending end is a unit step $e_G(t) = Eu_{-1}(t)$, with $Z_G = Z_c$. Determine and plot $v(x, t)$ versus time t at the sending end of the line.

SOLUTION From (13.2.12) with $Z_R = \dfrac{1}{sC_R}$ and $Z_G = Z_c$,

$$\Gamma_R(s) = \frac{\dfrac{1}{sC_R Z_c} - 1}{\dfrac{1}{sC_R Z_c} + 1} = \frac{-s + \dfrac{1}{Z_c C_R}}{s + \dfrac{1}{Z_c C_R}}$$

$$\Gamma_S(s) = \frac{1 - 1}{1 + 1} = 0$$

Then, from (13.2.10), with $E_G(s) = E/s$,

$$V(x,s) = \frac{E}{s}\left(\frac{1}{2}\right)\left[e^{-sx/v} + \left(\frac{-s + \dfrac{1}{Z_c C_R}}{s + \dfrac{1}{Z_c C_R}}\right) e^{s[(x/v)-2\tau]}\right]$$

$$= \frac{E}{2}\left[\frac{e^{-sx/v}}{s} + \frac{1}{s}\left(\frac{-s + \dfrac{1}{Z_c C_R}}{s + \dfrac{1}{Z_c C_R}}\right) e^{s[(x/v)-2\tau]}\right]$$

Using partial fraction expansion of the second term above,

$$V(x,s) = \frac{E}{2}\left[\frac{e^{-sx/v}}{s} + \left(\frac{1}{s} - \frac{2}{s + \dfrac{1}{Z_c C_R}}\right) e^{s[(x/v)-2\tau]}\right]$$

The inverse Laplace transform is

$$v(x,t) = \frac{E}{2}u_{-1}\left(t - \frac{x}{v}\right) + \frac{E}{2}[1 - 2e^{(-1/Z_c C_R)(t+x/v-2\tau)}]u_{-1}\left(t + \frac{x}{v} - 2\tau\right)$$

At the sending end, where $x = 0$,

$$v(0,t) = \frac{E}{2}u_{-1}(t) + \frac{E}{2}[1 - 2e^{(-1/Z_c C_R)(t-2\tau)}]u_{-1}(t - 2\tau)$$

$v(0,t)$ is plotted in Figure 13.6. As in Example 13.2, a forward traveling step voltage wave of E/2 volts is initiated at the sending end at $t = 0$. At $t = \tau$, when the forward traveling wave arrives at the receiving end, a backward traveling wave is initiated. The backward traveling voltage wave, an exponential with initial value $-E/2$, steady-state value $+E/2$, and time constant $Z_c C_R$, arrives at the sending end at $t = 2\tau$, where it is superimposed on the forward traveling wave. No additional waves are initiated, since the source impedance is matched to the line. In steady-state, the line and the capacitor at the receiving end are energized at E volts with zero current.

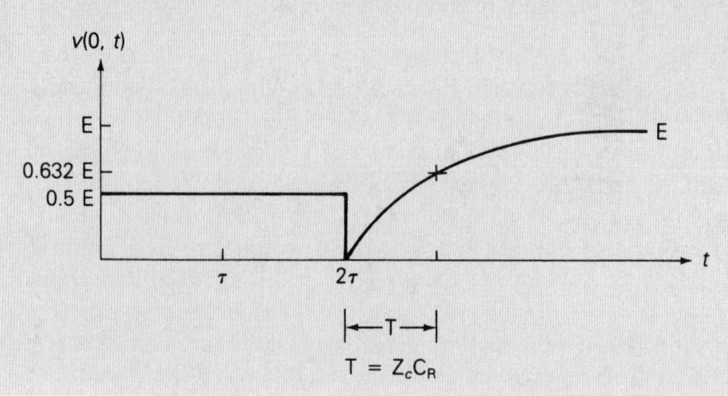

FIGURE 13.6 Voltage waveform for Example 13.3

The capacitor at the receiving end can also be viewed as a short circuit at the instant $t = \tau$, when the forward traveling wave arrives at the receiving end. For a short circuit at the receiving end, $\Gamma_R = -1$, and therefore the backward traveling voltage wavefront is $-E/2$, the negative of the forward traveling wave. However, in steady-state the capacitor is an open circuit, for which $\Gamma_R = +1$, and the steady-state backward traveling voltage wave equals the forward traveling voltage wave. ■

EXAMPLE 13.4 Single-phase lossless-line transients: step-voltage source with unmatched source resistance at sending end, unmatched resistive load at receiving end

At the receiving end, $Z_R = Z_c/3$. At the sending end, $e_G(t) = Eu_{-1}(t)$ and $Z_G = 2Z_c$. Determine and plot the voltage versus time at the center of the line.

SOLUTION From (13.2.12),

$$\Gamma_R = \frac{\frac{1}{3} - 1}{\frac{1}{3} + 1} = -\frac{1}{2} \qquad \Gamma_S = \frac{2 - 1}{2 + 1} = \frac{1}{3}$$

From (13.2.10), with $E_G(s) = E/s$,

$$V(x, s) = \frac{E}{s}\left(\frac{1}{3}\right)\frac{\left[e^{-sx/v} - \frac{1}{2}e^{s[(x/v)-2\tau]}\right]}{1 + \left(\frac{1}{6}e^{-2s\tau}\right)}$$

The preceding equation can be rewritten using the following geometric series:

$$\frac{1}{1+y} = 1 - y + y^2 - y^3 + y^4 - \cdots$$

with $y = \frac{1}{6}e^{-2s\tau}$,

$$V(x, s) = \frac{E}{3s}\left[e^{-sx/v} - \frac{1}{2}e^{s[(x/v)-2\tau]}\right] \times \left[1 - \frac{1}{6}e^{-2s\tau} + \frac{1}{36}e^{-4s\tau} - \frac{1}{216}e^{-6s\tau} + \cdots\right]$$

Multiplying the terms within the brackets,

$$V(x, s) = \frac{E}{3s}\left[e^{-sx/v} - \frac{1}{2}e^{s[(x/v)-2\tau]} - \frac{1}{6}e^{-s[(x/v)+2\tau]} + \frac{1}{12}e^{s[(x/v)-4\tau]} + \frac{1}{36}e^{-s[(x/v)+4\tau]} - \frac{1}{72}e^{s[(x/v)-6\tau]} + \cdots\right]$$

Taking the inverse Laplace transform,

$$v(x,t) = \frac{E}{3}\left[u_{-1}\left(t - \frac{x}{v}\right) - \frac{1}{2}u_{-1}\left(t + \frac{x}{v} - 2\tau\right) - \frac{1}{6}u_{-1}\left(t - \frac{x}{v} - 2\tau\right) + \frac{1}{12}u_{-1}\left(t + \frac{x}{v} - 4\tau\right) + \frac{1}{36}u_{-1}\left(t - \frac{x}{v} - 4\tau\right) - \frac{1}{72}u_{-1}\left(t + \frac{x}{v} - 6\tau\right)\cdots\right]$$

At the center of the line, where $x = l/2$,

$$v\left(\frac{l}{2}, t\right) = \frac{E}{3}\left[u_{-1}\left(t - \frac{\tau}{2}\right) - \frac{1}{2}u_{-1}\left(t - \frac{3\tau}{2}\right) - \frac{1}{6}u_{-1}\left(t - \frac{5\tau}{2}\right) + \frac{1}{12}u_{-1}\left(t - \frac{7\tau}{2}\right) + \frac{1}{36}u_{-1}\left(t - \frac{9\tau}{2}\right) - \frac{1}{72}u_{-1}\left(t - \frac{11\tau}{2}\right)\cdots\right]$$

$v(l/2, t)$ is plotted in Figure 13.7(a). Since neither the source nor the load is matched to the line, the voltage at any point along the line consists of an infinite series of forward and backward traveling waves. At the center of the line, the first forward traveling wave arrives at $t = \tau/2$; then a backward traveling wave arrives at $3\tau/2$, another forward traveling wave arrives at $5\tau/2$, another backward traveling wave at $7\tau/2$, and so on.

The steady-state voltage can be evaluated from the final value theorem. That is,

$$v_{ss}(x) = \lim_{t\to\infty} v(x,t) = \lim_{s\to 0} sV(x,s)$$

$$= \lim_{s\to 0}\left\{ s\left(\frac{E}{s}\right)\left(\frac{1}{3}\right)\frac{[e^{-sx/v} - \frac{1}{2}e^{s[(x/v)-2\tau]}]}{1 + \frac{1}{6}e^{-2s\tau}}\right\}$$

$$= E\left(\frac{1}{3}\right)\left(\frac{1 - \frac{1}{2}}{1 + \frac{1}{6}}\right) = \frac{E}{7}$$

The steady-state solution can also be evaluated from the circuit in Figure 13.7(b). Since there is no steady-state voltage drop across the lossless

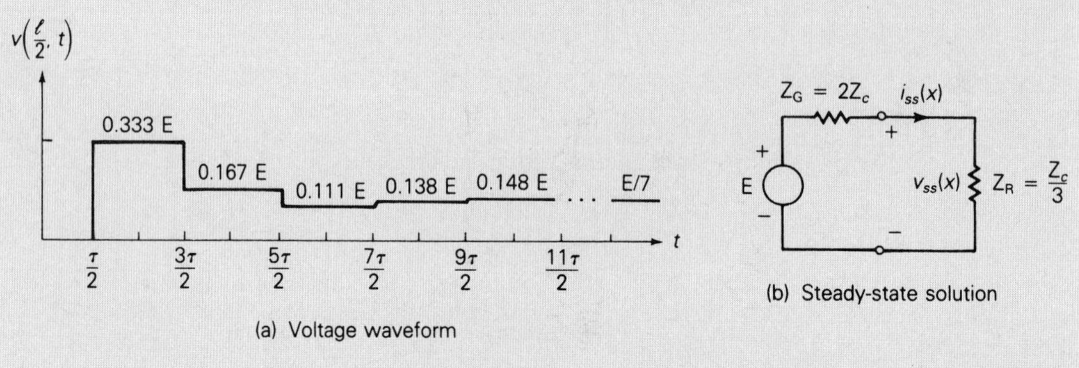

FIGURE 13.7 Example 13.4

line when a dc source is applied, the line can be eliminated, leaving only the source and load. The steady-state voltage is then, by voltage division,

$$v_{ss}(x) = E\left(\frac{Z_R}{Z_R + Z_G}\right) = E\left(\frac{\frac{1}{3}}{\frac{1}{3}+2}\right) = \frac{E}{7}$$ ■

13.3 BEWLEY LATTICE DIAGRAM

A lattice diagram developed by L. V. Bewley [2] conveniently organizes the reflections that occur during transmission-line transients. For the Bewley lattice diagram, shown in Figure 13.8, the vertical scale represents time and is scaled in units of τ, the transient time of the line. The horizontal scale represents line position x, and the diagonal lines represent traveling waves. Each reflection is determined by multiplying the incident wave arriving at an end by the reflection coefficient at that end. The voltage $v(x, t)$ at any point x and t on the diagram is determined by adding all the terms directly above that point.

FIGURE 13.8

Bewley lattice diagram for Example 13.5

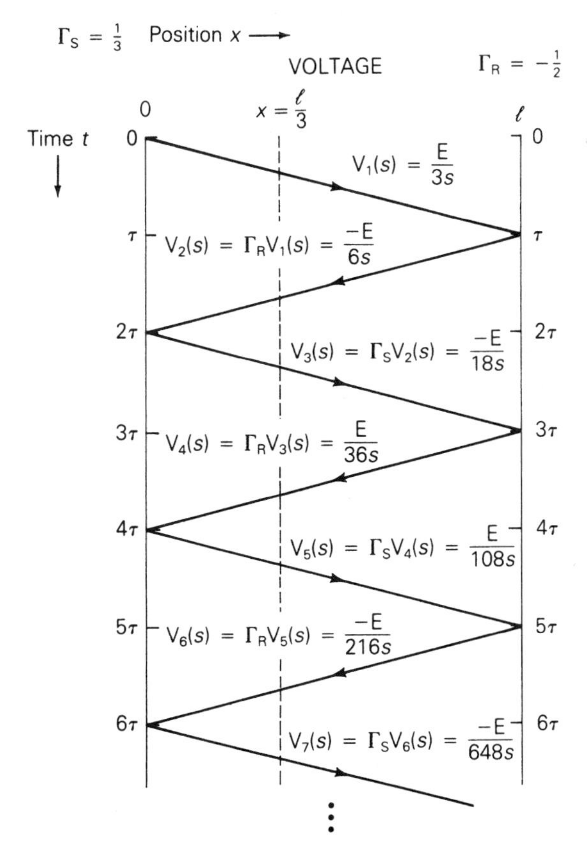

EXAMPLE 13.5 Lattice diagram: single-phase lossless line

For the line and terminations given in Example 13.4, draw the lattice diagram and plot $v(l/3, t)$ versus time t.

SOLUTION The lattice diagram is shown in Figure 13.8. At $t = 0$, the source voltage encounters the source impedance and the line characteristic impedance, and the first forward traveling wave is determined by voltage division:

$$\mathrm{V}_1(s) = \mathrm{E}_\mathrm{G}(s)\left[\frac{\mathrm{Z}_c}{\mathrm{Z}_c + \mathrm{Z}_\mathrm{G}}\right] = \frac{\mathrm{E}}{s}\left[\frac{1}{1+2}\right] = \frac{\mathrm{E}}{3s}$$

which is a step with magnitude (E/3) volts. The next traveling wave, a backward one, is $\mathrm{V}_2(s) = \Gamma_\mathrm{R}(s)\mathrm{V}_1(s) = (-\frac{1}{2})\mathrm{V}_1(s) = -\mathrm{E}/(6s)$, and the next wave, a forward one, is $\mathrm{V}_3(s) = \Gamma_\mathrm{s}(s)\mathrm{V}_2(s) = (\frac{1}{3})\mathrm{V}_2(s) = -\mathrm{E}/(18s)$. Subsequent waves are calculated in a similar manner.

The voltage at $x = l/3$ is determined by drawing a vertical line at $x = l/3$ on the lattice diagram, shown dashed in Figure 13.8. Starting at the top of the dashed line, where $t = 0$, and moving down, each voltage wave is added at the time it intersects the dashed line. The first wave v_1 arrives at $t = \tau/3$, the second v_2 arrives at $5\tau/3$, v_3 at $7\tau/3$, and so on. $v(l/3, t)$ is plotted in Figure 13.9.

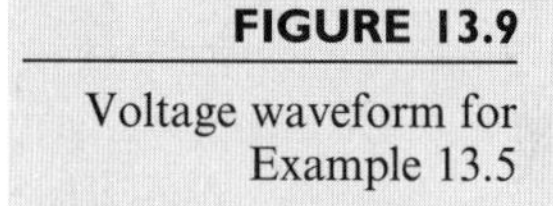

FIGURE 13.9 Voltage waveform for Example 13.5

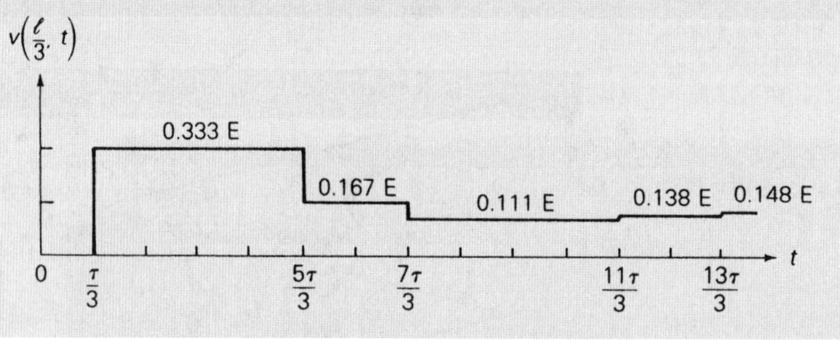

■

Figure 13.10 shows a forward traveling voltage wave V_A^+ arriving at the junction of two lossless lines A and B with characteristic impedances Z_A and Z_B, respectively. This could be, for example, the junction of an overhead line and a cable. When V_A^+ arrives at the junction, both a reflection V_A^- on line A and a refraction V_B^+ on line B will occur. Writing a KVL and KCL equation at the junction,

FIGURE 13.10 Junction of two single-phase lossless lines

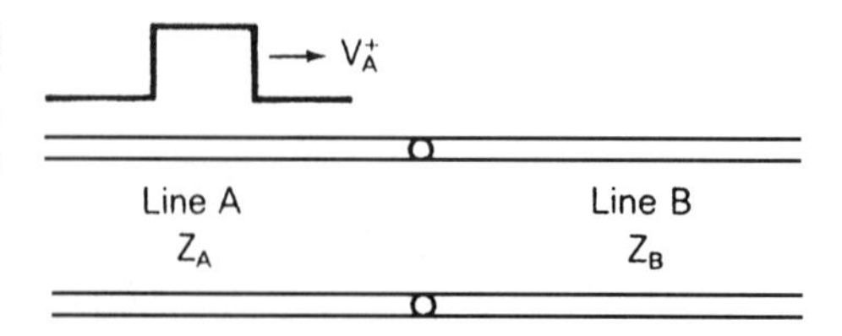

$$V_A^+ + V_A^- = V_B^+ \tag{13.3.1}$$

$$I_A^+ + I_A^- = I_B^+ \tag{13.3.2}$$

Recall that $I_A^+ = V_A^+/Z_A$, $I_A^- = -V_A^-/Z_A$, and $I_B^+ = V_B^+/Z_B$. Using these relations in (13.3.2),

$$\frac{V_A^+}{Z_A} - \frac{V_A^-}{Z_A} = \frac{V_B^+}{Z_B} \tag{13.3.3}$$

Solving (13.3.1) and (13.3.3) for V_A^- and V_B^+ in terms of V_A^+ yields

$$V_A^- = \Gamma_{AA} V_A^+ \tag{13.3.4}$$

where

$$\Gamma_{AA} = \frac{\dfrac{Z_B}{Z_A} - 1}{\dfrac{Z_B}{Z_A} + 1} \tag{13.3.5}$$

and

$$V_B^+ = \Gamma_{BA} V_A \tag{13.3.6}$$

where

$$\Gamma_{BA} = \frac{2\left(\dfrac{Z_B}{Z_A}\right)}{\dfrac{Z_B}{Z_A} + 1} \tag{13.3.7}$$

Note that Γ_{AA}, given by (13.3.5), is similar to Γ_R, given by (13.2.12), except that Z_B replaces Z_R. Thus, for waves arriving at the junction from line A, the "load" at the receiving end of line A is the characteristic impedance of line B.

EXAMPLE 13.6 Lattice diagram: overhead line connected to a cable, single-phase lossless lines

As shown in Figure 13.10, a single-phase lossless overhead line with $Z_A = 400\ \Omega$, $v_A = 3 \times 10^8$ m/s, and $l_A = 30$ km is connected to a single-phase lossless cable with $Z_B = 100\ \Omega$, $v_B = 2 \times 10^8$ m/s, and $l_B = 20$ km. At the sending end of line A, $e_g(t) = Eu_{-1}(t)$ and $Z_G = Z_A$. At the receiving end of line B, $Z_R = 2Z_B = 200\ \Omega$. Draw the lattice diagram for $0 \le t \le 0.6$ ms and plot the voltage at the junction versus time. The line and cable are initially unenergized.

SOLUTION From (13.2.13),

$$\tau_A = \frac{30 \times 10^3}{3 \times 10^8} = 0.1 \times 10^{-3} \quad \text{s} \qquad \tau_B = \frac{20 \times 10^3}{2 \times 10^8} = 0.1 \times 10^{-3} \quad \text{s}$$

From (13.2.12), with $Z_G = Z_A$ and $Z_R = 2Z_B$,

$$\Gamma_S = \frac{1-1}{1+1} = 0 \qquad \Gamma_R = \frac{2-1}{2+1} = \frac{1}{3}$$

From (13.3.5) and (13.3.6), the reflection and refraction coefficients for waves arriving at the junction from line A are

$$\left.\Gamma_{AA} = \frac{\frac{100}{400}-1}{\frac{100}{400}+1} = \frac{-3}{5} \qquad \Gamma_{BA} = \frac{2\frac{100}{400}}{\frac{100}{400}+1} = \frac{2}{5}\right\} \text{from line A}$$

Reversing A and B, the reflection and refraction coefficients for waves returning to the junction from line B are

$$\left.\Gamma_{BB} = \frac{\frac{400}{100}-1}{\frac{400}{100}+1} = \frac{3}{5} \qquad \Gamma_{AB} = \frac{2\frac{400}{100}}{\frac{400}{100}+1} = \frac{8}{5}\right\} \text{from line B}$$

The lattice diagram is shown in Figure 13.11. Using voltage division, the first forward traveling voltage wave is

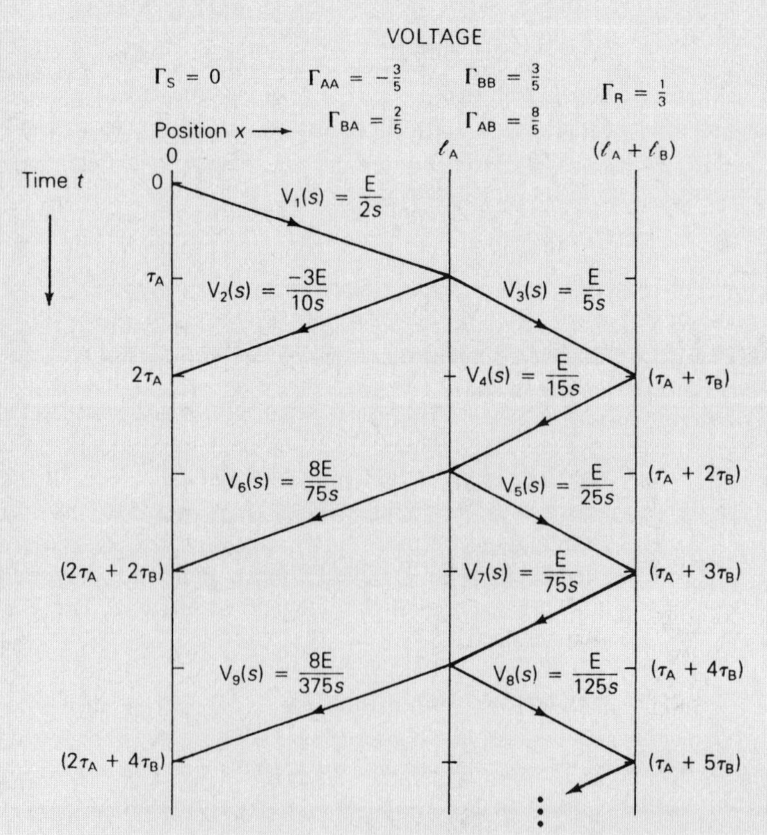

FIGURE 13.11 Lattice diagram for Example 13.6

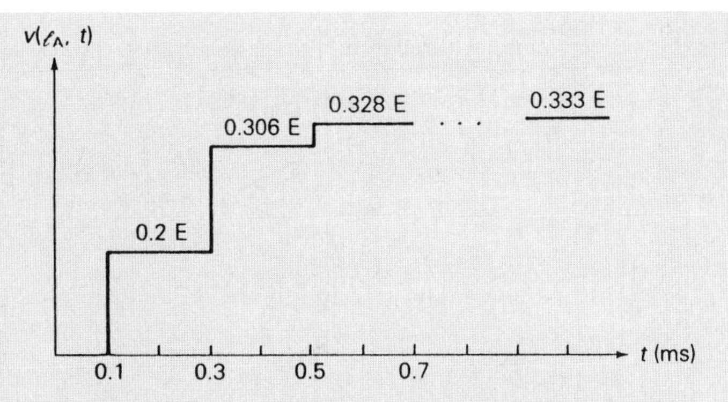

FIGURE 13.12 Junction voltage for Example 13.6

$$V_1(s) = E_G(s)\left(\frac{Z_A}{Z_A + Z_G}\right) = \frac{E}{s}\left(\frac{1}{2}\right) = \frac{E}{2s}$$

When v_1 arrives at the junction, a reflected wave v_2 and refracted wave v_3 are initiated. Using the reflection and refraction coefficients for line A,

$$V_2(s) = \Gamma_{AA}V_1(s) = \left(\frac{-3}{5}\right)\left(\frac{E}{2s}\right) = \frac{-3E}{10s}$$

$$V_3(s) = \Gamma_{BA}V_1(s) = \left(\frac{2}{5}\right)\left(\frac{E}{2s}\right) = \frac{E}{5s}$$

When v_2 arrives at the receiving end of line B, a reflected wave $V_4(s) = \Gamma_R V_3(s) = \frac{1}{3}(E/5s) = (E/15s)$ is initiated. When v_4 arrives at the junction, reflected wave v_5 and refracted wave v_6 are initiated. Using the reflection and refraction coefficients for line B,

$$V_5(s) = \Gamma_{BB}V_4(s) = \left(\frac{3}{5}\right)\left(\frac{E}{15s}\right) = \frac{E}{25s}$$

$$V_6(s) = \Gamma_{AB}V_4(s) = \left(\frac{8}{5}\right)\left(\frac{E}{15s}\right) = \frac{8E}{75s}$$

Subsequent reflections and refractions are calculated in a similar manner.

The voltage at the junction is determined by starting at $x = l_A$ at the top of the lattice diagram, where $t = 0$. Then, moving down the lattice diagram, voltage waves either just to the left or just to the right of the junction are added when they occur. For example, looking just to the right of the junction at $x = l_A^+$, the voltage wave v_3, a step of magnitude E/5 volts occurs at $t = \tau_A$. Then at $t = (\tau_A + 2\tau_B)$, two waves v_4 and v_5, which are steps of magnitude E/15 and E/25, are added to v_3. $v(l_A, t)$ is plotted in Figure 13.12.

The steady-state voltage is determined by removing the lossless lines and calculating the steady-state voltage across the receiving-end load:

$$v_{ss}(x) = E\left(\frac{Z_R}{Z_R + Z_G}\right) = E\left(\frac{200}{200 + 400}\right) = \frac{E}{3}$$

■

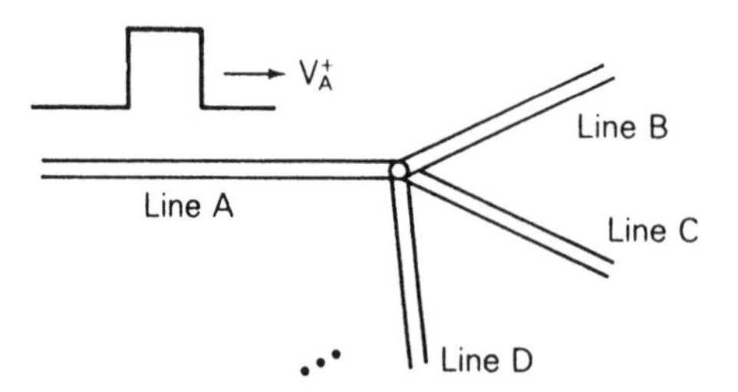

FIGURE 13.13 Junction of lossless lines A, B, C, D, and so on

The preceding analysis can be extended to the junction of more than two lossless lines, as shown in Figure 13.13. Writing a KVL and KCL equation at the junction for a voltage V_A^+ arriving at the junction from line A,

$$V_A^+ + V_A^- = V_B^+ = V_C^+ = V_D^+ = \cdots \tag{13.3.8}$$

$$I_A^+ + I_A^- = I_B^+ + I_C^+ + I_D^+ + \cdots \tag{13.3.9}$$

Using $I_A^+ = V_A^+/Z_A$, $I_A^- = -V_A^-/Z_A$, $I_B^+ = V_B^+/Z_B$, and so on in (13.3.9),

$$\frac{V_A^+}{Z_A} - \frac{V_A^-}{Z_A} = \frac{V_B^+}{Z_B} + \frac{V_C^+}{Z_C} + \frac{V_D^+}{Z_D} + \cdots \tag{13.3.10}$$

Equations (13.3.8) and (13.3.10) can be solved for V_A^+, V_B^+, V_C^+, V_D^+, and so on in terms of V_A^+. (See Problem 13.14.)

13.4 DISCRETE-TIME MODELS OF SINGLE-PHASE LOSSLESS LINES AND LUMPED RLC ELEMENTS

Our objective in this section is to develop discrete-time models of single-phase lossless lines and of lumped RLC elements suitable for computer calculation of transmission-line transients at discrete-time intervals $t = \Delta t$, $2\Delta t$, $3\Delta t$, and so on. The discrete-time models are presented as equivalent circuits consisting of lumped resistors and current sources. The current sources in the models represent the past history of the circuit—that is, the history at times $t - \Delta t$, $t - 2\Delta t$, and so on. After interconnecting the equivalent circuits of all the components in any given circuit, nodal equations can then be written for each discrete time. Discrete-time models, first developed by L. Bergeron [3], are presented first.

SINGLE-PHASE LOSSLESS LINE

From the general solution of a single-phase lossless line, given by (13.1.12) and (13.1.18), we obtain

FIGURE 13.14

Single-phase two-wire lossless line

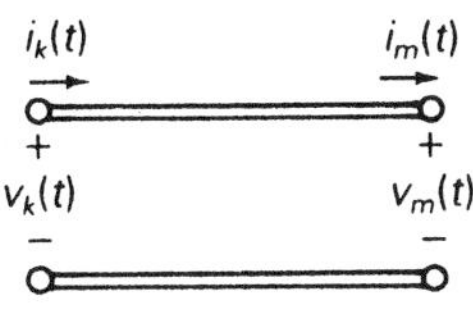

(a) Terminal variables

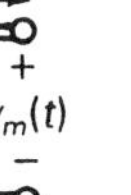

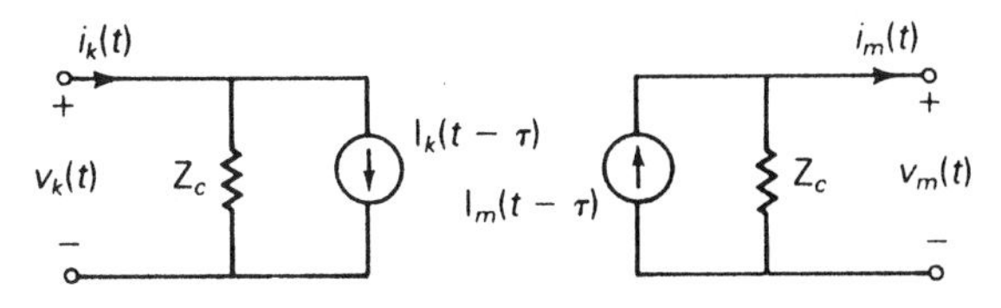

(b) Discrete-time equivalent circuit

$$v(x,t) + Z_c i(x,t) = 2v^+\left(t - \frac{x}{v}\right) \tag{13.4.1}$$

$$v(x,t) - Z_c i(x,t) = 2v^-\left(t + \frac{x}{v}\right) \tag{13.4.2}$$

In (13.4.1), the left side $(v + Z_c i)$ remains constant when the argument $(t - x/v)$ is constant. Therefore, to a fictitious observer traveling at velocity v in the positive x direction along the line, $(v + Z_c i)$ remains constant. If τ is the transit time from terminal k to terminal m of the line, the value of $(v + Z_c i)$ observed at time $(t - \tau)$ at terminal k must equal the value at time t at terminal m. That is,

$$v_k(t-\tau) + Z_c i_k(t-\tau) = v_m(t) + Z_c i_m(t) \tag{13.4.3}$$

where k and m denote terminals k and m, as shown in Figure 13.14(a).

Similarly, $(v - Z_c i)$ in (13.4.2) remains constant when $(t + x/v)$ is constant. To a fictitious observer traveling at velocity v in the negative x direction, $(v - Z_c i)$ remains constant. Therefore, the value of $(v - Z_c i)$ at time $(t - \tau)$ at terminal m must equal the value at time t at terminal k. That is,

$$v_m(t-\tau) - Z_c i_m(t-\tau) = v_k(t) - Z_c i_k(t) \tag{13.4.4}$$

Equation (13.4.3) is rewritten as

$$i_m(t) = I_m(t-\tau) - \frac{1}{Z_c} v_m(t) \tag{13.4.5}$$

where

$$I_m(t-\tau) = i_k(t-\tau) + \frac{1}{Z_c} v_k(t-\tau) \tag{13.4.6}$$

Similarly, (13.4.4) is rewritten as

$$i_k(t) = I_k(t-\tau) + \frac{1}{Z_c} v_k(t) \tag{13.4.7}$$

where

$$I_k(t-\tau) = i_m(t-\tau) - \frac{1}{Z_c} v_m(t-\tau) \tag{13.4.8}$$

Also, using (13.4.7) in (13.4.6),

$$\mathrm{I}_m(t-\tau) = \mathrm{I}_k(t-2\tau) + \frac{2}{\mathrm{Z}_c} v_k(t-\tau) \tag{13.4.9}$$

and using (13.4.5) in (13.4.8),

$$\mathrm{I}_k(t-\tau) = \mathrm{I}_m(t-2\tau) - \frac{2}{\mathrm{Z}_c} v_m(t-\tau) \tag{13.4.10}$$

Equations (13.4.5) and (13.4.7) are represented by the circuit shown in Figure 13.14(b). The current sources $\mathrm{I}_m(t-\tau)$ and $\mathrm{I}_k(t-\tau)$ shown in this figure, which are given by (13.4.9) and (13.4.10), represent the past history of the transmission line.

Note that in Figure 13.14(b) terminals k and m are not directly connected. The conditions at one terminal are "felt" indirectly at the other terminal after a delay of τ seconds.

LUMPED INDUCTANCE

As shown in Figure 13.15(a) for a constant lumped inductance L,

$$v(t) = \mathrm{L}\frac{di(t)}{dt} \tag{13.4.11}$$

Integrating this equation from time $(t-\Delta t)$ to t,

$$\int_{t-\Delta t}^{t} di(t) = \frac{1}{\mathrm{L}}\int_{t-\Delta t}^{t} v(t)\,dt \tag{13.4.12}$$

FIGURE 13.15

Lumped inductance

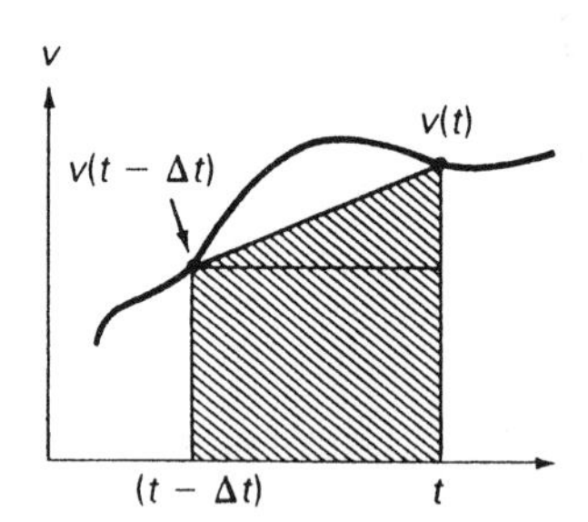

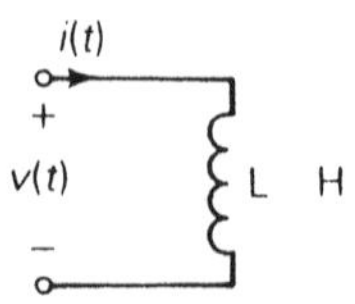

(a) Continuous time circuit

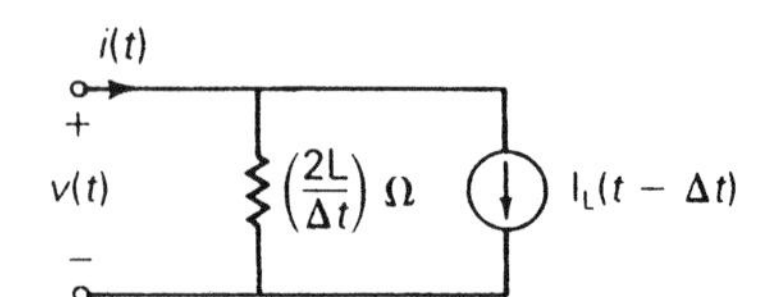

(b) Discrete-time circuit

Using the trapezoidal rule of integration,

$$i(t) - i(t - \Delta t) = \left(\frac{1}{L}\right)\left(\frac{\Delta t}{2}\right)[v(t) + v(t - \Delta t)]$$

Rearranging gives

$$i(t) = \frac{v(t)}{(2L/\Delta t)} + \left[i(t - \Delta t) + \frac{v(t - \Delta t)}{(2L/\Delta t)}\right]$$

or

$$i(t) = \frac{v(t)}{(2L/\Delta t)} + I_L(t - \Delta t) \tag{13.4.13}$$

where

$$I_L(t - \Delta t) = i(t - \Delta t) + \frac{v(t - \Delta t)}{(2L/\Delta t)} = I_L(t - 2\Delta t) + \frac{v(t - \Delta t)}{(L/\Delta t)} \tag{13.4.14}$$

Equations (13.4.13) and (13.4.14) are represented by the circuit shown in Figure 13.15(b). As shown, the inductor is replaced by a resistor with resistance $(2L/\Delta t)\ \Omega$. A current source $I_L(t - \Delta t)$ given by (13.4.14) is also included. $I_L(t - \Delta t)$ represents the past history of the inductor. Note that the trapezoidal rule introduces an error of the order $(\Delta t)^3$.

LUMPED CAPACITANCE

As shown in Figure 13.16(a) for a constant lumped capacitance C,

$$i(t) = C\frac{dv(t)}{dt} \tag{13.4.15}$$

Integrating from time $(t - \Delta t)$ to t,

$$\int_{t-\Delta t}^{t} dv(t) = \frac{1}{C}\int_{t-\Delta t}^{t} i(t)\,dt \tag{13.4.16}$$

Using the trapezoidal rule of integration,

$$v(t) - v(t - \Delta t) = \frac{1}{C}\left(\frac{\Delta t}{2}\right)[i(t) + i(t - \Delta t)]$$

FIGURE 13.16

Lumped capacitance

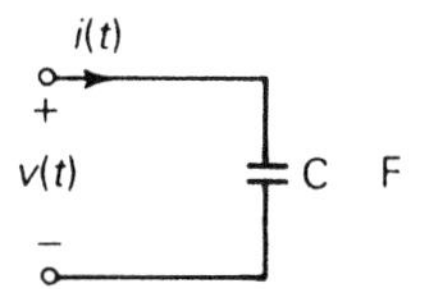

(a) Continuous time circuit

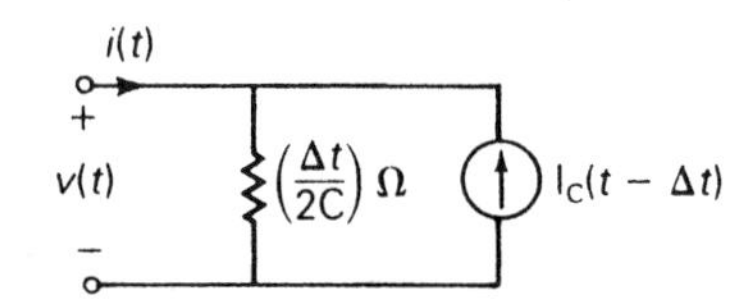

(b) Discrete-time circuit

Rearranging gives

$$i(t) = \frac{v(t)}{(\Delta t/2C)} - I_C(t - \Delta t) \tag{13.4.17}$$

where

$$I_C(t - \Delta t) = i(t - \Delta t) + \frac{v(t - \Delta t)}{(\Delta t/2C)} = -I_C(t - 2\Delta t) + \frac{v(t - \Delta t)}{(\Delta t/4C)} \tag{13.4.18}$$

Equations (13.4.17) and (13.4.18) are represented by the circuit in Figure 13.16(b). The capacitor is replaced by a resistor with resistance $(\Delta t/2C)$ Ω. A current source $I_C(t - \Delta t)$, which represents the capacitor's past history, is also included.

FIGURE 13.17

Lumped resistance

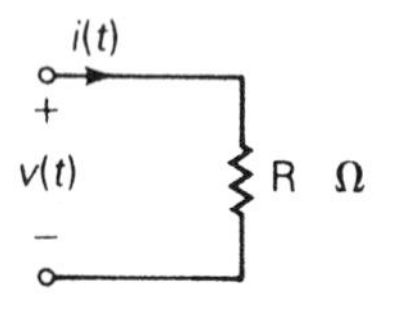

(a) Continuous time circuit

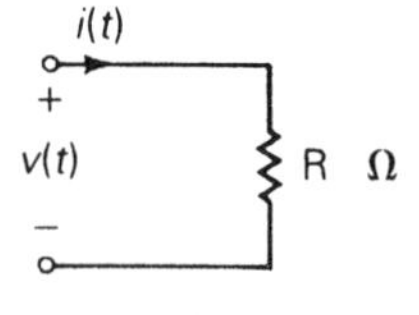

(b) Discrete-time circuit

LUMPED RESISTANCE

The discrete model of a constant lumped resistance R, as shown in Figure 13.17, is the same as the continuous model. That is,

$$v(t) = Ri(t) \tag{13.4.19}$$

NODAL EQUATIONS

A circuit consisting of single-phase lossless transmission lines and constant lumped RLC elements can be replaced by the equivalent circuits given in Figures 13.14(b), 13.15(b), 13.16(b), and 13.17(b). Then, writing nodal equations, the result is a set of linear algebraic equations that determine the bus voltages at each instant t.

EXAMPLE 13.7 Discrete-time equivalent circuit, single-phase lossless line transients, computer solution

For the circuit given in Example 13.3, replace the circuit elements by their discrete-time equivalent circuits and write the nodal equations that determine the sending-end and receiving-end voltages. Then, using a digital computer, compute the sending-end and receiving-end voltages for $0 \leqq t \leqq 9$ ms. For numerical calculations, assume $E = 100$ V, $Z_c = 400$ Ω, $C_R = 5$ μF, $\tau = 1.0$ ms, and $\Delta t = 0.1$ ms.

SOLUTION The discrete model is shown in Figure 13.18, where $v_k(t)$ represents the sending-end voltage $v(0, t)$ and $v_m(t)$ represents the receiving-end voltage $v(l, t)$. Also, the sending-end voltage source $e_G(t)$ in series with Z_G is converted to an equivalent current source in parallel with Z_G. Writing nodal equations for this circuit,

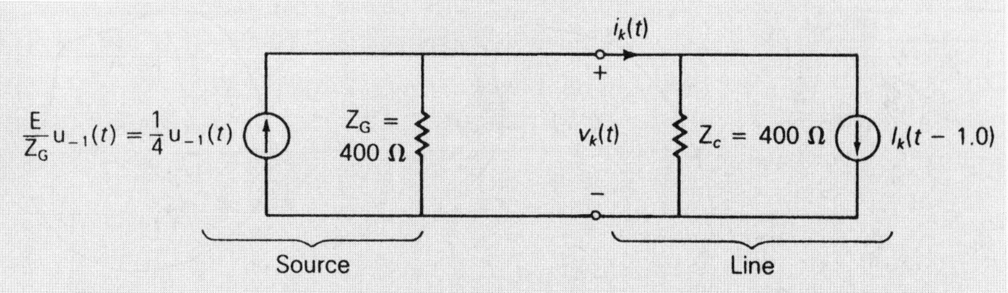

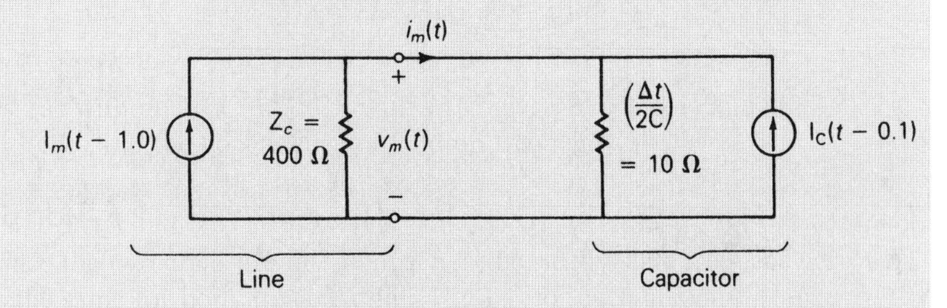

FIGURE 13.18

Discrete-time equivalent circuit for Example 13.7

$$\left[\begin{array}{c|c} \left(\frac{1}{400}+\frac{1}{400}\right) & 0 \\ \hline 0 & \left(\frac{1}{400}+\frac{1}{10}\right) \end{array}\right]\begin{bmatrix} v_k(t) \\ v_m(t) \end{bmatrix} = \begin{bmatrix} \frac{1}{4} - \mathrm{I}_k(t-1.0) \\ \mathrm{I}_m(t-1.0) + \mathrm{I_C}(t-0.1) \end{bmatrix}$$

Solving,

$$v_k(t) = 200\left[\tfrac{1}{4} - \mathrm{I}_k(t-1.0)\right] \tag{a}$$

$$v_m(t) = 9.75610[\mathrm{I}_m(t-1.0) + \mathrm{I_C}(t-0.1)] \tag{b}$$

The current sources in these equations are, from (13.4.9), (13.4.10), and (13.4.18), with the argument $(t-\tau)$ replaced by t,

$$\mathrm{I}_m(t) = \mathrm{I}_k(t-1.0) + \frac{2}{400}v_k(t) \tag{c}$$

$$\mathrm{I}_k(t) = \mathrm{I}_m(t-1.0) - \frac{2}{400}v_m(t) \tag{d}$$

$$\mathrm{I_C}(t) = -\mathrm{I_C}(t-0.1) + \frac{1}{5}v_m(t) \tag{e}$$

Equations (a) through (e) above are in a form suitable for digital computer solution. A scheme for iteratively computing v_k and v_m is as follows, starting at $t = 0$:

1. Compute $v_k(t)$ and $v_m(t)$ from equations (a) and (b).
2. Compute $\mathrm{I}_m(t)$, $\mathrm{I}_k(t)$, and $\mathrm{I_C}(t)$ from equations (c), (d), and (e). Store $\mathrm{I}_m(t)$ and $\mathrm{I}_k(t)$.
3. Change t to $(t+\Delta t) = (t+0.1)$ and return to (1) above.

FIGURE 13.19

Example 13.7

Output		
Time ms	VK Volts	VM Volts
0.00	50.00	0.00
0.20	50.00	0.00
0.40	50.00	0.00
0.60	50.00	0.00
0.80	50.00	0.00
1.00	50.00	2.44
1.20	50.00	11.73
1.40	50.00	20.13
1.60	50.00	27.73
1.80	50.00	34.61
2.00	2.44	40.83
2.20	11.73	46.46
2.40	20.13	51.56
2.60	27.73	56.17
2.80	34.61	60.34
3.00	40.83	64.12
3.20	46.46	67.53
3.40	51.56	70.62
3.60	56.17	73.42
3.80	60.34	75.95
4.00	64.12	78.24
4.20	67.53	80.31
4.40	70.62	82.18
4.60	73.42	83.88
4.80	75.95	85.41
5.00	78.24	86.80
5.20	80.31	88.06
5.40	82.18	89.20
5.60	83.88	90.22
5.80	85.41	91.15
6.00	86.80	92.00
6.20	88.06	92.76
6.40	89.20	93.45
6.60	90.22	94.07
6.80	91.15	94.64
7.00	92.00	95.15
7.20	92.76	95.61
7.40	93.45	96.03
7.60	94.07	96.40
7.80	94.64	96.75
8.00	95.15	97.06
8.20	95.61	97.34
8.40	96.03	97.59
8.60	96.40	97.82
8.80	96.75	98.03
9.00	97.06	98.22

Computer Program Listing

```
10 REM EXAMPLE 12.7
20 LPRINT "TIME VK VM"
30 LPRINT "ms Volts Volts"
40 IC=0
50 T=0
60 KPRINT=2
65 REM T IS TIME. KPRINT DETERMINES THE PRINTOUT INTERVAL.
70 REM LINES 110 to 210 COMPUTE EQS(a)–(e) FOR THE FIRST
80 REM TEN TIME STEPS (A TOTAL OF ONE ms) DURING WHICH
90 REM THE CURRENT SOURCES ON THE RIGHT HAND SIDE
100 REM OF THE EQUATIONS ARE ZERO. TEN VALUES OF
105 REM CURRENT SOURCES IK(J) AND IM(J) ARE STORED.
110 FOR J=1 TO 10
120 VK=200/4
130 VM=9.7561*IC
140 IM(J)=(2/400)*VK
150 IK(J)=(−2/400)*VM
160 IC=−IC+(1/5)*VM
170 Z=(J−1)/KPRINT
180 M=INT(Z)
190 IF M=Z THEN LPRINT USING "*** **"; T,VK,VM
200 T=T+.1
210 NEXT J
220 REM LINES 250 to 420 COMPUTE EQS(a)–(e) FOR TIME T
230 REM EQUAL TO AND GREATER THAN 1.0 ms. THE PAST TEN
240 REM VALUES OF IK(J) AND IM(J) ARE STORED
250 FOR J=1 TO 10
260 REM LINE 270 IS EQ(a).
270 VK=200*((1/4)−IK(J))
280 REM LINE 290 IS EQ(b).
290 VM=9.7561*(IM(J)+IC)
300 REM LINE 310 IS EQ(e).
310 IC=−IC+(1/5)*VM
320 REM LINES 330–360 ARE EQS (c) and (d).
330 C1=IK(J)+(2/400)*VK
340 C2=IM(J)−(2/400)*VM
350 IM(J)=C1
360 IK(J)=C2
370 Z=(J−1)/KPRINT
380 M=INT(Z)
390 IF M=Z THEN LPRINT USING "*** **"; T,VK,VM
400 T=T+.1
410 NEXT J
420 IF T<9.0 THEN GOTO 250
430 STOP
```

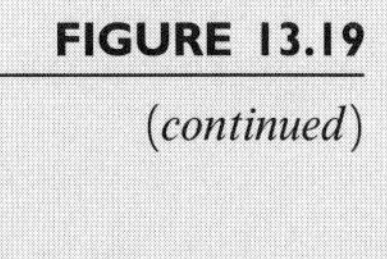

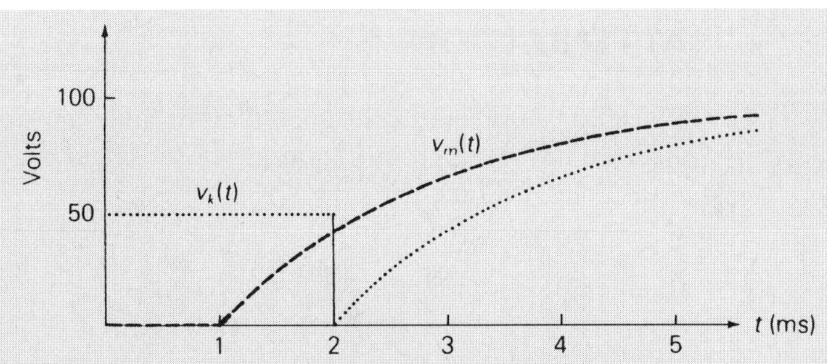

Note that since the transmission line and capacitor are unenergized for time t less than zero, the current sources $I_m(\ )$, $I_k(\ )$, and $I_C(\ )$ are zero whenever their arguments ( ) are negative. Note also from equations (a) through (e) that it is necessary to store the past ten values of $I_m(\ )$ and $I_k(\ )$.

A personal computer program written in BASIC that executes the above scheme and the computational results are shown in Figure 13.19. The plotted sending-end voltage $v_k(t)$ can be compared with the results of Example 13.3. ■

Example 13.7 can be generalized to compute bus voltages at discrete-time intervals for an arbitrary number of buses, single-phase lossless lines, and lumped RLC elements. When current sources instead of voltage sources are employed, the unknowns are all bus voltages, for which nodal equations $\mathbf{YV} = \mathbf{I}$ can be written at each discrete-time instant. Also, the dependent current sources in $\mathbf{I}$ are written in terms of bus voltages and current sources at prior times. For computational convenience, the time interval Δt can be chosen constant so that the bus admittance matrix $\mathbf{Y}$ is a constant real symmetric matrix as long as the RLC elements are constant.

13.5 LOSSY LINES

Transmission-line series resistance or shunt conductance causes the following:

1. Attenuation
2. Distortion
3. Power losses

We briefly discuss these effects as follows.

ATTENUATION

When constant series resistance R Ω/m and shunt conductance G S/m are included in the circuit of Figure 13.1 for a single-phase two-wire line, (13.1.3) and (13.1.4) become

$$\frac{\partial v(x,t)}{\partial x} = -\mathrm{R}i(x,t) - \mathrm{L}\frac{\partial i(x,t)}{\partial t} \tag{13.5.1}$$

$$\frac{\partial i(x,t)}{\partial x} = -\mathrm{G}v(x,t) - \mathrm{C}\frac{\partial v(x,t)}{\partial t} \tag{13.5.2}$$

Taking the Laplace transform of these, equations analogous to (13.1.7) and (13.1.8) are

$$\frac{d^2\mathrm{V}(x,s)}{dx^2} - \gamma^2(s)\mathrm{V}(x,s) = 0 \tag{13.5.3}$$

$$\frac{d^2\mathrm{I}(x,s)}{dx^2} - \gamma^2(s)\mathrm{I}(x,s) = 0 \tag{13.5.4}$$

where

$$\gamma(s) = \sqrt{(\mathrm{R}+s\mathrm{L})(\mathrm{G}+s\mathrm{C})} \tag{13.5.5}$$

The solution to these equations is

$$\mathrm{V}(x,s) = \mathrm{V}^+(s)e^{-\gamma(s)x} + \mathrm{V}^-(s)e^{+\gamma(s)x} \tag{13.5.6}$$

$$\mathrm{I}(x,s) = \mathrm{I}^+(s)e^{-\gamma(s)x} + \mathrm{I}^-(s)e^{+\gamma(s)x} \tag{13.5.7}$$

In general, it is impossible to obtain a closed form expression for $v(x,t)$ and $i(x,t)$, which are the inverse Laplace transforms of these equations. However, for the special case of a *distortionless* line, which has the property $\mathrm{R/L = G/C}$, the inverse Laplace transform can be obtained as follows. Rewrite (13.5.5) as

$$\gamma(s) = \sqrt{\mathrm{LC}[(s+\delta)^2 - \sigma^2]} \tag{13.5.8}$$

where

$$\delta = \frac{1}{2}\left(\frac{\mathrm{R}}{\mathrm{L}} + \frac{\mathrm{G}}{\mathrm{C}}\right) \tag{13.5.9}$$

$$\sigma = \frac{1}{2}\left(\frac{\mathrm{R}}{\mathrm{L}} - \frac{\mathrm{G}}{\mathrm{C}}\right) \tag{13.5.10}$$

For a distortionless line, $\sigma = 0$, $\delta = \mathrm{R/L}$, and (13.5.6) and (13.5.7) become

$$\mathrm{V}(x,s) = \mathrm{V}^+(s)e^{-\sqrt{\mathrm{LC}}[s+(\mathrm{R/L})]x} + \mathrm{V}^-(s)e^{+\sqrt{\mathrm{LC}}[s+(\mathrm{R/L})]x} \tag{13.5.11}$$

$$\mathrm{I}(x,s) = \mathrm{I}^+(s)e^{-\sqrt{\mathrm{LC}}[s+(\mathrm{R/L})]x} + \mathrm{I}^-(s)e^{+\sqrt{\mathrm{LC}}[s+(\mathrm{R/L})]x} \tag{13.5.12}$$

Using $v = 1/\sqrt{\mathrm{LC}}$ and $\sqrt{\mathrm{LC}}(\mathrm{R}/\mathrm{L}) = \sqrt{\mathrm{RG}} = \alpha$ for the distortionless line, the inverse transform of these equations is

$$v(x, t) = e^{-\alpha x} v^{+}\left(t - \frac{x}{v}\right) + e^{+\alpha x} v^{-}\left(t + \frac{x}{v}\right) \tag{13.5.13}$$

$$i(x, t) = e^{-\alpha x} i^{+}\left(t - \frac{x}{v}\right) + e^{\alpha x} i^{-}\left(t + \frac{x}{v}\right) \tag{13.5.14}$$

These voltage and current waves consist of forward and backward traveling waves similar to (13.1.12) and (13.1.13) for a lossless line. However, for the lossy distortionless line, the waves are attenuated versus x due to the $e^{\pm\alpha x}$ terms. Note that the attenuation term $\alpha = \sqrt{\mathrm{RG}}$ is constant. Also, the attenuated waves travel at constant velocity $v = 1/\sqrt{\mathrm{LC}}$. Therefore, waves traveling along the distortionless line do not change their shape; only their magnitudes are attenuated.

DISTORTION

For sinusoidal steady-state waves, the propagation constant $\gamma(j\omega)$ is, from (13.5.5), with $s = j\omega$

$$\gamma(j\omega) = \sqrt{(\mathrm{R} + j\omega\mathrm{L})(\mathrm{G} + j\omega\mathrm{C})} = \alpha + j\beta \tag{13.5.15}$$

For a lossless line, $\mathrm{R} = \mathrm{G} = 0$; therefore, $\alpha = 0$, $\beta = \omega\sqrt{\mathrm{LC}}$, and the phase velocity $v = \omega/\beta = 1/\sqrt{\mathrm{LC}}$ is constant. Thus, sinusoidal waves of all frequencies travel at constant velocity v without attenuation along a lossless line.

For a distortionless line $(\mathrm{R}/\mathrm{L}) = (\mathrm{G}/\mathrm{C})$, and $\gamma(j\omega)$ can be rewritten, using (13.5.8)–(13.5.10), as

$$\gamma(j\omega) = \sqrt{\mathrm{LC}\left(j\omega + \frac{\mathrm{R}}{\mathrm{L}}\right)^2} = \sqrt{\mathrm{LC}}\left(j\omega + \frac{\mathrm{R}}{\mathrm{L}}\right)$$

$$= \sqrt{\mathrm{RG}} + j\frac{\omega}{v} = \alpha + j\beta \tag{13.5.16}$$

Since $\alpha = \sqrt{\mathrm{RG}}$ and $v = 1/\sqrt{\mathrm{LC}}$ are constant, sinusoidal waves of all frequencies travel along the distortionless line at constant velocity with constant attenuation—that is, without distortion.

It can also be shown that for frequencies above 1 MHz, practical transmission lines with typical constants R, L, G, and C tend to be distortionless. Above 1 MHz, α and β can be approximated by

$$\alpha \simeq \frac{\mathrm{R}}{2}\sqrt{\frac{\mathrm{C}}{\mathrm{L}}} + \frac{\mathrm{G}}{2}\sqrt{\frac{\mathrm{L}}{\mathrm{C}}} \tag{13.5.17}$$

$$\beta \simeq \omega\sqrt{\mathrm{LC}} = \frac{\omega}{v} \tag{13.5.18}$$

Therefore, sinusoidal waves with frequencies above 1 MHz travel along a practical line undistorted at constant velocity $v = 1/\sqrt{\mathrm{LC}}$, with attenuation α given by (13.5.17).

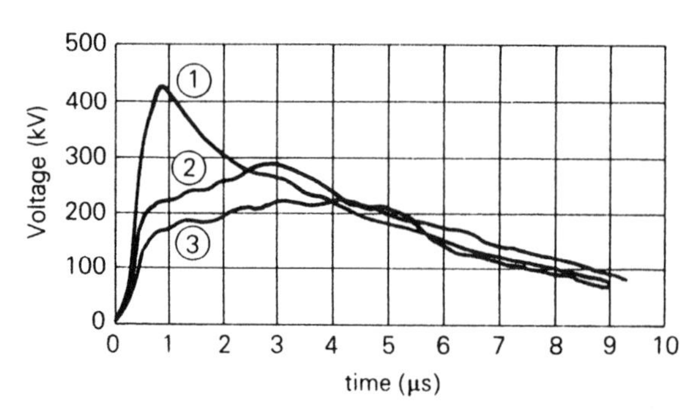

FIGURE 13.20

Distortion and attenuation of surges on a 132-kV overhead line [4] (H. M. Lacey, "The Lightning Protection of High-Voltage Overhead Transmission and Distribution Systems," Proceedings of the IEEE, 96 (1949), p. 287. © 1949 IEEE)

At frequencies below 1 MHz, these approximations do not hold, and lines are generally not distortionless. For typical transmission and distribution lines, (R/L) is much greater than (G/C) by a factor of 1000 or so. Therefore, the condition (R/L) = (G/C) for a distortionless line does not hold.

Figure 13.20 shows the effect of distortion and attenuation of voltage surges based on experiments on a 132-kV overhead transmission line [4]. The shapes of the surges at three points along the line are shown. Note how distortion reduces the front of the wave and builds up the tail as it travels along the line.

POWER LOSSES

Power losses are associated with series resistance R and shunt conductance G. When a current I flows along a line, I^2R losses occur, and when a voltage V appears across the conductors, V^2G losses occur. V^2G losses are primarily due to insulator leakage and corona for overhead lines, and to dielectric losses for cables. For practical lines operating at rated voltage and rated current, I^2R losses are much greater than V^2G losses.

As discussed above, the analysis of transients on single-phase two-wire lossy lines with constant parameters R, L, G, and C is complicated. The analysis becomes more complicated when skin effect is included, which means that R is not constant but frequency-dependent. Additional complications arise for a single-phase line consisting of one conductor with earth return, where Carson [5] has shown that both series resistance and inductance are frequency-dependent.

In view of these complications, the solution of transients on lossy lines is best handled via digital computation techniques. A single-phase line of length l can be approximated by a lossless line with half the total resistance $(Rl/2)$ Ω lumped in series with the line at both ends. For improved accuracy, the line can be divided into various line sections, and each section can be approximated by a lossless line section, with a series resistance lumped at both ends. Simulations have shown that accuracy does not significantly improve with more than two line sections.

FIGURE 13.21

Approximate model of a lossy line segment

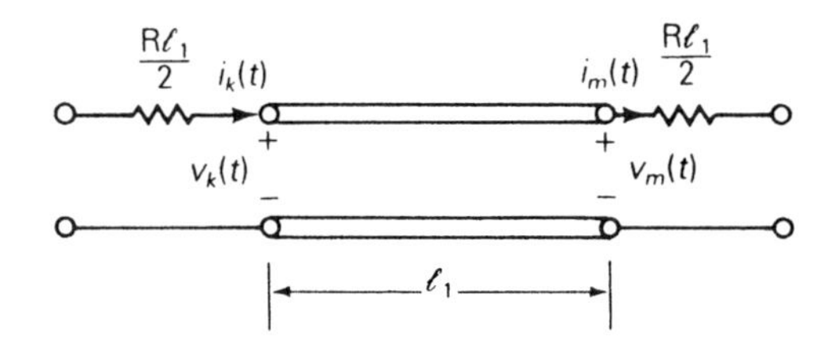

(a) Lossless line segment of length ℓ_1 with lumped line resistance

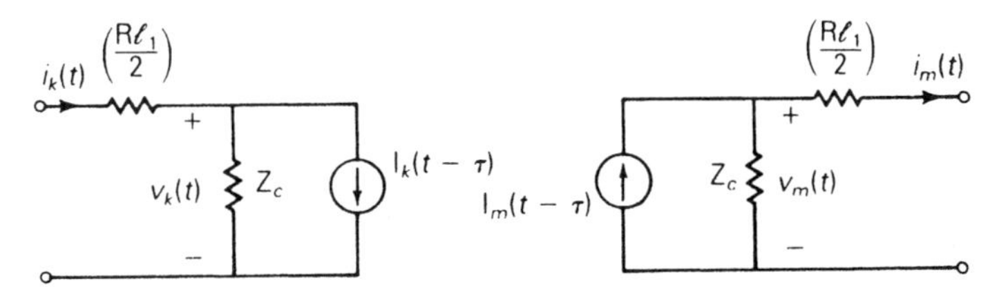

(b) Discrete-time model

Discrete-time equivalent circuits of a single-phase lossless line, Figure 13.14, together with a constant lumped resistance, Figure 13.17, can be used to approximate a lossy line section, as shown in Figure 13.21. Also, digital techniques for modeling frequency-dependent line parameters [6, 7] are available but we do not discuss them here.

13.6 MULTICONDUCTOR LINES

Up to now, we have considered transients on single-phase two-wire lines. For a transmission line with n conductors above a ground plane, waves travel in n "modes," where each mode has its own wave velocity and its own surge impedance. In this section we illustrate "model analysis" for a relatively simple three-phase line [8].

Given a three-phase, lossless, completely transposed line consisting of three conductors above a perfectly conducting ground plane, the transmission-line equations are

$$\frac{d\mathbf{V}(x,s)}{dx} = -s\mathbf{L}\mathbf{I}(x,s) \tag{13.6.1}$$

$$\frac{d\mathbf{I}(x,s)}{dx} = -s\mathbf{C}\mathbf{V}(x,s) \tag{13.6.2}$$

where

$$\mathbf{V}(x,s) = \begin{bmatrix} V_{ag}(x,s) \\ V_{bg}(x,s) \\ V_{cg}(x,s) \end{bmatrix} \qquad \mathbf{I}(x,s) = \begin{bmatrix} I_a(x,s) \\ I_b(x,s) \\ I_c(x,s) \end{bmatrix} \tag{13.6.3}$$

Equations (13.6.1) and (13.6.2) are identical to (13.1.5) and (13.1.6) except that scalar quantities are replaced by vector quantities. $\mathbf{V}(x,s)$ is the vector of line-to-ground voltages and $\mathbf{I}(x,s)$ is the vector of line currents. For a completely transposed line, the 3×3 inductance matrix $\mathbf{L}$ and capacitance matrix $\mathbf{C}$ are given by

$$\mathbf{L} = \begin{bmatrix} L_s & L_m & L_m \\ L_m & L_s & L_m \\ L_m & L_m & L_s \end{bmatrix} \quad \text{H/m} \tag{13.6.4}$$

$$\mathbf{C} = \begin{bmatrix} C_s & C_m & C_m \\ C_m & C_s & C_m \\ C_m & C_m & C_s \end{bmatrix} \quad \text{F/m} \tag{13.6.5}$$

For any given line configuration, $\mathbf{L}$ and $\mathbf{C}$ can be computed from the equations given in Sections 4.7 and 4.11. Note that L_s, L_m, and C_s are positive, whereas C_m is negative.

We now transform the phase quantities to modal quantities. First, we define

$$\begin{bmatrix} V_{ag}(x,s) \\ V_{bg}(x,s) \\ V_{cg}(x,s) \end{bmatrix} = \mathbf{T}_V \begin{bmatrix} V^0(x,s) \\ V^+(x,s) \\ V^-(x,s) \end{bmatrix} \tag{13.6.6}$$

$$\begin{bmatrix} I_a(x,s) \\ I_b(x,s) \\ I_c(x,s) \end{bmatrix} = \mathbf{T}_I \begin{bmatrix} I^0(x,s) \\ I^+(x,s) \\ I^-(x,s) \end{bmatrix} \tag{13.6.7}$$

$V^0(x,s)$, $V^+(x,s)$, and $V^-(x,s)$ are denoted *zero-mode*, *positive-mode*, and *negative-mode* voltages, respectively. Similarly, $I^0(x,s)$, $I^+(x,s)$, and $I^-(x,s)$ are *zero-*, *positive-*, and *negative-mode* currents. $\mathbf{T}_V$ and $\mathbf{T}_I$ are 3×3 constant transformation matrices, soon to be specified. Denoting $\mathbf{V}_m(x,s)$ and $\mathbf{I}_m(x,s)$ as the modal voltage and modal current vectors,

$$\mathbf{V}(x,s) = \mathbf{T}_V \mathbf{V}_m(x,s) \tag{13.6.8}$$

$$\mathbf{I}(x,s) = \mathbf{T}_I \mathbf{I}_m(x,s) \tag{13.6.9}$$

Using (13.6.8) and (13.6.9) in (13.6.1),

$$\mathbf{T}_V = \frac{d\mathbf{V}_m(x,s)}{dx} = -s\mathbf{L}\mathbf{T}_I\mathbf{I}_m(x,s)$$

or

$$\frac{d\mathbf{V}_m(x,s)}{dx} = -s(\mathbf{T}_v^{-1}\mathbf{L}\mathbf{T}_\mathrm{I})\mathbf{I}_m(x,s) \tag{13.6.10}$$

Similarly, using (13.6.8) and (13.6.9) in (13.6.2),

$$\frac{d\mathbf{I}_m(x,s)}{dx} = -s(\mathbf{T}_\mathrm{I}^{-1}\mathbf{C}\mathbf{T}_v)\mathbf{V}_m(x,s) \tag{13.6.11}$$

The objective of the modal transformation is to diagonalize the matrix products within the parentheses of (13.6.10) and (13.6.11), thereby decoupling these vector equations. For a three-phase completely transposed line, $\mathbf{T}_\mathrm{V}$ and $\mathbf{T}_\mathrm{I}$ are given by

$$\mathbf{T}_\mathrm{V} = \mathbf{T}_\mathrm{I} = \begin{bmatrix} 1 & 1 & 1 \\ 1 & -2 & 1 \\ 1 & 1 & -2 \end{bmatrix} \tag{13.6.12}$$

Also, the inverse transformation matrices are

$$\mathbf{T}_\mathrm{V}^{-1} = \mathbf{T}_\mathrm{I}^{-1} = \frac{1}{3}\begin{bmatrix} 1 & 1 & 1 \\ 1 & -1 & 0 \\ 1 & 0 & -1 \end{bmatrix} \tag{13.6.13}$$

Substituting (13.6.12), (13.6.13), (13.6.4), and (13.6.5) into (13.6.10) and (13.6.11) yields

$$\frac{d}{dx}\begin{bmatrix} \mathrm{V}^0(x,s) \\ \mathrm{V}^+(x,s) \\ \mathrm{V}^-(x,s) \end{bmatrix} = \begin{bmatrix} -s(\mathrm{L}_s + 2\mathrm{L}_m) & 0 & 0 \\ 0 & -s(\mathrm{L}_s - \mathrm{L}_m) & 0 \\ 0 & 0 & -s(\mathrm{L}_s - \mathrm{L}_m) \end{bmatrix} \times \begin{bmatrix} \mathrm{I}^0(x,s) \\ \mathrm{I}^+(x,s) \\ \mathrm{I}^-(x,s) \end{bmatrix} \tag{13.6.14}$$

$$\frac{d}{dx}\begin{bmatrix} \mathrm{I}^0(x,s) \\ \mathrm{I}^+(x,s) \\ \mathrm{I}^-(x,s) \end{bmatrix} = \begin{bmatrix} -s(\mathrm{C}_s + 2\mathrm{C}_m) & 0 & 0 \\ 0 & -s(\mathrm{C}_s - \mathrm{C}_m) & 0 \\ 0 & 0 & -s(\mathrm{C}_s - \mathrm{C}_m) \end{bmatrix} \times \begin{bmatrix} \mathrm{V}^0(x,s) \\ \mathrm{V}^+(x,s) \\ \mathrm{V}^-(x,s) \end{bmatrix} \tag{13.6.15}$$

From (13.6.14) and (13.6.15), the zero-mode equations are

$$\frac{d\mathrm{V}^0(x,s)}{dx} = -s(\mathrm{L}_s + 2\mathrm{L}_m)\mathrm{I}^0(x,s) \tag{13.6.16}$$

$$\frac{d\mathrm{I}^0(x,s)}{dx} = -s(\mathrm{C}_s + 2\mathrm{C}_m)\mathrm{V}^0(x,s) \tag{13.6.17}$$

These equations are identical in form to those of a two-wire lossless line, (13.1.5) and (13.1.6). By analogy, the zero-mode waves travel at velocity

$$v^0 = \frac{1}{\sqrt{(L_s + 2L_m)(C_s + 2C_m)}} \quad \text{m/s} \tag{13.6.18}$$

and the zero-mode surge impedance is

$$Z_c^0 = \sqrt{\frac{L_s + 2L_m}{C_s + 2C_m}} \quad \Omega \tag{13.6.19}$$

Similarly, the positive- and negative-mode velocities and surge impedances are

$$v^+ = v^- = \frac{1}{\sqrt{(L_s - L_m)(C_s - C_m)}} \quad \text{m/s} \tag{13.6.20}$$

$$Z_c^+ = Z_c^- = \sqrt{\frac{L_s - L_m}{C_s - C_m}} \quad \Omega \tag{13.6.21}$$

These equations can be extended to more than three conductors—for example, to a three-phase line with shield wires or to a double-circuit three-phase line. Although the details are more complicated, the modal transformation is straightforward. There is one mode for each conductor, and each mode has its own wave velocity and its own surge impedance.

The solution of transients on multiconductor lines is best handled via digital computer methods, and such programs are available [9, 10]. Digital techniques are also available to model the following effects:

1. Nonlinear and time-varying RLC elements [8]
2. Lossy lines with frequency-dependent line parameters [6, 7, 12]

13.7 POWER SYSTEM OVERVOLTAGES

Overvoltages encountered by power system equipment are of three types:

1. Lightning surges
2. Switching surges
3. Power frequency (50 or 60 Hz) overvoltages

LIGHTNING

Cloud-to-ground (CG) lightning is the greatest single cause of overhead transmission and distribution line outages. Data obtained over a 14-year period from electric utility companies in the United States and Canada and covering 25,000 miles or 40,000 km of transmission show that CG lightning accounted for about 26% of outages on 230-kV circuits and about 65% of outages on 345-kV circuits [13]. A similar study in Britain, also over a 14-year period, covering 50,000 faults on distribution lines shows that CG lightning accounted for 47% of outages on circuits up to and including 33 kV [14].

The electrical phenomena that occur within clouds leading to a lightning strike are complex and not totally understood. Several theories [15, 16, 17] generally agree, however, that charge separation occurs within clouds. Wilson [15] postulates that falling raindrops attract negative charges and therefore leave behind masses of positively charged air. The falling raindrops bring the negative charge to the bottom of the cloud, and upward air drafts carry the positively charged air and ice crystals to the top of the cloud, as shown in Figure 13.22. Negative charges at the bottom of the cloud induce a positively charged region, or "shadow," on the earth directly below the cloud. The electric field lines shown in Figure 13.22 originate from the positive charges and terminate at the negative charges.

When voltage gradients reach the breakdown strength of the humid air within the cloud, typically 5 to 15 kV/cm, an ionized path or downward *leader* moves from the cloud toward the earth. The leader progresses somewhat randomly along an irregular path, in steps. These leader steps, about 50 m long, move at a velocity of about 10^5 m/s. As a result of the opposite charge distribution under the cloud, another upward leader may rise to meet the downward leader. When the two leaders meet, a lightning discharge occurs, which neutralizes the charges.

The current involved in a CG lightning stroke typically rises to a peak value within 1 to 10 μs, and then diminishes to one-half the peak within 20 to 100 μs. The distribution of peak currents is shown in Figure 13.23 [20].

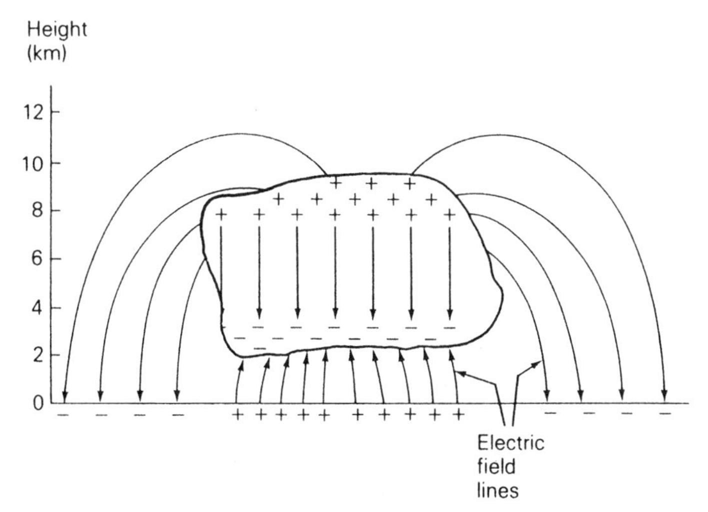

FIGURE 13.22 Postulation of charge separation within clouds [16] (G. B. Simpson and F. J. Scrase, "The Distribution of Electricity in Thunderclouds," Proceedings of the Royal Society A: Mathematical, Physical and Engineering Sciences, 161 (1937), p. 309)

FIGURE 13.23

Frequency of occurrence of lightning currents that exceed a given peak value [20] (IEEE Guide for the Application of Metal-Oxide Surge Arresters for Alternating-Current Systems, IEEE std. C62.22-1997 (New York: The Institute of Electrical and Electronics Engineers, http://standards.ieee.org/1998) © 1998 IEEE)

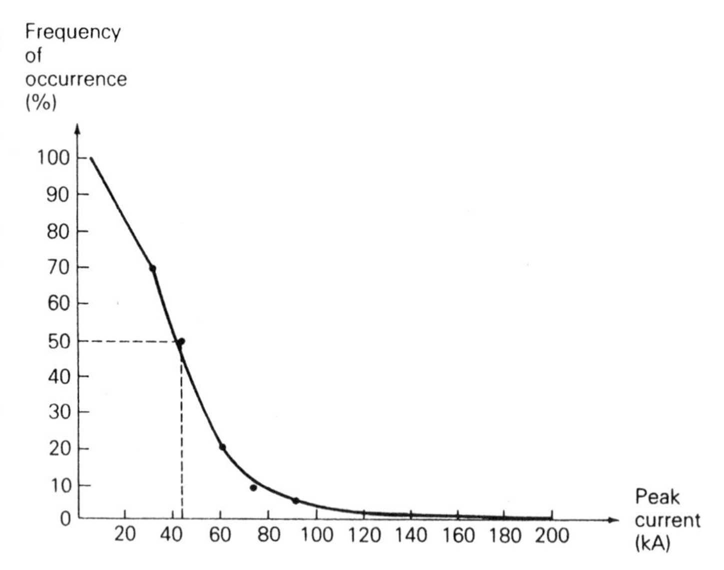

This curve represents the percentage of strokes that exceed a given peak current. For example, 50% of all strokes have a peak current greater than 45 kA. In extreme cases, the peak current can exceed 200 kA. Also, test results indicate that approximately 90% of all strokes are negative.

It has also been shown that what appears to the eye as a single flash of lightning is often the cumulative effect of many strokes. A typical flash consists of typically 3 to 5, and occasionally as many as 40, strokes, at intervals of 50 ms.

The U.S. National Lightning Detection Network® (NLDN), owned and operated by Global Atmospherics, Inc., is a system that senses the electromagnetic fields radiated by individual return strokes in CG flashes. As of 2001, the NLDN employed more than 100 ground-based sensors geographically distributed throughout the 48 contiguous United States. The sensors transmit lightning data to a network control center in Tucson, Arizona via a satellite communication system. Data from the remote sensors are recorded and processed in real time at the network control center to provide the time, location, polarity, and an estimate of the peak current in each return stroke. The real-time data are then sent back through the communications network for satellite broadcast dissemination to real-time users, all within 30–40 seconds of each CG lightning flash. Recorded data are also reprocessed off-line within a few days of acquisition and stored in a permanent database for access by users who do not require real-time information. NLDN's archive data library contains over 160 million flashes dating from 1989 [25, www.LightningStorm.com].

Figure 13.24 shows a lightning flash density contour map providing a representation of measured annual CG flash density detected by the NLDN

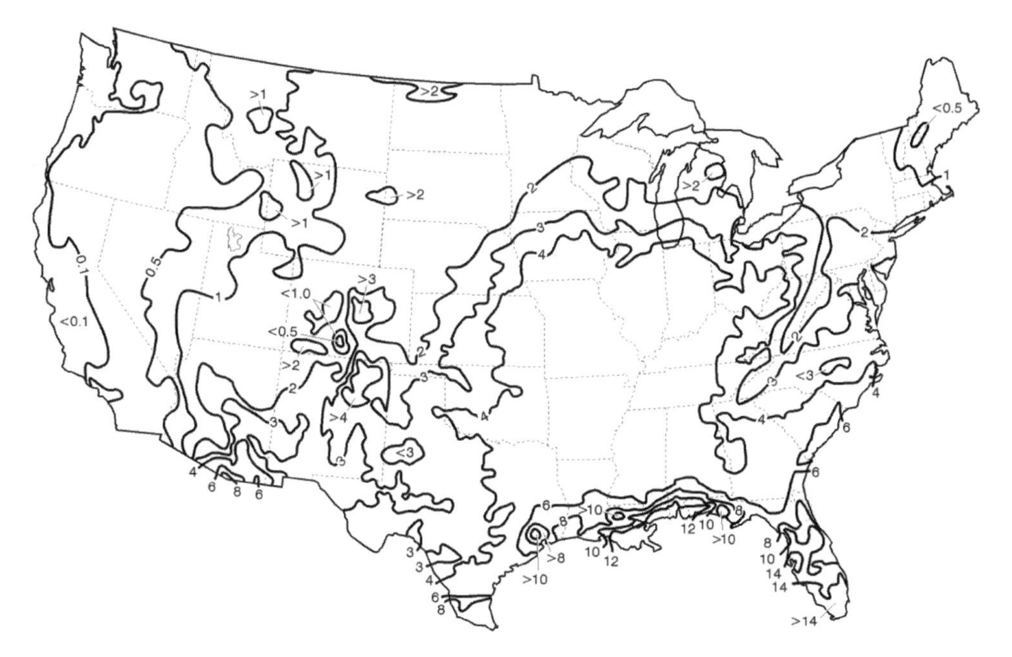

FIGURE 13.24 Lightning flash density contour map showing annual CG flash densities (#flashes/km^2/year) in the contiguous United States, as detected by the NLDN from 1989 to 1998. (Courtesy of Vaisala, Inc.)

from 1989 to 1998. As shown, average annual CG lightning flash densities range from about 0.1 flashes/km^2/year near the West Coast to more than 14 flashes/km^2/year in portions of the Florida peninsula. Figure 13.25 shows a high-resolution, 3 km $\times$ 3 km, CG flash density map in grid format.

Figure 13.26 shows an "asset exposure map" of all CG strikes in 1995 in a region that contains a 25-km 69-kV transmission line. This map provides an indication of the level of exposure to lightning within an exposure area that surrounds the transmission line. By combining this data with estimates of peak stroke currents and transmission line fault records, the lightning performance of the line and individual line segments can be quantified. Improvements in line design and line protection can also be evaluated.

Electric utilities use real-time lightning maps to monitor the approach of lightning storms, estimate their severity, and then either position repair crews in advance of storms, call them out, or hold them over as required. Utilities also use real-time lightning data together with on-line monitoring of circuit breakers, relays, and/or substation alarms to improve operations, minimize or avert damage, and speed up the restoration of their systems.

A typical transmission-line design goal is to have an average of less than 0.31 lightning outages per year per 100 km of transmission. For a given

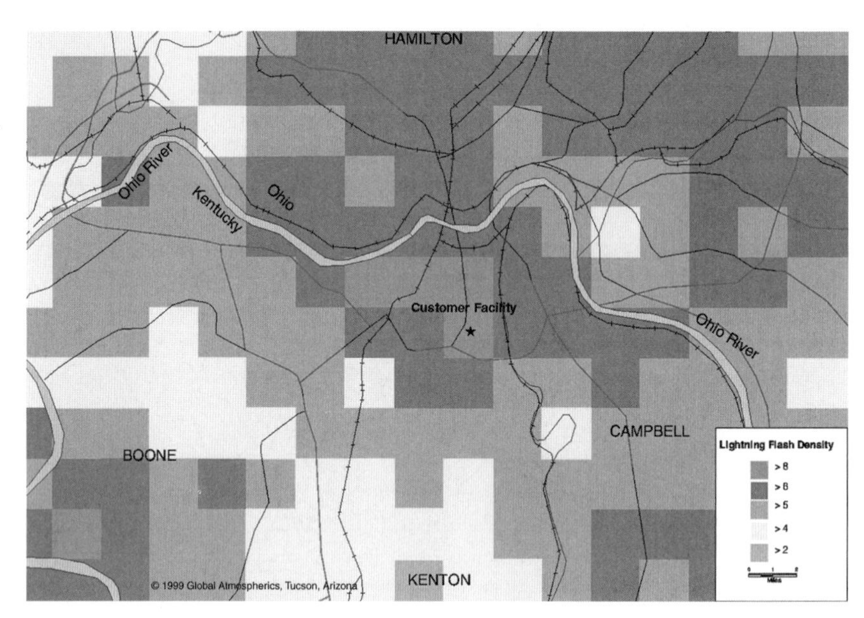

FIGURE 13.25 High-resolution CG flash density map in grid format (Courtesy of Vaisala, Inc.)

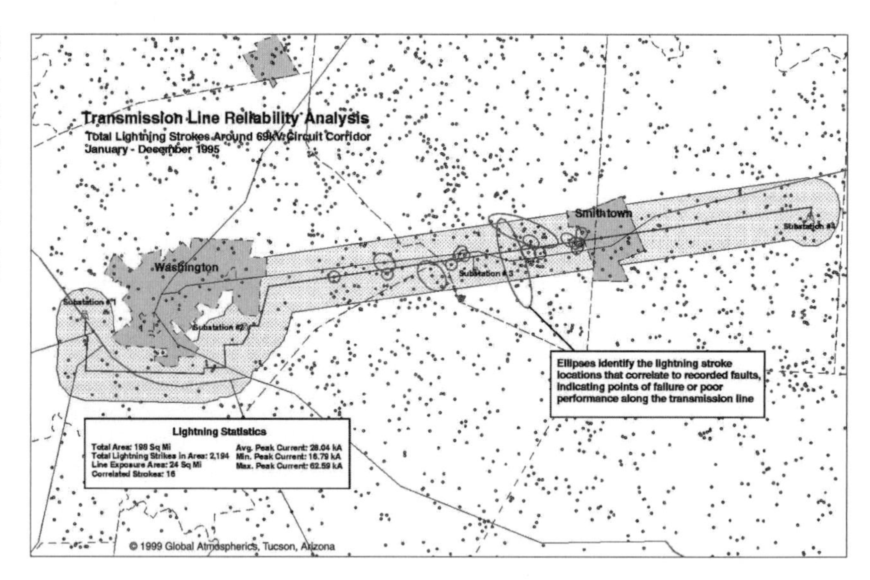

FIGURE 13.26 Asset exposure map showing CG strikes in 1995 in a region that contains a 15-mile (25-km) 69-kV transmission line (Courtesy of Vaisala, Inc.)

overhead line with a specified voltage rating, the following factors affect this design goal:

1. Tower height
2. Number and location of shield wires
3. Number of standard insulator discs per phase wire
4. Tower impedance and tower-to-ground impedance

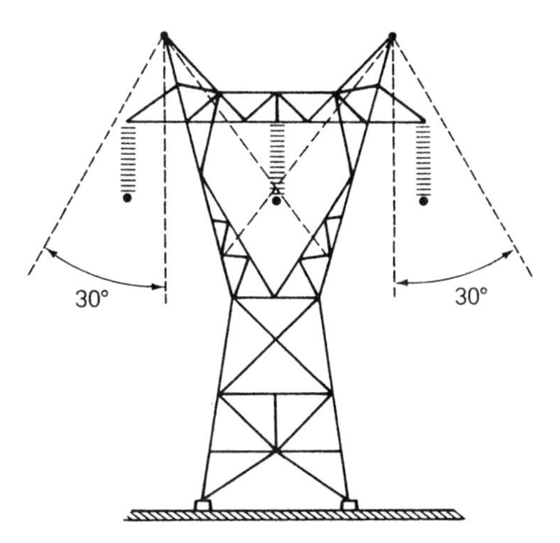

FIGURE 13.27
Effect of shield wires

It is well known that lightning strikes tall objects. Thus, shorter, H-frame structures are less susceptible to lightning strokes than taller, lattice towers. Also, shorter span lengths with more towers per kilometer can reduce the number of strikes.

Shield wires installed above phase conductors can effectively shield the phase conductors from direct lightning strokes. Figure 13.27 illustrates the effect of shield wires. Experience has shown that the chance of a direct hit to phase conductors located within $\pm 30°$ arcs beneath the shield wires is reduced by a factor of 1000 [18]. Some lightning strokes are, therefore, expected to hit these overhead shield wires. When this occurs, traveling voltage and current waves propagate in both directions along the shield wire that is hit. When a wave arrives at a tower, a reflected wave returns toward the point where the lightning hit, and two refracted waves occur. One refracted wave moves along the shield wire into the next span. Since the shield wire is electrically connected to the tower, the other refracted wave moves down the tower, its energy being harmlessly diverted to ground.

However, if the tower impedance or tower-to-ground impedance is too high, IZ voltages that are produced could exceed the breakdown strength of the insulator discs that hold the phase wires. The number of insulator discs per string (see Table 4.1) is selected to avoid insulator flashover. Also, tower impedances and tower footing resistances are designed to be as low as possible. If the inherent tower construction does not give a naturally low resistance to ground, driven ground rods can be employed. Sometimes buried conductors running under the line (called *counterpoise*) are employed.

SWITCHING SURGES

The magnitudes of overvoltages due to lightning surges are not significantly affected by the power system voltage. On the other hand, overvoltages due to switching surges are directly proportional to system voltage. Consequently,

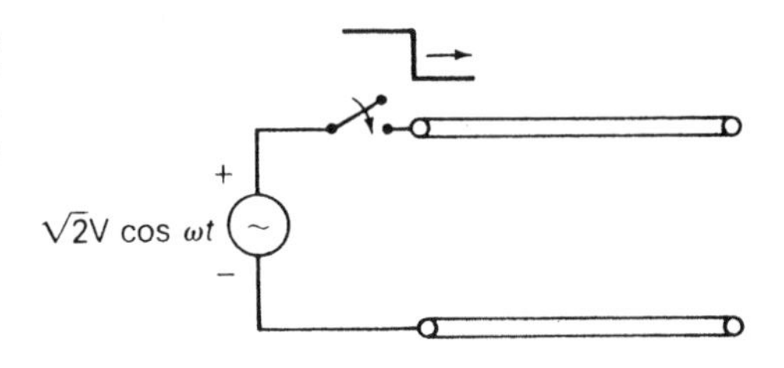

FIGURE 13.28 Energizing an open-circuited line

lightning surges are less important for EHV transmission above 345 kV and for UHV transmission, which has improved insulation. Switching surges become the limiting factor in insulation coordination for system voltages above 345 kV.

One of the simplest and largest overvoltages can occur when an open-circuited line is energized, as shown in Figure 13.28. Assume that the circuit breaker closes at the instant the sinusoidal source voltage has a peak value $\sqrt{2}$ V. Assuming zero source impedance, a forward traveling voltage wave of magnitude $\sqrt{2}$ V occurs. When this wave arrives at the open-circuited receiving end, where $\Gamma_R = +1$, the reflected voltage wave superimposed on the forward wave results in a maximum voltage of $2\sqrt{2}\ V = 2.83$ V. Even higher voltages can occur when a line is reclosed after momentary interruption.

In order to reduce overvoltages due to line energizing or reclosing, resistors are almost always preinserted in circuit breakers at 345 kV and above. Resistors ranging from 200 to 800 Ω are preinserted when EHV circuit breakers are closed, and subsequently bypassed. When a circuit breaker closes, the source voltage divides across the preinserted resistors and the line, thereby reducing the initial line voltage. When the resistors are shorted out, a new transient is initiated, but the maximum line voltage can be substantially reduced by careful design.

Dangerous overvoltages can also occur during a single line-to-ground fault on one phase of a transmission line. When such a fault occurs, a voltage equal and opposite to that on the faulted phase occurs at the instant of fault inception. Traveling waves are initiated on both the faulted phase and, due to capacitive coupling, the unfaulted phases. At the line ends, reflections are produced and are superimposed on the normal operating voltages of the unfaulted phases. Kimbark and Legate [19] show that a line-to-ground fault can create an overvoltage on an unfaulted phase as high as 2.1 times the peak line-to-neutral voltage of the three-phase line.

POWER FREQUENCY OVERVOLTAGES

Sustained overvoltages at the fundamental power frequency (60 Hz in the United States) or at higher harmonic frequencies (such as 120 Hz, 180 Hz, and so on) occur due to load rejection, to ferroresonance, or to permanent faults. These overvoltages are normally of long duration, seconds to minutes, and are weakly damped.

13.8

INSULATION COORDINATION

Insulation coordination is the process of correlating electric equipment insulation strength with protective device characteristics so that the equipment is protected against expected overvoltages. The selection of equipment insulation strength and the protected voltage level provided by protective devices depends on engineering judgment and cost.

As shown by the top curve in Figure 13.29, equipment insulation strength is a function of time. Equipment insulation can generally withstand high transient overvoltages only if they are of sufficiently short duration. However, determination of insulation strength is somewhat complicated. During repeated tests with identical voltage waveforms under identical conditions, equipment insulation may fail one test and withstand another.

For purposes of insulation testing, a standard impulse voltage wave, as shown in Figure 13.30, is defined. The impulse wave shape is specified by giving the time T_1 in microseconds for the voltage to reach its peak value and the time T_2 for the voltage to decay to one-half its peak. One standard wave

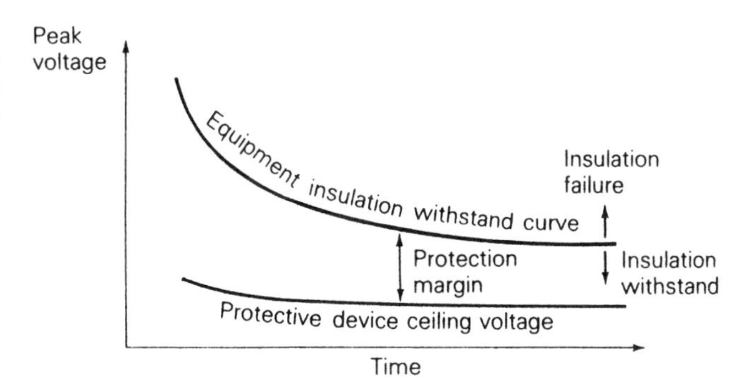

FIGURE 13.29 Equipment insulation strength

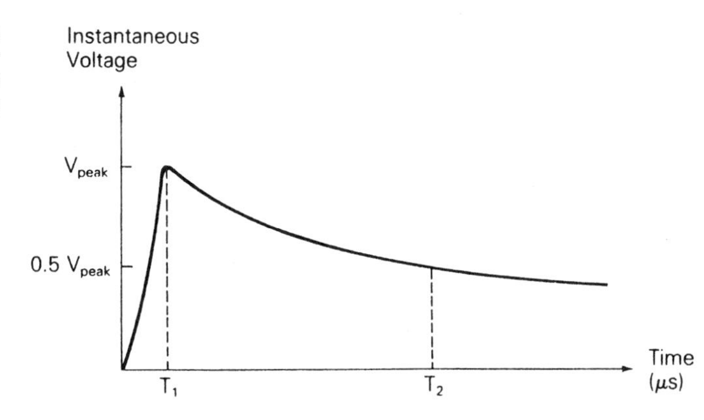

FIGURE 13.30 Standard impulse voltage waveform

is a 1.2×50 wave, which rises to a peak value at $T_1 = 1.2$ μs and decays to one-half its peak at $T_2 = 50$ μs.

Basic insulation level or BIL is defined as the peak value of the standard impulse voltage wave in Figure 13.30. Standard BILs adopted by the IEEE are shown in Table 13.1. Equipment conforming to these BILs must be capable of withstanding repeated applications of the standard waveform of positive or negative polarity without insulation failure. Also, these standard BILs apply to equipment regardless of how it is grounded. For nominal system voltages 115 kV and above, solidly grounded equipment with the reduced BILs shown in the table have been used.

BILs are often expressed in per-unit, where the base voltage is the maximum value of nominal line-to-ground system voltage. Consider for example a 345-kV system, for which the maximum value of nominal line-to-ground voltage is $\sqrt{2}(345/\sqrt{3}) = 281.7$ kV. The 1550-kV standard BIL shown in Table 13.1 is then $(1550/281.7) = 5.5$ per unit.

Note that overhead-transmission-line insulation, which is external insulation, is usually self-restoring. When a transmission-line insulator string flashes over, a short circuit occurs. After circuit breakers open to deenergize the line, the insulation of the string usually recovers, and the line can be rapidly reenergized. However, transformer insulation, which is internal, is not

TABLE 13.1 Standard and reduced basic insulation levels [18]

Nominal System Voltage kVrms	Standard BIL kV	Reduced BIL* kV
1.2	45	
2.5	60	
5.0	75	
8.7	95	
15	110	
23	150	
34.5	200	
46	250	
69	350	
92	450	
115	550	450
138	650	550
161	750	650
196	900	750
230	1050	825–900
287	1300	1000–1100
345	1550	1175–1300
500		1300–1800
765		1675–2300

* For solidly grounded systems

These BILs are based on 1.2×50 μs voltage waveforms. They apply to internal (or non-self-restoring) insulation such as transformer insulation, as well as to external (or self-restoring) insulation, such as transmission-line insulation, on a statistical basis.

(Westinghouse Electric Corporation, Electrical Transmission and Distribution Reference Book, 4th ed. (East Pittsburgh, PA: 1964).)

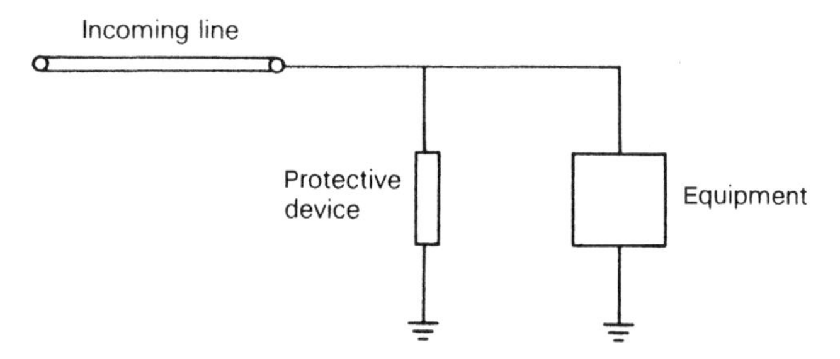

FIGURE 13.31

Single-line diagram of equipment and protective device

self-restoring. When transformer insulation fails, the transformer must be removed for repair or replaced.

To protect equipment such as a transformer against overvoltages higher than its BIL, a protective device, such as that shown in Figure 13.31, is employed. Such protective devices are generally connected in parallel with the equipment from each phase to ground. As shown in Figure 13.29, the function of the protective device is to maintain its voltage at a ceiling voltage below the BIL of the equipment it protects. The difference between the equipment breakdown voltage and the protective device ceiling voltage is the *protection margin.*

Protective devices should satisfy the following four criteria:

1. Provide a high or infinite impedance during normal system voltages, to minimize steady-state losses.
2. Provide a low impedance during surges, to limit voltage.
3. Dissipate or store the energy in the surge without damage to itself.
4. Return to open-circuit conditions after the passage of a surge.

One of the simplest protective devices is the rod gap, two metal rods with a fixed air gap, which is designed to spark over at specified overvoltages. Although it satisfies the first two protective device criteria, it dissipates very little energy and it cannot clear itself after arcing over.

A surge arrester, consisting of an air gap in series with a nonlinear silicon carbide resistor, satisfies all four criteria. The gap eliminates losses at normal voltages and arcs over during overvoltages. The resistor has the property that its resistance decreases sharply as the current through it increases, thereby limiting the voltage across the resistor to a specified ceiling. The resistor also dissipates the energy in the surge. Finally, following the passage of a surge, various forms of arc control quench the arc within the gap and restore the surge arrester to normal open-circuit conditions.

The "gapless" surge arrester, consisting of a nonlinear metal oxide resistor with no air gap, also satisfies all four criteria. At normal voltages the resistance is extremely high, limiting steady-state currents to microamperes and steady-state losses to a few watts. During surges, the resistance sharply decreases, thereby limiting overvoltage while dissipating surge energy. After

TABLE 13.2 Typical characteristics of station- and intermediate-class metal-oxide surge arresters [20]

Station Class									
Steady-State Operation: System Voltage and Arrester Ratings				Protective Levels: Range of Industry Maxima per Unit of MCOV			Durability Characteristics: IEEE Std C62.11-1993		
Max System Voltage L-L kV-rms[a]	Max System Voltage L-G kV-rms[a]	Min MCOV Rating kV-rms	Duty Cycle Ratings kV-rms	0.5 μs FOW Protective Level[b]	8/20 μs Protective Level[b]	Switching Surge Protective Level[c]	High Current Withstand Crest Amperes	Trans. Line Discharge Miles*	Pressure Relief kA rms (symmetrical)[d]
4.37	2.52	2.55	3	2.32–2.48	2.10–2.20	1.70–1.85	65 000	150	40–80
8.73	5.04	5.1	6–9	2.33–2.48	1.97–2.23	1.70–1.85	65 000	150	40–80
13.1	7.56	7.65	9–12	2.33–2.48	1.97–2.23	1.70–1.85	65 000	150	40–80
13.9	8.00	8.4	10–15	2.33–2.48	1.97–2.23	1.70–1.85	65 000	150	40–80
14.5	8.37	8.4	10–15	2.33–2.48	1.97–2.23	1.70–1.85	65 000	150	40–80
26.2	15.1	15.3	18–27	2.33–2.48	1.97–2.23	1.70–1.85	65 000	150	40–80
36.2	20.9	22	27–36	2.43–2.48	1.97–2.23	1.70–1.85	65 000	150	40–80
48.3	27.8	29	36–48	2.43–2.48	1.97–2.23	1.70–1.85	65 000	150	40–80
72.5	41.8	42	54–72	2.19–2.40	1.97–2.18	1.64–1.84	65 000	150	40–80
121	69.8	70	90–120	2.19–2.40	1.97–2.18	1.64–1.84	65 000	150	40–80
145	83.7	84	108–144	2.19–2.39	1.97–2.17	1.64–1.84	65 000	150	40–80
169	97.5	98	120–172	2.19–2.39	1.97–2.17	1.64–1.84	65 000	175	40–80
242	139	140	172–240	2.19–2.36	1.97–2.15	1.64–1.84	65 000	175	40–80
362	209	209	258–312	2.19–2.36	1.97–2.15	1.71–1.85	65 000	200	40–80
550	317	318	396–564	2.01–2.47	2.01–2.25	1.71–1.85	65 000	200	40–80
800	461	462	576–612	2.01–2.47	2.01–2.25	1.71–1.85	65 000	200	40–80
Intermediate class									
4.37–145	2.52–83.72	2.8–84	3–144	2.38–2.85	2.28–2.55	1.71–1.85	65 000	100	16.1[d]

[a]Voltage range A, ANSI C84.1-1989

[b]Equivalent front-of-wave protective level producing a voltage wave cresting in 0.5 μs. Protective level is maximum discharge voltage (DV) for 10 kA impulse current wave on arrester duty cycle rating through 312 kV, 15 kA for duty cycle ratings 396–564 kV and 20 kA for duty cycle ratings 576–612 kV, per IEEE Std C62.11-1993.

[c]Switching surge characteristics based on maximum switching surge classifying current (based on an impulse current wave with a time to actual crest of 45 μs to 60 μs) of 500 A on arrester duty cycle ratings 3–108 kV, 1000 A on duty cycle ratings 120–240 kV, and 2000 A on duty cycle ratings above 240 kV, per IEEE Std C62.11-1993.

[d]Test values for arresters with porcelain tops have not been standardized. Pressure relief classification is in 5 kA steps.

(From IEEE Guide for the Application of Metal-Oxide Surge Arresters for Alternating-Current Systems, IEEE std. C62.22-1997 (New York: The Institute of Electrical and Electronics Engineers, http://standards.ieee.org/1998) © 1998 IEEE)

*1 mile = 1.6 km

the surge passes, the resistance naturally returns to its original high value. One advantage of the gapless arrester is that its ceiling voltage is closer to its normal operating voltage than is the conventional arrester, thus permitting reduced BILs and potential savings in the capital cost of equipment insulation.

There are four classes of surge arresters: station, intermediate, distribution, and secondary. Station arresters, which have the heaviest construction, are designed for the greatest range of ratings and have the best protective characteristics. Intermediate arresters, which have moderate construction, are designed for systems with nominal voltages 138 kV and below. Distribution arresters are employed with lower-voltage transformers and lines, where there is a need for economy. Secondary arresters are used for nominal system voltages below 1000 V. A summary of the protective characteristics of station- and intermediate-class metal-oxide surge arresters is given in Table 13.2 [20]. This summary is based on manufacturers' catalog information.

Note that arrester currents due to lightning surges are generally less than the lightning currents shown in Figure 13.23. In the case of direct strokes to transmission-line phase conductors, traveling waves are set up in two directions from the point of the stroke. Flashover of line insulation diverts part of the lightning current from the arrester. Only in the case of a direct stroke to a phase conductor very near an arrester, where no line flashover occurs, does the arrester discharge the full lightning current. The probability of this occurrence can be significantly reduced by using overhead shield wires to shield transmission lines and substations. Recommended practice for substations with unshielded lines is to select an arrester discharge current of at least 20 kA (even higher if the isokeraunic level is above 40 thunderstorm days per year). For substations with shielded lines, lower arrester discharge currents, from 5 to 20 kA, have been found satisfactory in most situations [20].

EXAMPLE 13.8 Metal-oxide surge arrester selection

Consider the selection of a station-class metal-oxide surge arrester for a 345-kV system in which the maximum 60-Hz voltage under normal system conditions is 1.04 per unit. (a) Select a station-class arrester from Table 13.2 with a maximum continuous operating voltage (MCOV) rating that exceeds the 1.04 per-unit maximum 60-Hz voltage of the system under normal system conditions. (b) For the selected arrester, determine the protective margin for equipment in the system with a 1300-kV BIL, based on a 10-kA impulse current wave cresting in 0.5 μs.

SOLUTION (a) The maximum 60 Hz line-to-neutral voltage under normal system conditions is $1.04(345/\sqrt{3}) = 207$ kV. From Table 13.2, select a station-class surge arrester with a 209-kV MCOV. This is the lowest MCOV rating that exceeds the 207 kV providing the greatest protective margin as well as economy. (b) From Table 13.2 for the selected surge arrester, the maximum discharge voltage for a 10-kA impulse current wave cresting in 0.5 μs ranges from 2.19 to 2.36 in per-unit of MCOV, or 457–493 kV, depending on arrester manufacturer. Therefore, the protective margin ranges from (1300 – 493) = 807 kV to (1300 – 457) = 843 kV. ■

When selecting a metal-oxide surge arrester, it is important that the arrester MCOV exceeds the maximum 60-Hz system voltage (line-to-neutral) under normal conditions. In addition to considerations affecting the selection of arrester MCOV, metal-oxide surge arresters should also be selected to withstand temporary overvoltages in the system at the arrester location—for example, the voltage rise on unfaulted phases during line-to-ground faults. That is, the temporary overvoltage (TOV) capability of metal-oxide surge arresters should not be exceeded. Additional considerations in the selection of metal-oxide surge arresters are discussed in reference [22] (see www.cooperpower.com).

PROBLEMS

SECTION 13.2

13.1 From the results of Example 13.2, plot the voltage and current profiles along the line at times $\tau/2$, τ, and 2τ. That is, plot $v(x,\tau/2)$ and $i(x,\tau/2)$ versus x for $0 \leqslant x \leqslant l$; then plot $v(x,\tau)$, $i(x,\tau)$, $v(x,2\tau)$, and $i(x,2\tau)$ versus x.

13.2 Rework Example 13.2 if the source voltage at the sending end is a ramp, $e_G(t) = \mathrm{E}u_{-2}(t) = \mathrm{E}tu_{-1}(t)$, with $\mathrm{Z_G} = \mathrm{Z}_c$.

13.3 Referring to the single-phase two-wire lossless line shown in Figure 13.3, the receiving end is terminated by an inductor with $\mathrm{L_R}$ henries. The source voltage at the sending end is a step, $e_G(t) = \mathrm{E}u_{-1}(t)$ with $\mathrm{Z_G} = \mathrm{Z}_c$. Both the line and inductor are initially unenergized. Determine and plot the voltage at the center of the line $v(l/2,t)$ versus time t.

13.4 Rework Problem 13.3 if $\mathrm{Z_R} = \mathrm{Z}_c$ at the receiving end and the source voltage at the sending end is $e_G(t) = \mathrm{E}u_{-1}(t)$, with an inductive source impedance $\mathrm{Z_G}(s) = s\mathrm{L_G}$. Both the line and source inductor are initially unenergized.

13.5 Rework Example 13.4 with $\mathrm{Z_R} = 4\mathrm{Z}_c$ and $\mathrm{Z_G} = \mathrm{Z}_c/3$.

13.6 The single-phase, two-wire lossless line in Figure 13.3 has a series inductance $\mathrm{L} = (1/3) \times 10^{-6}$ H/m, a shunt capacitance $\mathrm{C} = (1/3) \times 10^{-10}$ F/m, and a 30-km line length. The source voltage at the sending end is a step $e_G(t) = 100u_{-1}(t)$ kV with $\mathrm{Z_G}(s) = 100\ \Omega$. The receiving-end load consists of a 100-Ω resistor in parallel with a 2-mH inductor. The line and load are initially unenergized. Determine (a) the characteristic impedance in ohms, the wave velocity in m/s, and the transit time in ms for this line; (b) the sending-and receiving-end voltage reflection coefficients in per-unit; (c) the Laplace transform of the receiving-end current, $I_R(s)$; and (d) the receiving-end current $i_R(t)$ as a function of time.

13.7 The single-phase, two-wire lossless line in Figure 13.3 has a series inductance $\mathrm{L} = 2 \times 10^{-6}$ H/m, a shunt capacitance $\mathrm{C} = 1.25 \times 10^{-11}$ F/m, and a 100-km line length. The source voltage at the sending end is a step $e_G(t) = 100u_{-1}(t)$ kV with a source impedance equal to the characteristic impedance of the line. The receiving-end load consists of a 100-mH inductor in series with a 1-μF capacitor. The line and load are initially unenergized. Determine (a) the characteristic impedance in Ω, the wave velocity in m/s, and the transit time in ms for this line; (b) the sending- and receiving-end voltage reflection coefficients in per-unit; (c) the receiving-end voltage $v_R(t)$ as a function of time; and (d) the steady-state receiving-end voltage.

13.8 The single-phase, two-wire lossless line in Figure 13.3 has a series inductance $L = 0.999 \times 10^{-6}$ H/m, a shunt capacitance $C = 1.112 \times 10^{-11}$ F/m, and a 60-km line length. The source voltage at the sending end is a ramp $e_G(t) = Etu_{-1}(t) = Eu_{-2}(t)$ kV with a source impedance equal to the characteristic impedance of the line. The receiving-end load consists of a 150-Ω resistor in parallel with a 1-μF capacitor. The line and load are initially unenergized. Determine (a) the characteristic impedance in Ω, the wave velocity in m/s, and the transit time in ms for this line; (b) the sending- and receiving-end voltage reflection coefficients in per-unit; (c) the Laplace transform of the sending-end voltage, $V_S(s)$; and (d) the sending-end voltage $v_S(t)$ as a function of time.

SECTION 13.3

13.9 Draw the Bewley lattice diagram for Problem 13.5, and plot $v(l/3, t)$ versus time t for $0 \leqslant t \leqslant 5\tau$. Also plot $v(x, 3\tau)$ versus x for $0 \leqslant x \leqslant l$.

13.10 Rework Problem 13.9 if the source voltage is a pulse of magnitude E and duration $\tau/10$; that is, $e_G(t) = E[u_{-1}(t) - u_{-1}(t - \tau/10)]$. $Z_R = 4Z_c$ and $Z_G = Z_c/3$ are the same as in Problem 13.9.

13.11 Rework Example 13.6 if the source impedance at the sending end of line A is $Z_G = Z_A/4 = 100$ Ω, and the receiving end of line B is short-circuited, $Z_R = 0$.

13.12 Rework Example 13.6 if the overhead line and cable are interchanged. That is, $Z_A = 100$ Ω, $v_A = 2 \times 10^8$ m/s, $l_A = 20$ km, $Z_B = 400$ Ω, $v_B = 3 \times 10^8$ m/s, and $l_B = 30$ km. The step voltage source $e_G(t) = Eu_{-1}(t)$ is applied to the sending end of line A with $Z_G = Z_A = 100$ Ω, and $Z_R = 2Z_B = 800$ Ω at the receiving end. Draw the lattice diagram for $0 \leqslant t \leqslant 0.6$ ms and plot the junction voltage versus time t.

12.13 As shown in Figure 13.32, a single-phase two-wire lossless line with $Z_c = 400$ Ω, $v = 3 \times 10^8$ m/s, and $l = 100$ km has a 400-Ω resistor, denoted R_J, installed across the center of the line, thereby dividing the line into two sections, A and B. The source voltage at the sending end is a pulse of magnitude 100 V and duration 0.1 ms. The source impedance is $Z_G = Z_c = 400$ Ω, and the receiving end of the line is

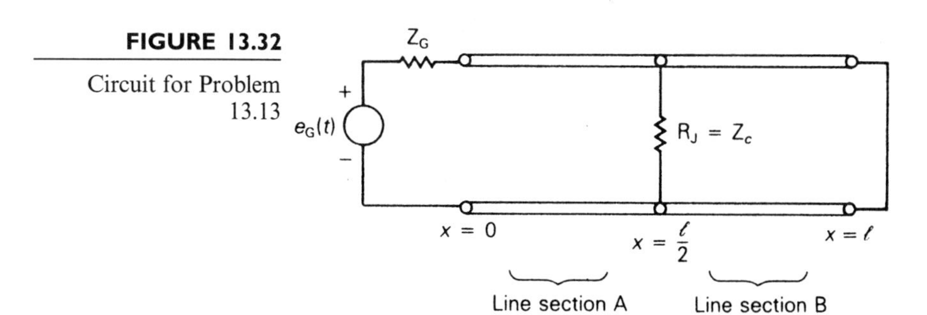

FIGURE 13.32

Circuit for Problem 13.13

short-circuited. (a) Show that for an incident voltage wave arriving at the center of the line from either line section, the voltage reflection and refraction coefficients are given by

$$\Gamma_{BB} = \Gamma_{AA} = \frac{\left(\frac{Z_{eq}}{Z_c}\right) - 1}{\left(\frac{Z_{eq}}{Z_c}\right) + 1} \qquad \Gamma_{AB} = \Gamma_{BA} = \frac{2\left(\frac{Z_{eq}}{Z_c}\right)}{\left(\frac{Z_{eq}}{Z_c}\right) + 1}$$

where

$$Z_{eq} = \frac{R_J Z_c}{R_J + Z_c}$$

(b) Draw the Bewley lattice diagram for $0 \leqslant t \leqslant 6\tau$. (c) Plot $v(l/2, t)$ versus time t for $0 \leqslant t \leqslant 6\tau$ and plot $v(x, 6\tau)$ versus x for $0 \leqslant x \leqslant l$.

13.14 The junction of four single-phase two-wire lossless lines, denoted A, B, C, and D, is shown in Figure 13.13. Consider a voltage wave v_A^+ arriving at the junction from line A. Using (13.3.8) and (13.3.9), determine the voltage reflection coefficient Γ_{AA} and the voltage refraction coefficients Γ_{BA}, Γ_{CA}, and Γ_{DA}.

13.15 Referring to Figure 13.3, the source voltage at the sending end is a step $e_G(t) = Eu_{-1}(t)$ with an inductive source impedance $Z_G(s) = sL_G$, where $L_G/Z_c = \tau/3$. At the receiving end, $Z_R = Z_c/4$. The line and source inductance are initially unenergized. (a) Draw the Bewley lattice diagram for $0 \leqslant t \leqslant 5\tau$. (b) Plot $v(l, t)$ versus time t for $0 \leqslant t \leqslant 5\tau$.

13.16 As shown in Figure 13.33, two identical, single-phase, two-wire, lossless lines are connected in parallel at both the sending and receiving ends. Each line has a 400-Ω characteristic impedance, 3×10^8 m/s velocity of propagation, and 100-km line length. The source voltage at the sending end is a 100-kV step with source impedance $Z_G = 100\ \Omega$. The receiving end is shorted ($Z_R = 0$). Both lines are initially unenergized. (a) Determine the first forward traveling voltage waves that start at time $t = 0$ and travel on each line toward the receiving end. (b) Determine the sending- and receiving-end voltage reflection coefficients in per-unit. (c) Draw the Bewley lattice diagram for $0 < t < 2.0$ ms. (d) Plot the voltage at the center of one line versus time t for $0 < t < 2.0$ ms.

FIGURE 13.33

Circuit for Problem 13.16

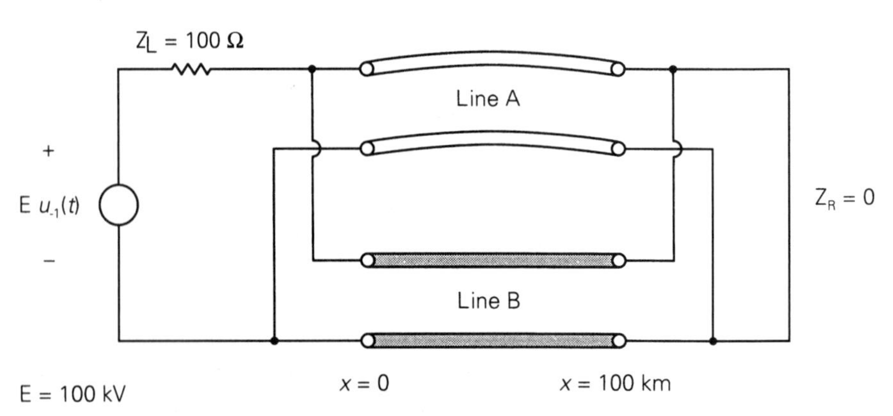

13.17 As shown in Figure 13.34, an ideal current source consisting of a 10-kA pulse with 50-μs duration is applied to the junction of a single-phase, lossless cable and a single-phase, lossless overhead line. The cable has a 200-Ω characteristic impedance, 2×10^8 m/s velocity of propagation, and 20-km length. The overhead line has a 300-Ω characteristic impedance, 3×10^8 m/s velocity of propagation, and 60-km length. The sending end of the cable is terminated by a 400-Ω resistor, and the receiving end of the overhead line is terminated by a 100-Ω resistor. Both the line and cable are initially unenergized. (a) Determine the voltage reflection coefficients Γ_S, Γ_R, Γ_{AA}, Γ_{AB}, Γ_{BA}, and Γ_{BB}. (b) Draw the Bewley lattice diagram for $0 < t < 0.8$ ms. (c) Determine and plot the voltage $v(0, t)$ at $x = 0$ versus time t for $0 < t < 0.8$ ms.

FIGURE 13.34

Circuit for Problem 13.17

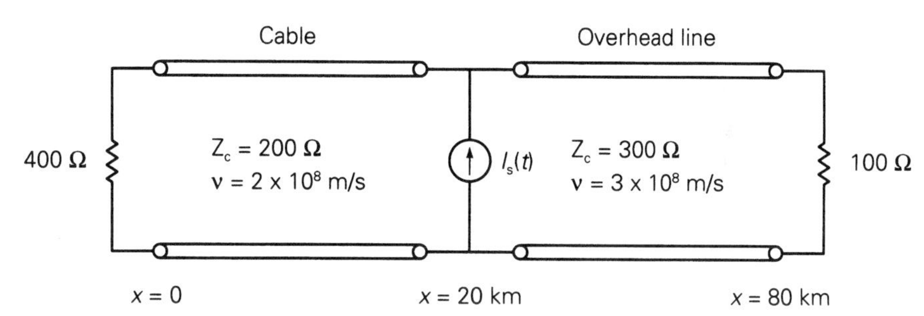

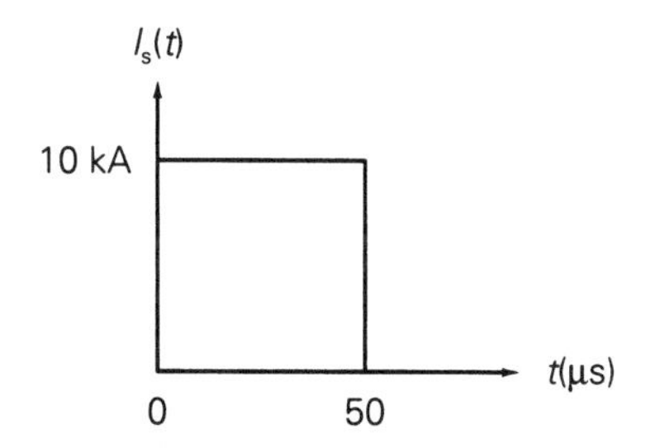

SECTION 13.4

13.18 For the circuit given in Problem 13.3, replace the circuit elements by their discrete-time equivalent circuits and write nodal equations in a form suitable for computer solution of the sending-end and receiving-end voltages. Give equations for all dependent sources. Assume $E = 1000$ V, $L_R = 10$ mH, $Z_c = 100$ Ω, $v = 2 \times 10^8$ m/s, $l = 40$ km, and $\Delta t = 0.02$ ms.

13.19 Repeat Problem 13.18 for the circuit given in Problem 13.13. Assume $\Delta t = 0.03333$ ms.

13.20 For the circuit given in Problem 13.7, replace the circuit elements by their discrete-time equivalent circuits. Use $\Delta t = 100$ μs $= 1 \times 10^{-4}$ s. Determine and show all resistance values on the discrete-time circuit. Write nodal equations for the discrete-time circuit, giving equations for all dependent sources. Then solve the nodal equations and determine the sending- and receiving-end voltages at the following times: $t = 100$, 200, 300, 400, 500, and 600 μs.

13.21 For the circuit given in Problem 13.8, replace the circuit elements by their discrete-time equivalent circuits. Use $\Delta t = 50\ \mu s = 5 \times 10^{-5}$ s and E = 100 kV. Determine and show all resistance values on the discrete-time circuit. Write nodal equations for the discrete-time circuit, giving equations for all dependent sources. Then solve the nodal equations and determine the sending- and receiving-end voltages at the following times: $t = 50$, 100, 150, 200, 250, and 300 μs.

SECTION 13.5

13.22 Rework Problem 13.18 for a lossy line with a constant series resistance R = 0.3 Ω/km. Lump half of the total resistance at each end of the line.

SECTION 13.8

13.23 Repeat Example 13.8 for a 115-kV system with a 1.08 per-unit maximum 60-Hz voltage under normal operating conditions and with a 450-kV BIL.

13.24 Select a station-class metal-oxide surge arrester from Table 13.2 for the high-voltage side of a three-phase 400 MVA, 345-kV Y/13.8-kV Δ transformer. The maximum 60-Hz operating voltage of the transformer under normal operating conditions is 1.10 per unit. The high-voltage windings of the transformer have a BIL of 1300 kV and a solidly grounded neutral. A minimum protective margin of 1.4 per unit based on a 10-kA impulse current wave cresting in 0.5 μs is required. (*Note*: Additional considerations for the selection of metal-oxide surge arresters are given in reference [22] (www.cooperpower.com).

CASE STUDY QUESTIONS

A. Why are circuit breakers and fuses ineffective in protecting against transient overvoltages due to lightning and switching surges?

B. Where are surge arresters located in power systems?

C. How does one select a surge arrester to protect specific equipment?

D. Which states in the U.S. have more than 1,000 MW of installed wind generation?

E. What are the positive impacts of wind generation on the reliability of power system grids? What are the negative impacts?

REFERENCES

1. A. Greenwood, *Electrical Transients in Power Systems*, 2d ed. (New York: Wiley Interscience, 1991).
2. L. V. Bewley, *Travelling Waves on Transmission Systems*, 2d ed. (New York: Wiley, 1951).
3. L. Bergeron, *Water Hammer in Hydraulics and Wave Surges in Electricity* (New York: Wiley, 1961).

4. H. M. Lacey, "The Lightning Protection of High-Voltage Overhead Transmission and Distribution Systems," *Proc. IEE, 96* (1949), p. 287.
5. J. R. Carson, "Wave Propagation in Overhead Wires with Ground Return," *Bell System Technical Journal 5* (1926), pp. 539–554.
6. W. S. Meyer and H. W. Dommel, "Numerical Modelling of Frequency-Dependent Transmission Line Parameters in an Electromagnetic Transients Program," *IEEE Transactions PAS*, vol. PAS-99 (September/October 1974), pp. 1401–1409.
7. A. Budner, "Introduction of Frequency-Dependent Line Parameters into an Electromagnetics Transients Program," *IEEE Transactions PAS*, vol. PAS-89 (January 1970), pp. 88–97.
8. D. E. Hedman, "Propagation on Overhead Transmission Lines: I—Theory of Modal Analysis and II—Earth Conduction Effects and Practical Results," *IEEE Transactions PAS* (March 1965), pp. 200–211.
9. H. W. Dommel, "A Method for Solving Transient Phenomena in Multiphase Systems," *Proceedings 2nd Power Systems Computation Conference*, Stockholm, 1966.
10. H. W. Dommel, "Digital Computer Solution of Electromagnetic Transients in Single- and Multiphase Networks," *IEEE Transactions PAS*, vol. PAS-88 (1969), pp. 388–399.
11. H. W. Dommel, "Nonlinear and Time-Varying Elements in Digital Simulation of Electromagnetic Transients," *IEEE Transactions PAS*, vol. PAS-90 (November/December 1971), pp. 2561–2567.
12. S. R. Naidu, *Transitorios Electromagnéticos em Sistemas de Poténcia*, Eletrobras/UFPb, Brazil, 1985.
13. "Report of Joint IEEE-EEI Committee on EHV Line Outages," *IEEE Transactions PAS, 86* (1967), p. 547.
14. R. A. W. Connor and R. A. Parkins, "Operations Statistics in the Management of Large Distribution System," *Proc. IEEE, 113* (1966), p. 1823.
15. C. T. R. Wilson, "Investigations on Lightning Discharges and on the Electrical Field of Thunderstorms," *Phil. Trans. Royal Soc.*, Series A, *221* (1920), p. 73.
16. G. B. Simpson and F. J. Scrase, "The Distribution of Electricity in Thunderclouds," *Proc. Royal Soc.*, Series A, *161* (1937), p. 309.
17. B. F. J. Schonland and H. Collens, "Progressive Lightning," *Proc. Royal Soc.*, Series A, *143* (1934), p. 654.
18. Westinghouse Electric Corporation, *Electrical Transmission and Distribution Reference Book*, 4th ed. (East Pittsburgh, PA: 1964).
19. E. W. Kimbark and A. C. Legate, "Fault Surge Versus Switching Surge, A Study of Transient Voltages Caused by Line to Ground Faults," *IEEE Transactions PAS, 87* (1968), p. 1762.
20. *IEEE Guide for the Application of Metal-Oxide Surge Arresters for Alternating-Current Systems*, IEEE std. C62.22-1997 (New York: The Institute of Electrical and Electronics Engineers, http://standards.ieee.org/1998).
21. C. Concordia, "The Transient Network Analyzer for Electric Power System Problems," Supplement to *CIGRE Committee No. 13 Report*, 1956.

22. *Varistar Type AZE Surge Arresters for Systems through 345 kV*, Cooper Power Systems Catalog 235-87 (Waukesha, WI: Cooper Power Systems, http://www.cooperpower.com, August 1997).

23. W. Grant, et al., "Change in the Air," *IEEE Power & Energy Magazine*, 7, 6 (November/December 2009), pp. 47–58.

24. W. R. Newcott, "Lightning, Nature's High-Voltage Spectacle," *National Geographic*, 184, 1 (July 1993), pp. 83–103.

25. K. L. Cummins, E. P. Krider, and M. D. Malone, "The U.S. National Lightning Detection Network TM and Applications of Cloud-to-Ground Lightning Data by Electric Power Utilities," *IEEE Transactions on Electromagnetic Compatibility*, 40, 4 (November 1998), pp. 465–480.

Distribution substation fed by two 22.9-kV (one overhead and one underground) lines through two 15-MVA, 22.9Δ/4.16Y kV distribution substation transformers. The transformers are located on either side of the switchgear building shown in the center. This substation feeds six 4.16-kV radial primary feeders through 5-kV, 1200-A vacuum circuit breakers (Courtesy of Danvers Electric)

14

POWER DISTRIBUTION

Major components of an electric power system are generation, transmission, and distribution. Distribution, including primary and secondary distribution, is that portion of a power system that runs from distribution substations to customer's service entrance equipment. In 2008 in the United States, distribution systems served approximately 138 million customers that consumed 3.7 trillion kWh [www.eia.doe.gov].

This chapter provides an overview of distribution. In Sections 14.1–14.3 we introduce the basic configurations and characteristics of distribution including primary and secondary distribution. In Sections 14.4 and 14.5 we discuss the application of transformers and capacitors in distribution systems. Then in Sections 14.6–14.9 we introduce distribution software, distribution reliability, distribution automation, and smart grids.

(J. D. Glover, "Electric Power Distribution," Encyclopedia of Energy Technology and The Environment, John Wiley & Sons, New York, NY, 1995. Reprinted with permission of John Wiley & Sons, Inc.)

CASE STUDY Utilities in North America and throughout the world are incorporating new technologies towards implementing the next-generation electricity grid known as the "intelligent grid" or "smart grid". A smart grid accommodates a wide variety of generation options, enables customers to interact with energy management systems to adjust their energy use, predicts looming failures and takes corrective actions to mitigate system problems in a self-healing manner. The move towards a smart grid started with the distribution system, including the introduction of advanced meter infrastructure (AMI), which provides utilities with two-way communication to meters at customers' premises and the ability to modify customers' service-level parameters. The next step is to implement distributed demand and control strategies that are integrated with AMI in a transition to the smart grid. The following article describes the evolution of smart grids and smart microgrids including their basic ingredients and topologies [19].

The Path of the Smart Grid

BY HASSAN FARHANGI

The utility industry across the world is trying to address numerous challenges, including generation diversification, optimal deployment of expensive assets, demand response, energy conservation, and reduction of the industry's overall carbon footprint. It is evident that such critical issues cannot be addressed within the confines of the existing electricity grid.

The existing electricity grid is unidirectional in nature. It converts only one-third of fuel energy into electricity, without recovering the waste heat. Almost 8% of its output is lost along its transmission lines, while 20% of its generation capacity exists to meet peak demand only (i.e., it is in use only 5% of the time). In addition to that, due to the hierarchical topology of its assets, the existing electricity grid suffers from domino-effect failures.

The next-generation electricity grid, known as the "smart grid" or "intelligent grid," is expected to address the major shortcomings of the existing grid. In essence, the smart grid needs to provide the utility companies with full visibility and pervasive control over their assets and services. The smart grid is required to be self-healing and resilient to system anomalies. And last but not least, the smart grid needs to empower its stakeholders to define and realize new ways of engaging with each other and performing energy transactions across the system.

("The Path of the Smart Grid," by Dr. Hassan Farhangi of BCIT. © 2010 IEEE. Reprinted, with permission, from IEEE Power & Energy Magazine, *8, 1 (January/February 2010), pp. 18–28)*

BASIC INGREDIENTS

To allow pervasive control and monitoring, the smart grid is emerging as a convergence of information technology and communication technology with power system engineering. Figure 1 depicts the

Existing Grid	Intelligent Grid
Electromechanical	Digital
One-Way Communication	Two-Way Communication
Centralized Generation	Distributed Generation
Hierarchical	Network
Few Sensors	Sensors Throughout
Blind	Self-Monitoring
Manual Restoration	Self-Healing
Failures and Blackouts	Adaptive and Islanding
Manual Check/Test	Remote Check/Test
Limited Control	Pervasive Control
Few Customer Choices	Many Customer Choices

Figure 1
The smart grid compared with the existing grid

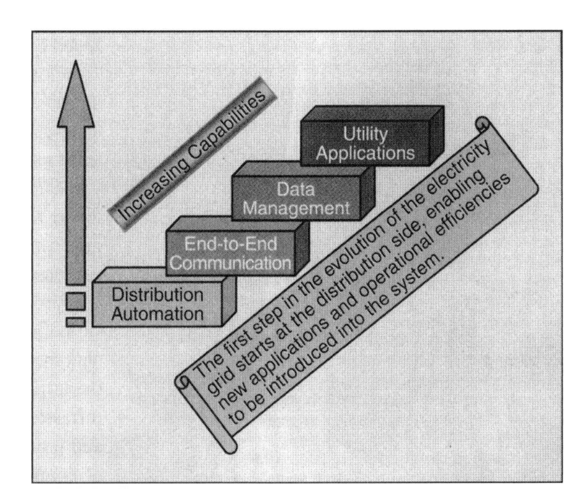

Figure 2
Utility-desired capabilities

salient features of the smart grid in comparison with the existing grid.

Given the fact that the roots of power system issues are typically found in the electrical distribution system, the point of departure for grid overhaul is firmly placed at the bottom of the chain. As Figure 2 demonstrates, utilities believe that investing in distribution automation will provide them with increasing capabilities over time.

Within the context of these new capabilities, communication and data management play an important role. These basic ingredients enable the utilities to place a layer of intelligence over their current and future infrastructure, thereby allowing the introduction of new applications and processes in their businesses. As Figure 3 depicts, the convergence of communication technology and information technology with power system engineering, assisted by an array of new approaches, technologies and applications, allows the existing grid to traverse the complex yet staged trajectory of architecture, protocols, and standards towards the smart grid.

SMART GRID DRIVERS

As the backbone of the power industry, the electricity grid is now the focus of assorted technological innovations. Utilities in North America and across the world are taking solid steps towards incorporating new technologies in many aspects of their operations and infrastructure. At the core of this transformation is the need to make more efficient use of current assets. Figure 4 shows a typical utility pyramid in which asset management is at the base of smart grid development. It is on this base that utilities build a foundation for the smart grid through a careful overhaul of their IT, communication, and circuit infrastructure.

As discussed, the organic growth of this well-designed layer of intelligence over utility assets enables the smart grid's fundamental applications to emerge. It is interesting to note that although the foundation of the smart grid is built on a lateral integration of these basic ingredients, true smart grid capabilities will be built on vertical integration of the upper-layer applications. As an example, a critical capability such as demand response may not be feasible without tight integration of smart meters and home area networks.

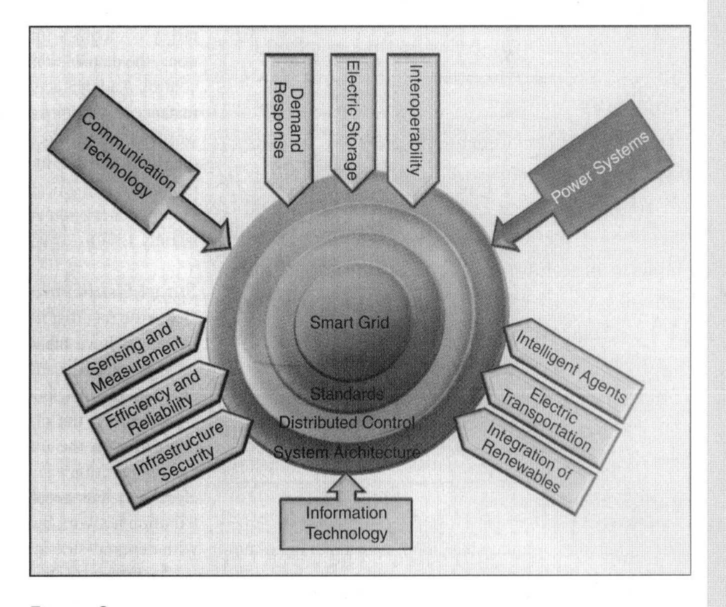

Figure 3
Basic smart grid ingredients (Gridwise Alliance)

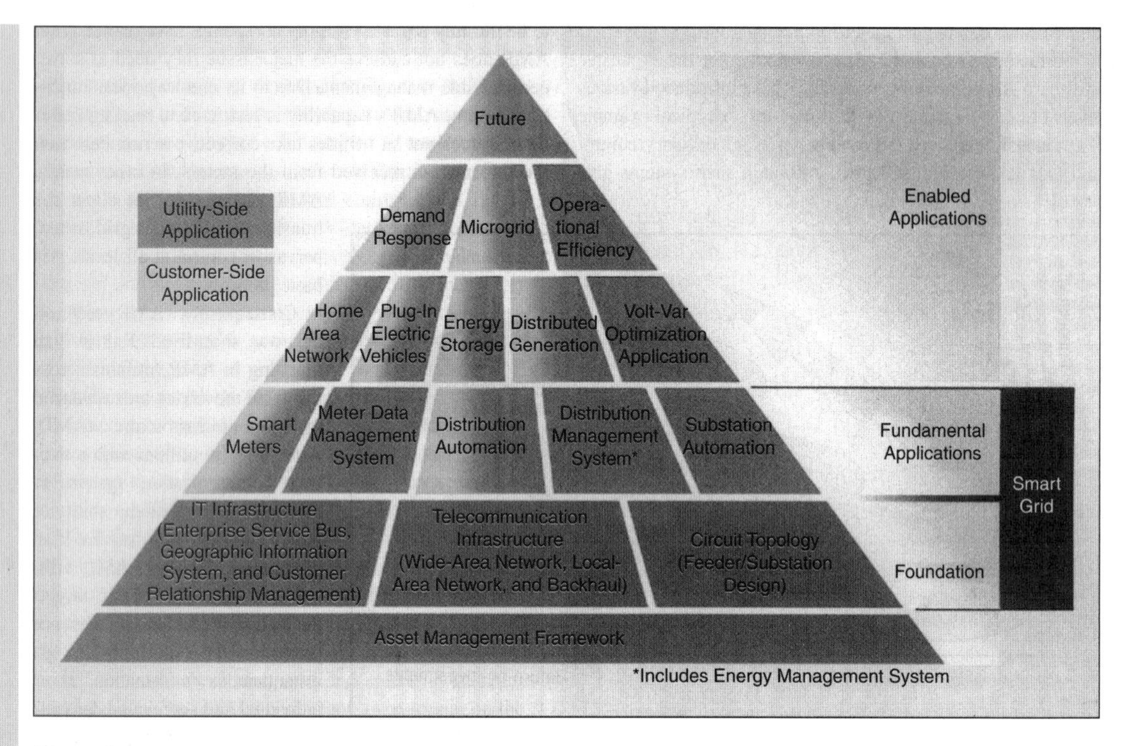

Figure 4
Smart grid pyramid (Courtesy of BC Hydro)

As such, one may argue that given the size and the value of utility assets, the emergence of the smart grid will be more likely to follow an evolutionary trajectory than to involve a drastic overhaul. The smart grid will therefore materialize through strategic implants of distributed control and monitoring systems within and alongside the existing electricity grid. The functional and technological growth of these embryos over time helps them emerge as large pockets of distributed intelligent systems across diverse geographies. This organic growth will allow the utilities to shift more of the old grid's load and functions onto the new grid and so to improve and enhance their critical services.

These smart grid embryos will facilitate the distributed generation and cogeneration of energy. They will also provide for the integration of alternative sources of energy and the management of a system's emissions and carbon footprint. And last but not least, they will enable utilities to make more efficient use of their existing assets through demand response, peak shaving, and service quality control.

The problem that most utility providers across the globe face, however, is how to get to where they need to be as soon as possible, at the minimum cost, and without jeopardizing the critical services they are currently providing. Moreover, utilities must decide which strategies and what road map they should pursue to ensure that they achieve the highest possible return on the required investments for such major undertakings.

As is the case with any new technology, the utilities in the developing world have a clear advantage over their counterparts in the developed world. The former have fewer legacy issues to grapple with and so may be able to leap forward without the

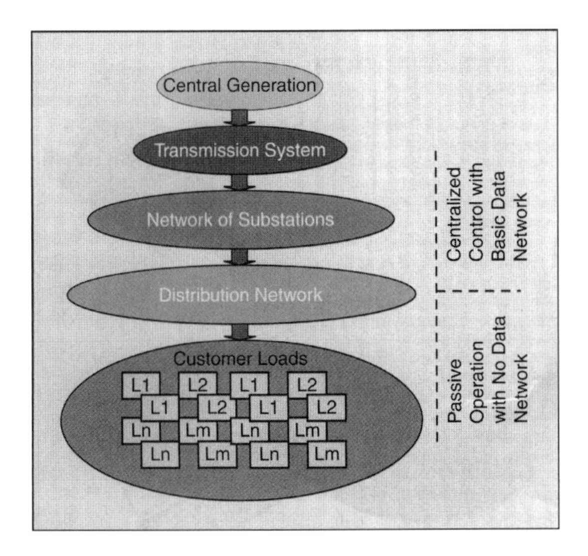

Figure 5
The existing grid

need for backward compatibility with their existing systems.

EVOLUTION OF THE SMART GRID

THE EXISTING GRID

The existing electricity grid is a product of rapid urbanization and infrastructure developments in various parts of the world in the past century. Though they exist in many differing geographies, the utility companies have generally adopted similar technologies. The growth of the electrical power system, however, has been influenced by economic, political, and geographic factors that are unique to each utility company.

Despite such differences, the basic topology of the existing electrical power system has remained unchanged. Since its inception, the power industry has operated with clear demarcations between its generation, transmission, and distribution subsystems and thus has shaped different levels of automation, evolution, and transformation in each silo.

As Figure 5 demonstrates, the existing electricity grid is a strictly hierarchical system in which power plants at the top of the chain ensure power delivery to customers' loads at the bottom of the chain. The system is essentially a one-way pipeline where the source has no real-time information about the service parameters of the termination points. The grid is therefore overengineered to withstand maximum anticipated peak demand across its aggregated load. And since this peak demand is an infrequent occurrence, the system is inherently inefficient.

Moreover, an unprecedented rise in demand for electrical power, coupled with lagging investments in the electrical power infrastructure, has decreased system stability. With the safe margins exhausted, any unforeseen surge in demand or anomalies across the distribution network causing component failures can trigger catastrophic blackouts.

To facilitate troubleshooting and upkeep of the expensive upstream assets, the utility companies have introduced various levels of command-and-control functions. A typical example is the widely deployed system known as supervisory control and data acquisition (SCADA). Although such systems give utility companies limited control over their upstream functions, the distribution network remains outside their real-time control. And the picture hardly varies all across the world. For instance, in North America, which has established one of the world's most advanced electrical power systems, less than a quarter of the distribution network is equipped with information and communications systems, and the distribution automation penetration at the system feeder level is estimated to be only 15% to 20%.

SMART GRID EVOLUTION

Given the fact that nearly 90% of all power outages and disturbances have their roots in the distribution network, the move towards the smart grid has to start at the bottom of the chain, in the distribution system.

Moreover, the rapid increase in the cost of fossil fuels, coupled with the inability of utility companies to expand their generation capacity in line

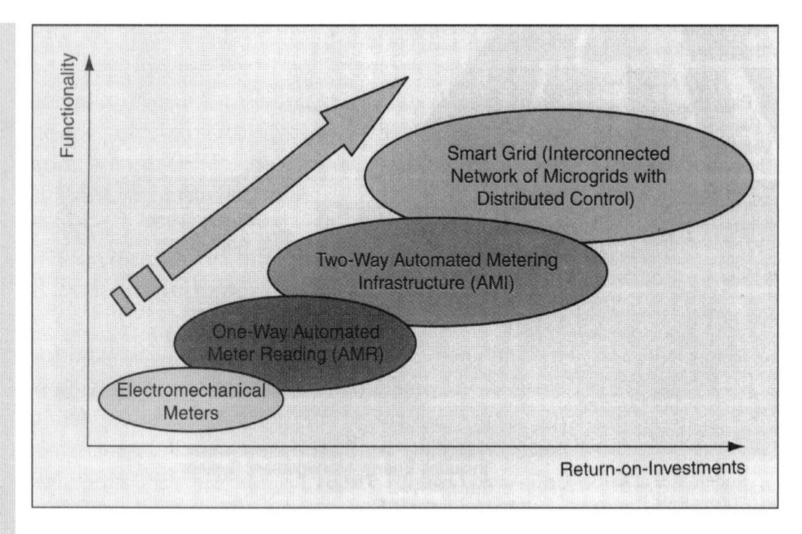

Figure 6
The evolution of the smart grid

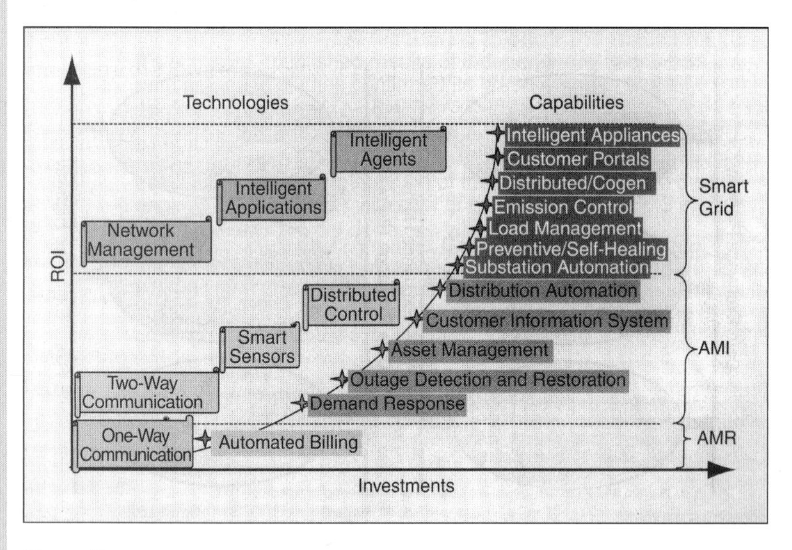

Figure 7
Smart grid return on investments

with the rising demand for electricity, has accelerated the need to modernize the distribution network by introducing technologies that can help with demand-side management and revenue protection.

As Figure 6 shows, the metering side of the distribution system has been the focus of most recent infrastructure investments. The earlier projects in this sector saw the introduction of automated meter reading (AMR) systems in the distribution network. AMR lets utilities read the consumption records, alarms, and status from customers' premises remotely.

As Figure 7 suggests, although AMR technology proved to be initially attractive, utility companies have realized that AMR does not address the major issue they need to solve: demand-side management. Due to its one-way communication system, AMR's capability is restricted to reading meter data. It does not let utilities take corrective action based on the information received from the meters. In other words, AMR systems do not allow the transition to the smart grid, where pervasive control at all levels is a basic premise.

Consequently, AMR technology was short-lived. Rather than investing in AMR, utilities across the world moved towards advanced metering infrastructure (AMI). AMI provides utilities with a two-way communication system to the meter, as well as the ability to modify customers' service-level parameters. Through AMI, utilities can meet their basic targets for load management and revenue protection. They not only can get instantaneous information about individual and aggregated demand, but they can also impose certain caps on consumption, as well as enact various revenue models to control their costs.

The emergence of AMI heralded a concerted move by stakeholders to further refine the ever-changing concepts around the smart grid. In fact, one of the major measurements that the utility companies apply in choosing among AMI technologies is whether or not they will be forward-compatible with their yet-to-be-realized smart grid's topologies and technologies.

TRANSITION TO THE SMART GRID

As the next logical step, the smart grid needs to leverage the AMI infrastructure and implement its distributed command-and-control strategies over the AMI backbone. The pervasive control and intelligence that embodies the smart grid has to reside across all geographies, components, and functions of the system. Distinguishing these three elements is significant, as it determines the topology of the smart grid and its constituent components.

SMART MICROGRIDS

The smart grid is the collection of all technologies, concepts, topologies, and approaches that allow the silo hierarchies of generation, transmission, and distribution to be replaced with an end-to-end, organically intelligent, fully integrated environment where the business processes, objectives, and needs of all stakeholders are supported by the efficient exchange of data, services, and transactions. A smart grid is therefore defined as a grid that accommodates a wide variety of generation options, e.g., central, distributed, intermittent, and mobile. It empowers consumers to interact with the energy management system to adjust their energy use and reduce their energy costs. A smart grid is also a self-healing system. It predicts looming failures and takes corrective action to avoid or mitigate system problems. A smart grid uses IT to continually optimize the use of its capital assets while minimizing operational and maintenance costs.

Mapping the above definitions to a practical architecture, one can readily see that the smart grid cannot and should not be a replacement for the existing electricity grid but a complement to it. In other words, the smart grid would and should coexist with the existing electricity grid, adding to its capabilities, functionalities, and capacities by means of an evolutionary path. This necessitates a topology for the smart grid that allows for organic growth, the inclusion of forward-looking technologies, and full backward compatibility with the existing legacy systems.

At its core, the smart grid is an ad hoc integration of complementary components, subsystems, and functions under the pervasive control of a highly intelligent and distributed command-and-control system. Furthermore, the organic growth and evolution of the smart grid is expected to come through the plug-and-play integration of certain basic structures called intelligent (or smart) microgrids. Microgrids are defined as interconnected networks of distributed energy systems (loads and resources) that can function whether they are connected to or separate from the electricity grid.

MICROGRID TOPOLOGY

A smart microgrid network that can operate in both grid-tied as well as islanded modes typically integrates the following seven components:

- It incorporates power plants capable of meeting local demand as well as feeding the unused energy back to the electricity grid. Such power plants are known as cogenerators and often use renewable sources of energy, such as wind, sun, and biomass. Some microgrids are equipped with thermal power plants capable of recovering the waste heat, which is an inherent by-product of fissile-based electricity generation. Called combined heat and power (CHP), these systems recycle the waste heat in the form of district cooling or heating in the immediate vicinity of the power plant.
- It services a variety of loads, including residential, office and industrial loads.

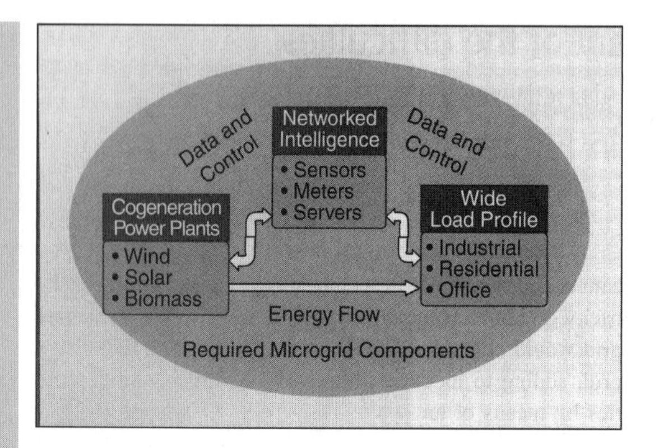

Figure 8
The topology of a smart microgrid

- It makes use of local and distributed power-storage capability to smooth out the intermittent performance of renewable energy sources.
- It incorporates smart meters and sensors capable of measuring a multitude of consumption parameters (e.g., active power, reactive power, voltage, current, demand, and so on) with acceptable precision and accuracy. Smart meters should be tamper-resistant and capable of soft connect and disconnect for load and service control.
- It incorporates a communication infrastructure that enables system components to exchange information and commands securely and reliably.
- It incorporates smart terminations, loads, and appliances capable of communicating their status and accepting commands to adjust and control their performance and service level based on user and/or utility requirements.
- It incorporates an intelligent core, composed of integrated networking, computing, and communication infrastructure elements, that appears to users in the form of energy management applications that allow command and control on all nodes of the network. These should be capable of identifying all terminations, querying them, exchanging data and commands with them, and managing the collected data for scheduled and/or on-demand transfer to the higher-level intelligence residing in the smart grid. Figure 8 depicts the topology of a smart microgrid.

SMART GRID TOPOLOGY

As Figure 9 shows, the smart grid is therefore expected to emerge as a well-planned plug-and-play integration of smart microgrids that will be interconnected through dedicated highways for command, data, and power exchange. The emergence of these smart microgrids and the degree of their interplay and integration will be a function of rapidly escalating smart grid capabilities and requirements. It is also expected that not all microgrids will be created equal. Depending on their diversity of load, the mix of primary energy sources, and the geography and economics at work in particular areas,

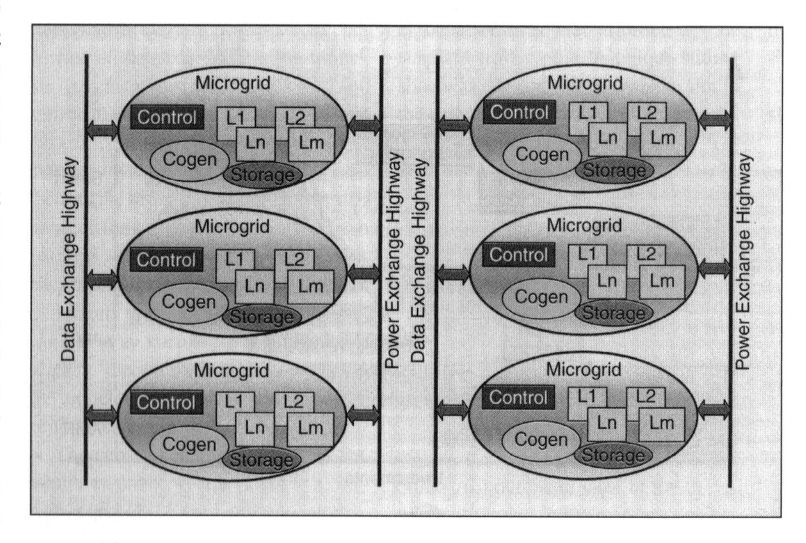

Figure 9
The smart grid of the future

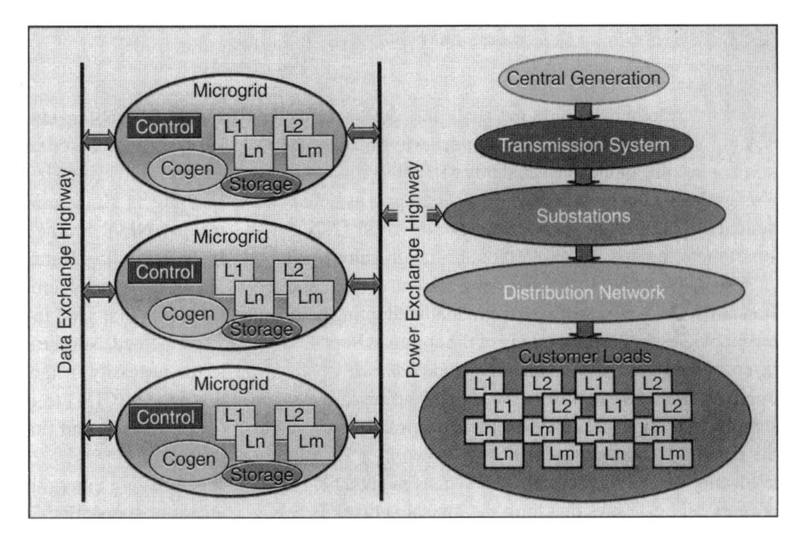

Figure 10
System topology for the smart grid in transition

among other factors, microgrids will be built with different capabilities, assets, and structures.

COEXISTENCE OF THE TWO GENERATIONS OF ELECTRICITY GRIDS

As discussed earlier, utilities require that the AMI systems now being implemented ensure an evolutionary path to the smart grid. The costs associated with AMI rollout are simply too high to permit an overhaul of the installed systems in preparation for an eventual transition to the smart grid.

As such, industry pundits believe that for the foreseeable future the old and the new grids will operate side by side, with functionality and load to be migrated gradually from the old system to the new one over time. And in the not too distant future, the smart grid will emerge as a system of organically integrated smart microgrids with pervasive visibility and command-and-control functions distributed across all levels.

The topology of the emerging grid will therefore resemble a hybrid solution, the core intelligence of which grows as a function of its maturity and extent. Figure 10 shows the topology of the smart grid in transition.

SMART GRID STANDARDS

Despite assurances from AMI technology providers, the utilities expect the transition from AMI to the smart grid to be far from a smooth ride. Many believe that major problems could surface when disparate systems, functions, and components begin to be integrated as part of a distributed command-and-control system. Most of these issues have their roots in the absence of the universally accepted interfaces, messaging and control protocols, and standards that would be required to ensure a common communication vocabulary among system components.

There are others who do not share this notion, however, arguing that given all the efforts under way in standardization bodies, the applicable standards will emerge to help with plug-and-play integration of various smart grid system components. Examples of such standards are ANSI C12.22 for smart metering and IEC 61850 for substation automation.

Moreover, to help with the development of the required standards, the power industry is gradually adopting different terminologies for the partitioning of the command-and-control layers of the smart grid. Examples include *home area network* or HAN (used to identify the network of communicating loads, sensors, and appliances beyond the smart meter and within the customer's premises); *local area network* or LAN (used to identify the network of integrated smart meters, field components, and gateways that form the logical network between distribution substations and a customer's premises); and, last but not least, *wide area network* or WAN (used to identify the network of upstream utility assets, including—but not limited to—power plants, distributed storage, substations, and so on).

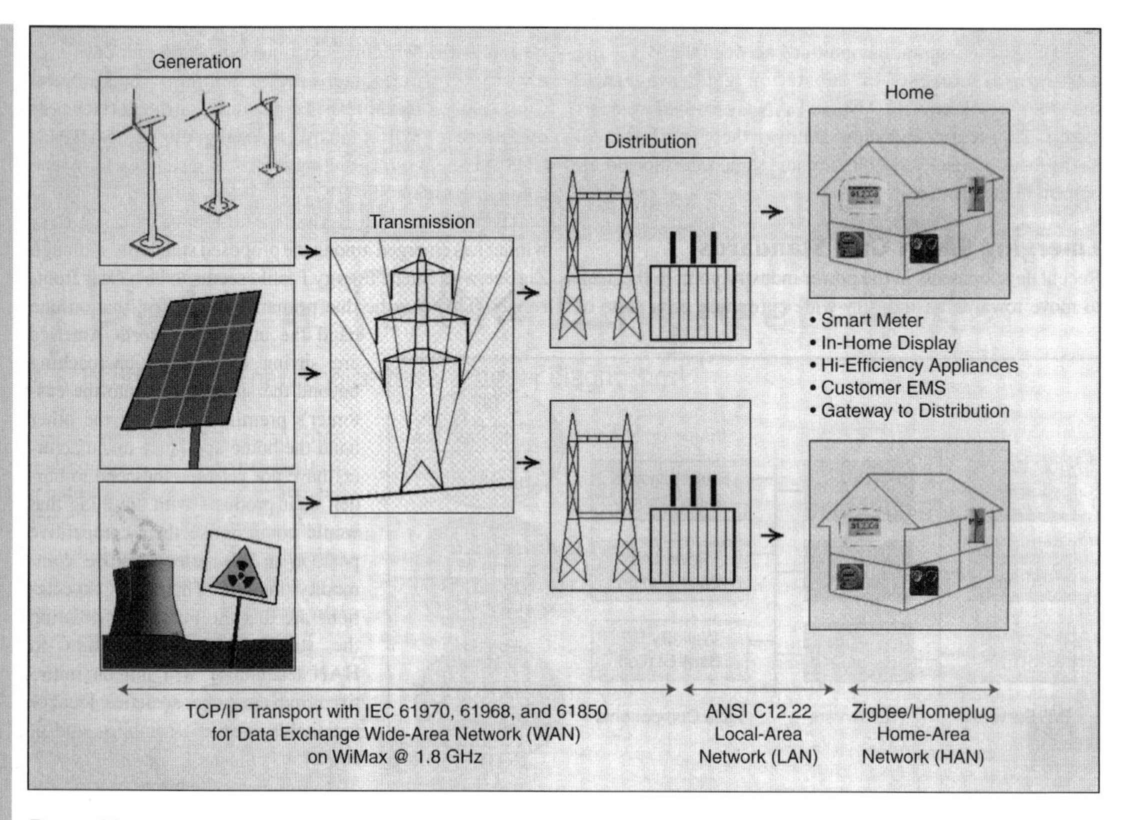

Figure 11
Emerging standards for the smart grid

As Figure 11 shows, the interface between the WAN and LAN worlds consists of substation gateways, while the interface between LAN and HAN is provided by smart meters. The security and vulnerability of these interfaces will be the focus of much technological and standardization development in the near future.

EMERGING SMART GRID STANDARDS

Recent developments in the power industry point to the need to move towards an industry-wide consensus on a suite of standards enabling end-to-end command and data exchange between various components of the smart grid. Focused efforts and leadership by NIST (United States National Institute of Standards and Technology) is yielding good results. NIST Framework and Roadmap for Smart Grid Interoperability Standards (Release 1.0, September 2009) identifies priority areas for standardization and a list of standards that need to be further refined, developed, and/or implemented. Similar efforts in Europe and elsewhere point to the necessity of the development of a common information model (CIM) to enable vertical and lateral integration of applications and functions within the smart grid. Among the list of proposed standards, IEC 61850 and its associate standards are emerging as favorites for WAN data communication, supporting TCP/IP, among other protocols, over fiber or a 1.8-GHz flavor of WiMax. In North America, ANSI C12.22, and its associated standards, is viewed as the favorite LAN standard, enabling a new generation of smart meters capable of communicating with their peers as well as with their corresponding substation gateways over a variety of wireless technologies. Similarly, the European Community's recently issued mandate for the development of Europe's AMI standard, replacing the aging DLMS/COSEM standard, is fueling efforts to develop a European counterpart for ANSI-C12.22.

The situation with HANs is a little murkier, as no clear winner has emerged among the proposed standards, although ZigBee with Smart Energy Profile seems to be a clear front-runner. This may be due primarily to the fact that on one hand the utilities in North America are shying away from encroaching beyond the smart meter into the customer's premises while on the other hand the home appliance manufacturers have not yet seen the need to burden their products with anything that would compromise their competitive position in this price-sensitive commodity market. Therefore, expectations are that the burden for creating the standardization momentum in HAN technology will fall on initiatives from consumer societies, local or national legislative assemblies, and/or concerned citizens.

In summary, the larger issue in the process of transitioning to the smart grid lies in the gradual rollout of a highly distributed and intelligent management system with enough flexibility and scalability to not only manage system growth but also to be open to the accommodation of ever-changing technologies in communications, IT, and power systems.

What would ensure a smooth transition from AMI to the smart grid would be the emergence of plug-and-play system components with embedded intelligence that could operate transparently in a variety of system integration and configuration scenarios. The embedded intelligence encapsulated in such components is often referred to with the term *intelligent agent.*

SMART GRID RESEARCH, DEVELOPMENT, AND DEMONSTRATION (RD&D)

Utility companies are fully cognizant of the difficulties involved in transitioning their infrastructure, organizations, and processes towards an uncertain future. The fact of the matter is that despite all the capabilities the smart grid promises to yield, the utilities, as providers of a critical service, still see as their primary concern keeping the lights on. Given the fact that utilities cannot and may not venture into adopting new technologies without exhaustive validation and qualification, one can readily see that one of the major difficulties utilities across the world are facing is the absence of near-real world RD&D capability to enable them develop, validate, and qualify technologies, applications, and solutions for their smart grid programs.

The problem most utility providers face is not the absence of technology. On the contrary, many disparate technologies have been developed by the industry (e.g., communication protocols, computing engines, sensors, algorithms, and models) to address utility applications and resolve potential issues within the smart grid.

The problem is that these new technologies have not yet been proven in the context of the utility providers' desired specifications, configurations, and architecture. Given the huge responsibility utilities have in connection with operating and maintaining their critical infrastructure, they cannot be expected to venture boldly and without proper preparation into new territories, new technologies, and new solutions. As such, utilities are in critical need of a near-real-world environment, with real loads, distribution gear, and diverse consumption profiles, to develop, test, and validate their required smart grid solutions. Such an environment would in essence constitute a smart microgrid.

Similar to a typical smart microgrid, an RD&D micro-grid will incorporate not only the three major components of generation, loads, and smart controls but also a flexible and highly programmable command-and-control overlay enabling engineers to develop, experiment with, and validate the utility's target requirements. Figure 12 depicts a programmable command-and-control overlay for an RD&D microgrid set up on the Burnaby campus of the British Columbia Institute of Technology (BCIT) in Vancouver, British Columbia, Canada.

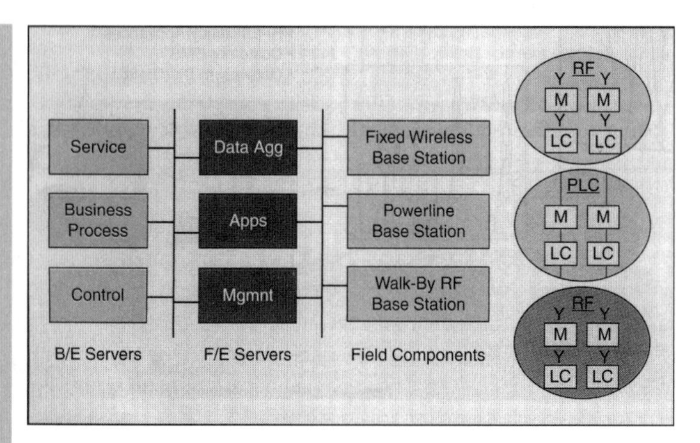

Figure 12
Command-and-control overlay of the BCIT RD&D microgrid

Sponsored by BC Hydro and funded jointly by the British Columbia government's Innovative Clean Energy (ICE) Fund and the Canadian government's Western Diversification Fund, BCIT's smart microgrid enables utility providers, technology providers, and researchers to work together to facilitate the commercialization of architectures, protocols, configurations, and models of the evolving smart grid. The ultimate goal is to chart a "path from lab to field" for innovative and cost-effective technologies and solutions for the evolving smart electricity grid.

In addition to a development environment, BCIT's smart microgrid is also a test bed where multitudes of smart grid components, technologies, and applications are integrated to qualify the merits of different solutions, showcase their capabilities, and accelerate the commercialization of technologies and solutions for the smart grid. As an example, Figure 13 shows how such an infrastructure may be programmed to enable utilities to develop, test, and validate their front-end and field capabilities in

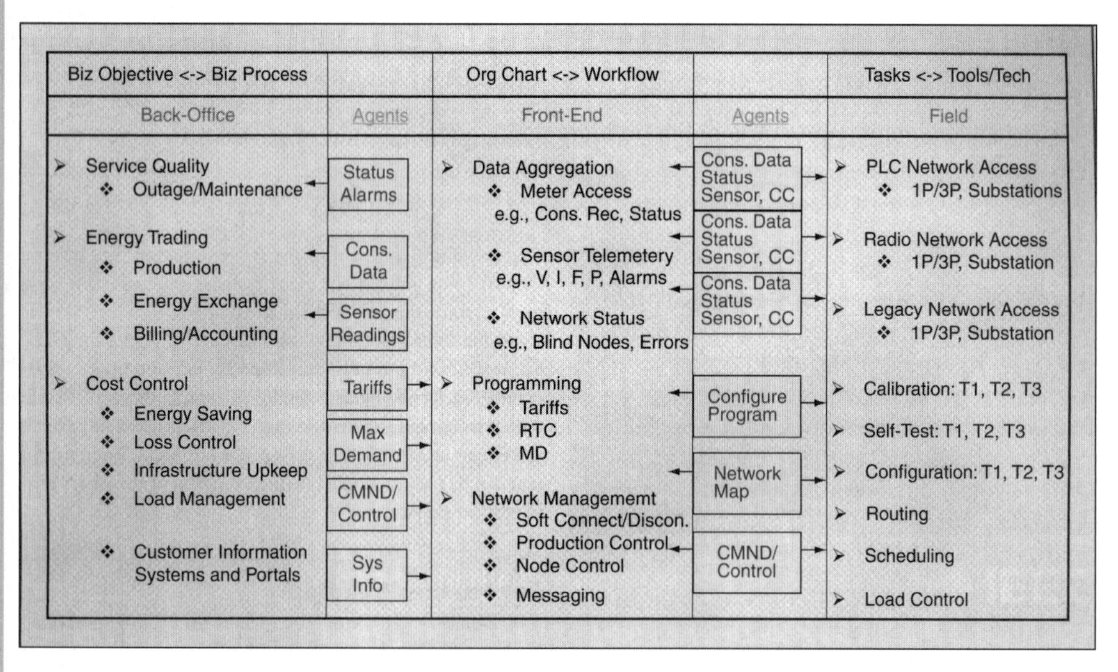

Figure 13
Process overlay of the BCIT RD&D microgrid

Figure 14
The BCIT smart grid control center (used with permission)

line with their already existing back-office business processes and tools.

Figure 14 shows BCIT's Smart Grid Development Lab, where powerful servers, protocol analyzers, routers, firewalls, and networking equipment are integrated with multiple base stations, gateways, and smart-metering installations to create an end-to-end smart grid control center (SGCC). Here, a variety of experiments, tests, and validation efforts can be programmed and carried out.

CONCLUSIONS

Exciting yet challenging times lie ahead. The electrical power industry is undergoing rapid change. The rising cost of energy, the mass electrification of everyday life, and climate change are the major drivers that will determine the speed at which such transformations will occur.

Regardless of how quickly various utilities embrace smart grid concepts, technologies, and systems, they all agree on the inevitability of this massive transformation. It is a move that will not only affect their business processes but also their organization and technologies.

At the same time, many research centers across the globe are working to ease this transition by developing the next-generation technologies required to realize the smart grid. As a member of Gridwise Alliance, BCIT is providing North American utilities with a state-of-the-art RD&D microgrid that can be used to accelerate the evolution of the smart grid in North America.

FOR FURTHER READING

S. J. Anders, "The emerging smart grid," Energy Policy Initiative Center, University of San Diego School of Law, Oct. 2007, pp. 4–8.

H. Farhangi, "Intelligent micro grid research at BCIT," *EnergyBiz Smart Grid Suppl.*, July 2008.

D. Moore and D. McDonnell, "Smart grid vision meets distribution utility reality," *Elect. Light Power,* pp. 1–6, Mar. 2007.

M. Smith, "Overview of federal R&D on microgrid technologies," in *Proc. Kythonos 2008 Symp. Microgrids,* June 2, pp. 2–8, 2008.

K. Moslehi, "Intelligent infrastructure for coordinated control of a self-healing power grid," in Proc. *IEEE Electrical Power System Conf. (EPEC'08)*, Vancouver, Canada, Oct. 2008, pp. 3–7.

K. Mauch and A. Foss, "Smart grid technology overview," *Natural Resources*, Canada, Sept. 2005 [Online]. Available: http://www.powerconnect.ca/newsevents/events/archive/2006/smartgrid/Smart%20Grid%20Technology%20Overview%20%20NRCanada%20%5BRead-Only%5D.pdf

H. Farhangi, "Intelligent micro grid research at BCIT," in Proc. *IEEE Electrical Power System Conf. (EPEC'08)*, Vancouver, Canada, Oct. 2008, pp. 1–7.

A. Vojdani, "Integration challenges of the smart grid—Enterprise integration of DR and meter data," in Proc. *IEEE Electrical Power System Conf. (EPEC'08)*, Vancouver, Canada, Oct. 2008, p. 21.

M. Amin and S. Wollenberg, "Toward a smart grid: Power delivery for the 21st century," *IEEE Power Energy Mag.*, vol. 3, no. 5, pp. 34–41, Sept.–Oct. 2005.

BIOGRAPHY

Hassan Farhangi is with the Technology Centre of British Columbia Institute of Technology.

14.1

INTRODUCTION TO DISTRIBUTION

Figure 14.1 shows the basic components of an electric power system [1–9]. Power plants convert energy from fuel (coal, gas, nuclear, oil, etc.) and from water, wind, or other forms into electric energy. Power plant generators, with typical ratings varying from 50 to 1300 MVA, are of three-phase construction, with three-phase armature windings embedded in the slots of stationary armatures. Generator terminal voltages, which are limited by material and insulation capabilities, range from a few kV for older and smaller units up to 20 kV for newer and larger units.

To reduce transmission energy losses, generator step-up (GSU) transformers at power plant substations increase voltage and decrease current.

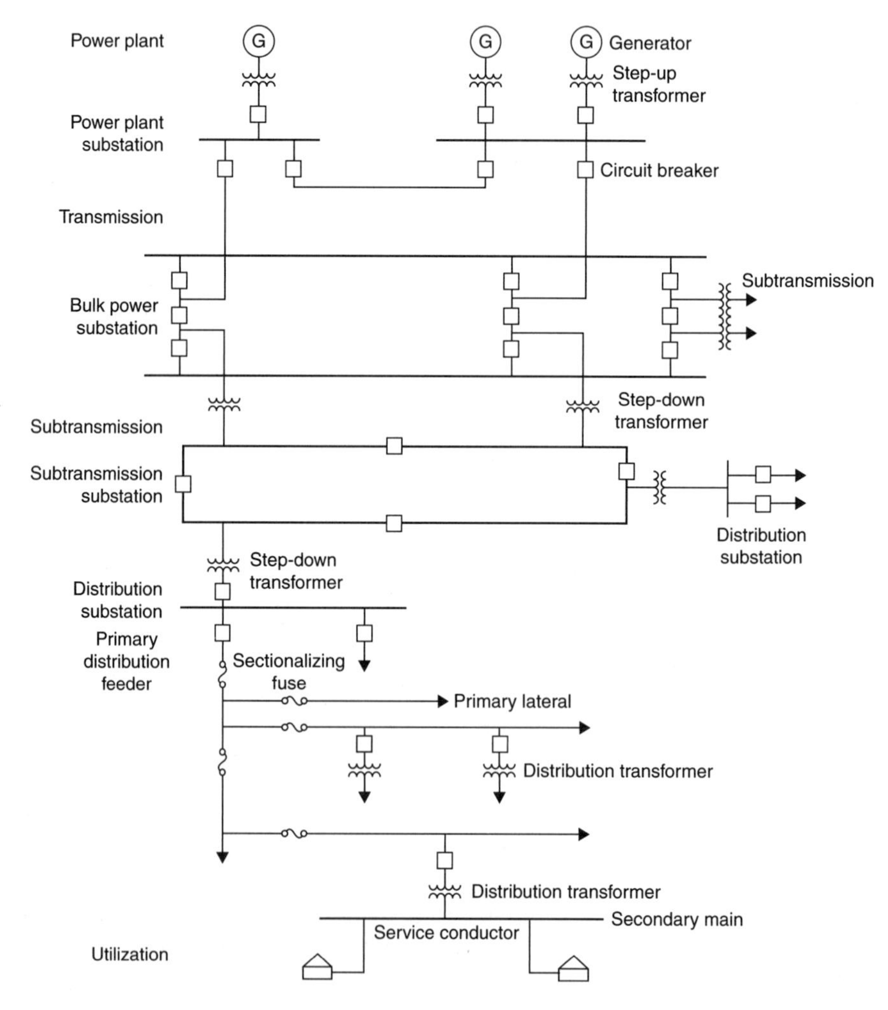

FIGURE 14.1 Basic components of an electric power system (J. D. Glover, "Electric Power Distribution," Encyclopedia of Energy Technology and The Environment, John Wiley & Sons, New York, NY, 1995. Reprinted with permission of John Wiley & Sons, Inc.)

Both the GSU transformers and the busses in these substations are protected by circuit breakers, surge arresters, and other protection equipment.

The transmission system serves three basic functions:

1. It delivers energy from generators to the system.

2. It provides for energy interchange among utilities.

3. It supplies energy to the subtransmission and distribution system.

The transmission system consists of a network of three-phase transmission lines and transmission substations, also called bulk power substations. Typical transmission voltages range from 230 up to 765 kV. Single-circuit three-phase ratings vary from 400 MVA at 230 kV up to 4000 MVA at 765 kV. In some cases, HVDC lines with solid-state converters are embedded in the transmission system as well as back-to-back ac-dc links.

The subtransmission system consists of step-down transformers, substations, and subtransmission lines that connect bulk power substations to distribution substations. In some cases, a subtransmission line may be tapped, usually through a circuit breaker, to supply a single-customer distribution load such as a large industrial plant. Typical subtransmission voltages range from 69 to 138 kV.

Distribution substations include step-down transformers (distribution substation transformers) that decrease subtransmission voltages to primary distribution voltages in the 2.2- to 46-kV range for local distribution. These transformers connect through associated circuit breaker and surge arrester protection to substation buses, which in turn connect through circuit breakers to three-phase primary distribution lines called distribution circuits or feeders. Each substation bus usually supplies several feeders. Typical distribution substation ratings vary from 15 MVA for older substations to 200 MVA or higher for newer installations. Distribution substations may also include equipment for regulating the primary voltage, such as load tap changers (LTCs) on the distribution substation transformers or separate voltage regulators.

Typical primary distribution feeder ratings include 4 MVA for 4.16 kV, 12 MVA for 13.8 kV, 20 MVA for 22.9 kV, and 30 MVA for 34.5-kV feeders. Feeders are usually segregated into several three-phase sections connected through sectionalizing fuses or switches. Each feeder section may have several single-phase laterals connected to it through fuses. Three-phase laterals may also be connected to the feeders through fuses or reclosers. Separate, dedicated primary feeders supply industrial or large commercial loads.

Feeders and laterals run along streets, as either overhead lines or underground cables, and supply distribution transformers that step the voltage down to the secondary distribution level (120 to 480 V). Distribution transformers, typically rated 5 to 5000 kVA, are installed on utility poles for overhead lines, and on pads at ground level or in vaults for underground cables. Distribution transformers are protected from overloads and faults by fuses or circuit breakers on the primary and/or the secondary side. From these transformers, energy flows through secondary mains and service conductors to supply single- or three-phase power directly to customer loads (residential, commercial, and light industrial).

Service conductors connect through meters, which determine kilowatthour consumption for customer billing purposes as well as other data for planning and operating purposes, to service panels located on customers' premises. Customers' service panels contain circuit breakers or fuses that connect to wiring that in turn supplies energy for utilization devices (lighting, appliances, motors, heating-ventilation-air conditioning, etc.).

Distribution of electric energy from distribution substations to meters at customers' premises has two parts:

1. Primary distribution, which distributes energy in the 2.2- to 46-kV range from distribution substations to distribution transformers, where the voltage is stepped down to customer utilization levels.
2. Secondary distribution, which distributes energy at customer utilization voltages of 120 to 480 V to meters at customers' premises.

14.2 PRIMARY DISTRIBUTION

Table 14.1 shows typical primary distribution voltages in the United States [1–9]. Primary voltages in the "15-kV class" predominate among U.S. utilities. The 2.5- and 5-kV classes are older primary voltages that are gradually being replaced by 15-kV class primaries. In some cases, higher 25- to 50-kV classes are used in new high-density load areas as well as in rural areas that have long feeders.

The three-phase, four-wire multigrounded primary system is the most widely used. Under balanced operating conditions, the voltage of each phase

TABLE 14.1 Typical Primary Distribution Voltages in the United States (J. D. Glover, "Electric Power Distribution," Encyclopedia of Energy Technology and The Environment, John Wiley & Sons, New York, NY, 1995. Reprinted with permission of John Wiley & Sons, Inc.)

Class, kV	Voltage, kV
2.5	2.4
5	4.16
8.66	7.2
15	12.47
	13.2
	13.8
25	22.9
	24.94
34.5	34.5
50	46

is equal in magnitude and 120° out of phase with each of the other two phases. The fourth wire in these Y-connected systems is used as a neutral for the primaries, or as a common neutral when both primaries and secondaries are present. Usually the windings of distribution substation transformers are Y-connected on the primary distribution side, with the neutral point grounded and connected to the common neutral wire. The neutral is also grounded at frequent intervals along the primary, at distribution transformers, and at customers' service entrances. Sometimes distribution substation transformers are grounded through an impedance (approximately one ohm) to limit short circuit currents and improve coordination of protective devices.

The three-wire delta primary system is also popular, although not as widely used as the four-wire multigrounded primary system. Three-wire delta primary systems are not being actively expanded. They are generally older and lower in voltage than the four-wire multigrounded type. They are also popular in industrial systems.

Rural areas with low-density loads are usually served by overhead primary lines with distribution transformers, fuses, switches, and other equipment mounted on poles. Urban areas with high-density loads are served by underground cable systems with distribution transformers and switchgear installed in underground vaults or in ground-level cabinets. There is also an increasing trend towards underground residential distribution (URD), particularly single-phase primaries serving residential areas. Underground cable systems are highly reliable and usually unaffected by weather. But the installation costs of underground distribution are significantly higher than overhead.

Primary distribution includes three basic systems:

1. Radial
2. Loop
3. Primary network systems

PRIMARY RADIAL SYSTEMS

The primary radial system, as illustrated in Figure 14.2, is a widely used, economical system often found in low-load-density areas [1, 3, 4, and 9]. It consists of separate three-phase feeder mains (or feeders) emanating from a distribution substation in a radial fashion, with each feeder serving a given geographical area. The photograph at the beginning of this chapter shows a distribution substation that supplies six radial feeders for a suburban residential area. A three-phase feeder main can be as short as a kilometer or two or as long as 30 km. Single-phase laterals (or branches) are usually connected to feeders through fuses, so that a fault on a branch can be cleared without interrupting the feeder. Single-phase laterals are connected to different phases of the feeder, so as to balance the loading on the three phases.

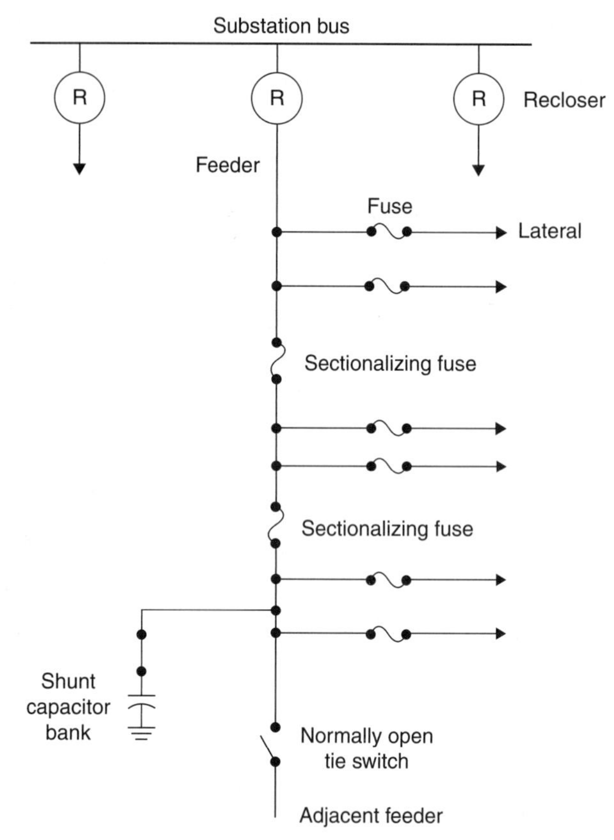

FIGURE 14.2 Primary radial system (J. D. Glover, "Electric Power Distribution," Encyclopedia of Energy Technology and The Environment, John Wiley & Sons, New York, NY, 1995. Reprinted with permission of John Wiley & Sons, Inc.)

To reduce the duration of interruptions, overhead feeders can be protected by automatic reclosing devices located at the distribution substation, at the first overhead pole, or at other locations along the feeder [11]. As an example, Figure 14.3 shows a pole-mount recloser for a 22.9-kV circuit. Studies have shown that the large majority of faults on overhead primaries are temporary, caused by lightning flashover of line insulators, momentary contact of two conductors, momentary bird or animal contact, or momentary tree limb contact. The recloser or circuit breaker with reclosing relays opens the circuit either "instantaneously" or with intentional time delay when a fault occurs, and then recloses after a short period of time. The recloser can repeat this open and reclose operation if the fault is still on the feeder. A popular reclosing sequence is two instantaneous openings (to clear temporary faults), followed by two delayed openings (allowing time for fuses to clear persistent downstream faults), followed by opening and lockout for persistent faults between the recloser and fuses. For safety purposes, the reclosing feature is bypassed during live line maintenance. Reclosing is not used on circuits that are primarily underground.

FIGURE 14.3

Pole-mount recloser for a three-phase 22.9-kV circuit. This recloser has an 800-A continuous current rating and a 16-kA interrupting rating. The 22.9-kV feeder is located near the top of the pole. There are two three-phase 4.16-kV circuits below the recloser. An antenna located below the 4.16-kV circuits is for remote control of the recloser from the dispatch center. A normally open bypass switch located on the top crossarm can be manually operated if the recloser fails to reclose (Courtesy of Danvers Electric)

To further reduce the duration and extent of customer interruptions, sectionalizing fuses are installed at selected intervals along radial feeders. In the case of a fault, one or more fuses blow to isolate the fault, and the unfaulted section upstream remains energized. In addition, normally open tie switches to adjacent feeders are incorporated, so that during emergencies, unfaulted sections of a feeder can be tied to the adjacent feeder. Spare capacity is often allocated to feeders to prevent overloads during such emergencies, or there may be enough diversity between loads on adjacent feeders to eliminate the need for spare capacity. Many utilities have also installed automatic fault locating equipment and remote controlled sectionalizers (controlled switches) at intervals along radial lines, so that faulted sections of a feeder can be isolated and unfaulted sections reenergized rapidly from a dispatch center, before the repair crew is sent out. Figure 14.4 shows a radio-controlled sectionalizing switch on a 22.9-kV circuit.

Shunt capacitor banks including fixed and switched banks are used on primary feeders to reduce voltage drop, reduce power losses, and improve power factor. Capacitors are typically switched off during the night for light loads and switched on during the day for heavy loads. Figure 14.5 shows a pole-mount switched capacitor bank. Computer programs are available to determine the number, size and location of capacitor banks that optimize voltage profile, power factor, and installation and operating costs. In some cases, voltage regulators are used on primary feeders.

FIGURE 14.4

S&C normally open radio-controlled sectionalizing switch on a 22.9-kV circuit. This switch has a 1200-A load-break capability (Courtesy of Danvers Electric)

One or more additional, independent feeders along separate routes may be provided for critical loads, such as hospitals that cannot tolerate long interruptions. Switching from the normal feeder to an alternate feeder can be done manually or automatically with circuit breakers and electrical interlocks to prevent the connection of a good feeder to a faulted feeder. Figure 14.6 shows a primary selective system, often used to supply concentrated loads over 300 kVA [3, 4, and 9]. There are two primary feeders with automatic switching in front (upstream) of the distribution transformer. In case of

FIGURE 14.5 Pole-mount three-phase 1800 kvar shunt capacitor bank for a 22.9-kV circuit. The capacitor bank, which is protected by 50-A type K fuses, is a switched bank (Courtesy of Danvers Electric)

feeder loss, automatic transfer to the other feeder is rapid and does not require fault locating before transfer.

PRIMARY LOOP SYSTEMS

The primary loop system, as illustrated in Figure 14.7 for overhead, is used where high service reliability is important [1, 3, 4 and 9]. The feeder loops around a load area and returns to the distribution substation, especially

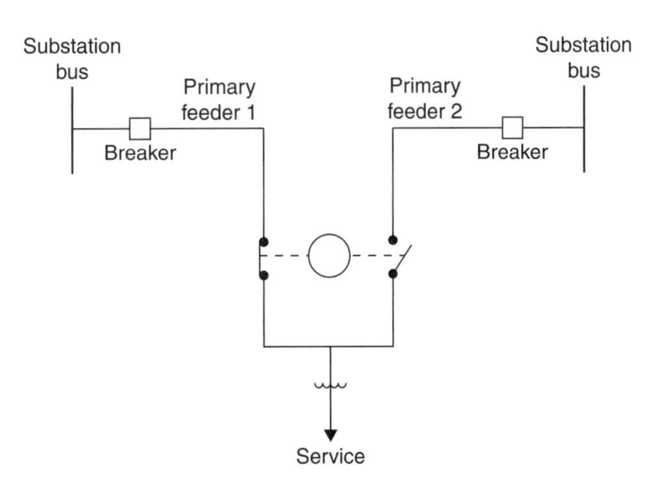

FIGURE 14.6 Primary selective system (J. D. Glover, "Electric Power Distribution," Encyclopedia of Energy Technology and The Environment, John Wiley & Sons, New York, NY, 1995. Reprinted with permission of John Wiley & Sons, Inc.)

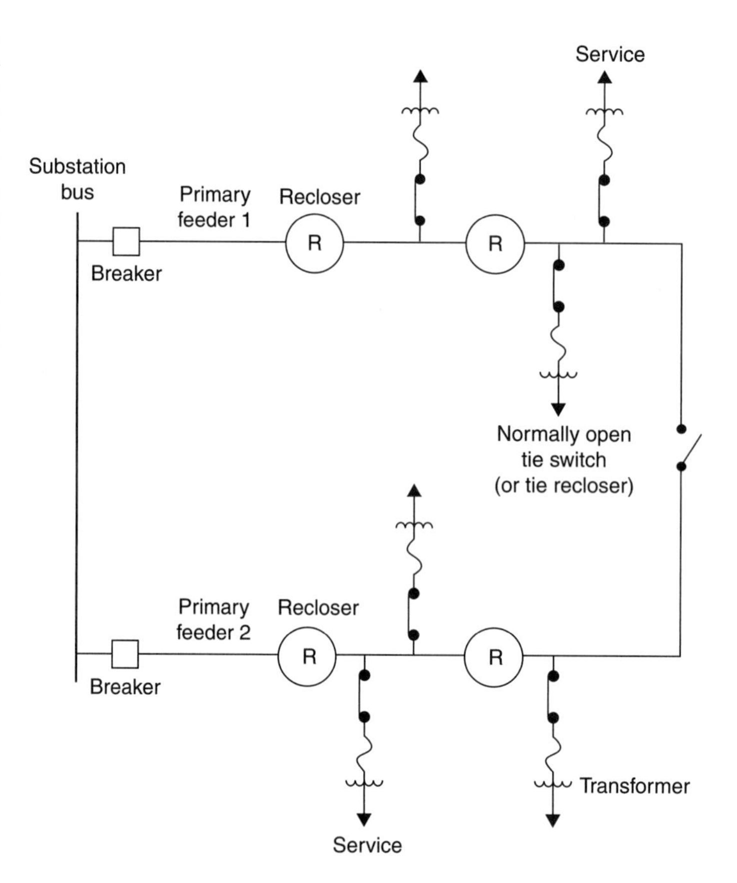

FIGURE 14.7 Overhead primary loop (J. D. Glover, "Electric Power Distribution," Encyclopedia of Energy Technology and The Environment, John Wiley & Sons, New York, NY, 1995. Reprinted with permission of John Wiley & Sons, Inc.)

providing two-way feed from the substation. The size of the feeder conductors, which are kept the same throughout the loop, is usually selected to carry the entire load connected to the loop, including future load growth. Reclosers and tie switches (sectionalizers) are used to reduce customer interruptions and isolate faulted sections of the loop. The loop is normally operated with the tie switch (or tie recloser) open. Power to a customer at any one time is supplied through a single path from the distribution substation, depending on the open/close status of the reclosers/sectionalizers. Each of the circuit breakers at the distribution substation can be connected to separate bus sections and fed from separate distribution substation transformers.

Figure 14.8 shows a typical primary loop for underground residential distribution (URD). The size of the cable, which is kept the same throughout the loop, is selected to carry the entire load, including future load growth. Underground primary feeder faults occur far less frequently than in overhead primaries, but are generally permanent. The duration of outages caused by primary feeder faults is the time to locate the fault and perform switching to isolate the fault and restore service. Fault locators at each distribution substation transformer help to reduce fault locating times.

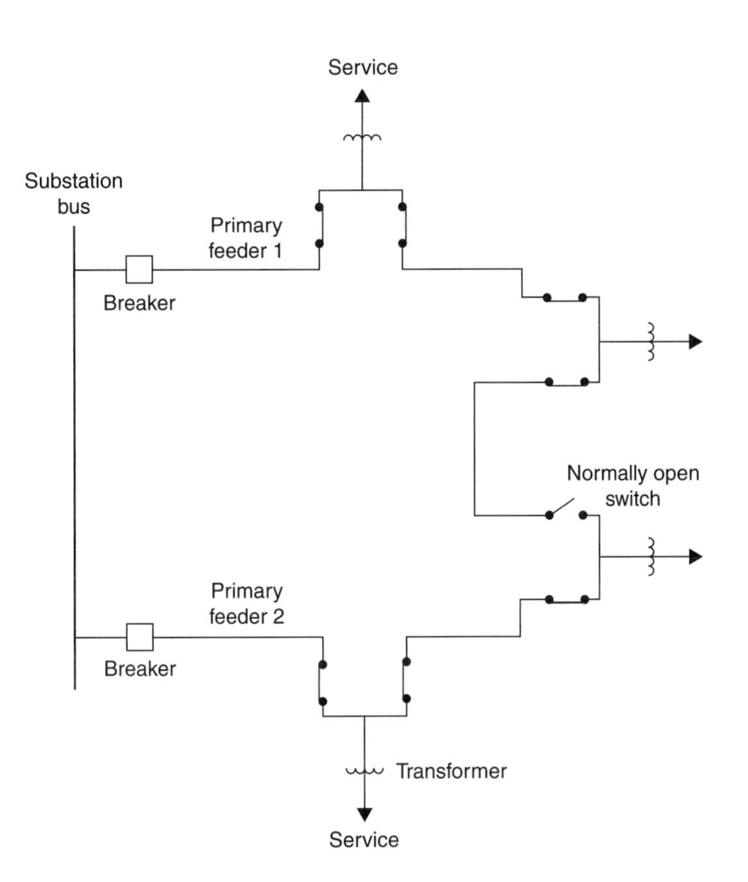

FIGURE 14.8 Underground primary loop (J. D. Glover, "Electric Power Distribution," Encyclopedia of Energy Technology and The Environment, John Wiley & Sons, New York, NY, 1995. Reprinted with permission of John Wiley & Sons, Inc.)

PRIMARY NETWORK SYSTEMS

Although the primary network system, as illustrated in Figure 14.9, provides higher service reliability and quality than a radial or loop system, only a few primary networks remain in operation in the United States today [1, 3, 4, and 9]. They are typically found in downtown areas of large cities with high load densities. The primary network consists of a grid of interconnected feeders supplied from a number of substations. Conventional distribution substations can be replaced by smaller, self-contained unit substations at selected network locations. Adequate voltage is maintained at utilization points by voltage regulators at distribution substations and by locating distribution transformers close to major load centers on the grid. However, it is difficult to maintain adequate voltage everywhere on the primary grid under various operating conditions. Faults on interconnected grid feeders are cleared by circuit breakers at distribution substations, and in some cases, by fuses on the primary grid. Radial primary feeders protected by circuit breakers or fuses can be tapped off the primary grid or connected directly at distribution substations.

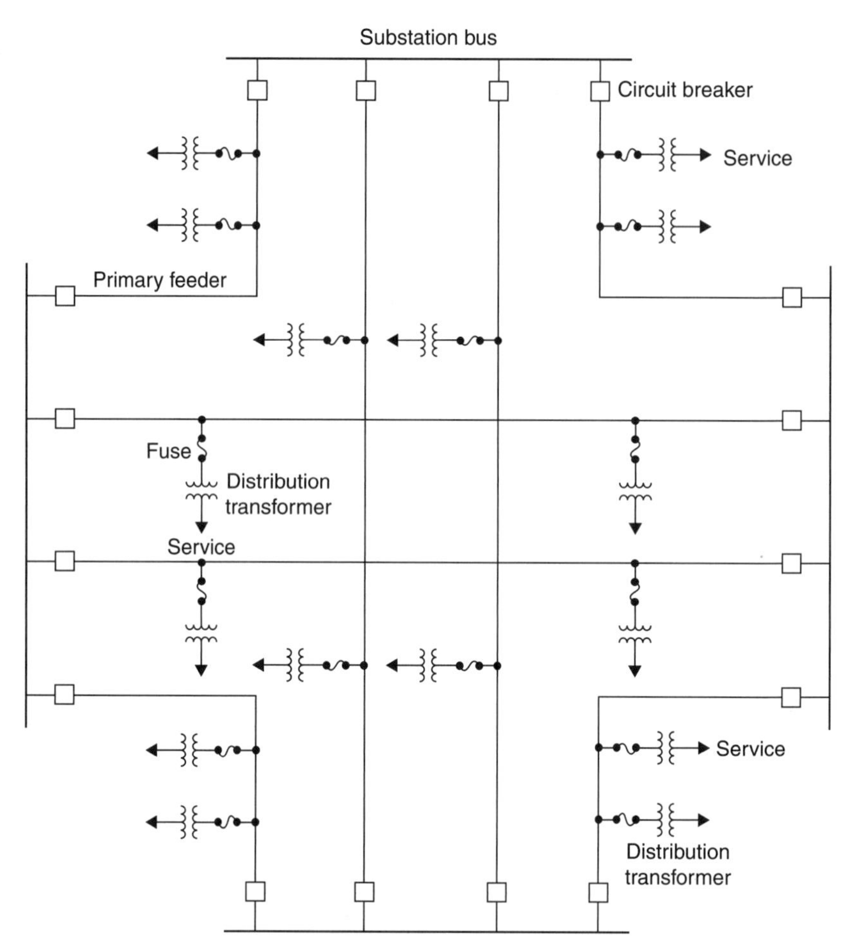

FIGURE 14.9 Primary network (J. D. Glover, "Electric Power Distribution," Encyclopedia of Energy Technology and The Environment, John Wiley & Sons, New York, NY, 1995. Reprinted with permission of John Wiley & Sons, Inc.)

14.3 SECONDARY DISTRIBUTION

Secondary distribution distributes energy at customer utilization voltages from distribution transformers up to meters at customers' premises. Table 14.2 shows typical secondary voltages and applications in the United States [1–9]. In residential areas, 120/240-V, single-phase, three-wire service is the most common, where lighting loads and outlets are supplied by 120-V, single-phase connections, and large household appliances such as electric ranges, clothes dryers, water heaters, and electric space heating are supplied by 240-V, single-phase connections. In urban areas serving high-density residential and commercial loads, 108Y/120-V, three-phase, four-wire service is common, where lighting, outlets, and small motor loads are supplied by 120-V, single-phase connections, and larger motor loads are supplied by 208-V, three-phase connections. In areas

TABLE 14.2

Typical Secondary Distribution Voltages in the United States (J. D. Glover, "Electric Power Distribution," Encyclopedia of Energy Technology and The Environment, John Wiley & Sons, New York, NY, 1995. Reprinted with permission of John Wiley & Sons, Inc.)

Voltage	# Phases	# Wires	Application
120/240 V	Single-phase	Three	Residential
208Y/120 V	Three-phase	Four	Residential/Commercial
480Y/277 V	Three-phase	Four	Commercial/Industrial/High Rise

with very high-density commercial and industrial loads as well as high-rise buildings, 480Y/277-V, three-phase, four-wire service is common, with fluorescent lighting supplied by 277-V, single-phase connections and motor loads supplied by 480-V, three-phase connections. Separate 120-V radial systems fed by small transformers from the 480-V system are used to supply outlets in various offices, retail stores, or rooms.

Figure 14.10 shows a typical residential customer voltage profile along a radial feeder. In accordance with ANSI standards, during normal conditions utilities in the United States are required to maintain customer voltage at the customer's service panel between 114 and 126 volts ($\pm 5\%$) based on a 120-V nominal secondary voltage. As shown in Figure 14.10, the first customer, closest to the substation, has the highest voltage and the last customer, furthest from the substation, has the lowest voltage. Proper distribution design dictates that the first customer's voltage is less than 126 V during light loads and the last customer's voltage is greater than 114 V during peak loads, so that all customers remain within 120 V $\pm$ 5% under all normal loading conditions. Load-tap-changing distribution substation transformers and voltage regulators (see Section 14.4) as well as shunt capacitors (see Section 14.5) are used to maintain customer voltages within ANSI limits.

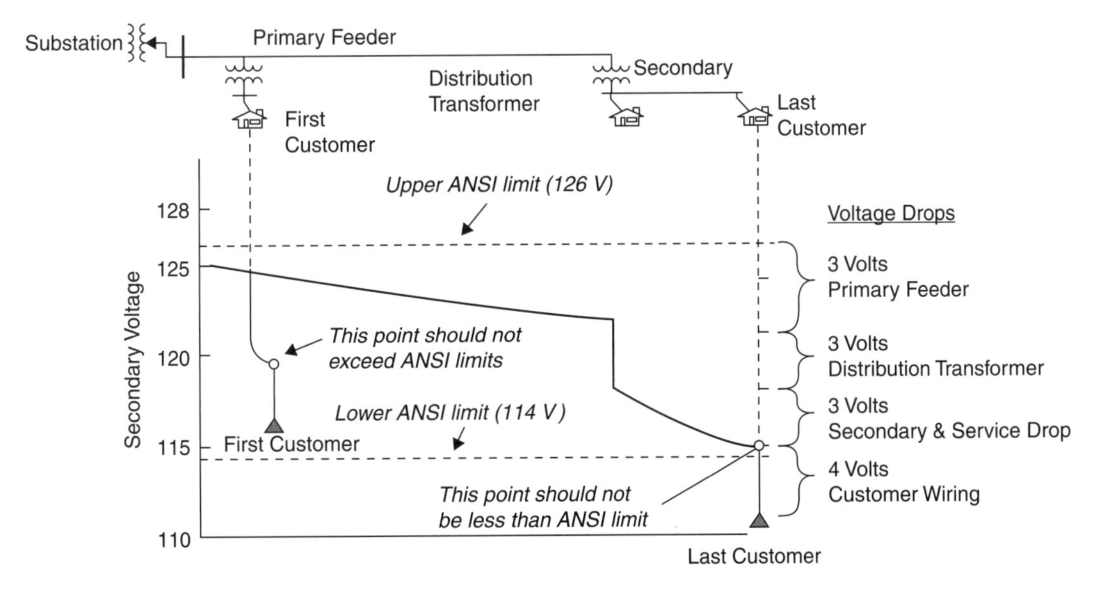

FIGURE 14.10 Typical residential customer voltage profile along a radial feeder, assuming no shunt capacitors or voltage regulators along the feeder

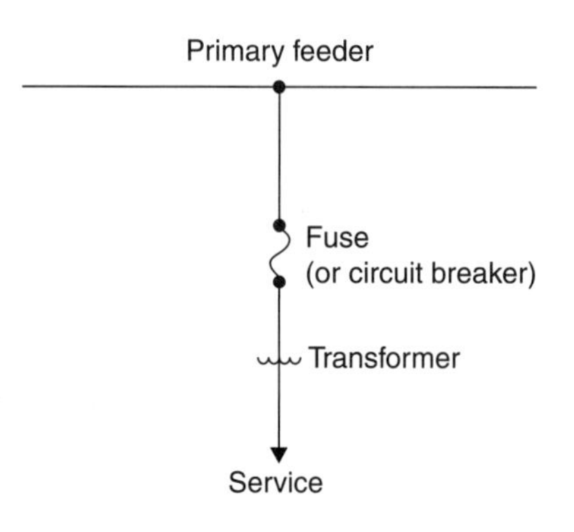

FIGURE 14.11 Individual distribution transformer supplying single-service secondary (J. D. Glover, "Electric Power Distribution," Encyclopedia of Energy Technology and The Environment, John Wiley & Sons, New York, NY, 1995. Reprinted with permission of John Wiley & Sons, Inc.)

There are four general types of secondary systems:

1. Individual distribution transformer per customer
2. Common secondary main
3. Secondary network
4. Spot network

INDIVIDUAL DISTRIBUTION TRANSFORMER PER CUSTOMER

Figure 14.11 shows an individual distribution transformer with a single service supplying one customer, which is common in rural areas where distances between customers are large and long secondary mains are impractical [3 and 4]. This type of system may also be used for a customer that has an unusually large load or for a customer that would otherwise have a low-voltage problem with a common secondary main. Although transformer installation costs and operating costs due to no-load losses are higher than those of other types of secondary systems, the installation costs of secondary mains are avoided.

COMMON SECONDARY MAIN

Figure 14.12 shows a primary feeder connected through one or more distribution transformers to a common secondary main with multiple services to a group of customers [3 and 4]. This type of secondary system takes advantage of diversity among customer demands that allows a smaller capacity of the transformer supplying a group compared to the sum of the capacities of individual transformers for each customer in the group. Also, the large transformer supplying a group can handle motor staring currents and other abrupt, load changes without severe voltage drops.

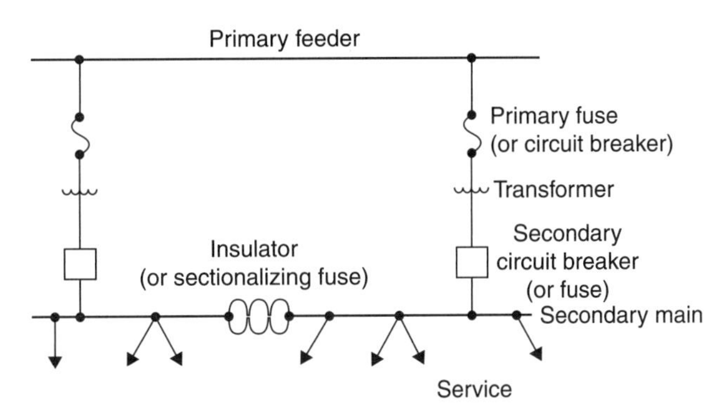

FIGURE 14.12 Common secondary main (J. D. Glover, "Electric Power Distribution," Encyclopedia of Energy Technology and The Environment, John Wiley & Sons, New York, NY, 1995. Reprinted with permission of John Wiley & Sons, Inc.)

In most cases, the common secondary main is divided into sections, where each section is fed by one distribution transformer and is also isolated from adjacent sections by insulators. In some cases, fuses are installed along a continuous secondary main, which results in banking of distribution transformers, also called banked secondaries.

SECONDARY NETWORK

Figure 14.13 shows a secondary network or secondary grid, which may be used to supply high-density load areas in downtown sections of cities, where

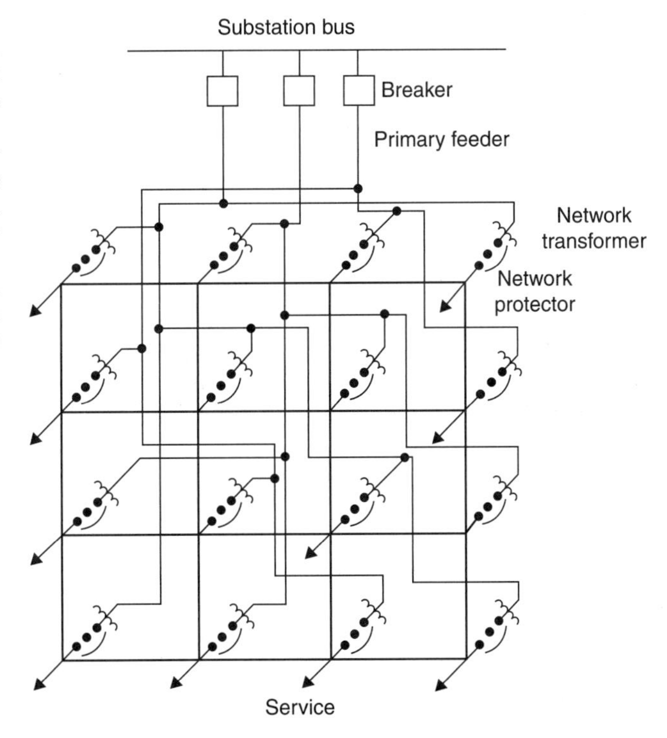

FIGURE 14.13 Secondary network (J. D. Glover, "Electric Power Distribution," Encyclopedia of Energy Technology and The Environment, John Wiley & Sons, New York, NY, 1995. Reprinted with permission of John Wiley & Sons, Inc.)

the highest degree of reliability is required and revenues justify grid costs [1, 3, 4, and 9]. The underground secondary network is supplied simultaneously by two or more primary feeders through network transformers. Most networks are supplied by three or more primary feeders with transformers that have spare capacity, so that the network can operate with two feeders out of service.

Secondary grids operate at either 208Y/120 or 480Y/277-V in the United States. Commonly used secondary cable sizes range from 4/0 to 500 kcmil (250 mm^2) AWG [5].

More than 260 cities in the United States have secondary networks [5]. New York City has the largest secondary network system in the United States with approximately 23,000 network transformers feeding various secondary networks and an online monitoring system that continuously monitors transformer loadings. Some of the secondary networks in New York City are fed by as many as 24 primary feeders operating in parallel [9].

Network transformers are protected by network protectors between the transformers and secondary mains. A network protector is an electrically operated low-voltage air circuit breaker with relays and auxiliary devices that automatically opens to disconnect the transformer from the network when the transformer or the primary feeder is faulted, or when there is a power flow reversal. The network protector also has the ability to close automatically when a feeder is energized [5]. Fuses may also be used for backup of network protectors.

In many cases especially on 208Y/120-V secondary networks, main protection of secondary cables has come from the ability of the cable system to "burn clear" with no fuse or other protective device. However, in many instances for 480Y/277-V secondary networks, this practice was not able to successfully burn clear, resulting in fires and considerable damage. As a solution, special fuses called cable limiters are commonly used at tie points in the secondary network to isolate faulted secondary cables. Cable limiters, which are designed with restricted sections of copper which act like a fuse, do not limit the magnitude of fault current like current limiting fuses. In high short circuit locations on the secondary network, current limiting fuses may be used instead of cable limiters.

In secondary network systems, a forced or scheduled outage of a primary feeder does not result in customer outages. Because the secondary mains provide parallel paths to customer loads, secondary cable failures usually do not result in customer outages either. Also, each network is designed to share the load equally among transformers and to handle large motor starting and other abrupt load changes without severe voltage drops.

SPOT NETWORK

Figure 14.14 shows a spot network consisting of a secondary network supplying a single, concentrated load such as a high-rise building or shopping center, where a high degree of reliability is required [1, 3, 4, and 9]. The secondary

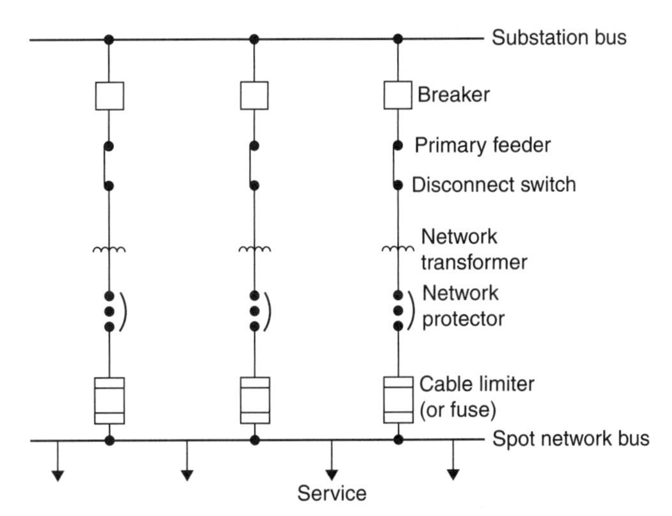

FIGURE 14.14 Secondary spot network (J. D. Glover, "Electric Power Distribution," Encyclopedia of Energy Technology and The Environment, John Wiley & Sons, New York, NY, 1995. Reprinted with permission of John Wiley & Sons, Inc.)

spot network bus is supplied simultaneously by two or more primary feeders through network transformers. In some cases, a spot network load as large as 25 MVA may be fed by up to six primary feeders. Most all spot networks in the United States operate at a 480Y/277-V secondary voltage [5]. Separate 120-V radial systems fed by small transformers from the 480-V system are used to supply outlets in various offices, retail stores, or rooms.

High service reliability and operating flexibility are achieved with a spot network fed by two or more primary feeders through network transformers. The secondary bus is continuously energized by all network transformers. Network protectors are used to automatically disconnect transformers from the spot network bus for transformer/feeder faults or for power-flow reversal, and cable limiters or fuses are used to protect against overloads and faults on secondary cables. Scheduled or forced outages of primary feeders occur without customer interruption or involvement. Spot networks also provide a very compact and reliable arrangement of components [5].

14.4 TRANSFORMERS IN DISTRIBUTION SYSTEMS

Transformers in distribution systems include distribution substation transformers and distribution transformers.

DISTRIBUTION SUBSTATION TRANSFORMERS

Distribution substation transformers come in a wide variety of ratings. Some of the typical characteristics of distribution substation transformers are given in Table 14.3.

TABLE 14.3

Typical Characteristics of Distribution Substation Transformers [5] (Power distribution engineering: fundamentals and applications by Burke. Copyright 1994 by TAYLOR & FRANCIS GROUP LLC - BOOKS. Reproduced with permission of TAYLOR & FRANCIS GROUP LLC - BOOKS in the format Textbook via Copyright Clearance Center)

Class, kV	Voltage, kV
Rating of High Voltage Winding	34.5 to 230 kV
Rating of Low Voltage Winding	2.4 to 46 kV
MVA Rating (OA)	2.5 to 75 MVA
Transformer Impedance	5 to 12 %
Number of Transformers in Substation	1 to 4
Loading	OA, OA/FA, OA/FA/FOA, OA/FA/FA
High Side Protection	Circuit Switches, Circuit Breakers, Fuses
Relay Protection	Overcurrent, Differential, Under-Frequency
Feeder Protection	Circuit Breakers, Reclosers

Distribution substation transformers usually contain mineral oil for insulating and cooling purposes (older transformers manufactured prior to 1978 originally contained askarels with high PCB content, but many of these have either been retired or re-classified as non-PCB transformers using perchloroethylene). In some units, an inert gas such as nitrogen fills the space above the oil, in order to keep moisture and air out of the oil, and the transformer tank is sealed. Some sealed transformers have a pressure relief diaphragm that is designed to rupture when the internal pressure exceeds a specified value, indicating possible deterioration of the insulation. Sealed transformers may also have a sudden pressure relay to either provide an alarm or de-energize the transformer when the internal pressure suddenly increases above a specified threshold [4].

Many distribution substation transformers have load tap changers (LTCs) that automatically regulate voltage levels based on loading conditions. Figure 14.15 shows a distribution substation transformer that has an internal LTC on the low-voltage side. Some distribution substations have distribution substation transformers with fixed taps and separate voltage regulators. A voltage regulator is basically an autotransformer with taps that automatically raise or lower voltage, operating in a similar way as LTCs on distribution substation transformers. Figure 14.16 shows a voltage regulator at a distribution substation. In addition to voltage regulators for distribution substations, there are also pole-mount voltage regulators that can be placed on feeders.

Some outdoor distribution substation transformers are equipped with a tank on the top of the transformer called a "conservator," in which expansion and contraction of the oil takes place. Condensation of moisture and formation of sludge occur within the conservator, which is also provided with a sump pump to draw off the moisture and sludge [4].

Distribution substation transformers have MVA ratings that indicate the continuous load that the transformers carry without exceeding a specified temperature rise of either 55°C (for older transformers) or 65°C (for newer transformers) above a specified ambient (typically 40°C). Also, distribution substation transformers are typically equipped with external radiators with fans and/or oil circulating pumps, in order to dissipate heat generated by

FIGURE 14.15 Three-phase 22.9 kVΔ/4.16 kVY distribution substation transformer rated 12 MVA OA/16 MVA FA1/20 MVA FA2. The transformer has fixed taps on the high-voltage side and an LTC on the low-voltage side (Courtesy of Danvers Electric)

copper and core losses. These transformers have multiple MVA ratings that include the following:

1. OA rating (passive convection with oil circulating pumps and fans off).
2. FA rating (with fans on but oil circulating pumps off).
3. FOA rating (with both fans and oil circulating pumps on). Some units, such as the one shown in Figure 14.15, may have two FA ratings, a

FIGURE 14.16 Voltage regulators at the 69/13.8 kV Lunenburg distribution substation, Lunenburg MA. The regulators are General Electric Type VR1 with ratings of 7.96 kV (line-to neutral), 437 A (Courtesy of Unitil Corporation)

lower FA rating with one of two sets of fans on, and a higher FA rating with both sets of fans on. Also, some units have water-cooled heat exchangers. The nameplate transformer impedance is usually given in percent using the OA rating as the base MVA [5].

EXAMPLE 14.1 Distribution Substation Transformer Rated Current and Short Circuit Current

A three-phase 230 kVΔ/34.5 kV Y distribution substation transformer rated 75 MVA OA/100 MVA FA/133 MVA FOA has a 7% impedance. (a) Determine the rated current on the low-voltage side of the transformer at its OA, FA, and FOA ratings. (b) Determine the per unit transformer impedance using a system base of 100 MVA and 34.5 kV on the low-voltage side of the transformer. (c) Calculate the short-circuit current on the low-voltage side of the transformer for a three-phase bolted fault on the low-voltage side. Assume that the prefault voltage is 34.5 kV.

SOLUTION

a. At the OA rating of 75 MVA,

$$I_{OA,L} = 75/(\sqrt{3} \times 34.5) = 7.372 \text{ kA per phase}$$

Similarly,

$$I_{FA,L} = 100/(\sqrt{3} \times 34.5) = 9.829 \text{ kA per phase}$$

$$I_{FOA,L} = 133/(\sqrt{3} \times 34.5) = 13.07 \text{ kA per phase}$$

b. The transformer impedance is 7% or 0.07 per unit based on the OA rating of 75 MVA. Using (3.3.11), the transformer per unit impedance on a 100 MVA system base is:

$$Z_{\text{puSystem Base}} = 0.07(100/75) = 0.09333 \text{ per unit}$$

c. For a three-phase bolted fault, using the transformer ratings as the base quantities,

$$\begin{aligned} Isc3\varphi &= 1.0/(0.07) = 14.286 \text{ per unit} \\ &= (14.286)(7.372) \\ &= 105.31 \text{ kA/phase} \end{aligned}$$

Note that in (c) above, the OA rating is used to calculate the short-circuit current, because the transformer manufacturer gives the per unit transformer impedance using the OA rating as the base quantity. ■

Most utilities have a planning and operating policy of loading distribution substation transformers within their nameplate OA/FA/FOA ratings during normal conditions, but possibly above their nameplate ratings during short-term emergency conditions. If one transformer has a scheduled or forced outage, the remaining transformer or transformers can continuously carry the entire substation load.

Typically there are two emergency loading criteria for distribution substation transformers:

1. A two-hour emergency rating, which gives time to perform switching operations and reduce loadings.
2. A longer-duration emergency rating (10 to 30 days), which gives time to replace a failed transformer with a spare that is in stock.

As one example, the distribution substation shown in the photograph at the beginning of this chapter has two transformers rated 9 MVA OA/12 MVA FA1/15 MVA FA2. The practice of the utility that owns the substation is to normally operate the substation at or below 15 MVA. As such, if there is a forced or scheduled outage of one transformer, the other transformer can supply all six 4.16-kV feeders without being loaded above its FA2 nameplate of 15 MVA. For this conservative operating practice, emergency transformer ratings above nameplate are not used.

Some utilities operate their distribution substation transformers above nameplate ratings during normal operating conditions, as well as during emergency conditions. ANSI/IEEE C-57.91-1995 entitled, *IEEE Guide for Loading Mineral-Oil-Immersed Transformers* identifies the risks of transformer loads in excess of nameplate rating and establishes limitations and guidelines, the application of which are intended minimize the risks to an acceptable limit [21, 22].

EXAMPLE 14.2 Distribution Substation Normal, Emergency, and Allowable Ratings

As shown in Figure 14.17, a distribution substation is served by two 138-kV sub-transmission lines, each connected to a 40 MVA (FOA nameplate rating) 138 kVΔ/12.5 kV Y distribution substation transformer, denoted TR1 and TR2. Both TR1 and TR2 are relatively new transformers with insulation systems designed for 65°C temperature rises under continuous loading conditions. Shunt capacitor banks are also installed at 12.5-kV bus 1 and bus 2. The utility that owns this substation has the following transformer loading criteria based on a percentage of nameplate rating:

1. 128% for normal summer loading.
2. 170% during a two-hour summer emergency.
3. 155% during a 30-day summer emergency.

(a) Assuming a 5% reduction for unequal transformer loadings, determine the summer "normal" rating of the substation. (b) Determine the "allowable" summer rating of the substation under the single-contingency loss of one transformer. (c) Determine the 30-day summer emergency rating of the substation under the single-contingency loss of one transformer.

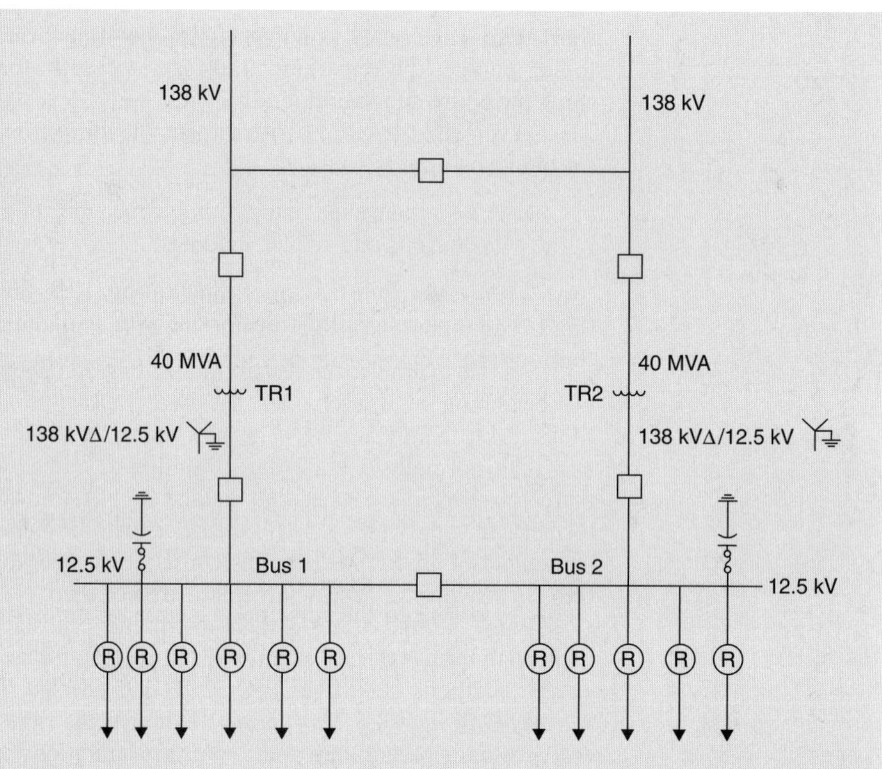

FIGURE 14.17 Distribution substation for Example 14.2 (J. D. Glover, "Electric Power Distribution," Encyclopedia of Energy Technology and The Environment, John Wiley & Sons, New York, NY, 1995. Reprinted with permission of John Wiley & Sons, Inc.)

SOLUTION

a. During normal operations, both transformers are in service. Using a 5% reduction to account for unequal transformer loadings, the summer normal substation rating is $1.28 \times (40 + 40) \times 0.95 = 97$ MVA. With both transformers in service, the substation can operate as high as 97 MVA without exceeding the summer normal rating of 128% or 51.2 MVA for each transformer.

b. The summer allowable substation rating, based on the single-contingency loss of one transformer, is $1.7 \times 40 = 68$ MVA. The transformer that remains in service is allowed to operate at 170% of its nameplate rating (68 MVA) for two hours, which gives time to perform switching operations to reduce the transformer loading to its 30-day summer emergency rating. Note that, even though the normal summer substation rating is 97 MVA, it is only allowed to operate up to 68 MVA, so that a transformer will not exceed its two-hour emergency rating in case the other transformer has an outage.

c. The 30-day summer emergency rating of the substation is $1.55 \times 40 = 62$ MVA. When one transformer has a permanent failure, the other can operate at 62 MVA for 30 days, which gives time to replace the failed transformer with a spare that is in stock. ■

DISTRIBUTION TRANSFORMERS

Distribution transformers connect the primary system (2.4 to 46 kV) to the secondary system (480 V and lower). Distribution transformers may be installed outdoors on overhead poles (pole-mount), outdoors at ground level on pads (padmount transformers), indoors within buildings, or underground in manholes and vaults.

Pole-mount transformers for overhead distribution are liquid-filled transformers that can be either single-phase or three-phase, depending on the load requirements and the primary supply configuration. Pole-mount distribution transformers may be manufactured as conventional transformers with no integral surge protection, overload protection, or short circuit-protection, or alternatively as completely self-protected (CSP) transformers.

For conventional pole-mount transformers, the protective devices are mounted external to the transformer. Typically a fuse cutout, which is a combination of a fuse and a switch, is installed adjacent to the conventional distribution transformer to disconnect it from the primary under overload conditions or an internal transformer failure. Similarly, a surge arrester is installed adjacent to the conventional transformer primary to protect it against transient overvoltages due to switching and lightning surges. Figure 14.18 shows three conventional single-phase pole-mount distribution transformers

FIGURE 14.18 Three conventional single-phase pole-mount 25-kVA transformers. The transformers are wired to form a three-phase bank rated 75 kVA, 4160VΔ–208/120 V grounded Y, which supplies secondary service for commercial customers. The transformers are supplied from a 4160-V primary through fused cutouts, with surge arresters mounted vertically on the sides of the transformer tanks (Courtesy of Danvers Electric)

that are wired to form a three-phase bank rated 75 kVA supplying 120/208 V overhead secondary service for commercial customers.

For CSP transformers, a primary fuse is located within the transformer tank. The surge arrester is mounted outside the tank, but connected to the primary bushing. Circuit breakers on the secondary side of CSP transformers provide protection from overloads and are coordinated with primary fuses.

Padmount transformers for underground distribution are liquid-filled or dry-type transformers that can be either single-phase or three-phase, outdoors or indoors. Single-phase padmount distribution transformers are typically designed for underground residential and commercial distribution systems where safety, reliability and aesthetics are especially important. Three-phase padmount distribution transformers are compact power centers usually for large commercial or industrial applications. Figure 14.19 shows a three-phase liquid-filled padmount transformer that supplies 480/277 V underground secondary service to an industrial plant. Dry-type padmount distribution transformers, whose insulation is solid (for example glass, silica, epoxy, or polyester resins) are primarily used where safety is a major concern, in close proximity to people such as at schools, hospitals, commercial buildings, and industrial plants, both indoors and outdoors.

Network transformers are large (300-to-2,500 kVA) liquid-filled, three-phase distribution transformers that are designed for use in underground vaults or in specially designed rooms within buildings to supply power to either secondary networks or spot networks. Their voltage ratings vary from 4.16-to-34.5 kVΔ or grounded Y for the high-voltage windings, and either 216 grounded Y/125 V or 480 grounded Y/277 V for the low-voltage windings. Network transformers are designed to be connected through network protectors that are integrally mounted on the transformer. Figure 14.20 on page 794 shows a network transformer from utility stock. Network transformers are built as either "vault type" (suitable for occasional submerged operation) or "subway type" (suitable for continuous submerged operation).

Table 14.4 on page 794 shows typical kVA ratings of distribution transformers. The kVA ratings of distribution transformers are based on the continuous load the transformers can carry without exceeding a specified temperature rise of either 55°C (for older transformers) or 65°C (for newer transformers above a specified ambient temperature (usually 40°C). When in service, distribution transformers are rarely loaded continuously at their rated kVA as they go through a daily load cycle. Oil-filled distribution transformers have a relatively long thermal time constant; that is, the load temperature rises slowly during load increases. As such, it is possible to load these transformers above their kVA ratings without compromising the life expectancy of the transformer. ANSI/EEE Std. C57.92-1981 is entitled IEEE Guide for Loading Mineral-Oil-Immersed Overhead and Pad-Mounted Distribution Transformers Rated 500 kVA and Less with 65°C or 55°C Average Winding Rise [23]. Table 14.5 on page 795 shows a typical loading guide, based on this standard.

(a)

(b)

FIGURE 14.19 Three-phase oil-filled padmount transformer shown with doors closed (a) and open (b). This padmount, rated 1,500 kVA OA, kVΔ–480/277 V grounded Y with internal fuses on the high-voltage side, supplies secondary service to an industrial plant (Courtesy of Danvers Electric)

FIGURE 14.20

General Electric 500 kVA, 13.8 kV delta-120/208V grounded Y network transformer from utility stock (Courtesy of Unitil Corporation)

Note that in accordance with Table 14.5, short-time loadings can be as high as 89% above the nameplate kVA rating for short durations. Also note that dry-type distribution transformers, which are not considered as rugged as liquid-filled units of the same rating, are not normally loaded above their kVA ratings.

TABLE 14.4

Standard Distribution Transformer kVA Ratings (J. D. Glover, "Electric Power Distribution," Encyclopedia of Energy Technology and The Environment, John Wiley & Sons, New York, NY, 1995. Reprinted with permission of John Wiley & Sons, Inc.)

Single-Phase	Three-Phase
kVA	kVA
5	30
10	45
15	75
25	112.5
38	150
50	225
75	300
100	500
167	750
250	1,000
333	1,500
500	2,500
	3,000
	3,750
	5,000

TABLE 14.5 Permissible Daily Short-Time Loading of Liquid-Filled Distribution Transformers Based on Normal Life Expectancy [5]

	Average Initial Load in Per Unit of Transformer Rating		
Period of Increased Loading, Hours	0.9	0.7	0.5
	Maximum Load in Per Unit of Transformer Rating		
0.5	1.59	1.77	1.89
1.0	1.40	1.54	1.60
2.0	1.24	1.33	1.37
4.0	1.12	1.17	1.19
8.0	1.06	1.08	1.08

14.5 SHUNT CAPACITORS IN DISTRIBUTION SYSTEMS

Loads in electric power systems consume real power (MW) and reactive power (Mvar). At power plants, many of which are located at long distances from load centers, real power is generated and reactive power may either be generated, such as during heavy load periods, or absorbed as during light load periods. Unlike real power (MW), the generation of reactive power (Mvar) at power plants and transmission of the reactive power over long distances to loads is not economically feasible. Shunt capacitors, however, are widely used in primary distribution to supply reactive power to loads. They draw leading currents that offset the lagging component of currents in inductive loads. Shunt capacitors provide an economical supply of reactive power to meet reactive power requirements of loads as well as transmission and distribution lines operating at lagging power factor. They can also reduce line losses and improve voltage regulation.

EXAMPLE 14.3 Shunt Capacitor Bank at End of Primary Feeder

Figure 14.21 shows a single-line diagram of a 13.8-kV primary feeder supplying power to a load at the end of the feeder. A shunt capacitor bank is located at the load bus. Assume that the voltage at the sending end of the feeder is 5% above rated and that the load is Y-connected with $R_{Load} = 20\ \Omega$/phase in parallel with load $jX_{Load} = j40\ \Omega$/phase. (a) With the shunt capacitor bank out of service, calculate the following: (1) line current; (2) voltage drop across the line; (3) load voltage; (4) real and reactive power delivered to the load; (5) load power factor; (6) real and reactive line losses; and (7) real power, reactive power, and apparent power delivered by the distribution substation. (b) The capacitor bank is Y connected with a reactance $X_C = 40\ \Omega$/phase. With the shunt

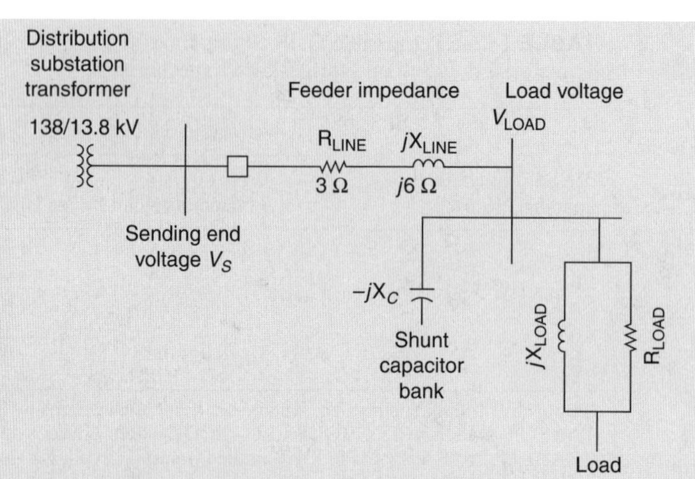

FIGURE 14.21 Single-line diagram of a primary feeder for Example 14.3 (J. D. Glover, "Electric Power Distribution," Encyclopedia of Energy Technology and The Environment, John Wiley & Sons, New York, NY, 1995. Reprinted with permission of John Wiley & Sons, Inc.)

capacitor bank in service, redo the calculations. Also calculate the reactive power supplied by the capacitor bank. (c) Compare the results of (a) and (b).

SOLUTION

a. Without the capacitor bank, the total impedance seen by the source is:

$$Z_{TOTAL} = R_{LINE} + jX_{LINE} + \frac{1}{\frac{1}{R_{LOAD}} + \frac{1}{jX_{LOAD}}}$$

$$Z_{TOTAL} = 3 + j6 + \frac{1}{\frac{1}{20} + \frac{1}{j40}}$$

$$Z_{TOTAL} = 3 + j6 + \frac{1}{0.0559\underline{/-26.57^\circ}}$$

$$= 3 + j6 + 17.89\underline{/26.56^\circ}$$

$$Z_{TOTAL} = 3 + j6 + 16 + j8 = 19 + j14$$

$$= 23.60\underline{/36.38^\circ}\ \Omega/\text{phase}$$

1. The line current is:

$$I_{LINE} = V_{SLN}/Z_{TOTAL} = \frac{1.05(13.8/\sqrt{3})\underline{/0^\circ}}{23.60\underline{/36.38^\circ}}$$

$$= 0.3545\underline{/-36.38^\circ}\ \text{kA/phase}$$

2. The voltage drop across the line is:

$$V_{DROP} = Z_{LINE}\, I_{LINE} = (3 + j6)(0.3545\underline{/-36.38^\circ})$$

$$= (6.708\underline{/63.43^\circ})(0.3545\underline{/-36.38^\circ})$$

$$= 2.378\underline{/27.05^\circ}\ \text{kV}$$

$$|V_{DROP}| = 2.378\ \text{kV}$$

3. The load voltage is:

$$V_{\text{LOAD}} = V_{\text{SLN}} - Z_{\text{LINE}} I_{\text{LINE}} = 1.05(13.8/\sqrt{3})\underline{/0^\circ} - 2.378\underline{/27.05^\circ}$$
$$= 8.366 - (2.117 + j1.081) = 6.249 - j1.081$$
$$= 6.342\underline{/-9.814^\circ}\ \text{kV}_{LN}$$
$$|V_{\text{LOAD}}| = 6.342\sqrt{3} = 10.98\ \text{kV}_{\text{LL}}$$

4. The real and reactive power delivered to the three-phase load is:

$$\text{P}_{\text{LOAD}3\varphi} = 3(\text{V}_{\text{LOADLN}})^2/\text{R}_{\text{LOAD}} = 3(6.342)^2/20 = 6.033\ \text{MW}$$
$$\text{Q}_{\text{LOAD}3\varphi} = 3(\text{V}_{\text{LOADLN}})^2/\text{X}_{\text{LOAD}} = 3(6.342)^2/40 = 3.017\ \text{Mvar}$$

5. The load power factor is:

$$\text{p.f.} = \cos[\tan^{-1}(\text{Q}/\text{P})]$$
$$= \cos[\tan^{-1}(3.017/6.033)]$$
$$= 0.89\ \text{lagging}$$

6. The real and reactive line losses are:

$$\text{P}_{\text{LINELOSS}3\varphi} = 3\ \text{I}_{\text{LINE}}{}^2\text{R}_{\text{LINE}} = 3(0.3545)^2(3) = 1.131\ \text{MW}$$
$$\text{Q}_{\text{LINELOSS}3\varphi} = 3\ \text{I}_{\text{LINE}}{}^2\text{X}_{\text{LINE}} = 3(0.3545)^2(6) = 2.262\ \text{Mvar}$$

7. The real power, reactive power, and apparent power delivered by the distribution substation are:

$$\text{P}_{\text{SOURCE}3\varphi} = \text{P}_{\text{LOAD}3\varphi} + \text{P}_{\text{LINELOSS}3\varphi} = 6.033 + 1.131 = 7.164\ \text{MW}$$
$$\text{Q}_{\text{SOURCE}3\varphi} = \text{Q}_{\text{LOAD}3\varphi} + \text{Q}_{\text{LINELOSS}3\varphi} = 3.017 + 2.262 = 5.279\ \text{Mvar}$$
$$\text{S}_{\text{SOURCE}3\varphi} = \sqrt{(7.164^2 + 5.279^2)} = 8.899\ \text{MVA}$$

b. With the capacitor bank in service, the total impedance seen by the source is:

$$Z_{\text{TOTAL}} = \text{R}_{\text{LINE}} + j\text{X}_{\text{LINE}} + \frac{1}{\frac{1}{\text{R}_{\text{LOAD}}} + \frac{1}{j\text{X}_{\text{LOAD}}} - \frac{1}{j\text{X}_C}}$$
$$Z_{\text{TOTAL}} = 3 + j6 + \frac{1}{\frac{1}{20} + \frac{1}{j40} - \frac{1}{j40}}$$
$$Z_{\text{TOTAL}} = 3 + j6 + \frac{1}{0.05} = 23 + j6 = 23.77\underline{/14.62^\circ}\ \Omega/\text{phase}$$

1. The line current is:

$$I_{\text{LINE}} = V_{\text{SLN}}/Z_{\text{TOTAL}} = \frac{1.05(13.8/\sqrt{3})/0^\circ}{23.77\underline{/14.62^\circ}}$$
$$= 0.3520\underline{/-14.62^\circ}\ \text{kA/phase}$$

2. The voltage drop across the line is:

$$V_{DROP} = Z_{LINE}\, I_{LINE} = (6.708\angle 63.43^\circ)(0.3520\angle -14.62^\circ)$$
$$= 2.361\angle 48.81^\circ \text{ kV}$$
$$|V_{DROP}| = 2.361 \text{ kV}$$

3. The load voltage is:

$$V_{LOAD} = V_{SLN} - Z_{LINE}\, I_{LINE}$$
$$= 1.05(13.8/\sqrt{3})\angle 0^\circ - 2.361\angle 48.81^\circ$$
$$= 8.366 - (1.555 + j1.778)$$
$$= 6.81 - j1.778$$
$$= 7.038\angle -14.62^\circ \text{ kV}_{LN}$$
$$|V_{LOAD}| = 7.038\sqrt{3} = 12.19 \text{ kV}_{LL}$$

4. The real and reactive power delivered to the three-phase load is:

$$P_{LOAD3\varphi} = 3(V_{LOADLN})^2/R_{LOAD} = 3(7.038)^2/20 = 7.430 \text{ MW}$$

$$Q_{LOAD3\varphi} = 3(V_{LOADLN})^2/X_{LOAD} = 3(7.038)^2/40 = 3.715 \text{ Mvar}$$

5. The load power factor is:

$$\text{p.f.} = \cos[\tan^{-1}(Q/P)]$$
$$= \cos[\tan^{-1}(3.715/7.430)]$$
$$= 0.89 \text{ lagging}$$

6. The real and reactive line losses are:

$$P_{LINELOSS3\varphi} = 3\ I_{LINE}^2\ R_{LINE} = 3(0.3520)^2(3) = 1.115 \text{ MW}$$

$$Q_{LINELOSS3\varphi} = 3\ I_{LINE}^2\ X_{LINE} = 3(0.3520)^2(6) = 2.230 \text{ Mvar}$$

7. The reactive power delivered by the shunt capacitor bank is:

$$Q_C = 3(V_{LOADLN})^2/X_C = 3(7.038)^2/40 = 3.715 \text{ Mvar}$$

8. The real power, reactive power, and apparent power delivered by the distribution substation are:

$$P_{SOURCE3\varphi} = P_{LOAD3\varphi} + P_{LINELOSS3\varphi} = 7.430 + 1.115 = 8.545 \text{ MW}$$

$$Q_{SOURCE3\varphi} = Q_{LOAD3\varphi} + Q_{LINELOSS3\varphi} - Q_C$$
$$= 3.715 + 2.230 - 3.715$$
$$= 2.230 \text{ Mvar}$$

$$S_{SOURCE3\varphi} = \sqrt{(8.545^2 + 2.230^2)} = 8.675 \text{ MVA}$$

c. Comparing the results of (a) and (b), with the shunt capacitor bank in service, the real power delivered to the load increases by 23% (from 6.033 to 7.430 MW) while at the same time:

- The line current decreases
- The real and reactive line losses decrease
- The voltage drop across the line decreases
- The reactive power delivered by the source decreases
- The load voltage increases

The above benefits are achieved by having the shunt capacitor bank (instead of the distribution substation) deliver reactive power to the load. ■

The location of a shunt capacitor bank along a primary feeder is important. If there were only one load on the feeder, the best location for the capacitor bank would be directly at the load, so as to minimize I^2R losses and voltage drops on the feeder. Note that if shunt capacitors were placed at the distribution substation, I^2R feeder losses and feeder voltage drops would not be reduced, because the total power including MW and Mvar would still have to be sent from the substation all the way to the load. Shunt capacitors at distribution substations, however, can be effective in reducing I^2R losses and voltage drops on the transmission or subtransmission lines that feed the distribution substations.

For a primary feeder that has a uniformly distributed load along the feeder, a common application is the "two-thirds" rule; that is, place 2/3 of the required reactive power 2/3 of the way down the feeder. Locating shunt capacitors 2/3 of the way down the feeder allows for good coordination between LTC distribution substation transformers or voltage regulators at the distribution substation. For other load distributions, computer software is available for optimal placement of shunt capacitor banks. We note that capacitors are rarely applied to secondary distribution systems due to their small economic advantage [3, 5].

During the daily load cycle, reactive power requirements change as a function of time. To meet the changing reactive power requirements, many utilities use a combination of fixed and switched capacitor banks. Fixed capacitor banks can be used to compensate for reactive power requirements at light loads, and switched capacitor banks can be added during heavy load conditions. The goal is to obtain a close-to-unity power factor throughout the day by switching capacitor banks on when needed and off when not needed. Methods of controlling switched capacitor banks include the following:

1. Voltage control
2. Current control
3. Var control
4. Time control
5. Temperature control
6. Radio dispatch/SCADA control

14.6

DISTRIBUTION SOFTWARE

Computer programs are available for planning, design, and operation of electric power distribution systems. Program functions include:

1. Arc flash hazard and fault analysis
2. Capacitor placement optimization
3. Circuit breaker duty
4. Conductor and conduit sizing—ampacity and temperature calculations
5. Database management
6. Demand management
7. Distribution reliability evaluation
8. Distribution short circuit calculations
9. Fault detection and location
10. Graphics for single-line diagrams and mapping systems
11. Harmonic analysis
12. Motor starting
13. Outage management
14. Power factor correction
15. Power flow/voltage drop computations
16. Power loss computations and costs of losses
17. Power quality and reliability
18. Relay and protective device coordination
19. Switching optimization
20. Tie capacity optimization
21. T & D modeling and analysis
22. Transformer sizing—load profile and life expectancy
23. Voltage/var optimization

Some of the vendors that offer distribution software packages are given as follows:

- ABB Network Control, Ltd., Switzerland
- ASPEN, San Mateo, CA
- Cooper Power Systems, Pittsburgh PA
- Cyme International, Burlington MA
- EDSA Corporation, Bloomfield MI
- Electrocon International Inc., Ann Arbor MI
- Operation Technology, Irvine CA
- Milsoft Utility Solutions Inc., Abilene TX
- RTDS Technologies, Winnipeg, Manitoba, CA

14.7

DISTRIBUTION RELIABILITY

Reliability in engineering applications, as defined in the *The Authoritative Dictionary of IEEE Standard Terms* (IEEE 100), is the probability that a device will function without failure over a specified time period or amount of usage. In the case of electric power distribution, reliability concerns have come from customers who want uninterrupted continuous power supplied to their facilities at minimum cost [11–17].

A typical goal for an electric utility is to have an overall average of one interruption of no more than two hours' duration per customer year. Given 8760 hours in a non-leap year, this goal corresponds to an Average Service Availability Index (ASAI) greater than or equal to 8758 service hours/8760 hours = 0.999772 = 99.9772%.

IEEE Standard 1366–2003 entitled, *IEEE Guide for Electric Power Distribution Reliability Indices,* defines the following distribution reliability indices [24]:

System Average Interruption Frequency Index (SAIFI):

$$\text{SAIFI} = \frac{\Sigma \text{ Total Number of Customers Interrupted}}{\text{Total Number of Customers Served}} \tag{14.7.1}$$

System Average Interruption Duration Index (SAIDI):

$$\text{SAIDI} = \frac{\Sigma \text{ Customer Interruption Duration}}{\text{Total Number of Customers Served}} \tag{14.7.2}$$

Customer Average Interruption Duration Index (CAIDI):

$$\text{CAIDI} = \frac{\Sigma \text{ Customer Interruption Duration}}{\text{Total Number of Customers Interrupted}} = \frac{\text{SAIDI}}{\text{SAIFI}} \tag{14.7.3}$$

Average Service Availability Index (ASAI):

$$\text{ASAI} = \frac{\text{Customer Hours Service Availability}}{\text{Customer Hours Service Demands}} \tag{14.7.4}$$

In accordance with IEEE Std. 1366–2003, when calculating the above reliability indices, momentary interruption events are not included. A momentary interruption event has an interruption duration that is limited to the time required to restore service by an interrupting device (including multiple reclosures of reclosers or circuit breakers). Switching operations must be completed within five minutes for a momentary interruption event. As such, customer interruption durations less than five minutes are excluded when

TABLE 14.6 Typical Values of Reliability Indices [24] (Based on IEEE Std. 1366–2003, IEEE Guide for Electric Power Distribution Reliability Indices, 2004)

SAIDI	SAIFI	CAIDI	ASAI
90 minutes/year	1.1 Interruptions/year	76 minutes/year	99.982%

calculating the reliability indices. IEEE Std. 1366-2003 also includes a method, when calculating reliability indices, for excluding major events, such as severe storms, for which the daily SAIDI exceeds a specified threshold.

The above formulas for reliability indices use customers out-of-service and customer-minutes out-of-service data. Electric utilities with outage management systems including geographical information systems (GIS) and customer information systems (CIS) are able to very accurately keep track of this data. Some utilities in the United States are required to report distribution reliability indices to state public service commissions, while other utilities may voluntarily report these indices to regional power associations. Typical values for these indices are given in Table 14.6 [24].

The following example uses outage data given in IEEE Std. 1366-2003 [24].

EXAMPLE 14.4 Distribution Reliability Indices

Table 14.7 gives 2010 annual outage data (sustained interruptions) from a utility's CIS database for one feeder. This feeder (denoted circuit 7075) serves 2,000 customers with a total load of 4 MW. Excluding momentary interruption events (less than five minutes duration) and major events, which are omitted from Table 14.7, calculate the SAIFI, SAIDI, CAIDI, and ASAI for this feeder.

SOLUTION

Using the outage data from Table 14.7 in (14.7.1)–(14.7.4):

$$\text{SAIFI} = \frac{200 + 600 + 25 + 90 + 700 + 1{,}500 + 100}{2{,}000}$$

$$= 1.6075 \text{ interruptions/year}$$

TABLE 14.7 Customer Outage Data for Example 14.4 [24] (Based on IEEE Std. 1366–2003, IEEE Guide for Electric Power Distribution Reliability Indices, 2004)

Outage Date	Time at Beginning of Outage	Outage Duration (minutes)	Circuit	Number of Customers Interrupted
3/17/2010	12:12:20	8.17	7075	200
5/5/2010	00:23:10	71.3	7075	600
6/12/2010	09:30:10	30.3	7075	25
8/20/2010	15:45:39	267.2	7075	90
8/31/2010	08:20:00	120	7075	700
9/03/2010	17:10:00	10	7075	1,500
10/07/2010	10:15:00	40	7075	100

$$\text{SAIDI} = \frac{(8.17 \times 200) + (71.3 \times 600) + (30.3 \times 25) + (267.2 \times 90) + (120 \times 700) + (10 \times 1{,}500) + (40 \times 100)}{2{,}000}$$

$$\text{SAIDI} = 86.110 \text{ minutes/year}$$

$$\text{CAIDI} = \frac{\text{SAIDI}}{\text{SAIFI}} = \frac{86.110}{1.6075} = 53.57 \text{ minutes/year}$$

$$\text{ASAI} = \frac{8760 \times 2000 - [(8.17 \times 200) + (71.3 \times 600) + (30.3 \times 25) + (267.2 \times 90) + (120 \times 700) + (10 \times 1{,}500) + (40 \times 100)]/60}{8760 \times 2000}$$

$$\text{ASAI} = 0.999836 = 99.9836\%$$

■

Table 14.8 lists basic outage reporting information recommended by an IEEE committee [15]. Table 14.9 lists generic and specific causes of outages, based on a U.S. Department of Energy study [16]. Many electric utilities routinely prepare distribution outage reports monthly, quarterly, and annually by town (municipality) or by district. The purposes of the reports are to monitor and evaluate distribution reliability, uncover weaknesses and potential problems, and make recommendations for improving reliability. These reports may include:

1. Frequency and duration reports, which provide data for the number of interruptions on distribution circuits together with power interrupted, average interruption duration, and causes.
2. Annual reports that sort outages according to cause of failure and according to circuit classification (for example, sort for each primary voltage; or sort for each conductor type including overhead open wire, overhead spacer cable, underground direct-burial, and cable in conduit).
3. Five- or ten-year trends for reliability indices, and outage trends for specific causes such as tree-contact outages for overhead distribution or dig-in outages for underground distribution.
4. Lists of problem circuits such as the 20 "worst" (lowest ASAI) circuits in a district, or all circuits with repeated outages during the reporting period.

TABLE 14.8

Basic Outage Reporting Information [15] (D. O. Koval and R. Billingon, "Evaluation of Distribution Circuit Reliability," Paper F77 067-2, IEEE Winter Power Meeting, New York, NY (January/February 1977). © 1977 IEEE)

1. Type, design, manufacturer, and other descriptions for classification purposes
2. Date of installation, location on system, length in case of a line
3. Mode of failure (short circuit, false operation, etc.)
4. Cause of failure (lightning, tree, etc.)
5. Times (both out of service and back in service, rather than outage duration alone), date, meteorological conditions when the failure occurred
6. Type of outage, forced or scheduled, transient or permanent (momentary or sustained)

TABLE 14.9 Generic and Specific Causes of Outages [16] (IEEE Committee Report, "List of Transmission and Distribution Components for Use in Outage Reporting and Reliability Calculations," IEEE Transactions on Power Apparatus and Systems, PAS 95, 54 (July/August 1976), pp. 1210–1215. © 1976 IEEE)

Weather	Miscellaneous	System Components	System Operation
Blizzard/snow	Airplane/helicopter	Electrical & mechanical:	System conditions:
Cold	Animal/bird/snake	Fuel supply	Stability
Flood	Vehicle:	Generating unit failure	High/low voltage
Heat	Automobile/truck	Transformer failure	High/low frequency
Hurricane	Crane	Switchgear failure	Line overload
Ice	Dig-in	Conductor failure	Transformer overload
Lightning	Fire/explosion	Tower, pole attachment	Unbalanced load
Rain	Sabotage/vandalism	Insulation failure:	Neighboring power system
		Transmission line	
Tornado	Tree	Substation	Public appeal:
Wind	Unknown	Surge arrester	Commercial & industrial
Other	Other	Cable failure	All customers
		Voltage control equipment:	Voltage Reduction:
		Voltage regulator	0–2% voltage reduction
		Automatic tap changer	Greater than 2–8% voltage reduction
		Capacitor	
		Reactor	Rotating Blackout
		Protection and control:	Utility personnel:
		Relay failure	System operator error
		Communication signal error	Power plant operator error
		Supervisory control error	
			Field operator error
			Maintenance error
			Other

Methods for improving distribution reliability include replacement of older distribution equipment, upgrades of problem circuits, crew staffing and training for fast responses to outages and rapid restoration of service, formal maintenance programs, and public awareness programs to reduce hazards in the vicinity of distribution equipment such as contractor dig-ins. Reliability evaluation has also become an important component of bid selections to procure new distribution equipment. Also, great strides in distribution reliability have come through distribution automation.

14.8 DISTRIBUTION AUTOMATION

Throughout its existence, the electric power industry has been a leader in the application of electric, electronic, and later computer technology for monitoring, control, and automation. Initially, this technology consisted of simple meters to show voltages and flows, and telephones to call the manned substations to do control operations. Yet as early as the 1950s supervisory control and the associated telemetering equipment was in widespread use with a 1955 AIEE (American Institute of Electrical Engineers) report [25] indicating

31% of transmission level stations (switching stations) had such control, and that the U.S. electric industry had more than 30,000 channel miles (48,000 channel km) for communication, and that continuous monitoring of watts, vars, and voltage was widespread. By the late 1960s increasingly sophisticated *energy management systems* (EMSs) were beginning to be deployed in electric utility control centers with applications that included automatic generation control, alarming, state estimation, on-line power flows, and contingency analysis. However, initially all of this monitoring, control, and automation was confined to the generators and the transmission level substations. Because of its larger number of devices and more diffuse nature, the costs of monitoring and automating the distribution system could not be justified.

The monitoring, control, and automation of the distribution system is known under the general rubric of *distribution automation* (DA). While prototype DA systems date back to the 1970s, it has only been in the last decade or so, as communication and computer costs have continued to decrease, that they have started to become widespread.

A primary reason for DA is to reduce the duration of customer outages. As was mentioned in Section 14.2, distribution systems are almost always radial, with sectionalizing fuses used to avoid having prolonged outages for most customers upstream from the fault location. Then sectionalizing switches can be used to further isolate the faulted area, and by closing normally open switches the unfaulted downstream sections of the feeder can be fed from adjacent feeders. DA can greatly reduce the time necessary to complete this process by either

1. providing the distribution system operators with the ability to remotely control the various sectionalizing switches, or

2. automating the entire process.

Real-time monitoring of the voltages and power flows is used to ensure there is sufficient capacity to pickup the unfaulted load on the adjacent feeders.

An example of this situation is illustrated in Figure 14.22, which can also be seen in PowerWorld Simulator by loading case Figure 14.22. This case represents a feeder system modeled using the primary loop approach from Section 14.2 with a nominal 13.8 kV feeder voltage. A total of 10 loads are represented, with each load classified as either primarily residential ('res'), primarily commercial ('com'), or industrial ('ind'); the bus name suffix indicates the type. The left side of the loop goes from the substation distribution bus 2 to bus 8ind; the right side of the loop goes from the substation distribution bus 3 to bus 13ind. Substation buses 2 and 3 are connected by a normally open bus tie breaker, while buses 8ind and 13ind are joined together by a normally open feeder segment to complete the loop. Each feeder line segment is assumed to use 336,400 26/7 ACSR line conductors and to be 0.6 miles (1 km) in length giving a per unit impedance (on a 100 MVA base) of $0.0964 + j0.1995$. The two 12 MVA 138/13.8 kV transformers have an impedance of $0.1 + j0.8$ per unit.

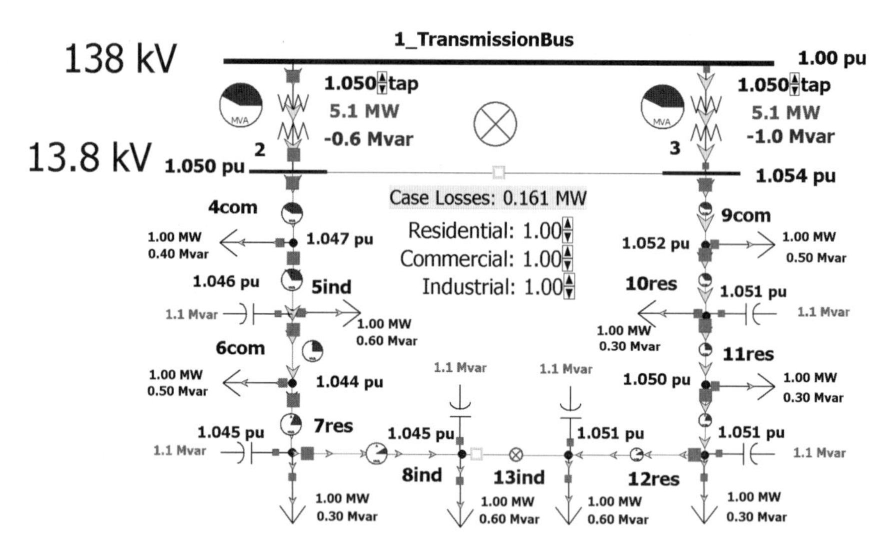

FIGURE 14.22 Primary loop feeder example, before fault

Assume a persistent fault occurs immediately downstream from the bus 3 breaker. This fault would be cleared by the bus 3 breaker, outaging all of the customers of the right branch feeder. Without DA, a line crew would need to be dispatched to locate the fault and then manually change the status of the appropriate sectionalizers to restore service to most customers on this feeder.

In contrast even with a simple DA, which just consisted of having the ability to remotely control the sectionalizers and monitor flow values/voltages, service could be more quickly restored to most customers on the feeder. This could be accomplished by first opening the sectionalizer downstream from bus 9com, then closing the sectionalizer between buses 8ind and 12res, then closing the distribution substation bustie breaker between buses 2 and 3 (after first balancing their taps to prevent circulating reactive power). This new configuration is shown in Figure 14.23.

Another important use for DA is the use of switched capacitor banks to minimize distribution losses and to better manage the customer voltage. Since the feeder load is continually changing, in order to maintain the desired feeder voltages and to minimize system losses, the status of the capacitors often needs to be changed. Without DA various techniques using only local information have been employed including temperature, current, voltage and reactive power sensors, or just simple timers. While these approaches are better than nothing, they all have limitations. By providing a more global view of the entire feeder, DA can greatly improve the situation.

To illustrate, again consider the Figure 14.22 case, which includes six 1.0 Mvar (nominal voltage) switched capacitors. The one-line display also shows the total system losses, and allows variation in the load multiplier for each of the three customer classes (residential, commercial, industrial) with a

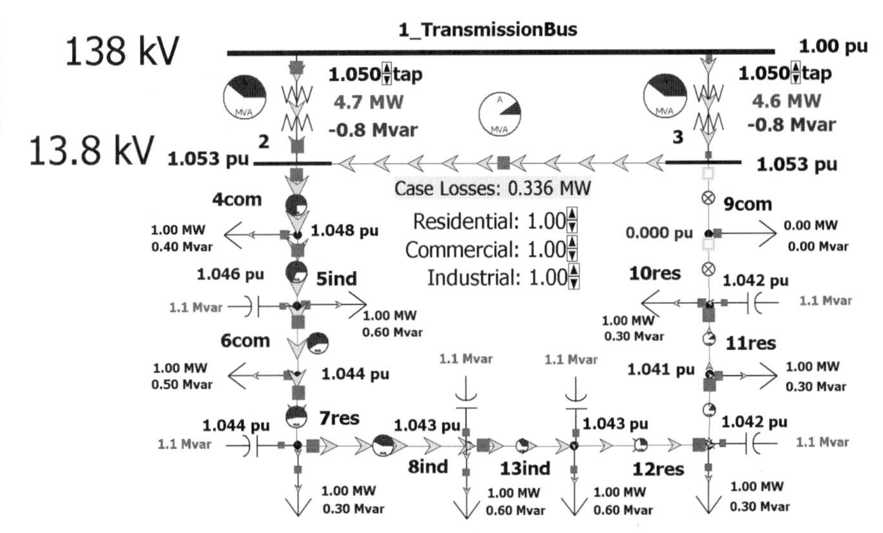

FIGURE 14.23 Primary loop feeder example, after fault and switching

typical value for each ranging between 0.5 and 1.25. Initially all of the capacitors are in-service with total losses of 0.161 MW. However by manually opening the capacitors at buses 7res and 12res the losses can be decreased modestly to 0.153 MW. Under lighter loading situations the loss reduction from capacitor optimization can be even more significant.

While the benefits of DA can be substantial, what had been holding back more widespread adoption of DA was the costs, both for the initial updated equipment installations, and the ongoing costs associated with maintaining the monitoring and control infrastructure such as communication costs. However many of these costs have continued to decrease resulting in more widespread adoption of DA technology.

14.9 SMART GRIDS

Over the last several years, the term "smart grid" has taken the electric power industry by storm, with its use being further cemented in the power industry lexicon with the launch of the *IEEE Transactions on Smart Grid Journal* in 2010. While a new word, the smart grid actually represents an evolutionary advancement on the technological innovation that has been present in the power industry since its inception in the 1880s. Such pervasive innovation over more than a century resulted in electrification being named the top engineering technology of the 20th century by the U.S. National Academy of Engineering in 2000. The smart grid represents a continuation of this application of advanced technology into the 21st century to take advantage of near ubiquitous computing and communication.

As defined in [29], "A smart grid uses digital technology to improve reliability, security and efficiency (both economic and energy) of the electric system from large generation, through the delivery system to electricity consumers and a growing number of distributed-generation and storage resources." Probably the best definition of the attributes of the smart grid is also given in [29] with its listing of six key characteristics:

1. Enables informed participation by customers;
2. Accommodates all generation and storage options;
3. Enables new procducts services, and markets;
4. Provides the power quality for the range of needs;
5. Optimizes asset utilization and operating efficiency;
6. Operates resiliently to disturbances, attacks and natural disasters.

While the smart grid covers large generation and high-voltage transmission, it is most germane to the distribution system and ultimately the end-use customer. The distribution system that, quoting from [28], "has traditionally been characterized as the most unglamorous component" of the power grid is suddenly front and center. Rather than just being a passive, radial conduit for power to flow from the networked transmission system, it will be the means for supporting a bi-directional flow of information and energy to customers who are no longer content to just receive a monthly electric bill. The large, new load potential of electric vehicles requires that the home electric meter and the distribution system become smarter, since the grid cannot reasonably accommodate charging a large number of car batteries as people return to their garages in the early evening when the remainder of the electric load is at peak demand. In many locations distributed energy resources, both fossil fuel-based and renewable, means new power flow patterns and continuing challenges for protection engineers.

Underlying the smart grid is the need for a trustworthy cyber infrastructure. As more smart grid technologies are deployed, the result will be a power grid increasingly dependent on communication and computing. Disruptions in this cyber infrastructure, either due to accidents, bugs, or deliberate attacks could result in wide scale blackouts.

PROBLEMS

SECTION 14.2

14.1 Are laterals on primary radial systems typically protected from short circuits? If so, how (by fuses, circuit breakers, or reclosers)?

14.2 What is the most common type of grounding on primary distribution systems?

14.3 What is the most common primary distribution voltage class in the United States?

14.4 Are reclosers used on: (a) overhead primary radial systems; (b) underground primary radial systems; (c) overhead primary loop systems; (d) underground primary loop systems? Why?

SECTION 14.3

14.5 What are the typical secondary distribution voltages in the United States?

14.6 What are the advantages of secondary networks? Name one disadvantage.

14.7 Using the internet, name three cities in the United States that have secondary network systems.

SECTION 14.4

14.8 A three-phase 138 kVΔ/13.8 kV Y distribution substation transformer rated 40 MVA OA/50 MVA FA/65MVA FOA has an 8% impedance. (a) Determine the rated current on the primary distribution side of the transformer at its OA, FA, and FOA ratings. (b) Determine the per unit transformer impedance using a system base of 100 MVA and 13.8 kV on the primary distribution side of the transformer. (c) Calculate the short-circuit current on the primary distribution side of the transformer for a three-phase bolted fault on the primary distribution side. Assume that the prefault voltage is 13.8 kV.

14.9 As shown in Figure 14.24, an urban distribution substation has one 30-MVA (FOA) and three 33.3 MVA (FOA), 138 kVΔ/12.5 kV Y transformers denoted TR1–TR4, which feed through circuit breakers to a ring bus. The transformers are older transformers designed for 55°C temperature rise. The ring bus contains eight bus-tie circuit breakers, two of which are normally open (NO), so as to separate the ring bus into two sections. TR1 and TR2 feed one section, and TR3 and TR4 feed the other section. Also, four capacitor banks, three banks rated at 6 Mvar and one at 9 Mvar, are connected to the ring bus. Twenty-four 12.5-kV underground primary feeders are served from the substation, 12 from each section. The utility that owns this substation has the following transformer summer loading criteria based on a percentage of nameplate rating:

1. 120% for normal summer loading.
2. 150% during a two-hour emergency.
3. 130% 30-day emergency loading.

Determine the following summer ratings of this substation: (a) the normal summer rating with all four transformers in service; (b) the allowable substation rating assuming the single-contingency loss of one transformer; and (c) the 30-day emergency rating under the single-contingency loss of one transformer. Assume that during a two-hour emergency, switching can be performed to reduce the total substation load by 10% and to approximately balance the loadings of the three transformers remaining in service. Assume a 5% reduction for unequal transformer loadings.

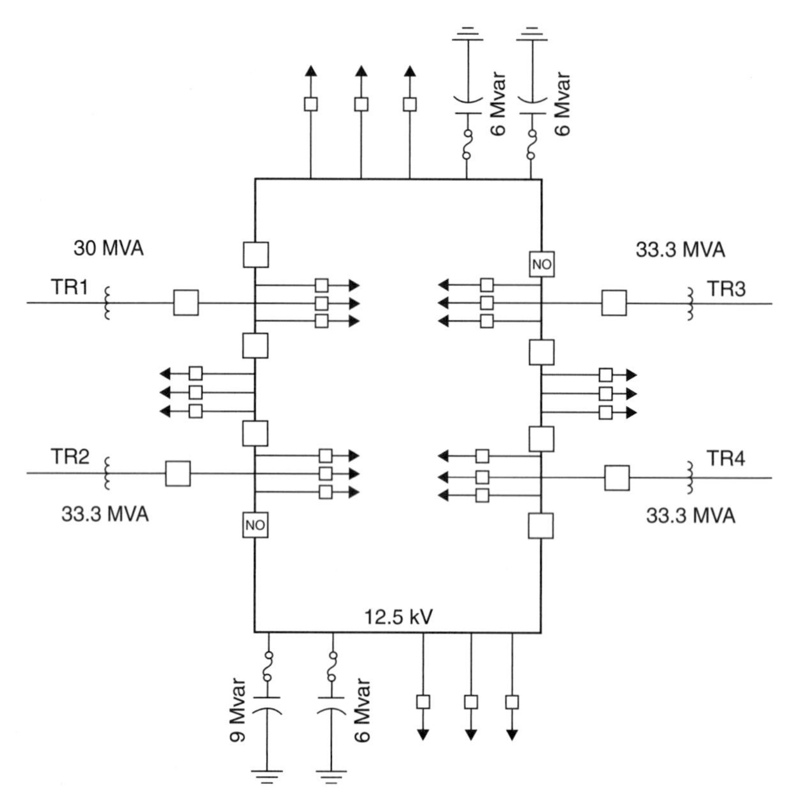

FIGURE 14.24 Distribution Substation for Problems 14.9 and 14.10

14.10 For the distribution substation given in Problem 14.9, assume that each of the four circuit breakers on the 12.5-kV side of the distribution substation transformers has a maximum continuous current of 2,000 A/phase during both normal and emergency conditions. Determine the summer allowable substation rating under the single-contingency loss of one transformer, based on not exceeding the maximum continuous current of these circuit breakers at 12.5-kV operating voltage. Assume a 5% reduction for unequal transformer loadings. Comparing the results of this problem with Problem 14.9, what limits the substation allowable rating, the circuit breakers or the transformers?

SECTION 14.5

14.11 (a) How many Mvar of shunt capacitors are required to increase the power factor on a 10 MVA load from 0.85 to 0.9 lagging? (b) How many Mvar of shunt capacitors are required to increase the power factor on a 10 MVA load from 0.90- to 0.95 lagging? (c) Which requires more reactive power, improving a low power-factor load or a high power-factor load?

14.12 Re-work Example 14.3 with $R_{Load} = 40\ \Omega$/phase, $X_{Load} = 60\ \Omega$/phase, and $X_C = 60\ \Omega$/phase.

SECTION 14.7

14.13 Table 14.10 gives 2010 annual outage data (sustained interruptions) from a utility's CIS database for feeder 8050. This feeder serves 4500 customers with a total load of 9 MW. Table 14.10 includes a major event that began on 11/04/2010 with 4000 customers out of service for approximately six day (10,053 minutes) due to an ice storm. Momentary interruption events (less than five minutes duration) are excluded from Table 14.10. Calculate the SAIFI, SAIDI, CAIDI, and ASAI for this feeder: (a) including the major event; and (b) excluding the major event.

TABLE 14.10 Customer Outage Data for Problem 14.13

Outage Date	Time at Beginning of Outage	Outage Duration (minutes)	Circuit	Number of Customers Interrupted
1/15/2010	01:24:20	14.4	8050	342
4/4/2010	08:14:20	151.2	8050	950
7/08/2010	07:15:46	89.8	8050	125
9/10/2010	15:45:39	654.6	8050	15
10/11/2010	07:40:59	32.7	8050	2,200
11/04/2010	22:30:00	10,053*	8050	4,000*
12/01/2010	14:18:07	40	8050	370

* Major event due to ice storm

14.14 Assume that a utility's system consists of two feeders: feeder 7075 serving 2000 customers and feeder 8050 serving 4500 customers. Annual outage data during 2010 is given in Table 14.6 and 14.10 for these feeders. Calculate the SAIFI, SAIDI, CAIDI, and ASAI for the system, excluding the major event.

SECTION 14.8

PW **14.15** Open PowerWorld Simulator case Problem 14_15 which represents a lower load scenario for the Figure 14.22 case. Determine the optimal status of the six switched shunts to minimize the system losses.

PW **14.16** Open PowerWorld Simulator case Problem 14_16 and note the case losses. Then close the bus tie breaker between buses 2 and 3. How do the losses change? How can the case be modified to reduce the system losses?

PW **14.17** Usually in power flow studies the load is treated as being independent of the bus voltage. That is, a constant power model is used. However, in reality the load usually has some voltage dependence, so if needed decreasing the feeder voltage magnitudes has a result of reducing the total system demand, at least temporarily. Open PowerWorld Simulator case Problem 14_17, which contains the Figure 14.22 system except the load model is set so 50% of the load is modeled as constant power, and 50% of the load is modeled as constant impedance (i.e., the load varies with the square of the bus voltage magnitude). By adjusting the tap positions for the two substation transformers and the capacitor banks, determine the operating configuration that minimizes the total load plus losses (shown on the display), with the constraint that all bus voltage must be at least 0.97 per unit.

SECTION 14.9

14.18 Select one of the smart grid characteristics from the list given in this section. Write a one page (or other instructor selected length) summary and analysis paper on a current news story that relates to this characteristic.

CASE STUDY QUESTIONS

A. What is a smart grid?

B. What provides the foundation for a smart grid?

C. Why is AMI technology preferred over AMR?

REFERENCES

1. J. D. Glover, "Electric Power Distribution," *Encyclopedia of Energy Technology and The Environment* (John Wiley & Sons, New York, NY, 1995).
2. D. G. Fink and H. W. Beaty, *Standard Handbook for Electrical Engineers*, 11th ed., (McGraw-Hill, New York, 1978), Sec 18.
3. T. Gonen, *Power Distribution Engineering* (Wiley, New York, 1986).
4. A. J. Pansini, *Electrical Distribution Engineering,* 2nd ed., The Fairmont Press (Liburn, GA., 1992).
5. J. J. Burke, *Power Distribution Engineering,* Marcel Dekker (New York, 1994).
6. Various co-workers, *Electric Distribution Systems Engineering Handbook, Ebasco Services Inc.,* 2nd ed., McGraw-Hill (New York, 1987).
7. Various co-workers, *Electrical Transmission & Distribution Reference Book,* Westinghouse Electric Corporation (Pittsburgh, 1964).
8. Various co-workers, *Distribution Systems Electric Utility Engineering Reference Book,* Vol. 3, Westinghouse Electric Corporation (Pittsburgh, 1965).
9. Various co-workers, *Underground Systems Reference Book,* Edison Electric Institute (New York, 1957).
10. R. Settembrini, R. Fisher, and N. Hudak, "Seven Distribution Systems: How Reliabilities Compare," *Electrical World,* 206, 5 (May 1992), pp. 41–45.
11. J. L. Blackburn, *Protective Relaying,* Marcel Dekker (New York, 1987).
12. R. Billinton, *Power System Reliability Evaluation,* Gordon and Breach (1988).
13. R. Billinton, R. N. Allan, and L. Salvaderi, *Applied Reliability Assessment in Electric Power Systems,* Institute of Electrical and Electronic Engineers (New York, 1991).
14. R. Billinton and J. E. Billinton, "Distribution Reliability Indices," *IEEE Transactions on Power Delivery,* 4, 1 (January 1989), pp. 561–568.
15. D. O. Koval and R. Billinton, "Evaluation of Distribution Circuit Reliability," Paper F77 067-2, *IEEE Winter Power Meeting* (New York, NY, January/February 1977).

16. IEEE Committee Report, "List of Transmission and Distribution Components for Use in Outage Reporting and Reliability Calculations," *IEEE Transactions on Power Apparatus and Systems*, PAS 95, 54 (July/August 1976), pp. 1210–1215.

17. U. S. Department of Energy, *The National Electric Reliability Study: Technical Reports,* DOE/EP-0003, (April 1981).

18. J. B. Bunch and co-workers, *Guidelines for Evaluating Distribution Automation,* EPRI-EL-3728, Project 2021-1, Electric Power Research Institute (Palo Alto, CA, November 1984).

19. T. Desmond, "Distribution Automation: What is it?, What does it do?," *Electrical World,* 206, 2 (February 1992), pp. 56 & 57.

20. H. Farhangi, "The Path of the Smart Grid," *IEEE Power & Energy Magazine,* 8, 1 (January/February 2010), pp. 18–28.

21. ANSI/IEEE Std. C57.91-1995, *IEEE Guide for Loading Mineral-Oil-Immersed Transformers*, Approved June 14, 1995.

22. ANSI/EEE Std. C57.92-1981, *IEEE Guide for Loading Mineral-Oil-Immersed Power Transformers Up to and including 100 MVA with 55°C or 65°C Average Winding Rise*, Approved January 12, 1981.

23. ANSI/EEE Std. C57.92-1981, *IEEE Guide for Loading Mineral-Oil-Immersed Overhead and Pad-Mounted Distribution Transformers Rated 500 kVA and Less with 65°C or 55°C Average Winding Rise*, Approved November 19, 1980.

24. IEEE Std. 1366-2003, *IEEE Guide for Electric Power Distribution Reliability Indices,* 2004.

25. AIEE Committee Report, "Supervisory Control and Associated Telemetering Equipment—A Survey of Current Practice," *AIEE Transactions Power Apparatus and Systems, Part III,* (January 1955), pp. 36–68.

26. Technical and System Requirements for Advance Distribution Automatin, EPRI Report 1010915, June 2004.

27. H. L. Willis, *Power Distribution Planning Reference Book*, 2nd Edition, CRC Press, Boca Raton, FL, 2004.

28. W. H. Kersting, Distribution System Modeling and Analysis, 2nd Edition, CRC Press, Boca Raton, FL, 2007.

29. *Smart Grid System Report,* U.S. DOE, July 2009.

APPENDIX

TABLE A.1

Typical average values of synchronous-machine constants

Constant (units)	Type	Symbol	Turbo-Generator (solid rotor)	Water-Wheel Generator (with dampers)	Synchro-nous Condenser	Synchro-nous Motor
Reactances (per unit)	Synchronous	X_d	1.1	1.15	1.80	1.20
		X_q	1.08	0.75	1.15	0.90
	Transient	X'_d	0.23	0.37	0.40	0.35
		X'_q	0.23	0.75	1.15	0.90
	Subtransient	X''_d	0.12	0.24	0.25	0.30
		X''_q	0.15	0.34	0.30	0.40
	Negative-sequence	X_2	0.13	0.29	0.27	0.35
	Zero-sequence	X_0	0.05	0.11	0.09	0.16
Resistances (per unit)	Positive-sequence	R (dc)	0.003	0.012	0.008	0.01
		R (ac)	0.005	0.012	0.008	0.01
	Negative-sequence	R_2	0.035	0.10	0.05	0.06
Time constants (seconds)	Transient	T'_{d0}	5.6	5.6	9.0	6.0
		T'_d	1.1	1.8	2.0	1.4
	Subtransient	$T''_d = T''_q$	0.035	0.035	0.035	0.036
	Armature	T_a	0.16	0.15	0.17	0.15

(Adapted from E. W. Kimbark, *Power System Stability: Synchronous Machines* (New York: Dover Publications, 1956/1968), Chap. 12)

TABLE A.2

Typical transformer leakage reactances

Rating of Highest Voltage Winding kV	BIL of Highest Voltage Winding kV	Leakage Reactance per unit*	
Distribution Transformers			
2.4	30	0.023–0.049	
4.8	60	0.023–0.049	
7.2	75	0.026–0.051	
12	95	0.026–0.051	
23	150	0.052–0.055	
34.5	200	0.052–0.055	
46	250	0.057–0.063	
69	350	0.065–0.067	
Power Transformers 10 MVA and Below			
8.7	110	0.050–0.058	
25	150	0.055–0.058	
34.5	200	0.060–0.065	
46	250	0.060–0.070	
69	350	0.070–0.075	
92	450	0.070–0.085	
115	550	0.075–0.100	
138	650	0.080–0.105	
161	750	0.085–0.011	
Power Transformers Above 10 MVA			
		Self-Cooled or Forced-Air-Cooled	Forced-Oil-Cooled
8.7	110	0.050–0.063	0.082–0.105
34.5	200	0.055–0.075	0.090–0.128
46	250	0.057–0.085	0.095–0.143
69	350	0.063–0.095	0.103–0.158
92	450	0.060–0.118	0.105–0.180
115	550	0.065–0.135	0.107–0.195
138	650	0.070–0.140	0.117–0.245
161	750	0.075–0.150	0.125–0.250
230	900	0.070–0.160	0.120–0.270
345	1300	0.080–0.170	0.130–0.280
500	1550	0.100–0.200	0.160–0.340
765		0.110–0.210	0.190–0.350

*Per-unit reactances are based on the transformer rating

TABLE A.3 Characteristics of copper conductors, hard drawn, 97.3% conductivity

Size of Conductor: Circular Mils	Size of Conductor: A.W.G. or B. & S.	Number of Strands	Diameter of Individual Strands (inches)	Outside Diameter (inches)	Breaking Strength (pounds)	Weight (pounds per mile)	Approx. Current Carrying Capacity* (amps)	Geometric Mean Radius at 60 Hz (feet)	r_a Resistance (Ohms per Conductor per Mile) 25°C (77°F) dc	25°C 25 Hz	25°C 50 Hz	25°C 60 Hz	50°C (122°F) dc	50°C 25 Hz	50°C 50 Hz	50°C 60 Hz	x_a Inductive Reactance (ohms per conductor per mile at 1 ft spacing) 25 Hz	x_a 50 Hz	x_a 60 Hz	x'_a Shunt Capacitive Reactance (megohms per conductor per mile at 1 ft spacing) 25 Hz	x'_a 50 Hz	x'_a 60 Hz
1 000 000	...	37	0.1644	1.151	43 830	16 300	1 300	0.0368	0.0585	0.0594	0.0620	0.0634	0.0640	0.0648	0.0672	0.0685	0.1666	0.333	0.400	0.216	0.1081	0.0901
900 000	...	37	0.1560	1.092	39 510	14 670	1 220	0.0349	0.0650	0.0658	0.0682	0.0695	0.0711	0.0718	0.0740	0.0752	0.1693	0.339	0.406	0.220	0.1100	0.0916
800 000	...	37	0.1470	1.029	35 120	13 040	1 130	0.0329	0.0731	0.0739	0.0760	0.0772	0.0800	0.0806	0.0826	0.0837	0.1722	0.344	0.413	0.224	0.1121	0.0934
750 000	...	37	0.1424	0.997	33 400	12 230	1 090	0.0319	0.0780	0.0787	0.0807	0.0818	0.0853	0.0859	0.0878	0.0888	0.1739	0.348	0.417	0.226	0.1132	0.0943
700 000	...	37	0.1375	0.963	31 170	11 410	1 040	0.0308	0.0836	0.0842	0.0861	0.0871	0.0914	0.0920	0.0937	0.0947	0.1759	0.352	0.422	0.229	0.1145	0.0954
600 000	...	37	0.1273	0.891	27 020	9 781	940	0.0285	0.0975	0.0981	0.0997	0.1006	0.1066	0.1071	0.1086	0.1095	0.1799	0.360	0.432	0.235	0.1173	0.0977
500 000	...	37	0.1162	0.814	22 510	8 151	840	0.0260	0.1170	0.1175	0.1188	0.1196	0.1280	0.1283	0.1296	0.1303	0.1845	0.369	0.443	0.241	0.1205	0.1004
500 000	...	19	0.1622	0.811	21 590	8 151	840	0.0256	0.1170	0.1175	0.1188	0.1196	0.1280	0.1283	0.1296	0.1303	0.1853	0.371	0.445	0.241	0.1206	0.1005
450 000	...	19	0.1539	0.770	19 750	7 336	780	0.0243	0.1300	0.1304	0.1316	0.1323	0.1422	0.1426	0.1437	0.1443	0.1879	0.376	0.451	0.245	0.1224	0.1020
400 000	...	19	0.1451	0.726	17 560	6 521	730	0.0229	0.1462	0.1466	0.1477	0.1484	0.1600	0.1603	0.1613	0.1619	0.1909	0.382	0.458	0.249	0.1245	0.1038
350 000	...	19	0.1357	0.679	15 590	5 706	670	0.0214	0.1671	0.1675	0.1684	0.1690	0.1828	0.1831	0.1840	0.1845	0.1943	0.389	0.466	0.254	0.1269	0.1058
350 000	...	12	0.1708	0.710	15 140	5 706	670	0.0225	0.1671	0.1675	0.1684	0.1690	0.1828	0.1831	0.1840	0.1845	0.1918	0.384	0.460	0.251	0.1253	0.1044
300 000	...	19	0.1257	0.629	13 510	4 891	610	0.01987	0.1950	0.1953	0.1961	0.1966	0.213	0.214	0.214	0.215	0.1982	0.396	0.476	0.259	0.1296	0.1080
300 000	...	12	0.1581	0.657	13 170	4 891	610	0.0208	0.1950	0.1953	0.1961	0.1966	0.213	0.214	0.214	0.215	0.1957	0.392	0.470	0.256	0.1281	0.1068
250 000	...	19	0.1147	0.574	11 360	4 076	540	0.01813	0.234	0.234	0.235	0.235	0.256	0.256	0.257	0.257	0.203	0.406	0.487	0.266	0.1329	0.1108
250 000	...	12	0.1443	0.600	11 130	4 076	540	0.01902	0.234	0.234	0.235	0.235	0.256	0.256	0.257	0.257	0.200	0.401	0.481	0.263	0.1313	0.1094
211 600	4/0	19	0.1055	0.528	9 617	3 450	480	0.01668	0.276	0.277	0.277	0.278	0.302	0.303	0.303	0.303	0.207	0.414	0.497	0.272	0.1359	0.1132
211 600	4/0	12	0.1328	0.552	9 483	3 450	490	0.01750	0.276	0.277	0.277	0.278	0.302	0.303	0.303	0.303	0.205	0.409	0.491	0.269	0.1343	0.1119
211 600	4/0	7	0.1739	0.522	9 154	3 450	480	0.01579	0.276	0.277	0.277	0.278	0.302	0.303	0.303	0.303	0.210	0.420	0.503	0.273	0.1363	0.1136
167 800	3/0	12	0.1183	0.492	7 556	2 736	420	0.01559	0.349	0.349	0.349	0.350	0.381	0.381	0.382	0.382	0.210	0.421	0.505	0.277	0.1384	0.1153
167 800	3/0	7	0.1548	0.464	7 366	2 736	420	0.01404	0.349	0.349	0.349	0.350	0.381	0.381	0.382	0.382	0.216	0.431	0.518	0.281	0.1405	0.1171
133 100	2/0	7	0.1379	0.414	5 926	2 170	360	0.01252	0.440	0.440	0.440	0.440	0.481	0.481	0.481	0.481	0.222	0.443	0.532	0.289	0.1445	0.1205
105 500	1/0	7	0.1228	0.368	4 752	1 720	310	0.01113	0.555	0.555	0.555	0.555	0.606	0.607	0.607	0.607	0.227	0.455	0.546	0.298	0.1488	0.1240
83 690	1	7	0.1093	0.328	3 804	1 364	270	0.00992	0.699	0.699	0.699	0.699	0.765				0.233	0.467	0.560	0.306	0.1528	0.1274
83 690	1	3	0.1670	0.360	3 620	1 351	270	0.01016	0.692	0.692	0.692	0.692	0.757				0.232	0.464	0.557	0.299	0.1495	0.1246
66 370	2	7	0.0974	0.292	3 045	1 082	230	0.00883	0.881	0.882	0.882	0.882	0.964				0.239	0.478	0.574	0.314	0.1570	0.1308
66 370	2	3	0.1487	0.320	2 913	1 071	240	0.00903	0.873				0.955				0.238	0.476	0.571	0.307	0.1537	0.1281
66 370	2	1	....	0.258	3 003	1 061	220	0.00836	0.864				0.945				0.242	0.484	0.581	0.323	0.1614	0.1345
52 630	3	7	0.0867	0.260	2 433	858	200	0.00787	1.112				1.216				0.245	0.490	0.588	0.322	0.1611	0.1343
52 630	3	3	0.1325	0.285	2 359	850	200	0.00805	1.101				1.204				0.244	0.488	0.585	0.316	0.1578	0.1315
52 630	3	1	....	0.229	2 439	841	190	0.00745	1.090		Same as dc		1.192		Same as dc		0.248	0.496	0.595	0.331	0.1656	0.1380
41 740	4	3	0.1180	0.254	1 879	674	180	0.00717	1.388				1.518				0.250	0.499	0.599	0.324	0.1619	0.1349
41 740	4	1	....	0.204	1 970	667	170	0.00663	1.374				1.503				0.254	0.507	0.609	0.339	0.1697	0.1415
33 100	5	3	0.1050	0.226	1 505	534	150	0.00638	1.750				1.914				0.256	0.511	0.613	0.332	0.1661	0.1384
33 100	5	1	....	0.1819	1 591	529	140	0.00590	1.733				1.895				0.260	0.519	0.623	0.348	0.1738	0.1449
26 250	6	3	0.0935	0.201	1 205	424	130	0.00568	2.21				2.41				0.262	0.523	0.628	0.341	0.1703	0.1419
26 250	6	1	....	0.1620	1 280	420	120	0.00526	2.18				2.39				0.265	0.531	0.637	0.356	0.1779	0.1483
20 820	7	1	....	0.1443	1 030	333	110	0.00468	2.75				3.01				0.271	0.542	0.651	0.364	0.1821	0.1517
16 510	8	1	....	0.1285	826	264	90	0.00417	3.47				3.80				0.277	0.554	0.665	0.372	0.1862	0.1552

*For conductor at 75°C, air at 25°C, wind 1.4 miles per hour (2 ft/sec), frequency = 60 Hz.

TABLE A.4 Characteristics of aluminum cable, steel, reinforced (Aluminum Company of America)—ACSR

Code Word	Circular Mils Aluminum	Aluminum		Aluminum Strand Diameter (inches)	Steel	Steel Strand Diameter (inches)	Outside Diameter (inches)	Copper Equivalent* Circular Mils or A.W.G.	Ultimate Strength (pounds)	Weight (pounds per mile)	Geometric Mean Radius at 60 Hz (feet)	Approx. Current Carrying Capacity† (amps)	r_a Resistance (Ohms per Conductor per Mile) 25°C (77°F) Small Currents dc	25 Hz	50 Hz	60 Hz	50°C (122°F) Current Approx. 75% Capacity‡ dc	25 Hz	50 Hz	60 Hz	x_a Inductive Reactance (ohms per conductor per mile at 1 ft spacing all currents) 60 Hz	x_a' Shunt Capacitive Reactance (megohms per conductor per mile at 1 ft spacing) 60 Hz
Joree	2 515 000	76	...	0.1819	19	0.0849	1.880		61 700		0.0621									0.0450	0.337	0.0755
Thrasher	2 312 000	76	...	0.1744	19	0.0814	1.802		57 300		0.0595									0.0482	0.342	0.0767
Kiwi	2 167 000	72	4	0.1735	7	0.1157	1.735		49 800		0.0570									0.0511	0.348	0.0778
Bluebird	2 156 000	84	4	0.1602	19	0.0961	1.762		60 300		0.0588									0.0505	0.344	0.0774
Chukar	1 781 000	84	4	0.1456	19	0.0874	1.602		51 000		0.0534									0.0598	0.355	0.0802
Falcon	1 590 000	54	3	0.1716	19	0.1030	1.545	1 000 000	56 000	10 777	0.0520	1 380	0.0587	0.0588	0.0590	0.0591	0.0646	0.0656	0.0675	0.0684	0.359	0.0814
Parrot	1 510 500	54	3	0.1673	19	0.1004	1.506	950 000	53 200	10 237	0.0507	1 340	0.0618	0.0619	0.0621	0.0622	0.0680	0.0690	0.0710	0.0720	0.362	0.0821
Plover	1 431 000	54	3	0.1628	19	0.0977	1.465	900 000	50 400	9 699	0.0493	1 300	0.0652	0.0653	0.0655	0.0656	0.0718	0.0729	0.0749	0.0760	0.365	0.0830
Martin	1 351 000	54	3	0.1582	19	0.0949	1.424	850 000	47 600	9 160	0.0479	1 250	0.0691	0.0692	0.0694	0.0695	0.0761	0.0771	0.0792	0.0803	0.369	0.0838
Pheasant	1 272 000	54	3	0.1535	19	0.0921	1.382	800 000	44 800	8 621	0.0465	1 200	0.0734	0.0735	0.0737	0.0738	0.0808	0.0819	0.0840	0.0851	0.372	0.0847
Grackle	1 192 500	54	3	0.1486	19	0.0892	1.338	750 000	43 100	8 082	0.0450	1 160	0.0783	0.0784	0.0786	0.0788	0.0862	0.0872	0.0894	0.0906	0.376	0.0857
Finch	1 113 000	54	3	0.1436	19	0.0862	1.293	700 000	40 200	7 544	0.0435	1 110	0.0839	0.0840	0.0842	0.0844	0.0924	0.0935	0.0957	0.0969	0.380	0.0867
Curlew	1 033 500	54	3	0.1384	7	0.1384	1.246	650 000	37 100	7 019	0.0420	1 060	0.0903	0.0905	0.0907	0.0909	0.0994	0.1005	0.1025	0.1035	0.385	0.0878
Cardinal	954 000	54	3	0.1329	7	0.1329	1.196	600 000	34 200	6 479	0.0403	1 010	0.0979	0.0980	0.0981	0.0982	0.1078	0.1088	0.1118	0.1128	0.390	0.0890
Canary	900 000	54	3	0.1291	7	0.1291	1.162	566 000	32 300	6 112	0.0391	970	0.104	0.104	0.104	0.104	0.1145	0.1155	0.1175	0.1185	0.393	0.0898
Crane	874 500	54	3	0.1273	7	0.1273	1.146	550 000	31 400	5 940	0.0386	950	0.107	0.107	0.107	0.108	0.1178	0.1188	0.1218	0.1228	0.395	0.0903
Condor	795 000	54	3	0.1214	7	0.1214	1.093	500 000	28 500	5 399	0.0368	900	0.117	0.118	0.118	0.119	0.1288	0.1308	0.1358	0.1378	0.401	0.0917
Drake	795 000	26	2	0.1749	7	0.1360	1.108	500 000	31 200	5 770	0.0375	900	0.117	0.117	0.117	0.117	0.1288	0.1288	0.1288	0.1288	0.399	0.0912
Mallard	795 000	30	2	0.1628	19	0.0977	1.140	500 000	38 400	6 517	0.0393	910	0.117	0.117	0.117	0.117	0.1288	0.1288	0.1288	0.1288	0.393	0.0904
Crow	715 500	54	3	0.1151	7	0.1151	1.036	450 000	26 300	4 859	0.0349	830	0.131	0.131	0.131	0.132	0.1442	0.1452	0.1472	0.1482	0.407	0.0932
Starling	715 500	26	2	0.1659	7	0.1290	1.051	450 000	28 100	5 193	0.0355	840	0.131	0.131	0.131	0.131	0.1442	0.1442	0.1442	0.1442	0.405	0.0928
Redwing	715 500	30	2	0.1544	19	0.0926	1.081	450 000	34 600	5 865	0.0372	840	0.131	0.131	0.131	0.131	0.1442	0.1442	0.1442	0.1442	0.399	0.0920
Flamingo	666 600	54	3	0.1111	7	0.1111	1.000	419 000	24 500	4 527	0.0337	800	0.140	0.140	0.141	0.141	0.1541	0.1571	0.1591	0.1601	0.412	0.0943
Rook	636 000	54	3	0.1085	7	0.1085	0.977	400 000	23 600	4 319	0.0329	770	0.147	0.147	0.148	0.148	0.1618	0.1638	0.1678	0.1688	0.414	0.0950
Grosbeak	636 000	26	2	0.1564	7	0.1216	0.990	400 000	25 000	4 616	0.0335	780	0.147	0.147	0.147	0.147	0.1618	0.1618	0.1618	0.1618	0.412	0.0946
Egret	636 000	30	2	0.1456	19	0.0874	1.019	400 000	31 500	5 213	0.0351	780	0.147	0.147	0.147	0.147	0.1618	0.1618	0.1618	0.1618	0.406	0.0937
Peacock	605 000	54	3	0.1059	7	0.1059	0.953	380 500	22 500	4 109	0.0321	750	0.154	0.155	0.155	0.155	0.1695	0.1715	0.1755	0.1775	0.417	0.0957
Squab	605 000	26	2	0.1525	7	0.1186	0.966	380 500	24 100	4 391	0.0327	760	0.154	0.154	0.154	0.154	0.1700	0.1720	0.1720	0.1720	0.415	0.0953
Dove	556 500	26	2	0.1463	7	0.1138	0.927	350 000	22 400	4 039	0.0313	730	0.168	0.168	0.168	0.168	0.1849	0.1859	0.1859	0.1859	0.420	0.0965
Eagle	556 500	30	2	0.1362	7	0.1362	0.953	350 000	27 200	4 588	0.0328	730	0.168	0.168	0.168	0.168	0.1849	0.1859	0.1859	0.1859	0.415	0.0957
Hawk	477 000	26	2	0.1355	7	0.1054	0.858	300 000	19 430	3 462	0.0290	670	0.196	0.196	0.196	0.196	0.216				0.430	0.0988
Hen	477 000	30	2	0.1261	7	0.1261	0.883	300 000	23 300	3 933	0.0304	670	0.196	0.196	0.196	0.196	0.216				0.424	0.0980
Ibis	397 500	26	2	0.1236	7	0.0961	0.783	250 000	16 190	2 885	0.0265	590	0.235		Same as dc		0.259		Same as dc		0.441	0.1015
Lark	397 500	30	2	0.1151	7	0.1151	0.806	250 000	19 980	3 277	0.0278	600	0.235				0.259				0.435	0.1006
Linnet	336 400	26	2	0.1138	7	0.0855	0.721	4/0	14 050	2 442	0.0244	530	0.278				0.306				0.451	0.1039
Oriole	336 400	30	2	0.1059	7	0.1059	0.741	4/0	17 040	2 774	0.0255	530	0.278				0.306				0.445	0.1032
Ostrich	300 000	26	2	0.1074	7	0.0835	0.680	188 700	12 650	2 178	0.0230	490	0.311				0.342				0.458	0.1057
Piper	300 000	30	2	0.1000	7	0.1000	0.700	188 700	15 430	2 473	0.0241	500	0.311				0.342				0.462	0.1049
Partridge	266 800	26	2	0.1013	7	0.0788	0.642	3/0	11 250	1 936	0.0217	460	0.350				0.385				0.465	0.1074

*Based on copper 97%, aluminum 61% conductivity.

†For conductor at 75°C, air at 25°C, wind 1.4 miles per hour (2 ft/sec), frequency = 60 Hz.

‡"Current Approx. 75% Capacity" is 75% of the "Approx. Current Carrying Capacity in Amps" and is approximately the current which will produce 50°C conductor temp. (25°C rise) with 25°C air temp., wind 1.4 miles per hour.

INDEX

PRINCIPAL UNITS USED IN MECHANICS

Quantity	International System (SI)			U.S. Customary System (USCS)		
	Unit	Symbol	Formula	Unit	Symbol	Formula
Acceleration (angular)	radian per second squared		rad/s^2	radian per second squared		rad/s^2
Acceleration (linear)	meter per second squared		m/s^2	foot per second squared		ft/s^2
Area	square meter		m^2	square foot		ft^2
Density (mass) (Specific mass)	kilogram per cubic meter		kg/m^3	slug per cubic foot		$slug/ft^3$
Density (weight) (Specific weight)	newton per cubic meter		N/m^3	pound per cubic foot	pcf	lb/ft^3
Energy; work	joule	J	N·m	foot-pound		ft-lb
Force	newton	N	$kg \cdot m/s^2$	pound	lb	(base unit)
Force per unit length (Intensity of force)	newton per meter		N/m	pound per foot		lb/ft
Frequency	hertz	Hz	s^{-1}	hertz	Hz	s^{-1}
Length	meter	m	(base unit)	foot	ft	(base unit)
Mass	kilogram	kg	(base unit)	slug		$lb\text{-}s^2/ft$
Moment of a force; torque	newton meter		N·m	pound-foot		lb-ft
Moment of inertia (area)	meter to fourth power		m^4	inch to fourth power		$in.^4$
Moment of inertia (mass)	kilogram meter squared		$kg \cdot m^2$	slug foot squared		$slug\text{-}ft^2$
Power	watt	W	J/s (N·m/s)	foot-pound per second		ft-lb/s
Pressure	pascal	Pa	N/m^2	pound per square foot	psf	lb/ft^2
Section modulus	meter to third power		m^3	inch to third power		$in.^3$
Stress	pascal	Pa	N/m^2	pound per square inch	psi	$lb/in.^2$
Time	second	s	(base unit)	second	s	(base unit)
Velocity (angular)	radian per second		rad/s	radian per second		rad/s
Velocity (linear)	meter per second		m/s	foot per second	fps	ft/s
Volume (liquids)	liter	L	$10^{-3}\ m^3$	gallon	gal.	$231\ in.^3$
Volume (solids)	cubic meter		m^3	cubic foot	cf	ft^3